Paris, 1958

Harish-Chandra

# Collected Papers II

## 1955 – 1958

*Editor*

Veeravalli Seshadri Varadarajan

Reprint of the 1984 Edition

 Springer

*Author*
Harish-Chandra (1923 Kanpur,
  India – 1983 Princeton, USA)

*Editor*
Veeravalli Seshadri Varadarajan
University of California
Los Angeles, USA

ISSN 2194-9875
ISBN 978-3-662-45446-6   (Softcover)
      978-0-387-90782-6   (Hardcover)
DOI 10.1007/978-3-662-45447-3
Springer Heidelberg New York Dordrecht London

Library of Congress Control Number: 2012954381

# Harish-Chandra
## Collected Papers

### Volume II
### (1955–1958)

Edited by V. S. Varadarajan

**Springer-Verlag Berlin Heidelberg GmbH**

Harish-Chandra
1923–1983

*Editor*

V. S. Varadarajan
Department of Mathematics
University of California
Los Angeles, CA 90024
U.S.A.

AMS Subject Classifications: 01A75, 22-XX

Library of Congress Cataloging in Publication Data
Harish-Chandra.
  Harish-Chandra collected papers.
  Contents: v. 1. 1944–1954—v. 2. 1955–1958—
[etc.]—v. 4. 1970–present.
  1. Mathematics—Collected works. I. Varadarajan,
V. S.  II. Title.
QA3.H294  1983        510        83-4828

Originally published by Springer-Verlag New York Berlin Heidelberg Tokyo in 1984
Softcover reprint of the hardcover 1st edition 1984



9  8  7  6  5  4  3  2  1

ISBN 978-1-4899-7409-9        ISBN 978-1-4899-7407-5 (eBook)
DOI 10.1007/978-1-4899-7407-5

# Table of Contents
# Volume II

# Bibliography of Harish-Chandra


[1944a]  (with Bhabha, H. J.) On the theory of point-particles. *Proc. Royal Soc. A.* **183**, 134–141.

[1944b]  On the removal of the infinite self-energies of point-particles. *Proc. Royal Soc. A.* **183**, 142–167.

[1945a]  On the scattering of scalar mesons. *Proc. Indian Acad. Sci. Sect. A.* **21**, 135–146.

[1945b]  Algebra of the Dirac-matrices. *Proc. Indian Acad. Sci. Sect. A.* **22**, 30–41.

[1946a]  (with Bhabha, H. J.) On the fields and equations of motion of point particles. *Proc. Royal Soc. A.* **185**, 250–268.

[1946b]  On the equations of motion of point particles. *Proc. Royal Soc. A.* **185**, 269–287.

[1946c]  A note on the $\sigma$-symbols. *Proc. Indian Acad. Sci. Sect. A.* **23**, 152–163.

[1946d]  The correspondence between the particle and the wave aspects of the meson and the photon. *Proc. Royal Soc. A.* **186**, 502–525.

[1947a]  On the algebra of the meson matrices. *Proc. Camb. Phil. Soc.* **43**, 414–421.

[1947b]  On relativistic wave equations. *Phys. Rev.* **71**, 793–805.

[1947c]  Equations for particles of higher spin. Report of an International Conference on Fundamental particles and Low temperatures held at the Cavendish Laboratory, Cambridge (1946). Vol. I, Fundamental Particles, 185–188. The Physical Society, London.

[1947d]  Infinite irreducible representations of the Lorentz group. *Proc. Royal Soc. A.* **189**, 372–401.

[1948a]  Relativistic equations for elementary particles. *Proc. Royal Soc. A.* **192**, 195–218.

[1948b]  Motion of an electron in the field of a magnetic pole. *Phys. Rev.* **74**, 883–887.

[1949a]  Faithful representations of Lie algebras. *Ann. of Math.* **50**, 68–76.

[1949b]  On representations of Lie algebras. *Ann. of Math.* **50**, 900–915.

[1950a]  On the radical of a Lie algebra. *Proc. Amer. Math. Soc.* **1**, 14–17.

[1950b]  On faithful representations of Lie groups. *Proc. Amer. Math. Soc.* **1**, 205–210.

[1950c]  Lie algebras and the Tannaka duality theorem. *Ann. of Math.* **51**, 299–330.

[1951a]  On some applications of the universal enveloping algebra of a semisimple Lie algebra. *Trans. Amer. Math. Soc.* **70**, 28–96.

[1951b]  Representations of semisimple Lie groups on a Banach space. *Proc. Nat. Acad. Sci. U.S.A.* **37**, 170–173.

## BIBLIOGRAPHY OF HARISH-CHANDRA

[1951c]  Representations of semisimple Lie groups. II. *Proc. Nat. Acad. Sci. U.S.A.* **37**, 362–365.

[1951d]  Representations of semisimple Lie groups. III. Characters. *Proc. Nat. Acad. Sci. U.S.A.* **37**, 366–369.

[1951e]  Representations of semisimple Lie groups. IV. *Proc. Nat. Acad. Sci. U.S.A.* **37**, 691–694.

[1951f]  Plancherel formula for complex semisimple Lie groups. *Proc. Nat. Acad. Sci. U.S.A.* **37**, 813–818.

[1952]  Plancherel formula for the $2 \times 2$ real unimodular group. *Proc. Nat. Acad. Sci. U.S.A.* **38**, 337–342.

[1953]  Representations of a semisimple Lie group on a Banach space. I. *Trans. Amer. Math. Soc.* **75**, 185–243.

[1954a]  Representations of semisimple Lie groups. II. *Trans. Amer. Math. Soc.* **76**, 26–65.

[1954b]  Representations of semisimple Lie groups. III. *Trans. Amer. Math. Soc.* **76**, 234–253.

[1954c]  The Plancherel formula for complex semisimple Lie groups. *Trans. Amer. Math. Soc.* **76**, 485–528.

[1954d]  On the Plancherel formula for the right $K$-invariant functions on a semisimple Lie group. *Proc. Nat. Acad. Sci. U.S.A.* **40**, 200–204.

[1954e]  Representations of semisimple Lie groups. V. *Proc. Nat. Acad. Sci. U.S.A.* **40**, 1076–1077.

[1954f]  Representations of semisimple Lie groups. VI. *Proc. Nat. Acad. Sci. U.S.A.* **40**, 1078–1080.

[1955a]  Integrable and square-integrable representations of a semi-simple Lie group. *Proc. Nat. Acad. Sci. U.S.A.* **41**, 314–317.

[1955b]  On the characters of a semisimple Lie group. *Bull. Amer. Math. Soc.* **61**, 389–396.

[1955c]  Representations of semisimple Lie groups. IV. *Amer. J. of Math.* **77**, 743–777.

[1956a]  Representations of semisimple Lie groups. V. *Amer. J. of Math.* **78**, 1–41.

[1956b]  Representations of semisimple Lie groups. VI. Integrable and square-integrable representations. *Amer. J. of Math.* **78**, 564–628.

[1956c]  The characters of semisimple Lie groups. *Trans. Amer. Math. Soc.* **83**, 98–163.

[1956d]  On a lemma of F. Bruhat. *J. Math. Pures. Appl.* (9) **35**, 203–210.

[1956e]  Invariant differential operators on a semisimple Lie algebra. *Proc. Nat. Acad. Sci. U.S.A.* **42**, 252–253.

[1956f]  A formula for semisimple Lie groups. *Proc. Nat. Acad. Sci. U.S.A.* **42**, 538–540.

[1957a]  Representations of semisimple Lie groups. *Proceedings of the International Congress of Mathematicians*, Amsterdam (1954). Vol. 1, 299–304. Erven P. Noordhoff N. V., Groningen, North Holland Publishing Company.

[1957b]  Differential operators on a semisimple Lie algebra. *Amer. J. of Math.* **79**, 87–120.

[1957c]  Fourier transforms on a semisimple Lie algebra. I. *Amer. J. of Math.* **79**, 193–257.

[1957d]  Fourier transforms on a semisimple Lie algebra. II. *Amer. J. of Math.* **79**, 653–686.

[1957e]  A formula for semisimple Lie groups. *Amer. J. of Math.* **79**, 733–760.

[1957f]  Spherical functions on a semisimple Lie group. *Proc. Nat. Acad. Sci. U.S.A.* **43**, 408–409.

[1958a]  Spherical functions on a semisimple Lie group. I. *Amer. J. of Math.* **80**, 241–310.

[1958b]  Spherical functions on a semisimple Lie group. II. *Amer. J. of Math.* **80**, 553–613.

[1959a]  Automorphic forms on a semisimple Lie group. *Proc. Nat. Acad. Sci. U.S.A.* **45**, 570–573.

[1959b]  Some results on differential equations and their applications. *Proc. Nat. Acad. Sci. U.S.A.* **45**, 1763–1764.

# BIBLIOGRAPHY OF HARISH-CHANDRA


[1960a]   Some results on differential equations.

[1960a′]   Supplement to "Some results on differential equations"

[1960b]   Differential equations and semisimple Lie groups.

[1961]   (with Borel, A.) Arithmetic subgroups of algebraic groups. *Bull. Amer. Math. Soc.* **67**, 579–583.

[1962]   (with Borel, A.) Arithmetic subgroups of algebraic groups. *Ann. of Math.* **75**, 485–535.

[1963]   Invariant eigendistributions on semisimple Lie groups. *Bull. Amer. Math. Soc.* **69**, 117–123.

[1964a]   Invariant distributions on Lie algebras. *Amer. J. of Math.* **86**, 271–309.

[1964b]   Invariant differential operators and distributions on a semi-simple Lie algebra. *Amer. J. of Math.* **86**, 534–564.

[1964c]   Some results on an invariant integral on a semisimple Lie algebra. *Ann. of Math.* **80**, 551–593.

[1965a]   Invariant eigendistributions on a semisimple Lie algebra. *Publ. Math. IHES* No. **27**, 5–54.

[1965b]   Invariant eigendistributions on a semisimple Lie group. *Trans. Amer. Math. Soc.* **119**, 457–508.

[1965c]   Discrete series for semisimple Lie groups. I. Construction of invariant eigendistributions. *Acta Math.* **113**, 241–318.

[1966a]   Two theorems on semisimple Lie groups. *Ann. of Math.* **83**, 74–128.

[1966b]   Discrete series for semisimple Lie groups. II. Explicit determination of the characters. *Acta. Math.* **116**, 1–111.

[1966c]   Harmonic analysis on semisimple Lie groups. Some recent advances in the basic sciences. Vol. 1 (1962, 1963, 1964), 35–40. *Belfer Graduate School of Science Annual Science Conference Proceedings.* Edited by A. Gelbart. Belfer Graduate School of Science, Yeshiva University, New York, New York.

[1967]   Characters of semisimple Lie groups. *Symposia on Theoretical Physics*, **4**, 137–142. (Lectures presented at the 1965 Third Anniversary Symposium of the Institute of Mathematical Sciences, Madras, India. Edited by A. Ramakrishnan.) Plenum Press, New York.

[1968a]   Harmonic analysis on semisimple Lie groups. *Proceedings of the International Congress of Mathematicians*, Moscow (1966), 89–94. "Mir", Moscow.

[1968b]   Automorphic forms on semisimple Lie groups. Notes by J. G. M. Mars. *Lecture Notes in Mathematics*, No. 62, Springer-Verlag, Berlin–Heidelberg–New York.

[1970a]   Some applications of the Schwartz space of a semisimple Lie group. *Lectures in Modern Analysis and Applications. II*, 1–7. Edited by C. T. Tamm. *Lecture Notes in Mathematics.* No. **140**, Springer-Verlag, Berlin–Heidelberg–New York.

[1970b]   Eisenstein series over finite fields. Functional analysis and related fields, 76–88. *Proceedings of a Conference in honor of Professor Marshall Stone held at the University of Chicago, May 1968.* Edited by F. E. Browder. Springer-Verlag, Berlin–Heidelberg–New York.

[1970c]   Harmonic analysis on semisimple Lie groups. *Bull. Amer. Math. Soc.* **76**, 529–551.

[1970d]   Harmonic analysis on reductive *p*-adic groups. Notes by G. van Dijk. *Lecture Notes in Mathematics*, No. **162**, Springer-Verlag, Berlin–Heidelberg–New York.

[1972]   On the theory of the Eisenstein integral. Conference on Harmonic Analysis, College Park, Maryland (1971), 123–149. Edited by D. Gulick and R. L. Lipsman. *Lecture Notes in Mathematics*, No. **266**, Springer-Verlag, Berlin–Heidelberg–New York.

[1973]   Harmonic analysis on reductive *p*-adic groups. Harmonic analysis on homogeneous spaces, 167–192. Edited by C. C. Moore. *Proceedings of Symposia in Pure Mathe-*

*matics*, Vol. **XXVI**, Amer. Math. Soc., Providence, R. I., U.S.A.

[1975]    Harmonic analysis on real reductive groups. I. The theory of the constant term. *J. of Functional Analysis* **19**, 104–204.

[1976a]    Harmonic analysis on real reductive groups. II. Wave packets in the Schwartz space. *Inventiones Math.* **36**, 1–55.

[1976b]    Harmonic analysis on real reductive groups. III. The Maass–Selberg relations and the Plancherel formula. *Ann. of Math.* **104**, 117–201.

[1977a]    The characters of reductive $p$-adic groups. *Contributions to Algebra*, 175–182. A collection of papers dedicated to Ellis Kolchin. Edited by H. Bass, P. J. Cassidy, and J. Kovacic. Academic Press, New York–San Francisco–London.

[1977b]    The Plancherel formula for reductive $p$-adic groups.

[1977b′]    Corrections to "The Plancherel formula for reductive $p$-adic groups"

[1978]    Admissible invariant distributions on reductive $p$-adic groups. Lie theories and their applications, 281–347. *Proceedings of the 1977 annual seminar of the Canadian Mathematical Congress*. Edited by W. Rossmann. Queen's papers in Pure and Applied Math. No. **48**. (Editors: A. J. Coleman and P. Ribenboim). Kingston, Ontario.

[1980]    A submersion principle and its applications. Geometry and Analysis, 95–102. Papers dedicated to the memory of V. K. Patodi, Indian Academy of Sciences, Bangalore, and the Tate Institute of Fundamental Research, Bombay.

[1983]    Supertempered distributions on real reductive groups. *Studies in Applied Mathematics, Advances in Mathematics, Supplementary Studies Series*, Vol. 8, Academic Press Inc., pp. 139–153, edited by Victor Guillemin.

# Abbreviations Used in the Bibliography

| | |
|---|---|
| Proc. Royal Soc. A. ... | Proceedings of the Royal Society, A... |
| Proc. Indian Acad. Sci. Sect. A | Proceedings of the Indian Academy of Sciences, Section A |
| Proc. Camb. Phil. Soc. | Proceedings of the Cambridge Philosophical Society |
| Phys. Rev. | The Physical Review |
| Ann. of Math. | Annals of Mathematics |
| Proc. Amer. Math. Soc. | Proceedings of the American Mathematical Society |
| Trans. Amer. Math. Soc. | Transactions of the American Mathematical Society |
| Proc. Nat. Acad. Sci. U.S.A. | Proceedings of the National Academy of Sciences of the United States of America |
| Bull. Amer. Math. Soc. | Bulletin of the American Mathematical Society |
| Amer. J. of Math. | American Journal of Mathematics |
| J. Math. Pures Appl. | Journal de Mathématiques Pures et Appliquées |
| Publ. Math. IHES | Publications Mathématiques, Institut des Hautes Études Scientifiques |
| Acta Math. | Acta Mathematica |
| J. Functional Analysis | Journal of Functional Analysis |
| Inventiones Math. | Inventiones Mathematicae |

# Preface

These volumes of the Collected Papers of Harish-Chandra are being brought out in response to a widespread feeling in the mathematical community that they would immensely benefit scholars and researchworkers, especially those in analysis, representation theory, arithmetic, mathematical physics, and other related areas. It is hoped that in addition to making his contributions more accessible by collecting them in one place, these volumes would help focus renewed attention on his ideas and methods as well as lend additional perspective to them.

The papers are arranged chronologically. Harish-Chandra is still an extraordinarily productive mathematician, and so I feel that there is no need to impose any structure on his output (past and current) other than that coming from the evolution of his own thinking. It was also decided to divide the papers into four volumes so that each volume would have a reasonable size. However, I have attempted to make the lines of division in such a way that each volume has a certain unity within itself.

It is impossible to mention all the individuals and organisations that helped me with this project. Nevertheless, I do want to record my special thanks to Phyllis Parris, for typing the two handwritten manuscripts on differential equations; to Walter Kaufmann-Bühler, for his constant enthusiasm and encouragement; and to Elise Oranges for her tireless work in producing the final copy. In addition I am extremely grateful to Roger Howe and to Nolan Wallach for agreeing to write commentaries on selected parts of Harish-Chandra's work, and doing it within the time frame that I gave them. Finally, I want to thank Mrs. Lily Harish-Chandra for providing me with the photographs of Harish-Chandra that appear in these volumes.

*Pacific Palisades, California*
*May, 1983*

V. S. VARADARAJAN

# Biographical Note

Harish-Chandra was born on October 11, 1923 in Kanpur, India. He received the M.Sc degree from Allahabad University in 1943 and spent the years 1943–1945 at the Indian Institute of Science, Bangalore, India. He went to Cambridge University in England as a graduate student of Dirac in 1945 and received his Ph.D from there in 1947. He was at Columbia University, New York, from 1950 to 1963. In 1963 he became a permanent member of the Institute for Advanced Study at Princeton, New Jersey, where he has been the IBM von Neumann Professor from 1970 onwards.

He was a Guggenheim Fellow from 1957 to 1958 and a Sloan Fellow from 1961 to 1963. He was elected a Fellow of the Royal Society in 1973 and a member of the National Academy of Sciences of the U.S.A. in 1981. He was awarded honorary doctorates by Delhi University in 1973 and by Yale University in 1981.

He and his wife live in Princeton. They have two daughters. Although he has not returned to India except for brief visits, he has always been profoundly concerned with India and Indian science. He remains a figure of great inspiration in his native land.

## Postscript

Harish-Chandra is no more. He suffered a fatal heart attack while out for a walk on the evening of Sunday, October 16, 1983. He survives in his work, which is a faithful reflection of his personality—lofty, intense, uncompromising.

*Princeton*                                          V. S. VARADARAJAN
*October 19, 1983*

# Introduction

V. S. Varadarajan

> *Bottom*: Masters, I am to discourse wonders:
> but ask me not what; for if I tell you, I am
> no true Athenian. I will tell you every-
> thing, right as it fell out.
>
> —Shakespeare, *A Midsummer Night's
> Dream*, Act Four, Scene II.

**§0.** The first scientific papers of Harish-Chandra were in theoretical physics. However, his interests shifted to mathematics rather early in his career. His prolonged and intense preoccupation with group representations began in the late 1940's when, at the suggestion of Dirac, he started investigating the infinite dimensional unitary representations of the Lorentz group [1947d]. Since then he has erected, almost singlehandedly, a monumental theory of harmonic analysis on reductive groups and their homogeneous spaces. His work is a profound synthesis of algebra, geometry, and analysis. The great force and continued resonance of his ideas have inspired a generation of mathematicians. The singlemindedness and courage with which he has pursued his goals, the beauty of his results, the power and originality of his methods, and the ultimate simplicity of his philosophy, as well as the conviction that sustains it, compel our admiration. There can be no doubt that his achievement is one of the greatest in mathematics in our time.

This introduction attempts to give an overview of this work. I have sketched, in an impressionistic way, the main lines of it in §2; the remaining sections contain additional detail on selected parts of his work so that one of his great achievements —the harmonic analysis of $L^2(G)$ for a real reductive group $G$—may come into clearer focus. I hope that this account, although incomplete and imperfect, may still be of value as a rough guide to the prospective reader.

**§1.** The early sources of motivation for representation theory and harmonic analysis were the classical theory of Fourier series and integrals and the theory of linear representations of finite groups. After the rise of topology and functional analysis in the early decades of this century, far-reaching generalizations of these classical themes began to emerge.

For a locally compact abelian group $G$ let $\hat{G}$ be the group of characters of $G$. $\hat{G}$ is also locally compact abelian and the Pontryagin–van Kampen theory pairs $G$ and $\hat{G}$ in a canonical duality. Given any function $f \in L^1(G)$ its Fourier transform is the function $\hat{f}$ on $\hat{G}$ given by

$$\hat{f}(\hat{x}) = \int_G f(x)\langle x, \hat{x}\rangle \, dx \qquad (\hat{x} \in \hat{G}, dx \text{ a Haar measure})$$

and $f \mapsto \hat{f}$ is a homomorphism of the convolution algebra $L^1(G)$ into the algebra of continuous functions on $\hat{G}$ that vanish at infinity. The Fourier transform can also be introduced on $L^2(G)$; it is then a unitary isomorphism of $L^2(G)$ with $L^2(\hat{G})$ where the Haar measure $d\hat{x}$ on $\hat{G}$ is uniquely determined by $dx$ (the measure dual to $dx$). In particular we have the *Plancherel formula*

$$\int_G |f(x)|^2 \, dx = \int_{\hat{G}} |\hat{f}(\hat{x})|^2 \, d\hat{x} \qquad (f \in L^2(G))$$

and the *inversion formula*

$$f(x) = \int_{\hat{G}} \hat{f}(\hat{x})\langle x, \hat{x}\rangle^{\mathrm{conj}} \, d\hat{x} \qquad (f \in L^1(G), \hat{f} \in L^1(\hat{G}))$$

which generalize the corresponding formulae from classical Fourier analysis on $\mathbb{R}^n$ ([W 1] [Lo] [Bl]).

The classical Poisson Summation formula can also be generalized to this setting. Let $\Gamma \subset G$ be a lattice in $G$, i.e., a discrete subgroup of $G$ such that $G/\Gamma$ is compact. Then $\Gamma^{\vee}$, the annihilator of $\Gamma$, is also a lattice in $\hat{G}$. If Fourier transforms are defined using the Haar measure that gives the measure 1 to $G/\Gamma$, then the Poisson formula takes the form

$$\sum_{\gamma \in \Gamma} f(\gamma) = \sum_{\gamma^{\vee} \in \Gamma^{\vee}} \hat{f}(\gamma^{\vee})$$

where $f$ and $\hat{f}$ are suitably restricted ([W 2]).

The Poisson formula has always been a powerful tool in number theory. The importance of Fourier analysis on general locally compact abelian groups became very clear with the work of Tate who, at the suggestion of Artin, used the general Poisson formula in the setting provided by the adele group of a number field to derive the analytic theory of the Hecke $L$-series attached to a Grössencharakter ([T] [W 2]).

Suppose now that $G$ is compact. Let $dx$ be the normalized Haar measure on $G$ so that $\int_G dx = 1$. Let $\hat{G}$ be the set of equivalence classes of irreducible (finite dimensional) unitary representations of $G$. The general facts concerning linear representations of finite groups extend essentially unchanged to $G$. Thus if $\pi \in \omega \in \hat{G}$ and $\phi, \psi$ are unit vectors in the representation space of $\pi$,

$$\int_G |(\pi(x)\phi, \psi)|^2 \, dx = \frac{1}{d(\omega)}$$

where $d(\omega)$ is the degree of $\omega$. Moreover, for $\omega \in \hat{G}$ let $\Theta_{\omega}$ be the corresponding character; then

$$\int_G \Theta_{\omega}\Theta_{\omega'}^{\mathrm{conj}} \, dx = \delta_{\omega\omega'} \qquad (\omega, \omega' \in \hat{G}).$$

The fundamental result is the Peter–Weyl theorem. It asserts the completeness of the irreducible representations and can be stated as the completeness formula

$$\int_G |f|^2 \, dx = \sum_{\omega \in \hat{G}} d(\omega) \|\pi_\omega(f)\|_2^2$$

where $f \in L^2(G)$, $\pi_\omega \in \omega$, and $\|\cdot\|_2$ is the Hilbert–Schmidt norm. The measure on $\hat{G}$, which assigns the mass $d(\omega)$ to $\omega$, may thus be viewed as the *Plancherel measure* of $G$.

Let $G$ be a compact Lie group. For $f \in L^2(G)$ let $\hat{f}$ be its Fourier transform, defined as the function on $\hat{G}$ given by

$$\hat{f}(\omega) = \int_G f \Theta_\omega^{\mathrm{conj}} \, dx \qquad (\omega \in \hat{G}).$$

The completeness formula can then be restated as

$$f(e) = \sum_{\omega \in \hat{G}} d(\omega) \hat{f}(\omega) \qquad (f \in C^\infty(G)).$$

In other words, we have the relation

$$\delta = \sum_{\omega \in \hat{G}} d(\omega) \Theta_\omega$$

in the space of *distributions* on the manifold $G$, $\delta$ being the delta function at the identity element of $G$.

It is natural to ask for an explicit determination of the irreducible representations of special groups such as the simple Lie groups which had been classified in the late 1890's by Killing and Élie Cartan. The work of Cartan and Hermann Weyl in the 1920's gave a complete solution of this problem. For Cartan it was a question of determining the irreducible finite dimensional representations of the (complex) simple *Lie algebras*; he obtained them as highest weight modules. Weyl had a more global view and worked with the *compact simple Lie groups*. He discovered beautiful formulae for the characters and dimensions of the irreducible representations. Moreover his discovery of the existence of a compact form of a complex simple Lie algebra ("unitarian trick") showed that the representation theory of a complex simple Lie algebra is essentially the same as that of the (simply connected) compact form associated to it, thus unifying his theory with that of Cartan ([We 2]; see also [V 1]).

The notion of distributions, which came into widespread use with the work of Laurent Schwartz [S], led to a deeper understanding of the notion of Fourier transform and to a great extension of the domain of its applicability. In the language of distributions the Plancherel formula for $\mathbb{R}^n$ became the "expansion" of the delta function at the origin in terms of the characters:

$$\delta = (2\pi)^{-n/2} \int_{\mathbb{R}^n} e(\lambda) \, d\lambda \qquad (e(\lambda)(x) = e^{i\lambda \cdot x}).$$

Suppose now that $G$ is a locally compact group, unimodular and second countable, which is neither compact nor abelian. Let $\hat{G}$ be the set of equivalence classes of its irreducible unitary representations (finite or infinite dimensional). There was no systematic theory of representations of such groups in the 1930's. But in Quantum Physics unitary representations of the various symmetry groups had already begun to play a prominent role. This is because the covariance properties of a system with a (connected) group of symmetries are expressed by means of a unitary representation of the group in question or an extension of it. In relativistic quantum physics this meant the inhomogeneous Lorentz group; its physically significant irreducible unitary representations (both single and double valued in the physicists' terminology) were classified by Wigner in a famous memoir in 1939 [Wi].

The situation changed completely in the 1940's. For, by then, the epoch-making papers of von Neumann (by himself and with Murray) on operator algebras in Hilbert spaces had appeared [MvN] [vN]. This theory provided the framework for the study of infinite dimensional unitary representations; in addition it revealed that new phenomena such as type II and type III algebras, direct integral decompositions, and so on, would have to be taken into account.

The systematic beginnings of unitary representation theory of general locally compact groups go back to 1943 when Gel'fand and Raikov proved that every such group had enough irreducible unitary representations to separate the points of the group [GR]. The subject developed rapidly in the late 1940's and early 1950's mainly at the hands of Godement, Mackey, Mautner, and Segal. Their point of view was functional analytic, and their main tool was the theory of operator algebras. Their work led, among other things, to a general decomposition theory of unitary representations, abstract Plancherel theorems, and to the recognition that a reasonable harmonic analysis can be expected only for type I groups, namely groups whose unitary representations were all of type I [Ma 1] [Ma 2] [Ma 3]. It was also in this period that the papers of Mackey on systems of imprimitivity and induced representations appeared, in which he extended the classical Frobenius theory to all separable locally compact groups. Mackey's work, which went far beyond what is suggested in these remarks, has proved to be very influential in the representation theory of nilpotent and solvable groups (cf. [AM] [Mo]).

These developments led to the following formulation of the central problem of harmonic analysis: given a separable locally compact group $G$ acting transitively on a space $X$ with an invariant measure $\mu$, and an irreducible unitary representation $\tau$ of the stabilizer in $G$ of a given point of $X$, decompose into irreducible constituents the representation of $G$ induced by $\tau$. If $\tau$ is the trivial representation, the induced representation is just the natural representation of $G$ in $L^2(X, \mu)$; if $G = X$ and the action is by left translations, the problem is the decomposition of the left regular representation of $G$. The decomposition may not be entirely discrete so that it has to be understood as a direct integral.

In this generality this problem is hopeless because the category of all separable locally compact groups is simply too vast for classical harmonic analysis to be anything more than a very uncertain guide. New examples were clearly needed, and it was natural to look to the simple Lie groups once again for this purpose. The papers of Gel'fand and Neumark [GN 1] and Bargmann [B], dealing with $SL(2,\mathbb{C})$ and $SL(2,\mathbb{R})$, and published independently of each other in 1947, mark the beginning of the subject of infinite dimensional representations of real semisimple Lie groups.

In [GN 1] Gel'fand and Neumark determined all the irreducible unitary representations of SL(2, $\mathbb{C}$) upto equivalence, discovered an explicit Plancherel formula, and proved that any unitary representation of SL(2, $\mathbb{C}$) could be decomposed as a direct integral of irreducible representations in a natural way. They discovered the surprising fact that this group had *additional* irreducible unitary representations which played no role in the harmonic analysis of $L^2(G)$. They called these the representations of the *supplementary series*. They followed this up with a remarkable and seminal series of papers in which they extended their theory to SL($n$, $\mathbb{C}$) and, to a lesser extent, to the other *complex classical groups*. This work, which was presented in detail in their beautiful monograph [GN 2], completely changed the perspective of the theory.

The group SL($n$, $\mathbb{C}$) acts transitively on a number of projective varieties such as the Grassmannians and the various flag spaces. The spaces of *holomorphic* sections of the complex analytic line bundles on these homogeneous spaces carry the finite dimensional complex analytic representations of SL($n$, $\mathbb{C}$). With remarkable insight Gel'fand and Neumark realized that the *smooth* sections of the *real analytic* line bundles on the same homogeneous spaces would transform (roughly speaking) according to the *infinite dimensional* irreducible representations of the group. These representations would be unitary under appropriate conditions. In this way, one had a single scheme for *all* irreducible representations, which contained the finite dimensional as well as the unitary ones as special cases. For example, $G = \mathrm{SL}(n, \mathbb{C})$ acts transitively on the space of all flags $(e_m)$ $(1 \leqslant m \leqslant n, e_1 \subset e_2 \subset \cdots \subset e_n, e_m$ is an $m$-dimensional linear subspace of $\mathbb{C}^n$). Then $X \approx G/B$ where $B$ is the subgroup of the upper triangular matrices. Gel'fand and Neumark proved that the unitary representations of $G$ induced by the unitary one dimensional representations of $B$, which they called the *principal series* of representations, are all irreducible, and that the regular representation of the group is a direct integral of them. They extended to SL($n$, $\mathbb{C}$) (as well as to the other complex classical groups) the construction of the supplementary series.

Although these representations are all infinite dimensional, Gel'fand and Neumark succeeded in associating *characters* to them. If $\xi$ is any one dimensional representation of $B$ and $\pi_\xi$ the representation of $G$ induced by $\xi$, the operator

$$\pi_\xi(x) = \int_G x(g)\pi_\xi(g)\, dg$$

is of trace class if $x$ is suitably restricted, and there is a (unique) class function $\theta_\xi$ on $G$ such that

$$\mathrm{tr}\big(\pi_\xi(x)\big) = \int_G x(g)\theta_\xi(g)\, dg.$$

It is natural to call $\theta_\xi$ the *character* of $\pi_\xi$. They evaluated $\theta_\xi$ on the subgroup $D$ of diagonal matrices and found for it a formula of the same general structure as Weyl's character formula. In terms of these characters they obtained the *Plancherel formula* for SL$(n, \mathbb{C})$ in the form

$$x(e) = \int_{\hat{D}} \hat{x}(\xi)\omega(\xi)\, d\xi \qquad (x \in C_c^\infty(G))$$

where

$$\hat{x}(\xi) = \text{trace}\big(\pi_\xi(x)\big) \qquad \text{(Fourier transform!)}$$

and $\omega(\xi)$ is an explicitly determined function $\geqslant 0$. If we rewrite this in the form

$$\delta = \int \omega(\xi)\theta_\xi\, d\xi,$$

its analogy with the Plancherel formula for $\mathbb{R}^n$ and compact Lie groups becomes clear; $\omega(\xi)\, d\xi$ is the Plancherel measure.

Let me now turn to the work of Bargmann which treated both SL$(2, \mathbb{C})$ and SL$(2, \mathbb{R})$. So far as SL$(2, \mathbb{C})$ was concerned the results were essentially the same as those in [GN 1]; however SL$(2, \mathbb{R})$ presented new phenomena. In addition to the principal and supplementary series, defined essentially as for the complex unimodular group, there now appeared a discretely parametrized series of irreducible unitary representations characterized by the remarkable fact that their matrix elements were already square integrable on the group manifold. This property meant that these representations would occur as *discrete direct summands* of the regular representation. The property of having square integrable matrix coefficients makes sense for irreducible unitary representations of arbitrary separable locally compact unimodular groups and is equivalent to their occurring as direct summands of the regular representation; they form the so-called *discrete series* of the group, and the set of the corresponding equivalence classes is denoted by $\hat{G}_d$. Their appearance in the case of SL$(2, \mathbb{R})$ meant that the decomposition of its regular representation had both discrete and continuous parts.

A highlight of the paper of Bargmann is the emergence of another new theme in our subject. He viewed the matrix elements of the irreducible representations as eigenfunctions of the Casimir differential operator defined on the group manifold (originally introduced by the physicist Casimir for SU(2)). In this way the problem of harmonic analysis in $L^2(G)$ became one of eigenfunction expansions for the Casimir operator. The matrix elements have simple transformation properties relative to the action (from both left and right) of the rotation subgroup of SL$(2, \mathbb{R})$; and so they are essentially determined by their values at the elements $\begin{pmatrix} e^t & 0 \\ 0 & e^{-t} \end{pmatrix}$ $(t > 0)$ of the group. Thus they may be viewed as functions on the half-line of positive reals; moreover the action of the Casimir operator coincides with that of a second order differential operator on this half-line. The spectral theory of such differential operators goes back to Hermann Weyl [We 1]. Bargmann used Weyl's theory to prove the completeness theorem, i.e., the fact that the linear space, spanned by the matrix elements of the discrete series and the "wave packets" of the matrix elements of the principal series, is dense in $L^2(G)$.

For the matrix elements of the discrete series Bargmann discovered remarkable generalizations of the Schur orthogonality relations. For the principal series the corresponding problem is the explicit determination of the Plancherel measure.

Bargmann found a beautiful expression for it in terms of the leading coefficients of the asymptotic expansions of the matrix elements.

The papers of Gel'fand and Neumark and Bargmann form a great watershed in our subject. Their work showed clearly, at least in retrospect, that to do harmonic analysis on real semisimple Lie groups it is absolutely necessary to operate at a level that allowed for a true interaction among the algebraic, geometric, and analytic properties of these groups.

This was the state of affairs in the early 1950's when Harish-Chandra entered the field. His methods combined in a deep and original manner the infinitesimal and global aspects of these problems, and, through an astonishing sequence of papers spanning almost three decades, led to a profound understanding of harmonic analysis of real reductive groups. Sometime in the 1960's he realized that his work contained deep analogies with the spectral theory of automorphic forms that was then being built through the efforts of Selberg, Gel'fand (and his collaborators), and Langlands. His exploration of these analogies has led in recent years to significant results in the harmonic analysis of reductive groups over local, global, and even finite fields, and has revealed the essential unity of the structure of Fourier analysis in these diverse contexts.


§2. Before starting with a description of Harish-Chandra's work we introduce a few definitions and some notation. $G$ is a real semisimple Lie group, connected and with finite center; $K$ is a maximal compact subgroup of $G$; and $\theta$, the involution that fixes $K$ pointwise; $\mathfrak{g} = \text{Lie}(G)$, $\mathfrak{k} = \text{Lie}(K)$. For any real Lie algebra $\mathfrak{m}$, $U(\mathfrak{m})$ is the universal enveloping algebra of the complexification $\mathfrak{m}_c$ of $\mathfrak{m}$.

Manifolds are smooth and second countable. They need not be connected but all the connected components will have the same dimension. For any manifold $M$, $C^\infty(M)$ (resp. $C_c^\infty(M)$) is the space of complex valued functions on $M$ that are smooth (resp. smooth with compact support). If $\omega$ is a volume element ( = exterior differential form of maximal degree which is nowhere zero) on $M$, then given any smooth differential operator $D$ on $M$ its transpose $D^t$ is the differential operator such that $\int_M Df \cdot g\omega = \int_M f \cdot D^t g\omega$ for all $f, g \in C_c^\infty(M)$. If $T$ is a distribution on $M$ the distribution $DT$ is defined by $\langle DT, f \rangle = \langle T, D^t f \rangle$ ($f \in C_c^\infty(M)$). Any function $F$ on $M$ that is locally integrable with respect to $\omega$ will be usually identified with the distribution $f \mapsto \int_M Ff\omega$. The actions of differential operators on smooth functions and distributions are compatible under this identification. If $M$ is a Lie group we shall always choose $\omega$ to be a left invariant volume element. In this case $U(\mathfrak{m}) = U(\mathfrak{m})^t$ is canonically isomorphic to the algebra of all left invariant differential operators on $M$; $X^t = - X$ for $X \in \mathfrak{m}$.

We write $\mathfrak{Z}$ for the center of $U(\mathfrak{g})$. Then $\mathfrak{Z}$ is canonically isomorphic to the algebra of all differential operators on $G$ that are both left and right invariant. $\mathfrak{Z}$ is a polynomial algebra in $l$ indeterminates where $l$ is the rank of $G$.

The centralizer in $G$ of a Cartan subalgebra of $\mathfrak{g}$ is called a Cartan subgroup. Cartan subgroups are parameter spaces for conjugacy classes, and so are absolutely fundamental objects in character theory which deals with class functions and distributions. They are abelian if $G$ is linear and are maximal tori if $G$ is compact. A compact Cartan subgroup is always connected and is in fact a torus. An element $x \in G$ is regular if its centralizer in $\mathfrak{g}$ is a Cartan subalgebra. If $l = rk(G)$, $t$ is an indeterminate and $D(x)$ is the coefficient of $t^l$ in $\det(\mathrm{Ad}(x) - 1 + t)$, $x$ is regular if and only if $D(x) \neq 0$. The set of regular points is denoted by $G'$; it is a dense open subset of $G$. $D$ is often called the discriminant of $G$.

Parabolic subgroups (psgrps) are normalizers in $G$ of parabolic subalgebras of $\mathfrak{g}$; a subalgebra $\mathfrak{p}$ of $\mathfrak{g}$ is parabolic if $\mathfrak{p}_c$ contains a Borel subalgebra of $\mathfrak{g}_c$. Any psgrp $P$ has a canonical decomposition $P = MAN \simeq M \times A \times N$. Here $N$ is the unipotent radical of $P$, namely the maximal normal subgroup of $P$ that consists entirely of unipotent elements; $M_1 = MA$ is reductive and is defined as $P \cap \theta(P)$, so that $P = M_1 N$ is essentially the Levi decomposition of $P$, while $A$ is the maximal $\mathbb{R}$-split* subgroup lying in the center of $M_1$. The subgroup $M$ is the intersection of the kernels of all the continuous homomorphisms of $M_1$ into the positive reals. The decomposition $P = MAN$ is called the Langlands decomposition of $P$. The parabolic subgroups are needed for defining the various series of irreducible unitary representations of $G$; moreover the formulation of many of the asymptotic questions in harmonic analysis depend in a fundamental way on the geometry of the set of all parabolic subgroups.

Harish-Chandra's early work was largely devoted to the development of algebraic methods for studying infinite dimensional representations of real semisimple Lie groups. These methods led him to discover the fundamental facts about unitary representations of $G$. Among these are the results that assert that $G$ is a type I group and that in every irreducible unitary representation of $G$ the irreducible representations of $K$ occur only with finite multiplicities.

Let us define a $(\mathfrak{g}, K)$-module to be a module $V$ for $\mathfrak{g}$ which has a compatible structure as a $K$-module such that $V$ is the algebraic sum of finite dimensional $K$-modules. $V$ is unitary if there is a positive definite scalar product $( \, , \, )$ for $V$ such that $(Xu, v) = -(u, Xv)$ for all $u, v \in V$, $X \in \mathfrak{g}$. One of Harish-Chandra's major discoveries was that the space $V(\pi)$ of $K$-finite vectors in an irreducible unitary representation $\pi$ is an irreducible $(\mathfrak{g}, K)$-module and that the assignment $\pi \mapsto V(\pi)$ induces a bijection of $\hat{G}$ with the set of isomorphism classes of irreducible $(\mathfrak{g}, K)$-modules that are unitary. He also proved that any irreducible $(\mathfrak{g}, K)$-module is of the form $V(\pi)$ for some irreducible representation $\pi$ acting on a Hilbert space and possessing an infinitesimal character, thus showing that in the notion of a $(\mathfrak{g}, K)$-module which is irreducible we have the correct formulation of the vague idea of an "arbitrary" irreducible representation of $G$.

In spite of the promise and power of the algebraic method he never pursued it beyond his initial investigations and was content to use his early work as the point of departure for the analytical theory.

The basic step in the analytical theory is the introduction of characters. The algebraic treatment had already shown that the multiplicities of the irreducible representations of $K$ that occur in an irreducible unitary representation of $G$ do not exceed a constant multiple of the degrees of the representations of $K$. It follows from this that for any function $f \in C_c^\infty(G)$ the operator

$$\pi(f) = \int_G f(x)\pi(x)\,dx$$

---

*This means that $\mathrm{Ad}(a)$ is diagonalizable over $\mathbb{R}$ for all $a \in A$.

is of trace class, and the linear functional

$$\Theta_\pi : f \mapsto \mathrm{tr}(\pi(f))$$

is a distribution on the group $G$. $\Theta_\pi$ depends only on the equivalence class $\omega$ of $\pi$; it is denoted by $\Theta_\omega$ and is called the character of $\pi$ (or $\omega$). It is an invariant distribution, i.e., a distribution that is invariant under all inner automorphisms of $G$. It determines $\omega$ completely. If $(\phi_i)_{i \geqslant 1}$ is an orthonormal basis of $K$-finite vectors for $\pi$ and $f_i$ is the matrix element defined by $\phi_i$, i.e., $f_i(x) = (\pi(x)\phi_i, \phi_i)$ $(x \in G)$,

$$\Theta_\omega = \sum_{i \geqslant 1} f_i,$$

this equation being understood in the sense of distributions, i.e., for all $u \in C_c^\infty(G)$,

$$\Theta_\omega(u) = \sum_{i \geqslant 1} \int_G f_i u \, dx.$$

Since any matrix element satisfies the differential equation

$$zf = \chi_\omega(z)f \qquad (z \in \mathfrak{Z})$$

where $\chi_\omega$ is the infinitesimal character of $\omega$, it follows that

$$z\Theta_\omega = \chi_\omega(z)\Theta_\omega \qquad (z \in \mathfrak{Z}).$$

Here we recall that the infinitesimal character of $\omega$ is the homomorphism according to which $\mathfrak{Z}$ acts (by Schur's lemma) on the space $V(\pi)$. In other words, $\Theta_\omega$ is an *invariant eigendistribution* on $G$.

Using the notion of characters one may formulate the problem of harmonic analysis on $G$ as follows (cf. the introductory remarks to [1957e] [1970c]): given any invariant distribution $T$ on $G$, express $T$ as a "linear combination" of the irreducible characters. For $f \in C_c^\infty(G)$ define its *Fourier transform* $\hat{f}$ to be the function on $\hat{G}$ given by

$$\hat{f}(\omega) = \Theta_\omega(f) \qquad (\omega \in \hat{G}).$$

The problem is then to define a "distribution" $\hat{T}$ on $\hat{G}$ such that

$$T(f) = \hat{T}(\hat{f}) \qquad (f \in C_c^\infty(G)).$$

$\hat{T}$ will be the "Fourier transform" of $T$. If $T = \delta$, the delta function at the identity element of $G$, the determination of $\hat{\delta}$ is the problem of finding the explicit Plancherel formula for $G$; $\hat{\delta}$ will have to be a nonnegative measure, the Plancherel measure. The derivation by Gel'fand and Neumark of the Plancherel formula for $SL(n,\mathbb{C})$, Harish-Chandra's generalization of this to all complex semisimple Lie groups [1951f] [1954c], and his treatment of the case of $SL(2,\mathbb{R})$ [1952], and the work of Gel'fand and Graev on $SL(n,\mathbb{R})$ [GG 1] [GG 2], may be regarded as initial evidence for this view.

A second approach to harmonic analysis is through *eigenfunction expansions*; this goes back to Bargmann as we saw in §1. Let $H \subset G$ be a closed unimodular subgroup, $X = G/H$, and let $\mathcal{H} = L^2(X)$ be the Hilbert space of functions square integrable with respect to the invariant measure on $X$. Let $\lambda$ be the natural representation of $G$ in $\mathcal{H}$. If $\mathcal{H}^\infty$ is the space of differentiable vectors in $\mathcal{H}$ for $\lambda$, we have a representation $a \mapsto \lambda(a)$ of $U(\mathfrak{g})$ in $\mathcal{K}^\infty$. If $z \in \mathfrak{Z}$ is Hermitian, i.e., $(z^{\mathrm{conj}})' = z$, the operator $\lambda(z)$ is essentially self adjoint on $\mathcal{K}^\infty$ and is invariant under $G$. So, in the first approximation, one may view the problem of decomposition of $\lambda$ as an eigenfunction expansion problem for the commuting algebra of differential operators $\lambda(z)$, $z \in \mathfrak{Z}$ (see the introductory remarks to [1968b]). The eigenspaces are parametrized by the points of the spectrum of the commutative algebra $\mathfrak{Z}$, i.e., by the homomorphisms $\chi\colon \mathfrak{Z} \to \mathbf{C}$. For a given $\chi$, the corresponding eigenspace would be a $G$-module which, typically, would be infinite dimensional. In view of the relationship between the eigenvalue problem and the representation theoretic question, we should expect that only those eigenfunctions will contribute to the spectral expansion in $L^2(X)$ whose $\chi$ are the infinitesimal characters of irreducible *unitary* representations of $G$ (not necessarily all of them either).

The basic example in this approach is the case $X = G$, viewed as a homogeneous space for $G$ (left action) or $G \times G$ (two-sided action). A variant of this in which the eigenspaces are finite dimensional is obtained as follows. Let $U$ be a finite dimensional Hilbert space which is a bimodule for $K$, i.e., which carries a unitary representation $\tau$ of $K \times K^{\vee}$, $K^{\vee}$ being the group opposite to $K$. The Hilbert space is now the space $L^2(G, \tau)$ of functions $f\colon G \to U$ which are square integrable and satisfy

$$f(k_1 x k_2) = k_1 f(x) k_2$$

for all $k_1, k_2 \in K$ and almost all $x \in G$. Such functions are called $\tau$-*spherical*; if $U = \mathbf{C}$ and $K \times K^{\vee}$ acts trivially, one speaks of *spherical* or *strictly spherical* functions. The terminology reminds us that we have a generalization of the classical theory of spherical harmonics. If $\chi$ is a homomorphism of $\mathfrak{Z}$ into $\mathbf{C}$ and $f \in L^2(G, \tau)$ is such that $zf = \chi(z)f$ for all $z \in \mathfrak{Z}$, $f$ is $C^\infty$ (even analytic), and the space of all such $f$ is of finite dimension (cf. [1960b]). One of Harish-Chandra's basic results is that the eigenfunctions needed for the $L^2$-expansion are precisely those that satisfy the so-called weak inequality ([1970c], p. 539).

The theory of automorphic forms gives another example of this point of view. The classical automorphic forms on the Poincaré half-plane may be interpreted as eigenfunctions for the Laplace–Beltrami operator. They can then be generalized to the context of nonanalytic forms and higher dimensional symmetric spaces, as was done by Maass [M] and Selberg [Se 1]. Following the work of Borel and Harish-Chandra [1962] on the structure of arithmetic subgroups of reductive groups and the associated homogeneous spaces, Gel'fand [G], Selberg [Se 1] [Se 2], and Langlands [L 1] constructed a general theory of automorphic forms. They approached the fundamental problem as the spectral theory of the invariant differential operators in $L^2(G/\Gamma)$, $G$ being the group of real points of a reductive group defined over $\mathbf{Q}$, and $\Gamma$ is the associated arithmetic subgroup of $G$. The classical Eisenstein series and its generalizations are examples of eigenfunctions of these differential operators. It is interesting to note that the homomorphisms $\chi$ in the spectrum of $\mathfrak{Z}$ that correspond to these eigenfunctions do not come from unitary representations, and so one must make an analytic continuation in $\chi$ in order to get the eigenfunctions needed for the spectral theory (cf. [1968b]).

Whichever point of view is adopted, it is not possible to go far without a good supply of irreducible unitary representations. The known results for complex groups and $SL(n,\mathbb{R})$ seemed to suggest that each conjugacy class of Cartan subgroups of $G$ gives rise to a "series" of irreducible unitary representations, and that the Plancherel measure is a *disjoint sum* of the measures contributed by each of these series (see [1957a] [1954e]). As a partial justification of this principle Harish-Chandra sketched in [1954e] a method of constructing a series of irreducible unitary representations associated to any Cartan subgroup $L$. Given $L$, one can select a parabolic subgroup $P = MAN$ such that $A$ is the maximal $\mathbb{R}$-split* component of $L$. If $\eta$ is an irreducible unitary representation of $M$ and $\nu \in \hat{A}$, the unitary representation $\pi_{\eta,\nu}$ of $G$ is defined as the representation induced by the representation $m \cdot a \cdot n \mapsto \nu(a)\eta(m)$ of $P$. The $\pi_{\eta,\nu}$ depend only on the conjugacy class of $L$ and are discretely decomposable into irreducible subrepresentations. If $L$ is chosen so that $\dim(A)$ is maximum, $P$ is a minimal parabolic subgroup, $M$ is compact, and what we have is called the *principal series*; for complex $G$ it *is* the principal series of Gel'fand and Neumark. Bruhat's work [Br 1] had shown that for minimal $P$ the $\pi_{\eta,\nu}$ are irreducible for $\nu$ in "general position." So one could hope that this would be so in the case of arbitrary $L$ also.

If $P$ is not minimal, $M$ is not compact and so we are still faced with the problem of determining $\hat{M}$. Harish-Chandra's remarkable idea, which is implicit but very clear in [1957a] and [1954e], was that if we restrict $\eta$ to vary always over the *discrete series* of $M$, the various series arising from a complete system of mutually nonconjugate Cartan subgroups would determine essentially disjoint parts of $\hat{G}$; and these representations would suffice to obtain the Plancherel formula for $G$. The discrete series is thus associated to the (unique) conjugacy class of compact Cartan subgroups, and the general discussion makes it clear that these may have to be constructed a priori and by quite different methods.

*Discrete series: beginnings.* Already in 1947 Godement had proved that the matrix elements of the discrete series satisfy orthogonality relations [Go]. In particular, one can associate to each $\omega \in \hat{G}_d$ a positive number $d(\omega)$, the *formal degree of* $\omega$, such that

$$\int_G |(\pi(x)\phi, \psi)|^2 \, dx = \frac{1}{d(\omega)}$$

where $\pi \in \omega$ and $\phi, \psi$ are arbitrary unit vectors in the representation space of $\pi$. Moreover if ${}^0L^2(G)$ is the closed linear span of the irreducible subspaces of $L^2(G)$ ( = the "discrete part" of $L^2(G)$), and $E$ is the orthogonal projection $L^2(G) \to {}^0L^2(G)$, then, for all $f \in L^1(G) \cap L^2(G)$,

$$\|Ef\|^2 = \sum_{\omega \in \hat{G}_d} d(\omega)\|\pi_\omega(f)\|_2^2$$

where $\pi_\omega \in \omega$ and $\|\cdot\|_2$ is the Hilbert-Schmidt norm; this shows that the Plancherel

---

*This means that $A$ is the maximal subgroup lying in the connected component of the identity of $L$ such that $\mathrm{Ad}(a)$ is diagonalizable over $\mathbb{R}$ for all $a \in A$.

measure assigns the mass $d(\omega)$ to the class $\omega$ [1956b]. These results are actually true for all second countable locally compact groups ([Ma 2], p. 640).

In a succession of beautiful papers Harish-Chandra added new evidence to the picture of harmonic analysis of $L^2(G)$ suggested above [1954f], [1955a], [1955c], [1956a], [1956b]. Assuming not only that $rk(G) = rk(K)$ but further that $G/K$ is Hermitian symmetric (this means that $G/K$ has a $G$-invariant complex structure) he constructed a whole family of discrete series representations. They were realized on certain Hilbert spaces of holomorphic functions on which $G$ acts naturally; they are the *holomorphic discrete series*. When $G/K$ is not Hermitian symmetric this method would give nothing; even in the Hermitian symmetric case it will not give the entire discrete series. Nevertheless the introductory remarks to [1955c] reveal his conviction that in these cases different methods would have to be invented.

These papers revealed profound analogies between the (holomorphic) discrete series and the finite dimensional representations. These representations were *highest weight modules* with respect to positive systems coming from the complex structure of $G/K$. Their formal degrees were given *exactly by Weyl's dimension formula* applied to their highest weights. Finally, their *characters made sense as point functions* on the set of regular points of a compact Cartan subgroup and *could be described there by precise analogues of the Weyl character formulae*.

These results were the first serious indication that the discrete series would be parametrized by the characters of the compact Cartan subgroup, and hence that the representations of the series associated to an arbitrary Cartan subgroup would be parametrized by the characters of that Cartan subgroup. One would thus have a natural map

$$\coprod_{1 \leqslant i \leqslant r} \hat{A}_i \to \hat{G}$$

where $A_1, A_2, \ldots, A_r$, is a complete system of mutually nonconjugate Cartan subgroups of $G$ (cf. the introductory remarks to [1957e]).

It is thus clear that by 1956 Harish-Chandra had essentially visualized the lines along which the Fourier analysis of $L^2(G)$ would have to be constructed. In retrospect it is really remarkable that he saw so clearly and so far; the clues available to him were certainly very meager.

*Discrete series: construction.* The determination of the discrete series turned out to be a prodigious undertaking ([1957b]–[1957e], [1964a]–[1966b]). His method, which was entirely novel, was roughly in three parts.

The first part is a direct and explicit construction of the invariant eigendistributions $\Theta_\xi$, parametrized by the (regular) characters $\xi$ of a compact Cartan subgroup $B \subset K$; these would ultimately be the characters of the representations of the discrete series. In the second part the question is that of showing that the Fourier coefficients (with respect to $K$) of the $\Theta_\xi$, which are clearly eigenfunctions for $\mathfrak{Z}$ that are $K$-finite, are in $L^2(G)$. The third part is concerned with the completeness of the $\Theta_\xi$; and the main result here is that these eigenfunctions and their translates span a dense subspace of $^0L^2(G)$.

Before the invariant eigendistributions could be constructed it is absolutely essential to understand the nature of their *singularities*. An initial exploration [1956c] had already shown that any invariant eigendistribution $\Theta$ on $G$ is an analytic function $F_\Theta$ on $G'$, the set of regular points of $G$. This function is locally integrable around the singular points also and so defines an invariant distribution $f \mapsto \int F_\Theta f \, dx$ ($f \in C_c^\infty(G)$) on $G$, also denoted by $F_\Theta$. Experience suggested that $\Theta = F_\Theta$ (cf. [1955b], although the statement there is somewhat guarded), and this actually turned out to be true in complete generality. This result, known as the *Regularity theorem*, lies at a great depth, and is the corner stone for the entire structure of harmonic analysis.

The function $F_\Theta$ will in general blow up in the neighborhood of a singular point of the group. This behavior however cannot be completely arbitrary because the differential equations satisfied by the characters remain valid in the neighborhood of the singular points also. Harish-Chandra's work revealed the true nature of the limitations on the behaviour of $F_\Theta$, namely that $F_\Theta$ should satisfy conditions of the following form: for suitable (explicitly determined) invariant differential operators $\nabla$ on $G'$, $\nabla F_\Theta$ should extend continuously to all of $G$.

The starting point of the construction of the $\Theta_\xi$ is now quite clear; one defines $\Theta_\xi$ at the regular points on the compact Cartan subgroup by the analogue of the Weyl character formula. But the problem of defining $\Theta_\xi$ on the other Cartan subgroups is nontrivial because an invariant eigendistribution is not uniquely determined by its values on $B' = B \cap G'$. Nevertheless Harish-Chandra discovered that there is uniqueness *if the distribution is suitably restricted at infinity on the group*. The condition is that the distributions should be tempered; this however is to anticipate later ideas. At this stage the condition had to be formulated in a more direct manner, and he did it by requiring that $|D|^{-1/2}$ be a *global majorant* for $\Theta_\xi$, $D$ being the discriminant of $G$. The existence problem, i.e., the construction of the $\Theta_\xi$, is however much deeper and depends in a fundamental way on *Fourier transforms on the Lie algebra of $G$*. Harish-Chandra proved that the $\Theta_\xi$ can be obtained by exponentiating from the Lie algebra the Fourier transforms of the invariant measures on appropriately chosen regular semisimple orbits [1965c].

The regularity theorem is the key that opens all the doors. If we look at the smooth, almost magical manner in which the theory flows once the regularity theorem is established, and also remember that there was very little in the known history of the subject to guide him in his prolonged struggle or to reassure him during moments of doubt, we have no choice except to place this among his greatest creative achievements.

The idea now is to use the method of integration over the conjugacy classes ( = orbits) to reduce the harmonic analysis of the discrete series to that on the compact Cartan subgroup. This is of course exactly the method used by Weyl in his famous completeness proof for the compact case. However the present situation is much more complicated, for one must take into account the contributions of the other Cartan subgroups. Before explaining Harish-Chandra's method of doing this I shall discuss in some detail the method of orbital integrals.

It was Gel'fand and Neumark who first realized that the theory of orbital integrals can be made the basis for a proof of the Plancherel formula [GN 2]. They treated $SL(n, \mathbb{C})$ by this method. In [1951f], [1954c] Harish-Chandra pushed this method through for all *complex* semisimple Lie groups. To motivate this let us first assume that $G$ is compact and let $T \subset G$ be a maximal torus. Using Weyl's formulae we can rewrite the Plancherel formula as a sum over $\hat{T}$:

$$f(e) = \sum_{\xi \in \hat{T}} \Omega(\xi) \int_G \Theta_\xi f \, dx \qquad (f \in C^\infty(G)).$$

Here $\Omega(\xi)$ is a polynomial function of $\xi$, namely a constant multiple of the product of the roots of a positive system; $\Theta_\xi$ is the Weyl character, given on $T$ by

$$\Delta(t)\Theta_\xi(t) = \sum_{s \in W} \varepsilon(s)(s\xi)(t) \qquad (t \in T)$$

where $W$ is the Weyl group and $\Delta$ is the Weyl denominator function. Integrating over the orbits and setting

$$F_f(t) = \Delta(t) \int_G f(xtx^{-1}) \, dx \qquad (t \in T)$$

we have

$$\int_G \Theta_\xi f \, dx = (-1)^p \hat{F}_f(\xi)$$

where $p$ is the number of positive roots and $\hat{F}_f$ is the Fourier transform of $F_f$ ($F_f$ is obviously in $C^\infty(T)$). So the Plancherel formula is equivalent to the assertion

$$f(e) = (-1)^p \sum_{\xi \in \hat{T}} \Omega(\xi) \hat{F}_f(\xi).$$

But, by Fourier transform theory on $T$, $(-1)^p \Omega(\xi) \hat{F}_f(\xi)$ is $\widehat{\Omega F_f}(\xi)$ where $\Omega$ *is now viewed as a constant coefficient differential operator on $T$*. So the Plancherel formula is equivalent to the relation

(L.F.) $$f(e) = (\Omega F_f)(e) \qquad (f \in C^\infty(G)).$$

We shall refer to (L.F.) as the *limit formula for the orbital integral.*

This derivation suggests a method for proving the Plancherel formula: first prove (L.F.) and then apply Fourier transform to it. It is remarkable that this method continues to work when $G$ is complex. The maximal torus is now replaced by a Cartan subgroup $L$. For any $f \in C_c^\infty(G)$, $F_{f,L}(h)$ will denote the integral of $f$ over the orbit of $h$:

$$F_{f,L}(h) = \Delta(h) \int_{G/L} f(xhx^{-1}) \, d\dot{x}$$

where $d\dot{x}$ is the invariant measure on $G/L$. Initially this makes sense only for $h$ regular, and in the regular subset of $L$, $F_{f,L}$ will be a smooth function. It can then be proved that $F_{f,L}$ extends to all of $L$ as an element of $C_c^\infty(L)$. The character formulae for the principal series now show that, with appropriate normalization,

$$\langle T_\xi, f \rangle = \hat{F}_{f,L}(\xi) \qquad (\xi \in \hat{L})$$

where $T_\xi$ is the distribution character of the principal series representation corresponding to the character $\xi$ of $L$. But now the limit formula

(L.F.) $$f(e) = (\Omega F_{f,L})(e) \qquad (f \in C_c^\infty(G))$$

holds, with same $\Omega$ as before, and the Fourier transformation of this relation gives the Plancherel formula

$$f(e) = \int_{\hat{L}} \hat{\Omega}(\xi)\langle T_\xi, f\rangle\, d\xi \qquad (f \in C_c^\infty(G))$$

where $\hat{\Omega}$ is the polynomial function which is the Fourier transform of the differential operator $\Omega$.

The set $\{e\}$ is a very complicated singular point in the space of conjugacy classes. In particular it is not the limit of orbits in general position, and so $f(e)$ is not the limit of $F_{f,L}(h)$ as $h \to e$. It is therefore striking that one can still recover $f(e)$ as the limit as $h \to e$ of *suitable derivatives* of $F_{f,L}(h)$.

It is natural to try to extend this method to all cases by establishing the limit formula and relating the Fourier transforms of the orbital integrals to the irreducible characters. However for arbitrary $G$ the orbital integrals and their derivatives have jumps at the singular points. Moreover there is more than one orbital integral to be considered since there is in general more than one conjugacy class of Cartan subgroups. Nevertheless calculations in special cases [1952] seemed to suggest that the Fourier transform of the orbital integral is very closely related to the irreducible characters, and so a limit formula should still represent a primitive version of the Plancherel formula (see the remarks at the beginning of [1957e]). In a series of papers in the 1950's ([1957b]–[1957e] and [1964c]) Harish-Chandra established the limit formula for the orbital integral associated to a *fundamental* Cartan subgroup, namely a Cartan subgroup whose maximal compact subgroup has the largest possible dimension (these form a single conjugacy class, and when $rk(G) = rk(K)$, fundamental means compact). In the limit formula

(L.F.) $$f(e) = (\Omega F_{f,L})(e) \qquad (L \text{ fundamental})$$

$F_{f,L}$ may still have jumps at singular points of $L$; *nevertheless the differential operator* $\Omega$, *which is still the product of roots of a positive system, kills all the jumps, i.e., $\Omega F_{f,L}$ extends continuously to all of $L$, so that the right side makes sense.*

The jumps of $F_{f,L}$ make the Fourier analysis of the relation (L.F.) a very delicate matter (see [1952]). These jumps are in fact the orbital integrals associated to Cartan subgroups which are "less compact" than $L$, suggesting that the jumps are caused by the contributions to $f$ from the representations associated to these less compact Cartan subgroups. Thus when $L$ is compact, one must work with matrix elements of the discrete series representations in order to get rid of the jumps. This can only be done by working with a space of rapidly decreasing functions instead of $C_c^\infty(G)$.

For this purpose Harish-Chandra introduced in [1966b] the Schwartz space $\mathcal{C}(G)$ of $G$. Overcoming many technical difficulties he proved in [1966b] that the theory of orbital integrals can be extended to the Schwartz space, that the limit formula (L.F.) is true for all $f$ in this space, and that the distributions $\Theta_\xi$ constructed earlier are *tempered*, with

$$\langle \Theta_\xi, f\rangle = \int_G \Theta_\xi f\, dx,$$

the integrals converging absolutely. Moreover he proved that *if $f$ is an eigenfunction* *for $\mathfrak{Z}$ in the Schwartz space, the integrals of $f$ on the regular non-elliptic orbits are all zero*; here we recall that an orbit is said to be *elliptic* if it meets a compact Cartan subgroup. If $G$ has no compact Cartan subgroup this means that all the orbital integrals of $f$ are zero; the limit formula would then imply that $f = 0$. We shall see presently that eigenfunctions of $\mathfrak{Z}$ in $L^2(G)$ are automatically in Schwartz space; so this argument already shows that the condition $rk(G) = rk(K)$ is necessary for the existence of the discrete series. If $rk(G) = rk(K)$, the same argument would still allow us to conclude that the orbital integral $F_{f, B}$ ($B \subset K$) has no jumps, i.e., it lies in $C^\infty(B)$. The Fourier transformation of the limit formula coupled with the formula for $\Theta_\xi$ on $B$ would show that the harmonic analysis of $f$ is entirely controlled by the $\Theta_\xi$. This is substantially Harish-Chandra's argument for the completeness of the $\Theta_\xi$.

To round out this picture it is necessary to know a method of determining when a given eigenfunction of $\mathfrak{Z}$ belongs to the Schwartz space. In [1966b] Harish-Chandra developed a powerful theory to describe the asymptotic behaviour of $\tau$-spherical eigenfunctions for $\mathfrak{Z}$ that satisfy the weak inequality. If $f$ is any $\tau$-spherical eigenfunction satisfying the weak inequality, Harish-Chandra was able to associate to $f$, at each point at infinity of the group, a so called *constant term*, that was the dominant term in the asymptotic development of $f$ there. These dominant terms are parametrized by the parabolic subgroups of $G$. If $P = MAN$ is a parabolic subgroup and $f_P$ is the associated constant term, $f_P$ is essentially a function of the same nature as $f$, but lives on the group $MA$; the difference between $f$ and $f_P$ has very rapid decay at infinity "in the direction of $P$." An important consequence of this theory is the result that $f$ lies in $L^2(G, \tau)$ *if and only if all its constant terms are zero; and that in this case, $f$ is in the Schwartz space.*

In order to apply this theory successfully two things have to be done. First, it must be shown that the Fourier coefficients of the $\Theta_\xi$ satisfy the weak inequality. This is technically rather subtle and is done in [1966a]. The second point is to show that all the constant terms of these Fourier coefficients are zero. Harish-Chandra's proof of this made use of a beautiful property of these constant terms, namely that they are *transitive* in a natural sense with respect to the partial ordering among the parabolic subgroups. If $f$ is a $\tau$-spherical eigenfunction built out of the Fourier coefficients of the $\Theta_\xi$, and $P \neq G$ is chosen so that $f_P \neq 0$ and $\dim(M)$ is minimal, the transitivity mentioned above would show that the restrictions of $f_P$ to $M$ are eigenfunctions in the Schwartz space of $M$ (actually linear combinations of such); so $M$ has to have a compact Cartan subgroup, implying that $P$ is associated to a Cartan subgroup of $G$ itself. This Cartan subgroup cannot be compact, and hence as the eigenhomomorphism to which $f$ belongs is regular and is associated to the compact Cartan subgroup, we have a contradiction.

These brief remarks cannot give an adequate idea of the pathbreaking nature of the papers [1965c] [1966b]. The audacity of their conceptions as well as the delicacy and craftsmanship that were needed in executing them, compel one to place them among the greatest works in Fourier analysis.

*The continuous spectrum: the theory of the Eisenstein integral and the explicit Plancherel formula.* With the discrete series completely determined Harish-Chandra turned to the problem of explicitly decomposing the regular representation of $G$. He went much farther than a mere $L^2$-theory and obtained a Fourier analysis of the

Schwartz space of $G$. The results were worked out in three long papers [1975], [1976a], and [1976b]; but the lectures [1970a], [1970c], and [1972] containing the announcements of these and many other results are interesting in themselves, for they give a bird's eye view of his entire approach.

Let $L$ be a Cartan subgroup written (canonically) as $T \cdot A$ where $T$ is compact and $A$ is connected and split over $\mathbb{R}$. Let $\mathcal{P}(A)$ be the finite set of parabolic subgroups $P = MAN$. Since $rk(M) = \dim(T)$, $\hat{M}_d \neq \phi$. For $\eta \in \omega \in \hat{M}_d$ and $\nu \in \mathfrak{a}^*$ ($\mathfrak{a} = \mathrm{Lie}(A)$), $\pi_{\eta,\nu}$ is the unitary representation of $G$ induced by the representation $m \cdot a \cdot n \mapsto e^{i\nu(\log a)} \cdot \eta(m)$ of $P$. The character $\Theta_{\omega,\nu}$ of $\pi_{\eta,\nu}$ is independent of $P$ ($M$ and $A$ are the same for all $P \in \mathcal{P}(A)$). Then (see [1976b])

(a)  $\pi_{\eta,\nu}$ is finitely decomposable; it is irreducible for regular $\nu$.
(b)  Let $W(A)$ be the Weyl group of $A$, i.e., the normalizer of $A$ modulo the centralizer of $A$; then $W(A)$ acts on $\hat{M}_d \times \mathfrak{a}^*$ and

$$\Theta_{s\omega,s\nu} = \Theta_{\omega,\nu} \qquad (s \in W(A)).$$

Actually results much stronger than (a) were proved by Harish-Chandra. For example, if $L$ is fundamental, *all* the $\pi_{\eta,\nu}$ are irreducible ([1976b]); special cases of this were known earlier, going all the way back to the irreducibility of the principal series of $SL(n,\mathbb{C})$ due to Gel'fand and Neumark.

One can now associate to $L$ the subspace $\mathcal{C}_L$ of $\mathcal{C}$ consisting of all functions that are orthogonal to the matrix elements of the representations $\pi_{\eta,\nu}$ attached to the Cartan subgroups that are not conjugate to $L$. $\mathcal{C}_L$ is closed in $\mathcal{C}$ and is invariant under translations. Central to Harish-Chandra's theory is the decomposition (Theorem 12, [1970c])

$$\mathcal{C} = \bigoplus_L \mathcal{C}_L \qquad (*)$$

where the sum is over a complete set of mutually nonconjugate Cartan subgroups; the sum is orthogonal and smooth (i.e., the projections are continuous in $\mathcal{C}$). It will turn out that $\mathcal{C}_L$ is exactly the space of "wave packets" of the matrix elements of the $\pi_{\eta,\nu}$ associated to $L$ (this can be formulated more precisely). Thus (*) is a beautiful and definitive formulation of the heuristic principle that "each Cartan subgroup makes a separate contribution to the Fourier analysis of $G$."

The splitting (*) allows one to work on each $\mathcal{C}_L$ separately. Let $\Omega_L = \hat{M}_d \times \mathfrak{a}^*$ and let us define, for $f \in \mathcal{C}$, its Fourier transform $\hat{f} = (\hat{f}_L)$ by

$$\hat{f}_L(\omega,\nu) = \int_G \Theta_{\omega,\nu}^{\mathrm{conj}} f \, dx \qquad ((\omega,\nu) \in \Omega_L).$$

Then $\hat{f}_L$ is $W(A)$-invariant, and we have an initial formulation of the Plancherel formula ([1972], Theorem 11): there is a unique $W(A)$-invariant continuous function $\mu_L$ on $\Omega_L$ ($\hat{M}_d$ is given the discrete topology) such that for all $f \in \mathcal{C}_L$,

$$f(e) = \sum_{\omega \in \hat{M}_d} d(\omega) \int_{\mathfrak{a}^*} \hat{f}_L(\omega : \nu) \mu_L(\omega : \nu) \, d\nu$$

the series converging absolutely. The function $\mu_L$ is $\geqslant 0$ and has other nice properties. For fixed $\omega$, it extends to a meromorphic function on $\mathfrak{a}_c^*$ that is holomorphic on $\mathfrak{a}^*$ and $>0$ at the regular points of $\mathfrak{a}^*$; moreover it is of at most polynomial growth on $\mathfrak{a}^*$.

The basic question now is of course the explicit calculation of $\mu_L$. Harish-Chandra's method for doing this was based on a remarkable *product formula* for $\mu_L$. To describe this, fix a parabolic subgroup $P \in \mathcal{P}(A)$; then one can associate in a canonical manner a set of pairs $(M_i, Q_i)$ $(1 \leq i \leq m)$ where $M_i$ is a reductive subgroup of $G$ with no split component and $Q_i = MA_iN_i$ is a parabolic subgroup of $M_i$ with $A_i \subset A$ and $\dim(A_i) = 1$. If $\mathfrak{a}_i = \mathrm{Lie}(A_i)$ and $\nu_i = \nu|_{\mathfrak{a}_i}$, then

$$\mu_L = c \prod_{1 \leq i \leq m} \mu_{L,i}, \qquad \mu_{L,i}(\omega : \nu) = \mu_{L,i}(\omega : \nu_i)$$

where $\mu_{L,i}$ is the $\mu$-function associated to $(M_i, Q_i)$ (see [1972], pp. 144–145 for a description of the $(M_i, Q_i)$); $c$ is a constant $>0$ independent of $\omega$ and $\nu$ and can be explicitly evaluated. This reduces the problem of explicit determination of $\mu_L$ to the case $(G, P)$ where $G$ has no split component, $P = MAN$, and $\dim(A) = 1$. If $rk(G) > rk(K)$, then $L$ is fundamental and the limit formula, by Fourier transformation on $L$, leads to the determination of $\mu_L$ as a *polynomial function* on $\hat{L}$; here we must remember that we are operating not on the whole Schwartz space but only on $\mathcal{C}_L$ so that the contributions of the other Cartan subgroups of $G$ do not enter the limit formula. Thus the analysis proceeds as if $G$ has a single conjugacy class of Cartan subgroups (see §24, [1976b]). If $rk(G) = rk(K)$, the treatment follows the model where $rk(G/K) = 1$ (cf. [1966a] and §36, [1976b]).

It is impossible to describe except in bare outline the architecture of this beautiful theory. I shall however try to compensate by discussing two important sources of motivation behind Harish-Chandra's treatment. The first is the harmonic analysis of *spherical functions* [1958a] [1958b]. Many of the crucial features that are typical of the continuous spectrum were discovered first in this context. The second source is the spectral theory of $L^2(G/\Gamma)$, due to Selberg [Se 2], Gel'fand and Piatetsky-Shapiro [G] [GP], and Langlands [L 1], $\Gamma$ being an arithmetic subgroup of $G$. Harish-Chandra's treatment of this in [1968b] is especially useful for this purpose.

The theory of spherical functions is essentially the harmonic analysis of $L^2(G/K)$. The fundamental objects are the elementary spherical functions $\varphi(\nu : x) = \varphi_\nu(x)$ and the transform that is defined by them. In [1958a] and [1958b] Harish-Chandra made the remarkable discovery that the $\varphi_\nu$, regarded as functions on $A$ (of the Iwasawa decomposition), look like a sum of plane waves at infinity on $A$; and that all of these have the same amplitude, in terms of which the Plancherel measure can be determined rather simply. More precisely,

$$d(a)\varphi(\nu : a) \sim \sum_{s \in W(A)} c_s(\nu) e^{is\nu(\log a)},$$

$d$ being a certain homomorphism of $A$ into the positive reals; the amplitudes satisfy

$$|c_s(\nu)| = |c(\nu)| \qquad (c = c_1)$$

and

$$|c(\nu)|^{-2} d\nu$$

is the Plancherel measure. Actually we have the relations

$$c_s(\nu) = c(s\nu)$$

which mirror the *functional equations*

$$\varphi_{s\nu} = \varphi_\nu$$

satisfied by the elementary spherical functions.

The computation of the spherical Plancherel measure thus reduces to determining the function $c$, which is a special case of the famous $c$-function of Harish-Chandra. It turned out that this function is meromorphic on $\mathfrak{a}_c^*$ and that it can be expressed as a product of the analogous $c$-functions associated to the real rank one subgroups of $G$ defined by the root system of $(G, A)$ (the Gindikin–Karpelevič product formula). Since the $c$-function can be explicitly evaluated for groups of real rank one, the story is complete.

For the harmonic analysis of arbitrary (not necessarily spherical) functions in the Schwartz space Harish-Chandra introduced the eigenfunctions that generalized the elementary spherical functions. Let $U$ be a finite dimensional Hilbert space carrying a unitary double representation $\tau$ of $K$. Let $\tau_M = \tau|_M$ and let $\mathfrak{L}$ be the space of $\tau_M$-spherical maps of $M$ into $U$ whose components (with respect to some basis of $U$) are in $\mathcal{C}(M) \cap {}^0L^2(M)$. Then $\dim(\mathfrak{L})$ is finite and for each $\psi \in \mathfrak{L}$, $\nu \in \mathfrak{a}_c^*$, and a given choice of $P$, a parabolic subgroup in $\mathscr{P}(A)$, Harish-Chandra defined an eigenfunction for $\mathfrak{Z}$ in $C^\infty(G, \tau)$, denoted by $E_{\psi, \nu}$ or $E_{P, \psi, \nu}$, called the *Eisenstein integral*. If $P$ is minimal, $U = \mathbf{C}$, $\tau$ is the trivial representation, and $\psi$ is 1, this is just the elementary spherical function (see [1972], p. 133).

The Eisenstein integrals satisfy the weak inequality, and so one can determine their behaviour at infinity on the group by applying the asymptotic theory of tempered eigenfunctions developed in [1966b] [1975]. If $Q$ is a parabolic subgroup in $\mathscr{P}(A)$ which may be different from $P$, then one finds that the constant term of the Eisenstein integral $E_{P, \psi, \nu}$ associated to $Q$ is of the form

$$\sum_{s \in W(A)} (\psi_{s, \nu})(m) e^{is\nu(\log a)}$$

where $\nu$ is a regular element of $\mathfrak{a}^*$ and $\psi_{s, \nu} \in \mathfrak{L}$. The maps

$$c_{Q|P}(s : \nu): \psi \mapsto \psi_{s, \nu}$$

are uniquely determined endomorphisms of $\mathfrak{L}$ and the relation

$$E_{P, \psi, \nu} \underset{Q}{\sim} \sum_{s \in W(A)} c_{Q|P}(s : \nu) \psi e^{is\nu}$$

expresses the fact that $E_{P, \psi, \nu}$ looks like a sum of plane waves at infinity in the direction of $Q$. This is clearly analogous to what happens in the spherical case.

Let us now fix $\omega$ and choose $\tau$ so that the subspace $\mathfrak{L}(\omega)$ of $\mathfrak{L}$ that "transforms according to $\omega$" is nonzero; this is always possible. It can then be shown that for regular $\nu$ $c_{Q|P}(1 : \nu)$ is a bijection of $\mathfrak{L}(\omega)$ onto itself. Let $c_{Q|P, \omega}(1 : \nu)$ be the restriction of $c_{Q|P}(1 : \nu)$ to $\mathfrak{L}(\omega)$. Then the fundamental result expressing the relationship between the Plancherel formula and the asymptotics of the Eisenstein integrals can be stated as follows (see [1976b], §13):

$$\mu_L(\omega:\nu)c_{Q|P,\omega}(1:\nu)^\dagger c_{Q|P,\omega}(1:\nu) = c \cdot id_\omega$$

where $id_\omega$ is the identity endomorphism of $\mathfrak{L}(\omega)$, $c > 0$ is a constant independent of $\omega$ and $\nu$ which can be explicitly calculated. One must also note that as $\mathfrak{L} \subset {}^0L(M, \tau_M)$, $\mathfrak{L}$ has a natural structure as a Hilbert space, and $\dagger$ is the adjoint operation in this structure.

The $c$-functions are closely related to the intertwining operators between the induced representations. The latter have natural product formulae which then lead to product formulae for the $c$-functions and thence to the $\mu$-functions. This is the origin of the product formula for the Plancherel measure.

A central role in Harish-Chandra's treatment of these questions is played by the functional equations of the Eisenstein integrals. The analogy with the theory of automorphic forms is most significant here, and so it may be useful to make a few remarks comparing the Eisenstein integral with the Eisenstein series. The starting point of this comparison is the following characterization of the "discrete part" of the Schwartz space of $G$, obtained by Harish-Chandra: if ${}^0\mathcal{C} = \mathcal{C}(G) \cap {}^0L^2(G)$, then an element $f$ in $\mathcal{C}(G)$ is in ${}^0\mathcal{C}$ if and only if

$$\int_N f(xn)\, dn = 0 \qquad (x \in G)$$

for all parabolic subgroups $P = MAN \neq G$ (Theorems 14 and 15, [1970c]). This is obviously analogous to the condition that defines a cusp form on $G/\Gamma$, so that it is natural to speak of the elements of ${}^0\mathcal{C}$ as *cusp forms* on $G$. Thus $\mathfrak{L}$ may be viewed as the space of $\tau_M$-spherical cusp forms on $M$; and the definition of the Eisenstein integral imitates that of the Eisenstein series with integration over $K$ replacing the summation over $\Gamma$ (cf. [1968b], p. 29). Both the characterization of ${}^0\mathcal{C}$ and the finite dimensionality of $\mathfrak{L}$ are deep lying consequences of the theory of the discrete spectrum.

The functional equations of the Eisenstein integrals involve the comparison of $E_{P,\psi,\nu}$ and $E_{Q,\psi',s\nu}$, and Harish-Chandra obtained them as a consequence of the remarkable principle asserting that *if just one summand is common to the constant terms of two such functions, they must be identical*. This principle follows from what he called the *Maass–Selberg relations* which may be described as follows: if $f$ is a $\tau$-spherical eigenfunction satisfying the weak inequality and suitable additional conditions (which are satisfied by the Eisenstein integrals), the intensities ( = norms in ${}^0L^2(M, \tau_M)$) of the plane waves that occur in its asymptotic forms along the parabolic subgroups in $\mathcal{P}(A)$ are all equal (cf. Theorem 14.1 and the results of §17, especially Lemmas 17.2 and 17.3 of [1976b]). The Maass–Selberg relations originally arose in the theory of automorphic forms ([1968b], Chap. IV, §2); the discovery and proof of their analogues in the present context is a highlight of [1976b].

*Representation theory of $\mathfrak{p}$-adic groups. The Lefschetz principle and the philosophy of cusp forms.* The beginnings of the representation theory of reductive $\mathfrak{p}$-adic groups go back to the late 1950's when Mautner's paper [Mau 1] on the representations of $GL(2, \Omega)$ and $PGL(2, \Omega)$ appeared, $\Omega$ being a local field (also called a $\mathfrak{p}$-adic field), i.e., a locally compact non-discrete field. It developed rapidly in the early 1960's at the hands of Bruhat, Gel'fand–Graev, Satake and others. Harish-Chandra became actively interested in this area in the mid 1960's; his ideas and results since then have proved to be of fundamental importance for the subject. I shall first try to motivate these developments.

Let $k$ be an algebraic number field. Then one can attach to $k$ the local fields $k_v$ which are its completions at the various places $v$, and the adele ring $\mathbf{A}$ which is a "restricted" direct product of the $k_v \cdot \mathbf{A}$ is an abelian ring which is locally compact in a natural topology; $k$ is diagonally imbedded in it as a discrete subring and $\mathbf{A}/k$ is compact. By the 1950's it had become a well understood principle in number theory that many arithmetic questions on $k$ could be studied over each $k_v$, and then by working over $\mathbf{A}$, the local data coming from the various $k_v$ could often be put together to reach global results. This is so in classfield theory, in the theory of simple algebras and Brauer groups, and in the arithmetic theory of quadratic forms [W 2].

After the development of the theory of linear algebraic groups in the 1950's it was realized that the formalism of adeles could be applied to any algebraic group defined over $k$. Let $\mathbf{G}$ be such a group, and for any $k$-algebra $R$, let $\mathbf{G}(R)$ be the group of $R$-points of $\mathbf{G}$. We then have the groups $G_v = \mathbf{G}(k_v)$ for each $v$ and the adele group $G_{\mathbf{A}} = \mathbf{G}(\mathbf{A})$. $G_v$ and $G_{\mathbf{A}}$ are locally compact, separable, and $G_k = \mathbf{G}(k)$ is (diagonally) imbedded naturally as a discrete subgroup of $G_{\mathbf{A}}$. With the availability of the Chevalley theory of reductive algebraic groups and the work of Borel and Harish-Chandra [1962] (see also [Bo 1]) on the structure of $G_{\mathbf{A}}/G_k$ it was natural to hope that one could begin studying the arithmetic aspects of the theory of reductive groups. A succession of closely related ideas and papers, especially those of Weil [W 3] [W 4] [W 5] and Langlands [L 1] [JL] (with Jacquet) [L 2] [L 3] led to an explosive development of this point of view in the 1960's. As a consequence, the supreme importance of the harmonic analysis of $L^2(G_{\mathbf{A}}/G_k)$ for arithmetic was recognized. For example, if $\mathbf{G} = GL(1)$, $G_{\mathbf{A}} = \mathbb{I}$ the idele group and $G_{\mathbf{A}}/G_k = \mathbb{I}/k^{\times}$ is the idele class group; a character of $\mathbb{I}/k^{\times}$ is essentially a Grössencharakter of Hecke. Tate's work (cf. §1) in 1947 had derived the Hecke theory of the associated $L$-series from the Poisson formula for $\mathbf{A}/k$. Also one knows from classfield theory that the characters of $\mathbb{I}/k^{\times}$ of finite order define (through Artin's reciprocity law) cyclic extensions of $k$. If $\mathbf{G} = GL(2)$, the modular forms on the Poincaré half plane which are eigenfunctions for the Hecke operators define (under suitable spectral conditions) irreducible representations of $G_{\mathbf{A}}$ that occur in $L^2(G_{\mathbf{A}}/G_k)$, showing that in the notion of an irreducible representation occurring in $L^2(G_{\mathbf{A}}/G_k)$ we have a far reaching generalization of the classical automorphic forms and Eisenstein series. The work of [L 3] suggested that for any integer $n \geqslant 1$ and any $n$-dimensional representation of the Galois group $\mathrm{Gal}(\bar{k}/k)$ ($\bar{k} = $ an algebraic closure of $k$) there would be associated an irreducible unitary representation occurring in $L^2(G_{\mathbf{A}}/G_k)$.

The irreducible unitary representations of $G_{\mathbf{A}}$ are tensor products of irreducible unitary representations of the local groups $G_v$. It is therefore natural to try to understand the representation theory of the groups $G_v$ as a first step. Of course these representations would reflect the arithmetic of the local fields $k_v$. For example, following [L 3], one would expect any $n$-dimensional representations of $\mathrm{Gal}(\bar{k}_v/k_v)$ to determine an irreducible unitary representation of $GL(n, k_v)$. This is a very rough sketch of some of the ideas that led to the great interest in the 1960's for constructing a theory of representations of groups $G = \mathbf{G}(\Omega)$ where $\Omega$ is a p-adic field and $\mathbf{G}$ is a reductive group defined over $\Omega$.

From the very beginning Harish-Chandra emphasized the striking similarities between the theory for real groups and the newly developing theory for the p-adic groups [1968a]. In [1970d] he formulated this as a broad heuristic principle, which he called the *Lefschetz principle*, asserting that whatever is true for real groups is also true for the p-adic ones. Its thrust was to use the rich theory of Fourier analysis on real reductive groups as a guide in the search for discovering the basic facts to be proved in the representation theory of the p-adic groups. In [1970d], as a first illustration of this principle, Harish-Chandra explored the theory of characters and orbital integrals on $G = \mathbf{G}(\Omega)$ as above. These investigations were completed in [1978] where he showed, assuming char $\Omega = 0$, that the theory of characters of $G$ resembles the theory for real groups to a remarkably close extent. Thus, characters of irreducible admissible representations are locally summable functions on $G$ which are locally constant on $G'$; and they can be obtained by exponentiating the Fourier transforms of the invariant measures on the (nilpotent) orbits in the Lie algebra. One should recall that in the real case such theorems came out of the profound study of the differential equations satisfied by the characters. In the present case there are no differential equations, and so it is quite surprising that such results are true. Harish-Chandra's work uses a beautiful theorem of Howe on invariant distributions on the Lie algebra of $G$ as a substitute for the differential equations.

The philosophy of cusp forms, which he had already introduced in [1970b] as the unifying principle for Fourier analysis on reductive groups is an even more striking illustration of the Lefschetz principle. In [1970b] he had illustrated it for reductive groups over finite fields, and it was the driving force in his work on the Fourier analysis of the Schwartz space of a real reductive group. He showed that the harmonic analysis of these groups can be worked out quite explicitly, based on the concept of cusp forms and parabolic induction. In [1973] and [1977a] he succeeded in pushing this method through for reductive p-adic groups. The cusp forms are once again (essentially) the matrix elements of the discrete series and this method led him to the theory of the Eisenstein integral, in particular, to the Maass–Selberg relations and Plancherel formula, in virtually complete analogy with the real case.

For groups over p-adic fields the discrete series is still not completely determined. Nevertheless, by taking Fourier analysis of reductive groups to this stage, Harish-Chandra has shown that its inner structure can be understood in terms of a small number of simple and yet general principles. This is a truly extraordinary accomplishment.


§3. *Orbital Integrals and Limit Formula.* I now wish to supplement this general survey with a little more detailed discussion of selected parts of Harish-Chandra's work. I shall begin with the theory of orbital integrals and the limit formula. The bulk of this work is contained in [1957b]–[1957e] and [1964c]; the extension of the theory to the Schwartz space of $G$ is treated in [1965b], [1966a], and [1966b].

We take $G$ to be the real form of a complex simply connected semi simple group. Let $L \subset G$ be a Cartan subgroup with $\mathfrak{h} = \mathrm{Lie}(L)$; $P$, a positive system of roots of $(\mathfrak{g}_c, \mathfrak{h}_c)$; and $\Delta$, the Weyl denominator on $L$. Define $\varpi$ and $\pi$ by $\varpi = \prod_{\alpha \in P} H_\alpha$, $\pi = \prod_{\alpha \in P} \alpha$. The orbital integral associated to $L$ is then defined for $f \in C_c^\infty(G)$, $h \in L' = L \cap G'$ by

$$F_{f,L}(h) = \varepsilon_R(h)\Delta(h)\int_{G/L = G^*} f(xhx^{-1})\, dx^* \tag{1}$$

where $\varepsilon_R(h)$ is a locally constant sign factor (§22, [1965b]). The analogous definition on $\mathfrak{g}$ is (see §5, [1964c])

$$\psi_{f,\mathfrak{h}}(H) = \varepsilon_R(H)\pi(H)\int_{G^*} f(H^x)\, dx^* \qquad (f \in C_c^\infty(\mathfrak{g})). \tag{2}$$

These are well defined because the regular orbits are closed, and it is easy to show that $F_{f,L}$ and $\psi_{f,\mathfrak{h}}$ are $C^\infty$ on $L'$ and $\mathfrak{h}'$, respectively.

The key to the entire theory is the system of differential equations satisfied by $F_f$ and $\psi_f$:

$$F_{zf,L} = \mu_{\mathfrak{g}/\mathfrak{h}}(z)F_{f,L} \qquad (z \in \mathfrak{Z}) \tag{3G}$$

$$\psi_{\partial(p)f,\mathfrak{h}} = \partial(p_\mathfrak{h})\psi_{f,\mathfrak{h}} \qquad (p \in I(\mathfrak{g})). \tag{3g}$$

Here, $I(\mathfrak{g})$ is the algebra of $G$-invariant elements in $S(\mathfrak{g}_c)$ and $p_\mathfrak{h}$ is the "restriction" of $p$ to $\mathfrak{h}$ while $\mu_{\mathfrak{g}/\mathfrak{h}}$ is as in §12 of [1965b]. To illustrate this let us examine how they control the growth of the orbital integrals near a singular point. Consider for instance $\psi_{f,\mathfrak{h}}$. The equations (3g) and the formula for integration on $\mathfrak{g}$ give the estimate

$$\int_{\mathfrak{h}'} |\pi||\partial(p_\mathfrak{h})\psi_{f,\mathfrak{h}}|\, d\mathfrak{h} \leqslant \int_{(\mathfrak{h}')^G} |\partial(p)f|\, d\mathfrak{g}. \tag{4}$$

From this one can conclude that all derivatives of $\psi_{f,\mathfrak{h}}$ are locally bounded, and in fact even more. The question is the often encountered one of obtaining point estimates in terms of $L^1$-norms of a function and its derivatives; but there is a technical complication because of the weight function $|\pi|$ which vanishes on $\mathfrak{h} \setminus \mathfrak{h}'$. In [1965b] (cf. Theorem 3) Harish-Chandra obtained a global estimate of the Sobolev type that can be used in the present situation. This can be formulated as follows (see [V 3], Proposition 7, p. 157). Let $V$ be a real finite dimensional vector space; $w = \prod_{1 \leqslant j \leqslant q} |\lambda_j|^{a_j}$ where $a_j > 0$, $\lambda_j \in V_c^*$, and $V' =$ the set where $w \neq 0$; let $S_0$ be a subalgebra of $S = S(V_c)$ such that $S$ is a finite module over $S_0$. Write $\mathfrak{U}^1(S_0 : w)$ (resp. $\mathfrak{U}^\infty$) for the Frechet space of all $g \in C^\infty(V')$ such that $\int |\partial(u)g|w\, dV < \infty$ (resp. $\sup|\partial(u)g| < \infty$) for all $u \in S_0$ (resp. $u \in S$). Then

$$\mathfrak{U}^1(S_0 : w) \subset \mathfrak{U}^\infty$$

and the inclusion is continuous. Since $\psi_{f,\mathfrak{h}} \in \mathfrak{U}^1(I(\mathfrak{h}):|\pi|)$ by (4), we must have $\psi_{f,\mathfrak{h}} \in \mathfrak{U}^\infty$. Actually, replacing $f$ by $qf$ where $q$ is an invariant polynomial on $\mathfrak{g}$, we find the much stronger result that $\psi_{f,\mathfrak{h}}$ lies in the Schwartz space $\mathcal{C}(\mathfrak{h}')$ of $\mathfrak{h}'$ and that the map $f \mapsto \psi_{f,\mathfrak{h}}$ is continuous with respect to the collection of seminorms of the form

$$f \mapsto \int_{(\mathfrak{h}')^G} |\partial(p)(qf)| \, d\mathfrak{g} \qquad (f \in C_c^\infty(\mathfrak{g})).$$

Since *these are all continuous on* $\mathcal{C}(\mathfrak{g})$, the orbital integral extends to a continuous map $\mathcal{C}(\mathfrak{g}) \to \mathcal{C}(\mathfrak{h}')$; in particular, the invariant measures on the orbits are all tempered. The original proof in [1957c] was essentially a variant of this type of argument.

The same argument works (up to a point) on the group also. The crucial case is when $L = B$ is compact and let us make this assumption. Let $\mathfrak{U}^\infty$ now be the space of all $g \in C^\infty(B')$ such that $\sup|ug| < \infty$ for all $u \in S(\mathfrak{b}_c)$ ($\mathfrak{b} = \mathrm{Lie}(B)$), regarded as a Frechet space in the usual way. Then the above argument will show that $F_{f,B} \in \mathfrak{U}^\infty$ and that $f \mapsto F_{f,B}$ will be continuous with respect to the topology induced on $C_c^\infty(G)$ by the seminorms

$$\nu_z : f \mapsto \int_{G_e} |zf| \, dx \qquad (z \in \mathfrak{Z})$$

where $G_e = (B')^G$ is the elliptic set. If one can show that

$$V(f) = \int_{G_e} |f| \, dx < \infty$$

for any $f \in \mathcal{C}(G)$ and that $V$ is a continuous seminorm on $\mathcal{C}(G)$, the continuity of the $\nu_z$ will follow, leading to the extension of the theory of orbital integrals to $\mathcal{C}(G)$. Let $\varphi_B$ be the characteristic function of the set $G_e$ and let

$$b(x) = \int_K \varphi_B(xk) \, dk \qquad (x \in G).$$

Since

$$\int_{G_e} |f| \, dx \leqslant \int_G |f| b \, dx$$

it is enough to prove that $b$ satisfies the weak inequality.

Harish-Chandra does this in [1966a] (Theorem 5). It is obvious that $b(x)$ is the measure of the set of all $k$ in $K$ for which $xk$ is elliptic. One can use the boundedness of the finite dimensional characters on $G_e$ to conclude that $b(x) \to 0$ when $x$ goes to infinity on $G$. For the sharper result needed here, one needs class functions with good growth properties on *all* Cartan subgroups. Harish-Chandra's proof depends on the existence of a locally summable class function $\Theta$ such that

(i)    $\Theta$ is a nonzero constant on $G_e$.

(ii)   If $L = T \cdot A$ is any Cartan subgroup, and $\mathfrak{z}$ is the centralizer of $A$ in $\mathfrak{g}$,

$$|\Theta(x)| \leqslant \mathrm{const.} |\det(1 - \mathrm{Ad}(x^{-1}))_{\mathfrak{g}/\mathfrak{z}}|^{-1/2} \qquad (x \in L').$$

(iii)   $\displaystyle\int_K \Theta(xk) \, dk = 0 \qquad (x \in G).$

Using his theory of the distributions $\Theta_\xi (\xi \in \hat{B})$ he shows that $\Theta = \sum_{s \in W} \varepsilon(s) \Theta_{s\rho}$ ($W$ = the Weyl group of $(\mathfrak{g}_c, \mathfrak{b}_c)$ and $\rho$ has the usual meaning), will have the required properties. $\Theta$ is essentially the character of the sum of the discrete series representations which have the same infinitesimal character as the trivial representation. It is interesting to observe that the same argument works in the $\mathfrak{p}$-adic case also; $\Theta$ may then be taken as the Steinberg character ([1973]).

The analysis of the jumps of the orbital integrals at the singular points comes down, through general arguments, to the case when the singular point is semi regular, and hence, to a calculation in $\mathfrak{sl}(2,\mathbb{R})$. Here there are two Cartan subalgebras $\mathfrak{h} = \mathbb{R} \cdot H$, $\mathfrak{b} = \mathbb{R} \cdot (X - Y)$ where

$$H = \begin{pmatrix} 1 & 0 \\ 0 & -1 \end{pmatrix}, \qquad X = \begin{pmatrix} 0 & 1 \\ 0 & 0 \end{pmatrix}, \qquad Y = \begin{pmatrix} 0 & 0 \\ 1 & 0 \end{pmatrix}.$$

The orbit $E_\theta$ of $\theta(X - Y)$ is a two-sheeted elliptic hyperboloid, and the orbit $H_t$ of $tH$ is a single sheeted hyperbolic hyperboloid; these are separated by $C$, the cone of nilpotents (light cone!), composed of two halves $C_\pm$ and the vertex $0$. $\psi_{f,\mathfrak{h}}$ is $C^\infty$ while $\psi_{f,\mathfrak{b}}$ is not. To see why, let $\theta \to 0+$, $t \to 0$. Then the $E_{\pm\theta}$ tend to $C_\pm$ while $H_t$ tends to $C$ (draw a figure). Thus

$$\lim_{\theta \to \pm 0} \psi_{f,\mathfrak{b}}(\pm\theta(X - Y)) = \pm\int_{C_\pm} f \, dC$$

$$\lim_{t \to 0+} \psi_{f,\mathfrak{b}}(tH) = \int_C f \, dC$$

giving the "jump formula"

$$\psi_{f,\mathfrak{b}}(0+) - \psi_{f,\mathfrak{b}}(0-) = \psi_{f,\mathfrak{h}}(0). \tag{5}$$

Generalized to arbitrary $\mathfrak{g}$, this is the basic relation that links the orbital integrals associated to different Cartan subalgebras ([1964c] Theorem 2); there is of course a corresponding version on the group. It shows that jumps arise only when *crossing root hyperplanes corresponding to singular imaginary roots*, and that for derivatives that are skew with respect to these roots, the jumps are $0$ (Theorem 1, [1964c]). The continuity of $\partial(\varpi)\psi_{f,\mathfrak{h}}$ and $\varpi F_{f,L}$ follow from this, as well as the vanishing of their values at $0$ and $e$ respectively if there are real roots, i.e., if $L$ is not fundamental. The formula (5) or rather its generalization, namely Theorem 2 of [1964c], implies that if $F_{f,\bar{L}} = 0$ for Cartan subgroups $\bar{L}$ "more split" than $L$, then $F_{f,L}$ is of class $C^\infty$ on $L$. We have already seen in §2 how this principle was decisive in the determination of the discrete spectrum. In particular, if $G$ is complex, more generally, if $L$ is "Iwasawa", $F_{f,L}$ is of class $C^\infty$; of course one can prove this by a direct evaluation of $F_{f,L}(h)$:

$$F_{f,L}(h) = d_P(h) \iint_{K \times N} f(khnk^{-1}) \, dk \, dn$$

$$\left( d_P(h) = |\det(\mathrm{Ad}(h)_\mathfrak{n})|^{1/2} \right).$$

Let us now turn to the limit formula. It is enough to prove it on the Lie algebra. Write

$$T(f) = \left( \partial(\varpi)\psi_{f,\mathfrak{h}} \right)(0) \qquad (f \in \mathcal{C}(\mathfrak{g})). \tag{6}$$

$T$ is a *tempered* invariant distribution on $\mathfrak{g}$ and the limit formula is the assertion that $T = c\delta$ where $c = c(\mathfrak{h})$ is a real constant which, for fundamental $\mathfrak{h}$, is $(-1)^q$ times a positive number, $q$ being the integer $1/2\{\dim(G/K) - rk(G) + rk(K)\}$.

If $\mathfrak{g}$ is complex or if it has a single conjugacy class of Cartan subalgebras, one can show that (cf. [1975], §§35–36) that

$$\psi_{\hat{f}, \mathfrak{h}} = c\hat{\psi}_{f, \mathfrak{h}} \qquad (f \in \mathcal{C}(\mathfrak{g})) \tag{7}$$

where $c$ is a constant $\neq 0$ (recall that $\psi_{f, \mathfrak{h}}$ is smooth and so lies in $\mathcal{C}(\mathfrak{h})$). The integration formula

$$\int_{\mathfrak{g}} u \, d\mathfrak{g} = \text{const.} \int_{\mathfrak{h}} \pi \psi_{u, \mathfrak{h}} \, d\mathfrak{h} \qquad (u \in \mathcal{C}(\mathfrak{g})),$$

on Fourier transformation, gives the limit formula ([1954c]). For arbitrary $\mathfrak{g}$ the relation (7) has to be suitably modified, and this requires a deeper study of the Fourier transforms $\hat{\mu}_H$ of the orbital measures $\mu_H \colon f \mapsto \int_{G_*} f(H^x) \, dx^*$. Since the invariant polynomials are constant on the orbits, the $\hat{\mu}_H$ are invariant eigendistributions:

$$\partial(p)\hat{\mu}_H = \tilde{p}(iH)\hat{\mu}_H \qquad (p \in I(\mathfrak{g})) \tag{8}$$

where $\tilde{p}$ is the polynomial corresponding to $p$. From the theory of invariant eigendistributions (see §4 below) it follows that $\hat{\mu}_H$ is an analytic function on $\mathfrak{g}'$, say $\hat{\mu}(H \colon \cdot)$, which can be studied in detail. If $\bar{\mathfrak{h}}$ is a Cartan subalgebra, $\bar{\mathfrak{h}}^+$ a connected component of $\bar{\mathfrak{h}}'$, $\overline{W}$ the Weyl group of $(\mathfrak{g}_c, \bar{\mathfrak{h}}_c)$, and $y$ is an element of the complex adjoint group taking $\mathfrak{h}_c$ to $\bar{\mathfrak{h}}_c$,

$$\pi(H)\bar{\pi}(\overline{H})\hat{\mu}(H \colon \overline{H}) = \sum_{s \in \overline{W}} c_s(H) e^{i\langle sH^y, \overline{H}\rangle} \tag{9}$$

for $H \in \mathfrak{h}'$, $\overline{H} \in \bar{\mathfrak{h}}^+$; the equations (3g) now imply that *the $c_s$ are locally constant on* $\mathfrak{h}'$ (Lemma 24, [1957c]). With this proviso, (9) is a generalization of (7) and leads directly to the conclusion that $\hat{T}$ is locally constant on $\mathfrak{g}'$. But $\text{supp}(T) \subset$ the set of nilpotents of $\mathfrak{g}$ so that $\hat{T}$ is $I(\mathfrak{g})$-finite. Since an invariant $I(\mathfrak{g})$-finite distribution on $\mathfrak{g}$, which is locally constant on $\mathfrak{g}'$, is constant on $\mathfrak{g}$, we have $\hat{T} = $ a constant. This argument (see §10, [1965a]) makes use of the regularity theorem on the Lie algebra; the earlier proof in [1957d] was more direct but was also more involved.

One should think of the $\hat{\mu}_H (H \in \mathfrak{h}')$ as the analogues of the irreducible characters of $G$ associated to $L$ by the inducing process. For example one can prove that if $\bar{\mathfrak{h}}$ is a Cartan subalgebra, $\hat{\mu}_H$ is $0$ on $(\bar{\mathfrak{h}}')^G$ if $\bar{\mathfrak{h}}$ is not conjugate to a Cartan subalgebra of the centralizer of the split component $\mathfrak{h}_\mathbf{R}$ of $\mathfrak{h}$; in particular this will be the case if $\dim(\bar{\mathfrak{h}}_\mathbf{R}) < \dim(\mathfrak{h}_\mathbf{R})$, i.e., $\bar{\mathfrak{h}}$ is "more compact" than $\mathfrak{h}$.

It remains to look at the case when $rk(G) = rk(K)$ and $\mathfrak{h} \subset \mathfrak{k}$. Harish-Chandra uses a fundamental solution $\Xi$, due to de Rham, of $\partial(\Omega)^{[n/2]}$ ($\Omega = $ the Casimir element, $n = \dim(\mathfrak{g})$), whose restriction to $\mathfrak{h}$ is a fundamental solution to $(-1)^q \partial(\Omega_\mathfrak{h})^{l/2}$, $l = \dim(\mathfrak{h})$. He now chooses $f \in \mathcal{C}(\mathfrak{g})$ such that

$$f(0) = 1, \qquad \psi_{f, \bar{\mathfrak{h}}} = 0 \quad \text{if } \bar{\mathfrak{h}} \text{ is not conjugate to } \mathfrak{h}. \tag{10}$$

Then in view of earlier remarks on the jumps of $\psi_f$, $\psi_{f, \mathfrak{h}}$ will be smooth and so will lie in $\mathcal{C}(\mathfrak{h})$; and the relation

$$1 = \int_{\mathfrak{g}} \Xi \cdot \left( \partial(\Omega)^{[n/2]} f \right) d\mathfrak{g}$$

will reduce to an integral on $\mathfrak{h}$ that gives the limit formula ([1964c]).

Thus everything comes down to the choice of $f$. Harish-Chandra takes $f = \hat{g}$ where $g \in C_c^\infty((\mathfrak{h}')^G)$ and $\int_{\mathfrak{g}} g \, d\mathfrak{g} = 1$; by our earlier remarks on $\hat{\mu}_H$, $\psi_{f,\bar{\mathfrak{h}}} = 0$ for $\bar{\mathfrak{h}}$ not conjugate to $\mathfrak{h}$.

The functions such as $f$ in the above proof are truly remarkable because their Fourier analysis is controlled completely by $\mathfrak{h}$. Their analogues on the group are clearly cusp forms. Since we do not have access to these until a substantial amount of Fourier analysis on the group is done, it is now clear why the proof of the limit formula takes place on the Lie algebra. Nevertheless the reader should note how closely this proof resembles the proof, on the group, of the separation of the discrete spectrum. I think the essential originality of Harish-Chandra's treatment of the limit formula was to have conceived of the transition to $\mathfrak{g}$, and to have realized that the theory of Fourier transforms on the Lie algebra could provide him with the "additive" analogues of the cusp forms that held the key to the solution of the problem.

§4. *The Regularity Theorem.* Announced in [1963], this is the central result in the long series of papers [1964a]–[1965b] on the structure of invariant eigendistributions. Harish-Chandra's approach was to prove it first on $\mathfrak{g}$ and then carry the result over to $G$. For mostly technical reasons he worked on the group with invariant distributions that were $\mathfrak{Z}$-finite (resp. $I(\mathfrak{g})$-finite on $\mathfrak{g}$). The essential problem is to investigate the structure of invariant distributions around semi simple points.

His main technique for doing this was the *method of descent*. Let $M$ be a manifold on which a Lie group $L$ acts, $m_0 \in M$, and let $E$ be a submanifold of $M$ which contains $m_0$ and is transversal to the orbit of $m_0$ at $m_0$. Then the descent mechanism is a method of reducing questions involving invariant distributions and differential operators defined on $M$ around $m_0$, to similar questions on $E$ around $m_0$. The classical example is the reduction of rotationally symmetric problems to problems on a half-line.

To set up the descent machinery (see [1964a]) we need volume elements $dE$, $dM$ on $E$ and $M$ respectively, with $dM$ invariant under $L$. Assume $\pi: L \times E \to M$ is submersive and let $M' = \pi(L \times E)$. Integrating on the fibers of $\pi$ with respect to the volume element $dL \, dE/dM$ will then give a "pull-back" $T \mapsto \beta_T$, taking distributions $T$ on $M'$ to distributions $\beta_T$ on $L \times E$. If $T$ is invariant, $\beta_T$ will be of the form $1 \otimes \sigma_T$, and $\sigma_T$ will be the *transfer* of $T$; the map $T \mapsto \sigma_T$ is injective. If $F$ is any $L$-invariant function and $F_E = F|_E$, $F$ is locally integrable on $M'$ if and only if $F_E$ is so on $E$; and $\sigma_F = F_E$. If $L$ is unimodular and $L_0$ is a closed subgroup of $L$ such that $E$ is $L_0$-stable and $dE$ is $L_0$-invariant, $\sigma_T$ is $L_0$-invariant. Suppose now that $D$ is an analytic invariant differential operator on $M'$. If $E$ is sufficiently small, we can choose an analytic differential operator $D'$ on $L \times E$, invariant under the action $x, (x', m) \mapsto (xx', m)$ of $L$, such that $D'$ and $D$ are $\pi$-related. The analytic differential operator

$\Delta(D)$ on $E$, obtained by retaining only the terms in $D'$ not involving differentiations with respect to $L$, is called a *radial component of $D$*. It is in general not unique because $D'$ is not unique; but it always satisfies

$$\sigma_{DT} = \Delta(D)\sigma_T. \tag{1}$$

If, however, $E$ is a local section at $m_0$ for the action of $L$, then $\Delta(D)$ is uniquely determined by the requirement

$$(DF)_E = \Delta(D)F_E \qquad \left(F \in (C^\infty(M'))^L\right) \tag{2}$$

and $D \mapsto \Delta(D)$ will be a homomorphism.

In the present context it is necessary to compute explicitly the radial components of $\mathfrak{Z}$ and $\partial(I(\mathfrak{g}))$ for the actions of $G$ on itself and on $\mathfrak{g}$, around semisimple points. Let a (resp. $X$) be a semisimple element of $G$ (resp. $\mathfrak{g}$); then its centralizer $Z$(resp. $\mathfrak{z}$) in $G$ (resp. $\mathfrak{g}$) is transversal to the orbit of a (resp. $X$). The radial components and distributions on $Z$ (resp. $\mathfrak{z}$) are then defined around a (resp. $X$) and invariant under $\Xi = Z^0$. If $a$ or $X$ is not central, $Z(X)$ will have lower dimension than $G(\mathfrak{g})$ and so induction on dimension is possible.

Harish-Chandra's main calculations may be summarized as follows.

(i) Here $G$ acts on $\mathfrak{g}$, $X$ is regular, and $\mathfrak{z} = \mathfrak{h}$ is a Cartan subalgebra; $M = \mathfrak{g}'$. We take $E = \mathfrak{h}'$ and find ([1957b] Lemma 8)

$$\Delta(\partial(p)) = \pi^{-1} \circ \partial(p_\mathfrak{h}) \circ \pi \qquad (p \in I(\mathfrak{g})). \tag{3}$$

Let $\mathcal{D}(\mathfrak{g})$ (resp. $\mathcal{D}(\mathfrak{h})$) be the algebra of differential operators with polynomial coefficients on $\mathfrak{g}$ (resp. $\mathfrak{h}$). In [1964b] Harish-Chandra proved, as a partial extension of (3), that for $D \in \mathcal{D}(\mathfrak{g})$, $\delta(D) = \pi \circ \Delta(D) \circ \pi^{-1}$ is in $\mathcal{D}(\mathfrak{h})$, invariant under the Weyl group, and

$$\delta(\hat{D}) = \widehat{\delta(D)} \quad (\wedge = \text{Fourier transform}). \tag{4}$$

(ii) $G$ acts on itself by conjugacy, $E = L'$ where $L$ is a Cartan subgroup, and $M = G'$. Then the radial components are unique, and ([1956c] Theorem 2)

$$\Delta(z) = \Delta^{-1} \circ \mu_{\mathfrak{g}/\mathfrak{h}}(z) \circ \Delta \qquad (z \in \mathfrak{Z}) \tag{5}$$

where $\Delta = \Delta_L$ is the Weyl denominator and $\mathfrak{h} = \text{Lie}(L)$.

(iii) $G$ acts on $\mathfrak{g}$; $X \in \mathfrak{g}$ semisimple. Let $\zeta(Z) = \det(\text{ad } Z)_{\mathfrak{g}/\mathfrak{z}}$; $\mathfrak{z}'$ is the subset of $\mathfrak{z}$ where $\zeta$ is not zero. The radial components are not unique; nevertheless, if $\Omega$ is an invariant open neighbourhood of $X$ in $\mathfrak{g}$ and $T$ an invariant distribution on $\Omega$,

$$\Delta(\partial(p))\sigma_T = |\zeta|^{-1/2} \circ \partial(p_\mathfrak{z}) \circ |\zeta|^{1/2}\sigma_T \qquad (p \in I(\mathfrak{g})) \tag{6}$$

thus generalizing (3) ([1964b] Theorem 2). In particular, $T$ is $I(\mathfrak{g})$-finite on $(\Omega \cap \mathfrak{z}')^G$ if and only if $|\zeta|^{1/2}\sigma_T$ is $I(\mathfrak{z})$-finite on $\Omega \cap \mathfrak{z}'$.

(iv) $G$ acts on itself; $a \in G$ is semisimple. Let $\nu(y) = \det\left(\text{Ad}((ay))^{-1} - 1\right)_{\mathfrak{g}/\mathfrak{z}}$ where $\mathfrak{z} = \text{Lie}(Z)$. Let $\Xi'$ be the subset of $\Xi$ where $\nu$ is nonzero. If $\Omega$ is a completely invariant (cf. [1963]) open neighbourhood of $a$ in $G$,

$$\Delta(z)\theta = \left(|\nu|^{-1/2} \circ \mu_{\mathfrak{g}/\mathfrak{z}}(z) \circ |\nu|^{1/2}\right)\theta \tag{7}$$

for all $\Xi$-invariant distributions $\theta$ on $\Omega_\Xi = \Xi' \cap (a^{-1}\Omega)$. In particular, an invariant distribution $T$ on $(a\Omega_{\Xi})^G$ is $\mathfrak{Z}$-finite if and only if $|\nu|^{1/2}\sigma_T$ is $\mathfrak{Z}_a$-finite on $\Omega_{\Xi}$; $\mathfrak{Z}_a = $ center of $U(\mathfrak{z})$ ([1965b], Lemma 22).

(v) Transfer from $\mathfrak{g}$ to $G$ via the exponential map. Let $z \mapsto p_z$ be the isomorphism (§12, [1965b]) of $\mathfrak{Z}$ with $I(\mathfrak{g})$ such that $\mu_{\mathfrak{g}/\mathfrak{h}}(z) = (p_z)_{\mathfrak{h}}$ for all Cartan subalgebras $\mathfrak{h} \subset \mathfrak{g}$. Define the invariant function $\xi = \xi_{\mathfrak{g}}$ on $\mathfrak{g}$ by

$$\xi(X) = |\det\{(e^{\operatorname{ad} X/2} - e^{-\operatorname{ad} X/2})/\operatorname{ad} X\}|^{1/2} \qquad (X \in \mathfrak{g}).$$

Then $dx = \xi^2 dX$ ($x = \exp X$) and $|\pi|\xi = |\Delta \circ \exp|$ on any Cartan subalgebra $\mathfrak{h}$. Let $U$ be a completely invariant open neighbourhood of $0$ in $\mathfrak{g}$ such that $U_G = \exp U$ is completely invariant open in $G$ and $\exp: U \to U_G$ is a diffeomorphism. We then have a $G$-equivariant pull back isomorphism $T \mapsto \tilde{T}$ of distributions from $U_G$ to $U$ such that at the level of locally summable functions

$$\tilde{F} = (F \circ \exp)\xi. \tag{8}$$

The main formula here is

$$(zT)^{\tilde{}} = \partial(p_z)\tilde{T} \tag{9}$$

for all invariant distributions $T$ (§14, [1965b]). Formula (9) implies that $T$ is $\mathfrak{Z}$-finite if and only if $\tilde{T}$ is $I(\mathfrak{g})$-finite.

The formulae (3) and (6) on the Lie algebra are proved essentially by direct calculation. But such a method cannot handle (7) or (9). It is easy in both cases to determine how the transferred differential operators act on smooth invariant functions; but their actions on invariant distributions are quite difficult to determine directly. Harish-Chandra reduced everything to calculations with invariant *smooth* functions by first proving the following remarkably general and beautiful theorem (Theorem 1, [1965b]): if $D$ is an invariant analytic differential operator on $\Omega$ such that $Df = 0$ for all invariant $f \in C^\infty(\Omega)$, then $DT = 0$ for all invariant distributions $T$ on $\Omega$. Its proof comes down, via descent, to an analogous result on $\mathfrak{g}$; the proof of this however uses the regularity theorem on $\mathfrak{g}$ ([1965a] Theorems 4 and 5; for an illuminating sketch of the argument see [1963]).

Fix a completely invariant open set $\Omega$ of $\mathfrak{g}$ and an invariant $I(\mathfrak{g})$-finite distribution $T$ on $\Omega$. Using (3) one finds that $T$ is an analytic function $F$ on $\Omega' = \Omega \cap \mathfrak{g}'$; $F$ as well as $\partial(p)F(p \in I(\mathfrak{g}))$ are locally summable on $\Omega$. The first step in proving the regularity theorem is to show that $T_F$, the distribution defined by $F$ on $\Omega$, is $I(\mathfrak{g})$-finite. If this is done, $\mathfrak{U} = T - T_F$ is invariant, $I(\mathfrak{g})$-finite, and supported on the singular set; the second step is to show that $0$ is the only such distribution.

For the first step it is a question of proving that $\partial(p)T_F = T_{\partial(p)F}$ for all $p \in I(\mathfrak{g})$. A formal argument (cf. [1957b], p. 99) reduces this to $p = \omega$, the Casimir operator. Let $J_0 = \partial(\omega)T_F - T_{\partial(\omega)F}$. $J_0$ can be expressed rather explicitly in terms of boundary values of the orbital integrals ([1965a] p. 15). Theorem 4 of [1964c] guarantees that such a distribution is zero on $\Omega$ if it is zero around the semi regular points of the noncompact type in $\Omega$. For $J_0$, this condition is equivalent, by descent, to the regularity theorem on $\mathfrak{s}\mathfrak{l}(2,\mathbb{R})$, which can be verified by direct calculation.

The second step, by (6) and induction on $\dim(\mathfrak{g})$, reduces to showing that $0$ is the only invariant $I(\mathfrak{g})$-finite distribution on $\mathfrak{g}$ supported by the set $\mathfrak{N}$ of nilpotents. The key to Harish-Chandra's proof of this (Theorems 4 and 5, [1964a]) is to view the space $\mathfrak{J}$ of invariant distributions supported by $\mathfrak{N}$ as a module for a Lie subalgebra

$\mathfrak{D} \subset \mathfrak{D}(\mathfrak{g})$ which is isomorphic to $\mathfrak{sl}(2,\mathbb{C})$ with a standard basis $\{H', X', Y'\}$ where $2Y' = \partial(\omega)$, $2X' = $ multiplication by $\tilde{\omega}$, and $H' = D + \frac{1}{2}\dim(\mathfrak{g})$ where $D$ is the Euler vector field. Using descent theory and the finiteness of $G \setminus \mathfrak{N}$ it is proved that $\mathfrak{T}$ is a weight module for $\mathfrak{D}$ none of whose weights ( $=$ eigenvalues of $H'$) is an integer $\geqslant 0$. This implies at once the much stronger result that 0 is the only $\partial(\omega)$-finite element of $\mathfrak{T}$. This completes the sketch of the proof of the regularity theorem on $\mathfrak{g}$; the formulae (7) and (9) then lead to the theorem on the group.

The behaviour around the singular points of the analytic functions defined by the invariant $\mathfrak{Z}$-finite (or $I(\mathfrak{g})$-finite) distributions can be studied in detail. Theorems 2, 3 and Lemma 25 of [1965a] do this on the Lie algebra while Lemmas 31, 34–37 of [1965b] treat the group situation. Let $L$ be a Cartan subgroup and $F$ the invariant locally summable function on $\Omega$ that defines a $\mathfrak{Z}$-finite distribution. If $\Phi_L = \Delta_L \cdot (F|_{\Omega \cap L'})$, then $\Phi_L$ extends analytically to $L'(R)$, the subset of $L$ where no real (global) root takes the value 1. Moreover, $\varpi_L \Phi_L = \Psi_L$ extends continuously on all of $\Omega \cap L$ and the system of functions $(\Psi_L)$ are compatible in the sense that

$$\Psi_{L_1} = \Psi_{L_2} \quad \text{on } L_1 \cap L_2 \cap \Omega. \tag{10}$$

These are generally known as the Harish-Chandra *matching conditions*; They are equivalent to the statement that $\nabla_G F$ extends to a continuous function on $\Omega$ (for the definition of $\nabla_G$ see Lemma 36 of [1965b]). Entirely analogous results are true on $\mathfrak{g}$.

§5. *Construction of the Invariant Eigendistributions* $\Theta_\xi$. In [1965c] the regularity theorem and the theory of Fourier transforms are used to construct the distributions $\Theta_\xi$ that will be the characters of the discrete series. Let $G$ be as in §3 with $rk(G) = rk(K)$; let $B \subset K$ be a Cartan subgroup, with $\mathfrak{b} = \mathrm{Lie}(B)$. We select a compact form $U$ of $G_c$ such that $K = U \cap G$ and $B \subset U$. $W$ and $W_c$ denote the respective Weyl groups of $(K, B)$ and $(U, B)$. $W_c$ operates on $B$ and the $\xi \in \hat{B}$ that have trivial stabilizer in $W_c$ are called regular. Theorem 3 of [1965c] then asserts that for any regular $\xi \in \hat{B}$ there is a unique invariant eigendistribution $\Theta_\xi$ on $G$ such that

    (a)  $|\Theta_\xi| \leqslant \text{const.}|D|^{-1/2}$   on $G'$
    (b)  $\Theta_\xi = \Delta^{-1} \Sigma_{s \in W} \varepsilon(s)(s\xi)$   on $B' = B \cap G'$.       (1)

The $\Theta_\xi$ are obtained by exponentiating suitable invariant eigendistributions on the Lie algebra. The condition (a) is essentially equivalent to the *temperedness* of these distributions on the Lie algebra; it thus allows Fourier transforms to be used for constructing them. In fact, if $\eta$ is the analogue of $D$ on $\mathfrak{g}(\eta = \pi^2$ on any Cartan subalgebra), the formula (8) of §4 shows that the conditions $|F| \leqslant \text{const.}|D|^{-1/2}$ and $|\tilde{F}| \leqslant \text{const.}|\eta|^{-1/2}$ are equivalent; the theory of orbital integrals can be used to show that the latter condition, for invariant eigendistributions with regular eigenvalues, is equivalent to temperedness. The fundamental result concerning the distributions on the Lie algebra is Theorem 2 of [1965c]: let $\lambda \in i\mathfrak{b}^*$ be regular; then there is a unique invariant distribution $T_\lambda = T_{\mathfrak{g}, \lambda}$ on $\mathfrak{g}$ such that

    (a)  $T_\lambda$ is tempered, $\partial(p)T_\lambda = p_\mathfrak{b}(\lambda)T_\lambda$     $(p \in I(\mathfrak{g}))$
    (b)  $T_\lambda = \pi^{-1} \Sigma_{s \in W} \varepsilon(s) e^{s\lambda}$   on $\mathfrak{b}' = \mathfrak{b} \cap \mathfrak{g}'$.

Let $\mathfrak{I}_\lambda$ be the space of all tempered invariant distributions $T$ on $\mathfrak{g}$ such that

$$\partial(p)T = p_{\mathfrak{b}}(\lambda)T \qquad (p \in I(\mathfrak{g})).$$

Let $A_\lambda$ be the space of all functions on $\mathfrak{b}$ of the form $\Sigma_{s \in W_c} c_s e^{s\lambda} (c_s \in \mathbb{C})$ which are $W$-skew. If $T \in \mathfrak{I}_\lambda$, $\pi \cdot (T|_{\mathfrak{b}'})$ belongs to $A_\lambda$ (Theorem 2, [1965a]), and the uniqueness part of the theorem above is the statement that the restriction map $R_{\mathfrak{b}}: T \mapsto \pi \cdot (T|_{\mathfrak{b}'})$ is injective. On the other hand, $-iH_\lambda$ is in $\mathfrak{b}$ and we have

$$(-iH_\lambda)^{G_c} \cap \mathfrak{g} = \coprod_{s \in W \setminus W_c} (-isH_\lambda)^G$$

showing that the Fourier transforms $\hat{\mu}_s (s \in W \setminus W_c)$ of the invariant measures on the $G$-orbits $(-isH_\lambda)^G$ are linearly independent elements of $\mathfrak{I}_\lambda$ (cf. (8) of §3). Since $\dim(A_\lambda) = [W_c : W]$, the injectivity of $R_{\mathfrak{b}}$ will imply that $\mathfrak{I}_\lambda$ is *spanned* by the $\hat{\mu}_s$ and that $R_{\mathfrak{b}}$ is an isomorphism of $\mathfrak{I}_\lambda$ with $A_\lambda$. This of course will prove Theorem 2 of [1965c].

The key result is thus the injectivity of $R_{\mathfrak{b}}$. To get a feeling for it consider $\mathfrak{g} = \mathfrak{sl}(2,\mathbb{R})$ with $\lambda \neq 0$ in $i\mathbb{R}$, $T \in \mathfrak{I}_\lambda$, $\partial(\omega)T = \lambda^2 T$ where $\omega = H^2 + 4YX$. Let $i\lambda > 0$. Then

$$i\theta T(\theta(X - Y)) = c_{\mathfrak{b}}^+ e^{\lambda\theta} - c_{\mathfrak{b}}^- e^{-\lambda\theta} \qquad (0 \neq \theta \in \mathbb{R})$$

$$tT(tH) = a_{\mathfrak{b}}^+ e^{-i\lambda t} - a_{\mathfrak{b}}^- e^{i\lambda t} \qquad (t > 0).$$

The matching conditions give

$$c_{\mathfrak{b}}^+ + c_{\mathfrak{b}}^- = a_{\mathfrak{b}}^+ + a_{\mathfrak{b}}^- \tag{2}$$

while the temperedness of $T$ implies that only bounded exponentials can appear in $tT(tH)$, so that $a_{\mathfrak{b}}^- = 0$. But then $a_{\mathfrak{b}}^+ = c_{\mathfrak{b}}^+ + c_{\mathfrak{b}}^-$, showing that $T$ is completely determined by $T|_{\mathfrak{b}'}$.

Such an argument works in general. In Lemma 28 of [1965c] Harish-Chandra proves that if $\mathfrak{h}$ is a Cartan subalgebra and $\mathfrak{h}^+$ is a component of $\mathfrak{h}'(R)$, one can choose Cartan subalgebras $\tilde{\mathfrak{h}}$ "adjacent" to $\mathfrak{h}$ (i.e., $\dim(\tilde{\mathfrak{h}}_\mathbf{R}) = \dim(\mathfrak{h}_\mathbf{R}) - 1)$ and components $\tilde{\mathfrak{h}}^+$ of $\tilde{\mathfrak{h}}'(R)$ such that (a) $Cl(\tilde{\mathfrak{h}}^+)$ and $Cl(\mathfrak{h}^+)$ interface along $\tilde{\mathfrak{h}} \cap \mathfrak{h}$ (which is a hyperplane in both $\mathfrak{h}$ and $\tilde{\mathfrak{h}}$); (b) the coefficients of the exponentials occurring in the formula for $T$ on $\mathfrak{h}^+$ and those coming from $\tilde{\mathfrak{h}}^+$ are related in a manner similar to (2); (c) only bounded exponentials occur. The proof that $T = 0$ on $\mathfrak{b}' \Rightarrow T = 0$ then goes by induction on $\dim(\mathfrak{h}_\mathbf{R})$. The regularity of $\lambda$ is decisive here.

For technical reasons it is necessary to prove the uniqueness even when $T$ is not everywhere defined. The domains $\Omega$ however cannot be arbitrary; roughly speaking $\Omega \cap \mathfrak{b}$ must be "unbounded in real directions." To make this precise note first that any semisimple $X \in \mathfrak{g}$ can be uniquely written as $X = X_e + X_h$ where (i) $X_h \in [\mathfrak{g}, \mathfrak{g}]$ (to allow for cases where $\mathfrak{g}$ is only reductive), $X_e$, $X_h$ are semisimple and $[X_e, X_h] = 0$; (ii) $\text{ad } X_e$ (resp. $\text{ad } X_h$) has only pure imaginary (resp. real) eigenvalues; $X_e$ is called the *elliptic component of* $X$. The uniqueness theorem is then true for the class $\mathfrak{E}(\mathfrak{g})$ of open sets $\Omega$ with the following properties:

(a)   $\Omega$ is completely invariant; if $X \in \Omega$, then $tX \in \Omega$ for $0 \leqslant t \leqslant 1$;

(b)   if $X \in \mathfrak{g}$ is semisimple, then $\Omega$ contains all semisimple elements $X'$ of $\mathfrak{g}$ such that $X_e' = X_e$ (in particular $X_e \in \Omega$).

Typical examples of sets in $\mathfrak{E}(\mathfrak{g})$ are $\mathfrak{g}(\varepsilon)$ $(\varepsilon > 0)$, the subset of all $X$ in $\mathfrak{g}$ such that $|\mathrm{Im}\,\lambda| < \varepsilon$ for any eigenvalue $\lambda$ of ad $X$. If $\mathfrak{z} \subset \mathfrak{g}$ is a reductive subalgebra containing $\mathfrak{b}$, then $\mathfrak{z} \cap \mathfrak{g}(\varepsilon) \in \mathfrak{E}(\mathfrak{z})$. These are the domains that occur in [1965c].

In lifting results to $G$ one can use the exponentiation only around *elliptic* elements of $G$; for, only these have centralizers in $\mathfrak{g}$ that admit Cartan subalgebras of compact type. For any $b \in B$ let $\mathfrak{z}_b$ be the centralizer of $b$ in $\mathfrak{g}$; and for any $\varepsilon > 0$, let

$$G_\varepsilon^{(b)} = \left(b\exp\!\left(\mathfrak{z}_b \cap \mathfrak{g}(\varepsilon)\right)\right)^G. \tag{4}$$

The sets $G_\varepsilon^{(b)}$, which are used by Harish-Chandra in the lifting process, have remarkable properties. They are of course completely invariant, open if $\varepsilon > 0$ is sufficiently small. Moreover, if $\varepsilon_b > 0$ are arbitrary numbers, $(b \in B)$,

$$G = \bigcup_{b \in B} G_{\varepsilon_b}^{(b)} \tag{5}$$

while for $b', b'' \in B$ and $\varepsilon', \varepsilon'' > 0$ (small enough),

$$G_{\varepsilon'}^{(b')} \cap G_{\varepsilon''}^{(b'')}$$

is the union of sets of the form $G_\varepsilon^{(b)}$. The uniqueness principle on the Lie algebras $\mathfrak{z}_b$ (in their localized version) will then give, in view of these properties, the uniqueness, not only on the whole group, but also for invariant $\mathfrak{Z}$-finite distributions satisfying (a) and (b) of (1), but defined only on $G_\varepsilon^{(b)}$. For the existence, a suitable linear combination of $T_{\mathfrak{z}_b, s\lambda}$ $(s \in W)$ will lift to a distribution $\Theta_\xi^{(b)}$ on $G_{\varepsilon_b}^{(b)}$ satisfying (a) and (b) of (1) on it if $\varepsilon_b > 0$ is sufficiently small; the compatibility of these on the overlaps of the $G_{\varepsilon_b}^{(b)}$ is a consequence of the strengthened form of the uniqueness mentioned above, and so, in view of (5), we would have constructed $\Theta_\xi$ on all of $G$ (cf. [V 3], Part II, §§2, 5).

The technique of using the matching conditions on the interfaces of adjacent Cartan subalgebras, which was used in the proof of uniqueness, has other applications. As an illustration, let us consider the distribution

$$\Theta_\xi^* = \sum_{s \in W_c} \varepsilon(s)\Theta_{s\xi}. \tag{6}$$

On $B$, $\Theta_\xi^*$ coincides, up to a constant factor, with the character of an irreducible finite dimensional representation of $U$. In particular, it is bounded and $W_c$-invariant. If $L$ is any other Cartan subgroup, and $W_{L,I}$ is the subgroup of the Weyl group of $(G_c, L_c)$ generated by the reflexions associated to the imaginary roots, the technique mentioned above can be used to prove by induction on $\dim(A)$ (where $A$ is the $\mathbb{R}$-split component of $L$), that $\Theta_\xi^*|_{L'}$ is $W_{L,I}$-invariant.[†] This will imply that

$$\sup_{L'} \left(|D_L|^{1/2}|\Theta_\xi^*|\right) < \infty \tag{7}$$

where

$$D_L(x) = \det\!\left(\left(Ad(x^{-1}) - 1\right)_{\mathfrak{g}/\mathfrak{z}}\right)$$

---

[†] It can be shown that $W_{L,I}$ operates on $L$ and stabilizes each component of $L'(R)$; see p. 307, [1965c].

($\mathfrak{z}$ = centralizer of $A$ in $\mathfrak{g}$) (§24, [1965c]). We have already seen in §3 the decisive role of the distributions $\Theta_\xi^*$ and the estimate (7) in the problem of extending the theory of orbital integrals to $\mathcal{C}(G)$.

**§6.** *Eigenfunction Asymptotics, Weak Inequality, and Analysis in the Schwartz Space.* I shall conclude this introduction with a few comments on the Schwartz space and Harish-Chandra's treatment of the asymptotic behaviour of the matrix elements of the irreducible unitary representations and their wave packets [1958a] [1958b] [1966b] [1975] [1976a]. Let $G = KA_0N_0$ be an Iwasawa decomposition with associated (minimal) parabolic subgroup $P_0 = M_0A_0N_0$; $A_0^+$, the positive chamber of $A_0$; $\mathfrak{a}_0 = \mathrm{Lie}(A_0)$ and $\rho_0 \in \mathfrak{a}_0^*$ defined as usual by $\rho_0(H) = \frac{1}{2}\mathrm{tr}(\mathrm{ad}\,H)_{\mathfrak{n}_0}$ where $\mathfrak{n}_0 = \mathrm{Lie}(N_0)$. We write $\sigma$ for the spherical function on $G$ such that $\sigma(a) = \|\log a\|\,(a \in A_0)$; $\sigma(x)$ is then the distance from $K$ to $xK$ in the Riemannian space $G/K$.

The concern in [1958a] [1958b] was with the elementary spherical functions $\varphi_\nu$. The basic estimate for them is (Theorem 3, [1958a]) the following

$$|\varphi_\nu(a)| \leqslant Ce^{-\rho_0(\log a)}(1 + \sigma(a))^m \qquad (a \in A_0^+, \nu \in \mathfrak{a}_0^*)$$

where $C > 0$, $m \geqslant 0$ are constants. In [1958b] it was proved that wave packets $f$ of the $\varphi_\nu$ satisfy, for each $m \geqslant 0$,

$$|f(a)| \leqslant C_m e^{-\rho_0(\log a)}(1 + \sigma(a))^{-m} \qquad (a \in A_0^+)$$

$C_m > 0$ being a constant. Since the Jacobian for the polar decomposition $G = K\,Cl(A_0^+)K$ behaves roughly as $e^{2\rho_0}$ on $A_0^+$, these estimates show that the $\varphi_\nu$ are in $L^{2+\varepsilon}(G)$ for every $\varepsilon > 0$ and that the wave packets are actually in $L^2(G)$. They motivate to some extent the following definition ([1970c], p. 539): a continuous function $f$ on $G$ is said to satisfy the *weak inequality* if there are constants $C > 0$, $m \geqslant 0$, such that

$$|f(k_1ak_2)| \leqslant Ce^{-\rho_0(\log a)}(1 + \sigma(a))^m \qquad (k_1, k_2 \in K,\ a \in A_0^+). \tag{1}$$

The Schwartz space $\mathcal{C}(G)$ is then defined as the space of all $f \in C^\infty(G)$ such that for $u, v \in U(\mathfrak{g})$ and any $m \geqslant 0$, the two-sided derivative $ufv$ of $f$ satisfies

$$|(ufv)(k_1ak_2)| \leqslant Ce^{-\rho_0(\log a)}(1 + \sigma(a))^{-m} \qquad (k_1, k_2 \in K,\ a \in A_0^+) \tag{2}$$

for some constant $C > 0$.

Actually, if we write $\varphi_0 = \Xi$, then

$$e^{-\rho_0(\log a)} \leqslant \Xi(a) \leqslant 1 \qquad (a \in A_0^+) \tag{3a}$$

so that in (1) and (2) we can replace $e^{-\rho_0(\log a)}$ by $\Xi(a)$. Since $\Xi$ has good properties from the point of view of harmonic analysis, this change improves the formal aspects of the theory also (cf. the proof that $\mathcal{C}(G)$ is a topological algebra under convolution, in §14, [1975]). In particular

$$\int_G \Xi^2(1 + \sigma)^{-m}\,dx < \infty \tag{3b}$$

if $m \gg 0$.

The space $\mathcal{C}(G)$ gives rise to a natural notion of tempered distributions on $G$. The fundamental nature of the weak inequality is then revealed by Theorem 14.1 of [1975]: for a $K$-finite $\mathfrak{Z}$-finite function on $G$, weak inequality is equivalent to temperedness. We also mention the corresponding results for class functions. For invariant $\mathfrak{Z}$-finite distributions $\Theta$ temperedness is equivalent to the estimate

$$|\Theta| \leqslant \text{const.}|D|^{-1/2}(1+\sigma)^m \qquad \text{(pointwise on } G') \tag{4}$$

for some $m \geqslant 0$, and then

$$\Theta(f) = \int_G \Theta f \, dx \qquad (f \in \mathcal{C}(G)) \tag{5}$$

the integral converging absolutely; the counterpart to (3b) is

$$\int_G |D|^{-1/2}\Xi(1+\sigma)^{-m} dx < \infty \tag{6}$$

for $m \gg 0$ (cf. [1975], §§10–14; see also the introductory remarks to [1970a]). In (4) the factor $(1+\sigma)^m$ can be dropped if $\Theta$ is an eigenfunction for a *regular* infinitesimal character, thereby elucidating the meaning of the condition $\sup|D|^{1/2}|\Theta_\xi| < \infty$ in the construction of the distributions $\Theta_\xi$. Temperedness (of the characters as well as the matrix elements) is the characteristic property of the series of representations $(\pi_{\eta,\nu})$ associated to the various Cartan subgroups of $G$.

In [1958b] Harish-Chandra discovered that the differential equations satisfied by the $\varphi_\nu$ become (roughly speaking), at infinity on $A_0$, differential equations with *constant coefficients*. This fact led to the result that $\varphi_\nu$ is asymptotically just a sum of plane waves. In [1966b] Harish-Chandra extended this method to handle all tempered $K$-finite $\mathfrak{Z}$-finite functions. Let $\tau$ be a finite dimensional double unitary representation of $K$ and let $\mathcal{C}(G, \tau)$ be the space of all $\tau$-spherical tempered $\mathfrak{Z}$-finite functions on $G$. Let $f \in \mathcal{C}(G,\tau)$. Fix a parabolic subgroup $P = MAN$. Then $G = K(MA)K$, and one can use the $\tau$-sphericity of $f$ to reduce the differential equations satisfied by $f$ on $G$ to a system of differential equations on $MA$; in the limit on $MA$, when $a \in A$ and $a \to {}_P\infty$ (this means that $a$ goes to infinity in such a way that for some $\varepsilon > 0$, $\alpha(\log a) \geqslant \varepsilon\sigma(a)$ for all roots $\alpha$ of $(P, A)$), the latter differential equations become the ones that define elements of $\mathcal{C}(MA, \tau_M)$. This suggests that $f$ may be approximated by an element of $\mathcal{C}(MA, \tau_M)$ at infinity on $MA$. That this is true is the content of Theorem 21.1 of [1975]; $f_P \in \mathcal{C}(MA, \tau_M)$ defined there is the constant term of $f$ along $P$. The fact that $f_P$ is a good approximation to $f$ in a suitable regime is contained in the following estimate (Lemma 23.4 [1975]):

$$|e^{\rho_0(\log a)}f(a) - e^{\rho_{00}(\log a)}f_P(a)| \leqslant C(1+\sigma(a))^m e^{-\kappa\rho_0(\log a)}; \tag{7}$$

here $\rho_{00}$ is the "$\rho_0$" of $MA$, $\kappa > 0$, $C > 0$, $m \geqslant 0$ are constants, and $a$ is to be restricted to regions of the following form ($\varepsilon > 0$ can be arbitrary)

$$A_0^+(P:\varepsilon) = \{a \in Cl(A_0^+)\,|\,\alpha(\log a) \geqslant \varepsilon\rho_0(\log a) \quad \text{for all roots } \alpha \text{ of } (P, A)\}. \tag{8}$$

We have already seen in §2 how decisive the theory of the constant term is in the treatment of the discrete spectrum. In particular the estimates (7) on sets $A_0^+(P:\varepsilon)$

show that for an $f \in \mathcal{Q}(G, \tau)$,

$$f \in L^2(G, \tau) \Leftrightarrow f \in \mathcal{C}(G, \tau) \Leftrightarrow f_P = 0 \quad \text{for all } P.$$

Let me mention now another illustration of the method of differential equations. Let $g$ be in $\mathcal{C}(G) \cap {}^0 L^2(G)$. We can write $g = \sum_j f_j$ where the $f_j$ are $K$-finite eigenfunctions, the sum being convergent in $L^2(G)$. The $f_j$ of course have vanishing constant terms and so decay exponentially in view of (7); by being careful in the analysis that led to (7) one can keep track of the dependence of (7) on the parameters of $f_j$ (the class $\omega \in \hat{G}_d$ to which $f_j$ belongs, and the $K$-types according to which $f_j$ transforms). The result is that the series $\sum_j f_j$ is convergent *in the topology of the Schwartz space*. On the other hand, the $f_j$, being $\mathfrak{Z}$-finite functions in $\mathcal{C}(G)$, are cusp forms (Theorem 18.1, [1975]) so that $g$ is a cusp form, showing that $\mathcal{C}(G) \cap {}^0 L^2(G)$ is precisely the space of cusp forms ([1970c], §8; [V 3], Part II, §16).

In the theory of the continuous spectrum one considers families of eigenfunctions, for instance the Eisenstein integrals. Harish-Chandra introduces the concept of ($\tau$-spherical) eigenfunctions $(\phi_\nu)$ of type II($\lambda$) associated to a Cartan subgroup $L = T \cdot A$ (§8, [1976a]) here $\lambda \in it^*$ ($t = \mathrm{Lie}(T)$) is regular, $\nu$ varies in $\mathfrak{a}^*$ and $\phi_\nu$ is smooth in $\nu$, satisfies the weak inequality in a uniform way, and satisfies the differential equation

$$z\phi_\nu = \mu_{\mathfrak{g}/\mathfrak{h}}(z)(\lambda + i\nu)\phi\nu \quad (z \in \mathfrak{Z}).$$

The fundamental idea now is to form wave packets

$$\phi_\alpha = \int_{\mathfrak{a}^*} \alpha(\nu)\phi_\nu \, d\nu \quad (\alpha \in \mathcal{C}(\mathfrak{a}^*)) \tag{9}$$

and to try to prove that $\phi_\alpha \in \mathcal{C}(G, \tau)$. This may not always be true even if $\alpha \in C_c^\infty(\mathfrak{a}^*)$. The point is that the constant terms $(\phi_{P, \nu})(P \in \mathcal{P}(A))$ are not eigenfunctions on $MA$ but only linear combinations of such (this is because the asymptotic form on $MA$ of the differential operator $z \in \mathfrak{Z}$ is $\mu_{\mathfrak{g}/\mathfrak{h}}(z)$, and $\mu_{\mathfrak{g}/\mathfrak{h}}(\mathfrak{Z}) \neq \mathfrak{Z}_{MA}$, the center of $U(\mathfrak{m} \oplus \mathfrak{a})$). For regular $\nu \in \mathfrak{a}^*$ one can write $\phi_{P, \nu}$ as $\sum_{s \in W(A)} \phi_{P, s, \nu}$; the $\phi_{P, s, \nu}$ will now be eigenfunctions (these statements have to be formulated with some care; see §§8–9 of [1976a]) but they may become singular in $\nu$ if $\nu$ approaches root hyperplanes in $\mathfrak{a}^*$ (for instance, in the spherical case, they are $c(s\nu)e^{is\nu}$). To overcome this, Harish-Chandra introduces the concept of eigenfunctions $(\phi_\nu)$ of type II'($\lambda$) in §9, loc.cit, and proves that if $(\phi_\nu)$ is of type II'($\lambda$), $\phi_\alpha \in \mathcal{C}(G, \tau)$ for any $\alpha \in \mathcal{C}(\mathfrak{a}^*)$, and $\alpha \mapsto \phi_\alpha$ is a continuous map (Theorem 13.1, loc.cit). The usefulness of this concept becomes clear from Theorem 9.1 of loc.cit asserting that if $(\phi_\nu)$ is of type II($\lambda$), $(\varpi(\lambda + i\nu)\phi_\nu)$ is of type II'($\lambda$). Theorem 11.1 of loc.cit actually gives a necessary and sufficient condition that $(\phi_\nu)$ which is of type II($\lambda$), is actually of type II'($\lambda$).

The wave packet theorem is proved by induction on $\dim(G)$. The basic point of course is to check that if $(\phi_\nu)$ is of type II'($\lambda$) on $G$, the $(\phi_{P, s, \nu})$ lead to eigenfunctions of type II'($\lambda$) on $MA$ (Lemma 9.1, [1976a]). If one had simply restricted oneself to Eisenstein integrals, the inductive step would have become much more complicated.

The wave packets of $E_{P,\psi,\nu}$ define *essentially* the inverse of the Fourier transform. Essentially, because we are using only $d\nu$ as the weight and not the Plancherel measure; or (equivalently), because we have not yet determined the true normalization of the $E_{P,\psi,\nu}$. To correct this one has to compute the Fourier transform of the wave packet $\phi_\alpha$ and relate it to $\alpha$. Theorem 13.2 of [1976a] does this. It is actually a generalization of the corresponding theorem 4 of [1958b] for spherical functions, and is based on the same circle of ideas. What it says is that if $\bar{P}$ is the parabolic subgroup opposite to $P$, the integral

$$\phi_\alpha^{\bar{P}}(m) = d_P(m)\int_{\bar{N}}\phi_\alpha(\bar{n}m)\,d\bar{n} \qquad (m \in MA)$$

can be evaluated by substituting for $\phi_\alpha$ the "truncated wave packets"

$$\phi_{P,\alpha}(m) = \int_{\mathfrak{a}^*}\alpha(\nu)\phi_{P,\nu}(m)\,d\nu \qquad (m \in MA)$$

(which are obtained by replacing $\phi_\nu$ by its constant term in the formation of the wave packet). If $(\phi_\nu)$ is the Eisenstein integral associated to a $Q \in \mathscr{P}(A)$, $\phi_{P,\nu}$ is expressible as a sum of plane waves $c_{P|Q}(s:\nu)e^{is\nu}$, and the integral of the truncated wave packet has an explicit formula in which the $c$-functions enter directly (Theorem 19.2, [1976b]). The theory of the $c$-functions and the functional equations of the Eisenstein integrals are now combined to derive the explicit Plancherel formula. I should mention however one technical difficulty in forming wave packets with the properly normalized weight function, namely, the control of the growth of the Plancherel measure at infinity on $\mathfrak{a}^*$; to overcome this one already needs the product formula and the explicit calculations in the case when $\dim(A)$ is equal to 1. I refer the reader to [1976b] for details.

## R\textsc{eferences}

[AM]  Auslander, L., and Moore, C. C., Unitary representations of solvable Lie groups. *Memoirs of the Amer. Math. Soc.* **62** (1966), 1–199.

[B]  Bargmann, V., Irreducible unitary representations of the Lorentz group, *Ann. of Math.* **48** (1947), 568–640.

[Bl]  Blattner, R. J., *General Background. Harmonic Analysis and Representations of Semisimple Lie Groups*. Edited by J. A. Wolf, M. Cahen, and M. De wilde. D. Reidel Publishing Company, Holland, 1980, pp. 1–67.

[Bo 1]  Borel, A., *Introduction aux groupes arithmétiques*. Hermann, Paris, 1969.

[Bo 2]  —, *Formes automorphes et séries de Dirichlet*. Springer Lecture Notes **514** (1976), pp. 183–222.

[Br 1]  Bruhat, F., Sur les représentations induites des groupes de Lie. *Bull. Soc. Math. France*, **84** (1956), 97–205.

[Br 2]  —, Sur les représentations des groupes classiques p-adiques, I, II. *Amer. J. of Math.* **83** (1961), 321–338, 343–368.

[G]  Gel'fand, I. M., Automorphic functions and the theory of representations. *Proc. Int. Cong. of Mathematicians, Stockholm, 1962*, pp. 74–85.

INTRODUCTION

[GG 1]  Gel'fand, I. M., and Graev, M. I., On a general method of decomposition of the regular representation of a Lie group into irreducible representations. *Dokl. Akad. Nauk. SSSR*, **92** (1953), 221–224.

[GG 2]  —, The analogue of Plancherel's theorem for real unimodular groups. *Dokl. Akad. Nauk, SSSR* **92** (1953), 461–464.

[GN 1]  Gel'fand, I. M., and Neumark, M. A., Unitary representations of the Lorentz group. *Izvestiya. Akad. Nauk. SSSR* **11** (1947), 411–504.

[GN 2]  —, Unitary representations of the classical groups. *Trudy Mat. Inst. Steklova*, **36** (1950), 1–288.

[GP]    Gel'fand, I. M., and Piatetsky-Shapiro, I. I., Unitary representations in the $G/\Gamma$ space, where $G$ is the group of real $n^{\text{th}}$ order matrices and $\Gamma$ is the subgroup of integral matrices. *Dokl. Akad. Nauk. SSSR* **147** (1962), 275–278.

[GR]    Gel'fand, I. M., and Raikov, D. A., Irreducible unitary representations of arbitrary locally bicompact groups. *Mat. Sbornik* (N.S) **13** (55) (1943), 301–316.

[Go]    Godement, R., Sur les relations d'orthogonalité de V. Bargmann. I.II. I: Résultats préliminaires, *C.R. Acad. Sci. Paris.* **225** (1947), 521–523; II: Démonstration génerale, *C.R. Acad. Sci. Paris*, **225** (1947), 657–659.

[JL]    Jacquet, H., and Langlands, R. P., *Automorphic Forms on* GL (2), Springer Lecture Notes **114** (1970).

[L 1]   Langlands, R. P., *On the Functional Equations Satisfied by Eisenstein Series*. Springer Lecture Notes **544** (1976).

[L 2]   —, *Euler Products*. Yale University Press, 1967.

[L 3]   —, *Problems in the Theory of Automorphic Forms. Lectures in Modern Analysis and Applications*, Springer Lecture Notes, **170** (1970), 18–86.

[Lo]    Loomis, L. H., *An Introduction to Abstract Harmonic Analysis*. Van Nostrand, New York, 1953.

[M]     Maass, H., Über eine neue Art von nichtanalytischen automorphen Funktionen. *Math. Ann.* **121** (1949), 141–183.

[Mau 1] Mautner, F. I., Spherical functions over p-adic fields, I. *Amer. J. of Math.*, **80** (1958), 441–457.

[Mau 2] —, Spherical functions over p-adic fields, II. *Amer. J. of Math.* **86** (1964), 171–200.

[Ma 1]  Mackey, G. W., *The Theory of Unitary Group Representations. Chicago Lectures in Mathematics*. The Chicago University Press, Chicago and London, 1976.

[Ma 2]  —, Infinite dimensional group representations. *Bull. Amer. Math. Soc.* **69** (1963), 628–686.

[Ma 3]  —, *Unitary Group Representations in Physics, Probability, and Number Theory*. Benjamin, 1978.

[Mo]    Moore, C. C., Representations of Solvable and nilpotent groups and harmonic analysis on nil and solvmanifolds. Harmonic analysis on homogeneous spaces. *Proceedings of Symposia in Pure Mathematics*, **XXVI**. Edited by C. C. Moore. *Amer. Math. Soc.* 1973, pp. 3–44.

[MvN]   Murray, F. J., and von Neumann, J., On rings of operators. I, II, IV. I: *Ann. of Math.* **37** (1936), 116–229; II: *Trans. Amer. Math. Soc.* **41** (1937), 208–248; IV: *Ann. of Math.* **44** (1943), 716–808.

[Se 1]  Selberg, A., Harmonic analysis and discontinuous groups in weakly symmetric Riemannian spaces with applications to Dirichlet series. *J. Indian Math. Soc.* **20** (1956), 47–87.

[Se 2]  —, Discontinuous groups and harmonic analysis. *Proc. Int. Cong. of Mathematicians, Stockholm* (1962), 177–189.

[S]     Schwartz, L., *Théorie des Distributions*, Hermann, Paris, 1973 (Nouvelle édition).

[T]        Tate, J., Fourier analysis in number fields and Hecke's Zeta functions. Thesis (Princeton), 1950. Reproduced in *Algebraic Number Theory*, Edited by J. W. S. Cassels and A. Frölich, Thompson Book Company Inc., Washington D.C., 1967.

[vN]       von Neumann, J., (a) On rings of operators III. *Ann. of Math.* **41** (1940), 94–161. (b) On rings of operators. Reduction theory. *Ann. of Math.* **50** (1949), 401–485.

[V 1]      Varadarajan, V. S., *Lie Groups, Lie Algebras, and Their Representations*. Prentice Hall, Englewood Cliffs, N.J., 1974.

[V 2]      —, The theory of characters and the discrete series for semisimple groups. Harmonic analysis on homogeneous spaces, *Proceedings of Symposia in Pure Mathematics*, **XXVI**. Edited by C. C. Moore. *Amer. Math. Soc.* 1973, pp. 45–99.

[V 3]      —, *Harmonic Analysis on Real Reductive Groups*. Springer Lecture Notes **576** (1976).

[W 1]      Weil, A., *L'integration dans les groupes topologiques et ses applications*. Hermann, Paris, 1940.

[W 2]      —, *Basic Number Theory*. Springer-Verlag, New York, 1967.

[W 3]      —, (a) Sur certains groupes d'opérateurs unitaires *Acta Math.* **111** (1964), 143–211. (b) Sur la formule de Siegel dans la théorie des groupes classiques. *Acta Math.* **113** (1965), 1–87.

[W 4]      —, Über die bestimmung Dirichletscher Reihen durch Funktionalgleichungen. *Math. Ann.* **168** (1967), 149–156.

[W 5]      —, *Automorphic Forms and Dirichlet Series*. Springer Lecture Notes **189** (1971).

[We 1]     Weyl, H., Über gewöhnliche Differentialgleichungen mit singularitäten und die zugehorigen Entwicklungen willkürlicher Funktionen. *Math. Ann.* **68** (1910), 220–269.

[We 2]     —, Theorie der Darstellung kontinuierlicher halbeinfacher Gruppen durch lineare Transformationen. I, II, III, und Nachtrag. I: *Math. Zeist.* **23** (1925), 271–309; II: *Math. Zeist.* **24** (1926), 328–376; III: *Math. Zeist.* **24** (1926), 377–395; Nachtrag: *Math. Zeist.* **24** (1926), 789–791.

[Wi]       Wigner, E. P., On unitary representations of the inhomogeneous Lorentz group. *Ann. Math.* **40** (1939), 149–204.

# Some Additional Aspects of Harish-Chandra's Work on Real Reductive Groups

Nolan R. Wallach

The main emphasis in the introduction to these volumes is the work of Harish-Chandra which led to the Plancherel theorem for reductive groups over local fields. This work did not historically follow a straight line and it included important results that are not directly related to the main theme. We include here a partial guide to some of these theorems and ideas. We label each paper to be discussed using the scheme of the Bibliography, e.g. [1951a].

[1951a] In this paper Harish-Chandra shows how to simultaneously construct a semisimple Lie algebra and all of its irreducible finite dimensional representations from its Cartan matrix. Recall that a Cartan matrix is an integral $l \times l$ matrix

$$A = (a_{ij})$$

with

(1)  $a_{ii} = 2, a_{ij} \leqslant 0, i \neq j$;
(2)  if $a_{ij} = 0$, then $a_{ji} = 0$;
(3)  $\det (A) \neq 0$;
(4)  the group generated by the transformations

$$s_i : x_j \mapsto x_j - a_{ji} x_i$$

is finite.

Starting with such a matrix $A$ he studies the algebra (either associative or Lie) with generators $X_i, Y_i, H_i, 1 \leqslant i \leqslant l$, and the relations

(R)
$$[H_i, H_j] = 0, \qquad [X_i, Y_j] = \delta_{ij} H_i$$
$$[H_i, X_j] = a_{ji} X_j, \qquad [H_i, Y_j] = -a_{ji} Y_j.$$

At first Harish-Chandra studies the associative algebra subject to (R). He later looks at the Lie algebra subject to (R) (the Lie subalgebra generated by the $X_i$, $Y_i$, $H_i$ in the associative algebra). Given a linear functional $\Lambda$ on the span of the $H_i$ with $\Lambda(H_i) = n_i$, $n_i \in \mathbb{Z}$, $n_i \geq 0$, he constructs a module for the algebra by taking a suitable quotient of what has come to be called a Verma module. He then takes its unique irreducible quotient (which we denote), $L(\Lambda)$. If $\mathfrak{g}'$ is the Lie algebra determined by the relations (R) and if $I = \{X \in \mathfrak{g}' \,|\, X$ acts by zero on every $L(\Lambda)\}$, then $\mathfrak{g} = \mathfrak{g}'/I$ defines the semisimple Lie algebra. The $L(\Lambda)$ are the finite dimensional irreducible representations of $\mathfrak{g}$. Parts of this argument are attributed by Harish-Chandra to Chevalley.

It is of some interest that the conditions (3), (4) are used only to guarantee that $\dim \mathfrak{g} < \infty$ and $\dim L(\Lambda) < \infty$. If the conditions (3) and (4) are dropped, the construction of Harish-Chandra gives the Kac-Moody Lie algebras and their standard modules.

**[1953]** In the first part of this paper Harish-Chandra initiates the algebraic theory of $(\mathfrak{g}, \mathfrak{k})$-modules (or Harish-Chandra modules). Here $\mathfrak{g}$ is a semisimple Lie algebra over the real numbers, $\theta$ is a Cartan involution of $\mathfrak{g}$, and $\mathfrak{k}$ is the fixed point algebra of $\theta$. A $(\mathfrak{g}, \mathfrak{k})$-module is a $\mathfrak{g}$-module, $M$, over $\mathbb{C}$ that splits into a (not necessarily finite) direct sum of irreducible $\mathfrak{k}$-modules. Harish-Chandra calls such modules quasi-semisimple for $\mathfrak{k}$. The basic theorem in the first part is Theorem 1 which may be stated as follows. Let $U(\mathfrak{g}_\mathbb{C}) \supset U(\mathfrak{k}_\mathbb{C})$ be the universal enveloping algebras of $\mathfrak{g}_\mathbb{C}$ and $\mathfrak{k}_\mathbb{C}$ respectively. Let $Z(\mathfrak{g}_\mathbb{C})$ be the center of $U(\mathfrak{g}_\mathbb{C})$. The theorem then says: if $W_1, W_2$ are finite dimensional semisimple $\mathfrak{k}_\mathbb{C}$-modules then

$$\mathrm{Hom}_{\mathfrak{k}}\left( W_1, U(\mathfrak{g}_\mathbb{C}) \underset{U(\mathfrak{k}_\mathbb{C})}{\otimes} W_2 \right)$$

is finitely generated as a $Z(\mathfrak{g}_\mathbb{C})$-module (the action of $Z(\mathfrak{g}_\mathbb{C})$ is $(z \cdot T)(v) = z \cdot T(v)$).

This theorem in particular implies that a $(\mathfrak{g}, \mathfrak{k})$-module $M$ that is finitely generated and is $Z(\mathfrak{g}_\mathbb{C})$-finite $(\dim Z(\mathfrak{g}_\mathbb{C}) \cdot m < \infty, m \in M)$ has finite $\mathfrak{k}$-multiplicities, i.e., is admissible. This implication is the starting point for the theory of admissible $(\mathfrak{g}, \mathfrak{k})$-modules. For the special case when $\mathfrak{g}$ is a complex Lie algebra looked upon as a real Lie algebra this result (as well as the basic idea of its proof) is in [1951a]. In [1951a] Harish-Chandra indicates that the result in this special case was suggested to him by Mautner.

In the second part of this paper Harish-Chandra initiates the theory of analytic vectors for Banach space representations of Lie groups. These vectors are called well-behaved in this paper. Let $(\pi, H)$ be a Banach representation of a connected Lie group $G$. Let $H^\omega$ be the space of analytic vectors. His main results are that $H^\omega$ is dense in $H$ and is $\mathfrak{g}$-invariant, and that the closure of a $\mathfrak{g}$-invariant subspace of $H^\omega$ is $G$-invariant.

Let $G$ be connected semisimple with finite center and let $K \subset G$ be the analytic subgroup of $G$ corresponding to $\mathfrak{k}$. Let $(\pi, H)$ be a Banach representation of $G$ such that if $v \in H^\omega$ $\dim Z(\mathfrak{g}_\mathbb{C})v < \infty$, and suppose that there exist $v_i \in H^\omega \cap H_K$, $1 \leq i \leq N$ such that $H$ is the closed linear span of $\pi(x)v_i$, $x \in G$, $1 \leq i \leq N$; here $H_K = \{v \in H \,|\, \pi(K) \cdot v$ spans a finite dimensional space$\}$. Then Theorem 1 and the theory of analytic vectors combine to prove

(1)  $H_K \subset H^\omega$; all the $K$-multiplicities in $H$ are finite.
(2)  $V \mapsto \mathrm{Cl}(V)$ is a bijection between the lattice of $\mathfrak{g}$-stable subspaces of $H_K$ and the lattice of $G$-stable closed subspaces of $H$.

These results are then used by him to prove that a connected semisimple Lie group is of type I in the sense of Murray and von Neumann.

**[1954a]** This paper contains the proof of his celebrated subquotient theorem. Let $G$ be linear, connected and semisimple. The theorem asserts that any irreducible $(\mathfrak{g}, \mathfrak{k})$-module that integrates to a $K$-module is isomorphic to a subquotient of the space of $K$-finite vectors of a representation (in general nonunitarily) induced from a one dimensional representation of an Iwasawa subgroup of $G$. He was later able to drop the linearity condition by using his results on differential equations [1960a].

**[1954c]** The main result of this paper is the Plancherel theorem for semisimple Lie groups defined over the field of complex numbers. Theorems 1 and 2 give the formula for the characters of the principal series representations for arbitrary connected semisimple Lie groups over $\mathbb{R}$, with finite center. The Plancherel theorem is based on this character computation and an integro-differential formula (Lemma 14) which generalizes earlier work of Gel'fand and Naimark. He generalized Lemma 14 to arbitrary semisimple Lie algebras in [1957c, d]; this provided one of the main steps in his proof of the Plancherel theorem for real reductive groups (see Varadarajan's introduction). Lemma 10 and its Corollary are important ingredients in the proofs of Lemma 14 and Theorem 2. He later used this result in his proof [1956d] of the Bruhat lemma.

**[1956b]** Section 7 of this paper is titled "A Digression on a Theorem of Cartan". Harish-Chandra considers the case $\mathfrak{g}_\mathbb{C}$ simple, and $\mathfrak{g}_\mathbb{C} = \mathfrak{k}_\mathbb{C} \oplus \mathfrak{p}_\mathbb{C}$ (complexified Cartan decomposition) with $\mathfrak{p}_\mathbb{C} = \mathfrak{p}^+ \oplus \mathfrak{p}^-$ a direct sum of two $\mathrm{Ad}\, K$-invariant nonzero subspaces. He had earlier shown [1955c] that there is $J \in \mathfrak{k}$ with $\mathrm{ad}\, J|_{\mathfrak{p}^+} = iI$, $\mathrm{ad}\, J|_{\mathfrak{p}^-} = -iI$, $\mathrm{ad}\, J|_{\mathfrak{k}} = 0$. Thus $[\mathfrak{p}^+, \mathfrak{p}^+] = [\mathfrak{p}^-, \mathfrak{p}^-] = 0$. Let $G$ be linear; then $G$ is contained in $G_\mathbb{C}$, a connected Lie group with Lie algebra $\mathfrak{g}_\mathbb{C}$. By going to a covering one may assume that $G_\mathbb{C}$ is simply connected. In $G_\mathbb{C}$ there is the connected parabolic subgroup

$$K_\mathbb{C} \exp \mathfrak{p}^+.$$

In [1956a] it is shown that

$$GK_\mathbb{C} \exp \mathfrak{p}^+ \subset (\exp \mathfrak{p}^-) K_\mathbb{C} (\exp \mathfrak{p}^+)$$

is an open subset, and that

$$G \cap (K_\mathbb{C} \exp \mathfrak{p}^+) = K.$$

One therefore has

$$GK_\mathbb{C} \exp \mathfrak{p}^+ = \exp(\Omega) K_\mathbb{C} \exp \mathfrak{p}^+$$

with $\Omega \subset \mathfrak{p}^-$ an open subset.

This gives rise to a diffeomorphism

$$\psi \colon G/K \to \Omega$$

given by

$$(\exp \psi(g)) K_{\mathrm{C}} \exp \mathfrak{p}^{+} = g K_{\mathrm{C}} \exp \mathfrak{p}^{+}$$

Using ad $J$ (defined above) one shows that $G/K$ has a $G$-invariant complex structure and it is easy to see that $\psi \colon G/K \to \Omega$ is complex analytic. Using his theory of strongly orthogonal roots and Lemma 20 Harish-Chandra shows that $\Omega$ is a bounded, symmetric domain in $\mathfrak{p}^{-}$. This gives a proof without going through classification of Cartan's basic theorem that every Hermitian symmetric domain is biholomorphic with a bounded symmetric domain. The map $\psi$ is now called the Harish-Chandra imbedding and is the starting point in the study of Hermitian symmetric domains.

In this paper there is also a very interesting argument which transfers the calculation of an integral on $G$ to the calculation of a corresponding integral on a compact form of $G_{\mathrm{C}}$ (see Lemma 28).

**[1960a]** This previously unpublished paper extends work in [1958a, b]. In this paper Harish-Chandra generalizes the classical theory of regular singular points and Frobenius's method to several variables. Most notable is his handling of what is now called the asymptotic expansion along the walls. These results were basic to the Langlands classification of irreducible $(\mathfrak{g}, \mathfrak{k})$-modules.

**[1962]** In this paper with A. Borel a general reduction theory of arithmetic subgroups of semisimple Lie groups over $\mathbb{R}$ is developed. Theorem 9.4 says that the volume of a fundamental domain for such a discrete subgroup is finite. Theorem 11.8 (which gives a positive answer to Godement's conjecture) is now called the Borel–Harish-Chandra criterion for compactness of a fundamental domain. It says that if $\Gamma \subset G$ is arithmetic then $G/\Gamma$ is compact if $\Gamma$ has no unipotent elements. This criterion is now the main method of proving compactness for arithmetic quotients.

**[1976b]** This paper is the culmination of Harish-Chandra's work on the Plancherel formula for real reductive groups. In addition to the completion of the proof of the Plancherel theorem this paper contains the completeness theorem for intertwining operators for cuspidal principal series representations (Theorem 37.1). This theorem is the basic ingredient in the analysis of the irreducibility of the cuspidal principal series representations.

**[1983]** The purpose of this paper is to study the basic ingredients that will be necessary for the spectral analysis of tempered, invariant distributions (for example the orbital integrals) on a real reductive group $G$. As in the case of the Plancherel theorem (which is concerned with the spectral analysis of the Dirac delta function at the identity of $G$) it is necessary to have the correct notion of the constant term. The theory of the constant term of a tempered, $Z(\mathfrak{g})$-finite, central distribution on $G$ is one of the main themes of this article.

Let $\Theta$ be a tempered, invariant, $Z(\mathfrak{g})$-finite distribution on $G$. Let $K \subset G$ be a maximal compact subgroup of $G$. Let $P$ be a proper parabolic subgroup of $G$ with standard Levi decomposition $P = MN$. The constant term of $\Theta$ along $P$, $\Theta_P$, should be a $Z(\mathfrak{m})$-finite tempered, central distribution on $M$ which is the limit in a suitable sense of $\Theta$ along the positive chamber relative to $P$ in $A$ ( = the standard split component of $M$). If $(\pi, H)$ is an irreducible tempered representation of $G$ and if $\Theta$ is the character of $\pi$ then there is a natural notion of $\Theta_P$ which we now describe. If $H_K$ is the space of $K$-finite vectors of $H$ then

$$H_K / \mathfrak{n} \cdot H_K$$

is an admissible, finitely generated, $(\mathfrak{m}, K \cap M)$-module. As a module for $\mathfrak{a}$, it splits into a direct sum of generalized weight spaces for $\mathfrak{a}$:

$$(H_K / \mathfrak{n} \cdot H_K)_\lambda.$$

Let $\rho$ be as usual the differential of the square root of the modular function, $\delta$, of $P$. Let $\bar{P} = \theta(P)$. Then $\Theta_{\bar{P}}$ is $\delta^{1/2}$ times the sum of the characters of the $(\mathfrak{m}, K \cap M)$-modules

$$(H_K / \mathfrak{n} \cdot H_K)_\lambda$$

with

$$\mathrm{Re}(\lambda - \rho, \alpha) = 0$$

for $\alpha$ a root of $(P, A)$. Harish-Chandra's definition in the general case is consistent with this definition for characters.

A tempered, invariant, $Z(\mathfrak{g})$-finite distribution $\Theta$ is called *super tempered* if $\Theta_P = 0$ for all proper parabolic subgroups $P$ of $G$. Thus the super tempered distributions are the analogues of the cusp forms of $G$. Using the results in [1960a] it is not difficult to see that if $\Theta$ is the character of an irreducible representation $\pi$ of $G$ which is tempered and unitary, then $\Theta$ is super tempered if and only if $\pi$ is square integrable. However, there are super tempered, invariant, eigendistributions that are not linear combinations of characters of discrete series representations. For example, if $(\pi, H)$ is the reducible unitary principle series representation of SL $(2, \mathbb{R})$, if $\pi^+$ and $\pi^-$ are the irreducible constituents of $\pi$, and if $\Theta^\pm$ are the characters of $\pi^\pm$, then $\Theta^+ - \Theta^-$ is super tempered, although neither $\Theta^+$ nor $\Theta^-$ is super tempered. One of the main results of this paper is that all super tempered eigendistributions that are not linear combinations of discrete series characters are given in terms of distributions generalizing $\Theta^+ - \Theta^-$ for SL $(2, \mathbb{R})$.

# The Work of Harish-Chandra on Reductive *p*-adic Groups

ROGER HOWE

Harish-Chandra's work on *p*-adic groups is broadly based on his experience with Lie groups. (In this survey, the term *Lie group* means an analytic group over **R**.) He himself made the connection quite explicit, formulating what he called the Lefschetz Principle and the Philosophy of Cusp Forms. His Lefschetz Principle [1970d] is a paraphrase for harmonic analysis of the rule of thumb of the same name from algebraic geometry. It asserts that the phenomena of representation theory for groups over *p*-adic fields, finite fields or adele rings are, suitably interpreted, essentially the same as for groups over **R**, i.e. Lie groups. In applying this principle the "suitable interpretation" is sometimes less than obvious; there are many respects in which *p*-adic harmonic analysis might seem rather different from real harmonic analysis. Nevertheless, it is a valuable principle, and there have been many examples where one theory will enlighten the other, or an attempt to reconcile apparent differences will lead to insight in both theories. For example, the cleanness of the theory of the constant term in the *p*-adic case led to a reworking of it for real groups [CM], [Wa]. It is a tribute to the robustness of Harish-Chandra's methods and to his tenacity that, despite serious obstacles including a lack of detailed understanding of the discrete series, characters and orbital integrals, he was able to reach a main goal, the Plancherel Formula [1977b].

The Philosophy of Cusp Forms is much more specific to harmonic analysis than is the Lefschetz Principle; it might be regarded as the primary concrete embodiment of that Principle in representation theory of reductive groups. It is most explicitly stated in [1970b] where it is traced back as far as a talk of Gel'fand at the Stockholm International Congress. It has become a basic feature of reductive harmonic analysis. In particular it is a unifying feature of Harish-Chandra's work.

The gist of the Philosophy of Cusp Forms is that the collection of all representations of a reductive group $G$ should be partitioned into disjoint classes, with each class being attached to a certain parabolic subgroup. (Or more precisely, to a class of associated parabolic subgroups. Recall that two parabolic subgroups are *associated* if their Levi components (the reductive components in their Levi decomposions) are conjugate.) The representations attached to $G$ itself are called *cuspidal* representations and the matrix coefficients of cuspidal representations are *cusp forms*. The representations associated to a proper parabolic subgroup $P$ are further associated to Weyl group orbits of cuspidal representations of $M$, the Levi component of $P$, and this further association is finite-to-one. Moreover, the representations of $G$ associated to a given cuspidal representation $\sigma$ of $M$ may be recovered as components of the induced representation

$$\mathrm{ind}_P^G\left(\sigma \otimes \delta_P^{1/2}\right)$$

where $\sigma$ is extended from $M$ to $P$ by letting it be trivial on the unipotent radical of $P$, and $\delta_P$ is the modular function of $P$. Finally, the method of passing from a representation of $G$ to its associated $P$ and representation $\sigma$ of $M$ is explicit; it is based on what Harish-Chandra calls the "theory of the constant term".

Thus the Philosophy of Cusp Forms provides an inductive strategy for understanding the representations of $G$. The total problem is divided into two parts: to determine the cusp forms and to analyze induced representations. Both problems are quite difficult and neither is solved in all cases, but the division has provided a fruitful method of attack and substantial progress has been made on each partial problem.

It should perhaps be pointed out that what one means by "cusp form" varies slightly according to context. For purposes of the Plancherel formula for real or $p$-adic groups, "cusp form" has an analytic meaning: a representation is cuspidal if and only if it is in the discrete series (modulo the center). For purposes of the theory of admissible representations of $p$-adic groups only supercuspidal representations are cuspidal, while in the Langlands classification [BW], [L2], [Vo] all representations which are tempered on the commutator subgroup play the part of cuspidal representations. The discrepancy between the cusp forms and the discrete spectrum is perhaps most vexed in the case of automorphic forms. All cusp forms in $L^2(G_A/G_k)$ (for notation see [Bo]) belong to the discrete spectrum, but the non-cuspidal or "residual discrete spectrum" is very rich and almost as poorly understood as the cusp forms. The variability of the notion of cusp form of course complicates the theory, but pays off amply in increased flexibility and applicability.

The determination of the discrete series of a semisimple Lie group was one of Harish-Chandra's great achievements. For semisimple $p$-adic groups this problem is still unsolved, except in some cases, and in fact it still seems quite far from solution. (See [MS] for some discussion of its current status.) An explicit description even of the representations of the compact group of norm units in a $p$-adic division algebra has not yet been given.

There are several reasons for the greater mystery of the discrete series, in the $p$-adic case. Recall that if $G$ is a connected semisimple Lie group, then [V]:

1) Up to conjugacy $G$ contains at most one compact Cartan subgroup $T$.
2) The discrete series of $G$ are parametrized in a natural way by the regular characters of $T$.

Some groups, including complex groups and $SL_n(\mathbf{R})$, $n \geqslant 3$, have no compact Cartan subgroup and therefore no discrete series. By contrast, all $p$-adic groups have compact Cartan subgroups and discrete series in abundance. Further, while it is roughly true that characters of the compact Cartan subgroups parametrize discrete series representations, there can be no such clean correspondence as for real groups. Some characters (even "regular" characters, although the definition of that term is itself problematic) will not correspond to representations. Further, some representations would seem to correspond to characters of several tori. And finally, there are some discrete series, analogous to the unipotent cuspidal representations constructed by Lusztig [Lu1] for groups over finite fields, which seem not to correspond to a character of any torus. Langlands [Bo], [L1] has conjectured that the discrete series should be parametrized by certain homomorphisms of the Weil group (of the local field over which the group is defined) into an appropriate group, called the $L$-group. This has been verified in some cases [K], [M], [He], [KM], [T]. But even the conjectures on how to describe the discrete series of $p$-adic groups are still undergoing refinement [Ar], [Lu2].

Because of the relative intractability of the determination of cuspidal representations for $p$-adic groups, Harish-Chandra along with most other workers in this area concentrated mainly on the second part of the cusp form program, namely, the connection between representations of a reductive group $G$ and its parabolic subgroups $P$. He also established what qualitative results he could about discrete series, and these were sufficient to yield a version of the Plancherel formula [1977b] quite analogous to, though somewhat less precise than, the real case. We will describe below his progress toward this goal in more detail.

Harish-Chandra's papers on $p$-adic groups are [1970d], [1973], [1977a], [1977b], [1978], [1980]. However, it should be noted that [1973] is only a summary of results, essentially an extended research announcement. The full proofs of the results in [1973] have not been published by Harish-Chandra. Instead they appear in Allan Silberger's book [Si], which is largely based on lectures of Harish-Chandra at the Institute for Advanced Study in the academic years 1971–1972 and 1972–1973. For understanding Harish-Chandra's work on $p$-adic groups, this book is an essential supplement to the papers appearing under his own name. Neither have the full proofs of some results stated in [1977b], [1978], or [1980] yet appeared.

The main topics treated in these papers are ones familiar from Harish-Chandra's work on real groups: the theory of characters, orbital integrals, the constant term, discrete series and the Schwartz space. All of these topics are mutually related via the Philosophy of Cusp Forms. In the real case all were subservient to the Plancherel Formula, but in the $p$-adic case neither character theory nor orbital integrals has been made as precise as in the real case, and eventually were bypassed by Harish-Chandra in his proof of the $p$-adic Plancherel Formula. We will discuss these topics in turn.

As a prelude to the more detailed discussion, a remark on technique is in order. Throughout his work on real groups, Harish-Chandra relies on the differential equations supplied by the universal enveloping algebra, especially its center. This fundamental avenue of approach is of course not available in the study of $p$-adic groups and replacing it is a major technical problem. For some purposes, the place of the differential equations is taken by certain finiteness theorems. Thus a result of Jacquet [J] forms the basis for the theory of the constant term, a result of Howe [Ho] is useful in character theory, and a finiteness result in [1977b] provides the final step in the Plancherel Theorem. We will discuss these in more detail below. But in fact, the lack of the differential equations has not been completely made good, and as was already noted, complete information about discrete series, character formulas, and orbital integrals is not yet available.

Character theory is the primary topic of [1977a], [1978], and [1980], and occupies the bulk of [1970d]. The paper [1977a] is an announcement of the results in [1978], which contains the most precise results on characters, and largely supersedes [1970d]. However, the paper [1978] is restricted to groups over fields of characteristic zero; in [1980] some of the results are extended to groups over fields of positive characteristic.

Before one can describe characters one must know in what sense they exist. For this, and much of the rest of representation theory for $p$-adic groups, a basic notion is that of *admissible representation*, first formalized in [JL]. Let $G$ be a locally compact topological group which is totally disconnected as topological space—a *t.d. group* in Harish-Chandra's terminology [1973]. This is the same as to say $G$ has a basis of neighborhoods of the identity consisting of open compact subgroups. All $p$-adic algebraic groups are t.d. groups. Let $\rho$ be a representation of $G$ on a vector space $V$. A vector $v \in V$ is called a *smooth vector* for $\rho$ if there is an open subgroup $K \subseteq G$ such that $v$ is invariant under $\rho(K)$, i.e., $\rho(k)v = v$ for $k \in K$. The set of smooth vectors for $\rho$ form a subspace $V^\infty \subseteq V$; this subspace $V^\infty$ is invariant by $G$. If $V$ is a complete locally convex topological vector space, and $\rho$ is a strongly continuous representation of $G$, then $V^\infty$ is a dense subspace of $V$.

For a given compact open group $K \subseteq G$, let $V^K$ denote the subspace of vectors in $V$ fixed by $\rho(K)$. Then $V^K \subseteq V^\infty$, and $V^\infty$ is a union of the $V^K$ as $K$ varies over all compact open subgroups of $G$. If $V = V^\infty$, then the representation $\rho$ is called *smooth*. If in addition all the spaces $V^K$ are finite dimensional, then $\rho$ is called *admissible*. A continuous representation of $\rho$ on a locally convex space whose associated smooth subrepresentation is admissible is also called admissible.

An admissible representation has a character in the following sense. Let $C_c^\infty(G)$ denote the space of locally constant, compactly supported, complex-valued functions on $G$. Then $C_c^\infty(G)$ is a convolution algebra under the usual definition of convolution on $G$. If $\rho$ is a smooth representation of $G$ on $V$, then we can define in an obvious and standard manner a representation, also denoted $\rho$, of $C_c^\infty(G)$ on $V$. If $\rho$ is an admissible representation, then $\rho(f)$ will be a finite rank operator. (If $f$ is invariant under left translation by the open subgroup $K \subseteq G$, then $\rho(f)(V) \subseteq V^K$.) Hence the trace of $\rho(f)$ will be well defined, and will depend linearly on $f$. Thus

$$\theta_\rho(f) = \operatorname{trace} \rho(f)$$

defines a linear functional on $C_c^\infty(G)$. By definition a linear functional on $C_c^\infty(G)$ is called a *distribution* on $G$. Thus we may say that an admissible representation has a character defined by a distribution on $G$—its "distribution character".

Thus the first question to ask concerning character theory of representations of $G$ is, are all reasonable (say unitary, though one can be more general) irreducible representations of $G$ admissible? The analogous question for Lie groups was one of the first issues resolved by Harish-Chandra [1953]. In the case of a reductive $p$-adic group the answer was much longer in coming. That the answer was "yes" was first established by Bernstein [Be]. A stronger result was proved for groups over fields of characteristic zero by Harish-Chandra in [1978], generalizing and extending [Ho]. Both results were based on [1970d] and in particular were resolutions of conjectures made in [1970d]. We will explain these conjectures.

The conjectures concern the supercuspidal representations, which are the most characteristic and the most mysterious phenomenon of representation theory on reductive $p$-adic groups.

Let $G$ be a locally compact group. Let $f$ be a function on $G$. Define the left and right translate $\lambda_g f$ and $\rho_g f$ of $f$ by $g$ in the usual way:

$$\lambda_g(f)(x) = f(g^{-1}x) \qquad \rho_g(f)(x) = f(xg) \qquad x, g \in G. \tag{1}$$

Let $H \subseteq G$ be a subgroup. We say $f$ is left (right) $H$-*invariant* if $\lambda_h(f) = f (\rho_h(f) = f)$ for $h \in H$. If $f$ is complex- or vector-valued we say $f$ is left (right) $H$-*finite* if the left translates $\lambda_h(f)$ (right translates $\rho_h(f)$) span a finite dimensional space. We say $f$ is left (right) compactly supported mod $H$ if there is a compact set $C \subseteq G$ such that $f = 0$ outside $HC$ (outside $CH$). If $H$ is central in $G$, we may omit the adjectives "left" and "right" in the definitions above.

Let $G$ be a reductive $p$-adic group—the rational points of a connected reductive algebraic group $\mathbf{G}$ defined over a $p$-adic field $\Omega$. Let $Z$ be the center of $G$. Let $P$ be a parabolic subgroup of $G$ and let $N$ be the unipotent radical of $P$. Let $\delta_P$ be the modular function of $P$. Let $f$ be a smooth complex- or vector-valued function on $G$, compactly supported mod $Z$. Then in analogy with the real case we can define a function $f^{(P)}$ on $P/N$ by the formula

$$f^{(P)}(m) = \delta_P(m)^{1/2} \int_N f(mn)\, dn \qquad m \in P.$$

Here $dn$ is Haar measure on $N$.

We say a function $f$ on $G$ is a *supercusp form* if

    i)   $f$ is smooth;
    ii)  $f$ is compactly supported mod $Z$;
    iii) $f$ is $Z$-finite; and
    iv) $(\lambda_g f)^{(P)} = 0$ for all proper parabolics $P \subseteq G$ and $g \in G$.

A *supercuspidal representation* is a representation whose matrix coefficients are supercusp forms. Let $\pi$ be a smooth representation of $G$ on the space $V$. Consider $T \in \mathrm{End}\, V$. We say $T$ is a *smooth operator* if there is an open subgroup $K \subseteq G$ such that

$$\pi(k)T = T = T_\pi(k) \qquad k \in K.$$

Let $\mathrm{End}^{\circ}(V)$ denote the space of smooth operators which also have finite rank. (Observe that if $\pi$ is admissible, then a smooth operator automatically has finite rank.) Then $\mathrm{End}^{\circ}(V)$ is a subalgebra of $V$, and is invariant by right and left multiplication by $\pi(g)$, for $g \in G$. Set

$$f_T(g) = \operatorname{trace} T\pi(g).$$

The function $f_T$ is called a *matrix coefficient* of $p$. It is easy to see that $f_T \in C^\infty(G)$, the space of locally constant functions on $G$. The space spanned by all matrix coefficients of all admissible representations is denoted $\mathcal{Q}(G)$. If $\pi$ is admissible and all matrix coefficients of $\pi$ are supercusp forms, then $\pi$ is called supercuspidal. It turns out ([1973] §6) that for $\pi$ to be supercuspidal, it is sufficient that its matrix coefficients be compactly supported mod $Z$.

Supercuspidal representations are obviously discrete series representations (mod $Z$). A general argument ([1970d], p. 6) using the Schur orthogonality relations for discrete series shows that all discrete series representations are admissible. In fact if $\pi$ is an irreducible square integrable representation (mod $Z$) of $G$ on a space $V$, and $V^K$ is the space of $\pi(K)$-invariant vectors in $V$ for an open subgroup $K \subseteq G$, one has

$$\dim V^K \leqslant \left( d(\pi)\mu(K) \right)^{-1}$$

where $d(\pi)$ is the formal degree of $\pi$ and $\mu(K)$ is the Haar measure of $K$. If $1_K$ denotes the trivial representation of $K$, then $\dim V^K$ is the multiplicity of the trivial representation of $K$ in $\pi$. It will be denoted $[\pi:1_K]$.

Harish-Chandra shows in [1970d], Part II, that the question of admissibility hinges on the possibility of bounding $[\pi:1_K]$ independently of $\pi$. He considers three possible facts that would do this. Observe that if $\pi$ is an irreducible admissible representation of $G$, then $\pi|Z$ is a multiple of some character $\chi_\pi = \chi$, called the *central character* of $\pi$. Consider the following statements.

i)    For each compact open subgroup $K$ there is a number $\delta(K)$ such that $[\pi:1_K] \leqslant \delta(K)$ for all supercuspidal irreducible representations $\pi$ of $G$.

ii)    For each compact open subgroup $K$ there is a number $m(K)$ such that, for any character $\chi$ of $Z$,

$$\sum_\pi [\pi:1_K] \leqslant m(K)$$

where $\pi$ ranges over all (equivalence classes of) irreducible supercuspidal representations of $G$ with central character $\chi$.

iii)    There is $\varepsilon > 0$ such that $d(\pi) > \varepsilon$ for all irreducible supercuspidal representations $\pi$ of $G$.

Obviously statement ii) implies i). The Schur orthogonality relations in fact imply that iii) implies ii). In [1970d], Part II, Harish-Chandra shows that statement i) implies that all irreducible unitary representations of $G$ are admissible. In [Be] Bernstein shows by a simple argument based on the structure of $C^\infty(G//K)$, the $K$ bi-invariant functions of compact support, that statement i) is true. In [1978] Harish-Chandra, as a consequence of a study of characters of supercuspidal representations, shows that if the ground field $\Omega$ has characteristic 0, then statement iii) is true. In fact, he shows that if Haar measure on $G$ is appropriately normalized, then the formal degrees $d(\pi)$ of supercuspidal representations are integers.

Once the issue of existence of characters is settled, one would like to know more precisely what they look like. For supercuspidal characters and when the ground field is of characteristic 0, Harish-Chandra already establishes an important qualitative result in [1970d]. Let $dg$ denote Haar measure on $G$, and let $\varphi$ be a complex-valued function on $G$ which is locally $L^1$, i.e. the integral

$$\int_C |\varphi|(g)\,dg$$

where $|\varphi|$ indicates the absolute value of $\varphi$ and $C$ is any open compact subset of $G$, is finite. Then $\varphi$ defines a distribution $D_\varphi$ on $G$ by the obvious formula

$$D_\varphi(f) = \int_G \varphi(g)f(g)\,dg \qquad f \in C_c^\infty(G).$$

If a distribution $D$ is equal to $D_\varphi$ for some $\varphi$ as above, we say $D$ is *locally $L^1$*. If the function $\varphi$ may be taken to be locally constant on some open set $S \subseteq G$, we say $D$ is *locally constant* on $S$. In [1970d], Part V, Harish-Chandra shows that the character of an irreducible supercuspidal representation is a locally constant function on the set $G'$ of regular elements of $G$. In Parts VII and VIII of [1970d] he goes on to show that if the ground field $\Omega$ has characteristic zero, then in fact the character of an irreducible supercuspidal representation is locally $L^1$ on $G$. These results are based on an integral formula ([1970d], Theorem 9) for the character of a supercuspidal representation and on a detailed analysis of the geometry of conjugacy classes, analogous to his work on real groups.

In [1978] Harish-Chandra extends these basic facts to all irreducible admissible representations. Furthermore he gives more precise information about the behavior of characters near singular points. These results make use of the exponential map from the Lie algebra $\mathfrak{g}$ of $G$ to $G$, and so are valid only when the ground field $\Omega$ has characteristic zero. They depend on a finiteness result, proved first for $GL_n$ in [Ho], and extended and generalized by Harish-Chandra in [1978]. This may be stated as follows. Let $L \subseteq \mathfrak{g}$ be a lattice, and let $C_c^\infty(\mathfrak{g}/L)$ be the functions on $\mathfrak{g}$ which are compactly supported and constant on cosets of $L$. Let $\mathrm{Ad}\,G$ denote the adjoint action of $G$ on $\mathfrak{g}$. Let $\omega \subseteq \mathfrak{g}$ be a compact set, let $\mathrm{Ad}\,G(\omega)$ denote the subset of $\mathfrak{g}$ swept out by the action of $\mathrm{Ad}\,G$ on $\omega$, and let $J(\omega)$ denote the space of distributions which are supported on $\mathrm{Ad}\,G(\omega)$ and which are invariant under $\mathrm{Ad}\,G$. Let $j_L J(\omega)$ denote the space of linear functionals on $C_c^\infty(\mathfrak{g}/L)$ obtained by restricting elements of $J(\omega)$ to $C_c^\infty(\mathfrak{g}/L)$. Then $j_L J(\omega)$ has finite dimension.

This finiteness result has two main consequences for characters. First, it gives a description of the possible singularities of characters near any point. It is in terms of Fourier transforms of invariant measures on nilpotent conjugacy classes. Let $x \in \mathfrak{g}$ be a nilpotent element, and let $\mathcal{O} = \mathcal{O}_x = \mathrm{Ad}\,G(x)$ denote the $G$-conjugacy class of $x$. There are only finitely many nilpotent conjugacy classes. On $\mathcal{O}$ there is supported a $G$-invariant measure $\nu_x$, unique up to multiples. Furthermore, a result of Deligne and Rao [Ra] says that the measure $\nu_x$ assigns finite mass to bounded subsets of $\mathcal{O}$; consequently $\nu_x$ defines a distribution on $\mathfrak{g}$.

We want to consider the Fourier transforms of the measures $\nu_x$. These may be defined in the usual way. If $f \in C_c^\infty(\mathfrak{g})$, its Fourier transform $\hat{f}$ is defined by

$$\hat{f}(x) = \int \hat{f}(x')\chi(B(x',x))\,dx'$$

where $B(\ ,\ )$ is an Ad $G$-invariant, non-degenerate symmetric bilinear form on $\mathfrak{g}$ (the Killing form if $\mathfrak{g}$ is semisimple), $\chi$ is a unitary character of the (additive group of the) base field $\Omega$, and $dx'$ is a Haar measure on $\mathfrak{g}$, normalized to make the map $f \to \hat{f}$ unitary. We can then extend $\hat{\ }$ to distributions by the recipe

$$\hat{D}(f) = D(\hat{f}) \qquad f \in C_c^\infty(\mathfrak{g}),\ D \in C_c^\infty(\mathfrak{g})^*. \tag{2}$$

Recall that $J(\omega)$ is the space of Ad $G$-invariant distributions supported on Ad $G(\omega)$ where $\omega \subseteq \mathfrak{g}$ is some compact set. Harish-Chandra in [1978] has described the Fourier transforms of elements of $J(\omega)$. To state his result we need the functions $\eta_\mathfrak{g}$ and $D_G$, familiar from the real theory, which measure how regular elements of $\mathfrak{g}$ or $G$ are.

Consider the polynomial $\det(t - \operatorname{ad} x)$ on $\mathfrak{g} \times \Omega$. Regard this as a polynomial in $t$ with coefficients depending on $x$. Then if $l$ is the rank of $\mathfrak{g}$ (i.e., the dimension of any Cartan subalgebra), the coefficient of $t^b$ is identically zero if $b < l$. The function $\eta_\mathfrak{g}$ is the coefficient of $t^l$. Thus

$$\det(t - \operatorname{ad} x) = t^l\eta_\mathfrak{g}(x) + t^{l+1}R(x,t). \tag{3}$$

The function $\eta_\mathfrak{g}$ may be described in another way, as follows. Let $x \in \mathfrak{g}$ be regular, let $\mathfrak{t} \subseteq \mathfrak{g}$ be the kernel of $\operatorname{ad} x$ and let $\mathfrak{t}^\perp$ be the orthogonal complement of $\mathfrak{t}$ with respect to the bilinear form $B$. Then

$$\eta_\mathfrak{g}(x) = \det\left(\operatorname{ad} x | \mathfrak{t}^\perp\right).$$

The function $D_G$ is the analogue on $G$ of $\eta_\mathfrak{g}$. The polynomial $\det(1 + t - \operatorname{Ad} g)$ also begins with the term $t^l$ and one writes

$$\det(1 + t - \operatorname{Ad} g) = t^l D_G(g) + t^{l+1}R'(g,t).$$

The functions $\eta_\mathfrak{g}$ and $D_G$ take values in $\Omega$, the ground field. Recall [W] that on $\Omega$ there is defined a natural absolute value

$$|\ \ |: \Omega \to \mathbf{R}.$$

We will need the real-valued functions gotten by taking absolute values of $\eta_\mathfrak{g}(x)$ or $D_G(x)$. Thus we write

$$|\eta_\mathfrak{g}|(x) = |\eta_\mathfrak{g}(x)| \qquad |D_G|(x) = |D_G(x)|.$$

Consider a distribution $D \in J(\omega)$. Harish-Chandra ([1978], Theorem 3) shows that $\hat{D}$ is locally $L^1$, locally constant on $\mathfrak{g}'$, the set of regular elements in $\mathfrak{g}$, and that $|\eta_\mathfrak{g}|^{1/2}\hat{D}$ is locally bounded on $\mathfrak{g}$. These results apply in particular to the Fourier transforms $\hat{\nu}_x$ of the invariant measures on nilpotent orbits.

Finally we can describe characters. Let $\theta_\pi$ be the character of an irreducible admissible representation $\pi$ of $G$. Then $\theta_\pi$ is a locally $L^1$ function, locally constant on $G'$, and the function $|D_G|^{1/2}\theta_\pi$ is locally bounded. Furthermore, if $\gamma \in G$ is a semisimple element, then the behavior of $\theta_\pi$ near $\gamma$ is as follows. Let $M$ be the centralizer of $\gamma$ in $G$, and $\mathfrak{m}$ the Lie algebra of $M$. Let $C$ be a small neighborhood of $0$ in $\mathfrak{m}$, on which the exponential map exp is defined. Then $\gamma \exp C$ is a neighborhood of $\gamma$ in $M$, and $\mathrm{Ad}\, G(\gamma \exp C)$ is an $\mathrm{Ad}\, G$-invariant neighborhood of $\gamma$ in $G$. If $C$ is chosen sufficiently small, then we may write

$$\theta_\pi(\gamma \exp Y) = \sum_\xi c_\xi \hat{\nu}_\xi(Y) \qquad Y \in C$$

where $\xi$ runs over the nilpotent conjugacy classes *in* $\mathfrak{m}$, and the $c_\xi$ are complex numbers depending on $\pi$ and $\gamma$. Roughly, we may say that locally $\theta_\pi$ is a linear combination of Fourier transforms of nilpotent orbits.

This result is analogous to the knowledge that an invariant eigendistribution on a real reductive group is a locally $L^1$ function, and is given locally on each Cartan subgroup by $D_G^{-1/2}$ times a linear combination of certain exponentials. What is still missing, except in special cases, are results as precise as Harish-Chandra's character formula for discrete series of real groups.

As was already stated, the above results apply when the ground field $\Omega$ has characteristic zero. When $\Omega$ is of positive characteristic, much less is known. However, in [1980] it is shown that in all characteristics characters are locally constant on the regular set. The techniques of this paper are different from those of [1973] or [1978]. The basic fact used is the submersiveness of a certain map ([1980], Theorem 1). Also, the following interesting fact is established. Let $K$ be a "good" maximal compact subgroup of $G$ in the sense of Bruhat and Tits [BT]. For an admissible representation $\pi$ of $G$ on a vector space $V$ and a regular element $x$ in $G$, set

$$T_x = \int_K \pi(kxk^{-1})\, dk.$$

Then $T_x \in \mathrm{End}^\circ(V)$, and in particular $T_x$ has finite rank. Then evidently $\theta_\pi(x) = \mathrm{tr}\, T_x$.

The study of the map $F_f$ or, in common current parlance, orbital integrals, is entwined with and in some sense dual to the study of characters, and currently is in roughly the same state. One has substantial knowledge of the qualitative behavior of $F_f$, but precise control analogous to the "jump formula" for real groups is still lacking. This would be provided by an explicit description of the "Shalika germs" which will be defined and discussed below.

First recall the definition of $F_f$. Fix a Cartan subgroup $A \subseteq G$, and fix an invariant measure $dg^*$ on $G/A$. Then for $f \in C_c^\infty(G)$ and $a \in A'$, where $A' = G' \cap A$ one sets

$$F_f(a) = |D_G(a)|^{1/2} \int_{G/A} f(gag^{-1})\, dg^*. \tag{4}$$

There is a close analogue of $F_f$ for the Lie algebra. If $\mathfrak{a}$ is the Lie algebra of $A$ and $x \in \mathfrak{a}$, put

$$\phi_f(x) = |\eta_\mathfrak{a}(x)|^{1/2} \int_{G/A} f(\mathrm{Ad}\, g(x))\, dg^* \qquad f \in C_c^\infty(\mathfrak{g}). \tag{5}$$

In formulas (4) and (5) the functions $D_G$ and $\eta_\mathfrak{a}$ are the ones defined above in (2) and (3).

Formulas (4) and (5) are very near parallels of the definitions of $F_f$ and $\phi_f$ for real groups. However, we should remark on one difference that has not been important up to now, but which needs clarification before these maps can be thoroughly understood. For real groups the function $D_G$ is a smooth function. The function $|D_G|^{1/2}$ is not smooth. However, on any Cartan subgroup $A$, there is a function $D_G^{A\ 1/2}$ such that $(D_G^{A\ 1/2})^2 = D_G$ on $A$ and which yields an optimal theory of orbital integrals. In other words, if one multiplies $|D_G|^{1/2}$ by appropriate phase factors, one obtains a better theory, in which the functions $F_f$ are as smooth as possible. It is the function $D_G^{A\ 1/2}$ rather than the positive $|D_G|^{1/2}$ which is used in Harish-Chandra's definition of $F_f$ for real groups. It would seem that some modification of $|D_G|^{1/2}$ by appropriate phases is also preferable in the $p$-adic case; but precisely how to do this has not yet been made clear.

The parts of Harish-Chandra's work devoted to $F_f$ on $p$-adic groups are [1970d], Part VI; [1973], §16; [1978] §§3, 4, 8, 9; and [1980], §5. In [1970d] it is already established that when the ground field $\Omega$ has characteristic 0, the function $F_f$ is bounded for any $f \in C_c^\infty(G)$. However, it is not shown that $F_f$ is locally constant. This is done in [1973], §16 (for the proofs see [Si]) by the same technique that establishes local constancy of characters. In fact, local constancy of $F_f$ is even established for $f$ belonging to Harish-Chandra's Schwartz space $\mathcal{C}(G)$ (see below for the definition of $\mathcal{C}(G)$); however local boundedness of $F_f$ for $f \in \mathcal{C}(G)$ is not proven. Local constancy of $F_f$ for $\Omega$ of positive characteristic, and $f \in \mathcal{C}(G)$ is proved in [1980], but boundedness not.

For real groups the behavior of $F_f(x)$ as $x$ approaches singular points is of much interest. This is also true for $p$-adic groups. The standard approach to the study of this limiting behavior is provided by an observation of Shalika [Sh]. Harish-Chandra's version of this result is Theorem 14 of [1978], which applies to the Lie algebra. As above, we let $\nu_x$ denote the invariant measures on the nilpotent $G$-conjugacy classes $\mathcal{O}_x$ in $\mathfrak{g}$. Fix a Cartan subalgebra $\mathfrak{a} \subseteq \mathfrak{g}$. Then there are functions $\Gamma_\mathcal{O}^\mathfrak{a} = \Gamma_\mathcal{O}$, for each nilpotent conjugacy class $\mathcal{O} = \mathcal{O}_x$ such that

$$\phi_f(a) - \sum \nu_x(f)\Gamma_\mathcal{O}(a)$$

vanishes in some neighborhood of the origin in $\mathfrak{a}$. The functions $\Gamma_\mathcal{O}$ are strictly speaking, only germs of functions, an equivalence class of functions equal in some neighborhood of 0; they are generally called "Shalika germs". However, they may be defined uniquely on all of $\mathfrak{a}$ by requiring them to satisfy an appropriate condition of homogeneity, namely

$$\Gamma_\mathcal{O}(t^2 a) = |t|^m \Gamma_\mathcal{O}(a) \qquad t \in \Omega^x, \ a \in \mathfrak{a}$$

where $m$ is an integer depending on $\mathcal{O}$. There are analogous germ expansions around non-zero singular points of $\mathfrak{a}$, and for $F_f$.

Clearly the functions $\Gamma_\mathcal{O}$ control the singularities of $\phi_f$. Considerable effort consequently has been devoted to determining them [Vi], [Ko], [Re1], [Re2], [Ro2]. However, they remain mysterious. Harish-Chandra shows in [1978], §9 that the $\Gamma_\mathcal{O}^\mathfrak{a}$ separate the nilpotent orbits $\mathcal{O}$ in the sense that the only linear combination $\Sigma c_\mathcal{O} \Gamma_\mathcal{O}^\mathfrak{a}$ which vanishes for all Cartan subalgebras $\mathfrak{a}$ is the trivial combination with all $c_\mathcal{O} = 0$. This is implied by his result ([1978], Theorem 10) that all invariant distributions annihilate any function $f \in C_c^\infty(\mathfrak{g})$ such that $\phi_f = 0$ for all Cartan subalgebras $\mathfrak{a}$. He also shows that $\Gamma_0^\mathfrak{a}$, the germ associated to the origin, is zero if $\mathfrak{a}$ is not an elliptic Cartan subalgebra, and on an elliptic Cartan is a constant times $|\eta_\mathfrak{a}|^{1/2}$. The value of the constant was conjectured by Harish-Chandra and determined by Rogawski [Ro1].

The third major topic in Harish-Chandra's work is the Plancherel Formula, and the associated analysis on $G$: the constant term, induced series of representations, wave packets, Eisenstein integrals, $c$-functions. For real groups, character theory and orbital integrals also contribute to the Plancherel Formula, but as explained above, for $p$-adic groups, they have not reached the state necessary for use in the Plancherel Formula. However certain technical aspects of the $p$-adic situation allow Harish-Chandra to complete his program without knowledge of the discrete series. The resulting Plancherel Formula is not as explicit as the one for real groups. But except for the lack of explicit knowledge of the discrete series, the Plancherel Formula for $p$-adic groups is quite parallel to that for real groups. Hence our discussion of it will be relatively brief, and will focus on the aspects which are particular to $p$-adic groups.

Since the theory of the constant term for real groups depends heavily on the differential equations supplied by the center of the enveloping algebra, its transference to $p$-adic groups requires new techniques. Harish-Chandra made a preliminary essay in [1970d], establishing the existence of the constant term contingent on a conjecture. However a firmer basis was provided shortly after by a result of Jacquet [J].

Let $\pi$ be an admissible representation of $G$ on the vector space $V$. Let $P \subseteq G$ be a parabolic subgroup with unipotent radical $N$. Let $V(N)$ be the subspace of $V$ spanned by vectors of the form $\pi(n)v - v$. Since $N$ is normalized by $P$, one can easily see that $V(N)$ is stable under $P$. Hence there is defined a representation of $P$ on $V_N = V/V(N)$. From the definition of $V(N)$ it is clear that $N$ acts trivially on $V_N$. Hence the action of $P$ on $V_N$ factors to an action $\pi_M$ of $P/N \simeq M$ on $V_N$. Jacquet showed (for $G = GL_n(\Omega)$, but with a general proof) that $V_N$ is also an admissible module for $M$. It turns out that $\pi$ is supercuspidal if and only if $V_N = \{0\}$.

Recall that $\mathcal{C}(G)$ is the space of all matrix coefficients of all admissible representations of $G$. Using Jacquet's theorem, Harish-Chandra establishes the existence of the constant term in a very strong form. Let $M$ now denote a Levi component of $P$, so that $M \simeq P/N$. Let $A$ denote the center of $M$. Let $\{\alpha_i\}$ be the simple roots of $A$ acting on $N$ by conjugation. Each $\alpha_i$ is a rational homomorphism

$$\alpha_i : A \to \Omega^\times.$$

For a given $t > 0$, set

$$A^+(t) = \{a \in A : |\alpha_i(a)| > t \text{ for each } \alpha_i\}.$$

As above let $\delta_P$ denote the modular function of $P$. Then Harish-Chandra shows ([1973], §6; see [Si], Chapter 2 for proofs) that given $f \in \mathcal{C}(G)$, there is a unique $f_P \in \mathcal{C}(M)$ such that, for any compact set $\omega \subseteq M$, one has

$$f(ma) = \delta_P(ma)^{1/2} f_P(ma) \qquad m \in \omega, \; a \in A^+(t)$$

for $t$ sufficiently large. Thus in the $p$-adic case one eventually has actual equality between $f$ and $f_P$, its *constant term along $P$*, and not merely an asymptotic relation.

Since $f_P$ is in $\mathcal{C}(M)$, it is in particular $A$-finite. The finite dimensional space spanned by the $A$-translates of $f_P$ will be spanned by generalized eigenspaces for $A$. If $Y$ is such an eigenspace, then there is a quasicharacter $\psi$ of $A$ such that $\lambda_a - \psi(a)$ is nilpotent on $Y$. (Here $\lambda_a$ is the left action of $A$, as in (1).) Harish-Chandra calls these characters $\psi$ the *exponents* of $f$ relative to the pair $(P, A)$. The $A$-finiteness of $f_P$ takes the place in many situations of the differential equations governing the constant term in the real case, and the exponents take the place of infinitesimal characters. To place limits on the elements of $\mathcal{C}(M)$ which can be constant terms of matrix coefficients of a fixed irreducible representation of $G$, Harish-Chandra develops a sharp form of the theory of intertwining operators for induced representations, pioneered by Bruhat for real groups [Br1].

With the constant term, Harish-Chandra can proceed toward the Plancherel Formula along very much the same road used in the real case. The theory of induced representations, of the Eisenstein integral, the $c$-functions, the Maass–Selberg relations, of wave packets and the Schwartz space proceed very much as in the real case. Already in [1973] he is able to state the Plancherel Formula for the wave packets constructed from a given series of induced representations (Theorem 34). In this Plancherel Formula, the Plancherel measure is determined by a product formula, as it is for real groups. However, it is not explicitly known. But luckily, it is not necessary to know it explicitly in the $p$-adic case. This is because any $p$-adic torus $A$, modulo an open compact subgroup, is discrete. Hence $\hat{A}^0$, the identity component of the Pontrjagin dual of $A$, is compact. It is a real torus—a product of circles. If $M$ is the Levi component of a parabolic subgroup $P$, and $A$ the center of $M$, then a series from which one builds wave packets consists of representations of the form $\pi \otimes \chi$ where $\pi$ is a square integrable (mod $A$) unitary representation of $M$ and $\chi \in \hat{A}^0$. In the real case, $\hat{A}^0$ is a vector group, and it is necessary to know that Plancherel measure on $\hat{A}^0$ grows only polynomially at $\infty$ in order to be able to control the Fourier transform. In fact, for real groups, Harish-Chandra gives an explicit expression for the Plancherel measure, and one can see directly that this expression has moderate growth. But since $\hat{A}^0$ is compact for $p$-adic groups, the growth of Plancherel measure is not an issue, and no explicit knowledge of it is necessary.

We will close this account by describing what is involved in the passage from the "local" Plancherel Formula of [1973] to the full formula announced in [1977b].

First the notion of constant term described above must be modified so that matrix coefficients of representations which are square integrable but not supercuspidal have zero constant term. This leads to the notion of the weak constant term.

Let $K$ be a good maximal compact subgroup of $G$. Let $\Xi$ be, as for real groups, the $K$-spherical matrix coefficient of the representation unitarily induced from the trivial representation of a minimal parabolic subgroup of $G$. Let $\sigma$ be the measure of slow growth on $G$; roughly $\sigma$ is the logarithm of $\Xi$ (see [1973]; §14). For each compact open $K_1 \subseteq K$, one defines $\mathcal{C}(G//K_1)$ to be the space of $K_1$ bi-invariant functions $f$ on $G$ such that

$$|f| \leqslant c(f, k)\Xi(1+\sigma)^{-k}$$

for an appropriate positive number $c(f, k)$ and for each positive number $k$. Then $\mathcal{C}(G)$, the *Schwartz space* of $G$ is the union over compact open $K_1$ of the spaces $\mathcal{C}(G//K_1)$.

Consider $\phi \in \mathcal{Q}(G)$. One says $\phi$ satisfies the *weak inequality* if for some constant $c$ and some integer $k$ one has the estimate

$$|\phi| \leq c\Xi(1 + \sigma)^k.$$

Such a $\phi$ is called *tempered*, and the subspace of $\mathcal{Q}(G)$ consisting of tempered matrix coefficients is denoted $\mathcal{Q}^w(G)$.

Let $f$ be in $\mathcal{Q}^w(G)$, and let $f_P$ be its constant term along the parabolic subgroup $P$. Write

$$f_P = \sum f_{P,\psi}$$

where $\psi$ runs over the exponents of $f$, and $f_{P,\psi}$ is the component of $f$ in the generalized $\psi$-eigenspace of the $A$ translates of $f$, $A$ being the center of a Levi component $M$ of $P$. The sum of the $f_{P,\psi}$ over those $\psi$ which are unitary is called the *weak constant term* of $f$ along $P$, and is denoted $f_P^w$. It is in $\mathcal{Q}^w(M)$. A *cusp form* on $G$ is an element of $\mathcal{Q}^w(G)$ such that the weak constant terms $(\lambda(g)f)_P^w$ of all left translates of $f$ are zero for all proper parabolic subgroups $P$ of $G$. The space of cusp forms on $G$ is denoted $^0\mathcal{Q}^w(G)$.

There is a notion of constant term for $\mathcal{C}(G)$ that is dual to that for $\mathcal{Q}^w(G)$. For $f \in \mathcal{C}(G)$ and $P$ a parabolic subgroup of $G$, set

$$f^{(P)}(m) = \delta_P(m)^{1/2} \int_N g(mn)\, dn \qquad m \in P/N$$

where $N$ is the unipotent radical of $P$. Let $^0\mathcal{C}(G)$ denote the subspace of $\mathcal{C}(G)$ consisting of those $f$ such that $(\lambda(g)f)^{(P)} = 0$ for all proper parabolic subgroups.

If $G$ were semisimple rather than reductive, then it is easy to see that $^0\mathcal{Q}^w(G) \subseteq {}^0\mathcal{C}(G)$. If $G$ possesses a non-compact center, then one must formulate the relation between $^0\mathcal{Q}^w$ and $^0\mathcal{C}$ in a more complicated way, but the situation is in essence the same. So to simplify the discussion we take $G$ semisimple. Then $^0\mathcal{Q}^w(G)$ consists of matrix coefficients of the discrete series, while $^0\mathcal{C}(G)$ is the subspace of $\mathcal{C}(G)$ which cannot be obtained by inducing tempered representations from proper parabolic subgroups. Hence to pass from the Plancherel Theorem of [1973], §17, to the Plancherel Formula for $G$, one needs to show that in fact $^0\mathcal{Q}^w(G) = {}^0\mathcal{C}(G)$. This essentially amounts to the statement that (for $G$ semisimple), the representation of $G$ on the span of the left translates of $f \in {}^0\mathcal{C}(G)$ is admissible, i.e., consists of finitely many irreducible summands. This is a consequence of Lemma 4 of [1977b]. A crucial step in the proof of Lemma 4 is Theorem 11, which states that if $\theta \in {}^0\mathcal{C}(G)$, then the map

$$f \rightarrow \int_{G/Z} dy^* \int_G f(x)\theta(yxy^{-1})\, dx$$

is well defined and yields a tempered distribution on $G$. Although the details of this result are not yet published, in flavor it reminds one of the analysis of supercuspidal characters in [1970d].

The form of Harish-Chandra's $p$-adic Plancherel Theorem is this. The space $\mathcal{C}(G)$ is the orthogonal direct sum of wave packets formed from series of representations induced unitarily from discrete series of (the Levi components of) parabolic subgroups of $P$. Moreover if two such series of induced representations yield the same subspace of $\mathcal{C}(G)$, then the parabolics from which they are induced are associate, and the representations of the Levi components are conjugate. To complete the analogy with real groups, one needs only to explicitly determine the discrete series; this is the outstanding problem left in $p$-adic representation theory.

## REFERENCES

[Ar]  J. Arthur, On some problems suggested by the trace formula, Special Year in Representation Theory, U. of Maryland, Nov. 1982, Springer Lecture Notes, to appear.

[Be]  I. N. Bernstein, All reductive $p$-adic groups are tame, *Fun. Anal. and App.* **8** (1974), 91–93.

[Bo]  A. Borel, Automorphic $L$-functions, in *Automorphic Forms, Representations, and L-functions, Proc. Sym. Pure Math.*, **XXXIII**, Part 2, Amer. Math. Soc. Providence, R.I., 1974, 27–63.

[BW]  A. Borel and N. Wallach, Continuous cohomology, discrete subgroups, and representations of reductive groups, *Ann. of Math. Studies* **94**, Princeton University Press, Princeton, 1980.

[Br1]  F. Bruhat, Sur les representations induites des groupes de Lie, *Bull. Soc. Math. France*, **84** (1956), 97–205.

[BT]  F. Bruhat and J. Tits, Groupes reductifs sur un corps local, I, Donnees radicielles valuees, *Pub. Math. I.H.E.S.* **41** (1972), 5–251.

[CM]  W. Casselman and D. Milicic, Asymptotic behavior of matrix coefficients of admissible representations, *Duke Math. J.* **49** (1982), 869–930.

[He]  G. Henniart, La conjecture de Langlands locale pour GL(3), *I.H.E.S. Notes* (1982).

[Ho]  R. Howe, The Fourier transform and germs of characters, *Math. Ann.* **208** (1974), 305–322.

[H]  H. Jacquet, Representations des groupes lineaires $p$-adiques, Theory of Group Representations and Harmonic Analysis (C.I.M.E., II, Ciclo, Montecatini terme, 1970) Edizione Cremonese, Roma, 1971, 119–220.

[JL]  H. Jacquet and R. Langlands, Automorphic forms on GL(2). *Lec. Notes Math.*, **260**, Springer-Verlag, Berlin, New York, 1972.

[Ko]  R. E. Kottwitz, Orbital integrals on GL(3), *Am. J. Math*, **102** (1980), 327–384.

[K]  P. Kutzko, The Langlands conjecture for $GL_2$ of a local field, *Ann. Math.* **112** (1980), 381–412.

[KM]  P. Kutzko and A. Moy, On the local Langlands Conjecture in prime dimension, preprint.

[L1]  R. Langlands, Problems in the theory of automorphic forms, *Lectures in Modern Analysis and Applications, Lecture Notes in Math.*, **170**, Springer-Verlag, New York, 1973, 18–86.

[L2]  R. Langlands, On the classification of irreducible representations of real reductive groups, *Notes, I.A.S.*, Princeton, 1973.

[Lu1]  G. Lusztig, Irreducible representations of finite classical groups, *Inv. Math.* **43** (1977), 125–175.

[Lu2] G. Lusztig, Some examples of square integrable representations of semisimple $p$-adic groups, preprint, *I.H.E.S.*, March 1982.

[M]   A. Moy, Local constants and the tame Langlands correspondence, Thesis, University of Chicago, 1982.

[MS]  A. Moy and P. Sally, Supercuspidal representations of $SL_n$ over a $p$-adic field: the tame case, preprint.

[Ra]  R. Rao, Orbital integrals in reductive groups, *Ann. of Math.* **96** (1972), 505–510.

[Re1] J. Repka, Shalika's germs for $p$-adic $GL(n)$: the leading term, preprint.

[Re2] J. Repka, Shalika's germs for $p$-adic $GL(n)$, II: the subregular term, preprint.

[Ro1] J. Rogawski, An application of the building to orbital integrals, *Comp. Math.* **42** (1981), 417–423.

[Ro2] J. Rogawski, Some remarks on Shalika germs, preprint.

[Sh]  J. Shalika, A theorem on semisimple $p$-adic groups, *Ann. Math.* **95** (1972), 226–242.

[Si]  A. Silberger, Introduction to Harmonic Analysis on Reductive $p$-adic Groups, *Math. Notes*, Princeton University Press, Princeton, N.J., 1979.

[T]   J. Tunnell, Report on the local Langlands conjecture for $GL_2$, *Proc. Symp. Pure Math.*, **XXXIII**, Part 2, Amer. Math. Soc., Providence, R.I., 1979, 135–138.

[V]   V. Varadarajan, Harmonic Analysis on Real Reductive Groups, *Lecture Notes in Math.*, **576**, Springer-Verlag, New York, 1976.

[Vi]  M. F. Vigneras, Caracterisation des integrales orbitales sur un groupe reductif $p$-adique, *J. Fac. Sci. Tokyo*, Sec. IA, **28**, (1982), 945–961.

[Vo]  D. Vogan, Representation of Real Reductive Lie Groups, *Progress in Math.*, **15**, Birkhauser, Boston, Basel, Stuttgart, 1981.

[Wa]  N. Wallach, Asymptotic expansions of generalized matrix entries, special year on representation theory, University of Maryland, Vol. I, *Lecture Notes in Mathematics*. Springer-Verlag, Berlin, Heidelberg, New York, Tokyo, 1983.

[W]   A. Weil, *Basic Number Theory*, 2nd ed., Grund. Math. Wiss., **144**, Springer-Verlag, Berlin, Heidelberg, New York, 1973.

# Permissions

Springer-Verlag would like to thank the original publishers of Harish-Chandra's papers for granting permissions to reprint specific papers in this collection. The following list contains the credit lines for those articles.

[1944a] Reprinted from *Proc. Royal Soc. A.* **183**, ©1944 by The Royal Society of London.
[1944b] Reprinted from *Proc. Royal Soc. A.* **183**, ©1944 by The Royal Society of London.
[1945a] Reprinted from *Proc. Indian Acad. Sci. Sect. A.* **21**, ©1945 by The Indian Academy of Sciences.
[1945b] Reprinted from *Proc. Indian Acad. Sci. Sect. A.* **22**, ©1945 by The Indian Academy of Sciences.
[1946a] Reprinted from *Proc. Royal Soc. A.* **185**, ©1946 by The Royal Society of London.
[1946b] Reprinted from *Proc. Royal Soc. A.* **185**, ©1946 by The Royal Society of London.
[1946c] Reprinted from *Proc. Indian Acad. Sci. Sect. A.* **23**, ©1946 by The Indian Academy of Sciences.
[1946d] Reprinted from *Proc. Royal Soc. A.* **186**, ©1946 by The Royal Society of London.
[1947a] Reprinted from *Proc. Camb. Phil. Soc.* **43**, ©1947 by Cambridge University Press.
[1947b] Reprinted from *Phys. Rev.*, **71**, ©1947 by The American Physical Society.
[1947c] Reprinted from Report of an International Conf. on Fundamental Particles and Low Temperatures held at the Cavendish Laboratory, Cambridge, (1946). Vol I, Fundamental Particles, ©1947 by The Institute of Physics.
[1947d] Reprinted from *Proc. Royal Soc. A.* **189**, ©1947 by The Royal Society of London.
[1948a] Reprinted from *Proc. Royal Soc. A.* **192**, ©1948 by The Royal Society of London.
[1948b] Reprinted from *Phys. Rev.* **74**, ©1948 by The American Physical Society.
[1949a] Reprinted from *Ann. of Math.* **50**, ©1949 by Princeton University Press.
[1949b] Reprinted from *Ann. of Math.* **50**, ©1949 by Princeton University Press.
[1950a] Reprinted from *Proc. Amer. Math. Soc.* **1**, ©1950 by The American Mathematical Society.
[1950b] Reprinted from *Proc. Amer. Math. Soc.* **1**, ©1950 by The American Mathematical Society.
[1950c] Reprinted from *Ann. of Math.* **51**, ©1950 by Princeton University Press.
[1951a] Reprinted from *Trans. Amer. Math. Soc.* **70**, ©1951 by The American Mathematical Society.

# Integrable and Square-Integrable Representations of a Semisimple Lie Group

HARISH-CHANDRA

DEPARTMENT OF MATHEMATICS, COLUMBIA UNIVERSITY

*Communicated by Paul A. Smith, March 11, 1955*

Let $G$ be a connected semisimple Lie group. We shall suppose for simplicity that the center of $G$ is finite. Let $\pi$ be an irreducible unitary representation of $G$ on a Hilbert space $\mathfrak{H}$. We say that $\pi$ is integrable (square-integrable) if there exists an element $\psi \neq 0$ in $\mathfrak{H}$ such that the function $(\psi, \pi(x)\psi)$ $(x \in G)$ is integrable (square-integrable) on $G$, with respect to the Haar measure. Assuming that the Haar measure $dx$ has been normalized in some way once for all and that $\pi$ is square-integrable, we denote by $d_\pi$ the positive constant given by the relation[1]

$$\int_G |(\psi, \pi(x)\psi)|^2 \, dx = \frac{1}{d_\pi},$$

where $\psi$ is any unit vector in $\mathfrak{H}$. Let $C_c^\infty(G)$ denote the set of all complex-valued functions on $G$ which are everywhere indefinitely differentiable and which vanish outside a compact set. Then the following result is an easy consequence of the Schur orthogonality[1] relations.

**Theorem 1.** *Let $\pi$ be an irreducible unitary representation of $G$ on $\mathfrak{H}$ which is square-integrable, and let $T_\pi$ denote the character[2] of $\pi$. Then, if $f \in C_c^\infty(G)$,*

$$T_\pi(f) = d_\pi \int_G dx \left( \int_G f(xyx^{-1})(\phi, \pi(y)\phi) \, dy \right),$$

*where $\phi$ is any unit vector in $\mathfrak{H}$.*

Now suppose $G$ has a Cartan subgroup $A$ which is compact. We extend $A$ to a maximal compact subgroup $K$. Let $\mathfrak{g}_0$ be the Lie algebra of $G$ and $\mathfrak{g}$ its complexification. We shall assume for simplicity that $G$ has a finite-dimensional faithful representation and therefore there exists a complex analytic group $G_c$ with the Lie algebra $\mathfrak{g}$ such that $G$ is the (real) analytic subgroup of $G_c$ corresponding to $\mathfrak{g}_0$. Let

$\mathfrak{h}_0$ and $\mathfrak{k}_0$ be the subalgebras of $\mathfrak{g}_0$ which correspond to $A$ and $K$, respectively. We define $\mathfrak{p}_0$, $\mathfrak{h}$, $\mathfrak{k}$, $\mathfrak{p}$ as in a previous note[3] and introduce a lexicographic order among roots (of $\mathfrak{g}$ with respect to $\mathfrak{h}$). For any root $\alpha$ we define $X_\alpha$, $X_{-\alpha}$ and $H_\alpha = [X_\alpha, X_{-\alpha}]$ as before,[3] so that $\alpha(H_\alpha) = 2$. Put $\mathfrak{n}_+ = \sum_{\alpha > 0} CX_\alpha$, $\mathfrak{n}^- = \sum_{\alpha > 0} CX_{-\alpha}$, where $C$ is the field of complex numbers and $\alpha$ runs over all positive roots. Then $\mathfrak{n}_+$, $\mathfrak{n}_-$ are subalgebras of $\mathfrak{g}$. Let $A_c$, $N_c^+$, $N_c^-$ be the complex analytic subgroups of $G_c$ corresponding to $\mathfrak{h}$, $\mathfrak{n}_+$, $\mathfrak{n}_-$, respectively. Then $G_c^0 = N_c^- A_c G$ is an open submanifold of $G_c$. If $\xi$ is a holomorphic character of $A_c$, we can choose a (unique) linear function $\Lambda$ on $\mathfrak{h}$ such that $\xi(\exp H) = e^{\Lambda(H)}$ ($H \in \mathfrak{h}$). We denote by $\mathfrak{H}_\Lambda$ the space of all holomorphic functions $\phi$ on $G_c^0$ such that

(i)      $\phi(naw) = \xi(a)\phi(w)$     $(n \in N_c^-, a \in A_c, w \in G_c^0)$,

(ii)     $\|\phi\|^2 = \int_G |\phi(x)|^2 \, dx < \infty$.

Then $\mathfrak{H}_\Lambda$ is a Hilbert space under the norm $\|\cdot\|$, and we get a representation $\pi_\Lambda$ of $G$ on $\mathfrak{H}_\Lambda$ if we put $(\pi_\Lambda(x)\phi)(y) = \phi(yx)$ ($\phi \in \mathfrak{H}_\Lambda$; $x, y \in G$). If $\mathfrak{H}_\Lambda \neq \{0\}$, $\pi_\Lambda$ is an irreducible unitary representation[4] which is square-integrable. Let $2\rho$ denote the sum of all positive roots. In case $G$ is simple and not compact, the following four conditions are both necessary and sufficient in order that $\mathfrak{H}_\Lambda \neq \{0\}$:

1.    The first Betti number of $G$ is 1.
2.    Every noncompact[3] positive root is totally positive.[3]
3.    $\Lambda(H_\alpha)$ is a nonnegative integer for every positive compact[3] root $\alpha$.
4.    $\Lambda(H_\beta) + \rho(H_\beta)$ is a negative integer for every positive noncompact[3] root $\beta$.

Let $2\rho_+$ denote the sum of all positive noncompact roots. Then, in the presence of the first three conditions, the following condition is sufficient to insure the integrability of $\pi$:

4'. $\Lambda(H_\beta) + \rho(H_\beta) < -2\rho_+(H_\beta) + 1$ for every positive noncompact root $\beta$. ·

From now on we shall suppose that $G$ is simple and not compact and that conditions 1 and 2 are fulfilled. We shall also assume that $G_c$ is simply connected. Then it is possible[3] to choose a fundamental system $(\alpha_0, \alpha_1, \ldots, \alpha_l)$ of positive roots such that $\alpha_0$ is noncompact, while $\alpha_1, \ldots, \alpha_l$ are all compact. Let $\Lambda_i$ denote the linear function on $\mathfrak{h}$ given by $\Lambda_i(H_{\alpha_j}) = \delta_{ij}$ ($0 \leqslant i, j \leqslant l$). Since $G_c$ is simply connected, there exists a holomorphic character $\xi_i$ of $A_c$ such that $\xi_i(\exp H) = e^{\Lambda_i(H)}$ ($H \in \mathfrak{h}$, $0 \leqslant i \leqslant l$). Moreover, from the theory of finite-dimensional representations of $G_c$ one can deduce the existence of a unique holomorphic function $g_i$ on $G_c$ such that $g_i(nan') = \xi_i(a)$ ($n \in N_c^-, a \in A_c, n' \in N_c^+$). Put $g = g_0^{m_0} g_1^{m_1} \cdots g_l^{m_l}$, where $m_0, \ldots, m_l$ are nonnegative integers, and let $g_y(x) = g(xy)$ ($x, y \in G_c$). Then, if $V$ is the vector space spanned over $C$ by the functions $g_y$ ($y \in G_c$) and $\pi$ is the representation of $G_c$ on $V$ given by $(\pi(y)f)(x) = f(xy)$ ($x, y \in G_c, f \in V$), $\pi$ is a finite-dimensional irreducible representation of $G_c$. The corresponding representation of $\mathfrak{g}$ is complex-linear, and its highest weight is $m_0\Lambda_0 + \cdots + m_l\Lambda_l$. On the other hand, suppose that $\Lambda$ is a linear function on $\mathfrak{h}$ satisfying conditions 3 and 4. Put $\lambda_i = \Lambda(H_{\alpha_i})$ ($0 \leqslant i \leqslant l$). Then $\lambda_0, \ldots, \lambda_l$ are all integers, and $\Lambda = \lambda_0\Lambda_0 + \cdots + \lambda_l\Lambda_l$. Moreover, $\lambda_0 < 0$, while $\lambda_1, \ldots, \lambda_l \geqslant 0$. Put $g_\Lambda = g_0^{\lambda_0} g_1^{\lambda_1} \cdots g_l^{\lambda_l}$. Then $g_\Lambda$ is a meromorphic function on $G_c$. However, it can be shown that $g_0$ is never zero on $G_c^0$, and so $g_\Lambda$ is holomorphic on $G_c^0$. Also, one can prove that $g_\Lambda \in \mathfrak{H}_\Lambda$, and therefore $\mathfrak{H}_\Lambda$ is the closure of the space spanned by the right translates of $g_\Lambda$ under $G$. Thus the analogy with the finite-dimensional case mentioned above is rather close.

Let $F$ be the set of all linear functions $\Lambda$ on $\mathfrak{h}$ which satisfy conditions 3 and 4. For every $\Lambda \in F$ we have defined a square-integrable representation $\pi_\Lambda$ above. Put $d_\Lambda = d_{\pi_\Lambda}$. Then, if $m$ is the number of totally positive roots, we have the following result.

**Theorem 2.** *It is possible to normalize the Haar measure of $G$ in such a way that*

$$d_\Lambda = (-1)^m \prod_{\alpha > 0} \left\{ \frac{\Lambda(H_\alpha) + \rho(H_\alpha)}{\rho(H_\alpha)} \right\}$$

*for every $\Lambda \in F$.*

The analogy with Weyl's formula[5] for the degree of an irreducible finite-dimensional representation in terms of its highest weight is obvious.

Let $\mathfrak{E}$ be the set of all equivalence classes of irreducible unitary representations of $G$. Let $\mathfrak{E}_0$ denote the subset of $\mathfrak{E}$ consisting of those classes $\omega$ which correspond to square-integrable representations. If $\omega \in \mathfrak{E}_0$ and $\pi \in \omega$, we put $d_\omega = d_\pi$. Also, define

$$N_\omega(f) = \left\| \int f(x) \pi(x) \, dx \right\|^2 \qquad (f \in C_c^\infty(G), \ \omega \in \mathfrak{E}),$$

where $\pi$ is any representation in $\omega$ and $\|A\|$ denotes the Hilbert-Schmidt norm[2] of an operator $A$. It follows from the work of von Neumann[6] and Mautner[7] and a previous result of mine[8] that there exists a *unique* positive measure $\mu$ on $\mathfrak{E}$ such that

$$\int_G |f(x)|^2 \, dx = \int_{\mathfrak{E}} N_\omega(f) \, d\mu$$

for all $f \in C_c^\infty(G)$. On the other hand, one can prove the following result:

**Theorem 3.** *The $\mu$-measure of a single point $\omega$ in $\mathfrak{E}_0$ is exactly $d_\omega$.*

For any $\Lambda \in F$ let $\omega_\Lambda$ denote the equivalence class of $\pi_\Lambda$. Then $\Lambda \to \omega_\Lambda$ is a one-one mapping of $F$ into $\mathfrak{E}_0$. We denote by $\mathfrak{E}_F$ the image of $F$ under this mapping. It is obvious from Theorems 2 and 3 that we now have an explicit formula for the restriction of the measure $\mu$ on $\mathfrak{E}_F$.

[1] See these PROCEEDINGS, **40**, 1076, 1954, Theorem 2; and R. Godement, *Compt. rend. Acad. sci.* (Paris), **225**, 657–659, 1947.

[2] See these PROCEEDINGS, **37**, 366–369, 1951.

[3] See *ibid.*, **40**, 1078, 1954, hereafter cited as "RVI."

[4]See RVI. The last statement of RVI is not correct. In order to rectify it, we have to replace $\pi_\xi$ there by the definition of $\pi_\Lambda$ given above in the present note.

[5]H. Weyl, *Math. Z.*, **24**, 328–395, 1925.

[6]J. von Neumann, *Ann. Math.*, **50**, 401–485, 1949.

[7]F. I. Mautner, *Ann. Math.*, **52**, 528–555, 1950.

[8]*Trans. Am. Math. Soc.*, **75**, 230, 1953, Theorem 7.

# ON THE CHARACTERS OF A SEMISIMPLE LIE GROUP

HARISH-CHANDRA

Let $G$ be a connected semisimple Lie group and let $Z$ denote its center. If $\pi$ is a representation [2c] of $G$ on a Hilbert space $\mathfrak{H}$ we consider the space $V$ consisting of all finite linear combinations of elements of the form

$$\int f(x)\pi(x)\psi dx \qquad\qquad (f \in C_c^\infty(G),\ \psi \in \mathfrak{H}),$$

where $dx$ is the Haar measure of $G$ and $C_c^\infty(G)$ is the set of all (complex-valued) functions on $G$ which are everywhere indefinitely differentiable and which vanish outside a compact set. $V$ is called the Gårding subspace of $\mathfrak{H}$. Let $R$ and $C$ be the fields of real and complex numbers respectively and $\mathfrak{g}_0$ the Lie algebra of $G$. We complexify $\mathfrak{g}_0$ to $\mathfrak{g}$ and denote by $\mathfrak{B}$ the universal enveloping algebra of $\mathfrak{g}$ [2a]. Then there exists a (uniquely determined) representation $\pi_V$ of $\mathfrak{B}$ on $V$ such that $\pi_V(X)\psi = \lim_{t\to 0} (1/t)\{\pi(\exp tX)\psi - \psi\}$ $(X \in \mathfrak{g}_0,\ \psi \in V,\ t \in R)$. Let $\mathfrak{Z}$ denote the center of $\mathfrak{B}$. We say that $\pi$ is quasi-simple if there exist homomorphisms $\eta$ and $\chi$ of $Z$ and $\mathfrak{Z}$ respectively into $C$ such that $\pi(\zeta)\phi = \eta(\zeta)\phi$, $\pi_V(z)\psi = \chi(z)\psi$ for all $\zeta \in Z$, $z \in \mathfrak{Z}$, $\phi \in \mathfrak{H}$ and $\psi \in V$. $\eta$ is then called the central character and $\chi$ the infinitesimal character of $\pi$. An irreducible unitary representation is automatically quasi-simple [5].

Let $A$ be a bounded linear operator on $\mathfrak{H}$. We say that $A$ is of the trace class or $A$ has a trace if for every complete orthonormal set $(\psi_j)_{j\in J}$ in $\mathfrak{H}$ the series[1] $\sum_{j\in J} (\psi_j, A\psi_j)$ converges absolutely and its sum is independent of the choice of the complete orthonormal set.[2] We call this sum the trace of $A$ and denote it by $\mathrm{Sp}\,A$. Now suppose $\pi$ is quasi-simple and irreducible. Then it can be shown (see [2e]) that for any $f \in C_c^\infty(G)$ the operator $\int f(x)\pi(x)dx$ is of the trace class. If we denote its trace by $T_\pi(f)$ we get a linear function $T_\pi$ on $C_c^\infty(G)$ which is actually a distribution (see [4; and 2e]). We call this distribution the character of $\pi$. Our object is to try to determine $T_\pi$.

---

An address delivered before the New York meeting of the Society on February 25, 1955 by invitation of the Committee to Select Hour Speakers for Eastern Sectional Meetings; received by the editors March 28, 1955.

[1] As usual $(\phi, \psi)$ denotes the scalar product of the two elements $\phi$ and $\psi$ in $\mathfrak{H}$.

[2] Actually it can be shown that this independence of the sum follows automatically from the absolute convergence of the series for every orthonormal base.

389

Let $x \to \text{Ad}(x)$ $(x \in G)$ denote the adjoint representation of $G$. If $\lambda$ is an indeterminate and $I$ is the identity mapping of $\mathfrak{g}_0$ we consider the characteristic polynomial $P_x(\lambda) = \det(\lambda I - \text{Ad}(x))$ of $\text{Ad}(x)$. Let $l$ be the highest integer such that $(\lambda - 1)^l$ divides $P_x(\lambda)$ for every $x \in G$. We expand $P_x(\lambda)$ in powers of $(\lambda - 1)$ and denote by $D(x)$ the coefficient of $(\lambda - 1)^l$ in this expansion. Then it is clear that $D$ is an analytic function on $G$ which, in view of our definition of $l$, cannot be identically zero. The integer $l$ is called the rank of $G$. Let $S$ denote the set of all $x \in G$ for which $D(x) = 0$. Then $S$ is a closed, nowhere dense subset of $G$ and its complement $G'$ is open and everywhere dense in $G$. An element $x \in G$ is called singular or regular according as $x \in S$ or $x \in G'$. Since $\text{Ad}(zx) = \text{Ad}(x)$ $(z \in Z)$ it is obvious that $ZS = S$ and $ZG' = G'$. Also since $P_{yxy^{-1}}(\lambda) = P_x(\lambda)$ $(x, y \in G)$, it follows that $D(yxy^{-1}) = D(x)$ and therefore $ySy^{-1} = S$, $yG'y^{-1} = G'$ $(y \in G)$.

When speaking of a real differentiable (or analytic) manifold $M$ let us agree to include the case when $M$ is not connected but the various connected components of $M$, which are all manifolds in the usual sense, have the same dimension. Under this definition every open subset $U$ of $M$ is again a manifold. Let $C_c^\infty(U)$ denote the set of all complex-valued functions on $M$ which are indefinitely differentiable and which vanish outside some compact subset of $U$. In particular suppose $U$ is an open subset of $G$ and $F$ is a (complex-valued) function on $U$. We say that $F$ is locally summable if it is summable on every compact subset of $U$ with respect to the Haar measure of $G$. $T$ being a distribution on $G$ we say that $T = F$ on $U$ if $F$ is locally summable and

$$T(f) = \int f(x) F(x) dx$$

for all $f \in C_c^\infty(U)$. Our main result may now be stated as follows.

THEOREM 1. *Let $\pi$ be an irreducible quasi-simple representation of $G$ on $\mathfrak{H}$ and let $T_\pi$ denote its character. Then there exists an analytic function $F_\pi$ on $G'$ such that $T_\pi = F_\pi$ on $G'$.*

Although in general $G'$ is not connected, there always exist a finite number of connected components $G_1, \cdots, G_r$ of $G'$ such that $ZG_i \cap ZG_j = \emptyset$ if $i \neq j$ and $G' = \bigcup_{i=1}^r ZG_i$. Moreover if $\eta_\pi$ is the central character of $\pi$ it is easy to show that $F_\pi(zx) = \eta_\pi(z) F_\pi(x)$ $(z \in Z, x \in G')$. Hence the knowledge of $F_\pi$ on $G_1 \cup G_2 \cup \cdots \cup G_r$ is sufficient to determine it completely. On the other hand it is possible to give examples in which $F_\pi$ vanishes everywhere on one of the components $G_i$ without being zero identically on $G'$. However in case $G$ is either a compact or a complex group, $G'$ is connected and then $F_\pi$ is completely determined by its restriction on any nonempty open subset of $G'$.

For any $x \in G$ and $f \in C_c^\infty(G)$ we define the function $f^x$ by $f^x(y) = f(x^{-1}yx)$ ($y \in G$). Then $f^x \in C_c^\infty(G)$ and $T_\pi(f^x) = T_\pi(f)$ (see [2e]). From this it follows that $F_\pi(xyx^{-1}) = F_\pi(y)$ ($x \in G$, $y \in G'$). Now let $A$ be a maximal abelian subgroup of $G$ which is not contained in $S$. (We call such a group a Cartan subgroup of $G$.) Obviously $A$ is closed in $G$. Let $A' = A \cap G'$ and $V = \bigcup_{x \in G} xA'x^{-1}$. Then $V$ is an open subset of $G'$ and it is clear from the above remarks that $F_\pi$ is determined completely on $V$ as soon as we know it on $A'$. Let $\mathfrak{h}_0$ denote the subalgebra of $\mathfrak{g}_0$ corresponding to $A$. Then the complexification $\mathfrak{h}$ of $\mathfrak{h}_0$ is a Cartan subalgebra of $\mathfrak{g}$. Let $W$ be the Weyl group (see [2b]) of $\mathfrak{g}$ with respect to $\mathfrak{h}$ so that $W$ is a finite group of nonsingular linear transformations of $\mathfrak{h}$. If $\Lambda$ is a linear function on $\mathfrak{h}$ and $s \in W$ we define the linear function $s\Lambda$ by the rule $s\Lambda(H) = \Lambda(s^{-1}H)$ ($H \in \mathfrak{h}$). Let $(H_1, \cdots, H_l)$ be a base for $\mathfrak{h}$ over $C$. A function $P$ on $\mathfrak{h}$ is called a polynomial function if there exists a polynomial $p(x_1, \cdots, x_l)$ in $l$ variables $(x_1, \cdots, x_l)$ with coefficients in $C$ such that $P(H) = p(a_1, \cdots, a_l)$ if $H = a_1 H_1 + \cdots + a_l H_l (a_i \in C)$. The degree of $p$ is also called the degree of $P$. Clearly these definitions are independent of the choice of the base $(H_1, \cdots, H_l)$.

THEOREM 2. *Let $\pi$ and $F_\pi$ be as in Theorem 1. Then there exists a linear function $\Lambda$ on $\mathfrak{h}$ with the following property. For any $a \in A'$ we can choose polynomial functions $p_s$ ($s \in W$) on $\mathfrak{h}$ such that*

$$F_\pi(a \exp H) = |D(a \exp H)|^{-1/2} \sum_{s \in W} p_s(H) e^{s\Lambda(H)}$$

*for all $H$ lying sufficiently near zero in $\mathfrak{h}_0$. $\Lambda$ is unique up to an operation of $W$ and if $N$ is the number of elements $\sigma$ in $W$ such that $\Lambda = \sigma\Lambda$, the degree of every $p_s$ is necessarily smaller than $N$.*

In particular if $s\Lambda \neq \Lambda$ except when $s$ is the unit element of $W$, $p_s$ must all be constants. (It is hardly necessary to point out the resemblance of the above formula to the one given by Weyl [6] for the irreducible characters of a compact semisimple Lie group. It should also be compared with the results of Gelfand and Naimark [7] on the unitary characters of the complex classical groups (see also [2f, p. 511])). Although $A'$ is not necessarily connected, it is possible to select a finite set $B_1, \cdots, B_r$ of its connected components such that $A' = \bigcup_{i=1}^r ZB_i$. Therefore in order to determine $F_\pi$ on $A'$, it is sufficient to know $\eta_\pi$ and the restrictions of $F_\pi$ on some nonempty open subsets of $B_1, \cdots, B_r$. Hence if $\Lambda$ is known, Theorem 2 gives us a formula for $F_\pi$ on $A'$ in terms of a finite number of undetermined constants. On the other hand we shall see presently that $\Lambda$ is completely determined (up to an operation of $W$) by the infinitesimal character $\chi_\pi$ of $\pi$.

Two Cartan subgroups $A_1$ and $A_2$ are said to be conjugate if $A_2 = xA_1x^{-1}$ for some $x \in G$. It is always possible to choose a finite number of distinct Cartan subgroups $A_1, \cdots, A_k$ such that every Cartan subgroup is conjugate to exactly one of these. Then if $A_i' = A_i \cap G'$ and $V_i = \bigcup_{x \in G} xA_i'x^{-1}$, $G'$ is the disjoint union of $V_1, \cdots, V_k$. This shows that if $\eta_\pi$ and $\chi_\pi$ are known, $F_\pi$ is completely determined in terms of a finite number of constants.[3]

Now we come to a brief outline of the proof. Let $M$ be a differentiable manifold, $Q$ a linear mapping of $C_c^\infty(M)$ into itself and $x_0$ a point in $M$. We say that $Q$ is a differential operator at $x_0$ if there exists a coordinate system $(t_1, \cdots, t_m)$ valid on an open neighborhood $U$ of $x_0$ and indefinitely differentiable functions $g_{i_1 i_2 \cdots i_p}$ on $U$ ($1 \leq i_1, \cdots,$ $i_p \leq m$, $0 \leq p \leq q$) such that if $f \in C_c^\infty(U)$, $Qf$ is zero outside $U$ and

$$Qf = \sum_{0 \leq p \leq q} \sum_{1 \leq i_1, \cdots, i_p \leq m} g_{i_1 i_2} \cdots {}_{i_p} \frac{\partial^p}{\partial t_{i_1} \cdots \partial t_{i_p}} f$$

on $U$. If $Q$ is a differential operator at every point in $M$ we say simply that it is a differential operator (on $M$). $T$ being a distribution on $M$ and $Q$ a differential operator we define a distribution $Q'T$ as follows:

$$(Q'T)(f) = T(Qf) \qquad\qquad (f \in C_c^\infty(M)).$$

In particular if $g$ is an indefinitely differentiable function on $M$ it defines a differential operator $Q: f \to gf$ ($f \in C_c^\infty(M)$). In this case we write $gT$ to denote $Q'T$ so that $(gT)(f) = T(gf)$. It is clear that the product of two differential operators is again a differential operator and therefore the differential operators form an algebra.

Coming back to $G$, we note that every $X \in \mathfrak{g}_0$ may be regarded as a differential operator on $G$ as follows:

$$(Xf)(x) = \left\{ \frac{d}{dt} f(x \exp tX) \right\}_{t=0} \qquad (f \in C_c^\infty(G),\, x \in G,\, t \in R).$$

Thus it is easy to see that $\mathfrak{B}$ may be identified in a natural way with a subalgebra of the algebra of differential operators on $G$. Then for any $b \in \mathfrak{B}$ we have a linear transformation $b'$ of the space of distribu-

---

[3] Actually it is possible to improve Theorems 1 and 2 and show that $T_\pi$ coincides with an analytic function on an open subset of $G$ which, in general, is larger than $G'$ and therefore has fewer connected components.

tions on $G$. Since $(b_1 b_2)' = b_2' b_1'$ $(b_1, b_2 \in \mathfrak{B})$ the mapping $b \to b'$ is an anti-representation of $\mathfrak{B}$. Let $\phi$ denote the anti-automorphism of $\mathfrak{B}$ over $C$ which is uniquely determined by the condition that $\phi(X) = -X$ $(X \in \mathfrak{g})$. Then $b \to (\phi(b))'$ $(b \in \mathfrak{B})$ is a representation of $\mathfrak{B}$. $T$ being any distribution on $G$ we now define $bT = (\phi(b))'T$ $(b \in \mathfrak{B})$. Then $(bT)(f) = T(\phi(b)f)$ $(f \in C_c^\infty(G))$. If $\chi_\pi$ is the infinitesimal character of $\pi$ and $z \in \mathfrak{Z}, f \in C_c^\infty(G)$, it is easy to see that

$$\int (\phi(z)f)(x)\pi(x)\psi dx = \pi_0(z)\left( \int f(x)\pi(x)\psi dx \right)$$

$$= \chi_\pi(z) \int f(x)\pi(x)\psi dx$$

for all $\psi \in \mathfrak{H}$. (Here $\pi_0$ is the representation of $\mathfrak{B}$ on the Gårding subspace of $\mathfrak{H}$.) From this it follows that $zT_\pi = \chi_\pi(z)T_\pi$ for all $z \in \mathfrak{Z}$. Hence $T_\pi$ is an eigen-distribution for each differential operator in $\mathfrak{Z}$.

On the other hand let $A$ be a Cartan subgroup of $G$. Put $A' = A \cap G'$ and $V = \bigcup_{x \in G} xA'x^{-1}$ as before. We regard $A'$ and $V$ as open submanifolds of $A$ and $G$ respectively. Let $G^*$ be the factor space $G/A$ consisting of cosets of the form $xA$ $(x \in G)$. If $h \in A$ and $x^* \in G^*$ we define $h^{x^*} = xhx^{-1}$ where $x$ is any element in the coset $x^*$. Let $dh$ and $dx^*$ respectively denote the Haar measure on $A$ and the invariant measure on $G^*$. Then we have the following lemma.

LEMMA 1. *There exists a distribution $\tau_\pi$ on $A'$ with the following property. If $f \in C_c^\infty(V)$, $T_\pi(f) = \tau_\pi(g)$ where $g$ is the function in $C_c^\infty(A')$ given by*

$$g(h) = |D(h)| \int_{G^*} f(h^{x^*})dx^* \qquad (h \in A').$$

Let $\mathfrak{h}_0$ be the Lie algebra of $A$. Any element $H \in \mathfrak{h}_0$ may be regarded as a differential operator on $A$ so that if $g \in C_c^\infty(A)$,

$$(Hg)(h) = \left\{ \frac{d}{dt} g(h \exp tH) \right\}_{t=0} \qquad (h \in A, t \in R).$$

Let $\mathfrak{h}$ be the subspace of $\mathfrak{g}$ spanned by $\mathfrak{h}_0$ over $C$ and $\mathfrak{U}$ the subalgebra of $\mathfrak{B}$ generated by $(1, \mathfrak{h})$. Then $\mathfrak{U}$ may be identified in a natural way with a subalgebra of the algebra of differential operators on $A$. For any distribution $\tau$ on $A'$ and $u \in \mathfrak{U}$ we define a distribution $u\tau$ on $A'$ as follows:

$$(u\tau)(g) = \tau(\phi(u)g) \qquad (g \in C_c^\infty(A')).$$

Here $\phi$ is the automorphism of $\mathfrak{U}$ given by $\phi(H) = -H$ $(H \in \mathfrak{h})$.

For every root $\alpha$ of $\mathfrak{g}$ (with respect to $\mathfrak{h}$), choose an element $X_\alpha \neq 0$ in $\mathfrak{g}$ such that $[H, X_\alpha] = \alpha(H)X_\alpha$ for all $H \in \mathfrak{h}$. We introduce some lexicographic order (see [2b]) in the set of all roots and denote by $P$ the set of positive roots under this order. Put $\mathfrak{n} = \sum_{\alpha \in P} CX_\alpha$. Then for every $z \in \mathfrak{Z}$ there exists a unique element $\gamma'(z) \in \mathfrak{U}$ such that $z - \gamma'(z) \in \mathfrak{W}\mathfrak{n}$ (see [2b, p. 72]). If $2\rho = \sum_{\alpha \in P} \alpha$ there exists a unique automorphism $\lambda$ of $\mathfrak{U}$ such that $\lambda(1) = 1$ and $\lambda(H) = H - \rho(H)$ $(H \in \mathfrak{h})$. We put $\gamma(z) = \lambda(\gamma'(z))$ $(z \in \mathfrak{Z})$. Let $W$ be the Weyl group of $\mathfrak{g}$ with respect to $\mathfrak{h}$. It is clear that every $s \in W$ defines an automorphism $u \to u^s$ of $\mathfrak{U}$ such that $1^s = 1$ and $H^s = sH$ $(H \in \mathfrak{h})$. An element $u \in \mathfrak{U}$ is called an invariant if $u = u^s$ for all $s \in W$. Let $J$ be the subalgebra of $\mathfrak{U}$ consisting of all invariants. Then (see [2b, Lemma 38]) the mapping $z \to \gamma(z)$ $(z \in \mathfrak{Z})$ defines an isomorphism of $\mathfrak{Z}$ onto $J$. Now let $\Lambda$ be a linear function on $\mathfrak{h}$. We can extend it uniquely to a homomorphism of $\mathfrak{U}$ into $C$ which takes the value 1 at 1. We agree to denote this extension also by $\Lambda$. Then as shown in [2b, Theorem 5] we can choose $\Lambda$ in such a way that $\chi_\pi(z) = \Lambda(\gamma(z))$ for all $z \in \mathfrak{Z}$. $\Lambda$ is determined up to an operation of $W$ by this condition.

Let $\Delta(h) = |D(h)|^{1/2}$ $(h \in A')$. Then $\Delta$ is an analytic function on $A'$ and therefore $\sigma_\pi = \Delta\tau_\pi$ is a well-defined distribution on $A'$. Now if we transform the differential equation $zT_\pi = \chi_\pi(z)T_\pi$ $(z \in \mathfrak{Z})$ for $T_\pi$ into a differential equation for $\sigma_\pi$ we get the following result which is one of the main steps of our argument.

LEMMA 2. $\tau_\pi$ *being as in Lemma 1 put* $\sigma_\pi = \Delta\tau_\pi$. *Then*

$$\gamma(z)\sigma_\pi = \chi_\pi(z)\sigma_\pi$$

*for every* $z \in \mathfrak{Z}$.

Now choose a linear function $\Lambda$ on $\mathfrak{h}$ such that $\chi_\pi(z) = \Lambda(\gamma(z))$ for $z \in \mathfrak{Z}$. Let $\zeta$ be an indeterminate and $u$ an element in $\mathfrak{U}$. Since $\mathfrak{U}$ is abelian we can consider the polynomial

$$\prod_{s \in W} (\zeta - u^s).$$

It is clear that every coefficient of this polynomial lies in $J$ and therefore if $w$ is the order of $W$, there exist uniquely determined elements $z_1(u), \cdots, z_w(u)$ in $\mathfrak{Z}$ such that

$$\prod_{s \in W}(\zeta - u^s) = \zeta^w + \gamma(z_1(u))\zeta^{w-1} + \gamma(z_2(u))\zeta^{w-2} + \cdots + \gamma(z_w(u)).$$

On replacing $\zeta$ by $u$ we immediately get the identity

$$u^w + u^{w-1}\gamma(z_1(u)) + u^{w-2}\gamma(z_2(u)) + \cdots + \gamma(z_w(u)) = 0$$

in $\mathfrak{U}$ if we recall that $\mathfrak{U}$ is abelian. Now apply the left side to $\sigma_\pi$ and use Lemma 2. Then

$$u^w \sigma_\pi + \chi_\pi(z_1(u)) u^{w-1} \sigma_\pi + \cdots + \chi_\pi(z_w(u)) \sigma_\pi = 0.$$

But $\chi_\pi(z_j(u)) = \Lambda(\gamma(z_j(u)))$, $1 \leq j \leq w$, and since $\Lambda$ is a homomorphism of $\mathfrak{U}$ into $C$ it is obvious that

$$\prod_{s \in W} (\zeta - \Lambda(u^s)) = \zeta^w + \Lambda(\gamma(z_1(u)))\zeta^{w-1} + \cdots + \Lambda(\gamma(z_w(u))).$$

Therefore the above differential equation for $\sigma_\pi$ may be written in the form

$$\prod_{s \in W} (u - \Lambda(u^s)) \sigma_\pi = 0.$$

However if $H_1, \cdots, H_l$ is a base for $\mathfrak{h}_0$ over $R$ and $\square = H_1^2 + \cdots + H_l^2$ it is obvious that the differential equation

$$\prod_{s \in W} (\square - \Lambda(\square^s)) \sigma_\pi = 0$$

is of the *elliptic* type (see Gårding [1]). Hence it follows from the work of Schwartz [4, p. 137] and John [3a, 3b] that $\sigma_\pi$ must be an analytic function on $A'$. Now if we take into account Theorem 5 of [2b] and the fact that $\prod_{s \in W} (u - \Lambda(u^s)) \sigma_\pi = 0$ for every $u \in \mathfrak{U}$, we get Theorem 2 without difficulty.

$\pi$ being any quasi-simple irreducible representation we denote by $T_\pi$, $\eta_\pi$, and $\chi_\pi$ respectively the character, the central character, and the infinitesimal character of $\pi$. Also we denote by $F_\pi$ the analytic function on $G'$ such that $T_\pi = F_\pi$ on $G'$. Let $T$ be a distribution on $G$. Since $D$ is an analytic function on $G$ the product $D^m T$ ($m \geq 0$) is a well-defined distribution. It is then possible to prove the following result.

LEMMA 3. *There exists an integer $m \geq 0$ with the following property. Suppose $\pi_1, \cdots, \pi_k$ is a finite set of quasi-simple irreducible representations and $c_1 F_{\pi_1} + \cdots + c_k F_{\pi_k} = 0$ ($c_i \in C$). Then if $T = c_1 T_{\pi_1} + \cdots + c_k T_{\pi_k}$, $D^m T = 0$.*

From this lemma one can deduce the following theorem.

THEOREM 3. *Let $\pi_0$ be a quasi-simple irreducible representation of $G$ such that $F_{\pi_0}$ is not identically zero. Then, apart from infinitesimal equivalence [2d, p. 230], there exist only a finite number of quasi-simple irreducible representations $\pi$ such that $F_\pi = F_{\pi_0}$.*

Let $\eta \neq 0$ and $\chi \neq 0$ respectively be given homomorphisms of $Z$ and $\mathfrak{Z}$ into $C$. Let $\omega$ denote the set of all quasi-simple irreducible representations $\pi$ of $G$ such that $\eta_\pi = \eta$ and $\chi_\pi = \chi$. As we have seen, it follows from Theorems 1 and 2 that the functions $F_\pi$ $(\pi \in \omega)$ span a finite-dimensional vector space over $C$. Hence if $\omega$ is not empty, we can choose a finite set of elements $\pi_1, \cdots, \pi_k$ in $\omega$ such that $F_{\pi_1}, \cdots, F_{\pi_k}$ form a base for this vector space. Then if $\pi$ is any representation in $\omega$, $F_\pi = c_1 F_{\pi_1} + \cdots + c_k F_{\pi_k}$ $(c_j \in C)$. If one could conclude from this equation that $T_\pi = c_1 T_{\pi_1} + \cdots + c_k T_{\pi_k}$ it would follow (see [2e, Theorem 6]) that $\pi$ is infinitesimally equivalent to some $\pi_j$ $(1 \leq j \leq k)$ and therefore, apart from infinitesimal equivalence, $\omega$ has only a finite number of representations. Therefore it is important to consider the following question.

*Let $\pi_1, \cdots, \pi_k$ be a finite set of quasi-simple irreducible representations of $G$ such that $c_1 F_{\pi_1} + \cdots + c_k F_{\pi_k} = 0$ $(c_i \in C)$. Then is it always true that $c_1 T_{\pi_1} + \cdots + c_k T_{\pi_k} = 0$?*

I believe the answer is yes but do not know how to prove it.

### References


1. L. Gårding, Math. Scand. vol. 1 (1953) pp. 55–72.
2a. Harish-Chandra, Ann. of Math. vol. 50 (1949) pp. 900–915.
2b. ———, Trans. Amer. Math. Soc. vol. 70 (1951) pp. 28–96.
2c. ———, Proc. Nat. Acad. Sci. U.S.A. vol. 37 (1951) pp. 366–369.
2d. ———, Trans. Amer. Math. Soc. vol. 75 (1953) pp. 185–243.
2e. ———, Trans. Amer. Math. Soc. vol. 76 (1954) pp. 234–253.
2f. ———, Trans. Amer. Math. Soc. vol. 76 (1954) pp. 485–528.
3a. F. John, Comm. Pure Appl. Math. vol. 3 (1950) pp. 273–304.
3b. ———, *Proceedings of the symposium on spectral theory and differential problems*, Stillwater, Okla., 1951, pp. 113–175.
4. L. Schwartz, *Théorie des distributions* I, Paris, Hermann, 1950.
5. I. E. Segal, Proc. Amer. Math. Soc. vol. 3 (1952) pp. 13–15.
6. H. Weyl, Math. Zeit. vol. 24 (1925) pp. 328–395.
7. I. M. Gelfand and M. A. Naimark, Trudi Mat. Inst. Steklova vol. 36 (1950)


Columbia University


Reprinted from
*Bull. Amer. Math. Soc.*
**61** (1955), 389–396

# REPRESENTATIONS OF SEMISIMPLE LIE GROUPS IV.*

By HARISH-CHANDRA.

**1. Introduction.** In an earlier paper of this series, I have given a method of constructing all irreducible quasi-simple representations of a connected semisimple group $G$ (at least in case $G$ has a finite-dimensional faithful representation) up to infinitesimal equivalence (see Theorem 4 of [5(d)]. However this method has one serious drawback. It can happen very well that the representation so obtained is infinitesimally unitary without being equivalent to a unitary representation. In order to make it unitary one has to introduce a new scalar product in a dense subspace of the representation space in some more or less artificial fashion and then complete this subspace with respect to the new norm. It would of course be much better if the unitary representations could be obtained in a more natural manner especially since it is often a rather difficult matter to decide whether a given representation is infinitesimally unitary or not. Suppose, for a moment, that $G$ is a linear group and $A$ is a Cartan subgroup of $G$. Then $A$ can be written as a direct product of two subgroups $A_-$ and $A_+$ of which the former is compact while the latter is isomorphic to a vector space. We say that $A$ is incompact if $A_+$ has the maximum possible dimension. Now in the above-mentioned construction it was quite essential that one should start from an incompact $A$. However the study of simple examples (such as that of the $2 \times 2$ real unimodular group [1, 5(e)]) is enough to convince one of the necessity of searching for alternative 'natural' methods of construction which start out from other types of Cartan subgroups. The work of Gelfand and Graev [4] on the Plancherel formula of the $n \times n$ real unimodular group points to the same conclusion. In a recent note [5(f)] I have given such a 'natural' mode of construction based on the concept of induced representations (see Mackey [6]), which is applicable to all Cartan subgroups. By a simple argument the main problem can be reduced to the case when $A$ is compact and therefore it is this particular case which requires intensive study. The object of the present paper is to deal with the algebraic aspects of this case, the function-theoretic aspects being left to a subsequent paper. Assuming that $G$ is simple and its first Betti number is 1, we shall

---

* Received July 8, 1955.

743

give a quite general method of constructing a series of irreducible unitary representations of $G$ parameterised by the characters of $A$. However if the first Betti number is zero (and $G$ is not compact) this method does not give anything at all [5(g)]. (It seems likely that this is not due to a fault of the method but rather reflects an essential difference in the two cases with regard to representation theory. It is noteworthy that a similar difference has been observed by E. Cartan [3(d)] in his theory of bounded homogeneous symmetric domains. There also exactly those simple groups come into play whose first Betti number is one.) The representations obtained in this manner bear a very close resemblance, in their algebraic structure, to finite-dimensional representations. As we shall see in another paper, they are also somewhat remarkable in so far as their matrix coefficients are square-integrable on the group [5(f), (g)]. In the last section we obtain a formula (see Corollary to Lemma 21) for the multiplicities of the various weights of these representations.

Some of the results of this paper have been announced in a short note [5(g)].

**2. Preliminary lemmas.** Let $R$ be the field of real numbers and $V$ a vector space over $R$ of finite dimension. We shall say that $V$ is ordered if there is given an order in $V$ satisfying the following condition. If $u, v$ are two elements in $V$ and $a$ is a positive real number then $u + v > 0$ and $au > 0$ whenever $u > 0$, $v > 0$. An element $u \, \varepsilon \, V$ will be called positive (with respect to the given order) if $u > 0$. Let $\mathfrak{F}$ be the space of all linear functions on $V$. Then there are two simple ways of introducing an order in $\mathfrak{F}$ which depend on the choice of a base either in $V$ or in $\mathfrak{F}$. Let $H_1, \cdots, H_l$ be a base for $V$. We define a function $\lambda \, \varepsilon \, \mathfrak{F}$ to be positive if $\lambda \neq 0$ and $\lambda(H_i) > 0$, $i$ being the least index such that $\lambda(H_i) \neq 0$. This is called the lexicographic order in $\mathfrak{F}$ corresponding to the base $(H_1, \cdots, H_l)$. Similarly let $(\lambda_1, \cdots, \lambda_l)$ be a base for $\mathfrak{F}$. Then every $\lambda \, \varepsilon \, \mathfrak{F}$ can be written as $\lambda = c_1\lambda_1 + \cdots + c_l\lambda_l \; (c_j \, \varepsilon \, R)$. We now define $\lambda$ to be positive if $\lambda \neq 0$ and $c_i > 0$ where $i$ is the least index such that $c_i \neq 0$. The order so defined is called the lexicographic order in $\mathfrak{F}$ corresponding to the base $(\lambda_1, \cdots, \lambda_l)$.

Let $\mathfrak{g}$ be a semisimple Lie algebra over the field $C$ of complex numbers and let $\mathfrak{h}$ be a Cartan[1] subalgebra of $\mathfrak{g}$. Let $\mathfrak{h}_R$ denote the subset of those elements of $\mathfrak{h}$ at which every root $\alpha$ (of $\mathfrak{g}$ with respect to $\mathfrak{h}$) takes a real value. Then $\mathfrak{h}_R$ is a vector space over $R$ and if $H_1, \cdots, H_l$ is a base for $\mathfrak{h}_R$

---

[1] We use the standard terminology of the theory of semisimple Lie algebras (see for example [5(b)]).

over $R$, it is also a base for $\mathfrak{h}$ over $C$. We say that a linear function $\lambda$ on $\mathfrak{h}$ is real if $\lambda(H)$ is real for $H \, \varepsilon \, \mathfrak{h}_R$. Then the set $\mathfrak{F}_R$ of all such real functions may be identified in the obvious way with the space of all linear functions on $\mathfrak{h}_R$. Hence corresponding to the base $(H_1, \cdots, H_l)$ we get a lexicographic order in $\mathfrak{F}_R$. Similarly if $\alpha_1, \cdots, \alpha_l$ are $l$ linearly independent roots they form a base for $\mathfrak{F}_R$ and we obtain the corresponding lexicographic order in $\mathfrak{F}_R$.

Let $\pi$ be a representation of $\mathfrak{g}$ on a vector space $V$ over $C$. A linear function $\Lambda$ on $\mathfrak{h}$ is called a weight of $\pi$ if there exists an element $\psi \neq 0$ in $V$ such that $\pi(H)\psi = \Lambda(H)\psi$ for all $H \, \varepsilon \, \mathfrak{h}$. A weight $\Lambda$ is called extreme if it is impossible to find a root $\alpha$ such that both $\Lambda + \alpha$ and $\Lambda - \alpha$ are also weights of $\pi$. It is well known (see Weyl [7]) that if $\dim V$ is finite every weight $\Lambda$ is real.

For each root $\alpha$ choose an element $X_\alpha \neq 0$ in $\mathfrak{g}$ such that $[H, X_\alpha] = \alpha(H)X_\alpha$ for all $H \, \varepsilon \, \mathfrak{h}$. Then $H_\alpha = [X_\alpha, X_{-\alpha}]$ lies in $\mathfrak{h}$ and we may suppose that $X_\alpha$ is so chosen that $\alpha(H_\alpha) = 2$. Let $\mathfrak{B}$ be the universal enveloping algebra of $\mathfrak{g}$. We identify representations of $\mathfrak{g}$ with those of $\mathfrak{B}$ in the usual way.

LEMMA 1. *Let $\pi$ be a representation of $\mathfrak{g}$ on $V$ and let $\psi \neq 0$ be an element belonging*[1] *to the weight $\Lambda$. Suppose $\alpha$ is a root such that $\pi(X_\alpha{}^r)\psi = \pi(X_{-\alpha}{}^r)\psi = 0$ for some integer $r \geq 1$. Then $\Lambda(H_\alpha)$ is an integer and $\Lambda - k\alpha$ is a weight of $\pi$ for all integers $k$ such that*

$$\min(0, \Lambda(H_\alpha)) \leq k \leq \max(0, \Lambda(H_\alpha)).$$

*In particular $\Lambda - \Lambda(H_\alpha)\alpha$ is a weight of $\pi$. Moreover if $\pi(X_\alpha)\psi = 0$, $\Lambda(H_\alpha)$ is equal to the smallest integer $r \geq 0$ such that $\pi(X_{-\alpha}{}^{r+1})\psi = 0$.*

The proof of the first two statements is substantially the same as that of Lemma 11 of [5(b)]. Now suppose $\pi(X_\alpha)\psi = 0$. Let $s$ be the least nonnegative integer such that $\pi(X_{-\alpha}{}^{s+1})\psi = 0$. Then if $\psi_{-1} = 0$, $\psi_0 = \psi$ and $\psi_k = \pi(X_{-\alpha}{}^k)\psi$ $(k \geq 1)$ the argument of Lemma 11 of [5(b)] shows that

$$\pi(X_\alpha)\psi_k = k(\Lambda(H_\alpha) - k + 1)\psi_{k-1} \qquad (k \geq 0).$$

Putting $k = s + 1$ we get $\Lambda(H_\alpha) = s$ and this proves the last part of assertion.

Let $\lambda$ be a linear function on $\mathfrak{h}$. We say that an element $b \, \varepsilon \, \mathfrak{B}$ is of rank $\lambda$ if $[H, b] = \lambda(H)b$ for all $H \, \varepsilon \, \mathfrak{h}$. Elements belonging to different ranks are linearly independent and if $\mathfrak{B}_\lambda$ is the set of all elements in $\mathfrak{B}$ of rank $\lambda$, $\mathfrak{B}$ is the direct sum of $\mathfrak{B}_\lambda$ for all $\lambda$ [5(b), Lemma 2]. By the component of $z$ of rank $\lambda$ $(z \, \varepsilon \, \mathfrak{B})$ we mean the component of $z$ in $\mathfrak{B}_\lambda$ in this direct sum decomposition.

Now suppose $\mathfrak{F}_R$ has been ordered in some way.

LEMMA 2. *Let $\pi$ be a representation of $\mathfrak{B}$ on a vector space $V$. Suppose $\Lambda$ is a linear function on $\mathfrak{h}$ and $\psi \neq 0$ a vector in $V$ such that:*

   (1)   *$\pi(X_\alpha)\psi = 0$ for every positive root $\alpha$;*

   (2)   *$\pi(H)\psi = \Lambda(H)\psi$ for all $H \varepsilon \mathfrak{h}$;*

   (3)   *$V = \pi(\mathfrak{B})\psi$.*

*Then if $\alpha_1, \cdots, \alpha_r$ are all the distinct positive roots of $\mathfrak{g}$ every weight of $\pi$ is of the form $\Lambda - (m_1\alpha_1 + \cdots + m_r\alpha_r)$ where $m_1, \cdots, m_r$ are nonnegative integers. Moreover if for each $\alpha_i$ we can find an integer $n_i \geqq 0$ such that $\pi(X_{-\alpha_i}{}^{n_i+1})\psi = 0$ then $\dim V$ is finite and $\pi$ is irreducible.*

Since $\mathfrak{g}$ is spanned by $(X_{\alpha_1}, \cdots, X_{\alpha_r}, \mathfrak{h}, X_{-\alpha_1}, \cdots, X_{-\alpha_r})$ it follows from Lemma 1 of [5(a)] and our assumptions above that $V$ is spanned by elements of the form $\pi(X_{-\alpha_1}{}^{m_1}X_{-\alpha_2}{}^{m_2} \cdots X_{-\alpha_r}{}^{m_r})\psi$. In view of Lemma 10 of [5(b)] our first assertion now follows immediately. Moreover it is clear that $C\psi$ is the space of all elements in $V$ which belong to the weight $\Lambda$.

Now suppose $\pi(X_{-\alpha_i}{}^{n_i+1})\psi = 0$, $1 \leqq i \leqq r$. Then we claim that $V$ is spanned by $\pi(X_{-\alpha_1}{}^{\nu_1} \cdots X_{-\alpha_r}{}^{\nu_r})\psi$, $(0 \leqq \nu_i \leqq n_i, 1 \leqq i \leqq r)$. For let $U$ be the space spanned by these elements. It would be sufficient to show that $\pi(X_{-\alpha_1}{}^{m_1} \cdots X_{-\alpha_r}{}^{m_r})\psi \varepsilon U$ for all $m_1, \cdots, m_r \geqq 0$. We shall do this by induction on $M = m_1 + \cdots + m_r$. If $M = 0$ the statement is obviously true and so we may suppose $M > 0$. Also we may assume that $m_i > n_i$ for some $i$. Let $j$ be the largest index such that $m_j > n_j$. If $j = r$, $\pi(X_{-\alpha_r}{}^{m_r})\psi = 0$ and so our statement is again true. Hence suppose $j < r$. Put

$$z = X_{-\alpha_1}{}^{m_1} \cdots X_{-\alpha_{j-1}}{}^{m_{j-1}}, \qquad w = X_{-\alpha_{j+1}}{}^{m_{j+1}} \cdots X_{-\alpha_r}{}^{m_r}$$

($z = 1$ if $j = 1$). Then $zX_{-\alpha_j}{}^{m_j}w = z[X_{-\alpha_j}{}^{m_j}, w] + zwX_{-\alpha_j}{}^{m_j}$ and since $m_j > n_j$, $\pi(X_{-\alpha_j}{}^{m_j})\psi = 0$. Therefore

$$\phi = \pi(X_{-\alpha_1}{}^{m_1} \cdots X_{-\alpha_r}{}^{m_r})\psi = \pi(z[X_{-\alpha_j}{}^{m_j}, w])\psi.$$

But $X_{-\alpha_1}, \cdots, X_{-\alpha_r}$ span a subalgebra of $\mathfrak{g}$ and therefore from Lemma 1 of [5(a)] $z[X_{-\alpha_j}{}^{m_j}, w]$ can be written as a linear combination of $X_{-\alpha_1}{}^{m_1'} \cdots X_{-\alpha_r}{}^{m_r'}$ with $m_1' + \cdots + m_r' < M$. Hence by our induction hypothesis $\phi \varepsilon U$. This proves that $U = V$ and therefore $\dim V$ is finite and $V$ is fully reducible under $\pi$. Hence there exists a subspace of $V'$ which is invariant and irreducible under $\pi(\mathfrak{B})$ and which contains a nonzero vector $\psi'$ belonging to the weight $\Lambda$. But then, as we have seen above $\psi' = c\psi$ $(c \varepsilon C)$ and therefore $V' = \pi(\mathfrak{B})\psi' = V$. This proves that $\pi$ is irreducible.

We shall need the following lemma due to E. Cartan [3(a)].

LEMMA 3 (Cartan). *Suppose $\pi$ is an irreducible finite-dimensional representation of $\mathfrak{g}$ and $\Lambda_0$ is its highest weight. Then if $W$ is the Weyl[1] group of $\mathfrak{g}$ (with respect to $\mathfrak{h}$) $s\Lambda_0$ ($s \varepsilon W$) are exactly all the extreme weights of $\pi$.*

We sketch the proof for the sake of completeness. It follows from the argument of Lemma 13 of [5(b)] that if $\Lambda$ is an extreme weight of $\pi$ so is also $s\Lambda$ for every $s \varepsilon W$. Since the highest weight is obviously an extreme weight, $s\Lambda_0$ ($s \varepsilon W$) are all extreme weights. Conversely suppose $\Lambda$ is an extreme weight of $\pi$. Let $\Lambda_1$ be the highest one among $s\Lambda$ ($s \varepsilon W$). Then $\Lambda_1$ is also an extreme weight and it would be enough to prove that $\Lambda_1 = \Lambda_0$. Let $\alpha$ be a positive root and $s_\alpha$ the corresponding Weyl reflexion. Then $s_\alpha \Lambda_1 = \Lambda_1 - \Lambda_1(H_\alpha)\alpha \leqq \Lambda_1$. Hence $\Lambda_1(H_\alpha) \geqq 0$. We claim that $\Lambda_1 + \alpha$ is not a weight of $\pi$. For if it were we could conclude from Lemma 1 (since $\pi(X_\alpha)$ and $\pi(X_{-\alpha})$ are both nilpotent) that $\Lambda_1 + \alpha - \nu\alpha$ is a weight for $0 \leqq \nu \leqq \Lambda_1(H_\alpha) + 2$. This would imply however that $\Lambda_1 + \alpha$ and $\Lambda_1 - \alpha$ are both weights, contradicting the fact that $\Lambda_1$ is extreme. Let $\psi \neq 0$ be an element in $V$ belonging to the weight $\Lambda_1$. Since $\Lambda_1 + \alpha$ is not a weight, $\pi(X_\alpha)\psi = 0$ and it follows from Lemma 2 that $\Lambda_1$ is the highest weight of $\pi$. Therefore $\Lambda_1 = \Lambda_0$ and Lemma 3 is proved.

A set $(\alpha_1, \cdots, \alpha_l)$ of roots of $\mathfrak{g}$ is called a fundamental system if it is a linearly independent set and if every root $\alpha$ can be written in the form $\alpha = d_1\alpha_1 + \cdots + d_l\alpha_l$ where $d_i$ are integers which are either all nonnegative or all nonpositive.

LEMMA[2] 4. *Let $\mathfrak{m}$ be a subalgebra of $\mathfrak{g}$ such that $\mathfrak{h} \subset \mathfrak{m}$ and for each root $\alpha$ either $X_\alpha$ or $X_{-\alpha}$ is in $\mathfrak{m}$. Then it is possible to choose a fundamental system $(\alpha_1, \cdots, \alpha_l)$ of roots such that $X_{\alpha_i} \varepsilon \mathfrak{m}$, $1 \leqq i \leqq l$.*

Let $\Gamma$ be the radical of $\mathfrak{m}$ and let $X \to X^*$ denote the natural mapping of $\mathfrak{m}$ on $\mathfrak{m}^* = \mathfrak{m}/\Gamma$. Then $\mathfrak{m}^*$ is a semisimple Lie algebra and $\mathfrak{h} = (\mathfrak{h} + \Gamma)/\Gamma$ is an abelian subalgebra of $\mathfrak{m}^*$. Let $X \to adX$ denote the adjoint representation of $\mathfrak{g}$. Then if $H \varepsilon \mathfrak{h}$, $adH$ is a semisimple[3] linear transformation of $\mathfrak{g}$. Hence if $ad^*H^*$ denotes the linear mapping of $\mathfrak{m}^*$ corresponding to $H^*$ under the adjoint representation of $\mathfrak{m}^*$, it is clear that $ad^*H^*$ is also semisimple. Therefore $\mathfrak{h}^*$ can be extended to a Cartan subalgebra $\mathfrak{h}_1^*$ of $\mathfrak{m}^*$. We introduce some lexicographic order among the roots of $\mathfrak{m}^*$ with respect to $\mathfrak{h}_1^*$. For every root $\gamma^*$ of $\mathfrak{m}^*$ choose an element $X_{\gamma^*}^* \neq 0$ in $\mathfrak{m}^*$ such that $[H^*, X_{\gamma^*}^*] = \gamma^*(H^*)X_{\gamma^*}^*$ for all $H^* \varepsilon \mathfrak{h}_1^*$. Let $\Gamma_1^* = \mathfrak{h}_1^* + \sum_{\gamma^* > 0} CX_{\gamma^*}$ and

---

[2] A very similar result has been obtained independently by A. Borel.

[3] This means that $\mathfrak{g}$ is fully reducible under $adH$.

let $\Gamma_1$ be the complete inverse image of $\Gamma_1^*$ in $\mathfrak{m}$. It is clear that $\Gamma_1^*$ is a solvable subalgebra of $\mathfrak{m}^*$. Hence $\Gamma_1$ is a solvable subalgebra of $\mathfrak{m}$ containing $\mathfrak{h} + \Gamma$.

Now let $\beta$ be any root of $\mathfrak{g}$. We claim that at least one of the two elements $X_\beta$ or $X_{-\beta}$ is in $\Gamma_1$. In order to prove this we may clearly assume that $X_\beta \varepsilon \mathfrak{m}$ but $X_\beta \notin \Gamma_1$. Then $[H, X_\beta] = \beta(H) X_\beta$ $(H \varepsilon \mathfrak{h})$ and by going over to $\mathfrak{m}^*$ we get $[H^*, X_\beta^*] = \beta(H) X_\beta^*$ $(H \varepsilon \mathfrak{h})$. Since $X_\beta \notin \Gamma_1$, $X_\beta^* \neq 0$ and so we can choose a root $\gamma^*$ of $\mathfrak{m}^*$ such that $\gamma^*(H^*) = \beta(H)$ for all $H \varepsilon \mathfrak{h}$. Define $\epsilon = 1$ or $-1$ according as $\gamma^*$ is positive or negative and put $Y^* = X_{\epsilon\gamma^*}^* \varepsilon \Gamma_1^*$. Then if $Y$ is any element in $\Gamma_1$ whose image in $\mathfrak{m}^*$ is $Y^*$, $Y \neq 0$ and

$$[H, Y] \equiv \epsilon\beta(H) Y \quad \mathrm{mod}\, \Gamma \qquad\qquad (H \varepsilon \mathfrak{h}).$$

However $\Gamma_1$ is invariant and fully reducible under $ad\, \mathfrak{h}$. Therefore we can select an element $Z \varepsilon \Gamma_1$ $(Z \neq 0)$ such that

$$[H, Z] = \epsilon\beta(H) Z.$$

But then $Z$ can differ from $X_{\epsilon\beta}$ only by a nonzero factor in $C$ and therefore $X_{\epsilon\beta} \varepsilon \Gamma_1 \subset \mathfrak{m}$. This proves our assertion.

Put $B(X, Y) = sp(adX adY)$ $(X, Y \varepsilon \mathfrak{g})$. For any linear function $\lambda$ on $\mathfrak{h}$ let $H_\lambda'$ denote the (unique) element in $\mathfrak{h}$ such that $B(H_\lambda', H) = \lambda(H)$ for all $H \varepsilon \mathfrak{h}$. Define $\langle \lambda, \mu \rangle = B(H_\lambda', H_\mu') = \lambda(H_\mu')$ for any two linear functions $\lambda, \mu$. Then it is well known [7] that if $\alpha$ is a root $H_\alpha' = c_\alpha H_\alpha$ where $c_\alpha$ is a *positive* real number. Let $\gamma_1, \cdots, \gamma_l$ be a fundamental system of roots of $\mathfrak{g}$. We introduce the lexicographic order (in $\mathfrak{F}_R$) corresponding to this system. Let $\Lambda_0$ be a linear function on $\mathfrak{h}$ such that $\Lambda_0(H_{\gamma_i})$ $1 \leq i \leq l$ are all positive integers. Now if $\alpha$ is a positive root $\alpha = d_1\gamma_1 + \cdots + d_l\gamma_l$ where $d_i$ are nonnegative integers and therefore $H_\alpha' = d_1 H_{\gamma_1}' + \cdots + d_l H_{\gamma_l}'$. Hence

$$c_\alpha \Lambda_0(H_\alpha) = \Lambda_0(H_\alpha') = \sum_{i=1}^{l} d_i \Lambda_0(H_{\gamma_i}') = \sum_i d_i c_{\gamma_i} \Lambda_0(H_{\gamma_i})$$

and in view of the fact that $c_\alpha, c_{\gamma_i}, \Lambda_0(H_{\gamma_i})$ are all positive and not all $d_i$ are zero we can conclude that $\langle \Lambda_0, \alpha \rangle$ and $\Lambda_0(H_\alpha)$ are both positive. Let $\pi$ be an irreducible finite-dimensional representation of $\mathfrak{g}$ with the highest weight $\Lambda_0$. (We know from Theorem 1 of [5(b)] that such a representation exists). Since $\Gamma_1$ is a solvable algebra, by Lie's theorem we can choose an element $\psi_0 \neq 0$ in the representation space $V$ and a linear function $\mu$ on $\Gamma_1$ such that $\pi(X)\psi_0 = \mu(X)\psi_0$ for all $X \varepsilon \Gamma_1$. Since $\mathfrak{h} \subset \Gamma_1$ it follows that $\mu$ coincides on $\mathfrak{h}$ with some weight $\Lambda$ of $\pi$. Let $\beta$ be a root such that $X_\beta \varepsilon \Gamma_1$.

Then $\psi_0$ is an eigenvalue of $\pi(X_\beta)$. But $\pi(X_\beta)$ and $\pi(X_{-\beta})$ are both nilpotent and therefore $\pi(X_\beta)\psi_0 = 0$ and $\pi(X_{-\beta}{}^r)\psi_0 = 0$ for $r$ sufficiently large. Hence it follows from Lemma 1 that $\Lambda(H_\beta) \geqq 0$ and $\Lambda(H_\beta) \neq 0$ unless $\pi(X_{-\beta})\psi_0 = 0$. Let $(\beta_1, \cdots, \beta_s)$ be all the distinct roots $\beta$ such that $X_\beta \varepsilon \Gamma_1$ and among these let $(\beta_1, \cdots, \beta_r)$ be all those for which $\Lambda(H_{\beta_i}) \neq 0$. Then $\Lambda(H_{\beta_i}) > 0$, $1 \leqq i \leqq r$ and in view of what we have proved above, $\mathfrak{g}$ is spanned by $\mathfrak{h}$, $X_{\beta_i}$ and $X_{-\beta_i}$, $1 \leqq i \leqq s$. Hence it follows from Lemma 1 of [5(a)] that $\pi(\mathfrak{B})\psi_0 = V$ is spanned by elements of the form $\pi(X_{-\beta_1}{}^{m_1} \cdots X_{-\beta_s}{}^{m_s})\psi_0$ where $m_1, \cdots, m_s$ are nonnegative integers. Since $\pi(X_{-\beta_j})\psi_0 = 0$, $r < j \leqq s$, $V$ is actually spanned by $\pi(X_{-\beta_1}{}^{m_1} \cdots X_{-\beta_r}{}^{m_r})\psi$. This proves that every weight of $\pi$ must be of the form $\Lambda - (m_1\beta_1 + \cdots + m_r\beta_r)$, $(m_i \geqq 0)$. We now claim that $\Lambda$ is an extreme weight of $\pi$. For otherwise let $\alpha$ be a root such that $\Lambda + \alpha$ and $\Lambda - \alpha$ are both weights. Then in view of the above result

$$\alpha = m_1\beta_1 + \cdots + m_r\beta_r, \qquad -\alpha = m_1'\beta_1 + \cdots + m_r'\beta_r$$

where $m_i, m_i' \geqq 0$. Hence $0 = (m_1 + m_1')\beta_1 + \cdots + (m_r + m_r')\beta_r$ and so

$$(m_1 + m_1')\langle\Lambda, \beta_1\rangle + (m_2 + m_2')\langle\Lambda, \beta_2\rangle + \cdots + (m_r + m_r')\langle\Lambda, \beta_r\rangle = 0.$$

But we have seen above that $\Lambda(H_{\beta_i}) > 0$ and therefore $\langle\Lambda, \beta_i\rangle > 0$, $i = 1, \cdots, r$. Hence the above equation implies that $m_i = m_i' = 0$, $1 \leqq i \leqq r$. This however is impossible since $\alpha \neq 0$. Therefore $\Lambda$ is an extreme weight and we can conclude from Lemma 3 that $\Lambda = s\Lambda_0$ for some $s \varepsilon W$.

We recall that $\langle\Lambda_0, \alpha\rangle \neq 0$ for every root $\alpha$. Hence $\langle\Lambda, \alpha\rangle = \langle\Lambda_0, s^{-1}\alpha\rangle \neq 0$ and therefore $\Lambda(H_\beta) > 0$ for every root $\beta$ such that $X_\beta$ lies in $\Gamma_1$. This shows in particular that for any root $\beta$ exactly one of the two elements $X_\beta$ and $X_{-\beta}$ is in $\Gamma_1$. Now put $\alpha_i = s\gamma_i$, $1 \leqq i \leqq l$. Then $(\alpha_1, \cdots, \alpha_l)$ is also a fundamental system of roots of $\mathfrak{g}$ and $\langle\Lambda, \alpha_i\rangle = \langle\Lambda_0, \gamma_i\rangle > 0$. Therefore $\Lambda(H_{\alpha_i}) > 0$ and hence $X_{\alpha_i} \varepsilon \Gamma_1 \subset \mathfrak{m}$. This proves the lemma.

COROLLARY 1. *Let $P$ be a set of roots of $\mathfrak{g}$ (with respect to $\mathfrak{h}$) satisfying the following two conditions.*

(1) *For any root $\beta$ exactly one of the two roots $\beta$ and $-\beta$ lies in $P$;*

(2) *If $\alpha$ and $\beta$ are in $P$ and $\alpha + \beta$ is a root then $\alpha + \beta \varepsilon P$.*

*Then it is possible to select roots $\alpha_i \varepsilon P$, $1 \leqq i \leqq l$ such that $(\alpha_1, \cdots, \alpha_l)$ is a fundamental system of roots and this system is unique (apart from a permutation of its elements). Moreover $P$ is exactly the set of positive roots in the lexicographic order corresponding to $(\alpha_1, \cdots, \alpha_l)$.*

It is clear that $\mathfrak{n} = \sum_{\beta \varepsilon P} CX_\beta$ is a subalgebra of $\mathfrak{g}$ and $\mathfrak{m} = \mathfrak{h} + \mathfrak{n}$ fulfils

9

all the conditions of Lemma 4.  Hence we can choose a fundamental system $(\alpha_1, \cdots, \alpha_l)$ such that $X_{\alpha_i} \varepsilon \mathfrak{m}$.  It is obvious that $\alpha_i \varepsilon P$.  Let $\beta$ be any positive root in the lexicographic order corresponding to this system.  Then $X_\beta$ is contained in the subalgebra of $\mathfrak{g}$ generated by $X_{\alpha_i}$, $1 \leq i \leq l$ (see Lemma 18 of [5(b)]).  Hence it follows from the second condition of our hypothesis that $X_\beta \varepsilon \mathfrak{n}$ and therefore $\beta \varepsilon P$.  So every positive root is in $P$ and hence in view of condition (1) no negative root can lie in $P$.  This proves that $P$ is exactly the set of all positive roots.  Finally it is obvious that if $\alpha$ is a root in $P$, it lies in the fundamental system $(\alpha_1, \cdots, \alpha_l)$ if and only if it cannot be written in the form $\alpha = \beta_1 + \cdots + \beta_r$ where $\beta_1, \cdots, \beta_r$ are all roots in $P$ and $r \geq 2$.  This shows that the system $(\alpha_1, \cdots, \alpha_l)$ is unique apart from order.

COROLLARY 2.  *Let $P$ be the set of all positive roots under any given order in $\mathfrak{F}_R$.  Then there exists a fundamental system $(\alpha_1, \cdots, \alpha_l)$ consisting of positive roots and this system is unique apart from a permutation.  Moreover $P$ is also the set of all positive roots in the lexicographic order corresponding to $(\alpha_1, \cdots, \alpha_l)$.*

For it is obvious that $P$ satisfies the two conditions of Corollary 1.

## 3.  Representations with an extreme vector.

DEFINITION.  *Let $\pi$ be a representation of $\mathfrak{B}$ on a vector space $V$.  An element $\psi$ in $V$ will be called an extreme vector of $\pi$ (with respect to $\mathfrak{h}$) if the following two conditions are fulfilled:*

(1)  *$\psi \neq 0$ and there exists a linear function $\Lambda$ on $\mathfrak{h}$ such that $\pi(H)\psi = \Lambda(H)\psi$ for all $H \varepsilon \mathfrak{h}$.*

(2)  *For any root $\beta$ at least one of the two elements $\pi(X_\beta)\psi$, $\pi(X_{-\beta})\psi$ is zero.*

Suppose $\psi$ is an extreme vector of $\pi$.  Let $\mathfrak{m}$ be the subalgebra of $\mathfrak{g}$ consisting of all elements $X \varepsilon \mathfrak{g}$ such that $\psi$ is an eigenvector of $\pi(X)$.  Then $\mathfrak{m}$ satisfies the conditions of Lemma 4.  If $\beta$ is a root such that $\pi(X_\beta)\psi \neq 0$, $\pi(X_\beta)\psi$ belongs to the weight $\Lambda + \beta$ and therefore $X_\beta \notin \mathfrak{m}$.  Hence if $X_\beta \varepsilon \mathfrak{m}$, $\pi(X_\beta)\psi = 0$.  Therefore from Lemma 4 it is possible to choose a fundamental system of roots $(\alpha_1, \cdots, \alpha_l)$ such that $\pi(X_{\alpha_i})\psi = 0$, $i = 1, \cdots, l$.  Then if we consider the lexicographic order corresponding to this system and set $\mathfrak{n} = \sum_{\beta>0} C X_\beta$, $\mathfrak{n} \subset \mathfrak{m}$ (Lemma 18 of [5(b)] and therefore $\pi(X_\beta)\psi = 0$ for all $\beta > 0$.  If $\beta_1, \cdots, \beta_r$ are all the distinct positive roots and we assume that $V = \pi(\mathfrak{B})\psi$, it follows from Lemma 2 that every weight of $\pi$ is of the form $\Lambda - (m_1\beta_1 + \cdots + m_r\beta_r)$, $(m_i \geqq 0)$ and therefore also of the form $\Lambda - (d_1\alpha_1 + \cdots + d_l\alpha_l)$ where $d_1, \cdots, d_l$ are nonnegative integers.

Conversely suppose we are already given an order in $\mathfrak{F}_R$ and $\psi$ is an extreme vector of the representation $\pi$. We say that $\psi$ is a *positive* extreme vector if $\pi(X_\beta)\psi = 0$ for all positive roots $\beta$. The above remarks show that any extreme vector is positive with respect to the lexicographic order corresponding to some fundamental system of roots.

LEMMA 5. *Let $\Lambda$ be a linear function on $\mathfrak{h}$. Then there exists at most one maximal left ideal $\mathfrak{M}$ in $\mathfrak{B}$ such that* [4]

$$\mathfrak{M} \supset \sum_{\beta>0} \mathfrak{B}X_\beta + \sum_{\beta>0} \mathfrak{B}(H_\beta - \Lambda(H_\beta))$$

*where $\beta$ runs over all positive roots.*

This is proved in exactly the same way as Lemma 1 of [5(b)].

COROLLARY. *Let $\pi_1$, $\pi_2$ be irreducible representations of $\mathfrak{B}$ on the vector spaces $V_1$, $V_2$ respectively. Suppose there exists an order in $\mathfrak{F}_R$ such that with respect to this order both $\pi_1$ and $\pi_2$ have positive extreme vectors belonging to the same weight $\Lambda$. Then $\pi_1$, $\pi_2$ are equivalent.*

Let $\psi_i$ be a positive vector of $\pi_i$ belonging to the weight $\Lambda$ $(i = 1, 2)$. Let $\mathfrak{M}_i$ be the set of all $b \, \varepsilon \, \mathfrak{B}$ such that $\pi_i(b)\psi_i = 0$. Since $\pi_i$ is irreducible $\mathfrak{M}_i$ is a maximal left ideal in $\mathfrak{B}$ and $\pi_i$ is equivalent to the natural representation (see [5(c), §3]) of $\mathfrak{B}$ on $\mathfrak{B}/\mathfrak{M}_i$. But $\mathfrak{M}_1 = \mathfrak{M}_2$ from Lemma 5 and so our assertion follows.

For reasons which we have explained briefly in Section 1 (see also [5(f)]), we shall now consider the following special case. Let $\mathfrak{g}_0$ be a semisimple Lie algebra over $R$. Henceforward I shall use the notation of my two previous papers [5(c), (d)]. Define $\mathfrak{k}_0$ and $\mathfrak{p}_0$ as in [5(c), §2] and let $\mathfrak{h}_0$ be a maximal abelian subalgebra of $\mathfrak{k}_0$. *We shall assume that $\mathfrak{h}_0$ is also maximal abelian in $\mathfrak{g}_0$.* Complexify $\mathfrak{g}_0$ to $\mathfrak{g}$ and let $\mathfrak{h}$, $\mathfrak{p}$, $\mathfrak{k}$ denote the subspaces of $\mathfrak{g}$ spanned over $C$ by $\mathfrak{h}_0$, $\mathfrak{p}_0$, $\mathfrak{k}_0$ respectively. Then $\mathfrak{h}$ is a Cartan subalgebra of $\mathfrak{g}$ and $[\mathfrak{h}, \mathfrak{k}] \subset \mathfrak{k}$, $[\mathfrak{h}, \mathfrak{p}] \subset \mathfrak{p}$. Therefore if $\beta$ is any root (of $\mathfrak{g}$ with respect to $\mathfrak{h}$) it is clear that $X_\beta$ lies either in $\mathfrak{k}$ or in $\mathfrak{p}$. We say that $\beta$ is compact or noncompact according as $X_\alpha \, \varepsilon \, \mathfrak{k}$ or $X_\alpha \, \varepsilon \, \mathfrak{p}$. Two roots $\alpha$ and $\beta$ are unlike if one of them is compact while the other is not. Otherwise they are like. Since $[\mathfrak{p}, \mathfrak{p}] \subset \mathfrak{k}$, $[\mathfrak{k}, \mathfrak{p}] \subset \mathfrak{p}$ and $[X_\alpha, X_{-\alpha}] = H_\alpha \, \varepsilon \, \mathfrak{h} \subset \mathfrak{k}$, it follows that $\alpha$ and $-\alpha$ are always like. If $\alpha, \beta$ and $\alpha + \beta$ are roots, it is known that

---

[4] That $\mathfrak{M}$ always exists follows from Lemma 10.

$[X_\alpha, X_\beta] = cX_{\alpha+\beta}$ where $c \neq 0$ $(c \, \varepsilon \, C)$. Therefore $\alpha + \beta$ is compact or not according as $\alpha, \beta$ are like or unlike.

Let $\pi$ be a representation of $\mathfrak{g}$ on a vector space $V$ with an extreme vector $\psi_0$ (with respect to $\mathfrak{h}$) belonging to the weight $\Lambda_0$. Introduce a lexicographic order so that $\pi(X_\beta)\psi_0 = 0$ for every positive root $\beta$. Let $P$ be the set of all positive roots and $P_k$ and $P_n$ respectively the subsets of $P$ consisting of all compact and noncompact roots in $P$. Then it is clear that

$$\mathfrak{k} = \mathfrak{h} + \sum_{\alpha \, \varepsilon \, P_k} (CX_\alpha + CX_{-\alpha}), \qquad \mathfrak{p} = \sum_{B \, \varepsilon \, P_n} (CX_\beta + CX_{-\beta}).$$

In accordance with our earlier notation [5(c)] we denote by $\mathfrak{X}$ the subalgebra of $\mathfrak{B}$ generated by $(1, \mathfrak{k})$. Since we are ultimately interested in the irreducible representations of the simply connected group corresponding to the algebra $\mathfrak{g}_0$, *we shall now make the further assumption that* $\dim \pi(\mathfrak{X})\psi_0$ *is finite and* $V = \pi(\mathfrak{B})\psi_0$.

Since $\mathfrak{X}$ is the universal enveloping algebra of $\mathfrak{k}$, we can identify representations of $\mathfrak{k}$ with those of $\mathfrak{X}$. Let $\sigma$ denote the representation of $\mathfrak{k}$ on $\mathfrak{g}$ given by $\sigma(X)Y = [X, Y]$ $(X \, \varepsilon \, \mathfrak{k}, Y \, \varepsilon \, \mathfrak{g})$.

DEFINITION. *A root $\beta$ of $\mathfrak{g}$ will be called totally positive if there exists a subset $P_\beta$ of $P$ such that* $\sigma(\mathfrak{X})X_\beta = \sum_{\gamma \, \varepsilon \, P_\beta} CX_\gamma$.

It is clear that every totally positive root is positive. Later (in Section 4) we shall give another equivalent definition of totally positive roots (see the corollary to Lemma 12). Let $P_+$ be the set of all totally positive roots and $Q$ the remaining set of positive roots. Our main object at present is to prove the following theorem.

THEOREM 1. *For every $\beta \, \varepsilon \, Q$, $\Lambda_0(H_\beta)$ is a nonnegative integer and it is the least integer $r \geq 0$ such that $\pi(X_{-\beta}{}^{r+1})\psi_0 = 0$.*

Let $Q'$ denote the set of all positive roots $\beta$ such that $\pi(X_{-\beta}{}^r)\psi_0 = 0$ for some $r \geq 1$. Since $\pi(X_\beta)\psi_0 = 0$ for $\beta \, \varepsilon \, P$, it follows from Lemma 1 that in order to prove the above theorem it is enough to show that $Q \subset Q'$. For this we need some lemmas.

LEMMA 6. *Suppose $\beta \, \varepsilon \, Q'$. Then for any $\psi \, \varepsilon \, V$ we can choose an integer $r \geq 1$ such that $\pi(X_\beta{}^r)\psi = \pi(X_{-\beta}{}^r)\psi = 0$.*

Let $\beta_1, \cdots, \beta_t$ be all the distinct positive roots. As we have already observed earlier $V = \pi(\mathfrak{B})\psi_0$ is then spanned by elements of the form $\pi(X_{-\beta_1}{}^{m_1} \cdots X_{-\beta_t}{}^{m_t})\psi_0$, $(m_j \geq 0)$. Hence it is enough to prove our statement for $\psi = \pi(x)\psi_0$ where $x = X_{-\beta_1}{}^{m_1} \cdots X_{-\beta_t}{}^{m_t} \, \varepsilon \, \mathfrak{B}$. Now if $X \, \varepsilon \, \mathfrak{g}$,

$Xx = xX + [X, x]$. Therefore if we denote by $adX$ the endomorphism $b \to [X, b]$ $(b \, \varepsilon \, \mathfrak{B})$ of $\mathfrak{B}$, it follows that (see pp. 40-41 of [5(b)])

$$X^p x = \sum_{0 \leq q \leq p} C_q{}^p \{ (adX)^{p-q} x \} X^q$$

for any integer $p \geq 0$. (Here $C_q{}^p = p! / (p-q)! q!$). However if $X = X_\beta$ or $X_{-\beta}$ the restriction of $adX$ on $\mathfrak{g}$ is nilpotent and from this it follows easily that $(adX) x^m = 0$ if $m$ is sufficiently large. Since $\beta \, \varepsilon \, Q'$, $\pi(X_\beta) \psi_0 = 0$ and we can choose an integer $s \geq 1$ such that $\pi(X_{-\beta}{}^s) \psi_0 = 0$. Then if we put $p = m + s$ we get

$$X^{m+s} = \sum_{0 \leq q \leq m+s} C_q{}^{m+s} \{ (adX) x^{m+s-q} \} X^q = \sum_{s < q \leq m+s} C_q{}^{m+s} \{ (adX) x^{m+s-q} \} X^q$$

since $(adX) x^m = 0$. On the other hand $\pi(X^q) \psi_0 = 0$ for $q \geq s$ and therefore $\pi(X^{m+s}) \psi = \pi(X x^{m+s}) \psi_0 = 0$. This proves our assertion.

COROLLARY. *Let $W'$ be the subgroup of the Weyl group $W$ generated by the Weyl reflexions $s_\beta$ corresponding to $\beta \, \varepsilon \, Q'$. Then if $\Lambda$ is a weight of $\pi$ and $s \, \varepsilon \, W'$, $s\Lambda$ is also a weight of $\pi$.*

Suppose $s = s_{\beta_1} s_{\beta_2} \cdots s_{\beta_r}$ where $\beta_i \, \varepsilon \, Q'$. If $r = 1$ the result follows immediately from Lemmas 6 and 1. Hence assume $r \geq 2$ and use induction on $r$. Then if $s' = s_{\beta_2} \cdots s_{\beta_r}$ it follows from our induction hypothesis that $s'\Lambda$ is a weight of $\pi$. Hence again by the above argument $s_{\beta_1}(s'\Lambda) = s\Lambda$ is also a weight of $\pi$.

LEMMA 7. *Let $\beta$ be a positive root. Suppose $s\beta < 0$ for some $s \, \varepsilon \, W'$. Then $\beta \, \varepsilon \, Q'$.*

Let $\beta_1, \cdots, \beta_l$ be a fundamental system of positive roots (see Corollary 2 to Lemma 4). Then we know from Lemma 2 that every weight of $\pi$ must be of the form $\Lambda_0 - (m_1 \beta_1 + \cdots + m_l \beta_l)$ where $m_i$ are nonnegative integers. Now suppose $\pi(X_{-\beta}{}^r) \psi_0 \neq 0$ for some positive integer. Then $\Lambda_0 - r\beta$ is a weight of $\pi$. Since $s \, \varepsilon \, W'$ it follows from the above corollary that both $s\Lambda_0$ and $s(\Lambda_0 - r\beta)$ are weights of $\pi$. Therefore

$$s\Lambda_0 = \Lambda_0 - (d_1 \beta_1 + \cdots + d_l \beta_l) \qquad\qquad (d_i \geq 0)$$

$$s\Lambda_0 - rs\beta = \Lambda_0 - (m_1 \beta_1 + \cdots + m_l \beta_l) \qquad\qquad (m_i \geq 0).$$

But $s\beta$ is a negative root and therefore $s\beta = -(e_1 \beta_1 + \cdots + e_l \beta_l)$ $(e_i \geq 0)$, $e_i$ being integers which are not all zero. Hence

$$-r(e_1 \beta_1 + \cdots + e_l \beta_l) = rs\beta = s\Lambda_0 - (s\Lambda_0 - rs\beta)$$

$$= (m_1 - d_1)\beta_1 + \cdots + (m_l - d_l)\beta_l$$

and therefore $m_i = d_i - re_i \geqq 0$. Put $d = \max d_i$. Then $d \geqq \max re_i \geqq r$. This proves that $\pi(X_{-\beta}^{d+1})\psi_0 = 0$ and so $\beta \, \varepsilon \, Q'$.

Notice that if $\alpha$ is a positive compact root $\alpha \, \varepsilon \, Q'$. For the dimension of $\pi(\mathfrak{X})\psi_0$ is finite and the nonzero vectors among $\pi(X_{-\alpha}^r)\psi_0$ $(r \geqq 0)$ are linearly independent since they belong to the distinct weights $\Lambda_0 - r\alpha$. Hence $\pi(X_{-\alpha}^r)\psi_0 = 0$ for $r$ sufficiently large. Therefore if $W_k$ is the subgroup of $W$ generated by the Weyl reflexions corresponding to the compact roots, $W_k \subset W'$.

Let $c_0$ be the center of $\mathfrak{k}_0$ and let $\mathfrak{k}_0' = [\mathfrak{k}_0, \mathfrak{k}_0]$. Then $\mathfrak{k}_0$ is the direct sum of $c_0$ and $\mathfrak{k}_0'$ (see $[5(c), \S 2]$) and since $\mathfrak{h}_0$ is maximal abelian in $\mathfrak{k}_0$, $c_0 \subset \mathfrak{h}_0$. Let $c$ and $\mathfrak{k}'$ be the respective complexifications of $c_0$ and $\mathfrak{k}_0'$ in $\mathfrak{k}$. Put $\mathfrak{h}' = \mathfrak{h} \cap \mathfrak{k}'$.

LEMMA 8. *Let $\beta$ be a positive root. If $\beta$ is zero on $c$, $s\beta < 0$ for some $s \, \varepsilon \, W_k$ and therefore $\beta \, \varepsilon \, Q'$.*

It is obvious that every compact root is identically zero on $c$. Since $\beta - s\beta$ $(s \, \varepsilon \, W_k)$ is a linear combination of compact roots it follows that if $\beta$ vanishes on $c$ the same holds for $s\beta$. Now consider $\lambda = \sum_{s \, \varepsilon \, W_k} s\beta$. Then $s_\alpha \lambda = \lambda$ and therefore $\lambda(H_\alpha) = 0$ for every compact root $\alpha$. Since $\mathfrak{h}'$ is a Cartan subalgebra of the semisimple algebra $\mathfrak{k}'$ (see $[5(c), \text{Lemma } 3]$) it follows that the elements $H_\alpha$ corresponding to all compact roots $\alpha$ span $\mathfrak{h}'$. Hence $\lambda = 0$ on $\mathfrak{h}'$. But as we have seen $s\beta$ $(s \, \varepsilon \, W_k)$ and therefore also $\lambda$ vanish on $c$. Hence $\lambda = 0$ and therefore $s\beta$ cannot all be positive. This proves that $s\beta < 0$ for some $s \, \varepsilon \, W_k$. But $W_k \subset W'$ and so we can conclude from Lemma 7 that $\beta \, \varepsilon \, Q'$.

We define the representation $\sigma$ of $\mathfrak{k}$ on $\mathfrak{g}$ as before.

LEMMA 9. *Let $\gamma$ be a root of $\mathfrak{g}$. Then $\sigma(\mathfrak{X})X_\gamma$ is irreducible under $\mathfrak{k}$.*

Since $\mathfrak{k}$ is reductive in $\mathfrak{g}$ $[5(c), \text{Lemma } 3]$ the representation $\sigma$ is semisimple. Also it is obvious that $\mathfrak{g}$ is fully reducible under $\sigma(\mathfrak{h})$ and $CX_\gamma$ is the space of all elements $Y \, \varepsilon \, \mathfrak{g}$ such that $\sigma(H)Y = \gamma(H)Y$ for every $H \, \varepsilon \, \mathfrak{h}$. Let $q$ be a subspace of $\sigma(\mathfrak{X})X_\gamma$ which is invariant under $\sigma(\mathfrak{k})$. Since $\sigma$ is semisimple there exists a subspace $q'$ complementary to $q$ in $\sigma(\mathfrak{X})X_\gamma$ which is also invariant under $\sigma(\mathfrak{k})$. Since $X_\gamma \, \varepsilon \, q + q'$, we can choose an element $Y \neq 0$ either in $q$ or $q'$ such that $\sigma(H)Y = \gamma(H)Y$ for all $H \, \varepsilon \, \mathfrak{h}$. Let $q_1 = q$ or $q'$ according as $Y$ is in $q$ or $q'$. Then as we have remarked above $CY = CX_\gamma$ and therefore $\sigma(\mathfrak{X})X_\gamma = \sigma(\mathfrak{X})Y \subset q_1$. This proves that $q_1 = \sigma(\mathfrak{X})X_\gamma$ and so $\sigma(\mathfrak{X})X_\gamma$ is irreducible.

COROLLARY. *If $\beta$ is any root such that $X_\beta \varepsilon \sigma(\mathfrak{X}) X_\gamma$ then $\beta - \gamma$ is identically zero on $\mathfrak{c}$.*

For if $H \varepsilon \mathfrak{c}$, $\sigma(H)$ lies in the center of $\sigma(\mathfrak{X})$. Hence in view of the irreducibility proved above, it follows from Schur's lemma that there exists a complex number $c$ such that $\sigma(H) Y = cY$ for all $Y \varepsilon \sigma(\mathfrak{X}) X_\gamma$. Hence in particular if $Y = X_\beta$, $cX_\beta = \sigma(H) X_\beta = \beta(H) X_\beta$. Therefore $c = \beta(H)$ and similarly $c = \gamma(H)$. Hence $\beta(H) - \gamma(H) = 0$ for every $H \varepsilon \mathfrak{c}$.

Now we come to the proof of Theorem 1. Let $\gamma_0$ be a positive root which is not in $Q'$. We have to show that $\gamma_0$ is totally positive. Since $P_k \subset Q'$, $\gamma_0$ is not compact and therefore $X_{\gamma_0} \varepsilon \mathfrak{p}$. Put $\mathfrak{p}_{\gamma_0} = \sigma(\mathfrak{X}) X_{\gamma_0}$. Then $\mathfrak{p}_{\gamma_0} \subset \mathfrak{p}$ and since $\mathfrak{p}$ is fully reducible under $\sigma(\mathfrak{h})$, $\mathfrak{p}_{\gamma_0}$ can be spanned by the elements $X_\beta$ corresponding to a certain number of noncompact roots $\beta$. Let $\beta_0$ and $\beta_1$ respectively be the highest and lowest among all those roots $\beta$ for which $X_\beta \varepsilon \mathfrak{p}_{\gamma_0}$. Then $\beta_0 \geqq \gamma_0 > 0$ and in order to prove that $\gamma_0$ is totally positive it would be enough to show that $\beta_1 > 0$. So suppose this is false and $\beta_1$ is negative. Since $\mathfrak{k} = \mathfrak{c} + \mathfrak{k}'$, it is obvious from Schur's lemma and Lemma 9 that $\mathfrak{p}_{\gamma_0}$ is irreducible under $\sigma(\mathfrak{k}')$. Let $\tau$ denote the irreducible representation of $\mathfrak{k}'$ on $\mathfrak{p}_{\gamma_0}$ defined by $\sigma$. Now $\mathfrak{h}'$ is a Cartan subalgebra of the semisimple algebra $\mathfrak{k}'$ and

$$\mathfrak{k}' = \mathfrak{h}' + \sum_{\alpha \varepsilon P_k} (CX_\alpha + CX_{-\alpha}).$$

Hence if we identify linear functions on $\mathfrak{h}'$ with those linear functions on $\mathfrak{h}$ which vanish on $\mathfrak{c}$, the roots of $\mathfrak{k}'$ (with respect to $\mathfrak{h}'$) coincide with the compact roots of $\mathfrak{g}$. For any root $\beta$ of $\mathfrak{g}$ let $\beta'$ denote the restriction of $\beta$ on $\mathfrak{h}'$. It is clear that $[X_\alpha, X_{\beta_0}] = [X_{-\alpha}, X_{\beta_1}] = 0$ for all positive compact roots $\alpha$. Therefore it follows from Lemma 2 that $\beta_0'$ is the highest and $\beta_1'$ the lowest weight (with respect to $\mathfrak{h}'$) of the irreducible representation $\tau$ of $\mathfrak{k}'$. Hence from Lemma 3, there exists an element $t$ in the Weyl group of $\mathfrak{k}'$ with respect to $\mathfrak{h}'$, such that $\beta_1' = t\beta_0'$. This means that we can choose $s \varepsilon W_k$ such that $\beta_1 - s\beta_0$ is identically zero on $\mathfrak{h}'$. However $\beta_0 - s\beta_0$ is a linear combination of compact roots and therefore it vanishes on $\mathfrak{c}$. On the other hand from the above corollary $\beta_1 - \beta_0$ is also zero on $\mathfrak{c}$. Therefore $\beta_1 - s\beta_0$ vanishes on both $\mathfrak{h}'$ and $\mathfrak{c}$ and so $\beta_1 = s\beta_0$. Since $\beta_0 > 0$ and $s\beta_0 = \beta_1 < 0$, we conclude from Lemma 7 that $\beta_0 \varepsilon Q'$.

Now $X_{\gamma_0} \varepsilon \mathfrak{p}_{\gamma_0}$ and $\gamma_0 \notin Q'$. Choose the highest $\beta$ such that $X_\beta \varepsilon \mathfrak{p}_{\gamma_0}$ and $\beta \notin Q'$. Then $\beta \geqq \gamma_0 > 0$ and $\beta \neq \beta_0$. We claim that $[X_\alpha, X_\beta] \neq 0$ for some positive compact root $\alpha$. For otherwise suppose $[X_\alpha, X_\beta] = 0$ for every $\alpha \varepsilon P_k$. Let $\alpha_1, \cdots, \alpha_r$ be all the distinct positive compact roots of $\mathfrak{g}$. Since $\mathfrak{p}_{\gamma_0}$ is irreducible under $\sigma(\mathfrak{k})$, it follows that $\sigma(\mathfrak{X})X_\beta = \mathfrak{p}_{\gamma_0}$. But then from Lemma 2 and the corollary to Lemma 9 we can conclude that

$$\beta_0 = \beta - (m_1\alpha_1 + \cdots + m_r\alpha_r) \qquad (m_i \geqq 0).$$

Since $\beta_0 \geqq \beta$ this implies that $m_1 = \cdots = m_r = 0$. However this is impossible as $\beta \neq \beta_0$. So let $\alpha$ be a positive compact root such that $[X_\alpha, X_\beta] \neq 0$. Then $\beta + \alpha$ is a root and $X_{\beta+\alpha} = c[X_\alpha, X_\beta] \, \varepsilon \, \mathfrak{p}_{\gamma_0}$ $(c \, \varepsilon \, C)$. Moreover $\beta + \alpha > \beta$ and therefore $\beta \, \varepsilon \, Q'$. Now $\beta + (\beta + \alpha)$ is not a root. For since $\beta + \alpha$ and $\beta$ are both noncompact $\beta + (\beta + \alpha)$, if it is a root, must be compact. On the other hand $\beta - (\beta + \alpha) = -\alpha$ is also a compact root. Since $2\beta = \beta + (\beta + \alpha) - \alpha$, $\beta$ is a linear combination of two compact roots and so it must be zero on c. Therefore from Lemma 8, $\beta \, \varepsilon \, Q'$. As this contradicts the definition of $\beta$, it follows that $\beta + (\beta + \alpha)$ is not a root while $\beta - (\beta + \alpha)$ is a root. Let $\nu$ be the largest integer $\geqq 0$ such that $\beta - \nu(\beta + \alpha)$ is a root. Then $\nu \geqq 1$ and

$$s_{\beta+\alpha}\beta = \beta - \nu(\beta + \alpha) \leqq \beta - (\beta + \alpha) = -\alpha < 0.$$

But $\beta + \alpha \, \varepsilon \, Q'$ and therefore from Lemma 7, $\beta \, \varepsilon \, Q'$ giving again a contradiction. This proves that $\beta_1 > 0$ and so $\gamma_0$ is totally positive.

**4. Another characterization of totally positive roots.** We shall now study the totally positive roots a little more closely. We call a root $\beta$ totally negative if $-\beta$ is totally positive.

LEMMA 10. *Let $\beta$ be a totally positive root. Then $s\beta$ $(s \, \varepsilon \, W_k)$ is also totally positive. Similarly if $\beta$ is totally negative the same is true of $s\beta$.*

Let $\alpha$ be a compact root and let $\gamma = s_\alpha\beta$. Then it is known (see Weyl [7]) that $X_\gamma$ can be written in the form $X_\gamma = c\sigma(Y^m)X_\beta$ where $c \, \varepsilon \, C$ and $Y = X_\alpha$ or $X_{-\alpha}$ and $m$ is a nonnegative integer. Therefore from Lemma 9, $\sigma(\mathfrak{X})X_\gamma = \sigma(\mathfrak{X})X_\beta$. This shows that if $\beta$ is totally positive the same holds for $s_\alpha\beta$ and therefore also for $s\beta$ $(s \, \varepsilon \, W_k)$. On the other hand if $\beta$ is totally negative $s(-\beta) = -s\beta$ $(s \, \varepsilon \, W_k)$ is totally positive and therefore $s\beta$ is totally negative.

COROLLARY 1. *If $\beta$ is totally positive it cannot be identically zero on c.*

Put $\lambda = \sum\limits_{s \, \varepsilon \, W_k} s\beta$. Then $s\beta$ are all totally positive and hence positive. Therefore $\lambda > 0$. But $s_\alpha\lambda = \lambda$ for every compact root $\alpha$ and so $\lambda$ is zero on $\mathfrak{h}'$. Hence $\lambda$ cannot be identically zero on c. But $\beta - s\beta$ $(s \, \varepsilon \, W_k)$ being a linear combination of compact roots is zero on c. This shows that if $w$

is the order of the group $W_k$, $\lambda$ coincides with $w\beta$ on c.  Therefore $\beta$ cannot vanish identically on c.

COROLLARY 2.  *Every totally positive root is noncompact.  If* $c = \{0\}$ *there are no totally positive roots.*

This is an immediate consequence of Corollary 1 above.

LEMMA 11.  *Let $\beta$ be a totally positive root and $\gamma$ a noncompact root which is not totally negative.  Then $\beta + \gamma$ is not a root.*

For otherwise suppose $\beta + \gamma$ is a root.  Since $\beta$ and $\gamma$ are both noncompact $\alpha = \beta + \gamma$ must be compact.  Then $-\gamma = \beta - \alpha$ and

$$X_{-\gamma} = c[X_{-\alpha}, X_\beta] \, \varepsilon \, \sigma(\mathfrak{X}) X_\beta \qquad\qquad (c \, \varepsilon \, C).$$

Hence from Lemma 9, $\sigma(\mathfrak{X}) X_{-\gamma} = \sigma(\mathfrak{X}) X_\beta$ and therefore since $\beta$ is totally positive the same must hold for $-\gamma$.  But this means that $\gamma$ is totally negative and so we get a contradiction.

Let $\theta$ denote the automorphism of $\mathfrak{g}$ of order 2 given by $\theta(X + Y) = X - Y$ $(X \, \varepsilon \, \mathfrak{k}, Y \, \varepsilon \, \mathfrak{p})$.  We define a linear mapping $\bar{\theta}$ of $\mathfrak{g}$ over $R$ by the rule $\bar{\theta}(cX) = \bar{c}\theta(X)$ $(c \, \varepsilon \, C, X \, \varepsilon \, \mathfrak{g}_0)$ where $\bar{c}$ is the complex conjugate of $c$.  Then $\bar{\theta}$ is the conjugation of $\mathfrak{g}$ with respect to its compact real form[5] $\mathfrak{u} = \mathfrak{k}_0 + (-1)^{\frac{1}{2}}\mathfrak{p}_0$ (see [5(c), §2]) and it is known (see Weyl [7]) that one can choose the elements $X_\gamma$ corresponding to the various roots $\gamma$ in such a way that $X_\gamma - X_{-\gamma}$ and $(-1)^{\frac{1}{2}}(X_\gamma + X_{-\gamma})$ both lie in $\mathfrak{u}$ (without disturbing our earlier condition that $\gamma(H_\gamma) = 2$ for $H_\gamma = [X_\gamma, X_{-\gamma}]$).  Assuming that this has been done, it follows immediately that $\bar{\theta}(X_\gamma) = -\epsilon_\gamma X_{-\gamma}$ where $\epsilon_\gamma = 1$ or $-1$ according as $\gamma$ is compact or not.

LEMMA 12.  *Let $\alpha_1, \cdots, \alpha_r$ be all the distinct positive compact roots and $\beta$ a totally positive root.  Then if $\gamma$ is any root of the form $\beta + m_1\alpha_1 + \cdots + m_r\alpha_r$ (where $m_1, \cdots, m_r$ are integers), $\gamma$ is totally positive and $X_\gamma \, \varepsilon \, \sigma(\mathfrak{X}) X_\beta$.*

We shall make use of an argument due to E. Cartan [3(c)].  Define the bilinear form $B(X, Y)$ on $\mathfrak{g}$ as before.  Then $-B(X, \bar{\theta}(X))$ $(X \, \varepsilon \, \mathfrak{g})$ is a positive definite Hermitian form on $\mathfrak{g}$.  Since $\beta$ is noncompact, $\mathfrak{p}_\beta = \sigma(\mathfrak{X}) X_\beta$ is a subspace of $\mathfrak{p}$.  Let $\mathfrak{q}$ denote the orthogonal complement[6]

---

[5] We fix once for all one square-root of $-1$ in $C$ and denote it by $(-1)^{\frac{1}{2}}$.

[6] I take this opportunity to correct a slip in one of my earlier papers [5(d), p. 29 bottom paragraph].  There the orthogonal complement is taken with respect to $B(X, Y)$ instead of $B(X, \bar{\theta}, Y))$.  This is a mistake which however is easy to rectify without making any substantial change in the argument.

of $\mathfrak{p}_\beta + \bar{\theta}(\mathfrak{p}_\beta)$ in $\mathfrak{p}$ (with respect to the above Hermitian form). Then we shall prove that $[\mathfrak{p}_\beta + \bar{\theta}(\mathfrak{p}_\beta), \mathfrak{q}] = \{0\}$. Since $\mathfrak{q} = \bar{\theta}(\mathfrak{q})$, it is enough to show that $[\mathfrak{p}_\beta, \mathfrak{q}] = \{0\}$. Let $X \, \varepsilon \, \mathfrak{p}_\beta$, $Y \, \varepsilon \, \mathfrak{q}$ and $Z \, \varepsilon \, \mathfrak{k}$. Then

$$B([X,Y], \bar{\theta}(Z)) = B(Y, [\bar{\theta}(Z), X]) = 0$$

since $[\bar{\theta}(Z), X]$ lies in $\mathfrak{p}_\beta$ and $\mathfrak{q}$ is orthogonal to $\bar{\theta}(\mathfrak{p}_\beta)$. But $[X,Y] \, \varepsilon \, [\mathfrak{p}, \mathfrak{p}] \subset \mathfrak{k}$ and so this shows that $[X,Y] = 0$ and our assertion is proved. Let $\mathfrak{k}_\beta$ be the centralizer of $\mathfrak{q}$ in $\mathfrak{k}$. Put $\mathfrak{g}_\beta = \mathfrak{p}_\beta + \bar{\theta}(\mathfrak{p}_\beta) + \mathfrak{k}_\beta$. We claim that $\mathfrak{g}_\beta$ is an ideal in $\mathfrak{g}$. Since $\mathfrak{p}_\beta + \bar{\theta}(\mathfrak{p}_\beta)$ is invariant under $\sigma(\mathfrak{k})$, the same holds for $\mathfrak{q}$ and $\mathfrak{k}_\beta$. Therefore $[\mathfrak{k}, \mathfrak{g}_\beta] \subset \mathfrak{g}_\beta$. Moreover $[\mathfrak{q}, \mathfrak{g}_\beta] = \{0\}$. Since $\mathfrak{g} = \mathfrak{k} + \mathfrak{q} + \mathfrak{p}_\beta + \bar{\theta}(\mathfrak{p}_\beta)$, it is enough to prove that $[\mathfrak{p}_\beta + \bar{\theta}(\mathfrak{p}_\beta), \mathfrak{g}_\beta] \subset \mathfrak{g}_\beta$. But $\mathfrak{g}_\beta = \mathfrak{p}_\beta + \bar{\theta}(\mathfrak{p}_\beta) + \mathfrak{k}_\beta$ and $\mathfrak{p}_\beta + \bar{\theta}(\mathfrak{p}_\beta)$ is invariant under $\sigma(\mathfrak{k})$. So it would be sufficient to show that if $X, Y$ are in $\mathfrak{p}_\beta + \bar{\theta}(\mathfrak{p}_\beta)$, $[X,Y]$ lies in $\mathfrak{k}_\beta$. However $[X,Y] \, \varepsilon \, \mathfrak{k}$ and if $Z \, \varepsilon \, \mathfrak{q}$

$$[[X,Y],Z] = [[X,Z],Y] + [X,[Y,Z]] = 0$$

since $[X,Z] = [Y,Z] = 0$. This proves that $[X,Y] \, \varepsilon \, \mathfrak{k}_\beta$ and therefore $\mathfrak{g}_\beta$ is an ideal in $\mathfrak{g}$. Let $\mathfrak{l}$ be the orthogonal complement of $\mathfrak{k}_\beta$ in $\mathfrak{k}$. Then $\mathfrak{g}' = \mathfrak{q} + \mathfrak{l}$ is clearly the orthogonal complement of $\mathfrak{g}_\beta$ in $\mathfrak{g}$. Let $X \, \varepsilon \, \mathfrak{g}$ and $Y \, \varepsilon \, \mathfrak{g}'$. Then if $Z \, \varepsilon \, \mathfrak{g}_\beta$,

$$B([X,Y], \bar{\theta}(Z)) = -B(Y, [X, \bar{\theta}(Z)]).$$

But

$$\bar{\theta}([X, \bar{\theta}(Z)]) = [\bar{\theta}(X), Z] \, \varepsilon \, \mathfrak{g}_\beta$$

and since $\mathfrak{g}'$ is orthogonal to $\mathfrak{g}_\beta$, $B(Y, [X, \bar{\theta}(Z)]) = 0$. Hence $[X,Y]$ being orthogonal to $\mathfrak{g}_\beta$ lies in $\mathfrak{g}'$. This shows that $\mathfrak{g}'$ is also an ideal in $\mathfrak{g}$. Hence $[\mathfrak{g}', \mathfrak{g}_\beta] \subset \mathfrak{g}' \cap \mathfrak{g}_\beta = \{0\}$. Moreover since $\mathfrak{q} = \bar{\theta}(\mathfrak{q})$ it follows that $\mathfrak{k}_\beta = \bar{\theta}(\mathfrak{k}_\beta)$, $\bar{\theta}(\mathfrak{g}_\beta) = \mathfrak{g}_\beta$ and $\bar{\theta}(\mathfrak{g}') = \mathfrak{g}'$. Now suppose $X \, \varepsilon \, \mathfrak{c}$. Since $\mathfrak{k} = \mathfrak{k}_\beta + \mathfrak{l}$, $X = Y + Z$ where $Y \, \varepsilon \, \mathfrak{k}_\beta$ and $Z \, \varepsilon \, \mathfrak{l}$. But $[\mathfrak{k}_\beta, \mathfrak{l}] = \{0\}$ and so it is obvious that both $Y$ and $Z$ are in $\mathfrak{c}$. This shows that $\mathfrak{c} = \mathfrak{c} \cap \mathfrak{k}_\beta + \mathfrak{c} \cap \mathfrak{l}$.

Now suppose $\gamma = \beta + m_1 \alpha_1 + \cdots + m_r \alpha_r$ is a root. Then $\gamma$ and $\beta$ coincide on $\mathfrak{c}$. Since $X_\beta \, \varepsilon \, \mathfrak{g}_\beta$, it commutes with $\mathfrak{l}$ and therefore $\beta$ vanishes identically on $\mathfrak{c} \cap \mathfrak{l}$. Hence the same holds for $\gamma$. Moreover since both $\mathfrak{g}_\beta$ and $\mathfrak{g}'$ are invariant under $\sigma(\mathfrak{h})$ it is clear that $X_\gamma$ lies either in $\mathfrak{g}_\beta$ or $\mathfrak{g}'$. We claim $X_\gamma \, \varepsilon \, \mathfrak{g}_\beta$. For otherwise if $X_\gamma \, \varepsilon \, \mathfrak{g}'$ it commutes with $\mathfrak{g}_\beta$ and therefore $\gamma$ vanishes identically on $\mathfrak{c} \cap \mathfrak{k}_\beta$. But we have already seen that $\gamma$ is zero on $\mathfrak{c} \cap \mathfrak{l}$. Therefore $\gamma$ and hence also $\beta$ vanishes identically on $\mathfrak{c}$. But since $\beta$ is totally positive this is impossible in view of Corollary 1 to Lemma 10. This proves that $X_\gamma \, \varepsilon \, \mathfrak{g}_\beta$. Now $\gamma$ cannot be a compact root since it coincides with $\beta$ on $\mathfrak{c}$ and therefore is not identically zero on $\mathfrak{c}$. Hence $X_\gamma \varepsilon \, \mathfrak{p} \cap \mathfrak{g}_\beta = \mathfrak{p}_\beta + \bar{\theta}(\mathfrak{p}_\beta)$. Since $\mathfrak{p}_\beta$ and $\bar{\theta}(\mathfrak{p}_\beta)$ are both invariant under $\sigma(\mathfrak{h})$ either $X_\gamma \varepsilon \, \mathfrak{p}_\beta$ or $X_\gamma \varepsilon \, \bar{\theta}(\mathfrak{p}_\beta)$. But $\bar{\theta}(X_\gamma) = -\epsilon_\gamma X_{-\gamma}$ and so either $X_\gamma$ or $X_{-\gamma}$ lies in $\mathfrak{p}_\beta$. Now if $X_{-\gamma} \varepsilon \, \mathfrak{p}_\beta$, it follows from the corollary to Lemma 9 that $\beta + \gamma$ is zero on $\mathfrak{c}$. But we have seen above that $\beta - \gamma$ is zero on $\mathfrak{c}$. Therefore $2\beta = (\beta + \gamma) + (\beta - \gamma)$ is also zero on $\mathfrak{c}$. Since $\beta$ is totally positive this is impossible (Corollary 1 to Lemma 10) and therefore $X_\gamma \varepsilon \, \mathfrak{p}_\beta$. Hence from Lemma 9 $\sigma(\mathfrak{X})X_\gamma = \sigma(\mathfrak{X})X_\beta$ and this proves that $\gamma$ is totally positive.

COROLLARY. *Let $\alpha_1, \cdots, \alpha_r$ be all the positive compact roots of $\mathfrak{g}$. Then a root $\beta$ is totally positive if and only if every root of the form $\beta + m_1\alpha_1 + \cdots + m_r\alpha_r$ (where $m_1, \cdots, m_r$ are integers) is positive.*

Suppose $\gamma = \beta + m_1\alpha_1 + \cdots + m_r\alpha_r$ is a root. If $\beta$ is totally positive, it follows from the above lemma that $\gamma$ is positive. Conversely suppose every root $\gamma$ of this form is positive. Then it is clear that $\beta \neq \pm \alpha_i$, $1 \leq i \leq r$ and therefore $\beta$ is noncompact and $\sigma(\mathfrak{X})X_\beta$ is spanned by the elements $X_\gamma$ corresponding to certain roots $\gamma$ of the above form. Since every such $\gamma$ is positive, it follows from our definition that $\beta$ is totally positive. The above corollary thus gives an alternative definition of total positivity.

**5. Some further results on totally positive roots.** First we recall a few well-known facts in the theory of semisimple Lie algebras. Let $\mathfrak{a}$ be an ideal in $\mathfrak{g}$ and $\beta$ a root of $\mathfrak{g}$. We say that $\beta$ is a root of $\mathfrak{a}$ if $X_\beta \varepsilon \, \mathfrak{a}$. Let $(\beta_1, \cdots, \beta_l)$ be a fundamental system of roots of $\mathfrak{g}$ and suppose among these $\beta_1, \cdots, \beta_m$ are all the roots of $\mathfrak{a}$. Then every root of $\mathfrak{a}$ is a linear combination of $\beta_1, \cdots, \beta_m$. Let $s_i$ denote the Weyl reflexion corresponding to $\beta_i$ and let $W_\mathfrak{a}$ be the subgroup of the Weyl group $W$ generated by $s_1, \cdots, s_m$. Then if $\beta$ is any root of $\mathfrak{a}$ the corresponding Weyl reflexion $s_\beta$ is contained in $W_\mathfrak{a}$.

Now as before let $P_+$ be the set of all totally positive roots. Put $\mathfrak{p}_+ = \sum_{\beta \varepsilon P_+} C X_\beta$ and $\mathfrak{p}_- = \sum_{\beta \varepsilon P_+} C X_{-\beta}$ so that $\mathfrak{p}_- = \bar{\theta}(\mathfrak{p}_+)$. Let $\mathfrak{q}$ be the orthogonal complement of $\mathfrak{p}_+ + \mathfrak{p}_-$ in $\mathfrak{p}$ (under the Hermitian form $B(X, \bar{\theta}(X))$). Then if $\mathfrak{l}$ is the centralizer of $\mathfrak{q}$ in $\mathfrak{k}$ we can show, as in the proof of Lemma 12, that $\mathfrak{g}_+ = \mathfrak{p}_+ + \mathfrak{p}_- + \mathfrak{l}$ is an ideal in $\mathfrak{g}$. Let $\mathfrak{g}'$ be orthogonal complement of $\mathfrak{g}_+$ in $\mathfrak{g}$. Then $\mathfrak{g}'$ is also an ideal in $\mathfrak{g}$ and $[\mathfrak{g}', \mathfrak{g}_+] = \{0\}$. It is obvious that $\mathfrak{g}' \cap \mathfrak{k}$ and $\mathfrak{g}' \cap \mathfrak{p}$ are the orthogonal complements of $\mathfrak{l}$ and $\mathfrak{p}_+ + \mathfrak{p}_-$ in $\mathfrak{k}$ and $\mathfrak{p}$ respectively and $\mathfrak{g}' = \mathfrak{g}' \cap \mathfrak{k} + \mathfrak{g}' \cap \mathfrak{p}$. Now if $\beta$ is a root, $X_\beta$ lies either in $\mathfrak{g}_+$ or $\mathfrak{g}'$. Since $\mathfrak{g}_+ \cap \mathfrak{p} = \mathfrak{p}_+ + \mathfrak{p}_-$, it follows that if $\beta$ is a noncompact root of $\mathfrak{g}_+$ it is either totally positive or totally negative.

Let $(\beta_1, \cdots, \beta_l)$ be a fundamental system of positive roots of $\mathfrak{g}$ (Corollary 2 to Lemma 4). If $\gamma$ is a positive root which is not contained in this fundamental system, then it is well known that $\gamma - \beta_i$ must also be a positive root for some $i$ $(1 \leqq i \leqq l)$. On the other hand $\beta_i - \beta$ can never be a positive root for any root $\beta > 0$. Hence a positive root is contained in the set $(\beta_1, \cdots, \beta_l)$ if and only if $\gamma - \beta$ is never a positive root for any root $\beta > 0$.

Now it follows from the definition of total positivity that $\mathfrak{p}_+$ is invariant under $\sigma(\mathfrak{k})$. Hence $\mathfrak{p}_+ = \sum_{\nu=1}^{q} \mathfrak{q}_\nu$ where $\mathfrak{q}_\nu$, $1 \leqq \nu \leqq q$, are nonzero subspaces of $\mathfrak{p}_+$ which are invariant and irreducible under $\sigma(\mathfrak{k})$ and the sum is direct. (If $\mathfrak{p}_+ = \{0\}$, $q = 0$). Let $\gamma_\nu$ be the lowest root such that $X_{\gamma_\nu} \varepsilon \mathfrak{q}_\nu$ and let $\alpha$ be a positive compact root. We claim $\gamma_\nu - \alpha$ is not a root. For otherwise it follows from Lemma 12 that $X_{\gamma_\nu - \alpha} \varepsilon \mathfrak{q}_\nu$ which however is impossible since $\gamma_\nu - \alpha < \gamma_\nu$. Now let $\beta$ be a noncompact positive root and suppose $\alpha = \gamma_\nu - \beta$ is a root. Since $\gamma_\nu$ and $\beta$ are both noncompact, $\alpha$ is compact and $\beta = \gamma_\nu - \alpha$. Hence by the above argument $\alpha$ cannot be positive. This proves that $\gamma_\nu - \beta$ can never be a positive root for any root $\beta > 0$. Therefore it follows from our criterion above that $\gamma_\nu$ is a root of the fundamental system $(\beta_1, \cdots, \beta_l)$.

Now suppose $(\beta_1, \cdots, \beta_r)$ are roots of $\mathfrak{g}_+$ while $(\beta_{r+1}, \cdots, \beta_l)$ are roots of $\mathfrak{g}'$ $(0 \leqq r \leqq l)$. Then if $\alpha$ is a positive compact root of $\mathfrak{g}_+$,

$$\alpha = m_1 \beta_1 + \cdots + m_r \beta_r \qquad\qquad (m_i \geqq 0).$$

We recall that if $\gamma$ is any compact root $\sum_{s \varepsilon W_k} s\gamma = 0$. Now suppose $\beta_1, \cdots, \beta_t$ are compact while $\beta_{t+1}, \cdots, \beta_r$ are noncompact (and hence totally positive). Then

$$0 = \sum_{s \varepsilon W_k} s\alpha = \sum_{i=1}^{r} m_i \sum_{s \varepsilon W_k} s\beta_i = \sum_{t < i \leqq r} m_i \sum_{s \varepsilon W_k} s\beta_i$$

But from Lemma 10 $s\beta_i$ $(i > t)$ is also totally positive. Since $m_i \geqq 0$ the above sum cannot be zero unless $m_i = 0$ $(t < i \leqq r)$. Therefore $\alpha = m_1 \beta_1 + \cdots + m_t \beta_t$. On the other hand let $\gamma$ be any positive noncompact root of $\mathfrak{g}_+$. Then $X_\gamma \varepsilon \mathfrak{q}_\nu = \sigma(\mathfrak{x}) X_{\gamma_\nu}$ for some $\nu$. Since $[\mathfrak{g}', \mathfrak{g}_+] = \{0\}$, it is clear that $\mathfrak{q}_\nu$ is irreducible under $\sigma(\mathfrak{k})$ and therefore $\gamma - \gamma_\nu$ is a linear combination of compact roots of $\mathfrak{g}_+$. Hence $\gamma = \gamma_\nu + m_1 \beta_1 + \cdots + m_t \beta_t$ where $m_1, \cdots, m_t$ are real numbers. But $\gamma_\nu$ is contained in the set $(\beta_1, \cdots, \beta_r)$ and since it is noncompact, $\gamma_\nu \neq \beta_i$. $1 \leqq i \leqq t$. Therefore since $\gamma > 0$ and $(\beta_1, \cdots, \beta_l)$ is a fundamental system, $m_1, \cdots, m_t$ must all be nonnegative integers. Thus we have obtained the following result.

**LEMMA 13.** *Let $(\beta_1, \cdots, \beta_l)$ be a fundamental system of positive roots of $\mathfrak{g}$. Let $(\alpha_1, \cdots, \alpha_r)$ be all the (distinct) compact roots of $\mathfrak{g}_+$ and $(\gamma_1, \cdots, \gamma_t)$ all the (distinct) noncompact roots of $\mathfrak{g}_+$ which occur in this fundamental system. Then every positive root $\gamma$ of $\mathfrak{g}_+$ can be written in the form*

$$\gamma = \mu_1 \gamma_1 + \cdots + \mu_t \gamma_t + m_1 \alpha_1 + \cdots + m_r \alpha_r$$

*where $\mu_i, m_j$ are all nonnegative integers. Moreover $\mu_1 + \cdots + \mu_t = 0$ or $1$ according as $\gamma$ is compact or not.*

**COROLLARY 1.** $t = \dim (\mathfrak{c} \cap \mathfrak{g}_+)$ *and* $r = \dim (\mathfrak{h}' \cap \mathfrak{g}_+)$.

For it is easy to see that $\mathfrak{h} = \mathfrak{h} \cap \mathfrak{g}_+ + \mathfrak{h} \cap \mathfrak{g}'$ and $\mathfrak{c} = \mathfrak{c} \cap \mathfrak{g}_+ + \mathfrak{c} \cap \mathfrak{g}'$. From this it follows that $r + t = \dim (\mathfrak{h} \cap \mathfrak{g}_+)$ and $r = \dim (\mathfrak{h}' \cap \mathfrak{g}_+)$. Since $\mathfrak{h} \cap \mathfrak{g}_+ = \mathfrak{h}' \cap \mathfrak{g}_+ + \mathfrak{c} \cap \mathfrak{g}_+$, $t = \dim (\mathfrak{c} \cap \mathfrak{g}_+)$.

**COROLLARY 2.** *If the algebra $\mathfrak{g}_0$ is simple, only the following two cases are possible.*

(1)  *There are no totally positive roots.*

(2)  *Every noncompact positive root is totally positive.*

*Suppose we are in case (2) and $\mathfrak{g}_0 \neq \mathfrak{k}_0$. Then there is exactly one noncompact root in the fundamental system $(\beta_1, \cdots, \beta_l)$ and $\dim_R \mathfrak{c}_0 = 1$.*

Let $\eta$ denote the conjugation of $\mathfrak{g}$ with respect to $\mathfrak{g}_0$ so that

$$\eta(X + (-1)^{\frac{1}{2}}Y) = X - (-1)^{\frac{1}{2}}Y \qquad\qquad (X, Y \,\varepsilon\, \mathfrak{g}_0).$$

Then $\eta = \bar{\theta} \circ \theta$ and since $\mathfrak{g}_+$ and $\mathfrak{g}'$ are invariant under both $\bar{\theta}$ and $\theta$ they are also invariant under $\eta$. Therefore $\mathfrak{g}_0 = \mathfrak{g}_0 \cap \mathfrak{g}_+ + \mathfrak{g}_0 \cap \mathfrak{g}'$ and $\mathfrak{g}_+, \mathfrak{g}'$ are spanned over $C$ by $\mathfrak{g}_0 \cap \mathfrak{g}_+$ and $\mathfrak{g}_0 \cap \mathfrak{g}'$ respectively. Hence it follows that if $\mathfrak{g}_0$ is simple either $\mathfrak{g}_+$ or $\mathfrak{g}'$ is $\{0\}$ and so we get the above two cases. Now suppose we are in case (2). Then every positive root is either compact or totally positive. Therefore if $\mathfrak{g}_0 \neq \mathfrak{k}_0$, totally positive roots must exist and so $t \geq 1$ in the notation of Lemma 13. On the other hand it follows from the simplicity of $\mathfrak{g}_0$ (see Cartan [3(e), p. 247]) that $\dim_R \mathfrak{c}_0 \leq 1$. Therefore from Corollary 1 above $t = \dim_R \mathfrak{c}_0 = 1$.

**LEMMA 14.** *Let $(\beta_1, \cdots, \beta_q)$ be all those roots in the fundamental system $(\beta_1, \cdots, \beta_l)$ (of Lemma 13) which are not totally positive and let $W'$ be the subgroup of the Weyl group $W$ generated by the Weyl reflexions* *corresponding to $\beta_i$, $1 \leq i \leq q$. Then if $\beta$ is any positive root of $\mathfrak{g}$ which is not totally positive we can find an index $i$ $(1 \leq i \leq q)$ and an element $s \, \varepsilon \, W'$ such that $\beta = s\beta_i$.*

For any root $\gamma$ let $s_\gamma$ denote the corresponding Weyl reflexion. Put $s_i = s_{\beta_i}$, $1 \leq i \leq l$. Then it is well known that $s_1, \cdots, s_l$ generate $W$. Now if some $\beta_i$ is a root of $\mathfrak{g}'$, it cannot be totally positive and therefore $i \leq q$. So we may suppose that $(\beta_1, \cdots, \beta_d)$ $(0 \leq d \leq q)$ are all the roots of $\mathfrak{g}'$ among $(\beta_1, \cdots, \beta_l)$. Since $[\mathfrak{g}_+, \mathfrak{g}'] = \{0\}$, it is easy to see that if $\gamma$ and $\gamma'$ are roots of $\mathfrak{g}_+$ and $\mathfrak{g}'$ respectively then $s_\gamma$ and $s_{\gamma'}$ commute and $s_\gamma \gamma' = \gamma'$, $s_{\gamma'} \gamma = \gamma$. Therefore in particular $s_i$ and $s_j$ commute if $i \leq d$ and $j > d$. Let $W''$ and $W_+$ be the subgroups of $W$ generated by $s_i$, $1 \leq i \leq d$ and $s_j$, $d < j \leq l$ respectively and let $\beta$ be a positive root which is not totally positive. First suppose $\beta$ is a root of $\mathfrak{g}'$. Then $\beta = s\beta_i$ for some $s \, \varepsilon \, W$ and some $i$ $(1 \leq i \leq l)$. But we can write $s^{-1} = uv$ where $u \, \varepsilon \, W''$ and $v \, \varepsilon \, W_+$. Then $v\beta = \beta$ and therefore $\beta_i = s^{-1}\beta = u\beta$. Since $u \, \varepsilon \, W''$, $u\beta - \beta$, and therefore also $\beta_i$, is a linear combination of roots of $\mathfrak{g}'$. This proves that $\beta_i$ is a root of $\mathfrak{g}'$ and therefore $i \leq d$. Since $u^{-1} \, \varepsilon \, W'' \subset W'$ and $\beta = u^{-1}\beta_i$, our assertion is true in this case. On the other hand if $\beta$ is a root of $\mathfrak{g}_+$, it must be compact. Now $\mathfrak{k}' \cap \mathfrak{g}_+$ being an ideal in the semisimple algebra $\mathfrak{k}'$, is semisimple and the compact roots of $\mathfrak{g}_+$ may be identified in the obvious manner with the roots of $\mathfrak{k}' \cap \mathfrak{g}_+$ with respect to $\mathfrak{h}' \cap \mathfrak{g}_+$. Then it follows from Lemma 13 that $(\beta_{d+1}, \cdots, \beta_q)$ is a fundamental system of roots of $\mathfrak{k}' \cap \mathfrak{g}_+$ and therefore $\beta = s\beta_j$ where $d < j \leq q$ and $s$ lies in the group generated by $(s_{d+1}, \cdots, s_q)$. This completes the proof of the lemma.

We shall now prove a result which will be needed in a subsequent paper.

LEMMA 15. *Let $\gamma$ be a totally positive root and $\alpha$ a compact root. Let $k$ and $k'$ respectively be the largest nonnegative integers such that $\gamma - k\alpha$ and $\gamma + k'\alpha$ are roots. Then $k + k' \leq 2$.*

Put $\gamma_0 = \gamma + k'\alpha$. Then it is known (see Weyl [7]) that $s_\alpha \gamma_0 = \gamma - k\alpha$ and $\gamma_0 - \nu\alpha$ is a root for an integer $\nu$ if and only if $0 \leq \nu \leq k + k' = \gamma_0(H_\alpha)$. So we have to prove that $\gamma_0(H_\alpha) \leq 2$. Hence let us suppose that $\gamma_0(H_\alpha) \geq 3$. Then $\gamma_0 - \alpha$ is a root and therefore also $\alpha - \gamma_0$. Hence $\alpha$ and $\alpha - \gamma_0$ are roots while $\alpha + \gamma_0$ is not a root. This implies (see Weyl [7]) that $s_{\gamma_0}\alpha \leq \alpha - \gamma_0$ or $\alpha(H_{\gamma_0}) \geq 1$ and therefore $\gamma_0(H_\alpha)\alpha(H_{\gamma_0}) \geq 3$. Now consider the root $\gamma' = s_{\gamma_0}s_\alpha\gamma_0$. It is easy to verify that $\gamma' = m\gamma_0 - \gamma_0(H_\alpha)\alpha$ where $m = \gamma_0(H_\alpha)\alpha(H_{\gamma_0}) - 1$. Since $\gamma$ is totally positive the same holds for $\gamma_0$ and $s_\alpha\gamma_0$ (corollary to Lemma 12) and therefore $\gamma' = s_{\gamma_0}s_\alpha\gamma_0$ is a root of $\mathfrak{g}_+$.

However $m \geq 2$ and so we get a contradiction with Lemma 13. This proves that $\gamma_0(H_\alpha) \leq 2$.

### 6. The converse of Theorem 1.

Assuming that we are given an order in $\mathfrak{F}_R$ we shall now pove the following converse of Theorem 1.

THEOREM 2. *Let* $\Lambda_0$ *be a linear function on* $\mathfrak{h}$ *such that* $\Lambda_0(H_\beta)$ *is a nonnegative integer for every positive root* $\beta$ *which is not totally positive. Then there exists a representation* $\pi$ *of* $\mathfrak{B}$ *on a vector space* $V$ *and a vector* $\psi \neq 0$ *in* $V$ *such that the following condition holds:*

(1)  $\pi(X_\beta)\psi = 0$ *for every positive root* $\beta$.

(2)  $\pi(H)\psi = \Lambda_0(H)\psi$ *for all* $H$ *in* $\mathfrak{h}$.

(3)  $\pi$ *is irreducible and* $\dim \pi(\mathfrak{X})\psi$ *is finite.*

*Moreover such a representation is unique up to equivalence.*

Choose a fundamental system $(\beta_1, \cdots, \beta_l)$ of positive roots and let $(\beta_1, \cdots, \beta_q)$ be all the roots in this system which are not totally positive. Then in the notation of Lemma 13, $(\gamma_1, \cdots, \gamma_t)$ are exactly the remaining roots of our fundamental system. Consider the left ideal

$$\mathfrak{M}_0 = \sum_{\beta \varepsilon P} \mathfrak{B}X_\beta + \sum_{\beta \varepsilon P} \mathfrak{B}(H_\beta - \Lambda_0(H_\beta)) + \sum_{1 \leq i \leq q} \mathfrak{B} Y_i^{\lambda_i+1}$$

in $\mathfrak{B}$. Here $P$ is the set of all positive roots, $\lambda_i = \Lambda_0(H_{\beta_i})$ and $Y_i = X_{-\beta_i}$, $1 \leq i \leq q$. First we claim that $\mathfrak{M}_0 \neq \mathfrak{B}$ (cf. Lemma 8 of [5(b)]). For otherwise if $\mathfrak{M}_0 = \mathfrak{B}$ we can choose $z_i \varepsilon \mathfrak{B}$ ($1 \leq i \leq q$) such that

$$1 \equiv \sum_{1 \leq i \leq q} z_i Y_i^{\lambda_i+1} \bmod \mathfrak{P}_{\Lambda_0}$$

where $\mathfrak{P}_{\Lambda_0} = \sum_{\beta \varepsilon P} \mathfrak{B}X_\beta + \sum_{\beta \varepsilon P} \mathfrak{B}(H_\beta - \Lambda(H_\beta))$. Since 1 is of rank zero (see Section 2 for the definition of rank) while $Y_i^{\lambda_i+1}$ is of rank $-(\lambda_i + 1)\beta_i$, we may assume (see the proof of Lemma 8 of [5(b)]) that $z_i$ is of rank $(\lambda_i + 1)\beta_i$. Put $X_i = X_{\beta_i}$, $Y_i = X_{-\beta_i}$ and $H_i = H_{\beta_i}$ ($1 \leq i \leq l$). Then $(1, X_i, H_i, Y_i)_{1 \leq i \leq l}$ generate $\mathfrak{B}$ (Lemma 18 of [5(b)]). Therefore if $\mathfrak{U}, \mathfrak{H}, \mathfrak{B}$ are the subalgebras of $\mathfrak{B}$ generated by $(1, X_1, \cdots, X_l)$, $(1, H_1, \cdots, H_l)$ and $(1, Y_1, \cdots, Y_l)$ respectively, it is easy to see that $\mathfrak{B} = \mathfrak{B}\mathfrak{H}\mathfrak{U}$ (see the proof of Lemma 2 of [5(b)]). However in view of the argument given in the proof of Lemma 8 of [5(b), p. 37] we may assume that $z_i \varepsilon \mathfrak{U}$. Since the product $X_{i_1}X_{i_2} \cdots X_{i_r}$ is of rank $\beta_{i_1} + \beta_{i_2} + \cdots \beta_{i_r}$ and since $\beta_1, \cdots, \beta_l$ are linearly independent, it follows that the only elements in $\mathfrak{U}$ of rank $(\lambda_i + 1)\beta_i$ are the numerical multiples of $X_i^{\lambda_i+1}$. Therefore $z_i = c_i X_i^{\lambda_i+1}$ ($c_i \varepsilon C$). But from Lemma 6 of [5(b)],

$$X_i^{\lambda_i+1}Y_i^{\lambda_i+1} = X_i^{\lambda_i}[X_i, Y_i^{\lambda_i+1}] + X_i^{\lambda_i}Y_i^{\lambda_i+1}X_i$$
$$= \lambda_i X_i^{\lambda_i}Y_i^{\lambda_i}(H_i - \lambda_i) + X_i^{\lambda_i}Y_i^{\lambda_i+1}X_i \equiv 0 \mod \mathfrak{P}_{\Lambda_0} \quad (1 \leq i \leq q).$$

Therefore $1 \equiv 0 \mod \mathfrak{P}_{\Lambda_0}$ and so $\mathfrak{B} = \mathfrak{P}_{\Lambda_0}$. But in view of the following lemma this is impossible and therefore $\mathfrak{M}_0 \neq \mathfrak{B}$.

LEMMA 16.   *Let $\Lambda$ be a linear function on $\mathfrak{h}$ and let $\mathfrak{P}_\Lambda$ be the left ideal in $\mathfrak{B}$ defined by $\mathfrak{P}_\Lambda = \sum_{\beta \epsilon P} \mathfrak{B}X_\beta + \sum_{\beta \epsilon P} \mathfrak{B}(H_\beta - \Lambda(H_\beta))$. Then $\mathfrak{P}_\Lambda \neq \mathfrak{B}$.*

The proof is similar to that of Lemma 5 of [5(b)]. Let us suppose $\mathfrak{P}_\Lambda = \mathfrak{B}$. Then if $\mathfrak{P} = \sum_{\beta \epsilon P} \mathfrak{B}X_\beta$, we can choose elements $u_\beta \epsilon \mathfrak{B}$ such that

$$1 \equiv \sum_{\beta \epsilon P} u_\beta(H_\beta - \Lambda(H_\beta)) \mod \mathfrak{P}.$$

Again since both 1 and $H_\beta - \Lambda(H_\beta)$ are of rank zero we may assume that $u_\beta$ is of rank zero. Then $u_\beta(H_\beta - \Lambda(H_\beta)) = (H_\beta - \Lambda(H_\beta))u_\beta$ and $u_\beta \epsilon \mathfrak{B} = \mathfrak{B}\mathfrak{H}\mathfrak{U}$. Since $u_\beta$ is of rank zero and we are interested only in the congruence mod $\mathfrak{P}$, we may assume that $u_\beta \epsilon \mathfrak{H}$. Hence

$$1 - \sum_{\beta \epsilon P} u_\beta(H_\beta - \Lambda(H_\beta)) \epsilon \mathfrak{P} \cap \mathfrak{H} = \{0\}$$

from Lemma 33 of [5(b)]. Now $\mathfrak{H}$ is the universal enveloping algebra of the abelian Lie algebra $\mathfrak{h}$ (Lemma 21 of [5(b)]). Therefore if $\rho$ is the representation of $\mathfrak{h}$ of degree 1 given by $\rho(H) = \Lambda(H)$ $(H \epsilon \mathfrak{h})$, $\rho$ can be extended uniquely to the representation of $\mathfrak{H}$ such that $\rho(1) = 1$. But since $1 = \sum_{\beta \epsilon P} u_\beta(H_\beta - \Lambda(H_\beta))$ it follows that

$$1 = \rho(1) = \sum_{\beta \epsilon P} \rho(u_\beta)\{\rho(H_\beta) - \Lambda(H_\beta)\} = 0$$

and so we get a contradiction. This proves the lemma.

Since $\mathfrak{M}_0 \neq \mathfrak{B}$, by Zorn's lemma[7] there exists a maximal left ideal $\mathfrak{M}$ in $\mathfrak{B}$ containing $\mathfrak{M}_0$. Let $b \to b^*$ denote the natural mapping and $\pi$ the natural representation of $\mathfrak{B}$ on $\mathfrak{B}^* = \mathfrak{B}/\mathfrak{M}$. Since $\mathfrak{M}$ is maximal, $\pi$ is irreducible. Moreover since $\mathfrak{M} \neq \mathfrak{B}$, $1^* \neq 0$ and $\pi(\mathfrak{M})1^* = \{0\}$. Hence if we put $V = \mathfrak{B}^*$ and $\psi = 1^*$ the first two conditions of Theorem 2 are fulfilled. So we have only to prove that $\dim \pi(\mathfrak{X})1^*$ is finite.

For any root $\beta$ let $s_\beta$ denote the Weyl reflexion corresponding to $\beta$ and let $W'$ be the subgroup of the Weyl group $W$ generated by $s_i = s_{\beta_i}$, $1 \leq i \leq q$. Since $X_i$ and $Y_i^{\lambda_i+1}$ $(1 \leq i \leq q)$ are in $\mathfrak{M}$ it follows easily (by the argument used in the proof of Lemma 6) that for any $b^* \epsilon \mathfrak{B}^*$ we can find an integer

---

[7] Since $\mathfrak{B}$ satisfies the ascending chain conditions for left ideal (see [2]) it is not necessary to invoke Zorn's lemma.

$r \geq 0$ such that $\pi(X_i^{r+1})b^* = \pi(Y_i^{r+1})b^* = 0$, $i = 1, \cdots, q$. Hence we can conclude from Lemma 1 that if $\Lambda$ is a weight of $\pi$ the same holds for $s\Lambda$ $(s \varepsilon W')$. Now let $\alpha$ be a positive compact root. Then $\alpha$ is not totally positive and therefore from Lemma 14, $\alpha = s\beta_i$ for some $s \varepsilon W'$ and some $i$ $(1 \leq i \leq q)$. Therefore $s_\alpha = ss_is^{-1} \varepsilon W'$. Now suppose $d$ is a positive integer such that $\pi(X_{-\alpha}^d)1^* \neq 0$. Then $\Lambda_0 - d\alpha$ is a weight of $\pi$ and hence $s_\alpha(\Lambda_0 - d\alpha)$ $= s_\alpha\Lambda_0 + d\alpha$ is also a weight of $\pi$. But then it follows from Lemma 2 that $\Lambda_0 - (s_\alpha\Lambda_0 + d\alpha) = (\Lambda_0(H_\alpha) - d)\alpha \geq 0$ in $\mathfrak{F}_R$ and therefore $d \leq \Lambda_0(H_\alpha)$. This shows that if $r > \Lambda_0(H_\alpha)$, $\pi(X_{-\alpha}^r)1^* = 0$. Hence if $\alpha_1, \cdots, \alpha_p$ are all the distinct positive compact roots of $\mathfrak{g}$, we can choose a positive integer $m$ such that $\pi(X_{-\alpha_i}^m)1^* = 0$ $(1 \leq i \leq p)$. From this we can conclude (see the proof of Lemma 2) that $\pi(\mathfrak{X})1^*$ is spanned by $\pi(X_{-\alpha_1}^{m_1} \cdots X_{-\alpha_p}^{m_p})1^*$, $0 \leq m_1, \cdots, m_p \leq m$. This proves that the dimension of $\pi(\mathfrak{X})1^*$ is finite. Since the last statement of Theorem 2 is an immediate consequence of the corollary to Lemma 5, the proof of this theorem is now complete.

We denote the representation $\pi$ of Theorem 2 by $\pi_{\Lambda_0}$. Then $\pi_{\Lambda_0}$ is uniquely determined by $\Lambda_0$ up to equivalence.

**7. Infinitesimally unitary representations.** For any given linear function $\Lambda$ on $\mathfrak{h}$, define the left ideal $\mathfrak{P}_\Lambda$ of $\mathfrak{B}$ as in Lemma 16. Then $\mathfrak{P}_\Lambda \neq \mathfrak{B}$ and therefore by Zorn's lemma[7] there exists a maximal left ideal in $\mathfrak{B}$ containing $\mathfrak{P}_\Lambda$. Moreover we know from Lemma 5 that this maximal left ideal is unique. We shall denote it by $\mathfrak{M}_\Lambda$ and the natural representation of $\mathfrak{B}$ on $\mathfrak{B}/\mathfrak{M}_\Lambda$ by $\pi_\Lambda$. In case $\Lambda$ satisfies the condition of Theorem 2, this definition of $\pi_\Lambda$ agrees with the one given in Section 6. We shall say that $\pi_\Lambda$ is *infinitesimally unitary* (see [5(c), p. 233]) if it is possible to define a scalar product $(a^*, b^*)$ in $\mathfrak{B}^* = \mathfrak{B}/\mathfrak{M}_\Lambda$ $(a^*, b^* \varepsilon \mathfrak{B}^*)$ such that under this product $\mathfrak{B}^*$ fulfills all the requirements of a Hilbert space except completeness and $(a^*, \pi_\Lambda(X)b^*) = -(\pi_\Lambda(X)a^*, b^*)$ for all $X \varepsilon \mathfrak{g}_0$ and $a^*, b^* \varepsilon \mathfrak{B}^*$. We shall now propose to investigate the conditions on $\Lambda$ under which $\pi_\Lambda$ is infinitesimally unitary.

Let $H, \cdots, H_l$ be a base for $\mathfrak{h}_0$ over $R$ and let $x_1, \cdots, x_l$ be $l$ independent (commutative) indeterminates. Let $\mathfrak{H}$ be the subalgebra of $\mathfrak{B}$ generated by $(1, \mathfrak{h})$. Since $\mathfrak{H}$ is the universal enveloping algebra of $\mathfrak{h}$, there exists a uniquely determined homomorphism $\xi_x$ of $\mathfrak{H}$ into $C[x] = C[x_1, \cdots, x_l]$ such that $\xi_x(H_i) = x_i$, $i = 1, \cdots, l$ and $\xi_x(1) = 1$. We shall now extend $\xi_x$ to a linear mapping of $\mathfrak{B}$ into $C[x]$ as follows. Put $\mathfrak{P} = \sum_{\beta \varepsilon P} \mathfrak{B}X_\beta$ as before and let $\mathfrak{U}$ and $\mathfrak{B}$ respectively be the subalgebras of $\mathfrak{B}$ generated by $(1, X_\beta)_{\beta \varepsilon P}$ and $(1, X_{-\beta})_{\beta \varepsilon P}$. Then as we have seen above $\mathfrak{B} = \mathfrak{B}\mathfrak{H}\mathfrak{U}$ and therefore it is

10

obvious that $\mathfrak{B} = \mathfrak{P} + \mathfrak{B}\mathfrak{H}$. Now every element of $\mathfrak{B}\mathfrak{H}$, which is of rank zero, must lie in $\mathfrak{H}$ and so for every $b \, \varepsilon \, \mathfrak{B}$ of rank zero we can choose an element $h \, \varepsilon \, \mathfrak{H}$ such that $b \equiv h \bmod \mathfrak{P}$. Moreover $h$ is unique since $\mathfrak{P} \cap \mathfrak{H} = \{0\}$ from Lemma 33 of [5(b)]. We define $\xi_x(b) = \xi_x(h)$. Now let $z$ be any element in $\mathfrak{B}$ and $z_0$ its component of rank zero (see Section 2). We extend $\xi_x$ on $\mathfrak{B}$ by setting $\xi_x(z) = \xi_x(z_0)$. If $\mathfrak{Z}$ is the center of $\mathfrak{B}$ and $\chi_x$ is the mapping of $\mathfrak{Z}$ into $C[x]$ as defined in [5(b), Lemma 36], it is obvious that $\xi_x(z) = \chi_x(z)$ $(z \, \varepsilon \, \mathfrak{Z})$.

LEMMA 17. *Let $z_1$, $z_2$ be two elements of $\mathfrak{B}$ and suppose at least one of them has rank zero. Then $\xi_x(z_1 z_2) = \xi_x(z_2 z_1) = \xi_x(z_1)\xi_x(z_2)$. Moreover $\xi_x(z) = \chi_x(z)$ if $z \, \varepsilon \, \mathfrak{Z}$.*

Suppose $z_1$ is of rank zero and $(z_2)_0$ is the component of $z_2$ of rank zero. Then $z_1(z_2)_0$ and $(z_2)_0 z_1$ are the components of rank zero of $z_1 z_2$ and $z_2 z_1$ respectively. Let $z_1 = h_1 + p_1$, $(z_2)_0 = h_2 + p_2$ where $h_1, h_2 \, \varepsilon \, \mathfrak{H}$ and $p_1, p_2 \, \varepsilon \, \mathfrak{P}$. Then $(z_2)_0 z_1 \equiv z_1(z_2)_0 \equiv h_1 h_2 \bmod \mathfrak{P}$ and this proves our assertion.

$\Lambda$ being any linear function on $\mathfrak{h}$, we denote by $\xi_\Lambda(b)$ $(b \, \varepsilon \, \mathfrak{B})$ the value of the polynomial $\xi_x(b)$ under the substitution $x_i = \Lambda(H_i)$, $1 \leq i \leq l$. Let $\eta$ be the conjugation of $\mathfrak{g}$ with respect to $\mathfrak{g}_0$. We can extend $-\eta$ uniquely to an anti-automorphism $\phi$ of $\mathfrak{B}$ over $R$. Then $\phi(ab) = \phi(ba)$ $(a, b \, \varepsilon \, \mathfrak{B})$ and $\phi(X) = -\eta(X)$ $(X \, \varepsilon \, \mathfrak{g})$.

LEMMA 18. *Let $\Lambda$ be a linear function on $\mathfrak{h}$. Then $\pi_\Lambda$ is infinitesimally unitary if and only if $\xi_\Lambda(\phi(z)z)$ is real and nonnegative for every $z \, \varepsilon \, \mathfrak{B}$.*

Suppose $\pi = \pi_\Lambda$ is infinitesimally unitary. Let $a \to a^*$ denote the natural mapping of $\mathfrak{B}$ on $\mathfrak{B}^* = \mathfrak{B}/\mathfrak{M}_\Lambda$ and $(a^*, b^*)$ the corresponding scalar product on $\mathfrak{B}^*$. We assume, as we may, that $(1^*, 1^*) = 1$. Then if $H \, \varepsilon \, \mathfrak{h}_0$, $(1^*, \pi(H)1^*) = -(\pi(H)1^*, 1^*)$ and therefore $\Lambda(H)$ is purely imaginary. Now suppose $z$ is an element of $\mathfrak{B}$ of rank $\lambda$ so that $[H, z] = \lambda(H)z$ $(H \, \varepsilon \, \mathfrak{h})$. Then if $H \, \varepsilon \, \mathfrak{h}_0$,

$$\lambda(H)(1^*, \pi(z)1^*) = (1^*, \pi([H, z])1^*)$$
$$= -(\pi(H)1^*, \pi(z)1^*) - (1^*, \pi(zH)1^*)$$
$$= \Lambda(H)(1^*, \pi(z)1^*) - \Lambda(H)(1^*, \pi(z)1^*) = 0$$

and so if $\lambda \neq 0$, $(1^*, \pi(z)1^*) = 0$. Therefore if $b$ is any element in $\mathfrak{B}$ and $b_0$ is its component of rank zero, $(1^*, \pi(b)1^*) = (1^*, \pi(b_0)1^*)$. Moreover if $b_0 \equiv h \bmod \mathfrak{P}$ $(h \, \varepsilon \, \mathfrak{H})$, it is obvious that

$$(1^*, \pi(b)1^*) = (1^*, \pi(h)1^*) = \xi_\Lambda(h) = \xi_\Lambda(b)$$

since $\pi(\mathfrak{P})1^* = \{0\}$. This proves that $(1^*, \pi(b)1^*) = \xi_\Lambda(b)$ for all $b \, \varepsilon \, \mathfrak{B}$. On the other hand $(\pi(X)a^*, b^*) = (a^*, \pi(\phi(X))b^*)$ $(a^*, b^* \, \varepsilon \, \mathfrak{B}, X \, \varepsilon \, \mathfrak{g})$ and so it is clear that $(\pi(z)a^*, b^*) = (a^*, \pi(\phi(z)b^*)$ $(z \, \varepsilon \, \mathfrak{B})$. Hence

$$\xi_\Lambda(\phi(z)z) = (1^*, \pi(\phi(z)z)1^*) = (\pi(z)1^*, \pi(z)1^*) \geqq 0.$$

This shows that our condition is necessary.

Conversely suppose $\xi_\Lambda(\phi(z)z)$ is real and nonnegative for all $z \, \varepsilon \, \mathfrak{B}$. Put $(w, z) = \xi_\Lambda(\phi(w)z)$ $(w, z \, \varepsilon \, \mathfrak{B})$ and $|z| = (z, z)^{\frac{1}{2}} \geqq 0$. Then a well-known elementary argument shows that $|(w, z)| \leqq |w| \, |z|$ and $|z + w| \leqq |z| + |w|$. Let $\mathfrak{M}$ be the set of all elements $z \, \varepsilon \, \mathfrak{B}$ such that $|z| = 0$. Since $|z_1 + z_2| \leqq |z_1| + |z_2|$, $\mathfrak{M}$ is a vector space. Moreover if $b, z, w \, \varepsilon \, \mathfrak{B}$ and $b' = \phi(b)$,

$$(w, bz) = \xi_\Lambda(\phi(w)bz) = \xi_\Lambda(\phi(b'w)z) = (b'w, z).$$

Hence if $|z| = 0$, $|(w, bz)| \leqq |b'w| \, |z| = 0$ and therefore in particular $|bz|^2 = 0$. This proves that $\mathfrak{M}$ is a left ideal. Let $b \to b^*$ denote the natural mapping and $\pi$ the natural representation of $\mathfrak{B}$ on $\mathfrak{B}^* = \mathfrak{B}/\mathfrak{M}$. It is clear that $(w, z)$ depends only on $w^*$ and $z^*$. Put $(w^*, z^*) = (w, z)$ and $|z^*| = |z|$. Then $|z^*| \neq 0$ unless $z^* = 0$ and if $X \, \varepsilon \, \mathfrak{g}_0$,

$$(w^*, \pi(X)z^*) = (w, Xz) = \xi_\Lambda(\phi(w)Xz) = -\xi_\Lambda(\phi(Xw)z)$$
$$= -(\pi(X)w^*, z^*) \qquad\qquad (w, z \, \varepsilon \, \mathfrak{B}).$$

Hence in order to prove the lemma it only remains to prove that $\mathfrak{M} = \mathfrak{M}_\Lambda$. It is clear that $\xi_\Lambda(z) = 0$ if $z \, \varepsilon \, \mathfrak{P}$ and therefore it follows from Lemma 17 that $\xi_\Lambda(\mathfrak{P}_\Lambda) = \{0\}$. But since $\mathfrak{P}_\Lambda$ is a left ideal this implies that $\mathfrak{M} \supset \mathfrak{P}_\Lambda$. Hence it would be enough to prove that $\mathfrak{M}$ is maximal. Since $\xi_\Lambda(1) = 1 \neq 0$, $\mathfrak{M} \neq \mathfrak{B}$. Now suppose $\mathfrak{M}'$ is a left ideal in $\mathfrak{B}$ such that $\mathfrak{B} \neq \mathfrak{M}'$ and $\mathfrak{M}' \supset \mathfrak{M}$. Then if $b' \, \varepsilon \, \mathfrak{M}'$ and $H \, \varepsilon \, \mathfrak{h}$

$$[H, b'] = Hb' - b'(H - \Lambda(H)) - \Lambda(H)b' \, \varepsilon \, \mathfrak{M}'$$

and from this we can conclude that all the components of $b'$ of various ranks also lie in $\mathfrak{M}'$. Let $(b')_0$ be its component of rank zero and choose $h \, \varepsilon \, \mathfrak{H}$ such that $(b')_0 \equiv h \bmod \mathfrak{P}$. Then it is obvious that $(b')_0 \equiv h \equiv \xi_\Lambda(h) \bmod \mathfrak{P}_\Lambda$. Since $(b')_0$ and $\mathfrak{P}_\Lambda$ are both contained in $\mathfrak{M}'$, $\xi_\Lambda(b') = \xi_\Lambda(h) \, \varepsilon \, \mathfrak{M}'$. However $\mathfrak{M}' \neq \mathfrak{B}$ and so this implies that $\xi_\Lambda(b') = 0$. Now put $b' = \phi(z')z'$ where $z'$ is any element in $\mathfrak{M}'$. Then $b' \, \varepsilon \, \mathfrak{M}'$ and therefore $(z', z') = \xi_\Lambda(\phi(z')z') = 0$. This shows that $z' \, \varepsilon \, \mathfrak{M}$ and hence $\mathfrak{M}' = \mathfrak{M}$. Our proof is now complete.

Notice that it follows automatically from the above condition on $\xi_\Lambda$ that $\dim \pi(\mathfrak{X})1^*$ is finite. For this it would be enough to verify (see Lemma 2) that if $\alpha$ is a positive compact root of $\mathfrak{g}$, $\pi(X_{-\alpha}{}^r)1^* = 0$ for some integer $r > 0$. But $\eta(X_\alpha) = -X_{-\alpha}$ (see Section 4) and therefore

$$\xi_\Lambda(\phi(X_{-\alpha}{}^r)X_{-\alpha}{}^r) = \xi_\Lambda(X_\alpha{}^r X_{-\alpha}{}^r) \geqq 0 \qquad (r \geqq 1).$$

Moreover we know from Lemma 6 of [5(b)] that

$$X_\alpha{}^r X_\alpha{}^{-r} = X_\alpha{}^{r-1}[X_\alpha, X_{-\alpha}{}^r] + X_\alpha{}^{r-1} X_{-\alpha}{}^r X_\alpha$$

$$\equiv r X_\alpha{}^{r-1} X_{-\alpha}{}^{r-1}(\Lambda(H_\alpha) - r + 1) \bmod \mathfrak{P}_\Lambda \qquad (r \geqq 1).$$

Therefore if $\lambda_\alpha = \Lambda(H_\alpha)$,

$$0 \leqq \xi_\Lambda(X_\alpha{}^r X_{-\alpha}{}^r) = r! \prod_{1 \leqq j \leqq r} (\lambda_\alpha - j + 1)$$

by induction on $r$. Putting $r = 1$ we conclude that $\lambda_\alpha \geqq 0$. Now let $r$ be the least integer such that $\lambda_\alpha - r + 1 \leqq 0$. Then the above product is negative unless $\lambda_\alpha = r - 1$. Hence we conclude that $\lambda_\alpha$ is a nonnegative integer and therefore $\xi_\Lambda(X_\alpha{}^r X_{-\alpha}{}^r) = 0$ for $r \geqq \lambda_\alpha + 1$. But since

$$\xi_\Lambda(X_\alpha{}^r X_{-\alpha}{}^r) = |\pi(X_{-\alpha}{}^r)1^*|^2,$$

$\pi(X_{-\alpha}{}^r)1^* = 0$ for $r \geqq \lambda_\alpha + 1$. This shows that $\pi$ then satisfies all the requirements of Theorem 1 and therefore $\Lambda(H_\beta)$ must be a nonnegative integer for all positive roots $\beta$ which are not totally positive.

Now suppose $\beta$ is a noncompact root. Then $\eta(X_{-\beta}) = X_\beta$ (see Section 4) and therefore

$$\xi_\Lambda(\phi(X_{-\beta}{}^r)X_{-\beta}{}^r) = (-1)^r \xi_\Lambda(X_\beta{}^r X_{-\beta}{}^r) = r! \prod_{1 \leqq j \leqq r} (j - 1 - \lambda_\beta) \quad (r \geqq 1)$$

where $\lambda_\beta = \Lambda(H_\beta)$. Since again this product must always be real and nonnegative, we conclude that $\lambda_\beta$ is real and $\leqq 0$. In particular if $\beta$ is not totally positive it follows, in view of our earlier remarks, that $\lambda_\beta = 0$. Thus we have obtained the following corollary.

COROLLARY 1. *In order that $\pi_\Lambda$ be infinitesimally unitary it is necessary that $\Lambda$ should satisfy the following conditions:*

(1)   $\Lambda(H_\alpha)$ *is a nonnegative integer for every positive compact root $\alpha$.*

(2)   $\Lambda(H_\beta) = 0$ *for every noncompact positive root $\beta$ which is not totally positive.*

(3)   $\Lambda(H_\gamma)$ *is real and $\leqq 0$ for every totally positive root $\gamma$.*

The space $\mathfrak{F}$ of linear functions on $\mathfrak{h}$, being a finite-dimensional vector space over $C$, has a natural topology.

COROLLARY 2. *The set of all linear functions $\Lambda$ on $\mathfrak{h}$ for which $\pi_\Lambda$ is infinitesimally unitary, is closed in $\mathfrak{F}$.*

For a fixed element $b \, \varepsilon \, \mathfrak{B}$, $\Lambda \to \xi_\Lambda(b)$ $(\Lambda \, \varepsilon \, \mathfrak{F})$ is obviously a polynomial function on $\mathfrak{F}$. Hence it is continuous and therefore the corollary follows from Lemma 18.

Define $\mathfrak{g}'$ as in the proof of Lemma 13.

LEMMA 19. *Suppose $\Lambda$ satisfies the conditions of Theorem 2 and $\Lambda(H_\beta) = 0$ for every positive root $\beta$ which is not totally positive. Then $\pi_\Lambda(X) = 0$ for all $X \, \varepsilon \, \mathfrak{g}'$.*

We use the notation of the proof of Lemma 13. Let $\alpha$ be a positive compact root of $\mathfrak{g}'$. Then we can choose a noncompact root $\beta$ of $\mathfrak{g}'$ such that $\beta + \alpha$ is a root. For otherwise $[X_\alpha, X_\beta] = 0$ for every noncompact root $\beta$ of $\mathfrak{g}'$. Since it is obvious that $\mathfrak{q} = \mathfrak{g}' \cap \mathfrak{p}$ is spanned by the elements $X_\beta$ corresponding to all noncompact roots $\beta$ of $\mathfrak{g}'$, $X_\alpha$ lies in the centralizer $\mathfrak{l}$ of $\mathfrak{q}$ in $\mathfrak{k}$. But $\mathfrak{l} \subset \mathfrak{g}_+$ and therefore $X_\alpha \, \varepsilon \, \mathfrak{g}_+ \cap \mathfrak{g}' = \{0\}$, which of course is false. So let $\beta$ be a noncompact root of $\mathfrak{g}'$ such that $\gamma = \beta + \alpha$ is a root. Since $\mathfrak{q} \cap (\mathfrak{p}_+ + \mathfrak{p}_-) = \{0\}$, no root of $\mathfrak{g}'$ can be totally positive or totally negative. Therefore $\Lambda(H_\beta) = \Lambda(H_\gamma) = 0$. But since $\alpha = \gamma - \beta$, $H_\alpha$ is a linear combination of $H_\beta$ and $H_\gamma$ and so $\Lambda(H_\alpha) = 0$. This shows that $\Lambda(H_\beta) = 0$ for every root $\beta$ of $\mathfrak{g}'$. Moreover if $\beta$ is a positive root of $\mathfrak{g}'$, it follows from Theorem 1 that $\pi_\Lambda(X_{-\beta})1^* = 0$ since $\beta$ is not totally positive. Therefore $\pi_\Lambda(\mathfrak{g}')1^* = \{0\}$ and since $\mathfrak{g}'$ is an ideal in $\mathfrak{g}$ and $\pi_\Lambda$ is irreducible, $\pi_\Lambda(\mathfrak{g}') = \{0\}$.

COROLLARY 1. *If $\pi_\Lambda$ is infinitesimally unitary $\pi_\Lambda(\mathfrak{g}') = \{0\}$.*

This is an immediate consequence of the above lemma and Corollary 1 to Lemma 18.

COROLLARY 2. *Suppose $\mathfrak{g}_0$ is simple and $\mathfrak{g}_0 \neq \mathfrak{k}_0$. Then if $\mathfrak{c}_0 = \{0\}$, $\pi_\Lambda$ cannot be infinitesimally unitary unless $\Lambda = 0$ in which case it is the zero representation of $\mathfrak{g}$ of degree 1.*

For then we know that $\mathfrak{g} = \mathfrak{g}'$ (see the proof of Corollary 2 to Lemma 13). Now apply Corollary 1 above.

In Corollary 1 to Lemma 18 we have obtained certain necessary conditions on $\Lambda$ in order that $\pi_\Lambda$ may be infinitesimally unitary. It is natural to ask how far these conditions are also sufficient. The following result gives a partial answer to this question.

THEOREM 3. *Let $2\rho$ denote the sum of all positive roots of $\mathfrak{g}$ and let $\Lambda$ be a linear function on $\mathfrak{h}$ satisfying the following conditions:*

(1) *$\Lambda(H_\alpha)$ is a nonnegative integer for every positive compact root $\alpha$.*

(2) *$\Lambda(H_\beta) = 0$ for every noncompact positive root $\beta$ which is not totally positive.*

(3) *$\Lambda(H_\gamma) + \rho(H_\gamma)$ is real and $\leqq 0$ for every totally positive root $\gamma$. Then $\pi_\Lambda$ is infinitesimally unitary.*

We leave the proof of this theorem to a subsequent paper (see however [5(h)]).

**8. Representations of the group.** Let $G$ be the simply connected Lie group with the Lie algebra $\mathfrak{g}_0$ and let $K$ be its analytic subgroup corresponding to $\mathfrak{k}_0$. Let $\Lambda$ be a linear function on $\mathfrak{h}$ which fulfills the conditions of Theorem 2. We shall now show that $\pi_\Lambda$ can be 'extended' to a representation of $G$ on a Hilbert space. Let $\Omega$ be the set of all equivalence classes of finite-dimensional irreducible representations of $K$. Henceforth we shall make free use of the terminology of [5(c)].

THEOREM 4. *Let $\Lambda$ be a linear function on $\mathfrak{h}$ such that $\Lambda(H_\beta)$ is a nonnegative integer for every positive root $\beta$ which is not totally positive. Then there exists an irreducible quasi-simple representation $\sigma$ of $G$ on a Hilbert space $\mathfrak{H}$ with the following property. Let $\mathfrak{H}_{\mathfrak{D}}$ ($\mathfrak{D} \varepsilon \Omega$) denote the subspace of $\mathfrak{H}$ consisting of all those elements which transform under $\sigma(K)$ according to $\mathfrak{D}$. Then the corresopnding representation $\sigma_0$ of $\mathfrak{B}$ defined on $\mathfrak{H}_0 = \sum_{\mathfrak{D} \varepsilon \Omega} \mathfrak{H}_{\mathfrak{D}}$ is equivalent to $\pi_\Lambda$.*

Put $\pi = \pi_\Lambda$ and define $\mathfrak{P}_\Lambda$ and $\mathfrak{M}_\Lambda$ as in Section 7. If $z$ is any element in $\mathfrak{B}$ of rank zero, $z \equiv \xi_\Lambda(z) \bmod \mathfrak{P}_\Lambda$ and therefore in particular if $z \varepsilon \mathfrak{Z}$, $z - \xi_\Lambda(z) \varepsilon \mathfrak{M}_\Lambda$. Moreover we know from Theorem 2 that $\dim \mathfrak{X}/\mathfrak{X} \cap \mathfrak{M}_\Lambda$ is finite and so we can conclude from Lemma 9 of [5(c)] that $\pi$ is a quasi-simple[8] irreducible representation of $\mathfrak{B}$.

Let $(\beta_1, \cdots, \beta_l)$ and $(\beta_1, \cdots, \beta_q)$ be defined as in the proof of Theorem 2 and let $\Lambda'$ be the linear function on $\mathfrak{h}$ given by $\Lambda'(H_{\beta_i}) = \Lambda(H_{\beta_i})$, $1 \leqq i \leqq q$ and $\Lambda'(H_{\beta_i}) = 0$, $q < i \leqq l$. Then from Theorem 1 of [5(b)], there exists a finite-dimensional irreducible representation $\rho$ of $\mathfrak{g}$ with the highest weight $\Lambda'$. Let $U$ be the representation space of $\rho$ and $\phi$ a nonzero vector in $U$ belonging to the weight $\Lambda'$. Then if $\alpha$ is any positive compact

---

[8] See [5(d), p. 56] for the definition of quasi-simplicity in this case.

root of $\mathfrak{g}$, $\rho(X_a)\phi = 0$ and $\rho(H_a)\phi = \Lambda'(H_a)\phi = \Lambda(H_a)\phi$ since $\alpha$ is a linear combination of $\beta_1, \cdots, \beta_q$ (Lemma 13). Similarly if $\psi \neq 0$ is a vector belonging to the weight $\Lambda$ of $\pi$, $\pi(X_a)\psi = 0$ and $\pi(H_a)\psi = \Lambda(H_a)\psi$. Let $\mathfrak{X}'$ be the subalgebra of $\mathfrak{B}$ generated by $(1\,\mathfrak{k}')$. Since $\mathfrak{k}'$ is semisimple we can apply Lemma 2 to $\mathfrak{X}'$ and conclude that the representations of $\mathfrak{k}'$ defined on $\pi(\mathfrak{X}')\psi$ and $\rho(\mathfrak{X}')\phi$ are both irreducible and therefore from Lemma 5 they are equivalent. Since $\rho$ is a finite-dimensional representation of $\mathfrak{g}$, the existence of $\sigma$ now follows from Theorem 4 of [5(d)].

COROLLARY. *Let* $\lambda$, $\mu$ *be two linear functions on* $\mathfrak{h}$. *Then if* $\pi_\lambda$ *and* $\pi_\mu$ *are both infinitesimally unitary the same holds for* $\pi_{\lambda+\mu}$.

Suppose $\pi_\lambda$ and $\pi_\mu$ are infinitesimally unitary. Then $\lambda$ and $\mu$ both satisfy the conditions of Theorem 2 (see Corollary 1 to Lemma 18) and therefore from Theorem 4 above and Theorem 9 of [5(c)] we can find two irreducible unitary representations $\sigma_\lambda$, $\sigma_\mu$ of $G$ on the Hilbert spaces $\mathfrak{H}_\lambda$, $\mathfrak{H}_\mu$ respectively such that the corresponding representations of $\mathfrak{B}$ defined on $\mathfrak{H}_\lambda{}^0 = \sum_{\mathfrak{D}\varepsilon\Omega} (\mathfrak{H}_\lambda)_\mathfrak{D}$ and $\mathfrak{H}_\mu{}^0 = \sum_{\mathfrak{D}\varepsilon\Omega} (\mathfrak{H}_\mu)_\mathfrak{D}$ are equivalent to $\pi_\lambda$ and $\pi_\mu$. We denote these representations of $\mathfrak{B}$ by $\sigma_\lambda{}^0$ and $\sigma_\mu{}^0$. Then it is clear from the proofs of Lemma 18 and Theorem 9 of [5(c)] that we can choose $\psi_\lambda \varepsilon \mathfrak{H}_\lambda{}^0$ and $\psi_\mu \varepsilon \mathfrak{H}_\mu{}^0$ such that

$$(\psi_\lambda, \sigma_\lambda{}^0(z)\psi_\lambda) = \xi_\lambda(z), \qquad (\psi_\mu, \sigma_\mu{}^0(z)\psi_\mu) = \xi_\mu(z)$$

for all $z \varepsilon \mathfrak{B}$. Let $\mathfrak{H}$ be the Hilbert space which is the Kronecker product of $\mathfrak{H}_\lambda$ and $\mathfrak{H}_\mu$ (in the sense of Hilbert space theory) and let $\psi = \psi_\lambda \times \psi_\mu$. Then if $\psi'$ and $\psi''$ are well-behaved under $\sigma_\lambda$ and $\sigma_\mu$ respectively it is obvious that $\psi' \times \psi''$ is well-behaved under the representation $\sigma = \sigma_\lambda \times \sigma_\mu$ of $G$ on $\mathfrak{H}$. Let $\sigma^0$ denote the representation of $\mathfrak{B}$ defined under $\sigma$ on the subspace $\mathfrak{H}^0$ consisting of all well-behaved elements in $\mathfrak{H}$. Then it is obvious that $\sigma^0(X)(\phi_1 \times \phi_2) = (\sigma_\lambda{}^0(X)\phi_1) \times \phi_2 + \phi_1 \times (\sigma_\mu{}^0(X)\phi_2)$ $(X \varepsilon \mathfrak{g}, \phi_1 \varepsilon \mathfrak{H}_\lambda{}^0, \phi_2 \varepsilon \mathfrak{H}_\mu{}^0)$ and therefore $\sigma^0(X_\beta)\psi = 0$ for all positive roots $\beta$ and $\sigma^0(H)\psi = (\lambda(H) + \mu(H))\psi$ if $H \varepsilon \mathfrak{h}$. Hence $\sigma^0(\mathfrak{P}_{\lambda+\mu})\psi = \{0\}$ in the notation of Section 7. On the other hand since $\sigma$ is a unitary representation of $G$,

$$(\phi', \sigma^0(X)\phi) = - (\sigma^0(X)\phi', \phi) \qquad (X \varepsilon \mathfrak{g}_0; \phi, \phi' \varepsilon \mathfrak{H}^0).$$

Therefore if $z$ is any element in $\mathfrak{B}$ of rank $\nu \neq 0$ it follows (see the proof of Lemma 18) that $(\psi, \sigma^0(z)\psi) = 0$. This shows that

$$(\psi, \sigma^0(b)\psi) = \xi_{\lambda+\mu}(b)|\psi|^2 \qquad (b \varepsilon \mathfrak{B})$$

and therefore

$$| \sigma^0(z)\psi |^2 = (\sigma^0(z)\psi, \sigma^0(z)\psi) = (\psi, \sigma^0(\phi(z)z)\psi) = \xi_{\lambda+\mu}(\phi(z)z) | \psi |^2$$

for any $z \varepsilon \mathfrak{B}$. Hence $\pi_{\lambda+\mu}$ is infinitesimally unitary from Lemma 18.

**9. An auxiliary result.** In this section we shall prove a result which will be required in another paper. Let $\Lambda_0$ be a linear function on $\mathfrak{h}$ satisfying the conditions of Theorem 3. As before let $P$ be the set of all positive roots and $P_k$ and $P_+$ respectively the subsets of $P$ consisting of all compact and all totally positive roots. Let $P'$ be the complement of $P_k \cup P_+$ in $P$. We know (Lemma 5) that there exists a unique left ideal $\mathfrak{Y}_{\Lambda_0}$ in $\mathfrak{X}$ containing

$$\sum_{a \varepsilon P_k} \mathfrak{X} X_a + \sum_{\gamma \varepsilon P} \mathfrak{X}(H_\gamma - \Lambda_0(H_\gamma)).$$

Our object now is to prove the following theorem.

THEOREM 5. *Let $\Lambda_0$ be a linear function on $\mathfrak{h}$ satisfying the conditions of Theorem 3. Then*

$$\mathfrak{M} = \sum_{\gamma \varepsilon P} \mathfrak{B} X_\gamma + \sum_{\gamma \varepsilon P'} \mathfrak{B} X_{-\gamma} + \mathfrak{B}\mathfrak{Y}_{\Lambda_0}$$

*is a maximal left ideal in $\mathfrak{B}$.*

It is obvious that $\Lambda_0$ fulfills the conditions of Theorem 2 and therefore we can consider the representation $\pi_{\Lambda_0}$ and choose a vector $\psi \neq 0$ in the representation space corresponding to Theorem 2. Define $\mathfrak{g}_+$ and $\mathfrak{g}'$ as in Section 5. Then we know from Lemma 19 that $\pi_{\Lambda_0}(\mathfrak{g}') = \{0\}$. Moreover it follows from Lemmas 2 and 5 (applied to $\mathfrak{k}$) that $\pi_{\Lambda_0}(\mathfrak{Y}_{\Lambda_0})\psi = \{0\}$. This shows that $\pi_{\Lambda_0}(\mathfrak{M})\psi = \{0\}$ and therefore $\mathfrak{M} \neq \mathfrak{B}$ since $\psi \neq 0$.

Let $\mathfrak{M}_1$ be the set of all elements $b \varepsilon \mathfrak{B}$ such that $\pi_{\Lambda_0}(b)\psi = 0$. Since $\mathfrak{Y}_{\Lambda_0}$ is a maximal left ideal in $\mathfrak{X}$, it is clear that $\mathfrak{M}_1 \cap \mathfrak{X} = \mathfrak{Y}_{\Lambda_0}$. Therefore in view of the fact that $\mathfrak{g}' \subset \mathfrak{M}_1$, we can conclude that $\mathfrak{g}' \cap \mathfrak{k} \subset \mathfrak{Y}_{\Lambda_0}$. On the other hand $\mathfrak{g}' = \sum_{\gamma \varepsilon P'} (C X_\gamma + C X_{-\gamma}) + \mathfrak{g}' \cap \mathfrak{k}$ and so it follows that $\mathfrak{g}' \subset \mathfrak{M}$.

Let $b \rightarrow b^*$ ($b \varepsilon \mathfrak{B}$) denote the natural mapping and $\pi$ the natural representation of $\mathfrak{B}$ on $\mathfrak{B}^* = \mathfrak{B}/\mathfrak{M}$. In order to prove that $\mathfrak{M}$ is a maximal left ideal in $\mathfrak{B}$ it is sufficient to show that $\pi$ is irreducible. Let $\mathfrak{B}_+$ be the subalgebra of $\mathfrak{B}$ generated by $(1, \mathfrak{g}_+)$. Since $\mathfrak{g}'$ is an ideal in $\mathfrak{g}$ and $\mathfrak{g}' \subset \mathfrak{M}$, it follows that $\pi(\mathfrak{g}') = \{0\}$ and therefore $\mathfrak{B}^* = \pi(\mathfrak{B}_+)1^*$. Hence it would be enough to prove that $\mathfrak{B}_+ \cap \mathfrak{M}$ is a maximal left ideal in $\mathfrak{B}_+$. Let $\mathfrak{X}_+$ be the subalgebra of $\mathfrak{X}$ generated by $(1, \mathfrak{k} \cap \mathfrak{g}_+)$. Since $\pi(\mathfrak{Y}_{\Lambda_0})1^* = \{0\}$ and since $\mathfrak{Y}_{\Lambda_0}$ is a maximal left ideal in $\mathfrak{X}$, it is clear that $\mathfrak{X}^* = \pi(\mathfrak{X})1^*$ is irreducible under $\pi(\mathfrak{X})$ and therefore also under $\pi(\mathfrak{X}_+)$. Hence $\mathfrak{Y}_+ = \mathfrak{X}_+ \cap \mathfrak{Y}_{\Lambda_0}$ is a maximal left ideal in $\mathfrak{X}_+$. Let $Q$ denote the set of all positive roots of $\mathfrak{g}_+$ (see Section 5). Then from Lemma 5, $\mathfrak{Y}_+$ must be the unique maximal left ideal in $\mathfrak{X}_+$ containing

$$\sum_{\gamma \varepsilon Q_k} \mathfrak{X}_+ X_\alpha + \sum_{\gamma \varepsilon Q} \mathfrak{X}_+ (H_\gamma - \Lambda_0(H_\gamma))$$

where $Q_k = Q \cap P_k$. Now if $\mathfrak{M}_+ = \sum_{\gamma \varepsilon Q} \mathfrak{B}_+ X_\gamma + \mathfrak{B}_+ \mathfrak{Y}_+$ it is obvious that $\mathfrak{B}_+ \cap \mathfrak{M} \supset \mathfrak{M}_+$ and therefore it is enough to prove that $\mathfrak{M}_+$ is a maximal left ideal in $\mathfrak{B}_+$. Let $2\rho_+ = \sum_{\gamma \varepsilon Q} \gamma$ and $2\rho = \sum_{\gamma \varepsilon P} \gamma$. If $\beta$ is a root in $\mathfrak{g}'$ and $\gamma$ a root in $\mathfrak{g}_+$ then $\beta(H_\gamma) = 0$. Hence it follows that $\rho(H_\gamma) = \rho_+(H_\gamma)$ for all $\gamma \varepsilon Q$. This shows (by comparison with the conditions of Theorem 3) that the problem of proving the maximality of $\mathfrak{M}_+$ in $\mathfrak{B}_+$ is exactly the same as our original problem if we replace $\mathfrak{g}$ by $\mathfrak{g}_+$. Therefore we may assume that $\mathfrak{g}' = \{0\}$ and $P = P_k \cup P_+$.

Let $\mathfrak{M}_1 \neq \mathfrak{M}$ be a left ideal in $\mathfrak{B}$ containing $\mathfrak{M}$. We have to show that $\mathfrak{M}_1 = \mathfrak{B}$. For any linear function $\Lambda$ on $\mathfrak{h}$, let $\mathfrak{B}_\Lambda^*$ denote the set of elements $b^* \varepsilon \mathfrak{B}^*$ such that $\pi(H)b^* = \Lambda(H)b^*$ $(H \varepsilon \mathfrak{h})$. Then it is clear that $\mathfrak{B}^* = \sum_\Lambda \mathfrak{B}_\Lambda^*$ and $\mathfrak{B}_\Lambda^* = \{0\}$ unless $\Lambda_0 - \Lambda$ is a linear combination of positive roots with coefficients which are all nonnegative integers (see the proof of Lemma 2). Moreover $\mathfrak{M}_1^* = \sum_\Lambda (\mathfrak{M}_1^* \cap \mathfrak{B}_\Lambda^*) \neq \{0\}$ since $\mathfrak{M}_1 \neq \mathfrak{M}$ (Lemma 1 of [5(b)]). Hence in view of the above remarks, there exists a highest weight $\Lambda_1$ such that $\mathfrak{M}_1^* \cap \mathfrak{B}_{\Lambda_1}^* \neq \{0\}$. Choose an element $\phi \neq 0$ in $\mathfrak{M}_1^* \cap \mathfrak{B}_{\Lambda_1}^*$. Then if $\gamma$ is a positive root $\mathfrak{M}_1^* \cap \mathfrak{B}_{\Lambda_1+\gamma}^* = \{0\}$ and therefore $\pi(X_\gamma)\phi = 0$. Now define $\chi_x(z)$ $(z \varepsilon \mathfrak{Z})$ as in Lemma 17 and for any linear function $\lambda$ on $\mathfrak{h}$ let $\chi_\lambda(z)$ denote the value of this polynomial under the substitution $x_i = \lambda(H_i)$ $(1 \leq i \leq l)$. (Here we are using the notation of Section 7). Then it is clear that $\pi(z)\phi = \chi_{\Lambda_1}(z)\phi$ for all $z \varepsilon \mathfrak{Z}$. Similarly $\pi(z)1^* = \chi_{\Lambda_0}(z)1^*$ $(z \varepsilon \mathfrak{Z})$ and therefore $\pi(z)b^* = \chi_{\Lambda_0}(z)b^*$ for every $b^* \varepsilon \mathfrak{B}^*$. Since $\phi \neq 0$, we conclude that $\chi_{\Lambda_0}(z) = \chi_{\Lambda_1}(z)$ for any $z \varepsilon \mathfrak{Z}$. But then if $W$ is the Weyl group of $\mathfrak{g}$ with respect to $\mathfrak{h}$, it follows from Theorem 5 of [5(b)] that $\Lambda_1 + \rho = s_0(\Lambda_0 + \rho)$ for some $s_0 \varepsilon W$. (Here $2\rho = \sum_{\gamma \varepsilon P} \gamma$).

Let $N_-$ denote the set of all totally negative roots (Section 4) and put $\bar{P} = P_k \cup N_-$. Since $\mathfrak{g}' = \{0\}$, it follows from Lemmas 11 and 12 that $\bar{P}$ satisfies the conditions of Corollary 1 to Lemma 4. Put $\Lambda_0' = \Lambda_0 + \rho$ and let $\mathfrak{F}_R(\bar{P})$ denote the set of those linear functions on $\mathfrak{h}$ which are linear combinations of roots in $\bar{P}$ with real nonnegative coefficients. Since

$\rho(H_\beta) > 0$ for every $\beta \varepsilon P$ (see Weyl [7]) it follows from our assumptions on $\Lambda_0$ that $\Lambda_0'(H_\gamma) \geqq 0$ for all $\gamma \varepsilon P$. As we shall see presently (Lemma 20 below), this implies that $\Lambda_0' - s\Lambda_0' \varepsilon \mathfrak{F}^+(\bar{P})$ for every $s \varepsilon W$. Let $(\beta_1, \cdots, \beta_l)$ be a fundamental system of roots in $P$. Suppose $(\beta_1, \cdots, \beta_t)$ are all the totally positive roots and $(\alpha_1, \cdots, \alpha_r)$ all the (distinct) compact roots among $(\beta_1, \cdots, \beta_l)$. Then we know from Lemma 13 that every totally negative root is of the form $-(\beta_i + m_1\alpha_1 + \cdots + m_r\alpha_r)$ where $1 \leqq i \leqq t$ and $m_1, \cdots, m_r \geqq 0$. Hence

$$\Lambda_0 - \Lambda_1 = (\Lambda_0 + \rho) - s_0(\Lambda_0 + \rho)$$
$$= -(\mu_1\beta_1 + \cdots + \mu_t\beta_t) + m_1'\alpha_1 + \cdots + m_r'\alpha_r$$

where $\mu_1, \cdots, \mu_t \geqq 0$. On the other hand $\mathfrak{B}_{\Lambda_1}^* \neq \{0\}$ and therefore, as we have seen above,

$$\Lambda_0 - \Lambda_1 = \nu_1\beta_1 + \cdots + \nu_t\beta_t + n_1\alpha_1 + \cdots + n_r\alpha_r$$

where $\nu_1, \cdots, \nu_t, n_1, \cdots, n_r \geqq 0$. Since $(\beta_1, \cdots, \beta_l)$ are linearly independent we conclude that $-\mu_i = \nu_i = 0$ $(1 \leqq i \leqq t)$ and therefore $\Lambda_0 - \Lambda_1$ is a linear combination of compact roots. Let $\gamma_1, \cdots, \gamma_p$ be all the (distinct) totally positive roots and $\delta_1, \cdots, \delta_q$ all the compact roots in $P$. Then $\mathfrak{B}^*$ is spanned by elements of the form $\pi(X_{-\gamma_1}^{\mu_1} \cdots X_{-\gamma_p}^{\mu_p} X_{-\delta_1}^{\lambda_1} \cdots X_{-\delta_q}^{\lambda_q})1^*$ where $\lambda_i$ and $\mu_i$ are nonnegative integers (see the proof of Lemma 2). Since $\phi \varepsilon \mathfrak{B}_{\Lambda_1}^*$, it must be a linear combination of those elements of the above form for which

$$\mu_1\gamma_1 + \cdots + \mu_p\gamma_p + \lambda_1\delta_1 + \cdots + \lambda_q\delta_q = \Lambda_0 - \Lambda_1.$$

However since $\Lambda_0 - \Lambda_1$ is a linear combination of compact roots it follows from Lemma 13 that $\mu_1 = \cdots = \mu_p = 0$ and therefore $\phi \varepsilon \pi(\mathfrak{X})1^* = \mathfrak{X}^*$. But $\pi(X_\alpha)\phi = 0$ $(\alpha \varepsilon P_k)$ and as we have seen above $\mathfrak{X}^*$ is irreducible under $\pi(\mathfrak{X})$. Therefore it follows from Lemma 2 (applied to $\mathfrak{X}$ and $\phi$) that $\Lambda_1 - \Lambda_0 \geqq 0$. But $\Lambda_0 - \Lambda_1 = \lambda_1\delta_1 + \cdots + \lambda_q\delta_q \geqq 0$. Hence $\Lambda_0 = \Lambda_1$ and $\lambda_1 = \cdots = \lambda_q = 0$. This shows that $\phi = c1^*$ $(c \varepsilon C)$ and since $\phi \neq 0$ it follows that $\mathfrak{M}_1^* \supset \pi(\mathfrak{B})\phi = \mathfrak{B}^*$ and therefore $\mathfrak{M}_1 = \mathfrak{B}$. Hence in order to complete the proof it remains to prove the following lemma.

LEMMA 20. *Let $\bar{P}_0$ be a fundamental system of roots of $\mathfrak{g}$ and let $\mathfrak{F}^+(\bar{P}_0)$ denote the set of all linear functions on $\mathfrak{h}$ which can be written as linear combinations of roots in $\bar{P}_0$ with real nonnegative coefficients. Then if $\Lambda \varepsilon \mathfrak{F}_R$ and $\Lambda(H_\beta) \geqq 0$ for all $\beta \varepsilon \bar{P}_0$, $\Lambda - s\Lambda \varepsilon \mathfrak{F}^+(\bar{P}_0)$ for every $s \varepsilon W$.*

It is enough to prove this lemma under the additional assumption that

$\Lambda(H_\beta)$ $(\beta \, \varepsilon \, P_0)$ are all rational since the general case then follows immediately by continuity. Moreover by multiplying $\Lambda$ by a suitable positive integer, the problem may be reduced to the case when $\Lambda(H_\beta)$ $(\beta \, \varepsilon \, P_0)$ are all nonnegative integers. But in this case we know (Theorem 1 of [5(b)] that there exists a finite-dimensional irreducible representation $\pi$ of $\mathfrak{g}$ such that $\Lambda$ is a weight of $\pi$ while $\Lambda + \gamma$ is not a weight for any root $\gamma \, \varepsilon \, \mathfrak{F}^+(P_0)$. Then it follows from Lemma 2 that if $\Lambda'$ is any other weight of $\pi$, $\Lambda - \Lambda' \, \varepsilon \, \mathfrak{F}^+(\bar{P}_0)$. Therefore since $s\Lambda$ is a weight for every $s \, \varepsilon \, W$, we get the result.

Let us keep to the notation of the proof of Theorem 5 and put $\mathfrak{p}_- = \sum_{\gamma \, \varepsilon \, P_+} CX_{-\gamma}$. Since $[\mathfrak{p}_-, \mathfrak{p}_-] = \{0\}$ (Lemma 11) the subalgebra $\mathfrak{P}_-$ of $\mathfrak{B}$ generated by $(1, \mathfrak{p}_-)$ is abelian. We consider the tensor product $\mathfrak{P}_- \times \mathfrak{X}^*$.

LEMMA 21. *Let $\Gamma$ denote the linear mapping $\mathfrak{P}_- \times \mathfrak{X}^*$ into $\mathfrak{B}^*$ such that $\Gamma(p \times x^*) = \pi(p)x^*$ $(p \, \varepsilon \, \mathfrak{P}_-, x^* \, \varepsilon \, \mathfrak{X}^*)$. Then $\Gamma$ is a linear isomorphism of $\mathfrak{P}_- \times \mathfrak{X}^*$ onto $\mathfrak{B}^*$.*

We have seen during the proof of Theorem 4 that $\mathfrak{B}^* = \pi(\mathfrak{B}_+)1^*$ and $\mathfrak{X}^* = \pi(\mathfrak{X}_+)1^*$. Also we proved that $\mathfrak{M}_+ = \sum_{\gamma \, \varepsilon \, Q} \mathfrak{B}X_\gamma + \mathfrak{B}_+Y_+$ is a maximal left ideal in $\mathfrak{B}_+$. Therefore $\mathfrak{B} = \mathfrak{B}_+ + \mathfrak{M}$, $\mathfrak{X} = \mathfrak{X}_+ + \mathfrak{Y}_{\Lambda_0}$ and $\mathfrak{M} \cap \mathfrak{B}_+ = \mathfrak{M}_+$. Put $\mathfrak{p}_+ = \sum_{\gamma \, \varepsilon \, P_+} CX_\gamma$ and let $\mathfrak{P}_+$ be the subalgebra of $\mathfrak{B}$ generated by $(1, \mathfrak{p}_+)$. Then if $\mathfrak{k}_+ = \mathfrak{k} \cap \mathfrak{g}_+$, $\mathfrak{g}_+ = \mathfrak{p}_+ + \mathfrak{p}_- + \mathfrak{k}_+$ and it follows from Lemma 12 of [5(c)] that $\mathfrak{B}_+ = \mathfrak{P}_-\mathfrak{P}_+\mathfrak{X}_+$. But $[\mathfrak{p}_+, \mathfrak{k}] \subset \mathfrak{p}_+ \subset \mathfrak{M}_+$ and therefore

$$\mathfrak{P}_+\mathfrak{X}_+ = \mathfrak{X}_+ + \mathfrak{P}_+\mathfrak{p}_+\mathfrak{X}_+ \subset \mathfrak{X}_+ + \mathfrak{M}_+.$$

Hence $\mathfrak{B}_+ = \mathfrak{P}_-\mathfrak{X}_+ + \mathfrak{M}_+$ and $\mathfrak{B}^* = \pi(\mathfrak{B}_+)1^* = \pi(\mathfrak{P}_-)\mathfrak{X}^* = \Gamma(\mathfrak{P}_- \times \mathfrak{X}^*)$. Moreover the mapping $(p \times p' \times x) \to pp'x$ $(p \, \varepsilon \, \mathfrak{P}_-, p' \, \varepsilon \, \mathfrak{P}_+, x \, \varepsilon \, \mathfrak{X}_+)$ defines a linear isomorphism $\mu$ of $\mathfrak{P}_- \times \mathfrak{P}_+ \times \mathfrak{X}_+$ onto $\mathfrak{B}_+$ [5(c), Lemma 12]. Therefore

$$\mathfrak{M}_+ = \mathfrak{B}_+\mathfrak{p}_+ + \mathfrak{B}_+\mathfrak{Y}_+ = \mathfrak{P}_-(\mathfrak{P}_+\mathfrak{p}_+)\mathfrak{X}_+ + \mathfrak{P}_-\mathfrak{P}_+\mathfrak{Y}_+$$
$$= \mu((\mathfrak{P}_- \times (\mathfrak{P}_+\mathfrak{p}_+) \times \mathfrak{X}_+) + (\mathfrak{P}_- \times \mathfrak{P}_+ \times \mathfrak{Y}_+)).$$

Hence by considering $\mu^{-1}(\mathfrak{M}_+) \cap \mu^{-1}(\mathfrak{P}_-\mathfrak{X}_+)$ it is obvious that $\mathfrak{P}_-\mathfrak{X}_+ \cap \mathfrak{M}_+ = \mathfrak{P}_-\mathfrak{Y}_+$ and this shows that the kernel of $\Gamma$ is zero. Therefore $\Gamma$ is an isomorphism of $\mathfrak{P}_- \times \mathfrak{X}^*$ onto $\mathfrak{B}^*$.

For any linear function $\Lambda$ on $\mathfrak{h}$ put $n(\Lambda) = \dim \mathfrak{B}_\Lambda^*$. We have seen above that $\mathfrak{X}^*$ is irreducible under $\pi(\mathfrak{k})$. Let $\Lambda_0, \Lambda_1, \cdots, \Lambda_r$ be all the distinct weights of the corresponding representation of $\mathfrak{k}$ and let $\mu_i$ denote the multiplicity of $\Lambda_i$ so that $\mu_i = \dim(\mathfrak{B}_{\Lambda_i}^* \cap \mathfrak{X}^*)$ $(0 \leq i \leq r)$. Then we get the following result from the above lemma.

COROLLARY. *Let* $(\gamma_1, \cdots, \gamma_q)$ *be all the distinct totally positive roots of* $\mathfrak{g}$. *Then if* $\Lambda$ *is any linear function on* $\mathfrak{h}$,

$$n(\Lambda) = \sum_{i=0}^{r} \mu_i n_i(\Lambda)$$

*where* $n_i(\Lambda)$ $(0 \leq i \leq r)$ *is the number of distinct sequences* $(m, \cdots, m_q)$ *of nonnegative integers such that* $\Lambda = \Lambda_i - (m_1\gamma_1 + \cdots + m_q\gamma_q)$.

Put $\mathfrak{X}_i^* = \mathfrak{B}_{\Lambda_i}^* \cap \mathfrak{X}^*$, $0 \leq i \leq r$. Then $\mathfrak{X}^*$ is the direct sum of $\mathfrak{X}_i^*$, $0 \leq i \leq r$ and we can choose a base $x_1^*, \cdots, x_d^*$ for $\mathfrak{X}^*$ which is composed of bases for the various $\mathfrak{X}_i^*$. Since $\mathfrak{P}_-$ is the universal enveloping algebra of the abelian Lie algebra $\mathfrak{p}_-$ and since $(X_{-\gamma_1}, \cdots, X_{-\gamma_q})$ is a base for $\mathfrak{p}_-$, it follows that the elements $X_{-\gamma_1}{}^{m_1} X_{-\gamma_q}{}^{m_2} \cdots X_{-\gamma_q}{}^{m_q}$ $(m_1, \cdots, m_q \geq 0)$ form a base for $\mathfrak{P}_-$. Hence it follows from the above lemma that $\pi(X_{-\gamma_1}{}^{m_1} \cdots X_{-\gamma_q}{}^{m_q}) x_j^*$ $(m_1, \cdots, m_q \geq 0, 1 \leq j \leq d)$ is a base for $\mathfrak{B}^*$. Moreover it is obvious that if $x^* \varepsilon \mathfrak{X}_i^*$ $(0 \leq i \leq r)$

$$\pi(X_{-\gamma_1}{}^{m_1} X_{-\gamma_2}{}^{m_2} \cdots X_{-\gamma_q}{}^{m_q}) x^* \varepsilon \mathfrak{B}_{\Lambda}{}^*$$

where $\Lambda = \Lambda_i - (m_1\gamma_1 + \cdots + m_q\gamma_q)$. Our assertion now follows immediately from these facts.

COLUMBIA UNIVERSITY,
NEW YORK, N. Y.

---

## REFERENCES.


[1] V. Bargmann, "Irreducible unitary representations of the Lorentz group," *Annals of Mathematics,* vol. 48 (1947), pp. 568-640.

[2] G. Birkhoff and P. M. Whitman, "Representation of Jordan and Lie algebras," *Transactions of the American Mathematical Society,* vol. 65 (1949), pp. 116-136.

[3] E. Cartan, (a) "Les groupes projectifs qui ne lassent invariante aucune multiplicité plane," *Bulletin de la Société Mathématique de France,* vol. 41 (1913), pp. 53-96.

———, (b) "Sur certaines formes riemanniennes remarquables des géométries à groupe fondamental simple," *Annales Scientifiques de l'École Normale Supérieure,* vol. 44 (1927), pp. 345-467.

———, (c) "Groupes simples clos et ouverts et géométrie riemannienne," *Journal de Mathématiques Pures et Appliquées,* 9th Series, vol. 8 (1929), pp. 1-33.

————, (d) "Sur les domaines bornés homogènes de l'espace de $n$ variables complexes," *Abhandlungen aus dem Mathematischen Seminar der Hamburgischen Universität*, vol. 11 (1935), pp. 116-162.

————, (e) "La topologie des espaces représentatifs des groupes de Lie," *L'Enseignement mathématique*, vol. 35 (1936), pp. 177-200.

[4] I. M. Gelfand and M. I. Graev, (a) "Unitarnye predstavleniia veshchestvennoi unimoduliarnoi gruppy (osnovnye nevyrozhdennye serii)," *Izvestia Akademiia Nauk SSSR*, vol. 17 (1953), pp. 189-248.

————, (b) "Analog formuly Plansherelia dlia veshchestvennykh poluprostykh grupp Li," *Doklady Akademiia Nauk SSSR*, N. S., vol. 92 (1953), pp. 461-464.

[5] Harish-Chandra, (a) "On representations of Lie algebras," *Annals of Mathematics*, vol. 50 (1949), pp. 900-915.

————, (b) "On some applications of the universal enveloping algebra of a semisimple Lie algebra," *Transactions of the American Mathematical Society*, vol. 70 (1951), pp. 28-96.

————, (c) "Representations of a semisimple Lie group on a Banach space I," *ibid.*, vol. 75 (1953), pp. 185-243.

————, (d) "Representations of semisimple Lie groups. II," *ibid.*, vol. 76 (1954), pp. 26-65.

————, (e) "Plancherel formula for the $2 \times 2$ real unimodular group," *Proceedings of the National Academy of Sciences*, vol. 38 (1952), pp. 337-342.

————, (f) "Representations of semisimple Lie groups. V," *ibid.*, vol. 40 (1954), p. 1076.

————, (g) "Representations of semisimple Lie groups. VI," *ibid.*, vol. 40 (1954), p. 1078.

————, (h) "Integrable and square-integrable representations of a semisimple Lie group," *ibid.*, vol. 41 (1955), p. 314.

[6] G. W. Mackey, "Induced representations of locally compact groups. I," *Annals of Mathematics*, vol. 55 (1952), pp. 101-139.

[7] H. Weyl, *The Structure and Representation of Continuous Groups*, The Institute for Advanced Study, Princeton, 1935.

Reprinted from
*Amer. J. of Math.*
77 (1955), 743–777

# REPRESENTATIONS OF SEMISIMPLE LIE GROUPS, V.*

By Harish-Chandra.

1. **Introduction.** Let $G$ be a connected semisimple Lie group and $A$ a Cartan subgroup of $G$. Under the assumption that the image of $A$ in the adjoint group of $G$ is compact, we have studied in detail [5(f)] certain irreducible representations of the Lie algebra of $G$ and seen that they can all be 'extended' to representations of the group (Theorem 4 of [5(f)]). In the present paper we shall obtain them directly as irreducible representations of $G$ on certain Hilbert spaces consisting of holomorphic functions on a suitable complex manifold.

It turns out that these representations are very intimately related with the finite-dimensional ones (Lemma 14) and the two have some striking similarities (Lemmas 6 and 12). Moreover, as we shall see in Theorem 3, under appropriate conditions some of these representations are unitary. In the last section we obtain a result on their characters. Some of the deeper analogies between these representations of $G$ and those of a compact smisimple group (see [5(e)]) will be discussed in another paper.

2. **Certain complex manifolds.** We keep to the notation of [5(f), §3]. Let $\mathfrak{g}_0$ be a semisimple Lie algebra over the field $R$ of real numbers. Define $\mathfrak{k}_0$ and $\mathfrak{p}_0$ as in [5(b), §2] and let $\mathfrak{h}_0$ be a maximal abelian subalgebra of $\mathfrak{k}_0$. *We shall assume that $\mathfrak{h}_0$ is also maximal abelian in* $\mathfrak{g}_0$ (see [5(f), §3]. Let $C$ denote the field of complex numbers. We complexify $\mathfrak{g}_0, \mathfrak{h}_0, \mathfrak{k}_0, \mathfrak{p}_0$ to $\mathfrak{g}, \mathfrak{h}, \mathfrak{k}, \mathfrak{p}$ respectively. Let $X \to adX$ $(X \varepsilon \mathfrak{g})$ denote the adjoint representation of $\mathfrak{g}$ and $\theta$ the automorphism of order 2 given by $\theta(X + Y) = X - Y$ $(X \varepsilon \mathfrak{k}, Y \varepsilon \mathfrak{p})$. Put[1] $\mathfrak{u} = \mathfrak{k}_0 + (-1)^{\frac{1}{2}}\mathfrak{p}_0$. Then $\mathfrak{u}$ is a compact real form of $\mathfrak{g}$. We denote by $\bar{\theta}$ and $\eta$ the conjugations of $\mathfrak{g}$ with respect to $\mathfrak{u}$ and $\mathfrak{g}_0$ respectively. Then

$$\bar{\theta}(X + (-1)^{\frac{1}{2}}Y) = X - (-1)^{\frac{1}{2}}Y \quad (X, Y \varepsilon \mathfrak{u}),$$
$$\eta(X' + (-1)^{\frac{1}{2}}Y') = X' - (-1)^{\frac{1}{2}}Y' \quad (X', Y' \varepsilon \mathfrak{g}_0)$$

and $\eta = \bar{\theta}\theta$.

We consider an arbitrary but fixed order (see [5(f), §2]) in the space

---

* Received July 8, 1955.

[1] We fix once for all a square-root of $-1$ in $C$ and denote it by $(-1)^{\frac{1}{2}}$.

1

$\mathfrak{F}_R$ of real linear functions[2] on $\mathfrak{h}$. Put[3] $\mathfrak{n}_+ = \sum_{\alpha>0} C X_\alpha$ and $\mathfrak{n}_- = \sum_{\alpha>0} C X_{-\alpha}$ (where $\alpha$ runs over all positive roots) and let $G_c$ be the simply connected complex analytic group with the Lie algebra $\mathfrak{g}$. We denote by $G_0$, $K_0$, $A_0$, $A_+$, $A_c$, $N_c^+$, $N_c$, $U$ the (real) analytic subgroups of $G_c$ corresponding to $\mathfrak{g}_0$, $\mathfrak{k}_0$, $\mathfrak{h}_0$, $(-1)^{\frac{1}{2}}\mathfrak{h}_0$, $\mathfrak{h}$, $\mathfrak{n}_+$, $\mathfrak{n}_-$, $\mathfrak{u}$ respectively. Then $K_0$, $A_0$ and $U$ are compact and $U$, $A_+$, $N_c^+$, $N_c^-$ are simply connected.[4] We now 'extend' $\theta$, $\bar{\theta}$ and $\eta$ to automorphisms of $G_c$. Since $\bar{\theta}(\mathfrak{n}_+) = \eta(\mathfrak{n}_+) = \mathfrak{n}_-$, it follows that $\bar{\theta}(N_c^+) = \eta(N_c^+) = N_c^-$. According to a theorem of Iwasawa [6] (see also Lemma 26 of [5(b)]) the mapping $(u, h, n) \to uhn$ $(u \, \varepsilon \, U, h \, \varepsilon \, A_+, n \, \varepsilon \, N_c^+)$ is a one-one regular analytic mapping of $U \times A_+ \times N_c^+$ onto $G_c$. We denote by $z \to Ad(z)$ $(z \, \varepsilon \, G_c)$ the adjoint representation of $G_c$.

Lemma 1. $N_c^- \cap A_+ N_c^+ = \{1\}$ and $N_c^- A_+ \cap N_c^+ = \{1\}$.

For suppose $an \, \varepsilon \, N_c^-$ $(a \, \varepsilon \, A_+, n \, \varepsilon \, N_c^+)$. Since $\mathfrak{n}_+$ is a nilpotent algebra, it is mapped onto $N_c^+$ under the exponential mapping (see Birkhoff [1]). Similarly for $\mathfrak{n}_-$. Hence we can choose $X \, \varepsilon \, \mathfrak{n}_+$ and $Y \, \varepsilon \, \mathfrak{n}_-$ such that $n = \exp X$, $an = \exp Y$. Now suppose $an \neq 1$. Then $Y \neq 0$ and there exists an element $H \, \varepsilon \, \mathfrak{h}$ such that $\exp(adY)H = H + Z$ where $Z \, \varepsilon \, \mathfrak{n}_-$ and $Z \neq 0$ (Lemma 8 of [5(d)]). Therefore it is obvious that

$$H + Z = \exp(adY)H = Ad(an)H = H + Z',$$

where $Z' \, \varepsilon \, \mathfrak{n}_+$. But since $\mathfrak{n}_+ \cap \mathfrak{n}_- = \{0\}$ it follows that $Z = Z' = 0$ and so we get a contradiction. This proves that $N_c^- \cap A_+ N_c^+ = \{1\}$. If we transform this result under the mapping $z \to \bar{\theta}(z^{-1})$ $(z \, \varepsilon \, G_c)$, we find that $N_c^- A_+ \cap N_c^+ = \{1\}$.

Corollary. $G_0 \cap A_+ N_c^+ = \{1\}$ and $N_c^- A_+ \cap G_0 = \{1\}$.

For suppose $x = an$ $(x \, \varepsilon \, G_0, a \, \varepsilon \, A_+, n \, \varepsilon \, N_c^+)$. Then $an = x = \eta(x) = a^{-1}\eta(n)$, since $\eta(H) = -H$ if $H \, \varepsilon \, (-1)^{\frac{1}{2}}\mathfrak{h}_0$. Therefore

$$\eta(n) = a^2 n \, \varepsilon \, N_c^- \cap A_+ N_c^+ = \{1\}$$

and so $a^2 = n = 1$. Since $A_+$ is abelian and simply connected this implies that $a = 1$ and therefore $x = 1$. The second assertion follows from the first under the mapping $z \to \eta(z^{-1})$ $(z \, \varepsilon \, G_c)$.

---

[2] A linear function on $\mathfrak{h}$ is called real if it takes real values on $(-1)^{\frac{1}{2}}\mathfrak{h}_0$ (see [5(f), § 2]).

[3] Any undefined terms or symbols should automatically be given the same meaning as in [5(f)].

[4] The proofs of these statements are well known and the necessary references can be found in [5(b)]. See in particular [3] and [7].

LEMMA 2. $G_0A_+N_c^+$ and $N_c^-A_+G_0$ are open in $G_c$.

Since $N_c^-A_+G_0$ is the image of $G_0A_+N_c^+$ under the topological mapping $z \to \eta(z^{-1})$ $(z \varepsilon G_c)$ it is enough to prove that $G_0A_+N_c^+$ is open. Let $\mathfrak{s}_+ = (-1)^{\frac{1}{2}}\mathfrak{h}_0 + \mathfrak{n}_+$. Then $\mathfrak{s}_+$ is the Lie algebra of $A_+N_c^+$ and so in view of the corollary $\mathfrak{g}_0 \cap \mathfrak{s}_+ = \{0\}$. Hence

$$\dim_R(\mathfrak{g}_0 + \mathfrak{s}_+) = \dim_R \mathfrak{g}_0 + \dim_R \mathfrak{s}_+.$$

On the other hand $\mathfrak{g}$ is the direct sum of $\mathfrak{h}$, $\mathfrak{n}_+$ and $\mathfrak{n}_-$ and therefore

$$\dim_R \mathfrak{g}_0 = \dim_C \mathfrak{g} = \dim_C \mathfrak{h} + \dim_C \mathfrak{n}_+ + \dim_C \mathfrak{n}_-$$

$$= \dim_R \mathfrak{h}_0 + 2\dim_C \mathfrak{n}_+ = \dim_R \mathfrak{h}_0 + \dim_R \mathfrak{n}_+ = \dim_R \mathfrak{s}_+.$$

This shows that $\dim_R(\mathfrak{g}_0 + \mathfrak{s}_+) = 2\dim_R \mathfrak{g}_0 = \dim_R \mathfrak{g}$, and therefore $\mathfrak{g}$ is the direct sum of $\mathfrak{g}_0$ and $\mathfrak{s}_+$. Our assertion now follows from Lemma 26 of [5(b)].

Since $A_+$ is simply connected, for every $a \varepsilon A$ there exists a unique element $H \varepsilon (-1)^{\frac{1}{2}}\mathfrak{h}_0$ such that $a = \exp H$. We denote this element by $\log a$. Also any element $z \varepsilon G_c$ can be written uniquely in the form $z = uhn$ $(u \varepsilon U,$ $h \varepsilon A_+,\ n \varepsilon N_c^+)$. We write $u(z)$ and $H(z)$ to denote $u$ and $\log h$ respectively. Then $z \to u(z)$ and $z \to H(z)$ are (real) analytic mappings of $G_c$ into $U$ and $(-1)^{\frac{1}{2}}\mathfrak{h}_0$ respectively.

LEMMA 3. Let $2\rho$ denote the sum of all the positive roots of $\mathfrak{g}$. Then if $dx$ is the Haar measure of $G_0$,

$$\int_{G_0} e^{-4\rho(H(x))} dx < \infty.$$

Let $\psi$ denote the mapping $x \to u(x)$ $(x \varepsilon G_0)$ of $G_0$ into $U$. It follows from the corollary to Lemma 1 that $\psi$ is univalent. Let $x = us$ $(x \varepsilon G_0, u \varepsilon U,$ $s \varepsilon A_+N_c^+)$. Then a simple calculation shows that if $X \varepsilon \mathfrak{g}_0$, $(d\psi)_x X = V$ where $V$ is the element of $\mathfrak{u}$ determined by the condition $V \equiv Ad(s)X \bmod \mathfrak{s}_+$. (Here $(d\psi)_x$ is the differential of $\psi$ at $x$ (see Chevalley [4]). Now we regard $\mathfrak{g}/\mathfrak{s}_+$ as a vector space over $R$ and denote by $D$ its endomorphism corresponding to $Ad(s)$. Since $\det(Ad(s)) = 1$, $\det D = \det(Ad(s^{-1})_{\mathfrak{s}_+}$ where $(Ad(s^{-1}))_{\mathfrak{s}_+}$ is the restriction of $Ad(s^{-1})$ on $\mathfrak{s}_+$. Then if $s = an$ $(a \varepsilon A_+, n \varepsilon N_c^+)$,

$$\det(Ad(s^{-1}))_{\mathfrak{s}_+} = \left| e^{-2\rho(\log a)} \right|^2 = e^{-4\rho(\log a)}$$

and therefore $\det D = e^{-4\rho(\log a)} = e^{-4\rho(H(x))}$. This calculation shows that

$$du = e^{-4\rho(H(x))} dx,$$

where $du$ and $dx$ are the Haar measures on $U$ and $G_0$ respectively. Hence

$$\int_{G_0} e^{-4\rho(H(x))} dx = \int_{\psi(G_0)} du \leqq \int_U du < \infty$$

because $U$ is compact.

Let $Q_+$ be the set of all totally positive roots of $\mathfrak{g}$ and $Q$ the remaining set of positive roots (see [5(f)]). Put $\mathfrak{p}_+ = \sum_{\beta \,\varepsilon\, Q_+} C X_\beta$, $\mathfrak{p}_- = \sum_{\beta \,\varepsilon\, Q_+} C X_{-\beta}$, $\mathfrak{p}' = \sum_{\beta \,\varepsilon\, Q} (C X_\beta + C X_{-\beta})$ and $\mathfrak{m} = \mathfrak{k} + \mathfrak{p}'$. Then $\mathfrak{p}_+$, $\mathfrak{p}_-$, $\mathfrak{m}$ are all subalgebras of $\mathfrak{g}$ which is a direct sum of these three. Moreover $\mathfrak{p}_+$ and $\mathfrak{p}_-$ are abelian and $[\mathfrak{m}, \mathfrak{p}_+] \subset \mathfrak{p}_+$, $[\mathfrak{m}, \mathfrak{p}_-] \subset \mathfrak{p}_-$ (see [5(f), §§ 4 and 5] for the proofs of these statements). Let $P_c^+, P_c^-$ and $M_c$ be the complex analytic subgroups of $G_c$ corresponding to $\mathfrak{p}_+$, $\mathfrak{p}_-$ and $\mathfrak{m}$ respectively.

LEMMA 4. *The mapping* $(q, m, p) \to qmp$ $(q \,\varepsilon\, P_c^-, m \,\varepsilon\, M_c, p \,\varepsilon\, P_c^+)$ *of* $P_c^- \times M_c \times P_c^+$ *into* $G_c$ *is univalent, holomorphic and regular.*

First we claim that $P_c^- M_c \cap P_c^+ = \{1\}$. For suppose $y \,\varepsilon\, P_c^- M_c \cap P_c^+$. Since $P_c^+$ is abelian, we can choose $Y \,\varepsilon\, \mathfrak{p}_+$ such that $y = \exp Y$. Also since $[\mathfrak{m}, \mathfrak{p}_-] \subset \mathfrak{p}_-$ and since $y \,\varepsilon\, P_c^- M_c$, it is clear that $Ad(y)\mathfrak{p}_- = \mathfrak{p}_-$. Now suppose $y \neq 1$ so that $Y \neq 0$. Then $Y = \sum_\gamma c_\gamma X_\gamma$ where $\gamma$ runs over all the totally positive roots and $c_\gamma \,\varepsilon\, C$. Let $\gamma_0$ be the lowest root such that $c_{\gamma_0} \neq 0$. Then $[Y, X_{-\gamma_0}] \equiv c_{\gamma_0} H_{\gamma_0} \bmod \mathfrak{n}_+$ and therefore it is obvious that

$$Ad(y) X_{-\gamma_0} \equiv X_{-\gamma_0} + c_{\gamma_0} H_{\gamma_0} \bmod \mathfrak{n}_+.$$

However $H_{\gamma_0} \notin \mathfrak{n}_+ + \mathfrak{n}_-$ and so it follows that $Ad(y) X_{-\gamma_0} \notin \mathfrak{p}_-$. Since this contradicts the fact that $Ad(y)\mathfrak{p}_- = \mathfrak{p}_-$ we must have $y = 1$ and therefore $P_c^- M_c \cap P_c^+ = \{1\}$. Taking the image of this equality under the mapping $z \to \eta(z^{-1})$ $(z \,\varepsilon\, G_c)$ we get $P_c^- \cap M_c P_c^+ = \{1\}$. Now if we make use of the fact that both $P_c^- M_c$ and $M_c P_c^+$ are subgroups of $G_c$, the univalence of our mapping follows straightaway. That it is holomorphic is an immediate consequence of the complex analyticity of $G_c$. Similarly the regularity follows from the relation $\mathfrak{g} = \mathfrak{p}_- + \mathfrak{m} + \mathfrak{p}_+$ (Lemma 26 of [5(b)]).

It is clear from Lemmas 2 and 4 that $G_0 A_+ N_c^+$, $N_c^- A_+ G_0$ and $P_c^- M_c P_c^+$ are open connected subsets of $G_c$ and therefore they may be regarded as open submanifolds of $G_c$.

LEMMA 5. $N_c^- A_+ G_0 A_+ N_c^+$ *is contained in* $P_c^- M_c P_c^+$.

Let $\exp \mathfrak{p}_0$ denote the set of all elements in $G_0$ of the form $\exp X$ $(X \,\varepsilon\, \mathfrak{p}_0)$. Then if $p \,\varepsilon\, \exp \mathfrak{p}_0$ and $p = uan$ $(u \,\varepsilon\, U, a \,\varepsilon\, A_+, n \,\varepsilon\, N_c^+)$,

$$p^{-1} = \theta(p) = ua^{-1}\bar{\theta}(n)$$

since $\bar\theta(u) = u$ and $\bar\theta(a) = a^{-1}$. Therefore

$$p^2 = \theta(n^{-1})a^2 n \,\varepsilon\, N_o^- A_+ N_o^+.$$

Now let $X$ be any element in $\mathfrak{p}_0$ and put $p = \exp(\tfrac{1}{2}X)$. Then

$$p^2 = \exp X \,\varepsilon\, N_o^- A_+ N_o^+$$

and therefore $\exp \mathfrak{p}_0 \subset N_o^- A_+ N_o^+$. On the other hand $\mathfrak{m} + \mathfrak{p}_+ \supset \mathfrak{h} + \mathfrak{n}_+$ and $\mathfrak{p}_- + \mathfrak{m} \supset \mathfrak{n}_- + \mathfrak{h}$. Therefore $A_+ N_o^+ \subset M_c P_c^+$ and $N_o^- A_+ \subset P_c^- M_c$ and so it follows that $\exp \mathfrak{p}_0 \subset P_c^- M_c P_c^+$. But $G_0 = K_0 (\exp \mathfrak{p}_0)$ (see Cartan [3] and Mostow [7]) and $K_0 P_c^- = P_c^- K_0$ since $[\mathfrak{k}, \mathfrak{p}_-] \subset \mathfrak{p}_-$. Moreover $K_0 \subset M_c$ and therefore $G_0 \subset P_c^- M_c P_c^+$. However we have seen already that $A_+ N_o^+ \subset M_c P_c^+$ and $N_o^- A_+ \subset P_c^- M_c$ and so it follows that $N_o^- A_+ G_0 A_+ N_o^+ \subset P_c^- M_c P_c^+$.

### 3. The simply connected covering manifold of $N_o^- A_+ G_0$.

Let $S$ denote the subgroup $N_o^- A_+$ of $G_c$. Then we have seen that $\bar W = S G_0$ is an open submanifold of $G_c$. Since $\mathfrak{g} = \mathfrak{n}_- + (-1)^{\frac{1}{2}}\mathfrak{h}_0 + \mathfrak{g}_0$ (see the proof of Lemma 2) the maping $(s, \bar x) \to s\bar x$ $(s\,\varepsilon\, S, \bar x\,\varepsilon\, G_0)$ of $S \times G_0$ into $\bar W$ is everywhere regular (Lemma 26 of [5(b)]) and therefore open. Also $S \cap G_0 = \{1\}$ (corollary to Lemma 1) and so this mapping is univalent. Hence it is a homeomorphism. Let $G$ be the simply connected covering group of $G_0$ and let $x \to \bar x$ $(x\,\varepsilon\, G)$ denote the natural homomorphism of $G$ onto $G_0$. Since $S$ is simply connected, $S \times G$ is a simply connected covering space of $S \times G_0$. Therefore we may also regard it as a covering space of $\bar W$ under the mapping $\nu: (s, x) \to s\bar x$ $(s\,\varepsilon\, S, x\,\varepsilon\, G)$. Since $\bar W$ is a complex manifold, we can introduce a complex structure in $S \times G$ in such a way that it becomes a covering manifold of $\bar W$ with respect to the mapping $\nu$. Let $W$ denote the complex manifold arising from $S \times G$ in this way. We identify $S$ and $G$ with subsets of $W$ under the topological mappings $s \to (s, 1)$ $(s\,\varepsilon\, S)$ and $x \to (1, x)$ $(x\,\varepsilon\, G)$. Then $S \cap G = (1, 1)$ is the common unit element of $S$ and $G$ which we shall denote by 1. Let $\bar W_l$ and $\bar W_r$ respectively be the set of all elements $\bar v$ and $\bar w$ in $\bar W$ such that $\bar v \bar W \subset \bar W$ and $\bar W \bar w \subset \bar W$. Put $W_l = \nu^{-1}(\bar W_l)$ and $W_r = \nu^{-1}(\bar W_r)$. Then if $z\,\varepsilon\, W_l$, the mapping $\bar l_z: \bar w \to \nu(z)\bar w$ $(\bar w\,\varepsilon\, \bar W)$ is obviously holomorphic on $\bar W$. Hence it is clear that there exists exactly one holomorphic mapping $l_z$ of $W$ into itself such that $\nu \circ l_z = \bar l_z \circ \nu$ and $l_z(1) = z$. Similarly if $z\,\varepsilon\, W_r$, there exists just one holomorphic mapping $r_z$ of $W$ into itself such that $\nu(r_z(w)) = \nu(w)\nu(z)$ $(w\,\varepsilon\, W)$ and $r_z(1) = z$. For convenience we shall write $l_z w$ and $r_v w$ instead of $l_z(w)$ and $r_v(w)$ respectively $(z\,\varepsilon\, W_l, v\,\varepsilon\, W_r, w\,\varepsilon\, W)$. Let $A$ be the analytic subgroup of $G$ corresponding to $\mathfrak{h}_0$. Then it is obvious that $S \subset W_l$, $G \subset W_r$ and $A \subset W_l \cap W_r$. Moreover if $u, v\,\varepsilon\, W_l$, $z = l_u v$ also lies in $W_l$ and $l_z = l_u l_v$. This shows that the multiplication in $W_l$, defined by the rule $uv = l_u v$ $(u, v \,\varepsilon\, W_l)$, is associative. Similarly we define an associative multiplication in $W_r$ by $zw = r_w z$ $(z, w \,\varepsilon\, W_r)$. It is easy to check that these two multiplications coincide on $W_r \cap W_l$.

We recall that $A_c$ is the complex analytic subgroup of $G_c$ corresponding to $\mathfrak{h}$. Put $\tilde{A}_c = v^{-1}(A_c)$. Since $A$ contains the center[5] of $G$, $\tilde{A}_c = A_+ \times A$ and so it is connected. Also $A_c \subset \bar{W}_l$ and therefore $\tilde{A}_c \subset W_l$. It is easy to verify that $\tilde{A}_c$ is a group (with respect to the multiplication defined in $W_l$) and actually it can be regarded as a covering group of $A_c$ under the mapping $a \to v(a)$ $(a \,\varepsilon\, \tilde{A}_c)$. Since $N_c^- \subset W_l$, the product $na$ $(n \,\varepsilon\, N_c^-, a \,\varepsilon\, \tilde{A}_c)$ is well defined in $W_l$.

**4. Holomorphic functions on $W$.** By a holomorphic character of $\tilde{A}_c$ we mean a holomorphic function $\xi \neq 0$ on $\tilde{A}_c$ such that $\xi(ab) = \xi(a)\xi(b)$ $(a, b \,\varepsilon\, \tilde{A}_c)$. $\xi$ being such a character, we denote by $\mathfrak{H}_\xi$ the space of all holomorphic functions $f$ on $W$ such that $f(l_{na} w) = f(w)\xi(a)$ $(n \,\varepsilon\, N_c^-,\ a \,\varepsilon\, \tilde{A}_c,\ w \,\varepsilon\, W)$. Our object now is to prove the following theorem.

THEOREM 1. $\mathfrak{H}_\xi = \{0\}$ *unless there exists a linear function* $\Lambda$ *on* $\mathfrak{h}$ *with the following two properties:* (1) $\xi(\exp H) = e^{\Lambda(H)}$ $(H \,\varepsilon\, \mathfrak{h})$ *and* (2) $\Lambda(H_\alpha)$ *is a nonnegative integer for every positive root which is not totally positive.*[3]

For any $\bar{x} \,\varepsilon\, G_0$ and $z \,\varepsilon\, G_c$ put $z^{\bar{x}} = \bar{x} z \bar{x}^{-1}$. If $y$ is a fixed element in $G$, $xyx^{-1}$ $(x \,\varepsilon\, G)$ depends only on $\bar{x}$ and so we may denote it by $y^{\bar{x}}$. Similarly if $w = (s, x) \,\varepsilon\, S \times G = W$ and $h \,\varepsilon\, A$, $r_{h^{-1}} l_h w = (s^h, x^h)$ and so for fixed $w$ it depends only on $h$. We shall denote it by $w^h$. It is clear from its definition that $w \to w^h$ $(w \,\varepsilon\, W)$ is a holomorphic mapping of $W$. $A_0$ being compact, we normalize its Haar measure $dh$ in such a way that $\displaystyle\int_{A_0} dh = 1$. We now need a few lemmas.

LEMMA 6. *There exists a function* $\psi \,\varepsilon\, \mathfrak{H}_\xi$ *such that*

$$\int_{A_0} \phi(r_x w^h)\, dh = \phi(x)\psi(w) \qquad\qquad (w \,\varepsilon\, W, x \,\varepsilon\, G)$$

*for every* $\phi \,\varepsilon\, \mathfrak{H}_\xi$. *This function is unique and* $\psi(1) = 1$ *if* $\mathfrak{H}_\xi \neq \{0\}$.

Let us first make some preliminary remarks. Every element $X$ in $\mathfrak{g}$ defines a right-invariant holomorphic infinitesimal transformation [4] $X'$ as

---

[5] For $A$ contains the center of $K$ (Weyl [9]) and $K$ contains the center of $G$ [7].

follows. $f$ being a function which is defined and holomorphic around some point[6] $z \, \varepsilon \, G_c$,

$$X'f(z) = \{df(\exp(-tX)z)/dt\}_{t=0}.$$

If $X, Y, Z$ are three elements in $\mathfrak{g}$ and $Z = [X, Y]$, it is easy to check that $Z'f = X'Y'f - Y'X'f$ in some neighborhood of $z$. Now since $\bar{W}$ is an open submanifold of $G_c$ and since the mapping $v$ of $W$ onto $\bar{W}$ is everywhere regular, there is exactly one holomorphic infinitesimal transformation on $W$ $v$-related to $X'$ [4, Chap. III, § V] which we denote again by $X'$. Let $V$ be an open set either in $W$ or $\bar{W}$ and let $\mathcal{E}$ be the space of all holomorphic functions on $V$. Then if we associate to each $X \, \varepsilon \, \mathfrak{g}$ the linear mapping $f \to X'f$ ($f \, \varepsilon \, \mathcal{E}$) of $\mathcal{E}$ into itself, we get a representation of $\mathfrak{g}$ on $\mathcal{E}$ which can be extended (uniquely) to a representation of the universal enveloping algebra $\mathfrak{B}$ of $\mathfrak{g}$. For any $b \, \varepsilon \, \mathfrak{B}$ we denote by $b'$ the corresponding operator on $\mathcal{E}$.

Let $w$ be a point in $G_c$ and $f$ a function which is defined and holomorphic in some neighbourhood of $w$. Then if $X \, \varepsilon \, \mathfrak{g}$ and $t$ is a complex variable, the function $F(t) = f(\exp(-tX)w)$ is defined and holomorphic around the origin in the complex plane and

$$\{(d^m/dt^m)F(t)\}_{t=0} = (X^m)'f(w).$$

Now let $(X_1, \cdots, X_n)$ be a base of $\mathfrak{g}$ over $C$. We put $X(z) = z_1 X_1 + \cdots + z_n X_n$ where $z_1, \cdots, z_n$ are complex numbers. Then the function $F(z) = f((\exp - X(z))w)$ is defined and holomorphic around the origin in $C^n$. Let $M = (m_1, \cdots, m_n)$ be any sequence of $n$ nonnegative integers. We write $z^M = z_1{}^{m_1} z_2{}^{m_2} \cdots z_n{}^{m_n}$, $M! = m_1! m_2! \cdots m_n!$, $|M| = m_1 + \cdots + m_n$ and

$$\partial^M F/\partial z^M = (\partial^{m_1 + \cdots + m_n}/\partial z_1{}^{m_1} \cdots \partial z_n{}^{m_n})F.$$

Then if $\delta$ is a sufficiently small positive number, $F(z)$ is defined and holomorphic for all $(z)$ such that $|z| = \max_i |z_i| < \delta$ and therefore

$$F(z) = \sum_M F(M) z^M/M! \qquad (|z| < \delta)$$

where $F(M) = (\partial^M F/\partial z^M)_0$, the suffix 0 denoting the value at the origin. Now replace $z_i$ by $tz_i$ where $t$ is a complex number and $|t| \leq 1$. Then

$$F(tz) = \sum_M F(M) t^{|M|} z^M/M!.$$

---

[6] Here $X'f(z)$ denotes the value of $X'f$ at $z$. A similar notation will be used in other cases as well.

On the other hand if $z = (z_1, \cdots, z_n)$ is fixed,

$$F(tz) = \sum_{m \geq 0} (t^m/m!)(X'(z))^m f(w)$$

for $|t|$ sufficiently small. (Here $X'(z) = (X(z))'$). This follows from the fact mentioned above that

$$\{(d^m/dt^m)F(tz)\}_{t=0} = (X'(z))^m f(w).$$

Hence comparing coefficients of powers of $t$ we get

$$(X'(z))^m f(w) = m! \sum_{m=|M|} F(M)z^M/M!.$$

Since this is true for all sufficiently small values of $|z|$, we can compare the coefficients of $z^M$ on both sides and conclude that $F(M) = X'(M)f(w)$, where $X(M)$ is the coefficient (in $\mathfrak{B}$) of $z^M$ in $(X(z))^m/m!$ $(m = |M|)$ and $X'(M) = (X(M))'$. This proves that

$$f((\exp - X(z))w) = \sum_M X'(M)f(w)z^M/M!$$

for all sufficiently small values of $|z|$.

Now we return to the lemma. Let $\phi$ and $x$ be fixed elements in $\mathfrak{H}_\xi$ and $G$ respectively. Put $f(w) = \phi(r_x w)$ $(w \,\varepsilon\, W)$. $\delta$ being a positive real number, let $Q_\delta$ denote the cube in $C^n$ consisting of all points $z$ with $|z| < \delta$. We assume that $\delta$ is so small that the following conditions are fulfilled: (1) $\exp(-X(z)) \,\varepsilon\, \bar{W}$ for $z \,\varepsilon\, Q_\delta$; (2) the mapping $z \to \exp(-X(z))$ is regular and therefore open on $Q_\delta$ and hence the set $\bar{V} = \exp Q_\delta$ is an open connected neighbourhood of 1 in $\bar{W}$; (3) $\bar{V}$ is evenly covered [4, Chap. II, § VI] under the mapping $v$ of $W$ onto $\bar{W}$. Let $V$ denote the component of 1 in $v^{-1}(\bar{V})$. Define a function $\bar{f}$ on $\bar{V}$ by the rule $\bar{f}(v(w)) = f(w)$ $(w \,\varepsilon\, V)$. Then $f$ is holomorphic on $\bar{V}$. For any $\bar{h} \,\varepsilon\, A_0$ we extend $Ad(\bar{h})$ to an automorphism of $\mathfrak{B}$ and put $b^{\bar{h}} = Ad(\bar{h})b$ $(b \,\varepsilon\, \mathfrak{B})$. Let $z_i(X)$ $i = 1, \cdots, n$ denote the coordinates of $X \,\varepsilon\, \mathfrak{g}$ with respect to the base $(X_1, \cdots, X_n)$. If $\epsilon$ is a positive number we denote by $\mathfrak{g}_\epsilon$ the set of all $X \,\varepsilon\, \mathfrak{g}$ such that $|z(X)| = \max_i |z_i(X)| < \epsilon$. Since $A_0$ is compact, we can choose $\epsilon$ so small that if $|z(X)| \leq \epsilon$, $|z(X^{\bar{h}})| \leq \delta/2$ for every $\bar{h} \,\varepsilon\, A_0$. Then since $\bar{f}(\exp(-X(z)))$ is holomorphic on $Q_\delta$, it follows from our result above that

$$\bar{f}(\exp(-X)) = \sum_{m \geq 0} (X^m)'\bar{f}(1)/m! \qquad\qquad (X \,\varepsilon\, \mathfrak{g}_\delta)$$

and therefore

$$\bar{f}(\exp(-X^{\bar{h}})) = \sum_{m \geq 0} (1/m!)((X^{\bar{h}})^m)'\bar{f}(1)/m! \qquad\qquad (X \,\varepsilon\, \mathfrak{g}_\epsilon, \bar{h} \,\varepsilon\, A_0).$$

For any $z \,\varepsilon\, Q_\delta$, let $w(z)$ denote the unique point in $V$ such that $v(w(z)) = \exp(-X(z))$. Then if $z(X)$ denotes the point $(z_1(X), \cdots, z_n(X))$ in $C^n$, it is clear that $w(z(X^{\hbar})) = (w(z(X)))^{\hbar}$ $(X \,\varepsilon\, \mathfrak{g}_c, \hbar \,\varepsilon\, A_0)$ and therefore

$$f((w(z))^{\hbar}) = \sum_M ((X(M))^{\hbar})' f(1) z^M / M! \qquad (|z| < \epsilon, \hbar \,\varepsilon\, A_0).$$

Moreover in view of our choice of $\epsilon$, for a fixed $z$ the convergence is uniform with respect to $\hbar$. Hence

$$\int_{A_0} f((w(z))^{\hbar}) d\hbar = \sum_M (z^M / M!) \int_{A_0} ((X(M))^{\hbar})' f(1) d\hbar \qquad (|z| < \epsilon).$$

Now let $q(M) = \int_{A_0} (X(M))^{\hbar} d\hbar$. Here the integral is well defined since the elements $(X(M))^{\hbar}$ span a finite-dimensional subspace of $\mathfrak{B}$. Then

$$\int_{A_0} f((w(z))^{\hbar}) d\hbar = \sum_M z^M q'(M) f(1) / M! \qquad (|z| < \epsilon).$$

It is obvious from the definition of $q(M)$ that $[H, q(M)] = 0$ if $H \,\varepsilon\, \mathfrak{h}$. Hence by the consideration of ranks (see §7 of [5(f)]) we see that there is exactly one element $h(M)$ in the subalgebra $\mathfrak{H}$ of $\mathfrak{B}$ generated by $(1, \mathfrak{h})$ such that $q(M) \equiv h(M)$ mod $\mathfrak{B}\mathfrak{n}_-$. On the other hand $f(l_n w) = f(w)$ if $n \,\varepsilon\, N_c^-$ and therefore it follows easily that $b'f(w) = 0$ if $b \,\varepsilon\, \mathfrak{B}\mathfrak{n}_-$ and $w \,\varepsilon\, W$. Moreover since $\tilde{A}_c$ is an abelian Lie group with the Lie algebra $\mathfrak{h}$, there exists a linear function $\Lambda$ on $\mathfrak{h}$ such that $\xi(\exp H) = e^{\Lambda(H)}$ $(H \,\varepsilon\, \mathfrak{h})$. Then $f(l_a w) = e^{\Lambda(H)} f(w)$ if $a = \exp H \,\varepsilon\, \tilde{A}_c$. We extend the mapping $H \to -\Lambda(H)$ $(H \,\varepsilon\, \mathfrak{h})$ to a (uniquely determined) homomorphism $\mu_\Lambda$ of $\mathfrak{H}$ into $C$ such that $\mu_\Lambda(1) = 1$. Then it is clear that $h'f(w) = \mu_\Lambda(h) f(w)$ for $h \,\varepsilon\, \mathfrak{H}$ and $w \,\varepsilon\, W$. Hence

$$q'(M) f(1) = h'(M) f(1) = \mu_\Lambda(h(M)) f(1),$$

and therefore

$$\int_{A_0} f((w(z))^{\hbar}) d\hbar = \sum_M z^M \mu_\Lambda(h(M)) f(1) / m! \qquad (|z| < \epsilon).$$

Now if we put $\Phi(x, w) = \int_{A_0} \phi(r_x w^{\hbar}) d\hbar$ and recall that $f(1) = \phi(x)$ we get

$$\Phi(x, w(z)) = \phi(x) \sum_M z^M \mu_\Lambda(h(M)) / M!$$

provided $|z| < \epsilon$. It is clear that the set of all points $w(z)$ $(|z| < \epsilon)$ is a neighbourhood of 1 in $W$. Hence from the principle of analytic continuation, there exists at most one holomorphic function $\psi$ on $W$ such that

$$\psi(w(z)) = \sum_M z^M \mu_\Lambda(h(M)) / M!$$

if $|z| < \epsilon$. On the other hand if $\mathfrak{H}_\xi \neq \{0\}$ we can choose a function $\phi_0 \neq 0$ in $\mathfrak{H}_\xi$. Since every element in $W$ is of the form $l_s x$ $(s \, \varepsilon \, S, x \, \varepsilon \, G)$, it is obvious that $\phi_0(x_0) \neq 0$ for some $x_0 \, \varepsilon \, G$. Moreover since $A_0$ is compact, the function

$$\Phi_0(x_0, w) = \int_{A_0} \phi_0(r_{x_0} w^h) \, d\bar{h} \qquad\qquad (w \, \varepsilon \, W)$$

is obviously holomorphic on $W$ and therefore the same holds for $\Phi_0(x_0, w)/\phi_0(x_0)$. But if we apply the above relation to $\phi_0$, we get

$$\Phi_0(x_0, w(z))/\phi_0(x_0) = \sum_M z^M \mu_\Lambda(h(M))/M! \qquad (|z| < \epsilon).$$

This shows that the function $\psi$ certainly exists (if $\mathfrak{H}_\xi \neq \{0\}$). On the other hand it is obvious that $\Phi_0(x_0, l_{na} w) = \xi(a)\Phi_0(x_0, w)$ $(n \, \varepsilon \, N_c^-, a \, \varepsilon \, \bar{A}_c, w \, \varepsilon \, W)$ and therefore $\psi \, \varepsilon \, \mathfrak{H}_\xi$. Furthermore since $\Phi(x, w)$ and $\phi(x)\psi(w)$ coincide on a neighbourhood of 1 in $W$ and since they are both holomorphic in $w$, they must coincide everywhere. This proves that

$$\int_{A_0} \phi(r_x w^h) \, d\bar{h} = \phi(x)\psi(w) \qquad\qquad (x \, \varepsilon \, G, w \, \varepsilon \, W),$$

and the uniqueness of $\psi$ is obvious from this formula. In particular if we put $\phi = \phi_0$, $x = x_0$ and $w = 1$, we get $\psi(1) = 1$. Finally if $\mathfrak{H}_\xi = \{0\}$, $\psi$ must also be zero and so $\psi$ is unique in any case. Thus the lemma is proved.

Let $\Lambda$ denote, as above, the linear function on $\mathfrak{h}$ such that $\xi(\exp H) = e^{\Lambda(H)}$ $(H \, \varepsilon \, \mathfrak{h})$.

LEMMA 7. *Let* $\phi \neq 0$ *be a function in* $\mathfrak{H}_\xi$. *Suppose there exists a linear function* $\Lambda'$ *on* $\mathfrak{h}$ *such that*

$$\phi(r_a w) = e^{\Lambda'(H)} \phi(w) \qquad\qquad (w \, \varepsilon \, W, H \, \varepsilon \, \mathfrak{h}_0)$$

*where* $a = \exp H \, \varepsilon \, A$. *Then* $\Lambda - \Lambda'$ *is a linear combination of positive roots with coefficients which are nonnegative integers.*

Let $V$ be an open connected neighbourhood of 1 in $W$ such that $V$ is mapped in a one-one fashion on $\bar{V} = \nu(V)$ under the mapping $\nu$. Define a function $\bar{\phi}$ on $\bar{V}$ by setting $\bar{\phi}(\nu(w)) = \phi(w)$ $(w \, \varepsilon \, V)$. Let $\alpha_1, \cdots, \alpha_r$ be all the distinct positive roots of $\mathfrak{g}$. We put $X_i = X_{\alpha_i}$ $1 \leq i \leq r$. Then $(X_1, \cdots, X_r)$ is a base for $\mathfrak{n}_+$ over $C$. Put $X(z) = z_1 X_1 + \cdots + z_r X_r$ (where the $z_i$'s are complex numbers) and for any positive $\epsilon$ let $\mathfrak{n}_+(\epsilon)$ denote the neighbourhood of zero in $\mathfrak{n}_+$ consisting of all $X(z)$ with $|z| = \max_i |z_i| < \epsilon$. We can choose $\epsilon$ so small that $\exp(-X(z)) \, \epsilon \, \bar{V}$ and

$$\bar{\phi}(\exp(-X(z))) = \sum_M X'(M)\bar{\phi}(1)z^M/M!$$

if $|z| < \epsilon$. (Here the notation is similar to what we used in the proof of Lemma 6. $M$ is a sequence $(m_1, \cdots, m_r)$ of $r$ nonnegative integers and $X(M)$ is the coefficient (in $\mathfrak{B}$) of $z^M$ in $(1/m!)(X(z))^m$ where $m = m_1 + \cdots + m_r$. Moreover $M! = m_1! \cdots m_r!$). Choose a positive real $\delta$ such that $(X(z))^{\hbar} \, \epsilon \, \mathfrak{n}_+(\epsilon)$ for all $\hbar \, \epsilon \, A_0$ if $|z| < \delta$. Then it is obvious that

$$\bar{\phi}(\exp(-X(z))^{\hbar}) = \sum_M ((X(M))^{\hbar})'\bar{\phi}(1)z^M/M! \qquad (|z| < \delta, \hbar \, \epsilon \, A_0).$$

However if $h = \exp H$ $(H \, \epsilon \, \mathfrak{h}_0)$, it is obvious that

$$\phi(w^{\hbar}) = \phi(l_h r_{h^{-1}} w) = \exp(\Lambda(H) - \Lambda'(H))\phi(w) \qquad (w \, \epsilon \, W).$$

Hence if $|z| < \delta$,

$$\bar{\phi}(\exp(-X(z))^{\hbar}) = \exp(\Lambda(H) - \Lambda'(H))\bar{\phi}(\exp(-X(z))).$$

On the other hand if $M = (m_1, \cdots, m_r)$, it is clear that

$$(X(M))^{\hbar} = e^{\alpha_M(H)}X(M),$$

where $\alpha_M = m_1\alpha_1 + \cdots + m_r\alpha_r$. Therefore

$$\exp(\Lambda(H) - \Lambda'(H)) \sum_M X'(M)\bar{\phi}(1)z^M/M!$$
$$= \sum_M e^{\alpha_M(H)}X'(M)\bar{\phi}(1)z^M/M!$$

for all $H \, \epsilon \, \mathfrak{h}_0$ and all $z = (z_1, \cdots, z_r)$ with $|z| < \delta$. Hence comparing coefficients of $z^M$ we get $X'(M)\bar{\phi}(1) = 0$ unless $\Lambda - \Lambda' = \alpha_M$. On the other hand $X'(M)\bar{\phi}(1)$ cannot be zero for all $M$. For otherwise $\bar{\phi}(\exp(-X(z)) = 0$ if $|z| < \delta$. But since $\mathfrak{g} = \mathfrak{n}_- + \mathfrak{h} + \mathfrak{n}_+$, the elements of the form $na\exp(-X(z))$ $(a \, \epsilon \, A_c, n \, \epsilon \, N_c^-, |z| < \delta)$ cover a neighbourhood of $1$ in $\bar{W}$. Since $\phi(l_{na}w) = \xi(a)\phi(w)$, it is obvious that

$$\bar{\phi}(na\exp(-X(z))) = \xi(a)\bar{\phi}(\exp(-X(z)) = 0$$

if $na\exp(-X(z)) \, \epsilon \, \bar{V}$ and $|z| < \delta$. This shows that $\bar{\phi}$ vanishes identically on a neighbourhood of $1$ in $\bar{V}$ and therefore $\phi$ is also zero on some neighbourhood of $1$ in $W$. But then $\phi$, being holomorphic, must be zero everywhere on $W$. Since this contradicts our hypothesis, $\Lambda - \Lambda' = \alpha_M$ for some $M$ and so the lemma is proved.

Every $X \, \epsilon \, \mathfrak{g}$ may be regarded as a (left-invariant) holomorphic infinitesimal transformation on $G_c$ [4, Chap. IV] and therefore also on its open submanifold $\bar{W}$. Then there is exactly one holomorphic infinitesimal transformation on $W$ which is $\nu$-related to $X$. We denote it also by $X$. Let $V$ be an open set either in $W$ or $\bar{W}$ and let $\mathcal{E}$ be the space of all holomorphic functions on $V$. Then these operations of $\mathfrak{g}$ define a representation of $\mathfrak{g}$ on $\mathcal{E}$ which may be extended uniquely to a representation of $\mathfrak{B}$. If $b \, \varepsilon \, \mathfrak{B}$ and $f \, \varepsilon \, \mathcal{E}$, we denote by $bf(w)$ the value of $bf$ at $w$ $(w \, \varepsilon \, V)$. Let $\mathfrak{X}$ be the subalgebra of $\mathfrak{B}$ generated by $(1, \mathfrak{k})$.

LEMMA 8. *Let $\psi$ be the function of Lemma 6. Then $H\psi = \Lambda(H)\psi$ for $H \, \varepsilon \, \mathfrak{h}$ and $X_\alpha\psi = 0$ for every positive root $\alpha$. Moreover the functions $b\psi$ $(b \, \varepsilon \, \mathfrak{X})$ span a finite-dimensional subspace of $\mathfrak{H}_\xi$.*

Since $r_u$ and $l_v$ $(u \, \varepsilon \, W_r, v \, \varepsilon \, W_l)$ commute, it is an easy matter to verify that if $f \, \varepsilon \, \mathfrak{H}_\xi$ and $X \, \varepsilon \, \mathfrak{g}$ then $Xf$ is also in $\mathfrak{H}_\xi$. Therefore we get a representation of $\mathfrak{B}$ on $\mathfrak{H}_\xi$. Now for any $\phi \, \varepsilon \, \mathfrak{H}_\xi$, consider the function

$$\Phi(x, w) = \phi(r_x w) \qquad\qquad (x \, \varepsilon \, G, w \, \varepsilon \, W)$$

on $G \times W$. Since $(x, w) \to \nu(r_x w) = \nu(w)\bar{x}$ is a (real) analytic mapping of $G \times W$ into $\bar{W}$, it is clear that $(x, w) \to r_x w$ is also an analytic mapping of $G \times W$ into $W$. If $W \, \varepsilon \, \mathfrak{g}_0$ it is obvious from the definition of $X\phi$ that

$$X\phi(w) = \{d\Phi(\exp tX, w)/dt\}_{t=0} \qquad\qquad (t \, \varepsilon \, R).$$

Moreover if $X, Y \, \varepsilon \, \mathfrak{g}_0$ and $Z = X + (-1)^{\frac{1}{2}}Y \, \varepsilon \, \mathfrak{g}$, $Z\phi = X\phi + (-1)^{\frac{1}{2}}(Y\phi)$. Therefore if $\Lambda'$ is a linear function on $\mathfrak{h}$, it follows from the above differential equation that $H\phi = \Lambda'(H)\phi$ for every $H \, \varepsilon \, \mathfrak{h}$ if and only if

$$\Phi(\exp H, w) = e^{\Lambda'(H)}\phi(w)$$

for all $H \, \varepsilon \, \mathfrak{h}_0$ and $w \, \varepsilon \, W$. In particular if we apply this criterion to $\psi$ and take into account the fact (which follows from Lemma 6) that $\psi(w^{\hbar}) = \psi(w)$ $(\hbar \, \varepsilon \, A_0, w \, \varepsilon \, W)$, we get $H\psi = \Lambda(H)\psi$ $(H \, \varepsilon \, \mathfrak{h})$. But then if $\phi = X_\alpha\psi$ it is clear that $H\phi = (\Lambda(H) + \alpha(H))\phi$ $(H \, \varepsilon \, \mathfrak{h})$ and therefore by the above criterion

$$\phi(r_h w) = \exp(\Lambda(H) + \alpha(H))\phi(w) \qquad\qquad (H \, \varepsilon \, \mathfrak{h}_0, w \, \varepsilon \, W)$$

where $h = \exp H \, \varepsilon \, A$. $\alpha$ being a positive root, we can conclude from Lemma 7 that $\phi = X_\alpha\psi = 0$.

To prove the last assertion we may assume that $\mathfrak{H}_\xi \neq \{0\}$. Let $K$ and $K'$ be the analytic subgroups of $G$ corresponding to $\mathfrak{k}_0$ and $\mathfrak{k}_0' = [\mathfrak{k}_0, \mathfrak{k}_0]$ respectively. Then $K'$ is compact and semisimple[4] and the function

$$\Psi(u, w) = \psi(r_u w) \qquad\qquad (u \, \varepsilon \, K', w \, \varepsilon \, W)$$

is continuous on $K' \times W$. Moreover $\Psi(1,1) = \psi(1) = 1$. Therefore from the Peter-Weyl Theorem for $K'$, there exists an irreducible character $\chi$ of $K'$ with the property that the function

$$\Psi'(u,w) = \int_{K'} \chi(v^{-1}) \Psi(uv,w) dv$$

is not identically zero on $K' \times W$. (Here $dv$ is the Haar measure on $K'$). Since $\Psi(uv,w) = \Psi(v, r_u w)$, it follows that $\Psi'(u,w) = \Psi'(1, r_u w)$. Therefore if $\psi'(w) = \Psi'(1,w)$ $(w \,\varepsilon\, W)$, it is obvious that $\psi' \,\varepsilon\, \mathfrak{H}_\xi$ and it is not zero. Now if $h \,\varepsilon\, A$,

$$\psi'(w^h) = \int_{K'} \chi(v^{-1}) \psi(r_v w^h) dv.$$

Moreover $K'$ is a normal subgroup of $K$ and it is clear that $\chi(h^{-1}v^{-1}h) = \chi(v^{-1})$. Therefore

$$\int_{K'} \chi(v^{-1}) \psi(r_v w^h) dv = \int_{K'} \chi(v^{-1}) \psi((r_v w)^h) dv = \int_{K'} \chi(v^{-1}) \psi(r_v w) dv = \psi'(w),$$

since $\psi(z^h) = \psi(z)$ $(z \,\varepsilon\, W)$. This shows that $\psi'(w^h) = \psi'(w)$ and therefore from Lemma 6,

$$\psi'(w) = \int_{A_0} \psi'(w^h) d\bar{h} = \psi'(1)\psi(w) \qquad (w \,\varepsilon\, W).$$

This proves that the function $\psi$ and $\psi'$ differ only by a constant factor which however cannot be zero since neither $\psi$ nor $\psi'$ is zero. For any fixed $u \,\varepsilon\, K'$ put $\psi_u'(w) = \Psi'(u,w) = \psi'(r_u w)$. Then it is clear from the definition of $\Psi'(u,w)$ that $\psi_u' \,\varepsilon\, \mathfrak{H}_\xi$ and the dimension of the subspace $V$ of $\mathfrak{H}_\xi$ spanned by all $\psi_u'$ $(u \,\varepsilon\, K')$ is finite. For any $\phi \,\varepsilon\, V$ define $\phi_u(w) = \phi(r_u w)$ $(u \,\varepsilon\, K', w \,\varepsilon\, W)$. Then if to each $u \,\varepsilon\, K'$ we associate the linear mapping $\sigma(u) : \phi \to \phi_u$ of $V$ into itself, we get a representation $\sigma$ of $K'$ on the finite-dimensional space $V$. We denote the corresponding representation of $\mathfrak{k}_0'$ also by $\sigma$. Since $\sigma(u)\phi(w) = \phi(r_u w)$ it follows immediately that $\sigma(X)\phi = X\phi$ $(X \,\varepsilon\, \mathfrak{k}_0', \phi \,\varepsilon\, V)$. On the other hand if $\mathfrak{c}_0$ is the center of $\mathfrak{k}_0$ and $h = \exp H \,\varepsilon\, A$ $(H \,\varepsilon\, \mathfrak{c}_0)$ it is clear that $\psi_u(r_h w) = \psi(r_h r_u w) = e^{\Lambda(H)}\psi(r_u w) = e^{\Lambda(H)}\psi_u(w)$. Since $V$ is spanned by the functions $\psi_u$, we can conclude that $\phi(r_h w) = e^{\Lambda(H)}\phi(w)$ for all $\phi \,\varepsilon\, V$. But as we have seen earlier this implies that $H\phi = \Lambda(H)\phi$ and therefore $V$ is invariant under the operations of $\mathfrak{k}_0' + \mathfrak{c}_0 = \mathfrak{k}_0$ and so also under those of $\mathfrak{k}$. Since $\psi \,\varepsilon\, V$, the last assertion of the Lemma follows immediately.

It is now obvious that Theorem 1 is a direct consequence of Lemma 8 and Theorem 1 of [5(f)].

**5. Representations on a Hilbert space of holomorphic functions.** We shall now prove a converse of Theorem 1. Since $\tilde{A}_c$ is the direct product of $A_+$ and $A$ and $A_+$ is simply connected, $\exp H = 1$ in $\tilde{A}_c$ $(H \varepsilon \mathfrak{h})$ if and only if $H \varepsilon \mathfrak{h}_0$ and $\exp H = 1$ in $A$. On the other hand if $D$ and $A'$ are the analytic subgroups of $A$ corresponding to[3] $c_0$ and $\mathfrak{h}_0' = \mathfrak{h}_0 \cap \mathfrak{k}_0'$, $A$ is the direct product of $D$ and $A'$ and $D$ is simply connected.[4] Therefore $\exp H \neq 1$ in $A$ unless $H \varepsilon \mathfrak{h}_0'$. Now suppose $H \varepsilon \mathfrak{h}_0'$. Since $A' \subset K'$ and $K'$ is compact, $\exp H = 1$ if and only if it lies in the kernel of every finite-dimensional irreducible representation of $K'$. But $K'$ is simply connected[4] and therefore there is a one-one correspondence between finite-dimensional representations of $K'$ and those of $\mathfrak{k}'$. So it is clear that $\exp H = 1$ if and only if $e^{\Lambda(H)} = 1$ for all weights $\Lambda$ (with respect to $\mathfrak{h}'$) of all finite-dimensional representations of $\mathfrak{k}'$. Since we may identify compact roots of $\mathfrak{g}$ with roots of $\mathfrak{k}'$, it follows[7] that a linear function $\Lambda$ on $\mathfrak{h}$ coincides on $\mathfrak{h}'$ with the weight of some finite-dimensional representation of $\mathfrak{k}'$, if and only if $\Lambda(H_\alpha)$ is an integer for every compact root $\alpha$. Therefore *in order that there should exist a (holomorphic) character $\xi$ of $\tilde{A}_c$ satisfying the equation $\xi(\exp H) = e^{\Lambda(H)}$ $(H \varepsilon \mathfrak{h})$ it is necessary and sufficient that $\Lambda(H_\alpha)$ should be an integer for every compact root $\alpha$.*

Now let $\mu$ and $\omega$ be two nonnegative (Haar) measurable functions on $G_0$. We assume that $\mu$ is not identically zero, $\omega$ is bounded on every compact set and $\mu(\bar{x}\bar{y}) \leqq \mu(\bar{x})\omega(\bar{y})$ $(\bar{x}, \bar{y} \varepsilon G_0)$. Then $\mu(\bar{y}) \leqq \mu(\bar{x})\omega(\bar{x}^{-1}\bar{y})$ and since $\mu$ is not identically zero it follows that $\mu(\bar{x}) \neq 0$. Moreover

$$\{\mu(\bar{x}\bar{y})\}^{-1} \leqq \{\mu(\bar{x})\}^{-1}\omega(\bar{y}^{-1})$$

and since $\omega(\bar{y})$ (and therefore also $\omega(\bar{y}^{-1})$) is bounded on every compact set, it follows from the above inequalities that the same holds for both $\mu$ and $1/\mu$. Hence if $d\bar{x}$ is the Haar measure on $G_0$, the two measures $\mu(\bar{x})d\bar{x}$ and $d\bar{x}$ are absolutely continuous with respect to each other.

Let $\Lambda$ be a *real*[2] linear function on $\mathfrak{h}$. We assume that $\Lambda(\dot{H_\alpha})$ is a nonnegative integer for every positive root $\alpha$ which is not totally positive. Then, as we have seen above, there exists a holomorphic character $\xi$ of $\tilde{A}_c$ such that $\xi(\exp H) = e^{\Lambda(H)}$ $(H \varepsilon \mathfrak{h})$. Let $f$ be a complex-valued function on $W$ such that $f(l_a w) = \xi(a)f(w)$ for all $a \varepsilon \tilde{A}_c$ and $w \varepsilon W$. If $Z$ is the center of $G$, $Z \subset A$[5] and therefore $f(l_z w) = \xi(z)f(w)$ $(z \varepsilon Z)$ and $|\xi(z)| = 1$ because $\Lambda$ is real. This shows that $|f(zx)| = |f(x)|$ $(z \varepsilon Z, x \varepsilon G)$ and so $|f(x)|$ depends only on $\bar{x}$. Hence if $|f(x)|$ happens to be a measurable function of $\bar{x}$,

---

[7] This can be deduced easily from Theorem 1 of [5(a)]. See also Weyl [9].

we can consider the integral $\int_{G_0} |f(x)|^2 \mu(\bar{x}) d\bar{x}$. Now let $\mathfrak{H}_\Lambda(\mu)$ denote the space of all holomorphic functions $f$ on $W$ satisfying the following two conditions:

$$(1) \qquad f(l_{na}w) = \xi(a)f(w) \qquad (n \, \varepsilon \, N_c^-, a \, \varepsilon \, \tilde{A}_c, w \, \varepsilon \, W)$$

$$(2) \qquad \| f \|^2 = \int_{G_0} |f(x)|^2 \mu(\bar{x}) d\bar{x} < \infty.$$

We put $\mathfrak{H} = \mathfrak{H}_\Lambda(\mu)$ for convenience. Then every $f \varepsilon \mathfrak{H}$ is certainly continuous on $G$. Therefore since the measures $d\bar{x}$ and $\mu(\bar{x})d\bar{x}$ are absolutely continuous with respect to each other, $\| f \|$ is positive unless $f$ vanishes identically on $G$. But then in view of condition (1) above, $f = 0$. Therefore in order to prove that $\mathfrak{H}$ is a Hilbert space, it only remains to prove that it is complete. After this has been done we intend to define a representation of $G$ on $\mathfrak{H}$ and prove that $\mathfrak{H} = \mathfrak{H}_\Lambda(\mu) \neq \{0\}$ for a suitable choice of $\mu$.

Let $x_0$ be a fixed element in $G$ and let $X_1, \cdots, X_n$ be a base for $\mathfrak{g}$ over $C$. If $X \varepsilon \mathfrak{g}$, we denote by $z_1(X), \cdots, z_n(X)$ the coordinates of $X$ with respect to this base and for any positive number $\epsilon$ we define (as before) $\mathfrak{g}_\epsilon$ to be the set of all $X \varepsilon \mathfrak{g}$ such that $|z(X)| = \max_i |z_i(X)| < \epsilon$. Now choose $\epsilon$ so small that the following conditions hold: (1) the mapping $X \to \exp(-X)$ of $\mathfrak{g}$ into $G_c$ is univalent and regular on $\mathfrak{g}_\epsilon$, (2) $\exp(-X)\bar{x}_0 \, \varepsilon \, \bar{W}$ for $X \varepsilon \mathfrak{g}_\epsilon$, (3) the set $\bar{V} = \exp(\mathfrak{g}_\epsilon)\bar{x}_0$ is evenly covered under the mapping $\nu$ of $W$ onto $\bar{W}$. Let $V$ denote the connected component of $x_0$ in $\nu^{-1}(\bar{V})$. Then $V$ and $\bar{V}$ are both open and $\nu$ defines a one-one regular holomorphic mapping of $V$ onto $\bar{V}$.

For any $X \varepsilon \mathfrak{g}$, consider the endomorphism

$$(1 - \exp(-adX))/adX = \sum_{m \geq 0} (-1)^m (adX)^m/(m+1)!$$

of the complex vector space $\mathfrak{g}$. We denote by $\Delta(X)$ its determinant. Then $\Delta(X)$ is clearly a holomorphic function on $\mathfrak{g}$ and a well-known computation [4, p. 157] shows that if $x^* = \exp(-X)\bar{x}_0$,

$$dx^* = d(x^*)^{-1} = |\Delta(X)|^2 |dX|^2 \qquad (X \varepsilon \mathfrak{g}_\epsilon)$$

where $dx^*$ is the Haar measure on $G_c$ and $|dX|^2$ is the Euclidean measure on $\mathfrak{g}$ (regarded as a vector space over $R$). Since $\Delta(0) = 1$, we may assume that $\epsilon$ is so small that $|\Delta(X)| \geq \frac{1}{2}$ on $\mathfrak{g}_\epsilon$.

On the other hand we know (see Corollary to Lemma 1 and Lemma 26 of [5(b)]) that $(n, a, \bar{x}) \to na\bar{x}$ $(n \, \varepsilon \, N_c^-, a \, \varepsilon \, A_+, \bar{x} \, \varepsilon \, G_0)$ is a one-one regular mapping of $N_c^- \times A_+ \times G_0$ onto $\bar{W}$ and an easy computation shows that

$$dx^* = e^{4\rho(\log a)} dn \, da \, d\bar{x},$$

where $x^* = na\bar{x}$ and $dx^*$, $dn$, $da$ are the (suitably normalised) Haar measures of $G_o$, $N_c^-$ and $A_+$ respectively. Let $V'$ be the set of all points $(n, a, \bar{x}) \varepsilon N_c^- \times A_+ \times G_0$ such that $na\bar{x} \varepsilon \bar{V}$ and let $V_1$, $V_2$, $V_3$ denote the projections of $V'$ on each of the factors $N_c^-$, $A_+$, $G_0$ respectively. Then they are all open and since the closure of $\mathfrak{g}_\epsilon$ is compact, it is clear that the closures of $\bar{V}, V', V_1, V_2, V_3$ are also all compact. We can choose open neighbourhoods $V_1', V_2', V_3'$ of $1, 1, \bar{x}_0$ in $N_c^-, A_+, G_0$ respectively such that $V_1'V_2'V_3' \subset \bar{V}$. Also choose a positive $\delta < \epsilon$ such that $\exp(-X)\bar{x}_0 \varepsilon V_1'V_2'V_3'$ if $X \varepsilon \mathfrak{g}_\delta$. If $f$ is any function in $\mathfrak{H}$, we denote by $\bar{f}$ the function on $\bar{V}$ given by $\bar{f}(\nu(w)) = f(w)$ $(w \varepsilon V)$. Then

$$\int_{\mathfrak{g}_\delta} |\bar{f}(\exp(-X)\bar{x}_0)|^2 \,|\,dX\,|^2 \leqq 4 \int_{\mathfrak{g}_\delta} |\bar{f}(\exp(-X)x_0)|^2 \,|\,\Delta(X)\,|^2 \,|\,dX\,|^2$$

$$\leqq 4 \int_{V_1 \times V_2 \times V_3} |\bar{f}(na\bar{x})|^2 \, e^{4\rho(\log a)} dn\,da\,d\bar{x} = M_1 \int_{V_3} |\bar{f}(\bar{x})|^2 \,d\bar{x}$$

where $M_1 = 4 \int_{V_1} \exp(2\Lambda(\log a) + 4\rho(\log a))\,da \int_{V_2} dn$. On the other hand since the closure of $V_3$ is compact, $1/\mu$ is bounded on $V_3$. Hence

$$\int_{V_3} |\bar{f}(\bar{x})|^2 \,d\bar{x} \leqq M_2 \int_{V_3} |\bar{f}(\bar{x})|^2 \mu(\bar{x})\,d\bar{x} \leqq M_2 \,\|f\|^2$$

where $M_2$ is an upper bound for $\mu^{-1}$ on $V_3$. This proves that

$$\int_{\mathfrak{g}_\delta} |\bar{f}(\exp(-X)\bar{x}_0)|^2 \,|\,dX\,|^2 \leqq M \,\|f\|^2$$

where $M = M_1M_2$. Since $\bar{f}(\exp(-X)\bar{x}_0)$ is a holomorphic function of $X$ on $\mathfrak{g}_\epsilon$, it follows from a classical argument (see Bochner and Martin [2, p. 117]) that if $\|f\|$ tends to zero, $\bar{f}(\exp(-X)\bar{x}_0)$ tends to zero uniformly on every compact subset of $\mathfrak{g}_\delta$. This proves that $\|f\| \to 0$ implies the uniform convergence of $f$ to zero on some neighbourhood of $x_0$ in $W$. Moreover since $x_0$ can be any point in $G$ and since $f(nax) = \xi(a)f(x)$ $(n \varepsilon N_c^-, a \varepsilon A_+, x \varepsilon G)$ it is clear that the same conclusion holds in some neighbourhood of any given point in $W$. Therefore if $f_m$ is a Cauchy sequence in $\mathfrak{H}$, it converges uniformly on every compact set in $W$ and hence the limit function $f$ is holomorphic on $W$ and clearly $f(l_{na}w) = \xi(a)f(w)$. Moreover it then follows by well-known elementary arguments that

$$\|f\|^2 = \int_{G_0} |f(x)|^2 \mu(\bar{x})\,d\bar{x} = \lim_{m \to \infty} \|f_m\|^2 < \infty$$

and $\|f - f_m\| \to 0$. This proves that $f$ lies in $\mathfrak{H}$ and $f_m$ converges to $f$ in $\mathfrak{H}$. Therefore $\mathfrak{H}$ is complete.

Now we define a representation $\pi$ of $G$ on $\mathfrak{H}$ as follows. If $f \varepsilon \mathfrak{H}$ and $y \varepsilon G$, $\pi(y)f$ is the function whose value at $w$ is $f(r_y w)$ $(w \varepsilon W)$. Since $\mu(\bar{x}\bar{y}^{-1}) \leqq \mu(\bar{x})\omega(\bar{y}^{-1})$,

$$\int_{G_0} |f(xy)|^2 \mu(\bar{x}) d\bar{x} \leqq \omega(\bar{y}^{-1}) \| f \|^2,$$

and so it follows easily that $\pi(y)f \varepsilon \mathfrak{H}$ and $\| \pi(y)f \|^2 \leqq \omega(\bar{y}^{-1}) \| f \|^2$. This shows moreover that the operators $\pi(y)$ remain uniformly bounded on any compact set. Now let $V = V^{-1}$ be any compact neighbourhood of 1 in $G$ and let $\bar{V}$ be its image in $G_0$ under the mapping $x \to \bar{x}$. For any given $\epsilon > 0$, we can choose a compact set $G_1$ in $G_0$ such that

$$\int_{{}^c G_1} |f(x)|^2 \mu(\bar{x}) d\bar{x} \leqq \epsilon^2.$$

(${}^c G_1$ is the complement of $G_1$ in $G_0$). Put $G_2 = G_1 \bar{V}$. Then $G_2$ is also compact and if $y \varepsilon V$, $({}^c G_2)\bar{y} \subset {}^c G_1$. Hence

$$\int_{{}^c G_2} |f(xy)|^2 \mu(\bar{x}) d\bar{x} \leqq \omega(\bar{y}^{-1}) \int_{({}^c G_1)} |f(x)|^2 \mu(\bar{x}) d\bar{x} \leqq M^2 \epsilon^2,$$

where $M^2$ is an upper bound for $\omega(\bar{y})$ on $\bar{V}$. Moreover

$$\| \pi(y)f - f \|^2 = \int_{G_2} |f(xy) - f(x)|^2 \mu(\bar{x}) d\bar{x} + \int_{{}^c G_2} |f(xy) - f(x)|^2 \mu(\bar{x}) d\bar{x}.$$

But

$$\int_{{}^c G_2} |f(xy) - f(x)|^2 \mu(\bar{x}) d\bar{x}$$

$$\leqq [(\int_{{}^c G_2} |f(xy)|^2 \mu(\bar{x}) d\bar{x})^{\frac{1}{2}} + (\int_{{}^c G_2} |f(x)|^2 \mu(\bar{x}) d\bar{x})^{\frac{1}{2}}]^2 \leqq \{(M+1)\epsilon\}^2,$$

and since $G_2$ is compact,

$$\operatorname*{Lim}_{y \to 1} \int_{G_2} |f(xy) - f(x)|^2 \mu(\bar{x}) d\bar{x} = 0.$$

As $\epsilon$ is arbitrary, this shows that $\operatorname*{Lim}_{y \to 1} \| \pi(y)f - f \| = 0$. Moreover it is obvious that $\pi(xy) = \pi(x)\pi(y)$ $(x, y \varepsilon G)$ and therefore $\pi$ is a representation [5(b), p. 201] of $G$ on $\mathfrak{H}$.

It is obvious that $\mathfrak{H}$ is contained in the space $\mathfrak{H}_\xi$ of Lemma 6.

Lemma 9. *If* $\mathfrak{H} = \mathfrak{H}_\Lambda(\mu) \neq \{0\}$, *the function* $\psi$ *of Lemma 6 lies in* $\mathfrak{H}$.

For if we choose $\phi_0 \neq 0$ in $\mathfrak{H}$, it is clear that $\phi_0(x_0) \neq 0$ for some $x_0 \varepsilon G_0$. Then it follows from Lemma 6 that

$$\int_{A_0} \phi_0(y^h x_0) d\bar{h} = \phi_0(x_0)\psi(y),$$

2

and therefore

$$|\phi_0(x_0)|^2 \int_{G_0} |\psi(y)|^2 \mu(\bar{y})\,d\bar{y} \leqq \int_{G_0} \mu(\bar{y})\,d\bar{y} \int_{A_0} |\phi_0(y^h x_0)|^2\,d\bar{h}$$

$$= \int_{A_0} d\bar{h} \int_{G_0} |\phi_0(y^h x_0)|^2 \mu(\bar{y})\,d\bar{y}.$$

But if $h \, \varepsilon \, A$,

$$\int_{G_0} |\phi_0(y^h x_0)|^2 \mu(\bar{y})\,d\bar{y} = \int_{G_0} |\phi_0(hyh^{-1}x_0)|^2 \mu(\bar{y})\,d\bar{y}$$

$$= \int_{G_0} |\phi_0(y)|^2 \mu(\bar{y}\bar{x}_0^{-1}\bar{h})\,d\bar{y} \leqq \omega(\bar{x}_0^{-1}\bar{h}) \|\phi_0\|^2,$$

since $|\xi(h)| = 1$. Therefore if $M$ is an upper bound for $\omega$ on the compact set $\bar{x}_0^{-1}A_0$, it follows that $|\phi_0(x_0)|^2 \|\psi\|^2 \leqq M \|\phi_0\|^2$. Since $\phi_0(x_0) \neq 0$, we conclude that $\|\psi\|^2 < \infty$ and therefore $\psi \, \varepsilon \, \mathfrak{H}$.

LEMMA 10. *Let $\phi$ be any element in $\mathfrak{H}$ which is well-behaved* [8] *under $\pi$. Then $\pi(X)\phi = X\phi$ $(X \, \varepsilon \, \mathfrak{g})$. Moreover if $\mathfrak{H} \neq \{0\}$, $\psi$ is well-behaved under $\pi$.*

We have seen that if $\phi$ tends to $\phi_0$ in $\mathfrak{H}$ then $\phi(w)$ tends to $\phi_0(w)$ uniformly on every compact set in $W$. In particular if $\phi$ is well-behaved under $\pi$ and $X \, \varepsilon \, \mathfrak{g}_0$, $(1/t)\{\pi(\exp tX)\phi - \phi\}$ $(t \, \varepsilon \, R)$ tends to $\pi(X)\phi$ in $\mathfrak{H}$ as $t \to 0$. On the other hand it is obvious that $(1/t)\{\pi(\exp tX)\phi - \phi\}$ tends to $X\phi$ uniformly on every compact set in $W$. Therefore $X\phi = \pi(X)\phi$ and by linearity this remains true if $X \, \varepsilon \, \mathfrak{g}$. Also we know that $\pi(z) = \xi(z)\pi(1)$ $(z \, \varepsilon \, Z)$. Therefore it follows from Theorem 4 of [5(b), p. 224] that the space of well-behaved elements is dense in $\mathfrak{H}$. On the other hand if $h \, \varepsilon \, A$, it is obvious that $\xi(h^{-1})\pi(h)$ depends only on $\bar{h}$ and so we may denote it by $\bar{\pi}(\bar{h})$. Then $\bar{h} \to \bar{\pi}(\bar{h})$ $(\bar{h} \, \varepsilon \, A_0)$ is a representation of $A_0$ on $\mathfrak{H}$ and

$$E = \int_{A_0} \bar{\pi}(\bar{h})\,d\bar{h}$$

is a bounded operator with $E^2 = E$. Moreover since $A_0$ is compact, it follows easily (see Lemma 29 of [5(b)]) that if $\phi$ is well-behaved under $\pi$, the same is true of $E\phi$. Moreover if we regard

$$E\phi = \int_{A_0} \bar{\pi}(\bar{h})\phi\,d\bar{h} \qquad\qquad (\phi \, \varepsilon \, \mathfrak{H})$$

as the limit of a sum in $\mathfrak{H}$, it is clear from the remarks on convergence made above that [6]

$$E\phi(w) = \int_{A_0} \bar{\pi}(\bar{h})\phi(w)\,d\bar{h} \qquad\qquad (w \, \varepsilon \, W).$$

---

[8] We use here and in the rest of this paper the terminology of [5(b)].

But if $h \,\varepsilon\, A$,

$$\bar{\pi}(h)\phi(w) = \xi(h^{-1})\pi(h)\phi(w) = \xi(h^{-1})\phi(r_h w) = \xi(h^{-1})\phi(l_h w^{\bar{a}}) = \phi(w^{\bar{a}})$$

where $a = h^{-1}$. Therefore, from Lemma 6,

$$E\phi(w) = \phi(1)\psi(w) \quad \text{or} \quad E\phi = \phi(1)\psi.$$

Now assume that $\mathfrak{H} \neq \{0\}$. Then from Lemma 9, $\psi \,\varepsilon\, \mathfrak{H}$ and so we can choose a sequence $\phi_m$ of well-behaved elements in $\mathfrak{H}$ such that $\phi_m \to \psi$ in $\mathfrak{H}$. Since $E$ is bounded

$$\phi_m(1)\psi = E\phi_m \to E\psi = \psi(1)\psi = \psi$$

from Lemma 6. Therefore $\phi_m(1) \neq 0$ if $m$ is sufficiently large and since $E\phi_m$ is well-behaved the same holds for $\psi = \{\phi_m(1)\}^{-1}E\phi_m$.

LEMMA 11. *Suppose $\mathfrak{H} \neq \{0\}$ and $\mathfrak{H}_1$ is the smallest closed subspace of $\mathfrak{H}$ containing $\psi$ which is invariant under $\pi(G)$. Then the representation of $G$ defined on $\mathfrak{H}_1$ under $\pi$ is irreducible and quasi-simple.*[8]

Let $\mathfrak{H}_2 \neq \{0\}$ be any closed invariant subspace of $\mathfrak{H}_1$. Choose $\phi \neq 0$ in $\mathfrak{H}_2$. Then as we have seen above $E\pi(x)\phi = \{\pi(x)\phi(1)\}\psi = \phi(x)\psi$ $(x \,\varepsilon\, G)$. Since $\phi \neq 0$ it is clear that $\phi(x) \neq 0$ for some $x$ and therefore $\psi \,\varepsilon\, \mathfrak{H}_2$. But this implies that $\mathfrak{H}_2 = \mathfrak{H}_1$ and therefore $\mathfrak{H}_2$ is irreducible.

Now let $z$ be any element in $\mathfrak{B}$ of rank[3] zero. Then it is clear that if $\phi$ is a well-behaved element in $\mathfrak{H}$,

$$\pi(h)\pi(z)\phi = \pi(z)\pi(h)\phi \qquad (h \,\varepsilon\, A),$$

and therefore $E\pi(z)\phi = \pi(z)E\phi$. In particular

$$E\pi(z)\psi = \pi(z)E\psi = \pi(z)\psi.$$

But we know that $E\phi = \phi(1)\psi$ for every $\phi \,\varepsilon\, \mathfrak{H}$. Therefore

$$\pi(z)\psi = E\pi(z)\psi = \chi(z)\psi,$$

where $\chi(z)$ is the value of $\pi(z)\psi$ at 1. Now if $z$ lies in the center of $\mathfrak{B}$, it follows that

$$\pi(z)\pi(x)\psi = \pi(x)\pi(z)\psi = \chi(z)\pi(x)\psi \qquad (x \,\varepsilon\, G).$$

Hence if $V$ is the subspace of $\mathfrak{H}_1$ spanned by $\pi(x)\psi$ $(x \,\varepsilon\, G)$, $\pi(z)\phi = \chi(z)\phi$ for all $\phi \,\varepsilon\, V$. Since $V$ consists of well-behaved elements and since $V$ is dense in $\mathfrak{H}_1$ the quasi-simplicity of the representation on $\mathfrak{H}_1$ now follows from [9] Lemma 32 of [5(b)].

LEMMA 12. *Suppose $\mathfrak{H} \neq \{0\}$ and $\pi$ is a unitary representation. Then $\mathfrak{H}$ is irreducible under $\pi$.*

In view of Lemma 11 it is enough to prove that $\mathfrak{H}_1 = \mathfrak{H}$. Let $\mathfrak{H}_2$ be the orthogonal complement of $\mathfrak{H}_1$ in $\mathfrak{H}$. Since $\pi$ is unitary, $\mathfrak{H}_2$ is invariant under $\pi$. Let $\phi$ be any element in $\mathfrak{H}_2$. Then we have seen that $E\pi(x)\phi = \phi(x)\psi$ $(x \varepsilon G)$. Therefore $\phi(x)\psi \varepsilon \mathfrak{H}_2 \cap \mathfrak{H}_1 = \{0\}$ and so $\phi(x) = 0$. This being true for every $x \varepsilon G$, it is clear that $\phi = 0$. This proves that $\mathfrak{H}_2 = \{0\}$ and therefore $\mathfrak{H}_1 = \mathfrak{H}$.

## 6. Computation of the function $\psi$.

We shall now determine the function $\psi$ explicitly. We know from Lemma 4 that $Q = P_c^- M_c P_c^+$ is an open submanifold of $G_c$ and there exists a (unique) holomorphic mapping $m$ of $Q$ into $M_c$ such that $q \varepsilon P_c^- m(q) P_c^+$ for every $q \varepsilon Q$. Let $\tilde{M}_c$ be the simply connected covering group of $M_c$ and $\gamma$ the natural homomorphism of $\tilde{M}_c$ onto $M_c$. Since $\bar{W} \subset Q$, $w \rightarrow m(\nu(w))$ is a holomorphic mapping of $W$ into $M_c$. But $W$ is simply connected and so there exists a (unique) holomorphic mapping $\tilde{m}$ of $W$ into $\tilde{M}_c$ such that $\tilde{m}(1) = 1$ and $\gamma \circ \tilde{m} = m \circ \nu$. Now if we use the notation of [5(f), §5] it is clear that $\mathfrak{m} = \mathfrak{g}' + \mathfrak{k}$. Hence if $\mathfrak{c}$ is the center of $\mathfrak{k}$, $\mathfrak{c}_+ = \mathfrak{c} \cap \mathfrak{g}_+$ is the center of $\mathfrak{m}$ and $\mathfrak{m}$ is the direct sum of $\mathfrak{c}_+$ and $\mathfrak{m}' = [\mathfrak{m}, \mathfrak{m}]$. Therefore $\tilde{M}_c$ being simply connected, there exists a (unique) holomorphic mapping $\Gamma$ of $W$ into $\mathfrak{c}_+$ such that $\tilde{m}(w) \varepsilon \tilde{M}_c' \exp \Gamma(w)$. Here $\tilde{M}_c'$ is the analytic subgroup of $\tilde{M}_c$ corresponding to $\mathfrak{m}'$.

LEMMA 13. *Let $w$ be any element in $W$. Then*

$$\Gamma(l_{na}w) = \Gamma(a) + \Gamma(w), \quad \Gamma(r_u w) = \Gamma(w) + \Gamma(u), \quad \Gamma(ux) = \Gamma(u) + \Gamma(x)$$

---

[9] It was pointed out to me by Dixmier that the proof of Lemma 33 of [5(b)] is incorrect. Since $\tilde{M}$ is dense in $\tilde{\mathfrak{H}}$ only in the weak topology, one cannot conclude that $|(\bar{\phi}, A\psi)| \leq |B|_N |\tilde{\phi}| |\psi|$ for all $\bar{\phi} \varepsilon \tilde{\mathfrak{H}}$ and $\psi \varepsilon M$, knowing it to be true for $\bar{\phi} \varepsilon \tilde{M}$ and $\psi \varepsilon M$. However suppose there exists a positive real number $a$ such that every element $\psi_0$ in $\tilde{\mathfrak{H}}$ can be approximated arbitrarily well (in the weak topology) by elements $\psi \varepsilon M$ with $|\psi| \leq a |\psi_0|$. Then it follows immediately that $|(\bar{\phi}, A\psi)| \leq a |B|_N |\tilde{\phi}| |\psi|$ for all $\bar{\phi} \varepsilon \tilde{\mathfrak{H}}$ and $\psi \varepsilon M$ and therefore $|A|_M \leq a |B|_N$. Hence, in particular, $A$ is bounded if $B$ is bounded. On the other hand if we use the notation of [5(b), pp. 226-227], it is clear that if $n$ is sufficiently large $|A_{f_n}\psi| \leq a |\psi|$ where $a = 2 \sup_{x \varepsilon \omega} |\pi(x)|$ and $\omega$ is a compact set outside which all $f_n$ are zero. From this it follows that $|\tilde{A}_{f_n}| \leq a$ and therefore the subspace $\tilde{V}$ of $\tilde{\mathfrak{H}}$ does have the above additional property. The proof of Lemma 32 of [5(b)] now goes through without any further modification.

*for $n \varepsilon N_c^-$, $a \varepsilon \tilde{A}_c$, $u \varepsilon K$ and $x \varepsilon G$. Moreover if $a = \exp H \varepsilon \tilde{A}_c$ $(H \varepsilon \mathfrak{H})$ and $H = H_+ + H'$ $(H_+ \varepsilon \mathfrak{c}_+, H' \varepsilon \mathfrak{h} \cap \mathfrak{m}')$ then $\Gamma(a) = H_+$.*

Since $[\mathfrak{m}, \mathfrak{p}_-] \subset \mathfrak{p}_-$ and $[\mathfrak{m}, \mathfrak{p}_+] \subset \mathfrak{p}_+$, it follows that $M_c Q = Q M_c = Q$ and $m(vq) = vm(q)$, $m(qv) = m(q)v$ $(v \varepsilon M_c, q \varepsilon Q)$. Moreover $\mathfrak{n}_- \subset \mathfrak{p}_- + \mathfrak{k}'$ $\subset \mathfrak{p}_- + \mathfrak{m}'$. Therefore $N_c^- \subset P_c^- \gamma(\tilde{M}_c')$. Similarly if $u \varepsilon K$, $m(q\bar{u}) = m(q)\bar{u}$ $(q \varepsilon Q)$ and $m(\bar{u}x) = \bar{u}m(x)$ $(x \varepsilon G)$. The first part of the lemma follows immediately from these facts. Now put $a_+ = \exp H_+$ and $a' = \exp H'$. Then $a = a_+ a'$ and therefore $\Gamma(a) = \Gamma(a_+) + \Gamma(a')$. But since $a' \varepsilon \tilde{M}_c'$, $\Gamma(a') = 0$. On the other hand it is obvious that $\Gamma(a_+) = H_+$ and so the lemma is proved.

By a complex representation of $G_c$ we mean a finite-dimensional representation such that the corresponding representation of $\mathfrak{g}$ is linear over $C$. Since $G_c$ is simply connected we may identify the finite-dimensional representations of $\mathfrak{g}$ with the complex representations of $G_c$. If $V$ is the representation space of such a representation $\pi$, it would be convenient to regard $V$ as a Hilbert space in such a way that $\pi$ becomes unitary on the subgroup $U$ corresponding to $\mathfrak{u} = \mathfrak{k}_0 + (-1)^{\frac{1}{2}}\mathfrak{p}_0$. Since $U$ is compact, this is always possible. Whenever we speak of a finite-dimensional representation of $\mathfrak{g}$ (or of a complex representation of $G_c$) we shall tacitly assume that such a Hilbert space structure has already been introduced in the representation space.

Now let $\Lambda$ be the linear function of Section 5 and let $\alpha_1, \cdots, \alpha_l$ be a fundamental system of positive roots of $\mathfrak{g}$ (see Corollary 2 to Lemma 4 of $[5(\mathrm{f})]$). We assume that $\alpha_1, \cdots, \alpha_t$ are all the totally positive roots among these. Define a linear function $\Lambda_0$ on $\mathfrak{h}$ by the conditions $\Lambda_0(H_{\alpha_i}) = 0$ $1 \leq i \leq t$ and $\Lambda_0(H_{\alpha_i}) = \Lambda(H_{\alpha_i})$ $(t < i \leq l)$. Then $\Lambda_0$ is a dominant integral function and therefore from Theorem 1 of $[5(\mathrm{a})]$ there exists an irreducible representation $\sigma$ of $\mathfrak{g}$ on the finite-dimensional space $V$ with the highest weight $\Lambda_0$. Let $\phi_0$ be a unit vector in $V$ belonging to the weight $\Lambda_0$ and put $\lambda = \Lambda - \Lambda_0$.

LEMMA 14. *Let $\xi$ be the holomorphic character of $\tilde{A}_c$ corresponding to $\Lambda$. Then the function $\psi$ of Lemma 6 is given by the formula*

$$\psi(w) = (\phi_0, \sigma(v(w))\phi_0)e^{\lambda(\Gamma(w))} \qquad (w \varepsilon W).$$

First observe that since $\sigma$ is unitary on $U$, the adjoint of the operator $\sigma(X)$ is $-\sigma(\bar{\theta}(X))$ $(X \varepsilon \mathfrak{g})$. Hence $\sigma(\bar{\theta}(z^{-1}))$ is the adjoint of $\sigma(z)$ $(z \varepsilon G_c)$. But $\bar{\theta}(N_c^-) = N_c^+$ and since $\phi_0$ belongs to the highest weight, $\sigma(n')\phi_0 = \phi_0$ if $n' \varepsilon N_c^+$. Moreover from Lemma 13, $\Gamma(l_n w) = \Gamma(w)$ $(n \varepsilon N_c^-)$. Therefore if $\psi_0$ denotes the expression on the right hand side of the above equation, $\psi_0(l_n w) = \psi_0(w)$ $(n \, \varepsilon \, N_c^-)$. Now let $a = \exp H \, \varepsilon \, \tilde{A}_c$ $(H \, \varepsilon \, \mathfrak{h})$. Then if $\bar{w} = \nu(w)$ $(w \, \varepsilon \, W)$,

$$(\phi_0, \sigma(\nu(l_a w))\phi_0) = (\sigma(\bar{\theta}\,(\bar{a}^{-1}))\phi_0, \sigma(\bar{w})\phi_0) = e^{\Lambda_0(H)}(\phi_0, \sigma(\bar{w})\phi_0)$$

since $\Lambda_0$ is real. On the other hand it follows from Lemma 13 that $\lambda(\Gamma(l_a w)) = \lambda(H_+) + \lambda(\Gamma(w))$ where $H = H_+ + H'$ $(H_+ \, \varepsilon \, \mathfrak{c}_+, H' \, \varepsilon \, \mathfrak{h} \cap \mathfrak{m}')$. Therefore

$$\psi_0(l_a w) = \exp(\Lambda_0(H) + \lambda(H_+))\psi_0(w).$$

But $\Lambda(H) = \Lambda(H_+) + \Lambda(H')$ and it is obvious from Lemma 13 of $[5(\mathrm{f})]$ that $H'$ is a linear combination of $H_{\alpha_i}$ $(t < i \leqq l)$ so that $\lambda(H') = 0$. Hence

$$\Lambda_0(H) + \lambda(H_+) = \Lambda_0(H) + \lambda(H) = \Lambda(H).$$

This proves that $\psi_0(l_a w) = \xi(a)\psi_0(w)$ and so $\psi_0 \, \varepsilon \, \mathfrak{H}_\xi$. Moreover if $a \, \varepsilon \, \Lambda$,

$$\psi_0(r_a w) \doteq (\phi_0, \sigma(\nu(w)\bar{a})\phi_0) \exp\left(\lambda(\Gamma(w)) + \lambda(\Gamma(a))\right) = \psi_0(w)\xi(a)$$

since $\sigma(\bar{a})\phi_0 e^{\lambda(\Gamma(a))} = \xi(a)\phi_0$. Hence

$$\psi_0(w^h) = \psi_0(l_h r_{h^{-1}} w) = \xi(h)\xi(h^{-1})\psi_0(w) = \psi_0(w) \qquad\qquad (h \, \varepsilon \, \Lambda)$$

and therefore $\psi_0(w) = \displaystyle\int_{\Lambda_0} \psi_0(w^h)d\bar{h} = \psi_0(1)\psi(w)$ from Lemma 6. But $\psi_0(1) = |\phi_0|^2 = 1$ and so $\psi_0 = \psi$.

We have still to show that the space $\mathfrak{H}_\Lambda(\mu) \neq \{0\}$ for a suitable choice of $\mu$. In order to do this we need some preliminary results on finite-dimensional representations of $\mathfrak{g}$ which we shall then apply to the above representation $\sigma$.

**7. Some results on finite-dimensional representations.** First we prove the following lemma.

LEMMA 15. *There exists an element $H \, \varepsilon \, \mathfrak{h}_0$ such that $\theta(X) = \exp(adH)X$ for all $X \, \varepsilon \, \mathfrak{g}$.*

Since $\theta(H) = H$ $(H \, \varepsilon \, \mathfrak{h})$ and $\theta^2(X) = X$ $(X \, \varepsilon \, \mathfrak{g})$, it is clear that $\theta(X_\alpha) = \pm X_\alpha$ for every root $\alpha$. If $(\alpha_1, \cdots, \alpha_l)$ is a fundamental system of roots we can choose $H \, \varepsilon \, \mathfrak{h}$ such that $\theta(X_{\alpha_i}) = e^{\alpha_i(H)}X_{\alpha_i}$ $1 \leqq i \leqq l$. It is obvious that $\alpha_1(H), \cdots, \alpha_l(H)$ are all purely imaginary and therefore $H \, \varepsilon \, \mathfrak{h}_0$. Moreover since $[X_{\alpha_i}, X_{-\alpha_i}] \, \varepsilon \, \mathfrak{h}$, it follows that $\theta(X_{-\alpha_i}) = e^{-\alpha_i(H)}X_{\alpha_i}$. Therefore $\theta$ and $\exp(adH)$ are two automorphisms of $\mathfrak{g}$ which coincide on $\mathfrak{h}$ and also at $X_{\alpha_i}$, $X_{-\alpha_i}$ $1 \leqq i \leqq l$. But $\mathfrak{g}$ is the smallest subalgebra of itself containing $\mathfrak{h}$ and $X_{\alpha_i}$, $X_{-\alpha_i}$ $1 \leqq i \leqq l$ (see Lemmas 18 and 19 of $[5(\mathrm{a})]$) and so $\theta = \exp(adH)$.

LEMMA 16. *Let $\pi$ be a representation of $\mathfrak{g}$ on a finite-dimensional space $V$. Then if $\phi$ is a vector belonging to some weight of $\pi$*

$$(\phi, \pi(z)\phi) = (\phi, \pi(\theta(z))\phi) \qquad (z \,\varepsilon\, G_c).$$

From Lemma 15 we can choose $h \,\varepsilon\, A_0$ such that $\theta(z) = hzh^{-1}$ for all $z \,\varepsilon\, G_c$. Then $(\phi, \pi(\theta(z))\phi) = (\pi(h^{-1})\phi, \pi(zh^{-1})\phi)$ since $\pi$ is unitary on $A_0$. But since $\phi$ belongs to a weight of $\pi$, $\pi(h^{-1})\phi = c\phi$ where $c$ is a unimodular complex number. Hence

$$(\phi, \pi(\theta(z))\phi) = |c|^2 (\phi, \pi(z)\phi) = (\phi, \pi(z)\phi).$$

Now $\phi$ being as above, we propose to study the growth of the function $F(X) = (\phi, \pi(\exp X)\phi)$ $(X \,\varepsilon\, \mathfrak{p}_0)$ at infinity. Let $X$ be a fixed element in $\mathfrak{p}_0$. Since $\theta(X) = -X$, $\pi(X)$ is a self-adjoint operator on $V$ and therefore we can choose an orthonormal base $(\phi_1, \cdots, \phi_d)$ for $V$ such that $\pi(X)\phi_i = \lambda_i \phi_i$ $(\lambda_i \,\varepsilon\, R,\ i = 1, \cdots, d)$. Then if $\phi = \sum_i a_i \phi_i$ $(a_i \,\varepsilon\, C)$,

$$F(X) = \sum_i |a_i|^2 e^{\lambda_i}.$$

Similarly $F(-X) = \sum_i |a_i|^2 e^{-\lambda_i}$. But since $\theta(X) = -X$, it follows from Lemma 16 that $F(X) = F(-X)$ and therefore $F(X) = \sum_i |a_i|^2 \cosh \lambda_i$. Now consider the function

$$c(t) = (\cosh t - 1)/t^2 = \sum_{n=1}^{\infty} t^{2n-2}/(2n)! \qquad (t \,\varepsilon\, R).$$

Then $c(t) = c(-t) \geqq 0$ and $c(t)$ increases with $t$ for $t \geqq 0$. Also

$$F(X) = \sum_i |a_i|^2 + \sum_i |a_i\lambda_i|^2 c(\lambda_i) = 1 + \sum_i |a_i\lambda_i|^2 c(|\lambda_i|)$$

if we assume that $|\phi| = 1$. On the other hand $|\pi(X)\phi|^2 = \sum_i |a_i\lambda_i|^2$ and therefore $d^{-\frac{1}{2}} |\pi(X)\phi| \leqq \max_i |a_i\lambda_i|$. Let $j$ be an index such that $|a_j\lambda_j| = \max_i |a_i\lambda_i|$. Then

$$F(X) \geqq 1 + |a_j\lambda_j|^2 c(|\lambda_j|) \geqq 1 + |a_j\lambda_j|^2 c(|a_j\lambda_j|)$$
$$\geqq 1 + d^{-1} |\pi(X)\phi|^2 c(d^{-\frac{1}{2}} |\pi(X)\phi|) = \cosh(d^{-\frac{1}{2}} |\pi(X)\phi|).$$

Therefore we have the following result.

LEMMA 17. *Let $\pi$ and $\phi$ be as in Lemma 16. Then if $|\phi| = 1$ and $X \,\varepsilon\, \mathfrak{p}_0$,*

$$(\phi, \pi(\exp X)\phi) \geqq \cosh(d^{-\frac{1}{2}} |\pi(X)\phi|)$$

*where $d = \dim V$.*

On the other hand we have the following result concerning $|\pi(X)\phi|$.

LEMMA 18. *Suppose $\phi$ belongs to the highest weight $\lambda$ of $\pi$. Then if $\lambda(H_\beta) \neq 0$ for every noncompact root $\beta$, $|\pi(X)\phi|^2$ $(X \,\varepsilon\, \mathfrak{p}_0)$ is a positive definite quadratic form on $\mathfrak{p}_0$.*

Let $X$ be an element in $\mathfrak{p}_0$. Then if we choose $X_\beta$, $X_{-\beta}$ as in $[5(\mathrm{f}), \S 4]$, we can write $X = \sum_\beta (c_\beta X_\beta + \bar{c}_\beta X_{-\beta})$ where $\beta$ runs over all noncompact positive roots and the bar denotes complex conjugate. Since $\phi$ belongs to the highest weight $\pi(X_\beta)\phi = 0$. Hence $\pi(X)\phi = \sum_\beta \bar{c}_\beta \pi(X_{-\beta})\phi$. On the other hand if $\lambda(H_\beta) \neq 0$ we know (Lemma 1 of $[5(\mathrm{f})]$) that $\pi(X_{-\beta})\phi \neq 0$ and it belongs to the weight $\lambda - \beta$. Since nonzero vectors belonging to distinct weights are linearly independent, we conclude that $\pi(X)\phi \neq 0$ unless $\bar{c}_\beta = 0$ for all $\beta$. This proves our assertion.

In Section 2 we have defined a (real) analytic mapping $z \to H(z)$ of $G_0$ into $(-1)^{\frac{1}{2}}\mathfrak{h}_0$ such that $z \,\varepsilon\, U(\exp H(z))N_c{}^+$ $(z \,\varepsilon\, G_0)$. We shall say that a linear function $\lambda$ on $\mathfrak{h}$ is *completely positive* if $\lambda(H_\alpha)$ is real and nonnegative for every positive root $\alpha$.

LEMMA 19. *If $\lambda$ is a completely positive linear function on $\mathfrak{h}$ then $\lambda(H(x)) \geqq 0$ for all $x \,\varepsilon\, G$. Moreover for any $X \,\varepsilon\, \mathfrak{p}_0$ the function $\lambda(H(\exp tX))$ $(t \,\varepsilon\, R)$ is non-decreasing for $t \geqq 0$. Finally $H(p) = H(p^{-1})$ if $p \,\varepsilon\, \exp \mathfrak{p}_0$.*

Let $\alpha_1, \cdots, \alpha_l$ be a fundamental system of positive roots. Define real linear functions $\Lambda_1, \cdots, \Lambda_l$ as follows: $\Lambda_i(H_{\alpha_j}) = \delta_{ij}$ $1 \leqq i, j \leqq l$ where $\delta_{ij} = 1$ or $0$ according as $i = j$ or not. Then $\lambda = \lambda_1 \Lambda_1 + \cdots + \lambda_l \Lambda_l$ where $\lambda_i = \Lambda(H_{\alpha_i})$ are nonnegative real numbers. Since $\Lambda_i$ are dominant integral functions (see $[5(\mathrm{a})$, p. 30$]$), it is clearly sufficient to prove the lemma under the assumption that $\lambda$ is such a function. Let $\pi$ be a finite-dimensional irreducible representation of $\mathfrak{g}$ on a space $V$ with the highest weight $\lambda$ $[5(\mathrm{a})$, Theorem 1$]$ and let $\phi$ be a unit vector in $V$ belonging to the weight $\lambda$. Then if $x = uhn$ $(x \,\varepsilon\, G_0, u \,\varepsilon\, U, h \,\varepsilon\, A_+, n \,\varepsilon\, N_c{}^+)$ it is clear that

$$|\pi(x)\phi| = |\pi(uh)\phi| = |\pi(h)\phi| = e^{\lambda(H(x))}.$$

On the other hand $x = vp$ $(v \,\varepsilon\, K_0, p \,\varepsilon\, \exp \mathfrak{p}_0)$ and therefore

$$|\pi(x)\phi|^2 = |\pi(p)\phi|^2 = (\phi, \pi(p^2)\phi),$$

since $\pi(p)$ is self-adjoint. But it follows from Lemma 17 that $(\phi, \pi(p^2)\phi) \geqq 1$ and therefore $e^{\lambda(H(x))} = |\pi(x)\phi| \geqq 1$. This proves that $\lambda(H(x)) \geqq 0$. Moreover if $X \,\varepsilon\, \mathfrak{p}$, $\pi(X)$ is self-adjoint and therefore

$$\exp 2\lambda(H(\exp tX)) = |\pi(\exp tX)\phi|^2$$
$$= (\phi, \pi(\exp 2tX)\phi) = (\phi, \pi(\exp(-2tX)\phi) \qquad (t \, \varepsilon \, R)$$

from Lemma 16. Hence

$$\exp 2\lambda(H(\exp tX)) = \tfrac{1}{2}(\phi, \pi(\exp 2tX)\phi) + \tfrac{1}{2}(\phi, \pi(\exp(-2tX)\phi)$$
$$= \sum_{n \geq 0} (2t)^{2n} |(\pi(X))^n\phi|^2/2n!.$$

Since $\lambda(H(\exp tX))$ is real and since only even powers of $t$ occur and all the coefficients of the series are nonnegative, the second assertion of the lemma is now obvious. Also it is clear that $\lambda(H(p)) = \lambda(H(p^{-1}))$ for $p \, \varepsilon \, \exp \mathfrak{p}_0$. This is true in particular for $\lambda = \Lambda_i \; i = 1, \cdots, l$. Since $\Lambda_1, \cdots, \Lambda_l$ is a base for the space of all linear functions on $\mathfrak{h}$, we conclude that $H(p) = H(p^{-1})$.

Let $B(X, Y) = sp(adX adY) \; (X, Y \, \varepsilon \, \mathfrak{g})$. Then $-B(\theta(X), X)$ is a positive definite Hermitian form on $\mathfrak{g}$ which we denote by $\|X\|^2$. We now regard $\mathfrak{g}$ as a Hilbert space under the corresponding norm $\|\cdot\|$. By combining Lemmas 17, 18 and 19 we can get the following stronger result.

LEMMA 20. *Suppose $\lambda$ is completely positive and $\lambda(H_\beta) \neq 0$ for every noncompact root $\beta$. Then there exists a positive real number $c$ such that $\lambda(H(\exp X)) \geq c\|X\|$ for all $X \, \varepsilon \, \mathfrak{p}_0$ lying outside some bounded set.*

First suppose $\lambda(H_{\alpha_i}) \; 1 \leq i \leq l$ are all integers so that $\lambda$ is a dominant integral function. We define $\pi$, $V$ and $\phi$ as in the proof of Lemma 19. Then from Lemma 17,

$$(\phi, \pi(\exp 2X)\phi) \geq \cosh(2d^{-\frac{1}{2}}|\pi(X)\phi|)$$

where $d = \dim V$. But we have seen above that if $p = \exp X$,

$$(\phi, \pi(\exp 2X)\phi) = |\pi(p)\phi|^2 = e^{2\lambda(H(p))}.$$

Hence

$$\lambda(H(p)) \geq \tfrac{1}{2}\log(\cosh(2d^{-\frac{1}{2}}|\pi(X)\phi|)) \geq d^{-\frac{1}{2}}|\pi(X)\phi| - \tfrac{1}{2}\log 2.$$

On the other hand we know from Lemma 18 that $|\pi(X)\phi|^2 \; (X \, \varepsilon \, \mathfrak{p}_0)$ is a positive definite quadratic form on $\mathfrak{p}_0$. Hence if $c_0$ is the least possible value of $|\pi(X)\phi|$ for all $X \, \varepsilon \, \mathfrak{p}_0$ with $\|X\| = 1$, $c_0 > 0$ and $|\pi(X)\phi| \geq c_0\|X\|$ $(X \, \varepsilon \, \mathfrak{p}_0)$. Therefore

$$\lambda(H(\exp X)) \geq c_0 d^{-\frac{1}{2}}\|X\| - \tfrac{1}{2}\log 2 \geq \tfrac{1}{2}c_0 d^{-\frac{1}{2}}\|X\|$$

provided $\|X\| \geq c_0^{-1}d^{\frac{1}{2}}\log 2$ $(X \, \varepsilon \, \mathfrak{p}_0)$. Hence we may take $c = \tfrac{1}{2}c_0 d^{-\frac{1}{2}}$.

Now we come to the general case. Define $\Lambda_1, \cdots, \Lambda_l$ as in the proof of Lemma 19. Then $\lambda = \lambda_1 \Lambda_1 + \cdots + \lambda_l \Lambda_l$ where $\lambda_i = \lambda(H_{\alpha_i})$ are nonnegative real numbers. Choose a positive number $r$ so large that if $\lambda_i \neq 0$, $r\lambda_i \geq 1$ $(1 \leq i \leq l)$ and put $\lambda_0 = a_1 \Lambda_1 + \cdots + a_l \Lambda_l$ where $a_i = 0$ or $1$ according as $\lambda_i = 0$ or not. Then $r\lambda - \lambda_0$ is completely positive and $\lambda_0$ is a dominant integral function. Let $\alpha$ be a positive root. Then $H_\alpha = b_1 H_{\alpha_1} + \cdots + b_l H_{\alpha_l}$ $(b_i \varepsilon R, b_i \geq 0)$. Suppose $b_i > 0$ $(1 \leq i \leq m)$ and $b_i = 0$ $(m < i \leq l)$. Then $\lambda(H_\alpha) = b_1 \lambda_1 + \cdots + b_m \lambda_m$ and $\lambda_0(H_\alpha) = b_1 a_1 + \cdots + b_m a_m$. Since $\lambda_i, a_i$ are all nonnegative and since $\lambda_i = 0$ if and only if $a_i = 0$, it is clear that $\lambda(H_\alpha) = 0$ if and only if $\lambda_0(H_\alpha) = 0$. Therefore $\lambda_0(H_\beta) \neq 0$ for any noncompact positive root $\beta$ and so, by the above proof, there exists a positive number $c'$ such that $\lambda_0(H(\exp X)) \geq c' \| X \|$ $(X \varepsilon \mathfrak{p}_0)$ provided that $\| X \|$ is sufficiently large. However since $r\lambda - \lambda_0$ is completely positive

$$r\lambda(H(\exp X)) \geq \lambda_0(H(\exp X))$$

(Lemma 19) and therefore

$$\lambda(H(\exp X)) \geq c' \| X \|/r \qquad\qquad (X \varepsilon \mathfrak{p}_0)$$

if $\| X \|$ is sufficiently large.

On the other hand it is quite easy to obtain a result in the opposite direction.

LEMMA 21. *Let $\lambda$ be a linear function on $\mathfrak{h}$. Then there exists a real number $c'$ such that $|\lambda(H(\exp X))| \leq c' \| X \|$ for all $X \varepsilon \mathfrak{p}_0$.*

Put $\lambda_i = \Lambda(H_{\alpha_i})$ $i = 1, 2, \cdots, l$. Then $\lambda = \lambda_1 \Lambda_1 + \cdots + \lambda_l \Lambda_l$ and $|\lambda(H)| \leq \sum_i |\lambda_i| |\Lambda_i(H)|$ $(H \varepsilon \mathfrak{h})$. Therefore it is obviously sufficient to prove the lemma under the assumption that $\lambda$ is a dominant integral function. Define $\pi$, $V$ and $\phi$ as above. Then if $p = \exp X$ $(X \varepsilon \mathfrak{p}_0)$,

$$e^{\lambda(H(p))} = |\pi(p)\phi| \leq |\pi(p)| \leq e^{|\pi(X)|}$$

where $|T|$ denotes, as usual, the bound of an operator $T$. Therefore from Lemma 19,

$$|\lambda(H(p))| = \lambda(H(p)) \leq |\pi(X)| \leq \{sp(\pi(X))^2\}^{\frac{1}{2}}$$

since $\pi(X)$ is self-adjoint. If $a'$ is the maximum value of $sp(\pi(X))^2$ for all $X \varepsilon \mathfrak{p}_0$ with $\| X \| = 1$, it is clear that $sp(\pi(X))^2 \leq a' \| X \|^2$ for every $X$ in $\mathfrak{p}_0$ and therefore $|\lambda(H(\exp X))| \leq c' \| X \|$ where $c' = (a')^{\frac{1}{2}}$.

We shall also need another similar result. Let $\rho$ be defined as in Lemma 3.

LEMMA 22. *Let $\pi$ be a representation of $\mathfrak{g}$ on a finite-dimensional space $V$. Then there exist positive real numbers $a$, $b$ such that*

$$|\pi(x^{-1})\phi| \leq be^{a\rho(H(x))}|\phi|$$

*for every $\phi \varepsilon V$ and $x \varepsilon G_0$.*

If $x = up$ $(u \varepsilon K_0, p \varepsilon \exp \mathfrak{p}_0)$, $\pi(x^{-1})\phi = \pi(p^{-1})\phi'$ where $\phi' = \pi(u)\phi$. Also $H(x) = H(p)$ and $|\phi'| = |\phi|$. So it is enough to consider the case when $u = 1$. Let $p = \exp X$ $(X \varepsilon \mathfrak{p}_0)$. Then

$$|\pi(p^{-1})\phi| \leq |\pi(p^{-1})||\phi| \leq e^{|\pi(X)|}|\phi|.$$

But as we have just seen above, there exists a real number $a_1$ such that $|\pi(X)| \leq a_1 \|X\|$ for all $X \varepsilon \mathfrak{p}_0$. On the other hand from Lemma 20 we can find positive real numbers $a_2$ and $M$ such that [10]

$$\rho(H(\exp X)) \geq a_2 \|X\|$$

for all $X \varepsilon \mathfrak{p}_0$ for which $\|X\| \geq M$. Let $b$ be an upper bound for $|\pi(\exp(-X))|$ when $\|X\| \leq M$ $(X \varepsilon \mathfrak{p}_0)$. Then if $a = a_1/a_2$ it is obvious that $|\pi(x^{-1})\phi| \leq be^{a\rho(H(x))}$ for all $\phi \varepsilon V$ and $x \varepsilon G$.

The following lemma describes some properties of the mapping $x \to \Gamma(x)$ $(x \varepsilon G)$.

LEMMA 23. *Let $\lambda$ be a real linear function on $\mathfrak{h}$. Then if $u \varepsilon K$, the real part of $\lambda(\Gamma(u))$ is zero. Moreover if $\lambda(H_\alpha) = 0$ for every positive root $\alpha$ which is not totally positive,*

$$\lambda(\Gamma(\exp X)) = 2\lambda(H(\exp \tfrac{1}{2}X)) \qquad (X \varepsilon \mathfrak{p}_0).$$

We know that $\mathfrak{h}$ is the direct sum of $\mathfrak{c}_+$ and $\mathfrak{h} \cap \mathfrak{m}'$ (in the notation of Section 6) and if $\alpha$ is a positive root which is not totally positive, $H_\alpha \varepsilon \mathfrak{h} \cap \mathfrak{m}'$. Without affecting its values on $\mathfrak{c}_+$ we can replace $\lambda$ by a linear function which vanishes identically on $\mathfrak{h} \cap \mathfrak{m}'$ and therefore assume that $\lambda(H_\alpha) = 0$ for all positive roots $\alpha$ which are not totally positive. Then $\lambda = \lambda_1 \Lambda_1 + \cdots + \lambda_t \Lambda_t$ (in the notation of the proof of Lemma 14) where $\lambda_1, \cdots, \lambda_t$ are real numbers. Obviously it would be enough to prove the lemma for $\lambda = \Lambda_i$ $i = 1, \cdots, t$. Hence we may assume that $\lambda$ is a dominant integral function. Let $\pi$ be an irreducible representation of $\mathfrak{g}$ on a finite-dimensional space $V$ with the highest weight $\lambda$ and let $\phi$ be a unit vector in $V$ belonging to the weight $\lambda$. Then if $\beta$ is a positive root which is not totally positive, $\lambda(H_\beta) = 0$ and therefore from Lemma 1 of [5(f)] $\pi(X_{-\beta})\phi = 0$. Hence it is clear that

---

[10] Here we have to make use of the fact (see Weyl [9]) that $\rho(H_\alpha) > 0$ for every positive root $\alpha$.

$\pi(\mathfrak{m}' + \mathfrak{p}_+)\phi = \{0\}$. On the other hand we have seen in Section 6 that $\bar{x} = qm(\bar{x})p$ $(x \,\varepsilon\, G)$ where $q \,\varepsilon\, P_c^-$, $p \,\varepsilon\, P_c^+$. Therefore

$$(\phi, \pi(\bar{x})\phi) = (\pi(\theta(q^{-1}))\phi, \pi(m(\bar{x}))\phi) = (\phi, \pi(m(\bar{x})\phi)$$

since $\bar{\theta}(q^{-1}) \,\varepsilon\, P_c^+$. Moreover since $\pi(\mathfrak{m}')\phi = \{0\}$, it is obvious that

$$\pi(m(\bar{x}))\phi = e^{\lambda(\Gamma(x))}\phi$$

and therefore $(\phi, \pi(\bar{x})\phi) = e^{\lambda(\Gamma(x))}$ $(x \,\varepsilon\, G)$. Now if $u \,\varepsilon\, K$, $\bar{u} \,\varepsilon\, M_c$ and therefore $\pi(\bar{u})\phi = e^{\lambda(\Gamma(u))}\phi$. But since $\pi(\bar{u})$ is unitary, $\lambda(\Gamma(u))$ must be purely imaginary. On the other hand if $x = p^2$ where $p = \exp\frac{1}{2}X$ $(X \,\varepsilon\, \mathfrak{p}_0)$, the above equation gives $|\pi(\bar{p})\phi|^2 = e^{\lambda(\Gamma(p^2))}$. But we have seen that $|\pi(\bar{p})\phi| = e^{\lambda(H(\bar{p}))}$. Hence

$$\lambda(\Gamma(\exp X)) - 2\lambda(H(\exp\tfrac{1}{2}X))$$

must be an integral multiple of $2\pi(-1)^{\frac{1}{2}}$. However it is a continuous function of $X$ and it is zero at $X = 0$. Therefore since $\mathfrak{p}_0$ is connected it must be everywhere zero on $\mathfrak{p}_0$. This proves the lemma.

Corollary. *If $u \,\varepsilon\, K$, $\Gamma(u)$ lies in $\mathfrak{h}_0$. Moreover*

$$\Gamma(\exp X) - \tfrac{1}{2}H(\exp\tfrac{1}{2}X) \,\varepsilon\, \mathfrak{h} \cap \mathfrak{m}'$$

*for all $X \,\varepsilon\, \mathfrak{p}_0$.*

This is an immediate consequence of the above lemma.

**8. Proof of the existence of representations.** In order to show that the space $\mathfrak{H}_\Lambda(\mu)$ of Lemma 9 is not zero for some $\mu$, it is enough to find a function $\mu$ on $G_0$, satisfying the conditions of Section 6 and such that

$$\int_{G_0} |\psi(x)|^2 \mu(\bar{x}) d\bar{x} < \infty$$

where $\psi$ is the function of Lemma 14. But $|\psi(x^{-1})| \leq |\sigma(\bar{x}^{-1})\phi_0| e^{\Re(\lambda(\Gamma(x^{-1})))}$, (in the notation of Lemma 14) where $\Re c$ denotes the real part of a complex number $c$. On the other hand we know from Lemma 29 that

$$|\sigma(\bar{x}^{-1})\phi_0| \leq b e^{a\rho(H(\bar{x}))} \qquad\qquad (x \,\varepsilon\, G)$$

for suitable real numbers $a$ and $b$. If $x = up$ $(u \,\varepsilon\, K, p = \exp X, X \,\varepsilon\, \mathfrak{p}_0)$ $\Gamma(x^{-1}) = \Gamma(p^{-1}) + \Gamma(u^{-1})$ (Lemma 13) and therefore

$$\Re(\lambda(\Gamma(x^{-1}))) = \lambda(\Gamma(p^{-1})) = 2\lambda(H(\exp(-\tfrac{1}{2}X))) = 2\lambda(H(\exp\tfrac{1}{2}X))$$

from Lemmas 23 and 19. It is obvious [10] that $r'\rho - 2\lambda$ is completely positive for a suitable real $r' \geq 0$. Then

$$2\lambda(H(\exp\tfrac{1}{2}X)) \leq r'\rho(H(\exp\tfrac{1}{2}X)) \leq r'\rho(H(\exp X)) = r'\rho(H(\bar{x}))$$

from Lemma 19. Hence $|\psi(x^{-1})| \leqq b e^{(a+r')\rho(H(\bar{x}))}$ $(x \varepsilon G)$. Now put $\mu(\bar{x}) = e^{-r\rho(H(\bar{x}^{-1}))}$ $(\bar{x} \varepsilon G_0)$ where $r \geqq 2a + 2r' + 4$. Then $\mu$ is a continuous function on $G_0$ which is everywhere positive and

$$\int_{G_0} |\psi(x)|^2 \mu(\bar{x}) d\bar{x} = \int_{G_0} |\psi(x^{-1})|^2 \mu(\bar{x}^{-1}) d\bar{x} \leqq b^2 \int_{G_0} e^{-4\rho(H(\bar{x}))} d\bar{x} < \infty$$

from Lemma 3. Moreover since $\rho$ is a dominant integral function (see Weyl [9]), there exists a finite-dimensional irreducible representation $\tau$ of $\mathfrak{g}$ with the highest weight $\rho$. Let $\zeta$ be a unit vector in the representation space belonging to the weight $\rho$. Then $|\tau(\bar{x}^{-1})\zeta| = e^{\rho(H(\bar{x}^{-1}))}$ $(\bar{x} \varepsilon G_0)$ and therefore, if $\bar{x}, \bar{y} \varepsilon G_0$,

$$e^{\rho(H(\bar{y}^{-1}\bar{x}^{-1}))} = |\tau(\bar{y}^{-1}\bar{x}^{-1})\zeta| \leqq |\tau(\bar{y}^{-1})| e^{\rho(H(\bar{x}^{-1}))}.$$

This proves that $\mu(\bar{x}\bar{y}) \leqq |\tau(\bar{y})|^r \mu(\bar{x})$ and since $|\tau(\bar{y})|$ is bounded on every compact set in $G_0$, $\mu$ fulfills all the conditions of Section 6. Hence $\psi \varepsilon \mathfrak{H}_\Lambda(\mu)$ and so $\mathfrak{H}_\Lambda(\mu) \neq \{0\}$.

We can now summarise our results in the following theorem.

THEOREM 2. *Let $\Lambda$ be a real linear function on $\mathfrak{h}$ such that $\Lambda(H_\alpha)$ is a nonnegative integer for every positive root $\alpha$ which is not totally positive. Let $\xi$ denote the character of $\bar{A}_c$ defined by $\xi(\exp H) = e^{\Lambda(H)}$ $(H \varepsilon \mathfrak{h})$ and let $\mu$ be a measurable function on $G_0$ which is everywhere positive and such that $\sup_{\bar{x} \varepsilon G_0} \mu(\bar{x}\bar{y})/\mu(\bar{x})$ $(\bar{y} \varepsilon G_0)$ remains bounded on every compact subset of $G_0$. Let $\mathfrak{H}_\Lambda(\mu)$ denote the Hilbert space of all holomorphic functions $f$ on $W$ which fulfill the following two conditions:*

$$(1) \quad f(l_{na}w) = \xi(a)f(w) \quad (a \varepsilon \bar{A}_c, n \varepsilon N_c^-, w \varepsilon W)$$

$$(2) \quad \|f\|^2 = \int_{G_0} |f(x)|^2 \mu(\bar{x}) d\bar{x} < \infty.$$

*Then if $\mathfrak{H}_\Lambda(\mu) \neq \{0\}$ there exists a unique function $\psi$ in $\mathfrak{H}_\Lambda(\mu)$ such that $\psi(1) = 1$ and $\psi(w^h) = \psi(w)$ for all $h \varepsilon \Lambda_0$ and $w \varepsilon W$. Put $\psi_x(w) = \psi(r_x w)$ $(x \varepsilon G, w \varepsilon W)$ and let $\mathfrak{H}$ denote the smallest closed subspace of $\mathfrak{H}_\Lambda(\mu)$ containing $\psi_x$ for all $x \varepsilon G$. Then we can define a quasi-simple irreducible representation $\pi$ of $G$ on $\mathfrak{H}$ by the rule*

$$\pi(x)\phi(w) = \phi(r_x w) \qquad\qquad (x \varepsilon G, \phi \varepsilon \mathfrak{H}, w \varepsilon W).$$

*This representation is infinitesimally equivalent to the representation $\pi_\Lambda$ of $[5(f), \S 6]$. Finally, it is always possible to choose $\mu$ in such a way that $\mathfrak{H}_\Lambda(\mu) \neq \{0\}$.*

The infinitesimal equivalence follows immediately from Lemmas 8 and 10, Theorem 2 of [5(f)] and Theorem 5 of [5(b)].

COROLLARY. *Suppose $\mu = 1$ and $\mathfrak{H}_\Lambda(\mu) \neq \{0\}$. Then $\mathfrak{H} = \mathfrak{H}_\Lambda(\mu)$, $\pi$ is unitary and*

$$(\psi, \pi(x)\psi) = \psi(x) \| \psi \|^2 \qquad\qquad (x \varepsilon G).$$

It is obvious that if $f \varepsilon \mathfrak{H}_\Lambda(\mu)$,

$$\int_{G_0} | f(xy) |^2 \, d\bar{x} = \| f \|^2 \qquad\qquad (y \varepsilon G),$$

and therefore from Lemma 12, $\mathfrak{H} = \mathfrak{H}_\Lambda(\mu)$ and $\pi$ is unitary. Moreover

$$\int_{A_0} \psi(y^h x) \, d\bar{h} = \psi(x)\psi(y) \qquad\qquad (x, y \varepsilon G)$$

from Lemma 6. Hence [11]

$$\psi(x) \| \psi \|^2 = \int_{G_0} \{ (\mathrm{conj}\, \psi(y)) \int_{A_0} \psi(y^h x) \, d\bar{h} \} d\bar{y}$$

$$= \int_{A_0} d\bar{h} \int_{G_0} (\mathrm{conj}\, \psi(y))\psi(yx) \, d\bar{y} = (\psi, \pi(x)\psi)$$

by Fubini's Theorem.


**9. Unitary representations.** Our next object is to look for unitary representations among those constructed above. However the result which we give below is not quite as strong as Theorem 3 of [5(f)] which was stated there without proof. We have to postpone its proof to another paper since it requires a deeper study of the representations.

THEOREM 3. *Let $(\alpha_1, \cdots, \alpha_l)$ be a fundamental system of positive roots and suppose $(\alpha_1, \cdots, \alpha_t)$ are all the totally positive roots in this system. Let $\lambda_i$ $(t < i \leqq l)$ be given nonnegative integers such that $\lambda_i = 0$ if $\alpha_i$ is not a root of[3] $\mathfrak{g}_+$. Then we can find a real number $c$ with the following property. If $\Lambda$ is a real linear function on $\mathfrak{h}$ and $\Lambda(H_{\alpha_i}) \leqq c$ $(1 \leqq i \leqq t)$, $\Lambda(H_{\alpha_i}) = \lambda_i$ $(t < i \leqq l)$, then[3] $\pi_\Lambda$ is infinitesimally unitary.*

Let $\Lambda$ be a real linear function such that $\Lambda(H_{\alpha_i}) = \lambda_i$ $(t < i \leqq l)$. Then $\Lambda(H_\alpha) = 0$ for every root $\alpha$ of $\mathfrak{g}'$ (see [5(f), §5]) and therefore $\pi_\Lambda(X) = 0$ if $X \varepsilon \mathfrak{g}'$ (Lemma 19 of [5(f)]). Since $\mathfrak{g}_0 = \mathfrak{g}_+ \cap \mathfrak{g}_0 + \mathfrak{g}' \cap \mathfrak{g}_0$ we can now restrict our attention to $\mathfrak{g}_+ \cap \mathfrak{g}_0$. Hence without any essential loss of generality, we may asume that $\mathfrak{g}' = \{0\}$ and so $\mathfrak{g} = \mathfrak{g}_+$. Then it follows

---

[11] conj $c$ denotes the complex conjugate of a number $c \varepsilon C$.

from Lemma 13 of $[5(f)]$ that $\alpha_i$ $(t < i \leq l)$ are all compact. Define a linear function $\lambda$ on $\mathfrak{h}$ by the conditions $\lambda(H_{\alpha_i}) = \Lambda(H_{\alpha_i})$ $1 \leq i \leq t$ and $\lambda(H_{\alpha_i}) = 0$ $t < i \leq l$. Then $\lambda$ is real and $\Lambda_0 = \Lambda - \lambda$ is a dominant integral function on $\mathfrak{h}$ and

$$\psi(x) = (\phi_0, \sigma(\bar{x})\phi_0)\, e^{\lambda(\Gamma(x))} \qquad (x \varepsilon G)$$

in the notation of Lemma 14. Now consider the integral

$$\int_{G_0} |\psi(x)|^2 \, d\bar{x}.$$

We know that

$$|\psi(x)| \leq |\sigma(\bar{x})\phi_0|\, e^{\Re(\lambda(\Gamma(x)))} = \exp(\Lambda_0(H(\bar{x})) + \Re(\lambda(\Gamma(x)))).$$

But if $x = u \exp X$ $(u \varepsilon K, X \varepsilon \mathfrak{p}_0)$

$$\Re(\lambda(\Gamma(x))) = \lambda(\Gamma(\exp X)) = 2\lambda(H(\exp \tfrac{1}{2}X))$$

from Lemmas 13 and 23. Now define $\Lambda_i$ $1 \leq i \leq l$ as in the proof of Lemma 19. Then $\lambda_0 = \Lambda_1 + \cdots + \Lambda_t$ is a completely positive linear function and it follows from Lemma 13 of $[5(f)]$ that $\lambda_0(H_\beta) = 1$ for every totally positive root. Since $\mathfrak{g} = \mathfrak{g}_+$, every noncompact positive root is totally positive and therefore Lemma 20 is applicable to $\lambda_0$. Hence there exists a positive number $a$ such that

$$\lambda_0(H(\exp X)) \geq a\,\|X\|,$$

for all $X \varepsilon \mathfrak{p}_0$ lying outside some bounded set. On the other hand from Lemma 21 we can find a real constant $b \geq 0$ such that

$$\Lambda_0(H(\exp X)) + 2\rho(H(\exp X)) \leq b\,\|X\| \qquad (X \varepsilon \mathfrak{p}_0).$$

Hence $\Lambda_0(H(\exp X)) + 2\rho(H(\exp X)) \leq (2b/a)\lambda_0(H(\exp \tfrac{1}{2}X))$, for all $X \varepsilon \mathfrak{p}_0$ with a sufficiently large value of $\|X\|$. We note that $b$ depends only on $\Lambda_0$ and therefore only on the integers $\lambda_i$ $(t < i \leq l)$. Now put $c = -b/a$. Then if $\Lambda(H_{\alpha_i}) \leq c$ $1 \leq i \leq t$, $-\lambda + c\lambda_0$ is completely positive and therefore from Lemma 19

$$-2\lambda(H(\exp \tfrac{1}{2}X)) \geq -2c\lambda_0(H(\exp \tfrac{1}{2}X)) \geq \Lambda_0(H(\exp X)) + 2\rho(H(\exp X)),$$

if $\|X\|$ is sufficiently large $(X \varepsilon \mathfrak{p}_0)$. Hence

$$\exp\{\Lambda_0(H(\exp X)) + 2\rho(H(\exp X)) + \lambda(\Gamma(\exp X))\}$$

is bounded on $\mathfrak{p}_0$ and if $M$ is an upper bound for it, it is obvious that

$$|\psi(x)| \leq M e^{-2\rho(H(\bar{x}))} \qquad (x \varepsilon G).$$

Therefore $\int_{G_0} |\psi(x)|^2 \, d\bar{x} \leq M^2 \int_{G_0} e^{-4\rho(H(\bar{x}))} d\bar{x} < \infty$ from Lemma 3. This proves that the space $\mathfrak{H}_\Lambda = \mathfrak{H}_\Lambda(1)$ (corresponding to the function $\mu = 1$) is not zero if $\Lambda(H_{\alpha_i}) \leq c \; (1 \leq i \leq t)$ and our assertion now follows immediately from Theorem 2 and its corollary.

**10. A result on characters.** Let $\Lambda_0$ be a linear function on $\mathfrak{h}$ satisfying the conditions of Theorem 2. We denote by $T_{\Lambda_0}$ the character $[5(c)]$ of the quasi-simple irreducible representation of $G$ defined in Theorem 2 corresponding to $\Lambda_0$. Since two infinitesimally equivalent quasi-simple irreducible representations have the same character (see $[5(c)]$, §7]), $T_{\Lambda_0}$ is independent of the choice of $\mu$ in Theorem 2 and so it is completely determined by $\Lambda_0$. We shall now try to obtain some information about this character under suitable assumptions on $\Lambda_0$.

Let $\mathfrak{H}$ be a Hilbert space and $Q$ a bounded operator on $\mathfrak{H}$. We say that $Q$ is *summable* if there exists a complete orthonormal set $(\psi_j)_{j \in J}$ in $\mathfrak{H}$ and a regular operator $B$ such that $\sum_{i,j \in J} |q_{ij}| < \infty$ where $q_{ij} = (\psi_i, BQB^{-1}\psi_j)$. Let $C_c^\infty(A)$ denote the set of all complex-valued functions on $A$ which are everywhere indefinitely differentiable and which vanish outside a compact set. Since $K$ is simply connected,[4] there exists (see Weyl [9]) an analytic function $\Delta_k$ on $A$ such that

$$\Delta_k(\exp H) = \prod_{\alpha \in P_k} \left( e^{\alpha(H)/2} - e^{-\alpha(H)/2} \right) \qquad (H \, \varepsilon \, \mathfrak{h}_0)$$

where $P_k$ is the set of all positive compact roots of $\mathfrak{g}$. As before we write $y^x = xyx^{-1} \; (x, y \, \varepsilon \, G)$. Let $dh$ and $d\bar{u}$ denote the Haar measures on $A$ and $K_0$ respectively. (We assume that $\int_{K_0} d\bar{u} = 1$). Then we have the following lemma.

LEMMA 24. *Let $\pi$ be a quasi-simple irreducible representation of $G$ on $\mathfrak{H}$. Then if $f \varepsilon C_c^\infty(A)$, the operator*

$$\int_A \int_{K_0} f(h) \Delta_k(h) \pi(h^{\bar{u}}) \, dh \, d\bar{u}$$

*is summable.*

We can choose a homomorphism $\eta$ of $K$ into $C$ such that $\eta(u^{-1})\pi(u)$ $(u \, \varepsilon \, K)$ depends only on $\bar{u}$ (see $[5(c)$, p. 249]). Hence if we put $\bar{\pi}(\bar{u}) = \eta(u^{-1})\pi(u)$ it is clear that $\bar{\pi}$ is a representation of the compact group $K_0$. Therefore we can find a regular operator $B$ such that $B\bar{\pi}(\bar{u})B^{-1}$ is unitary for every $\bar{u} \, \varepsilon \, K_0$. Hence, in view of our definition of the summa-

bility of an operator, it is clear that, without any loss of generality, we may
assume that $\bar{\pi}$ itself is a unitary representation. $\Omega$ being the set of all equivalence classes of irreducible finite-dimensional representations of $K$, we
denote, as usual, by $\mathfrak{H}_{\mathfrak{D}}$ ($\mathfrak{D}\,\varepsilon\,\Omega$) the subspace of $\mathfrak{H}$ consisting of those elements
which transform under $\pi(K)$ according to $\mathfrak{D}$. Then there exists an integer
$N$ such that $\dim \mathfrak{H}_{\mathfrak{D}} \leq N(d(\mathfrak{D}))^2$ ($\mathfrak{D}\,\varepsilon\,\Omega$) where $\dot{d}(\mathfrak{D})$ is the degree of any
representation in $\mathfrak{D}$ (see [5(c), Theorem 4]). Since $\bar{\pi}$ is unitary, the sub-
spaces $\mathfrak{H}_{\mathfrak{D}}$ are mutually orthogonal and so we can choose a complete ortho-
normal set $(\psi_j)_{j\,\varepsilon\,J}$ in $\mathfrak{H}$ such that each $\psi_j$ lies in some $\mathfrak{H}_{\mathfrak{D}}$ (see [5(a), Theorem
4]). Let $J(\mathfrak{D})$ denote the set of all $j\,\varepsilon\,J$ for which $\psi_j\,\varepsilon\,\mathfrak{H}_{\mathfrak{D}}$. Then $(\psi_j)_{j\,\varepsilon\,J(\mathfrak{D})}$
is an orthonormal base for $\mathfrak{H}_{\mathfrak{D}}$. Put $n(\mathfrak{D}) = (d(\mathfrak{D}))^{-1}\dim\mathfrak{H}_{\mathfrak{D}}$. Then $n(\mathfrak{D})$
is a nonnegative integer. Also if $E_{\mathfrak{D}}$ is the orthogonal projection of $\mathfrak{H}$ on
$\mathfrak{H}_{\mathfrak{D}}$, it follows from the Schur orthogonality relations on the compact group
$K_0$ that

$$E_{\mathfrak{D}}\Big(\int_{K_0}\pi(h^{\bar{u}})\,d\bar{u}\Big)E_{\mathfrak{D}} = \eta(h)E_{\mathfrak{D}}\Big(\int_{K_0}\bar{\pi}(\bar{u}h\bar{u}^{-1})\,d\bar{u}\Big)E_{\mathfrak{D}}$$

$$= d(\mathfrak{D})^{-1}\,sp(E_{\mathfrak{D}}\pi(h)E_{\mathfrak{D}})E_{\mathfrak{D}} = d(\mathfrak{D})^{-1}\zeta_{\mathfrak{D}}(h)E_{\mathfrak{D}} \qquad (h\,\varepsilon\,A)$$

where $\zeta_{\mathfrak{D}}$ is the character (on $K$) of the class $\mathfrak{D}$. This shows that if
$i, j\,\varepsilon\,J(\mathfrak{D})$,

$$(\psi_i, Q_f\psi_j) = d(\mathfrak{D})^{-1}\delta_{ij}\int_A f(h)\Delta_k(h)\zeta_{\mathfrak{D}}(h)\,dh$$

where $\delta_{ij}$ is the Kronecker symbol and $Q_f$ is the operator of our lemma.
Therefore if $\Omega_\pi$ is the set of all $\mathfrak{D}\,\varepsilon\,\Omega$ such that $\mathfrak{H}_{\mathfrak{D}} \neq \{0\}$,

$$\sum_{i,j\,\varepsilon\,J}|(\psi_i, Q_f\psi_j)| \leq N\sum_{\mathfrak{D}\,\varepsilon\,\Omega_\pi} d(\mathfrak{D})\Big|\int_A f(h)\Delta_k(h)\zeta_{\mathfrak{D}}(h)\,dh\Big|.$$

Let $\mathfrak{D}$ be a class in $\Omega_\pi$ and $\sigma$ a representation in $\mathfrak{D}$. We denote the corresponding representation of $\mathfrak{k}$ also by $\sigma$. Since[3] $\mathfrak{k} = \mathfrak{k}' + \mathfrak{c}$, it follows from
Schur's lemma that the representation space $V$ of $\sigma$ is irreducible under $\sigma(\mathfrak{k}')$.
Let $\sigma'$ denote the corresponding representation of $\mathfrak{k}'$ on $V$. Since $\mathfrak{h}$ is the
direct sum of $\mathfrak{c}$ and $\mathfrak{h}' = \mathfrak{h}\cap\mathfrak{k}'$ we may identify linear functions on $\mathfrak{h}'$ with
those linear functions on $\mathfrak{h}$ which vanish identically on $\mathfrak{c}$. Now $\mathfrak{k}'$ is semisimple and $\mathfrak{h}'$ is a Cartan subalgebra of $\mathfrak{k}'$ and it is clear that the roots of $\mathfrak{k}'$
with respect to $\mathfrak{h}'$ then coincide with the compact roots of $\mathfrak{g}$. Let $\lambda_{\mathfrak{D}}$ denote
the highest weight of $\sigma'$ (if we take $P_k$ as the set of positive roots of $\mathfrak{k}'$).
Then if $2\rho_k = \sum_{\alpha\,\varepsilon\,P_k}\alpha$ we know (see Weyl [9]) that

$$\Delta_k(\exp H)\zeta_{\mathfrak{D}}(\exp H) = \eta(\exp H)\sum_{s\,\varepsilon\,\mathfrak{w}_k}\epsilon(s)e^{s\lambda'_{\mathfrak{D}}(H)} \qquad (H\,\varepsilon\,\mathfrak{h}'\cap\mathfrak{k}_0)$$

3

where $\lambda'_{\mathfrak{D}} = \lambda_{\mathfrak{D}} + \rho_k$ and $\mathfrak{w}_k$ is the group generated by the Weyl reflexions $s_\alpha$ corresponding to $\alpha \, \varepsilon \, P_k$, and $\epsilon(s) = (-1)^r$ where $r$ is the number of negative roots among $s\alpha$ $(\alpha \, \varepsilon \, P_k)$. On the other hand the degrees of $\sigma$ and $\sigma'$ are the same and therefore (Weyl [9])

$$d(\mathfrak{D}) = \prod_{\alpha \varepsilon P_k} \lambda'_{\mathfrak{D}}(H_\alpha)/\rho_k(H_\alpha).$$

Now if we regard $\eta$ as a representation of $K$ of degree 1, it is obvious that $\eta(\exp X) = e^{\mu(X)}$ $(X \, \varepsilon \, \mathfrak{k}_0)$ where $\mu$ is a linear function on $\mathfrak{k}$ which vanishes identically on $\mathfrak{k}'$. Select a base $\Gamma_1, \cdots, \Gamma_r$ for $\mathfrak{c}_0 = \mathfrak{c} \cap \mathfrak{k}_0$ (over $R$) such that $\exp(t_1\Gamma_1 + \cdots + t_r\Gamma_r) = 1$ in $K_0$ $(t_1, \cdots, t_r \, \varepsilon \, R)$ if and only if $t_1, \cdots, t_r$ are all integers (see [5(c)], p. 239]). Then since $\eta(h^{-1})\sigma(h)$ $(h \, \varepsilon \, A)$ depends only on $\hbar$, it follows that $\sigma(\Gamma_i) = \mu(\Gamma_i) + 2\pi(-1)^{\frac{1}{2}}m_i$ $1 \leq i \leq r$ where $m_i$ are integers. We denote by $m_{\mathfrak{D}}$ the linear function on $\mathfrak{h}$ given by $m_{\mathfrak{D}}(H) = 0$ $(H \, \varepsilon \, \mathfrak{h}')$ and $m_{\mathfrak{D}}(\Gamma_i) = 2\pi(-1)^{\frac{1}{2}}m_i$ $1 \leq i \leq r$. Also put

$$|m_{\mathfrak{D}}| = (m_1^2 + \cdots + m_r^2 + 1)^{\frac{1}{2}}.$$

Then if $\nu_{\mathfrak{D}} = \lambda_{\mathfrak{D}} + m_{\mathfrak{D}} + \rho_k$,

$$\Delta_k(\exp H)\zeta_{\mathfrak{D}}(\exp H) = \eta(\exp H) \sum_{s \varepsilon \mathfrak{w}_k} \epsilon(s) e^{s\nu_{\mathfrak{D}}(H)} \qquad (H \, \varepsilon \, \mathfrak{h}_0)$$

and

$$d(\mathfrak{D}) = \prod_{\alpha \varepsilon P_k} \nu_{\mathfrak{D}}(H_\alpha)/\rho_k(H_\alpha)$$

since $sm_{\mathfrak{D}} = m_{\mathfrak{D}}$ $(s \, \varepsilon \, \mathfrak{w}_k)$ and $H_\alpha \, \varepsilon \, \mathfrak{h}'$ $(\alpha \, \varepsilon \, P_k)$.

Let $\mathfrak{U}$ be the subalgebra of $\mathfrak{B}$ generated by $(1, \mathfrak{h})$. We regard elements of $\mathfrak{U}$ as differential operators on $A$ in the usual way so that

$$(Hg)(h) = \{(d/dt)g(h \exp tH)\}_{t=0} \qquad (h \, \varepsilon \, A, H \, \varepsilon \, \mathfrak{h}_0, t \, \varepsilon \, R, g \, \varepsilon \, C_c^\infty(A)).$$

Since any element $s \, \varepsilon \, \mathfrak{w}_k$ permutes the compact roots among themselves, it follows that

$$\prod_{\alpha \varepsilon P_k} s\nu_{\mathfrak{D}}(H_\alpha) = \pm \prod_{\alpha \varepsilon P_k} \nu_{\mathfrak{D}}(H_\alpha)$$

and therefore it is obvious that there exists an element $z_0 \, \varepsilon \, \mathfrak{U}$ such that

$$z_0(e^{s\nu_{\mathfrak{D}}}) = d(\mathfrak{D})^2 \, |m_{\mathfrak{D}}|^2 \, e^{s\nu_{\mathfrak{D}}} \qquad (s \, \varepsilon \, \mathfrak{w}_k).$$

Here $e^{s\nu_{\mathfrak{D}}}$ is the function $g$ on $A$ given by [12] $g(\exp H) = e^{s\nu_{\mathfrak{D}}(H)}$ $(H \, \varepsilon \, \mathfrak{h}_0)$. Hence if $g_{\mathfrak{D}}(h) = \eta(h^{-1})\Delta_k(h)\zeta_{\mathfrak{D}}(h)$ it follows that $z_0 g_{\mathfrak{D}} = d(\mathfrak{D})^2 \, |m_{\mathfrak{D}}|^2 g_{\mathfrak{D}}$. Now let $\phi$ denote the automorphism of $\mathfrak{U}$ over $C$ such that $\phi(H) = -H$

---

[12] It follows easily from the discussion at the beginning of § 5 that such a function on $A$ actually exists.

and $\phi(1) = 1$ $(H \varepsilon \mathfrak{h})$. Then if $z = \phi(z_0)$, $f'(h) = f(h)\eta(h)$ $(h \varepsilon A)$ and $m$ is a positive integer, it is clear that

$$d(\mathfrak{D})^{2m} |m_{\mathfrak{D}}|^{2m} \int_A f'(h) g_{\mathfrak{D}}(h) dh$$

$$= \int_A f'(h) (z_0{}^m g_{\mathfrak{D}})(h) dh = \int_A (z^m f')(h) g_{\mathfrak{D}}(h) dh.$$

But $|e^{s\nu_{\mathfrak{D}}(H)}| = 1$ $(H \varepsilon \mathfrak{h}_0)$, hence $|g_{\mathfrak{D}}(h)| \leqq w$ $(h \varepsilon A)$ where $w$ is the order of $\mathfrak{w}_k$. Hence

$$\left| \int_A (z^m f')(h) g_{\mathfrak{D}}(h) dh \right| \leqq w \int_A |(z^m f')(h)| dh$$

and therefore

$$\sum_{\mathfrak{D} \varepsilon \Omega_\tau} d(\mathfrak{D}) \left| \int_A f(h) \Delta_k(h) \zeta_{\mathfrak{D}}(h) dh \right|$$

$$\leqq \sum_{\mathfrak{D} \varepsilon \Omega_\tau} d(\mathfrak{D})^{1-2m} |m_{\mathfrak{D}}|^{-2m} w \int_A |z^m f'(h)| dh.$$

However if $m$ is sufficiently large we know (see [5(c), p. 240]) that

$$\sum_{\mathfrak{D} \varepsilon \Omega_\tau} d(\mathfrak{D})^{1-2m} |m_{\mathfrak{D}}|^{-2m} < \infty$$

and this proves that $Q_f$ is summable.

Now every summable operator has a trace [5(c), Lemma 1]. Let $\tau_\pi(f)$ denote the trace of $Q_f$. Then the above proof shows that it is possible to choose a positive integer $m$ and a real constant $M$ such that

$$|\tau_\pi(f)| \leqq M \int_A |z^m f'(h)| dh$$

for all $f \varepsilon C_c^\infty(A)$. (Here $f' = f\eta$). This shows that $\tau_\pi$ is a distribution (see Schwartz [8]) on $A$ of finite order. Moreover

$$\tau_\pi(f) = \sum_{\mathfrak{D} \varepsilon \Omega_\tau} \sum_{j \varepsilon J(\mathfrak{D})} (\psi_j, Q_f \psi_j) = \sum_{\mathfrak{D} \varepsilon \Omega_\tau} n(\mathfrak{D}) \int_A f(h) \Delta_k(h) \zeta_{\mathfrak{D}}(h) dh$$

and this shows that $\tau_\pi$ does not change if we replace $\pi$ by another infinitesimally equivalent representation.

Now let $\Lambda_0$ be a real linear function on $\mathfrak{h}$ satisfying the following three conditions (cf. Theorem 3 of [5(f)]):

(1) $\Lambda_0(H_\alpha)$ is a nonnegative integer for every $\alpha \varepsilon P_k$.

(2) $\Lambda_0(H_\beta) = 0$ for every noncompact positive root which is not totally positive.

(3) $\quad \Lambda_0(H_\gamma) + \rho(H_\gamma) \leqq 0$ for every totally positive root $\gamma$. (Here $2\rho$ is the sum of all positive roots).

We consider the quasi-simple irreducible representation $\pi$ of $G$ defined in Theorem 2 corresponding to $\Lambda_0$. Put $\tau_{\Lambda_0} = \tau_\pi$ and let $\mathfrak{H}$ denote the representation space of $\pi$. Our object is to compute $\tau_{\Lambda_0}$. We shall see in another paper that $\tau_{\Lambda_0}$ is intimately related with $T_{\Lambda_0}$ (see however [5(g)]).

We keep to the notation of the proof of Lemma 24. By going over, if necessary, to an equivalent representation we may again assume that $\pi$ is unitary. For any linear function $\Lambda$ on $\mathfrak{h}$ let $\mathfrak{H}_\Lambda$ denote the set of all elements $\phi \varepsilon \mathfrak{H}$ such that $\pi(\exp H)\phi = e^{\Lambda(H)}\phi$ $(H \varepsilon \mathfrak{h}_0)$. Then it is clear that the subspaces $\mathfrak{H}_\Lambda$ are mutually orthogonal. Put $\mathfrak{H}^0 = \sum\limits_{\mathfrak{D} \varepsilon \Omega} \mathfrak{H}_\mathfrak{D}$ and $\mathfrak{H}_\Lambda{}^0 = \mathfrak{H}_\Lambda \cap \mathfrak{H}^0$. Since $\mathfrak{H}_\mathfrak{D}$ is finite-dimensional and fully reducible under $\pi(A)$, it is obvious that $\mathfrak{H}^0 = \sum\limits_\Lambda \mathfrak{H}_\Lambda{}^0$. Let $E_\Lambda$ denote the orthogonal projection of $\mathfrak{H}$ on $\mathfrak{H}_\Lambda$. Since $\mathfrak{H}^0$ is dense in $\mathfrak{H}$, $E_\Lambda\mathfrak{H}^0$ is dense in $\mathfrak{H}_\Lambda$. But it is clear that $E_\Lambda\mathfrak{H}^0 \subset \mathfrak{H}^0$ and therefore $E_\Lambda\mathfrak{H}^0 \subset \mathfrak{H}_\Lambda{}^0$. However $\dim \mathfrak{H}_\Lambda{}^0 < \infty$ (corollary to Lemma 21 of [5(f)]) and so it follows that $\mathfrak{H}_\Lambda = \mathfrak{H}_\Lambda{}^0$.

Let $P_+$ denote the set of all totally positive roots of $\mathfrak{g}$ and let $\mathfrak{h}_+$ be the open subset of $\mathfrak{h}$ consisting of all those element $H$ for which $|e^{\gamma(H)}| > 1$ for every $\gamma \varepsilon P_+$. Since every root takes pure imaginary values on $\mathfrak{h}_0$, it is obvious that $\mathfrak{h}_+ + \mathfrak{h}_0 \subset \mathfrak{h}$. Let $\sigma_\beta$ $(\beta \varepsilon P_+)$ denote the hyperplane in the real Euclidean space $\mathfrak{h}^* = (-1)^{\frac{1}{2}}\mathfrak{h}_0$ defined by the equation $\beta(H) = 0$. Consider the complement $\mathfrak{h}_1^*$ of $\bigcup\limits_{\beta \varepsilon P_+} \sigma_\beta$ in $\mathfrak{h}^*$. Let $(\alpha_1, \cdots, \alpha_l)$ be a fundamental system of positive roots and $H_0$ a point in $\mathfrak{h}$ such that $\alpha_i(H_0) = 1$ $1 \leqq i \leqq l$. Then $H_0 \varepsilon \mathfrak{h}_1^*$ and if $\mathfrak{h}_+^*$ is the connected component of $H_0$ in $\mathfrak{h}_1^*$, it is obvious that $\mathfrak{h}_+^* \subset \mathfrak{h}^* \cap \mathfrak{h}_+$. Conversely $\mathfrak{h}^* \cap \mathfrak{h}_+$ is a convex subset of $\mathfrak{h}_1^*$ and therefore it is connected. Hence $\mathfrak{h}_+^* = \mathfrak{h}^* \cap \mathfrak{h}_+$ and therefore $\mathfrak{h}_+ = \mathfrak{h}_0 + \mathfrak{h}_+^*$. In particular $tH_0 \varepsilon \mathfrak{h}_+^*$ for every $t > 0$ and so zero lies in the closure of $\mathfrak{h}_+$. Now consider the complex abelian Lie group $\tilde{A}_c$ defined in Section 3 and let $_+\tilde{A}_c$ denote the subset of those $h \varepsilon \tilde{A}_c$ which can be written in the form $h = \exp H$ $(H \varepsilon \mathfrak{h}_+)$. It is clear that $_+\tilde{A}_c$ is an open connected subset of $\tilde{A}_c$ whose closure contains 1. Also $_+\tilde{A}_c A = {}_+\tilde{A}_c$. We shall now define a bounded operator $\pi(h)$ on $\mathfrak{H}$ for every $h \varepsilon {}_+\tilde{A}_c$. Let $\mathfrak{F}_\pi$ be the set of those linear function $\Lambda$ on $\mathfrak{h}$ for which $\mathfrak{H}_\Lambda \neq \{0\}$. Then if $\exp H = 1$ in $A$ $(H \varepsilon \mathfrak{h}_0)$ it is obvious that $e^{\Lambda(H)} = 1$ and therefore, as we have seen in Section 5, there exists a holomorphic character $\xi_\Lambda$ of $\tilde{A}_c$ such that $\xi_\Lambda(\exp H) = e^{\Lambda(H)}$ $(H \varepsilon \mathfrak{h})$. Let $\psi$ be the element in $\mathfrak{H}$ corresponding to Theorem 2. Then $\psi \varepsilon \mathfrak{H}_{\Lambda_0}$ and $\mathfrak{H}^0 = \pi(\mathfrak{B})\psi$. Moreover if $V = \pi(\mathfrak{X})\psi$, $\dim V < \infty$ (Lemmas 8 and 10) and

$V$ is irreducible under $\pi(\mathfrak{k})$ (Lemma 2 of [5(f)]). Let $\mathfrak{D}_0$ denote the class [13] (in $\Omega_\pi$) and $\Lambda_0, \Lambda_1, \cdots, \Lambda_r$ all the (distinct) weights of this representation of $\mathfrak{k}$ on $V$. Then if $\gamma_1, \cdots, \gamma_q$ are all the (distinct) totally positive roots of $\mathfrak{g}$, every $\Lambda \varepsilon \mathfrak{F}_\pi$ can be written in the form

$$\Lambda = \Lambda_i - (m_1\gamma_1 + \cdots + m_q\gamma_q)$$

for some $i$ $(0 \leq i \leq r)$ and some non-negative integers $m_1, \cdots, m_q$ (see the corollary to Lemma 21 of [5(f)]). Therefore if $h \varepsilon_+\tilde{A}_c$

$$|\xi_\Lambda(h)| \leq \max_{0 \leq i \leq r} |\xi_{\Lambda_i}(h)| \qquad\qquad (\Lambda \varepsilon \mathfrak{F}_\pi).$$

Hence it is clear that the infinite sum $\sum\limits_{\Lambda \varepsilon \mathfrak{F}_\pi} \xi_\Lambda(h)E_\Lambda\phi$ converges in $\mathfrak{H}$ for any

$h \varepsilon_+\tilde{A}_c$ and $\phi \varepsilon \mathfrak{H}$. Let $\pi(h)\phi$ denote the limit of this sum. Then

$$|\pi(h)\phi|^2 \leq \max_{0 \leq i \leq r} |\xi_{\Lambda_i}(h)|^2 |\phi|^2$$

and therefore $\pi(h)$ is a bounded operator on $\mathfrak{H}$. It is obvious from its definition that if $h_+ \varepsilon_+\tilde{A}_c$ and $h \varepsilon A$, $\pi(hh_+) = \pi(h)\pi(h_+) = \pi(h_+)\pi(h)$.

Let $\zeta_{\mathfrak{D}_0}$ denote the character of the class $\mathfrak{D}_0$. Then if $d_i = \dim(V \cap \mathfrak{H}_{\Lambda_i})$ $0 \leq i \leq r$,

$$\zeta_{\mathfrak{D}_0}(h) = \sum_{0 \leq i \leq r} d_i\xi_{\Lambda_i}(h)$$

for any $h \varepsilon A$ and we can extend $\zeta_{\mathfrak{D}_0}$ to a holomorphic function on $\tilde{A}_c$ by means of the above formula. Let $\lambda$ be a linear function on $\mathfrak{h}$ such that $\lambda(H_\alpha)$ is an integer for every compact root $\alpha$. Then as we have seen in Section 5, there exists a holomorphic character $\xi_\lambda$ of $\tilde{A}_c$, such that $\xi_\lambda(\exp H) = e^{\lambda(H)}$ $(H \varepsilon \mathfrak{h})$. Put $\rho_+ = \frac{1}{2} \sum\limits_{\gamma \varepsilon P_+} \gamma$, $\rho_k = \frac{1}{2} \sum\limits_{\alpha \varepsilon P_k} \alpha$ and $\rho_0 = \rho_+ + \rho_k$. Then if $s_\alpha$ is the Weyl reflexion corresponding to a compact root $\alpha$, it follows from Lemma 10 of [5(f)] that $s_\alpha\rho_+ = \rho_+$ and therefore $\rho_+(H_\alpha) = 0$. Also if $\beta$ is any root, $\beta(H_\alpha)$ and $\rho_k(H_\alpha)$ are integers (see Weyl [9]). This shows that we can construct the corresponding characters $\xi_{\rho_+}, \xi_{\rho_k}, \xi_{\rho_0}, \xi_\beta$. Put

$$\Delta_k(h) = \xi_{\rho_k}(h) \prod_{\alpha \varepsilon P_k} \{1 - \xi_\alpha(h^{-1})\} = \prod_{\alpha \varepsilon P_k} (e^{\frac{1}{2}\alpha(H)} - e^{-\frac{1}{2}\alpha(H)}),$$

$$\Delta_+(h) = \xi_{\rho_+}(h) \prod_{\gamma \varepsilon P_+} \{1 - \xi_\gamma(h^{-1})\} = \prod_{\gamma \varepsilon P_+} \{e^{\frac{1}{2}\gamma(H)} - e^{-\frac{1}{2}\gamma(H)}\}$$

where $h = \exp H \varepsilon \tilde{A}_c$ $(H \varepsilon \mathfrak{h})$.

LEMMA 25. *If* $h \varepsilon_+\tilde{A}_c$ *the operator* $\pi(h)$ *is summable and*

$$sp\,\pi(h) = \{\Delta_+(h)\}^{-1}\xi_{\rho_+}(h)\zeta_{\mathfrak{D}_0}(h).$$

---

[13] As usual we identify finite-dimensional irreducible representations of $\mathfrak{k}$ with those of $K$.

*Moreover the series* $\sum_{\Lambda \in \mathfrak{F}_\pi} (\dim \mathfrak{H}_\Lambda) |\xi_\Lambda(h)|$ *converges uniformly on every compact subset of* $_+\tilde{A}_c$.

Let $h = \exp H$ $(H \in \mathfrak{h}_+)$. Then from the corollary to Lemma 21 of $[5(\mathrm{f})]$,

$$\sum_{\Lambda \in \mathfrak{F}_\pi} (\dim \mathfrak{H}_\Lambda) |\xi_\Lambda(h)|$$

$$= \sum_{0 \leq i \leq r} d_i \sum_{m_1, \cdots, m_q \geq 0} |\exp\{\Lambda_i(H) - m_1\gamma_1(H) - \cdots - m_q\gamma_q(H)\}|.$$

Since the real part of $\gamma_j(H)$ is positive, the series converges and clearly its convergence is uniform if $H$ remains in a compact subset of $\mathfrak{h}_+$. Moreover since $\pi(h)$ coincides with $\xi_\Lambda(h)E_\Lambda$ on $\mathfrak{H}_\Lambda$, it is now clear that $\pi(h)$ is summable and

$$sp\,\pi(h) = \sum_{\Lambda \in \mathfrak{F}_\pi} (\dim \mathfrak{H}_\Lambda)\xi_\Lambda(h)$$

$$= \sum_{0 \leq i \leq r} d_i \sum_{m_1, \cdots, m_q \geq 0} \exp(\Lambda_i(H) - m_1\gamma_1(H) - \cdots - m_q\gamma_q(H))$$

$$= \prod_{1 \leq j \leq q} \{1 - e^{-\gamma_j(H)}\}^{-1}\zeta_{\mathfrak{D}_0}(\exp H) = \{\Delta_+(h)\}^{-1}\xi_{\rho_+}(h)\zeta_{\mathfrak{D}_0}(h).$$

For any $f \in C_c^\infty(A)$ and $h_+ \in {}_+\tilde{A}_c$, put

$$Q_f(h_+) = \int_A f(h)\Delta_k(hh_+)\pi(hh_+)dh, \quad Q_f = \int_A \int_{K_0} f(h)\Delta_k(h)\pi(h^{\bar{u}})dh\,d\bar{u}.$$

LEMMA 26.   *If* $f \in C_c^\infty(A)$ *the operator* $Q_f(h_+)$ *is summable and*

$$\lim_{h_+ \to 1} sp\,Q_f(h_+) = sp\,Q_f \qquad\qquad (h_+ \in {}_+\tilde{A}_c).$$

Let $\Lambda \in \mathfrak{F}_\pi$. Then it is obvious that

$$Q_f(h_+)E_\Lambda = \{\int_A f(h)\Delta_k(hh_+)\xi_\Lambda(hh_+)dh\}E_\Lambda,$$

and therefore, in order to show that $Q_f(h_+)$ is summable, it is enough to prove that

$$\sum_{\Lambda \in \mathfrak{F}_\pi} (\dim \mathfrak{H}_\Lambda) |\int_A f(h)\Delta_k(hh_+)\xi_\Lambda(hh_+)dh| < \infty.$$

But this is obvious in view of the fact that

$$\sum_{\Lambda \in \mathfrak{F}_\pi} (\dim \mathfrak{H}_\Lambda) |\xi_\Lambda(hh_+)|$$

converges uniformly (with respect to $h$) on every compact subset of $A$ (Lemma 25).

Now we use the notation of the proof of Lemma 24. Since $Q_f(h_+)$ is summable,

$$sp\,Q_f(h_+) = \sum_{\mathfrak{D}\,\varepsilon\,\Omega_\pi} sp\,(E_\mathfrak{D}Q_f(h_+)E_\mathfrak{D}).$$

Then if we define $\eta$, $\mu$, $\nu_\mathfrak{D}$, $n(\mathfrak{D})$ and $m(\mathfrak{D})$ $(\mathfrak{D}\,\varepsilon\,\Omega_\pi)$ as before,

$$\eta(\exp H) = e^{\mu(H)},\, \Delta_k(\exp H)\zeta_\mathfrak{D}(\exp H) = e^{\mu(H)}\sum_{s\,\varepsilon\,\mathfrak{w}_k}\epsilon(s)e^{s\nu_\mathfrak{D}(H)} \qquad (H\,\varepsilon\,\mathfrak{h}_0)$$

and it is clear that we can extend $\eta$ and $\zeta_\mathfrak{D}$ to holomorphic functions on $\tilde{A}_c$ so that the above relations actually hold for all $H\,\varepsilon\,\mathfrak{h}$. Then if $\mathfrak{D}\,\varepsilon\,\Omega_\pi$,

$$sp\,(E_\mathfrak{D}Q_f(h_+)E_\mathfrak{D}) = n(\mathfrak{D})\int_A f(h)\Delta_k(hh_+)\zeta_\mathfrak{D}(hh_+)\,dh$$

and therefore

$$sp\,Q_f(h_+) = \sum_{\mathfrak{D}\,\varepsilon\,\Omega_\pi} n(\mathfrak{D})\int_A f(h)\Delta_k(hh_+)\zeta_\mathfrak{D}(hh_+)\,dh.$$

Now put $\nu'_\mathfrak{D} = \nu_\mathfrak{D} + \mu_0$ where $\mu_0$ is the restriction of $\mu$ on $\mathfrak{h}$,

$$g_\mathfrak{D}(h) = \eta(h^{-1})\Delta_k(h)\zeta_\mathfrak{D}(h) = \sum_{s\,\varepsilon\,\mathfrak{w}_k}\epsilon(s)\xi_{s\nu_\mathfrak{D}}(h) \qquad (h\,\varepsilon\,\tilde{A}_c)$$

and $f'(h) = f(h)\eta(h)$ $(h\,\varepsilon\,A)$. It is obvious from the definition of $\nu'_\mathfrak{D}$ that $\nu'_\mathfrak{D} - \rho_k\,\varepsilon\,\mathfrak{F}_\pi$ and therefore $s(\nu'_\mathfrak{D} - \rho_k)\,\varepsilon\,\mathfrak{F}_\pi$ for $s\,\varepsilon\,\mathfrak{w}_k$ (corollary to Lemma 6 of [5(f)]). Hence

$$|\xi_{s\nu_\mathfrak{D}}(h)| \leq \max_{\substack{0\leq i\leq r\\ s\,\varepsilon\,\mathfrak{w}_k}} |\xi_{\Lambda_i+s\rho_k}(h)|\,|\eta(h^{-1})| \qquad (h\,\varepsilon\,{}_+\tilde{A}_c).$$

We can now argue as in the proof of Lemma 24 and show that the series

$$\sum_{\mathfrak{D}\,\varepsilon\,\Omega_\pi} n(\mathfrak{D})\Big|\int_A f(h)\Delta_k(hh_+)\zeta_\mathfrak{D}(hh_+)\,dh\,\Big|$$

converges uniformly with respect to $h_+$ on $B\cap{}_+\tilde{A}_c$ where $B$ is a compact neighbourhood of 1 in $\tilde{A}_c$. Hence it follows that

$$\operatorname*{Lim}_{h_+\to 1} sp\,Q_f(h_+) = \sum_{\mathfrak{D}\,\varepsilon\,\Omega_\pi} n(\mathfrak{D})\operatorname*{Lim}_{h_+\to 1}\int_A f(h)\Delta_k(hh_+)\zeta_\mathfrak{D}(hh_+)\,dh$$

$$= \sum_{\mathfrak{D}\,\varepsilon\,\Omega_\pi} n(\mathfrak{D})\int_A f(h)\Delta_k(h)\zeta_\mathfrak{D}(h)\,dh = sp\,Q_f.$$

Now put

$$\Xi_{\Lambda_0} = \xi_{\rho_+}\zeta_{\mathfrak{D}_0}\Delta_k = \sum_{s\,\varepsilon\,\mathfrak{w}_k}\epsilon(s)\xi_{s(\Lambda_0+\rho_0)}.$$

COROLLARY 1.　$\tau_{\Lambda_0}(f) = \underset{h_+ \to 1}{\mathrm{Lim}} \int_A f(h)\,(\Xi_{\Lambda_0}(hh_+)/\Delta_+(hh_+))\,dh \quad (h_+ \,\varepsilon\, {}_+\bar{A}_0)$
for $f \varepsilon C_c^\infty(A)$.

This is an immediate consequence of Lemma 26 and the fact that $\tau_{\Lambda_0}(f) = spQ_f$.

COROLLARY [14] 2.　$\Delta_+\tau_{\Lambda_0} = \Xi_{\Lambda_0}$ on $A$.

Let $H_+$ be an element in $\mathfrak{h}_+{}^*$ so that $\gamma(H_+)$ is real and positive for every $\gamma \varepsilon P_+$. Then if $h_+(t) = \exp tH_+$ $(t > 0)$ we know from the above lemma that

$$\tau_{\Lambda_0}(\Delta_+ f) = \underset{t \to 0}{\mathrm{Lim}} \int_A \Delta_+(h)f(h)\{\Xi_{\Lambda_0}(hh_+(t))/\Delta_+(hh_+(t))\}dh$$

for $f \varepsilon C_o^\infty(A)$. Now we claim that there exists a constant $M$ such that

$$|\Delta_+(h)\Xi_{\Lambda_0}(hh_+(t))/\Delta_+(hh_+(t))| \leq M$$

for all $h \varepsilon A$ provided $t$ is sufficiently small and positive. This is seen as follows. If $\theta$ and $\epsilon$ are real numbers and $\epsilon \geq \frac{1}{2}$,

$$|(e^{(-1)^{\frac{1}{2}}\theta} - 1)/(\epsilon e^{(-1)^{\frac{1}{2}}\theta} - 1)|^2 = (2 - 2\cos\theta)/(1 + \epsilon^2 - 2\epsilon\cos\theta)$$
$$= 2(1 - \cos\theta)/\{(1 - \epsilon)^2 + 2\epsilon(1 - \cos\theta)\} \leq 1/\epsilon \leq 2.$$

Hence if $q$ is the number of totally positive roots of $\mathfrak{g}$, it is clear that

$$|\Delta_+(h)/\Delta_+(hh_+(t))| \leq 2^q$$

provided $t$ is sufficiently small and positive. Our assertion is now obvious. Therefore by Lebesgue's Theorem

$$\tau_{\Lambda_0}(\Delta_+ f) = \int_A f(h)\,\underset{t \to 0}{\mathrm{Lim}}\,\{\Delta_+(h)\Xi_{\Lambda_0}(hh_+(t))/\Delta_+(hh_+(t))\}dh.$$

But

$$\underset{t \to 0}{\mathrm{Lim}}\,\Delta_+(h)\Xi_{\Lambda_0}(hh_+(t))/\Delta_+(hh_+(t)) = \begin{cases} 0 \text{ if } \Delta_+(h) = 0 \\ \Xi_{\Lambda_0}(h) \text{ if } \Delta_+(h) \neq 0 \end{cases} \quad (h \,\varepsilon\, A)$$

and since the set of all $h \varepsilon A$ for which $\Delta_+(h) = 0$ is of Haar measure zero, we get

$$\tau_{\Lambda_0}(\Delta_+ f) = \int_A f(h)\Xi_{\Lambda_0}(h)\,dh.$$

This proves the corollary.

COLUMBIA UNIVERSITY.

---

[14] Here we use the standard terminology of the theory of distributions (Schwartz [8]).

## REFERENCES.

[1] G. Birkhoff, "Representability of Lie algebras and Lie groups by matrices," *Annals of Mathematics*, vol. 38 (1937), pp. 526-532.

[2] S. Bochner and W. T. Martin, *Several Complex Variables*, Princeton University Press, 1948.

[3] E. Cartan, "Sur certaines formes riemanniennes remarquables des géométries à groupe fondamental simple," *Annales Scientifiques de l'École Normale Supérieure*, vol. 44 (1927), pp. 345-467.

[4] C. Chevalley, *Theory of Lie Groups*, Princeton University Press, 1946.

[5] Harish-Chandra, (a) "On some applications of the universal enveloping algebra of a semisimple Lie algebra," *Transactions of the American Mathematical Society*, vol. 70 (1951), pp. 28-96.

———, (b) "Representations of a semisimple Lie group on a Banach space. I," *ibid.*, vol. 75 (1953), pp. 185-243.

———, (c) "Representations of semisimple Lie groups. III," *ibid.*, vol. 76 (1954), pp. 234-253.

———, (d) "The Plancherel formula for complex semisimple Lie groups," *ibid.*, vol. 76 (1954), pp. 485-528.

———, (e) "Integrable and square-integrable representations of a semisimple Lie group," *Proceedings of the National Academy of Sciences*, vol. 41 (1955), pp. 314-317.

———, (f) "Representations of semisimple Lie groups. IV," *American Journal of Mathematics* vol. 77 (1955) pp. 743-777.

———, (g) "On the characters of a semisimple Lie group," *Bulletin of the American Mathematical Society*, vol. 61 (1955), pp. 389-396.

[6] K. Iwasawa, "On some types of topological groups," *Annals of Mathematics*, vol. 50 (1949), pp. 507-557.

[7] G. D. Mostow, "A new proof of E. Cartan's theorem on the topology of semi-simple groups," *Bulletin of the American Mathematical Society*, vol. 55 (1949), pp. 969-980.

[8] L. Schwartz, *Théorie des distributions*, Paris, Hermann, 1950.

[9] H. Weyl, *The Structure and Representation of Continuous Groups*, The Institute for Advanced Study, Princeton, 1935.

Reprinted from
*Amer. J. of Math.*
**78** (1956), 1–41

# REPRESENTATIONS OF SEMISIMPLE LIE GROUPS VI.*

## Integrable and Square-Integrable Representations.

By Harish-Chandra.

---

**1. Introduction.** Let $G$ be a connected semisimple Lie group and let $Z$ denote its center. Then if $\pi$ is an irreducible unitary representation of $G$ on a Hilbert space $\mathfrak{H}$ we can find a unitary character $\eta_\pi$ of $Z$ such that $\pi(z) = \eta_\pi(z)\pi(1)$ for $z \varepsilon Z$. Let $x \to x^*$ denote the natural mapping of $G$ on $G^* = G/Z$. If $\phi, \psi \varepsilon \mathfrak{H}$, it is clear that $(\phi, \pi(x)\psi)(x \varepsilon G)$ depends only on $x^*$. Let $dx^*$ denote the Haar measure on $G^*$. We shall say that $\pi$ is square-integrable if there exists an element $\psi_0 \neq 0$ in such that

$$\int_{G^*} |(\psi_0, \pi(x)\psi_0)|^2 \, dx^* < \infty.$$

Similarly $\pi$ is said to be integrable if

$$\int_{G^*} |(\psi_0, \pi(x)\psi_0)| \, dx^* < \infty$$

for some $\psi_0 \neq 0$ in $\mathfrak{H}$. In this paper we intend to study in detail some examples of such representations.

Let $\pi$ be a square-integrable representation of $G$. Then, as shown by Godement [4(b)], the Schur orthogonality relations hold for $\pi$ and therefore there exists a positive constant $d_\pi$ such that

$$\int_{G^*} |(\phi, \pi(x)\psi)|^2 \, dx^* = d_\pi^{-1} |\phi|^2 |\psi|^2$$

for all $\phi, \psi \varepsilon \mathfrak{H}$. Naturally $d_\pi$ depends on the normalization of the measure $dx^*$ but once this has been fixed, $d_\pi$ can be considered as a function of $\pi$. In analogy with the case of compact groups we call $d_\pi$ the formal degree of $\pi$. If $Z$ is finite and $\omega$ is the equivalence class of $\pi$, $d_\pi$ is also equal to the mass of $\omega$ with respect to the Plancherel measure (see Section 5) just as in the compact case. Moreover for compact semisimple groups Weyl [11(a)] has given a formula for the degree of an irreducible representation in terms of its "highest weight." We shall se that substantially the same formula holds for the formal degree $d_\pi$ under suitable conditions. This is the principal result of this paper. It can be verified immediately in the case of the $2 \times 2$ real unimodular group by looking at the results obtained

---

* Received September 27, 1955.

by Bargmann [1, p. 634] by direct computation. This very simple case is, in a sense, fundamental and much of our argument will depend on the properties of this three-dimensional group.

This paper is divided into two parts. In Part I we obtain the Schur orthogonality relations and establish the connection between the formal degree and the Plancherel measure. We also obtain a formula for the character of a square-integrable representation which is quite similar to the corresponding formula in the compact case. Although the Schur orthogonality relations are now new, Godement's proof of them [4(b)] does not cover the case when $Z$ is infinite. (However it could perhaps be modified to include this case as well.) At any rate our method is quite different and it seems worthwhile to present it briefly even at the risk of some overlap with earlier work especially since the infinite case mentioned above is particularly important for us.

Part II is devoted to the proof of the analogue of Weyl's formula. This proof depends on a detailed comparison at each step between the compact and the non-compact cases and the entire argument is based on Lemma 22. In order to make this comparison we need some considerable algebraic preparation which consists of an intensive study of the root-structure of certain types of semisimple Lie algebras. As an incidental outcome of this study, we get in Section 7 a new proof of a theorem of E. Cartan [2(c), p. 145)] on the boundedness of certain complex domains. This proof, unlike that of Cartan, does not depend on the classification of simple groups.

The last two sections contain the proof of Lemma 22. Here I follow closely a method due to Weyl [11(a)] and Cartan [2(a)] and although the proof is somewhat long no new ideas are involved.

The results of this paper had been announced in a short note [5(d)].

## Part I.

**2. Preliminary lemmas.** Let $G$ be a connected semisimple Lie group and let $\mathfrak{g}_0$ denote its Lie algebra over the field $R$ of real numbers. Define $\mathfrak{k}_0$ as in [5(b)] and let $\mathfrak{c}_0$ be the center and $\mathfrak{k}_0' = [\mathfrak{k}_0, \mathfrak{k}_0]$ the derived algebra of $\mathfrak{k}_0$. Let $K$, $K'$ and $D$ denote the analytic subgroups of $G$ corresponding to $\mathfrak{k}_0$, $\mathfrak{k}_0'$ and $\mathfrak{c}_0$ respectively. We consider the space $C_c(G)$ of all (complex-valued) continuous functions on $G$ which vanish outside a compact set. For any two functions $f, g \in C_c(G)$ we define their convolution $f * g$ by

$$(f * g)(x) = \int_{G} f(y) g(y^{-1}x) dy \qquad (x \in G)$$

where $dy$ is the Haar measure on $G$. Under this operation $C_c(G)$ becomes an associative algebra. Let $\Omega$ denote the set of all equivalence classes of finite-dimensional irreducible representations of $K$ and let $\xi_{\mathfrak{D}}$ be the character (on $K$) of any class $\mathfrak{D} \varepsilon \Omega$. Choose a base $\Gamma_1, \cdots, \Gamma_r$ for $\mathfrak{c}_0$ over $R$ such that $\exp \Gamma_i \varepsilon D \cap Z$, $1 \leq i \leq r$. ($Z$ is the center of $G$.) This is possible since $D/D \cap Z$ is compact (see Mostow [9]). Let $\mathfrak{c}_1$ be the subset of $\mathfrak{c}_0$ consisting of all elements of the form $t_1 \Gamma_1 + \cdots + t_r \Gamma_r$ $(t_i \varepsilon R)$ with $|t_i| \leq \frac{1}{2}$, $1 \leq i \leq r$. Then $K_0 = K'(\exp \mathfrak{c}_1)$ is a compact subset of $K$. For any $f \varepsilon C_c(G)$ and $\mathfrak{D} \varepsilon \Omega$ we define two functions $_{\mathfrak{D}}f$ and $f_{\mathfrak{D}}$ in $C_c(G)$ as follows:

$$_{\mathfrak{D}}f(x) = d(\mathfrak{D}) \int_{K_0} \xi_{\mathfrak{D}}(u) f(ux) \, du, \qquad f_{\mathfrak{D}}(x) = d(\mathfrak{D}) \int_{K_0} f(xu) \xi_{\mathfrak{D}}(u) \, du$$

$(x \varepsilon G)$. Here $du$ is the Haar measure on $K$ normalized in such a way that $\int_{K_0} du = 1$. Also $d(\mathfrak{D})$ is the degree of any representation in $\mathfrak{D}$. Let $L(\mathfrak{D})$ $(\mathfrak{D} \varepsilon \Omega)$ denote the subspace of $C_c(G)$ consisting of all functions of the form $_{\mathfrak{D}}f$ $(f \varepsilon C_c(G))$. Since $_{\mathfrak{D}}(f * g) = (_{\mathfrak{D}}f) * g$ $(f, g \varepsilon C_c(G))$ it follows that $L(\mathfrak{D})$ is a right ideal in $C_c(G)$.

Let $\pi$ be a quasi-simple irreducible representation (see [5(b)] of $G$ on a Banach space $\mathfrak{H}$. For any $\mathfrak{D} \varepsilon \Omega$, let $\mathfrak{H}_{\mathfrak{D}}$ denote the subspace of $\mathfrak{H}$ consisting of all those elements which transform under $\pi(K)$ according to $\mathfrak{D}$. Let $\Omega_\pi$ be the set of all $\mathfrak{D} \varepsilon \Omega$ such that $\mathfrak{H}_{\mathfrak{D}} \neq 0$. Then $L_\pi = \sum_{\mathfrak{D} \varepsilon \Omega_\pi} L(\mathfrak{D})$ is a subalgebra of $C_c(G)$. Since $\pi$ is quasi-simple there exists a character $\eta_\pi$ of $Z$ such that $\pi(z) = \eta_\pi(z) \pi(1)$ $(z \varepsilon Z)$. We shall call $\eta_\pi$ *the central character of* $\pi$. For any $f \varepsilon C_c(G)$ let $\pi(f)$ denote the operator $\int_G f(x) \pi(x) \, dx$. Then $f \to \pi(f)$ defines a representation of $C_c(G)$ on $\mathfrak{H}$.

LEMMA 1. *The space* $\mathfrak{H}_0 = \sum_{\mathfrak{D} \varepsilon \Omega} \mathfrak{H}_{\mathfrak{D}}$ *is invariant and (algebraically) irreducible under* $\pi(L_\pi)$.

Let $E_{\mathfrak{D}}$ $(\mathfrak{D} \varepsilon \Omega)$ denote the canonical projection [5(b), p. 225] of $\mathfrak{H}$ on $\mathfrak{H}_{\mathfrak{D}}$. Then if $\mathfrak{D} \varepsilon \Omega_\pi$, one proves easily (see [5(c), p. 249]) that

$$E_{\mathfrak{D}} = d(\mathfrak{D}) \int_{K_0} \xi_{\mathfrak{D}}(u^{-1}) \pi(u) \, du$$

and therefore

$$E_{\mathfrak{D}} \pi(f) = \pi(_{\mathfrak{D}}f) \qquad\qquad (f \varepsilon C_c(G)).$$

On the other hand if $\mathfrak{D} \not\varepsilon \Omega_\pi$, $E_{\mathfrak{D}} = 0$. Hence it is clear that if $f \varepsilon L_\pi$, $\pi(f)$ maps $\mathfrak{H}$ into $\mathfrak{H}_0$. Let $\psi_0$ be any nonzero element in $\mathfrak{H}_0$. In order to prove the irreducibility of $\mathfrak{H}_0$ under $\pi(L_\pi)$ it would be enough to show that $\mathfrak{H}_\mathfrak{D} \subset \pi(L_\pi)\psi_0$ for all $\mathfrak{D} \varepsilon \Omega_\pi$. Suppose then that this is false for some $\mathfrak{D}$. Put $U = \mathfrak{H}_\mathfrak{D} \cap \pi(L_\pi)\psi_0$. Then $U \neq \mathfrak{H}_\mathfrak{D}$ and since $\dim \mathfrak{H}_\mathfrak{D}$ is finite [5(b)], there exists a linear function $\alpha \neq 0$ on $\mathfrak{H}_\mathfrak{D}$ which vanishes identically on $U$. Extend $\alpha$ to a continuous linear function on $\mathfrak{H}$ by setting $\alpha(\psi) = \alpha(E_\mathfrak{D}\psi)$ $(\psi \varepsilon \mathfrak{H})$. Since $\psi_0 \neq 0$ and $\pi$ is an irreducible representation of $G$, it is obvious that the continuous function $\alpha(\pi(x)\psi_0)$ $(x \varepsilon G)$ cannot be everywhere zero on $G$. Hence we can choose $f \varepsilon C_c(G)$ such that

$$\int f(x)\alpha(\pi(x)\psi_0)\,dx \neq 0.$$

Then

$$\alpha(\pi(_\mathfrak{D}f)\psi_0) = \alpha(E_\mathfrak{D}\pi(f)\psi_0) = \alpha(\pi(f)\psi_0) = \int f(x)\alpha(\pi(x)\psi_0)\,dx \neq 0.$$

But since $_\mathfrak{D}f \varepsilon L(D)$, $\pi(_\mathfrak{D}f)\psi_0 \varepsilon U$ and therefore $\alpha(\pi(_\mathfrak{D}f)\psi_0) = 0$. This contradiction proves the lemma.

COROLLARY. *For any* $\mathfrak{D} \varepsilon \Omega_\pi$, $\mathfrak{H}_\mathfrak{D}$ *is invariant and irreducible under* $\pi(L(\mathfrak{D}))$ *and the corresponding representation of* $L(\mathfrak{D})$ *on* $\mathfrak{H}_\mathfrak{D}$ *determines* $\pi$ *completely up to infinitesimal equivalence* [5(b), p. 230].

Since $E_\mathfrak{D}\pi(f) = \pi(_\mathfrak{D}f)$ $(f \varepsilon C_c(G))$ it is obvious that $\mathfrak{H}_\mathfrak{D}$ is invariant under $\pi(L(\mathfrak{D}))$. Let $\psi_0$ and $\psi$ be two elements in $\mathfrak{H}_\mathfrak{D}$ and suppose $\psi_0 \neq 0$. Then it follows from the above lemma that $\psi = \pi(f)\psi_0$ for some $f \varepsilon L_\pi$. But since $\psi \varepsilon \mathfrak{H}_\mathfrak{D}$,

$$\psi = E_\mathfrak{D}\psi = \pi(_\mathfrak{D}f)\psi_0.$$

However $_\mathfrak{D}f \varepsilon L(\mathfrak{D})$ and so the irreducibility is proved.

Let $\phi(x)$ $(x \varepsilon G)$ denote the trace of the restriction of $E_\mathfrak{D}\pi(x)E_\mathfrak{D}$ on $\mathfrak{H}_\mathfrak{D}$. Then if $f \varepsilon C_c(G)$, $\int f(x)\phi(x)\,dx$ is the trace of the restriction of $E_\mathfrak{D}\pi(f)E_\mathfrak{D}$ on $\mathfrak{H}_\mathfrak{D}$. But since $E_\mathfrak{D}\pi(f) = \pi(_\mathfrak{D}f)$, it follows that

$$\int f(x)\phi(x)\,dx = \int {}_\mathfrak{D}f(x)\phi(x)\,dx.$$

Hence if $\nu$ denotes the representation of $L(\mathfrak{D})$ on $\mathfrak{H}_\mathfrak{D}$

$$\int f(x)\phi(x)\,dx = \mathrm{Sp}(\nu(_\mathfrak{D}f)) \qquad\qquad (f \varepsilon C_c(G)).$$

$\phi$ being a continuous function, it is now obvious that the knowledge of the trace of $\nu$ determines it completely. From this our assertion follows (see [5(c), p. 235]).

**LEMMA 2.** *Let $\psi \neq 0$ be an element in $\mathfrak{H}$. Then $E_{\mathfrak{D}}\psi \neq 0$ for some $\mathfrak{D} \, \varepsilon \, \Omega_\pi$.*

Let $C_c^\infty(G)$ be the set of all functions $f \, \varepsilon \, C_c(G)$ which are everywhere indefinitely differentiable. Choose a neighborhood $V$ of 1 in $G$ such that $|\pi(x)\psi - \psi| \leq \frac{1}{2}|\psi|$ for $x \, \varepsilon \, V$. Since $K_0$ is compact, we can find another such neighborhood $V'$ with the property that $uV'u^{-1} \subset V$ for $u \, \varepsilon \, K_0$. Select a real-valued function $g \, \varepsilon \, C_c^\infty(G)$ such that $g \geq 0$, $\int g(x)dx = 1$ and $g = 0$ outside $V'$. Then if

$$f(x) = \int_{K_0} g(uxu^{-1})\,du \qquad\qquad (x \, \varepsilon \, G)$$

it is obvious that $f \, \varepsilon \, C_c^\infty(G)$, $f \geq 0$, $\int f(x)dx = 1$ and $f = 0$ outside $V$. Moreover since $K = K_0 Z$, one proves easily that $f(ux) = f(xu)$ $(u \, \varepsilon \, K, x \, \varepsilon \, G)$. Therefore $E_{\mathfrak{D}}\pi(f) = \pi(f)E_{\mathfrak{D}}$ and

$$|\pi(f)\psi - \psi| = \left|\int f(x)(\pi(x)\psi - \psi)dx\right| \leq \int f(x)|\pi(x)\psi - \psi|\,dx \leq \tfrac{1}{2}|\psi|.$$

Hence $\pi(f)\psi \neq 0$ and so it follows from Lemma 3 of [5(c)] that $\pi(f)E_{\mathfrak{D}}\psi = E_{\mathfrak{D}}\pi(f)\psi \neq 0$ for some $\mathfrak{D} \, \varepsilon \, \Omega_\pi$. This proves that $E_{\mathfrak{D}}\psi \neq 0$.

**COROLLARY.** *Let $\pi'$ be another quasi-simple irreducible representation of $G$ on a Banach space $\mathfrak{H}'$ and let $\psi' \neq 0$ be an element in $\mathfrak{H}'$. Suppose $\pi'(f)\psi' = 0$ whenever $\pi(f) = 0$ $(f \, \varepsilon \, C_c(G))$. Then $\pi$ and $\pi'$ are infinitesimally equivalent.*

Let $\eta$ be the central character of $\pi$. For any $z \, \varepsilon \, Z$ and $f \, \varepsilon \, C_c(G)$, define a function $_zf \, \varepsilon \, C_c(G)$ by $_zf(x) = f(z^{-1}x)$ $(x \, \varepsilon \, G)$. Then $\pi(_zf) = \eta(z)\pi(f)$ and therefore

$$\pi'(z)\pi'(f)\psi' = \pi'(_zf)\psi' = \eta(z)\pi'(f)\psi' \qquad\qquad (z \, \varepsilon \, Z, f \, \varepsilon \, C_c(G)).$$

Since elements of the form $\pi(f)\psi'$ $(f \, \varepsilon \, C_c(G))$ are dense in $\mathfrak{H}'$, we conclude that $\eta$ is also the central character of $\pi'$.

Now $Z \subset K$ and therefore, by Schur's lemma, there exists, for each $\mathfrak{D} \, \varepsilon \, \Omega$, a character $\eta_{\mathfrak{D}}$ of $Z$ such that $\sigma(z) = \eta_{\mathfrak{D}}(z)\sigma(1)$ $(z \, \varepsilon \, Z)$ for any representation $\sigma$ in $\mathfrak{D}$. Let $\Omega_0$ be the set of all those $\mathfrak{D}$ for which $\eta_{\mathfrak{D}} = \eta$. Then it is obvious that $\Omega_\pi \cup \Omega_{\pi'} \subset \Omega_0$ and if $E_{\mathfrak{D}}'$ is the canonical projection of $\mathfrak{H}'$ and $\mathfrak{H}_{\mathfrak{D}}'$,

$$E_{\mathfrak{D}} = d(\mathfrak{D})\int_{K_0} \xi_{\mathfrak{D}}(u^{-1})\pi(u)\,du. \qquad E_{\mathfrak{D}}' = d(\mathfrak{D})\int_{K_0} \xi_{\mathfrak{D}}(u^{-1})\pi'(u)\,du.$$

for all $\mathfrak{D} \,\varepsilon\, \Omega_0$ (see [5(c), p. 249]). Now, by the above lemma, we can choose $\mathfrak{D}_0 \,\varepsilon\, \Omega_{\pi'}$ such that $E_{\mathfrak{D}_0}'\psi' \neq 0$. Then $\pi'(f)E_{\mathfrak{D}_0}'\psi' \neq 0$ for some $f \,\varepsilon\, C_c(G)$. This implies that $\pi'(f_{\mathfrak{D}_0})\psi' \neq 0$ and therefore $\pi(f)E_{\mathfrak{D}_0} = \pi(f_{\mathfrak{D}_0}) \neq 0$. Hence $E_{\mathfrak{D}_0} \neq 0$ and so $\mathfrak{D}_0 \,\varepsilon\, \Omega_\pi \cap \Omega_{\pi'}$. Let $\nu$ and $\nu'$ be the corresponding representations of $L(\mathfrak{D}_0)$ on $\mathfrak{H}_{\mathfrak{D}_0}$ and $\mathfrak{H}_{\mathfrak{D}_0}'$ respectively. In view of the corollary to Lemma 1, it is sufficient to prove that $\nu$ and $\nu'$ are equivalent. But since they are both irreducible and finite-dimensional it would be enough to show that they have the same kernel. Let $\mathfrak{M}$ be the vernel of $\nu$ in $L(\mathfrak{D}_0)$. Then it is a maximal two-sided ideal in $L(\mathfrak{D}_0)$. If $g \,\varepsilon\, \mathfrak{M}$, $\pi(g_{\mathfrak{D}_0}) = \pi(g)E_{\mathfrak{D}_0} = 0$ and therefore $\pi'(g)E_{\mathfrak{D}_0}'\psi' = \pi'(g_{\mathfrak{D}_0})\psi' = 0$. This shows that $\nu'(\mathfrak{M})E_{\mathfrak{D}_0}'\psi' = 0$. Since $E_{\mathfrak{D}_0}'\psi' \neq 0$ and $\nu'$ is irreducible, it follows that the kernel of $\nu'$ must contain $\mathfrak{M}$ and therefore coincide with it. This completes the proof.

Let $Z_0$ be a fixed subgroup of $Z$ such that $Z/Z_0$ is finite and let $x \to x^*$ denote the natural mapping of $G$ and $G^* = G/Z_0$. Suppose $\alpha_1, \cdots, \alpha_r$ is a finite set of continuous linear functions on $\mathfrak{H}$ and $\psi_1, \cdots, \psi_r$ are certain given elements in $\mathfrak{H}$. Put

$$f(x) = \alpha_1(\pi(x)\psi_1) + \cdots + \alpha_r(\pi(x)\psi_r) \qquad (x \,\varepsilon\, G).$$

Then if the central character of $\pi$ is unitary, it is obvious that $|f(x)|$ depends only on on $x^*$. Let $dx^*$ denote the Haar measure on $G^*$.

LEMMA 3. *Assume that the function*

$$f(x) = \alpha_1(\pi(x)\psi_1) + \cdots + \alpha_r(\pi(x)\psi_r) \qquad (x \,\varepsilon\, G)$$

*is not identically zero on G. Then if the central character of $\pi$ is unitary and $\displaystyle\int_{G^*} |f(x)|^2 \, dx^* < \infty$, $\pi$ is infinitesimally equivalent to an irreducible unitary repersentation of G on a Hilbert space.*

Let $\eta_\pi$ denote the central character of $\pi$. Then $f(xz) = \eta_\pi(z)f(x)$ $(x \,\varepsilon\, G, z \,\varepsilon\, Z)$ and therefore it follows easily from the Peter-Weyl theorem for the compact group $K/D \cap Z$ that

$$\int_{K_0} |f(xu)|^2 \, du = \sum_{\mathfrak{D} \,\varepsilon\, \Omega_\pi} \int_{K_0} |f_{\mathfrak{D}}(xu)|^2 \, du.$$

Since $f \neq 0$, we can choose $\mathfrak{D}_0 \,\varepsilon\, \Omega_\pi$ such that $f_{\mathfrak{D}_0} \neq 0$. Then it is clear that

$$f_{\mathfrak{D}_0}(x) = \sum_{i=1}^{r} \alpha_i(\pi(x)E_{\mathfrak{D}_0}\psi_i)$$

and since

$$\int_{K_0} |f_{\mathfrak{D}_0}(xu)|^2 \, du \leqq \int_{K_0} |f(xu)|^2 \, du,$$

7

it follows that

$$\int |f_{\mathfrak{D}_0}(x)|^2\, dx^* \leqq \int |f(x)|^2\, dx^* < \infty.$$

Hence if we replace $\psi_i$ by $E_{\mathfrak{D}_0}\psi_i$ and $f$ by $f_{\mathfrak{D}_0}$, our problem is reduced to the case when the given elements of $\mathfrak{H}$ all lie in $\mathfrak{H}_{\mathfrak{D}_0}$. Moreover it is obvious that without loss of generality we may assume that $\psi_1, \cdots, \psi_r$ is a base for $\mathfrak{H}_{\mathfrak{D}_0}$ over the field $C$ of complex numbers. Since $L(\mathfrak{D}_0)$ is irreducible under $\pi(L(\mathfrak{D}_0))$ (Corollary to Lemma 1), it follows from Burnside's Theorem that

$$\pi(g_j)\psi_k = \delta_{jk}\psi_k \qquad\qquad 1 \leqq j, k \leqq r$$

for suitable elements $g_j \varepsilon L(\mathfrak{D}_0)$. ($\delta_{jk}$ is the Kronecker symbol.) Since $f \neq 0$ we may assume that $f_1(x) = \alpha_1(\pi(x)\psi_1)$ is not identically zero. Then

$$f_1(x) = \sum_{i=1}^{r} \alpha_i(\pi(x)\pi(g_1)\psi_i) = \int_G f(xy)g_1(y)\,dy$$

and therefore

$$|f_1(x)|^2 \leqq \int_\omega |f(xy)|^2\,dy \int_G |g_1(y)|^2\,dy$$

where $\omega$ is some compact subset of $G$ outside which $g_1$ is zero. This shows that $\int |f_1(x)|^2\,dx^* < \infty$ and since $f_1 \neq 0$, our problem is now reduced to the case when

$$f(x) = \alpha(\pi(x)\psi).$$

Here $\alpha$ is a continuous linear function on $\mathfrak{H}$ and $\psi \varepsilon \mathfrak{H}_{\mathfrak{D}_0}$.

Let $U'$ be the Hilbert space consisting of all measurable functions $g$ on $G$ such that (1) $g(xz) = g(x)\eta_\pi(z)$ $(x \varepsilon G, z \varepsilon Z)$ and (2) $\int_{G^*} |g(x)|^2 dx^* < \infty$. We define a representation $\sigma'$ of $G$ on $U'$ by $(\sigma'(y)g)(x) = g(xy)$ $(x, y \varepsilon G, g \varepsilon U')$. It is obvious that $f$ lies in $U'$. Let $U$ be the smallest closed subspace of $U'$ containing $f$ which is invariant under $\sigma'(G)$. We denote by $\sigma$ the representation of $G$ defined on $U$ under $\sigma'$. It is clear that $\sigma$ is unitary. We shall now show that $\pi$ is infinitesimally equivalent to $\sigma$.

For any $\mathfrak{D} \varepsilon \Omega$ let $F_{\mathfrak{D}}$ denote the corresponding canonical projection in $U$. Also, we denote the operator $\int g(x)\sigma(x)dx$ $(g \varepsilon C_c(G))$ by $\sigma(g)$. Since $f \varepsilon U_{\mathfrak{D}_0}$, it follows that $F_{\mathfrak{D}_0} \neq 0$ and therefore $F_{\mathfrak{D}_0}\sigma(g) = \sigma(_{\mathfrak{D}_0}g)$. Moreover it is obvious that elements of the form $\sigma(g)f$ $(g \varepsilon C_c(G))$ are dense in $U$. Therefore $\sigma(L(\mathfrak{D}_0))f$ is dense in $U_{\mathfrak{D}_0} = F_{\mathfrak{D}_0}U$. Now suppose $\pi(g)\psi = 0$ $(g \varepsilon C_c(G))$. Then

$$0 = \alpha(\pi(x)\pi(g)\psi) = \int \alpha(\pi(xy)\psi)g(y)\,dy = \int f(xy)g(y)\,dy$$

and therefore $\sigma(g)f = 0$. Hence if we can prove that $\sigma$ is irreducible (and therefore also quasi-simple (see [10(b)])), our assertion would follow from the Corollary to Lemma 2. However, in view of the above remarks, we can define a linear mapping $A$ of $\mathfrak{H}_{\mathfrak{D}_0}$ onto $\sigma(L(\mathfrak{D}_0))f$ such that

$$A(\pi(g)\psi) = \sigma(g)f \qquad (g \, \varepsilon \, L(\mathfrak{D}_0)).$$

Then $\dim \sigma(L(\mathfrak{D}_0))f \leqq \dim \mathfrak{H}_{\mathfrak{D}_0} < \infty$ and therefore since $\sigma(L(\mathfrak{D}_0))f$ is dense in $U_{\mathfrak{D}_0}, U_{\mathfrak{D}_0} = \sigma(L(\mathfrak{D}_0))f$. But $\mathfrak{H}_{\mathfrak{D}_0}$ is irreducible under $\pi(L(\mathfrak{D}_0))$ and so it follows from the existence of $A$ that the same is true for $U_{\mathfrak{D}_0}$ under $\sigma(L(\mathfrak{D}_0))$. Now suppose $V$ is a closed subspace of $U$ which is invariant under $\sigma(G)$ and let $W$ be the orthogonal complement of $V$ in $U$. In view of the above irreducibility either $V$ or $W$ must contain $U_{\mathfrak{D}_0}$. Suppose $V \supset U_{\mathfrak{D}_0}$. Then $f \varepsilon V$ and therefore $V = U$ from the definition of $U$. Similarly if $W \supset U_{\mathfrak{D}_0}, W = U$. This proves that $\sigma$ is an irreducible representation and so our lemma follows.

COROLLARY.[1] *Let $\pi$ be an irreducible unitary representation of $G$ on a Hilbert space $\mathfrak{H}$. Suppose there exist two elements $\phi_0 \neq 0$, $\psi_0 \neq 0$ in $\mathfrak{H}$ such that*

$$\int_{G^*} |(\phi_0, \pi(x)\psi_0)|^2 \, dx^* < \infty.$$

*Then there exists a positive real number $d_\pi$ such that*

$$\int_{G^*} |(\phi, \pi(x)\psi)|^2 \, dx^* = d_\pi^{-1} |\phi|^2 |\psi|^2$$

*for all $\phi, \psi$ in $\mathfrak{H}$.*

Let $V$ and $W$ respectively be the subspaces of $\mathfrak{H}$ consisting of all elements of the form $\pi(f)\phi_0$ and $\pi(f)\psi_0$ $(f \varepsilon C_c(G))$. Then $V$ and $W$ are both dense in $\mathfrak{H}$. Since $\pi$ is irreducible and unitary, it is quasi-simple [10(b)] and therefore $\dim \mathfrak{H}_{\mathfrak{D}} < \infty$ $(\mathfrak{D} \varepsilon \Omega)$. But if $\mathfrak{D} \varepsilon \Omega_\pi$, $E_{\mathfrak{D}}\pi(f) = \pi(_{\mathfrak{D}}f)$ $(f \varepsilon C_c(G))$ and hence $E_{\mathfrak{D}}V \subset V$. $V$ being dense in $\mathfrak{H}$, it follows that $E_{\mathfrak{D}}V$ is dense in $\mathfrak{H}_{\mathfrak{D}}$ and therefore $E_{\mathfrak{D}}V = \mathfrak{H}_{\mathfrak{D}}$. This shows that $\mathfrak{H}_0 = \sum_{\mathfrak{D}} \mathfrak{H}_{\mathfrak{D}} \subset V$. Similarly $\mathfrak{H}_0 \subset W$.

If $f, g \, \varepsilon \, C_c(G)$,

$$(\pi(f)\psi_0, \pi(x)\pi(g)\psi_0) = \int \int f(y)(\phi_0, \pi(y^{-1}xz)\psi_0)g(z)\,dy\,dz$$

---

[1] See Godement [4(b)].

and therefore it is obvious that

$$\int_{G^*} |(\phi, \pi(x)\psi)|^2 \, dx < \infty$$

for $\phi \, \varepsilon \, V$ and $\psi \, \varepsilon \, W$. Hence without loss of generality we may assume that $\phi_0, \psi_0 \, \varepsilon \, \mathfrak{H}_{\mathfrak{D}_0}$ for some $\mathfrak{D}_0 \, \varepsilon \, \Omega_\pi$. Let $\phi \neq 0$ be any element in $V$. Put $f_\phi(x) = (\phi, \pi(x)\psi_0)$ $(x \, \varepsilon \, G)$ and define $U'$ and $\sigma'$ as in the proof of Lemma 3. Let $U_\phi$ be the smallest closed subspace of $U'$ containing $f_\phi$ which is invariant under $\sigma'(G)$ and let $\sigma_\phi$ denote the corresponding representation of $G$ on $U_\phi$. Then as we have seen during the proof of Lemma 3, $\sigma_\phi$ is irreducible and quasi-simple. Also it is obvious that $\sigma_\phi(h)f_\phi$ $(h \, \varepsilon \, C_c(G))$ is the function $x \to (\phi, \pi(x)\pi(h)\psi_0)$ $(x \, \varepsilon \, G)$. Therefore it follows from the Corollary to Lemma 2 that $\pi$ and $\sigma_\phi$ are infinitesimally equivalent. Since they are both unitary they are equivalent [5(b), Theorem 8] and so there exists a unitary mapping $B_\phi$ of $\mathfrak{H}$ onto $U_\phi$ such that $B_\phi\pi(x) = \sigma_\phi(x)B_\phi$ $(x \, \varepsilon \, G)$. Moreover in view of what we have said above, there exists a linear mapping $A_\phi$ of $W$ into $U_\phi$ such that

$$A_\phi\pi(h)\psi_0 = \sigma_\phi(h)f_\phi \qquad\qquad (h \, \varepsilon \, C_c(G)).$$

Put $C_\phi = B_\phi^{-1}A_\phi$. Then $C_\phi$ is a linear mapping of $W$ into $\mathfrak{H}$ and $\pi(h)C_\phi = C_\phi\pi(h)$ $(h \, \varepsilon \, C_c(G))$. This holds in particular if $h \, \varepsilon \, L(\mathfrak{D})$ $(\mathfrak{D} \, \varepsilon \, \Omega_\pi)$ and therefore $C_\phi$ maps $\mathfrak{H}_{\mathfrak{D}}$ into itself. Now if we apply Schur's lemma to the finite-dimensional irreducible representation of $L(\mathfrak{D})$ on $\mathfrak{H}_{\mathfrak{D}}$, we can conclude from Lemma 1 that $C_\phi$ must be a scalar multiple of the identity on $\mathfrak{H}_0$. So there exists a complex number $c_\phi$ such that

$$\| \sigma_\phi(h)f_\phi \| = \| A_\phi\pi(h)\psi_0 \| = | c_\phi | \, \| B_\phi\pi(h)\psi_0 \| = | c_\phi | \, | \pi(h)\psi_0 |$$

if $h \, \varepsilon \, L_\pi$. (Here $\| \; \|$ denotes the norm in $U_\phi$.) But since

$$\| \sigma_\phi(h)f_\phi \|^2 = \int_{G^*} |(\phi, \pi(x)\pi(h)\psi_0)|^2 \, dx^*,$$

it follows that

$$\int_{G^*} |(\phi, \pi(x)\psi)|^2 \, dx^* = | c_\phi |^2 | \psi |^2$$

for all $\psi \, \varepsilon \, \mathfrak{H}_0$. If $\phi = 0$, we put $c_\phi = 0$ so that the above relation continues to hold in that case as well. Now suppose $\phi, \psi$ lie in $\mathfrak{H}_0 \subset V \cap W$. Then

$$\int |(\phi, \pi(x)\psi)|^2 \, dx^* = \int |(\psi, \pi(x)\phi)|^2 \, dx^*$$

and therefore

$$| c_\phi |^2 | \psi |^2 = | c_\psi |^2 | \phi |^2.$$

Hence we can find a real number $c$ such that $|c_\phi|^2 = c\,|\phi|^2$ for $\phi\,\varepsilon\,\mathfrak{H}_0$ and therefore

$$\int^{\boldsymbol{\cdot}} |(\phi, \pi(x)\psi)|^2\, dx^* = c\,|\phi|^2\,|\psi|^2 \qquad\qquad (\phi, \psi\,\varepsilon\,\mathfrak{H}_0).$$

Since $\mathfrak{H}_0$ is dense in $\mathfrak{H}$, it is obvious that $c$ is positive and an elementary argument shows that the above relation continues to hold for all $\phi, \psi\,\varepsilon\,\mathfrak{H}$. Now if we put $d_\pi = c^{-1}$ we get the assertion of the Corollary.

### 3.  The Schur orthogonality relations.

DEFINITION.  *Let $\pi$ be an irreducible unitary representation of $G$ on a Hilbert space $\mathfrak{H}$.  We say that $\pi$ is square-integrable, if there exist two elements $\phi_0 \neq 0$, $\psi_0 \neq 0$ in $\mathfrak{H}$ such that*

$$\int_{G^*} |(\phi_0, \pi(x)\psi_0)|^2\, dx^* < \infty.$$

*Similarly we say that $\pi$ is integrable if*

$$\int_{G^*} |(\phi_0, \pi(x)\psi_0)|\, dx^* < \infty$$

*for some nonzero elements $\phi_0$, $\psi_0$ in $\mathfrak{H}$.*

It is obvious that the above definitions do not depend on the choice of the subgroup $Z_0$ so long as $Z/Z_0$ is finite.  We have seen above (Corollary to Lemma 3) that if $\pi$ is square-integrable

$$\int_{G^*} |(\phi, \pi(x)\psi)|^2\, dx < \infty$$

for all $\phi, \psi\,\varepsilon\,\mathfrak{H}$.  In analogy with the case of compact groups, we shall call the number $d_\pi$ (of the Corollary to Lemma 3) the *formal degree* of $\pi$. Naturally $d_\pi$ depends on the choice of $Z_0$ and the normalization of the Haar measure of $G^*$.  However once these have been fixed, it is obvious that two equivalent square-integrable representations have the same formal degree.

The situation for integrable representations is somewhat similar.

LEMMA 4.  *Let $\pi$ be an integrable representation of $G$ on $\mathfrak{H}$.  Then if $\mathfrak{H}_0 = \sum_{\mathfrak{D}\,\varepsilon\,\Omega} \mathfrak{H}_{\mathfrak{D}}$,*

$$\int |(\phi, \pi(z)\psi)|\, dx^* < \infty$$

*for all $\phi$, $\psi$ in $\mathfrak{H}_0$.*

Choose nonzero elements $\phi_0$, $\psi_0$ in $\mathfrak{H}$ such that

$$\int |(\phi_0, \pi(x)\psi_0)|\, dx^* < \infty.$$

Then if $g, h \in C_c(G)$, it is easy to verify that the function $|(\pi(g)\phi_0, \pi(x)\pi(h)\psi_0)|$ is integrable on $G^*$. Let $V$ and $W$ be the set of all elements of the form $\pi(g)\phi_0$ and $\pi(g)\psi_0$ $(g \in C_c(G))$ respectively. Then, as we have seen during the proof of the Corollary to Lemma 3, $\mathfrak{H}_0 \subset V \cap W$ and therefore our assertion follows.

Let $\pi$ and $\pi'$ be two square-integrable representations of $G$ on the Hilbert spaces $\mathfrak{H}$ and $\mathfrak{H}'$ respectively. Then if their central characters coincide on $Z_0$, it is obvious that[2] $(\phi, \pi(x)\psi)\ \mathrm{conj}(\phi', \pi'(x)\psi')$ may be regarded as a function of $x^*$ on $G^*$ $(\phi, \psi \in \mathfrak{H}; \phi', \psi' \in \mathfrak{H}'; x \in G)$.

THEOREM 1 (The Schur orthogonality relations[1]). *If $\pi$ and $\pi'$ are not equivalent*

$$\int_{G^*} (\phi, \pi(x)\psi)\,\mathrm{conj}(\phi', \pi'(x)\psi')\, dx^* = 0$$

*for all $\phi, \psi \in \mathfrak{H}$ and $\phi', \psi' \in \mathfrak{H}'$. On the other hand if the two representations are equivalent under a unitary mapping $U$ of $\mathfrak{H}$ onto $\mathfrak{H}'$,*

$$\int_{G^*} (\phi, \pi(x)\psi)\,\mathrm{conj}(\phi', \pi'(x)\psi')\, dx^* = d_\pi^{-1}(U\phi, \phi')(\psi', U\psi)$$

$(\phi, \psi \in \mathfrak{H}; \phi'\psi' \in \mathfrak{H}')$ *where $d_\pi$ is the formal degree of $\pi$.*

Let $d_{\pi'}$ denote the formal degree of $\pi'$. Then it is obvious from the Corollary to Lemma 3 that

$$\int |(\phi, \pi(x)\psi)\,\mathrm{conj}(\phi', \pi'(x)\psi')|\, dx^* \leqq (d_\pi d_{\pi'})^{-1}\,|\phi|^2\,|\psi|^2\,|\phi'|^2\,|\psi'|^2.$$

Therefore for any given $\phi \in \mathfrak{H}$ and $\phi' \in \mathfrak{H}'$, there exists a bounded linear operator $A$ from $\mathfrak{H}$ to $\mathfrak{H}'$ such that

$$(\psi', A\psi) = \int (\phi, \pi(x)\psi)\,\mathrm{conj}(\phi', \pi'(x)\psi')\, dx^*$$

for all $\psi' \in \mathfrak{H}'$ and $\psi \in \mathfrak{H}$. It follows immediately from this relation that

$$(\psi', A\pi(x)\psi) = (\pi'(x^{-1})\psi', A\psi) \qquad\qquad (x \in G)$$

and therefore $A\pi(x) = \pi'(x)A$. In order to prove the first statement of the

---

[2] conj $c$ denotes the conjugate of a complex number $c$.

theorem it would be enough to show that if $A \neq 0$, $\pi$ and $\pi'$ are equivalent. So let us suppose $A \neq 0$. Choose $\psi$ in $\mathfrak{H}$ such that $A\psi \neq 0$. Then if $h \, \varepsilon \, C_c(G)$ it is clear that

$$A\pi(h)\psi = \pi'(h)A\psi$$

and therefore from the Corollary to Lemma 2, $\pi$ and $\pi'$ are infinitesimally equivalent. But since they are both unitary this implies that they are equivalent [5(b), Theorem 8].

In order to prove the second statement we may assume that $\pi' = \pi$ since

$$(\phi', \pi'(x)\psi') = (U^{-1}\phi', \pi(x)U^{-1}\psi')$$

in the general case. Hence if we keep to the above notation, $A$ is now a bounded linear operator on $\mathfrak{H}$, which commutes with $\pi(x)$ $(x \, \varepsilon \, G)$. Since $\pi$ is irreducible, $A$ must be a scalar multiple of the identity. Hence

$$\int (\phi, \pi(x)\psi)\operatorname{conj}(\phi', \pi(x)\psi')\,dx^* = c_{\phi, \phi'}(\psi', \psi)$$

where $c_{\phi, \phi'}$ is a complex number depending only on $\phi$ and $\phi'$. But obviously

$$\int (\phi, \pi(x)\psi)\operatorname{conj}(\phi', \pi(x)\psi')\,dx^* = \int (\psi', \pi(x)\phi')\operatorname{conj}(\psi, \pi(x)\phi)\,dx^*$$

and therefore

$$c_{\phi, \phi'}(\psi', \psi) = c_{\psi, \psi'}(\phi, \phi').$$

Choose $\psi' = \psi = \psi_0 \neq 0$ and put $c = (\psi_0, \psi_0)/|\psi_0|^2$. Then $c_{\phi, \phi'} = c(\phi', \phi)$ for all $\phi', \phi \, \varepsilon \, \mathfrak{H}$. In particular if we put $\phi = \phi' = \psi = \psi' = \psi_0$ we find that $c = d_\pi^{-1}$. Thus the theorem is proved.

**4. The character of a square-integrable representation.** Let $C_c^\infty(G)$ denote, as before, the subspace of $C_c(G)$ consisting of those functions which are indefinitely differentiable everywhere. For $x^* \, \varepsilon \, G^*$, we put $y^{x^*} = xyx^{-1}$ $(y \, \varepsilon \, G)$ where $x$ is any element in $G$ whose image in $G^*$ is $x^*$.

THEOREM 2. *Let $\pi$ be a square-integrable representation of $G$ on a Hilbert space and let $T_\pi$ denote the character [5(c)] of $\pi$. Then if $f \, \varepsilon \, C_c^\infty(G)$,*

$$T_\pi(f) = d_\pi \int_{G^*} dx^* \left\{ \int_G f(y^{x^*})(\phi, \pi(y)\phi)\,dy \right\}$$

*where $\phi$ is any unit vector in $\mathfrak{H}$ and $d_\pi$ is the formal degree of $\pi$.*

Let $Q$ denote the operator $\displaystyle\int_G f(y)\pi(y)\,dy$. We know [5(c), p. 243] that there exists a complete orthonormal set $\{\psi_j\}_{j \in J}$ in $\mathfrak{H}$ such that $\sum_{i,j \in J} |Q_{ij}| < \infty$ where $Q_{ij} = (\psi_i, Q\psi_j)$. Moreover $\pi(x^{-1})Q\pi(x)$ $(x \varepsilon G)$ depends only on $x^*$ and so we may denote it by $Q^{x^*}$. Then

$$(\phi, Q^{x^*}\phi) = (\pi(x)\phi, Q\pi(x)\phi) = \sum_i (\pi(x)\phi, \psi_i)(\psi_i, Q\pi(x)\phi)$$
$$= \sum_i \sum_j (\pi(x)\phi, \psi_i) Q_{ij}(\psi_j, \pi(x)\phi).$$

But if we make use of the Schwartz inequality and the Schur orthogonality relations, we get

$$\sum_{i,j} \int_{G^*} |(\pi(x)\phi, \psi_i) Q_{ij}(\psi_j, \pi(x)\phi)| \, dx^*$$
$$\leq \sum_{i,j} |Q_{ij}| \left\{ \int_{G^*} |(\pi(x)\phi, \psi_i)|^2 \, dx^* \int_{G^*} |(\psi_j, \, (x)\phi)|^2 \, dx^* \right\}^{\frac{1}{2}}$$
$$= d_\pi^{-1} \sum_{i,j} |Q_{ij}| < \infty$$

and therefore by Lebegue's Theorem the above series for $(\phi, Q^{x^*}\phi)$ may be integrated over $G^*$ term by term. Hence

$$\int_{G^*} (\phi, Q^{x^*}\phi) \, dx^* = \sum_i \sum_j Q_{ij} \int_{G^*} (\pi(x)\phi, \psi_i)(\psi_j, \pi(x)\phi) \, dx^*$$
$$= \sum_i \sum_j (Q_{ij}/d_\pi)(\psi_j, \psi_i) = d_\pi^{-1} \sum Q_{ii}$$
$$= d_\pi^{-1} \operatorname{Sp} Q = d_\pi^{-1} T_\pi(f).$$

But

$$(\phi, Q^{x^*}\phi) = \int_G f(y^{x^*})(\phi, \pi(y)\phi) \, dy$$

and so the theorem is proved.

It should be noticed that, in general, the double integral in the above theorem is not absolutely convergent and therefore the order of the two integrations cannot be interchanged.

**5. The discrete part of the Plancherel measure.** We shall assume in this section that $Z$ is finite and $Z_0 = \{1\}$. Let $\mathcal{E}$ denote the set of all equivalence classes of irreducible unitary representations of $G$. We consider the Hilbert space $L_2(G)$ consisting of all complex-valued functions on $G$ which are square-integrable with respect to the Haar measure. Let $\lambda$ denote the left regular representation of $G$ on $L_2(G)$ defined by

$$(\lambda(x)f)(y) = f(x^{-1}y) \qquad\qquad (x, y \varepsilon G; f \varepsilon L_2(G)).$$

Then $\lambda$ is unitary. We say that a class $\omega \varepsilon \mathcal{E}$ is discrete if there exists a closed subspace $\mathfrak{H} \neq 0$ of $L_2(G)$ which is invariant and irreducible under $\lambda(G)$ and such that the corresponding representation of $G$ on $\mathfrak{H}$ lies in $\omega$. It is known (see Godemont [4(a), Theorem 1]) that $\omega$ is discrete if and only if every representation in $\omega$ is square-integrable. Let $\mathcal{E}_0$ denote the set of all discrete classes in $\mathcal{E}$. If $\omega \varepsilon \mathcal{E}_0$, we denote by $d_\omega$ the formal degree of any representation in $\omega$.

For any $\omega \varepsilon \mathcal{E}$, let $T_\omega$ denote the character [5(c)] of any representation in $\omega$. Then it is known (see Segal [10(b)], Mautner [8] and [5(b), Theorem 7]) that there exists a unique positive measure $\mu$ on $\mathcal{E}$ such that

$$\int_G |f(x)|^2 \, dx = \int_{\mathcal{E}} T_\omega(\bar{f} * f) \, d\mu \qquad (f \varepsilon C_o(G))$$

where $\bar{f}(x) = \operatorname{conj} f(x^{-1})$ $(x \varepsilon G)$. We shall call $\mu$ the Plancherel measure on $\mathcal{E}$. First we prove the following simple lemma.

LEMMA 5. *Every single point $\omega_0$ in $\mathcal{E}$ is $\mu$-measurable.*

For any $f \varepsilon C_o(G)$ put $\|f\|_1 = \int_G |f(x)| \, dx$. Then under this norm $C_o(G)$ becomes a separable metric space. Let $\pi_0$ be a representation in $\omega_0$ and let $\mathfrak{M}_{\omega_0}$ denotes the set of all $f \varepsilon C_o(G)$ such that $\pi_0(f) = 0$. Then $\mathfrak{M}_{\omega_0}$ is also separable under the above metric and so we can select a sequence $\{\alpha_1, \alpha_2, \cdots\}$ in $\mathfrak{M}_{\omega_0}$ which is dense in $\mathfrak{M}_{\omega_0}$. Put $F_n(\omega) = T_\omega(\bar{\alpha}_n * \alpha_n)$ $(\omega \varepsilon \mathcal{E})$. Then $F_1, F_2, \cdots$ are all measurable functions on $\mathcal{E}$ and it follows from the Corollary to Lemma 2 that $\omega_0$ is the only point in $\mathcal{E}$ where they vanish simultaneously. From this the lemma follows immediately.

Our main object in this section is to prove the following theorem.

THEOREM 3. *If $\omega_0$ is a discrete class, $\mu(\omega_0) = d_{\omega_0}$.*

Define $\pi_0$, $\mathfrak{M}_{\omega_0}$ and $\{\alpha_n\}_{n \geq 1}$ as above and let $\psi_0$ be a nonzero vector in the representation space $\mathfrak{H}$ of $\pi_0$. Put $g(x) = (\pi_0(x)\psi_0, \psi_0)$. Then since $\omega_0$ is discrete, $g \varepsilon L_2(G)$. We first need the following lemma.

LEMMA 6. *There exists a sequence $g_1, g_2, \cdots$ in $C_o(G)$ and a subset $\mathcal{E}'$ of $\mathcal{E}$ satisfying the following conditions:*

(1)  *The complement of $\mathcal{E}'$ in $\mathcal{E}$ is of $\mu$-measure zero.*

(2)  $\operatorname*{Lim}_{n \to \infty} g_n = g$ *in* $L_2(G)$.

(3) *For every $\omega \varepsilon \mathcal{E}'$, $\underset{n \to \infty}{\mathrm{Lim}} T_\omega (\tilde{g}_n * g_n)$ exists and is finite and* [3]
$\underset{n \to \infty}{\mathrm{Lim}} T_\omega ((\alpha_m * g_n)\tilde{\ } * (\alpha * g_n)) = 0$ *for every* $m \geq 1$.

Since $C_c(G)$ is dense in $L_2(G)$ we can choose a sequence $\{h_1, h_2, \cdots\}$ in $C_c(G)$ such that $h_n \to g$ in $L_2(G)$. Then

$$\int_{\mathcal{E}} T_\omega(\tilde{h}_n * h_n) d\mu = \| h_n \|^2 \to \| g \|^2$$

where $\| \ \|$ denotes the norm in $L_2(G)$. Therefore by the Riesz-Fischer Theorem, there exists a subset $\mathcal{E}_0' \subset \mathcal{E}$ and a subsequence $\{h_n^{(0)}\}$ of $\{h_n\}$ such that [4] (1) $\mathcal{E} - \mathcal{E}_0'$ is of $\mu$-measure zero and (2) $\underset{n \to \infty}{\mathrm{Lim}} T(h_n^{(0)\tilde{\ }} * h_n^{(0)})$ exists and is finite for every $\omega \varepsilon \mathcal{E}_0'$. Now for each integer $r \geq 0$ we shall define a subsequence $\{h_n^{(r)}\}$ of $\{h_n^{(0)}\}$ and a subset $\mathcal{E}_r' \subset \mathcal{E}_0'$ such that (1) $\mathcal{E} - \mathcal{E}_r'$ is of $\mu$-measure zero and (2) $\underset{n \to \infty}{\mathrm{Lim}} T_\omega((\alpha_m * h_n^{(r)})\tilde{\ } * (\alpha_m * h_n^{(r)})) = 0$ for $\omega \varepsilon \mathcal{E}_r'$ and $1 \leq m \leq r$. This has already been done for $r = 0$. So assuming that $\{h_n^{(r)}\}$ and $\mathcal{E}_r'$ have been defined, we proceed to define $\{h_n^{(r+1)}\}$ and $\mathcal{E}_{r+1}'$ by induction. Suppose $\omega \varepsilon \mathcal{E}_r'$. Then $T_\omega(h_n^{(0)\tilde{\ }} * h_n^{(0)})$ converges to a finite limit and therefore if $\pi \varepsilon \omega$, $\pi(h_n^{(r)})$ converges with respect to the Hilbert-Schmidt (H. S.) norm to a bounded operator $A_\pi$. Now put $h_n' = \alpha_{r+1} * h_n^{(r)}$. Then $\pi(h_n') = \pi(\alpha_{r+1})\pi(h_n^{(r)})$ and since $\pi(\alpha_{r+1})$ is a bounded operator, it is obvious that $\pi(h_n')$ also converges to $\pi(\alpha_{r+1})A_\pi$ in the H. S. norm. On the other hand since $\| h_n^{(r)} - g \| \to 0$, $\alpha_{r+1} * h_n^{(r)} \to \alpha_{r+1} * g$ in $L_2(G)$. But

$$(\alpha_{r+1} * g)(x) = (\pi_0(x)\psi_0, \pi_0(\alpha_{r+1})\psi_0) = 0 \qquad (x \varepsilon G)$$

because $\alpha_{r+1} \varepsilon \mathfrak{M}_{\omega_0}$. Therefore $\alpha_{r+1} * h_n^{(r)} \to 0$ in $L_2(G)$ and so

$$\int_{\mathcal{E}} T_\omega((\alpha_{r+1} * h_n^{(r)})\tilde{\ } * (\alpha_{r+1} * h_n^{(r)})) d\mu \to 0$$

as $n \to \infty$. Hence by the Riesz-Fischer Theorem we can select a subsequence $\{h_n^{(r+1)}\}$ of $\{h_n^{(r)}\}$ and a subset $\mathcal{E}_{r+1} \subset \mathcal{E}$ such that (1) $\mathcal{E} - \mathcal{E}_{r+1}$ is of $\mu$-measure zero and (2) if $\omega \varepsilon \mathcal{E}_{r+1}$,

$$\underset{n \to \infty}{\mathrm{Lim}} T_\omega((\alpha_{r+1} * h_n^{(r+1)})\tilde{\ } * (\alpha_{r+1} * h_n^{(r+1)})) = 0.$$

Now put $\mathcal{E}_{r+1}' = \mathcal{E}_r' \cap \mathcal{E}_{r+1}$. Then $\mathcal{E}_{r+1}$ and $\{h_n^{(r+1)}\}$ satisfy all the required conditions. Hence if we take $\mathcal{E}' = \bigcap_{r=1}^{\infty} \mathcal{E}_r'$ and $g_n = h_n^{(n)}$ we get the assertion of the lemma.

---

[3] We write $f\tilde{\ }$ instead of $\tilde{f}$ ($f \varepsilon C_c(G)$) whenever it is convenient to do so.

[4] $\mathcal{E} - \mathcal{E}_0'$ is the complement of $\mathcal{E}_0'$ in $\mathcal{E}$.

Let us now come to the proof of the theorem. Let $\omega$ be a class in $\mathcal{E}'$ and $\pi$ a representation in $\omega$. Since $T_\omega(\tilde{g}_n * g_n)$ is convergent, the operators $\pi(g_n)$ converge to a limit in the H. S. norm. Let $A_\pi$ denote this limit. Since $\pi(\alpha_r)$ is a bounded operator $\pi(\alpha_r * g_n) = \pi(\alpha_r)\pi(g_n)$ also converges to $\pi(\alpha_r)A_\pi$ in the H. S. norm. However

$$\operatorname*{Lim}_{n \to \infty} T_\omega((\alpha_r * g_n)^\sim * (\alpha_r * g_n)) = 0$$

and therefore $\pi(\alpha_r)A_\pi = 0$ $(r \geq 1)$. Now suppose $\omega \neq \omega_0$. Then if $\psi$ lies in the representation space of $\pi, \pi(\alpha_r)A_\pi\psi = 0$. Since $\pi$ and $\pi_0$ are not equivalent (and therefore also not infinitesimally equivalent [5(b), Theorem 8]), it follows from the Corollary to Lemma 2 that $A_\pi\psi = 0$. This being true for every $\psi$, $A_\pi = 0$. But $\pi(g_n)$ tends to $A_\pi$ in the H. S. norm and so this shows that

$$\operatorname*{Lim}_{n \to \infty} T_\omega(\tilde{g}_n * g_n) = 0.$$

On the other hand $g_n \to g$ in $L_2(G)$ and therefore

$$\| g \|^2 = \operatorname*{Lim}_{n \to \infty} \int_{\mathcal{E}}^{\cdot} T_\omega(\tilde{g}_n * g_n)\, d\mu.$$

From this it follows that $\omega_0 \varepsilon \mathcal{E}'$. For otherwise, in view of what we have just said, $\operatorname*{Lim}_{n \to \infty} T_\omega(\tilde{g}_n * g_n) = 0$ for all $\omega \varepsilon \mathcal{E}'$ and since $\mathcal{E} - \mathcal{E}'$ is of $\mu$-measure zero we would have

$$\| g \|^2 = \operatorname*{Lim}_{n \to \infty} \int_{\mathcal{E}} T_\omega(\tilde{g}_n * g_n)\, d\mu = 0.$$

But this is false since $g$ is continuous and $g(1) = |\psi_0|^2 \neq 0$. Therefore $\omega_0 \varepsilon \mathcal{E}'$ and so $T_{\omega_0}(\tilde{g}_n * g_n)$ tends to a finite limit. This shows that

$$\| g \|^2 = \operatorname*{Lim}_{n \to \infty} \int_{\mathcal{E}'}^{\cdot} T_\omega(\tilde{g}_n * g_n)\, d\mu = \int_{\mathcal{E}'} \operatorname*{Lim}_{n \to \infty} T_\omega(\tilde{g}_n * g_n)\, d\mu$$

$$= \mu(\omega_0) \operatorname*{Lim}_{n \to \infty} T_{\omega_0}(\tilde{g}_n * g_n).$$

Now let [5] $\{\psi_1, \psi_2, \cdots\}$ be a complete orthonormal set in the representation space $\mathfrak{H}$ of $\pi_0$. Then if $A_n = \pi_0(g_n)$,

$$(\psi_i, A_n\psi_j) = \int_G g_n(x)(\psi_i, \pi_0(x)\psi_j)\, dx.$$

Since $\pi_0$ is square-integrable it follows that

$$\operatorname*{Lim}_{n \to \infty} (\psi_i, A_n\psi_j) = \int g(x)(\psi_i, \pi_0(x)\psi_j)\, dx = (\psi_i, \pi_0(g)\psi_j).$$

---

[5] It is not difficult to see that $\mathfrak{H}$ is separable.

But we have seen above that $A_n$ converges in the H. S. norm and so its limit must be $\pi_0(g)$. This proves that

$$\operatorname*{Lim}_{n \to \infty} T_{\omega_0}(\tilde{g}_n * g_n) = T_{\omega_0}(\tilde{g} * g)$$

and therefore

$$\| g \|^2 = \mu(\omega_0) T_{\omega_0}(\tilde{g} * g).$$

But

$$(\psi_i, \pi_0(g)\psi_j) = \int (\pi_0(x)\psi_0, \psi_0)(\psi_i, \pi_0(x)\psi_j) \, d\mathbf{x}$$

$$= d_{\omega_0}^{-1}(\psi_0, \psi_j)(\psi_i, \psi_0).$$

Hence

$$T_{\omega_0}(\tilde{g} * g) = \sum_{i,j} |(\psi_i, \pi_0(g)\psi_j)|^2 = d_{\omega_0}^{-2} \sum_{i,j} |(\psi_0, \psi_j)(\psi_i, \psi_0)|^2 = d_{\omega_0}^{-2} |\psi_0|^4.$$

On the other hand

$$\| g \|^2 = \int |(\pi(x)\psi_0, \psi_0)|^2 \, dx = d_{\omega_0}^{-1} |\psi_0|^4$$

and therefore $\mu(\omega_0) = d_{\omega_0}$.

COROLLARY 1. *Let $\pi$ be a square-integrable representation of $G$. Then if $f \varepsilon C_c(G)$,*

$$\| \int f(x)\pi(x) \, dx \|^2 \leq d_\pi^{-1} \int |f(x)|^2 \, dx$$

*where $\| A \|$ denotes the H. S. norm of an operator $A$.*

For $\int |f(x)|^2 \, dx = \int_{\mathcal{E}} T_\omega(\tilde{f} * f) \, d\mu \geq \mu(\omega_0) T_{\omega_0}(\tilde{f} * f)$ if $\omega_0$ is the class of $\pi$. But $\mu(\omega_0) = d_\pi$ and

$$T_{\omega_0}(\tilde{f} * f) = \| \int f(x)\pi(x) \, dx \|^2.$$

Hence the result

COROLLARY 2. *A class $\omega_0 \varepsilon \mathcal{E}$ is discrete if and only if $\mu(\omega_0) > 0$.*

We have seen that if $\omega_0$ is discrete $\mu(\omega_0) = d_{\omega_0}$ is positive. Conversely suppose $\mu(\omega_0)$ is positive. Let $\pi$ be a representation in $\omega_0$ and $\phi$ a unit vector in the representation space of $\pi$. Then if $f \varepsilon C_c(G)$,

$$|(\phi, \pi(f)\phi)|^2 \leq \| \int f(x)\pi(x) \, dx \|^2 = T_{\omega_0}(f * f).$$

On the other hand

$$\| f \|^2 = \int_{\mathcal{E}} T_\omega(\tilde{f} * f) \, d\mu \geq \mu(\omega_0) T_{\omega_0}(\tilde{f} * f).$$

Therefore

$$|(\phi, \pi(f)\phi)|^2 \leqq \mu(\omega_0)^{-1} \|f\|^2.$$

Let $L_1(G)$ denote, as usual, the space of all functions which are integrable on $G$. Then if $L = L_1(G) \cap L_2(G)$, it follows from the above inequality that

$$|(\phi, \pi(g)\phi)|^2 \leqq \mu(\omega_0)^{-1} \|g\|^2 \qquad (g \, \varepsilon \, L)$$

where $\pi(g) = \int g(x)\pi(x)\,dx$. $U$ being any compact neighborhood of 1 in $G$, we now define a function $g_U \, \varepsilon \, L$ as follows. $g_U(x) = (\pi(x)\phi, \phi)$ if $x \, \varepsilon \, U$ and $g_U(x) = 0$ otherwise. Then

$$(\phi, \pi(g_U)\phi) = \int g_U(x)(\phi, \pi(x)\phi)\,dx = \int_U |(\phi, \pi(x)\phi)|^2\,dx = \|g_U\|^2$$

and therefore

$$\|g_U\|^2 = \int_U |(\phi, \pi(x)\phi)|^2\,dx \leqq \mu(\omega_0)^{-1}.$$

Hence

$$\int |(\phi, \pi(x)\phi)|^2\,dx = \sup_U \int_U |(\phi, \pi(x)\phi)|^2\,dx \leqq \mu(\omega_0)^{-1}.$$

This proves that $\pi$ is square-integrable and therefore $\omega_0$ is discrete.

## Part II.

**6. Some algebraic results.** We shall now study in detail certain special representations which have been constructed in another paper [5(f)] and prove that, under suitable conditions, they are square-integrable or even integrable. Later (in Sections 9 and 10) we shall also obtain a formula for the formal degree of these representations.

Let $\mathfrak{h}_0$ be a maximal abelian subalgebra of $\mathfrak{k}_0$. In accordance with the assumption of [5(e), (f)] *we shall suppose that $\mathfrak{h}_0$ is also maximal abelian in $\mathfrak{g}_0$*. From now on we use the notation and the terminology of [5(e), §3] without further comment. Then $\mathfrak{h}$ is a Cartan subalgebra of $\mathfrak{g}$. Suppose an order has been introduced once for all in the space $\mathfrak{F}_R$ of real linear functions on $\mathfrak{h}$ (see [5(e), §2]) and $P$ is the set of all positive roots of $\mathfrak{g}$ (with respect to $\mathfrak{h}$) in this order. *We shall further assume that every non-compact root is totally positive.* Since we are now interested primarily in unitary representations, this assumption is justified in view of Corollary 1 to Lemma 19 of [5(e)].

Let I be a subalgebra of $\mathfrak{g}$. Suppose there exists a set $Q$ of totally positive roots such that

$$I = I \cap \mathfrak{k} + \sum_{\gamma \in Q} (CX_\gamma + CX_{-\gamma})$$

and let $\beta$ be the lowest root in $Q$. Then $X_\beta$, $X_{-\beta}$ and therefore also $H_\beta = [X_\beta, X_{-\beta}]$ are in I. Let $I_\beta$ denote the centralizer of $CH_\beta + CX_\beta + CX_{-\beta}$ in I. It is obvious that $I_\beta$ is invariant under $\theta$ and therefore

$$I_\beta = I_\beta \cap \mathfrak{k} + I_\beta \cap \mathfrak{p}.$$

LEMMA 7. $C(X_\beta + X_{-\beta}) + I_\beta \cap \mathfrak{p}$ *is exactly the set of all elements in* $I \cap \mathfrak{p}$ *which commute with* $X_\beta + X_{-\beta}$.

Let $Q'$ be the set of all roots in $Q$ other than $\beta$. Then if $X \varepsilon I \cap \mathfrak{p}$,

$$X = c_\beta' X_\beta + c_{-\beta}' X_{-\beta} + \sum_{\gamma \in Q'} (c_\gamma X + c_{-\gamma} X_{-\gamma})$$

where $c_\beta'$, $c_{-\beta}'$, $c_\gamma$, $c_{-\gamma}$ are complex numbers. Now

$$\mathfrak{g} = \mathfrak{h} + \sum_{\delta \in P} (CX_\delta + CX_{-\delta})$$

where the sum is direct. Hence it is clear that the component of $[X, X_\beta + X_{-\beta}]$ in $\mathfrak{h}$ is $(c_\beta' - c_{\beta}') H_\beta$. So if $X$ commutes with $X_\beta + X_{-\beta}$, $c_\beta' = c_{-\beta}'$ and therefore

$$Y = \sum_{\gamma \in Q} (c_\gamma X_\gamma + c_{-\gamma} X_{-\gamma})$$

also commutes with $(X_\beta + X_{-\beta})$. In order to prove the lemma it is enough to show that $Y \varepsilon I_\beta$. Let us suppose then that this is false. Define $c_\delta = 0$ for any root $\delta$ for which neither $\delta$ nor $-\delta$ is in $Q'$. Then it is obvious that there exists roots $\delta$ such that (1) $c_\delta \neq 0$ and (2) $X_\delta \not\varepsilon I_\beta$, for otherwise $Y$ would lie in $I_\beta$. Let $\delta_0$ be the highest such root. Since $[Y, X_\beta + X_{-\beta}] = 0$, it follows that $\delta_0 + \beta$ is not a root. However $X_{\delta_0} \not\varepsilon I_\beta$ and so $\delta_0 - \beta$ must be a root. The coefficient of $X_{\delta_0 - \beta}$ in $[Y, X_\beta]$ (with respect to the above decomposition of $\mathfrak{g}$ as a direct sum) is then clearly different from zero. This means that $\delta_0 - \beta = \gamma + \beta$ where $\gamma$ is some root with $c_\gamma \neq 0$. Hence $\gamma = \delta_0 - 2\beta$ is a root and $X_{\delta_0 - 2\beta} \varepsilon I$. Since $\delta_0$ and $\beta$ are both noncompact, $\alpha = \delta_0 - \beta$ is compact. Moreover $\beta$ being totally positive, $\delta_0 = \beta + \alpha$ and $2\beta - \delta_0 = \beta - \alpha$ are also totally positive (Lemma 12 of [5(e)]) and $X_{\beta + \alpha} = X_{\delta_0}$, $X_{\beta - \alpha} = X_{2\beta - \delta_0}$ are both in I. Therefore $\beta + \alpha$ and $\beta - \alpha$ are in $Q$. This however is impossible since $\beta$ is the lowest root in $Q$ and so the lemma is proved.

Let $Q_\beta$ be the set of all $\gamma \varepsilon Q$ such that $\gamma \neq \beta$ and neither $\gamma + \beta$ nor $\gamma - \beta$ is a root. Then it is obvious that

$$I_\beta = I_\beta \cap \mathfrak{k} + \sum_{\gamma \varepsilon Q_\beta} (CX_\gamma + CX_{-\gamma}).$$

Therefore $I_\beta$ satisfies the same condition as the one imposed above on $I$. Now we shall define a sequence $\mathfrak{g} = \mathfrak{g}_1 \supset \mathfrak{g}_2 \supset \mathfrak{g}_3 \supset \cdots$ of subalgebras of $\mathfrak{g}$ such that each $\mathfrak{g}_r$ satisfies this condition. The inductive definition is as follows. If $\mathfrak{g}_r \subset \mathfrak{k}$, $\mathfrak{g}_{r+1} = \mathfrak{g}_r$. Otherwise let $\beta$ be the lowest totally positive root such that $X_\beta \varepsilon \mathfrak{g}_r$. Then $\mathfrak{g}_{r+1}$ is the centralizer of $CH_\beta + CX_\beta + CX_{-\beta}$ in $\mathfrak{g}_r$. It is obvious that $\dim \mathfrak{g}_{r+1} < \dim \mathfrak{g}_r$ unless $\mathfrak{g}_r \subset \mathfrak{k}$ and therefore $\mathfrak{g}_r \subset \mathfrak{k}$ if $r$ is sufficiently large. Let $s \geq 0$ be the least integer such that $\mathfrak{g}_{s+1} \subset \mathfrak{k}$ and let $\gamma_r$ be the lowest totally positive root such that $X_{\gamma_r} \varepsilon \mathfrak{g}_r$ $(1 \leq r \leq s)$. One proves easily by induction on $r$ that if $X_\alpha \varepsilon \mathfrak{g}_r$ for some root $\alpha$ then $X_{-\alpha}$ is also in $\mathfrak{g}_r$.

LEMMA 8. $\gamma_i \pm \gamma_j$ $(1 \leq i < j \leq s)$ *is never a root or zero and the elements* $(X_{\gamma_i} + X_{-\gamma_i})$ $i = 1, 2, \cdots, s$ *span a maximal abelian subspace of* $\mathfrak{p}$ *over* $C$.

If $i < j$, $\mathfrak{g}_{i+1} \supset \mathfrak{g}_j$ and therefore $\mathfrak{g}_j$ commutes with $X_{\gamma_i}$ and $X_{-\gamma_i}$. Hence $\gamma_i \pm \gamma_j$ is not a root or zero. Let $\mathfrak{a}_\mathfrak{p}$ be the subspace of $\mathfrak{p}$ spanned by $(X_{\gamma_i} + X_{-\gamma_i})$ $i = 1, 2, \cdots, s$. Then $\mathfrak{a}_\mathfrak{p}$ is obviously abelian. Let $X$ be an element in $\mathfrak{p}$ which commutes with $\mathfrak{a}_\mathfrak{p}$. We have to show that $X \varepsilon \mathfrak{a}_\mathfrak{p}$. Suppose this is false. Then it is obvious that $X \notin \mathfrak{k} + \mathfrak{a}_\mathfrak{p}$. Since $\mathfrak{g}_{s+1} \subset \mathfrak{k}$, we can choose $r$ $(1 \leq r \leq s)$ such that $X \varepsilon \mathfrak{g}_r + \mathfrak{a}_\mathfrak{p}$ but $X \notin \mathfrak{g}_{r+1} + \mathfrak{a}_\mathfrak{p}$. Let $X = Y + Z$ $(Y \varepsilon \mathfrak{g}_r, Z \varepsilon \mathfrak{a}_\mathfrak{p})$. Since $X$ commutes with $X_{\gamma_r} + X_{-\gamma_r}$, the same holds for $Y$. Also $Y = X - Z \varepsilon \mathfrak{g}_r \cap \mathfrak{p}$. Therefore we conclude from Lemma 7 that

$$Y = c(X_{\gamma_r} + X_{-\gamma_r}) + Y_1$$

where $Y_1 \varepsilon \mathfrak{g}_{r+1} \cap \mathfrak{p}$ and $c \varepsilon C$. Then $Z_1 = Z + c(X_{\gamma_r} + X_{-\gamma_r})$ lies in $\mathfrak{a}_\mathfrak{p}$ and so

$$X = Y_1 + Z_1 \varepsilon \mathfrak{g}_{r+1} + \mathfrak{a}_\mathfrak{p}.$$

Since this contradicts the definition of $r$, the lemma follows.

COROLLARY. *Let* $\mathfrak{a}_{\mathfrak{p}_0} = \sum_{i=1}^{s} R(X_{\gamma_i} + X_{-\gamma_i})$. *Then* $\mathfrak{a}_{\mathfrak{p}_0} = \mathfrak{p}_0 \cap \mathfrak{a}_\mathfrak{p}$ *and therefore it is a maximal abelian subspace of* $\mathfrak{p}_0$.

We know that $\bar\theta(X_\gamma) = -X_{-\gamma}$ for any root $\gamma$ (see [5(e), § 4]. There is a small mistake on p. 757 of [5(e)]. In line 22 $\bar\theta$ should be replaced by $\eta$

which is the conjugation of $\mathfrak{g}$ with respect to $\mathfrak{g}_0$.). Hence $X_{\gamma_i} + X_{-\gamma_i} \varepsilon \mathfrak{p}_0$. Moreover if

$$X = \sum_{i=1}^{s} c(X_{\gamma_i} + X_{-\gamma_i}) \varepsilon \mathfrak{p}_0 \qquad (c_i \varepsilon C),$$

$X = -\tilde{\theta}(X)$ and therefore $c_i \varepsilon R$.

We now need some simple facts about a three dimensional Lie algebra.

LEMMA 9. *Let* $\mathfrak{l}$ *be the Lie algebra of dimension 3 spanned over $C$ by the elements $H, X, Y$ satisfying the following relations:*

$$[X, Y] = H, \qquad [H, X] = 2X, \qquad [H, Y] = -2Y.$$

*Let $\nu$ denote the automorphism of $\mathfrak{l}$ given by*

$$\nu(Z) = \exp \tfrac{\pi}{4} \mathrm{ad}(X - Y))Z \qquad (Z \varepsilon \mathfrak{l}).$$

*Then $\nu(H) = -(X + Y)$, $\nu(X + Y) = H$, $\nu(X - Y) = X - Y$. Moreover if $L$ is any complex analytic group with the Lie algebra $\mathfrak{l}$,*

$$\exp t(X + Y) = \exp (zY) \exp (\log(\cosh t)H) \exp zX \qquad (t \varepsilon C, \cosh l \neq 0)$$

*where* [6] $z = \tanh t$.

It is well known that $\mathfrak{l}$ is isomorphic to the Lie algebra of the group of all $2 \times 2$ complex matrices with determinant 1. Since this group is simply connected, it is enough to prove the above relations in it. Therefore we may identify $X, Y, H$ with matrices as follows:

$$X = \begin{pmatrix} 0 & 1 \\ 0 & 0 \end{pmatrix}, \quad Y = \begin{pmatrix} 0 & 0 \\ 1 & 0 \end{pmatrix}, \quad H = \begin{pmatrix} 1 & 0 \\ 0 & -1 \end{pmatrix}.$$

The required relations are now verified by a simple calculation.

Let $P_+$ be all the totally positive roots of $\mathfrak{g}$. Then $\gamma_i \varepsilon P_+$, $1 \leq i \leq s$. Consider the automorphism $\nu$ of $\mathfrak{g}$ given by $\nu = \exp \tfrac{\pi}{4} \mathrm{ad}(\sum_{j=1}^{s} (X_{\gamma_i} - X_{-\gamma_i}))$. It follows from Lemmas 8 and 9 that $\nu(X_{\gamma_j} + X_{-\gamma_j}) = H_{\gamma_j}$ and therefore $\nu(\mathfrak{a}_\mathfrak{p}) = \sum_{i=1}^{s} CH_{\gamma_i}$. This shows that $H_{\gamma_i}$ and therefore also $\gamma_i$ $(1 \leq i \leq s)$ are

---

[6] Our result is valid with any determination of the logarithm. But for the sake of definiteness let us make the following convention. Choose a fixed square root of $-1$ in $C$ and denote it by $(-1)^{\frac{1}{2}}$. Then if $z$ is a non-zero complex number

$$\log z = \log |z| + (-1)^{\frac{1}{2}} \phi$$

where $\log |z|$ and $\phi$ are real and $0 \leq \phi < 2\pi$. Hence in particular if $z$ is real and positive $\log z$ is real.

linearly independent. Let $\mathfrak{a}_{\mathfrak{k}}$ be the orthogonal complement of $\nu(\mathfrak{a}_\mathfrak{p})$ in $\mathfrak{h}$ with respect to the positive definition Hermitian form $-B(\bar{\theta}(X), X)$ ($X \varepsilon \mathfrak{g}$, see [5(e), §4]). Since $\bar{\theta}(H_\gamma) = -H_\gamma$ for every root $\gamma$, it is obvious that $B(H_{\gamma_i}, H) = 0$ and therefore $\gamma_i(H) = 0$ $1 \leq i \leq s$ if $H \varepsilon \mathfrak{a}_\mathfrak{k}$. This means that $X_{\gamma_i} - X_{-\gamma_i}$ $1 \leq i \leq s$ commute with $H$ and therefore $\nu(H) = H$ ($H \varepsilon \mathfrak{a}_\mathfrak{k}$). Hence if $\mathfrak{a} = \mathfrak{a}_\mathfrak{p} + \mathfrak{a}_\mathfrak{k}$, $\nu(\mathfrak{a}) = \nu(\mathfrak{a}_\mathfrak{p}) + \mathfrak{a}_\mathfrak{k} = \mathfrak{h}$. As $\nu$ is an automorphism, it follows that $\mathfrak{a}$ is a Cartan subalgebra of $\mathfrak{g}$.

Let $\alpha$, $\beta$ be two roots of $\mathfrak{g}$ and let $k$, $k'$ be the largest nonnegative integers such that $\beta - k\alpha$ and $\beta + k'\alpha$ are roots. Then it is known (see Weyl [11(b)]) that $\beta(H_\alpha) = k - k'$ and $\beta + r\alpha$ is a root or zero for an integer $r$ if and only if $-k \leq r \leq k'$ Moreover if $s_\alpha$ is the Weyl reflexion corresponding to $\alpha$, $s_\alpha(\beta + k'\alpha) = \beta - k\alpha$. These facts should be constantly borne in mind during the following discussion.

LEMMA 10. *If $\gamma, \delta \varepsilon P_+$, $\gamma(H_\delta) \geq 0$.*

For $\gamma + \delta$ is not a root (Lemma 11 of [5(e)]) and obviously it is not zero. Hence $\gamma(H_\delta) \geq 0$.

LEMMA 11. *Let $\alpha$ be any root such that $H_\alpha \varepsilon \mathfrak{a}_\mathfrak{k}$. Then $\alpha$ is compact and $\gamma_i \pm \alpha$ $(1 \leq i \leq s)$ can never be a root.*

Without loss of generality we may assume that $\alpha > 0$. If $\alpha$ is not compact, it must be totally positive and therefore $\alpha + \gamma_i$ is not a root [5(e), Lemma 11]. Since $H_\alpha \varepsilon \mathfrak{a}_\mathfrak{k}$, $\gamma_i(H_\alpha) = 0$ and so it follows that $\gamma_i - \alpha$ also cannot be a root or zero. Hence $X_\alpha$ commutes with $X_{\gamma_i}$, $X_{-\gamma_i}$ $1 \leq i \leq s$. This however is impossible since $\mathfrak{a}_\mathfrak{p}$ is maximal abelian in $\mathfrak{p}$. So $\alpha$ must be compact.

Now consider the sequence $\mathfrak{g} = \mathfrak{g}_1 \supset \mathfrak{g}_2 \supset \cdots$ introduced above. We shall prove that $X_\alpha \varepsilon \mathfrak{g}_r$ for every $r$. For otherwise choose the least $r \geq 0$ such that $X_\alpha \not\varepsilon \mathfrak{g}_{r+1}$. Since $\mathfrak{g}_1 = \mathfrak{g}$, $r \geq 1$ and $X_\alpha \varepsilon \mathfrak{g}_r$. In view of the fact that $X_\alpha \not\varepsilon \mathfrak{g}_{r+1}$, it is clear that either $\gamma_r + \alpha$ or $\gamma_r - \alpha$ is a root. But $\gamma_r$ is the lowest totally positive root $\gamma$ such that $X_\gamma \varepsilon \mathfrak{g}_r$. Since $X_\alpha \varepsilon \mathfrak{g}_r$, $X_{-\alpha}$ also lies in $\mathfrak{g}_r$ and so if $\gamma_r - \alpha$ were a root, $X_{\gamma_r - \alpha}$ would also lie in $\mathfrak{g}_r$. Since $\alpha$ is compact and positive, $\gamma_r - \alpha$ is also totally positive and $\gamma_r - \alpha < \gamma_r$. As this contradicts the definition of $\gamma_r$, we conclude that $\gamma_r - \alpha$ is not a root and therefore $\gamma_r + \alpha$ is a root. But this implies that $\gamma_r(H_\alpha) < 0$ which, in its turn, contradicts the fact $\gamma_r(H) = 0$ for all $H \varepsilon \mathfrak{a}_\mathfrak{k}$. Hence the lemma.

LEMMA 12. *Let $\alpha$ be a root. Then for any $i$ $(1 \leq i \leq s)$, $\gamma_i + \alpha$ and $\gamma_i - \alpha$ cannot both be roots.*

8

We may assume $\alpha > 0$. If $\alpha$ is noncompact $\gamma_i + \alpha$ cannot be a root [5(e), Lemma 11]. So we may assume that $\alpha$ is compact. Suppose then that for some $i$, $\gamma_i \pm \alpha$ are both roots. Then they are both totally positive and hence from Lemma 10,

$$\gamma_i(H_\delta) \pm \alpha(H_\delta) \geqq 0$$

for every $\delta \varepsilon P_+$ $(j \neq i)$. Then it follows from Lemma 8 that $\gamma_i(H_{\gamma_j}) = 0$ and therefore $\pm \alpha(H_{\gamma_j}) \geqq 0$. This means that $\alpha(H_{\gamma_j}) = 0$. On the other hand $\gamma_i + \alpha$, $\gamma_i$, $\gamma_i - \alpha$ are all roots and therefore it follows from Lemma 15 of $[5(e)]$ that $\gamma_i + 2\alpha$ and $\gamma_i - 2\alpha$ are not roots. Hence $\gamma_i(H_\alpha) = 0$. But this implies that $\alpha(H_{\gamma_i}) = 0$ and therefore $\alpha(H_{\gamma_j}) = 0$ $1 \leqq j \leqq s$. This however means that $H_\alpha \varepsilon \mathfrak{a}_f$ and so we get a contradiction with Lemma 11.

Let $\lambda$ and $\mu$ be two linear functions on $\mathfrak{h}$. We write $\lambda \sim \mu$ if $\lambda - \mu$ vanishes identically on $\nu(\mathfrak{a}_p) = \sum_{1 \leqq i \leqq s} CH_{\gamma_i}$.

LEMMA 13. *Let $\alpha$ be a positive compact root. Then there are only the following three mutually exclusive possibilities:*

(1)　$H_\alpha \varepsilon \mathfrak{a}_f$ *and therefore $\alpha \sim 0$ and $\gamma_i \pm \alpha$ $(1 \leqq i \leqq s)$ is never a root.*

(2)　*There exists a unique index $i$ $(1 \leqq i \leqq s)$ such that $\alpha + \frac{1}{2}\gamma_i \sim 0$.*

(3)　*There exists two unique indices $i$, $j$ $(1 \leqq i < j \leqq s)$ such that $\alpha \sim \frac{1}{2}(\gamma_j - \gamma_i)$.*

Since the first case is covered by Lemma 11, we may assume that $H_\alpha \notin \mathfrak{a}_f$. Then $\alpha(H_{\gamma_i}) \neq 0$ for some $i$ and therefore $X_\alpha \notin \mathfrak{g}_{s+1}$. Let $i$ be the least index $(1 \leqq i \leqq s)$ such that $X_\alpha \notin \mathfrak{g}_i$. Since $\alpha$ is positive, $\gamma_i + \alpha$ is a root while $\gamma_i - \alpha$ is not (see the proof of Lemma 11). Now suppose $\gamma_j + \epsilon\alpha$ is a root for some $j$ $(1 \leqq j \leqq s, \epsilon = \pm 1)$. If $j \neq i$ we claim $\epsilon = -1$. For otherwise suppose $\gamma_j + \alpha$ is a root. Then it follows from Lemmas 9 and 10 that $\alpha(H_{\gamma_i}) = \gamma_j(H_{\gamma_i}) + \alpha(H_{\gamma_i}) \geqq 0$. On the other hand since $\gamma_i + \alpha$ is a root while $\gamma_i - \alpha$ is not, it is clear that $\gamma_i(H_\alpha) < 0$ and therefore $\alpha(H_{\gamma_i}) < 0$. As this conflicts with our conclusion above, $\epsilon = -1$. So we have two cases. Either (1) $\gamma_j \pm \alpha$ is never a root for $j \neq i$ or (2) $\gamma_j - \alpha$ is a root for some $j \neq i$.

In the first case $\alpha(H_{\gamma_j}) = 0$ for all $j \neq i$. Moreover $\alpha$, $\alpha + \gamma_i$ are roots while $\alpha - \gamma_i$ is not. Since $\gamma_i$ and $\alpha + \gamma_i$ are both totally positive $\alpha + 2\gamma_i$ is not a root [5(e), Lemma 11]. Hence $\alpha(H_{\gamma_i}) = -1$. This shows that $\alpha(H_{\gamma_j}) + \frac{1}{2}\gamma_i(H_{\gamma_j}) = 0$ for all $j$ $(1 \leqq j \leqq s)$ and therefore $\alpha + \frac{1}{2}\gamma_i \sim 0$.

Now consider the second case. Let $j$ be the least index such that $\gamma_j - \alpha$ is a root. Then $j \neq i$ and in view of our definition of $i$, $j > i$. If $k$

is any index $(1 \leqq k \leqq s)$ other than $i$, $j$ we claim $\gamma_k \pm \alpha$ cannot be a root. We have already seen this for $\gamma_k + \alpha$. So suppose $\gamma_k - \alpha$ is a root. Then $\gamma_k(H_{\gamma_j}) - \alpha(H_{\gamma_j}) \geqq 0$ (Lemma 10) and therefore $- \alpha(H_{\gamma_j}) \geqq 0$ (Lemma 8). On the other hand $\alpha$, $\alpha - \gamma_j = - (\gamma_j - \alpha)$ are roots while $\alpha + \gamma_j$ is not. Therefore $\alpha(H_{\gamma_j}) > 0$ giving a contradiction. This proves that $\gamma_k \pm \alpha$ are never roots $(k \neq i, j)$. Moreover as we have seen above, $\alpha$, $\alpha + \gamma_i$ are roots while $\alpha - \gamma_i$ and $\alpha + 2\gamma_i$ are not and therefore $\alpha(H_{\gamma_i}) = - 1$. Similarly since $\alpha$, $\alpha - \gamma_j$ are roots while $\alpha + \gamma_j$, $\alpha + 2\gamma_j$ are not, $\alpha(H_{\gamma_j}) = 1$. Finally $\alpha(H_{\gamma_k}) = 0$ $(k \neq i, j)$ in view of our result above. Hence it is clear that $\alpha(H_{\gamma_k}) = \frac{1}{2}\gamma_j(H_{\gamma_k}) - \frac{1}{2}\gamma_i(H_{\gamma_k})$ $(1 \leqq k \leqq s)$ and therefore $\alpha \sim \frac{1}{2}(\gamma_j - \gamma_i)$. Moreover since $\gamma_k$ $(1 \leqq k \leqq s)$ vanish identically on $\mathfrak{a}_t$, it is clear that their restriction on $\nu(\mathfrak{a}_\mathfrak{p})$ are linearly independent. The uniqueness of the indices $i$ and $j$ in the second and third cases of our lemma and the mutual exclusiveness of the three possibilities are therefore obvious.

For any index $i$ $(1 \leqq i \leqq s)$ let $C_i$ denote the set of all compact roots $\alpha$ such that $\alpha + \frac{1}{2}\gamma_i \sim 0$. Similarly let $P_i$ denote the set of all totally positive roots $\gamma$ for which $\gamma \sim \frac{1}{2}\gamma_i$. If $\alpha \varepsilon C_i$, it follows from Lemma 13 that $- \alpha$ cannot be positive. This shows that $C_i$ consists of positive roots.

LEMMA 14. $\alpha \to \gamma_i + \alpha$ $(\alpha \varepsilon C_i)$ is a one-one mapping of $C_i$ onto $P_i$ $(1 \leqq i \leqq s)$.

For any root $\beta$ let $s_\beta$ denote the Weyl reflexion corresponding to $\beta$. Now if $\alpha \varepsilon C_i$, $\alpha \sim - \frac{1}{2}\gamma_i$ and therefore $\alpha(H_{\gamma_i}) = - 1$. Hence $s_{\gamma_i}\alpha = \alpha - \alpha(H_{\gamma_i})\gamma_i = \alpha + \gamma_i$ and so $\gamma_i + \alpha$ is a root which is obviously in $P_i$. Conversely if $\gamma \varepsilon P_i$, $\gamma \sim \frac{1}{2}\gamma_i$ and therefore $\gamma(H_{\gamma_i}) = 1$. Hence $s_{\gamma_i}\gamma = \gamma - \gamma_i$ is a root. But as $\gamma$ and $\gamma_i$ are both noncompact, $\alpha = \gamma - \gamma_i$ must be compact. Moreover $\alpha + \frac{1}{2}\gamma_i = \gamma - \frac{1}{2}\gamma_i \sim 0$ and therefore $\alpha \varepsilon C_i$. Since it is obvious from its definition that the mapping is one-one, the lemma is proved.

For any given pair of indices $i$, $j$ $(1 \leqq i < j \leqq s)$, let $C_{ij}$ denote the set of all compact roots $\alpha$ such that $\alpha \sim \frac{1}{2}(\gamma_j - \gamma_i)$. Similarly let $P_{ij}$ denote the set of all $\gamma \varepsilon P_+$ such that $\gamma \sim \frac{1}{2}(\gamma_j + \gamma_i)$. Again we conclude from Lemma 13 that every root in $C_{ij}$ is positive.

LEMMA 15. $\alpha \to \gamma_i + \alpha$ $(\alpha \varepsilon C_{ij})$ is a one-one mapping of $C_{ij}$ onto $P_{ij}$.

Let $\alpha \varepsilon C_{ij}$. Then $\alpha \sim \frac{1}{2}(\gamma_j - \gamma_i)$ and therefore $\alpha(H_{\gamma_i}) = - 1$. Hence $s_{\gamma_i}\alpha = \alpha + \gamma_i$ and so it is clear that $\gamma_i + \alpha \varepsilon P_{ij}$. Conversely if $\gamma \varepsilon P_{ij}$, $\gamma(H_{\gamma_i}) = 1$ and therefore $s_{\gamma_i}\gamma = \gamma - \gamma_i = \alpha$ (say). Then $\alpha$ is compact and $\alpha \sim \frac{1}{2}(\gamma_j - \gamma_i)$.

Let $C_0$ be the set of all positive roots $\alpha$ such that $\alpha \sim 0$. We know (Lemma 11) that every root in $C_0$ is compact.

LEMMA 16. *Let $P_0$ denote the set $(\gamma_1, \gamma_2, \cdots, \gamma_s)$. Then $P$ is the disjoint union of $C_0$, $C_i$, $C_{ij}$, $P_0$, $P_i$, $P_{ij}$ $(1 < i < j < s)$.*

Since $\gamma_1, \cdots, \gamma_s$ are linearly independent on $\nu(\mathfrak{a}_\mathfrak{p})$, it is obvious that these sets are all disjoint. Let $Q$ be their union. Then if $\gamma$ is any positive root, we have to show that $\gamma \varepsilon Q$. If $\gamma$ is compact, this follows from Lemma 13. So now suppose $\gamma \varepsilon P_+$. Since $\mathfrak{g}_{s+1} \subset \mathfrak{k}$, we can choose an index $i$ $(1 \leq i \leq s)$ such that $X_\gamma \varepsilon \mathfrak{g}_i$ but $X_\gamma \not\in \mathfrak{g}_{i+1}$. Moreover since $\gamma_i \varepsilon P_0 \subset Q$, we may assume that $\gamma \neq \gamma_i$. Then it follows from the definition of $\gamma_i$ that $\gamma > \gamma_i$. As both $\gamma$ and $\gamma_i$ are in $P_+$, $\gamma + \gamma_i$ is not a root [5(e), Lemma 11]. Therefore since $X_\gamma \not\in \mathfrak{g}_{i+1}$, $\alpha = \gamma - \gamma_i$ must be a root which is then obviously compact and positive. Therefore we can apply Lemma 13 to $\alpha$. Since $\gamma = \gamma_i + \alpha$ is a root, $\alpha \not\in C_0$ (Lemma 11). Hence either $\alpha \sim -\frac{1}{2}\gamma_j$ or $\alpha \sim \frac{1}{2}(\gamma_k - \gamma_j)$ for some $j$ or $(k, j)$ $(1 \leq j < k \leq s)$. In the first case $\gamma \sim \gamma_i - \frac{1}{2}\gamma_j$ and so[7] $\gamma(H_{\gamma_j}) = 2\delta_{ij} - 1$. But we know from Lemma 10, that $\gamma(H_{\gamma_j}) \geq 0$. Therefore $i = j$, $\gamma \sim \frac{1}{2}\gamma_i$ and $\gamma \varepsilon P_i$. In the second case $\gamma \sim \gamma_i + \frac{1}{2}(\gamma_k - \gamma_j)$ and $\gamma(H_{\gamma_j}) = \delta_{ij} - 1$ since $k \neq j$. Therefore again in view of the fact that $\gamma(H_{\gamma_j}) \geq 0$, we conclude that $i = j$ and hence $\gamma \varepsilon P_{jk}$. This shows that $\gamma \varepsilon Q$ and therefore $Q = P$.

LEMMA 17. *Let $\alpha$, $\beta$ be two roots such that $\alpha = \frac{1}{2}(\gamma_j - \gamma_i)$ and $\beta = \frac{1}{2}(\gamma_k - \gamma_j)$ $(1 \leq i, j, k \leq s)$. Then they are both compact if $k \neq i$, $\beta + \alpha = \frac{1}{2}(\gamma_k - \gamma_i)$ is a root.*

Consider the scalar product $\langle \alpha, \beta \rangle$ (see [5(e), §2]). Since $k \neq i$, it follows from Lemma 8 that $\langle \alpha, \beta \rangle = -\frac{1}{4}\langle \gamma_j, \gamma_j \rangle < 0$. Hence $\alpha(H_\beta) < 0$ and therefore $\alpha + \beta$ is a root. The compactness of $\alpha$ and $\beta$ is an immediate consequence of Lemma 16.

Let us say that $\gamma_i \prec \gamma_j$ $(1 \leq i, j \leq s)$ if $\frac{1}{2}(\gamma_j - \gamma_i)$ is a positive root. The above lemma shows that this relation is transitive and therefore it defines a partial order in the set $P_0$. It is obvious from the definition of $\gamma_i$ that $\gamma_i < \gamma_j$ if $i < j$. Hence $\gamma_i \prec \gamma_j$ implies $i < j$.

Let $r_i$ and $r_{ij}$ be the number of roots in $C_i$ and $C_{ij}$ $(1 \leq i < j \leq s)$ respectively. Then it follows from Lemmas 14 and 15 that these are also the number of roots in $P_i$ and $P_{ij}$ respectively. Put $2\rho_+ = \sum_{\beta \varepsilon P_+} \beta$. Then we have the following result.

---

[7] $\delta_{ij} = 1$ or $0$ according as $i = j$ or not.

LEMMA 18. $2\rho_+(H_{\gamma_i}) = 2 + r_i + \sum_{i<j\leq s} r_{ij} + \sum_{1\leq j<i} r_{ji}$ $(1\leq i\leq s)$.

Let $Q_i$ be the union of $P_i$, $P_{ij}$ $(i<j\leq s)$ and $P_{ji}$ $(1\leq j<i)$. Put $2\rho_i = \sum_{\gamma \varepsilon Q_i} \gamma$. Then since $\gamma_j(H_{\gamma_i}) = 2\delta_{ji}$, it is obvious that

$$2\rho_i(H_{\gamma_i}) = r_i + \sum_{i<j\leq s} r_{ij} + \sum_{1\leq j<i} r_{ji}.$$

Also if $\gamma$ is a totally positive root which does not lie in $Q_i$, it follows from Lemma 16 that $\gamma(H_{\gamma_i}) = 0$ unless $\gamma = \gamma_i$. Therefore since $\gamma_i(H_{\gamma_i}) = 2$,

$$2\rho_i(H_{\gamma_i}) = 2 + 2\rho_i(H_{\gamma_i})$$

and this gives the result.

LEMMA 19. $r_{ij}$ $(1\leq i<j\leq s)$ is even if and only if $\frac{1}{2}(\gamma_j - \gamma_i)$ is not a root.

Let $\theta'$ denote the automorphism $\nu\theta\nu^{-1}$ of $\mathfrak{g}$. Since $\theta(\mathfrak{a}) = \mathfrak{a}$ and $\mathfrak{h} = \nu(\mathfrak{a})$, it follows that $\theta'(\mathfrak{h}) = \mathfrak{h}$. Therefore if $\alpha$ is a root the linear function $H \to \alpha(\theta'H)$ $(H\varepsilon\mathfrak{h})$ is also a root. We denote it by $\theta'\alpha$. It is clear that $\theta'H = -H$ if $H\varepsilon\nu(\mathfrak{a}_p)$ and $\theta'H = H$ if $H\varepsilon\mathfrak{a}_f$. Hence $\alpha \neq -\theta'\alpha$ unless $\alpha$ vanishes identically on $\mathfrak{a}_f$. Now suppose $\alpha\varepsilon C_{ij}$ $(1\leq i<j\leq s)$. Then $\alpha \sim \frac{1}{2}(\gamma_j - \gamma_i)$ and therefore it is obvious that $\theta'\alpha \sim -\frac{1}{2}(\gamma_j - \gamma_i)$. In view of Lemma 16, this implies that $-\theta'\alpha \varepsilon C_{ij}$. Hence the mapping $\alpha \to -\theta'\alpha$ defines a permutation of order 2 in the set $C_{ij}$. Moreover $\alpha \neq -\theta'\alpha$ unless $\alpha = \frac{1}{2}(\gamma_j - \gamma_i)$. Therefore if we pair off $\alpha$ and $-\theta'\alpha$ together, it follows immediately that $r_{ij}$ is odd or even according as $\frac{1}{2}(\gamma_j - \gamma_i)$ is a root or not.

### 7. Digression on a theorem of Cartan. Put

$$\mathfrak{p}_+ = \sum_{\beta \varepsilon P_+} CX_\beta \quad \text{and} \quad \mathfrak{p}_- = \sum_{\beta \varepsilon P_+} CX_{-\beta}.$$

Then $\mathfrak{p}_+$, $\mathfrak{p}_-$ are abelian subalgebras of $\mathfrak{g}$ [5(e), Lemma 11] and $\mathfrak{g}$ is the direct sum of $\mathfrak{k}$, $\mathfrak{p}_+$ and $\mathfrak{p}_-$. Let $G_c$ denote the simply connected complex Lie group with the Lie algebra $\mathfrak{g}$ and let $\mathfrak{P}_c{}^+$, $K_c$, $\mathfrak{P}_c{}^-$ be its analytic subgroups corresponding to $\mathfrak{p}_+$, $\mathfrak{k}$, $\mathfrak{p}_-$ respectively. Also let $G_0$, $K_0$ be the real analytic subgroups of $G_c$ corresponding to $\mathfrak{g}_0$, $\mathfrak{k}_0$ respectively. Then $(q,k,p) \to qkp$ $(q\varepsilon\mathfrak{P}_c{}^-, k\varepsilon K_c, p\varepsilon\mathfrak{P}_c{}^+)$ is a one-one regular holomorphic mapping of the complex manifold $\mathfrak{P}_c{}^- \times K_c \times \mathfrak{P}_c{}^+$ into $G_c$ and $G_0$ is contained in $\mathfrak{P}_c{}^-K_c\mathfrak{P}_c{}^+$ [5(f), Lemmas 4 and 5].

LEMMA 20. *Let* $X = \sum_{i=1}^{s} t_i(X_{\gamma_i} + X_{-\gamma_i})$ $(t\varepsilon C)$. *Then*

$$\exp X = \exp Y \exp H \exp Z$$

*in $G_c$ where*[6]

$$Y = \sum_{i=1}^{s} (\tanh t_i) X_{-i}, \qquad Z = \sum_{i=1}^{s} (\tanh t_i) X_i \qquad H = \sum_{i=1}^{s} \log(\cosh t_i) H_i$$

*provided* $\cosh t_i \neq 0 \ 1 \leq i \leq s$.

This follows immediately from Lemmas 8 and 9.

Now if we put $(X, Y) = -B(\bar{\theta}(X), Y)$ and $\| X \| = (X, X)^{\frac{1}{2}}$ $(X, Y \in \mathfrak{g})$, $\mathfrak{g}$ becomes a finite-dimensional Hilbert space. Moreover since $\mathrm{ad}X$ is nilpotent for $X \in \mathfrak{p}_-$, it is easy to see that $X \to \exp X$ $(X \in \mathfrak{p}_-)$ is a one-one regular holomorphic mapping of $\mathfrak{p}_-$ onto $\mathfrak{P}_c^-$. Let $q \to \log q$ $(q \in \mathfrak{P}_c^-)$ denote its inverse. For $x \in G_0$, let $\zeta(x)$ denote the unique element in $\mathfrak{P}_c^-$ such that $x \in \zeta(x) K_c \mathfrak{P}_c^+$. Then we have the following result.

LEMMA 21. $\| \log \zeta(x) \|$ *remains bounded as $x$ varies in $G_0$.*

Let $\mathfrak{P}_0$ be the set of all elements in $G_0$ of the form $\exp X$ $(X \in \mathfrak{p}_0)$. Then it is known that $G_0 = K_0 \mathfrak{P}_0$ (see Cartan [2(b), p. 17], also Mostow [9]). Let $z \to \mathrm{Ad}(z)$ $(z \in G_c)$ denote the adjoint representation of $G_c$. It follows from the definition of $\bar{\theta}$ that if $k \in K_0$, $\mathrm{Ad}(k)$ is a unitary operator on $\mathfrak{g}$. Now $\mathfrak{a}_{\mathfrak{p}_0}$ is a maximal abelian subspace of $\mathfrak{p}_0$ (Lemma 8) and therefore $\mathfrak{p}_0 = \bigcup_{k \in K} \mathrm{Ad}(k) \mathfrak{a}_{\mathfrak{p}_0}$ (see Lemma 33 and also Cartan [2(a), p. 359]). Hence $G_0 = K_0 \mathfrak{A} K_0$ where $\mathfrak{A}$ is the analytic subgroup of $G_0$ corresponding to $\mathfrak{a}_{\mathfrak{p}_0}$. Moreover since $[\mathfrak{k}, \mathfrak{p}_-] \subset \mathfrak{p}_-$, it is obvious that $\zeta(kxk') = k\zeta(x)k^{-1}$ $(k, k' \in K_0, x \in G_0)$. Therefore if $x = kak'$ $(k, k \in K_0; a \in \mathfrak{A})$,

$$\log \zeta(x) = \mathrm{Ad}(k) (\log \zeta(a))$$

and so

$$\| \log \zeta(x) \| = \| \log \zeta(a) \|.$$

Now suppose $a = \exp X$ where $X = \sum_{i=1}^{s} t_i (X_{\gamma_i} + X_{-\gamma_i})$ $(t_i \in R)$. Then from Lemma 26,

$$\log \zeta(a) = \sum_{i=1}^{s} (\tanh t_i) X_{-\gamma_i}$$

and therefore

$$\| \log \zeta(a) \| \leq \sum_{i=1}^{s} \| X_{-\gamma_i} \|$$

since $| \tanh t | \leq 1$ for real $t$. Thus

$$\| \log \zeta(x) \| \leq \sum_{i=1}^{s} \| X_{-\gamma_i} \|$$

for all $x \in G_0$ and so the lemma is proved.

This result has the following significance in relation to the theory of bounded symmetric homogeneous domains of E. Cartan [2(c)]. We know that $G_0 K_c \mathfrak{P}_c^+$ is open in $\mathfrak{P}_c^- K_c \mathfrak{P}_c^+$ and $G_0 \cap (K_c \mathfrak{P}_{c+}) = K_0$ (see [5(f), §2]). Since $K_c \mathfrak{P}_c^+$ is a group and $\mathfrak{P}_c^- \cap (K_c \mathfrak{P}_c^+) = \{1\}$, we can identify $\mathfrak{P}_c^-$ with the factor space $(\mathfrak{P}_c^- K_c \mathfrak{P}_c^+)/K_c \mathfrak{P}_c^+$. In this way $G_0/K_0 = (G_0 K_c \mathfrak{P}_c^+)/K_c \mathfrak{P}_c^+$ becomes an open submanifold of $\mathfrak{P}_c^-$. The above lemma then shows that this submanifold is equivalent to a *bounded* domain in the complex Euclidean space $\mathfrak{p}_-$. This fact had previously been verified by Cartan [2(c)] by using the classification of all real simple groups and constructing the domain in each case separately.

**8. Transformation of certain integrals.** Let $(-1)^{\frac{1}{2}}$ denote a fixed square-root of $-1$ in $C$ and put $\mathfrak{u} = \mathfrak{k}_0 + (-1)^{\frac{1}{2}}\mathfrak{p}_0$. Then $\mathfrak{u}$ is a compact real form of $\mathfrak{g}$ (see [5(b), p. 187]). Let $\mathfrak{a}_{\mathfrak{p}_0}$ denote any (real) maximal abelian subspace of $\mathfrak{p}_0$. We denote by $\mathfrak{a}_\mathfrak{p}$ the complexification of $\mathfrak{a}_{\mathfrak{p}_0}$ in $\mathfrak{p}$. Define $G_c$, $G_0$, $K_0$ as in Section 7 and let $U$, $\mathfrak{A}$, $\mathfrak{A}^*$ and $\mathfrak{A}_c$ be the (real) analytic subgroups of $G_c$ corresponding to $\mathfrak{u}$, $\mathfrak{a}_{\mathfrak{p}_0}$, $(-1)^{\frac{1}{2}}\mathfrak{a}_{\mathfrak{p}_0}$ and $\mathfrak{a}_\mathfrak{p}$ respectively. Then $K_0$, $U$ and $\mathfrak{A}^*$ are compact (see §12 and [5(f), §2]). Put $\mathfrak{q} = [\mathfrak{g}, \mathfrak{a}_\mathfrak{p}]$. It is obvious that $\mathrm{Ad}(a)\mathfrak{q} = \mathfrak{q}$ for $a \varepsilon \mathfrak{A}_c$. We put

$$D(a) = \det(\mathrm{Ad}(a) - \mathrm{Ad}(a^{-1}))_\mathfrak{q} \qquad\qquad (a \varepsilon \mathfrak{A}_c)$$

where $(\mathrm{Ad}(a) - \mathrm{Ad}(a^{-1}))_\mathfrak{q}$ is the restriction of $\mathrm{Ad}(a) - \mathrm{Ad}(a^{-1})$ on $\mathfrak{q}$. Let $dx$, $dk$, $du$, $da$, $da^*$ denote the Haar measures on $G_0$, $K_0$ $U$, and $\mathfrak{A}^*$ respectively. We assume that

$$\int_{K_0} dk = \int_U du = 1.$$

On the other hand $da$ and $da^*$ are normalized as follows. The metric on $\mathfrak{g}$ (see Section 7) defines a Euclidean metric on the real vector space $\mathfrak{a}_{\mathfrak{p}_0}$ which is given by $\| H \|^2 = B(H, H)$ $(H \varepsilon \mathfrak{a}_{\mathfrak{p}_0})$. Let $dH$ denote the element of volume in $\mathfrak{a}_{\mathfrak{p}_0}$ corresponding to this Euclidean metric and put $e(H) = \exp(-1)^{\frac{1}{2}}H$. The mappings $H \to \exp H$ and $H \to e(H)$ $(H \varepsilon \mathfrak{a}_{\mathfrak{p}_0})$ define homomorphisms of the additive group $\mathfrak{a}_{\mathfrak{p}_0}$ onto $\mathfrak{A}$ and $\mathfrak{A}^*$ respectively and it is clear that these homomorphisms are local isomorphisms. Hence we can normalize the Haar measures $da$ and $da^*$ in such a way that $da = dH = da^*$ $(a = \exp H$ and $a^* = e(H)$, $H \varepsilon \mathfrak{a}_{\mathfrak{p}_0})$.

Let $\mathfrak{A}'$ and $\mathfrak{A}^{*\prime}$ be the sets of those points $a$ in $\mathfrak{A}$ and $\mathfrak{A}^*$ respectively where $D(a) \neq 0$. Then both $\mathfrak{A}'$ and $\mathfrak{A}^{*\prime}$ have only a finite number of connected components (see Section 12). Let $w$ and $w^*$ respectively denote their number. Moreover let $C_c(G_0)$ be the set of all continuous functions on $G_0$ which vanish outside a compact set.

**LEMMA 22.** *Let $g$ be a continuous function on $U$ and $B_0{}^*$ a connected component of $\mathfrak{A}^{*\prime}$. Then*

$$\int_U g(u)\,du \int_{\mathfrak{A}^*} |D(a^*)|^{\frac{1}{2}}\,da^* = w^* \int_{B_0{}^*} |D(a^*)|^{\frac{1}{2}}\,da^* \int_{K_0\times K_0} g(ka^*k')\,dk\,dk'.$$

*Moreover we can normalize the Haar measure $dx$ on $G_0$ in such a way that*

$$\int_{G_0} f(x)\,dx = w \int_{B_0} |D(a)|^{\frac{1}{2}}\,da \int_{K_0\times K_0} f(kak')\,dk\,dk'$$

*for all $f \varepsilon C_c(G_0)$ and every connected component $B_0$ of $\mathfrak{A}'$. This normalization of $dx$ and the numbers $w$, $w^*$ and $\int_{\mathfrak{A}^*} |D(a^*)|^{\frac{1}{2}}\,da^*$ are independent of the choice of $\mathfrak{a}_{r_0}$.*

Although the proof of this lemma is not difficult, due to some technical complications, it is rather long. Hence in order not to interrupt our main argument, we postpone it until Section 12.

Now we assume that $\mathfrak{a}_{p_0}$, $\mathfrak{a}_p$, $\mathfrak{a}_\mathfrak{k}$ and $\mathfrak{a}$ are defined as in Section 6 so that $\nu(\mathfrak{a}) = \mathfrak{h}$. Let $\Sigma$ be the set of all roots of $\mathfrak{g}$ with respect to $\mathfrak{a}$, which do not vanish identically on $\mathfrak{a}_p$. Then it is obvious that

$$|D(\exp H)| = \left| \prod_{\alpha\varepsilon\Sigma} (e^{\alpha(H)} - e^{-\alpha(H)}) \right| \qquad\qquad (H \varepsilon \mathfrak{a}_p).$$

Now every linear function $\lambda$ on $\mathfrak{h}$ defines a linear function $\lambda'$ on $\mathfrak{a}$ by the rule $\lambda'(H) = \lambda(\nu(H))$ $(H \varepsilon \mathfrak{a})$. Moreover since $\mathfrak{a} \cap \mathfrak{h} = \mathfrak{a}_\mathfrak{k}$ and $\nu(H) = H$ for $H \varepsilon \mathfrak{a}_\mathfrak{k}$, $\lambda$ and $\lambda'$ coincide on $\mathfrak{a} \cap \mathfrak{h}$. Finally since $\nu$ is an automorphism of $\mathfrak{g}$, it is obvious that $\lambda'$ is a root of $\mathfrak{g}$ with respect to $\mathfrak{a}$ if and only if $\lambda$ is a root with respect to $\mathfrak{h}$. Hence if we identify linear functions on $\mathfrak{h}$ with those on $\mathfrak{a}$ under the mapping $\lambda \to \lambda'$, the two sets of roots coincide. Then $\Sigma$ is exactly the set of those roots $\alpha$ for which $H_\alpha \not\varepsilon \mathfrak{a}_\mathfrak{k}$ (in the notation of Section 6). Let $Q$ be the set of those roots in $P$ which are not identically zero on $\nu(\mathfrak{a}_p)$. Then it follows from Lemma 16 that $Q$ is the disjoint union of $C_i$, $C_{ij}$, $P_0$, $P_i$, $P_{ij}$ $(1 \leqq i < j \leqq s)$. Moreover it is obvious that

$$|D(\exp H)|^{\frac{1}{2}} = \left| \prod_{\alpha\varepsilon Q} (e^{\alpha(H)} - e^{-\alpha(H)}) \right| \qquad\qquad (H \varepsilon \mathfrak{a}_p).$$

Now put

$$H = \sum_{i=1}^{s} t_i (X_{\gamma_i} + X_{-\gamma_i}) \qquad\qquad (t_i \varepsilon C).$$

Then

$$\nu(H) = \sum_{i=1}^{s} t_i H_{\gamma_i} \text{ and therefore}$$

$$\alpha(H) = \alpha(\nu(H)) = \sum_{i=1}^{s} t_i \alpha(H_{\gamma_i}) \qquad (\alpha \varepsilon Q).$$

Since $\gamma_i(H_{\gamma_j}) = 2\delta_{ij}$ $(1 \le i, j \le s)$ it is obvious (see Section 6) that

$$\begin{aligned}
\alpha(H) &= -t_i & \text{if} \quad &\alpha \varepsilon C_i, \\
\alpha(H) &= t_j - t_i & \text{if} \quad &\alpha \varepsilon C_{ij}, \\
\alpha(H) &= t_i & \text{if} \quad &\alpha \varepsilon P_i, \\
\alpha(H) &= t_j + t_i & \text{if} \quad &\alpha \varepsilon P_{ij} \qquad (1 \le i < j \le s).
\end{aligned}$$

Put

$$Q_1 = \bigcup_{1 \le i \le s} (C_i \cup P_i), \qquad Q_2 = \bigcup_{1 \le i < j \le s} (C_{ij} \cup P_{ij}),$$

Then in the notation of Section 6,

$$\left| \prod_{\alpha \varepsilon Q_1} (e^{\alpha(H)} - e^{-\alpha(H)}) \right| = \prod_{1 \le i \le s} 2^{2r_i} |\sinh t_i|^{2r_i},$$

$$\left| \prod_{\alpha \varepsilon P_0} (e^{\alpha(H)} - e^{-\alpha(H)}) \right| = \prod_{1 \le i \le s} |4 \sinh t_i \cosh t_i|.$$

Moreover since

$$\sinh(t_j - t_i) \sinh(t_j + t_i) = (\cosh t_j)^2 - (\cosh t_i)^2,$$

it follows that

$$\left| \prod_{\alpha \varepsilon Q_2} (e^{\alpha(H)} - e^{-\alpha(H)}) \right| = \prod_{1 \le i < j \le s} 2^{2r_{ij}} |(\cosh t_j)^2 - (\cosh t_i)^2|^{r_{ij}}.$$

Hence

$$|D(\exp H)|^{\frac{1}{2}} = \prod_{1 \le i \le s} 2^{2(r_i + 1)} |\sinh t_i|^{2r_i + 1} |\cosh t_i|$$

$$\times \prod_{1 \le i < j \le s} 2^{2r_{ij}} |(\cosh t_j)^2 - (\cosh t_i)^2|^{r_{ij}}.$$

Now suppose $\cosh t_i \ne 0$ $1 \le i \le s$ and put [6] $t_i' = \log(\cosh t_i)$.  Let

$$H' = \sum_{i=1}^{s} t_i H_{\gamma_i}$$

and consider the expression

$$\Delta(2H') = \prod_{\alpha \varepsilon C'} (e^{\alpha(H')} - e^{-\alpha(H')})$$

where $C'$ is the set of all positive compact roots which do not vanish identically on $\nu(\mathfrak{a}_\mathfrak{p})$. It follows from Lemma 16 that $C'$ is the disjoint union of $C_i$ and $C_{ij}$ $(1 \le i < j \le s)$.  Therefore it is clear that

$$\Delta(2H') = \prod_i \{\cosh t_i - (1/\cosh t_i)\} \prod_{i<j} \{(\cosh t_j/\cosh t_i) - (\cosh t_i/\cosh t_j)\}^{r_{ij}}$$

$$= \prod_i \{(\sinh t_i)^{2r_i}/(\cosh t_i)^{r_i}\} \prod_{i<j} \{(\cosh t_j)^2 - (\cosh t_i)^2\}^{r_{ij}}/(\cosh t_i \cosh t_j)^{r_{ij}}.$$

Hence

$$|D(\exp H)|^{\frac{1}{2}} = |\Delta(2H')| \{\prod_i 2^{2r_i+2} |\cosh t_i|^{r_i+1} |\sinh t_i|\}$$

$$\times \prod_{i<j} |4 \cosh t_i \cosh t_j|^{r_{ij}}.$$

But we know from Lemma 18 that

$$r_i + \sum_{i<j\leq s} r_{ij} + \sum_{1\leq j<i} r_{ji} = 2\rho_+(H_i) - 2$$

Therefore

$$\{\prod_i 2^{2r_i+2} |\cosh t_i|^{r_i+1}\} \prod_{i<j} |4 \cosh t_i \cosh t_j|^{r_{ij}}$$

$$= \prod_i 2^{2\rho_+(H\gamma_i)} |\cosh t_i|^{2\rho_+(H\gamma_i)-1} = 2^p \prod_i |\cosh t_i|^{2\rho_+(H\gamma_i)-1}$$

where $p = \sum_{1\leq i\leq s} 2\rho_+(H_{\gamma_i})$. Thus we have the following result.

$$|D(\exp H)|^{\frac{1}{2}} = 2^p |\Delta(2H')| \prod_i \{|\cosh t_i|^{2\rho_+(H\gamma_i)-1} |\sinh t_i|\}$$

We now introduce the partial order in the set $P_0 = (\gamma_1, \cdots, \gamma_s)$ as described at the end of Section 6. Let $\beta_1, \cdots, \beta_r$ be all the (distinct) minimal elements in $P_0$ under this order. For any $i$ $(1\leq i\leq r)$ consider the set $\sigma_i$ of all $\gamma \varepsilon P_0$ such that $\frac{1}{2}(\gamma - \beta_i)$ is a root. Then $\gamma$, $\gamma'$ are two distinct elements of $\sigma_i$, it follows from Lemma 17 that either $\gamma \prec \gamma'$ or $\gamma' \prec \gamma$. This shows that $\sigma_i$ is simply ordered. Therefore we may write it in the form

$$\beta_i = \beta_{i1} \prec \beta_{i2} \cdots \prec \beta_{is_i}.$$

Moreover if $i \neq j$, $\sigma_i$, $\sigma_j$ are disjoint $(1\leq i, j\leq r)$. For otherwise suppose $\gamma \varepsilon \sigma_i \cap \sigma_j$. Then $\frac{1}{2}(\gamma - \beta_i)$ and $\frac{1}{2}(\gamma - \beta_j)$ are both roots and therefore again from Lemma 17, $\frac{1}{2}(\beta_i - \beta_j)$ is a root. This however is impossible since both $\beta_i$ and $\beta_j$ are minimal in $P_0$. Thus $P_0$ is the disjoint union of $\sigma_1, \cdots, \sigma_r$. We put $t_{ij} = t_k$ if $\beta_{ij} = \gamma_k$ $(1\leq i\leq r, 1\leq j\leq s_i, 1\leq k\leq s)$. Let $\mathfrak{b}$ denote the subset of $\mathfrak{a}_{\mathfrak{p}_0}$ consisting of all elements $H = t_1 H_{\gamma_1} + \cdots + t_s H_{\gamma_s}$ $(t_j \varepsilon R)$ such that

$$0 < t_{i1} < t_{i2} < \cdots < t_{is_i} \qquad\qquad (1\leq i\leq r).$$

Similarly let $\mathfrak{b}^*$ denote the subset of $\mathfrak{b}$ consisting of those $H$ which satisfy the additional condition $t_{is_i} < \frac{\pi}{2}$ $(1\leq i\leq r)$. Then we have the following result.

LEMMA 23. *Put* $e(X) = \exp(-1)^{\frac{1}{2}}X$ $(X \,\varepsilon\, \mathfrak{g})$ *and*

$$\Delta(H) = \prod_{\alpha \,\varepsilon\, C'} (e^{\frac{1}{2}\alpha(H)} - e^{-\frac{1}{2}\alpha(H)}) \qquad\qquad (H \,\varepsilon\, \mathfrak{h}).$$

*For any* $H = \sum_i t_i(X_{\gamma_i} + X_{-\gamma_i})$ *in* $\mathfrak{b}$ *let* $H'$ *denote the element* [6]

$$H' = \log (\cosh t_1)H_{\gamma_1} + \cdots + \log (\cosh t_s)H_{\gamma_s}.$$

*in* $\mathfrak{h}$. *Similarly for any* $H = \sum_i t_i(X_{\gamma_i} + X_{-\gamma_i})$ *in* $\mathfrak{b}^*$, *let* $H^*$ *denote the element*

$$H^* = \log (\cos t_1)H_{\gamma_1} + \cdots + \log (\cos t_s)H_{\gamma_s}.$$

*Then*

$$| D(\exp H)|^{\frac{1}{2}} = 2^p \Delta(2H') \prod_i (\cosh t_i)^{2\rho_+(H\gamma_i)-1} \prod_i \sinh t_i \qquad (H \,\varepsilon\, \mathfrak{b})$$

*and*

$$| D(e(H))|^{\frac{1}{2}} = (-1)^p 2^p \Delta(2H^*) \prod_i (\cos t_i)^{2\rho_+(H\gamma_i)-1} \prod_i \sin t_i \qquad (H \,\varepsilon\, \mathfrak{b}^*).$$

If $H \,\varepsilon\, \mathfrak{b}$, $t_i > 0$ $(1 \leq i \leq s)$ and therefore $\sinh t_i > 0$. Hence in view of our earlier result, in this case it is enough to prove that $\Delta(2H')$ is real and positive. But we have already seen that

$$\Delta(2H') = \prod_i \{(\sinh t_i)^{2r_i}/(\cosh t_i)^{r_i}\}$$

$$\times \prod_{i<j} \{(\cosh t_j)^2 - (\cosh t_i)^2\}^{r_{ij}}/(\cosh t_i \cosh t_j)^{r_{ij}}.$$

Since $\cosh t$ is a positive increasing function of $t$ for $t > 0$, $\Delta(2H')$ has the same sign as

$$\eta = \prod_{1 \leq i < j \leq s} (t_j - t_i)^{r_{ij}}.$$

Now if $\frac{1}{2}(\gamma_j - \gamma_i)$ is not a root, we know from Lemma 19 that $r_{ij}$ is even. On the other hand if $\frac{1}{2}(\gamma_j - \gamma_i)$ is a root, $\gamma_i \prec \gamma_j$ and therefore, in view of the definition of $\mathfrak{b}$, $t_j - t_i > 0$. This shows that $\Delta(2H') \geq 0$ and so the first assertion of the lemma follows.

Now we come to the second case when $H \,\varepsilon\, \mathfrak{b}^*$. Since $\cosh((-1)^{\frac{1}{2}}t) = \cos t$ and $\sinh ((-1)^{\frac{1}{2}}t) = (-1)^{\frac{1}{2}} \sin t$ $(t \,\varepsilon\, R)$, it follows that

$$\Delta(2H^*) = \prod_i (-1)^{r_i}\{(\sin t_i)^{2r_i}/(\cos t_i)^{r_i}\}$$

$$\times \prod_{i<j} \{(\cos t_j)^2 - (\cos t_i)^2\}^{r_{ij}}/(\cos t_i \cos t_j)^{r_{ij}}$$

and therefore $\Delta(2H^*)$ is real. Moreover since $\cos t$ is a positive decreasing function of $t$ in the interval $0 < t < \frac{\pi}{2}$ it follows that $\Delta(2H^*)$ has the same sign as

$$\prod_i (-1)^{r_i} \prod_{i<j} (t_i - t_j)^{r_{ij}} = (-1)^q \eta$$

where $q = \sum_i r_i + \sum_{i<j} r_{ij}$. We have seen above that $\eta \geq 0$ on $\mathfrak{b}$ and therefore also on $\mathfrak{b}^*$ and from Lemma 18,

$$q = \sum_{i=1}^{s} \{ 2\rho_+(H_{\gamma_i}) - 2 \} = p - 2s.$$

Hence $(-1)^q = (-1)^p$. On the other hand $\cos t$ and $\sin t$ are both positive on the interval $0 < t < \frac{1}{2}$ and so the second statement of the lemma is an immediate consequence of our earlier expression for $|D|^{\frac{1}{2}}$.

Let $B$ and $B^*$ be the images in $\mathfrak{A}$ and $\mathfrak{A}^*$ of $\mathfrak{b}$ and $\mathfrak{b}^*$ under the mappings $H \to \exp H$ and $H \to e(H)$ respectively.

LEMMA 24. $B \cap \mathfrak{A}'$ *is both open and closed in* $\mathfrak{A}'$. *Similarly* $B^* \cap \mathfrak{A}^{*\prime}$ *is both open and closed in* $\mathfrak{A}^{*\prime}$.

Since $\mathfrak{b}$ and $\mathfrak{b}^*$ are obviously open in $\mathfrak{a}_{\mathfrak{p}_0}$ and since the mappings $H \to \exp H$ and $H \to e(H)$ are regular on $\mathfrak{a}_{\mathfrak{p}_0}$, it follows that $B$ and $B^*$ are open in $\mathfrak{A}$ and $\mathfrak{A}^*$ respectively. Let $\bar{\mathfrak{b}}$ and $\bar{\mathfrak{b}}^*$ respectively denote the closures of $\mathfrak{b}$ and $\mathfrak{b}^*$ in the real Euclidean space $\mathfrak{a}_{\mathfrak{p}_0}$. Then $\bar{\mathfrak{b}}^*$ is compact. Since $H \to \exp H$ is a topological mapping of $\mathfrak{a}_{\mathfrak{p}_0}$ onto $\mathfrak{A}$ (see Section 12), the image $\exp \bar{\mathfrak{b}}$ of $\bar{\mathfrak{b}}$ under this mapping is closed in $\mathfrak{A}$. Also $e(\bar{\mathfrak{b}}^*)$ is compact and therefore closed in $\mathfrak{A}^*$. Hence it is enough to prove that

$$(\exp \bar{\mathfrak{b}}) \cap \mathfrak{A}' = B \cap \mathfrak{A}', \qquad e(\bar{\mathfrak{b}}^*) \cap \mathfrak{A}^{*\prime} = B^* \cap \mathfrak{A}^{*\prime}.$$

Now let $H = \sum_{i=1}^{s} t_i (X_{\gamma_i} + X_{-\gamma_i})$ $(t_i \varepsilon R)$ be a point in $\bar{\mathfrak{b}}$. Then $0 \leq t_i$ and if $\gamma_i \prec \gamma_j$, $t_i \leq t_j$. On the other hand we have seen that

$$| D(\exp H) |^{\frac{1}{2}} = \prod_i 2^{2(r_i+1)} |\sinh t_i|^{2r_i+1} |\cosh t_i|$$
$$\times \prod_{i<j} 2^{2r_{ij}} |(\cosh t_j)^2 - (\cosh t_i)^2|^{r_{ij}}.$$

Hence if $D(\exp H) \neq 0$, $t_i \neq 0$ and $t_i \neq t_j$ if $r_{ij} > 0$ $(1 \leq i < j \leq s)$. This proves that in this case $H \varepsilon \mathfrak{b}$ and therefore $(\exp \bar{\mathfrak{b}}) \cap \mathfrak{A}' = B \cap \mathfrak{A}'$. Now suppose $H$ lies in $\bar{\mathfrak{b}}^*$. Then we have the additional conditions $0 \leq t_i \leq \frac{\pi}{2}$ $(i = 1, \cdots, s)$. Moreover again we know that

$$| D(e(H)) |^{\frac{1}{2}} = \prod_i 2^{2(r_i+1)} |\sin t_i|^{2r_i+1} |\cos t_i|$$
$$\times \prod_{i<j} 2^{2r_{ij}} |(\cos t_j)^2 - (\cos t_i)^2|^{r_{ij}}$$

and therefore if $D(e(H)) \neq 0$, $t_i \neq 0$, $\frac{\pi}{2}$ and $t_i \neq t_j$ whenever $r_{ij} > 0$

$(1 \leqq i < j \leqq s)$. This shows that $H \varepsilon \mathfrak{b}^*$ in this case and therefore

$$e(\bar{\mathfrak{b}}) \cap \mathfrak{A}^{*\prime} = B^* \cap \mathfrak{A}^{*\prime}.$$

COROLLARY. *Every connected component of $B \cap \mathfrak{A}'$ or $B^* \cap \mathfrak{A}^{*\prime}$ is also a connected component of $\mathfrak{A}'$ or $\mathfrak{A}^{*\prime}$ respectively. The number of connected components of $B \cap \mathfrak{A}'$ is the same as that of $B^* \cap \mathfrak{A}^{*\prime}$.*

The first statement is obvious from the above lemma. Now first we claim that the mapping $H \to e(H)$ is univalent on $\mathfrak{b}^*$. For suppose $e(H_1) = e(H_2)$ $(H_1, H_2 \varepsilon \mathfrak{b}^*)$. Then it is obvious that if $\alpha$ is any root in $\Sigma$, $\alpha(H_1) - \alpha(H_2)$ must be an integral multiple of $2\pi$. Hence in particular $\gamma_j(\nu(H_1)) - \gamma_j(\nu(H_2)) = 2n_j\pi$ $(1 \leqq j \leqq s)$ where $n_j$ is an integer. But since $H_1, H_2 \varepsilon \mathfrak{b}^*$, $0 < \gamma_j(\nu(H_i)) < \pi$ $i = 1, 2$ and therefore $n_j = 0$ $(1 \leqq j \leqq s)$. This however implies that $H_1 = H_2$. Since we already know that the mapping $H \to e(H)$ of $\mathfrak{b}^*$ onto $B^*$ is open and continuous it follows that it is topological.

Now let $\sigma_\alpha$ $(\alpha \varepsilon \Sigma)$ denote the hyperplane in $\mathfrak{a}_{\mathfrak{p}_0}$ consisting of all points $H$ such that $\alpha(H) = 0$. Let $\mathfrak{a}'_{\mathfrak{p}_0}$ denote the complement of $\bigcup\limits_{\alpha \varepsilon \Sigma} \sigma_\alpha$ in $\mathfrak{a}_{\mathfrak{p}_0}$. Then it is obvious that $D(\exp H) \neq 0$ $(H \varepsilon \mathfrak{a}_{\mathfrak{p}_0})$ if and only if $H \varepsilon \mathfrak{a}'_{\mathfrak{p}_0}$. Since the exponential mapping of $\mathfrak{a}_{\mathfrak{p}_0}$ onto $\mathfrak{A}$ is topological, $B \cap \mathfrak{A}'$ and $\mathfrak{b} \cap \mathfrak{a}'_{\mathfrak{p}_0}$ have the same number of components. Let $\mathfrak{b}_0$ be a connected component of $\mathfrak{b} \cap \mathfrak{a}'_{\mathfrak{p}_0}$. Then it is obvious that every root $\alpha \varepsilon \Sigma$ must keep constant sign on $\mathfrak{b}_0$. Therefore since $\mathfrak{b}$ is obviously a convex set, the same holds for $\mathfrak{b}_0$. Moreover if $H \varepsilon \mathfrak{b}_0$, it is obvious that the half-line consisting of all points $tH$ $(t > 0)$ lies entirely in $\mathfrak{b}_0$. But if $t$ is sufficiently small and positive $tH \varepsilon \mathfrak{b}^*$. This implies that $\mathfrak{b}_0$ contains some connected component of $\mathfrak{b}^* \cap \mathfrak{a}'_{\mathfrak{p}_0}$. Also if $H_1$, $H_2$ are two points in $\mathfrak{b}^*$ which both lie in $\mathfrak{b}_0$, the straight line-segment $J$ joining them is also contained in $\mathfrak{b}_0$. Hence $J \subset \mathfrak{a}'_{\mathfrak{p}_0}$. But since $\mathfrak{b}^*$ is obviously convex, $J \subset \mathfrak{b}^* \cap \mathfrak{a}'_{\mathfrak{p}_0}$. This shows that $\mathfrak{b}_0$ cannot contain two distinct connected components of $\mathfrak{b}^* \cap \mathfrak{a}'_{\mathfrak{p}_0}$ and therefore every component of $\mathfrak{b} \cap \mathfrak{a}'_{\mathfrak{p}_0}$ contains exactly one component of $\mathfrak{b}^* \cap \mathfrak{a}'_{\mathfrak{p}_0}$. Conversely it is obvious that every component of $\mathfrak{b}^* \cap \mathfrak{a}'_{\mathfrak{p}_0}$ is contained in exactly one component of $\mathfrak{b} \cap \mathfrak{a}'_{\mathfrak{p}_0}$. Hence these two sets have the same number of components.

On the other hand if $H = \sum\limits_{i=1}^{s} t_i(X_{\gamma_i} + X_{-\gamma_i})$ lies in $\mathfrak{b}^*$,

$$|D(e(H))|^{\frac{1}{2}} = \prod_i 2^{2(r_i+1)} |\sin t_i|^{2r_i+1} |\cos t_i|$$

$$\times \prod_{i<j} 2^{2r_{ij}} |(\cos t_j)^2 - (\cos t_i)^2|^{r_{ij}}.$$

Since $0 < t_i < \frac{\pi}{2}$, $D(e(H)) = 0$ if and only if $t_i = t_j$ for some pair of indices $i, j$ $(i < j)$ with $r_{ij} > 0$. But clearly this happens if and only if $\alpha(H) = 0$ for some $\alpha \varepsilon \Sigma$. Hence $D(e(H)) \neq 0$ if and only if $H \varepsilon \mathfrak{b}^* \cap \mathfrak{a}'_{\mathfrak{p}_0}$ and so the mapping $H \to e(H)$ maps $\mathfrak{b}^* \cap \mathfrak{a}'_{\mathfrak{p}_0}$ topologically onto $B^* \cap \mathfrak{A}^*$. Similarly $H \to \exp H$ $(H \varepsilon \mathfrak{b} \cap \mathfrak{a}'_{\mathfrak{p}_0})$ maps $\mathfrak{b} \cap \mathfrak{a}'_{\mathfrak{p}_0}$ topologically onto $B \cap \mathfrak{A}'$. Therefore $B \cap \mathfrak{A}'$, $\mathfrak{b} \cap \mathfrak{a}'_{\mathfrak{p}_0}$, $\mathfrak{b}^* \cap \mathfrak{a}'_{\mathfrak{p}_0}$, $B^* \cap \mathfrak{A}^*$ all have the same number of components.

**9. Application to representations of $G$.** Let $G$ be the simply connected covering group of $G_0$ and let $K$ be the analytic subgroup of $G$ corresponding to $\mathfrak{k}_0$. Define the complex manifold $W$ containing $G$ as in [5(f), §3] and the mapping $\Gamma$ of $G$ into the center $\mathfrak{c}$ of $\mathfrak{k}$ as in [5(f), §6]. Then $\Gamma(xu) = \Gamma(ux) = \Gamma(u) + \Gamma(x)$ $(u \varepsilon K, x \varepsilon G)$ [5(f), Lemma 13] and if $\lambda$ is a real linear function on $\mathfrak{h}$ (see [5(e), §2]) $\lambda(\Gamma(u))$ is pure imaginary [5(f), Lemma 23]. Let $\Lambda$ be a real linear function on $\mathfrak{h}$ such that $\Lambda(H_\alpha)$ is a non-negative integer for every positive compact root $\alpha$. Consider a fundamental system $(\alpha_1, \cdots, \alpha_l)$ of positive roots and suppose that $(\alpha_1, \cdots, \alpha_m)$ are all the totally positive roots in this system. Let $\Lambda_0$ denote the linear function on $\mathfrak{h}$ given by $\Lambda_0(H_{\alpha_i}) = 0$ $1 \leq i \leq m$ and $\Lambda_0(H_{\alpha_i}) = \Lambda(H_{\alpha_i})$ $m < i \leq l$. Then we can define (see [5(f), §6]) an irreducible complex representation $\sigma$ of $G_c$ on a finite-dimensional Hilbert space $V$ such that $\sigma$ is unitary on $U$ and its highest weight is $\Lambda_0$. Let $\phi_0$ be a unit vector in $V$ belonging to the highest weight $\Lambda_0$ and put $\lambda = \Lambda - \Lambda_0$. Then if $x \to \bar{x}$ $(x \varepsilon G)$ denotes the natural mapping of $G$ onto $G_0$, we consider the function

$$\psi_\Lambda(x) = (\phi_0, \sigma(\bar{x})\phi_0) e^{\lambda(\Gamma(x))} \qquad (x \varepsilon G)$$

on $G$. Since the center of $G$ lies in $K$ and $\Lambda$ is real, $|\psi_\Lambda(x)|$ depends only on $\bar{x}$. We propose to consider the integral (cf. [5(f), §9])

$$\int_{G_0} |\psi_\Lambda(x)|^2 \, d\bar{x}$$

where $d\bar{x}$ is the Haar measure on $G_0$. For any $a \varepsilon \mathfrak{A}$ let $\log a$ denote the unique element $H \varepsilon \mathfrak{a}_{\mathfrak{p}_0}$ such that $a = \exp H$. We put $\Gamma(a) = \Gamma(\exp' \log a)$ where $X \to \exp' X$ $(X \varepsilon \mathfrak{g}_0)$ is the exponential mapping of $\mathfrak{g}_0$ into $G$. Then if we normalize the various Haar measures in accordance with Lemma 22 and take into account the fact that $\lambda(\Gamma(a))$ is real for $a \varepsilon \mathfrak{A}$ [5(f), Lemma 23], it follows that

$$\int_{G_0} |\psi_\Lambda(x)|^2 \, d\bar{x} = w N^{-1} \int_B e^{2\lambda(\Gamma(a))} \, |D(a)|^{\frac{1}{2}} \, da \int_{K_0 \times K_0} (\phi_0, \sigma(kak')\phi_0)^2 \, dk \, dk'$$

where $B$ is defined[8] as in Lemma 24 and $N$ is the number of connected components of $B \cap \mathfrak{A}'$. Thus we are led to the integral

$$\int_{K_0 \times K_0} |(\phi_0, \sigma(kak')\phi_0)|^2 \, dk \cdot dk' \qquad (a \, \varepsilon \, B).$$

Let $H = \sum_i t_i(X_{\gamma_i} + X_{-\gamma_i}) \, \varepsilon \, \mathfrak{b} \ (t_i \, \varepsilon \, R)$ and let $a = \exp H$. Then we know from Lemma 20 that

$$a = \zeta h(a) \xi$$

where $\zeta \, \varepsilon \, \mathfrak{P}_c^-$, $\xi \, \varepsilon \, \mathfrak{P}_c^+$ (in the notation of Section 7) and

$$h(a) = \exp\left( \sum_{i=1}^{s} \log(\cosh t_i) H_i \right).$$

If we denote the corresponding representation of $\mathfrak{g}$ also by $\sigma$, it is obvious that $\sigma(\mathfrak{p}_+)\phi_0 = 0$ and therefore $\sigma(p)\phi_0 = \phi_0$ for $p \, \varepsilon \, \mathfrak{P}_c^+$. Since $k\mathfrak{P}_c^- k^{-1} = \mathfrak{P}_c^-$, $k\mathfrak{P}_c^+ k^{-1} = \mathfrak{P}_c^+ (k \, \varepsilon \, K_c)$ and $\tilde{\theta}(\mathfrak{P}_c^-) = \mathfrak{P}_c^+$, it follows that

$$(\phi_0, \sigma(kak')\phi_0) = (\phi_0, \sigma(kh(a)k')\phi_0).$$

Let $A_c$ be the complex analytic subgroup of $G_c$ corresponding to $\mathfrak{h}$. Then $h(a) \, \varepsilon \, A_c \subset K_c$ and

$$\int_{K_0 \times K_0} |(\phi_0, \sigma(kak')\phi_0)|^2 \, dk dk' = \int_{K_0 \times K_0} |(\phi_0, \sigma(kh(a)k')\phi_0)|^2 \, dk dk'.$$

On the other hand $\sigma(H)\phi_0 = \Lambda_0(H)\phi_0 \ (H \, \varepsilon \, \mathfrak{h})$ and $\sigma(X_\alpha)\phi_0 = 0$ for any positive root $\alpha$. Since this holds in particular for every positive compact root, it follows from Lemma 2 of [5(e)] (applied to $\mathfrak{k}$) that the subspace $V_0$ of $V$ spanned by $\sigma(k)\phi_0 \ (k \, \varepsilon \, K_c)$ is irreducible under $K_c$. Let $\sigma_0$ denote the corresponding representation of $K_c$ (and $\mathfrak{k}$) on $V_0$. Then obviously $\Lambda_0$ is the highest weight of $\sigma_0$. Now if we make use of the Schur orthogonality relations for the irreducible representation of the compact group $K_0$ on $V_0$, we find easily that

$$\int_{K_0 \times K_0} |(\phi_0, \sigma(kyk')\phi_0)|^2 \, dk \cdot dk' = (\dim V_0)^{-1} \int_{K_0} |\sigma(yk)\phi_0|^2 \, dk$$
$$= (\dim V_0)^{-2} \operatorname{Sp} \sigma_0(\tilde{\theta}(y)y)$$

if $y \, \varepsilon \, K_c$. Hence

$$\int_{K_0 \times K_0} |(\phi_0, \sigma(kak')\phi_0)|^2 \, dk dk' = (\dim V_0)^{-2} \chi_{\Lambda_0}(h(a))^2)$$

---

[8] Here we have to make use of the obvious fact that the complement of $\mathfrak{A}'$ in $\mathfrak{A}$ is of measure zero with respect to $da$.

where $\chi_{\Lambda_0}$ is the character of $\sigma_0$. But if $P_{\mathfrak{k}}$ is the set of all compact positive roots and $\rho_{\mathfrak{k}} = \frac{1}{2} \sum_{\alpha \in P_{\mathfrak{k}}} \alpha$, we know (see Weyl [11(a)]) that

$$\dim V_0 = \prod_{\alpha \in P_{\mathfrak{k}}} \{\Lambda_0(H_\alpha) + \rho_{\mathfrak{k}}(H_\alpha)/\rho_{\mathfrak{k}}(H_\alpha)\}$$
$$= \prod_{\alpha \in P_{\mathfrak{k}}} \{\Lambda(H_\alpha) + \rho_{\mathfrak{k}}(H_\alpha)/\rho_{\mathfrak{k}}(H_\alpha)\}$$

since $\lambda(H_\alpha) = 0$ $(\alpha \in P_{\mathfrak{k}})$ from Lemma 13 of [5(e)].

We now claim that

$$\lambda(\Gamma(a)) = \sum_{i=1}^{s} \log(\cosh t_i)\lambda(H_{\gamma_i})$$

where (in accordance with our convention) $\log(\cosh t_i)$ is real $(1 \leq i \leq s)$. Since both sides are linear in $\lambda$. it is enough to prove this under the assumption that $\lambda(H_{\alpha_i})$ $1 \leq i \leq m$ are all non-negative integers and $\lambda(H_{\alpha_i}) = 0$ $m < i \leq l$ (see the proof of Lemma 23 of [5(f)]). But then we can find an irreducible complex representation $\sigma'$ of $G_c$ on a finite-dimensional Hilbert space $V'$ with the highest weight $\lambda$ and assume that $\sigma'$ is unitary on $U$ (see [5(f), §6]). Let $\phi'$ be a unit vector in $V'$ belonging to the weight $\lambda$. Then, as we have seen during the proof of Lemma 23 of [5(f)],

$$(\phi', \sigma'(\tilde{x})\phi') = e^{\lambda(\Gamma(a))} \qquad (x \in G)$$

and therefore by the argument which we have already used above,

$$e^{\lambda(\Gamma(a))} = (\phi', \sigma'(a)\phi') = (\phi', \sigma'(h(a))\phi')$$
$$= \exp\left(\sum_{i=1}^{s} (\log \cosh t_i)\lambda(H_{\gamma_i})\right)$$

since $\phi'$ belongs to the weight $\lambda$. Our assertion now follows from the fact that both $\lambda(\Gamma(a))$ and $\sum_i (\log \cosh t_i)\lambda(H_{\gamma_i})$ are real.

If $b$ and $\mu$ are two real numbers and $b$ is positive we define $b^\mu$ in the usual way by $b^\mu = \exp(\mu \log b)$. Then the above result shows that

$$e^{\lambda(\Gamma(a))} = \prod_{i=1}^{s} (\cosh t_i)^{2\lambda(H_{\gamma_i})}$$

For any point $H = \sum_{i=1}^{s} t_i(X_{\gamma_i} + X_{-\gamma_i})$ $(t_i \in R)$ in $\mathfrak{a}_{\mathfrak{p}_0}$, we regard $(t_1, \cdots, t_s)$ as the coordinates of $H$ and denote by $dt$ the measure $dt_1 dt_2 \cdots dt_s$ on $\mathfrak{a}_{\mathfrak{p}_0}$. It is obvious that $dH = c\,dt$ where $c$ is a positive constant and $dH$ is the element of volume corresponding to the Euclidean metric on $\mathfrak{a}_{\mathfrak{p}_0}$ (see Section 8). Hence it follows from the above remarks that

$$\int_{G_0} |\psi_\Lambda(x)|^2 \, d\bar{x} = wN^{-1}(\dim V_0)^{-1} \int_B \chi_{\Lambda_0}((h(a))^2) e^{2\lambda(T(a))} |D(a)|^{\frac{1}{2}} \, da$$

$$= wcN^{-1}(\dim V_0)^{-2} \int_{\mathfrak{b}} \chi_{\Lambda_0}(\exp 2H') \prod_{i=1}^{s} (\cosh t_i)^{2\lambda(H\gamma_i)} |D(\exp H)^{\frac{1}{2}} \, dt$$

where $H = \sum_{i=1}^{s} t_i(X_{\gamma_i} + X_{-\gamma_i})$ and

$$H' = \sum_{i=1}^{s} (\log \cosh t_i) H_{\gamma_i} \tag{$t_i \varepsilon R$}$$

Let $\mathfrak{w}_{\mathfrak{f}}$ be the subgroup of the Weyl group (of $\mathfrak{g}$ with respect to $\mathfrak{h}$) generated by the Weyl reflexions $s_\alpha$ corresponding to $\alpha \varepsilon P_{\mathfrak{f}}$. For any $s \varepsilon \mathfrak{w}_{\mathfrak{f}}$ we define $\epsilon(s) = 1$ or $-1$ according as the permutation $\alpha \to s\alpha$ of the set of all compact roots $\alpha$, is even or odd. As in Lemma 16, let $C_0$ denote the set of those positive compact roots which vanish identically on $\nu(\mathfrak{a}_\mathfrak{p})$ and let $C'$ be the complement of $C_0$ in $P_{\mathfrak{f}}$. Then if $\rho_{\mathfrak{f}} = \frac{1}{2} \sum_{\alpha \varepsilon P_{\mathfrak{f}}} \alpha$ and $\rho_0 = \frac{1}{2} \sum_{\alpha \varepsilon C_0} \alpha$, we have the following result.

LEMMA 25. *Put* $\Lambda''_0 = \Lambda_0 + \rho_{\mathfrak{f}}$. *Then if $w_0$ is the order of the subgroup* $\mathfrak{w}_{\mathfrak{f}}^0$ *of* $\mathfrak{w}_{\mathfrak{f}}$ *generated by* $s_\alpha$ $(\alpha \varepsilon C_0)$,

$$\chi_{\Lambda_0}(\exp H) \prod_{\alpha \varepsilon C'} (e^{\frac{1}{2}\alpha(H)} - e^{-\frac{1}{2}\alpha(H)})$$

$$= \{w_0 \prod_{\alpha \varepsilon C_0} \rho_0(H_\alpha)\}^{-1} \sum_{s \varepsilon \mathfrak{w}_{\mathfrak{f}}} \epsilon(s) \{\prod_{\alpha \varepsilon C_0} s\Lambda''_0(H_\alpha)\} e^{s\Lambda''_0(H)}$$

*for all* $H \varepsilon \nu(\mathfrak{a}_\mathfrak{p})$.

For each $\alpha \varepsilon C'$ we define a holomorphic linear differential operator $D_\alpha$ of order one on the complex Euclidean space $\mathfrak{h}$ such that $D_\alpha \mu = \mu(H_\alpha)$ for every linear function $\mu$ on $\mathfrak{h}$. Then the operators $D_\alpha$ obviously commute with each other. Put $D = \prod_{\alpha \varepsilon C'} D_\alpha$. We know (see Weyl [11(a)]) that

$$\chi_{\Lambda_0}(\exp H \prod_{\alpha \varepsilon C'} (e^{\frac{1}{2}\alpha(H)} - e^{-\frac{1}{2}\alpha(H)}) \prod_{\alpha \varepsilon C_0} (e^{\frac{1}{2}\alpha(H)} - e^{-\frac{1}{2}\alpha(H)})$$

$$= \sum_{s \varepsilon \mathfrak{w}_{\mathfrak{f}}} \epsilon(s) e^{s\Lambda''_0(H)} \tag{$H \varepsilon \mathfrak{h}$}$$

Hence applying the differential operator $D$ to both sides and evaluating the result at a point $H$ in $\nu(\mathfrak{a}_\mathfrak{p})$, we find that

$$\chi_{\Lambda_0}(\exp H) \prod_{\alpha \varepsilon C'} (e^{\frac{1}{2}\alpha(H)} - e^{-\frac{1}{2}\alpha(H)})(D\Delta_0)_H$$

$$= \sum_{s \varepsilon \mathfrak{w}_{\mathfrak{f}}} \epsilon(s) \{\prod_{\alpha \varepsilon C_0} s\Lambda''_0(H_\alpha)\} e^{s\Lambda''_0(H)} \tag{$H \varepsilon \nu(\mathfrak{a}_\mathfrak{p})$}$$

9

since $\alpha(H) = 0$ of $\alpha \varepsilon C_0$ and $H \varepsilon \nu(\mathfrak{a}_\mathfrak{p})$. Here

$$\Delta_0 = \prod_{\alpha \varepsilon C_0} (e^{\alpha/2} - e^{-\alpha/2})$$

and $(D\Delta_0)_H$ denotes the value of $D\Delta_0$ at $H$. But it follows from well-known arguments (Weyl [11(a)]) that [9]

$$\Delta_0 = \sum_{s \varepsilon \mathfrak{w}_\mathfrak{f}^0} \epsilon_0(s) e^{s\rho_0}$$

where $\epsilon_0(s) = \pm 1$ and is determined by the rule

$$\prod_{\alpha \varepsilon C_0} (e^{\frac{1}{2}s\alpha} - e^{-\frac{1}{2}s\alpha}) = \epsilon_0(s) \Delta_0 \qquad\qquad (s \varepsilon \mathfrak{w}_\mathfrak{f}^0).$$

Hence

$$D\Delta_0 = \sum_{s \varepsilon \mathfrak{w}_\mathfrak{f}^0} \epsilon_0(s) \{ \prod_{\alpha \varepsilon C_0} s\rho_0(H_\alpha) \} e^{s\rho_0}.$$

But it is obvious that $\epsilon_0(s) = (-1)^q$ $(s \varepsilon \mathfrak{w}_\mathfrak{f}^0)$ if $q$ is the number of negative roots among $s\alpha$ $(\alpha \varepsilon C_0)$. Therefore

$$\prod_{\alpha \varepsilon C_0} s\rho_0(H_\alpha) = \epsilon_0(s) \prod_{\alpha \varepsilon C_0} \rho_0(H_\alpha)$$

and

$$(D\Delta_0)_H = w_0 \prod_{\alpha \varepsilon C_0} \rho_0(H_\alpha) \qquad\qquad (H \varepsilon \nu(\mathfrak{a}_\mathfrak{p}))$$

since $\rho_0(H) = 0$. Moreover if $\beta \varepsilon C_0$, not all the roots $s_\beta\alpha(\alpha \varepsilon C_0)$ are positive since $s_\beta\beta = -\beta$. This shows that $\sum_{\alpha \varepsilon C_0} s_\beta\alpha < \sum_{\alpha \varepsilon C_0} \alpha$ and therefore $s_\beta\rho_0 < \rho_0$. Since this implies that $\rho_0(H_\beta) > 0$, the lemma now follows.

Now again for any point $H = \sum_{i=1}^{s} t_i(X_{\gamma_i} + X_{-\gamma_i})$ $(t_i \varepsilon R)$ in $\mathfrak{b}$ put $H' = \sum_i \log(\cosh t_i) H_{\gamma_i}$. Then if $\rho = \frac{1}{2} \sum_{\alpha \varepsilon P} \alpha$ and $\Lambda' = \Lambda + \rho$, $\Lambda' = \Lambda''_0 + \lambda + \rho_+$. Moreover since $\lambda(H_\alpha) = 0$ for every compact root $\alpha$ [5(e), Lemma 13], it follows that $s\lambda = \lambda$ for all $s \varepsilon \mathfrak{w}_\mathfrak{f}$. Similarly it follows from Lemma 10 of [5(e)] that $s\rho_+ = \rho_+$ $(s \varepsilon \mathfrak{w}_\mathfrak{f})$ and therefore $\rho_+(H_\alpha) = 0$ if $\alpha \varepsilon P_\mathfrak{f}$. Hence we conclude from Lemmas 23 and 25 that

$$\chi_{\Lambda_0}(\exp 2H') \prod_{i=1}^{s} (\cosh t_i)^{2\lambda(H_{\gamma_i})} | D(\exp H) |^{\frac{1}{2}}$$

$$= 2^p \{ w_0 \prod_{\alpha \varepsilon C_0} \rho_0(H_\alpha) \}^{-1} \prod_i (\sinh t_i)$$

$$\times \sum_{\tau \varepsilon \mathfrak{w}_\mathfrak{f}} \epsilon(\tau) \{ \prod_{\alpha \varepsilon C_0} \tau\Lambda'(H_\alpha) \} \prod_i (\cosh t_i)^{2\tau\Lambda'(H_{\gamma_i})-1}.$$

---

[9] We have to apply here Weyl's result to the subalgebra $\mathfrak{a}_\mathfrak{f} + \sum_{\alpha \varepsilon C_0} (CX_\alpha + CX_{-\alpha})$ of $\mathfrak{g}$ which is obviously reductive [7].

Put $y_i = (\cosh t_i)^{-1}$. Then $\mathfrak{b}$ corresponds to the region defined by $0 < y_i < 1$ and $y_i > y_j$ if $\gamma_i \prec \gamma_j$ $(1 \leqq i < j \leqq s)$. Let us denote this region by $\mathfrak{b}_y$. Then if $dy = dy_1 dy_2 \cdots dy_s$, it is clear that

$$\int_{\mathfrak{b}} \chi_{\nu_0}(\exp 2H') e^{2\lambda(\Gamma(\exp H))} \mid D(\exp H) \mid^{\frac{1}{2}} dt$$

$$= 2^p \{ w_0 \prod_{a \in C_0} \rho_0(H_a) \}^{-1} \int_{\mathfrak{b}_y} \{ \sum_{\tau \in \mathfrak{w}_{\mathfrak{f}}} \epsilon(\tau) \, ( \prod_{a \in C_0} \tau \Lambda'(H_a) ) \prod_i y_i^{-2\tau\Lambda'(H\gamma_i)-1} \} dy.$$

Put $y_{ij} = (\cosh t_{ij})^{-1}$ $(1 \leqq i \leqq r, 1 \leqq j \leqq s_i)$ in the notation of Section 8. Then the region $\mathfrak{b}_y$ is defined by the inequalities

$$1 > y_{i1} > y_{i2} > \cdots > y_{is_i} > 0 \qquad\qquad (1 \leqq i \leqq r)$$

and an elementary computation shows that if $q_{ij}$ are *positive* real numbers

$$\int_{\mathfrak{b}_y} \prod_{i,j} y_{ij}^{q_{ij}-1} \, dy = \prod_{1 \leqq i \leqq r} \prod_{1 \leqq k \leqq s_i} ( \sum_{k \leqq j \leqq s_i} q_{ij})^{-1}.$$

We shall use this formula to determine the value of our integral.

Let $\mathfrak{F}$ denote the space of linear functions on $\mathfrak{h}$. By a rational function $\omega$ on $\mathfrak{F}$, we mean an element of the quotient field of the ring of polynomial functions on $\mathfrak{F}$ (see [5(b), p. 194]). We say that $\omega$ is defined at a point $\mu \varepsilon \mathfrak{F}$, if it can be written in the form $\omega = f/g$ where $f$ and $g$ are two polynomial functions on $\mathfrak{F}$ and $g(\mu) \neq 0$. It is obvious that in that case the ratio $f(\mu)/g(\mu)$ depends only on $\omega$ (and not on the choice of $f$ and $g$). This ratio is called the value of $\omega$ at $\mu$ and denoted by $\omega(\mu)$. We shall need the following simple lemma on rational functions.

LEMMA 26. *Let $f$ and $g \neq 0$ be two polynomial functions on $F$ and let $\omega = f/g$. Suppose there exists a base $\lambda_1, \cdots, \lambda_l$ for $\mathfrak{F}$ over $C$ and an integer $N_0$ with the following property. For any given integers $m_1, \cdots, m_l \geqq N_0$, $\omega(m_1\lambda_1 + \cdots + m_l\lambda_l) = 0$ provided $g(m_1\lambda_1 + \cdots + m_l\lambda_l) \neq 0$. Then $f = 0$.*

For otherwise $fg \neq 0$ and therefore we can choose integers $m_1, \cdots, m_l \geqq N_0$ such that $f(m_1\lambda_1 + \cdots + m_l\lambda_l) g(m_1\lambda_1 + \cdots + m_l\lambda_l) \neq 0$ (see Lemma 32 of [5(a)]). Then obviously $\omega(m_1\lambda_1 + \cdots + m_l\lambda_l) \neq 0$ and we get a contradiction.

COROLLARY. *Suppose $\omega_1$, $\omega_2$ are two rational functions on $\mathfrak{F}$ such that*

$$\omega_1(m_1\lambda_1 + \cdots + m_l\lambda_l) = \omega_2(m_1\lambda_1 + \cdots + m_l\lambda_l)$$

*for all integral values of $m_1, \cdots, m_l \geqq N_0$ whenever both sides are defined. Then $\omega_1 = \omega_2$.*

Let $\omega_1 = f_1/g_1$, $\omega_2 = f_2/g_2$ where $f_1$, $f_2$, $g_1$, $g_2$ are polynomial functions and $g_1 \neq 0$, $g_2 \neq 0$. Then $\omega_1 - \omega_2 = (g_2 f_1 - g_1 f_2)/g_1 g_2$. Since $\omega_1$ and $\omega_2$ are both defined at any point where $g_1 g_2$ does not vanish, it follows from the above lemma that $g_2 f_1 - g_1 f_2 = 0$ and therefore $\omega_1 = \omega_2$.

For any polynomial function $f$ and $\tau \varepsilon \mathfrak{w}_{\mathfrak{k}}$, we denote by $f^\tau$ the polynomial function $\mu \to f(\tau^{-1}\mu)$ $(\mu \varepsilon \mathfrak{F})$. Now define a polynomial function $F_0$ as follows:

$$F_0(\mu) = \prod_{1 \leq i \leq s} \prod_{\substack{1 \leq k \leq s_i \\ k \leq j \leq s_i}} \left( \sum q_{kj}(\mu) \right) \qquad (\mu \varepsilon \mathfrak{F})$$

where $q_{ij}(\mu) = \mu(H_{\gamma_p})$ if $\beta_{ij} = \gamma_p$ $(1 \leq p \leq s)$ in the notation of Section 8. Similarly let $f_0$, $g_\tau$ and $F_\tau$ $(\tau \varepsilon \mathfrak{w}_{\mathfrak{k}})$ be the polynomial functions given by

$$f_0(\mu) = \left\{ w_0 \prod_{a \varepsilon C_0} \rho_0(H_a) \right\} \prod_{a \varepsilon P_{\mathfrak{k}}} \left\{ (\mu(H_a) + \rho(H_a))/\rho(H_a) \right\}^2$$

$$g(\mu) = \prod_{a \varepsilon C_0} \left\{ \mu(\tau H_a) + \rho(\tau H_a) \right\}$$

$$F_\tau(\mu) = F_0{}^\tau(-2(\mu + \rho)) \qquad (\mu \varepsilon \mathfrak{F})$$

and put

$$\omega = 2^p \sum_{\tau \varepsilon \mathfrak{w}_{\mathfrak{k}}} \epsilon(\tau) g_\tau / f_0 F_\tau$$

where $p = 2 \sum_{1 \leq i \leq s} \rho_+(H_{\gamma_i})$. Then it follows from what we have proved above that

$$\int_{G_0} |\psi_\Lambda(x)|^2 \, d\bar{x} = 2^p wcN^{-1} \{f_0(\Lambda)\}^{-1} \int_{\mathfrak{b}_y} \left\{ \sum_{\tau \varepsilon \mathfrak{w}_{\mathfrak{k}}} \epsilon(\tau) g_\tau(\Lambda) \prod_i y_i^{-2\tau\Lambda'(H_{\gamma_i})-1} \right\} dy.$$

Now let us suppose that $\Lambda'(H_\gamma) < 0$ for every totally positive root $\gamma$. Then since any $\tau \varepsilon \mathfrak{w}_{\mathfrak{k}}$ permutes totally positive roots among themselves [5(e), Lemma 10], it follows that $\tau\Lambda'(H_\gamma) < 0$ $(\gamma \varepsilon P_+)$. Hence the above-mentioned formula is applicable to each term of the integral on the right and we see that the rational function $\omega$ is defined at $\Lambda$ and

$$\int_{G_0} |\psi_\Lambda(x)|^2 \, d\bar{x} = wcN^{-1}\omega(\Lambda).$$

Thus we have the following result.

LEMMA 27. *Let $\mathfrak{F}_G(P)$ denote the set of all linear functions on $\Lambda$ satisfying the following two conditions:*

(1)   $\Lambda$ *is real and* $\Lambda(H_a)$ *is a non-negative integer for every positive compact root* $\alpha$.

(2)   $\Lambda(H_\gamma) + \rho(H_\gamma) < 0$ *for every totally positive root* $\gamma$. *Then there*

*exists a uniquely determined rational function $\omega$ on $\mathfrak{F}$ which is defined every-where on $\mathfrak{F}_G(P)$ and such that*

$$\int_{G_0} |\psi_\Lambda(x)|^2 \, d\bar{x} = wcN^{-1}\omega(\Lambda)$$

*for all $\Lambda \varepsilon \mathfrak{F}_G(P)$.*

We have only to show that $\omega$ is unique. Let $(\alpha_1, \cdots, \alpha_l)$ be a funda-mental system of positive roots and suppose $(\alpha_1, \cdots, \alpha_m)$ are all the totally positive roots in this system. Define $l$ linear functions $\Lambda_i$ $(1 \leq i \leq l)$ on $\mathfrak{h}$ by the conditions $\Lambda_i(H_{\alpha_j}) = \delta_{ij}$ $1 \leq i, j \leq l$. Then if $\lambda_+ = \Lambda_1 + \cdots + \Lambda_m$, it follows from Lemma 13 of [5(e)] that $\lambda_+(H_\gamma) > 0$ for $\gamma \varepsilon P_+$ and $\lambda_+(H_\alpha) = 0$ for $\alpha \varepsilon P_{\mathfrak{f}}$. Hence it is obvious that we can find an integer $n \geq 2$ such that $\Lambda_i - n\lambda_+ \varepsilon \mathfrak{F}_G(P)$ $(1 \leq i \leq l)$. Therefore if $\lambda_i = \Lambda_i - n\lambda_+$, $(\lambda_1, \cdots, \lambda_l)$ is a base for $\mathfrak{F}$ over $C$ and $m_1\lambda_1 + \cdots + m_l\lambda_l \varepsilon \mathfrak{F}_G(P)$ for every set of positive integers $(m_1, \cdots, m_l)$. The uniqueness of $\omega$ now follows immediately from the Corollary to Lemma 26.

We shall now determine $\omega$ in another way. Let $\mathfrak{F}_U$ denote the set of all real linear functions $\Lambda$ on $\mathfrak{h}$ such that $\Lambda(H_\beta)$ is a non-negative integer for every positive root $\beta$. For a fixed $\Lambda \varepsilon \mathfrak{F}_U$, let $\sigma$ denote the irreducible complex representation (see [5(f), §6]) of $G_c$ on a finite-dimensional Hilbert space $V$ with the highest weight $\Lambda$. We assume that $\sigma$ is unitary on $U$. $\phi_0$, being a unit vector in $V$ belonging to the weight $\Lambda$, we consider the function

$$\psi_\Lambda^*(x) = (\phi_0, \sigma(x)\phi_0) \qquad (x \varepsilon G_c).$$

Since $V$ is irreducible under $\sigma(U)$, it follows from the Schur orthogonality relations for the compact group $U$, that

$$\int_U |\psi_\Lambda^*(u)|^2 \, du = (\dim V)^{-1}.$$

On the other hand if we normalize the various Haar measures according to Lemma 22 and put $\Lambda' = \Lambda + \rho$, we get

$$\int_U |\psi_\Lambda^*(u)|^2 \, du = w^*N^{-1} \Big( \int_{\mathfrak{A}^*} |D(a^*)|^{\frac{1}{2}} \, da^* \Big)^{-1}$$

$$\times \int_{B^*} |D(a^*)|^{\frac{1}{2}} \, da^* \int_{K_0 \times K_0} |\psi_\Lambda^*(ka^*k')|^2 \, dk \, dk'$$

where $B^*$ is defined as in Lemma 24 and we make use of the fact (see the Corollary to Lemma 24) that $B^*$ and $B$ both have $N$ connected components. Now let $H = \sum_i t_i(X_{\gamma_i} + X_{-\gamma_i}) \varepsilon \mathfrak{b}^*$ $(t_i \varepsilon R)$ and $a^* = e(H)$. Then if we put

$$H^* = \sum_i \log (\cos t_i) H_{\gamma_i}$$

and $h(a^*) = \exp H^* \, \varepsilon \, A_c$, we know from Lemma 20 that

$$a^* = \zeta h(a^*) \xi$$

where $\zeta \varepsilon \mathfrak{P}_c^-$ and $\xi \varepsilon \mathfrak{P}_c^+$. Therefore we conclude in the same way as before, that

$$(\phi_0, \sigma(ka^*k')\phi_0) = (\phi_0, \sigma(kh(a^*)k')\phi_0)$$

and therefore

$$\int_{K_0 \times K_0} |(\phi_0, \sigma(ka^*k')\phi_0)|^2 \, dk dk' = \int_{K_0 \times K_0} |(\phi_0, \sigma(kh(a^*)k')\phi_0)|^2 \, dk dk'.$$

On the other hand if $V_0$ is the subspace of $V$ spanned by $\sigma(k)\phi_0$ ($k \varepsilon K_c$), we prove exactly as before that the corresponding representation $\sigma_0$ of $K_c$ on $V_0$ is irreducible and has the highest weight $\Lambda$. Then if $\chi_\Lambda$ denotes the character of $\sigma_0$ we have

$$\int_{K_0 \times K_0} |(\phi_0, \sigma(ka^*k')\phi_0)|^2 \, dk dk' = (\dim V_0)^{-2} \chi_\Lambda((h(a^*))^2).$$

Therefore

$$\int_{B^*} |D(a^*)|^{\frac{1}{2}} da^* \int_{K_0 \times K_0} |\psi_\Lambda^*(ka^*k')|^2 \, dk dk'$$

$$= (\dim V_0)^{-2} \int_{B^*} \chi_\Lambda((h(a^*))^2)|D(a^*)|^{\frac{1}{2}} da^*$$

$$= (\dim V_0)^{-2} c \int_{\mathfrak{b}^*} \chi_\Lambda(\exp 2H^*)|D(e(H))|^{\frac{1}{2}} dt$$

where $c$ and $dt$ have the same meaning as before. But from Lemmas 23 and 25 it follows that[10]

$$\chi_\Lambda(\exp 2H^*)|D(e(H))|^{\frac{1}{2}}$$

$$= (-1)^p 2^p \{w_0 \prod_{\alpha \varepsilon C_0} \rho(H_\alpha)\}^{-1} \{ \prod_{1 \leq i \leq s} \sin t_i \}$$

$$\times \sum_{\tau \varepsilon \mathfrak{w}_t} \{\varepsilon(\tau) g_\tau(\Lambda) \prod_{1 \leq i \leq s} (\cos t_i)^{2\tau \Lambda'(H_{\gamma_i})-1}\}$$

where $\Lambda' = \Lambda + \rho$. Now put $y_i = \cos t_i$. Then $\mathfrak{b}^*$ corresponds to the region $\mathfrak{b}_y$ defined as before by $0 < y_i < 1$ and $y_i > y_j$ if $\gamma_i \prec \gamma_j$ ($1 \leq i < j \leq s$).

---

[10] The proof of Lemma 25 made no use of the fact that $\Lambda_0(H_\alpha) = 0$ for every totally positive root $\alpha$ lying in the fundamental system $(\alpha_1, \cdots, \alpha_l)$ of positive roots. Hence this lemma is applicable also to $\chi_\Lambda$.

Therefore if we use Weyl's formula [11(a)] for $\dim V_0$ and recall that $\rho(H_\alpha) = \rho_{\mathfrak{l}}(H_\alpha)$ $(\alpha \varepsilon P_{\mathfrak{l}})$, we get

$$(\dim V_0)^{-2} \int_{\mathfrak{b}^*} \chi_\Lambda(\exp 2H^*) \,|\, D(e(H))\,|^{\frac{1}{2}} \, dt$$

$$= (-1)^p \{2^p/f_0(\Lambda)\} \int_{\mathfrak{b}_y} \{ \sum_{\tau \varepsilon \mathfrak{w}_{\mathfrak{l}}} \epsilon(\tau) g_\tau(\Lambda) \prod_{1 \leqq i \leqq s} y_i{}^{2\tau\Lambda'(H\gamma_i)-1} \} \, dy.$$

We have seen that $\tau \varepsilon \mathfrak{w}_{\mathfrak{l}}$ permutes the totally positive roots among themselves and if $\gamma \varepsilon P_+$, $\Lambda'(H_\gamma) > 0$ since $\Lambda(H_\gamma) \geqq 0$ and $\rho(H_\gamma) > 0$ (see Weyl [11(b)]). Therefore $2\tau\Lambda'(H_{\gamma_i}) > 0$ and it follows that the right hand side of the above equation is

$$= (-1)^p 2^p \sum_{\tau \varepsilon \mathfrak{w}_{\mathfrak{l}}} \epsilon(\tau) g_\tau(\Lambda) \{f_0(\Lambda) F_0{}^\tau(2\Lambda')\}^{-1}.$$

Put $\Lambda^* = -(\Lambda + 2\rho)$. Then it is obvious from the definition of $F_0$, $f_0$, $F_\tau$ and $g_\tau$ $(\tau \varepsilon \mathfrak{w}_{\mathfrak{l}})$ that

$$F_\tau(\Lambda^*) = F_0{}^\tau(2\Lambda'), f_0(\Lambda^*) = f_0(\Lambda), g_\tau(\Lambda^*) = (-1)^{r_0} g_\tau(\Lambda)$$

where $r_0$ is the number of roots in $C_0$. Hence

$$\int_U |\,\psi_\Lambda{}^*(u)\,|^2 \, du = w^* c N^{-1} \left( \int_{\mathfrak{A}^*} |\, D(a^*)\,|^{\frac{1}{2}} \, da^* \right)^{-1} (-1)^{p+r_0} \omega(\Lambda^*).$$

But on the other hand by Weyl's formula [11(a)],

$$\int_U |\,\psi_\Lambda{}^*(u)\,|^2 \, du = (\dim V)^{-1} = \{ \prod_{\beta \varepsilon P} \Lambda'(H_\beta)/\rho(H_\beta) \}^{-1}$$

since $\Lambda$ is the highest weight of $\sigma$. This proves that

$$\omega(\Lambda^*) = (-1)^{p+r_0+m}(N/cw^*) \int_{\mathfrak{A}^*} |\, D(a^*)\,|^{\frac{1}{2}} \, da^*$$

$$\times \{ \prod_{\beta \varepsilon P} (\Lambda^*(H_\beta) + \rho(H_\beta))/\rho(H_\beta) \}^{-1}$$

where $m$ is the total number of positive roots. Let $\mathfrak{F}_U{}^*$ denote the set of all linear functions of the form $\Lambda^* = -(\Lambda + 2\rho)$ $(\Lambda \varepsilon \mathfrak{F}_U)$. Define the linear functions $\Lambda_i$ $1 \leqq i \leqq l$ as in the proof of Lemma 27. Then if $\lambda_1 = -(\Lambda_1 + 2\rho)$ and $\lambda_i = -\Lambda_i$ $2 \leqq i \leqq l$, $(\lambda_1, \cdots, \lambda_l)$ is a base for $\mathfrak{F}$ and $m_1\lambda_1 + \cdots + m_l\lambda_l \varepsilon \mathfrak{F}_U{}^*$ for positive integers $(m_1, \cdots, m_l)$. Therefore in view of the Corollary to Lemma 26, we can conclude that

$$\omega(\mu) = (-1)^{p+r_0+m}(N/cw^*) \int_{\mathfrak{A}^*} |\, D(a^*)\,|^{\frac{1}{2}} \, da^*$$

$$\times \{ \prod_{\beta \varepsilon P} (\mu(H_\beta) + \rho(H_\beta))/\rho(H_\beta) \}^{-1}$$

($\mu \varepsilon \mathfrak{F}$), both sides being defined and equal whenever at least one of them is defined. Hence in particular if $\Lambda \varepsilon \mathfrak{F}_G(P)$, it follows from Lemma 27 that

$$\int_{G_0} |\psi_\Lambda(x)|^2 \, d\bar{x} = (-1)^{p+r_0+m} (w/w^*) \int_{\mathfrak{A}^*} |D(a^*)|^{\frac{1}{2}} da^*$$
$$\times \{ \prod_{\beta \varepsilon P} \Lambda'(H_\beta)/\rho(H_\beta) \}^{-1}$$

where $\Lambda' = \Lambda + \rho$. On the other hand if $q$ is the number of totally positive roots, $\prod_{\beta \varepsilon P} \Lambda'(H_\beta)$ has obviously the same sign as $(-1)^q$. Therefore since the left side is positive and $\rho(H_\beta) > 0$ for every positive root $\beta$, $(-1)^{p+r_0+m} = (-1)^q$. Thus we have obtained the following result.

LEMMA 28. *If* $\Lambda \varepsilon \mathfrak{F}_G(P)$,

$$\int_{G_0} |\psi_\Lambda(x)|^2 \, d\bar{x} = (-1)^q (w/w^*) \int_{\mathfrak{A}^*} |D(a^*)|^{\frac{1}{2}} da^*$$
$$\times \{ \prod_{\beta \varepsilon P} (\Lambda(H_\beta) + \rho(H_\beta))/\rho(H_\beta) \}^{-1}.$$

Although our definition of $\mathfrak{a}_{p_0}$ depends on the order (which we have so far assumed to be fixed) on the space $\mathfrak{F}_R$ of real linear functions on $\mathfrak{h}$, we know from Lemma 22 that the above normalization of the measure $d\bar{x}$ and the numbers $w$, $w^*$, $\int_{\mathfrak{A}^*} |D(a^*)|^{\frac{1}{2}} da^*$ are actually independent of this order. Therefore if we normalize $d\bar{x}$ in such a way that

$$\int_{G_0} |\psi_\Lambda(x)|^2 \, d\bar{x} = | \prod_{\beta \varepsilon P} \Lambda(H_\beta) + \rho(H_\beta)/\rho(H_\beta)|^{-1}$$

for all $\Lambda \varepsilon \mathfrak{F}_G(P)$, this new normalization is also independent of the particular order in $\mathfrak{F}_R$. However since the definition of the function $\psi_\Lambda$ depends on this order, we shall now denote it by $\psi_\Lambda{}^P$. Then our result may be stated as follows.

THEOREM 4. *It is possible to normalize the Haar measure on* $d\bar{x}$ *on* $G_0$ *in such a way that the following condition is fulfilled. Let* $P$ *be the set of all positive roots under any given order* $\mathfrak{F}_R$ *and suppose every noncompact root in* $P$ *is totally positive. Let* $\rho = \frac{1}{2} \sum_{\beta \varepsilon P} \beta$ *and let* $\mathfrak{F}_G(P)$ *denote the set of all real linear functions* $\Lambda$ *on* $\mathfrak{h}$ *which satisfy the following two conditions:*

(1)  $\Lambda(H_\alpha)$ *is a non-negative integer for every positive compact root* $\alpha$.

(2)  $\Lambda(H_\beta) + \rho(H_\beta) < 0$ *for noncompact positive root* $\beta$.

*Then*

$$\int_{G_0} |\psi_\Lambda^P(x)|^2 \, d\bar{x} = |\prod_{\beta \epsilon P} \Lambda(H_\beta) + \rho(H_\beta)/\rho(H_\beta)|^{-1}$$

*for all* $\Lambda \epsilon \mathfrak{F}_G(P)$.

Now we return again to our fixed order in $\mathfrak{F}_R$. Let $\Lambda$ be a real linear function on $\mathfrak{h}$ such that $\Lambda(H_\alpha)$ is a non-negative integer for every compact positive root $\alpha$. Define $\psi(x)$ $(x \epsilon G)$ as in the beginning of this section.

LEMMA 29. $\int_{G_0} |\psi_\Lambda(x)|^2 \, d\bar{x} < \infty$ *if and only if* $\Lambda \epsilon \mathfrak{F}_G(P)$.

Define $\lambda_+$ as in the proof of Lemma 27. Then, as we have seen, $\lambda_+(H_\gamma) > 0$ for $\gamma \epsilon P_+$ and $\lambda_+(H_\alpha) = 0$ if $\alpha \epsilon P_\mathfrak{k}$. Put

$$t_0 = \max_{\gamma \epsilon P_+} \{\Lambda(H_\gamma) + \rho(H_\gamma)\}/\lambda_+(H_\gamma).$$

Then it is obvious that $\Lambda - t\lambda_+ \epsilon \mathfrak{F}_G(P)$ $(t \epsilon R)$ if and only if $t > t_0$ and therefore $t_0 \geqq 0$ if and only if $\Lambda \notin \mathfrak{F}_G(P)$. Moreover if $\Lambda_t = \Lambda - t\lambda_+$, it is clear that

$$\psi_{\Lambda_t}(x) = \psi_\Lambda(x) e^{-t\lambda_+(\Gamma(x))} \qquad (x \epsilon G).$$

But if $H = \sum_{1 \leqq i \leqq s} t_i(X_{\gamma_i} + X_{-\gamma_i})$ $(t_i \epsilon R)$ we have seen above that

$$\lambda_+(\Gamma(\exp H)) = \sum_{1 \leqq i \leqq s} (\log \cosh t_i)\lambda_+(H_{\gamma_i}).$$

Since $\cosh t_i \geqq 1$ and $\lambda_+(H_{\gamma_i}) \geqq 0$, it follows that $\lambda_+(\Gamma(a)) \geqq 0$ $(a \epsilon \mathfrak{A})$. Therefore it is obvious from Lemma 22, that

$$\int_{G_0} |\psi_\Lambda(x)|^2 \, d\bar{x} = \int_{G_0} |\psi_{\Lambda_t}(x) e^{t\lambda_+(\Gamma(x))}|^2 \, d\bar{x} \geqq \int_{G_0} |\psi_{\Lambda_t}(x)|^2 \, d\bar{x}$$

provided $t \geqq 0$. Now suppose $\Lambda \notin \mathfrak{F}_G(P)$ so that $t_0 \geqq 0$. Put $t = t_0 + \epsilon$ where $\epsilon$ is positive. Then, in view of Theorem 4, we get

$$\int_{G_0} \psi_\Lambda(x)|^2 \, d\bar{x} \geqq |\prod_{\beta \epsilon P} \{\Lambda(H_\beta) + \rho(H_\beta) - (t_0 + \epsilon)\lambda_+(H_\beta)\}/\rho(H_\beta)|^{-1}.$$

But it follows from the definition of $t_0$ that as $\epsilon \to 0$ the right side tends to infinity. Therefore

$$\int_{G_0} |\psi_\Lambda(x)|^2 \, d\bar{x} = \infty.$$

Conversely if $\Lambda \epsilon \mathfrak{F}_G(P)$ we know from Theorem 4 that

$$\int_{G_0} |\psi_\Lambda(x)|^2 \, d\bar{x} < \infty.$$

Thus the lemma is proved.

We shall now consider the question of the integrability of the function $\psi_\Lambda$. Let us recall that $\rho_+ = \tfrac{1}{2} \sum_{\beta \, \varepsilon \, P_+} \beta$.

LEMMA 30. *Let $\Lambda$ be a linear function on $\mathfrak{h}$ satisfying the following two conditions:*

*(1) $\Lambda(H_\alpha)$ is a non-negative integer for every compact positive root $\alpha$.*

*(2) $\Lambda(H_\beta) + \rho(H_\beta) < 1 - 2\rho_+(H_\beta)$ for every noncompact positive root $\beta$.*

*Then*

$$\int_{G_0} |\psi_\Lambda(x)| \, d\bar{x} < \infty.$$

We use the notation which was introduced at the beginning of this section. In view of Lemma 22, it is enough to prove that

$$\int_B |D(a)|^{\frac{1}{2}} da \int_{K_0 \times K_0} |(\phi_0, \sigma(kak')\phi_0)| \, e^{\lambda(\Gamma(a))} \, dk dk' < \infty.$$

But it follows from the Schwartz inequality that

$$\int_{K_0 \times K_0} |(\phi_0, \sigma(kak')\phi_0)| \, dk dk' \leq \{ \int_{K_0 \times K_0} |(\phi_0, \sigma(kak')\phi_0)|^2 \, dk dk' \}^{\frac{1}{2}}$$
$$= (\dim V_0)^{-1} \{ \mathrm{Sp}\, \sigma_0((h(a))^2) \}^{\frac{1}{2}}$$

as we have seen before. Since $\sigma_0(h(a))$ is obviously a positive definite self-adjoint transformation on $V_0$, it is clear that

$$\mathrm{Sp}\, \sigma_0((h(a))^2) \leq \{ \mathrm{Sp}\, \sigma_0(h(a)) \}^2.$$

Therefore

$$\int_{K_0 \times K_0} |(\phi_0, \sigma(kak')\phi_0)| \, dk dk' \leq (\dim V_0)^{-1} \chi_{\Lambda_0}(h(a)).$$

So it would be sufficient to prove that

$$\int_B \chi_{\Lambda_0}(h(a)) e^{\lambda(\Gamma(a))} |D(a)|^{\frac{1}{2}} da < \infty.$$

For any point

$$H = \sum_{1 \leq i \leq s} t_i(X_{\gamma_i} + X_{-\gamma_i}) \quad (t \, \varepsilon \, R) \text{ in } \mathfrak{h} \text{ we put } H' = \sum_{1 \leq i \leq s} (\log \cosh t_i) H_{\gamma_i}$$

as before. Then if $a = \exp H$, $h(a) = \exp H'$ and therefore it follows from Lemmas 23 and 25 that

$$\chi_{\Lambda_0}(h(a)) e^{\lambda(\Gamma(a))} |D(a)|^{\frac{1}{2}}$$

$$= c_1 \frac{\Delta(2H')}{\Delta(H')} \{ \sum_{\tau \, \varepsilon \, \mathfrak{w}_{\mathfrak{f}}} \epsilon(\tau) g_\tau(\Lambda) \prod_{i=1}^{s} (\cosh t_i)^{\tau\Lambda'(H\gamma_i)+\rho_+(H\gamma_i)-1} \} \prod_{i=1}^{s} (\sinh t_i)$$

where $\Lambda' = \Lambda + \rho$, $c_1$ is a positive constant and $g_\tau$ is defined as before. Now

$$\frac{\Delta(2H')}{\Delta(H')} = \prod_{\alpha \in C'} \left( e^{\frac{1}{2}\alpha(H')} + e^{-\frac{1}{2}\alpha(H')} \right)$$

and therefore if $r'$ is the number of roots in $C'$, it follows from Lemma 16 that

$$0 < \frac{\Delta(2H')}{\Delta(H')} \leq 2^{r'} \prod_{\alpha \in C'} e^{\frac{1}{2}|\alpha(H')|}$$

$$= 2^{r'} \prod_{i=1}^{s} (\cosh t_i)^{\frac{1}{2}r_i} \prod_{1 \leq i < j \leq s} (\cosh t_j / \cosh t_i)^{\frac{1}{2}r_{ij}}$$

in the notation of Section 6. (Here we have to make use of the fact that $t_j > t_i$ on $\mathfrak{b}$ if $r_{ij} > 0$ $(1 \leq i < j \leq s)$). But since

$$r_i - \sum_{i < j \leq s} r_{ij} + \sum_{1 \leq j < i} r_{ji} \leq 2\rho_+(H_{\gamma_i}) - 2$$

from Lemma 18, we conclude that

$$0 < \{\Delta(2H')/\Delta(H')\} \leq 2^{r'} \prod_{i=1}^{s} (\cosh t_i)^{\rho_+(H_{\gamma_i})-1}$$

and therefore if $c_2 = 2^{r'} c_1$,

$$\chi_{\Lambda_0}(h(a)) e^{\lambda(\Gamma(a))} |D(a)|^{\frac{1}{2}}$$

$$\leq c_2 \left\{ \sum_{\tau \in \mathfrak{w}_{\mathfrak{k}}} \epsilon(\tau) g_\tau(\Lambda) \prod_i (\cosh t_i)^{\tau\Lambda'(H_{\gamma_i})+2\rho_+(H_{\gamma_i})-2} \right\} \prod_i \sinh t_i.$$

Now put $y_i = (\cosh t_i)^{-1}$, $dy = dy_1 \cdots dy_s$ and define $\mathfrak{b}_y$ as before. Then

$$\int_{\mathfrak{b}} \prod_i (\cosh t_i)^{\tau\Lambda'(H_{\gamma_i})+2\rho_+(H_{\gamma_i})-2} \prod_i (\sinh t_i)\, dt$$

$$= \int_{\mathfrak{b}_y} \prod_i y_i^{-\tau\Lambda'(H_{\gamma_i})-2\rho_+(H_{\gamma_i})}\, dy \qquad (\tau \in \mathfrak{w}_{\mathfrak{k}}).$$

But we have seen earlier that $\tau\rho_+ = \rho_+$ and therefore

$$\tau\Lambda'(H_{\gamma_i}) + 2\rho_+(H_{\gamma_i}) = \Lambda'(H_\gamma) + 2\rho_+(H_\gamma)$$

where $\gamma = \tau^{-1}\gamma_i$. Hence it follows from our hypothesis on $\Lambda$ that

$$-\tau\Lambda'(H_{\gamma_i}) - 2\rho_+(H_{\gamma_i}) + 1 > 0$$

and therefore

$$\int_{\mathfrak{b}_y} \prod_i y_i^{-\tau\Lambda'(H_{\gamma_i})-2\rho_+(H_{\gamma_i})}\, dy < \infty.$$

This shows that

$$\int_B \chi_{\Lambda_0}(h(a)) e^{\lambda(\Gamma(a))} |D(a)|^{\frac{1}{2}}\, da < \infty$$

and so the lemma is proved.

**10. Integrable and square-integrable representations.** Suppose $\Lambda$ is a real linear function on $\mathfrak{h}$ which satisfies the first condition of Lemma 30. We define the Hilbert space $\mathfrak{H}_\Lambda$ as in Theorem 2 of [5(f)] corresponding to the function $\mu = 1$ on $G_0$. We know from Lemma 14 of [5(f)] and Lemma 29 that $\mathfrak{H}_\Lambda \neq 0$ if and only if $\Lambda \, \varepsilon \, \mathfrak{F}_G(P)$. So now let us assume that $\Lambda \, \varepsilon \, \mathfrak{F}_G(P)$ and let $\pi_\Lambda$ denote the representation of $G$ on $\mathfrak{H}_\Lambda$ (see [5(f), Theorem 2]). Then $\pi_\Lambda$ is irreducible and unitary and

$$(\psi_\Lambda, \pi_\Lambda(x)) = \psi_\Lambda(x) \| \psi_\Lambda \|^2 \qquad\qquad (x \, \varepsilon \, G)$$

from the Corollary to Theorem 2 of [5(f)]. (Here we use the usual notation for the norm and the scalar product in $\mathfrak{H}_\Lambda$). Since $\psi_\Lambda \neq 0$ and

$$\int_{G_0} | \psi_\Lambda(x) |^2 \, d\bar{x} = \| \psi_\Lambda \|^2 < \infty,$$

this prove that $\pi_\Lambda$ is square-integrable (see Section 3). Moreover if $\Lambda$ satisfies the second condition of Lemma 30, we prove in the same way that $\pi_\Lambda$ is integrable. Combining these results with Lemma 19 of [5(e)], we get Theorem 3 of [5(e)] which was stated there without proof.

It is clear from the definition of $\mathfrak{H}_\Lambda$ that every element $\phi \, \varepsilon \, \mathfrak{H}_\Lambda$ is compeletly determined by its restriction on $G$ and in fact

$$\| \phi \|^2 = \int_{G_0} | \phi(x) |^2 \, d\bar{x}.$$

Therefore since $\psi_\Lambda \, \varepsilon \, \mathfrak{H}_\Lambda$ and $\pi_\Lambda$ is irreducible, $\mathfrak{H}_\Lambda$ may be identified with the completion (under the above norm) of the space spanned by the right translates of the function $\psi_\Lambda$ under $G$. If we use the normalization of Theorem 4 for the Haar measure on $G_0$ and denote by $d_\Lambda$ the formal degree of $\pi$. (corresponding to the kernel $Z_0$ of the mapping $x \to \bar{x}$ of $G$ on $G_0$ (see Section 3)), it follows from Theorem 1 that

$$d_\Lambda^{-1} \| \psi_\Lambda \|^2 = \int_{G_0} | (\psi_\Lambda, \pi_\Lambda(x)\psi_\Lambda) |^2 \, d\bar{x} = \| \psi_\Lambda \|^4$$

**and therefore**

$$d_\Lambda = \| \psi_\Lambda \|^{-2} = \prod_{\beta \, \varepsilon \, P} | (\Lambda(H_\beta) + \rho(H_\beta))/\rho(H_\beta) |$$

from Theorem 4. This is the formula for the formal degree in terms of the highest weight $\Lambda$ of the representation. It is substantially the same as the corresponding formula of Weyl [11(a)] for a compact semisimple group.

**11. Similarity with finite-dimensional representations.** In order to simplify matters let us suppose in this section that $G$ is simple but not compact. Then if $(\alpha_0, \alpha_1, \cdots, \alpha_l)$ is a fundamental system of positive roots, we know (Corollary 2 to Lemma 13 of [5(e)]) that it contains exactly one noncompact root, which we may assume to be $\alpha_0$. Let $\Lambda_i$ $0 \leq i \leq l$ be the linear functions on $\mathfrak{h}$ given by $\Lambda_i(H_{\alpha_j}) = \delta_{ij}$ $0 \leq i, j \leq l$ and let $\sigma_i$ be an irreducible complex representation of $G_c$ (see [5(f), §6]) on a finite-dimensional Hilbert space $V_i$ with the highest weight $\Lambda_i$. We assume that $\sigma_i$ is unitary on $V$. Let $\phi_i$ be a unit vector in $V_i$ belonging to the weight $\Lambda_i$. Then if $\psi_i(z) = (\phi_i, \sigma_i(z)\phi_i)$ $(z \,\varepsilon\, G_c)$, it follows from Lemmas 6 and 14 of [5(f)] that $\psi_i(\bar{x}) = \psi_{\Lambda_i}(x)$ $(0 \leq i \leq l)$ and

$$\psi_0(\bar{x}) = \psi_{\Lambda_0}(x) = \exp(\Lambda_0(\Gamma(x))) \qquad (x \,\varepsilon\, G).$$

Hence in particular $\psi_{\Lambda_0}(x)$ is never zero. Now if $\Lambda$ is any linear function on $\mathfrak{h}$, it is obvious that $\Lambda = \lambda_0 \Lambda_0 + \cdots + \lambda_l \Lambda_l$ where $\lambda_i = \Lambda(H_{\alpha_i})$ $0 \leq i \leq l$. Therefore if $\Lambda \,\varepsilon\, \mathfrak{F}_G(P)$, $\lambda_1, \cdots, \lambda_l$ are all non-negative integers while $\lambda_0$ is a *negative* real number. Moreover it follows again from Lemma 6 of [5(f)] that

$$\psi_\Lambda(x) = e^{\lambda_0 \Lambda_0(\Gamma(x))} \psi_1^{\lambda_1}(\bar{x}) \cdots \psi_l^{\lambda_l}(\bar{x}) \qquad (x \,\varepsilon\, G).$$

In particular if $\lambda_0$ is also an integer, $\psi_\Lambda(x)$ depends only on $\bar{x}$ and so if we regard $\psi_\Lambda$ as a function on $G_0$, we have

$$\psi_\Lambda = \psi_0^{\lambda_0} \psi_1^{\lambda_1} \cdots \psi_l^{\lambda_l}.$$

On the other hand if $m_0, m_1, \cdots, m_l$ are non-negative integers,

$$\psi = \psi_0^{m_0} \psi_1^{m_1} \cdots \psi_l^{m_l}$$

is a holomorphic function on $G_c$ and again we can conclude from Lemma 6 of [5(f)] that

$$\psi(z) = (\phi, \sigma(z)\phi) \qquad (z \,\varepsilon\, G_c)$$

where $\sigma$ is the irreducible complex representation of $G_c$ on a finite-dimensional Hilbert space $V$ with the highest weight $m_0 \Lambda_0 + \cdots + m_l \Lambda_l$ and $\phi$ is a unit vector in $V$ belonging to this weight. (We assume of course that $\sigma$ is unitary on $U$). Then $\psi(u)$ $(u \,\varepsilon\, U)$ is a matrix coefficient of an irreducible representation of $U$ and therefore the space spanned by all the right translates of $\psi$ under $U$ is irreducible under the corresponding representation of $U$. Thus the similarity between this case and that of $\psi_\cdot$, $\mathfrak{H}_\Lambda$ and $\pi_\Lambda$ discussed above is now quite obvious.

**12. Proof of Lemma 22.** In order to prove Lemma 22, we shall use a method which has been extensively used before by Weyl [11(a)] and Cartan [2(a)]. It is no longer necessary to assume that $\mathfrak{h}_0$ is maximal abelian in $\mathfrak{g}_0$, since this assumption plays no role whatever in our proof. However we still consider $\mathfrak{g}$ as a Hilbert space under the norm $\|X\|^2 = -B(\bar{\theta}(X), X)$ $(X \varepsilon \mathfrak{g})$ and denote by $G_c$ a simply connected complex Lie group with the Lie algebra $\mathfrak{g}$. As before $G_0$, $K_0$ and $U$ are the (real) analytic subgroups of $G_c$ corresponding to $\mathfrak{g}_0$, $\mathfrak{k}_0$ and $\mathfrak{u} = \mathfrak{k}_0 + (-1)^{\frac{1}{2}}\mathfrak{p}_0$. We assume that $\int_{K_0} dk = 1$.

Let us introduce the following notation for the sake of convenience. Suppose $\mu$ and $\mu'$ are positive measures on two locally compact spaces $E$ and $E'$ respectively and $f$ is an open continuous mapping of $E$ into $E'$ which is locally one-one on $E$. Then we write $d\mu \underset{f}{\sim} d\mu'$ if there exists a positive constant $c$ with the following property. Let $U$ be an open set in $E$ such that $f$ is univalent on $U$. Then $\mu(U) = c\mu'(f(U))$. If it is clear from the context which mapping $f$ we have in mind, we write simply $d\mu \sim d\mu'$.

LEMMA [11] 31. $(k, X) \to k \exp X$ $(k \varepsilon K_0, X \varepsilon \mathfrak{p}_0)$ *is a one-one regular analytic mapping of* $K_0 \times \mathfrak{p}_0$ *onto* $G_0$.

Let $\phi$ denote this mapping. It is known (Cartan [2(b)], Mostow [9]) that $\phi$ is one-one and onto $G_0$. Also it is obviously analytic. Let $f$ be a function on $G_0$ which is defined and analytic around $x = k \exp X$. Then if $Y \varepsilon \mathfrak{p}_0$ and $Z \varepsilon \mathfrak{k}_0$,

$$\left\{ \frac{d}{dt} f(k \exp(X + tY)) \right\}_{t=0} = (Y'f)(x)$$

$$\left\{ \frac{d}{dt} f(k \exp tZ \exp X) \right\}_{t=0} = (Z'f)(x) \qquad (t \varepsilon R)$$

where (see Chevalley [3, p. 157])

$$Y' = \{(1 - \exp(-adX)))/adX\}Y$$

$$Z' = \exp(-adX)Z$$

and $(1 - \exp(-adX))/adX$ stands for the sum of the convergent series

$$\sum_{m \geq 0}^{\infty} (-1)^m (adX)^m/(m+1)!.$$

---

[11] The proofs of Lemmas 31 and 33, which we present here, are substantially the same as those of Cartan. However in view of later applications, we are interested not only in verifying that certain determinants are different from zero but also in computing their actual value. This compels us to reproduce the whole argument.

Let $T$ be the linear transformation of $\mathfrak{g}$ such that $TY = Y'$ and $TZ = Z'$. Since $(adX)\mathfrak{p} \subset \mathfrak{k}$ and $(adX)^2\mathfrak{p} \subset \mathfrak{p}$, it follows that

$$\exp(adX)Z' = Z$$

$$\exp(adX)Y' \equiv \{\sinh adX/adX\}Y \bmod k$$

where $(\sinh adX/adX) = s(X)$ is defined by the power series

$$\sum_{m \geq 0} (adX)^{2m}/(2m+1)!.$$

Since only even powers of $adX$ appear in this series $s(X)$ leaves $\mathfrak{p}$ invariant. Also $\tilde{\theta}(X) = -X$ and therefore $adX$ is self-adjoint. Hence $s(X)$ is a positive definite self-adjoint transformation. Therefore the same holds for its restriction $(s(X))_\mathfrak{p}$ on $\mathfrak{p}$. On the other hand it is obvious from the above congruence that

$$\det(\exp(adX)T) = \det((s(X))_\mathfrak{p}).$$

Since $G_c$ is semisimple $\det(\exp(adX)) = 1$ and therefore

$$\det T = \det(s(X))_\mathfrak{p}.$$

The right side is positive since $(s(X))_\mathfrak{p}$ is positive definite. This proves that $\det T \neq 0$ and therefore $\phi$ is regular at $(k, X)$.

Let $dX$ denote the element of volume in $p_0$ corresponding to the Euclidean metric $\|X\|$ $(X \varepsilon \mathfrak{p}_0)$.

COROLLARY. $dx \sim \det(\sinh adX/adX)_\mathfrak{p}\, dkdX$ $(x = k \exp X)$.

This follows immediately from our calculation above.

We now state a lemma which will be needed frequently.

LEMMA 32. *Let $M$ and $N$ be two manifolds of class $C^1$ satisfying the countability axioms. Then if $f$ is a differentiable mapping of $M$ into $N$, and $Q$ is a subset of $M$ then $\partial \mathrm{im}\, f(Q) \leq \partial \mathrm{im}\, Q$.*

Here we use the Brouwer-Urysohn-Menger definition[12] of dimension (Hurewicz and Wallman [6]). Although this result must be regarded as known, no easily accessible published proof of it seems to be available. Therefor we shall give a short proof in the Appendix (Section 13).

Let $\mathfrak{a}_{\mathfrak{p}_0}$ be a maximal abelian subspace of $\mathfrak{p}_0$. Then we have the following result[11] due to Cartan [2(a), p. 354].

---

[12] In order to distinguish it from the vector dimension of a complex vector space, we denote the topological dimension by "$\partial \mathrm{im}$."

**LEMMA 33.** $\mathfrak{p}_0 = \bigcup\limits_{k \,\varepsilon\, K_0} \mathrm{Ad}(k)\mathfrak{a}_{\mathfrak{p}_0}.$

Extend $\mathfrak{a}_{\mathfrak{p}_0}$ to a maximal abelian subalgebra $\mathfrak{a}_0$ of $\mathfrak{g}_0$ and let $\mathfrak{a}_{\mathfrak{p}}$ and $\mathfrak{a}$ be the complexifications of $\mathfrak{a}_{\mathfrak{p}_0}$ and $\mathfrak{a}_0$ respectively in $\mathfrak{g}$. Then $\mathfrak{a}$ is a Cartan subalgebra of $\mathfrak{g}$. We consider the set $\Sigma$ of all roots of $\mathfrak{g}$ (with respect to $\mathfrak{a}$) which do not vanish identically on $\mathfrak{a}_{\mathfrak{p}_0}$. For each $\alpha\,\varepsilon\,\Sigma$ let $\sigma_\alpha$ denote the hyperplane consisting of all points $H\,\varepsilon\,\mathfrak{a}_{\mathfrak{p}_0}$ such that $\alpha(H)=0$ and let $\mathfrak{a}'_{\mathfrak{p}_0}$ denote the complement of $\bigcup\limits_{\alpha\,\varepsilon\,\Sigma}\sigma_\alpha$ in $\mathfrak{a}_{\mathfrak{p}_0}$. Moreover let $M$ be the set of all $k\,\varepsilon\,K_0$ such that $\mathrm{Ad}(k)H=H$ for every $H\,\varepsilon\,\mathfrak{a}_{\mathfrak{p}_0}$. Obviously $M$ is a closed subgroup of $K_0$. Let $k\to\bar{k}$ $(k\,\varepsilon\,K_0)$ denote the natural mapping of $K_0$ on the factor space $\bar{K}_0 = K_0/M$ consisting of the cosets of the form $kM$. Consider the mapping $\phi$ of $\bar{K}_0 \times \mathfrak{a}_{\mathfrak{p}_0}$ into $\mathfrak{p}_0$ given by $\phi(\bar{k},H)=H^k=\mathrm{Ad}(k)H$ where $k$ is any element in the coset $\bar{k}$. $\phi$ is obviously analytic. Let $\mathfrak{m}_0$ be the centralizer of $\mathfrak{a}_{\mathfrak{p}_0}$ in $\mathfrak{k}_0$, so that $\mathfrak{m}_0$ is the Lie algebra of $M$. Let $f$ be a function on $\mathfrak{p}_0$ defined and analytic around $X=H_0^{k_0}$ $(H_0\,\varepsilon\,\mathfrak{a}_{\mathfrak{p}_0}, k_0\,\varepsilon\,K_0)$. Then if $Z\,\varepsilon\,\mathfrak{k}_0$ and $H\,\varepsilon\,\mathfrak{a}_{\mathfrak{p}_0}$,

$$\left\{\frac{d}{dt}f(\phi(\overline{k_0\exp tZ},H_0))\right\}_{t=0} = \left\{\frac{d}{dt}f(X-t\,\mathrm{Ad}(k_0)[H_0,Z])\right\}_{t=0}$$

$$\left\{\frac{d}{dt}f(\phi(\bar{k}_0,H_0+tH))\right\}_{t=0} = \left\{\frac{d}{dt}f(X+t\,\mathrm{Ad}(k_0)H)\right\}_{t=0}$$

$(t\,\varepsilon\,R)$. Now if $H_0\,\varepsilon\,\mathfrak{a}'_{\mathfrak{p}_0}$, $\dim([H_0,\mathfrak{k}])=\dim\mathfrak{k}-\dim\mathfrak{m}$ (where $\mathfrak{m}$ is the complexification of $\mathfrak{m}_0$). Also $\mathfrak{a}_{\mathfrak{p}}$ and $(\mathrm{ad}\,H_0)\mathfrak{k}$ are mutually orthogonal. Hence

$$\dim(\mathfrak{a}_{\mathfrak{p}}+(\mathrm{ad}\,H_0)\mathfrak{k})=\dim\mathfrak{a}_{\mathfrak{p}}+\dim\mathfrak{k}-\dim\mathfrak{m}=\dim\mathfrak{p}$$

from Lemma 4 of [5(b)]. This proves that

$$\mathrm{Ad}(k_0)(\mathfrak{a}_{\mathfrak{p}}+(\mathrm{ad}\,H_0)\mathfrak{k})=\mathfrak{p}.$$

Since $\dim(\bar{K}_0\times\mathfrak{a}_{\mathfrak{p}_0})=\dim\mathfrak{k}-\dim\mathfrak{m}+\dim\mathfrak{a}_{\mathfrak{p}}=\dim\mathfrak{p}=\dim_R\mathfrak{p}_0$, it follows that $\phi$ is regular on $\bar{K}_0\times\mathfrak{a}'_{\mathfrak{p}_0}$. Let $d\bar{k}$ denote the invariant measure on $\bar{K}_0$ such that $\int_{\bar{K}_0}d\bar{k}=1$. Similarly let $dH$ denote the element of volume in $\mathfrak{a}_{\mathfrak{p}_0}$ corresponding to the Euclidean metric $\|H\|$. Now introduce some (fixed) lexicographic order among roots and let $\Sigma^+$ be the set of positive roots in $\Sigma$ under this order. For each root $\alpha$ select an element $X_\alpha\neq 0$ in $\mathfrak{g}$ such that $[H,X_\alpha]=\alpha(H)X_\alpha$ for all $H\,\varepsilon\,\mathfrak{a}$. Let $\mathfrak{n}^+$ and $\mathfrak{n}^-$ respectively be the subspaces of $\mathfrak{g}$ spanned by $X_\alpha$ and $X_{-\alpha}$ $(\alpha\,\varepsilon\,\Sigma^+)$. Then (see the proof of Lemma 4 of [5(b)])

$$\mathfrak{g} = \mathfrak{n}^- + \mathfrak{a}_\mathfrak{p} + \mathfrak{m} + \mathfrak{n}^+ = \mathfrak{k} + \mathfrak{a}_\mathfrak{p} + \mathfrak{n}^+ = \mathfrak{p} + \mathfrak{m} + \mathfrak{n}^+$$

where the sums are all direct. Hence

$$\mathfrak{k}/\mathfrak{m} \cong \mathfrak{g}/(\mathfrak{m} + \mathfrak{a}_\mathfrak{p} + \mathfrak{n}^+) \cong \mathfrak{p}/\mathfrak{a}_\mathfrak{p}$$

and

$$\mathfrak{g}/(\mathfrak{m} + \mathfrak{a}_\mathfrak{p} + \mathfrak{n}^+) \cong \mathfrak{n}^-.$$

Therefore we may identify $\mathfrak{k}/\mathfrak{m}$ and $\mathfrak{p}/\mathfrak{a}_\mathfrak{p}$ with $\mathfrak{n}^-$ in a natural way. Now if $H \varepsilon \mathfrak{a}_\mathfrak{p}$, $\mathfrak{m}$ is contained in the kernel of the mapping $Z \to [H, Z]$ $(Z \varepsilon \mathfrak{k})$ and therefore by going to the residue classes we get a mapping of $\mathfrak{k}/\mathfrak{m}$ into $\mathfrak{p}/\mathfrak{a}_\mathfrak{p}$. In view of the above identification, this gives a linear transformation $T_H$ in $\mathfrak{n}^-$. It is clear that $T_H$ coincides with the restriction of $\mathrm{ad}\, H$ on $\mathfrak{n}^-$. Therefore in view of our calculation above we can conclude that

$$dX \sim |\det T_H|\, dH d\bar{k} = \prod_{\alpha \varepsilon \Sigma^+} |\alpha(H)|\, dH d\bar{k} \qquad (X = H^{\bar{k}})$$

$(H \varepsilon \mathfrak{a}'_{\mathfrak{p}_0}, \bar{k} \varepsilon \bar{K}_0)$. We shall need this result a little later.

Let $\alpha$ be a root in $\Sigma$. Since $\alpha$ takes real values on $\mathfrak{a}_{\mathfrak{p}_0}$, we can find an element $X \neq 0$ in $\mathfrak{g}_0$ such that $[H, X] = \alpha(H)X$ for all $H \varepsilon \mathfrak{a}_{\mathfrak{p}_0}$. We claim $X \notin \mathfrak{p}_0$. For otherwise since $[\mathfrak{p}_0, \mathfrak{p}_0] \subset \mathfrak{k}_0$, we get $[H, X] \varepsilon \mathfrak{k} \cap \mathfrak{p}_0 = 0$. This however is impossible since $X \neq 0$ and $\alpha(H) \neq 0$ for a suitable $H$ in $\mathfrak{a}_{\mathfrak{p}_0}$. Similarly we prove that $X \notin \mathfrak{k}_0$. Therefore $X = Y + Z$ $(Y \varepsilon \mathfrak{p}, Z \varepsilon \mathfrak{k}_0)$ and $Y \neq 0$, $Z \neq 0$. Then comparing components in $\mathfrak{k}_0$ and $\mathfrak{p}_0$ of both sides of the equation $[H, X] = \alpha(H)X$, we find that $[H, Y] = \alpha(H)Z$ and $[H, Z] = \alpha(H)Y$. Thus we have obtained the following result (see Cartan [2(a), p. 357]).

LEMMA 34. *For each $\alpha \varepsilon \Sigma$ we can choose nonzero elements $Y_\alpha$, $Z_\alpha$ in $\mathfrak{p}_0$ and $\mathfrak{k}_0$ respectively such that*

$$[H, Y_\alpha] = \alpha(H)Z_\alpha, \qquad [H, Z_\alpha] = \alpha(H)Y_\alpha$$

*for all $H \varepsilon \mathfrak{a}_\mathfrak{p}$.*

Let $M_\alpha$ $(\alpha \varepsilon \Sigma^+)$ denote the set of all elements $k \varepsilon K_0$ such that $\mathrm{Ad}(k)H = H$ for every $H \varepsilon \sigma_\alpha$. Then $M_\alpha$ is a compact subgroup of $K_0$ containing $M$. Moreover it is obvious that the element $Z_\alpha$ of the above lemma does not lie in $\mathfrak{m}_0$ while it is certainly contained in the Lie algebra of $M_\alpha$. Hence $\dim M_\alpha > \dim M$. Let $\phi_\alpha$ be the mapping of $(K_0/M_\alpha) \times \sigma_\alpha$ into $\mathfrak{p}_0$ defined by $\phi_\alpha(k^*, H) = \mathrm{Ad}(k)H$ where $k \varepsilon K_0$, $H \varepsilon \sigma_\alpha$ and $k^* = kM_\alpha$.

10

It is obvious that $\phi_\alpha$ is an analytic mapping and therefore it follows from Lemma 32 that

$$\partial\mathrm{im}\,\phi_\alpha((K_0/M_\alpha)\times\sigma_\alpha)\leqq\partial\mathrm{im}\,((K_0/M_\alpha)\times\sigma_\alpha)\leqq\dim_R\mathfrak{p}_0-2$$

since $\dim(K_0/M_\alpha)<\dim(K_0/M)$ and

$$\partial\mathrm{im}\,\sigma_\alpha=\dim_R\sigma_\alpha=\dim_R\mathfrak{a}_{\mathfrak{p}_0}-1.$$

On the other hand it is clear that

$$\phi_\alpha((K_0/M_\alpha)\times\sigma_\alpha)=\phi(\bar{K}_0\times\sigma_\alpha)$$

and therefore

$$\phi((K_0/M)\times(\bigcup_{\alpha\,\varepsilon\,\Sigma^+}\sigma_\alpha))=\bigcup_{\alpha\,\varepsilon\,\Sigma^+}\phi_\alpha((K_0/M_\alpha)\times\sigma_\alpha).$$

Hence if we denote this set by $\mathfrak{p}_s$,

$$\partial\mathrm{im}\,\mathfrak{p}_s\leqq\partial\mathrm{im}\,\mathfrak{p}_0-2.$$

Therefore the complement $^c\mathfrak{p}_s$ of $\mathfrak{p}_s$ in $\mathfrak{p}_0$ is connected (see Hurewicz and Wallman [6, p. 48, Theorem IV 4]). On the other hand let

$$\mathfrak{p}'_0=\phi(\bar{K}_0\times\mathfrak{a}'_{\mathfrak{p}_0}).$$

Since $\phi$ is regular and therefore open on $\bar{K}_0\times\mathfrak{a}'_{\mathfrak{p}_0}$, $\mathfrak{p}'_0$ is open in $\mathfrak{p}_0$. Let $r_0=\dim_R(\mathfrak{a}_{\mathfrak{p}_0}+\mathfrak{m}_0)$ and for any $X\,\varepsilon\,\mathfrak{p}_0$ let $r(X)$ denote the complex dimension of the centralizer of $X$ in $\mathfrak{g}$. Then it is obvious that $r(X)=r_0$ if $X\,\varepsilon\,\mathfrak{p}'_0$ while $r(X)>r_0$ if $X\,\varepsilon\,\mathfrak{p}_s$. Hence $\mathfrak{p}'\subset{}^c\mathfrak{p}_s$. Finally since $\bar{K}_0$ is compact, it follows immediately that $\mathfrak{p}'_0$ is closed in $^c\mathfrak{p}_s$. Therefore since $^c\mathfrak{p}_s$ is connected and $\mathfrak{p}'_0$ is not empty (because $\mathfrak{a}'_{\mathfrak{p}_0}\subset\mathfrak{p}'_0$) we conclude that $\mathfrak{p}'_0={}^c\mathfrak{p}_s$. Now let $\lambda$ be an indeterminate. Then it is clear that there exists a polynomial function $F$ on $\mathfrak{p}$ such that if $X\,\varepsilon\,\mathfrak{p}$, $F(X)$ is the coefficient of $\lambda^{r_0}$ in the characteristic polynomial of $\mathrm{ad}\,X$ (in $\lambda$). It is obvious that $r(X)>r_0$ if and only if $F(X)=0$. Hence $F$ is not identically zero and $\mathfrak{p}_s$ is contained in the set of zeros of $F$ on $\mathfrak{p}_0$. This shows that $^c\mathfrak{p}_s=\mathfrak{p}'_0$ is everywhere dense in $\mathfrak{p}_0$. But $\bar{K}_0$ is compact, and $\|H^k\|=\|H\|$ ($k\,\varepsilon\,\bar{K}$, $H\,\varepsilon\,\mathfrak{a}_{\mathfrak{p}_0}$) and so it follows that $\phi(\bar{K}_0\times\mathfrak{a}_{\mathfrak{p}_0})$ is closed in $\mathfrak{p}_0$. Therefore $\phi(\bar{K}_0\times\mathfrak{a}_{\mathfrak{p}_0})=\mathfrak{p}_0$. This completes the proof of Lemma 33.

Let $M'$ be the normalizer of $\mathfrak{a}_{\mathfrak{p}_0}$ in $K$. Then $M'$ is a closed subgroup of $K_0$ and $M$ is a normal subgroup of $M'$. If $H\,\varepsilon\,\mathfrak{a}_{\mathfrak{p}_0}$, $\mathrm{ad}\,H$ is self-adjoint and therefore $(\mathrm{ad}\,H)^2X=0$ implies $(\mathrm{ad}\,H)X=0$ ($X\,\varepsilon\,\mathfrak{g}$). From this it follows immediately that the Lie algebra of $M'$ is also $\mathfrak{m}_0$ and therefore $W=M'/M$

is discrete. But since it is also compact, $W$ must be finite. It operates as a group of linear transformations on $\mathfrak{a}_{\mathfrak{p}_0}$ as follows:

$$sH = \mathrm{Ad}(k)H \qquad\qquad (s \, \varepsilon \, W)$$

where $k$ is any element in $M'$ belonging to the coset $s$.

LEMMA 35. *Let $H$ be an element in $\mathfrak{a}'_{\mathfrak{p}_0}$. Then if $k \, \varepsilon \, K_0$ and $\mathrm{Ad}(k)H \, \varepsilon \, \mathfrak{a}_{\mathfrak{p}_0}$, $k$ must lie in $M'$.*

Since $H \, \varepsilon \, \mathfrak{a}'_{\mathfrak{p}_0}$, $\mathfrak{a}_{\mathfrak{p}_0}$ is exactly the centralizer of $H$ in $\mathfrak{p}_0$. Therefore $\mathrm{Ad}(k)\mathfrak{a}_{\mathfrak{p}_0}$ is the centralizer of $\mathrm{Ad}(k)H$ in $\mathfrak{p}_0$. But since $\mathrm{Ad}(k)H \, \varepsilon \, \mathfrak{a}_{\mathfrak{p}_0}$ and $\mathfrak{a}_{\mathfrak{p}_0}$ is abelian, $\mathfrak{a}_{\mathfrak{p}_0} \subset \mathrm{Ad}(k)\mathfrak{a}_{\mathfrak{p}_0}$. However $\mathfrak{a}_{\mathfrak{p}_0}$ and $\mathrm{Ad}(k)\mathfrak{a}_{\mathfrak{p}_0}$ obviously have the same dimension over $R$. Therefore $\mathfrak{a}_{\mathfrak{p}_0} = \mathrm{Ad}(k)\mathfrak{a}_{\mathfrak{p}_0}$ and so $k \, \varepsilon \, M'$.

$W$ also operates on $\bar{K}_0$ on the right as follows. Let $\bar{k} \, \varepsilon \, \bar{K}_0$ and $s \, \varepsilon \, W$. Since $M$ is a normal subgroup of $M'$, $\bar{k}s = km'M$ where $k$ and $m'$ are any two elements of $K_0$ and $M'$ respectively lying in the cosets $\bar{k}$ and $s$. It is obvious that $\phi(\bar{k}s, s^{-1}H) = \phi(\bar{k}, H)$ $(H \, \varepsilon \, \mathfrak{a}_{\mathfrak{p}_0})$. Conversely if $H \, \varepsilon \, \mathfrak{a}'_{\mathfrak{p}_0}$ and $\phi(\bar{k}, H) = \phi(\bar{k}_1, H_1)$ $(k, k_1 \, \varepsilon \, K_0, H_1 \, \varepsilon \, \mathfrak{a}_{\mathfrak{p}_0})$, it follows from Lemma 35 that $k_1^{-1}k \, \varepsilon \, M'$. This shows that $\bar{k}_1 = \bar{k}s$ and $H_1 = s^{-1}H$ for some $s \, \varepsilon \, W$. Then if $w_0$ is the order of the group $W$, it follows that every element in $\mathfrak{p}'_0$ has exactly $w_0$ distinct pre-images in $\bar{K}_0 \times \mathfrak{a}'_{\mathfrak{p}_0}$ under $\phi$. We have seen above that there exists a polynomial function $F \neq 0$ on $\mathfrak{p}_0$ such that $F(X) = 0$ $(X \, \varepsilon \, \mathfrak{p}_0)$ if and only if $X \notin \mathfrak{p}'_0$. This shows that the complement of $\mathfrak{p}'_0$ in $\mathfrak{p}_0$ is of Euclidean measure zero. Similarly it is obvious that the complement of $\mathfrak{a}'_{\mathfrak{p}_0}$ in $\mathfrak{a}_{\mathfrak{p}}$ is of Euclidean measure zero. Moreover we have seen above that

$$dX \sim \prod_{\alpha \, \varepsilon \, \Sigma^+} |\alpha(H)| \, dH d\bar{k} \, (X = H^{\bar{k}}, H \, \varepsilon \, \mathfrak{a}'_{\mathfrak{p}_0}, \bar{k} \, \varepsilon \, \bar{K}_0),$$

and so we have the following result.

LEMMA 36. *There exists a positive constant $c$ with the following property. If $f$ is any continuous function on $\mathfrak{p}_0$ vanishing outside a compact set,*

$$\int_{\mathfrak{p}_0} f(X)\,dX = c \int_{\bar{K}_0 \times \mathfrak{A}_{\mathfrak{p}_0}} f(H^{\bar{k}}) \prod_{\alpha \, \varepsilon \, \Sigma^+} |\alpha(H)| \, d\bar{k}dH.$$

On the other hand we have the following lemma.

LEMMA 37. *Let $\mathfrak{b}_0$ be a connected component of $\mathfrak{a}'_{\mathfrak{p}_0}$. Then*

$$\mathfrak{a}'_{\mathfrak{p}_0} = \bigcup_{s \, \varepsilon \, W} s\mathfrak{b}_0.$$

Let $\bar{\mathfrak{b}}_0$ denote the closure of $\mathfrak{b}_0$ in $\mathfrak{a}_{\mathfrak{p}_0}$. Since $\mathfrak{b}_0$ is obviously closed in $\mathfrak{a}'_{\mathfrak{p}_0}$, $\bar{\mathfrak{b}}_0 \cap \mathfrak{a}'_{\mathfrak{p}_0} = \mathfrak{b}_0$. Consider $\phi(\bar{K}_0 \times \mathfrak{b}_0) \subset \mathfrak{p}'_0$. Since $\phi$ is regular on $\bar{K}_0 \times \mathfrak{a}'_{\mathfrak{p}_0}$, $\phi(\bar{K}_0 \times \mathfrak{b}_0)$ is open. Moreover since $\bar{K}_0$ is compact it follows easily that $\phi(\bar{K}_0 \times \bar{\mathfrak{b}}_0)$ is closed in $\mathfrak{p}_0$. Hence $\phi(\bar{K}_0 \times \mathfrak{b}_0) = \phi(\bar{K}_0 \times \bar{\mathfrak{b}}_0) \cap \mathfrak{p}'_0$ is closed in $\mathfrak{p}'_0$. But we know that $\mathfrak{p}'_0$ is connected and therefore $\phi(\bar{K}_0 \times \mathfrak{b}_0) = \mathfrak{p}'_0$. Hence in particular if $H \, \varepsilon \, \mathfrak{a}'_{\mathfrak{p}_0}$, $H = \mathrm{Ad}(k)H_0$ for some $k \, \varepsilon \, K_0$ and $H_0 \, \varepsilon \, \mathfrak{b}_0$. But then it follows from Lemma 35 that $k \, \varepsilon \, M'$ and therefore $H = sH_0$ for some $s \, \varepsilon \, W$. This proves that $\mathfrak{a}'_{\mathfrak{p}_0} \subset \bigcup_{s \, \varepsilon \, W} s\mathfrak{b}_0$. Since the reverse inclusion is obvious we get the lemma.

COROLLARY. *$\mathfrak{a}'_{\mathfrak{p}_0}$ has only a finite number of connected components.*

Now as we have seen earlier, $\prod_{\alpha \, \varepsilon \, \Sigma^+} |\alpha(H)|^2$ $(H \, \varepsilon \, \mathfrak{a}_{\mathfrak{p}_0})$ is the determinant of the linear transformation in $\mathfrak{p}/\mathfrak{a}_{\mathfrak{p}}$ corresponding to $(adH)^2$. From this it follows that the value of this expression does not change if we replace $H$ by $sH$ $(s \, \varepsilon \, W)$. Therefore it is obvious from Lemma 37 that

$$\int_{K_0 \times \mathfrak{A}_{\mathfrak{p}_0}} f(\mathrm{Ad}(k)H) \prod_{\alpha \, \varepsilon \, \Sigma^+} |\alpha(H)| \, dk dH$$
$$= w \int_{K_0 \times \mathfrak{b}_0} f(\mathrm{Ad}(k)H) \prod_{\alpha \, \varepsilon \, \Sigma^+} |\alpha(H)| \, dk dH$$

where $w$ is the number of connected components of $\mathfrak{a}'_{\mathfrak{p}_0}$, $\mathfrak{b}_0$ is any such component and $f$ is a continuous function on $\mathfrak{p}_0$ vanishing outside a compact set. Combining this with Lemma 31 and its corollary, we obtain the following result.

LEMMA 38. *It is possible to normalize the Haar measure $dx$ on $G_0$ in such a way that the following condition is fulfilled. If $f$ is a continuous function on $G_0$ vanishing outside a compact set and $\mathfrak{b}_0$ is any connected component of $\mathfrak{a}'_{\mathfrak{p}_0}$,*

$$\int_{G_0} f(x) dx = \int_{K_0 \times \mathfrak{p}_0} f(k \exp X) \det(\sinh \mathrm{ad} X / \mathrm{ad} X)_\mathfrak{p} \, dk dX$$
$$= cw \int_{\mathfrak{b}_0} \prod_{\alpha \, \varepsilon \, \Sigma^+} |e^{\alpha(H)} - e^{-\alpha(H)}| \, dH \int_{K_0 \times K_0} f(k \exp H \, k') \, dk dk'$$

*where $w$ is the number of connected components of $\mathfrak{a}'_{\mathfrak{p}_0}$ and $c$ is a positive constant given by the relation*

$$\int_{\mathfrak{p}_0} e^{-\|X\|^2} \det(\sinh adX / adX)_\mathfrak{p} \, dX = c \int_{\mathfrak{A}_{\mathfrak{p}_0}} e^{-\|H\|^2} \prod_{\alpha \, \varepsilon \, \Sigma^+} |e^{\alpha(H)} - e^{-\alpha(H)}| \, dH.$$

For the proof we have only to notice the fact that

$$\det(\sinh adH / adH) = \prod_{\alpha \, \varepsilon \, \Sigma^+} (\sinh \alpha(H) / \alpha(H)) \qquad (H \, \varepsilon \, \mathfrak{a}_{\mathfrak{p}_0}).$$

The following lemma is also due to Cartan [2(a)].

LEMMA 39. *Let $'\mathfrak{a}_{\mathfrak{p}_0}$ be any maximal abelian subspace of $\mathfrak{p}_0$. Then*

$$'\mathfrak{a}_{\mathfrak{p}_0} = \mathrm{Ad}(k)\mathfrak{a}_{\mathfrak{p}_0}$$

*for some $k \varepsilon K_0$.*

Define $\mathfrak{a}'_{\mathfrak{p}_0}$ as above and choose $H_0 \varepsilon \mathfrak{a}'_{\mathfrak{p}_0}$. Then $\mathfrak{a}_{\mathfrak{p}_0}$ is exactly the centralizer of $H_0$ in $\mathfrak{p}_0$. If we apply Lemma 33 to $'\mathfrak{a}_{\mathfrak{p}_0}$ (instead of $\mathfrak{a}_{\mathfrak{p}_0}$), it follows that $H_0 \varepsilon \mathrm{Ad}(k^{-1})'\mathfrak{a}_{\mathfrak{p}_0}$ for some $k \varepsilon K_0$. Since $'\mathfrak{a}_{\mathfrak{p}_0}$ is abelian the same holds for $\mathrm{Ad}(k^{-1})'\mathfrak{a}_{\mathfrak{p}_0}$ and therefore $\mathrm{Ad}(k^{-1})'\mathfrak{a}_{\mathfrak{p}_0} \subset \mathfrak{a}_{\mathfrak{p}_0}$ or $'\mathfrak{a}_{\mathfrak{p}_0} \subset \mathrm{Ad}(k)\mathfrak{a}_{\mathfrak{p}_0}$. But then since $'\mathfrak{a}_{\mathfrak{p}_0}$ is maximal abelian in $\mathfrak{p}_0$ it is obvious that $'\mathfrak{a}_{\mathfrak{p}_0} = \mathrm{Ad}(k)\mathfrak{a}_{\mathfrak{p}_0}$.

Let $\mathfrak{A}$ be the analytic subgroup of $G_0$ corresponding to $\mathfrak{a}_{\mathfrak{p}_0}$. We conclude from Lemma 31 that $\mathfrak{A}$ is closed and $H \to \exp H$ ($H \varepsilon \mathfrak{a}_{\mathfrak{p}_0}$) is a topological mapping of $\mathfrak{a}_{\mathfrak{p}_0}$ onto $\mathfrak{A}$. Moreover as we have seen in Section 8,

$$| D(\exp H)| = \prod_{\alpha \varepsilon \Sigma} | e^{\alpha(H)} - e^{-\alpha(H)}|$$

$$= \prod_{\alpha \varepsilon \Sigma^+} | e^{\alpha(H)} - e^{-\alpha(H)}|^2 \qquad (H \varepsilon \mathfrak{a}_{\mathfrak{p}_0}).$$

Therefore if we define $\mathfrak{A}'$ as in Lemma 22, it is clear that $\mathfrak{A}' = \exp(\mathfrak{a}'_{\mathfrak{p}_0})$ and the number of connected component of $\mathfrak{A}'$ and $\mathfrak{a}'_{\mathfrak{p}_0}$ is the same. Moreover it follows from Lemma 39 without difficulty that the numbers $c$ and $w$ of Lemma 38 are independent of the choice of $\mathfrak{a}_{\mathfrak{p}_0}$. Therefore the second statement of Lemma 22 is now obvious.

It remains to prove the first part of Lemma 22. Let

$$e(X) = \exp(-1)^{\frac{1}{2}}X \varepsilon U \qquad (X \varepsilon \mathfrak{p}_0).$$

LEMMA 40. $\mathfrak{A}^* = e(\mathfrak{a}_{\mathfrak{p}_0})$ *is compact.*

Since $U$ is compact it is enough to prove that $\mathfrak{A}^*$ is closed in $U$. Let $\bar{\mathfrak{A}}^*$ denote the closure of $\mathfrak{A}^*$ in $U$. Obviously $\mathfrak{A}^*$ and therefore $\bar{\mathfrak{A}}^*$ is a connected Lie subgroup of $U$. $G_c$ being simply connected, we "extend" $\theta$ to a (complex) automorphism of $G_c$ so that $\theta(\exp X) = \exp \theta(X)$ ($X \varepsilon \mathfrak{g}$). Since $\theta(H) = -H$ ($H \varepsilon \mathfrak{a}_{\mathfrak{p}_0}$), it is clear that $\theta(a) = a^{-1}$ for $a \varepsilon \mathfrak{A}^*$ and therefore by continuity also for $a \varepsilon \bar{\mathfrak{A}}^*$. Hence if $X$ is an element in $\mathfrak{u}$ which lies in the Lie algebra of $\bar{\mathfrak{A}}^*$, $\theta(X) = -X$ and therefore $X \varepsilon \mathfrak{u} \cap \mathfrak{p} = (-1)^{\frac{1}{2}}\mathfrak{p}_0$. But since $\mathfrak{a}_{\mathfrak{p}_0}$ is maximal abelian in $\mathfrak{p}_0$, it is obvious that $X \varepsilon (-1)^{\frac{1}{2}}\mathfrak{a}_{\mathfrak{p}_0}$. This proves that $\mathfrak{A}^*$ and $\bar{\mathfrak{A}}^*$ have the same Lie algebra and therefore $\mathfrak{A}^* = \bar{\mathfrak{A}}^*$.

Define $\mathfrak{A}^{*\prime}$ as in Lemma 22. Then $e(H) \varepsilon \mathfrak{A}^{*\prime}$ ($H \varepsilon \mathfrak{a}_{\mathfrak{p}_0}$) if and only if

$$\prod_{\alpha \varepsilon \Sigma^+} | \sin \alpha(H)| \neq 0.$$

Put $U_* = U/K_0$ and $u_* = uK_0$ $(u \, \varepsilon \, U)$. We define an analytic mapping $\psi$ of $\bar{K}_0 \times \mathfrak{A}^*$ into $U_*$ as follows:

$$\psi(\bar{k}, a) = (ka)_* \qquad\qquad (\bar{k} \, \varepsilon \, \bar{K}, a \, \varepsilon \, \mathfrak{A}^*)$$

where $k$ is any element in the coset $\bar{k}$.

LEMMA 41. *$\psi$ is regular and open on $\bar{K}_0 \times \mathfrak{A}^{*\prime}$ and $\psi(\bar{K} \times \mathfrak{A}^*) = U_*$.*

Let $u$ be an element in $U$ and $f$ a function which is defined and analytic on some neighborhood of $u$ in $U$. Suppose $u = ka$ where $k \, \varepsilon \, K_0$ and $a \, \varepsilon \, \mathfrak{A}^*$. Then if $H \, \varepsilon \, (-1)^{\frac{1}{2}}\mathfrak{a}_{\mathfrak{p}_0}$ and $X \, \varepsilon \, k_0$,

$$\left\{ \frac{d}{dt} f(ka \exp tH) \right\}_{t=0} = (Hf)(u)$$

$$\left\{ \frac{d}{dt} f(k \exp tX \, a) \right\}_{t=0} = (X'f)(u)$$

where $X' = \mathrm{Ad}(a^{-1})X$. Let $T$ denote the linear mapping of $\mathfrak{k} + \mathfrak{a}_\mathfrak{p}$ into $\mathfrak{g}$ given by

$$H \to H \qquad\qquad (H \, \varepsilon \, \mathfrak{a}_\mathfrak{p})$$

$$X \to \mathrm{Ad}(a^{-1})X \qquad (X \, \varepsilon \, \mathfrak{k}).$$

Since $\mathfrak{m}$ lies in the kernel of $T$, we get a linear mapping $\bar{T}$ of $(\mathfrak{k} + \mathfrak{a}_\mathfrak{p})/\mathfrak{m}$ into $\mathfrak{g}/\mathfrak{k}$ by going over to the factor spaces. Now suppose $T(H + X) \, \varepsilon \, \mathfrak{k}$ $(H \, \varepsilon \, \mathfrak{a}_\mathfrak{p}, X \, \varepsilon \, \mathfrak{k})$. Then $H + \mathrm{Ad}(a^{-1})X \, \varepsilon \, \mathfrak{k}$. But

$$\mathrm{Ad}(a^{-1})X \equiv \tfrac{1}{2}(\mathrm{Ad}(a^{-1}) - \mathrm{Ad}(a))X \bmod \mathfrak{k}$$

and $(\mathrm{Ad}(a^{-1}) - \mathrm{Ad}(a))X \, \varepsilon \, \mathfrak{p}$. Hence

$$H + \tfrac{1}{2}(\mathrm{Ad}(a^{-1}) - \mathrm{Ad}(a))X \, \varepsilon \, \mathfrak{k} \cap \mathfrak{p} = 0.$$

However it is easy to check that $(\mathrm{Ad}(a^{-1}) - \mathrm{Ad}(a))\mathfrak{g}$ and $\mathfrak{a}_\mathfrak{p}$ are orthogonal. Therefore $H = 0$ and $\tfrac{1}{2}(Ad(a^{-1}) - Ad(a))X = 0$. Now if in particular $a \, \varepsilon \, \mathfrak{A}^{*\prime}$, this clearly implies that $X \, \varepsilon \, \mathfrak{m}$. This shows that if $a \, \varepsilon \, \mathfrak{A}^{*\prime}$, the kernel of $\bar{T}$ is zero. Moreover as we have seen before $\mathfrak{g} = \mathfrak{k} + \mathfrak{a}_\mathfrak{p} + \mathfrak{n}^+ = \mathfrak{p} + \mathfrak{m} + \mathfrak{n}^+$ and therefore

$$(\mathfrak{k} + \mathfrak{a}_\mathfrak{p})/\mathfrak{m} \cong \mathfrak{g}/(\mathfrak{m} + \mathfrak{n}^+) \cong \mathfrak{p} \cong \mathfrak{g}/\mathfrak{k}$$

Hence $(\mathfrak{k} + \mathfrak{a}_\mathfrak{p})/\mathfrak{m}$ and $\mathfrak{g}/\mathfrak{k}$ have the same dimension. This proves that $\bar{T}$ is an isomorphism of $(\mathfrak{k} + \mathfrak{a}_\mathfrak{p})/\mathfrak{m}$ onto $\mathfrak{g}/\mathfrak{k}$ and therefore $\psi$ is open and regular on $\bar{K}_0 \times \mathfrak{A}^{*\prime}$. Let $du_*$ and $d\bar{k}$ denote the invariant measures on the homogeneous spaces $U_*$ and $\bar{K}_0$ respectively. We assume that

$$\int_{U_*} du_* = \int_{\bar{K}_0} d\bar{k} = 1.$$

Since $\mathfrak{g} = \mathfrak{n}^- + \mathfrak{a}_\mathfrak{p} + \mathfrak{m} + \mathfrak{n}^+$,

$$\mathfrak{g}/(\mathfrak{m} + \mathfrak{n}^+) \cong \mathfrak{a}_\mathfrak{p} + \mathfrak{n}^-$$

and therefore if we identify $(\mathfrak{k} + \mathfrak{a}_\mathfrak{p})/\mathfrak{m}$ and $\mathfrak{g}/\mathfrak{k}$ with $\mathfrak{a}_\mathfrak{p} + \mathfrak{n}^-$ under the isomorphism indicated above, $\bar{T}$ may be regarded as a linear mapping of $\mathfrak{a}_\mathfrak{p} + \mathfrak{n}^-$ into itself. If $Z \,\varepsilon\, \mathfrak{n}^-$,

$$\mathrm{Ad}(a^{-1})(Z + \theta(Z)) \equiv \tfrac{1}{2}(\mathrm{Ad}(a^{-1}) - \mathrm{Ad}(a))(Z + \theta(Z))$$
$$\equiv Z' - \theta(Z') \bmod \mathfrak{k}$$

where $Z' = \tfrac{1}{2}(\mathrm{Ad}(a^{-1}) - \mathrm{Ad}(a))Z$. From this it follows easily that

$$\bar{T}(H + Z) = H + \tfrac{1}{2}(\mathrm{Ad}(a^{-1}) - \mathrm{Ad}(a))Z \qquad (H \,\varepsilon\, \mathfrak{a}_\mathfrak{p}, Z \,\varepsilon\, \mathfrak{n}^-)$$

and therefore

$$|\det \bar{T}| = 2^{-q}|D(a)|^{\frac{1}{2}}$$

where $q$ is the number of roots in $\Sigma^+$. This proves that

$$du_* \sim |D(a)|^{\frac{1}{2}} da\, d\bar{k} \qquad (u_* = \psi(\bar{k}, a))$$

if $\bar{k} \,\varepsilon\, \bar{K}_0$ and $a \,\varepsilon\, \mathfrak{A}^{*\prime}$ and $da$ denotes the Haar measure on $\mathfrak{A}^*$. We shall need this result presently.

For each $\alpha \,\varepsilon\, \Sigma^+$ we consider the character $\xi_\alpha$ of $\mathfrak{A}^*$ given by $\xi_\alpha(e(H)) = \exp((-1)^{\frac{1}{2}}\alpha(H))$ $(H \,\varepsilon\, \mathfrak{a}_{\mathfrak{p}_0})$. Then obviously $a \to \xi_\alpha(a^2)$ $(a \,\varepsilon\, \mathfrak{A}^*)$ is also a non-trivial character of $\mathfrak{A}^*$. Hence its kernel $\mathfrak{A}^*_\alpha$ is a closed subgroup of $\mathfrak{A}^*$ and $\dim \mathfrak{A}^*_\alpha = \dim \mathfrak{A}^* - 1$. Let $M_\alpha$ be the subgroup of $K_0$ consisting of all $k \,\varepsilon\, K_0$ such that $(ka)_* = a_*$ for every $a \,\varepsilon\, \mathfrak{A}^*_\alpha$. Obviously $M_\alpha$ is closed and it contains $M$. Now consider the elements $Y_\alpha$, $Z_\alpha$ of Lemma 34. If $H \,\varepsilon\, \mathfrak{a}_{\mathfrak{p}_0}$ it is obvious that

$$\mathrm{Ad}(e(H))Z_\alpha = \cos \alpha(H)Z_\alpha + (-1)^{\frac{1}{2}} \sin \alpha(H)Y_\alpha.$$

Moreover if $e(H) \,\varepsilon\, \mathfrak{A}^*_\alpha$, $\alpha(H)/\pi$ is an integer and therefore

$$\mathrm{Ad}(a)Z_\alpha = \pm Z_\alpha$$

for $a \,\varepsilon\, \mathfrak{A}^*_\alpha$. From this we conclude immediately that $Z_\alpha$ lies in the Lie algebra of $M_\alpha$. Since $Z_\alpha \,\not\varepsilon\, \mathfrak{m}_0$, $\dim M_\alpha > \dim M$. But $\psi(\bar{K}_0 \times \mathfrak{A}^*_\alpha)$ may be regarded as the image of $(K_0/M_\alpha) \times \mathfrak{A}^*_\alpha$ under an analytic mapping and therefore it follows from Lemma 32 that

$$\dim \psi(\bar{K}_0 \times \mathfrak{A}^*_a) \leqq \dim K_0/M_a + \dim \mathfrak{A}^*_a$$

$$\leqq \dim \bar{K}_0 + \dim \mathfrak{A}^* - 2 = \dim U_* - 2.$$

Hence if $V = \bigcup_{a \varepsilon \Sigma^+} \psi(\bar{K}_0 \times \mathfrak{A}^*_a)$, $V$ is compact and $\partial \text{im } V \leqq \dim U_* - 2$ and therefore (see Hurewicz and Wallman [6, p. 48]) the complement $^c V$ of $V$ in $U_*$ is connected. Moreover $^c V$ is a dense subset of $U_*$. Now consider

$$U'_* = {}^c V \cap \psi(\bar{K}_0 \times \mathfrak{A}^*).$$

Since $\bar{K}_0 \times \mathfrak{A}^*$ is compact, $U'_*$ is closed in $^c V$. On the other hand $V$ and $\psi(\bar{K}_0 \times \mathfrak{A}^{*\prime})$ cannot have any point in common (see Corollary 1 to Lemma 42 below) and therefore

$$U'_* = \psi(\bar{K}_0/\mathfrak{A}^{*\prime}).$$

Since $\psi$ is open on $\bar{K}_0 \times \mathfrak{A}^{*\prime}$, $U'_*$ is a nonempty open subset of $U_*$. Therefore $^c V$ being connected, we can conclude that $U'_* = {}^c V$. This shows that the compact set $\psi(\bar{K}_0 \times \mathfrak{A}^*)$ is dense in $U_*$. Hence $\psi(\bar{K}_0 \times \mathfrak{A}^*) = U_*$ and Lemma 41 is proved.

For any $a \varepsilon \mathfrak{A}^*$ and $s \varepsilon W$, define $a^s = kak^{-1}$ where $k$ is some element of $M'$ lying in the coset $s$. Then $a^s \varepsilon \mathfrak{A}^*$ and $|D(a^s)| = |D(a)|$. Hence $(\mathfrak{A}^{*\prime})^s = \mathfrak{A}^{*\prime}$.

LEMMA 42. *Suppose* $(ka_1)_* = (a_2)_*$ $(k \varepsilon K_0; a_1, a_2 \varepsilon \mathfrak{A}^*)$. *Then if* $a_1 \varepsilon \mathfrak{A}^{*\prime}$, $k$ *lies in* $M'$ *and* $a_2$ *in* $\mathfrak{A}^{*\prime}$.

For $ka_1 = a_2 k'$ where $k' \varepsilon K_0$. Applying the automorphism $\theta$ to both sides we get $ka_1^{-1} = a_2^{-1}k'$ and therefore $ka_1^2 k^{-1} = a_2^2$. Since $a_1 \varepsilon \mathfrak{A}^{*\prime}$, $\mathfrak{a}_{\mathfrak{p}_0}$ is exactly the set of all elements $H \varepsilon \mathfrak{p}_0$ such that $\text{Ad}(a_1^2)H = H$. Therefore it follows from the above relation that $\text{Ad}(k)\mathfrak{a}_{\mathfrak{p}_0} = \mathfrak{a}_{\mathfrak{p}_0}$ and so $k \varepsilon M'$. Also it is now clear that $a_2 \varepsilon \mathfrak{A}^{*\prime}$.

COROLLARY 1. *Suppose* $a \varepsilon \mathfrak{A}^{*\prime}$ *and* $\bar{k} \varepsilon \bar{K}_0$. *Then if* $w_0$ *is the order of* $W$, $\psi(\bar{k}, a)$ *has exactly* $w_0$ *distinct preimages in* $\bar{K}_0 \times \mathfrak{A}^*$ *namely* $(\bar{k}s^{-1}, a^s)$ $(s \varepsilon W)$.

Choose an element $k$ in the coset $\bar{k}$. Then if $\psi(\bar{k}_1, a_1) = \psi(\bar{k}, a)$ $(k_1 \varepsilon K_0, a_1 \varepsilon \mathfrak{A}^*)$, it follows from Lemma 42 that $k^{-1}k_1 \varepsilon M'$ and therefore if $s^{-1}$ denotes the corresponding element of $W$, $\bar{k}_1 = \bar{k}s^{-1}$ and $a_1 = a^s$.

COROLLARY 2. *There exists a positive number* $c$ *with the following property. If* $f$ *is any continuous function on* $U_*$,

$$\int_{U_*} f(u_*) du_* = c \int_{\mathfrak{A}^*} |D(a)|^2 da \int_{\bar{K}_0} f(\psi(\bar{k}, a)) d\bar{k}.$$

We have seen that $du_* \sim |D(a)|^{\frac{1}{2}} da dk$ if $u_* = \psi(k, a)$ and $k \varepsilon \bar{K}_0$, $a \varepsilon \mathfrak{A}^{*\prime}$. Now if we define $V = \bigcup_{a \varepsilon \Sigma^+} \psi(\bar{K}_0 \times \mathfrak{A}^*_a)$ as before, it follows from Lemma 45 (Section 13) that $V$ is a null set on $U_*$ with respect to the measure $du_*$. Similarly $\bigcup_{a \varepsilon \Sigma^+} \mathfrak{A}^*_a$ is a null set on $\mathfrak{A}^*$ with respect to the measure $da$. Since $\psi(\bar{K}_0 \times \mathfrak{A}^{*\prime}) = {}^cV$, our assertion follows from Corollary 1 above.

Put $J^* = K_0 \cap \mathfrak{A}^*$. Since $\mathfrak{k}_0 \cap (-1)^{\frac{1}{2}} \mathfrak{a}_{p_0} = 0$, $J^*$ is both discrete and compact and so it is finite. It is obvious that $(J^*)^s = J^*$ $(s \varepsilon W)$.

LEMMA 43. *Let $B^*_0$ be any connected component of $\mathfrak{A}^{*\prime}$. Then*

$$\mathfrak{A}^{*\prime} = \bigcup_{s \varepsilon W} (B^*_0)^s J^*.$$

*Hence $\mathfrak{A}^{*\prime}$ has only a finite number of connected components.*

Let $\bar{B}^*_0$ denote the closure of $B^*_0$ in $\mathfrak{A}^*$. Since $B^*_0$ is closed in $\mathfrak{A}^{*\prime}$, $B^*_0 = \bar{B}^*_0 \cap \mathfrak{A}^{*\prime}$. Now define $V$ and ${}^cV$ as above. Then

$$\psi(\bar{K}_0 \times B^*_0) = {}^cV \cap \psi(\bar{K}_0 \times \bar{B}^*_0)$$

is both open and closed in ${}^cV$ and therefore, ${}^cV$ being connected,

$${}^cV = \psi(\bar{K}_0 \times B^*_0) \quad \text{and} \quad U_* = \psi(\bar{K}_0 \times \bar{B}^*_0).$$

Now suppose $a \varepsilon \mathfrak{A}^{*\prime}$. Then we can find elements $b \varepsilon \bar{B}^*_0$ and $k \varepsilon K_0$ such that $a_* = (kb)_*$. Hence $(k^{-1}a)_* = b_*$ and therefore $k \varepsilon M'$ from Lemma 42. This means that $b_* = (a^s)_*$ for some $s \varepsilon W$ and $b^{-1} a^s \varepsilon K_0 \cap \mathfrak{A}^* = J^*$. Conversely if $z \varepsilon J^*$, $z^{-1} = \theta(z)$ and so $z^2 = 1$. Hence $|D(az)| = |D(a)|$ for any $a \varepsilon \mathfrak{A}^*$. This shows that $\bigcup_{s \varepsilon W} (B^*_0)^s J^*$ is contained in $\mathfrak{A}^{*\prime}$ and so the lemma follows.

COROLLARY. *Let $w^*$ be the number of connected components of $\mathfrak{A}^{*\prime}$. Then if $f$ is any continuous function on $U$,*

$$\int_U f(u) du \int_{\mathfrak{A}^*} |D(a)|^{\frac{1}{2}} da = w^* \int_{B_0^*} |D(a)|^{\frac{1}{2}} da \int_{K_0 \times K_0} f(kak') dk dk'.$$

Let $f^*$ denote the function on $U_*$ given by

$$f^*(u_*) = \int_{K_0} f(uk) dk \qquad (u \varepsilon U).$$

Then

$$\int_{U_*} f^*(u_*) du_* = \int_U f(u) du$$

and

$$\int_{K_0} f(kak') dk' = f^*((ka)_*).$$

Therefore

$$\int_{K_0 \times K_0} f(kak')\, dk dk' = \int_{K_0} f^*((ka_*)\, dk$$

and so it follows from Corollary 2 to Lemma 42 that

$$\int_U f(u)\, du = c \int_{\mathfrak{A}^*} |D(a)|^{\frac{1}{2}}\, da \int_{K_0 \times K_0} f(kak')\, dk dk'.$$

Now we can replace the integral on $\mathfrak{A}^*$ on the right by the corresponding integral on $\mathfrak{A}^{*'}$. We know that any component $B^*$ of $\mathfrak{A}^{*'}$ is of the form $(B^*_0)^s z$ $(s \varepsilon W, z \varepsilon J^*)$. Therefore if $m$ is any element of $M'$ in the coset $s$, it is clear that

$$\int_{B^*} |D(a)|^{\frac{1}{2}}\, da \int_{K_0 \times K_0} f(kak')\, dk dk'$$

$$= \int_{B_0^*} |D(a^s z)|^{\frac{1}{2}}\, da \int_{K_0 \times K_0} f(kmam^{-1}zk')\, dk dk'$$

$$= \int_{B_0^*} |D(a)|^{\frac{1}{2}}\, da \int_{K_0 \times K_0} f(kak')\, dk dk'$$

since $|D(a^s z)| = |D(a)|$. This shows that

$$\int_{\mathfrak{A}^*} |D(a)|^{\frac{1}{2}}\, da \int_{K_0 \times K_0} f(kak')\, dk dk' = w^* \int_{B_0^*} |D(a)|^{\frac{1}{2}}\, da \int_{K_0 \times K_0} f(kak')\, dk dk'.$$

In particular if we take the function $f = 1$, we get

$$1 = \int_U du = c \int_{\mathfrak{A}^*} |D(a)|^{\frac{1}{2}}\, da$$

and this gives our result.

Now if we normalize the Haar measure $da$ on $\mathfrak{A}^*$ in accordance with Lemma 22, it follows from Lemma 39 that the numbers $w^*$ and $\int_{\mathfrak{A}^*} |D(a)|^{\frac{1}{2}}\, da$ are independent of the choice of $\mathfrak{a}_{\mathfrak{p}_0}$. This completes the proof of Lemma 22.

**13. Appendix.** We shall now give a proof of Lemma 32. Let $m$ and $n$ be the dimensions of $M$ and $N$ respectively and let $(t_1, \cdots, t_m)$ be a coordinate system on $M$ which is valid on some open neighborhood $V$ of a point $x_0 \varepsilon M$. We assume $t_i(x_0) = 0$ $i = 1, \cdots, m$. For any positive integer $p$ let $R^p$ denote the Cartestian product of $R$ with itself $p$ times. It is obvious that we can chose a compact neighborhood $W$ of $x_0$ and a positive number $a$ such that $W \subset V$ and the mapping $x \to (t_1(x), \cdots, t_m(x))$ $(x \varepsilon W)$ maps $W$ topologically on the cube $|t_i| \leq a$ in $R^m$. Any set $W$ defined in this way

is called a cubic set in $M$ (with respect to the coordinate system $(t_1, \cdots, t_m)$).
Cubic sets in $N$ are defined similarly.

Since $M$ satisfies the countability axioms (see Chevalley [3]) we can
find a countable family $V_i$ $(i = 1, 2, \cdots)$ of cubic sets in $M$ such that the
following conditions hold: (1) $M = \bigcup_i V_i$ and (2) for each $i$, $f(V_i)$ is contained in some cubic set in $N$. Therefore, in view of the sum theorem of
dimension theory (Hurewicz and Wallman [6, p. 30]) it is enough to show
that $\partial \mathrm{im}\, f(V_i) \leqq m$ for each $i$. This however is an immediate consequence
of the lemma below.

We regard $R^p$ as an additive group in the usual way and if

$$a = (a_1, \cdots, a_p) \,\varepsilon\, R^p, \text{ we put } |a| = (\sum_{1 \leqq i \leqq p} a_i^2)^{\frac{1}{2}}.$$

Let $I$ be the unit interval $0 \leqq t \leqq 1$ in $R$ and let $I^m$ be the Cartesian product
of $I$ with itself $m$ times. Then $I^m \subset R^m$.

LEMMA 44. *Let $f$ be a mapping of $I^m$ into $R^n$ such that*

$$|f(b) - f(a)| \leqq c\,|b - a| \qquad\qquad (a, b \,\varepsilon\, I^m)$$

*where $c$ is a fixed number. Then $\partial \mathrm{im}\, f(I^m) \leqq m$.*

It is enough to prove that the Hausdorff $(m+1)$-measure of $f(I^m)$
is zero (see Hurewicz and Wallman [6, p. 104]). But this follows at once
from our hypothesis on $f$ and the fact that the Hausdorff $(m+1)$-measure
of $I^m$ is zero (Hurewicz and Wallman [6, p. 103]).

Let $\mu$ be a Borel measure on $N$. We say that $\mu$ is locally euclidean if
the following condition holds. For each $x_0$ in $N$ we can find a cubic set $W$
in $N$ with respect to a coordinate system $(t_1, \cdots, t_n)$ such that (1) $x_0$ lies
in the interior of $W$ and (2) $\mu$ is completely continuous (on $W$) with respect
to the Euclidean measure $dt = dt_1 \cdots dt_n$ on $W$.

LEMMA 45. *Let $N$ be a manifold satisfying the countability axioms
and let $\mu$ be a locally euclidean measure on $N$. Then if $Q$ is any subset of
$N$ with $\partial \mathrm{im}\, Q < \partial \mathrm{im}\, N$, $Q$ is a null set with respect to $\mu$.*

We can cover $N$ by a countable family of cubic sets $V_i$ $i = 1, 2, \cdots$.
For any fixed $i$, choose a coordinate system $(t_1, \cdots, t_n)$ valid on some open
neighborhood of $V_i$ such that $V_i$ is a cubic set with respect to this system.
Obviously it would be enough to prove that $Q \cap V_i$ is a null set with respect
to the measure $dt = dt_1 \cdots dt_n$. So we have to prove the following lemma.

LEMMA 46. *Let $Q$ be a subset of $I^n$ such that $\partial\text{im}\, Q < n$. Then $Q$ is a null set with respect to the Euclidean measure on $I^n$.*

Since $\partial\text{im}\, Q < n$, the Hausdorff $n$-measure of $Q$ is zero. Our assertion therefore follows immediately from the definition of the Hausdorff measure.

THE INSTITUTE FOR ADVANCED STUDY,
  PRINCETON, N. J.

COLUMBIA UNIVERSITY,
  NEW YORK, N. Y.

---

## REFERENCES.


[1] V. Bargmann, *Ann. of Math.*, vol. 48 (1947), pp. 568-640.

[2] E. Cartan, (a) *Ann. École Norm. Sup.*, vol. 44 (1927), pp. 345-467.
  (b) *J. Math. Pures Appl.*, vol. 8 (1929), pp. 1-33.
  (c) *Abh. Math. Seminar Hamburg*, vol. 11 (1925), pp. 116-162.

[3] C. Chevalley, *Theory of Lie groups*, Princeton University Press, 1946.

[4] R. Godement, *C. R. Acad. Sci. Paris*, vol. 225 (1947), (a) pp. 521-523, (b) pp. 657-659.

[5] Harish-Chandra. (a) *Trans. Amer. Math. Soc.*, vol. 70 (1951), pp. 28-96.
  (b)                vol. 75 (1953), pp. 185-243.
  (c)                vol. 76 (1953), pp. 234-253.
  (d) *Proc. Nat. Acad. Sci.*, vol. 41 (1955), pp. 314-317.
  (e) *Amer. Jour. Math.*, vol. 77 (1955), pp. 743-777.
  (f) *Amer. Jour. Math.*, vol. 78 (1956), pp. 1-41.

[6] W. Hurewicz and H. Wallman, *Dimension theory*, Princeton University Press, 1948.

[7] J. L. Koszul, *Bull. Soc. Math.*, vol. 78 (1950), pp. 65-127.

[8] F. I. Mautner, *Ann. of Math.*, vol. 52 (1950), pp. 528-556.

[9] G. D. Mostow, *Bull. Amer. Math. Soc.*, vol. 55 (1949), pp. 969-980.

[10] I. E. Segal, (a) *Ann. of Math.*, vol. 52 (1950), pp. 272-292.
  (b) *Proc. Amer. Math. Soc.*, vol. 3 (1952), pp. 13-15.

[11] H. Weyl, (a) *Math. Zeit.*, vol. 24 (1925), pp. 328-395.
  (b) **The structure and representations of continuous groups, Princeton,** The Institute for Advanced Study, 1935.



Reprinted from
*Amer. J. of Math.*
**78** (1956), 564–628

# THE CHARACTERS OF SEMISIMPLE LIE GROUPS

BY

HARISH-CHANDRA

1. **Introduction.** Let $\pi$ be a quasi-simple irreducible representation [6(c)] of a connected semisimple Lie group $G$. We denote by $C_c^\infty(G)$ the space of all complex-valued functions on $G$ which are indefinitely differentiable and which vanish outside a compact set. Then if $dx$ is the Haar measure on $G$, the operator

$$\pi(f) = \int_G f(x)\pi(x)dx \qquad (f \in C_c^\infty(G))$$

is of the trace class and the mapping $T_\pi : f \to sp(\pi(f))$ is a distribution on $G$ which is called the character of $\pi$ (see [6(d)]). In this paper we shall obtain some results on these characters. Let $l$ be the rank of $G$. We say that an element $x \in G$ is regular if $l$ is exactly the multiplicity of the eigenvalue 1 of the matrix which corresponds to $x$ in the adjoint representation of $G$. The regular elements form an open and dense subset $G'$ of $G$. We shall prove (Theorem 6) that $T_\pi$ coincides on $G'$ (in the sense of distribution theory [11]) with an analytic function $F_\pi$ which is given by a formula rather similar to that of Weyl [14(a)] for the compact case (see also [4] and [6(e), p. 511]). However it still remains an open question whether $F_\pi$ determines $T_\pi$ completely[1].

The central idea of our method is quite simple and can be explained as follows. Let $\mathfrak{g}$ be the complexification of the Lie algebra of $G$ and $\mathfrak{B}$ the universal enveloping algebra of $\mathfrak{g}$. Every element $b \in \mathfrak{B}$ can be regarded as a (left-invariant) differential operator on $G$ and so by duality it operates also on the space of distributions on $G$. Hence the distribution $bT_\pi$ is well defined. Let $\mathfrak{Z}$ be the center of $\mathfrak{B}$. Then it follows from our assumptions on $\pi$ that $T_\pi$ is an "eigen-distribution" of every operator $z \in \mathfrak{Z}$. Or, in other words, there exists a homomorphism $\chi_\pi$ of $\mathfrak{Z}$ into the field of complex numbers such that $zT_\pi = \chi_\pi(z)T_\pi$ $(z \in \mathfrak{Z})$. On the other hand, in view of its definition, $T_\pi(f^y) = T_\pi(f)$ $(y \in G, f \in C_c^\infty(G))$ where $f^y$ stands for the function $x \to f(y^{-1}xy)$ $(x \in G)$. (We express this property by saying that $T_\pi$ is an invariant distribution.) Now let $x_0$ be a regular element in $G$ and $Z_{x_0}$ the centralizer of $x_0$. Then we can choose a closed normal abelian subgroup $A$ of $Z_{x_0}$ such that $x_0 \in A$ and $Z_{x_0}/A$ is finite. Moreover if $B$ is a sufficiently small open neighborhood of $x_0$

---

Received by the editors September 28, 1955.

[1] If $F_\pi \neq 0$ one can prove that there is only a finite number of possibilities for $T_\pi$ (see [6(h), Theorem 3]).

98

in $A$, $V = \bigcup_{x \in G} xBx^{-1}$ is open in $G$. $T_\tau$ being invariant, it defines in a natural way a distribution $\tau_\tau$ on $B$ (regarded as an open submanifold of the Lie group $A$). Now if we transcribe the differential equations $zT_\tau = \chi_\tau(z) T_\tau$ $(z \in \mathfrak{Z})$ into the corresponding equations for $\tau_\tau$, we find that among the latter there are always some which are of the *elliptic* type (see for example Gårding [3] for the definition of ellipticity). Hence in view of the recent generaliations (see Schwartz [11, p. 136] and John [8]) of the classical theorem of Bernstein, one can conclude that $\tau_\tau$ is an analytic function on $V$. The precise form of this function can now be obtained by looking more closely at the differential equations for $\tau_\tau$.

The contents of this paper are as follows. §2 deals with preliminary results. In §3 we introduce the notion of a quasi-regular element of $G$. (For us the significance of this concept, which is less strict than that of a regular element—for example in a compact semisimple Lie group, every element is quasi-regular—lies in the fact that the above-mentioned results actually hold on the larger set of quasi-regular elements.) Then, after a brief discussion in §4 of some general properties of differential operators on Lie groups, we begin an intensive study of certain special types of operators. The results of this investigation (Theorems 1 and 2) permit us to prove that certain eigen-distributions of these operators are actually analytic functions (Theorem 3). The exact form of these functions is obtained in Theorem 4. These results are then applied to the theory of characters in §11. One particular case is considered in greater detail in §12 and there we find a rather striking connection (Theorem 8 and Lemma 44) between the characters of the finite- and the infinite-dimensional representations.

A short summary of the results of this paper has been published in [6(h)].

2. **Preliminaries.** Let $R$ and $C$ denote the fields of real and complex numbers respectively and $(-1)^{1/2}$ a fixed square-root of $-1$ in $C$. Let $\mathfrak{g}_0$ be a semisimple Lie algebra over $R$ and $\mathfrak{g}$ its complexification. We put $B(X, Y) = sp(\mathrm{ad}\, X\, \mathrm{ad}\, Y)$ $(X, Y \in \mathfrak{g})$ where $X \to \mathrm{ad}\, X$ is the adjoint representation of $\mathfrak{g}$. Let $\bar{\mathfrak{u}}$ be a real form of $\mathfrak{g}$. By the conjugation of $\mathfrak{g}$ with respect to $\bar{\mathfrak{u}}$ we mean the autmorphism $\bar{\theta}$ of $\mathfrak{g}$ over $R$ given by $\bar{\theta}(X + (-1)^{1/2}Y) = X - (-1)^{1/2}Y$ $(X, Y \in \bar{\mathfrak{u}})$. We note that if $\mathfrak{h}_0$ is a Cartan subalgebra of $\bar{\mathfrak{u}}$, its complexification $\mathfrak{h}$ is a Cartan subalgebra of $\mathfrak{g}$. The real form $\bar{\mathfrak{u}}$ is called compact if the quadratic form $B(X, X)$ is negative definite on $\bar{\mathfrak{u}}$. Let $\eta$ denote the conjugation of $\mathfrak{g}$ with respect to $\mathfrak{g}_0$. The following lemma has been proved by Mostow [10].

LEMMA 1 (MOSTOW). *Let $\bar{\mathfrak{h}}_0$ be a Cartan subalgebra of $\mathfrak{g}_0$ and $\bar{\mathfrak{h}}$ the complexification of $\bar{\mathfrak{h}}_0$ in $\mathfrak{g}$. Then there exists a compact real form $\bar{\mathfrak{u}}$ of $\mathfrak{g}$ with the following two properties:*
  1. $\eta(\bar{\mathfrak{u}}) = \bar{\mathfrak{u}}$,
  2. $\bar{\mathfrak{h}} \cap \bar{\mathfrak{u}}$ *is a Cartan subalgebra of $\bar{\mathfrak{u}}$.*

Now let $\mathfrak{u}$ be a fixed compact real form of $\mathfrak{g}$ such that $\eta(\mathfrak{u}) = \mathfrak{u}$ and let

$\mathfrak{k}_0 = \mathfrak{g}_0 \cap \mathfrak{u}$, $\mathfrak{p}_0 = \mathfrak{g}_0 \cap (-1)^{1/2}\mathfrak{u}$. Then $\mathfrak{g}_0 = \mathfrak{k}_0 + \mathfrak{p}_0$ and $\mathfrak{g} = \mathfrak{k} + \mathfrak{p}$ where $\mathfrak{k}$, $\mathfrak{p}$ are the complexifications of $\mathfrak{k}_0$, $\mathfrak{p}_0$ respectively in $\mathfrak{g}$ and both the sums are direct. Let $\bar{\theta}$ denote the conjugation of $\mathfrak{g}$ with respect to $\mathfrak{u}$ and put $\theta = \eta \circ \bar{\theta}$. Then $\theta$ is an automorphism of $\mathfrak{g}$ over $C$ and $\theta(X + Y) = X - Y$ ($X \in \mathfrak{k}$, $Y \in \mathfrak{p}$).

Let $G$ be a connected Lie group with the Lie algebra $\mathfrak{g}_0$ and let $x \to \mathrm{Ad}\,(x)$ denote the adjoint representation of $G$ on $\mathfrak{g}$. The following consequence of Lemma 1 was pointed out to me by Chevalley.

CoROLLARY. *Let $\bar{\mathfrak{h}}_0$ be a Cartan subalgebra of $\mathfrak{g}_0$. Then we can choose an element $x \in G$ such that if $\mathfrak{h}_0 = \mathrm{Ad}\,(x)\bar{\mathfrak{h}}_0$,*

$$\mathfrak{h}_0 = \mathfrak{h}_0 \cap \mathfrak{k}_0 + \mathfrak{h}_0 \cap \mathfrak{p}_0.$$

Choose $\bar{\mathfrak{u}}$ in accordance with Lemma 1 and let $\bar{\theta}$ denote the conjugation of $\mathfrak{g}$ with respect to $\bar{\mathfrak{u}}$. Put $\bar{\mathfrak{k}}_0 = \bar{\mathfrak{u}} \cap \mathfrak{g}_0$, $\bar{\mathfrak{p}}_0 = ((-1)^{1/2}\bar{\mathfrak{u}}) \cap \mathfrak{g}_0$. Then $\mathfrak{g}_0$ is the direct sum of $\bar{\mathfrak{k}}_0$ and $\bar{\mathfrak{p}}_0$ and it follows from a theorem of Cartan $[1(\mathrm{b})]$ that $\bar{\mathfrak{k}}_0 = \mathrm{Ad}\,(x^{-1})\mathfrak{k}_0$ for some $x \in K$. Now suppose $X \in \bar{\mathfrak{k}}_0$, $Y \in \bar{\mathfrak{p}}_0$. Then $B(X, Y) = B(\bar{\theta}(X), \bar{\theta}(Y)) = B(X, -Y)$ and therefore $X$ and $Y$ are orthogonal under the bilinear form $B$. But $B(X, X) < 0$ if $X \in \bar{\mathfrak{k}}_0$ ($X \neq 0$) and so it follows that $\bar{\mathfrak{p}}_0$ is exactly the set of elements of $\mathfrak{g}_0$ orthogonal to $\bar{\mathfrak{k}}_0$ under $B$. A similar relationship holds between $\mathfrak{k}_0$ and $\mathfrak{p}_0$ and therefore also between $\bar{\mathfrak{k}}_0 = \mathrm{Ad}\,(x^{-1})\mathfrak{k}_0$ and $\mathrm{Ad}\,(x^{-1})\mathfrak{p}_0$. This proves that $\mathrm{Ad}\,(x^{-1})\mathfrak{p}_0 = \bar{\mathfrak{p}}_0$. Now let $\mathfrak{h}_0 = \mathrm{Ad}\,(x)\bar{\mathfrak{h}}_0$. If $\mathfrak{h}$ is the complexification of $\mathfrak{h}_0$ in $\mathfrak{g}$, it follows from the second condition of Lemma 1 that $\bar{\mathfrak{h}}$ is also the complexification of $\bar{\mathfrak{h}} \cap \bar{\mathfrak{u}}$ and therefore $\bar{\theta}(\bar{\mathfrak{h}}) = \bar{\mathfrak{h}}$. But obviously $\bar{\theta}(\mathfrak{g}_0) = \mathfrak{g}_0$ and therefore

$$\bar{\theta}(\bar{\mathfrak{h}}_0) = \bar{\theta}(\bar{\mathfrak{h}} \cap \mathfrak{g}_0) = \bar{\mathfrak{h}} \cap \mathfrak{g}_0 = \bar{\mathfrak{h}}_0.$$

This shows that $\bar{\mathfrak{h}}_0 = \bar{\mathfrak{h}}_0 \cap \bar{\mathfrak{k}}_0 + \bar{\mathfrak{h}}_0 \cap \bar{\mathfrak{p}}_0$ and hence

$$\mathfrak{h}_0 = \mathfrak{h}_0 \cap \mathfrak{k}_0 + \mathfrak{h}_0 \cap \mathfrak{p}_0.$$

Let $\mathfrak{a}_0$ be a maximal abelian subspace of $\mathfrak{p}_0$. For each (real-valued) linear function $\lambda$ on $\mathfrak{a}_0$ let $\mathfrak{g}_{0,\lambda}$ be the set of all $X \in \mathfrak{g}_0$ such that $[H, X] = \lambda(H)X$ for all $H \in \mathfrak{a}_0$. Let $\Sigma$ be the set of those $\lambda \neq 0$ for which $\mathfrak{g}_{0,\lambda} \neq 0$. Then $\Sigma$ is a finite set. Let $K$ denote the analytic subgroup of $G$ corresponding to $\mathfrak{k}_0$.

LEMMA(²) 2. *Let $\mathfrak{h}_1$, $\mathfrak{h}_2$ be two Cartan subalgebras of $\mathfrak{g}_0$ both invariant under $\theta$ and such that $\mathfrak{h}_i \cap \mathfrak{p}_0 \subset \mathfrak{a}_0$ ($i = 1, 2$). Let $\Sigma_i$ be the set of those $\alpha \in \Sigma$ which vanish identically on $\mathfrak{h}_i \cap \mathfrak{p}_0$ ($i = 1, 2$). Then if $\Sigma_1 = \Sigma_2$ there exists an element $k \in K$ such that $\mathfrak{h}_2 = \mathrm{Ad}\,(k)\mathfrak{h}_1$.*

Choose an element $H_0 \in \mathfrak{h}_1 \cap \mathfrak{p}_0$ such that $\alpha(H_0) \neq 0$ ($\alpha \in \Sigma$) unless $\alpha \in \Sigma_1$. Also let $\mathfrak{a}_1$ denote the set of all those $H \in \mathfrak{a}_0$ for which $\alpha(H) = 0$ for every $\alpha \in \Sigma_1$. Obviously $\mathfrak{a}_1 \supset \mathfrak{h}_1 \cap \mathfrak{p}_0$. We shall now show that $\mathfrak{a}_1 = \mathfrak{h}_1 \cap \mathfrak{p}_0$. Let $\mathfrak{q}_0$ be

---

(²) This result had also been obtained independently by A. Borel.

the centralizer of $H_0$ in $\mathfrak{k}_0$. Then it is obvious that

$$\mathfrak{q}_0 \subset \mathfrak{g}_{0,0} + \sum_{\alpha \in \Sigma_1} \mathfrak{g}_{0,\alpha}$$

and therefore $[\mathfrak{q}_0, \mathfrak{a}_1] = \{0\}$. But it is clear that $\mathfrak{h}_1 \cap \mathfrak{k}_0 \subset \mathfrak{q}_0$ and so $\mathfrak{a}_1 + \mathfrak{h}_1 \cap \mathfrak{k}_0$ is an abelian algebra. However

$$\mathfrak{a}_1 + \mathfrak{h}_1 \cap \mathfrak{k}_0 \supset \mathfrak{h}_1 \cap \mathfrak{p}_0 + \mathfrak{h}_1 \cap \mathfrak{k}_0 = \mathfrak{h}_1.$$

Therefore since $\mathfrak{h}_1$ is maximal abelian in $\mathfrak{g}_0$, $\mathfrak{a}_1 = \mathfrak{h}_1 \cap \mathfrak{p}_0$.

This proves that $\mathfrak{h}_1 \cap \mathfrak{p}_0$ is completely determined by $\Sigma_1$. Hence if $\Sigma_1 = \Sigma_2$, $\mathfrak{h}_1 \cap \mathfrak{p}_0 = \mathfrak{h}_2 \cap \mathfrak{p}_0$. Then $\mathfrak{q}_0$, as defined above, is the centralizer in $\mathfrak{k}_0$ of $\mathfrak{h}_1 \cap \mathfrak{p}_0$ $= \mathfrak{h}_2 \cap \mathfrak{p}_0$. Replacing $G$, if necessary, by its adjoint group we may assume that $K$ is compact (see Mostow [10]). Then if $Q$ is the centralizer[3] of $\mathfrak{h}_1 \cap \mathfrak{p}_0$ in $K$, $Q$ being a closed subgroup of $K$, is compact. Also its Lie algebra is $\mathfrak{q}_0$. Since it is obvious that both $\mathfrak{h}_1 \cap \mathfrak{k}_0$ and $\mathfrak{h}_2 \cap \mathfrak{k}_0$ are maximal abelian subalgebras of $\mathfrak{q}_0$ it follows, in view of the compactness of $Q$, that $\mathfrak{h}_2 \cap \mathfrak{k}_0 = \mathrm{Ad}\,(k)(\mathfrak{h}_1 \cap \mathfrak{k}_0)$ for some $k \in Q$. But then

$$\mathrm{Ad}\,(k)\mathfrak{h}_1 = \mathrm{Ad}\,(k)(\mathfrak{h}_1 \cap \mathfrak{k}_0) + \mathrm{Ad}\,(k)(\mathfrak{h}_1 \cap \mathfrak{p}_0)$$

$$= \mathfrak{h}_2 \cap \mathfrak{k}_0 + \mathfrak{h}_1 \cap \mathfrak{p}_0 = \mathfrak{h}_2 \cap \mathfrak{k}_0 + \mathfrak{h}_2 \cap \mathfrak{p}_0 = \mathfrak{h}_2.$$

We say that two Cartan subalgebras $\mathfrak{h}_1$, $\mathfrak{h}_2$ are conjugate (or lie in the same conjugacy class) if $\mathfrak{h}_2 = \mathrm{Ad}\,(x)\mathfrak{h}_1$ for some $x \in G$.

COROLLARY. *We can select a finite number of Cartan subalgebras* $\mathfrak{h}_1, \cdots, \mathfrak{h}_r$ *of* $\mathfrak{g}_0$ *with the following two properties.* $\theta(\mathfrak{h}_i) = \mathfrak{h}_i$ $(i = 1, \cdots, r)$ *and if* $\mathfrak{h}_0$ *is any Cartan subalgebra of* $\mathfrak{g}_0$, $\mathfrak{h}_0$ *is conjugate to* $\mathfrak{h}_i$ *for exactly one* $i$.

The set $\Sigma$ being finite, there is only a finite number of possibilities for the subsets $\Sigma_1$, $\Sigma_2$ of Lemma 2. Hence if we combine the above results with the corollary to Lemma 1, our assertion follows from the fact (see Cartan [1(a)]) that any two maximal abelian subspaces of $\mathfrak{p}_0$ are conjugate under $K$.

Let $\lambda$ be an indeterminate. For any $x \in G$ consider the polynomial

$$D(x, \lambda) = \det \{(\lambda + 1)I - \mathrm{Ad}\,(x)\}$$

in $\lambda$. (Here $I$ is the identity mapping of $\mathfrak{g}$.) Let $\lambda^l$ be the highest power of $\lambda$ which divides $D(x, \lambda)$ for every $x$. Then $l$ is called the rank of $G$ (or of $\mathfrak{g}_0$) and, as is well known, every Cartan subalgebra of $\mathfrak{g}_0$ has (real) dimension $l$. Now clearly if $n = \dim \mathfrak{g}$,

$$D(x, \lambda) = \lambda^n + \sum_{r=l}^{n-1} (-1)^{n-r} D_r(x)\lambda^r$$

---

[3] Let $x$ and $X$ be two elements in $G$ and $\mathfrak{g}$ respectively. It is convenient to say that $x$ and $X$ commute if $\mathrm{Ad}(x)X = X$.

where $D_r$ $(l \leq r < n)$ are analytic functions on $G$ and $D_l$ is not identically zero. Let $S$ denote the set of zeros of $D_l$ on $G$. We call $S$ the singular set of $G$. Obviously it is a closed nowhere dense set. A point $x \in G$ is called singular or regular according as it lies in $S$ or not. If $z$ lies in the center $Z$ of $G$, $\mathrm{Ad}\,(zx) = \mathrm{Ad}(x)$ $(x \in G)$ and therefore $ZS = S$. Similarly $D(yxy^{-1},\,\lambda) = D(x,\,\lambda)$ $(x,\,y \in G)$ and therefore $ySy^{-1} = S$.

An element $X \in \mathfrak{g}$ is called regular if the multiplicity of the eigenvalue zero of $\mathrm{ad}\,X$ is exactly $l$. It is well known (see Weyl $[14(b)]$) that if $X$ is regular, its centralizer in $\mathfrak{g}$ is a Cartan subalgebra of $\mathfrak{g}$ and conversely every Cartan subalgebra contains a regular element. Let $A$ be a maximal abelian subgroup of $G$. We say that $A$ is a Cartan subgroup[4] of $\mathfrak{g}$ if its connected component of 1 is not contained in $S$. Obviously a Cartan subgroup is closed and hence it is a Lie subgroup of $G$.

Lemma 3.†$\,$*The Lie algebra of any Cartan subgroup of $G$ is a Cartan subalgebra of $\mathfrak{g}_0$. Conversely if $\mathfrak{h}_0$ is any Cartan subalgebra of $\mathfrak{g}_0$, there is exactly one Cartan subgroup $A$ of $G$ whose Lie algebra is $\mathfrak{h}_0$. If $H$ is a regular element in $\mathfrak{h}_0$, $A$ coincides with the centralizer[3] of $H$ in $G$.*

Corollary. *Let $Z_x$ denote the centralizer of a regular element $x$ in $G$. Then $Z_x$ contains exactly one Cartan subgroup $A$ of $G$. Moreover $A$ is normal in $Z_x$ and $Z_x/A$ is finite.*

Before proving the above lemma and its corollary, we need a few auxiliary results. Let $G_c$ be a complex analytic group with the Lie algebra $\mathfrak{g}$ and let $S_c$ denote the singular set of $G_c$. Then if $\mathfrak{h}$ is a Cartan subalgebra of $\mathfrak{g}$, we consider the complex analytic subgroup $A_c$ corresponding to $\mathfrak{h}$. Let $G_c'$ denote the complement of $S_c$ in $G_c$.

Lemma 4. $G_c'$ *is connected and* $G_c' \subset \bigcup_{z \in G_c} z A_c z^{-1}$. *Moreover if* $H_0$ *is a regular element in* $\mathfrak{h}$, *the centralizer of* $H_0$ *in* $G_c$ *coincides with* $A_c$.

Let $z \to \mathrm{Ad}\,(z)$ $(z \in G_c)$ denote the adjoint representation of $G_c$ on $\mathfrak{g}$. Then it is a complex representation (see $[6(g),\,\S 6]$). $\lambda$ being an indeterminate, consider the polynomial

$$D(z,\,\lambda) = \det\left\{(\lambda + 1)I - \mathrm{Ad}\,(z)\right\} \qquad\qquad (z \in G_c)$$

where $I$ is the identity mapping of $\mathfrak{g}$. The coefficients of this polynomial are holomorphic functions on $G_c$. $l$ being the rank of $\mathfrak{g}$, let $(-1)^l D_l(z)$ denote the coefficient of $\lambda^l$. Then $S_c$ is the set of zeros of the holomorphic function $D_l$. Since $D_l$ is not identically zero, $S_c$ is a complex-analytic subvariety of $G_c$ of one complex dimension (and therefore two real dimensions) less. Hence $G_c'$ is connected.

Put $V' = \bigcup_{z \in G_c} z A_c' z^{-1}$ where $A_c' = A_c \cap G_c'$. It is obvious that $V \subset G_c'$.

<hr>

[4] The definition of a Cartan subgroup given in $[6(h)]$ is not quite correct.

† See Lemma 8 of $[1964c]$ for the correct definition of a Cartan subgroup and the corrected version of the Lemma and Corollary — V.

Moreover if $a$ is an element in $A_c$ sufficiently near 1 but [†]different from 1, it is easily seen that $a$ is regular and therefore $V'$ is not empty. Therefore since $G'_c$ is connected, in order to prove the second statement of the lemma, it would be sufficient to show that $V'$ is both open and closed in $G'_c$. Let $G^*_c$ be the factor space $G_c/A_c$ consisting of all cosets of the form $zA_c$ $(z \in G_c)$. Put $a^{z*} = zaz^{-1}$ $(z \in G_c, a \in A_c)$ where $z \to z^*$ is the natural mapping of $G_c$ on $G^*_c$. Then an easy computation shows that the holomorphic mapping $\phi : (z^*, a) \to a^{z*} (z^* \in G^*_c, a \in A_c)$ of $G^*_c \times A_c$ into $G_c$ is everywhere regular on $G_c \times A'_c$ and therefore we conclude that $V' = \phi(G^*_c \times A'_c)$ is open in $G_c$. So now it remains to show that $V'$ is closed in $G'_c$. Introduce some lexicographic order among the roots of $\mathfrak{g}$ (with respect to $\mathfrak{h}$) and for each root $\alpha$ select an element $X_\alpha \neq 0$ such that $[H, X_\alpha] = \alpha(H)X_\alpha$ for every $H \in \mathfrak{h}$. Then $\mathfrak{n} = \sum_{\alpha > 0} CX_\alpha$ is a nilpotent subalgebra of $\mathfrak{g}$. Let $\mathfrak{h}_0$ be the real subspace of $\mathfrak{h}$ consisting of those elements where every root $\alpha$ takes a pure imaginary value. It is possible to choose (see Weyl [14(b)]) a compact real form $\bar{\mathfrak{u}}$ of $\mathfrak{g}$ which contains $\mathfrak{h}_0$. Let $U$ and $N$ be the (real) analytic subgroups of $G_c$ corresponding to $\bar{\mathfrak{u}}$ and $\mathfrak{n}$ respectively. Then $U$ is compact, $G_c = UA_cN$ and $A_cN$ is closed (see Iwasawa [7]). Hence

$$V = \bigcup_{u \in U} uA_cNu^{-1}$$

is also a closed set. Moreover it is obvious that

$$V \cap G'_c = \bigcup_{u \in U} u(A_cN)'u^{-1}$$

where $(A_cN)' = A_cN \cap G'_c$. Also we know (Lemmas 8 and 10 of [6(e)]) that

$$(A_cN)' = \bigcup_{n \in N} nA'_cn^{-1}$$

and therefore $V \cap G'_c = V'$. But since $G_c = UA_cN = UNA_c$, it follows that $V' \subset V$ and so $V' = V \cap G'_c$. This proves that $V'$ is closed in $G'_c$.

Now let $H_0$ be a regular element in $\mathfrak{h}$ and suppose Ad $(x)H_0 = H_0$ $(x \in G)$. Then $x = una$ $(u \in U, n \in N, a \in A_c)$ and so Ad $(n)H_0 = $ Ad $(u^{-1})H_0$. Since $\mathfrak{h}$ is the centralizer of $H_0$ in $\mathfrak{g}$, we conclude that Ad $(n)\mathfrak{h} = $ Ad $(u^{-1})\mathfrak{h}$. But Ad $(n)\mathfrak{h} \subset \mathfrak{h} + \mathfrak{n}$ and Ad $(u^{-1})\mathfrak{h}_0 \subset \bar{\mathfrak{u}}$. Moreover if $(\mathfrak{h} + \mathfrak{n}) \cap \bar{\mathfrak{u}} = \mathfrak{h}_1$, it is obvious that $[\mathfrak{h}_0, \mathfrak{h}_1] \subset \mathfrak{n} \cap \bar{\mathfrak{u}} = \{0\}$. Therefore since $\mathfrak{h}_0$ is maximal abelian in $\bar{\mathfrak{u}}$ it follows that $\mathfrak{h}_1 = \mathfrak{h}_0$. This shows that Ad $(u^{-1})\mathfrak{h}_0 = \mathfrak{h}_0$ and so Ad $(n)\mathfrak{h} = $ Ad $(u^{-1})\mathfrak{h} = \mathfrak{h}$. But from Lemma 8 of [6(e)], this implies that $n = 1$ and hence Ad $(u^{-1})H_0 = H_0$. Let $H_1, H_2$ be elements in $\mathfrak{h}_0$ such that $H_0 = H_1 + (-1)^{1/2}H_2$. Then it is obvious that Ad $(u^{-1})H_i = H_i$, $i = 1, 2$. Moreover since $H_0$ is regular, no root of $\mathfrak{g}$ (with respect to $\mathfrak{h}$) can vanish at both $H_1$ and $H_2$. Therefore we can choose $t \in R$ such that $H = H_1 + tH_2$ is a regular element in $\mathfrak{h}_0$. Let $A_0$ be the analytic subgroup of $U$ corresponding to $\mathfrak{h}_0$. Then since $U$ is compact and Ad $(u)H = H$, it follows (see Stiefel [12, Satz 7]) that $u \in A_0$. This shows that $x \in A_c$ and so the lemma is now proved completely.

---

$†$ $a$ should avoid the "root hyperplanes" — V.

COROLLARY. *Let $A_x$ denote the centralizer (in $G_c$) of a regular element $x$ in $G_c$. Then $A_x$ has only a finite number of connected components and its Lie algebra is a Cartan subalgebra of $\mathfrak{g}$.*

From Lemma 4 we can choose $z \in G_c$ such that $a = z^{-1}xz \in A_c$. It is obvious that $a$ is regular and therefore $\mathfrak{h}$ is the centralizer[3] of $a$ in $\mathfrak{g}$. Let $A^*$ be the normalizer of $A_c$ in $G_c$. It is clear that $zA_cz^{-1}$ is the connected component of 1 in $A_x$ and $A_x \subset zA^*z^{-1}$. Hence in order to prove the corollary it is sufficient to show that $A^*/A_c$ is finite. Let $y \in A^*$. Then $y = unb$ ($u \in U$, $n \in N$, $b \in A_c$) and Ad $(n)\mathfrak{h} =$ Ad $(u^{-1})\mathfrak{h}$. By the argument which we have already used in the proof of Lemma 4, it follows that $n = 1$ and so $u \in A^* \cap U$. This proves that $A^*/A_c \simeq (A^* \cap U)/(A_c \cap U)$. But since $\mathfrak{h}$ is its own normalizer in $\mathfrak{g}$, we conclude that $A^*/A_c$ is both discrete and compact. Therefore it is finite.

Now we come to Lemma 3 and its corollary. If $Z$ is the center of $G$, it is obviously permissible† for our proof to replace $G$ by $G/Z$. Hence we may assume that there exists a complex analytic group $G_c$ with the Lie algebra $\mathfrak{g}$ such that $G$ is the (real) analytic subgroup of $G_c$ corresponding to $\mathfrak{g}_0$. Then if $S_c$ and $G'_c$ are defined as above, it is obvious that $G \cap G'_c = G'$ and $G \cap S_c = S$. Let $A$ be a Cartan subgroup of $G$. We denote by $A_0$ the connected component of 1 and by $\mathfrak{h}_0$ the Lie algebra of $A$. Choose a regular element $a \in A_0$. $A$ being abelian, $a = \exp H$ for some $H \in \mathfrak{h}_0$ and it is obvious that $H$ is regular. Let $A_c$ and $\mathfrak{h}$ be the centralizers of $H$ in $G_c$ and $\mathfrak{g}$ respectively. From Lemma 4, $A_c$ is abelian and hence it is clear that $A = A_c \cap G$. Therefore $\mathfrak{h}_0 = \mathfrak{h} \cap \mathfrak{g}_0$ and this shows that $\mathfrak{h}_0$ is a Cartan subalgebra of $\mathfrak{g}_0$ and $A$ is the centralizer in $G$ of any regular element in $\mathfrak{h}_0$. Conversely let $\mathfrak{h}_0$ be a given Cartan subalgebra of $\mathfrak{g}_0$ and $H$ a regular element in $\mathfrak{h}_0$. Then if $A_c$ is the centralizer of $H$ in $G_c$, we know from Lemma 4 that $A_c$ is abelian. Hence it is obvious that $A = A_c \cap G$ is a maximal abelian subgroup of $G$. Moreover if $t$ is sufficiently small and positive $\exp tH$ is evidently regular. Therefore $A$ is a Cartan subgroup of $G$. All the statements of Lemma 3 are now obvious. However we still have to prove the corollary. $x$ being a regular element in $G$, let $\mathfrak{h}_x$ denote the centralizer[3] of $x$ in $\mathfrak{g}$ and $A_c$ the analytic subgroup of $G_c$ corresponding to $\mathfrak{h}_x$. Then if $A_x$ is the centralizer of $x$ in $G_c$, we have seen (corollary to Lemma 4) that $A_c$ is normal in $A_x$ and $A_x/A_c$ is finite. Therefore $(A_x \cap G)/(A_c \cap G)$ is also finite. But it is obvious that $Z_x = A_x \cap G$ and, as we saw above, $A = A_c \cap G$ is the Cartan subgroup of $G$ corresponding to $\mathfrak{h}_x \cap \mathfrak{g}_0$. Moreover since $\mathfrak{h}_x \cap \mathfrak{g}_0$ is the Lie algebra of $Z_x$, it is clear that $A$ is the only Cartan subgroup of $G$ contained in $Z_x$.

LEMMA 5. *Let $\mathfrak{h}_1, \cdots, \mathfrak{h}_r$ be defined as in the corollary to Lemma 2 and let $A_i$ be the Cartan subgroup of $G$ which corresponds to $\mathfrak{h}_i$ ($i = 1, \cdots, r$). Then*

$$G' \subset \bigcup_{i=1}^{r} \bigcup_{x \in G} xA_ix^{-1}.$$

--------

† See Lemma 8 of [1964c] — V.

Since $A_i \supset Z$ for every $i$, it is again permissible to replace $G$ by $G/Z$. Therefore, as in the proof of Lemma 3, we may assume that $G \subset G_c$. Let $y$ be a regular element in $G$ and let $\mathfrak{h}_y$ denote the centralizer of $y$ in $\mathfrak{g}$. Then as we have seen above $\mathfrak{h}_y \cap \mathfrak{g}_0$ is a Cartan subalgebra of $\mathfrak{g}_0$ and therefore $\mathfrak{h}_y \cap \mathfrak{g}_0 = \mathrm{Ad}\ (x)\mathfrak{h}_i$ for some $x \in G$ and some $i$ ($1 \leq i \leq r$). Now if $A_y$ is the Cartan subgroup of $G$ corresponding to $\mathfrak{h}_y \cap \mathfrak{g}_0$, it follows from Lemma 3 that $A_y = xA_i x^{-1}$. But since $A_y$ is the centralizer of $\mathfrak{h}_y$ in $G$ (see Lemma 3), $y \in xA_i x^{-1}$.

Two Cartan subgroups $A$ and $B$ of $G$ are said to be conjugate (or to lie in the same conjugacy class) if $B = xAx^{-1}$ for some $x \in G$.

LEMMA 6. *Every Cartan subgroup $A$ of $G$ is conjugate to $A_i$ for exactly one $i$* ($1 \leq i \leq r$).

This follows immediately from Lemma 3 and the corollary to Lemma 2.

Let $\mathfrak{h}_0$ be a Cartan subalgebra of $\mathfrak{g}_0$ such that $\theta(\mathfrak{h}_0) = \mathfrak{h}_0$. Let $A$ be the centralizer of $\mathfrak{h}_0$ in $G$ and $\overline{M}$ the connected component of 1 in the centralizer of $\mathfrak{h}_0 \cap \mathfrak{p}_0$ in $K$. We put $M = \overline{M}A$. Let $\mathfrak{m}_{\mathfrak{k}_0}$ be the centralizer of $\mathfrak{h}_0 \cap \mathfrak{p}_0$ in $\mathfrak{k}_0$. Then $\mathfrak{m}_{\mathfrak{k}_0}$ is the Lie algebra of $\overline{M}$ and it is obvious that $\mathrm{Ad}\ (a)\mathfrak{m}_{\mathfrak{k}_0} = \mathfrak{m}_{\mathfrak{k}_0}$ for $a \in A$. Hence $M$ is a group. Let $\overline{B}$ denote the centralizer of $\mathfrak{m}_{\mathfrak{k}_0}$ in $A \cap K$.

LEMMA 7. *Let $A_+$ denote the analytic subgroup of $G$ corresponding to $\mathfrak{h}_0 \cap \mathfrak{p}_0$. Then $A = (A \cap K)A_+$, $M \cap K = \overline{M}\,\overline{B}$ and $\overline{B}$ is the center of $M \cap K$. Moreover*

$$M \cap K = \bigcup_{m \in \overline{M}} m(A \cap K)m^{-1}$$

*and $A \cap K = \overline{A}\,\overline{B}$ where $\overline{A}$ is the connected component of 1 in $A \cap K$. Finally $M = \overline{M}\,\overline{B}A_+$ and $\overline{B}A_+$ is the center of $M$.*

Let $\exp \mathfrak{p}_0$ denote the set of all elements in $G$ of the form $\exp X$ ($X \in \mathfrak{p}_0$). Then we can "extend" $\theta$ to an automorphism of $G$ such that $\theta(kp) = kp^{-1}$ where $k \in K$ and $p \in \exp \mathfrak{p}_0$. Since $\theta(\mathfrak{h}_0) = \mathfrak{h}_0$, it follows that $\theta(A) = A$ and therefore if $a \in A$ and $a = kp$ ($k \in K$, $p \in \exp \mathfrak{p}_0$), $p^2 = \theta(a^{-1})a$ is also in $A$. But with regard to the positive definite quadratic form $-B(\theta(Y), Y)$ ($Y \in \mathfrak{g}_0$) on $\mathfrak{g}_0$, $\mathrm{ad}\ X$ ($X \in \mathfrak{p}_0$) is a symmetric operator and hence $\exp (\mathrm{ad}\ X)Y \neq Y$ ($Y \in \mathfrak{g}_0$) unless $[X, Y] = 0$. Therefore if $p = \exp X$ ($X \in \mathfrak{p}_0$) we can conclude that $(\mathrm{ad}\ X)\mathfrak{h}_0 = \{0\}$ and $X \in \mathfrak{h}_0 \cap \mathfrak{p}_0$. This proves that $p \in A_+$ and so $k \in A \cap K$. Now since $Z \subset \overline{B} \subset A \cap K$, it is clearly permissible to replace $G$ by $G/Z$ in proving the rest of the above lemma. So we may assume that $K$ and therefore also $A \cap K$ and $\overline{M}$ are compact. It is obvious that $\mathfrak{h}_0 \cap \mathfrak{k}_0$ is a maximal abelian subalgebra of $\mathfrak{m}_{\mathfrak{k}_0}$. Hence $\overline{M}$ being compact, it follows (see Weyl [14(b)]) that

$$\overline{M} = \bigcup_{m \in \overline{M}} m\overline{A}m^{-1}$$

and so $\overline{A}$ contains the center of $\overline{M}$. Now complexify $\mathfrak{h}_0$, $\mathfrak{m}_{\mathfrak{k}_0}$ to $\mathfrak{h}$, $\mathfrak{m}_{\mathfrak{k}}$ respectively and for every root $\alpha$ (of $\mathfrak{g}$ with respect to $\mathfrak{h}$) choose an element $X_\alpha \neq 0$ belonging to the root $\alpha$. Let $Q$ be the set of all roots $\alpha$ such that $X_\alpha \in \mathfrak{m}_{\mathfrak{k}}$. We can choose a fundamental system $(\alpha_1, \cdots, \alpha_r)$ of roots in $Q$ so that $\alpha_1, \cdots, \alpha_r$ are linearly independent and every root in $Q$ can be written in the form $d_1\alpha_1 + \cdots + d_r\alpha_r$ where $d_1, \cdots, d_r$ are integers which are either all nonnegative or all nonpositive. If $a \in A \cap K$, it is obvious that $\operatorname{Ad}(a)X_\alpha = c_\alpha X_\alpha$ $(\alpha \in Q)$ where $c_\alpha$ are unimodular complex numbers. Since $[X_\alpha, X_{-\alpha}] \in \mathfrak{h}$, $c_{-\alpha} = c_\alpha^{-1}$ $(\alpha \in Q)$. We can choose $H \in \mathfrak{h}_0 \cap \mathfrak{k}_0$ such that $c_{\alpha_i} = e^{\alpha_i(H)}$ $i = 1, \cdots, r$. Then $a_0 = \exp H \in \overline{A}$ and $\operatorname{Ad}(a_0^{-1}a)$ leaves $X_{\alpha_i}, X_{-\alpha_i}$ $(1 \le i \le r)$ fixed. But these elements together with $\mathfrak{h} \cap \mathfrak{k}$ generate the Lie algebra $\mathfrak{m}_{\mathfrak{k}}$, and so it follows that $a_0^{-1}a \in \overline{B}$. Hence $A \cap K = \overline{B}\overline{A}$. Moreover $\overline{A} \subset \overline{M}$ and therefore $M = \overline{M}A = \overline{M}\overline{B}A_+$ and $M \cap K = \overline{M}\overline{B}$. Since $\overline{A}$ contains the center of $\overline{M}$, it is now clear that $\overline{B}$ is the center of $\overline{M}\overline{B}$. Therefore

$$M \cap K = \overline{M}\overline{B} = \bigcup_{m \in \overline{M}} m(\overline{A}\overline{B})m^{-1} = \bigcup_{m \in \overline{M}} (A \cap K)m^{-1}.$$

On the other hand it is obvious that $\overline{B}A_+$ lies in the center of $M$. Conversely if $z$ is in the center of $M$, $z \in A$ since $A$ is maximal abelian. Hence $z = aa_+$ where $a \in A \cap K$ and $a_+ \in A_+$. Then $a$ is in the center of $M \cap K$ and therefore $a \in \overline{B}$. This completes the proof of the lemma.

Let $H_1, \cdots, H_p$ be a base for $\mathfrak{h}_0 \cap \mathfrak{p}_0$ over $R$. We extend it to a base $(H_1, \cdots, H_p, H_{p+1}, \cdots, H_l)$ for $\mathfrak{h}_0 \cap \mathfrak{p}_0 + (-1)^{1/2}(\mathfrak{h}_0 \cap \mathfrak{k}_0)$ over $R$ and introduce the lexicographic order among the roots corresponding to this base (see $[6(\mathrm{f}), \S 2]$). Let $P$ be the set of all positive roots under this order. Then $P$ is the union of three disjoint subsets $P_+, P_0, P_-$ defined as follows. $P_+$ is the set of those roots $\alpha \in P$ which do not vanish identically on $\mathfrak{h} \cap \mathfrak{p}$. If $\alpha$ is a positive root which is not in $P_+$, it is clear that $X_\alpha$ and $\theta(X_\alpha)$ both belong to $\alpha$ and therefore since $\theta^2$ is the identity, $\theta(X_\alpha) = \pm X_\alpha$. Then $\alpha$ lies in $P_0$ or $P_-$ according as $X_\alpha \in \mathfrak{p}$ or $X_\alpha \in \mathfrak{k}$. Put $\mathfrak{m} = \mathfrak{m}_{\mathfrak{k}} + \mathfrak{h} \cap \mathfrak{p}$ and

$$\mathfrak{q} = \sum_{\alpha \in P_+} (CX_\alpha + CX_{-\alpha}) + \sum_{\alpha \in P_0} (CX_\alpha + CX_{-\alpha}).$$

Then $\mathfrak{q}$ is exactly the set of those elements of $\mathfrak{g}$ which are orthogonal to $\mathfrak{m}$ with respect to the bilinear form $B(X, Y)$. Since $\mathfrak{m} \cap \mathfrak{g}_0$ is the Lie algebra of $M$, it follows that $\mathfrak{q}$ is invariant under $\operatorname{Ad}(m)$ $(m \in M)$. Let $I$ be the identity mapping of $\mathfrak{g}$ and $(\operatorname{Ad}(m) - I)_\mathfrak{q}$ the restriction of $\operatorname{Ad}(m) - I$ $(m \in M)$ on $\mathfrak{q}$. We denote by $'M$ the set of those elements $m \in M$ for which $\det(\operatorname{Ad}(m) - I)_\mathfrak{q} \ne 0$. Put $M' = M \cap G'$ and $A' = A \cap G'$. It is obvious that $'M \supset M'$.

LEMMA 8. $A'$ *is dense in* $A$.

For each root $\alpha$ we can define a character $\xi_\alpha$ of $A$ by $\operatorname{Ad}(a)X_\alpha = \xi_\alpha(a)X_\alpha$ $(a \in A)$. Select an element $H \in \mathfrak{h}_0$ such that $\alpha(H) \ne 0$ for every root $\alpha$. Then if $a \in A$, it is obvious that $\xi_\alpha(a \exp tH) = \xi_\alpha(a)e^{t\alpha(H)} \ne 1$ for every root $\alpha$, provided $t$ is a sufficiently small positive number. This proves our assertion.

COROLLARY. *Both* $'M$ *and* $M'$ *are open and dense in* $M$.

Since det $(\mathrm{Ad}\ (m) - I)_q$ is an analytic function of $m$ on $M$, the openness of $`M$ is obvious. Moreover $G'$ being open in $G$, $M'$ is open[5] in $M$. On the other hand

$$M = \bigcup_{m \in \bar{M}} mAm^{-1} \quad \text{and} \quad M' = \bigcup_{m \in \bar{M}} mA'm^{-1}.$$

Therefore since $A'$ is dense in $A$ and $`M \supset M'$ our assertion follows.

LEMMA 9. *There exists a finite number of connected components $A_1', \cdots, A_s'$ of $A'$ such that $A' = \bigcup_{i=1}^{s} ZA_i'$.*

For the purpose of this lemma, we can obviously replace $G$ by $G/Z$ and thus assume that $K$ is compact. Then $\bar{A}$ being defined as in Lemma 7, $(A \cap K)/\bar{A}$ is both discrete and compact and therefore it is finite. Put $A_0 = \bar{A}A_+$. Then it follows from Lemma 7 that every connected component of $A$ is of the form $a_0 A_0$ $(a_0 \in A \cap K)$ and so there is only a finite number of such components. Define the characters $\xi_\alpha$ of $A$ as above. Since $[X_\alpha, X_{-\alpha}] \in \mathfrak{h}$, it is obvious that $\xi_{-\alpha} = \xi_\alpha^{-1}$. Select complex numbers $c_\alpha$ such that $\xi_\alpha(a_0) = e^{c_\alpha}$ $(\alpha \in P)$. Since $A \cap K$ is compact, $c_\alpha$ are all purely imaginary. Let $r$ be any integer and $\alpha$ a root in $P$. Then by $\sigma_{\alpha,r}$ we denote the hyperplane in $\mathfrak{h}_0$ consisting of all elements $H$ such that $\alpha(H) = -c_\alpha + (-1)^{1/2}2\pi r$. It is clear that any compact subset of $\mathfrak{h}_0$ meets only a finite number of these hyperplanes. Let $\mathfrak{h}_0'$ be the complement in $\mathfrak{h}_0$ of the union of all these hyperplanes. Then every point in the real Euclidean space $\mathfrak{h}_0$ has a neighborhood $V$ such that $V \cap \mathfrak{h}_0'$ decomposes only into a finite number of connected components. From this we conclude immediately that if $\omega$ is a compact subset of $\mathfrak{h}_0$, there exists a neighborhood $V$ of $\omega$ for which $V \cap \mathfrak{h}_0'$ has only a finite number of connected components. Now $\bar{A}$ being compact, we can choose a compact set $\omega_{\mathfrak{t}_0}$ in $\mathfrak{h}_0 \cap \mathfrak{t}_0$ such that $\bar{A} \subset \exp \omega_{\mathfrak{t}_0}$. Put $\omega' = \mathfrak{h}_0' \cap (\mathfrak{h}_0 \cap \mathfrak{p}_0 + \omega_{\mathfrak{t}_0})$ and suppose $H' \in \omega'$. Then $H' = H + H_0$ where $H \in \mathfrak{h}_0 \cap \mathfrak{p}_0$, $H_0 \in \omega_{\mathfrak{t}_0}$. Consider $H_t' = tH + H_0$ $(t > 0)$. We claim $H_t' \in \omega'$. For otherwise $\alpha(H_t') = -c_\alpha + (-1)^{1/2}2\pi r$ for some $\alpha \in P$ and some integer $r$. But

$$\alpha(H_t') = t\alpha(H) + \alpha(H_0)$$

and $\alpha(H)$ is real while $\alpha(H_0)$ is purely imaginary. Therefore since $c_\alpha$ is also purely imaginary, the above relation is possible only if $\alpha(H) = 0$. But in that case

$$\alpha(H') = \alpha(H_t') = -c_\alpha + (-1)^{1/2}2\pi r$$

which however is impossible since $H' \in \mathfrak{h}_0'$. Let $V$ be a neighborhood of $\omega_{\mathfrak{t}_0}$ in $\mathfrak{h}_0$ such that $V \cap \mathfrak{h}_0'$ has only a finite number of connected components. Then the above result shows that every point of $\omega'$ can be joined to a point

---

[5] We note that $M$ is a closed subgroup of $G$ and therefore the natural topology on $M$ as a Lie group coincides with the one induced by $G$ (see Chevalley [2]). A similar statement holds for $A$.

in $V$ by a line lying entirely in $\omega'$. Therefore $\omega'$ also does not have more than a finite number of connected components. On the other hand it is obvious that the image of $\omega'$ under the mapping $H \to a_0 \exp H$ ($H \in \omega'$) is exactly $A' \cap a_0 A_0$. This proves that no component of $A$ contains more than a finite number of components of $A'$. Therefore since $A$ has only a finite number of components the same holds for $A'$.

COROLLARY. *There exists a finite number of connected components $G_1', \cdots, G_r'$ of $G'$ such that $G' = \bigcup_{i=1}^r Z G_i'$.*

This follows immediately from Lemmas 5 and 9.

LEMMA 10. *Let $A^*$ denote the normalizer of $A$ in $G$. Then $A^* = (A^* \cap K)A_+$, the factor group $A^*/A$ is finite and $a^* M a^{*-1} = M$ for any $a^* \in A^*$. Moreover if $x$ is an element in $G$ such that $xmx^{-1} \in M$ for some $m \in {}^\backprime M$, then $x \in A^* M$.*

Since $\theta(A) = A$ it is obvious that $\theta(A^*) = A^*$. Now suppose $kp \in A^*$ ($k \in K$, $p \in \exp \mathfrak{p}_0$). Then $p^2 = (\theta(kp))^{-1}kp \in A^*$ and therefore Ad $(p^2)\mathfrak{h}_0 = \mathfrak{h}_0$. But an element $H \in \mathfrak{h}_0$ belongs to $\mathfrak{p}_0$ (or $\mathfrak{k}_0$) if and only if all the eigenvalues of ad $H$ are real (or purely imaginary). This implies that Ad $(p^2)$ leaves both $\mathfrak{h}_0 \cap \mathfrak{p}_0$ and $\mathfrak{h}_0 \cap \mathfrak{k}_0$ invariant. Choose $Y \in \mathfrak{p}_0$ such that $p = \exp Y$. Since $[\mathfrak{p}_0, \mathfrak{p}_0] \subset \mathfrak{k}_0$ and $[\mathfrak{p}_0, \mathfrak{k}_0] \subset \mathfrak{p}_0$, we conclude that

$$\{ \text{Ad } (p^2) - \text{Ad } (p^{-2}) \} H = 0 \qquad (H \in \mathfrak{h}_0).$$

But as we have seen during the proof of Lemma 7, this is possible only if (ad $Y)H = 0$ for all $H \in \mathfrak{h}_0$. This means however that $Y \in \mathfrak{h}_0 \cap \mathfrak{p}_0$ and therefore $p \in A_+$ and $k \in A^* \cap K$. Moreover $\mathfrak{m}_{\mathfrak{k}_0}$ being the centralizer of $\mathfrak{h}_0 \cap \mathfrak{p}_0$ in $\mathfrak{k}_0$, it is obvious that Ad $(k)\mathfrak{m}_{\mathfrak{k}_0} = \mathfrak{m}_{\mathfrak{k}_0}$ and therefore $k \overline{M} k^{-1} = \overline{M}$. Since $M = \overline{M}A$, this proves that $a^* M a^{*-1} = M$ for any $a^* \in A^*$. Also $\mathfrak{h}_0$ being its own normalizer in $\mathfrak{g}_0$, it follows that $A^*$ and $A$ have the same connected component of 1. Therefore since $Z \subset A \cap K$ and $K/Z$ is compact, $A^*/A \simeq (A^* \cap K)/A \cap K$ is both discrete and compact and so it must be finite.

Now suppose $xmx^{-1} \in M$ for some $m \in {}^\backprime M$ and $x \in G$. Then $m = m_1 a m_1^{-1}$ where $a \in {}^\backprime M \cap A = {}^\backprime A$ and $m_1 \in M$. Similarly $xmx^{-1} = m_2 a' m_2^{-1}$ ($a' \in A$, $m_2 \in M$). Hence if $y = m_2^{-1} x m_1$, it is sufficient to prove that $y \in A^* M$. But $a \in y^{-1} A y$ and therefore it is obvious that Ad $(a)X = X$ if $X \in \text{Ad } (y^{-1})\mathfrak{h}_0$. On the other hand since $a \in {}^\backprime A$, Ad $(a)X \neq X$ ($X \in \mathfrak{g}_0$) unless $X \in \mathfrak{h}_0 \cap \mathfrak{p}_0 + \mathfrak{m}_{\mathfrak{k}_0} = \mathfrak{m}_0$. Therefore Ad $(y^{-1})\mathfrak{h}_0 \subset \mathfrak{m}_0$. Moreover Ad $(\overline{M})$ being compact, any two maximal abelian subalgebras of $\mathfrak{m}_0$ are conjugate under $\overline{M}$. Therefore we can select $m_3 \in \overline{M}$ such that Ad $(y^{-1})\mathfrak{h}_0 = \text{Ad } (m_3)\mathfrak{h}_0$. Then from Lemma 3, $ym_3 \in A^*$ and therefore $y \in A^* M$.

**3. Quasi-regular elements.** Let $Z_x$ and $\mathfrak{h}_x$ be the centralizers in $G$ and $\mathfrak{g}_0$ respectively of an element $x \in G$. We shall say that $x$ is quasi-regular if (1) $\mathfrak{h}_x$ contains a regular element and (2) there exists a closed normal abelian subgroup $A$ of $Z_x$ such that $Z_x/A$ is compact. It follows from the corollary to

Lemma 3 that every regular element is quasi-regular. Now suppose $x$ is quasi-regular and $H$ is a regular element in $\mathfrak{h}_x$. Then $x$ lies in the centralizer of $H$ in $G$ and therefore from Lemmas 3 and 6, $x$ is conjugate to an element of some $A_i$ $(1 \leq i \leq r)$. Thus we have obtained the following result.

LEMMA 11. *Let $'G$ be the set of all quasi-regular elements in $G$ and put $'A_i = {}'G \cap A_i$ $(1 \leq i \leq r)$ where $A_i$ have the same meaning as in Lemma 6. Then $'G = \bigcup_{i=1}^r \bigcup_{x \in G} x'A_i x^{-1}$.*

Let us now use the notation of Lemma 7 and define $'M$ as in the corollary to Lemma 8. Our object is to show that $'G \cap M = {}'M$. Suppose $a_0$ is a quasi-regular element in $A$. Define $\xi_\alpha$ $(\alpha \in P)$ as in the proof of Lemma 8 and let $\mathfrak{n}$ be the centralizer of $a_0$ in $\mathfrak{g}$. Then if $Q$ is the set of those roots $\alpha \in P$ for which $\xi_\alpha(a_0) = 1$, $\mathfrak{n} = \mathfrak{h} + \sum_{\alpha \in Q} (CX_\alpha + CX_{-\alpha})$. Moreover since $a_0$ lies in $G$, $\mathfrak{n}$ is the complexification of $\mathfrak{n}_0 = \mathfrak{n} \cap \mathfrak{g}_0$ and $\mathfrak{n}_0$ is the Lie algebra of $Z_{a_0}$. Now let $y \in Z_{a_0}$. Then it follows from the definition of quasi-regularity that all the eigenvalues of the restriction of Ad $(y)$ on $\mathfrak{n}$ are unimodular complex numbers. Therefore since $A \subset Z_{a_0}$, we conclude that $\left| \xi_\alpha(a) \right| = 1$ for all $a \in A$ and $\alpha \in Q$. This shows that every root in $Q$ takes only purely imaginary values on $\mathfrak{h}_0$. On the other hand since all roots take only real values on $\mathfrak{h}_0 \cap \mathfrak{p}_0$, it follows that $Q \subset P_0 \cup P_-$ and therefore $\theta(\mathfrak{n}) = \mathfrak{n}$. Furthermore it is obvious from the above expression for $\mathfrak{n}$ that it is reductive (see Koszul [9]) and so any abelian ideal of $\mathfrak{n}$ must be contained in $\mathfrak{h}$. Hence in view of the quasi-regularity of $a_0$, $\mathfrak{n}_0' = [\mathfrak{n}_0, \mathfrak{n}_0]$ must be a *compact semisimple* Lie algebra over $R$. On the other hand $\theta(\mathfrak{n}_0')$ $= \mathfrak{n}_0'$ and therefore $\mathfrak{n}_0' = \mathfrak{n}_0' \cap \mathfrak{k}_0 + \mathfrak{n}_0' \cap \mathfrak{p}_0$. But if $X \in \mathfrak{n}_0' \cap \mathfrak{p}_0$ all the eigenvalues of ad $X$ are real which, from the compactness and semisimplicity of $\mathfrak{n}_0'$, is impossible unless $X = 0$. This shows that $\mathfrak{n}_0' \subset \mathfrak{k}_0$ and therefore $\mathfrak{k} \supset [\mathfrak{h}, \mathfrak{n}]$ $\supset \sum_{\alpha \in Q} (CX_\alpha + CX_{-\alpha})$. Hence $Q \subset P_-$ and so $a_0 \in A \cap {}'M$. Conversely if $m \in {}'M$, $Z_m$ is contained in $A^*M$ and $A_+Z$ is a closed normal subgroup of $A^*M$ such that $A^*M/A_+Z$ is compact (see Lemma 10). Moreover since $m$ is conjugate to some element of $A$ (Lemma 7), the centralizer of $m$ in $\mathfrak{g}_0$ contains a regular element. This proves that $m$ is quasi-regular. Therefore $A \cap {}'G = A \cap {}'M$ and since $M = \bigcup_{m \in M} mAm^{-1}$, it follows that $'M = M \cap {}'G$. Thus we have the following lemma.

LEMMA 12. *$'M$ coincides with the set of quasi-regular elements in $M$.*

COROLLARY. *$'G$ is an open subset of $G$ and there exists a finite number of connected components $G_1, \cdots, G_N$ of $'G$ such that $'G = \bigcup_{i=1}^N ZG_i$.*

Consider the mapping $(x, m) \to xmx^{-1}$ $(x \in G, m \in {}'M)$. Since (Ad $(m^{-1})$ $- I)\mathfrak{q} = \mathfrak{q}$, every element in $\mathfrak{g}$ can be written in the form $X + (\text{Ad } (m^{-1}) - I) Y$ $(X \in \mathfrak{m}, Y \in \mathfrak{q})$. Hence this mapping is everywhere open and therefore $\bigcup_{x \in G} x'Mx^{-1}$ is an open set in $G$. But $'M = \bigcup_{m \in M} m'Am^{-1}$ where $'A = A \cap {}'G$. Hence $\bigcup_{x \in G} x'Ax^{-1}$ is an open set and we conclude from Lemma 11 that $'G$ is also open. Now $G'$ being the set of regular elements in $G$, $'G \supset G'$ and $G'$ is dense in $G$. Hence every connected component of $`G$ contains some connected component of $G'$. Define $G_i'\ 1 \leq i \leq r$ as in the corollary to Lemma 9 and select a connected component $G_i$ of $`G$ containing $G_i'$. Then $\bigcup_{i=1}^r ZG_i \supset G'$ and therefore every connected component of $`G$ must intersect and therefore coincide with $zG_i$ for some $z \in Z$ and some $i$. Hence $`G = \bigcup_{i=1}^r ZG_i$.

4. **Some general facts about differential operators.** Let $V$ be a locally connected topological space on each connected component of which there is defined the structure of a differentiable manifold of class $C^\infty$ (or of an analytic manifold) and dimension $n$. Under these conditions we say that $V$ is a differentiable (or analytic) manifold of dimension $n$. This definition differs from the usual one in not assuming that $V$ is connected. It has the advantage that every open subset $U$ of $V$ may now be regarded as an open submanifold of $V$. Let $C_c^\infty(U)$ denote the set of all complex-valued functions on $V$ which are everywhere differentiable and which vanish outside a compact set contained in $U$. Then $C_c^\infty(U)$ is a vector space over $C$ and a distribution $T$ on $U$ is a linear function on this vector space satisfying certain conditions of continuity (see Schwartz [11]). Let $D$ be a linear mapping of $C_c^\infty(V)$ into itself and $p$ a point in $V$. We say that $D$ is a differential operator at $p$ if there exists a co-ordinate system $(t_1, \cdots, t_n)$ valid on some open neighborhood $U$ of $p$ and having the following property. Let $\partial_i$ denote the operators $\partial/\partial t_i$ $(1 \leq i \leq n)$ of partial differentiation. We require that there should exist a finite set of indefinitely differentiable functions $g_{i_1 i_2 \cdots i_q}$ $(1 \leq i_1, \cdots, i_q \leq n, 0 \leq q \leq r)$ on $U$ such that for every $f \in C_c^\infty(U)$,

$$Df = \sum_{0 \leq q \leq r} \sum_{1 \leq i_1, \cdots, i_q \leq n} g_{i_1 i_2 \cdots i_q} \partial_{i_1} \partial_{i_2} \cdots \partial_{i_q} f$$

on $U$ and $Df = 0$ outside $U$. (In case $q = 0$, $\partial_{i_1}\partial_{i_2} \cdots \partial_{i_q} f$ should be interpreted to mean $f$.) If $D$ is a differential operator at every point of $V$, we say simply that it is a differential operator (on $V$). Clearly the differential operators form a subalgebra of the algebra of all endomorphisms of $C_c^\infty(V)$. In particular if $g$ is an indefinitely differentiable function on $V$ the mapping $f \to gf$ $(f \in C_c^\infty(V))$ is a differential operator which we again denote by $g$.

Let $C^\infty(V)$ denote the space of all indefinitely differentiable functions on $V$. Then any differential operator $D$ also defines an endomorphism of $C^\infty(V)$. In fact if $f \in C^\infty(V)$ and $p$ is a point in $V$, $(Df)(p) = (Dg)(p)$ where $g$ is any function in $C_c^\infty(V)$ such that $f - g$ is identically zero on some neighborhood of $p$. It would often be convenient to denote the value of $Df$ at $p$ by $f(p, D)$. $V$ being an analytic manifold, we say that $D$ is analytic if $Df$ is analytic at a point $p$ whenever $f$ is analytic at $p$ $(f \in C^\infty(V))$.

Now suppose $D_1, D_2$ are differential operators on two manifolds $V_1$ and $V_2$ respectively. Then $D = D_1 \times D_2$ is a differential operator on the product manifold $V_1 \times V_2$ (see Schwartz [11]). In fact if $f_1 \in C^\infty(V_1)$, $f_2 \in C^\infty(V_2)$ and $g(p_1, p_2) = f_1(p_1)f_2(p_2)(p_1 \in V_1, p_2 \in V_2)$ then

$$(Dg)(p_1, p_2) = f_1(p_1, D_1)f_2(p_2, D_2).$$

We shall sometimes denote the value of $Dh$ ($h \in C^\infty(V_1 \times V_2)$) at $(p_1, p_2)$ $\in V_1 \times V_2$ by $h(p_1, D_1; p_2, D_2)$.

Let $L$ be a Lie group and let $\mathfrak{l}_0$ denote its Lie algebra over $R$. Then if $L'$ is a nonempty open subset of $L$, every $X \in \mathfrak{l}_0$ defines a differential operator on $L'$. In fact if $f \in C_c^\infty(L')$, $Xf$ is given by $f(x, X) = \{ df(x \exp tX)/dt \}_{t=0}$ ($x \in L'$, $t \in R$). It is well known (see Chevalley [2]) that $[X, Y]f = X(Yf) - Y(Xf)$ ($X, Y \in \mathfrak{l}_0, f \in C_c^\infty(L')$) and therefore we get a representation of $\mathfrak{l}_0$ on $C_c^\infty(L')$. Let $\mathfrak{l}$ be the complexification of $\mathfrak{l}_0$ and $\mathfrak{L}$ the universal enveloping algebra of $\mathfrak{l}$ [6(a)]. We may extend this representation ( uniquely) to a representation of $\mathfrak{L}$ on $C_c^\infty(L')$ and therefore every element $b \in \mathfrak{L}$ defines a differential operator $f \to bf$ ($f \in C_c^\infty(L')$) on $L'$. Let $bf(x)$ denote the value of $bf$ at $x \in L'$.

LEMMA[6] 13. *Let $x_0$ be a point in $L'$ and $b$ an element in $\mathfrak{L}$. If $bf(x_0) = 0$ for every $f \in C_c^\infty(L')$ then $b = 0$.*

Let $X_1, \cdots, X_n$ be a base for $\mathfrak{l}_0$ over $R$. We can choose a positive number $\epsilon$ and an open neighborhood $U$ of $x_0$ in $L'$ with the following property. If $Q$ is the open cube consisting of all points $t = (t_1, \cdots, t_n)$ in $R^n$ with $|t| = \max_i |t_i| < \epsilon$, the mapping

$$t \to x_0 \exp X(t)$$

where $X(t) = t_1X_1 + \cdots + t_nX_n$, is a one-one regular mapping of $Q$ onto $U$. Then it is obvious that if $P(t)$ is any polynomial in $(t_1, \cdots, t_n)$ with complex coefficients, there exists a function $f_P \in C_c^\infty(L')$ such that $f_P(x_0 \exp X(t)) = P(t)$ for $|t| \leq \epsilon/2$. In particular $f_P$ is analytic at $x_0$ and its Taylor expansion

$$f_P(x_0 \exp X(t)) = \sum_{m \geq 0} \frac{1}{m!} (X(t))^m f(x_0)$$

is valid if $|t|$ is sufficiently small. Now $M$ being an ordered set $(m_1, \cdots, m_n)$ of $n$ non-negative integers, put $t^M = t_1^{m_1} \cdots t_n^{m_n}$, $|M| = m_1 + \cdots + m_n$ and let $X(M)$ denote the coefficient of $t^M$ in $(|M|!)^{-1}(X(t))^{|M|}$. Then

$$P(t) = f(x_0 \exp X(t)) = \sum_{m \geq 0} \sum_{|M| = m} t^M X(M)f_P(x_0)$$

provided $|t|$ is sufficiently small. Hence if $P_M$ is the coefficient of $t^M$ in $P(t)$, $P_M = X(M)f_P(x_0)$. On the other hand the elements $X(M)$ form a base for $\mathfrak{L}$ over $C$ (see [6(a)]). Therefore $b = \sum_M a(M)X(M)$ where $a(M) \in C$ and the sum is finite. If $b \neq 0$ we can choose $M_0$ such that $a(M_0) \neq 0$. Now put $P = t^{M_0}$. Then

$$bf_P(x_0) = a(M_0) \neq 0.$$

---

[6] This lemma is attributed to Schwartz by Godement [5].

This proves our assertion.

COROLLARY. *Let $D$ be a differential operator on $L'$ and $x_0$ a point in $L'$. Then there exists exactly one element $b \in \mathfrak{L}$ such that $f(x_0, D) = bf(x_0)$ for all $f \in C_c^\infty(L')$.*

We keep to the notation introduced above. Since $t_1, \cdots, t_n$ form a co-ordinate system on $U$ under the mapping $t \to x_0 \exp X(t)$ and

$$m_1! m_2! \cdots m_n! X(M) f(x_0) = \left\{ \frac{\partial^{m_1 + \cdots + m_n}}{\partial t_1^{m_1} \cdots \partial t_n^{m_n}} f(x_0 \exp X(t)) \right\}_{t=0}$$

for $M = (m_1, \cdots, m_n)$ and $f \in C_c^\infty(L')$, it follows from the definition of a differential operator that there exists an element $b = \sum_M a(M) X(M) \in \mathfrak{L}$ such that $f(x_0, D) = bf(x_0)$ for all $f \in C_c^\infty(L')$. On the other hand the uniqueness of $b$ is an immediate consequence of the above lemma.

We call the element $b$ of the above corollary the local expression of $D$ at $x_0$. Lemma 13 shows that the representation of $\mathfrak{L}$ on $C_c^\infty(L')$ is faithful and so we may identify elements of $\mathfrak{L}$ with the corresponding differential operators on $L'$. In this way $\mathfrak{L}$ becomes a subalgebra of the algebra of all differential operators on $L'$. Moreover if $X \in \mathfrak{l}_0$, $X$ is an analytic differential operator on $L'$ (see Chevalley [2]) and therefore the same holds for every $b \in \mathfrak{L}$.

Let $x \to \mathrm{Ad}\,(x)$ denote the adjoint representation of $L$ on $\mathfrak{l}$. We extend $\mathrm{Ad}\,(x)$ to an automorphism of $\mathfrak{L}$ over $C$ and write $a^x = \mathrm{Ad}\,(x)a$ $(a \in \mathfrak{L}, x \in L)$ whenever it is convenient to do so. Now suppose $L'$ is an invariant set which means that $xL'x^{-1} = L'$ for all $x \in L$. Then for any function $f$ on $L'$, define $f^x$ $(x \in L)$ to be the function $y \to f(x^{-1}yx)$ $(y \in L')$. We say that $f$ is invariant if $f = f^x$ for all $x \in L$. It is clear that $f \to f^x$ $(f \in C_c^\infty(L'))$ is a linear isomorphism of $C_c^\infty(L')$ onto itself and if $a \in \mathfrak{L}$, $f^x(y, a) = f(x^{-1}yx, \mathrm{Ad}\,(x^{-1})a)$ $(x \in L, y \in L')$. We shall always denote by $I_c^\infty(L')$ and $I^\infty(L')$ respectively the sets of invariant functions in $C_c^\infty(L')$ and $C^\infty(L')$. $D$ being any differential operator, the mapping $f \to (D(f^{x^{-1}}))^x$ $(f \in C_c^\infty(L'))$ is also a differential operator which we denote by $D^x$. In particular if $D \in \mathfrak{L}$, $D^x = \mathrm{Ad}\,(x)D$ and so this coincides with the definition of $a^x$ $(a \in \mathfrak{L})$ given above. We say that $D$ is invariant if $D^x = D$ for all $x \in L$. Let $D_y$ $(y \in L')$ denote the local expression of $D$ at $y$.

LEMMA 14. *If $D$ is a differential operator on $L'$,*

$$\mathrm{Ad}\,(x)D_y = (D^x)_{xyx^{-1}} \qquad\qquad (x \in L,\ y \in L').$$

*Therefore $D$ is invariant if and only if $D_{xyx^{-1}} = \mathrm{Ad}\,(x)D_y$ for all $x \in L$ and $y \in L'$.*

If $f \in C_c^\infty(L')$, it follows from the definition of $D^x$ that

$$f(xyx^{-1}, D^x) = f^{x^{-1}}(y, D) = f^{x^{-1}}(y, D_y)$$
$$= f(xyx^{-1}, \mathrm{Ad}\,(x)D_y).$$

Hence the lemma.

Similarly if $\tau$ is a distribution on $L'$, the mapping $f \to \tau(f^{x^{-1}})$ $(f \in C_c^\infty(L'))$ is also a distribution which we denote by $\tau^x$ $(x \in L)$. Again $\tau$ is said to be invariant if $\tau^x = \tau$ for all $x \in L$. Now suppose the group $L$ is unimodular so that its Haar measure $dx$ is both left- and right-invariant. Let $U$ be a (nonempty) open set in $L'$ and $F$ a complex-valued function on $U$. We say that $F$ is locally summable if it is measurable with respect to the Haar measure and $\int_\omega |F(x)| \, dx < \infty$ for any compact set $\omega$ contained in $U$. Corresponding to any such $F$, we define a distribution $\tau_F$ on $U$ as follows:

$$\tau_F(f) = \int f(x)F(x)dx \qquad\qquad (f \in C_c^\infty(U)).$$

If $\tau$ is a distribution (defined on some open set containing $U$) we say that $\tau = F$ on $U$ (or $\tau$ coincides with $F$ on $U$) if $\tau(f) = \tau_F(f)$ for every $f \in C_c^\infty(U)$. Let $D$ be a differential operator on $U$. Then there exists a uniquely determined differential operator $D^*$ on $U$ such that

$$\int (D^*f)g\,dx = \int f(Dg)dx \qquad\qquad (f, g \in C_c^\infty(U)).$$

$D^*$ is called the adjoint of $D$. The mapping $D \to D^*$ is obviously an anti-automorphism of order 2 of the algebra of all differential operators on $U$. In particular if $g \in C^\infty(U)$, $g^* = g$. Now if $\tau$ is a distribution and $D$ a differential operator on $U$, we define a distribution $D\tau$ by $(D\tau)(f) = \tau(D^*f)$ $(f \in C_c^\infty(U))$. In this way we get a representation of the algebra of differential operators on the space of distributions. $\tau$ is said to be an eigen-distribution of $D$ if $D\tau = c\tau$ where $c$ is a complex number. Let $\phi$ be the anti-automorphism of $\mathfrak{L}$ such that $\phi(X) = -X$ $(X \in \mathfrak{l})$. Then it is obvious that $X^* = \phi(X)$ if $X \in \mathfrak{l}$ and therefore $b^* = \phi(b)$ for all $b \in \mathfrak{L}$. Moreover if $U$ is an invariant set, it is easy to see that a differential operator $D$ on $U$ is invariant if and only if $D^*$ is invariant.

**5. Differential operators on $'M$.** Now let us return to the notation of Lemma 7 and define $'M$ as in the corollary to Lemma 8. Then $'M$ is an open submanifold of the Lie group $M$. Let $\mathfrak{B}$ be the universal enveloping algebra of $\mathfrak{g}$ and $\mathfrak{M}$ the subalgebra of $\mathfrak{B}$ generated by $(1, \mathfrak{m})$ where

$$\mathfrak{m} = \mathfrak{h} + \sum_{\alpha \in P-} (CX_\alpha + CX_{-\alpha}).$$

Then $\mathfrak{M}$ is the universal enveloping algebra of $\mathfrak{m}$ (see Lemma 21 of [6(b)]) and therefore $\mathfrak{B}$ and $\mathfrak{M}$ are certain algebras of differential operators on $G$ and $M$ respectively. For any $b \in \mathfrak{B}$ let $L_b$ and $R_b$ respectively denote the linear mappings $a \to ba$ and $a \to ab$ $(a \in \mathfrak{B})$ of $\mathfrak{B}$. $\phi$ being an indefinitely differentiable function on $G$, put $\Phi(x; m) = \phi(xmx^{-1})$ $(x \in G, m \in M)$. Then $\Phi \in C^\infty(G \times M)$

and if $X_1, \cdots, X_r \in \mathfrak{g}_0$ and $H_1, \cdots, H_s \in \mathfrak{m}_0 = \mathfrak{m} \cap \mathfrak{g}_0$, we have

$$\Phi(x, X_1 X_2 \cdots X_r; m, H_1 H_2 \cdots H_s)$$

$$= \left\{ \frac{\partial^{r+s}}{\partial u_1 \cdots \partial u_r \partial t_1 \cdots \partial t_s} \phi(x y(u) m m(t)(y(u))^{-1} x^{-1}) \right\}_{u_i = t_j = 0}$$

where $y(u) = \exp u_1 X_1 \cdots \exp u_r X_r$ and $m(t) = \exp t_1 H_1 \cdots \exp t_s H_s$ $(u_i, t_j \in R)$. Therefore, as a simple induction shows,

$$\Phi(x, X_1 X_2 \cdots X_r; m, H_1 H_2 \cdots H_s) = \phi(x m x^{-1}, w^x)$$

where

$$w = (L_{\mathrm{Ad}(m^{-1})X_1} - R_{X_1}) \cdots (L_{\mathrm{Ad}(m^{-1})X_r} - R_{X_r})(H_1 \cdots H_s).$$

This suggests the lemma below. Let $\mathfrak{B} \times \mathfrak{M}$ denote the tensor product of $\mathfrak{B}$ and $\mathfrak{M}$ and $S(\mathfrak{q})$ and $S(\mathfrak{g})$ the symmetric algebras over $\mathfrak{q}$ and $\mathfrak{g}$ respectively (see [6(c), §2]). Then $S(\mathfrak{q}) \subset S(\mathfrak{g})$. Put $\mathfrak{Q} = \lambda(S(\mathfrak{q}))$ where $\lambda$ is the canonical mapping [6(c), p. 192] of $S(\mathfrak{g})$ onto $\mathfrak{B}$. Then if $m \in M$, it is obvious that $\mathrm{Ad}(m) \mathfrak{Q} = \mathfrak{Q}$. Let $S_r(\mathfrak{g})$ be the space of those elements in $S(\mathfrak{g})$ which are homogeneous of degree $r$. We call an element $b \in \mathfrak{B}$ homogeneous of degree $r$ if $b \in \lambda(S_r(\mathfrak{g}))$ and of degree $\leq r$ if $b \in \sum_{0 \leq d \leq r} \lambda(S_d(\mathfrak{g}))$. Similarly we say that an element in $\mathfrak{B} \times \mathfrak{M}$ is of total degree $\leq r$ if it lies in the space $\sum_{d+e \leq r} \mathfrak{B}_d \times \mathfrak{M}_e$ where $\mathfrak{B}_d$ and $\mathfrak{M}_e$ respectively are the subspaces of $\mathfrak{B}$ and $\mathfrak{M}$ consisting of homogeneous elements of degrees $d$ and $e$.

$\quad$ LEMMA 15. *For each $m \in M$ there exists a unique linear mapping $\Gamma_m$ of $\mathfrak{B} \times \mathfrak{M}$ into $\mathfrak{B}$ such that $\Gamma_m(1 \times v) = v$ and*

$$\Gamma_m(X_1 X_2 \cdots X_r \times v) = (L_{\mathrm{Ad}(m^{-1})X_1} - R_{X_1}) \cdots (L_{\mathrm{Ad}(m^{-1})X_r} - R_{X_r})v$$

*for all $X_1, \cdots, X_r \in \mathfrak{g}$ and $v \in \mathfrak{M}$. Moreover if $m \in {}^\iota M$, $\Gamma_m$ defines a one-one mapping of $\mathfrak{Q} \times \mathfrak{M}$ onto $\mathfrak{B}$.*

The uniqueness of $\Gamma_m$ is obvious and so only the existence requires proof. Put

$$\gamma_m(X_1, \cdots, X_r; v) = L_{\mathrm{Ad}(m^{-1})X_1} - R_{X_1}) \cdots (L_{\mathrm{Ad}(m^{-1})X_r} - R_{X_r})v$$

for $m \in M$, $X_1, \cdots, X_r \in \mathfrak{g}$ and $v \in \mathfrak{M}$. If $Y_1, \cdots, Y_n$ is a base for $\mathfrak{g}_0$ over $R$, it would be enough to prove that if

$$\sum_{j=0}^{r} \sum_{1 \leq i_1, \cdots, i_j \leq n} c_{i_1, i_2 \cdots i_j} Y_{i_1} Y_{i_2} \cdots Y_{i_j} = 0 \qquad (c_{i_1 \cdots i_j} \in C)$$

then

$$\sum_{j=0}^{r} \sum_{1 \leq i_1, \cdots, i_j \leq n} c_{i_1 i_2 \cdots i_j} \gamma_m(Y_{i_1}, \cdots, Y_{i_j}; v) = 0.$$

Now $\phi$ and $\Phi$ being as above, we have seen that

$$\Phi(x, Y_{i_1} \cdots Y_{i_j}; m, \nu) = \phi(xmx^{-1}, b^x) \qquad (x \in G)$$

where

$$b = \gamma_m(Y_{i_1}, Y_{i_2}, \cdots, Y_{i_j}; \nu).$$

Hence we conclude that

$$0 = \sum_{j=0}^{r} \sum_{1 \leq i_1, \cdots, i_j \leq n} c_{i_1 \cdots i_j} \Phi(x, Y_{i_1} \cdots Y_{i_j}; m, \nu) = \phi(xmx^{-1}, a^x)$$

where

$$a = \sum_{j=0}^{r} \sum_{1 \leq i_1, \cdots, i_j \leq m} c_{i_1 \cdots i_j} \gamma_m(Y_{i_1}, \cdots, Y_{i}; \nu).$$

Putting $x=1$, we get $\phi(m, a)=0$. This being true for every $\phi \in C^\infty(G)$, it follows from Lemma 13 that $a=0$.

Now we come to the second statement of the lemma. Put

$$A(m) = (\mathrm{Ad}\,(m^{-1}) - I)_{\mathfrak{q}}$$

and suppose $m \in {}^\prime M$. Let $\mathfrak{B}_s$ denote the space of all elements in $\mathfrak{B}$ of degree $\leq s$ and put $\mathfrak{B}_{-1} = \{0\}$. Then $\mathfrak{B}_s$ mod $\mathfrak{B}_{s-1}$ ($s \geq 0$) is spanned by elements of the form $X_1 X_2 \cdots X_s$ where $X_1, \cdots, X_s \in \mathfrak{g}$. Moreover if $(i_1, i_2, \cdots, i_s)$ is any permutation of $(1, 2, \cdots, s)$,

$$X_1 X_2 \cdots X_s - X_{i_1} X_{i_2} \cdots X_{i_s} \in \mathfrak{B}_{s-1}$$

(see [6(a)]). Let $\mathfrak{Q}_d$ and $\mathfrak{M}_d$ be the sets of homogeneous elements of degree $d$ in $\mathfrak{Q}$ and $\mathfrak{M}$ respectively. We shall prove by induction on $s$ that

$$\sum_{d+e \leq s} \Gamma_m(\mathfrak{Q}_d \times \mathfrak{M}_e) = \mathfrak{B}_s.$$

If $Y_1, \cdots, Y_d \in \mathfrak{q}$ and $X_1, \cdots, X_e \in \mathfrak{m}$,

$$\Gamma_m(Y_1 \cdots Y_d \times X_1 \cdots X_e)$$
$$= (L_{\mathrm{Ad}(m^{-1})Y_1} - R_{Y_1}) \cdots (L_{\mathrm{Ad}(m^{-1})Y_d} - R_{Y_d})(X \cdots X_e).$$

Put $D_X = L_X - R_X$ ($X \in \mathfrak{g}$). Then since $D_X \mathfrak{B}_r \subset \mathfrak{B}_r$ ($r \geq 0$), it is clear that

$$\Gamma_m(Y_1 \cdots Y_d \times X_1 \cdots X_e)$$
$$= (L_{A(m)Y_1} + D_{Y_1}) \cdots (L_{A(m)Y_d} + D_{Y_d})(X_1 \cdots X_e)$$
$$\equiv (A(m)Y_1) \cdots (A(m)Y_d) X_1 \cdots X_e \bmod \mathfrak{B}_{d+e-1}.$$

This shows that $\Gamma_m(\mathfrak{Q}_d \times \mathfrak{M}_e) \subset \mathfrak{B}_s$ if $d+e \leq s$. Conversely since $m \in {}^\prime M$, $A(m)$ is nonsingular and so every element in $\mathfrak{q}$ can be written in the form $A(m)Y$ ($Y \in \mathfrak{q}$). But elements of the form $Z_1 \cdots Z_d X_1 X_2 \cdots X_e$ ($Z_1, \cdots, Z_d \in \mathfrak{q}$,

$X_1, \cdots, X_e \in \mathfrak{m}$) with $d+e=s$ span $\mathfrak{B}_s$ mod $\mathfrak{B}_{s-1}$ and the order of the factors is immaterial mod $\mathfrak{B}_{s-1}$. Therefore

$$\sum_{d+e=s} \Gamma_m(\mathfrak{Q}_d \times \mathfrak{M}_e) + \mathfrak{B}_{s-1} \supset \mathfrak{B}_s$$

and so our assertion followed by induction hypothesis. Moreover since $\mathfrak{g}$ is the direct sum of $\mathfrak{q}$ and $\mathfrak{m}$, $\mathfrak{B}_s$ and $\sum_{d+e \leqq s} \mathfrak{Q}_d \times \mathfrak{M}_e$ have the same dimension (see Lemma 12 of [6(c)]). Therefore $\Gamma_m$ must be one-one.

COROLLARY. *If $\nu_1, \nu_2 \in \mathfrak{M}$, then $\Gamma_m(\nu_1 \times \nu_2) \in \mathfrak{M}(m \in M)$ and*

$$\Gamma_m(b\nu_1 \times \nu_2) = \Gamma_m(b \times \Gamma_m(\nu_1 \times \nu_2)) \qquad (b \in \mathfrak{B}).$$

It would be enough to consider the case when $\nu_1 = X_1 X_2 \cdots X_d$ and $b = Z_1 Z_2 \cdots Z_r$, where $X_1, \cdot, X_d \in \mathfrak{m}$ and $Z_1, \cdots, Z_r \in \mathfrak{g}$. But then our assertions follow directly from the definition of $\Gamma_m$.

Let $f$ be a mapping of an analytic manifold $V$ into a vector space $U$. We say that $f$ is analytic if for every linear function $\mu$ on $U$ the function $p \to \mu(f(p))$ $(p \in V)$ is analytic on $V$. Put $d_+(m) = \det(\mathrm{Ad}\,(m^{-1}) - I)_\mathfrak{q}$ $(m \in M)$. (In case $\mathfrak{q} = \{0\}$, we have to put $d_+(m) = 1$.)

LEMMA 16. *For each $b \in \mathfrak{B}$ there exists a nonnegative integer $r$ and an analytic mapping $\gamma_b$ of $M$ into $\mathfrak{Q} \times \mathfrak{M}$ such that*

$$\Gamma_m(\gamma_b(m)) = (d_+(m))^r b \qquad (m \in M).$$

*Moreover if the degree of $b$ is $\leqq s$, the total degree of $\gamma_b(m)$ is also $\leqq s$.*

We shall use induction on $s$. It is obviously sufficient to consider the case when

$$b = Y_1 Y_2 \cdots Y_d X_1 X_2 \cdots X_e$$

where $Y_1, \cdots, Y_d \in \mathfrak{q}$, $X_1, \cdots, X_e \in \mathfrak{m}$ and $d+e=s$. Let $P(t, m)$ denote the characteristic polynomial of $A(m) = (\mathrm{Ad}\,(m^{-1}) - I)_\mathfrak{q}$ in the indeterminate $t$. Then if $q = \dim \mathfrak{q}$,

$$P(t, m) = t^q - \sigma_1(m)t^{q-1} + \cdots + (-1)^q \sigma_q(m)$$

where $\sigma_1, \cdots, \sigma_q$ are analytic functions on $M$ and $\sigma_q(m) = d_+(m)$. Put

$$B(m) = (-1)^{q+1}\{(A(m))^{q-1} - \sigma_1(m)(A(m))^{q-2} + \cdots + (-1)^{q-1}\sigma_{q-1}(m)I_\mathfrak{q}\}$$

where $I_\mathfrak{q}$ is the identity mapping of $\mathfrak{q}$. Then $B(m)A(m) = d_+(m)I_\mathfrak{q}$. Put $Z_i = B(m)Y_i$ so that $A(m)Z_i = d_+(m)Y_i$ $(i = 1, \cdots, d)$. Also let

$$Q(m) = (d!)^{-1} \sum Z_{i_1} Z_{i_2} \cdots Z_{i_d}$$

where the sum extends over the permutations $(i_1, \cdots, i_d)$ of $(1, 2, \cdots, d)$. Then it is clear that

$$\Gamma_m(Q(m) \times X_1 \cdots X_e) \equiv (d_+(m))^d Y_1 Y_2 \cdots Y_d X_1 \cdots X_e \bmod \mathfrak{B}_{s-1}.$$

Hence

$$b_0(m) = (d_+(m))^d b - \Gamma_m(Q(m) \times X_1 \cdots X_e) \in \mathfrak{B}_{s-1}$$

and in view of the definition of $\Gamma_m$ it is obvious that $m \to b_0(m)$ is an analytic mapping of $M$ into $\mathfrak{B}_{s-1}$. Therefore if $b_1, \cdots, b_N$ is a base for $\mathfrak{B}_{s-1}$,

$$b_0(m) = \sum_{i=1}^{N} a_i(m) b_i$$

where $a_i(m)$ are analytic functions on $M$. Now by induction hypothesis we can choose integers $r_i \geqq 0$ and analytic mappings $\gamma_{b_i}$ of $M$ into $\mathfrak{Q} \times \mathfrak{M}$ such that

$$\Gamma_m(\gamma_{b_i}(m)) = (d_+(m))^{r_i} b_i \qquad (i = 1, \cdots, N)$$

and the total degree of $\gamma_{b_i}(m)$ is $\leqq s-1$ $(m \in M)$. Put $r = \max_{1 \leq i \leq N} (d, r_i)$ and

$$\gamma_b(m) = (d_+(m))^{r-d} Q(m) \times X_1 \cdots X_e + \sum_{i=1}^{N} a_i(m)(d_+(m))^{r-r_i} \gamma_{b_i}(m).$$

Then $\gamma_b$ is an analytic mapping of $M$ into $\mathfrak{B}$,

$$\Gamma_m(\gamma_b(m)) = (d_+(m))^r b$$

and the total degree of $\gamma_b(m)$ is $\leqq s$ $(m \in M)$. This proves the lemma.

Now put $\mathfrak{Q}' = \sum_{d \geqq 1} \mathfrak{Q}_d$ and let $m \in {}^\prime M$. Then if $b \in \mathfrak{B}$, it follows from Lemma 15 that there is a unique element $\delta_m(b) \in \mathfrak{M}$ such that

$$b - \delta_m(b) = b - \Gamma_m(1 \times \delta_m(b)) \in \Gamma_m(\mathfrak{Q}' \times \mathfrak{M}).$$

Moreover in view of Lemma 16, $m \to \delta_m(b)$ is an analytic mapping of ${}^\prime M$ into $\mathfrak{M}$. Therefore it is obvious that there exists a (unique) analytic differential operator $\delta(b)$ on ${}^\prime M$ such that $\delta_m(b)$ is its local expression at any point $m \in {}^\prime M$.

LEMMA 17. *If $a \in \mathfrak{B}$ and $m \in M$,*

$$(\delta(a))^m = \delta(a^m).$$

For any $m \in M$, the mappings $q \to q^m$ and $\nu \to \nu^m$ $(q \in \mathfrak{Q}, \nu \in \mathfrak{M})$ are linear isomorphisms of $\mathfrak{Q}$ and $\mathfrak{M}$ respectively onto themselves. Hence there exists a linear isomorphism of $\mathfrak{Q} \times \mathfrak{M}$ onto itself which maps $q \times \nu$ onto $q^m \times \nu^m$. We denote it again by Ad $(m)$. On the other hand it is obvious from the definition of $\Gamma_{m_1}$ $(m_1 \in M)$ that

$$\mathrm{Ad}\,(m)(\Gamma_{m_1}(a \times \nu)) = \Gamma_{m m_1 m^{-1}}(a^m \times \nu^m)$$

for $a \in \mathfrak{B}$ and $\nu \in \mathfrak{M}$. So in particular if $m_1 \in {}^\prime M$,

$$a - \Gamma_{m_1}(1 \times \delta_{m_1}(a)) \in \Gamma_{m_1}(\mathfrak{Q}' \times \mathfrak{M})$$

and therefore

$$a^m - \mathrm{Ad}\ (m)(\Gamma_{m_1}(1 \times \delta_{m_1}(a))) \in \Gamma_{mm_1m^{-1}}(\mathfrak{Q}' \times \mathfrak{M})$$

since $\mathrm{Ad}(m)(\mathfrak{Q}' \times \mathfrak{M}) = \mathfrak{Q}' \times \mathfrak{M}$. This shows that

$$\delta_{mm_1m^{-1}}(a^m) = \mathrm{Ad}\ (m)(\delta_{m_1}(a))$$

and therefore it follows from Lemma 14 that $(\delta(a))^m = \delta(a^m)$.

COROLLARY. *If $a$ is an element in $\mathfrak{B}$ such that $a^m = a$ for all $m \in M$, then $\delta(a)$ is an invariant differential operator on $`M$.*

This is an immediate consequence of Lemma 17.

6. **The relationship between invariants of the Weyl group and certain differential operators.** Let $z$ be an element in the center of $\mathfrak{B}$. We shall now try to determine the differential operator $\delta(z)$ on $`M$. Let us first recall a few facts about reductive Lie algebras (see Koszul [9]). Let $\mathfrak{l}$ be a reductive Lie algebra over $C$ and let $\mathfrak{c}$ and $\mathfrak{l}'$ respectively denote the center and the derived algebra of $\mathfrak{l}$. Then $\mathfrak{l}'$ is semisimple and if $\Gamma$ is a Cartan subalgebra of $\mathfrak{l}$, it is obvious that $\Gamma = \mathfrak{c} + \Gamma'$ where $\Gamma' = \Gamma \cap \mathfrak{l}'$ is a Cartan subalgebra of $\mathfrak{l}'$. By a root of $\mathfrak{l}$ (with respect to $\Gamma$) we mean a root $\alpha$ of $\mathfrak{l}'$ (with respect to $\Gamma'$). We extend $\alpha$ to a linear function on $\Gamma$ by setting $\alpha(H) = 0$ $(H \in \mathfrak{c})$. Let $\mathfrak{w}$ be the Weyl group of $\mathfrak{l}'$ with respect to $\Gamma'$. If $s \in \mathfrak{w}$ we also extend $s$ to a linear transformation on $\Gamma$ by setting $sH = H$ $(H \in \mathfrak{c})$. For every root $\alpha$ choose an element $X_\alpha \neq 0$ such that $[H, X_\alpha] = \alpha(H)X_\alpha$ for all $H \in \Gamma$. Introduce some lexicrographic order among roots and let $\alpha_1, \cdots, \alpha_r$ be all the (distinct) positive roots under this order. Denote by $\mathfrak{L}$ the universal enveloping algebra of $\mathfrak{l}$ and by $\mathfrak{U}$ the subalgebra of $\mathfrak{L}$ generated by $(1, \Gamma)$. Let $u \to u^s$ $(u \in \mathfrak{U})$ denote the automorphism of $\mathfrak{U}$ which coincides with $s$ on $\Gamma(s \in \mathfrak{w})$.

LEMMA 18. *Let $\mathfrak{Z}$ be the center of $\mathfrak{L}$. Then for every $z \in \mathfrak{Z}$ there exists a unique element $\gamma'(z) \in \mathfrak{U}$ such that $z - \gamma'(z) \in \sum_{i=1}^{r} \mathfrak{L}X_{\alpha_i}$. Furthermore $z \to \gamma'(z)$ $(z \in \mathfrak{Z})$ is an algebraic isomorphism of $\mathfrak{Z}$ into $\mathfrak{U}$.*

If $\mathfrak{c} = \{0\}$ this is the same as Lemma 36 of [6(b)]. The proof in the general case also is practically the same and so we shall not repeat it here.

Let us say that an element $u \in \mathfrak{U}$ is an invariant (of $\mathfrak{w}$) if $u^s = u$ for all $s \in \mathfrak{w}$. Let $J$ denote the subalgebra of $\mathfrak{U}$ consisting of all invariants. Put $\rho = (\alpha_1 + \cdots + \alpha_r)/2$ and let $\lambda$ denote the automorphism of $\mathfrak{U}$ given by $\lambda(H) = H - \rho(H)(H \in \Gamma)$. Define $\gamma(z) = \lambda(\gamma'(z))$ $(z \in \mathfrak{Z})$.

LEMMA 19. *$z \to \gamma(z)$ $(z \in \mathfrak{Z})$ is an algebraic isomorphism of $\mathfrak{Z}$ onto $J$.*

If $\mathfrak{c} = \{0\}$ this follows from the results of [6(b), Part II, see equation (13) p. 71 and Lemmas 38 and 39]. In the general case let $\mathfrak{C}$, $\mathfrak{L}'$ and $\mathfrak{U}'$ be the subalgebras of $\mathfrak{L}$ generated by $(1, \mathfrak{c})$, $(1, \mathfrak{l}')$ and $(1, \Gamma')$ respectively. Then $\mathfrak{L} = \mathfrak{L}'\mathfrak{C}$, $\mathfrak{Z} = \mathfrak{Z}'\mathfrak{C}$ and $J = J'\mathfrak{C}$ where $\mathfrak{Z}'$ is the center of $\mathfrak{L}'$ and $J' = J \cap \mathfrak{U}'$. Since $\gamma(c) = c$ $(c \in \mathfrak{C})$ our assertion now follows from the corresponding result for $\mathfrak{l}'$.

For every automorphism $\tau$ of $\mathfrak{l}$ let $b \to b^\tau$ $(b \in \mathfrak{L})$ denote the corresponding automorphism of $\mathfrak{L}$.

LEMMA 20. *The mapping $\gamma$ defined above is independent of the lexicographic order of roots used in its definition. If $\tau$ is any automorphism of $\mathfrak{l}$ which leaves $\Gamma$ invariant, then $\gamma(z^\tau) = (\gamma(z))^\tau$ $(z \in \mathfrak{Z})$. Finally let $b \to b^*$ $(b \in \mathfrak{L})$ denote the anti-automorphism of $\mathfrak{L}$ which maps $X$ on $-X(X \in \mathfrak{l})$. Then $\gamma(z^*) = (\gamma(z))^*$ $(z \in \mathfrak{Z})$.*

Let $\beta$ be a linear function on $\Gamma$ and $\tau$ an automorphism of $\mathfrak{l}$ such that $\tau\Gamma = \Gamma$. Then we denote by $\tau\beta$ the function $H \to \beta(\tau^{-1}H)$ $(H \in \Gamma)$. If $\alpha$ is a root, it is obvious that $\tau\alpha$ is also a root and $\tau X_\alpha \in CX_{\tau\alpha}$. Now first suppose $\tau$ permutes the positive roots $(\alpha_1, \cdots, \alpha_r)$ among themselves. Then $\tau\rho = \rho$ and if $\nu$ is the automorphism of $\mathfrak{U}$ given by $\nu(H) = H + \rho(H)$ $(H \in \Gamma)$, $\nu$ commutes with $\tau$ (if the latter is regarded as an automorphism of $\mathfrak{U}$). Now let $z \in \mathfrak{Z}$. Then $z - \nu(\gamma(z)) \in \sum_{i=1}^r \mathfrak{L}X_{\alpha_i}$ from the definition of $\gamma$. Hence

$$z^\tau - (\nu(\gamma(z)))^\tau = z^\tau - \nu((\gamma(z))^\tau) \in \sum_{i=1}^r \mathfrak{L}\tau X_{\alpha_i} = \sum_{i=1}^r \mathfrak{L}X_{\alpha_i}.$$

Since $z^\tau \in \mathfrak{Z}$, it follows from Lemma 19 that $\nu(\gamma(z^\tau)) = \nu((\gamma(z))^\tau)$ and therefore $\gamma(z^\tau) = (\gamma(z))^\tau$. Now suppose $\tau$ is any automorphism of $\mathfrak{l}$ leaving $\Gamma$ invariant. Then $(\tau\alpha_1, \cdots, \tau\alpha_r)$ is the set of positive roots in a suitable lexicographic order and hence there exists[7] an element $s \in \mathfrak{w}$ such that $(s\alpha_1, \cdots, s\alpha_r)$ and $(\tau\alpha_1, \cdots, \tau\alpha_r)$ are the same apart from order. As is well known $s$ can be extended to an "inner" automorphism[8] of $\mathfrak{l}$. Then $z^s = z$ $(z \in \mathfrak{Z})$ and $\tau' = \tau s^{-1}$ permutes $(\alpha_1, \cdots, \alpha_r)$ among themselves. But then $z^\tau = z^{\tau'}$ and $(\gamma(z))^\tau = (\gamma(z))^{\tau'}$ since $(\gamma(z))^s = \gamma(z)$ $(z \in \mathfrak{Z})$. Hence in view of the above proof,

$$\gamma(z^\tau) = \gamma(z^{\tau'}) = (\gamma(z))^{\tau'} = (\gamma(z))^\tau.$$

Now put $\nu_\tau = \tau\nu\tau^{-1}$ ($\tau$ being regarded as an automorphism of $\mathfrak{U}$). Then $\nu_\tau(H) = H + \tau\rho(H)$ $(H \in \Gamma)$ and $\nu_\tau(\gamma(z^\tau)) = (\nu(\gamma(z)))^\tau$. This shows that

$$z^\tau - \nu_\tau(\gamma(z^\tau)) = (z - \nu(\gamma(z)))^\tau \in \sum_{i=1}^r \mathfrak{L}X_{\tau\alpha_i}.$$

Replacing $z$ by $z^{\tau^{-1}}$, we get

---

(7) This follows from the fact that one fundamental system of roots can be transformed into every other (apart from order) by an element of $\mathfrak{w}$ (see Weyl [14(b)] and [6(f), Corollary 2 to Lemma 4]).

(8) This means that there exists a transformation of the adjoint group of $\mathfrak{l}$ which coincides with $s$ on $\Gamma$ (see Weyl [14(b)]).

$$z - \nu_\tau(\gamma(z)) \in \sum_{i=1}^{r} \mathfrak{L} X_{\tau\alpha_i} \qquad (z \in \mathfrak{Z}).$$

Now $(\alpha_1', \cdots, \alpha_r')$ being the set of all positive roots in any other order, put $\rho' = (\alpha_1' + \cdots + \alpha_r')/2$ and let $\nu'$ be the automorphism of $\mathfrak{U}$ given by $\nu'(H) = H + \rho'(H)$ $(H \in \Gamma)$. Then in order to prove the first statement of the lemma, we have to show that

$$z - \nu'(\gamma(z)) \in \sum_{i=1}^{r} \mathfrak{L} X_{\alpha_i'}.$$

But, as has already been mentioned above, we can choose[7] an element $s \in \mathfrak{w}$ such that the two sets $(\alpha_1', \cdots, \alpha_r')$ and $(s\alpha_1, \cdots, s\alpha_r)$ are the same. Then $\rho' = s\rho$ and therefore $\nu' = \nu_s$ and

$$z - \nu'(\gamma(z)) = z - \nu_s(\gamma(z)) \in \sum_{i=1}^{r} \mathfrak{L} X_{s\alpha_i} = \sum_{i=1}^{r} \mathfrak{L} X_{\alpha_i'}.$$

Now we come to the last part. One proves without difficulty by using the notion of rank (see [6(b), p. 79]) that actually

$$z - \nu(\gamma(z)) \in \sum_{1 \leq i, j \leq r} X_{-\alpha_i} \mathfrak{L} X_{\alpha_j} \qquad (z \in \mathfrak{Z}).$$

Hence

$$z^* - (\nu(\gamma(z)))^* \in \sum_{1 \leq i, j \leq r} X_{\alpha_j} \mathfrak{L} X_{-\alpha_i}.$$

Define the automorphism $\nu^*$ of $\mathfrak{U}$ by $\nu^*(u) = (\nu(u^*))^*$ $(u \in \mathfrak{U})$. Then $\nu^*(H) = H - \rho(H)$ $(H \in \Gamma)$ and if we replace $z$ by $z^*$ in the above equation, we have

$$z - \nu^*((\gamma(z^*))^*) \in \sum_{1 \leq i, j \leq r} X_{\alpha_j} \mathfrak{L} X_{-\alpha_i}.$$

On the other hand it is obvious that under a suitable order $(-\alpha_1, \cdots, -\alpha_r)$ is the set of all positive roots. Since $-(\alpha_1 + \cdots + \alpha_r)/2 = -\rho$, it follows from what has been proved above that

$$z - \nu^*(\gamma(z)) \in \sum_{i=1}^{r} \mathfrak{L} X_{-\alpha_i}.$$

This shows that $\gamma(z) = (\gamma(z^*))^*$ and therefore $\gamma(z^*) = (\gamma(z))^*$ $(z \in \mathfrak{Z})$.

COROLLARY. *If $\tau$ is any anti-automorphism of $\mathfrak{L}$ such that $\mathfrak{l}^\tau = \mathfrak{l}$ and $\Gamma^\tau = \Gamma$, then $\gamma(z^\tau) = (\gamma(z))^\tau$ $(z \in \mathfrak{Z})$.*

The mapping $b \rightarrow (b^\tau)^*$ is obviously an automorphism of $\mathfrak{L}$ which leaves $\mathfrak{l}$ and $\Gamma$ invariant. Hence the corollary follows from the above lemma.

We shall now apply Lemma 19 to our problem. Let $\mathfrak{Z}$ and $\mathfrak{Z}_M$ denote the centers of $\mathfrak{B}$ and $\mathfrak{M}$ respectively and $\mathfrak{U}$ the subalgebra of $\mathfrak{B}$ generated by

$(1, \mathfrak{h})$. Let $W$ be the Weyl group of $\mathfrak{g}$ (with respect to $\mathfrak{h}$) and $W_-$ the subgroup generated by the Weyl reflexions $s_\alpha$ corresponding to $\alpha \in P_-$. Also denote by $J$ and $J_-$ respectively the subalgebras of $\mathfrak{U}$ consisting of all invariants of $W$ and $W_-$. It is obvious that $\mathfrak{m}$ is reductive and $\mathfrak{h}$ is a Cartan subalgebra of $\mathfrak{m}$. Then in accordance with Lemma 19 we have the isomorphisms $z \to \gamma(z)$ $(z \in \mathfrak{Z})$ and $w \to \gamma_-(w)$ $(w \in \mathfrak{Z}_M)$ of $\mathfrak{Z}$ and $\mathfrak{Z}_M$ onto $J$ and $J_-$ respectively. Since $J \subset J_-$, there exists an isomorphism $z \to \mu(z)$ of $\mathfrak{Z}$ into $\mathfrak{Z}_M$ such that $\gamma(z) = \gamma_-(\mu(z))$. The significance of $\mu$ is expressed by the theorem below. First let us recall that $d_+(m) = \det (\mathrm{Ad} (m^{-1}) - I)_\mathfrak{q} \neq 0$ if $m \in {}^\backprime M$ and therefore $|d_+|^{1/2}$ and $|d_+|^{-1/2}$ are analytic functions on ${}^\backprime M$. Moreover $\mu(z)$, being an element of $\mathfrak{M}$, is a differential operator on $M$.

THEOREM 1. *If $z \in \mathfrak{Z}$ and $\tau$ is an invariant distribution on ${}^\backprime M$, then*

$$\delta(z)\tau = |d_+|^{-1/2}\mu(z)(|d_+|^{1/2}\tau).$$

Let $D = |d_+|^{-1/2}\mu(z) \circ |d_+|^{1/2}$ where $\circ$ denotes the operator product. It is clear from Lemma 7 that $|d_+|^{-1/2}$, $\mu(z)$, $|d_+|^{1/2}$ are all invariant differential operators on ${}^\backprime M$. Therefore $D$ is also invariant. Also $\delta(z)$ is invariant from the corollary to Lemma 17. Put $\Theta = \delta(z) - D$. We have to show that $\tau(\Theta^* f) = 0$ for any $f \in C_c^\infty({}^\backprime M)$. Let $M^*$ be the analytic subgroup of the adjoint group of $\mathfrak{g}_0$ corresponding to $\mathfrak{m}_{\mathfrak{t}_0}$. Then $M^*$ is compact. Put $g^{m^*} = g^m$ $(g \in C_c^\infty({}^\backprime M)$, $m^* \in M^*)$ where $m$ is any element in $\overline{M}$ (see Lemma 7 for notation) whose image in $M^*$ is $m^*$. Moreover normalize the Haar measure $dm^*$ on $M^*$ in such a way that $\int_{M^*} dm^* = 1$. Then since $\tau$ and $\Theta^*$ are both invariant, it is clear that

$$\tau(\Theta^* f) = \tau((\Theta^* f)^{m^*}) = \tau(\Theta^*(f^{m^*})) = \tau(\Theta^* \bar{f})$$

where

$$\bar{f}(m) = \int_{M^*} f^{m^*}(m)\, dm^* \qquad\qquad (m \in {}^\backprime M).$$

Since $\bar{f} \in I_c^\infty({}^\backprime M)$, it would be enough to prove that $\Theta^* g = 0$ for every $g \in I_c^\infty({}^\backprime M)$. On the other hand if $dm$ is the Haar measure of $M$ and $f \in C_c^\infty({}^\backprime M)$, $g \in I_c^\infty({}^\backprime M)$, it is evident that

$$\int f\Theta^* g\, dm = \int \bar{f}\Theta^* g\, dm = \int (\Theta \bar{f}) g\, dm$$

where $\bar{f}$ is defined as above. Hence it is sufficient to prove the following lemma.

LEMMA 21. *If $f \in I_c^\infty({}^\backprime M)$, $\Theta f = 0$.*

$G'$ being the set of regular elements in $G$, put $A' = A \cap G'$. Since $\Theta f$

$\in I_c^\circ(`M)$ it would obviously be enough (see Lemmas 7 and 8) to show that $f(h, \Theta) = 0$ for $h \in A'$. For this we need some additional results which will be obtained in the next two sections.

**7. Differential operators on** $A'$. Put $\mathfrak{s} = \sum_{\alpha \in P}(C X_\alpha + C X_{-\alpha})$ and $\mathfrak{s}_- = \sum_{\alpha \in P_-}(C X_\alpha + C X_{-\alpha})$. Then $\mathfrak{g} = \mathfrak{s} + \mathfrak{h}$, $\mathfrak{m} = \mathfrak{s}_- + \mathfrak{h}$ and $\mathfrak{s} = \mathfrak{q} + \mathfrak{s}_-$ where all sums are direct. Let $S(\mathfrak{s})$ and $S(\mathfrak{s}_-)$ denote the symmetric algebras over $\mathfrak{s}$ and $\mathfrak{s}_-$ respectively and $\mathfrak{S}$ and $\mathfrak{S}_-$ their images in $\mathfrak{B}$ under the canonical mapping of $S(\mathfrak{g})$ onto $\mathfrak{B}$. Let $\mathfrak{S}_r$ be the space of homogeneous elements of degree $r$ in $\mathfrak{S}$. Put $\mathfrak{S}' = \sum_{r \geq 1} \mathfrak{S}_r$ and $\mathfrak{S}'_- = \mathfrak{S}_- \cap \mathfrak{S}'$. $\mathfrak{U}$ being the subalgebra of $\mathfrak{B}$ generated by $(1, \mathfrak{h})$, it is obvious that $\mathfrak{B} = \mathfrak{S}\mathfrak{U}$ and $\mathfrak{M} = \mathfrak{S}_-\mathfrak{U}$ (see [6(c), Lemma 12]). Let $\Gamma_m$ ($m \in M$) denote the mapping of Lemma 15. Notice that $A' \subset `M$ and $\mathfrak{U} \subset \mathfrak{M}$.

LEMMA 22. *If* $h \in A'$, $\Gamma_h$ *defines a one-one mapping* $\mathfrak{S} \times \mathfrak{U}$ *onto* $\mathfrak{B}$ *and also a one-one mapping of* $\mathfrak{S}_- \times \mathfrak{U}$ *onto* $\mathfrak{M}$.

The proof of these statements is exactly parallel to that of Lemma 9.
We note that $\mathfrak{B}\mathfrak{g} = \mathfrak{S}'\mathfrak{U} + \mathfrak{U}\mathfrak{h}$ and in view of the corollary to Lemma 15,

$$\Gamma_h(\mathfrak{U}\mathfrak{h} \times \mathfrak{U}) = \Gamma_h(\mathfrak{U} \times \Gamma_h(\mathfrak{h} \times \mathfrak{U})) = \{0\} \qquad (h \in A)$$

since $\Gamma_h(\mathfrak{h} \times \mathfrak{U}) = \{0\}$. Hence applying the same corollary once again, we get

$$\Gamma_h(\mathfrak{B}\mathfrak{g} \times \mathfrak{U}) = \Gamma_h(\mathfrak{S}'\mathfrak{U} \times \mathfrak{U}) \subset \Gamma_h(\mathfrak{S}' \times \mathfrak{U}).$$

This proves that

$$\Gamma_h(\mathfrak{S}' \times \mathfrak{U}) = \Gamma_h(\mathfrak{B}' \times \mathfrak{U}) \qquad (h \in A)$$

where $\mathfrak{B}' = \mathfrak{B}\mathfrak{g}$. Similarly $\Gamma_h(\mathfrak{S}'_- \times \mathfrak{U}) = \Gamma_h(\mathfrak{M}' \times \mathfrak{U})$ where $\mathfrak{M}' = \mathfrak{M}\mathfrak{m}$.

For each root $\alpha$ let $\xi_\alpha$ be the character of $A$ given by Ad $(h)X_\alpha = \xi_\alpha(h)X_\alpha$ ($h \in A$) and consider the multiplicative group $_0\Xi$ generated by all $\xi_\alpha$. Also let $_0\Xi_-$ be the subgroup of $_0\Xi$ generated by $\xi_\alpha$ ($\alpha \in P_-$). We denote by $\Xi$ the space of finite linear combinations of characters in $_0\Xi$ with coefficients in $C$. Obviously $\Xi$ is a ring under ordinary multiplication and the subspace $\Xi_-$ spanned by $_0\Xi_-$ is a subring of $\Xi$. Put

$$d(h) = \det (\text{Ad } (h^{-1}) - I)_\mathfrak{s}, \qquad d_-(h) = \det (\text{Ad } (h^{-1}) - I)_{\mathfrak{s}_-}$$

where the subscripts denote restrictions to the corresponding subspace.

LEMMA 23. *For each* $b \in \mathfrak{B}$ *we can choose elements* $v_1, \cdots, v_N \in \mathfrak{B} \times \mathfrak{U}$, $\xi_1, \cdots, \xi_N \in \Xi$ *and an integer* $r \geq 0$ *such that*

$$\Gamma_h\left(\sum_{i=1}^N \xi_i(h)v_i\right) = (d(h))^r b$$

*for all* $h \in A$.

Define $\mathfrak{B}_s$ ($s \geq -1$) as in the proof of Lemma 15 and suppose $b \in \mathfrak{B}_s$ ($s \geq 0$). We shall use induction on $s$. Let $X_1, \cdots, X_n$ be a base of $\mathfrak{g}$ consisting of the elements $X_\alpha$, $X_{-\alpha}$ $(\alpha \in P)$ and a base of $\mathfrak{h}$. Then it is obviously sufficient to consider the case when $b = Y_1 Y_2 \cdots Y_s$ where each $Y_i$ is equal to some $X_j$. Since the order of the factors is immaterial mod $\mathfrak{B}_{s-1}$, we may assume that $Y_j \in \mathfrak{h}$ for $t < j \leq s$ while $Y_j \notin \mathfrak{h}$ for $1 \leq j \leq t$. Also we may suppose that $t \geq 1$ for otherwise if $t = 0$, $b \in \mathfrak{U}$ and $\Gamma_h(1 \times b) = b$ $(h \in A)$. Put $Y_{t+1} \cdots Y_s = u \in \mathfrak{U}$ $(u = 1$ if $t = s)$ and let $\eta_i$ be the character of $A$ given by $\mathrm{Ad}\ (h) Y_i = \eta_i(h) Y_i$ $(1 \leq i \leq t)$. Then $\eta_i = \xi_\alpha$ for some root $\alpha$ and therefore $\eta_i(h) \neq 1$ if $h \in A'$. For any $j$ $(1 \leq j \leq t)$ put $y_j = Y_j \cdots Y_t$. Then[9]

$$
\begin{aligned}
\Gamma_h(y_j \times u) &= (L_{\mathrm{Ad}(h^{-1})Y_j} - R_{y_j})\Gamma_h(y_{j+1} \times u) \\
&= (\mathrm{Ad}\ (h^{-1})Y_j - Y_j)\Gamma_h(y_{j+1} \times u) + [Y_j, \Gamma_h(y_{j+1} \times u)] \\
&= (\eta_j(h^{-1}) - 1)Y_j\Gamma_h(y_{j+1} \times u) + [Y_j, \Gamma_h(y_{j+1} \times u)].
\end{aligned}
$$

Therefore it follows easily by induction on $t-j$ that for each $j$ we can choose elements $b_{ij} \in \mathfrak{B}_{s+t-j-1}$, $\zeta_{ij} \in \Xi$ $(1 \leq i \leq N_j)$ such that

$$
\Gamma_h(y_j \times u) = \prod_{j \leq i \leq t} (\eta_i(h^{-1}) - 1)y_j u + \sum_{i=1}^{N_j} \zeta_{ij}(h)b_{ij}.
$$

Putting $j = 1$ and $b_{i1} = b_i$, $\zeta_{i1} = \zeta_i$, $1 \leq i \leq N_1$,

$$
\Gamma_h(Y_1 \cdots Y_t \times u) = \prod_{i=1}^{t} (\eta_i(h^{-1}) - 1)b + \sum_{i=1}^{N_1} \zeta_i(h)b_i.
$$

Also it is obviously possible to find an integer $k \geq 0$ and $\zeta_0 \in \Xi$ such that

$$
\zeta_0(h) \prod_{1 \leq i \leq t} (\eta_i(h^{-1}) - 1) = (d(h))^k \qquad (h \in A).
$$

Hence if we put $v_0 = (Y_1 Y_2 \cdots Y_t) \times u$,

$$
\Gamma_h(\zeta_0(h)v_0) = (d(h))^k b + \sum_{j=1}^{N_1} \zeta_0(h)\zeta_j(h)b_j.
$$

Since $\Xi$ is a ring and $b_j \in \mathfrak{B}_{s-1}$ $(1 \leq j \leq N_1)$ our assertion now follows from the induction hypothesis.

If $b \in \mathfrak{B}$ and $h \in A'$, we conclude from Lemma 22 that there exists a *unique* element $\beta_h(b) \in \mathfrak{U}$ such that

$$
b - \beta_h(b) \in \Gamma_h(\mathfrak{S}' \times \mathfrak{U}) = \Gamma_h(\mathfrak{B}' \times \mathfrak{U}).
$$

Define $v_i$, $\xi_i$ and $r$ as in Lemma 23 and choose $u_i \in \mathfrak{U}$ such that $v_i - (1 \times u_i) \in \mathfrak{B}' \times \mathfrak{U}$. Then since $d(h)$ is never zero if $h \in A'$, it is clear that

$$
\beta_h(b) = (d(h))^{-r} \sum_{i=1}^{N} \xi_i(h)u_i \qquad (h \in A').
$$

Thus we have obtained the following result.

---

[9] If $a_1$, $a_2$ are two elements of an associative algebra, we write $[a_1, a_2] = a_1 a_2 - a_2 a_1$.

Corollary. *If $b \in \mathfrak{B}$, it is possible to choose elements $u_1, \cdots, u_N \in \mathfrak{U}$, $\xi_1, \cdots, \xi_N \in {}_0\Xi$ and an integer $r \geq 0$ such that*

$$\beta_h(b) = (d(h))^{-r} \sum_{i=1}^{N} \xi_i(h) u_i$$

*for all $h \in A'$.*

Now suppose $b \in \mathfrak{M}$ and $h \in A'$. Then from Lemma 22 we can choose $b_0 \in \mathfrak{U}$ such that $b - b_0 \in \Gamma_h(\mathfrak{S}'_- \times \mathfrak{U})$. But

$$\Gamma_h(\mathfrak{S}'_- \times \mathfrak{U}) \subset \Gamma_h(\mathfrak{B}' \times \mathfrak{U})$$

and so, in view of the uniqueness of $\beta_h(b)$, we conclude that $b_0 = \beta_h(b)$. Hence

$$b - \beta_h(b) \in \Gamma_h(\mathfrak{S}'_- \times \mathfrak{U}) = \Gamma_h(\mathfrak{M}' \times \mathfrak{U}).$$

The following lemma and corollary are proved in the same way as Lemma 23.

Lemma 24. *For each $b \in \mathfrak{M}$ we can choose elements $v_1, \cdots, v_N \in \mathfrak{M} \times \mathfrak{U}$, $\xi_1, \cdots, \xi_N \in \Xi_-$ and an integer $r \geq 0$ such that*

$$\Gamma_h \left( \sum_{i=1}^{N} \xi_i(h) v_i \right) = (d_-(h))^r b$$

*for all $h \in A$.*

Corollary. *If $b \in \mathfrak{M}$ it is possible to select elements $u_1, \cdots, u_N \in \mathfrak{U}$, $\xi_1, \cdots, \xi_N \in {}_0\Xi_-$ and an integer $r \geq 0$ such that*

$$\beta_h(b) = (d_-(h))^{-r} \sum_{i=1}^{N} \xi_i(h) u_i$$

*for all $h \in A'$.*

We now regard $A'$ as an open submanifold of the Lie group $A$. It follows from the above results that for each $b \in \mathfrak{B}$ there exists an analytic differential operator $\beta(b)$ on $A'$ whose local expression at $h \in A'$ is $\beta_h(b)$.

Lemma 25. *Suppose $b \in \mathfrak{M}$ and $f \in I^\infty('M)$. Then if $h \in A'$,*

$$f(h, b) = f(h, \beta_h(b)).$$

Since $b - \beta_h(b) \in \Gamma_h(\mathfrak{M}' \times \mathfrak{U})$, it would be enough to prove that $f(h, a) = 0$ if $a \in \Gamma_h(\mathfrak{M}' \times \mathfrak{U})$. It is clearly sufficient to consider the case when

$$a = \Gamma_h(X_1 \cdots X_r \times H_1 H_2 \cdots H_s)$$

where $X_1, \cdots, X_r \in \mathfrak{m} \cap \mathfrak{g}_0$ $(r \geq 1)$ and $H_1, \cdots, H_s \in \mathfrak{h}_0$ $(s \geq 0)$. Put $m(u) = \exp(u_1 X_1) \cdots \exp(u_r X_r)$, $h(t) = \exp(t_1 H_1 + \cdots + t_s H_s)$ $(u_i, t_j \in R)$. Then as we have seen in §5,

$$f(h, a) = \left\{ (\partial^{r+s}/\partial u_1 \cdots \partial u_r \partial t_1 \cdots \partial t_s) f(m(u) h h(t)(m(u))^{-1}) \right\}_{u_i = t_j = 0}.$$

But since $f$ is an invariant function and $r \geq 1$, the right-hand side is zero. This proves the lemma.

COROLLARY. *Let $g$ be the restriction on $A'$ of a function $f \in I^\infty('M)$. Then if $b \in \mathfrak{B}$, $\delta(b)f$ coincides with $\beta(b)g$ on $A'$.*

Let $h$ be a point in $A'$. Then $g(h, \beta(b)) = g(h, \beta_h(b)) = f(h, \beta_h(b))$ and $f(h, \delta(b)) = f(h, \delta_h(b))$. Therefore, in view of the above lemma, it would be enough to show that $\beta_h(\delta_h(b)) = \beta_h(b)$. But

$$b - \delta_h(b) \in \Gamma_h(\mathfrak{Q}' \times \mathfrak{M})$$

and $\mathfrak{M} = \Gamma_h(\mathfrak{S}_- \times \mathfrak{U})$ from Lemma 22. Therefore

$$\Gamma_h(\mathfrak{Q}' \times \mathfrak{M}) = \Gamma_h(\mathfrak{Q}' \times \Gamma_h(\mathfrak{S}_- \times \mathfrak{U}))$$
$$= \Gamma_h(\mathfrak{Q}'\mathfrak{S}_- \times \mathfrak{U}) \subset \cdot \Gamma_h(\mathfrak{B}' \times \mathfrak{U})$$

from the corollary to Lemma 15. On the other hand

$$\delta_h(b) - \beta_h(\delta_h(b)) \in \Gamma_h(\mathfrak{B}' \times \mathfrak{U})$$

and therefore

$$b - \beta_h(\delta_h(b)) \in \Gamma_h(\mathfrak{B}' \times \mathfrak{U}).$$

This proves that $\beta_h(b) = \beta_h(\delta_h(b))$.

8. **Determination of $\beta(z)$ in terms of the invariants.** As before (§6) let $W$ and $W_-$ respectively be the Weyl groups of $\mathfrak{g}$ and $\mathfrak{m}$ (relative to $\mathfrak{h}$) and let $J$ and $J_-$ be the corresponding subalgebras of invariants of $\mathfrak{U}$. In §6 we have defined the isomorphisms $\gamma$ and $\gamma_-$ of $\mathfrak{Z}$ and $\mathfrak{Z}_M$ onto $J$ and $J_-$ respectively. (We recall that $\mathfrak{Z}_M$ is the center of $\mathfrak{M}$.) The operators $\beta(z)$ and $\beta(w)$ can be expressed in terms of $\gamma(z)$ and $\gamma_-(w)$ ($z \in \mathfrak{Z}$, $w \in \mathfrak{Z}_M$) as follows.

THEOREM 2. *If $z \in \mathfrak{Z}$ and $w \in \mathfrak{Z}_M$ then*

$$\beta(z) = |d|^{-1/2} \gamma(z) \circ |d|^{1/2}, \qquad \beta(w) = |d_-|^{-1/2} \gamma_-(w) \circ |d_-|^{1/2}$$

*as differential operators on $A'$. (Here $\circ$ denotes operator product.)*

Assuming this for a moment, we shall now prove Lemma 21 and thus complete the proof of Theorem 1. Let $z \in \mathfrak{Z}$ and $f \in I_c^\infty('M)$. Then as we have seen in §6, it is enough to prove that $\Theta f$ is zero on $A'$. Let $g$ denote the restriction of $f$ on $A'$. Then by the corollary to Lemma 25 and Theorem 2, $\delta(z)f$ coincides on $A'$ with $\beta(z)g = |d|^{-1/2}\gamma(z)(|d|^{1/2}g)$. On the other hand $\gamma_-(\mu(z)) = \gamma(z)$. Therefore again from Theorem 2 and the corollary to Lemma 25, $|d_+|^{-1/2}\mu(z)(|d_+|^{1/2}f)$ coincides on $A'$ with $|d|^{-1/2}\gamma(z)(|d|^{1/2}g)$ since $d = d_+ d_-$. This proves that $\Theta f$ vanishes identically on $A'$.

We now come to the proof of Theorem 2. Put

$$\Phi = \beta(z) - |d|^{-1/2}\gamma(z) \circ |d|^{1/2},$$
$$\Psi = \beta(w) - |d_-|^{-1/2}\gamma_-(w) \circ |d_-|^{1/2}.$$

Then we have to show that $\Phi = \Psi = 0$. Put $G^* = \mathrm{Ad}\ (G) \simeq G/Z$ and $x^* = \mathrm{Ad}\ (x)$ $(x \in G)$. It is obvious from its definition (see Lemma 15) that $\Gamma_m$ depends only on $m^*$ $(m \in M)$ and therefore $\beta_h(b)$ $(b \in \mathfrak{B})$ only on $h^*$ $(h \in A')$. Since the same holds for $d(h)$ and $d_-(h)$, we can regard $\Phi$ and $\Psi$ as differential operators on $A'^*$ and it would clearly be enough to prove that $\Phi = \Psi = 0$ on $A'^*$. Let $G_0$ be any connected Lie group whose Lie algebra is $\mathfrak{g}_0$. Then the adjoint groups of $G$ and $G_0$ are the same and hence the above argument shows that it is permissible, for the proof of our theorem, to replace $G$ by $G_0$. Therefore if $G_c$ is the simply connected complex analytic group with the Lie algebra $\mathfrak{g}$, we may assume that $G$ is the real analytic subgroup of $G_c$ corresponding to $\mathfrak{g}_0$. Let $A_c$ and $M_c$ be the complex analytic subgroups which correspond to $\mathfrak{h}$ and $\mathfrak{m}$ respectively. Then we know from Lemmas 3 and 4 that $A \subset A_c$ and therefore (see Lemma 7) $M \subset M_c$. Now $\mathfrak{h}$, regarded as an additive group, is locally isomorphic to $A_c$ under the exponential mapping and so it is a complex Lie group whose Lie algebra is also $\mathfrak{h}$. Therefore elements of $\mathfrak{U}$ may now be considered as (holomorphic) differential operators on $\mathfrak{h}$. Let $\mathfrak{h}_1$ be the complete inverse image of $A$ (under the exponential mapping) in $\mathfrak{h}$. Then $\mathfrak{h}_1$ is closed and therefore it is a (real) Lie subgroup and its Lie algebra is $\mathfrak{h}_0$. Put

$$\Delta(H) = \prod_{\alpha \in P} (e^{\alpha(H)/2} - e^{-\alpha(H)/2}),$$
$$\Delta_-(H) = \prod_{\alpha \in P_-} (e^{\alpha(H)/2} - e^{-\alpha(H)/2}) \qquad\qquad (H \in \mathfrak{h}).$$

Then it is obvious that

$$d(\exp H) = (-1)^r \Delta(H)^2,$$
$$d_-(\exp H) = (-1)^s \Delta_-(H)^2 \qquad\qquad (H \in \mathfrak{h}_1)$$

where $r$ and $s$ are the numbers of roots in $P$ and $P_-$ respectively. For any function $\lambda$ on $\mathfrak{h}$ and $s \in W$, let $\lambda^s$ denote the function $H \to \lambda(s^{-1}H)$ $(H \in \mathfrak{h})$. (If $\lambda$ is a linear function we write $s\lambda$ instead of $\lambda^s$.) Then for any root $\alpha$, $s\alpha$ is also a root and therefore $\Delta^s = \epsilon(s)\Delta$ where $\epsilon(s) = \pm 1$. Similarly if $s \in W_-$, it is clear that $\Delta_-^s = \epsilon_-(s)\Delta_-$ where again $\epsilon_-(s) = \pm 1$. But if $s_\alpha$ is the Weyl reflexion corresponding to a root $\alpha \in P_-$, it is known (see Weyl [14(b)]) that $\epsilon(s_\alpha) = \epsilon_-(s_\alpha) = -1$. Therefore since $\epsilon$ and $\epsilon_-$ are both homomorphisms, it follows that $\epsilon(s) = \epsilon_-(s)$ for all $s \in W_-$. This shows in particular that if

$$\Delta_+(H) = \prod_{\alpha \in P_0 \cup P_+} (e^{\alpha(H)/2} - e^{-\alpha(H)/2}) \qquad\qquad (H \in \mathfrak{h}),$$

then $\Delta_+^s = \Delta_+$ for all $s \in W_-$. (We shall need this result a little later in §10.)

Consider the open submanifold $\mathfrak{h}'$ of $\mathfrak{h}$ consisting of those points $H$ where $\Delta(H)\neq 0$. Then $\mathfrak{h}_1' = \mathfrak{h}'\cap\mathfrak{h}_1$ is the complete inverse image of $A'$. We regard $\mathfrak{h}_1'$ as an open submanifold of the real Lie group $\mathfrak{h}_1$. It is obvious that corresponding to any differential operator $D$ on $A'$ there exists exactly one differential operator $\overline{D}$ on $\mathfrak{h}_1'$ with the following property. If $f\in C^{\infty}(A')$ and $\bar{f}(H)=f(\exp\ H)$ $(H\in\mathfrak{h}_1')$, then $\bar{f}(H,\ \overline{D})=f(\exp\ H,\ D)$. Moreover $\overline{D}=0$ if and only if $D=0$. Hence it would be sufficient to prove that $\overline{\Phi}=\overline{\Psi}=0$. It is clear that

$$\overline{\Phi} = \overline{\beta(z)} - \Delta^{-1}\gamma(z)\circ\Delta,$$

$$\overline{\Psi} = \overline{\beta(w)} - \Delta_-^{-1}\gamma_-(w)\circ\Delta_-.$$

Let $F$ and $F_-$ be the additive groups generated by the roots $\alpha$ in $P$ and $P_-$ respectively. Then it follows from the corollaries to Lemmas 23 and 24 that we can choose elements $\sigma_1,\cdots,\sigma_p\in F$, $\tau_1,\cdots,\tau_q\in F_-$, $u_1,\cdots,u_p$, $v_1,\cdots,v_q\in\mathfrak{U}$ and two integers $r,\ k\geq 0$ such that

$$\overline{\Phi} = \Delta^{-r}\sum_{i=1}^{p}e^{\sigma_i}u_i$$

$$\overline{\Psi} = \Delta_-^{-k}\sum_{j=1}^{q}e^{\tau_j}v_j.$$

Now suppose $\Phi\neq 0$. We may assume that $\sigma_1,\cdots,\sigma_p$ are all distinct and $u_i\neq 0\ 1\leq i\leq p$. Let $\mathfrak{F}_R$ be the space consisting of all real linear combinations of the roots $\alpha$. Then a lexicographic order is defined in $\mathfrak{F}_R$ (see §2) and hence $F$ and $F_-$ are ordered groups. Let $\sigma_1$ be the highest element among $(\sigma_1,\cdots,\sigma_p)$. Put $\rho=2^{-1}\sum_{\alpha\in P}\alpha$ and suppose $\nu$ is an irreducible representation of $\mathfrak{g}$ on a finite-dimensional space $V$ with the highest weight $\Lambda$. We denote the corresponding representations of $\mathfrak{B}$ and $G_c$ also by $\nu$. Let $\chi$ be the character of $\nu$ on $G_c$. Then if $\Lambda'=\Lambda+\rho$ one knows (see Weyl [14(a)]) that

$$\chi(\exp\ H)\Delta(H) = \sum_{s\in W}\epsilon(s)e^{s\Lambda'(H)} \qquad (H\in\mathfrak{h}).$$

On the other hand $\chi(xyx^{-1})=\chi(y)$ $(x,\ y\in G_c)$ and so as we have seen during the proof of Lemma 25, $\chi(h,\ a)=0$ if $h\in A$ and $a\in\Gamma_h(\mathfrak{B}'\times\mathfrak{U})$. This implies that $\chi(h,\ b)=\chi(h,\ \beta_h(b))(b\in\mathfrak{B},\ h\in A')$ and therefore if $\bar{\chi}(H)=\chi(\exp\ H)$ $(H\in\mathfrak{h})$ and $\beta_H(b)$ denotes the local expression of $\overline{\beta(b)}$ at $H\in\mathfrak{h}_1'$, it follows that

$$\bar{\chi}(H,\ \beta_H(z)) = \chi(\exp\ H,\ z).$$

Now $\sigma$ being a linear function on $\mathfrak{h}$, we may extend it (uniquely) to a homomorphism of $\mathfrak{U}$ into $C$ such that $\sigma(1)=1$. Then it is known (see [6(b), p. 73]) that $\nu(z)\phi=\Lambda'(\gamma(z))\phi$ for any $\phi\in V$ and therefore

$$z\chi = \Lambda'(\gamma(z))\chi.$$

In view of our result above, this implies that

$$\tilde{\chi}(H, \beta_H(z)) = \Lambda'(\gamma(z))\tilde{\chi}(H) \qquad (H \in \mathfrak{h}_1').$$

Put $\chi' = \tilde{\chi}\Delta$. Then since $\gamma(z) \in J$, it follows from the formula for $\chi'$ given above that

$$\chi'(H, \gamma(z)) = \Lambda'(\gamma(z))\chi'(H) \qquad (H \in \mathfrak{h}).$$

Therefore

$$\tilde{\chi}(H, \overline{\Phi}) = \tilde{\chi}(H, \beta_H(z)) - \Delta(H)^{-1}\chi'(H, \gamma(z)) = 0$$

for $H \in \mathfrak{h}_1'$. On the other hand let $\Lambda = \Lambda_1, \cdots, \Lambda_N$ be all the distinct weights of $\nu$ and let $n_i$ be the multiplicity of $\Lambda_i$ $(1 \leq i \leq N)$. Then $\Lambda_i \in \mathfrak{F}_R$ and $\Lambda_i < \Lambda$ $2 \leq i \leq N$ and

$$\tilde{\chi}(H) = \chi(\exp H) = \sum_{i=1}^{N} n_i e^{\Lambda_i(H)} \qquad (H \in \mathfrak{h}).$$

Therefore it follows from our formula for $\overline{\Phi}$ that

$$0 = \tilde{\chi}(H, \overline{\Phi}) = \Delta(H)^{-r} \sum_{i=1}^{p} \sum_{j=1}^{N} n_j \Lambda_j(u_i) \exp\left(\sigma_i(H) + \Lambda_j(H)\right)$$

for all $H \in \mathfrak{h}_1'$. But since $\mathfrak{h}_1'$ contains the nonempty open subset $\mathfrak{h}_0 \cap \mathfrak{h}'$ of $\mathfrak{h}_0$, it is obvious that a meromorphic function on $\mathfrak{h}$ cannot vanish everywhere on $\mathfrak{h}_1'$ unless it is identically zero. Therefore

$$\sum_{i=1}^{p} \sum_{j=1}^{N} n_j \Lambda_j(u_i) e^{\sigma_i + \Lambda_i} = 0.$$

But $\sigma_1 + \Lambda = \sigma_1 + \Lambda_1 > \sigma_i + \Lambda_j$ unless $i = j = 1$ and the exponentials of distinct linear functions are linearly independent (see for example Lemma 41 of [6(b)]). So it follows that $\Lambda(u_1) = 0$. On the other hand if $l = \dim_c \mathfrak{h}$, there exist $l$ linearly independent elements $\lambda_1, \cdots, \lambda_l \in \mathfrak{F}_R$ such that for any set $(m_1, \cdots, m_l)$ of non-negative integers $\Lambda_0 = m_1\lambda_1 + \cdots + m_l\lambda_l$ is the highest weight of some irreducible representation of $\mathfrak{g}$ (see [6(b), Theorem 1]). Therefore $\Lambda_0(u_1) = 0$ for all such $\Lambda_0$ and from this we can conclude (see Lemma 22 of [6(b)]) that $u_1 = 0$. However this contradicts our hypothesis and so $\overline{\Phi} = 0$.

Now we come to $\overline{\Psi}$. Again assuming that $\overline{\Psi} \neq 0$, we may suppose that $\tau_1 > \tau_2 > \cdots > \tau_q$ in $F_-$ and $v_j \neq 0$ $1 \leq j \leq q$. $\nu$, $\Lambda$ and $V$ being as above, let $\phi \neq 0$ be a vector in $V$ belonging to the weight $\Lambda$. Then if $U = \nu(\mathfrak{M})\phi$, we get a representation $\nu_1$ of $M_c$ on $U$ which is irreducible (see Lemma 2 of [6(f)]). Obviously $\Lambda$ is the highest weight of $\nu_1$. Hence if $\chi_1$ is the character of $\nu_1$ and $\Lambda'' = \Lambda + \rho_-$ (where $\rho_- = 2^{-1}\sum_{\alpha \in P_-} \alpha$), we have the formula (see Weyl [14(a)])

$$\chi_1(\exp H)\Delta_-(H) = \sum_{s \in W} \epsilon(s) e^{\Lambda''(sH)} \qquad (H \in \mathfrak{h}).$$

We can now apply the above argument to $\chi_1$ (instead of $\chi$) and conclude that $\Lambda(v_1) = 0$. Again this implies that $v_1 = 0$ and so we get a contradiction. Hence Theorem 2 is now proved.

9. **Eigen-distributions on** $`M$. Let $\tau$ be an invariant distribution on $`M$. Since $\delta(z)$ $(z \in \mathfrak{Z})$ is an invariant differential operator on $`M$ (see the corollary to Lemma 17), $\delta(z)\tau$ is also invariant. Therefore it follows from Theorem 1 that $\delta(z_1 z_2)\tau = \delta(z_1)\delta(z_2)\tau$ $(z_1, z_2 \in \mathfrak{Z})$. We shall say that $\tau$ is an eigen-distribution of $\mathfrak{Z}$ if it is an eigen-distribution of $\delta(z)$ for every $z \in \mathfrak{Z}$. In case this is so, it is clear that there exists a homomorphism $\chi$ of $\mathfrak{Z}$ into $C$ such that $\delta(z)\tau = \chi(z)\tau$ $(z \in \mathfrak{Z})$. The following theorem is the main result of the above theory.

THEOREM 3. *Let $\tau$ be an invariant distribution on $`M$. Then if it is an eigen-distribution of $\mathfrak{Z}$, it coincides with an analytic function on $`M$.*

Define $J$ and $J_-$ as in §8. Then $J_- \supset J$ and since $W$ is a finite group, it follows from the theory of invariants (see Weyl [14(c), p. 275]) that $J$ has a finite number of generators (over $C$) and $\mathfrak{U}$ is a finite module over $J$ (see [6(c), p. 200]). Therefore $J$ is a Noetherian ring and hence by a well-known theorem in algebra (see van der Waerden [13, vol. II, §99]) $J_-$ is also a finite module over $J$. Hence any element $w' \in J_-$ satisfies an equation of the form

$$w'^g + \gamma(z_1) w'^{g-1} + \cdots + \gamma(z_g) = 0$$

where $z_1, \cdots, z_g \in \mathfrak{Z}$ and $g$ is a positive integer. Now the notation being as in §8, let $w \in \mathfrak{Z}_M$ and put $w' = \gamma_-(w)$. Then by applying the isomorphism $\gamma_-^{-1}$ to the above equation, we get

$$w^g + \mu(z_1) w^{g-1} + \cdots + \mu(z_g) = 0.$$

Now put $\tau' = |d_+|^{1/2}\tau$. If $\chi$ is the homomorphism of $\mathfrak{Z}$ into $C$ corresponding to $\tau$, it follows from Theorem 1 that $\mu(z)\tau' = \chi(z)\tau'$ $(z \in \mathfrak{Z})$ and therefore since $\mathfrak{Z}_M$ is abelian,

$$w^g \tau' + \chi(z_1) w^{g-1}\tau' + \cdots + \chi(z_g)\tau' = 0.$$

Let $H_1, \cdots, H_p$ be a base for $\mathfrak{h}_0 \cap \mathfrak{p}$ over $R$. Since $-B(X, X)$ $(X \in \mathfrak{k}_0)$ is a positive-definite quadratic form on $\mathfrak{k}_0$, we can choose a base $X_1, \cdots, X_r$ for $\mathfrak{m}_{\mathfrak{k}_0}$ over $R$ such that $-B(X_i, X_j) = \delta_{ij}$ $(1 \leq i, j \leq r)$ where $\delta_{ij} = 1$ or $0$ according as $i = j$ or not. Put

$$\square = H_1^2 + \cdots + H_p^2 + X_1^2 + X_2^2 + \cdots + X_r^2.$$

It is well known that $X_1^2 + \cdots + X_r^2$ lies in $\mathfrak{Z}_M$ and therefore the same holds for $\square$. Hence in view of what we have said above, $\tau'$ satisfies an equation of the form

$$\square^g \tau' + c_1 \square^{g-1}\tau' + \cdots + c_g \tau' = 0$$

where $c_1, \cdots, c_g$ are complex numbers. But it is obvious from its definition that the analytic differential operator $\Box$ is of the *elliptic type* (see Gårding [3]) everywhere on $M$. The same therefore holds for

$$\Box^g + c_1\Box^{g-1} + \cdots + c_g$$

and so we can conclude from the work of Schwartz [11, p. 136] and John [8] (see also [3]) on the generalization of the classical Bernstein Theorem to distributions, that $\tau'$, and hence also $\tau$ are analytic functions on $`M$. This proves the theorem.

We shall now try to determine $\tau$. Put $`A = `M \cap A$. Then it is obvious from Lemma 7 that $`M = \bigcup_{m \in M} m`Am^{-1}$ and since $\tau$ is an invariant analytic function, it would be enough to know its restriction on $A' = A \cap G'$ (see Lemma 8). Let $\sigma'$ denote the restriction of $\tau' = |d_+|^{1/2}\tau$ on $A'$. Since $\mu(z)\tau' = \chi(z)\tau'$ $(z \in \mathfrak{Z})$, we conclude from the corollary to Lemma 25 that $\beta(\mu(z))\sigma' = \chi(z)\sigma'$. On the other hand

$$\beta(\mu(z)) = |d_-|^{-1/2}\gamma(z) \circ |d_-|^{1/2}$$

from Theorem 2. Therefore if $\sigma(h) = |d(h)|^{1/2}\tau(h)$ $(h \in A')$,

$$\gamma(z)\sigma = \chi(z)\sigma \qquad\qquad (z \in \mathfrak{Z}).$$

Let us now assume that $\tau \neq 0$ so that $\sigma \neq 0$ and $\chi(1) = 1$. Then from Theorem 5 of [6(b)], there exists a linear function $\Lambda$ on $\mathfrak{h}$ such that $\chi(z) = \Lambda(\gamma(z))$ for every $z \in \mathfrak{Z}$. Moreover the linear functions $s\Lambda$ $(s \in W)$ are the only ones which have this property so that $\Lambda$ is uniquely determined by $\chi$ up to an operation of $W$. Since $\gamma$ is an isomorphism of $\mathfrak{Z}$ onto $J$ we obtain the following differential equations for $\sigma$.

$$v\sigma = \Lambda(v)\sigma \qquad\qquad (v \in J).$$

Define $u^s$ $(u \in \mathfrak{U}, s \in W)$ as in §6. Then $\mathfrak{U}$ being abelian, we can, for any $u \in \mathfrak{U}$, consider the polynomial

$$\prod_{s \in W} (\zeta - u^s) = \zeta^w + v_1\zeta^{w-1} + \cdots + v_w$$

in the indeterminate $\zeta$ with coefficients in $J$. (Here $w$ is the order of $W$.) Replacing $\zeta$ by $u$, we obtain the identity

$$u^w + u^{w-1}v_1 + \cdots + v_w = 0$$

which when applied to $\sigma$ gives

$$u^w\sigma + \Lambda(v_1)u^{w-1}\sigma + \cdots + \Lambda(v_w)\sigma = 0.$$

On the other hand since $\Lambda$ is a homomorphism of $\mathfrak{U}$, it is obvious that

$$\prod_{s \in W} (\zeta - \Lambda(u^s)) = \zeta^w + \Lambda(v_1)\zeta^{w-1} + \cdots + \Lambda(v_w).$$

Therefore the above equation can be written in the form

$$\prod_{s \in W} (u - \Lambda(u^s))\sigma = 0.$$

Thus we have obtained the following result.

LEMMA 26. *Let* $\sigma(h) = \left| d(h) \right|^{1/2} \tau(h)$ $(h \in A')$ *where* $\tau$ *is the analytic function (on* '$M$*) of Theorem 3. Then there exists a linear function* $\Lambda$ *on* $\mathfrak{h}$ *such that*

$$v\sigma = \Lambda(v)\sigma$$

*for every* $v \in J$. *Moreover if* $\tau \neq 0$, $\Lambda$ *is unique up to an operation of* $W$ *and*

$$\prod_{s \in W} (u - \Lambda(u^s))\sigma = 0$$

*for all* $u \in \mathfrak{U}$.

In order to obtain more precise information about $\sigma$ from the above lemma we first need some simple facts about differential equations of a certain type. Let $I$ be an open interval on the real line $R$. We denote the variable on $R$ by $t$ and the differential operator $d/dt$ by $\partial$. Let $P(\zeta) \neq 0$ be a polynomial in an indeterminate $\zeta$ with complex coefficients.

LEMMA 27. *Let* $\xi$ *be an analytic function on* $I$ *such that* $P(\partial)\xi = 0$. *Then* $\xi(t) = p_1(t)e^{\lambda_1 t} + \cdots + p_s(t)e^{\lambda_s t}$ $(t \in I)$ *where* $p_1, \cdots, p_s$ *are polynomials (with complex coefficients) and* $\lambda_1, \cdots, \lambda_s$ *are all the distinct roots of the equation* $P(\zeta) = 0$. *If* $\mu_i$ *is the multiplicity of the root* $\lambda_i$ *of* $P$, *the degree of* $p_i$ *is less than* $\mu_i$ $(1 \leq i \leq s)$.

Although this lemma is undoubtedly known, we give a short proof for the sake of completeness. Without loss of generality we may assume that the origin lies in $I$. Let $r$ be the degree of $P$. First we claim that if $(\partial^j \xi)_{t=0} = 0$, $0 \leq j \leq r$, then $\xi = 0$. For otherwise if $t^m$ is the lowest power of $t$ appearing in the Taylor expansion $\xi(t) = a_0 t^m + a_1 t^{m+1} + \cdots$ around $t = 0$ $(a_i \in C, a_0 \neq 0)$, it is clear that $m > r$ and therefore the coefficient of $t^{m-r}$ in the Taylor expansion of $P(\partial)\xi$ is not zero. Hence it follows easily that any $r+1$ solutions of the differential equation $P(\partial)\Xi = 0$ which are analytic on $I$ must be linearly dependent over $C$. On the other hand it is easy to verify that the functions $t^q e^{\lambda_i t}$ $(0 \leq q < \mu_i, 1 \leq i \leq s)$ constitute a set of $r$ linearly independent solutions of this equation. Hence $\xi$ must be a linear combination of them.

COROLLARY. *Let* $U$ *be an open connected set in a Euclidean space* $E$ *of dimension* $l$ *over* $R$ *and let* $(t_1, \cdots, t_l)$ *denote the Cartesian coordinates in* $E$. *Put* $\partial_i = \partial/\partial t_i$ $(i = 1, \cdots, l)$ *and let* $P_1(\zeta), \cdots, P_l(\zeta)$ *be nonzero polynomials of degrees* $r_1, \cdots, r_l$ *respectively in the indeterminate* $\zeta$ *with coefficients in* $C$. *Consider an analytic function* $\xi$ *on* $U$ *such that* $P_i(\partial_i)\xi = 0$ $(1 \leq i \leq l)$. *Then* $\xi$ *is a linear combination with complex coefficients of functions of the form*

$$t_1^{m_1} t_2^{m_2} \cdots t_l^{m_l} \, e^{\lambda_1 t_1 + \cdots + \lambda_l t_l}$$

*where $m_1, \cdots, m_l$ are integers, $\lambda_i$ is a root of the equation $P_i(\zeta) = 0$ of multiplicity $\mu_i$ and $0 \leq m_i < \mu_i$ $(1 \leq i \leq l)$.*

Put $r = r_1 r_2 \cdots r_l$. It is easy to verify that the functions written above form a system of $r$ linearly independent solutions of our differential equations (see Lemma 41 of [6(b)]). Hence it is enough to prove that any $r+1$ analytic solutions are linearly dependent. Again we may assume that the origin lies in $U$. Then the equations $(\partial_1^{j_1} \partial_2^{j_2} \cdots \partial_l^{j_l} \xi)_{t_1 = \cdots = t_l = 0} = 0$, $0 \leq j_i < r_i$, $1 \leq i \leq l$ are exactly $r$ in number. Hence given $r+1$ analytic solutions $\xi_1, \cdots, \xi_{r+1}$, we can choose $c_i \in C$ $(1 \leq i \leq r+1)$, not all zero, such that $\xi = c_1 \xi_1 + \cdots + c_{r+1} \xi_{r+1}$ satisfies these $r$ conditions. We shall now prove that $\xi = 0$. Since $\xi$ is analytic it is enough to show that it is zero in some neighborhood of the origin. Hence it would be sufficient to prove the following lemma.

LEMMA 28. *Let $\epsilon$ be a positive real number and $U$ the cube $|t_i| < \epsilon$ $(1 \leq i \leq l)$ in $E$. Suppose $\xi$ is an analytic function on $U$ such that $P_i(\partial_i)\xi = 0$ $1 \leq i \leq l$ and*

$$(\partial_1^{j_1} \partial_2^{j_2} \cdots \partial_l^{j_l} \xi)_{t_1 = \cdots = t_l = 0} = 0$$

$0 \leq j_1 < r_1, \cdots, 0 \leq j_l < r_l$. *Then $\xi = 0$.*

We use induction on $l$. If $l = 1$ we have seen during the proof of Lemma 27 that our assertion is true. So assume $l \geq 2$ and let $\overline{U}$ denote the section of $U$ defined by $t_l = 0$. Let $\bar{\xi}_j$ be the restriction of $\xi_j = \partial_l^j \xi$ on $\overline{U}$ $(0 \leq j < r_l)$. Then it follows from our induction hypothesis that $\bar{\xi}_j$ is identically zero. For fixed $t_1, \cdots, t_{l-1}$ $(|t_i| < \epsilon, i = 1, \cdots, l-1)$ put $\eta(t) = \xi(t_1, \cdots, t_{l-1}, t)$ $(|t| < \epsilon)$. Then $(\partial^j \eta)_{t=0} = \bar{\xi}_j(t_1, \cdots, t_{l-1}) = 0$ and $P_l(\partial)\eta = 0$ where $\partial = d/dt$. Therefore $\eta = 0$ by the case $l = 1$ and so this proves that $\xi = 0$ on $U$.

After this preparation we return to our problem of determining $\sigma$. Let $h_0$ be a point in $A'$. We can choose an open connected neighborhood $U$ of zero in the real Euclidean space $\mathfrak{h}_0$ such that $h_0 \exp H \in A'$ for all $H \in U$. Then $F(H) = \sigma(h_0 \exp H)$ $(H \in U)$ is an analytic function on $U$. As in §8 we can regard elements of $\mathfrak{U}$ as differential operators on $\mathfrak{h}_0$ and therefore also on $U$. Then it follows from Lemma 26 that

$$\prod_{s \in W} (u - \Lambda(u^s)) F = 0 \qquad\qquad (u \in \mathfrak{U})$$

and

$$(v - \Lambda(v)) F = 0 \qquad\qquad (v \in J).$$

Let $H_1, \cdots, H_l$ be a base for $\mathfrak{h}_0$ over $R$ and let $(t_1, \cdots, t_l)$ denote the corresponding Cartesian coordinates in $\mathfrak{h}_0$. Then $\partial_i = H_i$ and if

$$P_i(\zeta) = \prod_{s \in W} (\zeta - s\Lambda(H_i)),$$

it is obvious that $P_i(\partial_i)F = P_i(H_i)F = 0$ $(1 \leq i \leq l)$. Moreover $s\Lambda(H_i)$ $(s \in W)$ are exactly the roots of the polynomial $P_i(\zeta)$. Hence it follows from the corollary to Lemma 27 that

$$F = p_1 e^{\lambda_1} + \cdots + p_r e^{\lambda_r}$$

where $p_1, \cdots, p_r$ are polynomial functions and $\lambda_1, \cdots, \lambda_r$ linear functions on $\mathfrak{h}$. We may assume that $\lambda_1, \cdots, \lambda_r$ are all distinct and $p_i \neq 0$ $(1 \leq i \leq r)$. It is obvious that $u e^\lambda = \lambda(u) e^\lambda$ $(u \in \mathfrak{U})$ for any linear function $\lambda$ on $\mathfrak{h}$. Hence if $v \in J$,

$$0 = (v - \Lambda(v))F = \sum_{i=1}^{r} p_i' e^{\lambda_i}$$

where $p_i' = (\lambda_i(v) - \Lambda(v))p_i + q_i$ and $q_i$ is a polynomial of lower degree than $p_i$. But the exponentials of distinct linear functions are known to be linearly independent over the ring of polynomial functions (see Lemma 41 of [6(b)]). Therefore $p_i' = 0$ and so

$$\lambda_i(v) = \Lambda(v) \qquad\qquad (1 \leq i \leq r).$$

This being true for every $v \in J$, it follows from Theorem 5 of [6(b)] that $\lambda_i = s_i \Lambda$ for some $s_i \in W$. Now let $W_\Lambda$ be the subgroup consisting of those elements $s \in W$ for which $s\Lambda = \Lambda$. Let $N$ be the order of $W_\Lambda$. We shall prove that the degree of each $p_i$ is smaller than $N$. For otherwise suppose the degree of $p_1$ is $n_1$ and $n_1 \geq N$. It is easily seen that we can choose an element $H_0 \in \mathfrak{h}_0$ such that $\Lambda(H_0) \neq s\Lambda(H_0)$ $(s \in W)$ unless $s \in W_\Lambda$ and the degree of the polynomial $p(t) = p_1(tH_0)$ in $t$ is $n_1$. Put $f(t) = F(tH_0)$. Then $f$ is an analytic function defined on some open interval $I$ in $R$ containing the origin. On the other hand, considering $H_0$ as a differential operator on $U$, we have the relation

$$F\left(tH_0, \prod_{s \in W} (H_0 - s\Lambda(H_0))\right) = 0 \qquad\qquad (t \in I).$$

From this it follows immediately that

$$\prod_{s \in W} (\partial - s\Lambda(H_0))f = 0$$

where $\partial = d/dt$. Put $p_i'(t) = p_i(tH_0)$ and $\lambda_i' = \lambda_i(H_0)$. Then

$$f(t) = \sum_{i=1}^{r} p_i'(t) e^{\lambda_i' t}.$$

Since $\lambda_1', \cdots, \lambda_r'$ are all distinct $e^{\lambda_1' t}, \cdots, e^{\lambda_r' t}$ are linearly independent over the field of rational functions of $t$ (Lemma 41 of [6(b)]). Moreover as we have seen $\lambda_1', \cdots, \lambda_r'$ are roots of the polynomial

$$\prod_{s \in W} (\zeta - s\Lambda(H_0))$$

all whose roots are of multiplicity $N$. Therefore it follows from Lemma 27 that the degree of $p(t) = p_1'(t)$ cannot exceed $N-1$. This contradiction proves our assertion.

We know from Lemma 7 that every element in $A$ can be written in the form $a \exp H$ where $a$ lies in the center of $M \cap K$ and $H \in \mathfrak{h}_0$. Let $\mathfrak{h}_1$ denote a nonempty open connected subset of $\mathfrak{h}_0$ such that $a \exp H \in {}'A$ for all $H \in \mathfrak{h}_1$. Since $A'$ is dense in $A$ (Lemma 8), we can choose an element $H_0 \in \mathfrak{h}_1$ such that $a \exp H_0 = h_0 \in A'$. Moreover $a$ being in the center of $M \cap K$, it is obvious that

$$\left| d_-(a \exp H) \right|^{1/2} = \left| d_-(\exp H) \right|^{1/2} = \epsilon \Delta_-(H)$$

if $H$ is sufficiently near $H_0$ in $\mathfrak{h}_1$. (Here $\epsilon = \pm 1$ and it is independent of $H$.) Hence if we apply the principle of analytic continuation to the function $\Delta_-(H)\tau(a \exp H)$ on $\mathfrak{h}_1$ we obtain the following theorem from our result above.

THEOREM 4. *Let $\tau \neq 0$ be an invariant analytic function on ${}'M$. Suppose $\tau$ is an eigenfunction of the operator $\delta(z)$ for every $z \in \mathfrak{Z}$. Then there exists a linear function $\Lambda$ on $\mathfrak{h}$ such that $\delta(z)\tau = \Lambda(\gamma(z))\tau$ $(z \in \mathfrak{Z})$ and $\Lambda$ is unique up to an operation of $W$. If $a$ is any point in the center of $M \cap K$ and $\mathfrak{h}_1$ a connected component of the set ${}'\mathfrak{h}_a$ of all points $H \in \mathfrak{h}_0$ for which $d_+(a \exp H) \neq 0$, we can select for each $s \in W$ a polynomial function $p_s$ on $\mathfrak{h}$ such that*

$$\Delta_-(H)\tau(a \exp H) = \left| d_+(a \exp H) \right|^{-1/2} \sum_{s \in W} p_s(H) e^{s\Lambda(H)} \qquad (H \in \mathfrak{h}_1).$$

*Let $W_\Lambda$ be the subgroup consisting of those elements $s \in W$ for which $s\Lambda = \Lambda$. Then under the assumption that $p_{s\sigma} = p_s$ $(s \in W, \sigma \in W_\Lambda)$, the polynomial functions $p_s$ are uniquely determined and their degrees are smaller than the order of $W_\Lambda$.*

The above formula is very similar to the one obtained by Weyl [14(a)] for the character of an irreducible finite-dimensional representation of $G$. Actually it is possible to improve our result in two ways. First of all we have not yet made full use of the differential equations $v\sigma = \Lambda(v)\sigma$ $(v \in J)$ of Lemma 26. Secondly the fact that $\tau$ defines an analytic function on ${}'A$ (and not merely on $A'$) allows us to obtain a little more information about $\Lambda$ and $p_s$. We shall however not pursue this matter here further.

10. **An important special case.** In this section we shall make the following two additional assumptions. (1) $\mathfrak{h}_0 \subset \mathfrak{k}_0$ and (2) $G$ has a faithful finite-dimensional representation. Then $G$ can be imbedded in a complex analytic group $G_c$ with the Lie algebra $\mathfrak{g}$ in such a way that the (real) analytic subgroup of $G_c$ corresponding to $\mathfrak{g}_0$ coincides with $G$. In this case $K$ is compact, $A_+ = \{1\}$ and $M = K$. Moreover if $A_0$ is the connected component of 1 in $A$, $A_0$ is a maximal connected abelian subgroup of $K$. But $K$ being compact, this implies [12] that $A_0$ is maximal abelian in $K$ and therefore $A = A_0$. Hence $A$

is connected. Moreover in this case there are no roots in the set $P_+$ so that $P = P_0 \cup P_-$.

Let $\mathfrak{u} = \mathfrak{k}_0 + (-1)^{1/2}\mathfrak{p}_0$ and let $U$ be the (real) analytic subgroup of $G_c$ corresponding to $\mathfrak{u}$. Then $U$ is compact and $K = G \cap U$ (see Cartan [1(a)] and Mostow [10]). As before let $W$ denote the Weyl group of $\mathfrak{g}$ with respect to $\mathfrak{h}$. For each $s \in W$ we can choose an element $u \in U$ such that Ad $(u)H = sH$ ($H \in \mathfrak{h}$) and therefore if we put $h^s = uhu^{-1}$ ($h \in A$), $h \to h^s$ defines an automorphism of $A$. For any function $f$ on $A$ we denote by $f^s$ ($s \in W$) the function $h \to f(h^{s^{-1}})$ ($h \in A$).

For each root $\alpha$ we have the character $\xi_\alpha$ of $A$ given by $\xi_\alpha (\exp H) = e^{\alpha(H)}$ ($H \in \mathfrak{h}_0$). As before (see §7) let $F$ and $_0\Xi$ be the additive and the multiplicative groups generated by $\alpha$ and $\xi_\alpha$ respectively ($\alpha \in P$). Then the mapping $\alpha \to \xi_\alpha$ ($\alpha \in P$) can be extended (uniquely) to a homomorphism $\lambda \to \xi_\lambda$ ($\lambda \in F$) of $F$ into $_0\Xi$. It is clear that if $\lambda \in F$ and $s \in W$, $s\lambda$ also lies in $F$. On the other hand if $\rho = 2^{-1} \sum_{\alpha \in P} \alpha$ and $s_\alpha$ is the Weyl reflexion corresponding to the root $\alpha$, it is known (see Weyl [14(b)]) that $s_\alpha\rho = \rho - k\alpha$ where $k$ is an integer. Hence $s_\alpha\rho - \rho \in F$ and from this it follows immediately that $s\rho - \rho \in F$ for every $s \in W$. Put $\xi_s = \xi_{s\rho-\rho}$ and

$$\Delta' = \prod_{\alpha \in P} (1 - \xi_\alpha^{-1}).$$

Then

$$\Delta'(\exp H) = \prod_{\alpha \in P} (1 - e^{-\alpha(H)}) = \Delta(H)e^{-\rho(H)} \qquad (H \in \mathfrak{h}_0)$$

if $\Delta(H)$ is defined as in §8. Therefore (see Weyl [14(b)])

$$\Delta' = \sum_{s \in W} \epsilon(s)\xi_s$$

and

$$(\Delta')^s = \epsilon(s)\xi_s^{-1}\Delta' \qquad (s \in W).$$

As usual let $I^\infty(U)$ denote the space of invariant functions in $C^\infty(U)$. Also we may regard elements of $\mathfrak{B}$ as differential operators on $U$.

LEMMA 29. *For any $f \in C^\infty(A)$ there exists a unique function $\phi_f \in I^\infty(U)$ such that*

$$\Delta'(h)\phi_f(h) = \sum_{s \in W} \epsilon(s)\xi_s(h)f^s(h) \qquad (h \in A).$$

*Moreover for any $b \in \mathfrak{B}$ we can select a finite set of elements $a_1, \cdots, a_r \in \mathfrak{U}$ such that*

$$\sup |b\phi_f| \leq \sup |a_1 f| + \cdots + \sup |a_r f|$$

*for every $f \in C^\infty(A)$.*

Since $U$ is compact, $U = \bigcup_{u \in U} uAu^{-1}$ and the uniqueness of $\phi_f$ is obvious. So only the existence requires proof. Let $\Theta$ denote the set of all characters of $A$. Put

$$c_\theta(f) = \int_A \theta(h^{-1})f(h)\,dh \qquad\qquad (f \in C^\infty(A),\, \theta \in \Theta),$$

the Haar measure $dh$ on $A$ being so normalized that $\int_A dh = 1$. Let $\Lambda_\theta$ denote the linear function on $\mathfrak{h}$ such that $\theta(\exp H) = e^{\Lambda_\theta(H)}$ ($H \in \mathfrak{h}_0$). Then if we agree to regard a linear function on $\mathfrak{h}$ as a homomorphism of $\mathfrak{U}$ into $C$ (see §§8 and 9), it is clear that $a\theta = \Lambda_\theta(a)\theta$ ($a \in \mathfrak{U}$). Choose a base $H_1, \cdots, H_l$ for $\mathfrak{h}_0$ over $R$ such that $\exp(t_1H_1 + \cdots + t_lH_l) = 1$ in $A$ if and only if $t_1, \cdots, t_l$ are all integers and put $a_0 = \{1 - (4\pi^2)^{-1}(H_1^2 + \cdots + H_l^2)\}^k \in \mathfrak{U}$ where $k$ is any integer greater than $l/2$. Then $\Lambda_\theta(H_i) = (-1)^{1/2}2\pi n_i$ where $n_i$ ($i = 1, \cdots, l$) are integers and therefore $\Lambda_\theta(a_0) = (1 + n_1^2 + \cdots + n_l^2)^k$. Hence if $M$ denotes the sum of the convergent series

$$\sum_{-\infty < n_1, \cdots, n_l < \infty} (1 + n_1^2 + \cdots + n_l^2)^{-(l+1)/2}$$

it is obvious that

$$\sum_{\theta \in \Theta} |c_\theta(f)| = \sum_{\theta \in \Theta} \Lambda_\theta(a_0)^{-1} \left| \int_A \theta(h^{-1})f(h, a_0)\,dh \right|$$

$$\leq M \sup |a_0 f| \qquad\qquad (f \in C^\infty(A)).$$

This proves that the series $\sum_{\theta \in \Theta} c_\theta(f)\theta(h)$ is absolutely convergent on $A$. Moreover as is well known, it converges to $f$ uniformly on $A$. Hence

$$\sum_{s \in W} \epsilon(s)\xi_s(h)f^s(h) = \sum_{\theta \in \Theta} c_\theta(f) \sum_{s \in W} \epsilon(s)\xi_s(h)\theta^s(h) \qquad (h \in A).$$

For every root $\alpha \in P$, put $H_\alpha = [X_\alpha, X_{-\alpha}]$. We may suppose that $X_\alpha$, $X_{-\alpha}$ are so normalized that $\alpha(H_\alpha) = 2$. Now define $d_\theta$ as follows:

$$\epsilon_\theta d_\theta = \prod_{\alpha \in P} \frac{\Lambda_\theta(H_\alpha) + \rho(H_\alpha)}{\rho(H_\alpha)} \qquad\qquad (\theta \in \Theta).$$

Here $\epsilon_\theta = 1$ if the right side (which is always real) is positive and $-1$ otherwise. Then it is known (see Weyl [14(b)]) that

$$\sum_{s \in W} \epsilon(s)e^{s\rho(H)}e^{s\Lambda_\theta(H)} = \begin{cases} \epsilon_\theta \chi_\theta(\exp H)\Delta(H) & \text{if } d_\theta \neq 0, \\ 0 & \text{if } d_\theta = 0 \end{cases}$$

for all $H \in \mathfrak{h}_0$. Here $\chi_\theta$ is the character of an irreducible representation of $U$ of degree $d_\theta$. Since $\Delta'(\exp H) = e^{-\rho(H)}\Delta(H)$ ($H \in \mathfrak{h}_0$), it follows that

$$\sum_{s \in W} \epsilon(s)\xi_s(h)f^s(h) = \sum_{\theta \in \Theta} c_\theta(f)\Delta'(h)\epsilon_\theta\chi_\theta(h) \qquad\qquad (h \in A)$$

if we agree to set $\chi_\theta = 0$ when $d_\theta = 0$. Now consider the series

$$\sum_{\theta \in \Theta} \epsilon_\theta c_\theta(f) \chi_\theta(u) \qquad\qquad (u \in U)$$

on $U$. We claim this series converges to a function in $C^\infty(U)$. Let $X_1, \cdots, X_r$ $\in \mathfrak{u}$ $(r \geq 0)$ and consider the series

$$\sum_{\theta \in \Theta} \epsilon_\theta c_\theta(f) \chi_\theta(u, X_1 X_2 \cdots X_r).$$

For any $\theta$ for which $d_\theta \neq 0$ let $\sigma_\theta$ denote the irreducible unitary representation of $U$ on a finite-dimensional Hilbert space $V$ corresponding to the character $\chi_\theta$. We denote the corresponding representations of $\mathfrak{g}$ and $\mathfrak{B}$ also by $\sigma_\theta$. Then

$$\left| \chi_\theta(u, X_1 \cdots X_r) \right| = \left| \operatorname{sp}\left(\sigma_\theta(u)\sigma_\theta(X_1 \cdots X_r)\right) \right| \qquad (u \in U).$$

Now for any linear transformation $T$ on $V$, let $T^*$ denote the adjoint of $T$ and put $\|T\| = \{\operatorname{sp}(TT^*)\}^{1/2}$. Then $\left|\operatorname{sp}(T_1 T_2)\right| \leq \|T_1\| \|T_2\|$ for any two linear transformations $T_1$ and $T_2$ and therefore

$$\left| \chi_\theta(u, X_1 X_2 \cdots X_r) \right| \leq d_\theta \|\sigma_\theta(X_1 \cdots X_r)\|$$

since $\|\sigma_\theta(u)\| = d_\theta$. On the other hand if $X \in \mathfrak{u}$, $-\sigma_\theta(X)$ is the adjoint of $\sigma_\theta(X)$ and therefore

$$\|\sigma_\theta(X_1 \cdots X_r)\|^2 = \operatorname{sp} \sigma_\theta(b)$$

where $b = (-1)^r(X_r X_{r-1} \cdots X_1)(X_1 X_2 \cdots X_r)$. On the other hand we can choose $z_1 \in \mathfrak{Z}$ such that $b$ can be written in the form

$$b = z_1 + \sum_{i=1}^{q} [b_i, b_i']$$

where $b_i, b_i' \in \mathfrak{B}$ (see Lemma 39 of $[6(\mathrm{b})]$). Then $\operatorname{sp} \sigma_\theta(b) = \operatorname{sp} \sigma_\theta(z_1)$. But in view of the above formula for $\chi_\theta$, it is clear that the highest weight of $\sigma_\theta$ must be of the form $s\Lambda_\theta' - \rho$ where $s \in W$ and $\Lambda_\theta' = \Lambda_\theta + \rho$. Therefore if $z = 1 + z_1$, it follows (see $[6(\mathrm{b}), \text{ p. } 73]$) that

$$\|\sigma_\theta(X_1 \cdots X_r)\| \leq 1 + \|\sigma_\theta(X_1 \cdots X_r)\|^2 = 1 + \operatorname{sp} \sigma_\theta(z_1) = 1 + d_\theta \Lambda_\theta'(\gamma(z_1))$$
$$\leq d_\theta \Lambda_\theta'(\gamma(z))$$

where $\gamma$ is defined as in §8. This shows that

$$\left| \chi_\theta(u, X_1 \cdots X_r) \right| \leq d_\theta^2 \Lambda_\theta'((\gamma z))$$

which remains true also if $d_\theta = 0$ since then $\chi_\theta = 0$. On the other hand

$$d_\theta^2 = \Lambda_\theta'(a)$$

where $a = \left\{ \prod_{\alpha \in P} \rho(H_\alpha) \right\}^{-2} \prod_{\alpha \in P} H_\alpha^2 \in \mathfrak{U}$. Therefore

$$\left| \chi_\theta(u, X_1 \cdots X_r) \right| \leqq \Lambda_\theta'(a\gamma(z))$$

and

$$\sum_{\theta \in \Theta} \left| \epsilon_\theta c_\theta(f) \chi_\theta(u, X_1 \cdots X_r) \right| \leqq \sum_{\theta \in \Theta} \left| c_\theta(f) \Lambda_\theta'(a\gamma(z)) \right|.$$

Let $\lambda$ denote the automorphism of $\mathfrak{U}$ over $C$ given by $\lambda(H) = H + \rho(H)$ ($H \in \mathfrak{h}$). Then if $a' = \lambda(a\gamma(z))$, it is clear that

$$a'\theta = \Lambda_\theta'(a\gamma(z))\theta \qquad\qquad (\theta \in \Theta).$$

Hence

$$c_\theta(f)\Lambda_\theta'(a\gamma(z)) = \int_A \theta(h, a')f(h^{-1})dh = \int_A \theta(h^{-1})f(h, a')dh$$

and therefore if $a'' = a_0 a'$,

$$\sum_{\theta \in \Theta} \left| \epsilon_\theta c_\theta(f) \chi_\theta(u, X_1 \cdots X_r) \right| \leqq \sum_{\theta \in \Theta} \left| \int \theta(h^{-1})f(h, a')dh \right| \leqq M \sup \left| a''f \right|$$

by our earlier result. Now if we introduce in $C^\infty(U)$ the topology which is usual in the theory of distributions (see Schwartz [11, p. 68]), it becomes a complete locally convex space and the above result shows that $\sum_{\theta \in \Theta} \epsilon_\theta c_\theta(f)\chi_\theta$ converges in $C^\infty(U)$ to some element $\phi_f$. This implies in particular the uniform convergence of $\sum_{\theta \in \Theta} \epsilon_\theta c_\theta(f)b\chi_\theta$ on $U$ to $b\phi_f$ ($b \in \mathfrak{B}$). Taking $b = 1$ we see that $\sum_{\theta \in \Theta} \epsilon_\theta c_\theta(f)\chi_\theta$ converges to $\phi_f$ uniformly on $U$. Therefore since the functions $\chi_\theta$ are all invariant, the same holds for $\phi_f$ and so $\phi_f \in I^\infty(U)$. Moreover

$$\Delta'(h)\phi_f(h) = \sum_{\theta \in \Theta} c_\theta(f)\epsilon_\theta \Delta'(h)\chi_\theta(h)$$

$$= \sum_{s \in W} \epsilon(s)\xi_s(h)f^s(h) \qquad\qquad (h \in A).$$

Finally it follows from the inequality obtained above that

$$\sup \left| b\phi_f \right| \leqq M \sup \left| a''f \right|$$

where $b = X_1 X_2 \cdots X_r$. Since every element in $\mathfrak{B}$ can be written as a linear combination of such elements $b$, the last statement of the lemma is now obvious.

COROLLARY 1. *Suppose $f$ is a finite linear combination of elements in $\Theta$ and $f^s = f$ ($s \in W$). Then there exists an invariant analytic function $\phi$ on $U$ which coincides with $f$ on $A$.*

For $c_\theta(f)$ being zero for all $\theta \in \Theta$ except a finite number,

$$\phi_f = \sum_{\theta \in \Theta} c_\theta(f)\epsilon_\theta \chi_\theta$$

is a finite linear combination of irreducible characters of $U$ and so it is analytic. Now since $f^s = f$ $(s \in W)$,

$$\sum_{s \in W} \epsilon(s) \xi_s f^s = f \sum_{s \in W} \epsilon(s) \xi_s = f\Delta'.$$

Hence $\Delta'(h)(f(h) - \phi_f(h)) = 0$ $(h \in A)$. But since $f$ and $\phi_f$ are both continuous we can conclude that $f(h) - \phi_f(h) = 0$ $(h \in A)$.

Now introduce on $C^\infty(A)$ and $C^\infty(K)$ the topologies used in the theory of distributions (see Schwartz [11, p. 66]). Obviously $I^\infty(K)$ is closed in $C^\infty(K)$.

COROLLARY 2. *For any $f \in C^\infty(A)$ there exists a unique function $F_f \in I^\infty(K)$ such that*

$$\Delta'(h)F_f(h) = \sum_{s \in W} \epsilon(s) \xi_s(h) f^s(h) \qquad (h \in A).$$

*Moreover $f \to F_f$ is a continuous linear mapping of $C^\infty(A)$ into $I^\infty(K)$.*

The uniqueness of $F_f$ follows from the fact that $K = \bigcup_{k \in K} kAk^{-1}$. On the other hand the restriction on $K$ of the function $\phi_f$ of Lemma 29 obviously satisfies the required conditions for $F_f$. The linearity of the mapping follows from the uniqueness of $F_f$ and the continuity from the last statement of Lemma 29.

COROLLARY 3. *Suppose $f$ is a finite linear combination of elements in $\Theta$ and $f^s = f$ $(s \in W_-)$. Then there exists an invariant analytic function on $K$ which coincides with $f$ on $A$.*

Since $K$ is compact and connected, it is obvious that the method of proof of Lemma 29 is applicable to the pair $(K, A)$ instead of $(U, A)$. The required result therefore follows from the Corollary 1 above.

Now in the present case $M = K$ and $\mathfrak{q} = \mathfrak{p}$. Hence $\mathfrak{X}$ being the subalgebra of $\mathfrak{B}$ generated by $(1, \mathfrak{k})$, $\mathfrak{M} = \mathfrak{X}$. Let $r$ be the number of roots in $P_0$. Then $\dim \mathfrak{p} = 2r$ since $(X_\alpha, X_{-\alpha})_{\alpha \in P_0}$ is a base for $\mathfrak{p}$ over $C$. Consider the polynomial

$$\prod_{\beta \in P_0} \{ t - (2 - \xi_\beta - \xi_\beta^{-1}) \} = t^r - \sigma_1 t^{r-1} + \cdots + (-1)^r \sigma_r$$

in the indeterminate $t$. $(\sigma_1, \cdots, \sigma_r$ are functions on $A$.) If $s \in W_-$ and $\beta \in P_0$, it is clear that either $s\beta$ or $-s\beta$ is in $P_0$. Therefore it is obvious that $\sigma_i^s = \sigma_i$ $(s \in W_-, 1 \leq i \leq r)$. Hence from Corollary 3 to Lemma 29 we can extend $\sigma_i$ to an invariant analytic function on $K$. We denote this extension again by $\sigma_i$. Let $S(k)$ denote the restriction of $(\mathrm{Ad}\ (k) - I)(\mathrm{Ad}\ (k^{-1}) - I)$ $(k \in K)$ on $\mathfrak{p}$. (Here $I$ is the identity mapping of $\mathfrak{g}$.) Then if we define $d_+(k)$ as in §5, we have the following result.

LEMMA 30. $\sigma_r(k) = d_+(k)$ *and*

$$(S(k))^r - \sigma_1(k)(S(k))^{r-1} + \cdots + (-1)^r\sigma_r(k)I_\mathfrak{p} = 0 \qquad (k \in K)$$

*where $I_\mathfrak{p}$ is the identity mapping of $\mathfrak{p}$.*

Since $K = \bigcup_{k \in K} kAk^{-1}$, it is obviously enough to prove these relations on $A$. But if $h \in A$, the eigenvalues of the restriction of Ad $(h)$ on $\mathfrak{p}$ are obviously $\xi_\beta(h)$ and $\xi_\beta(h^{-1})$ $(\beta \in P_0)$. Hence every eigenvalue of $S(h)$ is of the form

$$(\xi_\beta(h) - 1)(\xi_\beta(h^{-1}) - 1) = 2 - \xi_\beta(h) - \xi_\beta(h)^{-1} \qquad (\beta \in P_0).$$

Therefore if we replace $t$ by $S(h)$ in the polynomial

$$\prod_{\beta \in P_0} \{t - (2 - \xi_\beta(h) - \xi_\beta(h)^{-1})\} = t^r - \sigma_1(h)t^{r-1} + \cdots + (-1)^r\sigma_r(h)$$

we get zero (provided the last term is interpreted as $\sigma_r(h)I_\mathfrak{p}$). Moreover

$$\sigma_r(h) = \prod_{\beta \in P_0} (2 - \xi_\beta(h) - \xi_\beta(h)^{-1})$$
$$= \prod_{\beta \in P_0} (\xi_\beta(h) - 1)(\xi_\beta(h)^{-1} - 1) = d_+(h)$$

and so the lemma is proved. We note that $d_+$ is never negative on $A$ and hence also never on $K$.

Let $X_1, \cdots, X_n$ be a base for $\mathfrak{g}$ over $C$ and put $g_{ij} = B(X_i, X_j)$ $(1 \leq i, j \leq n)$ (see §2 for the definition of the bilinear form $B(X, Y)$ on $\mathfrak{g}$). Then since $\mathfrak{g}$ is semisimple, the matrix $(g_{ij})_{1 \leq i,j \leq n}$ is nonsingular. Let $(g^{ij})_{1 \leq i,j \leq n}$ denote its inverse. Put $\omega = \sum_{1 \leq i,j \leq n} g^{ij}X_iX_j$. Then it is well known that $\omega$ is independent of the choice of the base used in its definition and it lies in $\mathfrak{Z}$. In particular if $(X_1, \cdots, X_m)$ is a base for $\mathfrak{k}$ and $(X_{m+1}, \cdots, X_n)$ a base for $\mathfrak{p}$, $\omega = \omega_K + \omega_+$ where

$$\omega_K = \sum_{1 \leq i, j \leq m} g^{ij}X_iX_j,$$

and

$$\omega_+ = \sum_{m < i, j \leq n} g^{ij}X_iX_j.$$

Since $\omega^k = \omega$ $(k \in K)$, it follows easily that $\omega_K^k = \omega_K$ and $\omega_+^k = \omega_+$. Hence $\omega_K$ lies in the center $\mathfrak{Z}_K$ of $\mathfrak{X}$. We shall call $\omega$ and $\omega_K$ the Casimir operators of $\mathfrak{g}$ and $\mathfrak{k}$ respectively. Let us now use the notation of §5.

LEMMA 31. *There exists an analytic mapping $\gamma_\omega$ of $K$ into $\mathfrak{D} \times \mathfrak{X}$ such that*

$$\Gamma_k(\gamma_\omega(k)) = d_+(k)\omega \qquad (k \in K).$$

Put

$$T(k) = (-1)^{r-1}\{(S(k))^{r-1} - \sigma_1(k)(S(k))^{r-2} + \cdots + (-1)^{r-1}\sigma_{r-1}(k)I_\mathfrak{p}\}$$

where $S(k)$ is defined as in Lemma 30. Then

$$S(k)T(k) = T(k)S(k) = d_+(k)I_\mathfrak{p} \qquad\qquad (k \in K).$$

If $X_1, \cdots, X_q$ is a base for $\mathfrak{p}$, it is clear that

$$\omega_+ = \sum_{1 \le i, j \le q} g^{ij} X_i X_j$$

where $(g^{ij})_{1 \le i, j \le q}$ is the inverse of the matrix $g_{ij} = B(X_i, X_j)$ $1 \le i, j \le q$. Now $k$ being a fixed element in $K$, suppose $S'$, $T'$ are two linear transformations on $\mathfrak{p}$ which commute with the restriction of Ad $(k)$ on $\mathfrak{p}$. Put $Y_i = S'X_i$, $Z_i = T'X_i$, $X^i = \sum_{j=1}^q g^{ij} X_j$, $Y^i = S'X^i$ and $Z^i = T'X^i$ $(1 \le i \le q)$. It is obvious that

$$X = \sum_{i=1}^q B(X, X_i) X^i$$

for any $X \in \mathfrak{p}$ and therefore

$$\text{Ad } (k)X = \sum_{i=1}^q B(\text{Ad } (k)X, X_i) X^i = \sum_{i=1}^q B(X, \text{Ad } (k^{-1})X_i) X^i.$$

Hence

$$\text{Ad } (k)Y_i = S' \text{ Ad } (k)X_i = \sum_{j=1}^q B(X_i, \text{Ad } (k^{-1})X_j) S'X^j$$

and

$$\sum_{i=1}^q (\text{Ad } (k)Y_i)Z^i = \sum_{i,j} (S'X^j)(T'X^i) B(X_i, \text{Ad } (k^{-1})X_j)$$

$$= \sum_j (S'X^j)(T' \text{ Ad } (k^{-1})X_j) = \sum_j Y^j(\text{Ad } (k^{-1})Z_j).$$

This proves that

$$\sum_{1 \le i, j \le q} g^{ij}(\text{Ad } (k)Y_i)Z_j = \sum_{1 \le i, j \le q} g^{ij}Y_i(\text{Ad } (k^{-1})Z_j)$$

and therefore

$$\sum_{1 \le i, j \le q} g^{ij}[\text{Ad } (k)Y_i, Z_j] = \sum_{1 \le i, j \le q} g^{ij}[Y_i, \text{Ad } (k^{-1})Z_j].$$

Now put $T'(k) = \text{Ad } (k^{-1})T(k) - T(k)$, $X_j(k) = T(k)X_j$ and $X_j'(k) = T'(k)X_j$ $1 \le j \le q$. Then

$$\sum_{1 \le i, j \le q} g^{ij}\Gamma_k(X_i X_j(k) \times 1)$$

$$= \sum_{i,j} g^{ij}(\text{Ad } (k^{-1})X_i - X_i)(T'(k)X_j) + \sum_{i,j} g^{ij}[X_i, T'(k)X_j]$$

$$= \sum_{i,j} g^{ij}X_i\{(\text{Ad } (k) - I)T'(k)X_j\} + \sum_{i,j} g^{ij}[X_i, T'(k)X_j].$$

But $(\text{Ad } (k) - I)T'(k) = S(k)T(k) = d_+(k)I_\mathfrak{p}$. So if we put

$$b(k) = \frac{1}{2} \sum_{1 \leq i, j \leq q} g^{ij}(X_i X_j(k) + X_j(k)X_i),$$

it is clear that

$$\Gamma_k(b(k) \times 1) = d_+(k)\omega_+ + \sum_{i,j} g^{ij}[X_i, X_j'(k)] - \frac{1}{2} \sum_{i,j} g^{ij}\Gamma_k([X_i, X_j(k)] \times 1).$$

Put $Y(k) = \sum_{i,j} g^{ij}[X_i, X_j(k)]$ and $Y'(k) = \sum_{i,j} g^{ij}[X_i, X_j'(k)]$. It follows from our remarks above that $\Gamma_k(Y(k) \times 1) = \mathrm{Ad}\,(k^{-1})\,Y(k) - Y(k) = 0$. Moreover $Y'(k) \in [\mathfrak{p}, \mathfrak{p}] \subset \mathfrak{k}$. Hence if we set

$$\gamma_\omega(k) = b(k) \times 1 + d_+(k)(1 \times \omega_K) - 1 \times Y'(k),$$

$\Gamma_k(\gamma_\omega(k)) = d_+(k)\omega$. This proves the lemma.

On the other hand $`K$ being the open submanifold of $K$ consisting of those points $k$ for which $d_+(k) \neq 0$, we have a differential operator $\delta(\omega)$ defined on $`K$ (see §5). It follows from the definition of $\delta(\omega)$ that

$$d_+(k)\delta_k(\omega) = d_+(k)\omega_K - Y'(k) \qquad (k \in `K).$$

This means that there exists an analytic differential operator $D$ on $K$ which coincides with $d_+\delta(\omega)$ on $`K$.

LEMMA 32. *Let $D^*$ denote the adjoint of $D$. Then if $f \in I^\infty(K)$, $Df = D^*f$.*

In view of the corollary to Lemma 8, it is enough to show that $Df$ and $D^*f$ coincide on $K' = K \cap G'$. Hence, as we have seen during the proof of Theorem 1, it is sufficient to prove that

$$\int_K g(k)f(k, D)dk = \int_K g(k, D)f(k)dk$$

for every $g \in I_c^\infty(K')$. (Here $dk$ is the Haar measure on $K$.) But since $D$ is an invariant operator, $Df$ and $Dg$ are invariant functions. Hence (see Weyl [14(a)])

$$\int_K g(k)f(k, D)dk = \int_A |d_-(h)| g(h)f(h, D)dh$$

and

$$\int_K g(k, D)f(k)dk = \int_A |d_-(h)| g(h, D)f(h)dh$$

if $dk$ is suitably normalized. On the other hand since $d_+ = |d_+|$, we can conclude from Theorem 2 and the corollary to Lemma 25 that if $h \in A'$,

$$|d_-(h)| f(h, D) = f(h, D')$$

where $D'$ denotes the differential operator $|d|^{1/2}\gamma(\omega) \circ |d|^{1/2}$ on $A'$. Moreover

$(\gamma(\omega))^* = \gamma(\omega^*) = \gamma(\omega)$ from Lemma 20 and so $D'$ coincides with its adjoint on $A'$. Hence

$$\int_K g(k)f(k, D)dk = \int_A g(h)f(h, D')dh = \int_A g(h, D')f(h)dh$$
$$= \int_K g(k, D)f(k)dk.$$

Let $\gamma'(\omega)$ denote the image of $\gamma(\omega)$ under the automorphism of $\mathfrak{U}$ which maps $H$ on $H+\rho(H)$ $(H\in\mathfrak{h})$.

LEMMA 33. *Define the mapping $f\rightarrow F_f$ of $C^\infty(A)$ into $I^\infty(K)$ as in Corollary 2 to Lemma 29. Then*

$$DF_f = d_+ F_{\gamma'(\omega)f} \qquad\qquad (f\in C^\infty(A)).$$

It would be sufficient to prove that $DF_f$ and $d_+F_{\gamma'(\omega)f}$ coincide on $A'$. Let $g$ denote the restriction of $F_f$ on $A'$. Then

$$g(h) = \Delta'(h)^{-1}\sum_{s\in W}\epsilon(s)\xi_s(h)f^s(h) \qquad\qquad (h\in A')$$

and it follows from Theorem 2 and the corollary to Lemma 25 that $DF_f$ coincides on $A'$ with $D'g$ where $D' = d_+|d|^{-1/2}\gamma(\omega)\circ|d|^{1/2}$. Let $H_0$ be any point in $\mathfrak{h}_0$ such that $h_0 = \exp H_0 \in A'$ and let $V$ be an open connected neighborhood of $H_0$ in $\mathfrak{h}_0$ such that $\exp H\in A'$ for all $H\in V$. Put

$$g'(H) = \sum_{s\in W}\epsilon(s)e^{s\rho(H)}f^s(\exp H) \qquad\qquad (H\in V).$$

Then $g(\exp H)\Delta(H) = g'(H)$ $(H\in V)$ where $\Delta(H)$ is defined as in §8. Hence it is obvious that

$$g(\exp H, D') = d_+(\exp H)\Delta(H)^{-1}g'(H, \gamma(\omega)) \qquad\qquad (H\in V)$$

if we regard $\gamma(\omega)$ as a differential operator on $V$. On the other hand if $f'_s(H) = e^{s\rho(H)}f^s(\exp H)$ $(H\in\mathfrak{h}_0)$ it is clear that

$$f'_s(H, \gamma(\omega)) = f'_1(s^{-1}H, \gamma(\omega))$$

since $\gamma(\omega)$ is an invariant of $W$. Furthermore

$$f'_1(H, \gamma(\omega)) = e^{\rho(H)}f'(H, \gamma'(\omega)) \qquad\qquad (H\in\mathfrak{h}_0)$$

where $f'(H) = f(\exp H)$. Hence

$$g'(H, \gamma(\omega)) = \sum_{s\in W}\epsilon(s)e^{s\rho(H)}f'(s^{-1}H, \gamma'(\omega))$$
$$= \sum_{s\in W}\epsilon(s)e^{s\rho(H)}f_0^s(\exp H) \qquad\qquad (H\in V)$$

where $f_0 = \gamma'(\omega)f$ and therefore

$$g(\exp H, D') = d_+(\exp H)F_{\gamma'(\omega)f}(\exp H) \qquad (H \in V).$$

This proves that $DF_f$ coincides with $d_+F_{\gamma'(\omega)f}$ on some neighborhood of $h_0$ in $A'$. Since $h_0$ is an arbitrary point of $A'$ our assertion follows.

Let $\tau$ be a distribution on $K$. Then by Corollary 2 to Lemma 29, the mapping $\Phi: f \to \tau(d_+F_f)$ $(f \in C^\infty(A))$ is a distribution on $A$. The following theorem contains the main result of the present section.

THEOREM 5. *Suppose there exists a complex number $c$ such that $D\tau = cd_+\tau$. Then $\Phi$ coincides with an analytic function on $A$.*

If $f \in C^\infty(A)$, $D\tau(F_f) = \tau(DF_f) = \Phi(\gamma'(\omega)f)$ from Lemmas 32 and 33. But $D\tau(F_f) = c\tau(d_+F_f) = c\Phi(f)$. Hence

$$\Phi(\gamma'(\omega)f) = c\Phi(f) \qquad (f \in C^\infty(A)).$$

Now let $H_1, \cdots, H_l$ be a base for $\mathfrak{h}_0$ over $R$ such that $B(H_i, H_j) = -\delta_{ij}$ $1 \leq i, j \leq l$. Then if we normalize $X_\alpha, X_{-\alpha}$ $(\alpha \in P)$ in such a way that $B(X_\alpha, X_{-\alpha}) = 1$, it is clear that

$$\omega = -(H_1^2 + \cdots + H_l^2) + \sum_{\alpha \in P}(X_\alpha X_{-\alpha} + X_{-\alpha}X_\alpha).$$

Hence

$$\omega \equiv -(H_1^2 + \cdots + H_l^2) + \sum_{\alpha \in P} H_\alpha \bmod \sum_{\alpha \in P} \mathfrak{B}X_\alpha$$

where $H_\alpha = [X_\alpha, X_{-\alpha}]$. This proves (see Lemma 18) that

$$\gamma'(\omega) = -(H_1^2 + \cdots + H_l^2) + 2H_\rho$$

where $H_\rho = 2^{-1}\sum_{\alpha \in P} H_\alpha$ is characterized by the property that $B(H_\rho, H) = \rho(H)$ $(H \in \mathfrak{h})$. Put

$$\square = H_1^2 + \cdots + H_l^2 + 2H_\rho.$$

Then it is clear that $\Phi$ satisfies the differential equation

$$(\square + c)\Phi = 0.$$

Since the operator $\square$ is obviously of the elliptic type (see Gårding [3]) it follows [11; 8] that $\Phi$ is an analytic function on $A$.

We shall give an application of this theorem in §12.

**11. Application to representations.** We shall now apply the preceding theory to the problem of determining the characters of irreducible representations of $G$. Let $\pi$ be a representation of $G$ on a Hilbert space $\mathfrak{H}$ and let $V$ be the Gårding subspace of $\mathfrak{H}$ (see [6(c), p. 201]). We denote by $\Omega$ the set of all equivalence classes of finite-dimensional irreducible representations of $K$ and by $\mathfrak{H}_\mathfrak{D}$ $(\mathfrak{D} \in \Omega)$ the subspace consisting of those elements in $\mathfrak{H}$ which transform under $\pi(K)$ according to $\mathfrak{D}$. Let $\pi_0$ denote the representation of $\mathfrak{B}$ on $V$ corresponding to $\pi$ [6(c), p. 201] and $I$, $I_0$ the identity mappings of $\mathfrak{H}$ and $V$ respectively. We shall say that $\pi$ is quasi-simple if there exist([10])

(1) a homomorphism $\eta_\pi$ of $Z$ into the multiplicative group of complex numbers such that $\pi(\zeta) = \eta_\pi(\zeta)I$ $(\zeta \in Z)$,

(2) a homomorphism $\chi_\pi$ of $\mathfrak{Z}$ into $C$ such that $\pi_0(z) = \chi_\pi(z)I_0$ $(z \in \mathfrak{Z})$, and

(3) an integer $N$ such that dim $\mathfrak{H}_{\mathfrak{D}} \leq Nd(\mathfrak{D})^2$ for all $\mathfrak{D} \in \Omega$. $(d(\mathfrak{D})$ is the degree of any representation in $\mathfrak{D}$.)

$\eta_\pi$ and $\chi_\pi$ respectively are called the central and the infinitesimal characters of $\pi$. Let $Q$ be a bounded linear transformation on $\mathfrak{H}$. We say that $Q$ is summable if there exists a complete orthonormal set $(\psi_j)_{j \in J}$ in $\mathfrak{H}$ and a regular operator $A$ such that([11]) $\sum_{i,j \in J} |(\psi_i, AQA^{-1}\psi_j)| < \infty$. One knows [6(d), Lemma 1] that if $Q$ is summable and $A$, $B$ are bounded linear operators $AQB$ is of the trace class and sp $AQB = $ sp $BAQ = $ sp $QBA$. Hence in particular if $A$ is regular sp $AQA^{-1} = $ sp $Q$. Now $\pi$ being quasi-simple, consider the operator

$$Q_f = \int_G f(x)\pi(x)dx$$

where $dx$ is the Haar measure on $G$ and $f \in C_c^\infty(G)$. We know [6(e), §2] that $Q_f$ is summable and the mapping $T_\pi: f \to $ sp $Q_f$ $(f \in C_c^\infty(G))$ is a distribution on $G$. $T_\pi$ is called the character of $\pi$.

Let $'G$ be the set of quasi-regular elements in $G$ (see §3). In view of the corollary to Lemma 12, $'G$ may be considered as an open submanifold of $G$. Our next object is to prove the following theorem.

THEOREM 6. *Let $\pi$ be a quasi-simple representation of $G$ on $\mathfrak{H}$ and let $T_\pi$ be the character of $\pi$. Then $T_\pi$ coincides with an analytic function on $'G$.*

Since $T_\pi$ is an invariant distribution on $G$ (see [6(d)]), it coincides in the neighborhood of a point $x \in {}'G$ with an analytic function if and only if the same property holds at $yxy^{-1}$ $(y \in G)$. Hence, in view of Lemma 11, it is enough to prove this statement when $x \in {}'A = A \cap {}'G$ in the notation of §5. Let $dm$ denote the Haar measure on $M$.

LEMMA 34. *If $f \in C_c^\infty(G \times {}'M)$ the operator*

$$\int_{G \times {}'M} f(x, m)\pi(xmx^{-1})dxdm$$

*is summable. Moreover if $S_\pi(f)$ denotes its trace, the mapping*

---

([10]) Actually these conditions can be relaxed considerably without affecting our subsequent arguments. But it is hardly worthwhile to do so since they are already weak enough to be satisfied in all cases which are of interest to us.

([11]) As usual $(\phi, \psi)$ denotes the scalar product of two elements $\phi$ and $\psi$ in $\mathfrak{H}$ and $|\phi|$ the norm of $\phi$.

$$S_{\tau}: f \rightarrow S_{\tau}(f) \qquad\qquad (f \in C_c^{\infty}(G \times {}^{\backprime}M))$$

*is a distribution on* $G \times {}^{\backprime}M$.

We can regard $G \times M$ as a Lie group and $\mathfrak{B} \times \mathfrak{M}$ the enveloping algebra of the complexification of its Lie algebra. Then $G \times {}^{\backprime}M$ is an open invariant subset of $G \times M$. For any $b \in \mathfrak{B}$, let $D'(b)$ denote the differential operator on $G \times {}^{\backprime}M$ defined as follows. The local expression $(D'(b))_{x,m}$ of $D'(b)$ at $(x, m)$ $\in G \times {}^{\backprime}M$ is given by

$$(D'(b))_{x,m} = d_+(m)^{-r}\gamma_b(m)$$

in the notation of Lemma 16. Since $m \rightarrow \gamma_b(m)$ is an analytic mapping, it is clear that $D'(b)$ is in fact an analytic differential operator on $G \times {}^{\backprime}M$. On the other hand the mapping $b \rightarrow D'(b)$ is linear and for any given $b \in \mathfrak{B}$, we can find a finite set of linearly independent elements $b_1, \cdots, b_N \in \mathfrak{B}$ such that

$$\mathrm{Ad}\,(x^{-1})b = a_1(x)b_1 + \cdots + a_N(x)b_N$$

where $a_1, \cdots, a_N$ are analytic functions on $G$. Hence

$$(D'(b^{x^{-1}}))_{x,m} = \sum_{i=1}^{N} a_i(x)(D'(b_i))_{x,m}$$

and therefore there exists an analytic differential operator $D(b)$ on $G \times {}^{\backprime}M$ such that $(D(b))_{x,m} = (D'(b^{x^{-1}}))_{x,m}$ $(x \in G,\ m \in {}^{\backprime}M)$. The operator $D(b)$ has the following significance. $\phi$ being any function in $C^{\infty}(G)$, consider the integral

$$\int f(x, m)\phi(xmx^{-1})dxdm \qquad\qquad (f \in C_c^{\infty}(G \times {}^{\backprime}M)).$$

Put $\Phi(x, m) = \phi(xmx^{-1})$ $(x \in G,\ m \in {}^{\backprime}M)$. Then, as we have seen in §5,

$$\Phi(x, b; m, v) = \phi(xmx^{-1}, \mathrm{Ad}\,(x)(\Gamma_m(b \times v)))$$

for $b \in \mathfrak{B}$ and $v \in \mathfrak{M}$. Hence it is obvious that

$$\Phi(x, m, D_{x,m}(b)) = \phi(xmx^{-1}, b) \qquad\qquad (b \in \mathfrak{B})$$

and therefore

$$\int f(x, m)\phi(xmx^{-1}, b)dxdm = \int f(x, m, (D(b))^*)\phi(xmx^{-1})dxdm$$

where $(D(b))^*$ is the adjoint of $D(b)$.

Now in order to prove the summability of the operator

$$Q_f = \int f(x, m)\pi(xmx^{-1})dxdm \qquad\qquad (f \in C_c^{\infty}(G \times {}^{\backprime}M))$$

we proceed as follows. We can choose a homomorphism $\eta$ of $K$ into $C$ such that $\eta$ coincides with $\eta_\pi$ on a subgroup $Z_0$ of $Z$ of finite index (see $[6(d),$ p. 239]). Then the factor group $\overline{K} = K/Z_0$ is compact and we can define a representation $\bar{\pi}$ of $\overline{K}$ on $\mathfrak{H}$ by setting

$$\pi(\bar{u}) = \eta(u^{-1})\pi(u) \qquad\qquad (u \in K)$$

where $u \to \bar{u}$ is the natural mapping of $K$ on $\overline{K}$. Since $\overline{K}$ is compact, $\bar{\pi}$ is equivalent to a unitary representation. Hence in view of our definition of summability, it is sufficient to consider the case when $\bar{\pi}$ is unitary and therefore the spaces $\mathfrak{H}_{\mathfrak{D}}$ $(\mathfrak{D} \in \Omega)$ are mutually orthogonal (see $[6(d),$ p. 244]). For any $\phi \in C^\infty(G)$ define $\phi(b, x) = \phi(x, \operatorname{Ad}(x^{-1})b)$ $(x \in G, b \in \mathfrak{B})$ and let $\Omega_\pi$ be the set of those $\mathfrak{D} \in \Omega$ for which $\mathfrak{H}_{\mathfrak{D}} \neq \{0\}$. Then the following facts are known (see the proof of Lemma 3 of $[6(d)]$). If $z$ is an element lying in the center $\mathfrak{Z}_K$ of $\mathfrak{X}$, there exist, for each $\mathfrak{D} \in \Omega_\pi$, two complex numbers $n_1(\mathfrak{D}, z)$, $n_2(\mathfrak{D}, Z)$ with the following property. If $\psi_1 \in \mathfrak{H}_{\mathfrak{D}1}$, $\psi_2 \in \mathfrak{H}_{\mathfrak{D}2}$ $(\mathfrak{D}_1, \mathfrak{D}_2 \in \Omega_\pi)$ and $\phi(x) = (\psi_1, \pi(x)\psi_2)$ $(x \in G)$, then

$$\phi(z, x) = n_1(\mathfrak{D}_1, z)\phi(x), \qquad \phi(x, z) = \phi(x)n_2(\mathfrak{D}_2, z) \qquad (x \in G).$$

Moreover it is possible to choose $z_1, z_2 \in \mathfrak{Z}_K$ such that the numbers $n_1(\mathfrak{D}, z_1)$, $n_2(\mathfrak{D}, z_2)$ $(\mathfrak{D} \in \Omega_\pi)$ are all real and positive and

$$\sum_{\mathfrak{D} \in \Omega_\pi} \frac{d(\mathfrak{D})^2}{n_1(\mathfrak{D}, z_1)} < \infty, \qquad \sum_{\mathfrak{D} \in \Omega_\pi} \frac{d(\mathfrak{D})^2}{n_2(\mathfrak{D}, z_2)} < \infty.$$

Now let $(\psi_i)_{i \in J}$ be a complete orthonormal base for $\mathfrak{H}$ which is composed of bases for the various $\mathfrak{H}_{\mathfrak{D}}$ $(\mathfrak{D} \in \Omega_\pi)$. Let $J(\mathfrak{D})$ denote the set of those $j \in J$ for which $\psi_j \in \mathfrak{H}_{\mathfrak{D}}$. Then

$$\sum_{i, j \in J} |(\psi_i, Q_f \psi_j)| = \sum_{i, j \in J} \left| \int f(x, m)\phi_{ij}(xmx^{-1})dx\,dm \right|$$

where $\phi_{ij}(x) = (\psi_i, \pi(x)\psi_j)$ $(x \in G)$. On the other hand if $i \in J(\mathfrak{D}_1)$ and $j \in J(\mathfrak{D}_2)$,

$$\phi_{ij}(x, (\operatorname{Ad}(x^{-1})z_1)z_2) = \phi_{ij}(z_1, x, z_2) = n_1(\mathfrak{D}_1, z_1)n_2(\mathfrak{D}_2, z_2)\phi_{ij}(x).$$

Moreover we can find elements $b_1, \cdots, b_r \in \mathfrak{B}$ such that

$$(\operatorname{Ad}(x^{-1})z_1)z_2 = a_1(x)b_1 + \cdots + a_r(x)b_r \qquad\qquad (x \in G)$$

where $a_1, \cdots, a_r$ are analytic functions on $G$. Therefore in view of what we have seen above, it is obvious that there exists a differential operator $D$ on $G \times {}^\prime M$ such that

$$\int f(x, m, D)\phi(xmx^{-1})dx\,dm = \int f(x, m)\phi(xmx^{-1}, (\operatorname{Ad}(xm^{-1}x^{-1})z_1)z_2)dx\,dm$$

for all $f \in C_c^\infty(G \times {}^\prime M)$ and $\phi \in C^\infty(G)$. On the other hand, we can find compact sets $\omega_1$, $\omega_2$ in $G$ and $`M$ respectively such that a given $f$ is zero outside $\omega_1 \times \omega_2$. Let $\omega$ be the image of $\omega_1 \times \omega_2$ in $G$ under the mapping $(x, m) \to xmx^{-1}$. Then $\omega$ is compact and there exists a number $\mu$ such that $|\pi(x)| \leq \mu$ for $x \in \omega$. Then it is clear that

$$\left| \int f(x, m, D)\phi_{ij}(xmx^{-1})dxdm \right| \leq \mu \int |f(x, m, D)| \, dxdm$$

and therefore

$$\sum_{i,j \in J} |(\psi_i, Q_f\psi_j)| \leq \mu N^2 \int |f(x, m, D)| \, dxdm \sum_{\mathfrak{D}_1, \mathfrak{D}_2 \in \Omega_\pi} \frac{d(\mathfrak{D}_1)^2}{n_1(\mathfrak{D}_1, z_1)} \frac{d(\mathfrak{D}_2)^2}{n_2(\mathfrak{D}_2, z_2)}$$

since dim $\mathfrak{H}_\mathfrak{D} \leq Nd(\mathfrak{D})^2$ $(\mathfrak{D} \in \Omega_\pi)$. The series on the right being convergent, this proves that $Q_f$ is summable. Moreover the above result shows that for any compact set $\omega'$ in $G \times `M$, there exists a real number $\mu'$ such that

$$|\mathrm{sp}\, Q_f| \leq \mu' \int |f(x, m, D)| \, dxdm$$

where $f$ is any function in $C_c^\infty(G \times `M)$ which vanishes outside $\omega'$. Therefore the mapping $f \to \mathrm{sp}\, Q_f$ is a distribution and the lemma is proved.

Now suppose $\alpha \in C_c^\infty(G)$ and $\beta \in C_c^\infty(`M)$. We denote by $\alpha \times \beta$ the function $f$ in $C_c^\infty(G \times `M)$ given by $f(x, m) = \alpha(x)\beta(m)$ $(x \in G, m \in `M)$.

LEMMA 35. *There exists a distribution $\tau_\pi$ on $`M$ such that*

$$S_\pi(\alpha \times \beta) = \left( \int_G \alpha(x)dx \right) \tau_\pi(\beta) \qquad (\alpha \in C_c^\infty(G), \beta \in C_c^\infty(`M)).$$

Let $\lambda(y)\alpha$ $(y \in G)$ denote the function $x \to \alpha(y^{-1}x)$ $(x \in G)$. Similarly if $\sigma$ is a distribution on $G$, we denote by $\lambda(y)\sigma$ the distribution $\alpha \to \sigma(\lambda(y^{-1})\alpha)$ $(\alpha \in C_c^\infty(G))$. Let $\beta$ be a fixed element in $C_c^\infty(`M)$. Then it is obvious that the mapping

$$\sigma_\beta: \alpha \to S_\pi(\alpha \times \beta)$$

is a distribution on $G$. Moreover if

$$Q_\alpha = \int \alpha(x)\beta(m)\pi(xmx^{-1})dxdm,$$

it is clear that

$$Q_{\lambda(y)\alpha} = \pi(y)Q_\alpha\pi(y^{-1}).$$

Therefore since $Q_\alpha$ is summable,

$$\sigma_\beta(\alpha) = \mathrm{sp}\, Q_\alpha = \mathrm{sp}\, Q_{\lambda(y)\alpha} = \sigma_\beta(\lambda(y)\alpha).$$

This shows that $\lambda(y)\sigma_\beta = \sigma_\beta$ $(y \in G)$. On the other hand we have the following lemma.

LEMMA 36. *Let $\sigma$ be a distribution on $G$ such that $\lambda(x)\sigma = \sigma$ for all $x \in G$. Then there exists a complex number $c$ such that*

$$\sigma(\alpha) = c \int_G \alpha(x)dx$$

*for every $\alpha \in C_c^\infty(G)$.*

Assuming this for a moment, we can prove Lemma 35 as follows. Put $I(\alpha) = \int_G \alpha(x)dx$ $(\alpha \in C_c^\infty(G))$ and select a function $\alpha_0$ in $C_c^\infty(G)$ for which $I(\alpha_0) = 1$. Then it follows from Lemma 36 that

$$\sigma_\beta(\alpha) = \sigma_\beta(\alpha_0)I(\alpha) \qquad (\alpha \in C_c^\infty(G)).$$

But the mapping $\beta \rightarrow \sigma_\beta(\alpha_0) = S_\pi(\alpha_0 \times \beta)$ is obviously a distribution on ${}^\prime M$. Hence if we denote it by $\tau_\pi$,

$$S_\pi(\alpha \times \beta) = I(\alpha)\tau_\pi(\beta)$$

which proves our assertion.

Now in order to prove Lemma 36, we make use of the theory of right-invariant differential forms on $G$ (see Chevalley [2]). For each $X \in \mathfrak{g}_0$, we denote by $X'$ the right-invariant differential operator defined by

$$\alpha(x, X') = \left\{\frac{d}{dt}\alpha(\exp(-tX)x)\right\}_{t=0} \qquad (x \in G, \alpha \in C^\infty(G)).$$

$n$ being the dimension of $G$, let $\omega$ denote the invariant differential form of degree $n$ corresponding to the Haar measure $dx$. Then

$$\int_G \alpha\omega = I(\alpha) \qquad (\alpha \in C_c^\infty(G)).$$

Now suppose $I(\alpha) = 0$. Then by the de Rham Theorem for forms with compact support, there exists a differential form $\zeta$ of degree $n-1$ and class $C^\infty$ with compact support such that $\alpha\omega = d\zeta$. Let $\omega_1, \cdots, \omega_n$ be a base for right-invariant forms of degree 1. We may assume that $\omega = \omega_1 \wedge \omega_2 \wedge \cdots \wedge \omega_n$ where $\wedge$ denotes exterior product. Let $(X_1, \cdots, X_n)$ be a base for $\mathfrak{g}_0$ over $R$ such that $\omega_i(X_j') = \delta_{ij}$ $(1 \leq i, j \leq n)$. Then

$$\zeta = \sum_{i=1}^n \zeta_i \Omega_i$$

where $\zeta_i \in C_c^\infty(G)$ and $\Omega_i = \omega_1 \wedge \omega_2 \wedge \cdots \wedge \hat{\omega}_i \wedge \cdots \wedge \omega_n$, the circumflex signifying the deletion of the factor $\omega_i$ from the product. Then

$$\alpha\omega = d\zeta = \sum_{i=1}^{n} d\zeta_i \wedge \Omega_i + \sum_{i=1}^{n} \zeta_i d\Omega_i.$$

But $d\Omega_i = 0$ for otherwise $\omega$ would be the derivative of a right-invariant differential form and this is known to be impossible (see Koszul [9, Theorem 9.2]). On the other hand

$$d\zeta_i = \sum_{j=1}^{n} (X_j'\zeta_i)\omega_j$$

and therefore

$$\alpha\omega = \left\{ \sum_{j=1}^{n} (-1)^{i-1} X_j'\zeta_i \right\} \omega.$$

This shows that

$$\alpha = \sum_{j=1}^{n} (-1)^{i-1} X_j'\zeta_j.$$

Hence in order to prove that $I(\alpha)=0$ implies $\sigma(\alpha)=0$ $(\alpha\in C_c^\infty(G))$, it is sufficient to verify that $\sigma(X'\alpha)=0$ for every $X\in\mathfrak{g}_0$ and $\alpha\in C_c^\infty(G)$. Put

$$\alpha_t(x) = \alpha(\exp(-tX)x) \qquad\qquad (x\in G, t\in R).$$

Then if $b\in\mathfrak{B}$,

$$\alpha_t(x, b) = \alpha(\exp(-tX)x, b)$$

and from this it follows that if $\alpha' = X'\alpha$, $t^{-1}\{\alpha_t(x, b) - \alpha(x, b)\}$ converges uniformly on $G$ to $\alpha'(x, b)$ as $t$ tends to zero. This implies that

$$\sigma(\alpha') = \operatorname*{Lim}_{t\to 0} \frac{1}{t} (\sigma(\alpha_t) - \sigma(\alpha)) = 0$$

since $\alpha_t = \lambda(\exp tX)\alpha$. Hence $\sigma(X'\alpha)=0$ and our assertion is proved.

Select a function $\alpha_0 \in C_c^\infty(G)$ such that $I(\alpha_0)=1$. Then if $\alpha\in C_c^\infty(G)$ and $\alpha' = \alpha - I(\alpha)\alpha_0$, $I(\alpha')=0$. Hence $\sigma(\alpha')=0$ and therefore $\sigma(\alpha) = \sigma(\alpha_0)I(\alpha)$. This proves the lemma.

LEMMA 37. *The distribution $\tau_\pi$ on $'M$ is an invariant distribution.*

It is obvious that if $\alpha\in C_c^\infty(G)$, $\beta\in C_c^\infty('M)$ and $m\in M$,

$$S_\pi(\alpha \times \beta^m) = S_\pi(\alpha' \times \beta)$$

where $\alpha'(x) = \alpha(xm^{-1})$ $(x\in G)$. Hence if we choose $\alpha$ such that $\int_G \alpha(x)dx = 1$,

$$\tau_\pi(\beta^m) = S_\pi(\alpha \times \beta^m) = S_\pi(\alpha' \times \beta) = \tau_\pi(\beta)$$

and this shows that $\tau_\pi$ is invariant.

Define the differential operator $\delta(z)$ $(z \in \mathfrak{Z})$ on $`M$ as in Theorem 1.

LEMMA 38. $\delta(z)\tau_\pi = \chi_\pi(z)\tau_\pi$ $(z \in \mathfrak{Z})$ where $\chi_\pi$ is the infinitesimal character of $\pi$.

We use the notation of the proof of Lemma 34. Let $E_\mathfrak{D}$ denote the canonical projection $[6(c), p. 225]$ of $\mathfrak{H}$ on $\mathfrak{H}_\mathfrak{D}$ $(\mathfrak{D} \in \Omega_\pi)$. Then if $\phi_\mathfrak{D}(x) = \mathrm{sp}\,(E_\mathfrak{D}\pi(x)E_\mathfrak{D})$ $(x \in G)$ it is obvious from Lemma 34 that

$$Q(\alpha, \beta) = \int \alpha(x)\beta(m)\pi(xmx^{-1})dx\,dm \qquad (\alpha \in C_c^\infty(G),\ \beta \in C_c^\infty(`M))$$

is of the trace class and therefore (see $[6(d), \S2]$)

$$\mathrm{sp}\,Q(\alpha, \beta) = \sum_{\mathfrak{D} \in \Omega_\pi} \int \alpha(x)\beta(m)\phi_\mathfrak{D}(xmx^{-1})dx\,dm.$$

Also we know (see Lemmas 32 and 34 of $[6(c)]$) that $\phi_\mathfrak{D}$ is an analytic function on $G$ and

$$\phi_\mathfrak{D}(x, z) = \chi_\pi(z)\phi_\mathfrak{D}(x) \qquad (x \in G,\ z \in \mathfrak{Z}).$$

Put $\Phi_\mathfrak{D}(x; m) = \phi_\mathfrak{D}(xmx^{-1})$ $(x \in G,\ m \in `M)$. Then if $b \in \mathfrak{B}$ and $\nu \in \mathfrak{M}$, we have seen in §5 that

$$\Phi_\mathfrak{D}(x, b; m, \nu) = \phi_\mathfrak{D}(xmx^{-1},\ \mathrm{Ad}\,(x)(\Gamma_m(b \times \nu))).$$

On the other hand if $z$ is a fixed element in $\mathfrak{Z}$, we can, from Lemma 16, select $b_1, \cdots, b_r \in \mathfrak{Q}'$, $\nu_1, \cdots, \nu_r \in \mathfrak{M}$ and analytic functions $a_1, \cdots, a_r$ on $`M$ such that

$$\sum_{i=1}^r a_i(m)\Gamma_m(b_i \times \nu_i) + \Gamma_m(1 \times \delta_m(z)) = z.$$

Therefore since $z^x = z$,

$$\sum_{i=1}^r a_i(m)\Phi_\mathfrak{D}(x, b_i; m, \nu_i) + \Phi_\mathfrak{D}(x; m, \delta_m(z)) = \phi_\mathfrak{D}(xmx^{-1}, z) = \chi_\pi(z)\phi_\mathfrak{D}(xmx^{-1}).$$

Let $D_i$ and $\delta^*(z)$ respectively denote the adjoints of the differential operators $a_i(m)\nu_i$ and $\delta(z)$ on $`M$. If $b \to b^*$ $(b \in \mathfrak{B})$ is the anti-automorphism of $\mathfrak{B}$ which maps $X$ on $-X$ $(X \in \mathfrak{g})$, it is obvious that $b_i^*$ is the adjoint of the differential operator $b_i$ on $G$. Hence

$$\chi_\pi(z) \int \alpha(x)\beta(m)\phi_\mathfrak{D}(xmx^{-1})dx\,dm$$

$$= \sum_{i=1}^r \int \alpha(x, b_i^*)\beta(m, D_i)\Phi_\mathfrak{D}(x; m)dx\,dm + \int \alpha(x)\beta(m, \delta^*(z))\Phi_\mathfrak{D}(x; m)dx\,dm$$

and therefore it follows from Lemma 34 that

$$\chi_\pi(z) \, \text{sp} \, Q(\alpha, \beta) = \sum_{i=1}^{r} \text{sp} \, Q(b_i^*\alpha, D_i\beta) + \text{sp} \, Q(\alpha, \delta^*(z)\beta).$$

Now assume that $I(\alpha) = 1$ and apply Lemma 35. Then we get

$$\chi_\pi(z)\tau_\pi(\beta) = \sum_{i=1}^{r} I(b_i^*\alpha)\tau_\pi(D_i\beta) + \tau_\pi(\delta^*(z)\beta).$$

But since $b_i \in \mathfrak{Q}'$ the same holds for $b_i^*$. On the other hand if $X \in \mathfrak{g}_0$ and $f \in C_c^\infty(G)$, it is obvious that $I(f_t) = I(f)$ where $f_t(x) = f(x \exp tX)$ $(x \in G, t \in R)$. From this we conclude immediately that $I(Xf) = 0$. Hence $I(b_i^*\alpha) = 0$ $(1 \le i \le r)$ and therefore

$$\chi_\pi(z)\tau_\pi(\beta) = \tau_\pi(\delta^*(z)\beta) \qquad\qquad (\beta \in C_c^\infty(`M)).$$

This proves the statement of the lemma.

It now follows from Theorem 3 that $\tau_\pi$ is an analytic function on $`M$. In order to prove Theorem 6 we still have to investigate the relationship between the two distributions $T_\pi$ and $\tau_\pi$. Put $\mathfrak{q}_0 = \mathfrak{q} \cap \mathfrak{g}_0$. Then $\mathfrak{g}_0$ is the direct sum of $\mathfrak{q}_0$ and $\mathfrak{m}_0 = \mathfrak{m} \cap \mathfrak{g}_0$. Let $U'$ be an open neighborhood of 1 in $G$ such that if $U'' = (U')^{-1}U'$ then $U'' \cap (\tilde{A}M) \subset M$ where $\tilde{A}$ is the normalizer of $A$ in $G$. This is possible since $\tilde{A}M/M$ is a finite group (see Lemma 10). Let $G^* = G/M$ be the factor space consisting of all cosets of the form $xM$ $(x \in G)$ and let $\sigma$ denote the natural mapping $x \to x^*$ of $G$ on $G^*$. We can select an open neighborhood $U^*$ of $1^*$ in $G^*$ and a regular analytic mapping $\phi$ of $U^*$ into $U'$ such that $(\phi(x^*))^* = x^*$ and $\phi(1^*) = 1$ $(x^* \in U^*)$. Let $dx^*$ denote the invariant measure on $G^*$. Define an analytic mapping $\psi$ of $U^* \times `M$ into $G$ as follows: $\psi(x^*, m) = \phi(x^*)m(\phi(x^*))^{-1}$ $(x^* \in U^*, m \in `M)$. Suppose $\psi(x_1^*, m_1) = \psi(x_2^*, m_2)$ $(x_1^*, x_2^* \in U^*, m_1, m_2 \in `M)$. Then from Lemma 10

$$(\phi(x_1^*))^{-1}\phi(x_2^*) \in U'' \cap (\tilde{A}M) \subset M$$

and therefore $x_2^* = x_1^*$, $m_2 = m_1$. This shows that $\psi$ is univalent on $U^* \times `M$. For $Y \in \mathfrak{g}_0$ and $x^* \in U^*$ put[12] $Y_{x^*}^* = d\sigma_x Y$ where $x = \phi(x^*)$. Then if we identify the tangent space of $U^* \times `M$ at $(x^*, m)$ with $\mathfrak{g}_0$ under the linear isomorphism

$$Y + X \to Y_{x^*}^* \times X \qquad\qquad (Y \in \mathfrak{q}_0, X \in \mathfrak{m}_0),$$

a simple calculation shows that

$$d\psi_{x^*,m}(X + Y) = (\text{Ad} \, \phi(x^*))X + \text{Ad} \, \phi(x^*)\{\text{Ad} \, (m^{-1}) - I\}(d\phi_{x^*}Y_{x^*}^*).$$

(Here $I$ is the identity mapping of $\mathfrak{g}$.) Moreover $\sigma \circ \phi$ being the identity mapping on $U^*$,

---

[12] We follow here the terminology of Chevalley [2].

$$d\sigma_x(d\phi_{x^*}Y^*_{x^*} - Y) = 0$$

if $x = \phi(x^*)$ and therefore

$$d\phi_{x^*}Y^*_{x^*} \equiv Y \bmod \mathfrak{M}_0.$$

Therefore if $D_{x^*,m} = (\mathrm{Ad}\ \phi(x^*))^{-1}d\psi_{x^*,m}$,

$$D_{x^*,m}X = X \qquad\qquad (X \in \mathfrak{m}_0),$$
$$D_{x^*,m}Y \equiv \mathrm{Ad}\ (m^{-1})Y - Y \bmod \mathfrak{M}_0 \qquad\qquad (Y \in \mathfrak{q}_0)$$

and so it follows that

$$\det\ (d\psi_{x^*,m}) = \det D_{x^*,m} = d_+(m) \neq 0.$$

This proves that $\psi$ is everywhere regular on $U^* \times {}^{\backprime}M$. Hence it maps $U^* \times {}^{\backprime}M$ topologically on an openneighborhood $N$ of ${}^{\backprime}M$ in $G$. We regard $N$ as an open submanifold of $G$. Let $f \in C_c^\infty(N)$ and put

$$F(x^*, m) = f(\psi(x^*, m)) \qquad\qquad (x^* \in U^*, m \in {}^{\backprime}M).$$

Since $\psi$ defines an isomorphism of the two analytic manifolds $U^* \times {}^{\backprime}M$ and $N$, it is clear that $F \in C_c^\infty(U^* \times {}^{\backprime}M)$. Choose bases $(X_1, \cdots, X_p)$ and $(Y_1, \cdots, Y_q)$ for $\mathfrak{m}_0$ and $\mathfrak{q}_0$ respectively over $R$ and let $\omega^*$ and $\zeta$ denote the differential forms (of degrees $q$ and $p$) on $G^*$ and $M$ respectively corresponding to the measures $dx^*$ and $dm$. Then we may assume that

$$\omega^*_{x^*}(d\sigma_x Y_1, \cdots, d\sigma_x Y_q) = \zeta_m(X_1, \cdots, X_p) = 1 \qquad (x \in G, m \in M).$$

On the other hand if $\omega$ is the differential form on $G$ of degree $p+q$ corresponding to the measure $dx$, we may assume that $\omega_x(Y_1, \cdots, Y_q, X_1, \cdots, X_p) = 1$ $(x \in G)$. Now consider the form[12] $\omega' = (\delta\psi)\omega$ on $U^* \times {}^{\backprime}M$. Then it follows from the above result that

$$\omega'_{x^*,m}(Y_1, \cdots, Y_q, X_1, \cdots X_p)$$
$$= d_+(m)\omega_y(Y_1, \cdots, Y_q, X_1 X_1, \cdots X_p) = d_+(m)$$

where $y = \psi(x^*, m)$ $(x^* \in U^*, m \in {}^{\backprime}M)$ and this proves that

$$\omega'_{x^*,m} = d_+(m)\xi_{x^*,m}$$

where $\xi$ is the direct product of $\omega^*$ and $\zeta$ on $U^* \times {}^{\backprime}M$ so that

$$\xi_{x^*,m}(Y_1, \cdots, Y_q, X_1, \cdots, X_p)$$
$$= \omega^*_{x^*}((Y_1)^*_{x^*}, \cdots, (Y_q)^*_{x^*})\zeta_m(X_1, \cdots, X_p) = 1.$$

But this implies that

$$\int_N f(x)dx = \int_{U^* \times {}^{\backprime}M} F(x^*, m)\,|\,d_+(m)\,|\,dx^*dm.$$

Thus we have obtained the following result.

LEMMA 39. *If the measures $dx^*$, $dm$ and $dx$ on $G^*$, $M$ and $G$ respectively are suitably normalized,*

$$\int f(x)dx = \int_{U^* \times {}^{\backprime}M} f(\psi(x^*, m)) \, | \, d_+(m) \, | \, dx^* dm$$

*for every $f \in C_c^\infty(N)$.*

Now $\psi$ being univalent and regular on $U^* \times {}^{\backprime}M$, there exists an analytic function $F_\pi$ on $N$ such that $F_\pi(\psi(x^*, m)) = \tau_\pi(m)$ $(x^* \in U^*, m \in {}^{\backprime}M)$.

LEMMA 40. *Suppose the various measures have been normalized in accordance with Lemma 39. Then*

$$T_\pi(f) = \int f(x)F_\pi(x)dx$$

*for every $f \in C_c^\infty(N)$.*

For a given $f \in C_c^\infty(N)$ we can choose a compact subset $\omega^*$ of $U^*$ such that $f(\psi(x^*, m)) = 0$ $(x^* \in U^*, m \in {}^{\backprime}M)$ unless $x^* \in \omega^*$. Let $\alpha^*$ be a function in $C_c^\infty(U^*)$ which is identically equal to 1 on $\omega^*$. The mapping $(x^*, m) \to \phi(x^*)m$ is easily seen to be regular and univalent on $U^* \times M$. Therefore it defines an analytic isomorphism of $U^* \times M$ with the open submanifold $U = \phi(U^*)M$ of $G$. Select $\beta \in C_c^\infty(M)$ such that $\int_M \beta(m)dm = 1$. Then there exists an element $\alpha \in C_c^\infty(U)$ such that $\alpha(\phi(x^*)m) = \alpha^*(x^*)\beta(m)$ $(x^* \in U^*, m \in M)$. Now $g$ being any indefinitely differentiable function on $N$, put $f' = gf$. Then

$$\int_{G \times {}^{\backprime}M} \alpha(x)f'(xmx^{-1}) \, | \, d_+(m) \, | \, dxdm$$

$$= \int_{G^* \times M \times {}^{\backprime}M} \alpha^*(x^*)\beta(m_1)f'(\psi(x^*, m_1 m m_1^{-1})) \, | \, d_+(m) \, | \, dx^* dm_1 dm$$

since $dx = dx^* dm_1$ if $x = \phi(x^*)m_1$ $(x^* \in U^*, m_1 \in M)$. But $d_+(m_1^{-1} m m_1) = d_+(m)$ and therefore

$$\int_{{}^{\backprime}M} f'(\psi(x^*, m_1 m m_1^{-1})) \, | \, d_+(m) \, | \, dm = \int_{{}^{\backprime}M} f'(\psi(x^*, m)) \, | \, d_+(m) \, | \, dm.$$

Moreover $\int_M \beta(m_1)dm_1 = 1$ and $\alpha^* = 1$ on $\omega^*$. Hence

$$\int_{G \times {}^{\backprime}M} \alpha(x)f'(xmx^{-1}) \, | \, d_+(m) \, | \, dxdm = \int_{G^* \times {}^{\backprime}M} f'(\psi(x^*, m)) \, | \, d_+(m) \, | \, dx^* dm$$

$$= \int_G f'(x)dx$$

from Lemma 39. This proves that

$$\int f(x)g(x)dx = \int_{G\times`M} \alpha(x)f(xmx^{-1}) \, g(xmx^{-1}) \, \big| d_+(m) \big| \, dxdm$$

if $g\in C^\infty(N)$. On the other hand

$$T_\pi(f) = \sum_{\mathfrak{D}\in\Omega_\pi} \int f(x)\phi_\mathfrak{D}(x)dx$$

in the notation of the proof of Lemma 38. Therefore

$$T_\pi(f) = \sum_{\mathfrak{D}\in\Omega_\pi} \int \alpha(x)f(xmx^{-1})\phi_\mathfrak{D}(xmx^{-1}) \, \big| d_+(m) \big| \, dxdm.$$

Now put

$$f_1(x, m) = \alpha(x)f(xmx^{-1}) \big| d_+(m) \big| \qquad (x \in G, \, m \in `M).$$

Then $f_1\in C_c^\infty(G\times`M)$ and it follows from Lemma 34 that the right hand side of the above equation is equal to $S_\pi(f_1)$. On the other hand

$$S_\pi(f_1) = \int_{G\times`M} f_1(x, m)\tau_\pi(m)dxdm$$

from Lemma 35 (see Schwartz [11, p. 108]). Therefore

$$T_\pi(f) = \int_{G\times M} \alpha(x)f(xmx^{-1})\tau_\pi(m) \, \big| d_+(m) \big| \, dxdm = \int_G f(x)F_\pi(x)dx$$

from our result above. This proves Lemma 40 and since $N\supset`M\supset`A$, Theorem 6 follows.

For any quasi-simple representation $\pi$ of $G$ define $\eta_\pi$, $\chi_\pi$ and $T_\pi$ as above and let $F_\pi$ be the analytic function (on $`G$) which coincides with $T_\pi$ on $`G$. From the corollary to Lemma 12 we can choose a finite number of connected components $G_1, \cdots, G_N$ of $`G$ such that $`G=\bigcup_{i=1}^N ZG_i$. Then if $w$ is the order of the Weyl group of $\mathfrak{g}$ (with respect to any Cartan subalgebra $\mathfrak{h}$), we obtain the following result by combining Theorem 4 and Lemma 40.

THEOREM 7. *Suppose $\eta$ and $\chi$ respectively are given homomorphisms of $Z$ and $\mathfrak{Z}$ into $C$ and $\omega$ is a collection of quasi-simple representations $\pi$ of $G$ such that $\eta=\eta_\pi$ and $\chi=\chi_\pi$. Then the dimension of the linear space spanned over $C$ by the functions $F_\pi$ $(\pi\in\omega)$ cannot exceed $Nw$.*

Since

$$\int f(z^{-1}x)\pi(x)dx = \int f(x)\pi(zx)dx = \eta(z) \int f(x)\pi(x)dx \qquad (z \in Z)$$

and

$$T_\pi(f) = T_\pi(f^y) \qquad (y \in G)$$

for $f \in C_c^\infty(G)$ and $\pi \in \omega$, it is obvious that $F_\pi(zx) = \eta(z)F_\pi(x)$ and $F_\pi(yxy^{-1}) = F_\pi(x)$ $(x \in {}'G)$. Therefore if $V$ is the linear space spanned by $F_\pi$ $(\pi \in \omega)$, similar relations hold for every $F \in V$. Let $F_i$ denote the restriction on $G_i$ of a function $F$ in $V$. If $F$ vanishes identically on some nonempty open subset of $G_i$ then $F_i = 0$. Now the set $\bigcup_{x \in G} xG_ix^{-1}$ is connected since it is the image of $G \times G_i$ under the continuous mapping $(x, y) \to xyx^{-1}$. Hence it follows that $xG_ix^{-1} = G_i$ $(x \in G)$. Therefore from Lemma 11, $G_i$ must meet at least one Cartan subgroup $A_i$ of $G$ for which $\theta(A_i) = A_i$. Let $A_i'$ be the set of regular elements in $A_i$. By Lemma 8 we can select a nonempty open connected subset $B_i$ of $A_i'$ such that $B_i \subset A_i \cap G_i$. One proves without difficulty that $U_i = \bigcup_{x \in G} xB_ix^{-1}$ is open in $G$ and therefore if $F$ is everywhere zero on $B_i$, $F_i = 0$. Let $\overline{F}_i$ denote the restriction of $F$ on $B_i$ and let $\overline{V}_i$ be the linear space consisting of all $\overline{F}_i(F \in V)$. It follows from Lemma 38 and Theorem 4 that $\dim \overline{V}_i \leq w$. Now $\overline{V}$ being the direct sum of $\overline{V}_i$ $(1 \leq i \leq N)$, consider the linear mapping $F \to \overline{F} = (\overline{F}_1, \cdots, \overline{F}_N)$ of $V$ into $\overline{V}$. Since ${}'G = \bigcup_{i=1}^N ZG_i$, it is clear that $F = 0$ whenever $\overline{F} = 0$. Hence $\dim V \leq \dim \overline{V} \leq Nw$.

In view of Theorem 7 we can choose a finite number of representations $\pi_1, \cdots, \pi_r$ in $\omega$ such that $F_{\pi_1}, \cdots, F_{\pi_r}$ form a base for $V$. Then for any $\pi \in \omega$, $F_\pi = c_1 F_{\pi_1} + \cdots + c_r F_{\pi_r}$, where $c_1, \cdots, c_r$ are complex numbers. From this one would like to deduce that $T_\pi = c_1 T_{\pi_1} + \cdots + c_r T_{\pi_r}$. Although this conclusion seems to be in agreement with experience, I have not succeeded in constructing a proof of it. If true it has certain interesting consequences (see $[6(h)]$).

**12. Closer study of a special case.** We now make the two assumptions of §10 on $G$ and use the notation introduced there. Put

$$\Delta_-' = \prod_{\alpha \in P_-} (1 - \xi_\alpha^{-1})$$

and normalize the Haar measures $dk$ and $dh$ on $K$ and $A$ respectively in such a way that

$$\int_K dk = \int_A |\Delta_-'(h)|^2 dh = 1.$$

LEMMA 41. *Let $\pi$ be a quasi-simple representation of $G$. Then for any $f \in C^\infty(K)$ the operator*

$$\int_K f(k)\pi(k)dk$$

*is summable and if $\tau_\pi(f)$ denotes its trace, the mapping*

$$\tau_\pi : f \to \tau_\pi(f) \qquad\qquad (f \in C^\infty(K))$$

*is a distribution on $K$.*

This follows immediately if we apply the reasoning of $[6(d), \S 5]$ to the representation $k \to \pi(k)$ of $K$.

COROLLARY. *The distribution $\tau_\pi$ defined above coincides on ${}^\prime K$ with that of Lemma 35.*

For suppose $\alpha \in C_c^\infty(G)$ and $f \in C^\infty(K)$. Since $\int_K f(k)\pi(k)dk$ is summable, it follows without difficulty (see $[6(d), \text{ p. 237}]$) that

$$\int_{G \times K} \alpha(x)f(k)\pi(xkx^{-1})dxdk$$

is of the trace class and

$$\mathrm{sp}\left(\int_{G \times K} \alpha(x)f(k)\pi(xkx^{-1})dxdk\right) = \int_G \alpha(x)dx\ \mathrm{sp}\left(\int_K f(k)\pi(k)dk\right)$$

$$= \int_G \alpha(x)dx\ \tau_\pi(f).$$

Our assertion is a direct consequence of this fact.

On the other hand we have the following two lemmas.

LEMMA 42. *For any $f \in C^\infty(A)$ the operator*

$$\int_{A \times K} f(h)\Delta'_-(h)\pi(khk^{-1})dhdk$$

*is summable and if $\sigma_\pi(f)$ denotes its trace, the mapping*

$$\sigma_\pi : f \to \sigma_\pi(f) \qquad\qquad (f \in C^\infty(A))$$

*is a distribution on $A$.*

The proof of this is substantially the same as that of Lemma 24 of $[6(g)]$.

Now put $\rho_- = \sum_{\alpha \in P_-} \alpha$. Then if $s \in W_-$, $s\rho_- - \rho_-$ is a linear combination of roots in $P_-$ with integer coefficients (see Weyl $[14(b)]$). Hence $\xi_s^- = \xi_{s\rho_- - \rho_-}$ is a well-defined character of $A$ and

$$\Delta'_- = \sum_{s \in W_-} \epsilon(s)\xi_s^-.$$

Now if we apply the method of proof of Lemma 29 to the pair $(K, A)$ we obtain the following result.

LEMMA 43. *For any $f \in C^\infty(A)$ there exists a unique function $\phi_f \in I^\infty(K)$ such that*

$$\Delta'_-(h)\phi_f(h) = \sum_{s \in W_-} \epsilon(s)\xi_s^-(h)f^s(h) \qquad\qquad (h \in A).$$

COROLLARY. *$\sigma_\tau$ coincides on $`A$ with the analytic function $h \to \Delta'_-(h)\tau_\tau(h)$* $(h \in `A)$.

For suppose $f \in C_c^\infty(`A)$. Then it is obvious that $\phi_f \in I_c^\infty(`K)$ and

$$\int_K \phi_f(k)\pi(k)dk = \int_{A \times K} \phi_f(h) \, |\, \Delta'_-(h)\, |^2 \pi(khk^{-1})dhdk.$$

On the other hand if $h \in A$ and $s \in W_-$,

$$\int_K \pi(kh^s k^{-1})dk = \int_K \pi(khk^{-1})dk$$

and

$$\Delta'_-(h^s) = \epsilon(s)\xi_s^-(h^s)\Delta'_-(h).$$

Therefore[13]

$$\int_K \phi_f(k)\pi(k)dk = w_- \int_{A \times K} f(h)(\operatorname{conj} \Delta'_-(h))\pi(khk^{-1})dhdk$$

where $w_-$ is the order of $W_-$. On the other hand since $\tau_\tau$ coincides with an analytic function on $`K$,

$$\operatorname{sp}\left( \int_K \phi_f(k)\pi(k)dk \right) = \int_K \phi_f(k)\tau_\tau(k)dk = \int_A \phi_f(h) \, |\, \Delta'_-(h)\, |^2 \tau_\tau(h)dh.$$

Again since $\tau_\tau$ is an invariant function on $`K$ (Lemma 37), $\tau_\tau(h^s) = \tau_\tau(h)$ $(s \in W_-, h \in `A)$ and therefore

$$\int_A \phi_f(h) \, |\, \Delta'_-(h)\, |^2 \tau_\tau(h)dh = w_- \int_A f(h) \operatorname{conj} \Delta'_-(h)\tau_\tau(h)dh.$$

But

$$\operatorname{conj} \Delta'_- = \prod_{\alpha \in P_-} (1 - \xi_\alpha) = (-1)^r \xi_{2\rho_-} \Delta'_-$$

where $r$ is the number of roots in $P_-$. Hence replacing $f$ by $f\xi_{2\rho_-}^{-1}$, we obtain

$$\operatorname{sp}\left( \int_{A \times K} f(h)\Delta'_-(h)\pi(khk^{-1})dhdk \right) = \int f(h)(\Delta'_-(h)\tau_\tau(h)dh.$$

This proves our assertion.

We shall now try to obtain some information regarding the possible "singularities" of $\sigma_\tau$ on $A$. In an earlier paper [6(g), §10] we made a detailed study of certain examples and found that there one could always select a

---

[13] conj $c$ denotes the conjugate of a complex number $c$.

meromorphic function on $A$ so as to coincide with $\sigma_\pi$ on $`A$. The following result, although it is weaker, holds under quite general conditions. Define $\Delta'$ as in §10.

THEOREM 8. *Let $S_\pi$ be the distribution on $A$ given by*

$$S_\pi(f) = \mathrm{sp} \left\{ \int_{A \times K} |\Delta'(h)|^2 \sum_{s \in W} f^s(h) \pi(khk^{-1}) dh dk \right\} \qquad (f \in C^\infty(A)).$$

*Then $S_\pi$ coincides with an analytic function on $A$.*

That the above operator is actually summable and $S_\pi$ is a distribution follows immediately from Lemma 42. Moreover we have seen that if $g \in C^\infty(K)$ and $\alpha \in C_c^\infty(G)$,

$$\int \alpha(x) g(k) \pi(xkx^{-1}) dx dk$$

is of the trace class and

$$\mathrm{sp} \left( \int_{G \times K} \alpha(x) g(k) \pi(xkx^{-1}) dx dk \right) = \int_G \alpha(x) dx \, \tau_\pi(g).$$

Hence we conclude that

$$\int_G \alpha(x) dx \, \tau_\pi(g) = \sum_{\mathfrak{D} \in \Omega_\pi} \int_{G \times K} \alpha(x) g(k) \phi_\mathfrak{D}(xkx^{-1}) dx dk$$

the series being absolutely convergent. (Here the notation is the same as in the proof of Lemma 38.) Now choose $\alpha$ such that $\int \alpha(x) dx = 1$. Then

$$\tau_\pi(d_+ g) = \sum_{\mathfrak{D} \in \Omega_\pi} \int_{G \times K} \alpha(x) d_+(k) g(k) \phi_\mathfrak{D}(xkx^{-1}) dx dk$$

for $g \in C^\infty(K)$. On the other hand $\omega$ being the Casimir operator of $\mathfrak{g}$,

$$\phi_\mathfrak{D}(x, \omega) = \chi_\pi(\omega) \phi_\mathfrak{D}(x) \qquad (x \in G, \mathfrak{D} \in \Omega_\pi)$$

as we saw in §11. Therefore

$$\chi_\pi(\omega) \tau_\pi(d_+ g) = \sum_{\mathfrak{D} \in \Omega_\pi} \int_{G \times K} \alpha(x) d_+(k) g(k) \phi_\mathfrak{D}(xkx^{-1}, \omega) dx dk.$$

Now put

$$\Phi_\mathfrak{D}(x; k) = \phi_\mathfrak{D}(xkx^{-1}) \qquad (x \in G, k \in K, \mathfrak{D} \in \Omega_\pi)$$

and define $\gamma_\pi(k)$ and $D$ as in Lemmas 31 and 32. We have seen during the proof of Lemma 31 that

$$\gamma_\omega(k) = \sum_{i=1}^{r} a_i(k)(b_i \times 1) + 1 \times D_k \qquad (k \in K)$$

where $b_1, \cdots, b_r$ are homogeneous elements in $\mathfrak{Q}$ of degree 2, $a_1, \cdots, a_r$ are analytic functions on $K$ and $D_k$ is the local expression at $k$ of the differential operator $D$. Since $\omega^x = \omega$ $(x \in G)$ and $\Gamma_k(\gamma_\omega(k)) = d_+(k)\omega$ $(k \in K)$, we conclude (see §5) that

$$\sum_{i=1}^{r} a_i(k)\Phi_{\mathfrak{D}}(x, b_i; k) + \Phi_{\mathfrak{D}}(x; k, D_k) = d_+(k)\phi_{\mathfrak{D}}(xkx^{-1}, \omega)$$

and therefore

$$\int_{G \times K} \alpha(x)d_+(k)g(k)\phi_{\mathfrak{D}}(xkx^{-1}, \omega)dxdk$$

$$= \sum_{i=1}^{r} \int_{G \times K} \alpha(x, b_i^*)a_i(k)g(k)\phi_{\mathfrak{D}}(xkx^{-1})dxdk + \int_{G \times K} \alpha(x)g(k, D^*)\phi_{\mathfrak{D}}(xkx^{-1})dxdk$$

where $D^*$ is the adjoint of $D$ and $b_i^*$ is the image of $b_i$ under the anti-automorphism of $\mathfrak{B}$ which maps $X$ on $-X(X \in \mathfrak{g})$. Summing over all $\mathfrak{D} \in \Omega_r$, we get

$$\chi_\pi(\omega)\tau_\pi(d_+g) = \sum_{i=1}^{r} \int_G \alpha(x, b_i^*)dx\tau_\pi(a_ig) + \tau_\pi(D^*g).$$

But $b_i^*$ being homogeneous of degree 2 in $\mathfrak{B}$, it is obvious that

$$\int_G \alpha(x, b_i^*)dx = 0$$

and therefore

$$\tau_\pi(D^*g) = \chi_\pi(\omega)\tau_\pi(d_+g) \qquad (g \in C^\infty(K)).$$

This proves that $D\tau_\pi = \chi_\pi(\omega)d_+\tau_\pi$.

Now define the mapping $f \rightarrow F_f$ of $C^\infty(A)$ into $I^\infty(K)$ as in Corollary 2 to Lemma 29 and let $\Phi$ denote the distribution $f \rightarrow \tau_\pi(d_+F_f)$ $(f \in C^\infty(A))$. Then it follows from Theorem 5 that $\Phi$ coincides with an analytic function on $A$. On the other hand if $f \in C^\infty(A)$,

$$\int_K d_+(k)F_f(k)\pi(k)dk = \int_{A \times K} d_+(h)|\Delta'_-(h)|^2 F_f(h)\pi(khk^{-1})dhdk$$

and $d_+(h)|\Delta'_-(h)|^2 = |\Delta'(h)|^2$ $(h \in A)$. Now put

$$g = \sum_{s \in W} f^s.$$

Then

$$\Delta'(h)F_g(h) = \left( \sum_{s \in W} \epsilon(s)\xi_s(h) \right) g(h) = \Delta'(h)g(h) \qquad (h \in A)$$

and

$$\Phi(g) = \tau_\pi(d_+F_g) = \mathrm{sp}\left( \int_K d_+(k)F_g(k)\pi(k)dk \right)$$

$$= \mathrm{sp}\left( \int_{A \times K} |\Delta'(h)|^2 g(h)\pi(khk^{-1})dhdk \right).$$

Therefore if we put $\psi(h) = \sum_{s \in W} \Phi(h^s)$ $(h \in A)$, it is clear that

$$S_\pi(f) = \int_A f(h)\psi(h)dh \qquad (f \in C^\infty(A)).$$

This shows that $S_\pi$ coincides with $\psi$ on $A$.

The following lemma establishes a connection with finite-dimensional representations.

LEMMA 44. *Let $S_\pi$ be the analytic function on $A$ corresponding to Theorem 8. Then there exists an irreducible finite-dimensional representation $\sigma$ of $G$ and a complex number $c$ such that*

$$S_\pi(h) = c\,|\Delta'(h)|^2 T_\sigma(h) \qquad (h \in A).$$

*Here $T_\sigma$ is the character of $\sigma$.*

Let $A'$ be the set of regular elements in $A$. It is clear that if $h \in A'$, the same holds for $h^s$ $(s \in W)$. It follows from the corollary to Lemma 43 that

$$S_\pi(h) = |\Delta'(h)|^2 \bar\tau_\pi(h) \qquad (h \in A')$$

where

$$\bar\tau_\pi(h) = \sum_{s \in W} \tau_\pi(h^s) \qquad (h \in A').$$

Let $\mathfrak{h}_0'$ be the complete inverse image of $A'$ under the exponential mapping of $\mathfrak{h}_0$ into $A$ and let $V$ be a nonempty open connected subset of $\mathfrak{h}_0'$. Then from Theorem 4,

$$S_\pi(\exp H) = \Delta(H) \sum_{s \in W} p_s(H)e^{s\Lambda(H)} \qquad (H \in V)$$

where $\Lambda$ is a linear function and $p_s$ $(s \in W)$ are polynomial functions on $\mathfrak{h}$. ($\Delta(H)$ is defined as in §8.) But since both sides of this equation may be regarded as analytic functions on $\mathfrak{h}_0$, they must be equal everywhere. Moreover we may assume that $p_{st} = p_s$ if $t\Lambda = \Lambda$ $(s, t \in W)$. Then since $S_\pi(\exp tH) = S_\pi(\exp H)$ $(t \in W, H \in \mathfrak{h}_0)$ we conclude from Lemma 41 of $[6(\mathrm{b})]$ that

$$\epsilon(t)p_s(H) = p_{ts}(tH) \qquad (s, t \in W, H \in \mathfrak{h}_0).$$

Hence if we put $p = p_1$ and define $p^s$ by $p^s(H) = p(s^{-1}H)$ $(s \in W, H \in \mathfrak{h})$, it follows that $p_s = \epsilon(s)p^s$. Therefore

$$S_\pi(\exp H) = \Delta(H) \sum_{s \in W} \epsilon(s)p^s(H)e^{s\Lambda(H)} \qquad (H \in \mathfrak{h}_0).$$

Let $H_0$ be an element in $\mathfrak{h}_0$ such that $\exp H_0 = 1$. Then $e^{\alpha(H_0)} = 1$ for every root $\alpha$ and therefore $\Delta(H+H_0) = e^{-\rho(H_0)}\Delta(H)$ $(H \in \mathfrak{h})$. But since $S_\pi(\exp H) = S_\pi(\exp (H+H_0))$, we conclude that

$$p^s(H + H_0) \exp (s\Lambda(H_0) - \rho(H_0)) = p^s(H) \qquad (s \in W, H \in \mathfrak{h}_0).$$

In particular if $s = 1$,

$$p(H) = p(H + H_0) \exp (\Lambda(H_0) - \rho(H_0)).$$

Now we may suppose that $S_\pi \neq 0$ for otherwise the statement of the lemma is true trivially. Then $p \neq 0$ and if $p'$ is the homogeneous component of $p$ of the highest degree, it is obvious that

$$p' = p' \exp (\Lambda(H_0) - \rho(H_0))$$

so that $\exp (\Lambda(H_0) - \rho(H_0)) = 1$ and $p(H) = p(H+H_0)$. This means that $p$ can be regarded as a continuous function on $A$ and so it must be bounded. But a polynomial function cannot be bounded on $\mathfrak{h}_0$ unless it is a constant. Moreover it is clear that there exists a character $\theta$ of $A$ such that

$$\theta(\exp H) = \exp (\Lambda(H) - \rho(H)) \qquad (H \in \mathfrak{h}_0).$$

Therefore

$$S_\pi(h) = c' (\operatorname{conj} \Delta'(h)) \sum_{s \in W} \epsilon(s)\xi_s(h)\theta^s(h) \qquad (h \in A)$$

where $c'$ is a constant. But as we have seen during the proof of Lemma 29, there exists an irreducible finite-dimensional representation $\sigma$ of $G$ such that

$$\sum_{s \in W} \epsilon(s)\xi_s(h)\theta^s(h) = \epsilon\Delta'(h)T_\sigma(h) \qquad (h \in A)$$

where $T_\sigma$ is the character of $\sigma$ and $\epsilon = \pm 1$. Therefore if $c = \epsilon c'$,

$$S_\pi(h) = c \, | \Delta'(h) |^2 T_\sigma(h) \qquad (h \in A)$$

## References

1. E. Cartan, (a) Ann. École Norm. vol. 44 (1927) pp. 345–467.
    (b) J. Math Pures Appl. vol. 8 (1929) pp. 1–33.
2. C. Chevalley, *Theory of Lie groups*, Princeton University Press, 1946.
3. L. Gårding, Math. Scand. vol. 1 (1953) pp. 55–72.
4. I. M. Gelfand and M. A. Naimark, Trudi Mat. Inst. Steklova vol. 36 (1950).
5. R. Godement, Trans. Amer. Math. Soc. vol. 73 (1953) pp. 496–556.

6. Harish-Chandra, (a) Ann. of Math. vol. 50 (1949) pp. 900–915.
    (b) Trans. Amer. Math. Soc. vol. 70 (1951) pp. 28–96.
    (c) Trans. Amer. Math. Soc. vol. 75 (1953) pp. 185–243.
    (d) Trans. Amer. Math. Soc. vol. 76 (1954) pp. 234–253.
    (e) Trans. Amer. Math. Soc. vol. 76 (1954) pp. 485–528.
    (f) Amer. J. Math. vol. 77 (1955) pp. 743–777.
    (g) Amer. J. Math. vol. 78 (1956) pp. 1–41.
    (h) Bull. Amer. Math. Soc. vol. 61 (1955) pp. 389–396.
7. K. Iwasawa, Ann. of Math. vol. 50 (1949) pp. 507–557.
8. F. John, Proceedings of the Symposium on Spectral Theory and Differential Problems, Stillwater, Okla., 1951, pp. 113–175.
9. J. L. Koszul, Bull. Soc. Math. France vol. 78 (1950) pp. 65–127.
10. G D. Mostow, Bull. Amer. Math. Soc. vol. 55 (1949) pp. 969–980.
11. L. Schwartz, *Theorie des distributions* I, Paris, Hermann, 1950.
12. E Stiefel, Comment. Math. Helv. vol. 14 (1942) pp. 350–380.
13. B. L. van der Waerden, *Moderne Algebra*, Berlin, Springer, 1937.
14. H. Weyl, (a) Math. Zeit. vol. 24 (1925) pp. 328–395.
    (b) *The structure and representations of continuous groups*, Princeton, The Institute for Advanced Study, 1935.
    (c) *The classical groups*, Princeton University Press, 1939.

INSTITUTE FOR ADVANCED STUDY,
    PRINCETON, N. J.
COLUMBIA UNIVERSITY,
    NEW YORK, N. Y.

Reprinted from
*Trans. Amer. Math. Soc.*
**83** (1956), 98–163

*On a lemma of F. Bruhat;*

## By HARISH-CHANDRA.

In a recent Note [1(*b*)] F. Bruhat has given an interesting new proof of the irreducibility of certain unitary representations of semi-simple Lie groups. His proof is based on a certain remarkable property (*see* the lemma in [1(*a*)]) of such groups which, so far as I know, had not been noticed before. Bruhat actually verified it for the complex classical groups by direct computation, leaving the case of the exceptional groups open. The object of this paper is to give a general proof of this result which is independent of the classification and therefore is applicable to all semisimple Lie groups (whether real or complex). Another totally different proof (which is also general in the above sense) has been obtained by Chevalley [3]. Chevalley's proof is of an algebro-geometric nature and is not immediately applicable to real groups. On the other hand since it requires no restriction on the characteristic of the ground field, it has certain other applications [3].

Let G be a connected semisimple Lie group and $\mathfrak{g}$ its Lie algebra over the field R of real numbers. Let $\mathfrak{k}$ be the subalgebra of $\mathfrak{g}$ corresponding to a maximal compact subgroup of the adjoint group of G. We consider the orthogonal complement $\mathfrak{p}$ of $\mathfrak{k}$ in $\mathfrak{g}$ with respect to the bilinear form $B(X, Y) = \mathrm{sp}(\mathrm{ad}X\,\mathrm{ad}Y)$ $(X, Y \in \mathfrak{g})$ where $X \to \mathrm{ad}X$ is the adjoint representation of $\mathfrak{g}$. Then ($^1$) $\mathfrak{g}$ is the

---

($^1$) Most of the facts which we state here without proof are well known. In any case their proofs can be found in [4, § 2, and 9]. In order to see that present situation is identical with that considered in [4], one has to make use of a theorem of Cartan [2, p. 19] that any two maximal compact subgroups of a connected semisimple Lie group are conjugate.

direct sum of $\mathfrak{k}$ and $\mathfrak{p}$. Let $\mathfrak{a}$ be a maximal abelian subspace of $\mathfrak{p}$ and let $H_1, \ldots, H_p$ be a fixed base for $\mathfrak{a}$ over R. If $\lambda$ and $\mu$ are two linear functions on $\mathfrak{a}$ we say that $\lambda > \mu$ (or $\mu < \lambda$) if $\lambda \neq \mu$ and $\lambda(H_i) > \mu(H_i)$ where $i$ is the least index ($1 \leq i \leq p$) such that $\lambda(H_i) \neq \mu(H_i)$. Let $\mathfrak{g}_\lambda$ denote the set of all $X \in \mathfrak{g}$ such that $[H, X] = \lambda(H)X$ for every $H \in \mathfrak{a}$. We put $\mathfrak{n} = \sum_{\lambda > 0} \mathfrak{g}_\lambda$ and $\mathfrak{n}_- = \sum_{\lambda > 0} \mathfrak{g}_\lambda$. It is easy to see that $\mathfrak{n}$ and $\mathfrak{n}_-$ are nilpotent sub-algebras of $\mathfrak{g}$. Let K, A and N be the analytic subgroups of G corresponding to $\mathfrak{k}$, $\mathfrak{a}$ and $\mathfrak{n}$ respectively. They are all closed in G. Let M be the centralizer and M' the normalizer of A in K. Clearly M and M' are closed and since K contains the center Z of G, $Z \subset M$. Moreover M is a normal subgroup of M'. Let $\mathfrak{m}$ be the centralizer of $\mathfrak{a}$ in $\mathfrak{k}$. If $[X, \mathfrak{a}] \subset \mathfrak{a}$ for some $X \in \mathfrak{k}$, $(\mathrm{ad}\,H)^2 X = 0$ for every $H \in \mathfrak{a}$ and from this we conclude easily that $X \in \mathfrak{m}$. Therefore M and M' have the same Lie algebra $\mathfrak{m}$. Since $K/Z$ is compact and $Z \subset M$, it follows that $\mathfrak{w} = M'/M$ is both compact and discrete and so it is finite. By a theorem of Iwasawa [5] $(k, a, n) \to kan$ ($k \in K$, $a \in A$, $n \in N$) is a topological mapping of $K \times A \times N$ onto G. Therefore since $[\mathfrak{a}, \mathfrak{n}] \subset \mathfrak{n}$ it follows easily that $S = MAN$ is a closed subgroup of G. We shall consider double cosets of the form $SxS(x \in G)$. Since $M \subset S$, $S\sigma S = SxS$ where $x$ is any element in the coset $\sigma \in \mathfrak{w}$. We can now state Bruhat's result as follows.

Theorem. — *The mapping $\sigma \to S\sigma S (\sigma \in \mathfrak{w})$ is a one-one mapping of $\mathfrak{w}$ onto the set of all double cosets $SxS(x \in G)$.*

Let $\theta$ denote the automorphism (*see* Cartan [2] and Mostow [6]) of $\mathfrak{g}$ of order 2 given by $\theta(X+Y) = X - Y(X \in \mathfrak{k},\ Y \in \mathfrak{p})$. We consider the bilinear form

$$Q(X, Y) = -B(X, \theta(Y)) \qquad (X, Y \in \mathfrak{g})$$

on $\mathfrak{g}$. It is known that the corresponding quadratic form $Q(X, X)$ is positive definite and therefore it defines a Euclidean metric on $\mathfrak{g}$. Whenever we speak below of orthogonality (or other related concepts) in $\mathfrak{g}$, it is always with respect to this Euclidean metric. It is easy to verify that

$$Q([Z, X], Y) = Q(X, [\theta(Z), Y]) \qquad (X, Y, Z \in \mathfrak{g}).$$

Hence $\mathrm{ad}\, Z$ is a self-adjoint or a skew-adjoint transformation in $\mathfrak{g}$ according as $Z \in \mathfrak{p}$ or $Z \in \mathfrak{k}$. Hence it is obvious that the spaces $\mathfrak{a} + \mathfrak{m}$, $\mathfrak{n}$ and $\mathfrak{n}_-$ are mutually orthogonal and $\mathfrak{g}$ is their direct sum. If $Z \in \mathfrak{k}$, $\mathrm{ad}\, Z$ is a skew-adjoint transformation which commutes with $\theta$. Hence $\mathrm{Ad}(k)$ $(k \in K)$ is an orthogonal transformation which also commutes with $\theta$. [Here $x \to \mathrm{Ad}(x)$ $(x \in G)$ is the adjoint representation of $G$].

Let us call an element $H$ in $\mathfrak{a}$ regular if $\mathfrak{a} + \mathfrak{m}$ is exactly the centralizer of $H$ in $\mathfrak{g}$. We know [4, lemma 4] that regular elements exist. We shall first prove a few preliminary results.

LEMMA 1. — *Let* $H$ *and* $X$ *be two elements in* $\mathfrak{a} + \mathfrak{m}$ *and* $\mathfrak{n}$ *respectively. Then the characteristic polynomials of* $\mathrm{ad}\, H$ *and* $\mathrm{ad}(H + X)$ *are the same. Moreover if* $H$ *lies in* $\mathfrak{a}$ *and is regular,* $H + X = \mathrm{Ad}(n)H$ *for some* $n \in N$.

One proves without difficulty that if $\lambda$ and $\mu$ are two linear functions on $\mathfrak{a}$,

$$[\mathfrak{g}_\lambda, \mathfrak{g}_\mu] \subset \mathfrak{g}_{\lambda+\mu}.$$

Moreover $\mathfrak{g}_\lambda = \mathfrak{a} + \mathfrak{m}$ if $\lambda = 0$. These two facts should be borne in mind during the following discussion. Let $\lambda_1 < \lambda_2 < \ldots < \lambda_r$ be all the various linear functions $\lambda$ on $\mathfrak{a}$ for which $\mathfrak{g}_\lambda \neq \{0\}$. Put

$$\mathfrak{g}_i = \sum_{\lambda > \lambda_i} \mathfrak{g}_\lambda \qquad (1 \leq i \leq r)$$

and $\mathfrak{g}_0 = \mathfrak{g}$. Then $(^2)$ $\mathfrak{g}_0 = \mathfrak{g} > \mathfrak{g}_1 > \ldots > \mathfrak{g}_r = \{0\}$. Moreover each $\mathfrak{g}_i$ is invariant under $\mathrm{ad}\, H$ and $[\mathfrak{n}, \mathfrak{g}_i] \subset \mathfrak{g}_{i+1}$ $(0 \leq i < r)$. Therefore it is obvious that $\mathrm{ad}\, H$ and $\mathrm{ad}(H + X)$ define the same endomorphism of $\mathfrak{g}_i/\mathfrak{g}_{i+1}$ $(0 \leq i \leq r)$ and the first statement of the lemma is an immediate consequence of this fact. Now suppose $H$ is a regular element in $\mathfrak{a}$. Let $i$ be the greatest index $(1 \leq i \leq r + 1)$ such

---

$(^2)$ If $V$ is a subspace of a vector space $U$ we write $U > V$ if $U \neq V$.

that $X \in \mathfrak{g}_{i-1}$. We shall use induction on $r - i$ to prove our second assertion. We may assume that $i < r$ for otherwise $X = o$ and we can take $n = 1$. Then $X \notin \mathfrak{g}_i$ and we can choose $X' \in \mathfrak{g}_{\lambda_i}$ such that $X - X' \in \mathfrak{g}_i$. Since $X \in \mathfrak{n} = \sum_{\lambda > 0} \mathfrak{g}_\lambda$ and the component $X'$ of $X$ in $\mathfrak{g}_{\lambda_i}$ is not zero, it is obvious that $\lambda_i > o$ and therefore $X' \in \mathfrak{n}$. Moreover since $H$ is regular, $\lambda_i(H) \neq o$. Hence if $n_1 = \exp\{(\lambda_i(H))^{-1} X'\} \in N$, it is clear that

$$\operatorname{Ad}(n_1^{-1})\, H \equiv H - (\lambda_i(H))^{-1} \lfloor X', H \rfloor$$

$$\equiv H + X' \bmod \mathfrak{g}_i.$$

This shows that

$$(H + X) - \operatorname{Ad}(n_1^{-1})\, H \in \mathfrak{g}_i.$$

On the other hand $[\mathfrak{n}, \mathfrak{g}_i] \subset \mathfrak{g}_i$ and there fore $\operatorname{Ad}(n_1)\mathfrak{g}_i \subset \mathfrak{g}_i$. Hence

$$\operatorname{Ad}(n_1)\,(H + X) = H + X_1$$

where $X_1 \in \mathfrak{g}_i$. But then it follows from our induction hypothesis that $H + X_1 = \operatorname{Ad}(n_2)H$ for some $n_2 \in N$ and therefore $H + X = \operatorname{Ad}(n)H$ where $n = n_1^{-1} n_2$.

Lemma 2. — *Let* $Z \in \mathfrak{a} + \mathfrak{m} + \mathfrak{n}$. *Then if all the eigenvalues of* $\operatorname{ad} Z$ *are real* $Z \in \mathfrak{a} + \mathfrak{n}$ *and if* $\operatorname{ad} Z$ *is nilpotent* $Z \in \mathfrak{n}$.

Let $Z = H + Y + X$ ($H \in \mathfrak{a}$, $Y \in \mathfrak{m}$, $X \in \mathfrak{n}$). Then from lemma 1 $\operatorname{ad} Z$ and $\operatorname{ad}(H + Y)$ have the same charactéristic polynomial and hence the same eigenvalues. Moreover $\operatorname{ad} H$ commutes with $\operatorname{ad} Y$ and it is self-adjoint. Therefore all the eigenvalues of $\operatorname{ad} Y$ are also real. However $\operatorname{ad} Y$ is skew-adjoint and so this is possible only if $\operatorname{ad} Y = o$. But since $\mathfrak{g}$ is semisimple this implies that $Y = o$ and therefore $Z \in \mathfrak{a} + \mathfrak{n}$. Furthermore if $\operatorname{ad} Z$ is nilpotent all its eigenvalues are zero and so $Y = o$. Therefore $\operatorname{ad} H$ and $\operatorname{ad} Z$ have the same characteristic polynomial. This implies that $\operatorname{ad} H$ is nilpotent. But since it is self-adjoint it must be zero and so $H = o$. This proves that $Z = X \in \mathfrak{n}$.

Corollary. — *Suppose* $\operatorname{Ad}(x)H \in \mathfrak{a} + \mathfrak{m} + \mathfrak{n}$ *for some* $x \in G$ *and* $H \in \mathfrak{a}$. *Then* $\operatorname{Ad}(x)H \in \mathfrak{a} + \mathfrak{n}$. *Choose* $H_1 \in \mathfrak{a}$ *such that* $\operatorname{Ad}(x)H - H_1 \in \mathfrak{n}$. *Then if* $H$ *is regular the same holds for* $H_1$.

Let $H' = \mathrm{Ad}(x)H$. Then $\mathrm{ad}\,H' = \mathrm{Ad}(x)\,\mathrm{ad}\,H\,\mathrm{Ad}(x^{-1})$ and there fore $\mathrm{ad}\,H$ and $\mathrm{ad}\,H'$ have the same characteristic polynomial. Since $\mathrm{ad}\,H$ is self-adjoint all its eigenvalues are real and therefore the same is true of $\mathrm{ad}\,H'$. Hence our first assertion follows from lemma 2. Now suppose $H' = H_1 + X\,(X \in \mathfrak{n})$. Then from lemma 1, $\mathrm{ad}\,H'$ and $\mathrm{ad}\,H_1$ have the same characteristic polynomial. Therefore the multiplicity of the eigenvalue zero for $\mathrm{ad}\,H_1$ is the same as for $\mathrm{ad}\,H'$ or $\mathrm{ad}\,H$. Since $H$ is regular, its centralizer in $\mathfrak{g}$ is $\mathfrak{a} + \mathfrak{m}$. Therefore since $\mathrm{ad}\,H_1$ is self-adjoint and since $\mathfrak{a} + \mathfrak{m}$ is contained in the centralizer of $H_1$, it follows that $\mathfrak{a} + \mathfrak{m}$ is exactly this centralizer. Hence $H_1$ is regular.

LEMMA 3. — *Let $k$ be an element in* $\mathrm{K}$ *and* $H$ *a regular element in* $\mathfrak{a}$. *Then if* $\mathrm{Ad}(k)H \in \mathfrak{a}$, $k$ *lies in* $\mathrm{M}'$.

Since $H$ is regular, $\mathfrak{a}$ is exactly the centralizer of $H$ in $\mathfrak{p}$. Hence $\mathrm{Ad}(k)\mathfrak{a}$ is the centralizer of $H' = \mathrm{Ad}(k)H$ in $\mathrm{Ad}(k)\mathfrak{p} = \mathfrak{p}$. Therefore since $H' \in \mathfrak{a}$ et $\mathfrak{a}$ is abelian, $\mathfrak{a} \subset \mathrm{Ad}(k)\mathfrak{a}$. But $\mathfrak{a}$ and $\mathrm{Ad}(k)\mathfrak{a}$ have the same dimension and so $\mathfrak{a} = \mathrm{Ad}(k)\mathfrak{a}$. This proves that $k \in \mathrm{M}'$.

The following lemma is the main step of our proof.

LEMMA 4. — *For any $x \in \mathrm{G}$ we can choose an element $n \in \mathrm{N}$ such that*

$$\mathrm{Ad}(n)(\mathfrak{a} + \mathfrak{m}) \subset (\mathfrak{a} + \mathfrak{m} + \mathfrak{n}) \cap \{\mathrm{Ad}(x)(\mathfrak{a} + \mathfrak{m} + \mathfrak{n})\}.$$

Since $\mathrm{G} = \mathrm{KAN}$, $x = kt$ where $k \in \mathrm{K}$ and $t \in \mathrm{AN}$ and it is obvious that

$$\mathrm{Ad}(x)(\mathfrak{a} + \mathfrak{m} + \mathfrak{n}) = \mathrm{Ad}(k)(\mathfrak{a} + \mathfrak{m} + \mathfrak{n}).$$

Put

$$\Gamma = (\mathfrak{a} + \mathfrak{m} + \mathfrak{n}) \cap \{\mathrm{Ad}(k)(\mathfrak{a} + \mathfrak{m} + \mathfrak{n})\}$$

and suppose $X \in \Gamma \cap \mathfrak{n}$. Then $Y = \mathrm{Ad}(k^{-1})X$ lies in $\mathfrak{a} + \mathfrak{m} + \mathfrak{n}$ and since $\mathrm{ad}\,X$ is nilpotent the same hold for $\mathrm{ad}\,Y$. Hence from lemma 2, $Y \in \mathfrak{n}$ and therefore $X = \mathrm{Ad}(k)Y \in \mathfrak{n} \cap \mathrm{Ad}(k)\mathfrak{n}$. This proves that

$$\Gamma \cap \mathfrak{n} = \mathfrak{n} \cap \mathrm{Ad}(k)\mathfrak{n}.$$

On the other hand $\mathfrak{a} + \mathfrak{m} + \mathfrak{n}_-$ is the orthogonal complement of $\mathfrak{n}$ in $\mathfrak{g}$. Since $\mathrm{Ad}(k)$ is an orthogonal transformation, $\mathrm{Ad}(k)(\mathfrak{a} + \mathfrak{m} + \mathfrak{n}_-)$

is the orthogonal complement of $\mathrm{Ad}(k)\,\mathfrak{n}$ and therefore

$$V = (\mathfrak{a} + \mathfrak{m} + \mathfrak{n}_-) + \mathrm{Ad}(k)\,(\mathfrak{a} + \mathfrak{m} + \mathfrak{n}_-)$$

is the orthogonal complement of $\Gamma \cap \mathfrak{n}$. On the other hand $\theta$ commutes with $\mathrm{Ad}(k)$ and $\theta(\mathfrak{n}_-) = \mathfrak{n}$ since $\theta(H) = -H\,(H \in \mathfrak{a})$. Hence

$$\theta(V) = (\mathfrak{a} + \mathfrak{m} + \mathfrak{n}) + \mathrm{Ad}(k)\,(\mathfrak{a} + \mathfrak{m} + \mathfrak{n})$$

and

$$\dim V = \dim \theta(V) = 2 \dim (\mathfrak{a} + \mathfrak{m} + \mathfrak{n}) - \dim \Gamma.$$

Therefore

$$\begin{aligned}
\dim (\Gamma \cap \mathfrak{n}) &= \dim \mathfrak{g} - \dim V \\
&= \dim \Gamma + \dim \mathfrak{g} - 2 \dim (\mathfrak{a} + \mathfrak{m} + \mathfrak{n}) \\
&= \dim \Gamma - \dim (\mathfrak{a} + \mathfrak{m})
\end{aligned}$$

since

$$\dim \mathfrak{g} = \dim (\mathfrak{a} + \mathfrak{m}) + \dim \mathfrak{n} + \dim \mathfrak{n}_- = 2 \dim (\mathfrak{a} + \mathfrak{m} + \mathfrak{n}) - \dim (\mathfrak{a} + \mathfrak{m}).$$

This proves that

$$\dim (\Gamma/\Gamma \cap \mathfrak{n}) = \dim (\mathfrak{a} + \mathfrak{m}).$$

But on the other hand

$$\Gamma/\Gamma \cap \mathfrak{n} \simeq (\Gamma + \mathfrak{n})/\mathfrak{n} \subset (\mathfrak{a} + \mathfrak{m} + \mathfrak{n})/\mathfrak{n} \simeq \mathfrak{a} + \mathfrak{m}.$$

Hence we conclude that $\Gamma + \mathfrak{n} = \mathfrak{a} + \mathfrak{m} + \mathfrak{n}$. Now let $H$ be a regular element in $\mathfrak{a}$. Then $H + X \in \Gamma$ for some $X \in \mathfrak{n}$. From lemma 1 we can find an element $n \in N$ such that $H + X = \mathrm{Ad}(n)\,H$. Put $H_1 = H + X$ and $H_2 = \mathrm{Ad}(k^{-1})\,H_1$. Since $H_1 \in \Gamma$, $H_2 \in \mathfrak{a} + \mathfrak{m} + \mathfrak{n}$ and $H_2 = \mathrm{Ad}(k^{-1}n)\,H$. Therefore it follows from the corollary to lemma 2 that $H_2 = H_3 + Y$ where $H_3$ is a regular element in $\mathfrak{a}$ and $Y \in \mathfrak{n}$. Mereover it follows from lemma 1 that $H_2 = \mathrm{Ad}(n_1)\,H_3$ for some $n_1 \in N$. This proves that

$$\mathrm{Ad}(n)\,H = \mathrm{Ad}(kn_1)\,H_3.$$

Since $H$ and $H_3$ are regular their centralizers are both equal to $\mathfrak{a} + \mathfrak{m}$ and therefore

$$\mathrm{Ad}(n)\,(\mathfrak{a} + \mathfrak{m}) = \mathrm{Ad}(kn_1)\,(\mathfrak{a} + \mathfrak{m}).$$

However it is obvious that

$$\mathrm{Ad}(n')\,(\mathfrak{a} + \mathfrak{m}) \subset \mathfrak{a} + \mathfrak{m} + \mathfrak{n}$$

for any $n' \in N$.   Therefore

$$\mathrm{Ad}(n)\,(\mathfrak{a} + \mathfrak{m}) \subset \Gamma$$

and the lemma is proved.

Now we come to the proof of the theorem.   Let $x$ be any element in G.   We first have to find an element $k \in M'$ such that $SxS = SkS$. From lemma 4 we can choose $n \in N$ such that

$$\mathrm{Ad}(n)\,(\mathfrak{a} + \mathfrak{m}) \subset \mathrm{Ad}(x)\,(\mathfrak{a} + \mathfrak{m} + \mathfrak{n}).$$

Then if $y = n^{-1}x$, $SxS = SyS$ and

$$\mathrm{Ad}(y^{-1})\,(\mathfrak{a} + \mathfrak{m}) \subset \mathfrak{a} + \mathfrak{m} + \mathfrak{n}.$$

Let H be a regular element in $\mathfrak{a}$.   Then it follows from the corollary to lemma 2 that $\mathrm{Ad}(y^{-1}) H = H_1 + X$ where $H_1$ is a regular element in $\mathfrak{a}$ and $X \in \mathfrak{n}$.   Moreover from lemma 1 we can choose $n_1 \in N$ such that $H_1 + X = \mathrm{Ad}(n_1) H_1$.   Then if $z = y n_1$, $SxS = SyS = SzS$ and $H = \mathrm{Ad}(z) H_1$.   Let $z = kt$ where $k \in K$ and $t \in AN$.   Then $\mathrm{Ad}(k^{-1}) H = \mathrm{Ad}(t) H_1$.   However $\mathrm{Ad}(k^{-1}) H \in \mathfrak{p}$ and $\mathrm{Ad}(t) H_1 \in \mathfrak{a} + \mathfrak{n}$. Hence

$$\mathrm{Ad}(k^{-1}) H \in \mathfrak{p} \cap (\mathfrak{a} + \mathfrak{n}).$$

On the other hand if $Y \in \mathfrak{p} \cap (\mathfrak{a} + \mathfrak{n})$,

$$Y = - \theta(Y) \in \theta(\mathfrak{a} + \mathfrak{n}) = \mathfrak{a} + \mathfrak{n}_-.$$

But since $\mathfrak{n}_-$ is orthogonal to $\mathfrak{a} + \mathfrak{n}$,

$$(\mathfrak{a} + \mathfrak{n}) \cap (\mathfrak{a} + \mathfrak{n}_-) = \mathfrak{a}.$$

This shows that $\mathfrak{p} \cap (\mathfrak{a} + \mathfrak{n}) = \mathfrak{a}$ and therefore $\mathrm{Ad}(k^{-1}) H \in \mathfrak{a}$.   In view of lemma 3 we can now conclude that $k \in M'$.   Moreover since $t \in N \subset S$,

$$SxS = SzS = SkS.$$

On the other and suppose $k_1$, $k_2$ are two elements in M' such that $Sk_1 S = Sk_2 S$.   Then we can choose $s$, $t' \in S$ such that $sk_1 = k_2 t'$. It is obvious that $S = MAN = NAM$ and therefore $s$ can be written in the forme $s = nam\,(n \in N,\ a \in A,\ m \in M)$.   Then

$$sk_1 = n k_1 k_1^{-1}\,(am)\,k_1.$$

Since $k_1 \in M'$, $k_1^{-1} a k_1 \in A$ and therefore $k_1^{-1}(am) k_1 \in AM \subset S$. Hence $nk_1 = k_2 t$ for some $t \in S$. Let H be any element in $\mathfrak{a}$. In view view of the fact that $t \in S = \text{NAM}$ it is clear that $\text{Ad}(t) H = H + X$ where $X \in \mathfrak{n}$. Put $H_1 = \text{Ad}(k_1) H$, $H_2 = \text{Ad}(k_2) H$. Since $k_1$, $k_2 \in M'$, $H_1$, $H_2$ lie in $\mathfrak{a}$. Moreover

$$\text{Ad}(n) H_1 = \text{Ad}(nk_1) H = \text{Ad}(k_2 t) H = H_2 + \text{Ad}(k_2) X.$$

On the other hand it is obvious that $\text{Ad}(n) H_1 = H_1 + X_1$ where $X_1 \in \mathfrak{n}$. Hence

$$H_1 - H_2 = - X_1 + \text{Ad}(k_2) X.$$

But $\mathfrak{m} + \mathfrak{a}$ is invariant under $\text{Ad}(k_2)$ since $k_2 \in M'$. Moreover $\mathfrak{n} + \mathfrak{n}_-$ is the orthogonal complement of $\mathfrak{a} + \mathfrak{m}$ in $\mathfrak{g}$ and so it is also invariant under $\text{Ad}(k_2)$. Therefore

$$H_1 - H_2 \in \mathfrak{a} \cap (\mathfrak{n} + \mathfrak{n}_-) = \{ o \}.$$

This shows that $\text{Ad}(k_1) H = \text{Ad}(k_2) H$ for every $H \in \mathfrak{a}$ and so $k_1^{-1} k_2 \in M$. Therefore $k_1$ and $k_2$ define the same element of $\mathfrak{w}$. The proof of the theorem is now complete.


## REFERENCES


[1] F. Bruhat, *C. R. Acad. Sc.*, t. 238, 1954 : (*a*) p. 437-439; (*b*) p. 550-553.
[2] E. Cartan, *J. Math. pures et appl.*, t. 8, 1929, p. 1-33.
[3] C. Chevalley, *Tôhoku Math. J.*, 1955 (in print).
[4] Harish-Chandra, *Trans. Amer. Math. Soc.*, t. 75, 1953, p. 185-243.
[5] K. Iwasawa, *Ann. Math.*, t. 50, 1949, p. 507-557.
[6] G. D. Mostow, *Bull. Amer. Math. Soc.*, t. 55, 1949, p 969-980.



Reprinted from
*J. Math. Pures. Appl.*
(9) **35** (1956), 203–210

# Invariant Differential Operators on a Semisimple Lie Algebra

Harish-Chandra

INSTITUTE FOR ADVANCED STUDY
*Communicated by M. Morse, March 1, 1956*

Let $R$ and $C$ be the fields of real and complex numbers, respectively, and $E_0$ a vector space over $R$ of finite dimension. Then, $E$ being the complexification of $E_0$, we consider the symmetric algebra $S(E)$ and the algebra $Q(E)$ of polynomial functions on $E$. We regard $E_0$ as a differentiable manifold and, for any differential operator $D$ and indefinitely differentiable function $f$, denote by $f(X; D)$ the value of $Df$ at $X \in E_0$. Corresponding to any $X$ in $E_0$ we define the differential operator $\partial(X)$ by

$$ f(Y; \partial(X)) = \left\{ \frac{d}{dt} f(Y + tX) \right\}_{t=0} \quad (Y \in E_0,\ t \in R). $$

Then $\partial$ can be extended uniquely to an isomorphism of $S(E)$ into the algebra of differential operators on $E_0$. Let $U$ be an open subset of $E_0$. By $\mathcal{C}(U)$ we mean the space of all functions $f$ on $U$ of class $C^\infty$ such that $\tau(q, \partial(p); f) = \sup_{X \in U} |q(X)f(X; \partial(p))| < \infty$ for all $q \in Q(E)$ and $p \in S(E)$. We define a topology in $\mathcal{C}(U)$ by means of the collection of these seminorms $\tau(q, \partial(p))$ $(q \in Q(E),$ $p \in S(E))$.

Let $\mathfrak{g}_0$ be a semisimple Lie algebra over $R$ and $\mathfrak{h}_0$ a Cartan subalgebra of $\mathfrak{g}_0$. Complexify $\mathfrak{g}_0$, $\mathfrak{h}_0$ to $\mathfrak{g}$ and $\mathfrak{h}$, respectively. We can identify[1] $S(\mathfrak{g})$ and $Q(\mathfrak{g})$ by means of the fundamental bilinear form $B(X, Y) = \mathrm{sp}(\mathrm{ad}\, X\, \mathrm{ad}\, Y)(X, Y \in \mathfrak{g})$ on $\mathfrak{g}$. Let $G_0$ be the (connected) adjoint group of $\mathfrak{g}_0$. A function $f$ on $\mathfrak{g}_0$ will be called invariant if $f(xX) = f(X)$ for all $x \in G_0$ and $X \in \mathfrak{g}_0$. Let $I(\mathfrak{g})$ be the subalgebra of $S(\mathfrak{g})$ consisting of invariant polynomial functions. Similarly, let $I(\mathfrak{h})$ denote the algebra of those elements in $S(\mathfrak{h})$ which are invariant under the Weyl group $W$ (of $\mathfrak{g}$ with respect to $\mathfrak{h}$). For any $p \in I(\mathfrak{g})$, let $\bar{p}$ denote the restriction of the polynomial function $p$ on $\mathfrak{h}$. Then $p \to \bar{p}$ is an isomorphism of $I(\mathfrak{g})$ onto $I(\mathfrak{h})$. Let $P$ denote the set of all positive roots (of $\mathfrak{g}$ with respect to $\mathfrak{h}$) under some fixed order. Put $\pi = \prod_{\alpha \in P} \alpha$. Then $\pi \in S(\mathfrak{h})$. Notice that if $q \in S(\mathfrak{h})$, $\partial(q)$ is a differential operator on $\mathfrak{h}_0$.

**Theorem**[2] **1.** *Suppose that $f$ is an invariant function on $\mathfrak{g}_0$ of class $C^\infty$. Then, if*

$p \in I(\mathfrak{g})$,

$$\pi(H)f(H; \partial(p)) = g(H; \partial(\bar{p})) \qquad (H \in \mathfrak{h}_0),$$

*where g is the function on $\mathfrak{h}_0$ given by $g(H) = \pi(H)f(H)$.*

Let $dx$ denote the Haar measure of $G_0$.

**Corollary.** *Suppose that $G_0$ is compact. Then*[3]

$$\pi(H)\pi(H')\int_{G_0} \exp B(xH, H')\, dx = c \sum_{s \in W} \varepsilon(s)\exp B(sH, H')$$

*for all $H, H'$ in $\mathfrak{h}$. Here c is a constant which is easily determined by the condition* $\int_{G_0} dx = 1$.

Let $A_0$ be the Cartan subgroup of $G_0$ corresponding to $\mathfrak{h}_0$, and let $dx^*$ denote the invariant measure on the factor space $G^* = G_0/A_0$. Define $x^*H = xH$ $(H \in \mathfrak{h})$, where $x$ is any element in the coset $x^* \in G^*$. Let $\mathfrak{h}_0'$ be the set of those elements $H \in \mathfrak{h}_0$, where $\pi(H) \neq 0$. Then, if $f \in \mathcal{C}(\mathfrak{g}_0)$, the integral

$$F_f(H) = \pi(H)\int_{G^*} f(x^*H)\, dx^*$$

is convergent for $H \in \mathfrak{h}_0'$, and $F_f$ is of class $C^\infty$ on $\mathfrak{h}_0'$. Moreover, it follows from Theorem 1 that $F_{\partial(p)f} = \partial(\bar{p})F_f (p \in I(\mathfrak{g}))$. This relation, in its turn, implies the following result.

**Theorem 2.** *The mapping $f \to F_f$ is a continuous mapping of $\mathcal{C}(\mathfrak{g}_0)$ into $\mathcal{C}(\mathfrak{h}_0')$.*

For any $f \in \mathcal{C}(\mathfrak{g}_0)$, let $\tilde{f}$ denote the Fourier transform of $f$, so that

$$\tilde{f}(X) = \int_{\mathfrak{g}_0} \exp(iB(X, Y))f(Y)\, dY \qquad (X \in \mathfrak{g}_0),$$

where $dY$ is the (suitably normalized) Euclidean measure on $\mathfrak{g}_0$. Then it is possible to obtain interesting relations between $F_f$ and $F_{\tilde{f}}$. For example, if $G_0$ is either compact or complex,

$$F_{\tilde{f}}(H) = \int_{\mathfrak{h}_0} \exp(iB(H, H'))F_f(H')\, dH' \qquad (H \in \mathfrak{h}_0')$$

where $dH'$ is the (suitably normalized) Euclidean measure on $\mathfrak{h}_0$. Similar but more complicated relations hold in other cases as well, under suitable restrictions on $f$.

If $G_0$ is a complex semisimple group, the above relation between $F_f$ and $F_{\tilde{f}}$ can be obtained directly without much difficulty. This, in fact, is the starting point of the proof of Theorem 1.

---

[1] See *Trans. Am. Math. Soc.*, **75**, 194, 1953.

[2] This theorem should be compared with Lemma 2 of *Bull. Am. Math. Soc.*, **61**, 394, 1955.

[3] $\varepsilon(s) = 1$ or $-1$, according as $s$ takes an even or odd number of positive roots into negative roots.

# A Formula for Semisimple Lie Groups

HARISH-CHANDRA

INSTITUTE FOR ADVANCED STUDY

*Communicated by Marston Morse, June 26, 1956*

In order to save space we shall follow strictly the notation of an earlier paper.[1] Define $\mathfrak{k}_0$, $\mathfrak{p}_0$ as usual,[2] and suppose that $\mathfrak{h}_0 = \mathfrak{h}_0 \cap \mathfrak{k}_0 + \mathfrak{h}_0 \cap \mathfrak{p}_0$. Then we say that $\mathfrak{h}_0$ is fundamental if $\mathfrak{h}_0 \cap \mathfrak{k}_0$ is maximal Abelian in $\mathfrak{k}_0$. Any two fundamental Cartan subalgebras of $\mathfrak{g}_0$ are conjugate under $G_0$.

**Theorem 1.** *Let $\mathfrak{h}_1$ be a connected component of $\mathfrak{h}_0'$. Then there exists a real number $c$ such that*

$$\lim_{H \to 0} F_f(H; \partial(\pi)) = cf(0) \qquad (H \in \mathfrak{h}_1)$$

*for every $f \in \mathcal{C}(\mathfrak{g}_0)$. Moreover, $c = 0$ if $\mathfrak{h}_0$ is not fundamental.*

**Theorem 2.** *Suppose that $\mathfrak{h}_0$ is fundamental and $\mathfrak{h}_1, \ldots, \mathfrak{h}_r$, are all the distinct connected components of $\mathfrak{h}_0'$. Let $c_j$ be the real number of Theorem 1 corresponding to $\mathfrak{h}_j$ $(1 \leqslant j \leqslant r)$. Then $c_1 + c_2 + \cdots + c_r \neq 0$.*

The following result plays an essential role in the proofs of the above theorems. Let $dH$ denote the Euclidean measure on $\mathfrak{h}_0$ and $dk$ the Haar measure on the compact analytic subgroup $K_0$ of $G_0$ corresponding to $\mathfrak{k}_0$.

**Lemma 1.** *Assume that $\mathfrak{h}_0 \subset \mathfrak{k}_0$, and for any $g \in \mathcal{C}(\mathfrak{h}_0)$ put*

$$\hat{g}(X) = \int_{K_0 \times \mathfrak{h}_0} \exp(iB(X, kH))\pi(H)^2 \sum_{s \in W} g(sH)\, dk\, dH \qquad (X \in \mathfrak{g}_0).$$

*Then the integral*

$$F_{\hat{g}}(H) = \pi(H)\int_{G^*} \hat{g}(x^*H)\, dx^*$$

*is convergent for $H \in \mathfrak{h}'_0$, and there exists a constant $c \neq 0$ such that*

$$\sum_{s \in W} \epsilon(s) F_{\hat{g}}(sH') = c \int_{\mathfrak{h}_0} \sum_{s \in W} \epsilon(s) \exp(iB(H', sH)) \pi(H) g(H) \, dH$$

*for all $H' \in \mathfrak{h}'_0$ and $g \in \mathcal{C}(\mathfrak{h}_0)$.*

Let $G$ be any connected Lie group whose Lie algebra is $\mathfrak{g}_0$, and let $A$ be the Cartan subgroup of $G$ corresponding to $\mathfrak{h}_0$.

**Lemma 2.** *Let $\Xi$ be the centralizer in $G$ of an element $a_0 \in A$. Denote the natural mapping of $G$ on $\bar{G} = G/\Xi$ by $x \to \bar{x}$ $(x \in G)$. Then we can find a neighborhood $B$ of $a_0$ in $A$ with the following property. Given any compact set $\omega$ in $G$, there exists a compact set $\bar{\Omega}$ in $\bar{G}$ satisfying the condition that if $xax^{-1} \in \omega$ for some $a \in B$ and $x \in G$, then $\bar{x} \in \bar{\Omega}$.*

By combining this with Theorem 2 of Paper 1, we can obtain certain results on $G$ as follows. $\lambda$ being an indeterminate, let $D(x)$ $(x \in G)$ denote the coefficient of $\lambda^l$ in $\det\{(\lambda + 1)I - \mathrm{Ad}(x)\}$, where $l$ is the rank and $I$ the identity mapping of $\mathfrak{g}$. Let $A'$ be the set of those points $a \in A$ where $D(a) \neq 0$. We regard $A'$ as an open submanifold of the Lie group $A$. Note that $G/A = G_0/A_0 = G^*$, and put $a^{x^*} = xax^{-1}$ $(a \in A, x \in G)$, where $x^*$ denotes the coset $xA$ in $G^*$. Let $C_c^\infty(G)$ be the set of all indefinitely differentiable functions on $G$ which vanish outside a compact set. Then the integral

$$\phi_f(a) = |D(a)|^{1/2} \int_{G^*} f(a^{x^*}) \, dx^*$$

converges for $f \in C_c^\infty(G)$ and $a \in A'$, and $\phi_f$ is of class $C^\infty$ on $A'$. Let $\mathfrak{B}$ be the universal enveloping algebra of $\mathfrak{g}$ and $\mathfrak{H}$ the subalgebra of $\mathfrak{B}$ generated by $(1, \mathfrak{h})$. Then elements of $\mathfrak{B}$ and $\mathfrak{H}$ can be regarded as differential operators on $G$ and $A$, respectively.[3] Let $\mathfrak{Z}$ denote the center of $\mathfrak{B}$ and $\gamma$ the isomorphism of $\mathfrak{Z}$ into $\mathfrak{H}$ described in Paper 2 (p. 394). Then one can show that $\phi_{zf} = \gamma(z)\phi_f$ $(f \in C_c^\infty(G)$, $z \in \mathfrak{Z})$. Let us call a subset of $G$ *bounded* if its closure is compact. Consider the space $\mathcal{C}_0(A')$ of all functions $g$ on $A'$ of class $C^\infty$ satisfying the following two conditions: (1) $g$ is zero (on $A'$) outside some bounded set; (2) $\tau_v(g) = \sup_{a \in A'}|g(a; v)| < \infty$ for every $v \in \mathfrak{H}$. We topologize $C_c^\infty(G)$ in the usual way[4] and $\mathcal{C}_0(A')$ by means of the seminorms $\tau_v$ $(v \in \mathfrak{H})$.

**Theorem 3.** *$\phi_f \in \mathcal{C}_0(A')$ for $f \in C_c^\infty(G)$ and $f \to \phi_f$ is a continuous mapping of $C_c^\infty(G)$ into $\mathcal{C}_0(A')$. Moreover, for any compact set $\omega$ in $G$ we can find a bounded subset $B$ of $A'$ with the following property. If the carrier of $f$ is contained in $\omega$, then $\phi_f = 0$ outside $B$.*

**Corollary.** *Let $da$ denote the Haar measure on $A$. Then if $g$ is a measurable function on $A$ which is bounded on every compact set, the mapping $T_g: f \to \int_A g\phi_f \, da$ $(f \in C_c^\infty(G))$ is a distribution on $G$.*

The distributions of the form $T_g$, where $g$ is a character of $A$, are closely related to the characters of $G$ (see Paper 2).

$S(\mathfrak{h})$ is isomorphic to $\mathfrak{H}$ under a mapping which preserves every element of $\mathfrak{h}$, and so for any $q \in S(\mathfrak{h})$ we get a differential operator on $A$, which we shall denote by $\partial(q)$. Then Theorems 1 and 2 have the following analogues on $G$.

**Theorem 4.** *Let $A_1$ be a connected component of $A'$ whose closure contains 1. Then there exists a constant $c$ such that*

$$\lim_{a \to 1} \phi_f(a; \partial(\pi)) = cf(1) \qquad (a \in A_1)$$

*for all $f \in C_c^\infty(G)$. Moreover, $c = 0$ if $\mathfrak{h}_0$ is not fundamental.*

**Theorem 5.** *Now suppose that $\mathfrak{h}_0$ is fundamental and $A_1, \ldots, A_r$ are all the distinct components of $A'$ whose closures contain 1. Let $c_j$ be the constant of Theorem 4 corresponding to $A_j$ $(1 \leqslant j \leqslant r)$. Then not every $c_j$ can be zero.*

Theorem 5 gives a simple formula for $f(1)$ in terms of $\phi_f$. In case $G$ is compact, the proof of this formula is quite trivial. Moreover, I had proved it earlier[5] for (1) a complex semisimple group and (2) the $2 \times 2$ real unimodular group. More recently Gelfand and Graev[6] had extended it to the $n \times n$ real unimodular group.

[1] These PROCEEDINGS, **42**, 252–253, 1956. This will be referred to as "Paper 1."
[2] See *Trans. Am. Math. Soc.*, **75**, 187, 1953.
[3] See *Bull. Am. Math. Soc.*, **61**, 389–396, 1955. This will be referred to as "Paper 2."
[4] See L. Schwartz, *Théorie des distributions*, Vol. 1 (Paris: Hermann & Cie, 1950).
[5] See *Trans. Am. Math. Soc.*, **76**, 514, 1954, and these PROCEEDINGS, **38**, 339, 1952.
[6] *Doklady Akad. Nauk S.S.S.R.*, N.S., **92**, 461–464, 1953.

# REPRESENTATIONS OF SEMISIMPLE LIE GROUPS

HARISH-CHANDRA

Let $G$ be a Lie group and $\mathfrak{H}$ a Banach space. A representation $\pi$ of $G$ on $\mathfrak{H}$ is a mapping which assigns to every element $x$ in $G$ a bounded linear operator $\pi(x)$ on $\mathfrak{H}$ such that the following two conditions are fulfilled: (1) $\pi(xy) = \pi(x)\,\pi(y)$ $(x, y \in G)$, $\pi(1) = I$ and (2) the mapping $(x, \psi) \to \pi(x)\psi$ of $G \times \mathfrak{H}$ into $\mathfrak{H}$ is continuous. (Here 1 is the unit element of $G$ and $I$ is the unit operator.) In particular if $\mathfrak{H}$ is a Hilbert space and $\pi(x)$ is a unitary operator for every $x \in G$, we say that $\pi$ is a unitary representation. In the study of finite-dimensional representations the success of the infinitesimal method is well known. Our object is to make this method applicable also to the infinite-dimensional case.

First we discuss a few preliminary notions. Let $f$ be a mapping of some neighborhood of the origin on the real line into the Banach space $\mathfrak{H}$. We say $f$ is analytic at zero if it is possible to write it in the form

$$f(t) = \sum_{n \geq 0} \psi_n t^n$$

where $\psi_n \in \mathfrak{H}$ and the equality means that the series on the right converges in $\mathfrak{H}$ to $f(t)$ for all values of $t$ sufficiently near zero. The generalisation to the case of several real variables is obvious and so it is clear what is meant by an analytic mapping of a (real) analytic manifold $M$ into $\mathfrak{H}$. Now let $\psi$ be a fixed element in $\mathfrak{H}$ and consider the mapping $x \to \pi(x)\psi$ of $G$ into $\mathfrak{H}$. We shall say that $\psi$ is well-behaved (under $\pi$) if this mapping is analytic. Let $W$ be the space of all well-behaved elements in $\mathfrak{H}$. Then if $\mathfrak{g}_0$ is the Lie algebra of $G$ one can prove that the limit

$$\lim_{t \to 0} \frac{1}{t} \{\pi(\exp tX)\psi - \psi\}$$

exists for every $X \in \mathfrak{g}_0$ and $\psi \in W$ and again lies in $W$. If we denote it by $\pi_W(X)\psi$ we get a linear mapping $\pi_W(X)$ of $W$ into itself. It is easy to verify that

$$\pi_W([X, Y]) = \pi_W(X)\pi_W(Y) - \pi_W(Y)\pi_W(X) \qquad (X,\ Y \in \mathfrak{g}_0)$$

and therefore this defines a representation $\pi_W$ of $\mathfrak{g}_0$ on $W$. Let $\mathfrak{g}$ denote the complexification of $\mathfrak{g}_0$. We extend $\pi_W$ to a representation of $\mathfrak{g}$ by linearity. The importance of the space $W$ derives from the following result:

*Let $U$ be a subspace of $W$ which is invariant under $\pi_W(\mathfrak{g})$. Then its closure $\overline{U}$ is invariant under $\pi(G)$.*

The next step is to prove that $W$ is dense in $\mathfrak{H}$. This can be done for a fairly large class of Lie groups. However we are primarily interested in the semisimple case. So let us suppose from now on that $G$ is connected and semisimple. Let $Z$ denote the center of $G$. We shall say that $\pi$ is a permissible representation if it maps $Z$ into scalar multiples of the unit operator. Then one can show that if $\pi$ is permissible $W$ is dense in $\mathfrak{H}$.

Let $X_1, \ldots, X_n$ be a base for $\mathfrak{g}$ over the field $C$ of complex numbers and let $[X_i, X_j] = \sum_{k=1}^{n} c_{ij}^k X_k$, $1 \leq i, j \leq n$ $(c_{ij}^k \in C)$. Consider the associative algebra $\mathfrak{B}$ generated by $(1, X_1, \ldots, X_n)$ which is free except for the relations $X_i X_j - X_j X_i = \sum_{k=1}^{n} c_{ij}^k X_k$. Then $\mathfrak{B} \supset \mathfrak{g}$ and it is independent of the particular base $(X_1, \ldots, X_n)$ used in its construction. $\mathfrak{B}$ is called the universal enveloping algebra of $\mathfrak{g}$. It is obvious that every representation of $\mathfrak{g}$ can be extended uniquely to a representation of $\mathfrak{B}$ and so there is a natural $1-1$ correspondence between representations of $\mathfrak{B}$ and $\mathfrak{g}$. We shall denote two such corresponding representations usually by the same letter.

Now $\pi_W$, being a representation of $\mathfrak{g}$, may also be regarded as a representation of $\mathfrak{B}$ on $W$. Let $\mathfrak{Z}$ be the center of $\mathfrak{B}$. We shall say that the representation $\pi$ of $G$ is quasi-simple if $\pi(x)$ and $\pi_W(z)$ are both scalar multiples of the unit operator for $x \in Z$ and $z \in \mathfrak{Z}$. Let $\chi$ be the homomorphism of $\mathfrak{Z}$ into $C$ such that $\pi_W(z) = \chi(z)\pi_W(1)$ $(z \in \mathfrak{Z})$. We call $\chi$ the infinitesimal character of $\pi$.

Let $K$ be the complete inverse image in $G$ of a maximal compact subgroup of the adjoint group of $G$. Then $K$ is connected and it contains $Z$. Let $\Omega$ denote the set of all equivalence-classes of finite-dimensional irreducible representations of $K$. We denote by $\mathfrak{H}_{\mathfrak{D}}$ $(\mathfrak{D} \in \Omega)$ the subspace consisting of all those elements $\psi \in \mathfrak{H}$ which transform [1] under $\pi(K)$ according to the class $\mathfrak{D}$. Let $W_{\mathfrak{D}} = \mathfrak{H}_{\mathfrak{D}} \cap W$. One can prove that if $\pi$ is permissible the space [2] $\mathfrak{H}_0 = \sum_{\mathfrak{D} \in \Omega} W_{\mathfrak{D}}$ is dense in $\mathfrak{H}$. Choose an element $\psi \neq 0$ in $\mathfrak{H}_0$ and let $U$ be the smallest closed subspace of $\mathfrak{H}$ which is invariant under $\pi(G)$ and which contains $\psi$. Then if $\pi$ is quasi-simple dim $(U \cap \mathfrak{H}_{\mathfrak{D}}) < \infty$ for every $\mathfrak{D} \in \Omega$. In particular if $\pi$ is an irreducible representation $U = \mathfrak{H}$ and therefore dim $\mathfrak{H}_{\mathfrak{D}}$ is finite.

Now suppose $\mathfrak{H}$ is a Hilbert space and $\pi$ is a quasi-simple unitary representation. Then one can deduce, from the above result, the existence of a closed

---

[1] This means that the linear space spanned by $\pi(u)\psi$ $(u \in K)$ is of finite dimension and the representation of $K$ defined on it is a direct sum of representations of class $\mathfrak{D}$.

[2] $\sum_{\mathfrak{D} \in \Omega} W_{\mathfrak{D}}$ consists of all finite linear combinations of elements in $\bigcup_{\mathfrak{D} \in \Omega} W_{\mathfrak{D}}$.

subspace $U \neq \{0\}$ which is invariant and irreducible under $\pi(G)$. This fact has the following significance in relation to the theory of factors of Murray and von Neumann [6]. A unitary representation $\pi$ is called a factor representation if the weakly closed algebra generated by $\pi(G)$ is a factor. It is not difficult to prove that every factor representation is quasi-simple [8] and so the above remarks are applicable to it. Hence if $\pi$ is a factor representation $\mathfrak{H}$ contains a minimal closed invariant subspace and therefore from a well-known result of Murray and von Neumann [6] this factor must be of type $I$. This proves that only factors of type $I$ can arise from unitary representations of a semisimple group.

Now we return to the case where $\mathfrak{H}$ is a Banach space and assume that $\pi$ is quasi-simple and irreducible. Then we have seen above that $\dim \mathfrak{H}_{\mathfrak{D}} < \infty$ $(\mathfrak{D} \in \Omega)$. Actually one can prove that

$$\dim \mathfrak{H}_{\mathfrak{D}} \leqq N\big(d(\mathfrak{D})\big)^2 \quad (\mathfrak{D} \in \Omega)$$

where $d(\mathfrak{D})$ is the degree of any representation in class $\mathfrak{D}$ and $N$ is an integer independent of $\mathfrak{D}$. This estimate can be improved under suitable assumptions on $\pi$ or $G$. For example if we assume that $G$ has a faithful finite-dimensional representation, the factor $N$ can be dropped (see [5(a)] Theorem 4) and [4]). The above inequality has some interesting consequences. Suppose $\mathfrak{H}$ is a Hilbert space and $f(x)$ is a complex-valued square-integrable function on $G$ which vanishes outside a compact set. Then it follows that the operator $\int_G f(x)\pi(x)dx$

$(dx$ is the element of Haar measure on $G$) is of the Hilbert-Schmidt class. This fact had been observed by Gelfand and Naimark [3] in their study of representations of the complex classical groups.

One can also define a character of $\pi$ as follows. We say that a bounded operator $A$ on $\mathfrak{H}$ has a trace (or is of the trace class) if for every complete orthonormal set $\psi_i$ $(i = 1, 2, \ldots)$ in $\mathfrak{H}$ the series $\sum_{i \geq 1} (\psi_i, A\psi_i)$ is convergent and its sum is independent of the particular choice of this set. This sum is then called the trace of $A$ and we shall denote it by $\mathrm{sp}\, A$. Let $C_c^\infty(G)$ denote the class of all complex-valued functions on $G$ which are everywhere indefinitely differentiable and which vanish outside a compact set. Then if $\pi$ is a quasi-simple irreducible representation of $G$ on $\mathfrak{H}$ we consider the operator $\int f(x)\,\pi(x)dx$ $(f \in C_c^\infty(G))$. One proves that it has a trace which we denote by $T_\pi(f)$. The mapping $T_\pi : f \to T_\pi(f)$ is clearly a linear function on $C_c^\infty(G)$. It follows from the inequality $\dim \mathfrak{H}_{\mathfrak{D}} \leqq N(d(\mathfrak{D}))^2$ that it is actually a distribution in the sense of L. Schwartz [9]. We call this distribution the character of $\pi$. Equivalent representations have the same character. Conversely two irreducible unitary representations [1]) are equivalent if their characters are equal. The existence of these characters had been noticed by Gelfand and Naimark [3] for certain irreducible unitary representations of complex classical groups.

We now come to the Plancherel formula. Let $\mathfrak{E}$ be the set of all equivalence classes of irreducible unitary representations of $G$ and let $C_c(G)$ denote the set of all complex-valued continuous functions on $G$ which vanish outside a compact set. Choose any class $\omega \in \mathfrak{E}$ and a representation $\pi$ in $\omega$. Then we have seen that the operator $\int f(x)\pi(x)dx$ is of the Hilbert-Schmidt class. It is clear that its Hilbert-Schmidt norm depends only on the class $\omega$ of $\pi$. We denote the square of this norm by $N_\omega(f)$. Then in analogy with the Plancherel formula for locally compact abelian groups or the Peter—Weyl theorem for compact groups, we would like to obtain a positive measure $d\omega$ on $\mathfrak{E}$ such that

$$\int\limits_G |f(x)|^2\,dx = \int\limits_{\mathfrak{E}} N_\omega(f)d\omega$$

for all $f \in C_c(G)$. In view of the fact that any factor arising from a unitary representation of $G$ must always be of type $I$, it follows from the reduction theory of von Neumann [7] that such a measure $d\omega$ exists and is unique. However the problem of computing it and relating it to the structure of $G$ still remains. In case $G$ is a *complex* semisimple group it is possible to determine this measure explicitly. The main reason why we have to assume that $G$ be complex, is that in that case all Cartan subgroups of $G$ are conjugate. Since this, in general, is not true for real groups the corresponding problem for them is much more complicated. There each conjugacy class of Cartan subgroups seems to make a separate contribution to the Plancherel formula, so that the total measure $d\omega$ is the sum of the measures contributed by these various conjugacy classes. For example the group of $2 \times 2$ real unimodular matrices has two such classes and corresponding to these we get the continuous and the discrete series of representations in the Plancherel formula (see [1] and [5(b)]). Recently Gelfand and Graev [2] have made a beautiful application of the method of analytic continuation of Riesz for solving certain hyperbolic differential equations (see [9 p. 50]), towards the determination of $d\omega$. By this powerful method they are able to handle the case of $n \times n$ real unimodular group in detail and their results confirm the above picture. As a partial explanation of this apparently intimate connection between the conjugacy classes of Cartan subgroups and the various series' of represen-

---

[1]) Every irreducible unitary representation is automatically quasi-simple [8].

tations which appear in the Plancherel formula we shall sketch here a general method of associating, in a natural way, certain irreducible unitary representations of $G$ to each such conjugacy class. Let $A$ be a given Cartan subgroup of $G$. An easy argument shows that the really important case is when $A \subset K$ and $G$ is simple. (The general case can be reduced to this without much trouble.) Moreover in order to avoid inessential complication, let us assume that $G$ has a faithful finite-dimensional representation. Then we can regard $G$ as the real analytic subgroup corresponding to $\mathfrak{g}_0$, of a complex analytic group $G_c$ with the Lie algebra $\mathfrak{g}$. Let $\mathfrak{h}_0$ be the Lie algebra of $A$ and $\mathfrak{h}$ its complexification in $\mathfrak{g}$. Then it is possible to choose a nilpotent subalgebra $\mathfrak{n}$ of $\mathfrak{g}$ such that $\mathfrak{g} = \mathfrak{g}_0 + \mathfrak{h} + \mathfrak{n}$. Let $A_c$ and $N_c$ be the complex analytic subgroups of $G_c$ corresponding to $\mathfrak{h}$ and $\mathfrak{n}$ respectively. Then $G_c^0 = GA_cN_c$ is an open subset of $G_c$ and therefore it may be regarded as a complex manifold. Let $\xi$ be a complex-analytic character of $A_c$. Consider the space $\mathfrak{H}_\xi$ of all holomorphic functions $\varphi$ on $G_c^0$ such that $\varphi(xan) = \varphi(x)\,\xi(a)$ $(x \in G_c^0,\ a \in A_c,\ n \in N_c)$ and

$$\| \varphi \|^2 = \int_G | \varphi(x) |^2 dx < \infty.$$

It is not difficult to show that $\mathfrak{H}_\xi$ is complete under the norm $\| \varphi \|$ and so it is a Hilbert space. Now if $\mathfrak{H}_\xi \neq \{0\}$ we can define a unitary representation $\pi_\xi$ of $G$ on it as follows:

$$(\pi_\xi(x)\varphi)(y) = \varphi(x^{-1}y) \quad (x \in G,\ y \in G_c^0).$$

It can be proved that this representation, which is obviously unitary, is irreducible and square-integrale i.e.

$$\int_G |(\varphi, \pi_\xi(x)\varphi)|^2\, dx < \infty$$

for every $\varphi \in \mathfrak{H}_\xi$. (Here we have used the usual notation for the scalar product in $\mathfrak{H}_\xi$.) One can prove that $\mathfrak{H}_\xi = \{0\}$ unless the first Betti number of $G$ is 1 or what is equivalent, unless the center of $K$ is non-discrete. (We assume naturally that $G$ is not compact). On the other hand if this condition is fulfilled $\mathfrak{H}_\xi \neq \{0\}$ provided $\xi$ is suitably chosen.

As we have already mentioned above, the general case can be reduced to the one considered above and so corresponding to each Cartan subgroup $A$ we get a series of unitary irreducible representations of $G$.


### References

[1]  V. Bargmann, Ann. of Math. vol. 48 (1947) pp. 568—640.

[2]  I. M. Gelfand and M. I. Graev, Doklady Akad. Nauk SSSR, N.S. (a) vol. 92 (1953) pp. 221—224, (b) vol. 92 (1953) pp. 461—464.

[3]  I. M. GELFAND and M. A. NAIMARK, Trudi Mat. Inst. Steklova vol. 36 (1950).
[4]  R. GODEMENT, Trans. Amer. Math. Soc. vol. 73 (1952) pp. 496—566.
[5]  HARISH-CHANDRA, (a) Trans. Amer. Math. Soc. vol. 76 (1954) pp. 26—65, (b) Proc. Nat. Acad. Sci. U.S.A. vol. 38 (1952) pp. 337—342.
[6]  F. J. MURRAY and J. VON NEUMANN, Ann. of Math. vol. 37 (1936) pp. 116—229.
[7]  J. VON NEUMANN, Ann. of Math. vol. 50 (1949) pp. 401—485.
[8]  I. E. SEGAL, Proc. Amer. Math. Soc. vol. 3 (1952) pp. 13—15.
[9]  L. SCHWARTZ, Théorie des distributions vol. 1 Paris, Hermann, 1950.

COLUMBIA UNIVERSITY, NEW YORK, N.Y., U.S.A.

Reprinted from
*Proceedings of the International
Congress of Mathematicians* 1954
Amsterdam, September 2–September 9
Volume I

# DIFFERENTIAL OPERATORS ON A SEMISIMPLE LIE ALGEBRA.*

By Harish-Chandra.

1. **Introduction.** The object of this paper is to present in as concise and coherent a form as possible, certain results on differential operators which will be needed in two subsequent papers for application to the theory of Fourier transforms on a semisimple Lie algebra $\mathfrak{g}_0$. When $\mathfrak{g}_0$ is compact, this theory is not difficult (see Section 5) although some of the results obtained here (e. g. **Theorems** 2 and 3) seem to be new. Therefore the main application of our results will be to the case when $\mathfrak{g}_0$ is noncompact. There it will appear that one can obtain certain formulas of a more or less analytic (as against algebraic) nature, which bear a great resemblance to their counterparts in the compact case (see [4(i)]). However, in view of their non-algebraic character, it would seem rather unlikely that the main burden of their proof could be transferred to the compact case by some formal device such as the "unitarian trick." For instance, let us consider the following example. Let $\mathfrak{g}$ the the Lie algebra of all $2 \times 2$ complex matrices of trace zero and $G_c$ its (connected) complex adjoint group. Let $\mathfrak{g}_0$ and $\mathfrak{u}$ be the real subalgebras of $\mathfrak{g}$ consisting of the real and the skew-hermitian matrices respectively and $G, U$ the corresponding real analytic subgroups of $G_c$. Then $\mathfrak{g}_0 \cap \mathfrak{u} = \mathfrak{h}_0$, where $\mathfrak{h}_0$ is the vector space spanned over the real field $R$ by $H = \begin{pmatrix} 0 & 1 \\ -1 & 0 \end{pmatrix}$. Let $f, g$ be functions of class $C^\infty$ on $\mathfrak{g}_0$ and $\mathfrak{u}$ respectively which vanish outside a compact set. Put

$$\phi_f(t) = t \int_G f(txH)\,dx, \qquad \psi_g(t) = t \int_U g(tuH)\,du \qquad (t \in R)$$

where $dx$ and $du$ are the Haar measures on $G$ and $U$ respectively. It is not difficult to show that $\phi_f$ is well defined and of class $C^\infty$ for all $t \neq 0$ and one can prove that

$$\operatorname*{Lim}_{t \to 0} ((d/dt)\phi_f) = c_1 f(0), \qquad \operatorname*{Lim}_{t \to 0} ((d/dt)\psi_g) = c_2 g(0)$$

where $c_1$, $c_2$ are nonzero constants independent of $f$ and $g$. The proof of this formula is quite trivial for $g$ and follows from the fact that

$$\operatorname*{Lim}_{t \to 0} \int_U g(tuH)\,du = g(0),$$

---

* Received July 30, 1956.

87

if $\int_U du = 1$. But a similar statement about $f$ is actually false and a much more delicate investigation of $\phi_f$ is required. Nevertheless the resemblance in the two cases is so striking that one cannot give up the feeling that they both must be governed by some common principle. However a little closer examination reveals important differences. For example if we replace $H$ by

$$H' = \begin{pmatrix} 1 & 0 \\ 0 & -1 \end{pmatrix} \text{ and define}$$

$$\phi_f'(t) = t \int_G f(txH')\,dx,$$

we get $\underset{t \to 0}{\mathrm{Lim}}\,((d/dt)\phi_f') = 0$. On the other hand in the compact case $H$ is

conjugate to $H'' = \begin{pmatrix} (-1)^{\frac{1}{2}} & 0 \\ 0 & -(-1)^{\frac{1}{2}} \end{pmatrix}$ under $U$ and therefore nothing new

is obtained if we replace $H$ by $H''$. This shows that the situation in the noncompact case is considerably more complicated.

Returning to the general case, let $\mathfrak{g}$ be a semisimple Lie algebra over the complex field $C$. We shall consider the algebra $\mathfrak{D}(\mathfrak{g})$ of polynomial differential operators on $\mathfrak{g}$ (see Section 2 for the precise definition). Let $\mathfrak{g}_0$ and $\mathfrak{u}$ be two real forms of $\mathfrak{g}$ and suppose $\mathfrak{u}$ is compact. It is clear that the complex group $G_c$ of $\mathfrak{g}$ operates on $\mathfrak{D}(\mathfrak{g})$. Hence we can consider the subalgebra $\mathfrak{J}'$ consisting of those elements in $\mathfrak{D}(\mathfrak{g})$ which are invariant under $G_c$. The structure of $\mathfrak{J}'$ is, of course, completely determined by $\mathfrak{u}$. On the other hand, elements of $\mathfrak{J}'$ can also be regarded as differential operators on $\mathfrak{g}_0$ and therefore the internal structure of $\mathfrak{J}'$ is reflected in the differential equations on $\mathfrak{g}_0$ and this provides the required link from $\mathfrak{u}$ to $\mathfrak{g}_0$. For instance let $\mathfrak{g}, \mathfrak{g}_0$ and $\mathfrak{u}$ be as in the above example. Choose complex coordinates $x, y, z$ in $\mathfrak{g}$ such that the corresponding matrix is given by $\begin{pmatrix} z & x \\ y & -z \end{pmatrix}$. Then if the differential operator.

$$D = \partial^2/\partial z^2 + 4\partial^2/\partial x \partial y$$

is interpreted suitably on $\mathfrak{g}_0$ and $\mathfrak{u}$, one finds that $\phi_{Df} = d^2\phi_f/dt^2$ and $\psi_{Dg} = -d^2\psi_g/dt^2$. This is a special case of a result (Lemma 15) which will be proved in its full generality in another paper (see however [4(h)]).

Theorem 1 (Section 3) is the central result of this paper. In Section 5 we give its applications to the case of a compact semisimple Lie algebra. This discussion also enables us to obtain some algebraic results (e. g. Lemma 18) which are valid in the general case as well. Section 8 is devoted to the detailed study of a special case which will have an important significance for later applications. Theorem 5 contains the main result of this section.

**2. Preliminary definitions.** Let $R$ and $C$ be the fields of real and complex numbers respectively and $E_0$ a vector space over $R$ of finite dimension. We consider the symmetric algebra [4(c), p. 191] $S(E)$ over the complexification $E$ of $E_0$. For any $X \in E_0$ let $\partial(X)$ denote the differential operator on $E_0$ given by [1]

$$f(Y; \partial(X)) = \{(d/dt)f(Y + tX)\}_{t=0} \qquad (Y \in E_0, f \in C^\infty(E_0), t \in R).$$

Then it is obvious that the mapping $X \to \partial(X)$ can be extended uniquely to an algebraic isomorphism of $S(E)$ (over $C$) into the algebra of differential operators on $E_0$. Thus for any $p \in S(E)$, we get a differential operator $\partial(p)$ on $E$. Now suppose there is given a real nondegenerate symmetric bilinear form $B(X, Y)$ $(X, Y \in E_0)$ on $E_0$. We extend this form on $E$ by linearity (over $C$) and then use it to identify $E$ with its dual. In this identification an element $X \in E$ corresponds to the linear function $Y \to B(X, Y)$ $(Y \in E)$ and therefore $S(E)$ can now be regarded as the algebra of all polynomial functions on $E$. In accordance with this intepretation we shall often refer to elements of $S(E)$ as polynomials on $E$.

Let $\mathcal{E}$ be the algebra of all differential operators on $E_0$. Then $\mathcal{E} \supset C^\infty(E_0)$, and therefore $S(E)$ and $\partial(S(E))$ are both subalgebras of $\mathcal{E}$. Let $\mathcal{D}(E)$ denote the subalgebra of $\mathcal{E}$ generated by $S(E) \cup \partial(S(E))$. The elements of $\mathcal{D}(E)$ will be called polynomial differential operators on $E_0$. But, whenever convenient, we can regard an element of $\mathcal{D}(E)$ also as a *holomorphic* differential operator on the *complex* manifold $E$. Let $H$ be a group of nonsingular linear mappings of $E$ into itself which preserve $B$ so that $B(xX, xY) = B(X, Y)$ $(X, Y \in E; x \in H)$. Then for any $x \in H$, we shall define an automorphism $d \to d^x$ $(d \in \mathcal{D}(E))$ of $\mathcal{D}(E)$. In order to do this we have to define $(d^x f)(X_0)$, where $f$ is a holomorphic function on some neighborhood of $X_0$ in $E$. If $g$ is a function defined on a subset $V$ of $E$, let $g^x$ denote the function on $xV$ given by $g^x(X) = g(x^{-1}X)$ $(X \in xV)$. If $V$ is open and $g$ is holomorphic on $V$, it is clear that $g^x$ is holomorphic on $xV$. We now define $(d^x f)(X_0) = (df^{x^{-1}})^x(X_0)$. It is easy to check that the operator $d^x$ so defined, is actually a holomorphic differential operator and $d \to d^x$ is a homomorphism of $\mathcal{D}(E)$ into the algebra of all holomorphic differential operators on $E$. Moreover if $p \in S(E)$ it is obvious that $p^x$ is the

---

[1] $M$ being a differentiable manifold (which need not be connected [4(e), (f)]), we denote by $C^\infty(M)$ the space of all (complex-valued) functions on $M$ of class $C^\infty$ and by $C_c(M)$ the space of continuous functions on $M$ which vanish outside some sompact set. If $D$ is a differential operator on $M$ and $f \in C^\infty(M), f(p; D)$ denotes the value of $Df$ at a point $p$ in $M$ (see [4(e), §4]). Moreover $C_c^\infty(M) = C_c(M) \cap C^\infty(M)$.

polynomial function $X \to p(x^{-1}X)$ $(X \in E)$ and similarly $(\partial(X))^x = \partial(xX)$. Therefore $(p\partial(q))^x = p^x\partial(q^x)$ $(p, q \in S(E))$. Since every element in $\mathfrak{D}(E)$ is evidently a sum of elements of the form $p\partial(q)$, it follows that $d^x \in \mathfrak{D}(E)$. Finally, the mapping $d \to d^x$ has the inverse $d \to d^{x^{-1}}$ and so it is an automorphism of $\mathfrak{D}(E)$.

Let $S_r(E)$ $(r \geqq 0)$ denote the subspace of $S(E)$ spanned by elements of the form $X_1 X_2 \cdots X_r$ where $X_i \in E$. (In case $r = 0$, $S_r(E) = C$.) As usual a polynomial on $E$ is said to be homogeneous of degree $r$ if it lies in $S_r(E)$. For any two polynomials $p, q$ let $\langle p, q \rangle$ denote the value of $\partial(p)q$ at zero. It is obvious that if $p$ and $q$ are homogeneous, $\langle p, q \rangle = 0$ unless their degrees are equal. Notice that $\langle X, Y \rangle = B(X, Y)$ $(X, Y \in E)$ and it follows from the symmetry of $B$ that $\langle p, q \rangle = \langle q, p \rangle$. In fact, let $(X_1, \cdots, X_n)$ be a base for $E$ such that[2] $B(X_i, X_j) = \delta_{ij}$. Then if $p = \sum_{m_i \geqq 0} a(m_1, \cdots, m_n) X_1^{m_1} X_2^{m_2} \cdots X_n^{m_n}$ $(a(m_1, \cdots, m_n) \in C)$, it is obvious that

$$\langle X_1^{m_1} \cdots X_n^{m_n}, p \rangle = m_1! \cdots m_n! \, a(m_1, \cdots, m_n).$$

Hence

$$\langle q, p \rangle = \sum m_1! \cdots m_n! \, a(m_1, \cdots, m_n) b(m_1, \cdots, m_n)$$

if $q = \sum b(m_1, \cdots, m_n) X_1^{m_1} \cdots X_n^{m_n}$. This also shows that $\langle p, q \rangle = 0$, for all $q \in S(E)$ implies $p = 0$. Moreover $(\partial(p)q)^x = \partial(p^x)q^x$ and therefore $\langle p^x, q^x \rangle = \langle p, q \rangle$ $(x \in H)$.

Let $U$ be a nonempty open subset of $E_0$. We shall denote by $\mathscr{B}(U)$ the class of all functions $f \in C^\infty(U)$ such that

$$\nu_d(f) = \sup_{X \in U} |f(X; d)| < \infty$$

for every $d \in \mathfrak{D}(E)$. We topologise $\mathscr{B}(U)$ by means of the seminorms $\nu_d$ $(d \in \mathfrak{D}(E))$ and in this way $\mathscr{B}(U)$ becomes a locally convex space. If $D$ is a differential operator on $U$ and $X$ a point in $U$, it is obvious that there exists a unique element $p \in S(E)$ such that $f(X; D) = f(X; \partial(p))$ for every $f \in C^\infty(U)$. *We call $\partial(p)$ the local expression of $D$ at $X$. It will usually be denoted by $D_X$.* Let $H_0$ be a group of nonsingular linear mappings of $E_0$ into itself. Then for every function $f$ on $U$ and $x \in H_0$, let $f^x$ denote the function on $U$ given by $f^x(X) = f(x^{-1}X)$ $(X \in U)$. Then corresponding to any differential operator $D$ on $U$, we define a new differential operator $D^x$ on $U$ by $D^x f = (Df^{x^{-1}})^x$ $(f \in C^\infty(U))$. One proves without difficulty that $(D_X)^x = (D^x)_{xX}$ $(X \in U)$. $D$ is said to be invariant under $H_0$ if $D^x = D$ for all $x \in H_0$.

---

[2] $\delta_{ij}$ is the usual Kronecker symbol. It is equal to 1 or 0 according as $i = j$ or not.

Select a fixed square root of $-1$ in $C$ and denote it by $(-1)^{\frac{1}{2}}$. Also let $dX$ denote the Euclidean measure on $E_0$ (normalized in some way). Then for any $f \in \mathscr{E}(E_0)$, the Fourier transform $\tilde{f}$ of $f$ is defined as follows.

$$\tilde{f}(Y) = \int_{E_0} \exp((-1)^{\frac{1}{2}} B(Y,X)) f(X) dX \qquad (Y \in E_0).$$

It is well known [8(I), p. 105] that $\tilde{f}$ also lies in $\mathscr{E}(E_0)$ and in fact $f \to \tilde{f}$ is a topological mapping of $\mathscr{E}(E_0)$ onto itself. Moreover the measure $dX$ can be so normalized that

$$f(Y) = \int_{E_0} \exp\{-(-1)^{\frac{1}{2}} B(Y,X)\} \tilde{f}(X) dX \qquad (Y \in E_0)$$

for every $f \in \mathscr{E}(E_0)$. *This normalization will be called the regular normalization.*

LEMMA 1. *There exist a unique automorphism* [3] $d \to \hat{d}$ *of* $\mathfrak{D}(E)$ *such that* $\hat{X} = -(-1)^{\frac{1}{2}} \partial(X)$ *and* $(\partial(X))^{\hat{}} = -(-1)^{\frac{1}{2}} X$ $(X \in E)$. *Moreover* $(df)^{\tilde{}} = \hat{d}\tilde{f}$ *for any* $d \in \mathfrak{D}(E)$ *and* $f \in \mathscr{E}(E_0)$.

Since $\mathfrak{D}(E)$ is generated by $(1, E + \partial(E))$, the uniqueness is obvious. The rest follows immediately from the theory of Fourier transforms [8(I), p. 105].

COROLLARY. *For any* $d$ *in* $\mathfrak{D}(E)$, *we can select another element* $d' \in \mathfrak{D}(E)$ *such that*

$$\int_{E_0} |\, d\tilde{f}\,|\, dX \leqq \nu_{d'}(f) \qquad (f \in \mathscr{E}(E_0)).$$

Choose a polynomial $p \in S(E)$ such that $p$ is never zero on $E_0$ and $c = \int_{E_0} |p|^{-1} dX < \infty$. Then if $\phi \in \mathscr{E}(E_0)$,

$$\nu_1(\tilde{\phi}) = \sup_{X \in E_0} |\tilde{\phi}(X)| \leqq \int_{E_0} |\phi|\, dX \leqq c\nu_1(p\phi).$$

Now it follows from the above lemma that $pd = \hat{d}_1$ for some $d_1 \in \mathfrak{D}(E)$. Hence

$$\int_{E_0} |\, d\tilde{f}\,|\, dX \leqq c\nu_1(pd\tilde{f}) = c\nu_1((d_1 f)^{\tilde{}}) \leqq c^2 \nu(pd_1 f) \qquad (f \in \mathscr{E}(E_0))$$

and so we can take $d' = c^2 p d_1$.

---

[3] We shall write $d^{\wedge}$, $f^{\sim}$ respectively instead of $\hat{d}$, $\tilde{f}$ whenever it is convenient to do so.

The adjoint of a differential operator $D$ on $E_0$ is, by definition, the unique differential operator $D^*$ such that

$$\int_{E_0} f(X;D)g(X)\,dX = \int_{E_0} f(X)g(X;D^*)\,dX$$

for all $f, g \in C_c^\infty(E_0)$. The mapping $D \to D^*$ is an anti-automorphism of the algebra of all differential operators on $E_0$ and $(D^*)^* = D$. In particular if $p$ is a homogeneous element in $S(E)$ of degree $m$, it is obvious that $p^* = p$ and $(\partial(p))^* = (-1)^m \partial(p)$. This shows that $\mathfrak{D}(E))^* = \mathfrak{D}(E)$ and so we get the following result.

LEMMA 2. *There exists a unique anti-automorphism* $d \to d^*$ *of* $\mathfrak{D}(E)$ *such that* $p^* = p$ *and* $(\partial(p))^* = (-1)^m \partial(p)$ *for any homogeneous polynomial in* $S(E)$ *of degree* $m$.

Again the uniqueness follows from the fact that $\mathfrak{D}(E)$ is generated by $S(E) \cup \partial(S(E))$.

$U$ and $H_0$ being as before, let $\tau$ denote a distribution on $U$ [8(I)] and $dX$ the Euclidean measure on $E_0$. Then, if $F$ is a measurable and locally summable function on $U$, we say that[4] $\tau = F$ if

$$\tau(f) = \int_U fF\,dX$$

for every $f \in C_c^\infty(U)$. Let $V$ be a subset of $C_c^\infty(U)$. Then by the $C_c^\infty$-topology on $V$, we mean the topology induced under the usual topology of $C_c^\infty(U)$ [8(I), p. 67]. Similarly if $V'$ is a subset of $\mathscr{E}(U)$, the $\mathscr{E}$-topology on $V'$ is the one induced on it by $\mathscr{E}(U)$. A distribution $\tau$ on $E_0$ will be called a $\mathscr{E}$-distribution if it is continuous under the $\mathscr{E}$-topology on $C_c^\infty(E_0)$. In case this is so, $\tau$ can be extended uniquely to a continuous function on $\mathscr{E}(E_0)$ [8(II, pp. 93-94]. If $D$ is a differential operator and $\tau$ a distribution on $U$, the mapping $f \to \tau(D^*f)$ $(f \in C_c^\infty(U))$, where $D^*$ is the adjoint of $D$, is also a distribution and it will be denoted by $D\tau$. If $\tau$ is continuous with respect to the $\mathscr{E}$-topology, the same holds for $D\tau$ provided $D \in \mathfrak{D}(E)$. $\tau$ is said to be invariant under $H_0$ if $\tau(f^x) = \tau(f)$ for all $f \in C_c^\infty(U)$ and $x \in H_0$. More generally, a $\mathscr{E}$-distribution $\tau$ on $U$ is a continuous linear mapping of $\mathscr{E}(U)$ into $C$. If $D \in \mathfrak{D}(E)$, the mapping $D\tau : f \to \tau(D^*f)$ $(f \in \mathscr{E}(U))$ is also a $\mathscr{E}$-distribution on $U$.

We now introduce a notation which will be used rather frequently. Let $M_1$, $M_2$ be two manifolds and $f$ a function on $M_1 \times M_2$ of class $C^\infty$.

---

[4] Of course this definition depends on the normalization of $dX$.

Then if $m_i \in M_i$ and $D_i$ is a differential operator on $M_i$ $(i = 1, 2)$, we shall denote the value of $(D_1 \times D_2)f$ at $(m_1, m_2)$ by $f(m_1; D_1 : m_2; D_2)$. Moreover $D_i$ will be suppressed in this notation whenever it is 1. For example $f(m_1 : m_2; D_2) = f(m_1; 1 : m_2; D_2)$.

Now suppose $F$ is the complexification in $E$ of a vector subspace $F_0$ of $E_0$. Then $S(F) \subset S(E)$. Let $B_F$ denote the restriction of the bilinear form $B$ on $F$ and suppose $B_F$ is nondegenerate. Then we can identify $F$ with its dual under $B_F$ and thus regard $S(F)$ as the algebra of all polynomial functions on $F$. For any function $f$ on $E_0$, let $\bar{f}$ denote its restriction on $F_0$. Then if $f \in S(E)$, it is clear that $\bar{f}$ is a polynomial function[5] on $F_0$ and therefore it lies in $S(F)$. Moreover if $f \in S(F)$, $f = \bar{f}$ in $S(E)$. Similarly $\partial(S(F)) \subset \partial(S(E))$ and therefore $\mathfrak{D}(F) \subset \mathfrak{D}(E)$. Elements of $\mathfrak{D}(F)$ can therefore be regarded as differential operators either on $F_0$ or $E_0$. But in view of the relation $\overline{df} = d\bar{f}$ $(f \in C^\infty(E_0), d \in \mathfrak{D}(F))$, it is not necessary to distinguish between the two meanings. Also it follows from Lemma 2 that the adjoint operation in $\mathfrak{D}(E)$ coincides on $\mathfrak{D}(F)$ with the corresponding operation in $\mathfrak{D}(F)$. The same obviously holds for the scalar product $\langle p, q \rangle$ $(p, q \in S(F))$.

**3. Invariant differential operators.** Let $\mathfrak{g}_0$ be a semisimple Lie algebra over $R$ and $\mathfrak{h}_0$ a Cartan subalgebra of $\mathfrak{g}_0$. We denote the corresponding complexifications by $\mathfrak{g}$ and $\mathfrak{h}$ respectively. Then if we take $E_0 = \mathfrak{g}_0$, $F_0 = \mathfrak{h}_0$ and[6] $B(X, Y) = \mathrm{sp}(\mathrm{ad}\, X\, \mathrm{ad}\, Y)$ $(X, Y \in \mathfrak{g}_0)$, all the above conditions are fulfilled. Let $G$ be the (connected) adjoint group of $\mathfrak{g}_0$. Then every $x \in G$ can be regarded as a nonsingular linear transformation of $\mathfrak{g}$ which preserves $B$ and so it defines an automorphism $D \to D^x$ $(D \in \mathfrak{D}(E))$ of $\mathfrak{D}(E)$. We call a function $f$ of $\mathfrak{g}_0$ invariant (under $G$) if $f^x = f$ $(x \in G)$. Let $I(\mathfrak{g})$ be the set of invariant elements in $S(\mathfrak{g})$. Then, if $f$ is an invariant function in $C^\infty(\mathfrak{g}_0)$ and $p \in I(\mathfrak{g})$, it is obvious that $\partial(p)f$ is also invariant. We would like to express $\overline{\partial(p)f}$ in terms of $\partial(\bar{p})$ and $\bar{f}$ (where the bar denotes restriction on $\mathfrak{h}_0$). Let $\alpha_1, \cdots, \alpha_r$ be all the distinct positive roots of $\mathfrak{g}$ (with respect to $\mathfrak{h}$) under some fixed order (see [4($\mathfrak{g}$), §2]). Then $\pi = \alpha_1 \alpha_2 \cdots \alpha_r$ is a polynomial function on $\mathfrak{h}$. We intend to prove that $\pi \overline{\partial(p)f} = \partial(\bar{p})(\pi\bar{f})$.

First we introduce some notation. Let $U$ be an open subset of $\mathfrak{g}_0$.

---

[5] We agree to identify a polynomial function on $E$ with its restriction on $E_0$. Similarly for $F$.

[6] $X \to \mathrm{ad}\, X$ $(X \in \mathfrak{g})$ denotes the adjoint representation of $\mathfrak{g}$.

Select open subsets $\mathfrak{g}_1$, $G_1$ of $\mathfrak{g}_0$ and $G$ respectively such that $xX \in U$ for $x \in G_1$, $X \in \mathfrak{g}_1$. For any $f \in C^\infty(U)$, consider the function

$$g(x:X) = f(xX) \qquad (x \in G_1, X \in \mathfrak{g}_1)$$

on $G_1 \times \mathfrak{g}_1$. If $d$ and $D$ are differential operators on $G_1$ and $\mathfrak{g}_1$ respectively, we put $f(x;d:X;D) = g(x;d:X;D)$. (Again we suppress $d$ or $D$ whenever it is 1.) Let $\mathfrak{B}$ be the universal enveloping algebra of $\mathfrak{g}$. Then elements of $\mathfrak{B}$ can be regarded as left-invariant differential operators on $G$ [4(e), §4]. All products of the form $X_1{}^{m_1} \cdots X_r{}^{m_r}$ $(X_i \in \mathfrak{g})$ are meant to be in $S(\mathfrak{g})$ unless it is explicitly stated otherwise. Now suppose $Y \in \mathfrak{g}_1$, $p \in S(\mathfrak{g})$ and $f \in C^\infty(U)$. Then

$$f(x:Y;\partial(p)) = f(xY;\partial(p^x)) \qquad (x \in G_1)$$

and therefore if $X \in \mathfrak{g}_0$,

$$f(x;X:Y;\partial(p)) = \{(d/dt)f(x_tY;\partial(p^{x_t}))\}_{t=0} \qquad (t \in R)$$

where $x_t = x \exp tX$. For any $Z \in \mathfrak{g}$, let $d_Z$ denote the derivation of $S(\mathfrak{g})$ which coincides with $\operatorname{ad} Z$ on $\mathfrak{g}$. It is clear that

$$\{(d/dt)f(x_tY;\partial(p^{x_t}))\}_{t=0}$$
$$= \{(d/dt)f(x_tY;\partial(p^x))\}_{t=0} + \{(d/dt)f(xY;\partial(p^{x_t}))\}_{t=0}$$

and

$$f(x_tY;\partial(p^x)) = f(x:z_tY;\partial(p)), \ f(xY;\partial(p^{x_t})) = f(x:Y;\partial(p^{z_t})),$$

where $z_t = \exp tX$. Hence

$$f(x;X:Y;\partial(p)) = f(x:Y;\partial((d_XY)p + d_Xp)).$$

For any $q \in S(\mathfrak{g})$, let $L_q$ denote the linear mapping $p \to qp$ $(p \in S(\mathfrak{g}))$ of $S(\mathfrak{g})$. Then we have the following result.

LEMMA 3. *Let* $f \in C^\infty(U)$, $X_1, \cdots, X_r \in \mathfrak{g}$ *and* $p \in S(\mathfrak{g})$. *Moreover let* $\cdot$ *denote multiplication in* $\mathfrak{B}$. *Then if* $Y \in \mathfrak{g}_1$,

$$f(x;X_1 \cdot X_2 \cdots \cdots X_r:Y;\partial(p)) = f(x:Y;\partial(q)) \qquad (x \in G_1),$$

*where* [7]

$$q = (L_{[X_1,Y]} + d_{X_1})(L_{[X_2,Y]} + d_{X_2}) \cdots (L_{[X_r,Y]} + d_{X_r})p.$$

For $r = 1$ this has been proved above. So now suppose $r > 1$ and use induction. Put

$$q' = (L_{[X_2,Y]} + d_{X_2}) \cdots (L_{[X_r,Y]} + d_{X_r})p.$$

------

[7] $[X, Y] = d_XY$ $(X, Y \in \mathfrak{g})$ as usual.

Then by induction hypothesis

$$f(x';X_2 \cdot \cdots \cdot X_r:Y;\partial(p)) = f(x':Y;\partial(q')) \qquad (x' \in G_1).$$

Hence $f(x;X_1 \cdot X_2 \cdot \cdots \cdot X_r:Y;\partial(p))$

$$= \{(d/dt)f(x\exp tX_1;X_2 \cdot \cdots \cdot X_r:Y;\partial(p))\}_{t=0}$$

$$= \{(d/dt)f(x\exp tX_1:Y;\partial(q'))\}_{t=0}$$

$$= f(x;X_1:Y;\partial(q')) = f(x:Y;\partial((L_{[X_1,Y]}+d_{X_1})q')))$$

$= f(x:Y;\partial(q))$.  This proves the lemma.

Now consider the tensor product $\mathfrak{B} \mathbf{X} S(\mathfrak{g})$.  The above lemma shows[8] **that for every $Y \in \mathfrak{g}_0$ there exists a unique linear mapping $\Gamma_Y$ of $\mathfrak{B} \mathbf{X} S(\mathfrak{g})$** into $S(\mathfrak{g})$ such that $\Gamma_Y(1 \mathbf{X} p) = p$ and

$$\Gamma_Y(X_1 \cdot X_2 \cdot \cdots \cdot X_r \mathbf{X} p) = (L_{[X_1,Y]}+d_{X_1}) \cdots (L_{[X_r,Y]}+d_{X_r})p$$

$(X_1, \cdots, X_r \in \mathfrak{g}; p \in S(\mathfrak{g}))$.  For any root $\alpha$ (of $\mathfrak{g}$ with respect to $\mathfrak{h}$), select an element $X_\alpha \neq 0$ in $\mathfrak{g}$ such that $[H,X_\alpha] = \alpha(H)X_\alpha$ for all $H \in \mathfrak{h}$. Put $\mathfrak{s} = \sum_\alpha CX_\alpha$ where $\alpha$ runs over all the roots.  Then the symmetric algebra $S(\mathfrak{s})$ over $\mathfrak{s}$ is a subalgebra of $S(\mathfrak{g})$.  Let $\lambda$ denote the canonical mapping [4(c), p. 192] of $S(\mathfrak{g})$ into $\mathfrak{B}$ and put $\mathfrak{S} = \lambda(S(\mathfrak{s}))$.

**LEMMA 4.**  *Let $H_0$ be an element in $\mathfrak{h}_0$ such that $\pi(H_0) \neq 0$.  Then $\Gamma_{H_0}$ defines a one-one mapping of $\mathfrak{S} \mathbf{X} S(\mathfrak{h})$ onto $S(\mathfrak{g})$.*

Let $S_d(\mathfrak{g})$, $S_d(\mathfrak{s})$ and $S_d(\mathfrak{h})$ denote the spaces of homogeneous elements of degree $d$ in $S(\mathfrak{g})$, $S(\mathfrak{s})$ and $S(\mathfrak{h})$ respectively and put $\mathfrak{S}_d = \lambda(S_d(\mathfrak{s}))$. We shall prove by induction on $r$ that

$$\sum_{d+e \leqq r} \Gamma_{H_0}(\mathfrak{S}_d \mathbf{X} S_e(\mathfrak{h})) = \sum_{s \leqq r} S_s(\mathfrak{g}).$$

This is obvious if $r = 0$.  So now suppose $r \geqq 1$.  It is clear that the left side is contained in the right side.  Hence in view of our induction hypothesis it is enough to prove that

$$\sum_{d+e \leqq r} \Gamma_{H_0}(\mathfrak{S}_d \mathbf{X} S_e(\mathfrak{h})) + \mathscr{S}_{r-1} \supset S_r(\mathfrak{g}),$$

where $\mathscr{S}_k = \sum_{s \leqq k} S_s(\mathfrak{g})$.  Since $\mathfrak{g} = \mathfrak{h} + \mathfrak{s}$, $S_r(\mathfrak{g})$ is spanned by elements of the form $Z_1 Z_2 \cdots Z_d H_1 H_2 \cdots H_e$ ($Z_i \in \mathfrak{s}$, $H_j \in \mathfrak{h}$), where $d + e = r$.  Moreover the restriction of $\mathrm{ad}\, H_0$ on $\mathfrak{s}$ is nonsingular since $\pi(H_0) \neq 0$.  Therefore we can select $Y_i \in \mathfrak{s}$ such that $[H_0, Y_i] = -Z_i$ $1 \leqq i \leqq d$.  Then

---

$$\Gamma_{H_0}(Y_1 \cdot Y_2 \cdots \cdots Y_d \mathbf{X} H_1 \cdots H_e) = (L_{Z_1} + d_{Y_1}) \cdots (L_{Z_d} + d_{Y_d})(H_1 \cdots H_e)$$
$$\equiv Z_1 Z_2 \cdots Z_d H_1 \cdots H_e \bmod \mathscr{S}_{r-1}.$$

On the other hand $Y_1 \cdot Y_2 \cdots \cdots Y_d - \lambda(Y_1 Y_2 \cdots Y_d) \in \lambda(\mathscr{S}_{d-1})$ (see [4(a)]) and $\Gamma_{H_0}(\lambda(\mathscr{S}_{d-1}) \mathbf{X} H_1 \cdots H_e) \subset \mathscr{S}_{d+e-1} = \mathscr{S}_{r-1}$. Hence it follows that

$$\Gamma_{H_0}(\lambda(Y_1 \cdots Y_d) \mathbf{X} H_1 H_2 \cdots H_e) - (Z_1 Z_2 \cdots Z_d H_1 \cdots H_e) \in \mathscr{S}_{r-1}$$

and this proves our assertion. Now the spaces $\mathscr{S}_r$ and $\sum\limits_{d+e \leq r} \mathfrak{S}_d \mathbf{X} S_e(\mathfrak{h})$ have the same dimension from Lemma 12 of [4(c)]. Therefore $\Gamma_{H_0}$ must be one-one.

Let $U$ and $V$ be two vector spaces over $C$ and suppose the dimension of $U$ is finite. A mapping $f$ of $U$ into $V$ will be called a polynomial mapping if (1) the subspace of $V$ spanned by $f(U)$ is of finite dimension and (2) for every linear function $\mu$ on $V$, $u \to \mu(f(u))$ $(u \in U)$ is a polynomial function on $U$.

LEMMA 5. *For every $p \in S(\mathfrak{g})$ there exists an integer $r \geq 0$ and a polynomial mapping $\gamma_p$ of $\mathfrak{h}$ into $\mathfrak{S} \mathbf{X} S(\mathfrak{h})$ such that $\Gamma_H(\gamma_p(H)) = \pi(H)^r p$ $(H \in \mathfrak{h}_0)$.*

This is proved in substantially the same way as Lemma 16 of [4(e)]. Now put $\mathfrak{S}' = \sum\limits_{d \geq 1} \mathfrak{S}_d$ and let $\mathfrak{h}_0'$ denote the set of all $H \in \mathfrak{h}_0$ where $\pi(H) \neq 0$. It follows from Lemma 4 that for any $H \in \mathfrak{h}_0'$ and $p \in S(\mathfrak{g})$, there exists a unique element $\beta_H(p) \in S(\mathfrak{h})$ such that $p - \beta_H(p) \in \Gamma_H(\mathfrak{S}' \mathbf{X} S(\mathfrak{h}))$. Hence we can define a mapping $\delta_{H}'(H \in \mathfrak{h}_0')$ of $\partial(S(\mathfrak{g}))$ into $\partial(S(\mathfrak{h}))$ by $\delta_H'(\partial(p)) = \partial(\beta_H(p))$ $(p \in S(\mathfrak{g}))$. We extend this to a mapping of $\mathfrak{D}(\mathfrak{g})$ into $\partial(S(\mathfrak{g}))$ by setting $\delta_H'(D) = \delta_H'(D_H)$ $(D \in \mathfrak{D}(\mathfrak{g}))$ where $D_H$ is the local expression of $D$ at $H$. Then it follows from Lemma 5 that there exists an integer $m \geq 0$ and a polynomial mapping $\gamma$ of $\mathfrak{h}$ into $\partial(S(\mathfrak{h}))$ such that $\delta_H'(D) = \pi(H)^{-m}\gamma(H)$ $(H \in \mathfrak{h}_0')$. This shows that there exists a differential operator $\delta'(D)$ on $\mathfrak{h}_0'$ whose local expression at any point $H \in \mathfrak{h}_0'$ coincides with $\delta_H'(D)$. The significance of $\delta'(D)$ is given by the following lemma.

LEMMA 6. *Let $U$ be an open neighborhood in $\mathfrak{g}_0$ of a point $H \in \mathfrak{h}_0'$. Suppose $f$ is a function in $C^\infty(U)$ and $f(xX) = f(X)$ whenever $(x, X)$ is sufficiently near $(1, H)$ in $G \times \mathfrak{g}_0$. Then if $\bar{f}$ denotes the restriction of $f$ on $U \cap \mathfrak{h}_0'$, $f(H; D) = \bar{f}(H; \delta'(D))$ for $D \in \mathfrak{D}(\mathfrak{g})$.*

Select $p \in S(\mathfrak{g})$ such that $D_H = \partial(p)$. Then we know that

$$p = \Gamma_H(1 \mathbf{X} \beta_H(p) + \sum_{i=1}^{r} s_i \mathbf{X} h_i)$$

where $s_i \in \mathfrak{S}'$ and $h_i \in S(\mathfrak{h})$. Therefore from Lemma 3,

$$f(x:H\,;\partial(p)) = f(x:H\,;\partial(\beta_H(p))) + \sum_{i=1}^{r} f(x\,;s_i:H\,;\partial(h_i)),$$

if $x$ lies sufficiently near 1 in $G$. But it is obvious from our assumption on $f$ that $f(x\,;b:X) = 0$ $(b \in \mathfrak{B}\mathfrak{g})$ if $(x,X)$ is sufficiently near $(1,H)$ in $G \times \mathfrak{g}_0$. Hence $f(x:H\,;\partial(p)) = f(x:H\,;\partial(\beta_H(p)))$. But $\partial(p) = D_H$ and $\partial(\beta_H(p)) = \delta_H'(D_H) = \delta_H'(D)$. Therefore putting $x = 1$, we get $f(H\,;D) = f(H\,;\delta_H'(D))$, which is equivalent to the assertion of the lemma.

Let $\mathfrak{S}'$ be the subalgebra consisting of those elements $D \in \mathfrak{D}(\mathfrak{g})$ which are invariant under $G$.

LEMMA 7. *The mapping* $D \to \delta'(D)$ $(D \in \mathfrak{S}')$ *is a homomorphism of* $\mathfrak{S}'$ *into the algebra of all differential operators on* $\mathfrak{h}_0'$.

Let $H_0$ be a point in $\mathfrak{h}_0'$ and $A$ the Cartan subgroup of $G$ (see [4(e), §2]) corresponding to $\mathfrak{h}_0$. Let $x \to x^*$ denote the natural mapping of $G$ on $G^* = G/A$ and put $x^*H = xH$ $(x \in G, H \in \mathfrak{h}_0)$. Select open connected neighborhoods $U$ and $V$ of $H_0$ and 1 in $\mathfrak{h}_0'$ and $G$ respectively. Let $V^*$ be the image of $V$ in $G^*$. Then the mapping $\phi : (x^*, H) \to x^*H$ of $V^* \times U$ into $\mathfrak{g}_0$ is easily seen to be regular (see [4(d), p. 501]). Therefore since the two manifolds $V^* \times U$ and $\mathfrak{g}_0$ have the same dimension, $N = \phi(V^* \times U)$ is an open submanifold of $\mathfrak{g}_0$. Moreover it follows that if $V$ and $U$ are sufficiently small, $\phi$ is univalent and therefore it defines an analytic isomorphism of $V^* \times U$ onto $N$. For any $g \in C^\infty(U)$, let $f_g$ denote the function in $C^\infty(N)$ given by $f_g(\phi(x^*, H)) = g(H)$ $(x^* \in V^*, H \in U)$. Then $f_g(xH) = g(H)$ $(x \in V, H \in U)$. Now suppose $D_1, D_2 \in \mathfrak{S}'$ and put $F = D_2 f_g$. Since $D_2$ is invariant under $G$, it is clear that Lemma 6 is applicable to $F$ and therefore $F(H\,;D_1) = \bar{F}(H\,;\delta'(D_1))$ $(H \in U)$. On the other hand by applying the same lemma to $f_g$ we get

$$\bar{F}(H) = f_g(H\,;D_2) = g(H\,;\delta'(D_2)) \qquad (H \in U).$$

Therefore, $f_g(H\,;D_1D_2) = F(H\,;D_1) = \bar{F}(H\,;\delta'(D_1)) = g(H\,;\delta'(D_1)\delta'(D_2))$. But again from Lemma 6, $f_g(H\,;D_1D_2) = g(H\,;\delta'(D_1D_2))$. Therefore, $g(H\,;\delta'(D_1D_2)) = g(H\,;\delta'(D_1)\delta'(D_2))$ and this proves that $\delta'(D_1D_2) = \delta'(D_1)\delta'(D_2)$ on $U$. But since $U$ is a neighborhood of an arbitrary point $H_0 \in \mathfrak{h}_0'$, we conclude that $\delta'(D_1D_2) = \delta'(D_1)\delta'(D_2)$ on $\mathfrak{h}_0'$.

We recall that $I(\mathfrak{g}) = S(\mathfrak{g}) \cap \mathfrak{S}'$. Let $\mathfrak{S}(\mathfrak{g})$ be the subalgebra of $\mathfrak{S}'$ generated by $I(\mathfrak{g}) \cup \partial(I(\mathfrak{g}))$. We shall now determine $\delta'(D)$ for $D \in \mathfrak{S}(\mathfrak{g})$. Obviously, if $p \in I(\mathfrak{g})$, $\delta'(p) = \bar{p}$, where $\bar{p}$ is the restriction of $p$ on $\mathfrak{h}$.

7

Therefore in view of Lemma 7, it is sufficient to determine $\delta'(D)$ for $D \in \partial(I(\mathfrak{g}))$.

LEMMA 8.  *If $p \in I(\mathfrak{g})$ then[9] $\delta'(\partial(p)) = \pi^{-1}\partial(\bar{p}) \circ \pi$.*

Define a polynomial function $\omega$ on $\mathfrak{g}$ by $\omega(X) = \mathrm{sp}(\mathrm{ad}\, X)^2$ $(X \in \mathfrak{g})$. (We shall call $\omega$ the Casimir polynomial of $\mathfrak{g}$.)  Then $\omega \in I(\mathfrak{g})$ and we shall first compute $\delta'(\omega)$.  Let $P$ be the set of all positive roots.  For any $\alpha \in P$, normalize the elements $X_\alpha$, $X_{-\alpha}$ in such a way that $B(X_\alpha, X_{-\alpha}) = 1$ and put $H_\alpha = [X_\alpha, X_{-\alpha}]$.  Then $B(H_\alpha, H) = \alpha(H)$ $(H \in \mathfrak{h})$ and therefore $\alpha = H_\alpha$ under our identification of $\mathfrak{h}$ with its dual.  Now consider the polynomial function

$$\xi = \omega - \bar{\omega} - 2 \sum_{\alpha \in P} X_\alpha X_{-\alpha}.$$

If $X = H + \sum_{\alpha \in P} (c_\alpha X_\alpha + c_{-\alpha} X_{-\alpha})$ $(H \in \mathfrak{h}, c_\alpha, c_{-\alpha} \in C)$, it is clear that

$$\omega(X) = \mathrm{sp}(\mathrm{ad}\, X)^2 = \mathrm{sp}(\mathrm{ad}\, H)^2 + 2 \sum_{\alpha \in P} c_{-\alpha} c_\alpha.$$

Therefore, since $B(X, X_\beta) = c_{-\beta}$ for any root $\beta$, it follows that $\xi(X) = 0$. Hence $\omega = \bar{\omega} + 2 \sum_{\alpha \in P} X_\alpha X_{-\alpha}$.  Now if $H \in \mathfrak{h}_0'$,

$$\Gamma_{H_0}((X_\alpha \cdot X_{-\alpha}) \mathbf{X} 1) = (L_{[X_\alpha, H_0]} + d_{X_\alpha})[X_{-\alpha}, H] = \alpha(H_0)X_\alpha X_{-\alpha} - \alpha(H_0)H_\alpha$$

and $X_\alpha \cdot X_{-\alpha} = \tfrac{1}{2}(X_\alpha \cdot X_{-\alpha} + X_{-\alpha} \cdot X_\alpha) + \tfrac{1}{2}H_\alpha$ in $\mathfrak{B}$.  Hence

$$\Gamma_{H_0}(\sum_{\alpha \in P} \alpha(H_0)^{-2}(X_\alpha \cdot X_{-\alpha} + X_{-\alpha} \cdot X_\alpha)\mathbf{X} 1 + (1 \mathbf{X} \bar{\omega}) + 2 \sum_{\alpha \in P} \alpha(H_0)^{-1}(1 \mathbf{X} H_\alpha))$$

$= \bar{\omega} + 2 \sum_{\alpha \in P} X_\alpha X_{-\alpha} = \omega$.  This proves that

$$\delta_{H_0}'(\partial(\omega)) = \partial(\bar{\omega}) + 2 \sum_{\alpha \in P} \alpha(H_0)^{-1}\partial(H_\alpha)$$

since $X_\alpha \cdot X_{-\alpha} + X_{-\alpha} \cdot X_\alpha \in \mathfrak{S}'$.  On the other hand let us compute $\pi^{-1}\partial(\bar{\omega}) \circ \pi$.  For any two differential operators $D_1$, $D_2$ put $\{D_1, D_2\} = D_1 \circ D_2 - D_2 \circ D_1$.  Then, if $\alpha_1, \cdots, \alpha_r$ are all the distinct positive roots,

$$\{\partial(\bar{\omega}), \pi\} = \sum_{k=1}^{r} \alpha_1 \alpha_2 \cdots \alpha_{k-1}\{\partial(\bar{\omega}), \alpha_k\} \circ (\alpha_{k+1} \cdots \alpha_r).$$

Moreover a simple calculation shows[10] that $\{\partial(\bar{\omega}), \alpha\} = 2\partial(H_\alpha)$.  Therefore

$$\partial(\bar{\omega}) \circ \pi = \pi\partial(\bar{\omega}) + 2 \sum_{k=1}^{r} \alpha_1 \alpha_2 \cdots \alpha_{k-1} \partial(H_{\alpha_k}) \circ (\alpha_{k+1} \cdots \alpha_r).$$

---

[9] Whenever it is necessary to avoid confusion, we denote the product of two differential operators $D_1$, $D_2$ by $D_1 \circ D_2$.

[10] This is best done by making use of a base $H_1, \cdots, H_l$ for $\mathfrak{h}$ such that $B(H_i, H_j) = \delta_{ij}$ $(1 \leqq i, j \leqq l)$.

But $\{\partial(H_\alpha),\beta\} = \beta(H_\alpha)$ for any two roots $\alpha,\beta$, and so we can conclude that

$$\partial(\overline{\omega}) \circ \pi = \pi\partial(\overline{\omega}) + 2\sum_{k=1}^{r} \alpha_k^{-1}\partial(H_{\alpha_k}) + q,$$

where $q \in S(\mathfrak{h})$. Applying both sides of the above equation to the constant function 1, we find that $q = \partial(\overline{\omega})\pi$. But we shall see presently (Lemma 9 and the Corollary to Lemma 10) that $\partial(\overline{\omega})\pi = 0$. Therefore by comparing this with our earlier expression for $\delta_{H_0}{}'(\partial(\omega))$, it follows that $\delta'(\partial(\omega)) = \pi^{-1}\partial(\overline{\omega}) \circ \pi$. This proves our assertion in case $p = \omega$.

Now let $\mu$ be the derivation of $\mathfrak{D}(\mathfrak{g})$ given by $\mu D = \frac{1}{2}\{\partial(\omega), D\}$ $(D \in \mathfrak{D}(\mathfrak{g}))$. A direct computation shows that $\mu X = \partial(X)$ $(X \in \mathfrak{g})$. We shall now prove by induction on $m$ that $\mu^m(X_1 X_2 \cdots X_m) = m!\,\partial(X_1 X_2 \cdots X_m)$ for $X_1, \cdots, X_m \in \mathfrak{g}$. Since $\mu^2 X = \mu\partial(X) = 0$ $(X \in \mathfrak{g})$, it follows from the Leibnitz rule for derivations that

$$\mu^m(X_1 X_2 \cdots X_m) = (\mu^m(X_1 \cdots X_m)) \circ X_m + m\mu^{m-1}(X_1 \cdots X_{m-1}) \circ \mu X_m.$$

But $\mu^{m-1}(X_1 \cdots X_{m-1}) = (m-1)!\,\partial(X_1 X_2 \cdots X_{m-1})$ by induction hypothesis. Therefore $\mu^m(X_1 \cdots X_{m-1}) = 0$ and

$$\mu^m(X_1 X_2 \cdots X_m) = m!\,\partial(X_1 \cdots X_{m-1}) \circ \mu X_m = m!\,\partial(X_1 \cdots X_{m-1}X_m).$$

This shows that if $q$ is a homogeneous element in $S(\mathfrak{g})$ of degree $m$, then $\mu^m q = m!\,\partial(q)$.

On the other hand let us put $\bar{\mu}\xi = \frac{1}{2}\{\delta'(\omega),\xi\}$ for any differential operator $\xi$ on $\mathfrak{h}_0'$. Then $\bar{\mu}$ is a derivation of the algebra of differential operators on $\mathfrak{h}_0'$ and it follows from Lemma 7 that $\delta'(\mu D) = \bar{\mu}\delta'(D)$ for $D \in \mathfrak{J}'$. Let $p$ be a homogeneous element in $I(\mathfrak{g})$ of degree $m$. Since $\mathfrak{J}'$ is obviously stable under $\mu$, we conclude that

$$m!\,\delta'(\partial(p)) = \delta'(\mu^m p) = \bar{\mu}^m\delta'(p) = \bar{\mu}^m\bar{p}.$$

Now let $\mathfrak{A}$ be an associative algebra and for any $a \in \mathfrak{A}$, let $d_a$ denote the derivation of $\mathfrak{A}$ defined by $d_a b = \frac{1}{2}(ab - ba)$ $(b \in \mathfrak{A})$. Then one proves easily by induction on $k$ that

$$d_a{}^k b = 2^{-k}\sum_{0 \le r \le k} C_r{}^k(-1)^r a^{k-r}ba^r \qquad (k \ge 0).$$

We now suppose that $\mathfrak{A}$ has a unit element, $c$ is an element in $\mathfrak{A}$ with an inverse $c^{-1}$ and $a' = c^{-1}ac$. Then if $b \in \mathfrak{A}$ and it commutes with $c$, it is clear that

$$d_{a'}{}^k b = c^{-1}(d_a{}^k b)c \qquad (k \ge 0).$$

Let us now take $\mathfrak{A}$ to be the algebra of all differential operators on $\mathfrak{h}_0'$. Then if $a = \partial(\bar{\omega})$, $b = \bar{p}$ and $c = \pi$, all the above conditions are fulfilled and $a' = \pi^{-1}\partial(\bar{\omega}) \circ \pi = \delta'(\omega)$ as we have seen above. Moreover if $q$ is any homogeneous element in $S(\mathfrak{h})$ of degree $k$, we prove in the same way as above that $d_a{}^k q = k!\,\partial(q)$. Therefore

$$\mu^m \bar{p} = d_{a'}{}^m \bar{p} = \pi^{-1}(d_a{}^m \bar{p}) \circ \pi = m!\,\pi^{-1}\partial(\bar{p}) \circ \pi$$

and this proves that $\delta'(\partial(p)) = \pi^{-1}\partial(\bar{p}) \circ \pi$ for any homogeneous element $p \in I(\mathfrak{g})$. But it is obvious that $I(\mathfrak{g}) = \sum_{m \geq 0} I(\mathfrak{g}) \cap S_m(\mathfrak{g})$ and so the assertion of the lemma follows.

Let $W$ denote the Weyl group of $\mathfrak{g}$ with respect to $\mathfrak{h}$ (see [4(b), p. 29]). Then $W$ is a group of nonsingular linear mappings of $\mathfrak{h}$ and therefore (see Section 2) it operates on $\mathfrak{D}(\mathfrak{h})$. Let $\mathfrak{J}(\mathfrak{h})$ denote the subalgebra of those elements in $\mathfrak{D}(\mathfrak{h})$ which are invariant under $W$.

THEOREM 1. *There exists a unique homomorphism $\delta$ of $\mathfrak{J}(\mathfrak{g})$ into $\mathfrak{J}(\mathfrak{h})$ such that $\delta(p) = p$ and $\delta(\partial(p)) = \partial(\bar{p})$ ($p \in I(\mathfrak{g})$). Moreover $\delta'(D) = \pi^{-1}\delta(D) \circ \pi$ on $\mathfrak{h}_0'$ for any $D \in \mathfrak{J}(\mathfrak{g})$.*

Since $\mathfrak{J}(\mathfrak{g})$ is generated by $I(\mathfrak{g}) \cup \partial(I(\mathfrak{g}))$, the uniqueness of $\delta$ is obvious. Let $\mathfrak{H}'$ denote the algebra of all differential operators on $\mathfrak{h}_0'$. It is clear that we can regard $\mathfrak{D}(\mathfrak{h})$ as a subalgebra of $\mathfrak{H}'$. Put $\delta(D) = \pi\delta'(D) \circ \pi^{-1}$. Then it follows from Lemma 7, that $\delta$ is a homomorphism of $\mathfrak{J}(\mathfrak{g})$ into $\mathfrak{H}'$. Moreover $\delta(p) = \bar{p}$, $\delta(\partial(p)) = \partial(\bar{p})$ ($p \in I(\mathfrak{g})$) from Lemma 8 and $\bar{p}$ is invariant under $W$ (see Lemma 9 below). Therefore it is obvious that $\delta$ maps $\mathfrak{J}(\mathfrak{g})$ into $\mathfrak{J}(\mathfrak{h})$ and so the theorem is proved.

**4. Invariants of the Weyl group.** An element $q \in S(\mathfrak{h})$ is said to be invariant (under $W$) if $q^s = q$ for all $s \in W$. Let $I(\mathfrak{g})$ be the subalgebra of $S(\mathfrak{h})$ consisting of all invariants. The following result is due to Chevalley (see [1, exposé 19, p. 10]).

LEMMA 9 (Chevalley). *If $p \in I(\mathfrak{g})$, then $\bar{p} \in I(\mathfrak{h})$, and $p \to \bar{p}$ is an isomorphism of $I(\mathfrak{g})$ onto $I(\mathfrak{h})$.*

For any root $\alpha$ let $s_\alpha$ denote the Weyl reflexion corresponding to $\alpha$. An element $q \in S(\mathfrak{h})$ will be called skew (or skew-invariant) if $q^{s_\alpha} = -q$ for every root $\alpha$. It is known (see Weyl [9]) that $\pi$ is skew.

LEMMA 10. *If $q$ is a skew-invariant in $S(\mathfrak{h})$, then $q = \pi q_0$ for some $q_0 \in I(\mathfrak{h})$.*

For any root $\alpha$, let $\sigma_\alpha$ denote the hyperplane in $\mathfrak{h}$ given by the equation $\alpha = 0$. Then if $p \in S(\mathfrak{h})$, it follows from the definition of $s_\alpha$ that $p^{s_\alpha} - p$ vanishes everywhere on $\sigma_\alpha$. Therefore $p^{s_\alpha} - p$ is divisible by $\alpha$ in the polynomial ring $S(\mathfrak{h})$. This shows that $q = -\frac{1}{2}(q^{s_\alpha} - q)$ is divisible by $\alpha$. Now the hyperplanes $\sigma_\alpha$ corresponding to distinct positive roots $\alpha$ are distinct. Therefore since $S(\mathfrak{h})$ is a unique factorization domain, $q$ is divisible by $\prod_{\alpha \in P} \alpha = \pi$. This means that $q = \pi q_0$, where $q_0 \in S(\mathfrak{h})$. But since both $\pi$ and $q$ are skew, it follows that $q_0^{s_\alpha} = q_0$ $(\alpha \in P)$. Therefore since $W$ is generated by the reflexions $s_\alpha$ $(\alpha \in P)$, we conclude that $q_0 \in I(\mathfrak{h})$.

COROLLARY. *If $q$ is a homogeneous element in $I(\mathfrak{h})$ of positive degree, then $\partial(q)\pi = 0$.*

Let $r$ be the degree of $\pi$. Then it is obvious that $\partial(q)\pi$ is homogeneous and its degree is lower than $r$. On the other hand, since $q \in I(\mathfrak{h})$, it is clear that $\partial(q)\pi$ is a skew-invariant. Therefore in view of the above lemma, we conclude that $\partial(q)\pi = 0$.

Put $S = S(\mathfrak{h})$ and $J = I(\mathfrak{h})$ and let $J_+$ be the ideal in $J$ spanned by all homogeneous elements of $J$ of positive degree. Since $W$ is a finite group generated by reflexions. it follows from a result of Chevalley [2] that the dimension of the vector space $S/J_+S$ is equal to the order $w$ of $W$. Moreover if $Q_S$ and $Q_J$ are the quotient fields of $S$ and $J$ respectively, then $[Q_S : Q_J] = w$ where $[Q_S : Q_J]$ denotes, as usual, the degree of the field-extension $Q_S/Q_J$. It is obvious that if an element of $S$ lies in $J$, the same holds for all its homogeneous components. Hence we can select homogeneous elements $v_1, \cdots, v_w \in S$ such that $S = V + J_+S$, where $V$ is the linear space spanned over $C$ by $(v_1, \cdots, v_w)$.

LEMMA 11. *Every element $v$ of $S(\mathfrak{h})$ can be written exactly in one way in the form $v = u_1 v_1 + \cdots + u_w v_w$, where $u_i \in I(\mathfrak{h})$ $(1 \leqq i \leqq w)$.*

It follows easily by induction on $m$ that $S = JV + J_+^m S$ $(m \geqq 1)$. Let $v$ be any homogeneous element in $S$ of degree $d$ and let $d_i$ be the degree of $v_i$. Choose $m > d$ and select elements $u_i', \cdots, u_w' \in J$ such that $v - (u_1' v_1 + \cdots + u_w' v_w) \in J_+^m S$. Let $u_i$ denote the homogeneous component of $u_i'$ of degree $d - d_i$. Then, by considering the homogeneous component of of degree $d$ in the above relation, it is clear that $v - (u_1 v_1 + \cdots + u_w v_w) = 0$. This proves that $S = Jv_1 + \cdots + Jv_w$ and now it follows easily that $Q_S = Q_J S = \sum_{i=1}^{w} Q_J v_i$. However $[Q_S : Q_J] = w$ and so $v_1, \cdots, v_w$ must be linearly independent over $Q_J$. This proves that an element $v \in S$ can be written in exactly one way in the form $v = u_1 v_1 + \cdots + u_w v_w$ $(u_i \in J)$.

We shall now make use of the above lemma to obtain some results on the solutions of a certain system of differential equations which will be important for our applications.

LEMMA 12. *Let $\tau$ be a distribution on an open subset $U$ of $\mathfrak{h}_0$. Suppose for each $q \in I(\mathfrak{h})$, there exists a complex number $c_q$ such that $\partial(q)\tau = c_q\tau$. Then $\tau$ coincides with an analytic function on $U$.*

Let $H_1, \cdots, H_l$ be a base for $\mathfrak{h}_0$ over $R$. Put $\Omega = H_1^2 + H_2^2 + \cdots + H_l^2 \in S(\mathfrak{h})$ and consider the polynomial

$$\prod_{s \in W} (\zeta - \Omega^s) = \zeta^w + q_1\zeta^{w-1} + \cdots + q_w$$

in the indeterminants $\zeta$ with coefficients $q_i \in I(\mathfrak{h})$. Then

$$\Omega^w + \Omega^{w-1}q_1 + \cdots + q_w = \prod_{s \in W} (\Omega - \Omega^s) = 0$$

and therefore $\{\partial(\Omega)^w + c_1\partial(\Omega)^{w-1} + \cdots + c_w\}\tau = 0$, where $c_i = c_{q_i}$ $(1 \leq i \leq w)$. However the above differential equation is obviously of the elliptic type (see [3] for the definition of ellipticity). Therefore (see Schwartz [8(1), p. 136] and John [5]), $\tau$ is an analytic function on $U$.

LEMMA 13. *Define $(v_1, \cdots, v_w)$ as in Lemma 11 and let $\phi$ be an analytic function on a nonempty open connected subset $U$ of $\mathfrak{h}_0$. Suppose $\phi$ is an eigenfunction of $\partial(u)$ for every $u \in I(\mathfrak{h})$. Then if $\partial(v_i)\phi$ $(1 \leq i \leq w)$ all vanish simultaneously at some point of $U$, $\phi$ must be identically zero.*

Suppose $\phi(H_0; \partial(v_i)) = 0$ $(1 \leq i \leq w)$ for some $H_0 \in U$. For any $v \in S(\mathfrak{h})$, we can select $u_1, \cdots, u_w \in I(\mathfrak{h})$ such that $v = \sum_{i=1}^{w} u_i v_i$ (Lemma 11). But then $\partial(u_i)\phi = c_i\phi$ $(c_i \in C)$ and therefore

$$\phi(H_0; \partial(v)) = \sum_{i=1}^{w} c_i\phi(H_0; \partial(v_i)) = 0.$$

This shows that all derivatives of $\phi$ vanish at $H_0$ and so $\phi$, being an analytic function, must be identically zero.

COROLLARY. *Suppose $H'$ is a point in $\mathfrak{h}$ such that $\pi(H') \neq 0$ and $\phi$ an analytic function on $U$ satisfying the system of differential equations*

$$\partial(q)\phi = q(H')\phi \qquad\qquad (q \in I(\mathfrak{h})).$$

*Then there exist constants $c_s$ $(s \in W)$ such that*

$$\phi(H) = \sum_{s \in W} c_s \exp B(H, sH') \qquad\qquad (H \in U).$$

*Moreover they are uniquely determined.*

It follows from Lemma 4 of [4(e)] that the $w$ points $sH'$ $(s \in W)$ are all distinct. Put $\phi_s(H) = \exp B(H, sH')$ $(H \in U)$. Then $\phi_s$ $(s \in W)$ are linearly independent analytic functions on $U$ (see Lemma 41 of [4(b)]). Moreover since $\partial(H)\phi_s = B(H, sH')\phi_s$ $(H \in \mathfrak{h})$, it is clear that $\partial(q)\phi_s = q(H')\phi_s$ for $q \in I(\mathfrak{h})$. Therefore $\phi_s$ is a solution of our system of differential equations. Let $E$ be the vector space (over $C$) consisting of all analytic solutions of this system. It is obviously enough to prove that $\phi_s$ $(s \in W)$ form a base for $E$ or, what is equivalent, that $\dim E \leq w$. Choose a point $H_0 \in U$. Assuming that $\dim E > w$, we can select $\phi \neq 0$ in $E$ satisfying the $w$ linear conditions $\phi(H_0; \partial(v_i)) = 0$ $1 \leq i \leq w$. But we know from Lemma 13 that this is impossible.

The following lemma will be required in another paper.

LEMMA 14. *Suppose $\phi$ is an analytic function on $U$ such that $\partial(q)\phi = 0$ for all homogeneous elements $q \in I(\mathfrak{h})$ of positive degree. Then there exists a polynomial function $p \in S(\mathfrak{h})$ such that $\phi = p$ on $U$.*

For any $H \in \mathfrak{h}$, consider the polynomial

$$\prod_{s \in W} (\zeta - sH) = \zeta^w + q_1\zeta^{w-1} + \cdots + q_w$$

in the **indeterminate** $\zeta$ with coefficients $q_i \in I(\mathfrak{h})$. Then $H^w + H^{w-1}q_1 + \cdots + q_w = 0$; and therefore $\partial(H^w + H^{w-1}q_1 + \cdots + q_w)\phi = 0$. But since $q_i$ are all homogeneous and of positive degree, $\partial(q_i)\phi = 0$ $(1 \leq i \leq w)$ and therefore $\partial(H^w)\phi = 0$. Hence if $H_1, \cdots, H_l$ is a base for $\mathfrak{h}_0$ over $R$, $\partial(H_i{}^w)\phi = 0$ $1 \leq i \leq l$. Obviously, this implies that $\phi$ is a polynomial of degree $\leq lw$.

**5. Applications to the compact case.** In this section we shall assume that $G$ is compact. Let $dx$ denote the normalized Haar measure of $G$, so that $\int_G dx = 1$. For any $f \in C^\infty(\mathfrak{g}_0)$, let $\phi_f$ denote the function on $\mathfrak{h}_0$ given by

$$\phi_f(H) = \pi(H) \int_G f(xH)dx \qquad (H \in \mathfrak{h}_0).$$

It is obvious that $\phi_f \in C^\infty(\mathfrak{h}_0)$. Define $\delta$ as in Theorem 1.

LEMMA 15. $\phi_{\xi f} = \delta(\xi)\phi_f$ *for* $\xi \in \mathfrak{J}(\mathfrak{g})$ *and* $f \in C^\infty(\mathfrak{g}_0)$.

We use the notation of Section 3. It follows from Lemmas 3 and 5 that we can select elements $s_i \in \mathfrak{S}'$, $h_i \in S(\mathfrak{h})$ and analytic functions $a_i$ $(1 \leqq i \leqq m)$ on $\mathfrak{h}_0'$ such that

$$f(xH;\xi) = f(x:H;\delta'(\xi)) + \sum_{i=1}^{m} a_i(H)f(x;s_i:H;\partial(h_i))$$

for $H \in \mathfrak{h}_0'$.  But it is obvious that $\int_G f(x;b:H)dx = 0$ $(H \in \mathfrak{h}_0)$ for $b \in \mathfrak{B}\mathfrak{g}$.  Hence it follows from Theorem 1 that

$$\phi_{\xi f}(H) = \pi(H) \int_G f(x:H;\delta'(\xi))dx = \phi_f(H;\delta(\xi)) \qquad (H \in \mathfrak{h}_0').$$

However, $\phi_{\xi f}$ and $\delta(\xi)\phi_f$ are both in $C^\infty(\mathfrak{h}_0)$ and $\mathfrak{h}_0'$ is dense in $\mathfrak{h}_0$.  Therefore $\phi_{\xi f} = \delta(\xi)\phi_f$.

LEMMA 16.    $\phi_f(0;\delta(\pi)) = \langle \pi, \pi \rangle f(0)$   for all   $f \in C^\infty(\mathfrak{g}_0)$.

Put $g(H) = \int f(xH)dx$ $(H \in \mathfrak{h}_0)$.  Since $\pi$ is a homogeneous polynomial, it is obvious that the local expression of $\partial(\pi) \circ \pi$ at the origin is exactly $\langle \pi, \pi \rangle$.  Hence the value of $\partial(\pi)\phi_f$ at zero is $\langle \pi, \pi \rangle g(0) = \langle \pi, \pi \rangle f(0)$.

As usual, we define $\epsilon(s) = 1$ or $-1$ according as $\pi^s = \pi$ or $-\pi$ $(s \in W)$.

THEOREM 2.

$$\pi(H)\pi(H') \int_G \exp B(xH, H')dx = w^{-1}\langle \pi, \pi \rangle \sum_{s \in W} \epsilon(s) \exp B(sH, H')$$

for all $H, H' \in \mathfrak{h}$.

Select an element $H' \in \mathfrak{h}$ such that $\pi(H') \neq 0$, and put $f(X) = \exp B(X, H')$ $(X \in \mathfrak{g}_0)$.  Then

$$\phi_f(H) = \pi(H) \int_G \exp B(xH, H')dx$$

and therefore, from Theorem 1 and Lemma 15, $\phi_{\partial(p)f} = \partial(\bar{p})\phi_f$ $(p \in I(\mathfrak{g}))$. But it is clear that $\partial(Y)f = B(Y, H')f$ $(Y \in \mathfrak{g})$, and so $\partial(q) = q(H')f$ for $q \in S(\mathfrak{y})$.  Hence $\partial(\bar{p})\phi_f = p(H')\phi_f$ $(p \in I(\mathfrak{g}))$.  But in view of Lemma 9, this implies that $\partial(q)\phi_f = q(H')\phi_f$ for all $q \in I(\mathfrak{h})$, and we conclude from the Corollary to Lemma 13 that

$$\phi_f(H) = \sum_{s \in W} c_s \exp B(sH, H') \qquad (H \in \mathfrak{h}_0),$$

where $c_s$ $(s \in W)$ are constants.  On the other hand, since $G$ is compact, we can select, for each $s \in W$, an element $x_s \in G$ such that $x_s H = sH$ $(H \in \mathfrak{h}_0)$.

Therefore it is obvious that $s$ maps $\mathfrak{h}_0$ into itself and $\phi_f(sH) = \epsilon(s)\phi_f(H)$, and so

$$\phi_f(H) = w^{-1}\sum_{s \in W}\epsilon(s)\phi_f(sH) = w^{-1}c\sum_{s \in W}\epsilon(s)\exp B(sH, H') \qquad (H \in \mathfrak{h}_0),$$

where $c = \sum_{s \in W}\epsilon(s)c_s$. On the other hand, $\phi_f(0;\partial(\pi)) = \langle\pi,\pi\rangle f(0) = \langle\pi,\pi\rangle$ from Lemma 16. Put $\psi_s(H) = \exp B(sH, H')$ $(H \in \mathfrak{h}_0)$. Then $\partial(q)\psi_s = q(s^{-1}H')\psi_s$ $(q \in S(\mathfrak{h}))$, and therefore $\partial(\pi)\psi_s = \epsilon(s)\pi(H')\psi_s$. Hence

$$\langle\pi,\pi\rangle = \phi_f(0;\partial(\pi)) = w^{-1}c\pi(H')w = c\pi(H'),$$

and so $c = \langle\pi,\pi\rangle/\pi(H')$. This proves that

$$\pi(H)\pi(H')\int_G \exp B(x, H, H')dx = w^{-1}\langle\pi,\pi\rangle\sum_{s \in W}\epsilon(s)\exp B(sH, H')$$

for $H \in \mathfrak{h}_0$ and $H' \in \mathfrak{h}$ provided $\pi(H') \neq 0$. But since both sides can obviously be considered as holomorphic functions on $\mathfrak{h} \times \mathfrak{h}$, it follows that they must be equal for all $H, H' \in \mathfrak{h}$.

Let $dX$ and $dH$ denote the Euclidean measures on $\mathfrak{g}_0$ and $\mathfrak{h}_0$, respectively. We may assume (see Lemma 9 of [4(d)]) that they are so normalized that

$$\int_{\mathfrak{g}_0}f(X)dX = \int_{\mathfrak{h}_0}|\pi(H)|^2\,dH\int_G f(xH)dx \qquad (f \in C_c(\mathfrak{g}))$$

For any $f \in \mathcal{B}(\mathfrak{g}_0)$ and $g \in \mathcal{B}(\mathfrak{h}_0)$, let $\tilde{f}$ and $\tilde{g}$ denote the corresponding Fourier transforms, so that

$$\tilde{f}(Y) = \int_{\mathfrak{g}_0}\exp\{(-1)^{\frac{1}{2}}B(Y, X)\}f(X)dX,$$

$$\tilde{g}(H') = \int_{\mathfrak{h}_0}\exp\{(-1)^{\frac{1}{2}}B(H', H)\}g(H)dH$$

$(Y \in \mathfrak{g}_0, H' \in \mathfrak{h}_0)$. Then we have the following result.

THEOREM 3. *If* $f \in \mathcal{B}(\mathfrak{g}_0)$, *the corresponding function* $\phi_f$ *lies in* $\mathcal{B}(\mathfrak{h}_0)$ *and* $f \to \phi_f$ *is a continuous mapping of* $\mathcal{B}(\mathfrak{g}_0)$ *into* $\mathcal{B}(\mathfrak{h}_0)$. *Moreover* [3] $\phi_{\tilde{f}} = (-1)^r\langle\pi,\pi\rangle(\phi_f)^{\tilde{}}$, *where* $r$ *is the number of positive roots.*

It is seen without difficulty that for any $D \in \mathfrak{D}(\mathfrak{g})$, we can find a finite number of elements $D_1, \cdots, D_m \in \mathfrak{D}(\mathfrak{g})$ and analytic functions $a_1, \cdots, a_m$ on $G$ such that $D^x = a_1(x)D_1 + \cdots + a_m(x)D_m$ $(x \in G)$. Now suppose $f \in \mathcal{B}(\mathfrak{g}_0)$ and $d \in \mathfrak{D}(\mathfrak{h})$. Then it is clear that

$$\phi_f(H;d) = \int_G f(xH;D^x)dx \qquad (H \in \mathfrak{h}_0)$$

where $D = d \circ \pi$. Since $G$ is compact, we can select a positive number $c$ such that $|a_i(x)| \leqq c$ $(1 \leqq i \leqq m, x \in G)$. Hence

$$|\phi_f(H;d)| \leqq \sum_{i=1}^{m} c \sup_{X \in \mathfrak{g}_0} |f(X;D_i)|,$$

and this proves that $\phi_f \in \mathcal{B}(\mathfrak{h}_0)$ and the mapping $f \to \phi_f$ of $\mathcal{B}(\mathfrak{g}_0)$ into $\mathcal{B}(\mathfrak{h}_0)$ is continuous.

Put $\eta(X:Y) = \exp\{(-1)^{\frac{1}{2}} B(X,Y)\}$ $(X, Y \in \mathfrak{g}_0)$. Then if $f \in \mathcal{B}(\mathfrak{g}_0)$,

$$\phi_{\tilde{f}}(H) = \pi(H) \int_G \tilde{f}(xH)dx = \pi(H) \int_G dx \int_{\mathfrak{g}_0} \eta(xH:Y)f(Y)dY \quad (H \in \mathfrak{h}_0).$$

But

$$\int_{\mathfrak{g}_0} \eta(xH:Y)f(Y)dY = \int_{G \times \mathfrak{h}_0} \eta(xH:yH')\,|\pi(H')|^2 f(yH')dy\,dH'$$

$$= (-1)^r \int_{G \times \mathfrak{h}_0} \eta(xH:yH')\pi(H')^2 f(yH')dy\,dH'$$

since every root takes only pure imaginary values on $\mathfrak{h}_0$. Therefore we conclude from Theorem 2 that

$$\phi_{\tilde{f}}(H) = w^{-1}\langle\pi,\pi\rangle(-1)^r \sum_{s \in W} \epsilon(s) \int_{G \times \mathfrak{h}_0} \pi(H')\eta(sH:H')f(yH')dy\,dH'$$

$$= (-1)^r w^{-1}\langle\pi,\pi\rangle \sum_{s \in W} \epsilon(s) \int_{\mathfrak{h}_0} \eta(sH:H')\phi_f(H')dH'.$$

But we have seen that $\phi_f(sH') = \epsilon(s)\phi_f(H')$ $(s \in W)$. Hence

$$\int_{\mathfrak{h}_0} \eta(sH:H')\phi_f(H')dH' = \int_{\mathfrak{h}_0} \eta(H:H')\phi_f(sH')dH'$$

$$= \epsilon(s) \int_{\mathfrak{h}_0} \eta(H:H')\phi_f(H')dH'.$$

This proves that $\phi_{\tilde{f}}(H) = (-1)^r\langle\pi,\pi\rangle \int \eta(H:H')\phi_f(H')dH'$ $(H \in \mathfrak{h}_0)$, and therefore $\phi_{\tilde{f}} = (-1)^r\langle\pi,\pi\rangle(\phi_f)^{\tilde{}}$.

In another paper we shall obtain a suitable analogue of the above theorem for the case when $G$ is not compact.

Let $J(\mathfrak{g}_0)$ and $J(\mathfrak{h}_0)$ denote the sets of all functions in $\mathcal{B}(\mathfrak{g}_0)$ and $\mathcal{B}(\mathfrak{h}_0)$ which are invariant under $G$ and $W$ respectively. For any function $f$ on $\mathfrak{g}_0$, let $\tilde{f}$ denote its restriction on $\mathfrak{h}_0$.

THEOREM 4. *If* $f \in J(\mathfrak{g}_0)$, *then* $\tilde{f} \in J(\mathfrak{h}_0)$ *and* $f \to \tilde{f}$ *is a topological mapping of* $J(\mathfrak{g}_0)$ *onto* $J(\mathfrak{h}_0)$.

Put $\eta'(X:Y) = \exp\{-(-1)^{\frac{1}{2}} B(X,Y)\}$ $(X,Y \in \mathfrak{g}_0)$. Then if $p \in S(\mathfrak{g})$, it is clear that $\eta'(X;\partial(p):Y) = \hat{p}(Y)\eta'(X:Y)$ where $\hat{p}$ is the polynomial function $Z \to p(-(-1)^{\frac{1}{2}}Z)$ on $\mathfrak{g}$. Now we assume that the normalization of the Euclidean measure $dH$ on $\mathfrak{h}_0$ is regular and put

$$\psi_g(X) = \int_{\mathfrak{h}_0} \xi(X:H)\pi(H)\tilde{g}(H;\partial(\pi))dH \qquad (g \in J(\mathfrak{h}_0), X \in \mathfrak{g}_0),$$

where $\xi(X:Y) = \int_G \eta'(xX:Y)dx$ $(X,Y \in \mathfrak{g}_0)$. Then it follows from Theorem 2 that if $H' \in \mathfrak{h}_0$,

$$\pi(H')\psi_g(H') = c \int_{\mathfrak{h}_0} \sum_{s \in W} \epsilon(s)\eta'(H:sH')\tilde{g}(H;\partial(\pi))dH,$$

where $c = w^{-1}\langle \pi,\pi \rangle$. Now it is clear that [11]

$$\int_{\mathfrak{h}_0} \eta'(H:sH')\tilde{g}(H;\partial(\pi))dH = (-1)^r \int_{\mathfrak{h}_0} \eta'(H;\partial(\pi):sH')\tilde{g}(H)dH$$

$$= (-1)^r \hat{\pi}(sH') \int \eta'(H:sH')\tilde{g}(H)dH = (-1)^{r/2}\epsilon(s)\pi(H')g(sH')$$

$(s \in W)$. Therefore, since $g$ is invariant under $W$, we get

$$\pi(H')\psi_g(H') = (-1)^{r/2} wc\pi(H')g(H') \qquad (H' \in \mathfrak{h}_0).$$

But as $\bar{\psi}_g$ and $g$ are both of class $C^\infty$, this implies that $\bar{\psi}_g = c'g$, where $c' = (-1)^{r/2}\langle \pi,\pi \rangle$. It is obvious from its definition that $\psi_g \in C^\infty(\mathfrak{g}_0)$ and it is invariant under $G$. Moreover, since $\mathfrak{g}_0 = \bigcup_{x \in G} x\mathfrak{h}_0$, it is obvious that $\psi_g$ is the only invariant function on $\mathfrak{g}_0$ which coincides with $c'g$ on $\mathfrak{h}$. Therefore $q\psi_g = \psi_{\tilde{q}g}$ if $q \in I(\mathfrak{g})$.

Put

$$v_D(F) = \sup_{X \in \mathfrak{g}_0} |F(X;D)|, \qquad \sigma_d(\phi) = \sup_{H \in \mathfrak{h}_0} |\phi(H;d)|$$

for $F \in C^\infty(\mathfrak{g}_0)$, $\phi \in C^\infty(\mathfrak{h})$, $D \in \mathfrak{D}(\mathfrak{g})$ and $d \in \mathfrak{D}(\mathfrak{h})$. We claim that for any $D \in \mathfrak{D}(\mathfrak{g})$, we can select a finite number of elements $d_1, \cdots, d_N \in \mathfrak{D}(\mathfrak{h})$ such that

$$v_D(\psi_g) \leqq \sum_{i=1}^{N} \sigma_{d_i}(g)$$

for $g \in J(\mathfrak{h}_0)$. It is enough to prove this for $D = p\partial(q)$ $(p,q \in S(\mathfrak{g}))$. We shall use induction on the degree of $q$ which we denote by $m$. If $m = 0$, we may assume that $D = p$. Since $G$ is compact, the Casimir polynomial $\omega$ is a negative-definite quadratic form on $\mathfrak{g}_0$, and so it is clear that we can

---

[11] $(-1)^{k/2} = ((-1)^{\frac{1}{2}})^k$ for any integer $k$.

select $p_0 \in I(\mathfrak{g})$ such that $p_0$ takes only real nonnegative values on $\mathfrak{g}_0$ and $|p(X)| \leqq p_0(X)$ $(X \in \mathfrak{g}_0)$. Then $p_0 \psi_g = \psi_{\bar{p}_0 g}$, and therefore $\nu_D(\psi_g) \leqq \nu_{p_0}(\psi_g)$ $= \nu_1(\psi_{\bar{p}_0 g})$. From Lemma 1 and its Corollary, there exist elements $p_1 \in S(\mathfrak{h})$ and $d \in \mathfrak{D}(\mathfrak{h})$ such that $(\bar{p}_0 \phi)^{\tilde{}} = \partial(p_1)\tilde{\phi}$ and

$$\int_{\mathfrak{h}_0} |\pi \partial(\pi p_1)\tilde{\phi}| \, dH \leqq \sigma_d(\phi)$$

for $\phi \in \mathscr{B}(\mathfrak{h}_0)$. Therefore

$$\nu_D(\psi_g) \leqq \nu_1(\psi_{\bar{p}_0 g}) \leqq \int |\pi \partial(\pi p_1)\tilde{g}| \, dH \leqq \sigma_d(g) \qquad (g \in J(\mathfrak{h}_0))$$

since $|\xi(X:Y)| \leqq 1$ $(X, Y \in \mathfrak{g}_0)$. Now suppose $m \geqq 1$. Again choose $p_0 \in I(\mathfrak{g})$ such that $|p(X)| \leqq p_0(X)$ for $X \in \mathfrak{g}_0$. Then if $D_0 = p_0 \partial(q)$, it is obvious that $\nu_D(\psi_g) \leqq \nu_{D_0}(\psi_g)$. Also, it is clear that

$$D_0 = \partial(q) \circ p_0 + \sum_{i=1}^{k} p_i \partial(q_i),$$

where $k$ is some integer, $p_i, q_i \in S(\mathfrak{g})$ and the degree of each $q_i$ is smaller than $m$. In view of the induction hypothesis, it is sufficient to prove our statement for $\Delta = \partial(q) \circ p_0$. But $\Delta \psi_g = \partial(q)\psi_{g_0}$, where $g_0 = \bar{p}_0 g$, and

$$\xi(X;\partial(q):Y) = \int_G \hat{q}(xY)\eta'(X:xY)\, dx \qquad (X, Y \in \mathfrak{g}_0).$$

We again select $q_0 \in I(\mathfrak{g})$ such that $|\hat{q}(X)| \leqq q_0(X)$ for $X \in \mathfrak{g}_0$. Then $|\xi(X;\partial(q):Y)| \leqq q_0(Y)$ $(X, Y \in \mathfrak{g}_0)$, and therefore

$$|\psi_{g_0}(X;\partial(q))| \leqq \int_{\mathfrak{h}_0} |\tilde{q}_0 \pi \partial(\pi)\tilde{g}_0| \, dH.$$

Choose $p_1 \in S(\mathfrak{h})$ and $d \in \mathfrak{D}(\mathfrak{h})$ such that $(\bar{p}_0 \phi)^{\tilde{}} = \partial(p_1)\tilde{\phi}$ and

$$\int_{\mathfrak{h}_0} |\tilde{q}_0 \pi \partial(\pi p_1)\tilde{\phi}| \, dH \leqq \sigma_d(\phi)$$

for $\phi \in \mathscr{B}(\mathfrak{h})$. Then $\nu_\Delta(\psi_g) = \nu_{\partial(q)}(\psi_{g_0}) \leqq \sigma_d(g)$ $(g \in J(\mathfrak{h}_0))$.

The above result shows that if $g \in J(\mathfrak{h}_0)$, then $\psi_g$ lies in $J(\mathfrak{g}_0)$ and [12] $\psi' : g \to (c')^{-1}\psi_g$ is a continuous mapping of $J(\mathfrak{h}_0)$ into $J(\mathfrak{g}_0)$. On the other hand, it is obvious that if $f \in J(\mathfrak{g}_0)$, then $\tilde{f} \in J(\mathfrak{h}_0)$ and the mapping $\tau : f \to \tilde{f}$ of $J(\mathfrak{g}_0)$ into $J(\mathfrak{h}_0)$ is continuous and univalent. Since $\tau(\psi_g) = c'g$, it follows that $\psi_{\tilde{f}} = c'f$ and therefore these two mappings $\tau$ and $\psi'$ are inverses of each other. Hence the theorem.

The following result will be needed in Section 8.

---

[12] We shall prove in Section 6 that $\langle \pi, \pi \rangle \neq 0$ (see the Corollary to Lemma 18).

LEMMA 17. *For any* $f \in \mathcal{B}(\mathfrak{h}_0)$ *there exists a unique element* $\Psi_f$ *in* $J(\mathfrak{h}_0)$ *such that*

$$\pi(H)\Psi_f(H) = \sum_{s \in W} \epsilon(s) f(sH) \qquad (H \in \mathfrak{h}_0).$$

*Moreover* $f \to \Psi_f$ *is a continuous mapping of* $\mathcal{B}(\mathfrak{h}_0)$ *into* $J(\mathfrak{g}_0)$.

Put

$$\Psi_f(X) = c_1 \int_{\mathfrak{h}_0} \xi(X:H)\pi(H) \sum_{s \in W} \epsilon(s)\tilde{f}(sH)dH \qquad (f \in \mathcal{B}(\mathfrak{h}_0), X \in \mathfrak{g}_0),$$

where $\tilde{f}$ is the Fourier transform of $f$ and [12] $c_1 = \langle \pi, \pi \rangle^{-1}$. Then it follows from Theorem 2 that

$$\pi(H')\Psi_f(H') = \sum_{s \in W} \epsilon(s) f(sH') \qquad (H' \in \mathfrak{h}_0).$$

Furthermore, we show exactly as above that $\Psi_f \in J(\mathfrak{g}_0)$ and the mapping $f \to \Psi_f$ is continuous. Finally, since $\mathfrak{g}_0 = \bigcup_{x \in G} x\mathfrak{h}_0$, every invariant function in $C^\infty(\mathfrak{g}_0)$ is uniquely determined by its restriction on $\mathfrak{g}_0'$. This proves the uniqueness of $\Psi_f$.

**6. Some further results.** We now discard the assumption about the compactness of $G$. Define $\delta$ as in Theorem 1, and, for $\xi \in \mathfrak{D}(\mathfrak{g})$ and $\eta \in \mathfrak{D}(\mathfrak{h})$, let $\xi_0$ and $\eta_0$ denote their respective local expressions at zero. It is clear that if $\xi \in \mathfrak{F}(\mathfrak{g})$, then $\xi_0 \in \partial(I(\mathfrak{g}))$.

LEMMA 18. $(\partial(\pi) \circ \delta(\xi))_0 = (\partial(\pi) \circ \delta(\xi_0))_0$ *for every* $\xi \in \mathfrak{F}(\mathfrak{g})$.

It is known (see Mostow [7]) that we can find a compact real form $\mathfrak{u}$ of $\mathfrak{g}$ such that $\mathfrak{h} \cap \mathfrak{u}$ is a Cartan subalgebra of $\mathfrak{u}$. Let $G_c$ be the (connected) complex adjoint group of $\mathfrak{g}$ and $U$ the (connected) adjoint group of $\mathfrak{u}$. Then both $G$ and $U$ are real analytic subgroups of $G_c$ corresponding to the real subalgebras $\mathfrak{g}_0$ and $\mathfrak{u}$ of $\mathfrak{g}$. Then (see Section 2) $G_c$ operates on $\mathfrak{D}(\mathfrak{g})$, and it is clear that $I(\mathfrak{g})$ is exactly the algebra of those polynomials $p$ on $\mathfrak{g}$ which are invariant under $G_c$. Therefore the two algebras $\mathfrak{F}(\mathfrak{g})$ and $\mathfrak{F}(\mathfrak{h})$ are the same whether we start from the real form $\mathfrak{g}_0$ or $\mathfrak{u}$. Let $du$ denote the normalized Haar measure of $U$, and for any $f \in C^\infty(\mathfrak{u})$ define

$$\phi_f(H) = \pi(H) \int_U f(uH)du \qquad (H \in \mathfrak{h} \cap \mathfrak{u})$$

as in Section 5. In order to prove the lemma, let us first suppose that $\xi_0 = 0$. Then from Lemma 16, $\phi_{\xi f}(0; \partial(\pi)) = 0$ for all $f \in C^\infty(\mathfrak{u})$. But we know (Lemma 15) that $\phi_{\xi f} = \delta(\xi)\phi_f$. This proves that $\partial(\pi)(\delta(\xi)\phi_f)$ has the value zero at the origin. Now suppose $f \in I(\mathfrak{g})$. Then $\phi_f = \pi \bar{f}$, where $\bar{f}$ is the restriction of $f$ on $\mathfrak{h} \cap \mathfrak{u}$. Hence we conclude from Lemma 9 that $(\partial(\pi) \circ \delta(\xi))(\pi p)$ takes the value zero at the origin for every $p \in I(\mathfrak{h})$. Select $q \in S(\mathfrak{h})$ such that $(\partial(\pi) \circ \delta(\xi))_0 = \partial(q)$. Since $\delta(\xi)$ is invariant under $W$, it is obvious that $q^s = \epsilon(s)q$ $(s \in W)$. We claim $q = 0$. For otherwise we could select $p_1 \in S(\mathfrak{h})$ such that $\langle q, p_1 \rangle \neq 0$. But $\langle q, p_1 \rangle = \langle q^s, p_1{}^s \rangle = \epsilon(s)\langle q, p_1{}^s \rangle$ $(s \in W)$. Hence $\langle q, p_2 \rangle = \langle q, p_1 \rangle \neq 0$, where $p_2 = w^{-1} \sum_{s \in W} \epsilon(s)p_1{}^s$. However it follows from Lemma 10 that $p_2 = \pi p$ for some $p \in I(\mathfrak{h})$, and therefore $\langle q, \pi p \rangle \neq 0$. On the other hand, in view of the definition of $q$, $\langle q, \pi p \rangle$ is exactly the value of $(\partial(\pi) \circ \delta(\xi))(\pi p)$ at the origin, which, as we have seen above, must be zero. This contradiction proves that $q = 0$, and therefore $(\delta(\pi) \circ \delta(\xi))_0 = 0$. The general case is reduced to the one discussed above by replacing $\xi$ by $\xi - \xi_0$.

Let $\bar{\theta}$ denote the conjugation of $\mathfrak{g}$ with respect to $\mathfrak{u}$, and $p \to p_*$ $(p \in S(\mathfrak{g}))$ the automorphism of $S(\mathfrak{g})$ over $R$ which coincides with $-\bar{\theta}$ on $\mathfrak{g}$. Select a base $X_1, \cdots, X_n$ for $\mathfrak{u}$ over $R$ such that $B(X_i, X_j) = -\delta_{ij}$, $1 \leq i, j \leq n$. Then if

$$p = \sum_{m_i \geq 0} a(m_1, \cdots, m_n) X_1{}^{m_1} \cdots X_n{}^{m_n} \qquad (a(m_1, \cdots, m_n) \in C),$$

it is clear that $\langle p_*, p \rangle = \sum_{m_i \geq 0} m_1! \cdots m_n! \, |a(m_1, \cdots, m_n)|^2 > 0$ unless $p = 0$. On the other hand, since every root $\alpha$ takes only pure imaginary values on $\mathfrak{h} \cap \mathfrak{u}$, it is obvious that $\pi_* = \pi$. Therefore $\langle \pi^k, \pi^k \rangle$ is a positive real number for every integer $k \geq 0$. Moreover, since $\pi^2 \in I(\mathfrak{h})$, there exists (Lemma 9) a unique element $\eta \in I(\mathfrak{g})$ such that $\bar{\eta} = \pi^2$. It is clear that $\eta_*$ is also in $I(\mathfrak{g})$ and $\bar{\eta}_* = (\pi^2)_* = \pi^2 = \bar{\eta}$. Hence we conclude from Lemma 9 that $\eta_* = \eta$, and therefore $\langle \eta^k, \eta^k \rangle > 0$ for any integer $k \geq 0$. Put $\xi = \partial(\eta^k) \circ \eta^k$ for some fixed $k$. It is obvious that $\xi_0 = \langle \eta^k, \eta^k \rangle$, and therefore we get the following Corollary from Lemma 18.

COROLLARY. *Put* $\xi = \partial(\eta^k) \circ \eta^k$ *for some integer* $k \geq 0$. *Then*

$$(\partial(\pi) \circ \delta(\xi))_0 = \langle \eta^k, \eta^k \rangle \partial(\pi).$$

*Moreover,* $\langle \pi^k, \pi^k \rangle$ *and* $\langle \eta^k, \eta^k \rangle$ *are positive real numbers.*

We shall now prove an algebraic result which will be needed in another paper. As before let $\omega$ denote the Casimir polynomial of $\mathfrak{g}$ and define a differential operator $\sigma$ on $\mathfrak{g}_0$ as follows: $f(X; \sigma) = f(X; \partial(X))$ $(X \in \mathfrak{g}_0, f \in C^\infty(\mathfrak{g}_0))$. It is obvious that $\sigma \in \mathfrak{D}(\mathfrak{g})$.

LEMMA 19. *For any integer* $k \geq 1$,

$$\partial(\omega^k) \circ \omega^k \equiv 4^k k! (\tfrac{1}{2}n + \sigma + k - 1)(\tfrac{1}{2}n + \sigma + k - 2) \cdots (\tfrac{1}{2}n + \sigma)$$
$$\mod \omega \mathfrak{D}(\mathfrak{g}),$$

*where* $n = \dim_C \mathfrak{g}$.

Define $\{D_1, D_2\} = D_1 D_2 - D_2 D_1$ as before for any two differential operators $D_1, D_2$. Then a simple calculation shows that $\{\partial(\omega), \omega\} = 2n + 4\sigma$. Moreover, if $p$ is a homogeneous polynomial in $S(\mathfrak{g})$ of degree $m$, it is easily seen that $\sigma p = mp$, and therefore it is clear that $\partial(\omega)\sigma = (\sigma + 2)\partial(\omega)$. Hence

$$\{\partial(\omega^k), \omega\} = \sum_{0 \le m < k} \partial(\omega^{k-1-m})(2n + 4\sigma)\partial(\omega^m)$$
$$= \sum_{0 \le m < k} (2n + 4(\sigma + 2k - 2 - 2m))\partial(\omega^{k-1})$$
$$= 4k(\tfrac{1}{2}n + \sigma + k - 1)\partial(\omega^{k-1}).$$

Therefore

$$\partial(\omega^k) \circ \omega^k \equiv \{\partial(\omega^k), \omega\} \circ \omega^{k-1} \equiv 4k(\tfrac{1}{2}n + \sigma + k - 1)\partial(\omega^{k-1}) \circ \omega^{k-1} \mod \omega \mathfrak{D}(\mathfrak{g}),$$

and our assertion now follows by induction.

This lemma has the following significance for our later applications. Let $t$ denote a real variable.

COROLLARY. *For any* $f \in C_c^\infty(\mathfrak{g}_0)$, *put* $f_k = \partial(\omega^k)(\omega^k f)$ $(k \ge 0)$. *Then if* $X \ne 0$ *is an element in* $\mathfrak{g}_0$ *such that* $\omega(X) = 0$, *we have* [13]

$$\int_0^\infty f_k(tX)dt = 4^k k! \{\Gamma(\tfrac{1}{2}n + k - 1)/\Gamma(\tfrac{1}{2}n - 1)\} \int_0^\infty f(tX)dt$$

*for all integers* $k \ge 0$.

It is obvious that $g(tX; \sigma) = tg(tX; \partial(X)) = t(d/dt)g(tX)$ for any $g \in C_c^\infty(\mathfrak{g}_0)$. Hence

$$\int_0^\infty g(tX; \sigma + 1)dt = \int_0^\infty (d/dt)(tg(tX))dt = 0.$$

On the other hand, it follows from Lemma 19 that

$$f_k = c_k f + (\sigma + 1)g_1 + \omega g_2,$$

where $g_1, g_2 \in C_c^\infty(\mathfrak{g})$ and $c_k = 4^k k! \Gamma(\tfrac{1}{2}n + k - 1)/\Gamma(\tfrac{1}{2}n - 1)$. Therefore, since $\omega(tX) = t^2 \omega(X) = 0$, we conclude that

$$\int_0^\infty f_k(tX)dt = c_k \int_0^\infty f(tX)dt.$$

---

[13] Here $\Gamma$ stands for the classical Gamma function.

**7. Differential operators on a reductive subalgebra.** Let $\mathfrak{m}_0$ be a Lie subalgebra of $\mathfrak{g}_0$. We assume that $\mathfrak{m}_0 \supset \mathfrak{h}_0$ and $\mathfrak{m}_0$ is reductive in $\mathfrak{g}_0$ (see Koszul [6]). Then we can select a subspace $\mathfrak{q}_0$ of $\mathfrak{g}_0$ such that $[\mathfrak{m}_0, \mathfrak{q}_0] \subset \mathfrak{q}_0$ and $\mathfrak{g}_0$ is the direct sum of $\mathfrak{m}_0$ and $\mathfrak{q}_0$. Let $\mathfrak{m}, \mathfrak{q}$ be the complexification of $\mathfrak{m}_0, \mathfrak{q}_0$, respectively, in $\mathfrak{g}$ and let $Q$ be the set of all roots $\alpha \in P$ for which $X_\alpha \in \mathfrak{m}$. Then it is clear that $\mathfrak{m} = \mathfrak{h} + \sum_{\alpha \in Q} (CX_\alpha + CX_{-\alpha})$ and $\mathfrak{q} = \sum_{\alpha \in Q'} (CX_\alpha + CX_{-\alpha})$, where $Q'$ is the complement of $Q$ in $P$. Since the restriction of the bilinear form $B$ on $\mathfrak{m}$ is obviously nondegenerate, we can apply the formalism described in Section 2 to $E_0 = \mathfrak{g}_0$ and $F_0 = \mathfrak{m}_0$. For any $X \in \mathfrak{m}$, let $\zeta(X)$ denote the determinant of the restriction of $\mathrm{ad}\, X$ on $\mathfrak{q}$. (If $\mathfrak{q} = \{0\}$, we define $\zeta(X)$ to be 1.) Then $\zeta$, being a polynomial function on $\mathfrak{m}$, lies in $S(\mathfrak{m})$. Moreover, $\zeta$ coincides on $\mathfrak{h}$ with $(-1)^s \prod_{\alpha \in Q'} \alpha^2$, where $s$ is the number of roots in $Q'$. Hence $\zeta \neq 0$. Let $\mathfrak{m}_0'$ denote the set of all $X \in \mathfrak{m}_0$ where $\zeta(X) \neq 0$. Then $\mathfrak{m}_0'$ is an open dense subset of $\mathfrak{m}_0$ and $\mathfrak{h}_0' \subset \mathfrak{m}_0'$. Put $\mathfrak{Q} = \lambda(S(\mathfrak{q}))$, where $\lambda$ is the canonical mapping of $S(\mathfrak{g})$ into $\mathfrak{B}$. The next two lemmas are proved in the same way as Lemmas 4 and 5.

LEMMA 20. *If $X \in \mathfrak{m}_0'$, $\Gamma_X$ defines a one-one mapping of $\mathfrak{Q} \mathbf{X} S(\mathfrak{m})$ onto $S(\mathfrak{g})$.*

LEMMA 21. *For every $p \in S(\mathfrak{g})$, there exists an integer $k \geq 0$ and a polynomial mapping $\gamma_p$ of $\mathfrak{m}$ into $\mathfrak{Q} \mathbf{X} S(\mathfrak{m})$ such that $\Gamma_X(\gamma_p(X)) = \zeta(X)^k p$ $(X \in \mathfrak{m}_0)$.*

For any $x \in G$, let $b \to b^x$ $(b \in \mathfrak{B})$ denote the unique antomorphism of $\mathfrak{B}$ which coincides with $x$ on $\mathfrak{g}$. Then we also get an automorphism $\mu \to \mu^x$ of $\mathfrak{B} \mathbf{X} S(\mathfrak{g})$ given by $(b \mathbf{X} p)^x = b^x \mathbf{X} p^x$ $(b \in \mathfrak{B}, p \in S(\mathfrak{g}))$. It is obvious from the definition of $\Gamma_Y$ $(Y \in \mathfrak{g}_0)$ that $(\Gamma_Y(\mu))^x = \Gamma_{xY}(\mu^x)$ $(\mu \in \mathfrak{B} \mathbf{X} S(\mathfrak{g}))$. Let $M$ be the analytic subgroup of $G$ corresponding to $\mathfrak{m}_0$. It is obvious that $\zeta^m = \zeta$, and therefore $m\mathfrak{m}_0' = \mathfrak{m}_0'$ $(m \in M)$. For any $p \in S(\mathfrak{g})$, let $\beta_p(X)$ $(X \in \mathfrak{m}_0')$ denote the unique element in $\mathfrak{Q} \mathbf{X} S(\mathfrak{m})$ (see Lemma 20) such that $\Gamma_X(\beta_p(X)) = p$. Then if $m \in M$,

$$p^m = (\Gamma_X(\beta_p(X)))^m = \Gamma_{mX}(\beta_p(X)^m) \qquad (X \in \mathfrak{m}_0'),$$

and this shows that $\beta_{p^m}(mX) = (\beta_p(X))^m$. Thus we have obtained the following result.

LEMMA 22. *For any $p \in S(\mathfrak{g})$ and $X \in \mathfrak{m}_0'$, let $\beta_p$ denote the unique* *element in* $\mathfrak{Q} \mathbf{X} S(\mathfrak{m})$ *such that* $\Gamma_X(\beta_p(X)) = p$. *Then* $\beta_{p^m}(mX) = (\beta_p(X))^m$ $(m \in M)$.

Put $\mathfrak{Q}' = \sum_{d \geqq 1} \lambda(S_d(\mathfrak{q}))$, where $S_d(\mathfrak{q})$ is the space of homogeneous elements in $S(\mathfrak{q})$ of degree $d$. Let $\mu_X(p)$ $(X \in \mathfrak{m}_0', p \in S(\mathfrak{g}))$ denote the unique element in $S(\mathfrak{m})$ such that $\beta_p(X) - (1 \mathbf{X} \mu_X(p)) \in \mathfrak{Q}' \mathbf{X} S(\mathfrak{m})$. Then it follows from Lemma 21 that there exists a differential operator $\Delta(p)$ on $\mathfrak{m}_0'$ whose local expression at $X$ coincides with $\partial(\mu_X(p))$. On the other hand, since $\zeta$ is nowhere zero on $\mathfrak{m}_0'$, $|\zeta|^{\frac{1}{2}}$ and $|\zeta|^{-\frac{1}{2}}$ are analytic functions on $\mathfrak{m}_0'$, and so, for any $q \in S(\mathfrak{m})$, $|\zeta|^{-\frac{1}{2}} \partial(q) \circ |\zeta|^{\frac{1}{2}}$ is a well-defined differential operator on $\mathfrak{m}_0'$. Let $p_\mathfrak{m}$ denote the restriction of a polynomial $p \in S(\mathfrak{g})$ on $\mathfrak{m}$. Then $p_\mathfrak{m} \in S(\mathfrak{m})$ and $\Delta(p) - |\zeta|^{-\frac{1}{2}} \partial(p_\mathfrak{m}) \circ |\zeta|^{\frac{1}{2}}$ is a differential operator on $\mathfrak{m}_0'$. We shall denote its local expression at $X \in \mathfrak{m}_0'$ in the usual way by $(\Delta(p) - |\zeta|^{-\frac{1}{2}} \partial(p_\mathfrak{m}) \circ |\zeta|^{\frac{1}{2}})_X$. Let $\mathfrak{M}$ be the subalgebra of $\mathfrak{B}$ generated by $(1, \mathfrak{m})$. We put $\mathfrak{M}' = \mathfrak{M}\mathfrak{m}$.

LEMMA 23. *If* $p \in I(\mathfrak{g})$ *and* $X \in \bigcup_{m \in M} m\mathfrak{h}_0'$, *then*

$$(\Delta(p) - |\zeta|^{-\frac{1}{2}} \partial(p_\mathfrak{m}) \circ |\zeta|^{\frac{1}{2}})_X \in \partial(\Gamma_X(\mathfrak{M}' \mathbf{X} S(\mathfrak{m}))).$$

Let $D = \Delta(p) - |\zeta|^{-\frac{1}{2}} \partial(p_\mathfrak{m}) \circ |\zeta|^{\frac{1}{2}}$. Since both $\zeta$ and $p_\mathfrak{m}$ are invariant under $M$, it follows from Lemma 22 that $D_{mX} = (D_X)^m$ $(m \in M, X \in \mathfrak{m}_0')$. Hence it would be sufficient to prove that $D_H$ lies in $\partial(\Gamma_H(\mathfrak{M}' \mathbf{X} S(\mathfrak{m})))$ for $H \in \mathfrak{h}_0'$. Since $\mathfrak{m}_0$ is reductive, it is clear that Lemma 4 can be applied to $\mathfrak{m}_0$ in place of $\mathfrak{g}_0$. We now use the notation of Section 3 and put $\mathfrak{S}_\mathfrak{m} = \mathfrak{S} \cap \mathfrak{M}$ and $\mathfrak{S}_\mathfrak{m}' = \mathfrak{S}' \cap \mathfrak{M}$. Also fix an element $H_0 \in \mathfrak{h}_0'$. Since $S(\hat{\mathfrak{s}}) \cap S(\mathfrak{m})$ $= S(\hat{\mathfrak{s}} \cap \mathfrak{m})$ and the canonical mapping $\lambda$ of $S(\mathfrak{g})$ onto $\mathfrak{B}$ is a linear isomorphism, it follows that $\lambda(S(\hat{\mathfrak{s}} \cap \mathfrak{m})) = \mathfrak{S}_\mathfrak{m}$ and $\mathfrak{S}_\mathfrak{m}' = \mathfrak{S} \cap \mathfrak{M}'$. Therefore from Lemma 4 (applied to $\mathfrak{m}_0$), $\Gamma_{H_0}$ defines a one-one mapping of $\mathfrak{S}_\mathfrak{m} \mathbf{X} S(\mathfrak{h})$ into $S(\mathfrak{m})$. Moreover $\zeta(H)$ is real and equal to $\pm \prod_{\alpha \in Q'} (\alpha(H))^2$ if $H \in \mathfrak{h}_0$. Therefore by applying Lemmas 7 and 8 to $\mathfrak{m}_0$ (instead of $\mathfrak{g}_0$), we conclude that

$$(|\zeta|^{-\frac{1}{2}} \partial(p_\mathfrak{m}) \circ |\zeta|^{\frac{1}{2}})_{H_0} - (\pi^{-1} \partial(\bar{p}) \circ \pi)_{H_0} \in \partial(\Gamma_{H_0}(\mathfrak{S}_\mathfrak{m}' \mathbf{X} S(\mathfrak{h}))),$$

where $\bar{p}$ is the restriction of $p$ (and therefore also of $p_\mathfrak{m}$) on $\mathfrak{h}$. On the other hand, it follows from the definition of $\Delta(p)$ that

$$(\Delta(p))_{H_0} - \partial(p) \in \partial(\Gamma_{H_0}(\mathfrak{Q}' \mathbf{X} S(\mathfrak{m}))).$$

Therefore

$$D_{H_0} \equiv \partial(p) - (\pi^{-1} \partial(\bar{p}) \circ \pi)_{H_0} \equiv 0 \quad \mod \partial(\Gamma_{H_0}(\mathfrak{S}' \mathbf{X} S(\mathfrak{m})))$$

8

from Lemma 8. Now it is an immediate consequence of the defintion of $\Gamma_Y$ $(Y \in \mathfrak{g}_0)$ that $\Gamma_Y(b_1 b_2 \mathbf{X} q) = \Gamma_Y(b_1 \mathbf{X} \Gamma_Y(b_2 \mathbf{X} q))$ $(b_1, b_2 \in \mathfrak{B}, q \in S(\mathfrak{g}))$. Therefore, since $S(\mathfrak{m}) = \Gamma_{H_0}(\mathfrak{S}_\mathfrak{m} \mathbf{X} S(\mathfrak{h}))$, it follows that $\Gamma_{H_0}(\mathfrak{S}' \mathbf{X} S(\mathfrak{m}))$ $\subset \Gamma_{H_0}(\mathfrak{B}' \mathbf{X} S(\mathfrak{h}))$, where $\mathfrak{B}' = \mathfrak{B}\mathfrak{g}$. But $\mathfrak{B}' = \mathfrak{S}' + \mathfrak{B}\mathfrak{h}$ and $\Gamma_{H_0}(\mathfrak{B}\mathfrak{h} \mathbf{X} S(\mathfrak{h}))$ $= \{0\}$. Therefore $\Gamma_{H_0}(\mathfrak{B}' \mathbf{X} S(\mathfrak{h})) = \Gamma_{H_0}(\mathfrak{S}' \mathbf{X} S(\mathfrak{h}))$. This proves that $\Gamma_{H_0}(\mathfrak{S}' \mathbf{X} S(\mathfrak{m})) = \Gamma_{H_0}(\mathfrak{S}' \mathbf{X} S(\mathfrak{h}))$, and hence $D_{H_0} \in \partial(S(\mathfrak{m}) \cap \Gamma_{H_0}(\mathfrak{S}' \mathbf{X} S(\mathfrak{h})))$. But, as we have seen, $S(\mathfrak{m}) = \Gamma_{H_0}(\mathfrak{S}_\mathfrak{m} \mathbf{X} S(\mathfrak{h}))$ and $\Gamma_{H_0}$ is univalent on $\mathfrak{S} \mathbf{X} S(\mathfrak{h})$ (Lemma 4). Therefore

$$D_{H_0} \in \partial(\Gamma_{H_0}(\mathfrak{S}_\mathfrak{m}' \mathbf{X} S(\mathfrak{h}))) \subset \partial(\Gamma_{H_0}(\mathfrak{M}' \mathbf{X} S(\mathfrak{m}))),$$

and so the lemma is proved.

COROLLARY. *If* $H \in \mathfrak{h}_0'$ *then*

$$(\Delta(p))_H - (\pi^{-1} \partial(\bar{p}) \circ \pi)_H \in \partial(\Gamma_H(\mathfrak{M}' \mathbf{X} S(\mathfrak{h})))$$

*for* $p \in I(\mathfrak{g})$.

For, in view of what we have seen above, the left side lies in

$$\partial(S(\mathfrak{m}) \cap \Gamma_H(\mathfrak{S}' \mathbf{X} S(\mathfrak{m}))) \quad \text{and} \quad S(\mathfrak{m}) \cap \Gamma_H(\mathfrak{S}' \mathbf{X} S(\mathfrak{m})) \subset \Gamma_H(\mathfrak{M}' \mathbf{X} S(\mathfrak{h})).$$

**8. Closer study of a special case.** Define $\mathfrak{k}_0$ and $\mathfrak{p}_0$ as in $[4(c), §2]$ adn let $\mathfrak{k}, \mathfrak{p}$ denote their respective complexifications in $\mathfrak{g}$. *Throughout this section we shall assume that* $\mathfrak{h}_0 \subset \mathfrak{k}_0$. Then the theory developed in Section 7 can be applied to $\mathfrak{m}_0 = \mathfrak{k}_0$. In this case $\mathfrak{q}_0 = \mathfrak{p}_0$ and $\zeta(X) = \det(\operatorname{ad} X)_\mathfrak{p}$ $(X \in \mathfrak{k})$, where $(\operatorname{ad} X)_\mathfrak{p}$ is the restriction of $\operatorname{ad} X$ on $\mathfrak{p}$. Put $\mathfrak{P} = \lambda(S(\mathfrak{p}))$. Then we intend to prove the following result concerning the Casimir polynomial $\omega$ of $\mathfrak{g}$.

LEMMA [14] 24. *There exists a unique polynomial mapping* $\gamma_\omega$ *of* $\mathfrak{k}$ *into* $\mathfrak{P} \mathbf{X} S(\mathfrak{k})$ *such that* $\Gamma_X(\gamma_\omega(X)) = \zeta(X)\omega$ $(X \in \mathfrak{k}_0)$. *Moreover* $(\gamma_\omega(X))^k = \gamma_\omega(kX)$ *for* $k \in K$ *and* $X \in \mathfrak{k}$.

The case $\mathfrak{q}_0 = \mathfrak{k}_0$ being trivial, we assume that $\mathfrak{p}_0 \neq 0$. Let $P_-$ and $P_0$ be the sets of all compact and all noncompact roots in $P$, respectively (see $[4(g), \text{p. } 751]$). Then if $r$ is the number of roots in $P_0$, $\dim \mathfrak{p} = 2r$. Let $t$ be an indeterminate and consider the polynomial

$$\prod_{\beta \in P_0} (t - \beta^2) = t^r + q_1 t^{r-1} + \cdots + q_r$$

with coefficients $q_i \in S(\mathfrak{h})$. Let $W_\mathfrak{k}$ be the subgroup of the Weyl group $W$

---

[14] The results of this section are quite similar to those of $[4(e), §10]$.

generated by the reflexions $s_\alpha$ corresponding to $\alpha \in P_-$. Then, since $\mathfrak{k}$ is reductive (Lemma 5 of [4(c)]), $W_\mathfrak{k}$ can also be regarded as the Weyl group of $\mathfrak{k}$ with respect to $\mathfrak{h}$ (see [4(e), §6]). Let $K$ denote the analytic subgroup of $G$ corresponding to $\mathfrak{k}_0$ and $I(\mathfrak{k})$ the set of those $p \in S(\mathfrak{k})$ which are invariant under $K$. Similarly, let $I_\mathfrak{k}(\mathfrak{h})$ be the set of all elements in $S(\mathfrak{h})$ which are invariant under $W_\mathfrak{k}$. Then Chevalley's Theorem (Lemma 9) is obviously applicable to $(\mathfrak{k}, \mathfrak{h})$ in place of $(\mathfrak{g}, \mathfrak{h})$. Hence $p \to \bar{p}$ $(p \in I(\mathfrak{k}))$ is an isomorphism of $I(\mathfrak{k})$ onto $I_\mathfrak{k}(\mathfrak{h})$. Now if $\beta$ is a noncompact root and $s \in W_\mathfrak{k}$, then $s\beta$ is also noncompact (see the proof of Lemma 10 of [4(g)]). From this it follows that $q_i \in I_\mathfrak{k}(\mathfrak{h})$, $1 \leq i \leq r$. Hence we can select $\sigma_i \in I(\mathfrak{k})$ such that $q_i = \bar{\sigma}_i$. For any $X \in \mathfrak{k}$, put $S(X) = (\operatorname{ad} X)_\mathfrak{p}{}^2$. Then we shall first prove the following lemma.

LEMMA 25. $\sigma_r = \zeta$ and

$$(S(X))^r + \sigma_1(X)(S(X))^{r-1} + \cdots + \sigma_r(X)I_\mathfrak{p} = 0 \qquad (X \in \mathfrak{k}).$$

(*Here $I_\mathfrak{p}$ is the identity mapping of $\mathfrak{p}$.*)

It is clear that $\zeta \in I(\mathfrak{k})$ and $\bar{\zeta} = (-1)^r \prod_{\beta \in P_0} \beta^2 = q_r$. Therefore, in view of the above *isomorphism* of $I(\mathfrak{k})$ onto $I_\mathfrak{k}(\mathfrak{h})$, we can conclude that $\zeta = \sigma_r$. Moreover $\tau : X \to \{(S(X))^r + \sigma_1(X)(S(X))^{r-1} + \cdots + \sigma_r(X)I_\mathfrak{p}\}$ $(X \in \mathfrak{k})$ is obviously a polynomial mapping of $\mathfrak{k}$ into the space of endomorphisms of $\mathfrak{p}$, and therefore it would be sufficient to prove that $\tau$ maps $\mathfrak{k}_0$ into zero. Moreover, since $K$ is compact [7], $\mathfrak{k}_0 = \bigcup_{k \in K} k\mathfrak{h}_0$, and so it would clearly be enough to show that $\tau$ maps $\mathfrak{h}_0$ into zero. But if $H \in \mathfrak{h}$, $S(H)$ is semisimple and all its eigenvalues are of the form $\beta(H)^2$ $(\beta \in P_0)$. Therefore

$$\tau(H) = \prod_{\beta \in P_0} (S(H) - \beta(H)^2 I_\mathfrak{p}) = 0,$$

and this proves our assertion.

Now put

$$T(X) = -\{(S(X))^{r-1} + \sigma_1(X)(S(X))^{r-2} + \cdots + \sigma_{r-1}(X)I_\mathfrak{p}\} \qquad (X \in \mathfrak{k}),$$

so that $S(X)T(X) = T(X)S(X) = \zeta(X)I_\mathfrak{p}$. Select a base $Y_1, \cdots, Y_p$ $(p = 2r)$ for $\mathfrak{p}$ over $C$. Since $\mathfrak{k}$ and $\mathfrak{p}$ are orthogonal under $B$, $B$ is nondegenerate on $\mathfrak{p}$. Hence we can choose $Y_i' \in \mathfrak{p}$ $(1 \leq i \leq p)$ such that $B(Y_i', Y_j) = \delta_{ij}$ $(1 \leq i, j \leq p)$. Let $\omega_\mathfrak{k}$ and $\omega_\mathfrak{p}$ denote the restrictions of $\omega$ on $\mathfrak{k}$ and $\mathfrak{p}$, respectively. Then in view of the above-mentioned orthogonality of $\mathfrak{k}$ and $\mathfrak{p}$, $\omega = \omega_\mathfrak{k} + \omega_\mathfrak{p}$. Moreover, it is easy to see that $\omega_\mathfrak{p} = \sum_{1 \leq i \leq p} Y_i Y_i'$.

Now fix an element $X \in \mathfrak{k}$ and let $S', T'$ be two linear transformations of $\mathfrak{p}$ which commute with $(\operatorname{ad} X)_{\mathfrak{p}}$. Put $U_i = S'Y_i$, $V_i = T'Y_i$, $U_i' = S'Y_i'$, $V_i' = T'Y_i'$ $(1 \leq i \leq p)$. Note that $Z = \sum\limits_{1 \leq i \leq p} B(Z, Y_i')Y_i = \sum\limits_{1 \leq i \leq p} B(Z, Y_i)Y_i'$ $(Z \in \mathfrak{p})$, and therefore $(\operatorname{ad} X)Z = -\sum\limits_i B(Z, (\operatorname{ad} X)Y_i')Y_i$. Hence

$$(\operatorname{ad} X)U_i = S'(\operatorname{ad} X)Y_i = -\sum_j B(Y_i, (\operatorname{ad} X)Y_j')U_j,$$

$$(\operatorname{ad} X)V_i' = T'(\operatorname{ad} X)Y_i' = \sum_j B((\operatorname{ad} X)Y_i', Y_j)V_j',$$

and so if we consider the tensor product $\mathfrak{p} \mathbf{X} \mathfrak{p}$, it follows that

$$\sum_i ((\operatorname{ad} X)U_i) \mathbf{X} V_i' = -\sum_{i,j} B(Y_i, (\operatorname{ad} X)Y_j')U_j \mathbf{X} V_i' = -\sum_j U_j \mathbf{X} (\operatorname{ad} X)V_j'.$$

Obviously, this implies that

$$\sum_i [X, U_i]V_i' = -\sum_i U_i[X, V_i'],$$

$$\sum_i [[X, U_i], V_i'] = -\sum_i [U_i, [X, V_i']].$$

Now put $Y_j(X) = T(X)Y_j$ $(1 \leq j \leq p)$ so that $S(X)Y_j(X) = \zeta(X)Y_j$. Then if $U, V \in \mathfrak{p}$ and $X \in \mathfrak{k}_0$,

$$\Gamma_X(U \cdot V \mathbf{X} 1) = (L_{[U,X]} + d_U)[V, X] = [X, U][X, V] - [U, [X, V]]$$

in the notation of Section 3. Hence

$$\sum_{1 \leq j \leq p} \Gamma_X(Y_j' \cdot Y_j(X) \mathbf{X} 1) = \sum_j [X, Y_j'][X, Y_j(X)] - \sum_j [Y_j', [X, Y_j(X)]].$$

But since $(\operatorname{ad} X)_{\mathfrak{p}}T(X)$ commutes with $(\operatorname{ad} X)_{\mathfrak{p}}$, it follows from our result above that

$$\sum_j [X, Y_j'][X, Y_j(X)] = -\sum_j Y_j'(S(X)Y_j(X)) = -\zeta(X)\omega_{\mathfrak{p}},$$

and therefore

$$\sum_{1 \leq j \leq p} \Gamma_X(Y_j' \cdot Y_j(X) \mathbf{X} 1) = -\zeta(X)\omega_{\mathfrak{p}} - \sum_j [Y_j', [X, Y_j(X)]].$$

Now put $b(X) = -\tfrac{1}{2}\sum\limits_i (Y_i' \cdot Y_i(X) + Y_i(X) \cdot Y_i')$ $(X \in \mathfrak{k})$. Then $b(X) \in \mathfrak{P}$ and $b(X) = -\sum\limits_j Y_j' \cdot Y_j(X) + \tfrac{1}{2}\sum\limits_j [Y_j', Y_j(X)]$. Moreover, it follows from our result above that

$$\sum_j [(\operatorname{ad} X)Y_j', Y_j(X)] = -\sum_j [Y_j', (\operatorname{ad} X)Y_j(X)],$$

and therefore $\Gamma_X(b(X) \mathbf{X} 1) = \zeta(X)\omega_{\mathfrak{p}} + \sum_j [Y_j', [X, Y_j(X)]]$ $(X \in \mathfrak{k}_0)$. Put $Z(X) = \sum_j [Y_j', [X, Y_j(X)]]$ and

$$\gamma_\omega(X) = b(X) \mathbf{X} 1 + \zeta(X)(1 \mathbf{X} \omega_{\mathfrak{k}}) - (1 \mathbf{X} Z(X)) \qquad (X \in \mathfrak{k}).$$

Then $Z(X) \in [\mathfrak{p}, \mathfrak{p}] \subset \mathfrak{k}$, and therefore $\gamma_\omega(X) \in \mathfrak{P} \mathbf{X} S(\mathfrak{k})$. Moreover,

$$\Gamma_X(\gamma_\omega(X)) = \zeta(X)\omega_{\mathfrak{p}} + \zeta(X)\omega_{\mathfrak{k}} = \zeta(X)\omega \qquad (X \in \mathfrak{k}_0)$$

and it is clear from its definition that $\gamma_\omega$ is a polynomial mapping of $\mathfrak{k}$ into $\mathfrak{P} \mathbf{X} S(\mathfrak{k})$. We still have to prove its uniqueness.

Let $\mathfrak{k}_0'$ be the set of all points $X \in \mathfrak{k}_0$ where $\zeta(X) \neq 0$. It follows from Lemma 20 that $\gamma_\omega$ is completely determined on $\mathfrak{k}_0'$ by the condition $\Gamma_X(\gamma_\omega(X)) = \zeta(X)\omega$. But it is obvious that two polynomial mappings of $\mathfrak{k}$ must be identical if they coincide on some nonempty open subset of $\mathfrak{k}_0$. This proves that $\gamma_\omega$ is unique.

For a fixed $k \in K$, consider the mapping $\gamma_\omega': X \to (\gamma_\omega(kX))^{k^{-1}}$ $(X \in \mathfrak{k})$. It is clear that $\gamma_\omega'$ is a polynomial mapping which, in view of Lemma 22, must coincide with $\gamma_\omega$ on $\mathfrak{k}_0'$. Hence $\gamma_\omega' = \gamma_\omega$, and therefore $(\gamma_\omega(X))^k = \gamma_\omega(kX)$ $(k \in K, X \in \mathfrak{k})$. This completes the proof of Lemma 24.

We can now define a differential operator $D$ on $\mathfrak{k}_0$ by

$$D_X = \zeta(X)\partial(\omega_{\mathfrak{k}}) - \partial(Z(X)) \qquad (X \in \mathfrak{k}_0)$$

in the above notation. It is clear that $D \in \mathfrak{D}(\mathfrak{k})$ and it follows from Lemma 24 that $D$ is invariant under $K$. Finally, it is obvious from the defintion of $\Delta(\omega)$ (in Lemma 23) that $D = \zeta\Delta(\omega)$ on $\mathfrak{k}_0'$. Let $D^*$ denote the adjoint of $D$.

LEMMA 26. *Let $I^\infty(\mathfrak{k}_0)$ denote the set of all functions $f \in C^\infty(\mathfrak{k}_0)$ which are invariant under $K$. Then $Df = D^*f$ for $f \in I^\infty(\mathfrak{k}_0)$.*

Let $f$ be a fixed element in $I^\infty(\mathfrak{k}_0)$. We have to prove that $Df - D^*f = 0$. Since $D$ is invariant under $K$, the same holds for $D^*$, and therefore $Df - D^*f$ lies in $I^\infty(\mathfrak{k}_0)$. Hence, in view of the fact that $\mathfrak{k}_0 = \bigcup_{k \in K} (k\mathfrak{h}_0)$, it would be enough to prove that $Df - D^*f$ is zero on $\mathfrak{h}_0'$. Let $H_0$ be any point in $\mathfrak{h}_0'$ and select an open connected neighborhood $U$ of $H_0$ in $\mathfrak{h}_0'$. Then $V = \bigcup_{k \in K} (kU)$ is an open connected neighborhood of $H_0$ in $\mathfrak{k}_0$. Let $dX$ denote the Euclidean measure on $\mathfrak{k}_0$. It is obvious that if $g \in C_c^\infty(V)$,

$$\int_V (Df - D^*f)g \, dX = \int_V (Df - D^*f)g_1 \, dX,$$

where $g_1(X) = \int_K g(kX)\,dk$ $(X \in V)$ and $dk$ is the normalized Haar measure on $K$. Therefore, in order to prove that $Df - D^*f$ is zero on $V$, it is sufficient to show that

$$\int (Df)\phi\,dX = \int f(D\phi)\,dX$$

for every $\phi \in I^\infty(\mathfrak{k}_0)$ which vanishes outside some compact set in $V$. Fix such a function $\phi$. Now $V \subset \mathfrak{k}_0'$ and therefore $D = \zeta\Delta(\omega)$ on $V$. Moreover since $\zeta$ takes only nonzero real values on the connected set $V$, it must keep constant sign. Hence it follows from Lemma 23 that

$$\psi(X; D) = \epsilon\psi(X; |\zeta|^{\frac{1}{2}}\partial(\omega_{\mathfrak{k}}) \circ |\zeta|^{\frac{1}{2}}) \qquad (X \in V, \psi \in I^\infty(\mathfrak{k}_0)),$$

where $\epsilon = 1$ or $-1$ according as $\zeta$ is positive or negative on $V$. Therefore, if $f' = |\zeta|^{\frac{1}{2}}f$ and $\phi' = |\zeta|^{\frac{1}{2}}\phi$, we have only to show that

$$\int (\partial(\omega_{\mathfrak{k}})f')\phi'\,dX = \int f'(\partial(\omega_{\mathfrak{k}})\phi')\,dX.$$

But, $\omega_{\mathfrak{k}}$ being a homogeneous polynomial of degree 2, $\partial(\omega_{\mathfrak{k}})$ is self-adjoint. Moreover, $\phi' \in C_o^\infty(V)$ and $f'$ is of class $C^\infty$ on $V$. Therefore the above relation is obvious, and so the lemma is proved.

Since every root $\alpha$ takes only pure imaginary values on $\mathfrak{h}_0$, it is clear that the corresponding Weyl reflexion $s_\alpha$ maps $\mathfrak{h}_0$ into itself. Hence $s\mathfrak{h}_0 = \mathfrak{h}_0$ for any $s \in W$. Put $J(\mathfrak{k}_0) = I^\infty(\mathfrak{k}_0) \cap \mathscr{B}(\mathfrak{k}_0)$ and consider the $\mathscr{B}$-topology on $J(\mathfrak{k}_0)$.

LEMMA 27. *For any $f \in \mathscr{B}(\mathfrak{h}_0)$, there exists exactly one element $\psi_f \in I^\infty(\mathfrak{k}_0)$ such that $\pi(H)\psi_f(H) = \sum\limits_{s \in W} \epsilon(s)f(sH)$ $(H \in \mathfrak{h}_0)$. Furthermore, $\psi_f$ lies in $J(\mathfrak{k}_0)$ and $f \to \psi_f$ is a continuous mapping of $\mathscr{B}(\mathfrak{h}_0)$ into $J(\mathfrak{k}_0)$.*

Since $\mathfrak{k}_0 = \bigcup\limits_{k \in K} k\mathfrak{h}_0$, a function in $I^\infty(\mathfrak{k}_0)$ is completely determined by its restriction on $\mathfrak{h}_0'$. Therefore the uniqueness of $\psi_f$ is obvious. Now put $\mathfrak{u} = \mathfrak{k}_0 + (-1)^{\frac{1}{2}}\mathfrak{p}_0$. Then $\mathfrak{u}$ is a compact real form of $\mathfrak{g}$. Let $U$ be the (connected) adjoint group of $\mathfrak{u}$ and let $J(\mathfrak{u})$ denote the set of all functions in $\mathscr{B}(\mathfrak{u})$ which are invariant under $U$. We consider the $\mathscr{B}$-topology on $J(\mathfrak{u})$. Then from Lemma 17, there exists, for each $f \in \mathscr{B}(\mathfrak{h}_0)$, a unique element $\Psi_f \in J(\mathfrak{u})$ such that

$$\pi(H)\Psi_f(H) = \sum_{s \in W} \epsilon(s)f(sH) \qquad (H \in \mathfrak{h}_0)$$

and $f \to \Psi_f$ is a continuous mapping of $\mathscr{E}(\mathfrak{h}_0)$ into $J(\mathfrak{u})$. Let $\sigma$ denote the mapping of $\mathscr{E}(\mathfrak{u})$ into $\mathscr{E}(\mathfrak{k}_0)$ which assigns to any function its restriction on $\mathfrak{k}_0$. It is obvious that $\sigma$ is continuous and it maps $J(\mathfrak{u})$ into $J(\mathfrak{k}_0)$. Therefore $f \to \psi_f = \sigma \Psi_f$ is a continuous mapping of $\mathscr{E}(\mathfrak{h}_0)$ into $J(\mathfrak{k}_0)$ and $\psi_f$ satisfies the condition of the lemma.

Put $I_c^\infty(\mathfrak{k}_0) = I^\infty(\mathfrak{k}_0) \cap C_c^\infty(\mathfrak{k}_0)$, and consider $C_c^\infty(\mathfrak{h}_0)$ and $I_c^\infty(\mathfrak{k}_0)$ under the $C_c^\infty$-topology. If $\Omega$ is any compact set in $\mathfrak{h}_0$ and $f$ a function in $C_c^\infty(\mathfrak{h}_0)$ whose carrier lies in $\Omega$, then it is obvious that the carrier of $\psi_f$ is contained in $\Omega' = \bigcup_{k \in K} k\Omega$. Since $\Omega'$ is compact, it follows that $\psi_f \in I_c^\infty(\mathfrak{k}_0)$ and $f \to \psi_f$ $(f \in C_c^\infty(\mathfrak{h}_0))$ is a continuous mapping of $C_c^\infty(\mathfrak{h}_0)$ into $I_c^\infty(\mathfrak{k}_0)$.

LEMMA 28. $D\psi_f = \zeta \psi_{\partial(\bar\omega)f}$ for all $f \in \mathscr{E}(\mathfrak{h}_0)$.

It is sufficient to prove that $D\psi_f - \zeta \psi_{\partial(\bar\omega)f}$ is zero on $\mathfrak{h}_0'$. Select a point $H_0 \in \mathfrak{h}_0'$. We know that $D = \zeta \Delta(\omega)$ on $\mathfrak{k}_0'$, and therefore it follows from the Corollary of Lemma 23 that

$$\psi_f(H_0; D) = \zeta(H_0)\psi_f(H_0; \Delta(\omega)) = \zeta(H_0)\pi(H_0)^{-1}g(H_0; \partial(\bar\omega)),$$

where $g$ is the function on $\mathfrak{h}_0$ given by

$$g(H) = \pi(H)\psi_f(H) = \sum_{s \in W} \epsilon(s)f^s(H) \qquad\qquad (H \in \mathfrak{h}_0).$$

On the other hand, $(\partial(\bar\omega)f)^s = \partial(\bar\omega)f^s$ $(s \in W)$ since $\bar\omega \in I(\mathfrak{h})$, and therefore $\partial(\bar\omega)g = \sum_{s \in W} \epsilon(s)(\partial(\bar\omega)f)^s$. This proves that

$$g(H; \partial(\bar\omega)) = \pi(H)\psi_{\partial(\bar\omega)f}(H) \qquad\qquad (H \in \mathfrak{h}_0),$$

and therefore $\psi_f(H_0; D) = \zeta(H_0)\psi_{\partial(\bar\omega)f}(H_0)$.

Let $\tau$ be a distribution on $\mathfrak{k}_0$ which is invariant under $K$. Then we define a distribution $\Phi$ on $\mathfrak{h}_0$ by $\Phi(f) = \tau(\zeta\psi_f)$ $(f \in C_c^\infty(\mathfrak{h}_0))$. Notice that if $\tau$ is a $\mathscr{E}$-distribution, the same holds for $\Phi$. The following theorem is the main result of this section.

THEOREM 5.[15] *Suppose there exists a complex number $c$ such that $D\tau = c\zeta\tau$. Then $\Phi$ coincides with an analytic function on $\mathfrak{h}_0$.*

Let $f$ be a function in $C_c^\infty(\mathfrak{h}_0)$. Then $\tau(D^*\psi_f) = \tau(D\psi_f) = \Phi(\partial(\bar\omega)f)$ from Lemmas 26 and 28. But $\tau(D^*\psi_f) = c\tau(\zeta\psi_f) = c\Phi(f)$ since $D\tau = c\zeta\tau$. Therefore $\Phi$ satisfies the differential equation $\partial(\bar\omega)\Phi - c\Phi = 0$. But since $\mathfrak{h}_0 \subset \mathfrak{k}_0$, $\bar\omega$ is a negative-definite quadratic form on $\mathfrak{h}_0$, and so this equation

---

[15] Compare this with Theorem 5 of [4(e)].

is of the elliptic type.  Hence we can conclude (see [8(I), p. 136] and [5]) that $\Phi$ is an analytic function on $\mathfrak{h}_0$.

We shall see in another paper that the above theorem plays an essential role in the theory of Fourier transforms on a noncompact semisimple Lie algebra.

Institute for Advanced Study,
Columbia University.

## REFERENCES.

[1] P. Cartier, Séminaire "Sophus Lie," 1954-55, Ecole Normale Supérieure, Paris.

[2] C. Chevalley, *Amer. Jour. Math.*, vol. 77 (1955), pp. 778-782.

[3] L. Gårding, *Math. Scand.*, vol. 1 (1953), pp. 55-72.

[4] Harish-Chandra, (a) *Ann. of Math.*, vol. 50 (1949), pp. 900-915.
    (b) *Trans. Amer. Math. Soc.*, vol. 70 (1951), pp. 28-96.
    (c) " " " " vol. 75 (1953), pp. 185-243.
    (d) " " " " vol. 76 (1954), pp. 485-528.
    (e) " " " " vol. 83 (1956), pp. 98-163.
    (f) *Bull. Amer. Math. Soc.*, vol. 61 (1955), pp. 389-396.
    (g) *Amer. Jour. Math.*, vol. 77 (1955), pp. 743-777.
    (h) *Proc. Nat. Acad. Sci., U.S.A.*, vol. 42 (1956), pp. 252-253.
    (i) " " " " " vol. 42 (1956), pp. 538-540.

[5] F. John, *Proceedings of the symposium on spectral theory and differential problems*, Stillwater, Okla. (1951), pp. 113-175.

[6] J. L. Koszul, *Bull. Soc. Math. France*, vol. 78 (1950), pp. 65-127.

[7] G. D. Mostow, *Bull. Amer. Math. Soc.*, vol. 55 (1949), pp. 969-980.

[8] L. Schwartz, *Théorie des distributions*, I (1950), II (1951), Paris, Hermann.

[9] H. Weyl, *The structure and representations of continuous groups*, Princeton, The Institute for Advanced Study, 1935.

Reprinted from
*Amer. J. of Math.*
**79** (1957), 87–120

# FOURIER TRANSFORMS ON A SEMISIMPLE LIE ALGEBRA I.*

By Harish-Chandra.

---

**1. Introduction.** Let $\mathfrak{g}_0$ be a real semisimple Lie algebra and $\mathfrak{h}_0$ a Cartan subalgebra of $\mathfrak{g}_0$. Consider the connected adjoint group $G$ of $\mathfrak{g}_0$ and its Cartan subgroup $A$ corresponding to $\mathfrak{h}_0$ (see $[2(e), \S2]$). Put $G^* = G/A$ and $x^*H = xH$ ($x \in G, H \in \mathfrak{h}_0$), where $x \to x^*$ is the natural mapping of $G$ on $G^*$. Let $dx^*$ denote the invariant measure on $G^*$. Define $\pi$ as in $[2(h), \S3]$ and let $\mathfrak{h}_0'$ be the set of those points $H \in \mathfrak{h}_0$ where $\pi(H) \neq 0$. Then we shall show that the integral

$$\phi_f(H) = \pi(H) \int_{G^*} f(x^*H) \, dx^* \qquad (H \in \mathfrak{h}_0')$$

is convergent for $f \in \mathscr{B}(\mathfrak{g}_0)$ (see $[2(h), \S2]$ for the notation used here and below.). Moreover, if $\bar{f}$ denotes the Fourier transform of $f$, there exists a rather simple relationship between $\phi_{\bar{f}}$ and $\phi_f$ (see Lemma 24). This fact is the appropriate generalization to the noncompact case of Theorem 3 of $[2(h)]$. Let $\mathfrak{h}_1$ be a connected component of $\mathfrak{h}_0'$. One of our principal objects is to show that

$$\operatorname*{Lim}_{H \to 0} \phi_f(H; \partial(\pi)) = cf(0) \qquad (H \in \mathfrak{h}_1, f \in \mathscr{B}(\mathfrak{g}_0)),$$

where $c$ is a complex number independent of $f$. (This corresponds to Lemma 16 of $[2(h)]$ in the compact case). Actually, it would turn out that $c = 0$ very often. Nevertheless, $\mathfrak{h}_0$ and $\mathfrak{h}_1$ can always be so chosen that $c \neq 0$. (The proof of these statements is rather long and will be completed only in the next paper of this series.) As we shall see in another paper, these results are intimately related to the Plancherel formula for any connected Lie group which is locally isomorphic to $G$. In the special case of the $n \times n$ real unimodular group, the above expression for $f(0)$ in terms of $\phi_f$ has been obtained by Gelfand and Graev [1].

Let $\mathfrak{k}_0$ be the subalgebra of $\mathfrak{g}_0$ corresponding to a maximal compact subgroup $K$ of $G$. We shall see that for most of our proofs, the crucial case is when $\mathfrak{h}_0 \subset \mathfrak{k}_0$. Lemma 5 contains the fundamental inequalities which

---

* Received July 30, 1956.

193

allow us to establish the main properties of $\phi_f$ (see Theorem 2) in this case. In Section 5, we discard the above assumption on $\mathfrak{h}_0$ and extend these results to the general case. The most important one among them is the relation $\phi_{\partial(p)f} = \partial(\bar{p})\phi_f$ $(p \in I(\mathfrak{g}), f \in \mathscr{C}(\mathfrak{g}_0))$ which, in its turn, follows from Theorem 1 of [2(h)]. On the other hand, these differential equations give us the above-mentioned results concerning Fourier transforms (see Section 6).

Select $\mathfrak{h}_0$ and $\mathfrak{h}_1$ as above, and put

$$T'(f) = \lim_{H \to 0} \phi_f(H; \partial(\pi)) \qquad\qquad (H \in \mathfrak{h}_1, f \in \mathscr{C}(\mathfrak{g}_0)).$$

Then $T'$ is a $\mathscr{C}$-distribution on $\mathfrak{g}_0$ and we have to show that $T' = c$, where $c$ is a constant. In Section 7, we shall obtain the weaker result that $T'$ coincides with a constant on each connected component of the set of regular elements in $\mathfrak{g}_0$. The problem is thus reduced (see the corollary to Lemma 29) to proving that the various constants so obtained are all equal. This requires a deeper investigation which will be carried out in another paper.

In Section 8, we define the notion of a fundamental Cartan subalgebra and prove that any two such algebras are conjugate under $G$. Moreover, we show that

$$\lim_{H \to 0} \phi_f(H; \partial(\pi)) = 0 \qquad\qquad (H \in \mathfrak{h}_1, f \in \mathscr{C}(\mathfrak{g}_0))$$

if $\mathfrak{h}_0$ is not fundamental. Thus it only remains to study the case when $\mathfrak{h}_0$ is fundamental. In particular, this is so if $\mathfrak{h}_0 \subset \mathfrak{k}_0$, and so we again make this assumption. The principal result of Section 9 is contained in Theorem 5, and its proof is based on Theorem 5 of [2(h)]. The rest of this paper is devoted to proving that the constant $c_0$ of Theorem 5 is not zero. For this, we have to make use of some (unpublished) work of de Rham on the elementary solutions of a certain type of differential operators. (Similar results also appear to have been obtained by Gelfand and Graev [1].) We apply de Rham's formula to the operator $\partial(\omega)$, where $\omega$ is the Casimir polynomial of $\mathfrak{g}_0$. This enables us to prove (see Lemma 41) the existence of a connected component $\mathfrak{h}_1$ of $\mathfrak{h}_0'$ and of a function $g \in \mathscr{C}(\mathfrak{g}_0)$ such that

$$\lim_{H \to 0} \phi_g(H; \partial(\pi)) \neq 0 \qquad\qquad (H \in \mathfrak{h}_1).$$

We shall need this fact in our study of the general case when $\mathfrak{h}_0$ is fundamental but the corresponding Cartan subgroup is not necessarily compact.

The main results of Theorem 3 have been announced in a short note [2(i)] (see also [2(j)]).

**2. A preliminary result.** Let $R$ and $C$ be the fields of real and complex numbers, respectively, $E_0$ a vector space over $R$ of finite dimension and $F_0$ a subspace of $E_0$. Let $E$ be the complexification of $E_0$ and $F$ the complexification of $F_0$ in $E$. We identify (real-valued) linear functions on $E_0$ with their linear extensions on $E$. Let $E'$ be the space of all such functions. Then elements of $E'$ will be called *real* linear functions on $E$. Let $F'$ be the corresponding space for $F$. Then $E'$ and $F'$ are vector spaces over $R$. For any $\lambda \in E'$, let $\bar{\lambda}$ denote the restriction of $\lambda$ on $F$. Then $\lambda \to \bar{\lambda}$ is a linear mapping of $E'$ into $F'$ whose kernel we denote by $E_1'$. Suppose we have introduced orders $[2(\mathrm{f}), \S 2]$ in $E'$ and $F'$. Then these orders are said to be compatible if $\bar{\lambda} > 0$ implies $\lambda > 0$ $(\lambda \in E')$. Given any order in $F'$, we can always define an order on $E'$ compatible with it as follows. Select any order on $E_1'$ and call an element $\lambda \in E'$ positive if either $\bar{\lambda}$ is positive in $F'$, or $\lambda \in E_1'$ and it is positive in the order chosen in $E_1'$.

Let $\mathfrak{g}_0$ be a semisimple Lie algebra over $R$. Define $\mathfrak{k}_0$ and $\mathfrak{p}_0$ as in $[2(\mathrm{c}), \S 2]$ and let $\mathfrak{a}_{\mathfrak{p}_0}$ be a maximal abelian subspace of $\mathfrak{p}_0$. We extend it to a Cartan subalgebra $\mathfrak{a}_0$ of $\mathfrak{g}_0$. Then $\mathfrak{a}_0 = \mathfrak{a}_{\mathfrak{p}_0} + \mathfrak{a}_{\mathfrak{k}_0}$ where $\mathfrak{a}_{\mathfrak{k}_0} = \mathfrak{a}_0 \cap \mathfrak{k}_0$ (see $[2(\mathrm{c}), \mathrm{p}.\ 188]$). Complexify $\mathfrak{g}_0$ to $\mathfrak{g}$ and let $\mathfrak{k}$, $\mathfrak{p}$, $\mathfrak{a}_\mathfrak{p}$, $\mathfrak{a}$, $\mathfrak{a}_\mathfrak{k}$ denote the complexifications of $\mathfrak{k}_0$, $\mathfrak{p}_0$, $\mathfrak{a}_{\mathfrak{p}_0}$, $\mathfrak{a}_0$, $\mathfrak{a}_{\mathfrak{k}_0}$, respectively, in $\mathfrak{g}_0$. Now take[1] $E_0 = \mathfrak{a}_{\mathfrak{p}_0} + (-1)^{\frac{1}{2}} \mathfrak{a}_{\mathfrak{k}_0}$ and $F_0 = \mathfrak{a}_{\mathfrak{p}_0}$ in the above set up, and select compatible orders in the spaces of real linear functions on $\mathfrak{a}$ and $\mathfrak{a}_\mathfrak{p}$, respectively. Let $(\alpha_1, \cdots, \alpha_l)$ be a fundamental system $[2(\mathrm{f}), \S 2]$ of positive roots (of $\mathfrak{g}$ with respect to $\mathfrak{a}$) in our order. Then if $p = \dim \mathfrak{a}_\mathfrak{p} = \dim_R \mathfrak{a}_{\mathfrak{p}_0}$, we have the following result.

LEMMA 1. *It is possible to select $p$ roots, say $(\alpha_1, \cdots, \alpha_p)$, among $(\alpha_1, \cdots, \alpha_l)$ with the following two properties.*[2]

(1)   *$\bar{\alpha}_1, \cdots, \bar{\alpha}_p$ are linearly independent.*

(2)   *If $\alpha$ is any root, $\bar{\alpha} = m_1 \bar{\alpha}_1 + \cdots + m_p \bar{\alpha}_p$ where $m_i$ are integers which are either all nonnegative or all nonpositive.*

Let $Q$ be the set of all positive roots of $\mathfrak{g}$ in our order and $Q_+$ the subset consisting of those $\alpha \in Q$ for which $\bar{\alpha} \neq 0$. Select a minimal subset $(\alpha_1, \cdots, \alpha_q)$ (say) of our fundamental system such that, for every $\alpha \in Q_+$, $\bar{\alpha}$ can be written in the form $\bar{\alpha} = m_1 \bar{\alpha}_1 + \cdots + m_q \bar{\alpha}_q$, where $m_i$ are all nonnegative integers. Then it is enough to prove that $\bar{\alpha}_1, \cdots, \bar{\alpha}_q$ are linearly independent. This requires a few lemmas.

---

[1] We select once for all a square root of $-1$ in $C$ and denote by $(-1)^{k/2}$ its $k$-th power for any integer $k$.

[2] The bar denotes restriction on $\mathfrak{a}_\mathfrak{p}$.

LEMMA 2. *If $i, j$ are two distinct indices $(1 \leq i, j \leq q)$, then $\bar{\alpha}_i - \bar{\alpha}_j \neq 0$ or $\bar{\alpha}$ for any $\alpha \in Q_+$.*

Without loss of generality, we may assume that $i = 1$, $j = 2$. In view of the minimality of our system $(\alpha_1, \cdots, \alpha_q)$, it is obvious that $\bar{\alpha}_1 - \bar{\alpha}_2 \neq 0$. So let us now suppose that $\bar{\alpha}_1 = \bar{\alpha} + \bar{\alpha}_2$ for some $\alpha \in Q_+$. Then $\bar{\alpha}_1 = (m_1 \bar{\alpha}_1 + \cdots + m_q \bar{\alpha}_q) + \bar{\alpha}_2$ where $m_1, \cdots, m_q$ are nonnegative integers. Hence

$$(m_1 - 1)\bar{\alpha}_1 + (m_2 + 1)\bar{\alpha}_2 + \sum_{3 \leq k \leq q} m_k \bar{\alpha}_k = 0.$$

It is clear that $m_1 = 0$, for otherwise, since $\bar{\alpha}_k > 0$ $(1 \leq k \leq q)$, the left side would be positive. Therefore

$$\bar{\alpha}_1 = (m_2 + 1)\bar{\alpha}_2 + \sum_{3 \leq k \leq q} m_k \bar{\alpha}_k,$$

and we get a contradiction with the minimality of the system $(\alpha_1, \cdots, \alpha_q)$.

Let $B$ denote the fundamental bilinear form on $\mathfrak{g}$ [2 (c), §2] and $\bar{\theta}$ the conjugation with respect to the compact real form $\mathfrak{u} = \mathfrak{k}_0 + (-1)^{\frac{1}{2}}\mathfrak{p}_0$. Then $\| X \|^2 = -B(X, \bar{\theta}(X))$ $(X \in \mathfrak{g})$ is a positive-definite Hermitian form on $\mathfrak{g}$. Since the restriction of $B$ on $\mathfrak{a}_{\mathfrak{p}_0}$ is real and nondegenerate, for each real linear function $\lambda$ on $\mathfrak{a}_\mathfrak{p}$, there exists a unique element $\Gamma_\lambda \in \mathfrak{a}_{\mathfrak{p}_0}$ such that $B(\Gamma, \Gamma_\lambda) = \lambda(\Gamma)$ for all $\Gamma \in \mathfrak{a}_\mathfrak{p}$. Define $\theta$ as in [2 (c), §2].

LEMMA 3. *Suppose $\lambda$ is a real linear function on $\mathfrak{a}_\mathfrak{p}$ and $X$ an element in $\mathfrak{g}_0$ such that $[\Gamma, X] = \lambda(\Gamma)X$ for all $\Gamma \in \mathfrak{a}_\mathfrak{p}$. Then $[\theta(X), X] = \| X \|^2 \Gamma_\lambda$.*

Let $X = Y + Z$ $(Y \in \mathfrak{p}_0, Z \in \mathfrak{k}_0)$. Since $[\mathfrak{p}, \mathfrak{p}] \subset \mathfrak{k}$ and $[\mathfrak{p}, \mathfrak{k}] \subset \mathfrak{p}$, it follows that $[\Gamma, Y] = \lambda(\Gamma)Z$, $[\Gamma, Z] = \lambda(\Gamma)Y$ $(\Gamma \in \mathfrak{a}_\mathfrak{p})$. Moreover, $[\theta(X), X] = 2[Z, Y]$, and so it is clear that $[\theta(X), X]$ lies in $\mathfrak{p}_0$ and it commutes with every element in $\mathfrak{a}_{\mathfrak{p}_0}$. Therefore, since $\mathfrak{a}_{\mathfrak{p}_0}$ is maximal abelian in $\mathfrak{p}_0$, $[\theta(X), X] \in \mathfrak{a}_{\mathfrak{p}_0}$. Finally, if $\Gamma \in \mathfrak{a}_\mathfrak{p}$,

$$B(\Gamma, [\theta(X), X]) = -B([\Gamma, X], \theta(X)) = -\lambda(\Gamma)B(X, \theta(X)) = \lambda(\Gamma)\| X \|^2,$$

and this proves that $[\theta(X), X] = \| X \|^2 \Gamma_\lambda$.

For any two linear functions $\lambda$ and $\mu$ on $\mathfrak{a}_\mathfrak{p}$, put $\langle \lambda, \mu \rangle = \lambda(\Gamma_\mu) = B(\Gamma_\lambda, \Gamma_\mu)$ and $| \lambda | = \langle \lambda, \lambda \rangle^{\frac{1}{2}} \geq 0$. Then if $\lambda \neq 0$, $| \lambda | = \| \Gamma_\lambda \| > 0$. Let $\Sigma$ be the set of all linear functions $\lambda$ on $\mathfrak{a}_\mathfrak{p}$ for which there exists an element $X_\lambda \neq 0$ in $\mathfrak{g}_0$ with the property that $[\Gamma, X_\lambda] = \lambda(\Gamma)X_\lambda$ for all $\Gamma \in \mathfrak{a}_\mathfrak{p}$. It is obvious that $\Sigma$ consists of zero and $\pm \bar{\alpha}$ $(\alpha \in Q_+)$.

LEMMA 4. *Suppose $\lambda, \mu \in \Sigma$ and $\lambda - \mu \notin \Sigma$. Then $\langle \lambda, \mu \rangle \leq 0$.*

Select nonzero elements $X_\lambda$, $X_\mu$ in $\mathfrak{g}_0$ as above. We may assume that $\|X_\mu\| = 1$. It follows from our assumption that neither $\lambda$ nor $\mu$ is zero. Put $X = 2^{\frac{1}{2}} |\mu|^{-1} X_\mu$, $Y = \theta(X)$ and $\Gamma_0 = 2|\mu|^{-2} \Gamma_\mu$. Then $[Y, X] = \Gamma_0$, $[\Gamma_0, X] = 2X$, $[\Gamma_0, Y] = -2Y$. Hence $I_0 = RX + RY + R\Gamma_0$ is a three-dimensional Lie algebra and $Z \to \operatorname{ad} Z$ $(Z \in I_0)$ is a representation of $I_0$ on $\mathfrak{g}_0$. Now $(\operatorname{ad} \Gamma_0) X_\lambda = \lambda(\Gamma_0) X_\lambda$ and $(\operatorname{ad} Y) X_\lambda = 0$ since $\lambda - \mu \notin \Sigma$. Therefore it follows immediately from the theory of finite-dimensional representations of $I_0$ (see [2(a), Lemma 1]) that $\lambda(\Gamma_0) \le 0$. But $\lambda(\Gamma_0) = 2|\mu|^{-2} \lambda(\Gamma_\mu)$, and so we conclude that $\langle \lambda, \mu \rangle \le 0$.

COROLLARY. *If $i \ne j$ $(1 \le i, j \le q)$ then $\langle \tilde\alpha_i, \tilde\alpha_j \rangle \le 0$.*

This is an immediate consequence of Lemmas 2 and 4.

Now we can prove that $\tilde\alpha_1, \cdots, \tilde\alpha_q$ are linearly independent. For otherwise, since they are real, we could select $a_i \in R$ $(1 \le i \le q)$ not all zero such that $a_1 \tilde\alpha_1 + \cdots + a_q \tilde\alpha_q = 0$. We may suppose that $a_1, \cdots, a_k > 0$ while $a_{k+1}, \cdots, a_q \le 0$. Then if $b_j = -a_j$ $(k < j \le q)$,

$$a_1 \tilde\alpha_1 + \cdots + a_k \tilde\alpha_k = b_{k+1} \tilde\alpha_{k+1} + \cdots + b_q \tilde\alpha_q = \lambda \quad \text{(say)}.$$

Therefore

$$|\lambda|^2 = \sum_{1 \le i \le k} \sum_{k < j \le q} a_i b_j \langle \tilde\alpha_i, \tilde\alpha_j \rangle \le 0$$

from the above corollary. This implies that $\lambda = a_1 \tilde\alpha_1 + \cdots + a_k \tilde\alpha_k = 0$ which, however, is impossible since $\tilde\alpha_1, \cdots, \tilde\alpha_k$ are all positive and $a_1, \cdots, a_k > 0$. This completes the proof of Lemma 1.

## 3. Some inequalities.

**3. Some inequalities.** The positive-definite Hermitian form

$$\|X\|^2 = -B(X, \theta(X)) \qquad\qquad (X \in \mathfrak{g})$$

defines the structure of a Hilbert space on $\mathfrak{g}$. For any linear transformation $T$ in $\mathfrak{g}$, put $\|T\|^2 = \operatorname{sp}(T^*T)$, where $T^*$ is the adjoint of $T$ with respect to this structure. It is clear that $(\operatorname{ad} X)^* = -\operatorname{ad}(\bar\theta(X))$ $(X \in \mathfrak{g})$, and therefore $\|\operatorname{ad} X\| = \|X\|$. Let $G$ be the connected adjoint group of $\mathfrak{g}_0$ and $K$ the analytic subgroup of $G$ corresponding to $\mathfrak{k}_0$. Then every element $k \in K$ is a unitary transformation on $\mathfrak{g}$, and therefore $\|k\| = 1$.

Let $\mathfrak{h}_0$ be a Cartan subalgebra of $\mathfrak{g}_0$. *Until the end of Section 4 we shall assume that $\mathfrak{h}_0 \subset \mathfrak{k}_0$.* Let $\mathfrak{h}$ denote the complexification of $\mathfrak{h}_0$ in $\mathfrak{g}$ and $P$ the set of all positive roots (under some arbitrary but fixed order) of $\mathfrak{g}$ with respect to $\mathfrak{h}$. Let $P_-$ and $P_0$ be the subsets of $P$ consisting of the compact and the noncompact roots, respectively (see [2(f), p. 751]). Put $\pi_0 = \prod_{\alpha \in P_0} \alpha$. Then $\pi_0$ is a polynomial function on $\mathfrak{h}$.

LEMMA 5. *Suppose* $x \in G$ *and* $H \in \mathfrak{h}_0$. *Then*

$$\| xH + \theta(xH) \|^2 = 2 \| H \|^2 + 2 \| xH \|^2.$$

*Moreover, there exist integers* $m, q \geqq 0$ *and a positive number* $c$ *such that*

$$\| x \| \, | \pi_0(H) |^q \leqq c \| xH + \theta(xH) \|^m$$

*for all* $x \in G$ *and* $H \in \mathfrak{h}_0$.

$K$ being compact [5], any maximal abelian subalgebra of $\mathfrak{k}_0$ is conjugate to $\mathfrak{h}_0$ under $K$. Hence we can select $k \in K$ such that $k(\mathfrak{a}_{\mathfrak{k}_0}) \subset \mathfrak{h}_0$. Then $k(\mathfrak{a}_{\mathfrak{p}_0})$ is also a maximal abelian subspace of $\mathfrak{p}_0$, and therefore, by replacing $\mathfrak{a}_0$ by $k\mathfrak{a}_0$, we may assume that $\mathfrak{a}_{\mathfrak{k}_0} \subset \mathfrak{h}_0$. Let $A_\mathfrak{p}$ be the analytic subgroup of $G$ corresponding to $\mathfrak{a}_{\mathfrak{p}_0}$. Then $G = KA_\mathfrak{p}K$ (see [2(g), §12]), and every element in $K$ is a unitary transformation on $\mathfrak{g}$ which commutes with $\theta$. Therefore for the proof of our lemma, it is obviously enough to consider the case when $x = ak$ $(a \in A_\mathfrak{p}, k \in K)$. Fix an element $H \in \mathfrak{h}_0$, and put $kH = X$. Then $xH = aX$ and $\theta(xH) = a^{-1}X$. Hence $\| xH + \theta(xH) \| = \| aX + a^{-1}X \|$. But since $a$ is self-adjoint, it is obvious that $\| aX + a^{-1}X \|^2 = \| aX \|^2 + \| a^{-1}X \|^2 + 2 \| X \|^2$. Moreover, $\| a^{-1}X \| = \| \theta(aX) \| = \| aX \|$, and therefore

$$\| xH + \theta(xH) \|^2 = 2 \| X \|^2 + 2 \| aX \|^2 = 2 \| H \|^2 + 2 \| xH \|^2.$$

This proves the first assertion. The proof of the second statement is, however, more complicated.

Define $Q$ and $Q_+$ as in Section 2, and for any root $\beta$ of $\mathfrak{g}$ with respect to $\mathfrak{a}$, choose an element $X_\beta \neq 0$ in $\mathfrak{g}$ such that $[\Gamma, X_\beta] = \beta(\Gamma).X_\beta$ for all $\Gamma \in \mathfrak{a}$. Then the function $\theta\beta : \Gamma \to \beta(\theta\Gamma)$ is also a root.[3] Let $Q_-$ be the complement of $Q_+$ in $Q$. Then if $\beta \in Q_-$, $\theta\beta = \beta$ and $X_\beta, X_{-\beta} \in \mathfrak{k}$. On the other hand $\theta\beta < 0$ if $\beta \in Q_+$. Put $\mathfrak{q} = \sum_{\beta \in Q_+} CX_\beta$, $\mathfrak{q}^- = \sum_{\beta \in Q_-} CX_{-\beta}$, and let $\mathfrak{l}$ be the centralizer of $\mathfrak{a}_\mathfrak{p}$ in $\mathfrak{k}$. Then $\mathfrak{l} \supset \mathfrak{a}_\mathfrak{k}$ and

$$\mathfrak{g} = \mathfrak{q}^- + \mathfrak{l} + \mathfrak{a}_\mathfrak{p} + \mathfrak{q} = \mathfrak{k} + \mathfrak{a}_\mathfrak{p} + \mathfrak{q} = \mathfrak{p} + \mathfrak{l} + \mathfrak{q},$$

where all the sums are direct.[3] Therefore

$$\dim \mathfrak{k} = \dim \mathfrak{g} - \dim \mathfrak{a}_\mathfrak{p} - \dim \mathfrak{q} = \dim \mathfrak{l} + \dim \mathfrak{q}.$$

Define a linear mapping $\phi$ of $\mathfrak{l} + \mathfrak{q}$ into $\mathfrak{k}$ by $\phi(X) = X + \theta(X)$ $(X \in \mathfrak{l} + \mathfrak{q})$. Since $(\mathfrak{l} + \mathfrak{q}) \cap \mathfrak{p} = \{0\}$, the kernel of $\phi$ is zero. Therefore since $\mathfrak{k}$ and $\mathfrak{l} + \mathfrak{q}$ have the same dimension, $\phi$ is an isomorphism of $\mathfrak{l} + \mathfrak{q}$ onto $\mathfrak{k}$.

---

[3] All the facts stated here are well known (see [2(c), p. 188]).

Now if $\Gamma \in \mathfrak{a}_{\mathfrak{p}_0}$, $\operatorname{ad} \Gamma$ is a self-adjoint operator on $\mathfrak{g}$, and its eigenvalues are zero and $\pm \beta(\Gamma)$ $(\beta \in Q_+)$. Therefore

$$\| (\operatorname{ad} \Gamma)^2 Z \| \geqq \lambda(\Gamma) \| (\operatorname{ad} \Gamma) Z \| \qquad (Z \in \mathfrak{g}, \Gamma \in \mathfrak{a}_{\mathfrak{p}_0}),$$

where $\lambda(\Gamma) = \min |\beta(\Gamma)|$ and $\beta$ runs over all roots in $Q_+$ for which $\beta(\Gamma) \neq 0$. (In case $\Gamma = 0$, we put $\lambda(\Gamma) = 0$.)

Now select[4] $\alpha_1, \cdots, \alpha_p \in Q_+$ according to Lemma 1, and choose $\Gamma_i \in \mathfrak{a}_{\mathfrak{p}_0}$ $(1 \leqq i \leqq p)$ such that[5] $\alpha_i(\Gamma_j) = \delta_{ij}$ $(1 \leqq i, j \leqq p)$. Then $(\Gamma_1, \cdots, \Gamma_p)$ is a base for $\mathfrak{a}_{\mathfrak{p}_0}$ over $R$, and it is clear that $\lambda(\Gamma_i) \geqq |\alpha_i(\Gamma_i)| = 1$. Therefore

$$\| (\operatorname{ad} \Gamma_i)^2 Z \| \geqq \| (\operatorname{ad} \Gamma_i) Z \| \qquad (Z \in \mathfrak{g}, 1 \leqq i \leqq p).$$

Let $Y$ be an element in $\mathfrak{q}_0 = \mathfrak{q} \cap \mathfrak{g}_0$. Then $\operatorname{ad} Y$ is nilpotent and $B(Y, Y) = B(\theta(Y), \theta(Y)) = 0$. Therefore

$$\| \phi(Y) \|^2 = -2B(Y, \theta(Y)) = 2 \| Y \|^2.$$

Hence if $\mathfrak{l}_0 = \mathfrak{l} \cap \mathfrak{g}_0$ and $X \in \mathfrak{l}_0 + \mathfrak{q}_0$,

$$\| \phi((\operatorname{ad} \Gamma_i)^2 X) \|^2 = 2 \| (\operatorname{ad} \Gamma_i)^2 X \|^2$$

since $(\operatorname{ad} \Gamma_i)^2 X \in \mathfrak{q}_0$. Moreover, $(\operatorname{ad} \Gamma_i)^2$ commutes with $\theta$. Therefore

$$\| \phi((\operatorname{ad} \Gamma_i)^2 X) \| = \| (\operatorname{ad} \Gamma_i)^2 \phi(X) \| \geqq \| (\operatorname{ad} \Gamma_i) \phi(X) \|$$

as we have seen above. Hence

$$\| (\operatorname{ad} \Gamma_i)^2 X \| \geqq 2^{-\frac{1}{2}} \| (\operatorname{ad} \Gamma_i) \phi(X) \| \qquad (X \in \mathfrak{l}_0 + \mathfrak{q}_0, 1 \leqq i \leqq p).$$

On the other hand, we know (see Iwasawa [3]) that $\mathfrak{g}_0$ is the direct sum of $\mathfrak{k}_0$, $\mathfrak{a}_{\mathfrak{p}_0}$ and $\mathfrak{q}_0$, so that

$$\dim_R \mathfrak{q}_0 = \dim_R \mathfrak{g}_0 - \dim_R \mathfrak{k}_0 - \dim_R \mathfrak{a}_{\mathfrak{p}_0}$$
$$= \dim \mathfrak{g} - \dim \mathfrak{k} - \dim \mathfrak{a}_\mathfrak{p} = \dim \mathfrak{q}.$$

Therefore $\dim_R (\mathfrak{l}_0 + \mathfrak{q}_0) = \dim \mathfrak{l} + \dim \mathfrak{q} = \dim \mathfrak{k} = \dim_R \mathfrak{k}_0$. Since $\phi(\mathfrak{l}_0 + \mathfrak{q}_0)$ is obviously contained in $\mathfrak{g}_0$, it follows that $\phi$ defines a linear isomorphism of $\mathfrak{l}_0 + \mathfrak{q}_0$ onto $\mathfrak{k}_0$. Let $\psi$ denote the inverse of this isomorphism. Then the above result may be written in the form

$$\| (\operatorname{ad} \Gamma_i)^2 \psi(X) \| \geqq 2^{-\frac{1}{2}} \| (\operatorname{ad} X) \Gamma_i \| \qquad (X \in \mathfrak{k}_0, 1 \leqq i \leqq p).$$

Let $(\operatorname{ad} X)_\mathfrak{p}$ denote the restriction of $\operatorname{ad} X$ on $\mathfrak{p}$ $(X \in \mathfrak{k}_0)$, and put $\mu(X) = |\gamma|$,

---

[4] All our lemmas are true trivially when $\mathfrak{g}_0 = \mathfrak{k}_0$. So whenever it is convenient for the proof, we shall assume that $\mathfrak{p}_0 \neq \{0\}$.

[5] $\delta_{ij} = 1$ or $0$ according as $i = j$ or $i \neq j$.

where $\gamma$ is an eigenvalue of $(\operatorname{ad} X)_\mathfrak{p}$ with the smallest possible absolute value. Then since $\operatorname{ad} X$ is skew-adjoint, it is obvious that $\| (\operatorname{ad} X) Y \| \geq \mu(X) \| Y \|$ $(Y \in \mathfrak{p})$.  Therefore

$$\prod_{i=1}^{p} \| (\operatorname{ad} \Gamma_i)^2 \psi(X) \| \geq 2^{-p/2} \prod_{1 \leq i \leq p} \| (\operatorname{ad} X) \Gamma_i \| \geq 2^{-p/2} \mu(X)^p \prod_{1 \leq i \leq p} \| \Gamma_i \|.$$

Moreover it is obvious that $\mu(kX) = \mu(X)$ $(k \in K, X \in \mathfrak{k}_0)$.  Hence

$$\prod_{1 \leq i \leq p} \| (\operatorname{ad} \Gamma_i)^2 \psi(kH) \| \geq c_0 \mu(H)^p \qquad (H \in \mathfrak{h}_0, k \in K)$$

where $c_0 = 2^{-p/2} \prod_{1 \leq i \leq p} \| \Gamma_i \|$ is a positive constant.  On the other hand, it follows from the definition of $\mu(H)$ that $\mu(H) = \min_{\beta \in P_0} | \beta(H) |$ $(H \in \mathfrak{h}_0)$.  Therefore

$$\prod_{1 \leq i \leq p} \| (\operatorname{ad} \Gamma_i)^2 \psi(kH) \| \geq c_0 \min_{\beta \in P_0} | \beta(H) |^p \qquad (H \in \mathfrak{h}_0, k \quad K).$$

Now observe that $| \beta(H) | \leq \| H \|$ for any $\beta \in P$.  Hence if $q_0$ is the number of roots in $P_0$,

$$\| H \|^{q_0 - 1} \min_{\beta \in P_0} | \beta(H) | \geq \prod_{\beta \in P_0} | \beta(H) | = | \pi_0(H) |,$$

and therefore

$$\| H \|^{p(q_0 - 1)} \prod_{1 \leq i \leq p} \| (\operatorname{ad} \Gamma_i)^2 \psi(kH) \| \geq c_0 | \pi_0(H) |^p \qquad (H \in \mathfrak{h}_0, k \in K).$$

Let $\mathfrak{a}_{\mathfrak{p}_0}^+$ be the set of all $\Gamma \in \mathfrak{a}_{\mathfrak{p}_0}$ such that $\alpha(\Gamma) \geq 0$ for every $\alpha \in Q$. It is obvious that $\Gamma = t_1 \Gamma_1 + \cdots + t_p \Gamma_p$ $(t_i \in R)$ lies in $\mathfrak{a}_{\mathfrak{p}_0}^+$ if and only if $t_1, \cdots, t_p \geq 0$.  Let $A_\mathfrak{p}^+$ be the image of $\mathfrak{a}_{\mathfrak{p}_0}^+$ in $A_\mathfrak{p}$ under the exponential mapping.  Then it is known (see $[2(g)$, Lemma 38]) that $G = K A_\mathfrak{p}^+ K$. Hence in the proof of the second assertion of our lemma, we can assume that $x \in A_\mathfrak{p}^+ K$.  Select a maximal set $(\alpha_1, \cdots, \alpha_r)$ $(r \geq p)$ of roots in $Q_+$ such that $\bar{\alpha}_1, \cdots, \bar{\alpha}_r$ are all distinct.  Then it is obvious that $\mathfrak{q} = \sum_{1 \leq j \leq r} \mathfrak{q}_j$, where $\mathfrak{q}_j$ is the set of all $Y \in \mathfrak{q}$ for which $[\Gamma, Y] = \alpha_j(\Gamma) Y (\Gamma \in \mathfrak{a}_\mathfrak{p})$.  Moreover, $\operatorname{ad} \Gamma$ is self-adjoint for $\Gamma \in \mathfrak{a}_{\mathfrak{p}_0}$, and so it is clear that $\mathfrak{q}_j$ $(1 \leq j \leq r)$ are mutually orthogonal.  Similarly $\mathfrak{l}$, $\mathfrak{q}$ and $\mathfrak{q}^-$ are orthogonal to each other. Now suppose $x = ak$ $(a \in A_\mathfrak{p}^+, k \in K)$ and $H$ is a fixed element in $\mathfrak{h}_0$ such that $\pi_0(H) \neq 0$.  Then if $\psi(kH) = L + X$ $(L \in \mathfrak{l}_0, X \in \mathfrak{q}_0)$, we have $kH = 2L + X + \theta(X)$.  Hence

$$\| xH + \theta(xH) \|^2 = 2 \| H \|^2 + 2 \| xH \|^2$$

$$= 16 \| L \|^2 + 4 \| X \|^2 + 2 \| aX \|^2 + 2 \| a^{-1} X \|^2.$$

Put $\nu = \max_{\alpha, i} | \alpha(\Gamma_i) |$ for $\alpha \in Q$ and $1 \leq i \leq p$.  Then since no eigenvalue of ad $\Gamma_i$ can exceed $\nu$ in absolute value, $\| (\operatorname{ad} \Gamma_i)^2 Z \| \leq \nu^2 \| Z \|$ $(Z \in \mathfrak{g}, 1 \leq i \leq p)$. Therefore if $a = \exp \Gamma$ $(\Gamma \in \mathfrak{a}_{\mathfrak{p}_0}{}^+)$ and $X_j$ is the orthogonal projection of $X$ in $\mathfrak{q}_j$, we have

$$\| aX \|^2 \geq \nu^{-4} \| (\operatorname{ad} \Gamma_i)^2 aX \|^2 = \nu^{-4} \sum_{1 \leq j \leq r} e^{2\alpha_j(\Gamma)} \alpha_j(\Gamma_i)^4 \| X_j \|^2 \qquad (1 \leq i \leq p).$$

Now we claim that

$$e^{\alpha(\Gamma)} \alpha(\Gamma_i)^2 \geq e^{\alpha_i(\Gamma)} \alpha(\Gamma_i)^2$$

for $\alpha \in Q_+$. This is obvious if $\alpha(\Gamma_i) = 0$, and so let us suppose that $\alpha(\Gamma_i) \neq 0$. We know that $\bar{\alpha} = \nu_1 \bar{\alpha}_1 + \cdots + \nu_p \bar{\alpha}_p$, where $\nu_j = \alpha(\Gamma_j)$ $(1 \leq j \leq p)$ are non-negative integers. Therefore, since $\Gamma \in \mathfrak{a}_{\mathfrak{p}_0}{}^+$ and $\nu_i = \alpha(\Gamma_i) \geq 1$, we conclude that $\alpha(\Gamma) \geq \nu_i \alpha_i(\Gamma) \geq \alpha_i(\Gamma)$, and our inequality follows. This shows that

$$\| aX \|^2 \geq \nu^{-4} \sum_{1 \leq j \leq r} e^{2\alpha_i(\Gamma)} \alpha_j(\Gamma_i)^4 \| X_j \|^2 = \nu^{-4} e^{2\alpha_i(\Gamma)} \| (\operatorname{ad} \Gamma_i)^2 X \|^2 \qquad (1 \leq i \leq p),$$

and therefore

$$\| xH + \theta(xH) \|^p \geq \| aX \|^p \geq \nu^{-2p} e^{\gamma(\Gamma)} \prod_{1 \leq i \leq p} \| (\operatorname{ad} \Gamma_i)^2 X \|,$$

where $\gamma = \alpha_1 + \alpha_2 + \cdots + \alpha_p$. On the other hand,

$$\| x \|^2 = \| a \|^2 = \operatorname{sp} a^2 = l + \sum_{\alpha \in Q} (e^{2\alpha(\Gamma)} + e^{-2\alpha(\Gamma)}),$$

where $l = \dim \mathfrak{a}$. Moreover, it is obvious that $\alpha(\Gamma) \leq \nu \gamma(\Gamma)$ for any $\alpha \in Q$. Therefore if $n = \dim \mathfrak{g}$, $\| x \|^2 \leq n e^{2\nu\gamma(\Gamma)}$. Also $(\operatorname{ad} \Gamma_i)^2 \psi(kH) = (\operatorname{ad} \Gamma_i)^2 X$. Hence it follows from our earlier result and the first assertion of the lemma that

$$\| x \| \leq n^{\frac{1}{2}} \nu^{2p\nu} \| xH + \theta(xH) \|^{p\nu} \{ \prod_{1 \leq i \leq p} \| (\operatorname{ad} \Gamma_i)^2 X \| \}^{-\nu}$$

$$\leq n^{\frac{1}{2}} \nu^{2p\nu} \| xH + \theta(xH) \|^{p\nu} \{ c_0 \, | \pi_0(H) |^p \, \| H \|^{-p(q_0-1)} \}^{-\nu}$$

$$\leq n^{\frac{1}{2}} \nu^{2p\nu} \| xH + \theta(xH) \|^{p\nu} \{ c_0 \, | \pi_0(H) |^p \, \| xH + \theta(xH) \|^{-p(q_0-1)} \}^{-\nu}.$$

So by taking $q = p\nu$, $m = p\nu q_0$ and $c = n^{\frac{1}{2}} \nu^{2p\nu} c_0^{-\nu}$, we get the second statement of the lemma.

LEMMA 6. *Let $\lambda$ be any real linear function on $\mathfrak{a}_\mathfrak{p}$. Then there exists an integer $m \geq 0$ such that*

$$e^{\lambda(\Gamma)} \leq \| \exp \Gamma \|^m$$

*for $\Gamma \in \mathfrak{a}_{\mathfrak{p}_0}{}^+$.*

It is clear that [4]

$$\| \exp \Gamma \|^2 \geq \sum_{\alpha \in Q_+} e^{2\alpha(\Gamma)} \geq r \exp(2 r^{-1} \sum_{\alpha \in Q_+} \alpha(\Gamma)) \qquad (\Gamma \in \mathfrak{a}_{\mathfrak{p}_0}{}^+),$$

where $r$ is the number of roots in $Q_+$. Moreover, we can obviously choose a positive integer $s$ such that

$$s(\alpha_1(\Gamma) + \alpha_2(\Gamma) + \cdots + \alpha_p(\Gamma)) \geqq \lambda(\Gamma)$$

for $\Gamma \in \mathfrak{a}_{p_0}{}^+$. Therefore

$$\| \exp \Gamma \|^{rs} \geqq \exp(s \sum_{\alpha \in Q_+} \alpha(\Gamma)) \geqq \exp \lambda(\Gamma) \qquad (\Gamma \in \mathfrak{a}_{p_0}{}^+),$$

and this proves the lemma.

COROLLARY 1. *Let $dx$ denote the Haar measure of $G$. Then there exists an integer $s \geqq 0$ such that*

$$\int_G \| x \|^{-s}\, dx < \infty.$$

Let $dk$ and $d\Gamma$, respectively, denote the Haar measure on $K$ and the Euclidean measure on $\mathfrak{a}_{p_0}$. Then it is known (see [2(g), Lemma 38]) that one can normalize $dx$ in such a way that

$$\int_G g(x)\, dx = \int_{K \times \mathfrak{a}_{p_0}{}^+ \times K} g(k(\exp \Gamma)k')\mu(\Gamma)\, dk\, d\Gamma\, dk' \qquad (g \in C_c(G)).$$

Here $C_c(G)$ is the set of all continuous functions on $G$ which vanish outside a compact set and

$$\mu(\Gamma) = \prod_{\alpha \in Q_+} (e^{\alpha(\Gamma)} - e^{-\alpha(\Gamma)}).$$

It is obvious that

$$\mu(\Gamma) \leqq \exp(\sum_{\alpha \in Q_+} \alpha(\Gamma)) \qquad (\Gamma \in \mathfrak{a}_{p_0}{}^+).$$

Select a real linar function $\lambda$ on $\mathfrak{a}_p$ such that $\int_{\mathfrak{a}_{p_0}{}^+} e^{-\lambda(\Gamma)}\, d\Gamma < \infty$. Then from Lemma 6,

$$\exp(\lambda(\Gamma) + \sum_{\alpha \in Q_+} \alpha(\Gamma)) \leqq \| \exp \Gamma \|^s \qquad (\Gamma \in \mathfrak{a}_{p_0}{}^+)$$

for some integer $s$. Hence

$$\mu(\Gamma) \| \exp \Gamma \|^{-s} \leqq e^{-\lambda(\Gamma)},$$

and therefore

$$\int_G \| x \|^{-s}\, dx = \int_{\mathfrak{a}_{p_0}{}^+} \| \exp \Gamma \|^{-s} \mu(\Gamma)\, d\Gamma \leqq \int_{\mathfrak{a}_{p_0}{}^+} e^{-\lambda(\Gamma)}\, d\Gamma < \infty$$

if we assume that $\int_K dk = 1$.

Let $\sigma$ be a representation of $G$ on a finite-dimensional Hilbert space $V$.

For any linear transformation $T$ on $V$, put $\|T\| = \{\mathrm{sp}(T^*T)\}^{\frac{1}{2}}$, where $T^*$ is the adjoint of $T$.

COROLLARY 2. *We can find an integer $m \geq 0$ and a positive number $c$ such that $\|\sigma(x)\| \leq c\|x\|^m$ for all $x \in G$.*

Let us denote the corresponding representation of $\mathfrak{g}$ also by $\sigma$. If $S$ is a nonsingular linear transformation on $V$, we know that $\|S\sigma(x)S^{-1}\| \leq \|S\|\|\sigma(x)\|\|S^{-1}\|$, and therefore it is permissible, for our proof, to replace $\sigma$ by an equivalent representation on $V$. Since $\mathfrak{u} = \mathfrak{k}_0 + (-1)^{\frac{1}{2}}\mathfrak{p}_0$ is a compact real form of $\mathfrak{g}$, we may, by going over to an equivalent representation, assume that $\sigma(X)$ is skew-adjoint for $X \in \mathfrak{u}$. Then $\sigma(k)$ is unitary and $\sigma(a)$ self-adjoint ($k \in K, a \in A_\mathfrak{p}$). Therefore, since $G = KA_\mathfrak{p}^+K$, it is enough to consider the case when $x \in A_\mathfrak{p}^+$. Let $d = \dim V$ and $\Lambda_1, \cdots, \Lambda_r$ be all the distinct weight of $\sigma$ (with respect to $\mathfrak{a}$). Then $\Lambda_i$ are all real linear functions on $\mathfrak{a}$ and so from Lemma 6, we can choose an integer $m \geq 0$ such that $e^{\Lambda_i(\Gamma)} \leq \|\exp \Gamma\|^m$ ($1 \leq i \leq r$) for $\Gamma \in \mathfrak{a}_{\mathfrak{p}_0}^+$. This proves that

$$\|\sigma(a)\|^2 \leq d \max_i e^{2\Lambda_i(\Gamma)} \leq d\|a\|^{2m}$$

if $a = \exp \Gamma$ ($\Gamma \in \mathfrak{a}_{\mathfrak{p}_0}^+$).

**4. Applications of the above inequalities.** Let $\mathscr{S}_0$ denote the set of all continuous functions $f$ on $\mathfrak{g}_0$ such that

$$\nu_m(f) = \sup_{X \in \mathfrak{g}_0} |f(X)|(1 + \|X + \theta(X)\|)^m < \infty$$

for every integer $m \geq 0$. Also let $\mathfrak{h}_0''$ denote the set of those points $H \in \mathfrak{h}_0$ where $\pi_0(H) \neq 0$.

LEMMA 7. *Let $m$ be a given integer $\geq 0$. Then we can find integers $q, m_1 \geq 0$ and a positive number $c$, such that*

$$(1 + \|H\|)^{q_1}|\pi_0(H)|^q \int_G \|x\|^m |f(xH)|\, dx \leq c\nu_{m_1+q_1}(f)$$

*for $H \in \mathfrak{h}_0''$, $f \in \mathscr{S}_0$ and any integer $q_1 \geq 0$. Moreover if $\Omega$ is a compact subset of $G \times \mathfrak{h}_0'' \times G$ and $f \in \mathscr{S}_0$, the integral*

$$\int_G \|zxy\|^m |f(zxyH)|\, dx$$

*converges uniformly with respect to $(z, H, y) \in \Omega$.*

Choose an integer $s \geq 0$ such that $\int_G \|x\|^{-s}\, dx < \infty$ (Corollary 1 to

Lemma 6). Then from Lemma 5, we can select integers $q, m_1 \geqq 0$ and a positive number $c_1$ with the property that

$$|\pi_0(H)|^q \| x \|^{m+s} \leqq c_1 \| xH + \theta(xH) \|^{m_1}$$

for all $H \in \mathfrak{h}_0$ and $x \in G$. Then since $\| H \| \leqq \| xH + \theta(xH) \|$ (Lemma 5),

$$(1 + \| H \|)^{q_1} |\pi_0(H)|^q \| x \|^m |f(xH)| \leqq c_1 \nu_{m_1+q_1}(f) \| x \|^{-s},$$

and therefore if we put

$$c = c_1 \int \| x \|^{-s} \, dx,$$

our first assertion follows. Now suppose $(z, H, y) \in \Omega$ and $f \in \mathcal{S}_0$. Then

$$\| zxy \|^m |f(zxyH)| \leqq c_1 |\pi_0(H)|^{-q} \nu_{m_1}(f) \| zxy \|^{-s}.$$

But $\| x \| = \| z^{-1}(zxy)y^{-1} \| \leqq \| z^{-1} \| \| zxy \| \| y^{-1} \|$. Therefore

$$\| zxy \| \geqq \| z^{-1} \|^{-1} \| x \| \| y^{-1} \|^{-1}$$

and

$$\| zxy \|^m |f(zxyH)| \leqq c_1 |\pi_0(H)|^{-q} \nu_{m_1}(f) \| z^{-1} \|^s \| y^{-1} \|^s \| x \|^{-s}$$

It is obvious that $|\pi_0(H)|^{-q} \| z^{-1} \|^s \| y^{-1} \|^s$ remains bounded on the compact set $\Omega$, and so our statement about uniform convergence follows from the fact that $\int \| x \|^{-s} \, dx < \infty$.

From now on we shall make use of the notation and terminology of $[2(\mathrm{h})]$ without further comment. Let $\mathcal{S}$ be the set of all $f \in C^\infty(\mathfrak{g}_0)$ such that $Df \in \mathcal{S}_0$ for every $D \in \partial(S(\mathfrak{g}))$. Let $\mathfrak{B}$ be the universal enveloping algebra of $\mathfrak{g}$, and define $\Gamma_X$ $(X \in \mathfrak{g}_0)$ as in $[2(\mathrm{h}), \S 3]$. Then we have seen (Lemma 3 of $[2(\mathrm{h})]$) that

$$g(x; b : X; \partial(\xi)) = g(x : X; \partial(\Gamma_X(b \mathbf{X} \xi))),$$

where $g \in C^\infty(\mathfrak{g}_0)$, $b \in \mathfrak{B}$, $\xi \in S(\mathfrak{g})$, $x \in G$ and $X \in \mathfrak{g}_0$. Suppose $b$ and $\xi$ are fixed. Then it is clear, from the definition of $\Gamma_X$, that we can select a finite number of linearly independent elements $\xi_1, \cdots, \xi_r \in S(\mathfrak{g})$ and also $p_1, \cdots, p_r \in S(\mathfrak{g})$ such that

$$\Gamma_X(b \mathbf{X} \xi) = p_1(X)\xi_1 + \cdots + p_r(X)\xi_r \qquad (X \in \mathfrak{g}_0).$$

Therefore

$$g(x; b : X; \partial(\xi)) = \sum_{1 \leqq i \leqq r} p_i(X) g(x : X; \partial(\xi_i)).$$

On the other hand, if $\eta \in S(\mathfrak{g})$,

$$g(x : X; \partial(\eta)) = g(xX; \partial(\eta^x)).$$

Therefore

$$g(x:X;\partial(\xi_i)) = g(xX;\partial(\xi_i^x)) \qquad (1 \le i \le r),$$

and again we can select a finite number of linearly independent elements $\eta_j$ $(1 \le j \le s)$ in $S(\mathfrak{g})$ such that

$$\xi_i^x = \sum_{1 \le j \le s} a_{ij}(x)\eta_j \qquad (1 \le i \le r),$$

where $a_{ij}$ are analytic functions on $G$. Moreover, we know from Corollary 2 to Lemma 6 that $|a_{ij}(x)| \le a\,\|x\|^m$ $(x \in G)$, where $m$ is a suitable non-negative integer and $a$ a positive constant. Then

$$|g(x;b:X;\partial(\xi))| \le a\,\|x\|^m \sum_{i,j} |p_i(X)|\,|g(xX;\partial(\eta_j))|.$$

Now suppose $g \in \mathscr{S}$. Then $\partial(\eta_j)g \in \mathscr{S}_0$, and it follows from Lemma 7, that

$$\int \|xy\|^m\,|g(xyH;\partial(\eta_j))|\,dx$$

converges uniformly with respect to $(y,H)$ on every compact subset $\Omega$ of $G \times \mathfrak{h}_0''$. Thus we have obtained the following result.

LEMMA 8. *Suppose $f \in \mathscr{S}$, $b \in \mathfrak{B}$ and $\xi \in S(\mathfrak{g})$. Then if $\Omega$ is a compact subset of $G \times \mathfrak{h}_0''$, the integral*

$$\int_G |f(xy;b:H;\partial(\xi))|\,dx$$

*converges uniformly with respect to $(y,H) \in \Omega$.*

Put $\mathfrak{B}' = \mathfrak{B}\mathfrak{g}$.

COROLLARY 1. *The notation being as above,*

$$\int_G f(x;b:H;\partial(\xi))\,dx = 0$$

*if $b \in \mathfrak{B}'$.*

For a fixed $H \in \mathfrak{h}_0''$, put

$$F(y) = \int_G f(xy:H;\partial(\xi))\,dx \qquad (y \in G).$$

Then it follows from Lemma 8 that $F$ is of class $C^\infty$ on $G$ and

$$F(y;b) = \int_G f(xy;b:H;\partial(\xi))\,dx$$

for $b \in \mathfrak{B}$. But since the above integral for $F(y)$ converges absolutely, it follows from the right invariance of the Haar measure of $G$, that $F(y)$ is actually independent of $y$. Therefore $F(y;b) = 0$ if $b \in \mathfrak{B}'$, and our assertion follows by putting $y = 1$.

COROLLARY 2. *Put* $\phi_f'(H) = \int_G f(xH)dx$ *for* $f \in \mathfrak{S}$ *and* $H \in \mathfrak{h}_0''$. *Then* $\phi_f'$ *is of class* $C^\infty$ *on* $\mathfrak{h}_0''$ *and if* $v \in S(\mathfrak{h})$,

$$\phi_f'(H;\partial(v)) = \int_G f(x:H;\partial(v))dx \qquad (H \in \mathfrak{h}_0'').$$

*This is an immediate consequence of the uniform convergence of*

$$\int_G |f(x:H;\partial(v))|\,dx \qquad (v \in S(\mathfrak{h}))$$

*with respect to* $H$ *on each compact subset of* $\mathfrak{h}_0''$.

LEMMA 9. *For given* $b \in \mathfrak{B}$ *and* $\xi \in S(\mathfrak{g})$, *we can select integers* $q, M \geqq 0$ *and a finite number of elements* $d_1, \cdots, d_r \in \partial(S(\mathfrak{g}))$ *such that*

$$|\pi_0(H)|^q(1 + \|H\|)^{q_1} \int |f(x;b:H;\partial(\xi))|\,dx \leqq \sum_{1 \leqq i \leqq r} \nu_{M+q_1}(d_i f) \quad (H \in \mathfrak{h}_0'')$$

*for* $f \in \mathfrak{S}$ *and any integer* $q_1 \geqq 0$.

Let us keep to the notation of the proof of Lemma 8. Select $q$, $m_1$ and $c$ corresponding to Lemma 7. Since $p_i$ are polynomials, we can select an integer $q_2 \geqq 0$ and a positive number $c_1$ such that

$$|p_i(X)| \leqq c_1(1 + \|X\|)^{q_2} \qquad (1 \leqq i \leqq r)$$

for all $X \in \mathfrak{g}$. Then if $q_1 \geqq 0$, $f \in \mathfrak{S}$ and $H \in \mathfrak{h}_0''$,

$$(1 + \|H\|)^{q_1}|\pi_0(H)|^q \int |f(x;b:H;\partial(\xi))|\,dx$$

$$\leqq |\pi_0(H)|^q(1 + \|H\|)^{q_1+q_2} ac_1 r \sum_j \int \|x\|^m |f(xH;\partial(\eta_j))|\,dx$$

$$\leqq acc_1 r \sum_j \nu_{m_1+q_1+q_2}(\partial(\eta_j)f).$$

The assertion of the lemma is now obvious.

Define $\phi_f'$ $(f \in \mathfrak{S})$ as in Corollary 2 to Lemma 8.

COROLLARY. *For each* $d \in \mathfrak{D}(\mathfrak{h})$, *there exist integers* $q, m \geqq 0$ *and a finite number of elements* $d_1, \cdots, d_r \in \partial(S(\mathfrak{g}))$ *such that*

$$|\pi_0(H)|^q(1 + \|H\|)^{q_1}|\phi_f'(H;d)| \leqq \sum_{1 \leqq i \leqq r} \nu_{m+q_1}(d_i f) \qquad (H \in \mathfrak{h}_0'')$$

*for* $f \in \mathfrak{S}$ *and any integer* $q_1 \geqq 0$.

It is obviously enough to prove this when $d$ is of the form $u\partial(v)$ $(u, v \in S(\mathfrak{g}))$. Select an integer $q_2 \geq 0$ and a positive number $c$ such that $|u(H)| \leq c(1 + \|H\|)^{q_2}$ $(H \in \mathfrak{h})$. Then

$$|\phi_f'(H; d)| \leq c(1 + \|H\|)^{q_2} \int |f(x : H; \partial(v))| \, dx$$

from Corollary 2 to Lemma 8, and the desired result now follows from Lemma 9.

We recall that $I(\mathfrak{g})$ is the algebra of those polynomials in $S(\mathfrak{g})$ which are invariant under $G$, and for any $p \in S(\mathfrak{g})$, $\bar{p}$ denotes the restriction of $p$ on $\mathfrak{h}$. Furthermore, $\pi = \prod_{\alpha \in P} \alpha$.

THEOREM 1. *For any $f \in \mathscr{E}$, put*

$$\phi_f(H) = \pi(H) \int_G f(xH) \, dx \qquad (H \in \mathfrak{h}_0'').$$

*Then $\phi_f$ is a function of class $C^\infty$ on $\mathfrak{h}_0''$ and*

$$\phi_{\partial(\xi)f} = \partial(\bar{\xi}) \phi_f \qquad (\xi \in I(\mathfrak{g})).$$

The first statement follows from Corollary 2 to Lemma 8. Let $\mathfrak{h}_0'$ be the set of those points $H \in \mathfrak{h}_0$ where $\pi(H) \neq 0$. Then $\mathfrak{h}_0'$ is a dense subset of $\mathfrak{h}_0''$, and so for a fixed $\xi \in I(\mathfrak{g})$, it would be enough to prove that $\phi_{\partial(\xi)f} - \partial(\bar{\xi})\phi_f$ is zero on $\mathfrak{h}_0'$. Fix an element $H \in \mathfrak{h}_0'$, and put $d = \partial(\bar{\xi}) \circ \pi$. Then $d \in \mathfrak{D}(\mathfrak{h})$, and we know from Lemma 8 of [2(h)] that

$$f(xH; \partial(\xi)) = \pi(H)^{-1}f(x : H; d_H) + \sum_{1 \leq i \leq r} f(x; b_i : H; \partial(v_i)) \qquad (x \in G),$$

where $b_i \in \mathfrak{B}'$, $v_i \in S(\mathfrak{h})$ $(1 \leq i \leq r)$ and $d_H$ is the local expression of $d$ at $H$. Therefore it follows from Lemma 8 and its corollaries that

$$\phi_{\partial(\xi)f}(H) = \pi(H) \int f(xH; \partial(\xi)) \, dx = \int f(x : H; d_H) \, dx$$
$$= \phi_f'(H; d) = \phi_f(H; \partial(\bar{\xi})).$$

This proves the theorem.

For any integer $m \geq 0$ and $d \in \partial(S(\mathfrak{g}))$, consider the function $v_{m,d}$: $f \to v_m(df)$ $(f \in \mathscr{E})$ on $\mathscr{E}$. Obviously, it is a seminorm, and we can define a topology on $\mathscr{E}$ by means of the collection of all such seminorms $v_{m,d}$. In this way $\mathscr{E}$ becomes a locally convex space. Since $\mathfrak{h}_0''$ is an open subset of $\mathfrak{h}_0$, the space $\mathscr{E}(\mathfrak{h}_0'')$ can be defined as usual (see [2(h), §2]).

THEOREM 2. *For any $f$ in $\mathscr{E}$, $\phi_f$ lies in $\mathscr{E}(\mathfrak{h}_0'')$ and $f \to \phi_f$ is a continuous mapping of $\mathscr{E}$ into $\mathscr{E}(\mathfrak{h}_0'')$.*

Let $U$ be a subset of $\mathfrak{h}_0''$. We shall say that $U$ has property (A) if for every $v \in S(\mathfrak{h})$, we can select an integer $M \geq 0$, a real constant $c \geq 1$ and a finite number of elements $d_1, \cdots, d_r \in \partial(S(\mathfrak{g}))$ such that

$$(A) \qquad (1 + \| H \|)^{q'} | \phi_f(H; \partial(v)) | \leq c^{q'} \sum_{1 \leq j \leq r} \nu_{M+q'}(d_j f)$$

for $H \in U$, $f \in \mathcal{S}$ and all integers $q' \geq 0$. Clearly the union of a finite number of sets with property (A) again has the same property. Similarly (A) holds for the closure of $U$ in $\mathfrak{h}_0''$ whenever it holds for $U$. Let $w$ be the order of the Weyl group $W$ (of $\mathfrak{g}$ with respect to $\mathfrak{h}$). From Lemma 11 of $[2(h)]$, we can select homogeneous elements $v_1, \cdots, v_w \in S(\mathfrak{h})$, such that $S(\mathfrak{h}) = \sum_{1 \leq i \leq w} I(\mathfrak{h}) v_i$. For any $v \in S(\mathfrak{h})$, let $\gamma_i(v)$ denote the unique elements in $I(\mathfrak{g})$ such that $v = \sum_{1 \leq i \leq w} \overline{\gamma_i(v)} v_i$ (see $[2(h),$ Lemmas 9 and 11$]$), and put $D_i(v) = \partial(\gamma_i(v))$. Then if $f \in \mathcal{S}$, we know from Theorem 1 that

$$\partial(v)\phi_f = \sum_{1 \leq i \leq w} \partial(v_i)\phi_{D_i(v)f},$$

and therefore

$$| \phi_f(H; \partial(v)) | \leq \sum_{1 \leq i \leq w} | \phi_{D_i(v)f}(H; \partial(v_i)) | \qquad\qquad (H \in \mathfrak{h}_0'').$$

This shows that in order to prove that $U$ has property (A), it is enough to check this property for $v = v_1, v_2, \cdots, v_w$ only. Finally, it is obvious that Theorem 2 would be proved if we can show that $\mathfrak{h}_0''$ has property (A).

Since $\mathfrak{h}_0'$ is a dense subset of $\mathfrak{h}_0''$ and it has only a finite number of connected components, it is enough to prove that each connected component of $\mathfrak{h}_0'$ has property (A). Let $\mathfrak{h}_+$ be a connected component of $\mathfrak{h}_0'$ and $P'$ the set of those roots $\alpha$ (of $\mathfrak{g}$ with respect to $\mathfrak{h}$) for which $-(-1)^{\frac{1}{2}}\alpha$ takes only positive values on $\mathfrak{h}_+$. Then we can select a fundamental system $(\alpha_1, \cdots, \alpha_l)$ of roots in $P'$ (see $[2(f),$ p. 749$]$). Choose a base $H_1, \cdots, H_l$ for $\mathfrak{h}_0$ over $R$ such that $\alpha_i(H_j) = (-1)^{\frac{1}{2}}\delta_{ij}$ $(1 \leq i, j \leq l)$, and let $(t_1, \cdots, t_l)$ denote the system of Cartesian coordinates in $\mathfrak{h}_0$ corresponding to this base. By a monomial we shall mean a function $T$ on $\mathfrak{h}_0$ of the form $T = t_1^{q_1} t_2^{q_2} \cdots t_l^{q_l}$, where $q_1, \cdots, q_l$ are nonnegative integers. $q_i$ is called the exponent of $t_i$ in $T$. In particular the monomial whose exponents are all zero is 1. We introduce a lexicographic order in the set of all monomials as follows. Suppose $T = t_1^{q_1} \cdots t_l^{q_l}$, $T' = t_1^{q_1'} \cdots t_l^{q_l'}$ are two distinct monomials. Then we say that $T \prec T'$ (or $T' \succ T$) if $q_i < q_i'$ where $i$ is the least index $(1 \leq i \leq l)$ such that $q_i \neq q_i'$. It is obvious that every nonempty set of monomials has a lowest element. Furthermore, observe that every monomial takes only positive values on $\mathfrak{h}_+$. We say that a monomial $T$ has property

(A′), if we can choose an integer $M \geq 0$, a real number $c \geq 1$ and a finite set of elements $d_1, \cdots, d_r \in \partial(S(\mathfrak{g}))$ such that

$$\text{(A′)} \qquad T(H)(1 + \| H \|)^{q'} \, | \, \phi_f(H; \partial(v_i)) | \leq c^q \sum_{1 \leq j \leq r} \nu_{M+q'}(d_j f) \qquad (1 \leq i \leq w)$$

for all $f \in \mathscr{S}$, $H \in \mathfrak{h}_+$ and $q' \geq 0$. In order to prove that $\mathfrak{h}_+$ has property (A), it is enough to show that the monomial 1 has property (A′).

First of all we claim that monomials with property (A′) actually do exist. If $\alpha \in P'$, we know that $\alpha = \sum_{1 \leq i \leq l} m_i \alpha_i$, where the $m_i$ are nonnegative integers. Select an index $i$ such that $m_i \geq 1$. Then it is clear that $| \alpha(H) | \geq t_i(H)$ $(H \in \mathfrak{h}_+)$. This shows that there exists a monomial $T_0$ such that $| \pi_0(H) | \geq T_0(H)$ for $H \in \mathfrak{h}_+$. Moreover, it follows easily from the Corollary to Lemma 9 that $T_0{}^q$ has property (A′) for a suitable integer $q \geq 0$.

Now let $T$ be the lowest monomial with property (A′). We have to prove that $T = 1$. So let us assume that $T \neq 1$, and let $m$ be the least index $(1 \leq m \leq l)$ such that the exponent $q$ of $t_m$ in $T$ is positive. Put $t = t_m$. Then $T = t^q T_1$, where $T_1$ is a monomial in which the exponents of $t_i$ are zero for $1 \leq i \leq m$. Write $\partial' = \partial(H_m) = \partial/\partial t$, and let

$$H_m v_i = \sum_{1 \leq i \leq w} p_{ij} v_j \qquad (1 \leq i \leq w).$$

where $p_{ij} \in I(\mathfrak{g})$. Fix an element $f \in \mathscr{S}$. Then

$$\partial'(\partial(v_i)\phi_f) = \sum_j \partial(v_j)\phi_{\partial(p_{ij})f}.$$

Set $\phi_i = T_1 \partial(v_i)\phi_f$ and $\psi_i = T_1 \sum_{1 \leq i \leq w} \partial(v_j)\phi_{\partial(p_{ij})f}$ $(1 \leq i \leq w)$. Then $\partial'\phi_i = \psi_i$. Moreover, if $M, c, d_1, \cdots, d_r$ are selected in accordance with (A′) for $T = t^q T_1$, it is clear that

$$t^q \mu^{q'} \, | \, \psi_i \, | \leq c^{q'} a_{q'}, \quad t^q \mu^{q'} \, | \, \phi_i \, | \leq c^{q'} b_{q'} \qquad (1 \leq i \leq w, q' \geq 0)$$

on $\mathfrak{h}_+$. Here $\mu(H) = 1 + \| H \|$ $(H \in \mathfrak{h}_0)$ and

$$a_{q'} = \sum_{1 \leq j \leq r} \sum_{1 \leq i, k \leq w} \nu_{M+q'}(d_j \partial(p_{ik})f),$$

$$b_{q'} = \sum_{1 \leq j \leq r} \nu_{M+q'}(d_j f).$$

Let $\tau_+, \tau_-$ denote the subsets of $\mathfrak{h}_+$ defined by the conditions $t \geq 1$ and $0 < t \leq 1$, respectively. For any $H \in \mathfrak{h}_+$, let $H_0$ denote the unique point of intersection of the line $H + sH_m$ $(s \in R)$ with the "hyperplane" $\sigma = \tau_+ \cap \tau_-$. Put $\bar\mu(H) = \mu(H_0)$ $(H \in \mathfrak{h}_+)$. Then if $H \in \tau_-$,

$$\mu(H) \leq 1 + \| H_0 \| + \| H_m \| \leq (1 + \| H_0 \|)c_1 = c_1 \bar\mu(H),$$

2

where $c_1 = 1 + \| H_m \|$. Similarly $\bar{\mu}(H) \leqq c_1 \mu(H)$. Now fix an integer $q' \geqq 0$. Then

$$| \psi_i | \leqq c^{q'} a_{q'} \mu^{-q'} t^{-q} \leqq (cc_1)^{q'} a_{q'} (\bar{\mu})^{-q'} t^{-q}$$

on $\tau_-$. Suppose $H$ is a point in $\tau_-$ and $H_0$ the corresponding point on $\sigma$. Then by integrating the differential equation $\partial' \phi_i = \psi_i$ along the line segment $t_k = t_k(H)$ $(k \neq m, 1 \leqq k \leqq l)$, $t(H) \leqq t = t_m \leqq 1$, we find that

$$| \phi_i(H_0) - \phi_i(H) | = \Big| \int_{t(H)}^1 \psi_i(H_0 - (1-s)H_m) ds \Big| \leqq (cc_1)^{q'} a_{q'} (\bar{\mu}(H))^{-q'} \int_{t(H)}^1 s^{-q} ds$$

since $\bar{\mu}$ is constant on this segment. Now put

$$\eta(H) = \begin{cases} \{t(H)^{-q+1} - 1\}/(q-1) & \text{if } q \geqq 2, \\ |\log t(H)| & \text{if } q = 1, \end{cases}$$

for $H \in \tau_-$. Then

$$| \phi_i(H) | \leqq | \phi_i(H_0) | + (cc_1)^{q'} a_{q'} (\bar{\mu}(H))^{-q'} \eta(H) \qquad (H \in \tau_-).$$

On the other hand, since $t(H_0) = 1$,

$$\mu(H_0)^{q'} | \phi_i(H_0) | \leqq c^{q'} b_{q'}.$$

Moreover, $\mu(H_0) = \bar{\mu}(H)$, and therefore

$$| \phi_i | \leqq \{c^{q'} b_{q'} + (cc_1)^{q'} a_{q'} \eta\} (\bar{\mu})^{-q'}$$

$$\leqq \{(cc_1)^{q'} b_{q'} + (cc_1^2)^{q'} a_{q'} \eta\} \mu^{-q'}$$

on $\tau_-$. Hence

$$t^{q-1} \mu^{q'} | \phi_i | \leqq (cc_1^2)^{q'} \{b_{q'} + a_{q'} \eta t^{q-1}\}$$

on $\tau_-$. On the other hand,

$$t^{q-1} \mu^{q'} | \phi_i | \leqq t^q \mu^{q'} | \phi_i | \leqq c^{q'} b_{q'} \leqq (cc_1^2)^{q'} b_{q'}$$

on $\tau_+$. Now first suppose $q \geqq 2$. Then $\eta t^{q-1} \leqq 1$ on $\tau_-$. Hence

$$t^{q-1} \mu^{q'} | \phi_i | \leqq (cc_1^2)^{q'} \{b_{q'} + a_{q'}\} \qquad (1 \leqq i \leqq w)$$

on $\mathfrak{h}_+$. Then if we put $T_2 = t^{q-1} T_1$, the above inequality can be written as follows:

$$T_2 \mu^{q'} | \partial(v_i)\phi_f | \leqq (cc_1^2)^{q'} \sum_{1 \leqq j \leqq r} \{v_{M+q'}(d_j f) + \sum_{1 \leqq h, k \leqq w} v_{M+q'}(d_j \partial(p_{hk})f)\}$$

$(1 \leqq i \leqq w)$. Therefore since $f$ and $q'$ are arbitrary, it is clear that $T_2$ has property (A'). However, $T_2$ is lower than $T$, and so we get a contradiction with the definition of $T$. Hence we conclude that the case $q \geqq 2$ is impossible, and so $q = 1$. But then

$$\mu^{q'} | \phi_i | \leqq (cc_1^2)^{q'} \{b_{q'} + a_{q'} | \log t |\}$$

on $\tau_-$. Put $c_2 = cc_1{}^2$, $r' = r(w^2 + 1)$, and let $d_j'$ $(1 \leq j \leq r')$ denote the $r'$ differential operators $d_j \partial(p_{hk})$ and $d_j$ $(1 \leq j \leq r, 1 \leq h, k \leq w)$. Then since $f$ and $q'$ were arbitrary, the above result shows that

$$T_1 \mu^{q'} \,|\, \partial(v_i)\phi_g \,| \leq c_2{}^{q'} \{ \sum_{1 \leq j \leq r'} v_{M+q'}(d_j'g) \}(1 + |\log t|)$$

on $\tau_-$ for any $g \in \mathscr{S}$ and $q' \geq 0$. Applying this result to $\partial(p_{hk})f$, we obtain

$$\mu^{q'} \,|\, \psi_i \,| \leq c_2{}^{q'} a'_{q'}(1 + |\log t|) \qquad\qquad (1 \leq i \leq w, q' \geq 0)$$

on $\tau_-$, where

$$a'_{q'} = \sum_{1 \leq j \leq r'} \sum_{1 \leq h, k \leq w} v_{M+q'}(d_j'\partial(p_{hk})f).$$

Again, we find by integration that if $H \in \tau_-$,

$$|\phi_i(H)| \leq |\phi_i(H_0)| + \int_{t(H)}^{1} |\psi_i(H_0 - (1-s)H_m)|\, ds$$
$$\leq \{c^{q'}b_{q'} + 2(c_1 c_2)^{q'} a'_{q'}\}(\mu(H))^{-q'},$$

since $\displaystyle\int_0^1 (1 + |\log s|)\, ds = 2$. Therefore

$$\mu^{q'} \,|\, \phi_i \,| \leq c_3{}^{q'}(b_{q'} + a'_{q'})$$

on $\tau_-$, where $c_3 = 2c_1{}^2 c_2$. On the other hand, since $q = 1$,

$$\mu^{q'} \,|\, \phi_i \,| \leq t\mu^{q'} \,|\, \phi_i \,| \leq c^{q'}b_{q'} \leq c_3{}^{q'}b_{q'}$$

on $\tau_+$. Therefore

$$\mu^{q'} T_1 \,|\, \partial(v_i)\phi_f \,| \leq c_3{}^{q'} \{ \sum_{1 \leq j \leq r'} \sum_{1 \leq h, k \leq w} v_{M+q'}(d_j'\partial(p_{hk})f) + \sum_{1 \leq j \leq r} v_{M+q'}(d_jf) \}$$

on $\mathfrak{h}_+$ for $q' \geq 0$ and $1 \leq i \leq w$. Again, since $f$ is an arbitrary element of $\mathscr{S}$ and $c_3$ is independent of $f$ and $q'$, it is clear that the monomial $T_1$ has property (A$'$), and so we get a contradiction as before. This completes the proof of Theorem 2.

**5. The general case.** We shall now discard the assumption that $\mathfrak{h}_0 \subset \mathfrak{k}_0$ and try to extend Theorems 1 and 2 of Section 4 to the general case. But first we need some preliminary work.

As before let $\bar\theta$ denote the conjugation of $\mathfrak{g}$ with respect to its compact real form $\mathfrak{u} = \mathfrak{k}_0 + (-1)^{\frac{1}{2}}\mathfrak{p}_0$.

LEMMA 10. *Let* $\mathfrak{m}$ *be a subalgebra of* $\mathfrak{g}$ *such that* $\bar\theta(\mathfrak{m}) = \mathfrak{m}$. *Then* $\mathfrak{m}$ *is reductive in* $\mathfrak{g}$ [4].

It follows from our hypothesis that $\mathfrak{m}$ is the complexification of the subalgebra $\mathfrak{m}_0 = \mathfrak{m} \cap \mathfrak{u}$. But $\mathfrak{u}$ being compact, $\mathfrak{m}_0$ is reductive in $\mathfrak{u}$ and so $\mathfrak{m}$ is reductive in $\mathfrak{g}$.

Let $\mathfrak{h}_0$ be a Cartan subalgebra of $\mathfrak{g}_0$ such that $\theta(\mathfrak{h}_0) = \mathfrak{h}_0$. For any (real-valued) linear function $\lambda$ on $\mathfrak{h}_{\mathfrak{p}_0} = \mathfrak{h}_0 \cap \mathfrak{p}_0$, let $\mathfrak{g}_{0\lambda}$ denote the space of those $X \in \mathfrak{g}_0$ for which $[H, X] = \lambda(H)X$ for all $H \in \mathfrak{h}_{\mathfrak{p}_0}$. Let $\mathfrak{m}_0$ be the orthogonal complement[6] of $\mathfrak{h}_{\mathfrak{p}_0}$ in $\mathfrak{g}_{00}$. Introduce an order [2(f), §2] in the space of linear functions on $\mathfrak{h}_{\mathfrak{p}_0}$ and put $\mathfrak{n}_0 = \sum_{\lambda>0} \mathfrak{g}_{0\lambda}$. Obviously, $\mathfrak{m}_0$ and $\mathfrak{n}_0$ are subalgebras of $\mathfrak{g}_0$ and $\mathfrak{n}_0$ is nilpotent. Moreover, it follows from its definition that $\theta(\mathfrak{m}_0) = \mathfrak{m}_0$. Therefore, from Lemma 10, $\mathfrak{m}_0$ is reductive in $\mathfrak{g}_0$.

Let $G$ be a connected Lie group with the Lie algebra $\mathfrak{g}_0$, and let $K$, $N$ and $A_\mathfrak{p}$ be the analytic subgroups of $G$ corresponding to $\mathfrak{k}_0$, $\mathfrak{n}_0$ and $\mathfrak{h}_{\mathfrak{p}_0}$, respectively. Put $\mathfrak{m}_{\mathfrak{p}_0} = \mathfrak{m}_0 \cap \mathfrak{p}_0$.

LEMMA 11. *The mapping* $(k, X, a, n) \to k(\exp X)an$ $(k \in K, X \in \mathfrak{m}_{\mathfrak{p}_0},$ $a \in A_\mathfrak{p}, n \in N)$ *is a one-one regular analytic mapping of* $K \times \mathfrak{m}_{\mathfrak{p}_0} \times A_\mathfrak{p} \times N$ *onto* $G$.

Let $M$ be the analytic subgroup of $G$ corresponding to $\mathfrak{m}_0$. Since $\theta(\mathfrak{m}_0) = \mathfrak{m}_0$, it follows that $\mathfrak{m}_0 = \mathfrak{m}_0 \cap \mathfrak{k}_0 + \mathfrak{m}_{\mathfrak{p}_0}$. Moreover, $[\mathfrak{m}_0, \mathfrak{n}_0] \subset \mathfrak{n}_0$, and therefore $\mathfrak{s}_0 = \mathfrak{m}_0 + \mathfrak{h}_{\mathfrak{p}_0} + \mathfrak{n}_0$ is a Lie algebra and $\mathfrak{n}_0$ is an ideal in $\mathfrak{s}_0$. Hence $S = MA_\mathfrak{p}N$ is a group and $N$ is normal in $S$. Let $M_\mathfrak{p}$ denote the subset of $M$ consisting of elements of the form $\exp X$ $(X \in \mathfrak{m}_{\mathfrak{p}_0})$. First we shall show that $G = KM_\mathfrak{p}A_\mathfrak{p}N$. Extend $\mathfrak{h}_{\mathfrak{p}_0}$ to a maximal abelian subspace $\mathfrak{a}_{\mathfrak{p}_0}$ of $\mathfrak{p}_0$. Let $\bar{\mathfrak{F}}$ and $\mathfrak{F}$ be the spaces of (real-valued) linear functions on $\mathfrak{a}_{\mathfrak{p}_0}$ and $\mathfrak{h}_{\mathfrak{p}_0}$, respectively. Select an order on $\bar{\mathfrak{F}}$ compatible with the one already chosen on $\mathfrak{F}$ (see Section 2). Let $\bar{\mathfrak{g}}_{0\lambda}$ $(\lambda \in \bar{\mathfrak{F}})$ denote the set of those $X \in \mathfrak{g}_0$ for which $[H, X] = \lambda(H)X$ for all $H \in \mathfrak{a}_{\mathfrak{p}_0}$. Then if $\bar{\mathfrak{n}}_0 = \sum_{\lambda>0} \bar{\mathfrak{g}}_{0\lambda}$, it is clear that $\bar{\mathfrak{n}}_0$ is a nilpotent subalgebra of $\mathfrak{g}_0$ containing $\mathfrak{n}_0$. Let $A_+$ and $\bar{N}$ be the analytic subgroups of $G$ corresponding to $\mathfrak{a}_{\mathfrak{p}_0}$ and $\bar{\mathfrak{n}}_0$, respectively. Then Iwasawa [3] has shown that $G = KA_+\bar{N}$. Moreover, in view of the above compatibility, it is clear that $\mathfrak{a}_{\mathfrak{p}_0} + \bar{\mathfrak{n}}_0 \subset \mathfrak{g}_{00} + \mathfrak{n}_0 = \mathfrak{s}_0$. Hence $A_+\bar{N} \subset S$. Therefore $G = KA_+\bar{N} = KMA_\mathfrak{p}N$. Now we claim that $M = M_K M_\mathfrak{p}$, where $M_K = M \cap K$. This is seen as follows. Since $\mathfrak{h}_0$ is maximal abelian in $\mathfrak{g}_0$, it follows that $\mathfrak{h}_{\mathfrak{k}_0} = \mathfrak{h}_0 \cap \mathfrak{k}_0$ is maximal abelian in $\mathfrak{m}_0$. Therefore the center of $\mathfrak{m}_0$ is contained in $\mathfrak{h}_{\mathfrak{k}_0}$, and, since $\mathfrak{m}_0$ is reductive, $\mathfrak{m}_0' = [\mathfrak{m}_0, \mathfrak{m}_0]$ is the orthogonal complement of this center in $\mathfrak{m}_0$. This shows that $\mathfrak{m}_0' \supset \mathfrak{m}_{\mathfrak{p}_0}$.

---

[6] Whenever we speak of orthogonality or related concepts, they are to be understood with respect to the positive-definite Hermitian form $\| X \|^2$ $(X \in \mathfrak{g})$ on $\mathfrak{g}$.

Moreover $\mathfrak{m}_0$, and therefore also $\mathfrak{m}_0'$, is invariant under $\theta$, so that $\mathfrak{m}_0'$ $= \mathfrak{m}_0' \cap \mathfrak{k}_0 + \mathfrak{m}_{\mathfrak{p}_0}$. This means that $\mathfrak{m}_0' \cap \mathfrak{k}_0 + (-1)^{\frac{1}{2}}\mathfrak{m}_{\mathfrak{p}_0}$ is a semisimple subalgebra of $\mathfrak{u}$, and therefore it is compact. Then if $M'$ and $M_K'$, respectively, are the analytic subgroups of $G$ corresponding to $\mathfrak{m}_0'$ and $\mathfrak{m}_0' \cap \mathfrak{k}_0$, we can conclude (see Mostow [5]) that $M' = M_K'M_{\mathfrak{p}}$. Now let $c_0$ be the center of $\mathfrak{m}_0$ and $D$ the corresponding analytic subgroup of $M$. Then $c_0 \subset \mathfrak{m}_0 \cap \mathfrak{k}_0$, and therefore $D \subset M \cap K$ and $M = DM'$. This proves that $M = M_K M_{\mathfrak{p}}$, and hence $G = KM_{\mathfrak{p}}A_{\mathfrak{p}}N$.

Now in order to prove that our mapping is one-one, we shall first show that $M \cap A_{\mathfrak{p}}N = \{1\}$. Suppose $m = a \exp X$ $(m \in M, a \in A_{\mathfrak{p}}, X \in \mathfrak{n}_0)$. Then if [7] $b \in A_{\mathfrak{p}}$,

$$m = bmb^{-1} = a \exp \mathrm{Ad}(b)X.$$

Since the exponential mapping is well known to be one-one on $\bar{\mathfrak{n}}_0$, it follows that $X = \mathrm{Ad}(b)X$ for all $b \in A_{\mathfrak{p}}$. This implies that $[H, X] = 0$ $(H \in \mathfrak{h}_{\mathfrak{p}_0})$, and hence $X \in \mathfrak{n}_0 \cap (\mathfrak{h}_{\mathfrak{p}_0} + \mathfrak{m}_0) = \{0\}$. Therefore $m = a$. Let $\exp \mathfrak{p}_0$ denote the set of all elements of the form $\exp Y$ $(Y \in \mathfrak{p}_0)$. Then every element in $G$ can be written *uniquely* in the form $kp$ $(k \in K, p \in \exp \mathfrak{p}_0$, see Mostow [5]). On the other hand $M = M_K M_{\mathfrak{p}}$, and therefore if $m = kp$, we can conclude that $k \in M_K$, $p \in M_{\mathfrak{p}}$. Moreover, $a \in A_{\mathfrak{p}} \subset \exp \mathfrak{p}_0$. Therefore $k = 1$ and $p \in M_{\mathfrak{p}} \cap A_{\mathfrak{p}}$. But again, since the exponential mapping is one-one on $\mathfrak{p}_0$, $p = \exp Y$, where $Y \in \mathfrak{m}_{\mathfrak{p}_0} \cap \mathfrak{h}_{\mathfrak{p}_0} = \{0\}$. This proves that $m = 1$. Now suppose $k \in K \cap S$. Then $k = man$ $(m \in M, a \in A_{\mathfrak{p}}, n \in N)$, and $\mathrm{Ad}(k)$ leaves $\mathfrak{n}_0$ invariant. But $\mathrm{Ad}(k) = \theta \mathrm{Ad}(k)\theta^{-1}$, and therefore $\mathrm{Ad}(k)$ also leaves $\theta(\mathfrak{n}_0) = \sum_{\mu < 0} \mathfrak{g}_{0\mu}$ invariant. This implies that $\mathrm{Ad}(n) = \mathrm{Ad}(a^{-1}m^{-1}k)$ leaves $\theta(\mathfrak{n}_0)$ invariant. However, this is possible only if $n = 1$ as follows from Lemma 13 below. Hence $k = ma$. Now let $m = m_1 p$, where $m_1 \in M_K, p \in M_{\mathfrak{p}}$. Since $\mathfrak{h}_{\mathfrak{p}_0}$ and $\mathfrak{m}_0$ commute, $m_1^{-1}k = pa \in K \cap (\exp \mathfrak{p}_0) = \{1\}$, and so $a = p^{-1}$. But $\mathfrak{m}_{\mathfrak{p}_0} \cap \mathfrak{h}_{\mathfrak{p}_0} = \{0\}$ and so $M_{\mathfrak{p}} \cap A_{\mathfrak{p}} = \{1\}$. Therefore $a = p^{-1} = 1$, and this proves that $k = m_1 \in M_K$. Hence $K \cap S = M_K$. Let us now assume that $k_1 m_1 a_1 n_1 = k_2 m_2 a_2 n_2$ $(k_i \in K, m_i \in M_{\mathfrak{p}}, a_i \in A_{\mathfrak{p}}, n_i \in N, i = 1, 2)$. Put $m_i a_i n_i = s_i$. Then $k_1^{-1} k_2 \in K \cap S = M_K$, and therefore $k_2 = k_1 m$, where $m \in M_K$. Then $m_1^{-1}(mm_2) \in M \cap A_{\mathfrak{p}}N = \{1\}$. Hence $mm_2 = m_1$. But since both $m_1, m_2 \in M_{\mathfrak{p}} \subset \exp \mathfrak{p}_0$ and $m \in M_K \subset K$, it follows that $m = 1$ and $m_2 = m_1$. This proves that $k_1 = k_2$. Also one knows (see Iwasawa [3]) that the mapping $(a, n) \to an$ $(a \in A_+, n \in \bar{N})$ is one-one. Hence $a_1 = a_2$, $n_1 = n_2$, and this proves that the mapping of Lemma 11 is one-one.

---

[7] $x \to \mathrm{Ad}(x)$ $(x \in G)$ denotes, as usual, the adjoint representation of $G$.

The proof of the regularity is straightforward and we shall only sketch it here. Let $\phi$ denote our mapping. In order to prove that $\phi$ is regular at $(k, X, a, n)$ $(k \in K, X \in \mathfrak{m}_{\mathfrak{p}_0}, a \in A_\mathfrak{p}, n \in N)$, we have to verify that the following linear mapping $D$ of $\mathfrak{g}_0$ is nonsingular.[8]

$$DZ = Z \quad (Z \in \mathfrak{n}_0), \qquad DH = \mathrm{Ad}(n^{-1})H \quad (H \in \mathfrak{h}_{\mathfrak{p}_0}),$$

$$DY = \mathrm{Ad}(n^{-1}a^{-1})\{(1 - \exp(-\operatorname{ad} X))/\operatorname{ad} X\}Y \qquad (Y \in \mathfrak{m}_{\mathfrak{p}_0}),$$

$$DU = \mathrm{Ad}(n^{-1}a^{-1})\exp(-\operatorname{ad} X)U \qquad (U \in \mathfrak{k}_0).$$

Put $D' = \exp(\operatorname{ad} X)\mathrm{Ad}(an)D$. Then since $\mathfrak{g}_0$ is semisimple, it is clear that $\det D = \det D'$. Moreover, if $I$ is the identity mapping of $\mathfrak{g}_0$, $\mathrm{Ad}(n) - I$ is nilpotent, and so the restriction of $\mathrm{Ad}(n)$ on $\mathfrak{n}_0$ has determinant 1. Similarly, since $X \in \mathfrak{m}_{\mathfrak{p}_0} \subset \mathfrak{m}_0'$, the restriction of $\exp(\operatorname{ad} X)$ on $\mathfrak{h}_{\mathfrak{p}_0} + \mathfrak{n}_0$ also has determinant 1. Hence if $D_1$ is the linear mapping of $\mathfrak{g}_0/(\mathfrak{h}_{\mathfrak{p}_0} + \mathfrak{n}_0)$ defined by $D'$, $\det D = \det \mathrm{Ad}_{\mathfrak{n}_0}(a) \det D_1$, where $\mathrm{Ad}_{\mathfrak{n}_0}(a)$ is the restriction of $\mathrm{Ad}(a)$ on $\mathfrak{n}_0$. But $D_1$ leaves every element in $\mathfrak{k}_0 \bmod (\mathfrak{h}_{\mathfrak{p}_0} + \mathfrak{n}_0)$ fixed. Hence it defines a linear mapping $D_2$ of $\mathfrak{g}_0/(\mathfrak{k}_0 + \mathfrak{h}_{\mathfrak{p}_0} + \mathfrak{n}_0)$ and $\det D_1 = \det D_2$. On the other hand, it is clear that

$$\mathfrak{g}_0 = \mathfrak{g}_{00} + \sum_{\lambda > 0} \mathfrak{g}_{0\lambda} + \sum_{\mu < 0} \mathfrak{g}_{0\mu} = \mathfrak{h}_{\mathfrak{p}_0} + \mathfrak{m}_0 + \mathfrak{n}_0 + \theta(\mathfrak{n}_0).$$

Moreover, if $Z \in \mathfrak{n}_0$, then $\theta(Z) = (\theta(Z) + Z) - Z \in \mathfrak{k}_0 + \mathfrak{n}_0$, and so $\mathfrak{n}_0 + \theta(\mathfrak{n}_0) \subset \mathfrak{k}_0 + \mathfrak{n}_0$. Therefore $\mathfrak{g}_0 = \mathfrak{h}_{\mathfrak{p}_0} + \mathfrak{m}_{\mathfrak{p}_0} + \mathfrak{k}_0 + \mathfrak{n}_0$. We claim this sum is direct. For suppose $U \in \mathfrak{k}_0 \cap (\mathfrak{h}_{\mathfrak{p}_0} + \mathfrak{m}_{\mathfrak{p}_0} + \mathfrak{n}_0)$. Then

$$U = \theta(U) \in (\mathfrak{h}_{\mathfrak{p}_0} + \mathfrak{m}_{\mathfrak{p}_0} + \mathfrak{n}_0) \cap (\mathfrak{h}_{\mathfrak{p}_0} + \mathfrak{m}_{\mathfrak{p}_0} + \theta(\mathfrak{n}_0)) = \mathfrak{h}_{\mathfrak{p}_0} + \mathfrak{m}_{\mathfrak{p}_0}.$$

Hence $U \in \mathfrak{k}_0 \cap (\mathfrak{h}_{\mathfrak{r}_0} + \mathfrak{m}_{\mathfrak{p}_0}) \subset \mathfrak{k}_0 \cap \mathfrak{p}_0 = \{0\}$. Therefore we may identify $\mathfrak{g}_0/(\mathfrak{k}_0 + \mathfrak{h}_{\mathfrak{p}_0} + \mathfrak{n}_0)$ with $\mathfrak{m}_{\mathfrak{p}_0}$ in the obvious way. If $Y \in \mathfrak{m}_{\mathfrak{p}_0}$,

$$D'Y = \{(\exp(\operatorname{ad} X) - 1)/\operatorname{ad} X\}Y \equiv (\sinh \operatorname{ad} X/\operatorname{ad} X)Y \bmod \mathfrak{k}_0.$$

Therefore it is clear that

$$\det D = \det(\sinh \operatorname{ad} X/\operatorname{ad} X)_{\mathfrak{m}_{\mathfrak{p}_0}} \det \mathrm{Ad}_{\mathfrak{n}_0}(a)$$

(where the suffix $\mathfrak{m}_{\mathfrak{p}_0}$ denotes restriction on $\mathfrak{m}_{\mathfrak{p}_0}$). Since $\mathrm{Ad}(a)$ and $\sinh \operatorname{ad} X/\operatorname{ad} X$ are positive-definite self-adjoint operators on $\mathfrak{g}$, it follows that $\det D > 0$, and therefore $\phi$ is regular. In fact, the following lemma is an immediate consequence of the above computation.

---

[8] $\{(1 - \exp(-\operatorname{ad} X))/\operatorname{ad} X\}$ and $\sinh \operatorname{ad} X/\operatorname{ad} X$ are defined in the usual way by means of their convergent power series.

LEMMA 12. *Let $dk$, $da$, $dn$ and $dx$ denote the Haar measures on $K$, $A_\mathfrak{p}$, $N$ and $G$, respectively, and $dX$ the Euclidean measure on $\mathfrak{m}_{\mathfrak{p}_0}$. Then it is possible to normalize them in such a way that*

$$dx = \det(\sinh\,\mathrm{ad}\,X/\mathrm{ad}\,X)_{\mathfrak{m}_{\mathfrak{p}_0}} \det \mathrm{Ad}_{\mathfrak{n}_0}(a)\,dk\,da\,dn\,dX,$$

*where $x = k(\exp X)an$ $(k \in K, X \in \mathfrak{m}_{\mathfrak{p}_0}, a \in A_{\mathfrak{p}_0}, n \in N)$.*

The following corollary is now obvious from Lemmas 11 and 12.

COROLLARY. *The mapping $(k, X, n, a) \to k(\exp X)na$ is also a one-one regular mapping of $K \times \mathfrak{m}_{\mathfrak{p}_0} \times N \times A_\mathfrak{p}$ onto $G$, and if $x = k(\exp X)na$,*

$$dx = \det(\sinh\,\mathrm{ad}\,X/\mathrm{ad}\,X)_{\mathfrak{m}_{\mathfrak{p}_0}}\,dk\,dX\,dn\,da$$

*with the same normalization of the measures as in Lemma 12.*

In order to complete the proof of Lemma 11, we still have to prove the following result.

LEMMA 13. *Let $X$ be an element in $\mathfrak{n}_0$ such that $\theta(\mathfrak{n}_0)$ is invariant under $\mathrm{Ad}(\exp X)$. Then $X = 0$.*

Let $\lambda_1 < \lambda_2 < \cdots < \lambda_r$ be all the positive elements $\lambda \in \mathfrak{F}$ such that $\mathfrak{g}_{0\lambda} \neq \{0\}$. Then if $X \neq 0$, let $i$ be the largest index $(1 \leq i \leq r)$ such that $X \in \sum_{j \geq i} \mathfrak{g}_{0\lambda_j}$. Then $X = X_i + X_{i+1} + \cdots + X_r$, where $X_j \in \mathfrak{g}_{0\lambda_j}$ $(1 \leq j \leq r)$ and $X_i \neq 0$. Now it is obvious that if $Y \in \mathfrak{g}_{0\lambda_i}$,

$$\exp(\mathrm{ad}\,X)\theta(Y) \equiv \theta(Y) + [X_i, \theta(Y)] \bmod \mathfrak{n}_0$$

and $[X_i, \theta(Y)] \in \mathfrak{g}_{00} = \mathfrak{h}_{\mathfrak{p}_0} + \mathfrak{m}_0$. However the left side lies in $\theta(\mathfrak{n}_0)$ by hypothesis. Therefore, since the sum $\mathfrak{h}_{\mathfrak{p}_0} + \mathfrak{m}_0 + \mathfrak{n}_0 + \theta(\mathfrak{n}_0)$ is direct, we conclude that $[X_i, \theta(Y)] = 0$. Hence if $H \in \mathfrak{h}_{\mathfrak{p}_0}$,

$$\lambda_i(H)B(X_i, \theta(Y)) = B([H, X_i], \theta(Y)) = B(H, [X_i, \theta(Y)]) = 0.$$

Since $\lambda_i \neq 0$, we can select $H$ such that $\lambda_i(H) \neq 0$, and thus obtain $B(X_i, \theta(Y)) = 0$. Now take $Y = X_i$. Then we get $\| X_i \|^2 = 0$, and so $X_i = 0$. This contradiction proves the lemma.

Select an order on the space of (real-valued) linear functions on $\mathfrak{h}_{\mathfrak{p}_0} + (-1)^{\frac{1}{2}}\mathfrak{h}_{\mathfrak{t}_0}$ compatible with the one already defined in $\mathfrak{F}$. Complexify $\mathfrak{h}_0$, $\mathfrak{m}_{\mathfrak{p}_0}$, $\mathfrak{n}_0$, etc., to $\mathfrak{h}$, $\mathfrak{m}_\mathfrak{p}$, $\mathfrak{n}$, respectively. Let $P$ be the set of all positive roots (of $\mathfrak{g}$ with respect to $\mathfrak{h}$) under our order. For each root $\alpha$, select an element $X_\alpha \neq 0$ in $\mathfrak{g}$ such that $[H, X_\alpha] = \alpha(H)X_\alpha$ for all $H \in \mathfrak{h}$. Let $P_+$ be the set of those $\alpha \in P$ which are not identically zero on $\mathfrak{h}_\mathfrak{p}$. The remaining roots in $P$ are divided into two disjoint classes $P_0$ and $P_-$ as follows. If $\alpha$ is a positive root which is not in $P_+$, then $X_\alpha$ lies either in $\mathfrak{p}$ or in $\mathfrak{k}$ (see [2(e), §2]), and, by definition, $\alpha$ is in $P_0$ or $P_-$ accordingly. Then $P$ is the disjoint union of $P_+$, $P_0$ and $P_-$, and it is obvious that $\mathfrak{n} = \sum_{\alpha \in P_+} CX_\alpha$. Define polynomial functions $\pi_+$, $\pi_0$ and $\pi_-$ on $\mathfrak{h}$ by $\pi_+ = \prod_{\alpha \in P_+} \alpha$, $\pi_0 = \prod_{\beta \in P_0} \alpha$, $\pi_- = \prod_{\alpha \in P_-} \alpha$, and put $\pi = \pi_+ \pi_0 \pi_-$.

We shall now assume that $G$ is the (connected) adjoint group of $\mathfrak{g}_0$. Let $A$ be the Cartan subgroup [2(e), §2] of $G$ corresponding to $\mathfrak{h}_0$. Then $A = A_K A_\mathfrak{p}$, where $A_K = A \cap K$ (Lemma 7 of [2(e)]). Let $x \to x^*$ denote the natural mapping of $G$ on the factor space $G^* = G/A$. We put $x^* H = xH$ for $x \in G$ and $H \in \mathfrak{h}$. Since $K$ is compact, we can normalize $dk$ in such a way that $\int_K dk = 1$. Let $dm$ denote the Haar measure on $M$ and $dx^*$ the invariant measure on $G^*$.

LEMMA 14. *It is possible to normalize $dm$ and $dx^*$ in such a way that*

$$\int_{G^*} f(x^*) \, dx^* = \int_{K \times M \times N} f((kmn)^*) \, dk \, dm \, dn$$

*for $f \in C_c(G^*)$.*

It is clear that if the Haar measure $db$ on $A$ is suitably normalized, then

$$\int g(x) \, dx = \int_{G^*} dx^* \int_A g(xb) \, db \qquad (g \in C_c(G)).$$

On the other hand, from the Corollary to Lemma 12,

$$\int g(x) \, dx = \int g(k \exp Xna) s(X) \, dk \, dX \, dn \, da,$$

where $s(X) = \det(\sinh \operatorname{ad} X / \operatorname{ad} X)_{\mathfrak{m}_{\mathfrak{p}_0}}$ $(X \in \mathfrak{m}_{\mathfrak{p}_0})$. Normalize the Haar measure $dm'$ on $M_K = M \cap K$ in such a way that $\int dm' = 1$. Then it is well known (see [2(g), §12]) that $dm$ can be so normalized that $dm = s(X) \, dm' \, dX$, where $m = m' \exp X$ $(m' \in M_K, X \in \mathfrak{m}_{\mathfrak{p}_0})$. Therefore it is obvious that

$$\int g(k \exp Xna) s(X) \, dk \, dX \, dn \, da = \int g(kmna) \, dk \, dm \, dn \, da.$$

On the other hand, $db = c \, da' \, da$ $(b = a'a, a' \in A_K, a \in A_\mathfrak{p})$, where $c$ is a positive constant and $da'$ is the Haar measure on $A_K$ $(\int da' = 1)$. Therefore, since $a' M a'^{-1} = M$, $a' N a'^{-1} = N$ $(a' \in A_K)$, it is clear that

$$c \int g(kmna)\,dk\,dm\,dn\,da = \int_{K\times M\times N} dk\,dm\,dn \int_A g(kmnb)\,db.$$

Hence

$$c \int_{G^*} dx^* \int_A g(xb)\,db = \int_{K\times M\times N} dk\,dm\,dn \int_A g(kmnb)\,db$$

for $g \in C_o(G)$. Now suppose $f \in C_c(G^*)$. Then we can select (see Weil [7, p. 43]) $g \in C_c(G)$ such that

$$f(x^*) = \int_A g(xb)\,db \qquad\qquad (x \in G).$$

Hence

$$\int_{G^*} f(x^*)\,dx^* = c^{-1} \int_{K\times M\times N} f((kmn)^*)\,dk\,dm\,dn,$$

and this proves our assertion.

COROLLARY. *With the normalizations of Lemma 14 we also have*

$$\int_{G^*} f(x^*)\,dx^* = \int_{K\times M\times N} f((knm)^*)\,dk\,dn\,dm$$

*for $f \in C_c(G^*)$.*

Since $[\mathfrak{m}_0, \mathfrak{n}_0] \subset \mathfrak{n}_0$, it follows that $mNm^{-1} = N$ $(m \in M)$. Therefore, in order to prove the corollary, it is sufficient to show that $|\det \mathrm{Ad}_\mathfrak{n}(m)| = 1$ $(m \in M)$, where $\mathrm{Ad}_\mathfrak{n}(m)$ is the restriction of $\mathrm{Ad}(m)$ on $\mathfrak{n}$. Let $m = m_1 m_2$ $(m_1 \in M_K, m_2 \in M_\mathfrak{p})$. Since $M_K$ is compact and $\mathfrak{m}_{\mathfrak{p}_0} \subset [\mathfrak{m}_0, \mathfrak{m}_0]$, it follows that $|\det \mathrm{Ad}_\mathfrak{n}(m_1)| = \det \mathrm{Ad}_\mathfrak{n}(m_2) = 1$, and therefore

$$|\det \mathrm{Ad}_\mathfrak{n}(m)| = |\det \mathrm{Ad}_\mathfrak{n}(m_1) \det \mathrm{Ad}_\mathfrak{n}(m_2)| = 1.$$

Put $\mathfrak{l}_0 = \mathfrak{h}_{\mathfrak{p}_0} + \mathfrak{m}_0$. Then $\mathfrak{l}_0$ is a Lie algebra which, by Lemma 10, is reductive in $\mathfrak{g}_0$. Let $\mathfrak{l}$ denote the complexification of $\mathfrak{l}_0$ in $\mathfrak{g}$ and $\zeta(L)$ $(L \in \mathfrak{l})$ the determinant of the restriction of $\mathrm{ad}\,L$ on $\mathfrak{n}$. It is obvious that $\zeta$ is a polynomial function on $\mathfrak{l}$ which takes only real values on $\mathfrak{l}_0$.

LEMMA 15. *It is possible to normalize the Euclidean measure $dZ$ on $\mathfrak{n}_0$ so that the following condition is fulfilled. If $L$ is an element in $\mathfrak{l}_0$ such that $\zeta(L) \neq 0$, then*

$$\int_N f(nL)\,dn = |\zeta(L)|^{-1} \int_{\mathfrak{n}_0} f(L+Z)\,dZ \qquad (f \in C_c(\mathfrak{l}_0 + \mathfrak{n}_0)).$$

COROLLARY. *Let $H$ be an element in $\mathfrak{h}_0$ such that $\pi_+(H) \neq 0$. Then*

$$\int |f(x^*H)|\,dx^* = |\zeta(H)|^{-1} \int |f(k(mH+Z))|\,dk\,dm\,dZ$$

*for $f \in C_c(\mathfrak{g}_0)$.*

It is clear that $\zeta(mH) = \zeta(H) = \pi_+(H)$ $(m \in M)$. Our corollary is therefore an immediate consequence of Lemma 15 and the Corollary of Lemma 14. On the other hand, since $[I_0, \mathfrak{n}_0] \subset \mathfrak{n}_0$, it is clear that $nL - L \in \mathfrak{n}_0$ for $n \in N$ and $L \in I_0$. Therefore Lemma 15 follows from the result below.

LEMMA 16. *Let $L$ be an element in $I_0$ such that $\zeta(L) \neq 0$. Then the mapping $Z \to \exp(\mathrm{ad}\, Z)L - L$ $(Z \in \mathfrak{n}_0)$ is a one-one regular mapping of $\mathfrak{n}_0$ onto itself.*

Let $\phi$ denote this mapping. Then in order to show that $\phi$ is regular at a point $Z \in \mathfrak{n}_0$, we have to consider the endomorphism $D$ of $\mathfrak{n}_0$ given by

$$DY = - (\mathrm{ad}\, L)\{(1 - \exp(-\mathrm{ad}\, Z))/\mathrm{ad}\, Z\}Y \qquad (Y \in \mathfrak{n}_0)$$

and prove that $\det D \neq 0$. But since $\mathrm{ad}\, Z$ is nilpotent, it is clear that

$$|\det D| = |\zeta(L)| \neq 0,$$

and so $\phi$ is everywhere regular. Moreover, it is well known that $Z \to \exp Z$ is a one-one mapping of $\mathfrak{n}_0$ onto $N$. Therefore if $\exp(\mathrm{ad}\, Z)L = L$ $(Z \in \mathfrak{n}_0)$, it follows that $\exp(t\,\mathrm{ad}\, L)Z = Z$ $(t \in R)$, and therefore $[L, Z] = 0$. But since $\zeta(L) \neq 0$, this implies that $Z = 0$. This proves that $nL \neq L$ $(n \in N)$ unless $n = 1$, and therefore $\phi(Z_1) = \phi(Z_2)$ $(Z_1, Z_2 \in \mathfrak{n}_0)$ unless $Z_1 = Z_2$. Hence $\phi$ is univalent.

We have still to prove that $\phi(\mathfrak{n}_0) = \mathfrak{n}_0$. Since $\mathfrak{n}_0$ is connected, it would be enough to show that $\phi(\mathfrak{n}_0)$ is closed in $\mathfrak{n}_0$. This however is an immediate consequence of the following statement. *Suppose $Z$ varies in $\mathfrak{n}$ in such a way that $\|\exp(\mathrm{ad}\, Z)L - L\|$ remains bounded. Then $\|Z\|$ also remains bounded.* This is proved as follows. Since $CL + Z$ is a solvable Lie algebra over the complex field, we can choose a base $Z_1, \cdots, Z_r$ for $\mathfrak{n}$ over $C$ and complex numbers $c_1, \cdots, c_r$ such that $[L, Z_i] \equiv c_i Z_i \bmod \sum_{k>i} CZ_k$ and $[Z_i, Z_j] \in \sum_{k>j} CZ_k$ $(1 \leq i \leq j \leq r)$. Moreover $c_i \neq 0$ since $\zeta(L) \neq 0$. Let $t_1, \cdots, t_r$ denote the complex Cartesian coordinates in $\mathfrak{n}$ corresponding to this base. Then it is clear that for any integer $j$ $(1 \leq j \leq r)$, there exists a polynomial $p_j$ in $(j - 1)$ variables $\tau_1, \cdots, \tau_{j-1}$ with complex coefficients such that

$$t_j((\exp \mathrm{ad}\, Z)L - L) = - c_j t_j(Z) + p_j(t_1(Z), \cdots, t_{j-1}(Z))$$

for $Z \in \mathfrak{n}$. Hence it follows by induction on $j$ that if $\|(\exp \mathrm{ad}\, Z)L - L\|$ remains bounded, the same holds for $|t_j(Z)|$. This proves our statement and therefore also the lemma.

In view of the fact that $|\det D| = |\zeta(L)|$, Lemma 15 is now obvious

Let $q_0$ be any subalgebra of $g_0$ such that $\theta(q_0) = q_0$, and let $q$ be the complexification of $q_0$ in $g$. It is obvious that the fundamental bilinear form $B$ is nondegenerate on $q$, and we know that $q$ is reductive in $g$ (Lemma 10). For any $p \in S(g)$, let $p_q$ denote the restriction of $p$ on $q$. Moreover, if $q_0'$ is a nonempty open subset of $q_0$, consider the space $\mathscr{S}(q_0')$ of functions $f \in C^\infty(q_0')$ such that

$$\nu_m(df; q_0') = \sup_{X \in q_0'} |f(X; d)|(1 + \|X + \theta(X)\|)^m < \infty$$

for every $d \in \partial(S(q))$ and every integer $m \geq 0$. We define a topology in $\mathscr{S}(q_0')$ by means of these seminorms $f \to \nu_m(df; q_0')$ $(m \geq 0, d \in \partial(S(q)))$. In this way, $\mathscr{S}(q_0')$ becomes a locally convex space. Similarly, put $\tau_m(df; q_0') = \sup_{X \in q_0'} |f(X; d)|(1 + \|X\|)^m$ for $d \in \mathfrak{D}(q)$ and $f \in \mathscr{C}(q_0')$. We note that

$$\nu_m(df; q_0') \leq 2^m \tau_m(df; q_0') \qquad (d \in \partial(S(q))$$

since $\|X + \theta(X)\| \leq \|X\| + \|\theta(X)\| = 2\|X\|$ $(X \in g_0)$.

LEMMA 17. *For any* $f \in \mathscr{S}(g_0)$, *let* $g_f$ *denote the function on* $\mathfrak{l}_0$ *given by*

$$g_f(L) = \int_{\mathfrak{n}_0} f(L + Z) dZ \qquad (L \in \mathfrak{l}_0).$$

*Then* $g_f \in \mathscr{S}(\mathfrak{l}_0)$ *and* $f \to g_f$ *is a continuous mapping of* $\mathscr{S}(g_0)$ *into* $\mathscr{S}(\mathfrak{l}_0)$. *Moreover* $g_{\partial(p)f} = \partial(p_\mathfrak{l})g_f$ *for* $p \in I(g)$. *Finally, if* $f \in \mathscr{C}(g_0)$ *then* $g_f \in \mathscr{C}(\mathfrak{l}_0)$ *and* $f \to g_f$ $(f \in \mathscr{C}(g_0))$ *is a continuous mapping of* $\mathscr{C}(g_0)$ *into* $\mathscr{C}(\mathfrak{l}_0)$.

Notice that $\mathfrak{l}$, $\mathfrak{n}$ and $\theta(\mathfrak{n})$ are mutually orthogonal subspaces of $g$. Hence $\|L + Z + \theta(L + Z)\|^2 = \|L + \theta(L)\|^2 + 2\|Z\|^2$ $(L \in \mathfrak{l}_0, Z \in \mathfrak{n}_0)$. Hence if $d \in \partial(S(\mathfrak{l}))$,

$$(1 + \|Z\|)^m |f(L + Z; d)| \leq (1 + \|L + Z + \theta(L + Z)\|)^m |f(L + Z; d)|$$
$$\leq \nu_m(df; g_0).$$

We can obviously choose $m$ so large that $\int (1 + \|Z\|)^{-m} dZ < \infty$. This shows that

$$\int_{\mathfrak{n}_0} |f(L + Z; d)| \, dZ \leq \nu_m(df; g_0) \int (1 + \|Z\|)^{-m} dZ.$$

Hence the integral

$$\int_{\mathfrak{n}_0} f(L + Z; d) dZ$$

converges uniformly with respect to $L \in \mathfrak{l}_0$, and therefore $g_f$ is of class $C^\infty$ and

$$g_f(L;d) = \int_{\mathfrak{n}_0} f(L+Z;d)\,dZ.$$

Moreover, it is obvious that

$$\nu_{m_1}(dg_f;\mathfrak{l}_0) \leqq \nu_{m+m_1}(df;\mathfrak{g}_0) \int_{\mathfrak{n}_0} (1+\|Z\|)^{-m}dZ \qquad (m_1 \geqq 0).$$

Therefore $g_f \in \mathscr{S}(\mathfrak{l}_0)$ and the mapping $f \to g_f$ of $\mathscr{S}(\mathfrak{g}_0)$ into $\mathscr{S}(\mathfrak{l}_0)$ is continuous.

Now let $f$ be a fixed element in $\mathscr{S}(\mathfrak{g}_0)$ and $\Omega$ a compact subset in $\mathfrak{g}_0$. For any $d \in \partial(S(\mathfrak{g}))$ and $Y \in \Omega$, consider the integral,

$$\int_{\mathfrak{n}_0} |f(Y+Z;d)|\,dZ.$$

Then

$$\|Z\| \leqq \|Z+\theta(Z)\| \leqq \|Z+Y+\theta(Z+Y)\|$$
$$+ \|Y+\theta(Y)\| \leqq \|X+\theta(X)\| + 2a \qquad (Y \in \Omega),$$

where $X = Y+Z$ and $a = \sup_{Y \in \Omega} \|Y\|$. Hence

$$(1+\|Z\|)^m |f(Y+Z;d)| \leqq (1+2a)^m \nu_m(df;\mathfrak{g}_0),$$

and this proves that the above integral converges uniformly with respect to $Y \in \Omega$. Now put

$$g(Y) = \int_{\mathfrak{n}_0} f(Y+Z)\,dZ \qquad (Y \in \mathfrak{g}_0).$$

Then it follows from the above result that $g \in C^\infty(\mathfrak{g}_0)$ and

$$g(Y;d) = \int_{\mathfrak{n}_0} f(Y+Z;d)\,dZ \qquad (d \in \partial(S(\mathfrak{g})), Y \in \mathfrak{g}_0).$$

But it is clear that $g(Y+Z) = g(Y)$ $(Z \in \mathfrak{n}_0)$. Therefore $dg = 0$ for $d \in \partial(S(\mathfrak{g})\mathfrak{n})$. Therefore, in order to prove the second statement of the lemma, it is enough to show that $p - p_\mathfrak{l} \in S(\mathfrak{g})\mathfrak{n}$ for $p \in I(\mathfrak{g})$. This will be done in Lemma 18 below.

Now suppose $f \in \mathscr{C}(\mathfrak{g}_0)$. Then since $\max(\|L\|, \|Z\|) \leqq \|L+Z\|$ $(L \in \mathfrak{l}_0, Z \in \mathfrak{n}_0)$,

$$(1+\|Z\|)^m (1+\|L\|)^{m_1} |f(L+Z;d)| \leqq \tau_{m+m_1}(df;\mathfrak{g}_0)$$

for $d \in \partial(S(\mathfrak{l}))$ and $m, m_1 \geqq 0$. Hence

$$\tau_{m_1}(dg_f;\mathfrak{l}_0) \leqq \tau_{m+m_1}(df;\mathfrak{g}_0) \int_{\mathfrak{n}_0} (1+\|Z\|)^{-m}dZ,$$

and this proves that $g_f \in \mathscr{B}(I_0)$ and $f \to g_f$ is a continuous mapping of $\mathscr{B}(\mathfrak{g}_0)$ into $\mathscr{B}(I_0)$.

For any $X \in \mathfrak{g}$, let $d_X$ denote the derivation of $S(\mathfrak{g})$ which coincides with $\operatorname{ad} X$ on $\mathfrak{g}$.

LEMMA 18. *Let $p$ be an element in $S(\mathfrak{g})$ such that $d_H p = 0$ for $H \in \mathfrak{h}_\mathfrak{p}$. Then $p - p_I \in S(\mathfrak{g})\mathfrak{n}$.*

Let $\alpha_1, \cdots, \alpha_r$ be all the distinct roots in $P_+$. Then $X_{\alpha_1}, \cdots, X_{\alpha_r}$ is a base for $\mathfrak{n}$ over $C$. Let $\mathfrak{M}$ be the set of all monomials (in $S(\mathfrak{g})$) of the form $X_{\alpha_1}{}^{m_1} \cdots X_{\alpha_r}{}^{m_r}$, where $m_1, \cdots, m_r$ are nonnegative integers. For any monomial $M = X_{\alpha_1}{}^{m_1} \cdots X_{\alpha}{}^{m_r}$, put $\beta_M = m_1 \alpha_1 + \cdots + m_r \alpha_r$ and extend $\theta$ to an automorphism of $S(\mathfrak{g})$. Then it is clear that $d_H \theta(M) = -\beta_M(H)\theta(M)$ $(H \in \mathfrak{h}_\mathfrak{p})$. Since $\mathfrak{g}$ is the direct sum of $I$, $\mathfrak{n}$ and $\theta(\mathfrak{n})$, it follows that $S(\mathfrak{g})$ is the direct sum of $S(I)S(\theta(\mathfrak{n}))$ and $S(\mathfrak{g})\mathfrak{n}$. Hence for any $q \in S(\mathfrak{g})$, there exist unique elements $q_M \in S(I)$ $(M \in \mathfrak{M})$ such that

$$q - \sum_M q_M \theta(M) \in S(\mathfrak{g})\mathfrak{n}.$$

Now apply $d_H$ $(H \in \mathfrak{h}_\mathfrak{p})$ to this relation. Then

$$d_H q + \sum_M q_M \beta_M(H)\theta(M) \in S(\mathfrak{g})\mathfrak{n}$$

since $[\mathfrak{h}_\mathfrak{p}, I] = \{0\}$. In particular, if $d_H q = 0$, we get

$$\sum_M q_M \beta_M(H)\theta(M) \in S(\mathfrak{g})\mathfrak{n}.$$

But as we have observed above, $(S(I)S(\theta(\mathfrak{n}))) \cap (S(\mathfrak{g})\mathfrak{n}) = \{0\}$. Therefore we conclude that $q_M \beta_M(H) = 0$ for every $M \in \mathfrak{M}$. On the other hand, if $M \neq 1$, it is obvious that the restriction of $\beta_M$ on $\mathfrak{h}_\mathfrak{p}$ is positive in our order, and so we can select $H \in \mathfrak{h}_\mathfrak{p}$ such that $\beta_M(H) \neq 0$. Therefore if $q = p$, we conclude that $p_M = 0$ for $M \neq 1$. This means that $p - p_1 \in S(\mathfrak{g})\mathfrak{n}$. But since $p_1 \in S(I)$ and $I$ and $\mathfrak{n} + \theta(\mathfrak{n})$ are mutually orthogonal, it follows that $p_I = (p_1)_I = p_1$. Hence $p - p_I \in S(\mathfrak{g})\mathfrak{n}$.

COROLLARY. *Let $p$ be an element in $S(\mathfrak{g})$ such that $d_H p = 0$ for all $H \in \mathfrak{h}_\mathfrak{p}$. Then $\partial(p_I)g_f = g_{\partial(p)f}$ $(f \in \mathscr{B}(\mathfrak{g}_0))$ in the notation of Lemma 17.*

This follows immediately from Lemma 18.

Put $\pi_\mathfrak{m} = \pi_0 \pi_-$ and let $\mathfrak{h}_0'$ be the set of all points $H \in \mathfrak{h}_0$ where $\pi(H) \neq 0$. We know that $I_0 = \mathfrak{h}_{\mathfrak{p}_0} + \mathfrak{c}_0 + \mathfrak{m}_0'$, where $\mathfrak{c}_0$ is the center and $\mathfrak{m}_0'$ the derived algebra of $\mathfrak{m}_0$. Moreover, $\mathfrak{m}_0'$ is semisimple and $\mathfrak{m}_0' = \mathfrak{m}_0' \cap \mathfrak{k}_0 + \mathfrak{m}_{\mathfrak{p}_0}$. Finally, $\mathfrak{h}_{I_0} = \mathfrak{h}_0 \cap \mathfrak{k}_0$ is maximal abelian in $\mathfrak{m}_0$, and therefore $\mathfrak{h}_{I_0} \cap \mathfrak{m}_0'$ is a Cartan subalgebra of $\mathfrak{m}_0'$. Let $D$ and $M'$ be the analytic subgroups of $M$

corresponding to $c_0$ and $m_0'$. Then $D \subset K$ and $M = DM'$. Let $d\gamma$, $dm'$ denote the Haar measures on $D$ and $M'$, respectively. We normalize them in such a way that $\int_D d\gamma = 1$ and

$$\int_M f(m)\,dm = \int_{M' \times D} f(m'\gamma)\,dm'd\gamma$$

for $f \in C_c(M)$. Now if $X \in c_0$ and $Y, Z \in m_0$, then $B(X, [Y, Z]) = B([X, Y], Z)$ $= 0$, and from this it follows easily that $c_0$ and $m_0'$ are mutually orthogonal.[9] Moreover $m_0$ is orthogonal to $\mathfrak{h}_{\mathfrak{p}_0}$ by definition. Hence $\mathfrak{h}_{\mathfrak{p}_0} + c_0$ is orthogonal to $m_0'$. It is clear that $X_\alpha \in m$ for $\alpha \in P_0 \cup P_-$. Therefore since $[\mathfrak{h}_\mathfrak{p} + c, m]$ $= \{0\}$, we conclude that $\alpha$ vanishes identically on $\mathfrak{h}_\mathfrak{p} + c$. This proves that $\pi_m(H) = \pi_m(H_-)$ $(H \in \mathfrak{h})$, where $H_-$ is the orthogonal projection of $H$ in $\mathfrak{h} \cap m'$.

For any $g \in \mathscr{S}(I_0)$ and $X \in \mathfrak{h}_{\mathfrak{p}_0} + c_0$, let $g_X$ denote the function on $m_0'$ given by $g_X(Y) = g(X + Y)$ $(Y \in m_0')$. Then if $d \in \partial(S(m'))$, it is clear that $(dg)_X = dg_X$. Moreover since $m_0'$ and $\mathfrak{h}_{\mathfrak{p}_0} + c_0$ are mutually orthogonal, we have

$$\nu_q(dg_X; m_0') \leqq \nu_q(dg; I_0) \qquad\qquad (q \geqq 0).$$

This proves that $g_X \in \mathscr{S}(m_0')$. Furthermore,

$$(1 + \| X + \theta(X) \|)^{q_1}(1 + \| Y + \theta(Y) \|)^{q_2} \leqq (1 + \| X + Y + \theta(X + Y) \|)^{q_1 + q_2}$$

$(X \in \mathfrak{h}_{\mathfrak{p}_0} + c_0; Y \in m_0'; q_1, q_2 \geqq 0)$. Therefore

$$\sup_{X \in \mathfrak{h}_{\mathfrak{p}_0} + c_0} (1 + \| X + \theta(X) \|)^{q_1} \nu_{q_2}(dg_X; m_0') \leqq \nu_{q_1 + q_2}(dg; I_0).$$

Let $\mathfrak{h}_0''$ be the set of those points $H \in \mathfrak{h}_0$ where $\pi_0(H) \neq 0$. Then for $H \in \mathfrak{h}_0''$ and $g \in \mathscr{S}(I_0)$, consider the integral

$$\pi_m(H) \int_M g(mH)\,dm = \pi_m(H_-) \int_{M'} g_{H_0}(m'H_-)\,dm',$$

where $H_0$ and $H_-$, respectively, are the orthogonal projections of $H$ in $\mathfrak{h}_{\mathfrak{p}_0} + c_0$ and $\mathfrak{h}_0 \cap m_0'$. The results of Section 4 are applicable to $m_0'$ since $\mathfrak{h}_{I_0} \cap m_0'$ is maximal abelian in $m_0'$. Therefore since $g_{H_0} \in \mathscr{S}(m_0')$ and $\pi_0(H_-) = \pi_0(H)$ $\neq 0$, the above integral is well defined. We denote its value by $\psi_g(H)$. It follows easily from Lemma 9 (applied to $m_0'$) that $\psi_g$ is of class $C^\infty$ on $\mathfrak{h}_0''$ and $\psi_g(H; d) = \psi_{dg}(H)$ for $d \in \partial(S(\mathfrak{h}_\mathfrak{p} + c))$. Now suppose $d_1 \in \partial(S(\mathfrak{h}_\mathfrak{p} + c))$ and $d_2 \in \partial(S(\mathfrak{h} \cap m'))$. Then $\psi_g(H; d_1d_2) = \psi_{d_1g}(H; d_2)$. Moreover, as we have seen during the proof of Theorem 2, we can choose an integer $N \geqq 0$,

---

[9] Cf. footnote 6.

a real constant $c \geq 1$ and a finite number of elements $d_1', \cdots, d_r' \in \partial(S(\mathfrak{m}'))$ such that

$$(1 + \| H_- \|)^{q'} | \psi_f(H; d_2) | \leq c^{q'} \sum_{1 \leq j \leq r} \nu_{N+q'}(d_j' f_{H_0}; \mathfrak{m}_0')$$

for any $f \in \mathscr{S}(I_0)$, $H \in \mathfrak{h}_0''$ and $q' \geq 0$. Therefore

$$(1 + \| H + \theta(H) \|)^{q'} | \psi_g(H; d_1 d_2) |$$
$$\leq (1 + \| H_0 + \theta(H_0) \|)^{q'} (1 + 2 \| H_- \|)^{q'} | \psi_{d_1 g}(H; d_2) |$$
$$\leq (2c)^{q'} (1 + \| H_0 + \theta(H_0) \|)^{q'} \sum_{1 \leq j \leq r} \nu_{N+q'}(d_j'(d_1 g)_{H_0}; \mathfrak{m}_0')$$
$$\leq (2c)^{q'} \sum_{1 \leq j \leq r} \nu_{N+2q'}(d_j' d_1 g; I_0)$$

since $d_j'(d_1 g)_{H_0} = (d_j' d_1 g)_{H_0}$. This proves that $\psi_g \in \mathscr{S}(\mathfrak{h}_0'')$ and the mapping $g \to \psi_g$ of $\mathscr{S}(I_0)$ into $\mathscr{S}(\mathfrak{h}_0'')$ is continuous. Let $I(I)$ denote the set of all elements in $S(I)$ which are invariant under $M$. Then it follows from Theorem 1 (applied to $\mathfrak{m}_0'$) that $\psi_{\partial(p)g} = \partial(\bar{p}) \psi_g$ ($p \in I(I), g \in \mathscr{S}(I_0)$). Finally, suppose $g \in \mathscr{E}(I_0)$. Then

$$(1 + \| H \|)^{q'} | \psi_g(H; d_1 d_2) |$$
$$\leq (1 + \| H_0 \|)^{q'} (1 + \| H_- \|)^{q'} | \psi_{d_1 g}(H; d_2) |$$
$$\leq c^{q'} (1 + \| H_0 \|)^{q'} \sum_{1 \leq j \leq r} \nu_{N+q'}(d_j'(d_1 g)_{H_0}; \mathfrak{m}_0')$$
$$\leq 2^{N+q'} c^{q'} \sum_{1 \leq j \leq r} \tau_{N+q'}(d_j' d_1 g; I_0)$$

since

$$(1 + \| H_0 \|)^{q'} \tau_{N+q'}((d_j' d_1 g)_{H_0}; \mathfrak{m}_0') \leq \tau_{N+2q'}(d_j' d_1 g; I_0).$$

This shows that in this case $\psi_g \in \mathscr{E}(\mathfrak{h}_0'')$ and $g \to \psi_g$ ($g \in \mathscr{E}(I_0)$) is a continuous mapping of $\mathscr{E}(I_0)$ into $\mathscr{E}(\mathfrak{h}_0'')$. Thus we have proved the following result.

LEMMA 19. *For any* $g \in \mathscr{S}(I_0)$, *let* $\psi_g$ *denote the function on* $\mathfrak{h}_0''$ *defined by*

$$\psi_g(H) = \pi_{\mathfrak{m}}(H) \int_M g(mH) \, dm \qquad (H \in \mathfrak{h}_0'').$$

*Then* $\psi_g \in \mathscr{S}(\mathfrak{h}_0'')$ *and* $g \to \psi_g$ *is a continuous mapping of* $\mathscr{S}(I_0)$ *into* $\mathscr{S}(\mathfrak{h}_0'')$. *Moreover,* $\psi_{\partial(p)g} = \partial(\bar{p}) \psi_g$ *for* $p \in I(I)$ *and* $g \in \mathscr{S}(I_0)$. *Finally, if* $g \in \mathscr{E}(I_0)$, *then* $\psi_g \in \mathscr{E}(\mathfrak{h}_0'')$ *and the mapping* $g \to \psi_g$ ($g \in \mathscr{E}(I_0)$) *of* $\mathscr{E}(I_0)$ *into* $\mathscr{E}(\mathfrak{h}_0'')$ *is also continuous.*

Now for $f \in \mathscr{S}(\mathfrak{g}_0)$ and $H \in \mathfrak{h}_0'$, put

$$\phi_f(H) = \pi(H) \int_{G^*} f(x^* H) \, dx^*.$$

That this integral is well defined is seen as follows. Put

$$f_1(X) = \int_K |f(kX)|\, dk \qquad\qquad (X \in \mathfrak{g}_0).$$

Then

$$\int_{G^*} |f(x^*H)|\, dx^* = |\zeta(H)|^{-1} \int_{M \times \mathfrak{n}_0} f_1(mH + Z)\, dm\, dZ$$

from the corollary to Lemma 15. Select an integer $r$ such that

$$c = \int_{\mathfrak{n}_0} (1 + \|Z\|)^{-r} dZ < \infty.$$

Then if $g(X) = \int_{\mathfrak{n}_0} f_1(X + Z)\, dZ \quad (X \in \mathfrak{l}_0)$,

$$(1 + \|X + \theta(X)\|)^q g(X) \leq c\nu_{q+r}(f; \mathfrak{g}_0)$$

for any integer $q \geq 0$. Hence by applying Lemma 7 to $\mathfrak{m}_0'$, we conclude that $\int_M g(mH)\, dm < \infty$. This proves that the above integral for $\phi_f$ is convergent. For any $f \in \mathscr{S}(\mathfrak{g}_0)$, let $\bar{f}$ denote the function

$$\bar{f}(X) = \int_K f(kX)\, dk \qquad\qquad (X \in \mathfrak{g}_0).$$

Since $\theta(kX) = k\theta(X)$, it is easy to see that $\bar{f} \in \mathscr{S}(\mathfrak{g}_0)$ and $f \to \bar{f}$ is a continuous mapping of $\mathscr{S}(\mathfrak{g}_0)$ into itself. (Similarly if $f \in \mathscr{B}(\mathfrak{g}_0)$, one proves easily that $\bar{f} \in \mathscr{B}(\mathfrak{g}_0)$ and $f \to \bar{f}$ $(f \in \mathscr{B}(\mathfrak{g}_0))$ is a continuous mapping of $\mathscr{B}(\mathfrak{g}_0)$ into itself). Moreover, the above calculation shows that

$$\phi_f(H) = \pi(H)|\pi_+(H)|^{-1} \int_{M \times \mathfrak{n}_0} \bar{f}(mH + Z)\, dm\, dZ$$

$$= \epsilon_+(H)\pi_{\mathfrak{m}}(H) \int_{M \times \mathfrak{n}_0} \bar{f}(mH + Z)\, dm\, dZ \qquad (f \in \mathscr{S}(\mathfrak{g}_0)),$$

where $\epsilon_+(H) = 1$ or $-1$ according as $\pi_+(H)$ is positive [10] or negative. Put

$$\psi_f(H) = \pi_{\mathfrak{m}}(H) \int_{M \times \mathfrak{n}_0} \bar{f}(mH + Z)\, dm\, dZ \qquad (H \in \mathfrak{h}_0'', f \in \mathscr{S}(\mathfrak{g}_0)).$$

It follows from Lemmas 17 and 19 that $\psi_f \in \mathscr{S}(\mathfrak{h}_0'')$ and $f \to \psi_f$ is a continuous mapping of $\mathscr{S}(\mathfrak{g}_0)$ into $\mathscr{S}(\mathfrak{h}_0'')$. Define $g_f$ as in Lemma 17. Then

$$\psi_f(H) = \pi_{\mathfrak{m}}(H) \int_M g_f(mH)\, dm \qquad (f \in \mathscr{S}(\mathfrak{g}_0), H \in \mathfrak{h}_0'').$$

---

[10] Notice that $\pi_+(H)$ is real since it is equal to the determinant of the restriction of ad $H$ on $\mathfrak{n}_0$

Now suppose $p \in I(\mathfrak{g})$. If $f_1 = \partial(p)f$, it is obvious that $\bar{f}_1 = \partial(p)\bar{f}$. Hence

$$\psi_{\partial(p)f}(H) = \pi_\mathfrak{m}(H) \int_M g_{\bar{f}}(mH; \partial(p_1)) \, dm = \psi_f(H; \partial(\bar{p}))$$

from Lemmas 17 and 19. Finally, if $f \in \mathscr{C}(\mathfrak{g}_0)$, we conclude from these two lemmas that $\psi_f \in \mathscr{C}(\mathfrak{h}_0'')$ and the mapping $f \to \psi_f$ ($f \in \mathscr{C}(\mathfrak{g}_0)$) of $\mathscr{C}(\mathfrak{g}_0)$ into $\mathscr{C}(\mathfrak{h}_0'')$ is continuous. Thus we have obtained the following theorem.

THEOREM 3. *For any* $f \in \mathscr{S}(\mathfrak{g}_0)$, *the integral*

$$\phi_f(H) = \pi(H) \int_{G^*} f(x^*H) \, dx^*$$

*is absolutely convergent for* $H \in \mathfrak{h}_0'$ *and the function* $\phi_f$, *so defined, lies in* $\mathscr{S}(\mathfrak{h}_0')$. *Moreover,* $f \to \phi_f$ *is a continuous mapping of* $\mathscr{S}(\mathfrak{g}_0)$ *into* $\mathscr{S}(\mathfrak{h}_0')$, *and*

$$\phi_{\partial(p)f} = \partial(\bar{p})\phi_f$$

*for* $p \in I(\mathfrak{g})$. *Let* $\mathfrak{h}_0''$ *be the set of those points* $H \in \mathfrak{h}_0$ *where* $\prod_{\alpha \in P_0} \alpha(H) \neq 0$ *and let* $\epsilon_+(H)$ *denote the sign of* $\pi_+(H)$ ($H \in \mathfrak{h}_0'$). *Then* $\epsilon_+\phi_f$ ($f \in \mathscr{S}(\mathfrak{g})$) *can be extended to a function of class* $C^\infty$ *on* $\mathfrak{h}_0''$. *Finally,* $\phi_f \in \mathscr{C}(\mathfrak{h}_0')$ *if* $f \in \mathscr{C}(\mathfrak{g}_0)$, *and the mapping* $f \to \phi_f$ *of* $\mathscr{C}(\mathfrak{g}_0)$ *into* $\mathscr{C}(\mathfrak{h}_0')$ *is also continuous.*

Define the homomorphism $\delta$ of $\mathfrak{J}(\mathfrak{g})$ into $\mathfrak{J}(\mathfrak{h})$ as in Theorem 1 of [2(h)].

COROLLARY. *If* $\xi \in \mathfrak{J}(\mathfrak{g})$ *and* $f \in \mathscr{C}(\mathfrak{g}_0)$, *then* $\phi_{\xi f} = \delta(\xi)\phi_f$.

From the above theorem, this is true if $\xi \in \partial(I(\mathfrak{g}))$. Since $\mathfrak{J}(\mathfrak{g})$ is generated by $I(\mathfrak{g}) \cup \partial(I(\mathfrak{g}))$, it is enough to verify the above statement for $\xi \in I(\mathfrak{g})$. But then $\xi f \in \mathscr{C}(\mathfrak{g}_0)$, and it is obvious from the definition of $\phi_f$ that

$$\phi_{\xi f} = \bar{\xi}\phi_f = \delta(\xi)\phi_f.$$

**6. Fourier transforms.** Let $dX$ denote the regular Euclidean measure on $\mathfrak{g}_0$ (see [2(h),§2]) and for any $f \in \mathscr{C}(\mathfrak{g}_0)$, let $\bar{f}$ denote the Fourier transform of $f$ so that

$$\bar{f}(Y) = \int_{\mathfrak{g}_0} \exp\{(-1)^{\frac{1}{2}}B(Y,X)\} f(X) \, dX \qquad (Y \in \mathfrak{g}_0).$$

Fix an element $H_0 \in \mathfrak{h}_0'$ and consider the mapping $T_{H_0}: f \to \phi_{\bar{f}}(H_0)$ ($f \in \mathscr{C}(\mathfrak{g}_0)$). It follows from Theorem 3 that $T_{H_0}$ is a $\mathscr{C}$-distribution. We intend to study this distribution more closely. For any $p \in S(\mathfrak{g})$, let $\hat{p}$ denote the

3

polynomial given by $\hat{p}(X) = p(-(-1)^{\frac{1}{2}}X)$ $(X \in \mathfrak{g})$. Then (see Lemma 1 of [2(h)]) $(\partial(p)f)^{\tilde{}} = \hat{p}\tilde{f}$ and $(pf)^{\tilde{}} = \partial(\hat{p})\tilde{f}$. Moreover, if $p \in I(\mathfrak{g})$, the same holds for $\hat{p}$. Therefore it follows from Theorem 3 and its corollary that

$$T_{H_0}(\partial(p)f) = \hat{p}(H_0)T_{H_0}(f) \qquad (f \in \mathscr{E}(\mathfrak{g}_0), p \in I(\mathfrak{g})),$$

and this proves that $\partial(p)T_{H_0} = p((-1)^{\frac{1}{2}}H_0)T_{H_0}$ for $p \in I(\mathfrak{g})$. Thus we have obtained the following lemma.

LEMMA 20. *For any $p \in S(\mathfrak{g})$, define the polynomial $p'$ by $p'(X)$ $= p((-1)^{\frac{1}{2}}X)$ $(X \in \mathfrak{g})$. Then $\partial(p)T_{H_0} = p'(H_0)T_{H_0}$ for $p \in I(\mathfrak{g})$.*

Let $l$ be the rank and $n$ the dimension of $\mathfrak{g}$. Then if $\lambda$ is an indeterminate and $I$ the identity mapping of $\mathfrak{g}$,

$$\det(\lambda I - \operatorname{ad} X) = \lambda^n + d_1(X)\lambda^{n-1} + d_2(X)\lambda^{n-2} + \cdots + d_{n-l}(X)\lambda^l \quad (X \in \mathfrak{g}),$$

where $d_j \in I(\mathfrak{g})$ and $d_{n-l} \neq 0$. It is obvious that the polynomials $d_j$ take only real values on $\mathfrak{g}_0$ and $d_{n-l}(H) = (-1)^r \pi(H)^2$ $(H \in \mathfrak{h})$, where $r = \frac{1}{2}(n-l)$ is the number of positive roots. Put $\eta = (-1)^r d_{n-l}$. Then (see Lemma 9 of [2(h)]) $\eta$ is the unique element in $I(\mathfrak{g})$ which coincides with $\pi^2$ on $\mathfrak{h}$. Moreover, as we have already observed above, $\eta$ takes only real values on $\mathfrak{g}_0$. *We call a point $X \in \mathfrak{g}$ singular or regular according as $\eta(X)$ is zero or not.* Let $\mathfrak{g}_0'$ be the set of all regular elements in $\mathfrak{g}_0$. Then $\mathfrak{g}_0'$ is an open dense subset of $\mathfrak{g}_0$. Our object is to show that $T_{H_0}$ coincides with an analytic function on $\mathfrak{g}_0'$.

Let $\bar{H}_0$ be a regular element in $\mathfrak{g}_0$ and $\bar{\mathfrak{h}}_0$ the centralizer of $\bar{H}_0$ in $\mathfrak{g}_0$. Then $\bar{\mathfrak{h}}_0$ is a Cartan subalgebra of $\mathfrak{g}_0$ (see Weyl [8]). Let $\bar{A}$ be the Cartan subgroup of $G$ corresponding to $\bar{\mathfrak{h}}_0$. Put $\bar{G} = G/\bar{A}$ and let $x \to \bar{x}$ denote the natural mapping of $G$ on $\bar{G}$. Complexify $\bar{\mathfrak{h}}_0$ to $\bar{\mathfrak{h}}$, and let $P$ be the set of all positive roots of $\mathfrak{g}$ with respect to $\bar{\mathfrak{h}}$ under some fixed order. Put $\bar{\pi} = \prod_{\alpha \in P} \alpha$, $\bar{\mathfrak{h}}_0' = \bar{\mathfrak{h}}_0 \cap \mathfrak{g}_0'$ and $\bar{x}\bar{H} = x\bar{H}$ $(x \in G, \bar{H} \in \bar{\mathfrak{h}})$. It is well known (see [2(d), p. 501]) that the mapping $\psi \colon (\bar{x}, \bar{H}) \to \bar{x}\bar{H}$ of $\bar{G} \times \bar{\mathfrak{h}}_0$ into $\mathfrak{g}_0$ is everywhere regular on $\bar{G} \times \bar{\mathfrak{h}}_0'$. Let $\tilde{A}$ be the normalizer of $\bar{\mathfrak{h}}_0$ in $G$. Then $\tilde{A}/\bar{A}$ can be regarded (see [2(e), §2]) as a subgroup of the Weyl group $\bar{W}$ of $\mathfrak{g}$ with respect $\bar{\mathfrak{h}}$, and therefore it is finite. Since $\bar{A}$ is the centralizer of $\bar{H}_0$ in $G$ (Lemma 3 of [2(e)]), it is clear that we can select an open connected neighborhood $V$ of $\bar{H}_0$ in $\bar{\mathfrak{h}}_0'$ such that $\psi$ is univalent on $\bar{G} \times V$. Put $\mathfrak{g}_1 = \psi(\bar{G} \times V)$. Then $\mathfrak{g}_1$ is an open connected neighborhood of $\bar{H}_0$ in $\mathfrak{g}_0'$ and $\psi$ defines an analytic isomorphism of $\bar{G} \times V$ with $\mathfrak{g}_1$. Let $d\bar{x}$ denote the invariant measure on $\bar{G}$ and $d\bar{H}$ the regular Euclidean measure on $\bar{\mathfrak{h}}_0$. Then a simple calculation shows (see [2(d), p. 501]) that $d\bar{x}$ can be so normalized that

$$dX = |\eta(\bar{H})|\, d\bar{x}\, d\bar{H} = |\bar{\pi}(\bar{H})|^2\, d\bar{x}\, d\bar{H} \qquad (X = \bar{x}\bar{H})$$

for $(\bar{x}, \bar{H}) \in \bar{G} \times V$.

Now let $T$ be a distribution on $\mathfrak{g}_0'$ which is invariant under $G$ (see [2(h), §2]). For any $g \in C_c^\infty(\bar{G} \times V)$, let $F_g$ denote the function in $C_c^\infty(\mathfrak{g}_1)$ given by $F_g(\bar{x}\bar{H}) = g(\bar{x}, \bar{H})$ $(\bar{x} \in \bar{G}, \bar{H} \in V)$. Then we can define a distribution $\tau$ on $\bar{G} \times V$ by $\tau(g) = T(F_g)$ $(g \in C_c^\infty(\bar{G} \times V))$. For $\alpha \in C_c^\infty(\bar{G})$ and $\beta \in C_c^\infty(V)$, let $\alpha \times \beta$ denote the function $g$ in $C_c^\infty(\bar{G} \times V)$ given by $g(\bar{x}, \bar{H}) = \alpha(\bar{x})\beta(\bar{H})$ $(\bar{x} \in \bar{G}, \bar{H} \in V)$.

LEMMA 21. *There exists a distribution $S$ on $V$ such that $\tau(\alpha \times \beta)$*

$$= \int \alpha(\bar{x})\, d\bar{x}\, S(\beta)\ \text{for all}\ \alpha \in C_c^\infty(\bar{G})\ \text{and}\ \beta \in C_c^\infty(V).$$

Let $dx$ and $da$ denote the Haar measures of $G$ and $\bar{A}$, respectively. We assume that $da$ is so normalized that

$$\int g(x)\, dx = \int_{\bar{G}} d\bar{x} \int_{\bar{A}} g(xa)\, da \qquad (g \in C_c(G)).$$

For any $f \in C_c^\infty(G \times V)$, let $\bar{f}$ denote the function in $C_c^\infty(\bar{G} \times V)$ given by

$$\bar{f}(\bar{x}, \bar{H}) = \int_{\bar{A}} f(xa, \bar{H})\, da.$$

Then we can define a distribution $\tau'$ on $G \times V$ by $\tau'(f) = \tau(\bar{f})$ $(f \in C_c^\infty(G \times V))$. For any $\alpha \in C_c^\infty(G)$, define $\alpha' \in C_c^\infty(\bar{G})$ by

$$\alpha'(\bar{x}) = \int_{\bar{A}} \alpha(xa)\, da \qquad (x \in G).$$

Then one sees without difficulty that $\alpha \to \alpha'$ is a continuous mapping of $C_c^\infty(G)$ onto $C_c^\infty(\bar{G})$. Hence if we denote the function $(x, \bar{H}) \to \alpha(x)\beta(\bar{H})$ on $G \times V$ by $\alpha \times \beta$ $(\alpha \in C_c^\infty(G),\ \beta \in C_c^\infty(V))$, it is enough to prove the existence of a distribution $S$ on $V$ such that

$$\tau'(\alpha \times \beta) = \int \alpha(x)\, dx\, S(\beta) \qquad (\alpha \in C_c^\infty(G),\ \beta \in C_c^\infty(V)).$$

Now fix $\beta$. Then $\sigma_\beta : \alpha \to \tau'(\alpha \times \beta)$ is a distribution on $G$. Moreover, since $T$ is invariant under $G$, it is clear that $\sigma_\beta$ is invariant under left translations of $G$. Therefore it follows from Lemma 36 of [2(e)], that

$$\tau'(\alpha \times \beta) = S(\beta) \int_G \alpha(x)\, dx,$$

where $S(\beta)$ is a complex number independent of $\alpha$. Now select $\alpha_0 \in C_c^\infty(G)$ such that $\int_G \alpha_0(x)\,dx = 1$. Then since $\beta \to \tau'(\alpha_0 \times \beta) = S(\beta)$ $(\beta \in C_c^\infty(V))$ is obviously a distribution on $V$, our assertion follows.

Since $\eta(\bar{H}) = \bar{\pi}(\bar{H})^2$ $(\bar{H} \in \bar{\mathfrak{h}})$, $\bar{\pi}$ is nowhere zero on $V$. Hence $\sigma = (\bar{\pi})^{-1}S$ is also a distribution on $V$.

LEMMA 22. *Let $p$ be a homogeneous element in $I(\mathfrak{g})$ and let $\bar{p}$ denote its restriction on $\mathfrak{h}$. Suppose $\partial(p)T = cT$ $(c \in C)$. Then $\partial(\bar{p})\sigma = c\sigma$.*

Select $\alpha_0 \in C_c^\infty(G)$ such that $\int \alpha_0(x)\,dx = 1$, and, for any $\beta \in C_c^\infty(V)$, let $F_\beta$ denote the function in $C_c^\infty(\mathfrak{g}_1)$ given by

$$F_\beta(\bar{x}\bar{H}) = \beta(\bar{H})\alpha_0'(\bar{x}) \qquad (\bar{x} \in \bar{G},\, \bar{H} \in V),$$

where $\alpha_0'(\bar{x}) = \int_{\bar{A}} \alpha_0(xa)\,da$ $(x \in G)$. We put $F_\beta(x:\bar{H}) = F_\beta(x\bar{H})$ $(x \in G,\, \bar{H} \in V)$ and use the notation of Lemma 3 of $[2(\mathrm{h})]$. Then it follows from Lemmas 5 and 6 of $[2(\mathrm{h})]$, that

$$F_\beta(x\bar{H};\partial(p)) = \bar{\pi}(\bar{H})^{-1}F_\beta\,(x:\bar{H};\partial(\bar{p})\circ\bar{\pi})$$
$$+ \sum_{1 \leq i \leq N} a_i(\bar{H})F_\beta(x;b_i:\bar{H};\partial(\bar{u}_i))$$

$(x \in G,\, \bar{H} \in V)$, where $b_i \in \mathfrak{Bg},\, \bar{u}_i \in S(\bar{\mathfrak{h}})$ and the $a_i$ are analytic functions on $V$ $(1 \leq i \leq N)$. Moreover, in view of Lemma 22 of $[2(\mathrm{h})]$, we can assume that $b_i{}^a = b_i$ $(a \in \bar{A})$. On the other hand,

$$F_\beta(x:\bar{H}) = \int \alpha_0(xa)\,da\,\beta(\bar{H}).$$

Therefore

$$F_\beta(\bar{x}\bar{H};\partial(p)) = \alpha_0'(\bar{x})\bar{\pi}(\bar{H})^{-1}\beta(\bar{H};\partial(\bar{p})\circ\bar{\pi})$$

$$+ \sum_{1 \leq i \leq N} a_i(\bar{H}) \int \alpha_0(xa;b_i)\,da\,\beta(\bar{H};\partial(\bar{u}_i)).$$

But

$$\int d\bar{x} \int \alpha_0(xa;b_i)\,da = \int \alpha_0(x;b_i)\,dx = 0$$

since $b_i \in \mathfrak{Bg}$. Therefore

$$T(\partial(p)F_\beta) = \sigma(\partial(\bar{p})(\bar{\pi}\beta)).$$

Let $m$ be the degree of $p$. Then

$$T(\partial(p)F_\beta) = (-1)^m cT(F_\beta) = (-1)^m c\sigma(\bar{\pi}\beta),$$

and this shows that

$$\sigma(\partial(\ddot{p})(\bar{\pi}\beta)) = (-1)^m c\sigma(\bar{\pi}\beta).$$

But since $\bar{\pi}$ is nowhere zero on $V$, this implies that $\partial(\bar{p})\sigma = c\sigma$.

COROLLARY. *Suppose $T$ is an eigen-distribution of $\partial(p)$ for every $p \in I(\mathfrak{g})$. Then $\sigma$ coincides with an analytic function on $V$.*

This follows from Lemma 12 of [2(h)] and the above result.

We shall now apply this corollary to the distribution $T = T_{H_0}$ of Lemma 20. Let $G_c$ be the (connected) complex adjoint group of $\mathfrak{g}$. Since any two Cartan subalgebras of $\mathfrak{g}$ are conjugate under $G_c$ (see [2(e), Lemma 4]), we can select $y \in G_c$ such that $y\mathfrak{h} = \bar{\mathfrak{h}}$. Then if $p \in I(\mathfrak{g})$, $p^y = p$, and therefore $p'(H_0) = p((-1)^{\frac{1}{2}}yH_0)$. In particular, $\eta(yH_0) = \eta(H_0) \neq 0$, and so $\bar{\pi}((-1)^{\frac{1}{2}}yH_0) \neq 0$. Therefore it follows from Lemma 20 and the Corollary of Lemma 13 of [2(h)], that

$$\sigma(\bar{H}) = \sum_{s \in \bar{W}} c_s \exp\{(-1)^{\frac{1}{2}}B(s\bar{H}, yH_0)\} \qquad (\bar{H} \in V),$$

where the $c_s$ are certain complex numbers. Moreover, we can conclude from Lemma 21 (see Schwartz [6(I), p. 108]) that

$$T_{H_0}(f) = \int_{\bar{G} \times V} f(\bar{x}\bar{H})\,\bar{\pi}(\bar{H})\,\sigma(\bar{H})\,d\bar{x}\,d\bar{H} \qquad (f \in C_c^\infty(\mathfrak{g}_1)).$$

Now put

$$\bar{\phi}_f(\bar{H}) = \bar{\pi}(\bar{H}) \int_{\bar{G}} f(\bar{x}\bar{H})\,d\bar{x} \qquad (\bar{H} \in \bar{\mathfrak{h}}_0').$$

Then the above relation becomes

$$\phi_f(H_0) = \sum_{s \in \bar{W}} c_s(H_0) \int \bar{\phi}_f(\bar{H}) \exp\{(-1)^{\frac{1}{2}}B(s\bar{H} \cdot yH_0)\}\,d\bar{H} \qquad (H_0 \in \mathfrak{h}_0')$$

for $f \in C_c^\infty(\mathfrak{g}_1)$. (Since the numbers $c_s$ may depend on $H_0$, we have now written them as $c_s(H_0)$.) Let $\mathfrak{h}_1$ be a connected component of $\mathfrak{h}_0'$. We shall prove that the $c_s$ are actually constant on $\mathfrak{h}_1$. Let $H_0$ be a fixed point in $\mathfrak{h}_1$ and $U$ a sufficiently small open connected neighborhood of $H_0$ in $\mathfrak{h}_1$. It is enough to prove that the $c_s$ are constant on $U$. Since $\eta(yH_0) = \eta(H_0) \neq 0$, it is clear that $syH_0 \neq yH_0$ for any $s \neq 1$ in $\bar{W}$ (see Lemma 4 of [2(e)]). Hence we can choose polynomials $q_s \in S(\bar{\mathfrak{h}})$ $(s \in \bar{W})$ such that

$$q_s(-(-1)^{\frac{1}{2}}t^{-1}yH_0) = \begin{cases} 0 & \text{if } s \neq t \\ 1 & \text{if } s = t \end{cases} \qquad (t \in \bar{W}).$$

Furthermore, select $\beta_s \in C_c^\infty(V)$ such that

$$\int_V \exp\{(-1)^{\frac{1}{2}} B(yH_0, s\bar{H})\} \beta_s(\bar{H}) d\bar{H} = 1 \qquad (s \in \bar{W}).$$

Fix an element $\alpha \in C_c^\infty(\bar{G})$ such that $\int \alpha(\bar{x}) d\bar{x} = 1$, and define $f_s \in C_c^\infty(\mathfrak{g}_1)$ by $f_s(\bar{x}\bar{H}) = \alpha(\bar{x})\beta_s(\bar{H}; \partial(q_s))\bar{\pi}(\bar{H})^{-1} (\bar{x} \in \bar{G}, \bar{H} \in V)$. Then if $H \in U$,

$$\phi_{f_s}(H) = \sum_{t \in \bar{W}} c_t(H) \int \beta_s(\bar{H}; \partial(q_s)) \exp\{(-1)^{\frac{1}{2}} B(t\bar{H}, yH)\} d\bar{H}$$

$$= \sum_{t \in \bar{W}} c_t(H) a_{s,t}(H) \qquad (s \in \bar{W}),$$

where

$$a_{s,t}(H) = q_s(-(-1)^{\frac{1}{2}} t^{-1} yH) \int \beta_s(\bar{H}) \exp\{(-1)^{\frac{1}{2}} B(t\bar{H}, yH)\} d\bar{H}.$$

But it is clear that $a_{s,t}$ are analytic functions on $U$ and

$$a_{s,t}(H_0) = \begin{cases} 1 & \text{if } s = t, \\ 0 & \text{if } s \neq t. \end{cases}$$

Therefore if $U$ is sufficiently small, the determinant of the matrix $(a_{s,t})_{s,t \in \bar{W}}$ is nowhere zero on $U$. Let $(b_{s,t})_{s,t \in \bar{W}}$ denote the inverse matrix. Then the $b_{s,t}$ are also analytic functions on $U$ and

$$c_t(H) = \sum_{s \in \bar{W}} b_{t,s}(H) \phi_{f_s}(H) \qquad (H \in U)$$

Therefore since $\phi_{f_s}$ are of class $C^\infty$ on $U$ (Theorem 3), the same holds for $c_t$.

Put

$$\sigma(H : \bar{H}) = \sum_{s \in \bar{W}} c_s(H) \exp\{(-1)^{\frac{1}{2}} B(yH, s\bar{H})\} \qquad (H \in U, \bar{H} \in V).$$

Then $\sigma$ is of class $C^\infty$ on $U \times V$. On the other hand, $\partial(p)\bar{f} = (p'f)^{\tilde{}}$ $(p \in S(\mathfrak{g}), f \in C_c^\infty(\mathfrak{g}))$, and therefore if $p \in I(\mathfrak{g})$,

$$\phi_{(p'f)}{}^{\tilde{}} = \partial(\bar{p})\phi_{\bar{f}}$$

from Theorem 3. Select $\alpha$ as above and define $f \in C_c^\infty(\mathfrak{g}_1)$ by $f(\bar{x}\bar{H}) = \alpha(\bar{x})\beta(\bar{H})\bar{\pi}(\bar{H})^{-1}$ $(\beta \in C_c^\infty(V))$. Then $\bar{\phi}_f(\bar{H}) = \beta(\bar{H})$ and $\bar{\phi}_{p'f}(\bar{H}) = p'(\bar{H})\beta(\bar{H}) (\bar{H} \in V)$, and so the above equation becomes

$$\int p'(\bar{H})\beta(\bar{H})\sigma(H : \bar{H}) d\bar{H} = \int \beta(\bar{H})\sigma(H; \partial(\bar{p}) : \bar{H}) d\bar{H} \qquad (H \in U).$$

As this is true for every $\beta \in C_c^\infty(V)$, we conclude that

$$\sigma(H; \partial(\bar{p}) : \bar{H}) = p'(\bar{H})\sigma(H : \bar{H}) \qquad (p \in I(\mathfrak{g}))$$

for $H \in U$ and $\bar{H} \in V$. Thus it is enough to prove the following result.

LEMMA 23. *Suppose $c_s$ ($s \in \bar{W}$) are functions of class $C^\infty$ on $U$ and*

$$\sigma(H:\bar{H}) = \sum_{s \in \bar{W}} c_s(H)\exp\{(-1)^{\frac{1}{2}}B(yH, s\bar{H})\} \qquad (H \in U, \bar{H} \in V).$$

*Then if $\sigma(H;\partial(\bar{p}):\bar{H}) = p'(\bar{H})\sigma(H:\bar{H})$ for all $p \in I(\mathfrak{g})$ and $(H, \bar{H}) \in U \times V$, it follows that each $c_s$ is constant on $U$.*

Fix $\bar{H}$ in $V$ for a moment, and let $\mu_s$ denote the linear function $H \to (-1)^{\frac{1}{2}}B(s\bar{H}, yH)$ $(H \in \mathfrak{h})$ on $\mathfrak{h}$. Since $\bar{\pi}(\bar{H}) \neq 0$, the functions $\mu_s$ ($s \in W$) are all distinct. Therefore $e^{\mu_s}$ are linearly independent over $S(\mathfrak{h})$ (Lemma 41 of [2(b)]). This means that the coefficients of $e^{\mu_s}$ in $\sigma(H;\partial(\bar{p}):\bar{H}) - p'(\bar{H})\sigma(H:\bar{H})$ must all be zero. But it is obvious that for a fixed $s$, this coefficient is of the form

$$c_s(H;\partial(\bar{p})) + \sum_{1 \leq i \leq N} q_i(\bar{H})b_i(H) - p'(\bar{H})c_s(H),$$

where the $q_i$ are certain homogeneous elements in $S(\bar{\mathfrak{h}})$ of positive degree and the $b_i$ are functions of class $C^\infty$ on $U$. Since $\bar{H}$ was an arbitrary element of $V$, we conclude that

$$c_s(H;\partial(\bar{p})) = 0 \qquad (s \in \bar{W}, H \in U)$$

for all homogeneous elements $p \in I(\mathfrak{g})$ of positive degree. But then it follows from Lemmas 12 and 14 of [2(h)] that $c_s$ is a polynomial function. Now again fix $\bar{H} \in V$. Then, in view of the equation $\sigma(H;\partial(\bar{p}):\bar{H}) = p'(\bar{H})\sigma(H:\bar{H})$ $(p \in I(\mathfrak{g}))$, we conclude from the Corollary of Lemma 13 of [2(h)] that

$$\sigma(H:\bar{H}) = \sum_{s \in W} b_s e^{\mu_s(H)} \qquad (H \in U),$$

where the $b_s$ are certain complex numbers independent of $H$. However the $e^{\mu_s}$ are linearly independent over $S(\mathfrak{h})$, and therefore $c_s = b_s$ ($s \in \bar{W}$). This proves that the $c_s$ are constant on $U$.

Thus we have obtained the following result which should be compared with Theorem 3 of [2(h)].

LEMMA 24. *Let $\mathfrak{h}_1$ be a connected component of $\mathfrak{h}_0'$. Then there exist complex numbers $c_s$ ($s \in \bar{W}$) such that*

$$\phi_{\bar{f}}(H) = \sum_{s \in \bar{W}} c_s \int \bar{\phi}_f(\bar{H})\exp\{(-1)^{\frac{1}{2}}B(s\bar{H}, yH)\}d\bar{H} \qquad (H \in \mathfrak{h}_1)$$

*for all $f \in C_c^\infty(\mathfrak{g}_1)$.*

For any $X \in \mathfrak{g}_0'$, put $\epsilon(X) = 1$ or $-1$ according as $\eta(X)$ is positive or negative. Then it is clear that for a fixed $H \in \mathfrak{h}_1$, there exists an analytic function $F_H$ on $\mathfrak{g}_1$ such that

$$F_H(\bar{x}\bar{H}) = \epsilon(\bar{H})\bar{\pi}(\bar{H})^{-1} \sum_{s \in \bar{W}} c_s \exp\{(-1)^{\frac{1}{2}} B(s\bar{H}, yH)\}$$

$(\bar{x} \in \bar{G}, \bar{H} \in V)$ and

$$\phi_{\bar{f}}(H) = \int f(X) F_H(X) dX \qquad\qquad (f \in C_c^{\infty}(\mathfrak{g}_1)).$$

(Compare this formula for $F_H$ with Theorem 2 of [2(h)]). This shows that the distribution $f \to \phi_{\bar{f}}(H)$ coincides on $\mathfrak{g}_1$ with $F_H$. Thus we have proved the following theorem.

THEOREM 4. *For a fixed* $H \in \mathfrak{h}_0'$ *consider the* $\mathscr{E}$-*distribution* $T_H : f \to \phi_{\bar{f}}(H)$ $(f \in \mathscr{E}(\mathfrak{g}_0))$ *on* $\mathfrak{g}_0$. *Then* $T_H$ *coincides on* $\mathfrak{g}_0'$ *with an analytic function* $F_H$.

Since $T_H$ is invariant under $G$, the same holds for $F_H$. The exact form of $F_H$ has already been obtained above.

**7. The distributions $T$ and $T'$.** Let $H_1, \cdots, H_l$ be a base for $\mathfrak{h}_0$ over $R$. It is clear that there exists a positive constant $c$ such that if $H = t_1 H_1 + \cdots + t_l H_l$ $(t_i \in R)$, then $|t_i| \leq c \|H\|$ $(1 \leq i \leq l)$. Now suppose $\mathfrak{h}_1$ is a connected component of $\mathfrak{h}_0'$ and $f \in \mathscr{E}(\mathfrak{h}_1)$. Then $\partial(H)f = \sum_{1 \leq i \leq l} t_i \partial(H_i)f$ and therefore

$$\nu(\partial(H)f) \leq c \|H\| \sum_{1 \leq i \leq l} \nu(\partial(H_i)f) \qquad\qquad (H \in \mathfrak{h}_0),$$

where $\nu(g) = \sup_{H \in \mathfrak{h}_1} |g(H)|$ $(g \in \mathscr{E}(\mathfrak{h}_1))$. Suppose $H, H'$ are two points in $\mathfrak{h}_1$ and the straight line-segment joining them lies entirely in $\mathfrak{h}_1$. Then if $f \in \mathscr{E}(\mathfrak{h}_1)$.

$$f(H') - f(H) = \int_0^1 (d/dt)f(H + t(H' - H))dt,$$

and therefore

$$|f(H') - f(H)| \leq \nu(\partial(H' - H)f) \leq c \| H' - H \| \sum_{1 \leq i \leq l} \nu(\partial(H_i)f).$$

But in view of Lemma 43 of the Appendix (§12), we can conclude that these inequalities are actually valid for any two points $H, H' \in \mathfrak{h}_1$, and so we have the following result.

LEMMA 25. *Let* $\mathrm{Cl}(\mathfrak{h}_1)$ *denote the closure of* $\mathfrak{h}_1$ *in* $\mathfrak{h}_0$. *Then every* $f \in \mathscr{E}(\mathfrak{h}_1)$ *can be extended to a continuous function on* $\mathrm{Cl}(\mathfrak{h}_1)$.

For any $\phi \in \mathscr{B}(\mathfrak{h}_0')$ and $\xi \in \mathfrak{D}(\mathfrak{h})$, put

$$\underset{\mathfrak{h}_1}{\operatorname{Lim}} \phi(0) = \underset{H \to 0}{\operatorname{Lim}} \phi(H), \qquad \underset{\mathfrak{h}_1}{\operatorname{Lim}} \phi(0;\xi) = \underset{H \to 0}{\operatorname{Lim}} \phi(H;\xi) \qquad (H \in \mathfrak{h}_1).$$

It is clear from the above corollary that these limits exist and

$$\left| \underset{\mathfrak{h}_1}{\operatorname{Lim}} \phi(0;\xi) \right| \leqq \nu(\xi\phi).$$

Therefore for a fixed $\xi$, the mapping $\phi \to \underset{\mathfrak{h}_1}{\operatorname{Lim}} \phi(0;\xi)$ is continuous on $\mathscr{B}(\mathfrak{h}_0')$. Let $\xi_0$ denote the local expression of $\xi$ at zero. Then if $\xi_0 = 0$, it is obvious that $\underset{\mathfrak{h}_1}{\operatorname{Lim}} \phi(0;\xi) = 0$. Hence

$$\underset{\mathfrak{h}_1}{\operatorname{Lim}} \phi(0;\xi) = \underset{\mathfrak{h}_1}{\operatorname{Lim}} \phi(0;\xi_0)$$

for $\phi \in \mathscr{B}(\mathfrak{h}_0')$ and $\xi \in \mathfrak{D}(\mathfrak{h})$.

We know from Theorem 3 that if $f \in \mathscr{B}(\mathfrak{g}_0)$, then $\phi_f \in \mathscr{B}(\mathfrak{h}_0')$. Hence we can consider the mapping $T : f \to \underset{\mathfrak{h}_1}{\operatorname{Lim}} \phi_f(0;\partial(\pi))$ $(f \in \mathscr{B}(\mathfrak{g}_0))$. It follows immediately from Theorem 3 and what we have said above, that $T$ is a $\mathscr{B}$-distribution.

LEMMA 26. *Define the polynomial $\eta \in I(\mathfrak{g})$ as in Section 6. Then*

$$\eta^k \partial(\eta^k) T = \langle \eta^k, \eta^k \rangle T$$

*for any integer $k \geqq 0$.*

Put $\xi = \partial(\eta^k) \circ \eta^k$. Then we know from the Corollary of Theorem 3 that $\phi_{\xi f} = \delta(\xi)\phi_f$ $(f \in \mathscr{B}(\mathfrak{g}_0))$, and therefore

$$T(\xi f) = \underset{\mathfrak{h}_1}{\operatorname{Lim}} \phi_f(0;\partial(\pi) \circ \delta(\xi)) = \langle \eta^k, \eta^k \rangle T(f)$$

from the Corollary of Lemma 18 of $[2(\mathrm{h})]$. Our assertion now follows from the fact that the degree of $\eta$ is even.

It is one of our principal objects to prove that $T(f) = cf(0)$ $(f \in \mathscr{B}(\mathfrak{g}_0)$, where $c$ is a complex number independent of $f$. However, the proof of this fact is rather long and difficult, and we shall not be able to complete it in this paper. We consider first the Fourier transform $T'$ of $T$ given by $T'(f) = T(\bar{f})$ $(f \in \mathscr{B}(\mathfrak{g}_0))$. Then $T'$ is also a $\mathscr{B}$-distribution on $\mathfrak{g}_0$, and the above-mentioned fact about $T$ is obviously equivalent to the statement that $T'$ coincides on $\mathfrak{g}_0$ with a constant. This we shall first prove in a weaker form. As before, let $\mathfrak{g}_0'$ denote the set of regular elements in $\mathfrak{g}_0$.

LEMMA 27. *Let $T'$ be the $\mathscr{B}$-distribution on $\mathfrak{g}_0$ defined by*

$$T'(f) = \underset{\mathfrak{h}_1}{\operatorname{Lim}} \phi_{\bar{f}}(0;\partial(\pi)) \qquad\qquad (f \in \mathscr{B}(\mathfrak{g}_0)).$$

*Then on each connected component of $\mathfrak{g}_0'$, $T'$ coincides with a constant.*

It is obviously sufficient to prove that if $\bar{H}_0 \in \mathfrak{g}_0'$, then $T'$ coincides on some open neighborhood of $\bar{H}_0$ in $\mathfrak{g}_0'$ with a constant. We use the notation of the proof of Lemma 24. Then

$$\phi_{\bar{f}}(H) = \sum_{s \in \bar{W}} c_s \int_V \bar{\phi}_f(\bar{H}) \exp\{(-1)^{\frac{1}{2}} B(s\bar{H}, yH)\} d\bar{H} \qquad (H \in \mathfrak{h}_1)$$

for $f \in C_c^\infty(\mathfrak{g}_1)$. Hence

$$\phi_{\bar{f}}(H; \partial(\pi)) = \sum_{s \in \bar{W}} c_s (-1)^{r/2} \int_V \bar{\phi}_f(\bar{H}) \pi(y^{-1} s\bar{H}) \exp\{(-1)^{\frac{1}{2}} B(s\bar{H}, yH)\} d\bar{H}$$

where $r$ is the degree of $\pi$. This implies that

$$T'(f) = \sum_{s \in \bar{W}} c_s (-1)^{r/2} \int_V \bar{\phi}_f(\bar{H}) \pi(y^{-1} s\bar{H}) d\bar{H}.$$

But $\pi(y^{-1} s\bar{H})^2 = \eta(y^{-1} s\bar{H}) = \eta(\bar{H}) = \bar{\pi}(\bar{H})^2$. Therefore it is obvious that $\epsilon_s = \pi(y^{-1} s\bar{H}) / \bar{\pi}(\bar{H})$ is constant for $\bar{H} \in V$. Hence

$$T'(f) = c' \int_V \bar{\pi}(\bar{H}) \bar{\phi}_f(\bar{H}) d\bar{H},$$

where $c' = \sum_{s \in \bar{W}} (-1)^{r/2} c_s \epsilon_s$. Now put $\epsilon = 1$ or $-1$ according as $\eta$ is positive or negative on $\mathfrak{g}_1$. Then

$$\int_V \bar{\pi}(\bar{H}) \bar{\phi}_f(\bar{H}) d\bar{H} = \epsilon \int f(X) dX \qquad (f \in C_c^\infty(\mathfrak{g}_1)),$$

and therefore

$$T'(f) = c \int f(X) dX \qquad (f \in C_c^\infty(\mathfrak{g}_1))$$

if $c = c'\epsilon$. This proves that $T'$ coincides with $c$ on $\mathfrak{g}_1$.

Lemma 28. $\mathfrak{g}_0'$ has only a finite number of connected components.

From the corollary of Lemma 2 of [2(e)], we can select a finite number of Cartan subalgebras $\mathfrak{h}_{(i)}$ $(1 \leq i \leq N)$ such that every Cartan subalgebra of $\mathfrak{g}_0$ is conjugate to some $\mathfrak{h}_{(i)}$ under $G$. Put $\mathfrak{h}_{(i)}' = \mathfrak{h}_{(i)} \cap \mathfrak{g}_0'$ and $V_i = \bigcup_{x \in G} x \mathfrak{h}_{(i)}'$. Let $X$ be an element $\mathfrak{g}_0'$, and let $\mathfrak{h}_X$ denote the centralizer of $X$ in $\mathfrak{g}_0$. Then since $X$ is regular, $\mathfrak{h}_X$ is a Cartan subalgebra. Hence $\mathfrak{h}_X = x\mathfrak{h}_{(i)}$ for some $x \in G$ and some $i$ $(1 \leq i \leq N)$. Therefore $X \in \mathfrak{h}_X \cap \mathfrak{g}_0' \subset V_i$, and this proves that $\mathfrak{g}_0' = V_1 \cup V_2 \cup \cdots \cup V_N$. On the other hand, we know that $\mathfrak{h}_0'$ (and therefore, similarly $\mathfrak{h}_{(i)}'$) has only a finite number of connected components. Since $G$ is connected, it is therefore obvious that the same holds for each $V_i$. But this implies the statement of the lemma.

LEMMA 29.  *Let $\mathfrak{g}_1, \mathfrak{g}_2, \cdots, \mathfrak{g}_N$ be all the distinct connected components of $\mathfrak{g}_0'$ and $c_i$ the constant such that $T' = c_i$ on $\mathfrak{g}_i$ $(1 \leqq i \leqq N)$.  Let $\Omega$ be an open subset of $\mathfrak{g}_0$ whose closure is compact.  Then there exists an integer $k \geqq 0$ such that*

$$\langle \eta^k, \eta^k \rangle T'(f) = \sum_{1 \leqq i \leqq N} c_i \int_{\mathfrak{g}_i} \eta(X)^k f(X; \partial(\eta^k)) dX$$

*for all $f \in C_c^\infty(\Omega)$.*

Let $\tau'$ denote the distribution on $\mathfrak{g}_0$ given by

$$\tau'(g) = T'(g) - \sum_{1 \leqq i \leqq N} c_i \int_{\mathfrak{g}_i} g(X) dX \qquad (g \in C_c^\infty(\mathfrak{g}_0)).$$

Then it is clear that the carrier of $\tau'$ is contained in the set of singular points of $\mathfrak{g}_0$.  For any $f \in C_c^\infty(\Omega)$, let $\tau_f$ denote the distribution on $R$ defined as follows:

$$\tau_f(\beta) = \tau'(\beta(\eta)f) \qquad (\beta \in C_c^\infty(R)),$$

where $\beta(\eta)$ is the function on $\mathfrak{g}_0$ whose value at $X \in \mathfrak{g}_0$ is $\beta(\eta(X))$.  It is clear that no point in $R$ other than zero, can lie in the carrier of $\tau_f$.  Moreover, since the closure of $\Omega$ is compact, $\tau'$ is of finite order on $\Omega$ (see Schwartz [6(I) pp. 82-83]).  Hence there exists an integer $k$ (independent of $f$) such that the order of $\tau_f$ (on $R$) is $< k$.  This implies that $\tau'(\eta^k f) = 0$ for all $f \in C_c^\infty(\Omega)$.

For any $p \in S(\mathfrak{g})$ define $\hat{p}$ as in the proof of Lemma 20.  Then obviously, $\hat{\eta}$ coincides on $\mathfrak{h}$ with $(-1)^r \pi^2$ (where $r$ is the degree of $\pi$).  Therefore it follows from Lemma 9 of $[2(h)]$ that $\hat{\eta} = (-1)^r \eta$, and so

$$(\eta^k \partial(\eta^k)g)^\sim = \partial(\eta^k)(\eta^k \tilde{g}) \qquad (g \in \mathscr{B}(\mathfrak{g}_0))$$

from Lemma 1 of $[2(h)]$.  Hence we conclude from Lemma 26 that

$$T'(\eta^k \partial(\eta^k)g) = \langle \eta^k, \eta^k \rangle T'(g).$$

Now suppose $f \in C_c^\infty(\Omega)$.  Then $\tau'(\eta^k \partial(\eta^k)f) = 0$, and therefore

$$\langle \eta^k, \eta^k \rangle T'(f) = T'(\eta^k \partial(\eta^k)f) = \sum_{1 \leqq i \leqq N} c_i \int_{\mathfrak{g}_i} \eta(X)^k f(X; \partial(\eta^k)) dX.$$

COROLLARY.  *If $c_1, c_2, \cdots, c_N$ are all equal then $T' = c_1$ on $\mathfrak{g}_0$.*

It is sufficient to prove that $T' = c_1$ on $\Omega$.  But if $f \in C_c^\infty(\Omega)$, we know from the above lemma that

$$\langle \eta^k, \eta^k \rangle T'(f) = c_1 \int_{\mathfrak{g}_0} \eta(X)^k f(X; \partial(\eta^k)) dX$$

since $c_1 = c_2 = \cdots = c_N$. Moreover, since $\eta$ is a homogeneous polynomial of even degree, $\partial(\eta^k)$ is self-adjoint, and therefore

$$\int_{\mathfrak{g}_0} \eta(X)^k f(X; \partial(\eta^k)) \, dX = \langle \eta^k, \eta^k \rangle \int_{\mathfrak{g}_0} f(X) \, dX.$$

On the other hand, $\langle \eta^k, \eta^k \rangle \neq 0$ (see the corollary to Lemma 18 of [2(h)]). Therefore

$$T'(f) = c_1 \int_{\mathfrak{g}_0} f(X) \, dX,$$

and this proves that $T' = c_1$ on $\Omega$.

In view of this corollary, our main task now is to prove that the constants $c_1, \cdots, c_N$ of Lemma 29 are all equal. This will be done in another paper. However we note the following result for later use.

LEMMA 30. *Let $\xi$ be any homogeneous polynomial in $I(\mathfrak{g})$ of positive degree. Then $\partial(\xi) T' = 0$.*

Define $\xi'$ as in Lemma 20. Then if $f \in \mathcal{B}(\mathfrak{g}_0)$.

$$T'(\partial(\xi)f) = T'((\partial(\xi)f)^{\tilde{}}) = (-1)^m T(\xi' \tilde{f}),$$

where $m$ is the degree of $\xi$. Since $\xi' \in I(\mathfrak{g})$, it follows from Theorem 3 that $\phi_{\xi' \tilde{f}} = \bar{\xi}' \phi_{\tilde{f}}$, where $\bar{\xi}'$ is the restriction of $\xi'$ on $\mathfrak{h}_0$. Therefore

$$T(\xi' \tilde{f}) = \lim_{\mathfrak{h}_1} \phi_{\tilde{f}}(0; \partial(\pi) \circ \bar{\xi}') = \lim_{\mathfrak{h}_1} \phi_{\tilde{f}}(0; d),$$

where $d$ is the local expression of $\partial(\pi) \circ \bar{\xi}'$ at zero. But since $m \geq 1$, it follows from Lemma 18 of [2(h)] that $d = 0$. This proves that $\partial(\xi) T' = 0$.

## 8. Fundamental Cartan subalgebras.

Let $\omega$ be the Casimir polynomial of $\mathfrak{g}$ and $\Gamma_0$ a Cartan subalgebra of $\mathfrak{g}_0$. Put $l_-(\Gamma_0) = \sup_{\Gamma_1} (\dim_R \Gamma_1)$, where $\Gamma_1$ runs over all linear subspaces of $\Gamma_0$ on which $\omega$ is negative-definite. Moreover let $l_- = \sup_{\Gamma_0} l_-(\Gamma_0)$, where $\Gamma_0$ runs over all Cartan subalgebras of $\mathfrak{g}_0$. We say that $\Gamma_0$ is *fundamental* if $l_-(\Gamma_0) = l_-$. It is obvious that $l_-(x\Gamma_0) = l_-(\Gamma_0)$ $(x \in G)$. Hence if $\Gamma_0$ is fundamental, the same holds for any Cartan subalgebra conjugate to $\Gamma_0$ under $G$.

LEMMA 31. $l_- = \operatorname{rank} \mathfrak{k}$.

Let $\Gamma_0$ be a fundamental Cartan subalgebra of $\mathfrak{g}_0$. By going over to a conjugate subalgebra, we may assume (see the Corollary of Lemma 1 of [2(e)]) that $\theta(\Gamma_0) = \Gamma_0$. Then since $\omega$ is positive-definite on $\mathfrak{p}_0$ and negative-definite on $\mathfrak{k}_0$, it is clear that $l_- = l_-(\Gamma_0) = \dim_R(\Gamma_0 \cap \mathfrak{k}_0)$.   Hence $l_- \leqq \operatorname{rank} \mathfrak{k}$. Conversely, let $\mathfrak{a}_{\mathfrak{k}_0}$ be a maximal abelian subalgebra of $\mathfrak{k}_0$.   We extend it to a Cartan subalgebra $\mathfrak{a}_0$ of $\mathfrak{g}_0$.   Then

$$l_- \geqq l_-(\mathfrak{a}_0) \geqq \dim_R \mathfrak{a}_{\mathfrak{k}_0} = \operatorname{rank} \mathfrak{k}.$$

This proves that $l_- = \operatorname{rank} \mathfrak{k}$.

COROLLARY.   *Let $\Gamma_0$ be a Cartan subalgebra of $\mathfrak{g}_0$ such that $\theta(\Gamma_0) = \Gamma_0$. Then $\Gamma_0$ is fundamental if and only if $\Gamma_0 \cap \mathfrak{k}_0$ is maximal abelian in $\mathfrak{k}_0$.*

For it is obvious that $l_-(\Gamma_0) = \dim_R(\Gamma_0 \cap \mathfrak{k}_0)$, and therefore from the above lemma, $l_-(\Gamma_0) = l_-$ if and only if $\Gamma_0 \cap \mathfrak{k}_0$ is maximal abelian in $\mathfrak{k}_0$.

LEMMA 32.   *Let $\Gamma_{\mathfrak{k}_0}$ be a maximal abelian subalgebra of $\mathfrak{k}_0$ and $\Gamma_0$ the centralizer of $\Gamma_{\mathfrak{k}_0}$ in $\mathfrak{g}_0$.   Then $\Gamma_0$ is a Cartan subalgebra of $\mathfrak{g}_0$ and $\theta(\Gamma_0) = \Gamma_0$.*

Since $\theta(\Gamma_{\mathfrak{k}_0}) = \Gamma_{\mathfrak{k}_0}$, it is obvious that $\theta(\Gamma_0) = \Gamma_0$.   Let $X, Y \in \Gamma_{\mathfrak{p}_0} = \Gamma_0 \cap \mathfrak{p}_0$. Then $Z = [X, Y] \in \Gamma_0 \cap \mathfrak{k}_0 = \Gamma_{\mathfrak{k}_0}$ and $\|Z\|^2 = -B(Z, Z) = B(Y, [X, Z]) = 0$ since $[X, Z] = 0$.   This proves that $Z = 0$, and therefore $\Gamma_0 = \Gamma_{\mathfrak{k}_0} + \Gamma_{\mathfrak{p}_0}$ is abelian.   Let $\mathfrak{a}_0$ be any Cartan subalgebra of $\mathfrak{g}_0$ containing $\Gamma_{\mathfrak{k}_0}$.   Then $\mathfrak{a}_0 \subset \Gamma_0$. But since $\mathfrak{a}_0$ is maximal abelian in $\mathfrak{g}_0$, it follows that $\mathfrak{a}_0 = \Gamma_0$.   This proves our assertion.

COROLLARY.   *Any two fundamental Cartan subalgebras of $\mathfrak{g}_0$ are conjugate under $G$.*

Let $\Gamma_1$, $\Gamma_2$ be two fundamental Cartan subalgebras of $\mathfrak{g}_0$.   By replacing them by conjugate subalgebras, we may assume that $\theta(\Gamma_i) = \Gamma_i$ $(i = 1, 2)$. Then from the Corollary to Lemma 31, $\Gamma_i \cap \mathfrak{k}_0$ is maximal abelian in $\mathfrak{k}_0$. However, since $K$ is compact, any two maximal abelian subalgebras of $\mathfrak{k}_0$ are conjugate under $K$.   Hence $\Gamma_2 \cap \mathfrak{k}_0 = (k\Gamma_1) \cap \mathfrak{k}_0$ for some $k \in K$.   But then it follows from Lemma 32 that $\Gamma_2 = k\Gamma_1$.

LEMMA 33.   *Let $\Gamma_0$ be a Cartan subalgebra of $\mathfrak{g}_0$ and $\Gamma$ its complexification in $\mathfrak{g}$.   Then the following two conditions on $\Gamma_0$ are equivalent.*

(1)   $\Gamma$ *is not fundamental.*

(2)   *There exists a root $\alpha$ (of $\mathfrak{g}$ with respect to $\Gamma$) which takes only real values on $\Gamma_0$.*

Again we may assume that $\theta(\Gamma_0) = \Gamma_0$.   Since every root $\alpha$ takes only pure imaginary values on $\Gamma_0 \cap \mathfrak{k}_0$, condition (2) is obviously equivalent to the following:

(3)   *There exists a root $\alpha$ which vanishes identically on $\Gamma \cap \mathfrak{k}$.*

Now if $\Gamma_0$ is fundamental, $\Gamma$ is exactly the centralizer of $\Gamma \cap \mathfrak{k}$ in $\mathfrak{g}$ (Lemma 32), and therefore no root can vanish identically on $\Gamma \cap \mathfrak{k}$. This proves that (3) implies (1). Conversely, suppose (3) is false. Then no root can vanish identically on $\Gamma \cap \mathfrak{k}$, and therefore it is obvious that $\Gamma$ is exactly the centralizer of $\Gamma \cap \mathfrak{k}$ in $\mathfrak{g}$. This proves that $\Gamma_0 \cap \mathfrak{k}_0$ is maximal abelian in $\mathfrak{k}_0$, and so $\Gamma_0$ is fundamental. Hence (1) and (3), and therefore also (1) and (2), are equivalent.

We now return to the notation of Section 5. Let $Q_+$ be the set of those $\alpha \in P$ which take only real values on $\mathfrak{h}_0$. Put $Q' = Q_+ \cup P_-$, and let $W_{Q'}$ be the subgroup of the Weyl group $W$ (of $\mathfrak{g}$ with respect to $\mathfrak{h}$) generated by the Weyl reflexions $s_\alpha$ corresponding to $\alpha \in Q'$.

LEMMA 34. *Suppose $s \in W_{Q'}$. Then $s\mathfrak{h}_0 = \mathfrak{h}_0$ and*

$$\phi_f(sH) = \epsilon(s)\phi_f(H)$$

*for $f \in \mathscr{B}(\mathfrak{g}_0)$ and $H \in \mathfrak{h}_0'$.*

Let $A'$ be the normalizer of $A$ in $G$. Then for every $y \in A'$, we get a transformation $\tau_y$ of $G^* = G/A$ onto itself given by $\tau_y(x^*) = (xy)^*$ $(x \in G)$. Since $A$ is of finite index in $A'$ (see $[2(e), \S2]$), $\tau_y{}^k$ is the identity for some integer $k \geq 1$, and from this it follows immediately that the measure $dx^*$ on $G^*$ is invariant under $\tau_y$.

Now from Corollary 2 to Lemma 46 of the Appendix (Section 12) we can choose $y \in G$ such that $yH = sH$ for all $H \in \mathfrak{h}$. Then it is clear that $s\mathfrak{h}_0 = \mathfrak{h}_0$ and $y \in A'$. Therefore if $f \in \mathscr{B}(\mathfrak{g}_0)$,

$$\phi_f(sH) = \pi(sH) \int_{G^*} f(x^*yH)dx^* = \epsilon(s)\pi(H) \int_{G^*} f(x^*H)dx^* = \epsilon(s)\phi_f(H)$$

for $H \in \mathfrak{h}'$.

COROLLARY. *If $\mathfrak{h}_0$ is not fundamental, the distribution $T$ of Lemma 26 is zero.*

We know from Lemma 33 that $Q_+$ is not empty. Let $V_1$ denote the closure of $\mathfrak{h}_1$ in $\mathfrak{h}_0$. It follows from Lemma 45 of the Appendix ($\S12$) that we can choose a point $H_0 \in V_1$ and a root $\alpha \in Q_+$ such that $\alpha(H_0) = 0$ while $\beta(H_0) \neq 0$ for any positive root $\beta \neq \alpha$. It is obvious that $tH_0 \in V_1$ for any positive number $t$. Let $s_\alpha$ denote the Weyl reflexion corresponding to $\alpha$ and for a fixed $f \in \mathscr{B}(\mathfrak{g}_0)$, put

$$\psi_f(H) = \epsilon_+(H)\phi_f(H) \qquad\qquad (H \in \mathfrak{h}_0')$$

in the notation of Theorem 3. Since $\mathfrak{h}_\mathfrak{k}$ and $\mathfrak{h}_\mathfrak{p}$ are orthogonal, it is clear that $s_\alpha \beta = \beta$ for any $\beta \in P_0 \cup P_-$, and therefore $\pi_+^{s_\alpha} = -\pi_+$. This shows that $\epsilon_+(s_\alpha H) = -\epsilon_+(H)$, and hence $\psi_f(s_\alpha H) = \psi_f(H)$ from Lemma 34.

Now we extend $\psi_f$ to a function of class $C^\infty$ on $\mathfrak{h}_0''$ (see Theorem 3). Then since $\pi^{s_\alpha} = -\pi$, it is obvious that

$$\psi_f(s_\alpha H; \partial(\pi)) = -\psi_f(H; \partial(\pi)) \qquad\qquad (H \in \mathfrak{h}_0'').$$

Therefore if $\sigma_\alpha$ is the hyperplane in $\mathfrak{h}_0$ consisting of those $H \in \mathfrak{h}_0$ where $\alpha(H) = 0$, it is clear that $\partial(\pi)\psi_f$ is zero on $\sigma_\alpha \cap \mathfrak{h}_0''$. Select a sequence $\{t_k\}_{k \geq 1}$ of positive numbers such that $\underset{k \to \infty}{\mathrm{Lim}}\, t_k = 0$. Then $t_k H_0 \in V_1 \cap \sigma_\alpha \cap \mathfrak{h}_0''$, and so we can choose $H_k \in \mathfrak{h}_1$ such that $\| t_k H_0 - H_k \| \leq 2^{-k}$ and $|\psi_f(H_k; \partial(\pi))| \leq 2^{-k}$. Then $H_k \to 0$ and $T(f) = \underset{k \to \infty}{\mathrm{Lim}}\, \phi_f(H_k; \partial(\pi))$. But

$$|\phi_f(H_k; \partial(\pi))| = |\psi_f(H_k; \partial(\pi))| \leq 2^{-k},$$

and therefore $T(f) = 0$. This proves that $T = 0$.

We shall see in another paper that if $\mathfrak{h}_0$ is fundamental and $\mathfrak{h}_1$ is suitably chosen, then $T \neq 0$. For a special case, this follows from Lemma 41 of Section 11.


**9.  Closer study of a special case.** *In this section we shall assume that* $\mathfrak{h}_0 \subset \mathfrak{k}_0$ and therefore rank $\mathfrak{g} =$ rank $\mathfrak{k}$. Let $dY$ denote the Euclidean measure on $\mathfrak{k}_0$, and, for any $g \in \mathscr{E}(\mathfrak{k}_0)$, put

$$\hat{g}(X) = \int_{\mathfrak{k}_0} \exp\{(-1)^{\frac{1}{2}} B(X, Y)\} g(Y)\, dY \qquad\qquad (X \in \mathfrak{g}_0).$$

It is clear that $\hat{g}(Y + Z) = \hat{g}(Y)$ $(Y \in \mathfrak{k}_0, Z \in \mathfrak{p}_0)$ and $\hat{g}$ coincides on $\mathfrak{k}_0$ with the Fourier transform of $g$. Hence

$$\hat{g}(Y + Z; \partial(p))(1 + \| Y + Z + \theta(Y + Z) \|)^m = \hat{g}(Y; \partial(p_\mathfrak{k}))(1 + 2 \| Y \|)^m.$$

for $p \in S(\mathfrak{g})$ and $m \geq 0$. (Here $p_\mathfrak{k}$ is the restriction of $p$ on $\mathfrak{k}$). This shows that $\hat{g} \in \mathscr{A}(\mathfrak{g}_0)$ and $g \to \hat{g}$ is a continuous mapping of $\mathscr{E}(\mathfrak{k}_0)$ into $\mathscr{A}(\mathfrak{g}_0)$. Let $\bar{\mathfrak{h}}_0$ be any Cartan subalgebra of $\mathfrak{g}_0$ such that $\theta(\bar{\mathfrak{h}}_0) = \bar{\mathfrak{h}}_0$. We use the notation of Section 6 and define $\bar{\mathfrak{h}}_0'$, $\bar{\pi}$, $\bar{W}$, $\bar{G}$, $d\bar{x}$, etc. as before. Put

$$\bar{\phi}_f(\bar{H}) = \bar{\pi}(\bar{H}) \int_{\bar{G}} f(\bar{x}\bar{H})\, d\bar{x} \qquad\qquad (f \in \mathscr{A}(\mathfrak{g}_0), \bar{H} \in \bar{\mathfrak{h}}_0').$$

Then we know from Theorem 3 that $\bar{\phi}_f \in \mathscr{A}(\bar{\mathfrak{h}}_0')$ and $f \to \bar{\phi}_f$ is a continuous mapping of $\mathscr{A}(\mathfrak{g}_0)$ into $\mathscr{A}(\bar{\mathfrak{h}}_0')$. Fix a point $\bar{H}_0$ in $\bar{\mathfrak{h}}_0'$, and put $\sigma_{\bar{H}_0}(g) = \bar{\phi}_{\hat{g}}(\bar{H}_0)$ $(g \in \mathscr{E}(\mathfrak{k}_0))$. Then it is obvious that $\sigma_{\bar{H}_0}$ is a $\mathscr{E}$-distribution on $\mathfrak{k}_0$. Define $D$ and $\zeta$ as in Theorem 5 of [2(h)] and $p'$ as in Lemma 20 $(p \in S(\mathfrak{g}))$.

**LEMMA 35.** $D\sigma_{\bar{H}_0} = \omega'(\bar{H}_0)\zeta\sigma_{\bar{H}_0}$ *where* $\omega$ *is the Casimir polynomial of* $\mathfrak{g}$.

For any $f \in \mathscr{S}(\mathfrak{g}_0)$ and $\alpha \in C_c^\infty(G)$, put

$$f_\alpha(X) = \int_G \alpha(x)f(x^{-1}X)\,dx \qquad\qquad (X \in \mathfrak{g}_0).$$

Then since

$$\int_{\bar{G}} |f(\bar{x}\bar{H})|\,d\bar{x} < \infty \qquad\qquad (\bar{H} \in \bar{\mathfrak{h}}_0')$$

(Theorem 3), it follows by Fubini's theorem that

$$\bar{\pi}(\bar{H}) \int_{\bar{G}} f_\alpha(\bar{x}\bar{H})\,d\bar{x} = \int_G \alpha(x)\,dx\,\bar{\phi}_f(\bar{H}).$$

Hence

$$\bar{\pi}(\bar{H}_0) \int \hat{g}_\alpha(\bar{x}\bar{H}_0)\,d\bar{x} = \int_G \alpha(x)\,dx\,\sigma_{\bar{H}_0}(g)$$

for $g \in \mathscr{B}(\mathfrak{k}_0)$ and $\alpha \in C_c^\infty(G)$. Now select $\alpha_0 \in C_c^\infty(G)$ such that $\int_G \alpha_0(x)\,dx = 1$. Then

$$\sigma_{\bar{H}_0}(g) = \bar{\pi}(\bar{H}_0) \int \hat{g}_{\alpha_0}(\bar{x}\bar{H}_0)\,d\bar{x} \qquad\qquad (g \in \mathscr{B}(\mathfrak{k}_0)),$$

and since $\omega' \in I(\mathfrak{g})$, it is obvious that

$$\omega'(\bar{H}_0)\sigma_{\bar{H}_0}(g) = \bar{\pi}(\bar{H}_0) \int_G \omega'(\bar{x}\bar{H}_0)\hat{g}_{\alpha_0}(\bar{x}\bar{H}_0)\,d\bar{x}.$$

Put

$$\Lambda_X(x:Y) = \exp\{(-1)^{\frac{1}{2}}B(X,xY)\} \qquad\qquad (x \in G,\, Y \in \mathfrak{k}_0)$$

for a fixed $X \in \mathfrak{g}_0$. From Lemma 24 of [2(h)] we know that $\Gamma_Y(\gamma_\omega(Y)) = \zeta(Y)\omega$ $(Y \in \mathfrak{k}_0)$ and

$$\gamma_\omega(Y) = \sum_{1 \le i \le r} q_i(Y)(b_i \mathbf{X} 1) + 1 \mathbf{X} D_Y \qquad\qquad (Y \in \mathfrak{k}_0).$$

Here $D_Y$ is the local expression of $D$ at $Y$, the $q_i$ are polynomial functions on $\mathfrak{k}$ and the $b_i$ are homogeneous elements in $\mathfrak{P}$ of order 2 $(1 \le i \le r)$. Let $\lambda_X(Z) = \exp\{(-1)^{\frac{1}{2}}B(X,Z)\}$ $(Z \in \mathfrak{g}_0)$ for a fixed $X \in \mathfrak{g}_0$. Then from Lemma 3 of [2(h)] it is clear that

$$\zeta(Y)\lambda_X(xY;\partial(\omega)) = \sum_{1 \le i \le r} q_i(Y)\Lambda_X(x;b_i:Y) + \Lambda_X(x:Y;D)$$

and therefore

$$\int_{G \times \mathfrak{k}_0} \alpha_0(x)\zeta(Y)\lambda_X(xY;\partial(\omega))g(Y)\,dx\,dY$$

$$= \sum_{1 \le i \le r} \int_{G \times \mathfrak{k}_0} \alpha_0(x;b_i{}^*)q_i(Y)g(Y)\lambda_X(xY)\,dx\,dY$$

$$+ \int_{G \times \mathfrak{k}_0} \alpha_0(x)g(Y;D^*)\lambda_X(xY)\,dx\,dY,$$

where the star denotes the adjoint of a differential operator. But

$$\lambda_X(Z;\theta(\omega)) = \omega'(X)\lambda_X(Z) \qquad\qquad (Z \in \mathfrak{g}_0).$$

Therefore [11] if $g_i = q_i g$,

$$\omega'(\zeta g)\hat{\,}_{\mathfrak{a}_0} = \sum_{1 \leq i \leq r} (\hat{g}_i)_{b_i \cdot \mathfrak{a}_0} + (D^*g)\hat{\,}_{\mathfrak{a}_0}.$$

Moreover since the $b_i$ are homogenous of degree 2, it is obvious that $b_i = b_i^* \in \mathfrak{B}\mathfrak{g}$ and so $\int \alpha_0(x;b_i^*)dx = 0$. So we conclude that

$$\omega'(\bar{H}_0)\sigma_{\bar{H}_0}(\zeta g) = \sigma_{\bar{H}_0}(D^*g),$$

and this proves that $D\sigma_{\bar{H}_0} = \omega'(\bar{H}_0)\zeta\sigma_{\bar{H}_0}$.

Define the mapping $f \to \psi_f$ of $\mathcal{B}(\mathfrak{h}_0)$ into $J(\mathfrak{k}_0)$ as in Lemma 27 of [2(h)] and put $\Phi_{\bar{H}_0}(f) = \sigma_{\bar{H}_0}(\zeta\psi_f)$ $(f \in \mathcal{B}(\mathfrak{h}_0))$. Then $\Phi_{\bar{H}_0}$ is a $\mathcal{B}$-distribution, which, from Theorem 5 of [2(h)], coincides with an analytic function on $\mathfrak{h}_0$. Our first object is to show that $\Phi_{\bar{H}_0} = 0$ if $\bar{\mathfrak{h}}_0$ is not conjugate to $\mathfrak{h}_0$ under $G$. For this we need a few lemmas.

LEMMA 36.　*For any $f \in C_c^\infty(G \times \mathfrak{h}_0')$, put*

$$\hat{f}(X) = \int_{G \times \mathfrak{h}_0'} \exp\{(-1)^{\frac{1}{2}}B(X,xH)\}f(x,H)\,dx\,dH \qquad (X \in \mathfrak{g}_0),$$

*where $dH$ is the regular Euclidean measure on $\mathfrak{h}_0'$. Then $\hat{f} \in \mathcal{B}(\mathfrak{g}_0)$ and $f \to \hat{f}$ is a continuous mapping of $C_c^\infty(G \times \mathfrak{h}_0')$ into $\mathcal{B}(\mathfrak{g}_0)$.*

Put $G^* = G/A$ as before, and let $H_0$ be any point in $\mathfrak{h}_0'$. Then we can select (see Section 6) an open neighborhood $U$ of $H_0$ in $\mathfrak{h}_0'$ such that the mapping $\phi : (x^*, H) \to x^*H$ of $G^* \times U$ into $\mathfrak{g}_0$ is univalent and regular and, if the Euclidean measure $dX$ on $\mathfrak{g}_0$ is suitably normalized, $dX = |\pi(H)|^2 dx^* dH$ $(X = x^*H, x^* \in G^*, H \in U)$. Then $N = \phi(G^* \times U)$ is an open subset of $\mathfrak{g}_0$ and $\phi$ defines an analytic isomorphism of $G^* \times U$ with $N$. Therefore, for any $f \in C_c^\infty(G \times U)$, we can define a function $F_f \in C_c^\infty(N)$ by

$$F_f(x^*H) = |\pi(H)|^2 \int_A f(xa,H)\,da \qquad (x \in G, H \in U).$$

Here $da$ is the Haar measure on $A$ normalized so that

$$\int_G g(x)\,dx = \int_{G^*} dx^* \int_A g(xa)\,da \qquad (g \in C_c(G)).$$

---

[11] Here $(h)\hat{\,}_{\mathfrak{a}} = (\hat{h})_{\mathfrak{a}}$ for $h \in \mathcal{B}(\mathfrak{k}_0)$ and $a \in C_c^\infty(G)$.

4

It is now clear that

$$\hat{f}(X') = \int_{\mathfrak{g}_0} \exp\{(-1)^{\frac{1}{2}} B(X', X)\} F_f(X) dX \qquad (X' \in \mathfrak{g}_0),$$

and therefore $\hat{f}$ is the Fourier transform of $F_f$. This proves that $\hat{f} \in \mathscr{E}(\mathfrak{g}_0)$. Moreover, since $\phi$ is an analytic isomorphism, the mapping $f \to F_f$ of $C_c^\infty(G \times U)$ into $C_c^\infty(N)$ is obviously continuous, and so the same holds for the mapping $f \to \hat{f}$ of $C_c^\infty(G \times U)$ into $\mathscr{E}(\mathfrak{g}_0)$. The statement of the lemma now follows from the fact that any compact subset of $\mathfrak{h}_0'$ can be covered by a finite number of sets of the type $U$.

For a fixed $\bar{H} \in \bar{\mathfrak{h}}_0'$, put

$$T_{\bar{H}}(g) = \bar{\pi}(\bar{H}) \int_{\bar{G}} g(\bar{x}\bar{H}) d\bar{x} \qquad (g \in \mathscr{E}(\mathfrak{g}_0)).$$

Then from Theorem 3, $T_{\bar{H}}$ is a $\mathscr{E}$-distribution on $\mathfrak{g}_0$. Put $\Phi_f(\bar{H}) = T_{\bar{H}}(\hat{f})$ for $f \in C_c^\infty(G \times \mathfrak{h}_0')$. Then it follows from Lemma 36 that for fixed $\bar{H}$ the mapping $f \to \Phi_f(\bar{H})$ is a distribution on $G \times \mathfrak{h}_0'$. If $\alpha \in C_c^\infty(G)$ and $\beta \in C_c^\infty(\mathfrak{h}_0')$, we denote by $\alpha \times \beta$ the function $\gamma \in C_c^\infty(G \times \mathfrak{h}_0')$ given by $\gamma(x, H) = \alpha(x)\beta(H)$ $(x \in G, H \in \mathfrak{h}_0')$.

LEMMA 37. *For any $\bar{H} \in \bar{\mathfrak{h}}_0'$, there exists a distribution $\tau_{\bar{H}}'$ on $\mathfrak{h}_0'$ such that*

$$\Phi_{\alpha \times \beta}(\bar{H}) = \int_G \alpha(x) dx \, \tau_{\bar{H}}'(\beta)$$

*for $\alpha \in C_c^\infty(G)$ and $\beta \in C_c^\infty(\mathfrak{h}_0')$.*

It is obvious that for a fixed $\beta$, the mapping $\alpha \to \Phi_{\alpha \times \beta}(\bar{H})$ is a distribution on $G$ which is invariant under left translations. Therefore from Lemma 36 of [2(e)], there exists a constant $c_\beta$ such that

$$\Phi_{\alpha \times \beta}(\bar{H}) = c_\beta \int_G \alpha(x) dx \qquad (\alpha \in C_c^\infty(G)).$$

Select $\alpha_0 \in C_c^\infty(G)$ such that $\int_G \alpha_0(x) dx = 1$. Then $c_\beta = \Phi_{\alpha_0 \times \beta}(\bar{H})$. But it is evident that the mapping $\tau_{\bar{H}}' : \beta \to \Phi_{\alpha_0 \times \beta}(\bar{H})$ $(\beta \in C_c^\infty(\mathfrak{h}_0'))$ is a distribution on $\mathfrak{h}_0'$. Hence

$$\Phi_{\alpha \times \beta}(\bar{H}) = \int_G \alpha(x) dx \, \tau_{\bar{H}}'(\beta) \qquad (\alpha \in C_c^\infty(G), \beta \in C_c^\infty(\mathfrak{h}_0')).$$

Put $\tau_{\bar{H}} = \pi \tau_{\bar{H}}'$ $(\bar{H} \in \bar{\mathfrak{h}}_0')$.

LEMMA 38. *For any* $p \in I(\mathfrak{g})$ *and* $\bar{H} \in \bar{\mathfrak{h}}_0'$,

$$\partial(\bar{p})\tau_{\bar{H}} = p'(\bar{H})\tau_{\bar{H}}$$

*where* $\bar{p}$ *denotes the restriction of* $p$ *on* $\mathfrak{h}$.

We use the notation of the proof of Lemma 35. It is obvious that

$$\lambda_X(Z;\partial(q)) = q'(X)\lambda_X(Z) \qquad\qquad (X, Z \in \mathfrak{g}_0; q \in S(\mathfrak{g})).$$

Select $\alpha \in C_c^\infty(G)$ such that $\int \alpha(x)dx = 1$, and fix $\beta \in C_c^\infty(\mathfrak{h}_0')$. Then if

if $p \in I(\mathfrak{g})$, it follows from Lemmas 5 and 8 of [2(h)] that

$$p'(X)\lambda_X(xH) = \lambda_X(xH;\partial(p))$$

$$= \sum_{1 \leq i \leq N} a_i(H)\Lambda_X(x;b_i:H;\partial(u_i)) + \pi(H)^{-1}\Lambda_X(x:H;\partial(\bar{p})\circ\pi)$$

for $x \in G$, $H \in \mathfrak{h}_0'$ and $X \in \mathfrak{g}_0$. Here $u_i \in S(\mathfrak{h})$, $b_i \in \mathfrak{B}\mathfrak{g}$ and the $a_i$ are analytic functions on $\mathfrak{h}_0'$. Hence if $\gamma = \alpha \times \beta$,

$$p'(X)\hat{\gamma}(X) = \sum_{1 \leq i \leq N} \int \alpha_i(x)\beta_i(H)\lambda_X(xH)dxdH + \int \alpha(x)\beta'(H)\lambda_X(xH)dxdH$$

where $\alpha_i = b_i^*\alpha$, $\beta_i = (\partial(u_i))^*(a_i\beta)$ $(1 \leq i \leq N)$ and $\beta' = (\pi^{-1}\partial(\bar{p})\circ\pi)^*\beta$. (As usual the star denotes the adjoint of a differential operator). But then

$$\int \alpha_i(x)dx = \int \alpha(x;b_i^*)dx = 0$$

and so it follows from Lemma 37 that

$$T_{\bar{H}}(p'\hat{\gamma}) = \tau_{\bar{H}}'(\beta').$$

Moreover $p' \in I(\mathfrak{g})$ and therefore it is obvious that

$$T_{\bar{H}}(p'\gamma) = p'(\bar{H})T_{\bar{H}}(\hat{\gamma}) = p'(\bar{H})\tau_{\bar{H}}'(\beta).$$

This proves that

$$\tau_{\bar{H}}'(\beta') = p'(\bar{H})\tau_{\bar{H}}'(\beta).$$

But $\beta' = (\pi^{-1}\partial(\bar{p})\circ\pi)^*\beta$. Hence if we replace $\beta$ by $\pi\beta$, we get

$$\tau_{\bar{H}}((\partial(\bar{p}))^*\beta) = p'(\bar{H})\tau_{\bar{H}}(\beta),$$

and so the lemma is proved.

Now normalize the Euclidean measure $dY$ on $\mathfrak{k}_0$ in such a way that

$$\int_{\mathfrak{k}_0} g(Y)dY = \int_{\mathfrak{h}_0} |\pi_-(H)|^2 dH \int_K g(kH)dK \qquad (g \in C_c(\mathfrak{k}_0)).$$

Here $\pi_- = \prod\limits_{\alpha \in P_-} \alpha$ and $dk$ is the Haar measure on $K$ $(\int_K dk = 1)$. It is easily seen that $\zeta(H)|\pi_-(H)|^2 = (-1)^r \pi(H)^2$ $(H \in \mathfrak{h}_0)$, where $r$ is the degree of $\pi$. Therefore it follows that if $f \in \mathscr{B}(\mathfrak{h}_0)$ and $X \in \mathfrak{g}_0$, then

$$\int_{\mathfrak{t}_0} \exp\{(-1)^{\frac{1}{2}} B(X,Y)\} \zeta(Y) \psi_f(Y) dY$$
$$= (-1)^r \int_{K \times \mathfrak{h}_0} \exp\{(-1)^{\frac{1}{2}} B(X, kH)\} \pi(H) \sum_{s \in W} \epsilon(s) f(sH) \, dk \, dH.$$

Now select $\beta \in C_c^\infty(\mathfrak{h}_0')$ and $\alpha \in C_c^\infty(G)$ $(\int \alpha(x) dx = 1)$, and put [12]

$$\alpha_1(x) = \int_K \alpha(xk) dk, \qquad \beta_1 = \pi \sum_{s \in W} \epsilon(s) \beta^s$$

and $\gamma = \alpha_1 \times \beta_1$. Then $T_{\bar{H}}(\hat{\gamma}) = \Phi_{\alpha_1 \times \beta_1}(\bar{H}) = \tau_{\bar{H}}'(\beta_1)$ since $\int \alpha_1(x) dx = 1$. On the other hand, if $g = \zeta \psi_\beta$,

$$\hat{\gamma}(X) = \int \exp\{(-1)^{\frac{1}{2}} B(X, xkH)\} \alpha(x) \beta_1(H) \, dx \, dk \, dH$$
$$= (-1)^r \hat{g}_\alpha(X) \qquad\qquad (X \in \mathfrak{g}_0)$$

in the notation of the proof of Lemma 35. Therefore

$$\tau_{\bar{H}}'(\beta_1) = T_{\bar{H}}(\hat{\gamma}) = (-1)^r \sigma_{\bar{H}}(g) = (-1)^r \Phi_{\bar{H}}(\beta),$$

and so we get the following result.

LEMMA 39. *If the Euclidean measure on $\mathfrak{t}_0$ is suitably normalized, we have the relation*

$$\Phi_{\bar{H}}(\beta) = (-1)^r \sum_{s \in W} \epsilon(s) \tau_{\bar{H}}(\beta^s)$$

*for all* $\beta \in C_c^\infty(\mathfrak{h}_0')$ *and* $\bar{H} \in \bar{\mathfrak{h}}_0'$.

COROLLARY 1. $\partial(\bar{p}) \Phi_{\bar{H}} = p'(\bar{H}) \Phi_{\bar{H}}$ *for* $p \in I(\mathfrak{g})$ *and* $\bar{H} \in \bar{\mathfrak{h}}_0'$.

We know from Lemma 9 of [2(h)] that $\bar{p}$ is invariant under $W$. Therefore

$$(\partial(\bar{p}))^* \beta^s = ((\partial(\bar{p}))^* \beta)^s \qquad\qquad (s \in W, \beta \in C_c^\infty(\mathfrak{h}_0'))$$

and

$$\Phi_{\bar{H}}((\partial(\bar{p}))^* \beta) = (-1)^r \sum_{s \in W} \epsilon(s) \tau_{\bar{H}}((\partial(\bar{p}))^* \beta^s)$$
$$= (-1)^r p'(\bar{H}) \sum_{s \in W} \epsilon(s) \tau_{\bar{H}}(\beta^s) = p'(\bar{H}) \Phi_{\bar{H}}(\beta)$$

---

[12] As usual $\beta^s$ $(s \in W)$ is defined by $\beta^s(H) = \beta(s^{-1}H)$ $(H \in \mathfrak{h}_0)$.

from Lemma 38. This proves that $\partial(\bar{p})\Phi_{\bar{H}} - p'(\bar{H})\Phi_{\bar{H}}$ is zero on $\mathfrak{h}_0'$. But we have seen that $\Phi_{\bar{H}}$ is an analytic function on $\mathfrak{h}_0$. Therefore $\partial(\bar{p})\Phi_{\bar{H}} = p'(\bar{H})\Phi_{\bar{H}}$ on $\mathfrak{h}_0$.

COROLLARY 2. *Suppose $\mathfrak{h}_0$ is not conjugate to $\bar{\mathfrak{h}}_0$ under $G$. Then $\Phi_{\bar{H}} = 0$ for every $\bar{H} \in \bar{\mathfrak{h}}_0'$.*

Fix $\bar{H} \in \bar{\mathfrak{h}}_0'$ and consider the $\mathscr{C}$-distribution $\tilde{\Phi} : f \to \Phi_{\bar{H}}(f')$ $(f \in \mathscr{C}(\mathfrak{h}_0))$, where

$$f'(H') = \int_{\mathfrak{h}_0} \exp\{-(-1)^{\frac{1}{2}} B(H', H)\} f(H)\, dH \qquad (H' \in \mathfrak{h}_0).$$

Then it follows from Corollary 1 above that $\bar{p}\tilde{\Phi} = p(\bar{H})\tilde{\Phi}$ $(p \in I(\mathfrak{g}))$, and therefore if $H_0$ is any point lying in the carrier of $\tilde{\Phi}$, it is obvious that $p(H_0) = p(\bar{H})$ for all $p \in I(\mathfrak{g})$. Let $\lambda$ be an indeterminate and $I$ the identity mapping of $\mathfrak{g}_0$. Then

$$\det(\lambda I - \operatorname{ad} X) = \sum_{l \le m \le n} \lambda^m p_m(X) \qquad (X \in \mathfrak{g}),$$

where $p_m \in I(\mathfrak{g})$. Let $\bar{\mathfrak{h}}$ be the complexification of $\bar{\mathfrak{h}}_0$ in $\mathfrak{g}$ and $\bar{P}$ the set of all positive roots (under some fixed order) of $\mathfrak{g}$ with respect to $\bar{\mathfrak{h}}$. Then

$$\det(\lambda I - \operatorname{ad} H_0) = (-1)^r \lambda^l \prod_{\alpha \in P} (\lambda - \alpha(H_0))^2$$

and similarly,

$$\det(\lambda I - \operatorname{ad} \bar{H}) = (-1)^r \lambda^l \prod_{\bar{\alpha} \in \bar{P}} (\lambda - \bar{\alpha}(\bar{H}))^2.$$

On the other hand, we know from the Corollary to Lemma 32 that $\bar{\mathfrak{h}}_0$ is not a fundamental Cartan subalgebra of $\mathfrak{g}_0$. Therefore from Lemma 33, there exists a root $\bar{\alpha}_0 \in \bar{P}$ which takes only real values on $\bar{\mathfrak{h}}_0$. Then $\bar{\alpha}_0(\bar{H}) \ne 0$ since $\bar{H} \in \bar{\mathfrak{h}}_0'$. On the other hand, every root $\alpha \in P$ takes only pure imaginary values on $\mathfrak{h}_0$ since $\mathfrak{h}_0 \subset \mathfrak{k}_0$. This proves that

$$\det(\lambda I - \operatorname{ad} H_0) \ne \det(\lambda I - \operatorname{ad} \bar{H}),$$

and therefore $p_m(H_0) \ne p_m(\bar{H})$ for some $m$ $(l \le m \le n)$. This shows that the carrier of $\tilde{\Phi}$ is empty, and therefore since $\tilde{\Phi}$ is the Fourier transform of $\Phi_{\bar{H}}$ (see Schwartz [6(II)]), we conclude that $\Phi_{\bar{H}} = 0$.

Let us now consider the case when $\bar{\mathfrak{h}}_0 = \mathfrak{h}_0$ and denote by $\Phi(H_0 : H)$ the value of the analytic function $\Phi_{H_0}$ at $H$ $(H_0 \in \mathfrak{h}_0', H \in \mathfrak{h}_0)$.

LEMMA 40. *There exists a locally constant function [13] $c$ on $\mathfrak{h}_0'$ such that*

$$\Phi(H_0 : H) = c(H_0) \sum_{s \in W} \epsilon(s) \exp\{(-1)^{\frac{1}{2}} B(sH_0, H)\}$$

*for $H_0 \in \mathfrak{h}_0'$ and $H \in \mathfrak{h}_0$.*

---

[13] This means that $c$ is constant on some neighborhood of every point in $\mathfrak{h}_0'$ and therefore also on each connected component of $\mathfrak{h}_0'$.

Fix $H_0 \in \mathfrak{h}_0'$. Then it follows from Corollary 1 of Lemma 39 and the Corollary to Lemma 13 of [2(h)] that

$$\Phi(H_0:H) = \sum_{s \in W} c_s \exp\{(-1)^{\frac{1}{2}} B(sH_0, H)\} \qquad (H \in \mathfrak{h}_0),$$

where the $c_s$ are uniquely determined constants. On the other hand, it is obvious from the definition of $\Phi_{H_0}$ that $\Phi_{H_0}(\beta^s) = \epsilon(s)\Phi_{H_0}(\beta)$ $(\beta \in \mathscr{B}(\mathfrak{h}_0)$, $s \in W)$. Hence $\Phi(H_0:sH) = \epsilon(s)\Phi(H_0:H)$. Therefore, in view of the uniqueness of $c_s$, we conclude that $\epsilon(s)c_s$ $(s \in W)$ are all equal. This proves that

$$\Phi(H_0:H) = c(H_0) \sum_{s \in W} \epsilon(s) \exp\{(-1)^{\frac{1}{2}} B(sH_0, H)\} \qquad (H_0 \in \mathfrak{h}_0', H \in \mathfrak{h}_0),$$

where $c$ is a function on $\mathfrak{h}_0'$.

Now select $\alpha \in C_c^{\infty}(G)$ such that $\int \alpha(x)\,dx = 1$, and for $\beta \in C_c^{\infty}(\mathfrak{h}_0')$, put $\gamma = \alpha \times \pi\beta$. Then $\tau_{H_0}(\beta) = T_{H_0}(\hat{\gamma}) = \phi_{\hat{\gamma}}(H_0)$ in the notations of Lemma 36 and Theorem 3. On the other hand,

$$\Phi_{H_0}(\beta) = (-1)^r \sum_{s \in W} \epsilon(s)\tau_{H_0}(\beta^s)$$

from Lemma 39. Therefore

$$c(H) \sum_{s \in W} \epsilon(s)\tilde{\beta}(sH) = (-1)^r \sum_{s \in W} \epsilon(s)\phi_{\hat{\gamma}_s}(H)$$

$(H \in \mathfrak{h}_0', \beta \in C_c^{\infty}(\mathfrak{h}_0'))$, where $\tilde{\beta}$ is the Fourier transform of $\beta$ and $\gamma_s = \alpha \times \pi\beta^s$. Since $C_c^{\infty}(\mathfrak{h}_0')$ is dense in $\mathscr{B}(\mathfrak{h}_0)$ under the norm

$$\|\beta_1\| = \int_{\mathfrak{h}_0} |\beta_1(H)|\,dH \qquad (\beta_1 \in \mathscr{B}(\mathfrak{h}_0)),$$

it follows without difficulty that, for a given $H_0 \in \mathfrak{h}_0'$, we can select $\beta \in C_c^{\infty}(\mathfrak{h}_0')$ such that $\sum_{s \in W} \epsilon(s)\tilde{\beta}(sH_0) \neq 0$. Hence, in view of the fact that $\hat{\gamma}_s \in \mathscr{B}(\mathfrak{g}_0)$ (Lemma 36) and $\phi_g$ is of class $C^{\infty}$ on $\mathfrak{h}_0'$ for $g \in \mathscr{B}(\mathfrak{g}_0)$ (Theorem 3), we conclude that $c$ is of class $C^{\infty}$ on $\mathfrak{h}_0'$. Now suppose $p \in I(\mathfrak{g})$. Then it is evident that

$$\gamma(X;\partial(p)) = \int \exp\{(-1)^{\frac{1}{2}} B(X, xH)\}\alpha(x)p'(H)\pi(H)\beta(H)\,dx\,dH \qquad (X \in \mathfrak{g}_0).$$

Hence if $\beta' = \bar{p}'\beta$ and $\gamma' = \alpha \times \pi\beta'$, we get $\partial(p)\hat{\gamma} = \hat{\gamma}'$, and therefore

$$\tau_{H_0}(\beta') = \phi_{\partial(p)\hat{\gamma}}(H_0) = \phi_{\hat{\gamma}}(H_0;\partial(\bar{p}))$$

from Theorem 3. Moreover $(\beta')^s = \bar{p}'\beta^s$ $(s \in W)$ since $\bar{p}'$ is invariant under $W$. Therefore

$$\Phi_{H_0}(\beta') = (-1)^r \sum_{s \in W} \epsilon(s) \tau_{H_0}((\beta')^s) = (-1)^r \sum_{s \in W} \epsilon(s) \phi_{\hat{\gamma}_s}(H_0; \partial(\bar{p}))$$

$$= \int_{\mathfrak{h}_0} \Phi(H_0; \partial(\bar{p}) : H) \beta(H) dH.$$

Since $\beta$ is an arbitrary element in $C_c^\infty(\mathfrak{h}_0')$, we conclude that

$$\Phi(H_0; \partial(\bar{p}) : H) = p'(H)\Phi(H_0 : H) \qquad (H_0 \in \mathfrak{h}_0', H \in \mathfrak{h}_0).$$

But now it follows from Lemma 23 that $c$ is locally constant.

CoROLLARY. *There exists a complex number $c_0$ such that*

$$\sum_{s \in W} \Phi(sH' : H) = c_0 \sum_{s \in W} \epsilon(s) \exp\{(-1)^{\frac{1}{2}} B(sH', H)\}$$

*for all $H' \in \mathfrak{h}_0'$ and $H \in \mathfrak{h}_0$.*

Let $\mathfrak{h}_1$ be a connected component of $\mathfrak{h}_0'$. Then (see **Lemma 44** of the Appendix, Section 12) $\mathfrak{h}_0' = \bigcup_{s \in W} s\mathfrak{h}_1$, and so it is evident that $c_0 = \sum_{s \in W} c(sH')$ $(H' \in \mathfrak{h}_0')$ is independent of $H'$. From this our assertion follows **immediately**.

The main results of this section can now be summarized in the following theorem.

THEOREM 5. *Let $\mathfrak{h}_0$ and $\bar{\mathfrak{h}}_0$ be two Cartan subalgebras of $\mathfrak{g}_0$. We assume that $\mathfrak{h}_0 \subset \mathfrak{k}_0$ and* [14] *$\theta(\bar{\mathfrak{h}}_0) = \bar{\mathfrak{h}}_0$. For any $f \in \mathscr{B}(\mathfrak{h}_0)$, put*

$$\hat{f}(X) = \int_{K \times \mathfrak{h}_0} \exp\{(-1)^{\frac{1}{2}} B(X, kH)\} \pi(H)^2 \sum_{s \in W} f(sH) dk dH \qquad (X \in \mathfrak{g}_0).$$

*Then the integral*

$$\phi_{\hat{f}}(\bar{H}) = \bar{\pi}(\bar{H}) \int_{\bar{G}} \hat{f}(\bar{x}\bar{H}) d\bar{x}$$

*is convergent for $\bar{H} \in \bar{\mathfrak{h}}_0'$. Moreover in case $\bar{\mathfrak{h}}_0$ is not conjugate to $\mathfrak{h}_0$ under $G$, $\phi_{\hat{f}} = 0$ on $\bar{\mathfrak{h}}_0'$. On the other hand, if $\bar{\mathfrak{h}}_0 = \mathfrak{h}_0$, there exists a complex number $c_0$ such that*

$$\sum_{s \in W} \epsilon(s) \phi_{\hat{f}}(sH') = c_0 \int_{\mathfrak{h}_0} \sum_{s \in W} \epsilon(s) \exp\{(-1)^{\frac{1}{2}} B(H', sH)\} \pi(H) f(H) dH$$

*for $H' \in \mathfrak{h}_0'$ and $f \in \mathscr{B}(\mathfrak{h}_0)$.*

---

[14] In view of the corollary to Lemma 1 of [2(e)], our assumption that $\theta(\bar{\mathfrak{h}}_0) = \bar{\mathfrak{h}}_0$ is obviously unnecessary. We make it here only for convenience.

It is evident that $\phi_{\hat{f}}(\bar{H}) = (-1)^r \Phi_{\bar{H}}(\pi f)$ $(f \in \mathcal{B}(\mathfrak{h}_0),\ \bar{H} \in \bar{\mathfrak{h}}_0')$ in our earlier notation. Therefore, from Corollary 2 to Lemma 39, $\phi_{\hat{f}} = 0$ if $\bar{\mathfrak{h}}_0$ is not conjugate to $\mathfrak{h}_0$. On the other hand, if $\bar{\mathfrak{h}}_0 = \mathfrak{h}_0$, our assertion follows from the Corollary to Lemma 40. We shall prove in Section 11 that $c_0 \neq 0$.

**10. Some work of de Rham.** Let $E$ be a Euclidean space over $R$ of dimension $n \geqq 3$, and let $p, q$ be two nonnegative integers such that $p + q = n$. Let $(x_1, \cdots, x_n)$ be a Cartesian system of coordinates on $E$ and put $y_i = x_{p+i}$ $(1 \leqq i \leqq q)$. We consider the differential operator

$$\square = \frac{\partial^2}{\partial x_1{}^2} + \cdots + \frac{\partial^2}{\partial x_p{}^2} - \left( \frac{\partial^2}{\partial y_1{}^2} + \cdots + \frac{\partial^2}{\partial y_q{}^2} \right)$$

on $E$. Let $dX$ denote the Euclidean measure on $E$ so normalized that $dX = dx_1 \cdots dx_n$. Then every locally summable function $\xi$ on $E$ defines, as usual, a distribution

$$f \to \int_E f\xi \, dX \qquad\qquad (f \in C_c^\infty(E))$$

which will also be denoted by $\xi$. Let $\delta$ denote the Dirac distribution given by $\delta(f) = f(0)$ and put $u = x_1{}^2 + \cdots + x_p{}^2 - (y_1{}^2 + \cdots + y_q{}^2)$. Then de Rham has shown [15] that there exists a function $\xi$ on $R$ with the following properties. Put $\Xi(X) = \xi(u(X))$ $(X \in E)$. Then $\Xi$ is a locally summable function on $E$ and

$$\square^{[n/2]}\Xi = \delta$$

in the sense of distribution theory. (Here $[t]$ denotes the largest integer not exceeding $t \in R$). The actual expressions for $\xi$ in the various possible cases are given below in terms of the Heaviside function $Y(t)$ which is 0 or 1 according as $t \leqq 0$ or $t > 0$ $(t \in R)$. There are four cases.[16]

(1)    If $n \equiv p \equiv 1 \bmod 2$,

$$\xi(t) = (-1)^{(p-1)/2} \{ \pi^{\frac{1}{2}(n-2)} 4^{n-2} \Gamma((n-2)/2) \}^{-1} Y(t) t^{-\frac{1}{2}}.$$

(2)    If $n \equiv q \equiv 1 \bmod 2$,

$$\xi(t) = (-1)^{p/2} \{ \pi^{\frac{1}{2}(n-2)} 4^{n-2} \Gamma((n-2)/2) \}^{-1} Y(-t) |t|^{-\frac{1}{2}}.$$

(3)    If $n \equiv p \equiv 0 \bmod 2$,

$$\xi(t) = (-1)^{(p-2)/2} \{ (4\pi)^{n/2} \Gamma(n/2) \}^{-1} \log |t|.$$

---

[15] I am grateful to Professor de Rham for communicating his results to me, which, so far as I know, have not yet been published.

[16] Here $\Gamma$ stands for the classical Gamma function.

(4)   If $p \equiv q \equiv 1 \bmod 2$,

$$\xi(t) = (-1)^{(p-1)/2} \{ (4\pi)^{(n-2)/2} \Gamma((n-2)/2) \}^{-1} Y(t).$$

Put $P = x_1^2 + \cdots + x_p^2$, $Q = y_1^2 + \cdots, y_q^2$. Then a simple calculation shows that

$$c = \int_E | \Xi | (1 + P)^{-m} (1 + Q)^{-m} \, dX < \infty$$

if $m$ is a sufficiently large positive integer. Hence

$$\int | \Xi f | \, dX \leqq c \sup_{X \in E} |(1 + P(X))^m (1 + Q(X))^m f(X)|$$

for $f \in \mathscr{B}(E)$. This proves that $\Xi$ can be regarded as a $\mathscr{B}$-distribution. Therefore, since $\square$ is obviously self-adjoint, we get

$$f(0) = \int_E \Xi \square^{[n/2]} f \, dX$$

for any $f \in \mathscr{B}(E)$.

**11.  An application of the above formula.**  Let $\omega$ be the Casimir polynomial of $\mathfrak{g}$ and put $\square = \partial(\omega)$. Then it is clear from the results of Section 10 that there exists a function $\xi$ on $R$ such that $\Xi(X) = \xi(\omega(X))$ $(X \in \mathfrak{g}_0)$ is a locally summable function on $\mathfrak{g}_0$ and

$$f(0) = \int_{\mathfrak{g}_0} \Xi \partial(\omega^m) f \, dX \qquad (f \in \mathscr{B}(\mathfrak{g}_0)).$$

Here $m = \lceil n/2 \rceil$, $n = \dim_R \mathfrak{g}_0$ and $dX$ is the regular Euclidean measure on $\mathfrak{g}_0$. (Since $\mathfrak{g}_0$ is semisimple it is obvious that $n \geqq 3$). We shall apply this formula to prove that the constant $c_0$ of Theorem 4 is not zero. So let us assume that $\mathfrak{h}_0$ is a Cartan subalgebra of $\mathfrak{g}_0$ which is contained in $\mathfrak{k}_0$. Select a finite set of Cartan subalgebras $\mathfrak{h}_0 = \mathfrak{h}_1, \mathfrak{h}_2, \cdots, \mathfrak{h}_N$ of $\mathfrak{g}_0$ such that $\theta(\mathfrak{h}_i) = \mathfrak{h}_i$ $(1 \leqq i \leqq N)$ and every Cartan subalgebra of $\mathfrak{g}_0$ is conjugate under $G$ to $\mathfrak{h}_i$ for exactly one $i$ $(1 \leqq i \leqq N)$. We know (see the Corollary to Lemma 2 of [2(e)]) that this is possible. Let $(\mathfrak{h}_i)_c$ denote the complexification of $\mathfrak{h}_i$ in $\mathfrak{g}$ and $P_i$ the set of all positive roots of $\mathfrak{g}$ with respect to $(\mathfrak{h}_i)_c$ under some fixed order. (We assume that $P_1 = P$.) Put $\pi_i = \prod_{\alpha \in P_i} \alpha$ and let $W_i$ denote the Weyl group of $\mathfrak{g}$ with respect to $(\mathfrak{h}_i)_c$. Consider the Cartan subgroup $A_i$ of $G$ corresponding to $\mathfrak{h}_i$ and the normalizer $A_i'$ of $A_i$

in $G$. Then [17] the factor group $W_i' = A_i'/A_i$ can be regarded as a subgroup of $W_i$. Put $G_i^* = G/A_i$, and define $x^*H$ $(x^* \in G_i^*, H \in \mathfrak{h}_i)$ in the usual way. We have seen in Section 8 that $W_i'$ operates on $G_i^*$ on the right. Let $\mathfrak{g}_0'$ be the set of regular elements of $\mathfrak{g}_0$, and put $\mathfrak{h}_i' = \mathfrak{g}_0' \cap \mathfrak{h}_i$. Consider the mapping $\psi_i : (x^*, H) \to x^*H$ of $G_i^* \times \mathfrak{h}_i'$ into $\mathfrak{g}_0'$, and put $\psi_i(G_i^* \times \mathfrak{h}_i') = \mathfrak{g}_i$. Then $\psi_i$ is everywhere regular on $G_i^* \times \mathfrak{h}_i'$, and therefore $\mathfrak{g}_i$ is open in $\mathfrak{g}_0$ and if $w_i'$ is the order of $W_i'$, any element $x^*H$ $(x^* \in G_i^*, H \in \mathfrak{h}_i')$ of $\mathfrak{g}_i$ has exactly $w_i'$ distinct pre-images under $\psi_i$, namely, $(x^*s, s^{-1}H_i)$ $(s \in W')$. Let $d_iH$ denote the regular Euclidean measure on $\mathfrak{h}_i$. Then the invariant measure $d_ix^*$ on $G_i^*$ can be so normalized that

$$\int_{\mathfrak{g}_i} g(X)\,dX = \int_{G_i^* \times \mathfrak{h}_i'} |\pi_i(H)|^2 g(x^*H)\,d_ix^*\,d_iH \qquad (g \in C_c(\mathfrak{g}_i)).$$

Since no two among $\mathfrak{h}_1, \cdots, \mathfrak{h}_N$ are conjugate under $G$, it is clear that the sets $\mathfrak{g}_1, \cdots, \mathfrak{g}_N$ are disjoint. Moreover, if $X$ is any regular element in $\mathfrak{g}_0$, its centralizer $\mathfrak{h}_X$ (in $\mathfrak{g}_0$) in a Cartan subalgebra of $\mathfrak{g}_0$ which must be conjugate to some $\mathfrak{h}_i$. This proves that $\mathfrak{g}_0' = \bigcup_{1 \leq i \leq N} \mathfrak{g}_i$, and therefore

$$\int_{\mathfrak{g}_0} g(X)\,dX = \int_{\mathfrak{g}_0'} g(X)\,dX = \sum_{1 \leq i \leq N} \int_{G_i^* \times \mathfrak{h}_i'} |\pi_i(H)|^2 g(x^*H)\,d_ix^*\,d_iH$$
$$(g \in C_c(\mathfrak{g}_0)).$$

Now for any $f \in \mathcal{B}(\mathfrak{g}_0)$, put

$$\phi_f^{(i)}(H) = \pi_i(H) \int_{G_i^*} f(x^*H)\,d_ix^* \qquad (H \in \mathfrak{h}_i').$$

From Theorem 3, we know that this integral is convergent, and it follows from Fubini's Theorem that [18]

$$\int_{\mathfrak{g}_0} \Xi f\,dX = \sum_{1 \leq i \leq N} \int_{\mathfrak{h}_i'} (\mathrm{conj}\,\pi_i(H))\,\Xi(H)\,\phi_f^{(i)}(H)\,d_iH \qquad (f \in \mathcal{B}(\mathfrak{g}_0)).$$

Now put $G_1^* = G^*$, $d_1x^* = dx^*$, $d_1H = dH$ and $\pi_1 = \pi$ as before and select two real-valued nonnegative functions $\alpha_0$ and $\beta_0$ in $C_c^\infty(G^*)$ and $C_c^\infty(\mathfrak{h}_0')$, respectively, such that

$$\int_{G^*} \alpha_0(x^*)\,dx^* = \int_{\mathfrak{h}_0} |\pi(H)|^2 \beta_0(H)\,dH = 1.$$

---

[17] All the facts stated here have already been seen in Sections 6 and 8.
[18] We denote the complex conjugate of $c$ by $\mathrm{conj}\,c$ $(c \in C)$.

Put $\alpha(x^*) = (w_1')^{-1} \sum_{s \in W_1'} \alpha_0(x^*s)$ and [12] $\beta = w^{-1} \sum_{s \in W} \beta_0{}^s$, where $W = W_1$ and $w$ is the order of $W$. We have seen in Section 8 that the measure $dx^*$ is invariant under the transformation $x^* \to x^*s$ of $G^*$ for any fixed $s \in W_1'$. Hence

$$\int_{G^*} \alpha(x^*)\,dx^* = \int_{\mathfrak{h}_0} |\pi(H)|^2 \beta(H)\,dH = 1,$$

and it is obvious that there exists a function $F \in C_c^\infty(\mathfrak{g}_1)$ such that $F(x^*H) = \alpha(x^*)\beta(H)$ $(x^* \in G^*, H \in \mathfrak{h}_0')$. Consider its Fourier transform $\tilde{F}$ and, for any fixed $i$ $(1 \leq i \leq N)$, put $\mathfrak{h}_i = \tilde{\mathfrak{h}}_0$. Then since

$$\tilde{F}(X') = \int \exp\{(-1)^{\frac{1}{2}} B(X', X)\} F(X)\,dX$$

$$= \int_{G^* \times \mathfrak{h}_0'} \exp\{(-1)^{\frac{1}{2}} B(X', x^*H)\}\alpha(x^*) |\pi(H)|^2 \beta(H)\,dx^*dH,$$

it is obvious from Lemma 37 that

$$\phi_{\tilde{F}}^{(i)}(\bar{H}) = (-1)^r \tau_{\bar{H}}(\pi\beta) \qquad (\bar{H} \in \dot{\mathfrak{h}}_0')$$

in the notation of Section 9. Moreover $(\pi\beta)^s = \epsilon(s)\pi\beta$ $(s \in W)$, and so it follows from Lemma 39 that

$$\tau_{\bar{H}}(\pi\beta) = w^{-1}(-1)^r \Phi_{\bar{H}}(\pi\beta).$$

Therefore if $i \neq 1$, $\phi_{\tilde{F}}^{(i)} = 0$ from Corollary 2 to Lemma 39. Now suppose $i = 1$. Then if we write $\phi_{\tilde{F}}$ instead of $\phi_{\tilde{F}}^{(1)}$, we find from the Corollary to Lemma 40 that

$$\sum_{s \in W} \epsilon(s)\phi_{\tilde{F}}(sH') = c_0 \int_{\mathfrak{h}_0} \exp\{(-1)^{\frac{1}{2}} B(H', H)\}\pi(H)\beta(H)\,dH \quad (H' \in \mathfrak{h}_0')$$

since $(\pi\beta)^s = \epsilon(s)\pi\beta$ $(s \in W)$. On the other hand, $\tilde{F} \in \mathcal{B}(\mathfrak{g}_0)$, and therefore

$$\tilde{F}(0) = \int_{\mathfrak{g}_0} \Xi\partial(\omega^m)\tilde{F}\,dX = \sum_{1 \leq i \leq N} \int_{\mathfrak{h}_0'} (\mathrm{conj}\,\pi_i(H))\phi_{F_m}^{(i)}(H)\Xi(H)\,dH,$$

where $F_m = \partial(\omega^m)\tilde{F}$. But from Theorem 3,

$$\phi_{F_m}^{(i)} = \partial(\omega_i{}^m)\phi_{\tilde{F}}^{(i)}$$

where $\omega_i$ is the restriction of $\omega$ on $\mathfrak{h}_i$. Therefore $\phi_{F_m}^{(i)} = 0$ if $i \neq 1$, and hence

$$\tilde{F}(0) = (-1)^r \int_{\mathfrak{h}_0'} \pi(H)\phi_{\tilde{F}}(H; \partial(\overline{\omega}^m))\Xi(H)\,dH,$$

where $\bar{\omega} = \omega_1$. But since $\Xi(H) = \xi(\omega(H))$ $(H \in \mathfrak{h}_0')$ and $\bar{\omega}^s = \bar{\omega}$ $(s \in W)$, it is obvious that $\Xi(sH) = \Xi(H)$. Moreover, $\pi^s = \epsilon(s)\pi$, and so we conclude that

$$\tilde{F}(0) = (-1)^r w^{-1} \int_{\mathfrak{h}_0'} \pi(H)\phi'(H;\partial(\bar{\omega}^m))\Xi(H)dH,$$

where $\phi'(H) = \sum_{s \in W} \epsilon(s)\phi_{\tilde{F}}(sH)$ $(H \in \mathfrak{h}_0')$. On the other hand,

$$\tilde{F}(0) = \int_{\mathfrak{g}_0} F(X)dX = \int_{G^* \times \mathfrak{h}_0'} F(x^*H)|\pi(H)|^2 dx^*dH$$

$$= \int \alpha(x^*)dx^* \int \beta(H)|\pi(H)|^2 dH = 1.$$

Therefore $\phi'$ cannot be identically zero on $\mathfrak{h}_0'$, and this proves that $c_0 \neq 0$.

LEMMA 41. *Suppose $\mathfrak{h}_0$ is a Cartan subalgebra of $\mathfrak{g}_0$ which is contained in $\mathfrak{k}_0$. Let $\mathfrak{h}_{(i)}$ $(1 \leq i \leq q)$ be all the distinct connected components of $\mathfrak{h}_0' = \mathfrak{h}_0 \cap \mathfrak{g}_0'$. Then there exists a function $g \in \mathcal{B}(\mathfrak{g}_0)$ such that*

$$\sum_{1 \leq i \leq q} \mathop{\mathrm{Lim}}_{\mathfrak{h}_{(i)}} \phi_g(0;\partial(\pi)) \neq 0$$

*in the notation of Section 7. Moreover the constant $c_0$ of Theorem 5 is not zero.*

Put $g = \tilde{F}$. Then since $\pi^s = \epsilon(s)\pi$ $(s \in W)$, it is clear that

$$\phi'(H';\partial(\pi)) = \sum_{s \in W} \phi_g(sH';\partial(\pi))$$

$$= c_0(-1)^{r/2} \int_{\mathfrak{h}_0} \exp\{(-1)^{\frac{1}{2}} B(H',H)\}\pi(H)^2\beta(H)dH$$
$$(H' \in \mathfrak{h}_0'),$$

and therefore

$$\mathop{\mathrm{Lim}}_{H' \to 0} \sum_{s \in W} \phi_g(sH';\partial(\pi)) = c_0(-1)^{r/2} \int_{\mathfrak{h}_0} \pi(H)^2\beta(H)dH = c_0(-1)^{r/2} \neq 0.$$

Since every $s \in W$ permutes $\mathfrak{h}_{(1)}, \cdots, \mathfrak{h}_{(q)}$ among themselves, the statement of the lemma is now obvious.

## 12. Appendix.

LEMMA 42. *Let $U$ be an open convex subset of a real Euclidean space $E$ of dimension $l$, and let $\sigma_1, \cdots, \sigma_N$ be a finite number of vector subspaces of $E$ of dimension $\leq l-2$. Let $U'$ denote the complement in $U$ of the set $U \cap (\bigcup_{1 \leq i \leq N} \sigma_i)$. Then if $X_1, X_2$ are any two points in $U'$, we can select*

$X_2' \in U'$ *arbitrarily near* $X_2$ *such that the straight line-segment joining* $X_1$ *and* $X_2'$ *lies entirely in* $U'$.

By replacing $\sigma_i$ by a larger linear subspace, if necessary, we may clearly assume that $\dim \sigma_i = l - 2$ $(1 \leq i \leq N)$. For each $i$, select two linear functions $\alpha_i$, $\beta_i$ on $E$ such that $\sigma_i$ is exactly the set of those $X \in E$ where $\alpha_i(X) = \beta_i(X) = 0$. Obviously, $\alpha_i$ and $\beta_i$ are linearly independent for every $i$. Let $E'$ denote the complement of $\bigcup_{1 \leq i \leq N} \sigma_i$ in $E$. If $Y_1, Y_2$ are two points in $E'$, it is evident that the line-segment joining $Y_1, Y_2$ cannot meet $\sigma_i$ unless $\{\alpha_i(Y_1)/\beta_i(Y_1)\} = \{\alpha_i(Y_2)/\beta_i(Y_2)\}$. Now suppose $X_1, X_2$ are two points in $U'$ and $\Sigma$ is the set of those $i$ $(1 \leq i \leq N)$ for which $\{\alpha_i(X_1)/\beta_i(X_1)\} = \{\alpha_i(X_2)/\beta_i(X_2)\}$. For the proof of the lemma, we may obviously assume that $\Sigma$ is not empty. Then since $\alpha_j, \beta_j$ are linearly independent for every $j$, we can select $X \in E$ such that

$$\alpha_i(X)\beta_i(X_1) - \beta_i(X)\alpha_i(X_1) \neq 0$$

for all $i \in \Sigma$. Then if $\epsilon$ is a sufficiently small positive number and $X_2' = X_2 + \epsilon X$, it is clear that $X_2' \in U'$ and

$$\alpha_j(X_2')\beta_j(X_1) - \beta_j(X_2')\alpha_j(X_1) \neq 0 \qquad (1 \leq j \leq N).$$

This proves that the line-segment joining $X_1$ and $X_2'$ lies in $U \cap E' = U'$.

COROLLARY. *$U'$ is connected.*

Now we use the notation of Section 6. Define $\mathfrak{h}_0'$ and $\mathfrak{h}_1$ as in Lemma 24 and let $Q$ be the set of those roots $\alpha \in P$ for which $\alpha^2$ takes only real values on $\mathfrak{h}_0$. Since all roots take only real values on $(-1)^{\frac{1}{2}}\mathfrak{h}_{\mathfrak{k}_0} + \mathfrak{h}_{\mathfrak{p}_0}$, it is obvious that a positive root lies in $Q$ if and only if it vanishes identically either on $\mathfrak{h}_{\mathfrak{k}_0}$ or on $\mathfrak{h}_{\mathfrak{p}_0}$. Let $P'$ denote the complement of $Q$ in $P$, and, for any $\alpha \in P$, let $\sigma_\alpha$ denote the set of those $H \in \mathfrak{h}_0$ where $\alpha(H) = 0$. Then $\sigma_\alpha$ is a linear subspace of $\mathfrak{h}_0$, and if $l = \dim_R \mathfrak{h}_0$, it is clear that $\dim_R \sigma_\alpha = l - 1$ or $l - 2$ according as $\alpha$ lies in $Q$ or $P'$.

For any $\alpha \in Q$, define $\alpha^* = \alpha$ or $(-1)^{\frac{1}{2}}\alpha$ according as $\alpha$ takes only real or only pure imaginary values on $\mathfrak{h}_0$. Then since $\mathfrak{h}_1$ is connected, $\alpha^*$ must keep constant sign on $\mathfrak{h}_1$. Put $\alpha' = \epsilon_\alpha \alpha^*$, where $\epsilon_\alpha = 1$ or $-1$ according as $\alpha^*$ is positive or negative on $\mathfrak{h}_1$. Let $\mathfrak{h}_2$ be the set of all $H \in \mathfrak{h}_0$ where $\alpha'(H) > 0$ $(\alpha \in Q)$. Clearly, $\mathfrak{h}_2$ is an open convex subset of $\mathfrak{h}_0$ and $\mathfrak{h}_2 \cap \mathfrak{h}_0'$ is the complement in $\mathfrak{h}_2$ of the set $\mathfrak{h}_2 \cap (\bigcup_{\alpha \in P'} \sigma_\alpha)$. Therefore it follows from the Corollary of Lemma 42 that $\mathfrak{h}_2 \cap \mathfrak{h}_0'$ is connected. On the other hand, it is obvious that $\mathfrak{h}_2 \cap \mathfrak{h}_0' \supset \mathfrak{h}_1$. Therefore, since $\mathfrak{h}_1$ is a maximal connected subset of $\mathfrak{h}_0'$, we conclude that $\mathfrak{h}_1 = \mathfrak{h}_2 \cap \mathfrak{h}_0'$. So the following result is an immediate consequence of Lemma 42.

LEMMA 43. *If $H_1, H_2$ are any two points in $\mathfrak{h}_1$, we can select $H_2' \in \mathfrak{h}_1$ arbitrarily near $H_2$ such that the straight line-segment joining $H_1$ and $H_2'$ lies entirely in $\mathfrak{h}_1$.*

*Remark.* Let $V_1$ be the closure of $\mathfrak{h}_1$ in $\mathfrak{h}_0$. Since $\mathfrak{h}_1 = \mathfrak{h}_2 \cap \mathfrak{h}_0'$ is dense in $\mathfrak{h}_2$, it is obvious that $V_1$ is the set of those points $H \in \mathfrak{h}_0$ where $\alpha'(H) \geqq 0$ $(\alpha \in Q)$. Moreover, $V_1 \cap \mathfrak{h}_0' = \mathfrak{h}_2 \cap \mathfrak{h}_0' = \mathfrak{h}_1$. We shall need these facts a little later.

For a linear function $\lambda$ on $\mathfrak{h}$ and $s \in W$, define the linear function $s\lambda$ by $(s\lambda)(H) = \lambda(s^{-1}H)$ $(H \in \mathfrak{h})$. Then if $\alpha$ is a root, the same holds for $s\alpha$. Let $W_R$ be the subgroup consisting of those elements in $W$ which map $\mathfrak{h}_0$ onto itself. Then if $\alpha^2$ is real on $\mathfrak{h}_0$, the same obviously holds for $(s\alpha)^2$ $(s \in W_R)$. This means that $s$ permutes $\{\sigma_\alpha\}_{\alpha \in Q}$ and $\{\sigma_\alpha\}_{\alpha \in P'}$ separately among themselves. Let $s_\alpha$ denote the Weyl reflexion corresponding to any root $\alpha$. Then $s_\alpha H = H - 2\{\alpha(H)/\alpha(H_\alpha)\}H_\alpha$ $(H \in \mathfrak{h})$, where $H_\alpha$ is the element in $\mathfrak{h}$ determined by the condition $B(H, H_\alpha) = \alpha(H)$ $(H \in \mathfrak{h})$. If $\alpha \in Q$, it is clear that $\alpha(H)H_\alpha \in \mathfrak{h}_0$ for $H \in \mathfrak{h}_0$, and therefore $s_\alpha \in W_R$. Let $W_Q$ be the subgroup of $W$ generated by $s_\alpha$ $(\alpha \in Q)$. Then $W_Q \subset W_R$.

LEMMA 44. $\mathfrak{h}_0' = \bigcup_{s \in W_Q} s\mathfrak{h}_1.$

Let $U$ be the complement in $\mathfrak{h}_0$ of the union of $\sigma_\alpha \cap \sigma_\beta$ $(\alpha, \beta \in P, \alpha \neq \beta)$. Since every root takes only real values on $\mathfrak{h}_{\mathfrak{p}_0} + (-1)^{\frac{1}{2}}\mathfrak{h}_{\mathfrak{k}_0}$, it is obvious that [18] $\alpha(\theta(H)) = -\operatorname{conj}\alpha(H)$ $(H \in \mathfrak{h}_0, \alpha \in P)$. Therefore [19] $\alpha = \pm \theta\alpha$ if and only if $\alpha \in Q$. Hence it follows from the definition of $U$ that no root in $P'$ can vanish anywhere on $U$. Moreover, it is obvious that $sU = U$ $(s \in W_R)$, and from the Corollary to Lemma 42, $U$ is connected. Let $V_1$ denote the closure of $\mathfrak{h}_1$ in $\mathfrak{h}_0$, and put $V = \bigcup_{s \in W_Q} sV_1$. We shall first prove that $V = \mathfrak{h}_0$.

Since $U$ is connected and dense in $\mathfrak{h}_0$, it would be enough to show that $V \cap U$ is open. But $V \cap U = \bigcup_{s \in W_Q} s(V_1 \cap U)$. Therefore it would be sufficient to prove that any point $H_0 \in V_1 \cap U$, lies in the interior of $V \cap U$.

---

[19] For any linear function $\lambda$ on $\mathfrak{h}$, $\theta\lambda$ denotes the function $H \to \lambda(\theta(H))$ $(H \in \mathfrak{h})$.

Since $\mathfrak{h}_1 \subset V \cap U$ and $\mathfrak{h}_1$ is open, we may suppose that $H_0 \notin \mathfrak{h}_1$. Then, in view of the definition of $U$, there is exactly one root $\alpha_0 \in P$ such that $H_0 \in \sigma_{\alpha_0}$ and moreover $\alpha_0 \in Q$. Therefore $\dim \sigma_{\alpha_0} = l - 1$ and $s_{\alpha_0}$ is the reflexion in the hyperplane $\sigma_{\alpha_0}$. Select a positive number $\epsilon$, and let $N$ be the set of those points $H \in \mathfrak{h}_0$ where $\alpha_0'(H) \geqq 0$ and $\| H - H_0 \| \leqq \epsilon$. Then if $\epsilon$ is sufficiently small, it is clear that $\alpha'(H) \geqq \tfrac{1}{2}\alpha'(H_0)$ for any $H \in N$ and $\alpha$ in $Q$. Therefore $N \subset V_1$ (see the remark after Lemma 43). But it is obvious that $N \cup s_{\alpha_0} N$ consists of all points $H \in \mathfrak{h}_0$ for which $\| H - H_0 \| \leqq \epsilon$, and therefore, since $U$ is open, $H_0$ lies in the interior of $V \cap U$. This proves that $V = \mathfrak{h}_0$.

Now we have seen above that $V_1 \cap \mathfrak{h}_0' = \mathfrak{h}_1$. Therefore

$$\mathfrak{h}_0' = V \cap \mathfrak{h}_0' = \bigcup_{s \in W_Q} s(V_1 \cap \mathfrak{h}_0') = \bigcup_{s \in W_Q} s\mathfrak{h}_1.$$

Thus the lemma is proved.

Let $Q_+$ be the set of those roots $\alpha \in Q$ which take only real values on $\mathfrak{h}_0$. It is obvious that a positive root $\beta$ lies in $Q_+$ if and only if $H_\beta \in \mathfrak{h}_{\mathfrak{p}_0}$. Hence $Q_+ \subset P_+$ and $Q = Q_+ \cup P_0 \cup P_-$ in the notation of Section 5.

LEMMA 45. *If $Q_+$ is not empty, there exists a root $\alpha \in Q_+$ such that $V_1 \cap U \cap \sigma_\alpha$ is not empty.*

For otherwise suppose $V_1 \cap U \cap \sigma_\alpha = \emptyset$ for every $\alpha \in Q_+$. Let $W_1$ be the subgroup of $W_Q$ generated by $s_\beta$ for $\beta \in P_0 \cup P_-$. Then if we use the notation of the proof of Lemma 44, it is clear that $H_0 \in V_1 \cap U \cap \sigma_{\alpha_0}$ and therefore $\alpha_0 \in P_0 \cup P_-$. Hence it follows immediately that $\bigcup_{s \in W_1} s(V_1 \cap U)$ is open, and therefore $\mathfrak{h}_0 = \bigcup_{s \in W_1} sV_1$ and $\mathfrak{h}_0' = \bigcup_{s \in W_1} s\mathfrak{h}_1$. Now select a root $\alpha \in Q_+$. Then $\alpha$ keeps constant sign on $\mathfrak{h}_1$. Moreover since $\mathfrak{h}_\mathfrak{p}$ and $\mathfrak{h}_\mathfrak{t}$ are orthogonal, it is obvious that $s\alpha = \alpha$ for all $s \in W_1$. Hence $\alpha$ keeps constant sign on $\mathfrak{h}_0' = \bigcup_{s \in W_1} s\mathfrak{h}_1$. As this is evidently false, our assertion follows.

LEMMA 46. *For any $\alpha \in P$, we can select $X_\alpha, X_{-\alpha}$ in such a way that $\bar{\theta}(X_\alpha) = - X_{-\alpha}$ and $[X_\alpha, X_{-\alpha}] = \{2/\alpha(H_\alpha)\}H_\alpha$. Moreover, if $\alpha \in Q_+$, we can assume that $X_\alpha, X_{-\alpha} \in \mathfrak{g}_0$.*

Since $\operatorname{conj} \alpha(H) = - \alpha(\theta(H))$ $(H \in \mathfrak{h}_0)$, it follows that $[H, \bar{\theta}(X_\alpha)] = - \alpha(H)\bar{\theta}(X_\alpha)$ $(H \in \mathfrak{h}_0)$, and therefore we can assume that $X_{-\alpha} = - \bar{\theta}(X_\alpha)$.

Then $B(X_a, X_{-a}) = -B(X_a, \bar{\theta}(X_a)) = \|X_a\|^2$. Now normalize $X_a$ in such a way that $\|X_a\|^2 = 2/\alpha(H_a)$. Then if $H \in \mathfrak{h}$,

$$B([X_a, X_{-a}], H) = B(X_{-a}, [H, X_a]) = \alpha(H)B(X_{-a}, X_a) = \{2/\alpha(H_a)\}\alpha(H),$$

and this shows that $[X_a, X_{-a}] = \{2/\alpha(H_a)\}H_a$. Now suppose $\alpha \in Q_+$. Then if $\zeta$ denotes the conjugation of $\mathfrak{g}$ with respect to $\mathfrak{g}_0$, it is clear that $[H, \zeta(X_a)] = \alpha(H)\zeta(X_a)$ $(H \in \mathfrak{h}_0)$. Moreover $\|\zeta(X)\| = \|X\|$ for any $X \in \mathfrak{g}$. Hence $\zeta(X_a) = c^2 X_a$, where $c$ is a unimodular complex number. Therefore, replacing $X_a$ by $cX_a$, we can assume that $\zeta(X_a) = X_a$ without disturbing the value of $\|X_a\|$. Then $X_a \in \mathfrak{g}_0$ and $X_{-a} = -\bar{\theta}(X_a) = -\theta(X_a)$ is also in $\mathfrak{g}_0$.

COROLLARY 1. *For a fixed* $\alpha \in P$, *select* $X_a, X_{-a}$ *according to the above lemma and put* $\mathfrak{l} = CH_a + CX_a + CX_{-a}$. *Then*

$$\mathfrak{l}_0 = \mathfrak{l} \cap \mathfrak{g}_0 = \begin{cases} RH_a + RX_a + RX_{-a} \text{ if } \alpha \in Q_+, \\ R(-1)^{\frac{1}{2}}H_a + R(-1)^{\frac{1}{2}}(X_a - X_{-a}) + R(X_a + X_{-a}) \text{ if } \alpha \in P_0, \\ R(-1)^{\frac{1}{2}}H_a + R(X_a - X_{-a}) + R(-1)^{\frac{1}{2}}(X_a + X_{-a}) \text{ if } \alpha \in P_-. \end{cases}$$

If $\alpha \in Q_+$, our assertion is obvious. Now suppose $\alpha \in P_0$. Then $X_a \in \mathfrak{p}$, and therefore

$$X_{-a} = -\bar{\theta}(X_a) = \zeta(X_a).$$

Therefore $X_a + X_{-a}$ and $(-1)^{\frac{1}{2}}(X_a - X_{-a})$ are in $\mathfrak{l}_0$ in this case. Similarly, $X_a \in \mathfrak{k}$ and $X_{-a} = -\bar{\theta}(X_a) = -\zeta(X_a)$ if $\alpha \in P_-$. Therefore $X_a - X_{-a}$ and $(-1)^{\frac{1}{2}}(X_a + X_{-a})$ are in $\mathfrak{l}_0$ in this case. Moreover, it is obvious that $\dim_R \mathfrak{l}_0 \leq \dim_C \mathfrak{l} \leq 3$, and so the above statements follow.

Put $H' = 2\alpha(H_a)^{-1}H_a$, $X' = X_a$, $Y' = X_{-a}$. Then $[H', X'] = 2X'$, $[H', Y'] = -2Y'$, $[X', Y'] = H'$, and a simple calculation shows that $\exp(\frac{\pi}{2}\mathrm{ad}(X' - Y'))H' = -H'$. On the other hand, it is obvious that $\exp(\frac{\pi}{2}\mathrm{ad}(X' - Y'))H = H$ if $\alpha(H) = 0$ $(H \in \mathfrak{h})$. Therefore

$$\exp(\tfrac{\pi}{2}\mathrm{ad}(X' - Y'))H = s_a H \qquad\qquad (H \in \mathfrak{h}).$$

COROLLARY 2. *If* $\alpha \in Q_+ \cup P_-$ *there exists an element* $x \in K$ *such that* $xH = s_a H$ *for all* $H \in \mathfrak{h}$.

This follows from the fact that $X' - Y' \in \mathfrak{k}_0$ if $\alpha \in Q_+ \cup P_-$.

INSTITUTE FOR ADVANCED STUDY,
COLUMBIA UNIVERSITY.

---

# REFERENCES.

[1] I. M. Gelfand and M. I. Graev, *Doklady Akad. Nauk. SSSR N. S.*, vol. 92 (1953), pp. 461-464.

[2] Harish-Chandra, (a) *Ann. of Math.*, vol. 51 (1950), pp. 299-330.

        (b) *Trans. Amer. Math. Soc.*, vol. 70 (1951), pp. 28-96.

        (c)   "     "     "     " vol. 75 (1953), pp. 185-243.

        (d)   "     "     "     " vol. 76 (1954), pp. 485-528.

        (e)   "     "     "     " vol. 83 (1956), pp. 98-163.

        (f) *Amer. Jour. Math.*, vol. 77 (1955), pp. 743-777.

        (g)   "     "     " vol. 78 (1956), pp. 564-628.

        (h)   "     "     " vol. 79 (1957), pp. 87-120.

        (i) *Proc. Nat. Acad. Sci. U.S.A.*, vol. 42 (1956), pp. 252-253.

        (j)   "     "     "     "     " vol. 42 (1956), pp. 538-540.

[3] K. Iwasawa, *Ann. of Math.*, vol. 50 (1949), pp. 507-557.

[4] J. L. Koszul, *Bull. Soc. Math. France*, vol. 78 (1950), pp. 65-127.

[5] G. D. Mostow, *Bull. Amer. Math. Soc.*, vol. 55 (1949), pp. 969-980.

[6] L. Schwartz, *Théorie des distributions*, I (1950), II (1951), Paris, Hermann.

[7] A. Weil, *L'integration dans les groupes topologiques et ses applications*, Paris, Hermann, 1940.

[8] H. Weyl, *The structure and representations of continuous groups*, Princeton, The Institute for Advanced Study, 1935.

Reprinted from
*Amer. J. of Math.*
**79** (1957), 193–257

# FOURIER TRANSFORMS ON A SEMISIMPLE LIE ALGEBRA II.*

By Harish-Chandra.

**1. Introduction.** The purpose of this paper is to complete the proofs of two results (Theorems 2 and 3) which have already been mentioned in $[4(e), \S 1]$. For the first one, the problem reduces to showing that the constants of the Corollary to Lemma 29 of $[4(e)]$ are all equal. It is obvious that this can be done only by taking the singular elements of $g_0$ into account. However, as we shall see, it is enough to consider only the best among these, namely, the semiregular ones (see Section 2). Let $g_0'$ be the set of all regular elements in $g_0$. We first show that a given semiregular element $H_0$ of $g_0$ can lie in the closure of at most three distinct connected components of $g_0'$. Let $I_0$ be the real semisimple Lie algebra of dimension 3 spanned by the elements $H$, $X$, $Y$ satisfying the relations $[H, X] = 2X$, $[H, Y] = -2Y$, $[X, Y] = H$. Then zero is the only semiregular element in $I_0$ and certain computations on $g_0$ around $H_0$ can be reduced to similar computations on $I_0$ around zero (see Lemma 8 and its Corollary). Now the distribution $T'$ of $[4(e), \S 7]$ satisfies certain differential equations $[4(e),$ Lemma 30]. If we transcribe these equations on $I_0$, they become simple enough to be handled directly (see Lemma 12). In this way, one proves that $T'$ coincides with a constant on some neighborhood of $H_0$. Theorem 2 now follows from the fact that the union of regular and semiregular elements of $g_0$ is connected (Lemma 13). Once Theorem 2 is proved, Theorem 3 can be obtained without much difficulty from Lemma 41 of $[4(e)]$.

For the convenience of the reader all calculations on $I_0$ have been collected together in the Appendix (Section 6).

The main results of this paper have been announced in a short note $[4(f)]$.

**2. Transformation of certain integrals.** We keep to the notation of $[4(d), (e)]$. Call an element $X \in g$ *semiregular* if (1) $\text{ad} X$ is semisimple and (2) the centralizer of $X$ in $g$ is of dimension $l + 2$ (where $l = \text{rank } g$). Let $\mathfrak{h}_0$ be a Cartan subalgebra of $g_0$ such that $\theta(\mathfrak{h}_0) = \mathfrak{h}_0$ and let $H_0$ be a semiregular element in $\mathfrak{h}_0$. Define $P$, $P_+$, $P_0$ and $P_-$ as in Section 5 of $[4(e)]$.

---

* Received July 30, 1956.

653

Then since $H_0$ is semiregular, it is obvious that $\alpha_0(H_0) = 0$ for exactly one root $\alpha_0 \in P$, and the centralizer of $H_0$ in $\mathfrak{g}$ is $\mathfrak{z} = \mathfrak{h} + CX_{\alpha_0} + CX_{-\alpha_0}$. Let $Q$ be the set of those roots $\alpha \in P$ for which $\alpha^2$ takes only real values on $\mathfrak{h}_0$. We claim $\alpha_0 \in Q$. For otherwise, suppose $\alpha_0 \notin Q$ and put $\beta = -\theta\alpha_0$. Then $\beta$ is also a root and $\beta(H) = \mathrm{conj}\,\alpha_0(H)$ $(H \in \mathfrak{h}_0)$. This implies that $\beta(H_0) = 0$, and therefore $X_\beta \in \mathfrak{z}$. On the other hand $\beta \neq \pm\alpha_0$ since $\alpha_0 \notin Q$. Therefore it is obvious that $X_\beta \notin \mathfrak{z}$, and we get a contradiction. Hence $\alpha_0 \in Q$. Moreover, it is clear that $\bar{\theta}(\mathfrak{z}) = \mathfrak{z}$, and therefore (see [4(e), Lemma 10]) $\mathfrak{z}$ is reductive in $\mathfrak{g}$ and $\mathfrak{z} = \sigma + \mathfrak{l}$, where $\sigma$ is the center and $\mathfrak{l} = [\mathfrak{z}, \mathfrak{z}]$ $= CH_{\alpha_0} + CX_{\alpha_0} + CX_{-\alpha_0}$ the derived algebra of $\mathfrak{z}$. Obviously $\sigma$ is the set of all $H \in \mathfrak{h}$, where $\alpha_0(H) = 0$. Put $\sigma_0 = \sigma \cap \mathfrak{h}_0$. Then $\mathfrak{z}_0 = \mathfrak{z} \cap \mathfrak{g}_0 = \sigma_0 + \mathfrak{l}_0$, where $\mathfrak{l}_0 = \mathfrak{l} \cap \mathfrak{g}_0$. Now $\dim_R \mathfrak{z}_0 = \dim \mathfrak{z} = l + 2$ and $\dim_R \sigma_0 = l - 1$. Hence $\dim_R \mathfrak{l}_0 = 3$, and this shows that $\mathfrak{l}_0$ is a real form of the complex semisimple Lie algebra $\mathfrak{l}$. Put $H_{\alpha_0}' = 2\alpha_0(H_{\alpha_0})^{-1}H_{\alpha_0}$ and define $Q_+$ as in [4(e), § 12]. We select $X_{\alpha_0}$ and $X_{-\alpha_0}$ in accordance with Corollary 1 of Lemma 46 of [4(e)]. Then $\mathfrak{l}_0 = RH_{\alpha_0}' + RX_{\alpha_0} + RX_{-\alpha_0}$ if $\alpha_0 \in Q_+$ and

$$\mathfrak{l}_0 = R(-1)^{\frac{1}{2}}H_{\alpha_0}' + R(X_{\alpha_0} - X_{-\alpha_0}) + R(-1)^{\frac{1}{2}}(X_{\alpha_0} + X_{-\alpha_0}) \ \text{ if } \ \alpha_0 \in P_-.$$

In the latter case, $\mathfrak{l}_0$ is compact (see Section 6). Finally,

$$\mathfrak{l}_0 = R(-1)^{\frac{1}{2}}H_{\alpha_0}' + R(-1)^{\frac{1}{2}}(X_{\alpha_0} - X_{-\alpha_0}) + R(X_{\alpha_0} + X_{-\alpha_0}) \ \text{ if } \ \alpha_0 \in P_0.$$

Let $\mathfrak{l}'$ be the set of those points in $\mathfrak{l}$ which are regular (with respect to $\mathfrak{l}$). Put $\mathfrak{l}_0' = \mathfrak{l}' \cap \mathfrak{l}_0$.

Let $\mathfrak{q}$ be the orthogonal complement of $\mathfrak{z}$ in $\mathfrak{g}$. Then $\mathfrak{q} = \sum\limits_{\substack{\alpha \in P \\ \alpha \neq \alpha_0}} (CX_\alpha + CX_{-\alpha})$ and $[\mathfrak{z}, \mathfrak{q}] \subset \mathfrak{q}$. Let $\zeta(Z)$ $(Z \in \mathfrak{z})$ denote the determinant of the restriction of $\mathrm{ad}\,Z$ on $\mathfrak{q}$. Then $\zeta$ is a polynomial function on $\mathfrak{z}$ which takes only real values on $\mathfrak{z}_0$ and $\zeta(H) = (-1)^{r-1} \prod\limits_{\substack{\alpha \in P \\ \alpha \neq \alpha_0}} \alpha(H)^2$ $(H \in \mathfrak{h})$, where $r$ is the number of roots in $P$. This shows that $\zeta(H_0) \neq 0$. Let $\mathfrak{z}_0'$ be the set of those points $Z \in \mathfrak{z}_0$ where $\zeta(Z) \neq 0$. Select an open connected neighborhood $U$ of $H_0$ in $\mathfrak{z}_0'$ and consider the factor space $G^* = G/\Xi$, where $G$ is the (connected) adjoint group of $\mathfrak{g}_0$ and $\Xi$ is the centralizer of $H_0$ in $G$. We denote by $x \to x^*$ $(x \in G)$ the natural mapping $G$ on $G^*$. Choose an open connected neighborhood $V^*$ of $1^*$ in $G^*$. Then if $V^*$ is sufficiently small, we can define (see Chevalley [2, p. 111]) an analytic mapping $\phi$ of $V^*$ into $G$ with the following properties: (1) $(\phi(x^*))^* = x^*$ $(x^* \in V^*)$ and (2) $\phi$ is regular on $V^*$. Put $V = \phi(V^*)$ and consider the mapping $\psi : (x^*, Z) \to \phi(x^*)Z$ $(x^* \in V^*, Z \in U)$ of $V^* \times U$ into $\mathfrak{g}_0$. We shall prove that $\psi$ is regular. For

$Y \in \mathfrak{g}_0$ and $X^* \in V^*$, put[1] $Y_{x^*}{}^* = d\tau_x(Y)$, where $x = \phi(x^*)$ and $\tau$ is the natural mapping of $G$ on $G^*$. Identify the tangent space of $V^* \times U$ at $(x^*, Z)$ with $\mathfrak{g}_0$ under the linear isomorphism

$$Y + X \leftrightarrow Y_{x^*}{}^* \times X \qquad\qquad (Y \in \mathfrak{q}_0, X \in \mathfrak{z}_0).$$

Then a simple calculation shows that

$$d\psi_{x^*, Z}(Y + X) = \phi(x^*)X - \phi(x^*)\, \mathrm{ad}\, Z(d\phi_{x^*} Y_{x^*}{}^*).$$

But since $\tau\phi$ is the identity mapping on $V^*$, $d\tau_x(d\phi_{x^*} Y_{x^*}{}^* - Y) = 0$ if $x = \phi(x^*)$, and therefore $d\phi_{x^*} Y_{x^*}{}^* \equiv Y \bmod \mathfrak{z}_0$. Hence if $D_{x^*, Z} = (\phi(x^*))^{-1} d\psi_{x^*, Z}$, it is clear that

$$D_{x^*, Z}X = X \quad (X \in \mathfrak{z}_0), \qquad D_{x^*, Z}Y \equiv -(\mathrm{ad}\, Z)Y \quad (Y \in \mathfrak{q}_0),$$

and so we conclude that $|\det(d\psi_{x^*, Z})| = |\det D_{x^*, Z}| = |\zeta(Z)| \neq 0$, since $U \subset \mathfrak{z}_0'$. This proves that $\psi$ is everywhere regular on $V^* \times U$. Therefore, since the dimensions of the two manifolds $V^* \times U$ and $\mathfrak{g}_0$ are equal, it follows that $\psi$ is also open and univalent provided $V^*$ and $U$ are sufficiently small. Put $N = \psi(V^* \times U)$. Then $N$ is an open connected neighborhood of $H_0$ in $\mathfrak{g}_0$ and $\psi$ defines an analytic isomorphism of $V^* \times U$ with $N$.

Since $\mathfrak{z}_0$ is reductive, the group $\Xi$ is unimodular, and therefore (see Weil [8, p. 45]) there exists an invariant measure $dx^*$ on $G^*$. Let $dX$ and $dZ$ denote the Euclidean measures on $\mathfrak{g}_0$ and $\mathfrak{z}_0$, respectively. Then the above calculation shows that $dx^*$ can be so normalized that

$$\int_N f(X)\, dX = \int_{V^* \times U} f(\psi(x^*, Z)) |\zeta(Z)|\, dx^* dZ \qquad (f \in C_c(N)).$$

Select once for all an element $a_0 \in C_c^\infty(V^*)$ such that $\int a_0(x^*)\, dx^* = 1$, and for any $\gamma \in C_c^\infty(U)$ define $f_\gamma \in C_c^\infty(N)$ by $f_\gamma(\psi(x^*, Z)) = a_0(x^*)\gamma(Z)$ $(x^* \in V^*, Z \in U)$. Then the following result is obvious.

LEMMA 1. *Let* '$U$ *be any open subset of* $U$ *and put* '$N = \psi(V^* \times 'U)$. *Then*

$$\int_{'N} f_\gamma(X)\, dX = \int_{'U} \gamma |\zeta|\, dZ \ \text{for all}\ \gamma \in C_c^\infty('U).$$

On the other hand, it is clear that $\sigma_0$ and $\mathfrak{l}_0$ are mutually orthogonal. Let $U_\sigma$ and $U_{\mathfrak{l}}$, respectively, denote the orthogonal projections of $U$ in $\sigma_0$ and $\mathfrak{l}_0$. Then by replacing $U$, if necessary, by a smaller open connected

---

[1] We follow here the terminology of Chevalley [2]. The present computation is very similar to the one performed during the proof of Lemma 39 of [4(c)].

neighborhood of $H_0$ in $\mathfrak{z}_0'$, we can assume that (1) $U = U_\sigma + U_{\mathfrak{l}}$, (2) $U_\sigma$ is convex and $\alpha(H) \neq 0$ for any $H \in U_\sigma$ and $\alpha \neq \alpha_0$ in $P$, (3) if $Z \in U_{\mathfrak{l}}$ then $tZ$ $(0 \leq t \leq 1)$ is also in $U_{\mathfrak{l}}$. Define the polynomial $\eta$ as in [4(e), §6] and let $\mathfrak{g}_0'$ be the set of all regular elements in $\mathfrak{g}_0$. Put $U' = U \cap \mathfrak{g}_0'$ and $U_{\mathfrak{l}}' = U_{\mathfrak{l}} \cap \mathfrak{l}_0'$. Then we claim that $U' = U_\sigma + U_{\mathfrak{l}}'$. This is seen as follows. Let $\lambda$ be an indeterminate and $I$ the identity maping of $\mathfrak{g}$. Then it is obvious that

$$\det(\lambda I - \operatorname{ad} Z) = \det(\lambda I - \operatorname{ad} Z)_\mathfrak{q} \det(\lambda I - \operatorname{ad} Z)_\mathfrak{z} \qquad (Z \in \mathfrak{z}),$$

where the subscripts signify restrictions on the corresponding subspaces. Now suppose $Z \in U$ and $Z = H + Y$, where $H \in U_\sigma$ and $Y \in U_{\mathfrak{l}}$. Then it is evident that

$$\det(\lambda I - \operatorname{ad} Z)_\mathfrak{z} = \lambda^{l-1} \det(\lambda I - \operatorname{ad} Y)_{\mathfrak{l}},$$

and therefore $\det(\lambda I - \operatorname{ad} Z) = \lambda^{l-1} \det(\lambda I - \operatorname{ad} Z)_\mathfrak{q} \det(\lambda I - \operatorname{ad} Y)_{\mathfrak{l}}$. The coefficient of $\lambda^l$ on the left side is $(-1)^r \eta(Z)$, where $r$ is the number of roots in $P$. On the other hand, the constant term in $\det(\lambda I - \operatorname{ad} Z)_\mathfrak{q}$ is $(-1)^q \zeta(Z)$, where $q = \dim \mathfrak{q}$. Moreover, since $\mathfrak{l}$ is of rank 1, $\lambda$ divides $\det(\lambda I - \operatorname{ad} Y)_{\mathfrak{l}}$, and the coefficient of $\lambda$ in it is $-\frac{1}{2}\Omega(Y)$, where $\Omega$ is the Casimir polynomial of $\mathfrak{l}$ (see Section 6). This means that

$$(-1)^r \eta(Z) = \tfrac{1}{2}(-1)^{q+1} \zeta(Z)\Omega(Y).$$

Therefore, since $\zeta$ is never zero on $U$, $Z \in U'$ if and only if $Y \in \mathfrak{l}_0'$, and this proves that $U' = U_\sigma + U_{\mathfrak{l}}'$. In view of the above condition imposed on $U_{\mathfrak{l}}$, it follows (see Appendix, §6) that $U_{\mathfrak{l}}'$ is connected or it has exactly three connected components according as $\mathfrak{l}_0$ is compact or not. Put $N' = N \cap \mathfrak{g}_0'$. Then $N' = \psi(V^* \times U')$, and therefore we get the following result.

LEMMA 2. *If $\alpha_0 \in P_-$ then $\mathfrak{l}_0$ is compact and $N \cap \mathfrak{g}_0'$ is connected. On the other hand, if $\alpha_0 \in Q_+ \cup P_0$, $\mathfrak{l}_0$ is not compact and $N \cap \mathfrak{g}_0'$ has exactly three connected components.*

*From now on we shall assume that $\alpha_0 \in Q_+ \cup P_0$.* Call an element $Z \in \mathfrak{l}_0'$ hyperbolic or elliptic according as $\Omega(Z)$ is positive or negative. Let $\mathfrak{l}_i$ $(i = 1, 2, 3)$ denote the three connected components of $\mathfrak{l}_0'$. We assume that $\Omega$ is positive on $\mathfrak{l}_1$ and negative on $\mathfrak{l}_2$ and $\mathfrak{l}_3$ (see Appendix, §6). Put

$$H' = H_{\alpha_0}', \; X' = X_{\alpha_0}, \; Y' = X_{-\alpha_0} \text{ if } \alpha_0 \in Q_+ \text{ and } H' = (-1)^{\frac{1}{2}}(X_{\alpha_0} - X_{-\alpha_0}),$$

$$X' = \tfrac{1}{2}\{X_{\alpha_0} + X_{-\alpha_0} + (-1)^{\frac{1}{2}}H_{\alpha_0}'\}, \quad Y' = \tfrac{1}{2}\{X_{\alpha_0} + X_{-\alpha_0} - (-1)^{\frac{1}{2}}H_{\alpha_0}'\}$$

if $\alpha_0 \in P_0$. Then $\mathfrak{l}_0 = RH' + RX' + RY'$ and $[H', X'] = 2X'$ $[H', Y']$

$=-2Y'$, $[X', Y'] = H'$. Moreover $H'$ is hyperbolic while $X' - Y'$ is elliptic. We shall suppose that $X' - Y' \in I_2$. Let $d\mu_\sigma$ and $d\mu_I$ denote the Euclidean measures on $\sigma_0$ and $I_0$, respectively. Every element in $I_0$ can be written in the form $tH' + xX' + yY'$ $(t, x, y \in R)$. We normalize $d\mu_I$ in such a way that [2] $d\mu_I = (2\pi)^{-1} dt\, dx\, dy$. Then since $\mathfrak{z}_0$ is the orthogonal sum of $\sigma_0$ and $I_0$, we can assume that $dZ = d\mu_\sigma d\mu_I$. Put

$$\mathfrak{a}_0 = \sigma_0 + RH' \quad \text{and} \quad \mathfrak{b}_0 = \sigma_0 + R(X' - Y').$$

Then $\mathfrak{a}_0$ and $\mathfrak{b}_0$ are Cartan subalgebras of $\mathfrak{z}_0$ (and also of $\mathfrak{g}_0$) which are invariant under $\theta$. Let $L$ be the analytic subgroup of $G$ corresponding to $I_0$, and let $K_1$ be the connected component of $1$ in $K \cap L$. Since $I_0 \cap \mathfrak{k}_0 = R(X' - Y')$, $K_1$ is the one-parameter subgroup corresponding to $X' - Y'$. Moreover $K_1$ is compact. For any $f \in C_c^\infty(\mathfrak{z}_0)$, put

$$\Phi_f(t, H_\sigma) = \int_{-\infty}^{\infty} \bar{f}(H_\sigma + tH' + xX')\, dx \qquad (t > 0),$$

$$\Psi_f(\phi, H_\sigma) = |\phi| \int_0^{\infty} \bar{f}(H_\sigma + \phi(e^t X' - e^{-t} Y'))(e^t - e^{-t})\, dt \quad (\phi \in R, \phi \neq 0)$$

for $H_\sigma \in \sigma_0$. Here $\bar{f}(Z) = \int_{K_1} f(k_1 z)\, dk_1$ $(Z \in \mathfrak{z}_0)$ and $dk_1$ is the normalized Haar measure on $K_1$ so that $\int_{K_1} dk_1 = 1$. Let $R^+$ and $R^-$ denote the sets of all positive and negative real numbers, respectively, and put $R' = R^+ \cup R^-$. Also let $\mathfrak{a}_1$, $\mathfrak{b}_2$, $\mathfrak{b}_3$, respectively, denote the sets of all elements of the form $H_\sigma + tH'$, $H_\sigma + t(X' - Y')$ and $H_\sigma - t(X' - Y')$ $(t \in R^+)$. Put $\mathfrak{b}_0' = \mathfrak{b}_2 \cup \mathfrak{b}_3$. We can also regard $\Phi_f$ and $\Psi_f$ as functions on $\mathfrak{a}_1$ and $\mathfrak{b}_0'$, respectively, as follows:

$$\Phi_f(H_\sigma + tH') = \Phi_f(t, H_\sigma), \qquad \Psi_f(H_\sigma + \phi(X' - Y')) = \Psi_f(\phi, H_\sigma)$$

$(H_\sigma \in \sigma_0, t \in R^+, \phi \in R')$. Let $\mathfrak{a}, \mathfrak{b}$ be the complexifications of $\mathfrak{a}_0, \mathfrak{b}_0$ in $\mathfrak{g}$. Consider the complex analytic subgroup $L_c$ corresponding to $I$ in the (connected) complex adjoint group $G_c$ of $\mathfrak{g}$. Define two elements $v_\mathfrak{a}$ and $v_\mathfrak{b}$ in $L_c$ as follows. $v_\mathfrak{a} = 1$, $v_\mathfrak{b} = \exp\{(-1)^{\frac{1}{2}}(\pi/4)\mathrm{ad}(X_{\alpha_0} + X_{-\alpha_0})\}$ if [2] $\alpha_0 \in Q_+$ and $v_\mathfrak{a} = \exp\{-(-1)^{\frac{1}{2}}(\pi/4)\mathrm{ad}(X_{\alpha_0} + X_{-\alpha_0})\}$, $v_\mathfrak{b} = 1$ if $\alpha_0 \in P_0$. Then (see §6) $H' = v_\mathfrak{a} H_{\alpha_0}'$, $(X' - Y') = (-1)^{\frac{1}{2}} v_\mathfrak{b} H_{\alpha_0}'$, and therefore $v_\mathfrak{a} \mathfrak{h} = \mathfrak{a}$, $v_\mathfrak{b} \mathfrak{h} = \mathfrak{b}$. For any root $\alpha$ of $\mathfrak{g}$ with respect to $\mathfrak{h}$, we define a root $\alpha_\mathfrak{a}$ with respect to $\mathfrak{a}$ by $\alpha_\mathfrak{a}(H) = \alpha(v_\mathfrak{a}^{-1} H)$ $(H \in \mathfrak{a})$. Similarly, define a root $\alpha_\mathfrak{b}$ with respect to $\mathfrak{b}$ by $\alpha_\mathfrak{b}(H) = \alpha(v_\mathfrak{b}^{-1} H)$ $(H \in \mathfrak{b})$. We shall say that $\alpha_\mathfrak{a}$ and $\alpha_\mathfrak{b}$ correspond to $\alpha$

---

[2] Here $\pi$ denotes, as usual, the smallest positive root of the equation $\sin t = 0$.

14

under $\nu_\mathfrak{a}$ and $\nu_\mathfrak{b}$, respectively. Put $\tau = (\alpha_0)_\mathfrak{a}$ and $\lambda = (\alpha_0)_\mathfrak{b}$. Then $\tau(H') = 2$ and $\lambda(X' - Y') = 2(-1)^{\frac{1}{2}}$. Let $P_\mathfrak{a}$ and $P_\mathfrak{b}$ be the sets of roots with respect to $\mathfrak{a}$ and $\mathfrak{b}$ corresponding to $P$ under $\nu_\mathfrak{a}$ and $\nu_\mathfrak{b}$, respectively. Define the Euclidean measures $d\mu_\mathfrak{a}$, $d\mu_\mathfrak{b}$ on $\mathfrak{a}$, $\mathfrak{b}$ by $d\mu_\mathfrak{a} = d\mu_\sigma dt$, $d\mu_\mathfrak{b} = d\mu_\sigma d\phi$ corresponding to $H = H_\sigma + tH'$ and $\bar{H} = H_\sigma + \phi(X' - Y')$ $(H \in \mathfrak{a}_0;\ \bar{H} \in \mathfrak{b}_0;\ H_\sigma \in \sigma_0;\ t, \phi \in R)$. Put $\mathfrak{z}_i = \sigma_0 + \mathfrak{l}_i$ $(i = 1, 2, 3)$.

LEMMA 3. *For any $f \in C_c^\infty(\mathfrak{z}_0)$ define $\Phi_f$, $\Psi_f$ as above. Then $\Phi_f$ and $\Psi_f$ lie in $\mathscr{E}(\mathfrak{a}_1)$ and $\mathscr{E}(\mathfrak{b}_0')$, respectively, and*

$$\int_{\mathfrak{z}_1} f(Z)\,dZ = \tfrac{1}{2} \int_{\mathfrak{a}_1} \tau \Phi_f \, d\mu_\mathfrak{a}, \qquad \int_{\mathfrak{z}_i} f(Z)\,dZ = \tfrac{1}{2} \int_{\mathfrak{b}_i} |\lambda|\, \Psi_f \, d\mu_\mathfrak{b} \qquad (i = 2, 3).$$

Since $\sigma$ is the center of $\mathfrak{z}$, it follows without difficulty that $\partial(p)\Phi_f = \Phi_{\partial(p)f}$, $\partial(p)\Psi_f = \Psi_{\partial(p)f}$ for $p \in S(\sigma)$ (see [4(d), §2] for notation). Therefore since the carrier of $f$ is compact, our statement follows from the work of the Appendix, especially Lemma 16.

Let $\Xi'$ be the normalizer of $\sigma_0$ in $G$ and $\Xi_0$ the analytic subgroup of $G$ corresponding to $\mathfrak{z}_0$.

LEMMA 4. *$\Xi'$ is also the normalizer of $\mathfrak{z}_0$ in $G$ and every element of $\Xi'$ leaves $\mathfrak{l}_0$ invariant. Moreover, $\Xi' \supset \Xi \supset \Xi_0$ and $\Xi_0$ is of finite index in $\Xi'$.*

Since $\sigma_0$ is the center of $\mathfrak{z}_0$ and $\mathfrak{z}_0$ the centralizer of $\sigma_0$ in $\mathfrak{g}_0$, it is obvious that $\Xi'$ is also the normalizer of $\mathfrak{z}_0$ in $G$. Moreover, since $\mathfrak{l}_0$ is the derived algebra of $\mathfrak{z}_0$, $x\mathfrak{l}_0 = \mathfrak{l}_0$ for $x \in \Xi'$. Finally, since $\mathfrak{z}_0$ is the centralizer of $H_0$ in $\mathfrak{g}_0$, it is clear that $\Xi_0 \subset \Xi \subset \Xi'$. The finiteness of $\Xi'/\Xi_0$ is a consequence of Lemma 15 of the Appendix (Section 6).

We shall now show that if $U$ is sufficiently small, $xU \cap U = \varnothing$ $(x \in G)$ unless $x \in \Xi$. This is done as follows. Since $S(\mathfrak{l}) \subset S(\mathfrak{g})$, we can regard $\Omega$ as a polynomial function on $\mathfrak{g}$. But then, since $\sigma$ and $\mathfrak{l}$ are orthogonal, $\Omega(H + Z) = \Omega(Z)$ $(H \in \sigma, Z \in \mathfrak{l})$, and therefore $\Omega(H_0) = 0$. Choose a positive number $\delta$. Then we may assume that $U$ is so small that $|\Omega(Z)| < \delta^2 |\alpha(H_\sigma(Z))|^2$ $(Z \in U)$ for all $\alpha \neq \alpha_0$ in $P$. Here $H_\sigma(Z)$ denotes the orthogonal projection of $Z$ in $\sigma$ for any $Z \in \mathfrak{z}$. Since $[\sigma, \mathfrak{l}] = \{0\}$, it is clear that $H_\sigma(yZ) = H_\sigma(Z)$ $(y \in L_c, Z \in \mathfrak{z})$. Hence

$$\min_{\substack{\alpha \neq \alpha_0 \\ \alpha \in P}} |\alpha(H_\sigma(Z))| = \min_{\substack{\alpha \neq \tau \\ \alpha \in P_\mathfrak{a}}} |\alpha(H_\sigma(Z))| = \min_{\substack{\alpha \neq \lambda \\ \alpha \in P_\mathfrak{b}}} |\alpha(H_\sigma(Z))|.$$

Moreover, if $Z \in \mathfrak{a}$, $\Omega(Z) = 2\tau(Z)^2$ (see Section 6). Let $\tilde{U} = \bigcup_{y \in L} yU$. Then it follows that if $Z \in \tilde{U} \cap \mathfrak{a}_0$

$$|\tau(Z)| < \delta\, |\alpha(H_\sigma(Z))| \quad \text{for all } \alpha \neq \tau \text{ in } P_\mathfrak{a}.$$

Similarly, if $Z \in \bar{U} \cap \mathfrak{b}_0$, then $|\lambda(Z)| < \delta |\alpha(H_\sigma(Z))|$ for all $\alpha \neq \lambda$ in $P_\mathfrak{b}$. Now suppose $\delta$ is so small that $\delta \leq \frac{1}{2}$ and

$$\delta |\alpha(H')| \leq 1 \quad (\alpha \in P_\mathfrak{a}), \qquad \delta |\beta(X' - Y')| \leq 1 \quad (\beta \in P_\mathfrak{b}).$$

Then since $Z = H_\sigma(Z) + \frac{1}{2}\tau(Z)H'$ for $Z \in \bar{U} \cap \mathfrak{a}_0$, it follows that $\alpha(Z) = \alpha(H_\sigma(Z)) + \frac{1}{2}\tau(Z)\alpha(H')$ for $\alpha \in P_\mathfrak{a}$. Therefore if $\alpha \neq \tau$,

$$|\alpha(Z)| \geq 2\delta |\alpha(Z)| \geq 2\delta\{|\alpha(H_\sigma(Z))| - \frac{1}{2}|\tau(Z)||\alpha(H')|\}$$
$$> (2 - \delta |\alpha(H')|)|\tau(Z)| \geq |\tau(Z)|.$$

This proves that $|\alpha(Z)| > |\tau(Z)|$ for $Z \in \bar{U} \cap \mathfrak{a}_0$ and $\alpha \neq \tau$ in $P_\mathfrak{a}$. Similarly, $|\alpha(Z)| > |\lambda(Z)|$ $(\alpha \in P_\mathfrak{b}, \alpha \neq \lambda)$ for $Z \in \bar{U} \cap \mathfrak{b}_0$. Moreover, we have the following result.

LEMMA 5. *The two Cartan subalgebras $\mathfrak{a}_0$ and $\mathfrak{b}_0$ are not conjugate under $G$.*

Since $\theta(\sigma_0) = \sigma_0$, $H' \in \mathfrak{p}_0$ and $X' - Y' \in \mathfrak{k}_0$ (see [4(e), §12]), it is clear that $\mathfrak{a}_0 \cap \mathfrak{k}_0 = \sigma_0 \cap \mathfrak{k}_0$ while $\mathfrak{b}_0 \cap \mathfrak{k}_0 = \sigma_0 \cap \mathfrak{k}_0 + R(X' - Y')$. Hence $l_-(\mathfrak{b}_0) = l_-(\mathfrak{a}_0) + 1$ in the notation of [4(e), §8]. Therefore $\mathfrak{a}_0$ and $\mathfrak{b}_0$ cannot be conjugate.

Put $U_i = U \cap \mathfrak{z}_i$ $(i = 1, 2, 3)$.

COROLLARY. *Suppose $Z_1 \in U_1$ and $Z_2 \in U_2 \cup U_3$. Then $Z_1$ and $Z_2$ cannot be conjugate under $G$.*

For otherwise suppose $Z_2 = xZ_1$ $(x \in G)$. Since $U_1 \subset \sigma_0 + I_1$, we can select (see Section 6) $y_1 \in L$ such that $Z_1' = y_1 Z_1 \in \mathfrak{a}_1$. Similarly, we can choose $y_2 \in L$ such that $Z_2' = y_2 Z_2 \in \mathfrak{b}_0'$. Obviously $Z_1, Z_2$ are regular in $\mathfrak{g}_0$, and therefore the same holds for $Z_1', Z_2'$. Hence $\mathfrak{a}_0$ and $\mathfrak{b}_0$, respectively, are the centralizers of $Z_1'$ and $Z_2'$ in $\mathfrak{g}_0$. Moreover, $Z_2' = x'Z_1'$ $(x' = y_2 x y_1^{-1})$, and therefore $\mathfrak{b}_0 = x'\mathfrak{a}_0$. As this contradicts Lemma 5, the corollary follows.

Now suppose $xZ \in U$ for some $x \in G$ and $Z \in U$. First, let us assume that $Z \in U_1$. Then obviously $xZ$ is regular in $\mathfrak{g}_0$, and so it follows from the Corollary to Lemma 5 that $xZ \in U_1$. Hence $y_1 Z, y_2 xZ$ are in $U_\sigma + R^+H'$ for some $y_1, y_2 \in L$. Put $Z_1 = y_1 Z$ and $x_1 = y_2 x y_1^{-1}$. Then since $\mathfrak{a}_0$ is the centralizer in $\mathfrak{g}_0$ both of $Z_1$ and $x_1 Z_1$, we conclude that $x_1 \mathfrak{a}_0 = \mathfrak{a}_0$. Let $W_\mathfrak{a}$ be the Weyl group of $\mathfrak{g}$ with respect to $\mathfrak{a}$. Then $x_1$ defines an element $s \in W_\mathfrak{a}$. Put $s\tau = \tau'$ and $Z_2 = x_1 Z_1$. It is clear that $\tau'(Z_2) = \tau(Z_1)$. On the other hand, since $Z_1 \in \bar{U} \cap \mathfrak{a}_0$, $|\tau(Z_1)| = \min_{\alpha \in P_\mathfrak{a}} |\alpha(Z_1)|$. Therefore $|\tau'(Z_2)| = \min_{\alpha \in P_\mathfrak{a}} |\alpha(Z_2)|$. But since $Z_2$ is also in $\bar{U} \cap \mathfrak{a}_0$, we know that $|\tau(Z_2)|$

$< |\alpha(Z_2)|$ for any root $\alpha \neq \tau$ in $P_\alpha$. Hence $\tau' = \pm \tau$. Now $\sigma_0$ is the set of all $H \in \mathfrak{a}_0$ where $\tau(H) = 0$. Therefore $x_1\sigma_0 = s\sigma_0 = \sigma_0$. Since $L$ is contained in $\Xi$, this proves that $x \in \Xi'$.

Similarly, if $Z \in U_2 \cup U_3$, we can choose $y_1 \in L$ such that $Z_1 = y_1 Z \in U_\sigma + R'(X' - Y')$. Again it follows from the Corollary to Lemma 5 that $xZ \in U_2 \cup U_3$, and therefore $y_2 xZ \in U_\sigma + R'(X' - Y')$ for some $y_2 \in L$. Hence $x_1\mathfrak{b}_0 = \mathfrak{b}_0$ ($x_1 = y_2 x y_1^{-1}$) and the corresponding element $s$ of the Weyl group $W_\mathfrak{b}$ (of $\mathfrak{g}$ with respect to $\mathfrak{b}$) maps $\lambda$ into $\pm \lambda$, and therefore again $x \in \Xi'$.

Finally suppose $Z$ is singular. For any $X \in \mathfrak{g}_0$, let $\mathfrak{n}_X$ denote the set of all $Y \in \mathfrak{g}_0$ such that $(\operatorname{ad} X)^q Y = 0$ for some integer $q \geq 1$. We call $\mathfrak{n}_X$ the null space of $X$. It is obvious that if $X$ is a singular element in $U$, $\mathfrak{n}_X = \mathfrak{z}_0$. Hence $\mathfrak{z}_0$ is the null space of both $Z$ and $xZ$, and therefore $x\mathfrak{z}_0 = \mathfrak{z}_0$. In view of Lemma 4 this implies again that $x \in \Xi'$.

Now from Lemma 4, we can select a finite number of elements $u_0 = 1$, $u_1, \cdots, u_M \in \Xi'$ such that the cosets $u_i \Xi_0$ ($0 \leq i \leq M$) are distinct and their union is $\Xi'$. Suppose $u_i \in \Xi$ for $0 \leq i \leq m$ and $u_i \notin \Xi$ for $m < i \leq M$. Then $u_i H_0 \neq H_0$ ($m < i \leq M$), and we may assume that $U_\sigma$ is so small that $U_\sigma \cap u_i U_\sigma = \emptyset$ ($m < i \leq M$). Select $i$ such that $x \in u_i \Xi_0$. Then we claim that $i \leq m$. For otherwise, suppose $i > m$. Since $U = U_\sigma + U_\mathrm{I}$, it follows that $xU = xU_\sigma + xU_\mathrm{I}$. On the other hand, from Lemma 4, $xU_\sigma$ and $xU_\mathrm{I}$ are contained in $\sigma_0$ and $\mathrm{I}_0$, respectively, and since $\sigma$ is the center of $\mathfrak{z}$, $xU_\sigma = u_i U_\sigma$, and therefore $U_\sigma \cap xU_\sigma = \emptyset$. But obviously, this implies that $U \cap xU = \emptyset$, and we get a contradiction with our assumption that $xZ \in U \cap xU$. Hence $i \leq m$ and therefore $x \in \Xi$.

Put $\bar{U}_\sigma = \bigcap_{1 \leq i \leq m} u_i U_\sigma$, $\bar{U}_\mathrm{I} = \bigcup_{\xi \in \Xi} \xi U_\mathrm{I}$ and $\bar{U} = \bar{U}_\sigma + \bar{U}_\mathrm{I}$. Since $U_\sigma$ is convex, the same holds for $\bar{U}_\sigma$. Moreover, in view of its definition, $\Omega$ is invariant under any automorphism of $\mathrm{I}$. Hence if $Z \in \bar{U}_\mathrm{I}$, it is clear that $|\Omega(Z)| < \delta^2 |\alpha(H_\sigma(Z))|^2$ for all roots $\alpha \neq \alpha_0$ in $P$. Therefore it follows from the above proof that $x\bar{U} \cap \bar{U} = \emptyset$ ($x \in G$) unless $x \in \Xi$.

LEMMA 6. *It is possible to select $U$ in Lemma 1 so that it satisfies the following two further conditions.*

(1) $\quad \xi U = U$ ($\xi \in \Xi$),

(2) $\quad U \cap xU = \emptyset$ *for any $x \in G$ which is not in $\Xi$.*

As we have already seen, $\bar{U}$ satisfies the above two conditions. The univalence of the mapping $(x^*, Z) \to \phi(x^*)Z$ of $V^* \times \bar{U}$ into $\mathfrak{g}_0$ is an immediate consequence of the second condition. Since the earlier conditions imposed on $U$ (see p. 656) are obviously fulfilled by $\bar{U}$, the statement of the lemma follows.

Let $A$ and $B$ be the Cartan subgroups of $G$ corresponding to $\mathfrak{a}_0$ and $\mathfrak{b}_0$, respectively. Since $L$ is a normal subgroup of $\Xi'$ (see Lemma 4), $AL$ and $BL$ are subgroups of $\Xi$ and $AL = LA$, $BL = LB$. Let $[\Xi : LB]$ denote the index of $LB$ in $\Xi$.

LEMMA 7.  $\Xi = LA$ and $[\Xi : LB] \leq 2$. Let $s_\lambda$ denote the Weyl reflexion in $W_\mathfrak{b}$ corresponding to the root $\lambda \in P_\mathfrak{b}$. Then $[\Xi : LB] = 2$ if and only if there exists an element $\xi \in \Xi$ which coincides with $s_\lambda$ on $\mathfrak{b}$.

Let $x$ be any element in $\Xi$. Then $x\mathfrak{l}_0 = \mathfrak{l}_0$ from Lemma 4, and therefore $\Omega(xH') = \Omega(H')$, $\Omega(x(X' - Y')) = \Omega(X' - Y')$. This shows that $xH' \in \mathfrak{l}_1$ and $x(X' - Y') \in \mathfrak{l}_2 \cup \mathfrak{l}_3$. Hence we can select $y_1, y_2 \in L$ such that $y_1^{-1}xH' \in R^+H'$ and $y_2^{-1}x(X' - Y') \in R'(X' - Y')$. But since $\mathrm{sp}(\mathrm{ad}(zX))^2 = \mathrm{sp}(\mathrm{ad}\,X)^2$ $(z \in G, X \in \mathfrak{g})$ this implies that $y_2^{-1}xH' = H'$ and $y_2^{-1}x(X' - Y') = \pm(X' - Y')$. Now if $t$ is sufficiently small and positive, $H_0 + tH'$ is nonsingular and so its centralizer in $G$ is $A$ (see [4(c), Lemma 3]). Hence $y_1^{-1}x \in A$, and so $x \in LA$. This proves that $\Xi = LA$. On the other hand, choose, if possible, an element $\xi \in \Xi$ such that $\xi(X' - Y') = -(X' - Y')$. In case $y_2^{-1}x(X' - Y') = X' - Y'$, the above argument applied to $H_0 + t(X' - Y')$ $(t \in R')$ shows that $x \in LB$. So now suppose $y_2^{-1}x(X' - Y') = -(X' - Y')$. Then $\xi^{-1}y_2^{-1}x(X' - Y') = (X' - Y')$, and therefore $\xi^{-1}y_2^{-1}x \in B$. This proves that $x \in L\xi B = \xi LB$ in this case. Hence $[\Xi : LB] \leq 2$. Moreover, it is obvious that $[\Xi : LB] = 1$ if $\xi$ does not exist. So let us assume that $\xi$ exists. We claim $\xi \notin LB$. For otherwise, $\xi = yb$ $(y \in L, b \in B)$, and therefore $y(X' - Y') = \xi(X' - Y') = -(X' - Y')$. But this is impossible (see Section 6), and therefore $[\Xi : LB] = 2$ in this case. Since $\xi\sigma = \sigma$, it is clear that $\xi\mathfrak{b} = \mathfrak{b}$. Let $s$ be the element of $W_\mathfrak{b}$ corresponding to $\xi$. Then

$$s(H_0 + t(X' - Y')) = \xi(H_0 + t(X' - Y'))$$
$$= H_0 - t(X' - Y') = s_\lambda(H_0 + t(X' - Y')) \quad (t \in R).$$

But if $t$ is suitably chosen, $H_0 + t(X' - Y')$ is regular in $\mathfrak{g}_0$. Then since $s^{-1}s_\lambda$ leaves it fixed, we can conclude (see [4(c), Lemma 4]) that $s = s_\lambda$. This completes the proof of the lemma.

Now select $U$ once for all according to Lemma 6.

COROLLARY.  Let $x$ be an element in $G$ such that $x(\mathfrak{a}_1 \cap U)$ meets $\mathfrak{a}_1 \cap U$. Then $x \in A$. Similarly, if $y(\mathfrak{b}_i \cap U)$ meets $\mathfrak{b}_i \cap U$ $(y \in G, i = 2, 3)$, then $y \in B$.

Select $H_1 \in \mathfrak{a}_1 \cap U$ such that $xH_1 \in \mathfrak{a}_1 \cap U$. Then since $\mathfrak{a}_0$ is the centralizer in $\mathfrak{g}_0$ of both $H_1$ and $H_2 = xH_1$, $\mathfrak{a}_0 = x\mathfrak{a}_0$. Hence $x$ defines an element $s \in W_\mathfrak{a}$. But $x\sigma = \sigma$, since $x \in \Xi$ from Lemma 6. Therefore it is clear that $s\tau = \pm \tau$. But $\tau(s^{-1}H_2) = \tau(H_1) > 0$ and $H_2 \in \mathfrak{a}_1$. So we conclude that $s\tau = \tau$, and therefore $sH' = H'$. This proves that $x(H_0 + tH') = H_0 + tH'$ $(t \in R)$, and therefore from [4(c), Lemma 3] $x \in A$. The proof in the other case is quite similar.

Define $N = \psi(V^* \times U)$ as in Lemma 1, and put $U_i = U \cap \mathfrak{z}_i$, $N_i = \psi(V^* \times U_i)$ $(i = 1, 2, 3)$. Moreover, let $\mathfrak{b}^+ = \mathfrak{b}_2$, $\mathfrak{b}^- = \mathfrak{b}_3$, $N^+ = N_2$, $N^- = N_3$ if $[\Xi : LB] = 1$ and $\mathfrak{b}^+ = \mathfrak{b}^- = \mathfrak{b}_2 \cup \mathfrak{b}_3$, $N^+ = N^- = N_2 \cup N_3$ if $[\Xi : LB] = 2$. Let $p$ be any polynomial in $S(\mathfrak{g})$ which is invariant under $L$. Then we denote by $p_\mathfrak{a}$ and $p_\mathfrak{b}$ its restrictions on $\mathfrak{a}$ and $\mathfrak{b}$, respectively. The following lemma contains the main result of this section.

LEMMA 8. *For any* $\gamma \in C_c^\infty(U)$, *define* $f_\gamma \in C_c^\infty(N)$ *as in Lemma 1. Then*

$$\int_{N_1} p\partial(q) f_\gamma \, dX = \tfrac{1}{2} \int_{\mathfrak{a}_1} \tau p_\mathfrak{a} \, |\zeta_\mathfrak{a}|^{\frac{1}{2}} \partial(q_\mathfrak{a}) \Phi_{|\zeta|^{\frac{1}{2}}\gamma} d\mu_\mathfrak{a},$$

$$\int_{N^\pm} p\partial(q) f_\gamma \, dX = \tfrac{1}{2} \int_{\mathfrak{b}^\pm} |\lambda| \, p_\mathfrak{b} \, |\zeta_\mathfrak{b}|^{\frac{1}{2}} \partial(q_\mathfrak{b}) \Psi_{|\zeta|^{\frac{1}{2}}\gamma} \, d\mu_\mathfrak{b}$$

*for* $p, q \in I(\mathfrak{g})$ *and* $\gamma \in C_c^\infty(U)$.

If $U_\sigma$ is the orthogonal projection of $U$ in $\sigma_0$, $U \cap \mathfrak{a}_1 \subset U_\sigma + R_1H' \subset \mathfrak{a}_0 \cap \mathfrak{g}_0'$ where $R_1$ is an open interval in $R^+$. Therefore, since $U_\sigma$ is connected, $U \cap \mathfrak{a}_1$ is contained in a connected component of $\mathfrak{a}_0 \cap \mathfrak{g}_0'$. Put $\pi_\mathfrak{a} = \prod_{\alpha \in P\mathfrak{a}} \alpha$. Then $\pi_\mathfrak{a}^2$ takes only real values on $\mathfrak{a}_0$, and so we can select a complex number $\epsilon$ such that $\epsilon^4 = 1$ and $\epsilon\pi_\mathfrak{a}$ is everywhere positive on $U \cap \mathfrak{a}_1$. Since $U_1 \subset U_\sigma + I_1$, it is obvious that $N_1 \subset \bigcup_{x \in G} x(\mathfrak{a}_1 \cap U)$. We claim $N_1 = N \cap \bigcup_{x \in G} x(\mathfrak{a}_1 \cap U)$. For suppose $zH \in N$ for some $z \in G$ and $H \in \mathfrak{a}_1 \cap U$. Then $zH = \phi(x^*)Z$ for some $x^* \in V^*$ and $Z \in U$. It follows from the Corollary to Lemma 5 that $Z \in U_1$ and therefore $Z = yH_1$ $(y \in L, H_1 \in \mathfrak{a}_1 \cap U)$. But then we conclude from Lemma 6 that $z^{-1}\phi(x^*)y \in \Xi$, and therefore $z \in \phi(x^*)\Xi$. Since $\Xi H \subset U_1$ (by the Corollary to Lemma 5), it follows that $zH \in \phi(x^*)U_1 \subset N_1$, and this proves our assertion. *Moreover, the above argument shows that if* $zH \in N$ $(z \in G, H \in \mathfrak{a}_1 \cap U)$, *then* $z^* \in V^*$. We shall need this fact presently.

Put $\bar{G} = G/A$ and define $\bar{x}H = xH$ $(x \in G, H \in \mathfrak{a}_0)$, where $x \to \bar{x}$ is the natural mapping of $G$ onto $\bar{G}$. Then in view of the above result and the

Corollary to Lemma 7, we can normalize the invariant measure $d\bar{x}$ on $\bar{G}$ in such a way (see [4(a), p. 501]) that

$$\int_{N_1} g(X)\,dX = \int_{\mathfrak{a}_1 \cap U} (\epsilon\pi_\mathfrak{a}(H))^2 d\mu_\mathfrak{a} \int_{\bar{G}} g(\bar{x}H)\,d\bar{x} \qquad (g \in C_c(N_1)).$$

Put $\bar{\Xi} = \Xi/A$ and let $d\bar{\xi}$ denote the invariant measure on $\bar{\Xi}$. Then it is clear that if $d\bar{\xi}$ is suitably normalized,

$$\int_{\bar{G}} g(\bar{x}H)\,d\bar{x} = \int_{G^*} dx^* \int_{\bar{\Xi}} g(x(\bar{\xi}H))\,d\bar{\xi} \qquad (g \in C_c^\infty(\mathfrak{g}_0), H \in \mathfrak{a}_0 \cap \mathfrak{g}_0').$$

Now put $\psi_g(H) = \pi_\mathfrak{a}(H) \int_{\bar{G}} g(\bar{x}H)\,d\bar{x}$ $(H \in \mathfrak{a}_0 \cap \mathfrak{g}_0')$ for $g \in C_c^\infty(\mathfrak{g}_0)$. (It follows from Theorem 3 of [4(e)] that the above integral is convergent.) Then if $g = f_\gamma$ $(\gamma \in C_c^\infty(U))$,

$$\psi_{f\gamma}(H) = \pi_\mathfrak{a}(H) \int_{G^*} dx^* \int_{\bar{G}} f_\gamma(x(\bar{\xi}H))\,d\bar{\xi} \qquad (H \in \mathfrak{a}_1 \cap U).$$

We have seen above that if $zH \in N$ for some $z \in G$ and $H \in \mathfrak{a}_1 \cap U$, then $z^* \in V^*$. Therefore

$$\psi_{f\gamma}(H) = \pi_\mathfrak{a}(H) \int_{V^*} dx^* \int_{\bar{G}} f_\gamma(\phi(x^*)(\bar{\xi}H))\,d\bar{\xi} \qquad (H \in \mathfrak{a}_1 \cap U).$$

But $f_\gamma(\phi(x^*)Z) = a_0(x^*)\gamma(Z)$ $(x^* \in V^*, Z \in U)$ from the definition of $f_\gamma$. Therefore, since $U$ is invariant under $\Xi$, we get

$$\psi_{f\gamma}(H) = \pi_\mathfrak{a}(H) \int_{V^*} a_0(x^*)\,dx^* \int_{\bar{\Xi}} \gamma(\bar{\xi}H)\,d\bar{\xi}$$
$$= \pi_\mathfrak{a}(H) \int_{\bar{\Xi}} \gamma(\bar{\xi}H)\,d\bar{\xi} \qquad (H \in \mathfrak{a}_1 \cap U).$$

Moreover, since $\Xi = LA$ it is obvious (see Section 6) that

$$\tau(H) \int_{\bar{\Xi}} \gamma(\bar{\xi}H)\,d\bar{\xi} = c\Phi_\gamma(H) \qquad (H \in \mathfrak{a}_1),$$

where $c$ is a positive number independent of $H$ or $\gamma$. Therefore

$$\psi_{f\gamma}(H) = c\pi_\mathfrak{a}'(H)\Phi_\gamma(H) \qquad (\gamma \in C_c^\infty(U), H \in \mathfrak{a}_1 \cap U),$$

where $\pi_\mathfrak{a}'$ is the product of all roots $\alpha \neq \tau$ in $P_\mathfrak{a}$. On the other hand, $\zeta$ is invariant under $L$ and since $\tau$ is everywhere positive on $\mathfrak{a}_1$, it is clear that $|\zeta_\mathfrak{a}|^{\frac{1}{2}} = \epsilon\pi_\mathfrak{a}'$ on $\mathfrak{a}_1 \cap U$. Hence $\psi_{f\gamma} = c\epsilon^{-1}\Phi_{|\zeta|^{\frac{1}{2}}\gamma}$ on $\mathfrak{a}_1 \cap U$. Now suppose $p, q \in I(\mathfrak{g})$. Then

$$\int_{N_1} p(X) f_\gamma(X;\partial(q))\,dX = \int_{\mathfrak{a}_1 \cap U} (\epsilon \pi_\mathfrak{a}(H))^2 p(H)\,d\mu_\mathfrak{a} \int_{\bar{G}} f_\gamma(\bar{x}H;\partial(q))\,d\bar{x}$$

$$= \epsilon^2 \int_{\mathfrak{a}_1 \cap U} \pi_\mathfrak{a}(H) p(H) \psi_{f\gamma}(H;\partial(q_\mathfrak{a}))\,d\mu_\mathfrak{a}$$

from Theorem 3 of [4(e)]. But

$$\epsilon^2 \pi_\mathfrak{a} \partial(q_\mathfrak{a})\psi_{f\gamma} = c\epsilon\pi_\mathfrak{a}'\tau\partial(q_\mathfrak{a})(\Phi_{|\zeta|\frac12\gamma}) = c\,|\,\zeta_\mathfrak{a}\,|^{\frac12}\tau\partial(q_\mathfrak{a})(\Phi_{|\zeta|\frac12\gamma})$$

on $\mathfrak{a}_1 \cap U$. Therefore since $\Phi_{|\zeta|\frac12\gamma}$ is zero on $\mathfrak{a}_1$ outside $\mathfrak{a}_1 \cap U$, it follows that

$$\int_{N_1} p(X) f_\gamma(X;\partial(q))\,dX = c\int_{\mathfrak{a}_1} |\,\zeta_\mathfrak{a}\,|^{\frac12}\tau p_\mathfrak{a}\partial(q_\mathfrak{a})\Phi_{|\zeta|\frac12\gamma}\,d\mu_\mathfrak{a}.$$

In order to determine $c$, we put $p = q = 1$. Then

$$\int_{N_1} f_\gamma\,dX = c\int_{\mathfrak{a}_1} |\,\zeta_\mathfrak{a}\,|^{\frac12}\tau\Phi_{|\zeta|\frac12\gamma}\,d\mu_\mathfrak{a} = c\int_{\mathfrak{a}_1} \tau\Phi_{|\zeta|\gamma}\,d\mu_\mathfrak{a} = 2c\int_{U_1} |\,\zeta\,|\gamma\,dZ$$

from Lemma 3. Therefore we conclude from Lemma 1 that $c = \frac12$.

Now we come to the second formula. Since $U_i \subset U_\sigma + I_i$ $(i = 2, 3)$, we prove as before that $N_i \subset \bigcup_{x \in G} x(\mathfrak{b}_i \cap U)$. Moreover, by the same argument as used above, we show that $N_2 \cup N_3 = N \cap (\bigcup_{x \in G} x(\mathfrak{b}_0' \cap U))$. Now first suppose $[\Xi : LB] = 1$. Then no point in $N_2$ can be conjugate (under $G$) to a point in $N_3$. For otherwise, we could choose $H_i \in \mathfrak{b}_i \cap U$ $(i = 2, 3)$ and $x \in G$ such that $xH_2 = H_3$. But as we have seen during the proof of Lemma 7, this implies that $x$ coincides with $s_\lambda$ on $\mathfrak{b}$, which is impossible since $[\Xi : LB] = 1$. Hence $N_i \cap (\bigcup_{x \in G} x(\mathfrak{b}_j \cap U))$ is empty if $i = 2, j = 3$ or $i = 3$, $j = 2$. This shows that $N_i = N \cap \bigcup_{x \in G} x(\mathfrak{b}_i \cap U)$ $(i = 2, 3)$ in this case, and the proof of the second formula now proceeds in the same way as that of the first.

So let us now assume that $[\Xi : LB] = 2$ and select an element $z \in \Xi$ which coincides with $s_\lambda$ on $\mathfrak{b}$. Then $z\mathfrak{b}_2 = \mathfrak{b}_3$ and therefore $N_2 \cup N_3 = N \cap (\bigcup_{x \in G} x(\mathfrak{b}_2 \cap U))$. Define $\pi_\mathfrak{b} = \prod_{\alpha \in P_\mathfrak{b}} \alpha$ and choose, as before, a complex number $\epsilon$ such that $\epsilon^4 = 1$ and $\epsilon\pi_\mathfrak{b}^2$ is everywhere positive on $U \cap \mathfrak{b}_2$. We use our earlier notation except that now $\bar{G} = G/B$, $\bar{\Xi} = \Xi/B$ and

$$\psi_g(H) = \pi_\mathfrak{b}(H)\int_{\bar{G}} g(\bar{x}H)\,d\bar{x} \qquad (H \in \mathfrak{b}_0 \cap \mathfrak{g}_0', g \in C_c^\infty(\mathfrak{g}_0)).$$

Then again

$$\psi_{f\gamma}(H) = \pi_\mathfrak{b}(H)\int_{\bar{\Xi}} \gamma(\bar{\xi}H)\,d\bar{\xi} \qquad (H \in \mathfrak{b}_2 \cap \mathfrak{g}_0', \gamma \in C_c^\infty(U)).$$

Let $\bar{L} = (LB)/B$. Then since $\Xi = LB \cup (LB)z$, it follows that

$$\int_{\bar{\Xi}} \gamma(\bar{\xi}H)\,d\bar{\xi} = \int_{\bar{L}} \gamma(\bar{\xi}H)\,d\bar{\xi} + \int_{\bar{L}} \gamma(\bar{\xi}(s_\lambda H))\,d\bar{\xi} \qquad (H \in \mathfrak{b}_2).$$

Moreover it is clear (see Section 6) that there exists a positive number $c$ such that

$$|\lambda(H)| \int_{\bar{L}} \gamma(\bar{\xi}H)\,d\bar{\xi} = c\Psi_\gamma(H) \qquad (H \in \mathfrak{b}_2 \cup \mathfrak{b}_3, \gamma \in C_c^\infty(U)).$$

For any function $g$ on $\mathfrak{b}_2 \cup \mathfrak{b}_3$, let $g^{s\lambda}$ denote the function $H \to g(s_\lambda H)$ $(H \in \mathfrak{b}_2 \cup \mathfrak{b}_3)$. Then since $\epsilon\pi_\mathfrak{b} = |\pi_\mathfrak{b}| = |\zeta_\mathfrak{b}|^{\frac{1}{2}}|\lambda|$ on $\mathfrak{b}_2 \cap U$, we conclude that

$$\psi_{f\gamma} = c\epsilon^{-1}\{\Psi_{|\zeta|^{\frac{1}{2}}\gamma} + (\Psi_{|\zeta|^{\frac{1}{2}}\gamma})^{s\lambda}\}$$

on $\mathfrak{b}_2 \cap U$. Now suppose $p, q \in I(\mathfrak{g})$. Then since

$$N^+ = N_2 \cup N_3 = N \cap \bigcup_{x \in G} x(\mathfrak{b}_2 \cap U),$$

it follows that

$$\int_{N^+} p(X)f_\gamma(X;\partial(q))\,dX = \int_{\mathfrak{b}_2 \cap U} (\epsilon\pi_\mathfrak{b}(H))^2 p(H)\,d\mu_\mathfrak{b} \int_{\bar{G}} f_\gamma(\bar{x}H;\partial(q))\,d\bar{x}$$

$$= \epsilon^2 \int_{\mathfrak{b}_2 \cap U} \pi_\mathfrak{b}(H)p(H)\psi_{f\gamma}(H;\partial(q_\mathfrak{b}))\,d\mu_\mathfrak{b} \text{ as before.  Furthermore,}$$

$$\epsilon^2\pi_\mathfrak{b}\partial(q_\mathfrak{b})\psi_{f\gamma} = c|\zeta_\mathfrak{b}|^{\frac{1}{2}}|\lambda|\partial(q_\mathfrak{b})\{\Psi_{|\zeta|^{\frac{1}{2}}\gamma} + (\Psi_{|\zeta|^{\frac{1}{2}}\gamma})^{s\lambda}\}$$

on $\mathfrak{b}_2 \cap U$. Therefore, since $\mathfrak{b}^+ = \mathfrak{b}_2 \cup \mathfrak{b}_3$, it is clear that

$$\int_{N^+} p\partial(q)f_\gamma\,dX = c\int_{\mathfrak{b}^+} |\zeta_\mathfrak{b}|^{\frac{1}{2}}|\lambda|\,p_\mathfrak{b}\partial(q_\mathfrak{b})\Psi_{|\zeta|^{\frac{1}{2}}\gamma}\,d\mu_\mathfrak{b}.$$

In order to determine $c$, we again put $p = q = 1$. Then

$$\int_{N^+} f_\gamma\,dX = c\int_{\mathfrak{b}^+} |\lambda|\,\Psi_{|\zeta|\gamma}\,d\mu_\mathfrak{b} = 2c\int_{U_2 \cup U_3} |\zeta|\,\gamma\,dZ$$

from Lemma 3. Therefore $c = \frac{1}{2}$ from Lemma 1. This completes the proof of Lemma 8.

Define the polynomial $\eta \in I(\mathfrak{g})$ as in [4(e), §6] and put $\epsilon_\mathfrak{a} = |\pi_\mathfrak{a}|\pi_\mathfrak{a}^{-1}$, $\epsilon_\mathfrak{b} = |\pi_\mathfrak{b}|\pi_\mathfrak{b}^{-1}$ on $\mathfrak{a}_1 \cap U$ and $\mathfrak{b}_0' \cap U$, respectively. Then as we have seen above, $\epsilon_\mathfrak{a}$ and $\epsilon_\mathfrak{b}$ are constant on $\mathfrak{a}_1 \cap U$ and $\mathfrak{b}_i \cap U$ $(i = 2, 3)$, respectively. Moreover since $\pi_\mathfrak{b}^{s\lambda} = -\pi_\mathfrak{b}$, it is clear that $\epsilon_\mathfrak{b}(H_2) = -\epsilon_\mathfrak{b}(H_3)$ if $H_i \in \mathfrak{b}_i \cap U$ $(i = 2, 3)$.

COROLLARY. *Let $k$ be any integer $\geq 0$. Then for any $\xi \in I(\mathfrak{g})$,*

$$\int_{N_1} \eta^k \partial(\eta^k \xi) f_\gamma \, dX = \tfrac{1}{2} \int_{\mathfrak{a}_1} \epsilon_\mathfrak{a} \pi_\mathfrak{a}^{2k+1} \partial(\pi_\mathfrak{a}^{2k} \xi_\mathfrak{a}) \Phi_{|\zeta|^{\frac{1}{2}}\gamma} d\mu_\mathfrak{a},$$

$$\int_{N^\pm} \eta^k \partial(\eta^k \xi) f_\gamma \, dX = \tfrac{1}{2} \int_{\mathfrak{b}^\pm} \epsilon_\mathfrak{b} \pi_\mathfrak{b}^{2k+1} \partial(\pi_\mathfrak{b}^{2k} \xi_\mathfrak{b}) \Psi_{|\zeta|^{\frac{1}{2}}\gamma} d\mu_\mathfrak{b}$$

*for $\gamma \in C_c^\infty(U)$.*

This follows immediately from Lemma 8 if we observe that $\eta_\mathfrak{a} = \pi_\mathfrak{a}^2$, $\eta_\mathfrak{b} = \pi_\mathfrak{b}^2$ and $\tau \mid \zeta_\mathfrak{a} \mid^{\frac{1}{2}} = \mid \pi_\mathfrak{a} \mid = \epsilon_\mathfrak{a}\pi_\mathfrak{a}$, $\mid \lambda \mid \mid \zeta_\mathfrak{b} \mid^{\frac{1}{2}} = \mid \pi_\mathfrak{b} \mid = \epsilon_\mathfrak{b}\pi_\mathfrak{b}$ on $\mathfrak{a}_1 \cap U$ and $\mathfrak{b}_0' \cap U$, respectively.

**3. Proof of Theorem 1.** Since $\mathfrak{h} = \sigma + CH_{\alpha_0}$, it is clear that every polynomial in $S(\mathfrak{h})$ can be written uniquely in the form $p = \sum_{k \geq 0} p_k \alpha_0^k$ where $p_k \in S(\sigma)$. Let $s_0$ denote the Weyl reflexion in $\mathfrak{h}$ corresponding to $\alpha_0$. Then since $H_{\alpha_0}$ is orthogonal to $\sigma$, it is obvious that $p_k^{s_0} = p_k$ and $s_0\alpha_0 = -\alpha_0$. Therefore if $p^{s_0} = p$, we can conclude that $p_k = 0$ if $k$ is odd. Now let $\xi$ be a homogeneous element in $I(\mathfrak{g})$ of positive degree and $r$ an integer $\geq 0$. Then if $\bar{\xi}$ is the restriction of $\xi$ on $\mathfrak{h}$, it follows that (see [4(d), Lemma 9]) that

$$\bar{\xi}\pi^{2r} = \sum_{k \geq 0} \alpha_0^{2r+2k} q_k,$$

where $q_k \in S(\sigma)$. Similarly,

$$\pi^{2r+1} = \sum_{m \geq 0} \alpha_0^{2r+1+2m} q_m',$$

where again $q_m' \in S(\sigma)$. Put

$$p_k = \sum_{m \geq 0} \langle \alpha_0^{2r+1+2m}, \alpha_0^{2r+1+2m} \rangle \partial(q_{m+1+k}) q_m' \qquad (k \geq 0)$$

in the notation of [4(d), §2]. Let $d^0(p)$ denote the degree of any polynomial $p \in S(\mathfrak{g})$. Then it is obvious that $q_k$ and $q_m'$ are homogeneous and

$$d^0(q_k) = d^0(\xi) + 2r d^0(\pi) - (2r + 2k),$$

$$d^0(q_m') = (2r + 1) d^0(\pi) - (2r + 1 + 2m).$$

Therefore $p_k$ is also homogeneous and

$$d^0(p_k) = d^0(\pi) - d^0(\xi) + 2k + 1 \qquad (k \geq 0).$$

LEMMA 9. *For any given integer $r \geq 0$, we can choose a homogeneous element $\xi \in I(\mathfrak{g})$ of positive degree such that the corresponding polynomial $p_k \neq 0$ for some $k$.*

Put $\epsilon = 1$ or $(-1)^{\frac{1}{2}}$ according as $\alpha_0$ lies in $Q_+$ or in $P_0$. Then $\alpha^* = \epsilon \alpha_0$ takes only real values on $\mathfrak{h}_0$. Let $\mathfrak{h}_+$ denote the set of all $H \in \mathfrak{h}_0$ where $\alpha^*(H) > 0$. Similarly, let $\mathfrak{h}_-$ be the set of those $H \in \mathfrak{h}_0$ where $\alpha^*(H) < 0$. We call $\mathfrak{h}_+$ and $\mathfrak{h}_-$, respectively, the upper and lower halves of $\mathfrak{h}_0$.

Let $d\mu$ denote the Euclidean measure on $\mathfrak{h}_0$ and $\mathcal{E}$ the distribution on $\mathfrak{h}_0$ given by

$$\mathcal{E}(g) = \int_{\mathfrak{h}_+} g \, d\mu \qquad\qquad (g \in C_c^\infty(\mathfrak{h}_0)).$$

Put $\mathcal{E}_r = \partial(\pi^{2r})(\pi^{2r+1}\mathcal{E})$. We shall first prove the following result.

LEMMA 10. *There exists a homogenous element $\bar{\xi} \in I(\mathfrak{h})$ of positive degree such that $H_0$ lies in the carrier of $\partial(\bar{\xi})\mathcal{E}_r$.*

Let $\square$ denote the polynomial function on $\mathfrak{h}$ defined by $\square(H)$ ' $= -B(H, \theta(H))$ $(H \in \mathfrak{h})$. Let $W$ be the Weyl group of $\mathfrak{g}$ with respect to $\mathfrak{h}$, and consider the polynomial

$$\prod_{s \in W} (t - \square^s) = t^w + q_1 t^{w-1} + \cdots + q_w \qquad (q_i \in S(\mathfrak{h}))$$

in the indeterminate $t$ with coefficients in $S(\mathfrak{h})$. It is obvious that $q_i$ $(1 \leq i \leq w)$ are homogeneous elements in $I(\mathfrak{h})$ of positive degree. Replacing $t$ by $\square$ we get,

$$\square^w + \square^{w-1} q_1 + \cdots + q_w = \prod_{s \in W} (\square - \square^s) = 0.$$

Now if our assertion is false, we can choose an open connected neighborhood $V$ of $H_0$ in $\mathfrak{h}_0$ such that $\partial(q_i)\mathcal{E}_r = 0$ on $V$ $(1 \leq i \leq w)$. Therefore since

$$\partial(\square^w) + \partial(\square^{w-1})\partial(q_1) + \cdots + \partial(q_w) = 0,$$

we conclude that $\partial(\square^w)\mathcal{E}_r = 0$ on $V$. But $\square$ is obviously a positive-definite quadratic form in $\mathfrak{h}_0$, and therefore this differential equation is of the *elliptic type* (see Garding [3]). Hence it follows (see [7, p. 136] and [6]) that $\mathcal{E}_r$ must coincide on $V$ with an analytic function. But $\mathcal{E} = 0$ on $\mathfrak{h}_-$, and therefore $\mathcal{E}_r = 0$ on $V \cap \mathfrak{h}_-$. Moreover, $V \cap \mathfrak{h}_-$ is open and not empty since $\alpha_0(H_0) = 0$. Therefore we conclude that $\mathcal{E}_r = 0$ on $V$. On the other hand, $V \cap \mathfrak{h}_+$ is also not empty and $\mathcal{E} = 1$ on $V \cap \mathfrak{h}_+$. Therefore $\partial(\pi)\mathcal{E}_r = \partial(\pi^{2r+1})\pi^{2r+1} = \langle \pi^{2r+1}, \pi^{2r+1} \rangle$ on $V \cap \mathfrak{h}_+$. But we know from the Corollary to Lemma 18 of [4(d)] that $\langle \pi^{2r+1}, \pi^{2r+1} \rangle$ is a positive constant. This contradiction proves the lemma.

Now in order to prove Lemma 9, we select (see Lemma 9 of [4(d)]) a homogeneous element $\xi \in I(\mathfrak{g})$ of positive degree such that its restriction $\bar{\xi}$

on $\mathfrak{h}$ fulfills the condition of Lemma 10. It is obvious that $\partial(H)\mathcal{E} = 0$ for $H \in \sigma$. Moreover,

$$\mathcal{E}(\partial(\alpha_0)(\alpha_0 g)) = \int_{\mathfrak{h}_+} \partial(\alpha_0)(\alpha_0 g)\, d\mu = 0 \qquad (g \in C_c^\infty(\mathfrak{h}_0)),$$

and therefore $\alpha_0 \partial(\alpha_0)\mathcal{E} = 0$. We claim that

$$\partial(\alpha_0^k)(\alpha_0^m \mathcal{E}) = (\partial(\alpha_0^k)\alpha_0^m)\mathcal{E} \qquad (m \geq k \geq 0).$$

This is obviously true for $k = 0$. So we assume $k \geq 1$ and use induction on $k$. Then $\partial(\alpha_0^{k-1})(\alpha_0^m \mathcal{E}) = (\partial(\alpha_0^{k-1})\alpha_0^m)\mathcal{E}$, and therefore

$$\partial(\alpha_0^k)(\alpha_0^m\mathcal{E}) = (\partial(\alpha_0^k)\alpha_0^m)\mathcal{E} + (\partial(\alpha_0^{k-1})\alpha_0^m)\partial(\alpha_0)\mathcal{E} = (\partial(\alpha_0^k)\alpha_0^m)\mathcal{E}$$

since $m > k - 1$ and $\alpha_0 \partial(\alpha_0)\mathcal{E} = 0$. Hence if $k > m$,

$$\partial(\alpha_0^k)(\alpha_0^m\mathcal{E}) = \partial(\alpha_0^{k-m})\{(\partial(\alpha_0^m)\alpha_0^m)\mathcal{E}\} = \langle \alpha_0^m, \alpha_0^m \rangle \partial(\alpha_0^{k-m})\mathcal{E}.$$

Now consider $\partial(\bar{\xi})\mathcal{E}_r$. Then if

$$\bar{\xi}\pi^{2r} = \sum_{k \geq 0} \alpha_0^{2r+2k} q_k, \qquad \pi^{2r+1} = \sum_{m \geq 0} \alpha_0^{2r+1+2m} q_m'$$

$(q_k, q_m' \in S(\sigma))$ as before, it is clear that

$$\partial(\bar{\xi})\mathcal{E}_r = \partial(\bar{\xi}\pi^{2r})(\pi^{2r+1}\mathcal{E}) = \sum_{k,m \geq 0} \partial(\alpha_0^{2r+2k} q_k)(\alpha_0^{2r+1+2m} q_m'\mathcal{E})$$

$$= \sum_{k,m \geq 0} (\partial(q_k)q_m')\partial(\alpha_0^{2r+2k})(\alpha_0^{2r+1+2m}\mathcal{E})$$

$$= \sum_{m \geq k \geq 0} (\partial(q_k)q_m')(\partial(\alpha_0^{2r+2k})\alpha_0^{2r+1+2m})\mathcal{E}$$

$$+ \sum_{k > m \geq 0} (\partial(q_k)q_m')\langle \alpha_0^{2r+1+2m}, \alpha_0^{2r+1+2m}\rangle \partial(\alpha_0^{2k-2m-1})\mathcal{E}$$

since $\sigma$ and $H_{\alpha_0}$ are orthogonal. Now the first sum is obviously equal to $(\partial(\bar{\xi}\pi^{2r})\pi^{2r+1})\mathcal{E}$. But $\partial(\bar{\xi}\pi^{2r})\pi^{2r+1}$ is a homogeneous polynomial in $S(\mathfrak{h})$ of degree $d^0(\pi) - d^0(\xi)$ which is skew-invariant under W. Therefore it follows from Lemma 10 of $[4(\mathrm{d})]$ that it is zero. Put $\delta^{(k)} = \partial(\alpha_0^{k+1})\mathcal{E}$ $(k \geq 0)$. Then

$$\partial(\bar{\xi})\mathcal{E}_r = \sum_{k \geq 0} p_k \delta^{(2k)},$$

where the polynomials $p_k \in S(\sigma)$ are defined as before by

$$p_k = \sum_{m \geq 0} (\partial(q_{m+1+k})q_m')\langle \alpha_0^{2r+1+2m}, \alpha_0^{2r+1+2m}\rangle.$$

On the other hand, since $\partial(\bar{\xi})\mathcal{E}_r \neq 0$, not every $p_k$ can be zero. This completes the proof of Lemma 9.

Now fix $r$ and $\xi$ once for all as in Lemma 9 and let $k_0$ be an integer such that $p_{k_0} \neq 0$. As before, let $d\mu_\sigma$ denote the Euclidean measure on $\sigma_0$.

LEMMA 11. *If $U_\sigma$ is any open neighborhood of $H_0$ in $\sigma_0$, we can select a function $f \in C_c^\infty(U_\sigma)$ such that*

$$\int_{U_\sigma} f p_k \, d\mu_\sigma = \begin{cases} 0 & \text{if } k \neq k_0, \\ 1 & \text{if } k = k_0 \end{cases} \qquad (k \geqq 0).$$

Let $k_1 < k_2 < \cdots < k_m$ be all the values of $k$ for which $p_k \neq 0$. Then as we have seen $p_{k_i}$ is a homogeneous polynomial in $S(\sigma)$ of degree $d^0(\pi) - d^0(\xi) + 2k_i + 1$ $(1 \leqq i \leqq m)$. Therefore, regarded as functions on $U_\sigma, p_{k_1}, \cdots, p_{k_m}$ are linearly independent over $C$, and our assertion follows from Lemma 20 of the Appendix (Section 6).

Now let $\mathfrak{a}_1 = \sigma_0 + R^+ H'$ and $\mathfrak{b}_0' = \sigma + R'(X' - Y')$ as before. We define the $\mathcal{B}$-distributions (see [4(d), §2]) $\mathcal{E}_\mathfrak{a}$ and $\mathcal{E}_\mathfrak{b}^\pm$ on $\mathfrak{a}_1$ and $\mathfrak{b}_0'$, respectively, as follows:

$$\mathcal{E}_\mathfrak{a}(g_1) = \int_{\mathfrak{a}_1} g_1 \, d\mu_\mathfrak{a} \qquad (g_1 \in \mathcal{B}(\mathfrak{a}_1)),$$

$$\mathcal{E}_\mathfrak{b}^+(g_2) = \int_{\mathfrak{b}_2} g_2 \, d\mu_\mathfrak{b}, \qquad \mathcal{E}_\mathfrak{b}^-(g_2) = \int_{\mathfrak{b}_3} g_2 \, d\mu_\mathfrak{b} \qquad (g_2 \in \mathcal{B}(\mathfrak{b}_0')).$$

Let $\delta_\mathfrak{a}^{(k)} = \partial(\tau^{k+1}) \mathcal{E}_\mathfrak{a}$ and $^\pm\delta_\mathfrak{b}^{(k)} = \partial(\lambda^{k+1}) \mathcal{E}_\mathfrak{b}^\pm$ $(k \geqq 0)$. For any automorphism $y$ of $\mathfrak{g}$ and $p \in S(\mathfrak{g})$, let $p^y$ denote the polynomial function $X \to p(y^{-1}X)$ $(X \in \mathfrak{g})$. Then it is clear that $\pi_\mathfrak{a} = \pi^{y\mathfrak{a}}$, $\pi_\mathfrak{b} = \pi^{y\mathfrak{b}}$ and $p^{y\mathfrak{a}} = p^{y\mathfrak{b}} = p$ for $p \in S(\sigma)$. Moreover, $(\alpha_0)^{y\mathfrak{a}} = \tau$, $(\alpha_0)^{y\mathfrak{b}} = \lambda$ and $(\bar{\xi})^{y\mathfrak{a}} = \xi_\mathfrak{a}$, $(\bar{\xi})^{y\mathfrak{b}} = \xi_\mathfrak{b}$ in the notation of Lemma 8. Therefore

$$\xi_\mathfrak{a} \pi_\mathfrak{a}^{2r} = \sum_{k \geqq 0} \tau^{2r+2k} q_k, \qquad \pi_\mathfrak{a}^{2r+1} = \sum_{m \geqq 0} \tau^{2r+1+2m} q_m'$$

$$\xi_\mathfrak{b} \pi_\mathfrak{b}^{2r} = \sum_{k \geqq 0} \lambda^{2r+2k} q_k, \qquad \pi_\mathfrak{b}^{2r+1} = \sum_{m \geqq 0} \lambda^{2r+1+2m} q_m'.$$

Moreover, it again follows without difficulty that $\partial(H)\mathcal{E}_\mathfrak{a} = 0$, $\partial(H)\mathcal{E}_\mathfrak{b}^\pm = 0$ for $H \in \sigma$ and $\tau\partial(\tau)\mathcal{E}_\mathfrak{a} = 0$, $\lambda\partial(\lambda)\mathcal{E}_\mathfrak{b}^\pm = 0$. Therefore we conclude in the same way as before, that

$$\partial(\xi_\mathfrak{a}\pi_\mathfrak{a}^{2r})(\pi_\mathfrak{a}^{2r+1}\mathcal{E}_\mathfrak{a}) = \sum_{k \geqq 0} p_k \delta_\mathfrak{a}^{(2k)}$$

$$\partial(\xi_\mathfrak{b}\pi_\mathfrak{b}^{2r})(\pi_\mathfrak{b}^{2r+1}\mathcal{E}_\mathfrak{b}) = \sum_{k \geqq 0} p_k {}^\pm\delta_\mathfrak{b}^{(2k)}.$$

Now $U$ being defined as in Lemma 8, let $U_\sigma$ and $U_\mathfrak{l}$ denote the orthogonal projections of $U$ in $\sigma_0$ and $\mathfrak{l}_0$, respectively. Select $g_0 \in C_c^\infty(U_\sigma)$ according to Lemma 11, and for any $\beta \in C_c^\infty(U_\mathfrak{l})$, define $\gamma_\beta \in C_c^\infty(U)$ by

$$\gamma_\beta(H_\sigma + Z) = g_0(H_\sigma)\beta(Z) | \zeta(H_\sigma + Z) |^{-\frac{1}{2}} \qquad (H_\sigma \in U_\sigma, Z \in U_\mathfrak{l}).$$

Moreover, let $F_\beta = f_{\gamma_\beta}$ (in the notation of Lemmas 1 and 8) and

$$\phi_\beta(t) = \int_{-\infty}^{\infty} \bar\beta(tH' + xX')\,dx \qquad\qquad (t \in R^+)$$

$$\psi_\beta(s) = |s| \int_0^{\infty} \bar\beta(s(e^t X' - e^{-t}Y'))(e^t - e^{-t})\,dt \qquad\qquad (s \in R'),$$

where [2] $\bar\beta(Z) = \pi^{-1} \int_0^{\pi} \beta(u_t Z)\,dt$ and $u_t = \exp(t\,\mathrm{ad}(X' - Y'))$. Put $\phi_\beta(0)$ $= \mathrm{Lim}_{t \to 0} \phi_\beta(t)\ (t \in R^+)$, $\psi_\beta^+(0) = \mathrm{Lim}_{s \to 0} \psi_\beta(s)\ (s \in R^+)$ and $\psi_\beta^-(0) = \mathrm{Lim}_{s \to 0} \psi_\beta(s)$ $(s \in R^-)$. All these limits exist (see Section 6). Also let $\pi'$ denote the product of all roots $\alpha \ne \alpha_0$ in $P$. Then $\pi'(H_0) \ne 0$. We set $\epsilon = |\pi'(H_0)|/\pi'(H_0)$.

LEMMA 12. *Let* $q = d^0(\xi)$ *and* $a = \epsilon(-1)^q \{\alpha_0(H_{a_0})\}^{2k_0+1} 2^{k_0-2}$, *where* $k_0$ *has the same meaning as in Lemma 11. Then for any* $\beta \in C_c^\infty(U_I)$,

$$\int_{N_1} \eta^r \partial(\eta^r \xi) F_\beta\,dX = a\phi_{\beta_0}(0),$$

*where* $\beta_0 = \partial(\Omega^{k_0})\beta$. *Similarly,*

$$\int_{N^\pm} \eta^r \partial(\eta^r \xi) F_\beta\,dX = -a\psi_{\beta_0}^{\pm}(0)$$

*if* $[\Xi : LB] = 1$, *and, if* $[\Xi : LB] = 2$,

$$\int_{N_2 \cup N_3} \eta^r \partial(\eta^r \xi) F_\beta\,dX = -a\{\psi_{\beta_0}^+(0) + \psi_{\beta_0}^-(0)\}.$$

Define $\epsilon_\mathfrak{a}$ and $\epsilon_\mathfrak{b}$ on $\mathfrak{a}_1 \cap U$ and $\mathfrak{b}_0' \cap U$ as in the Corollary to Lemma 8. Then $\epsilon_\mathfrak{a}$ is constant on $\mathfrak{a}_1 \cap U$ and $\epsilon_\mathfrak{b}$ on $\mathfrak{b}_i \cap U$ $(i = 2, 3)$, and it is obvious that $\pi_\mathfrak{a}'(H_0) = \pi_\mathfrak{b}'(H_0) = \pi'(H_0) \ne 0$, where $\pi_\mathfrak{a}' = (\pi')^{\nu_\mathfrak{a}}$, $\pi_\mathfrak{b}' = (\pi')^{\nu_\mathfrak{b}}$. Therefore since $\pi_\mathfrak{a} = \tau \pi_\mathfrak{a}'$, by making $H$ tend to $H_0$ $(H \in \mathfrak{a}_1 \cap U)$, we conclude immediately that $\epsilon_\mathfrak{a} = \epsilon$. Similarly, $\epsilon_\mathfrak{b} = -(-1)^{\frac{1}{2}}\epsilon$ and $(-1)^{\frac{1}{2}}\epsilon$ on $\mathfrak{b}_2$ and $\mathfrak{b}_3$, respectively. Hence it follows from the Corollary to Lemma 8 that

$$\int_{N_1} \eta^r \partial(\eta^r \xi) F_\beta\,dX = \tfrac{1}{2}\epsilon \mathcal{E}_\mathfrak{a}(\pi_\mathfrak{a}^{2r+1}\partial(\pi_\mathfrak{a}^{2r}\xi_\mathfrak{a})\Phi_{|\zeta|^{\frac{1}{2}}\gamma_\beta})$$

$$= \tfrac{1}{2}\epsilon(-1)^q \sum_{k \geq 0} \delta_\mathfrak{a}^{(2k)}(p_k \Phi_{|\zeta|^{\frac{1}{2}}\gamma_\beta}) = -\tfrac{1}{2}\epsilon(-1)^q \sum_{k \geq 0} \mathcal{E}_\mathfrak{a}(p_k \partial(\tau^{2k+1})\Phi_{|\zeta|^{\frac{1}{2}}\gamma_\beta})$$

in view of our result above. But it is clear that

$$\Phi_{|\zeta|^{\frac{1}{2}}\gamma_\beta}(H_\sigma + tH') = g_0(H_\sigma)\psi_\beta(t) \qquad\qquad (H_\sigma \in U_\sigma, t \in R^+).$$

Let $H_\tau$ denote the element in $\mathfrak{a}$ such that $B(H, H_\tau) = \tau(H)$ for all $H \in \mathfrak{a}$. Then it is clear that $\tau(H_\tau) = \alpha_0(H_{a_0}) = \langle \alpha_0, \alpha_0 \rangle$, and hence $H_\tau = \tfrac{1}{2}\langle \alpha_0, \alpha_0 \rangle H'$. Therefore

$$\mathcal{E}_{\mathfrak{a}}\left(p_k\partial(\tau^{2k+1})\Phi_{|\zeta|\frac{1}{2}\gamma_\beta}\right)$$

$$= \langle\alpha_0,\alpha_0\rangle^{2k+1}2^{-(2k+1)}\int_{U_\sigma}p_kg_0d\mu_\sigma\int_0^\infty\left(d^{2k+1}\phi_\beta/dt^{2k+1}\right)dt.$$

On the other hand, we know from Lemma 17 (Appendix, Section 6) that

$$\phi_{\partial(\Omega^k)\beta}=2^{-3k}d^{2k}\phi_\beta/dt^{2k}.$$

So it follows from the definition of $g_0$ that

$$\int_{N_1}\eta^r\partial(\eta^r\xi)F_\beta\,dX=\epsilon\langle\alpha_0,\alpha_0\rangle^{2k_0+1}2^{k_0-2}(-1)^q\phi_{\beta_0}(0),$$

and this proves the first formula.

Now suppose $[\Xi:LB]=1$. Then it follows again from the Corollary to Lemma 8 that

$$\int_{N^\pm}\eta^r\partial(\eta^r\xi)F_\beta dX=\mp\tfrac{1}{2}\epsilon(-1)^{\frac{1}{2}}\mathcal{E}_{\mathfrak{b}}^\pm\left(\pi_{\mathfrak{b}}^{2r+1}\partial(\pi_{\mathfrak{b}}^{2r}\xi_{\mathfrak{b}})\Psi_{|\zeta|\frac{1}{2}\gamma_\beta}\right)$$

$$=\mp\tfrac{1}{2}\epsilon(-1)^{q+\frac{1}{2}}\sum_{k\geqq0}{}^\pm\delta_{\mathfrak{b}}^{(2k)}\left(p_k\Psi_{|\zeta|\frac{1}{2}\gamma_\beta}\right)=\pm\tfrac{1}{2}\epsilon(-1)^{q+\frac{1}{2}}\sum_{k\geqq0}\mathcal{E}_{\mathfrak{b}}^\pm\left(p_k\partial(\lambda^{2k+1})\Psi_{|\zeta|\frac{1}{2}\gamma_\beta}\right).$$

Let $H_\lambda$ denote the element in $\mathfrak{b}$ such that $B(H,H_\lambda)=\lambda(H)$ $(H\in\mathfrak{b})$. Then $\lambda(H_\lambda)=\alpha_0(H_{\alpha_0})=\langle\alpha_0,\alpha_0\rangle$ and $H_\lambda=\nu_{\mathfrak{b}}H_{\alpha_0}=-\tfrac{1}{2}(-1)^{\frac{1}{2}}\langle\alpha_0,\alpha_0\rangle(X'-Y')$. Therefore

$$\mathcal{E}_{\mathfrak{b}}^+\left(p_k\partial(\lambda^{2k+1})\Psi_{|\zeta|\frac{1}{2}\gamma_\beta}\right)$$

$$= \langle\alpha_0,\alpha_0\rangle^{2k+1}(-1)^{k-\frac{1}{2}}2^{-(2k+1)}\int_{U_\sigma}p_kg_0d\mu_\sigma\int_0^\infty\left(d^{2k+1}\psi_\beta/dt^{2k+1}\right)dt$$

and

$$\mathcal{E}_{\mathfrak{b}}^-\left(p_k\partial(\lambda^{2k+1})\Psi_{|\zeta|\frac{1}{2}\gamma_\beta}\right)$$

$$= \langle\alpha_0,\alpha_0\rangle^{2k+1}(-1)^{k-\frac{1}{2}}2^{-(2k+1)}\int_{U_\sigma}p_kg_0d\mu_\sigma\int_{-\infty}^0\left(d^{2k+1}\psi_\beta/dt^{2k+1}\right)dt.$$

On the other hand,

$$\psi_{\partial(\Omega^k)\beta}=(-1)^k2^{-3k}d^{2k}\psi_\beta/dt^{2k}$$

from Lemma 17 (Section 6). Therefore

$$\mathcal{E}_{\mathfrak{b}}^\pm\left(p_k\partial(\lambda^{2k+1})\Psi_{|\zeta|\frac{1}{2}\gamma_\beta}\right)$$

$$= \pm\langle\alpha_0,\alpha_0\rangle^{2k+1}(-1)^{\frac{1}{2}}2^{k-1}\psi_{\partial(\Omega^k)\beta}^\pm(0)\int_{U_\sigma}p_kg_0d\mu_\sigma,$$

and by taking into account the definition of $g_0$ we get,

$$\int_{N^\pm}\eta^r\partial(\eta^r\xi)F_\beta\,dX=-\epsilon\langle\alpha_0,\alpha_0\rangle^{2k_0+1}(-1)^q2^{k_0-2}\psi_{\beta_0}^\pm(0).$$

Finally, suppose $[\Xi : LB] = 2$. Then again from the Corollary to Lemma 8,

$$\int_{N_2 \cup N_3}^{\bullet} \eta^r \partial(\eta^r \xi) F_\beta \, dX = -\tfrac{1}{2}\epsilon(-1)^{\frac{1}{2}}\mathcal{E}_\mathfrak{d}(\pi_\mathfrak{d}^{2r+1}\partial(\pi_\mathfrak{d}^{2r}\xi_\mathfrak{d})\Psi_{|\zeta|^{\frac{1}{2}}\gamma_\beta}),$$

where $\mathcal{E}_\mathfrak{d} = \mathcal{E}_\mathfrak{d}^+ - \mathcal{E}_\mathfrak{d}^-$, and we conclude from the above caluculation that

$$\int_{N_2 \cup N_3}^{\bullet} \eta^r \partial(\eta^r \xi) F_\beta \, dX = -\epsilon \langle \alpha_0, \alpha_0 \rangle^{2k_0+1} (-1)^q 2^{k_0-2}[\psi_{\beta_0}^+(0) + \psi_{\beta_0}^-(0)].$$

This completes the proof of Lemma 12.

We are now in a position to prove the main result of this section.

THEOREM 1. *Let $r$ and $\xi$ be defined as in Lemma 9 and let $c_1$, $c_2$, $c_3$ be three complex numbers. We assume that $c_2 = c_3$ in case $[\Xi : LB] = 2$. Let $N_0$ be an open neighborhood of $H_0$ in $N$ and let $T$ be the distribution on $N_0$ defined as follows:*

$$T(f) = \sum_{1 \leq i \leq 3} c_i \int_{N_i}^{\bullet} \eta^r \partial(\eta^r) f \, dX \qquad (f \in C_c^\infty(N_0)).$$

*Then $\partial(\xi)T = 0$ implies that $c_1 = c_2 = c_3$.*

Let $_0U$ be an open neighborhood of $H_0$ in $U$ and $_0V^*$ an open neighborhood of $1^*$ in $V^*$ (see Lemma 1 for the definition of $V^*$). We can choose $_0V^*$ and $_0U$ so small that $\phi(x^*)Z \in N_0$ for $x^* \in {}_0V^*$, $Z \in {}_0U$. Let $_0U_\sigma$ and $_0U_\mathfrak{l}$ be the orthogonal projections of $_0U$ in $\sigma_0$ and $\mathfrak{l}_0$, respectively. We may further assume that $_0U = {}_0U_\sigma + {}_0U_\mathfrak{l}$. Moreover, we can suppose that the carriers of the functions $a_0$ and $g_0$ (used in the definitions of $f_\gamma$ and $F_\beta$ in Lemmas 1 and 12, respectively) are contained in $_0V^*$ and $_0U_\sigma$, respectively. Then it is clear that $F_\beta \in C_c^\infty(N_0)$ for any $\beta \in C_c^\infty(_0U_\mathfrak{l})$, and therefore $T(\partial(\xi)F_\beta) = 0$ by our hypothesis. On the other hand, it follows from Lemma 12 that

$$T(\partial(\xi)F_\beta) = a\{c_1\phi_{\partial(\Omega^{k_0})\beta}(0) - c_2\psi^+_{\partial(\Omega^{k_0})\beta}(0) - c_3\psi^-_{\partial(\Omega^{k_0})\beta}(0)\}$$

since $c_2 = c_3$ when $[\Xi : LB] = 2$. Now put $\beta = \Omega^{k_0}\gamma$ ($\gamma \in C_c^\infty(_0U_\mathfrak{l})$) and $\gamma_0 = \partial(\Omega^{k_0})(\Omega^{k_0}\gamma)$. Then it is clear that

$$c_1\phi_{\gamma_0}(0) - c_2\psi_{\gamma_0}^+(0) - c_3\psi_{\gamma_0}^-(0) = 0$$

for all $\gamma \in C_c^\infty(_0U_\mathfrak{l})$. On the other hand if $c = 2^{2k_0}k_0!\Gamma(k_0 + \tfrac{1}{2})/\Gamma(\tfrac{1}{2})$, it follows from the Corollary of Lemma 18 (Section 6) that $\phi_{\gamma_0}(0) = c\phi_\gamma(0)$, $\psi_{\gamma_0}^\pm(0) = c\psi_\gamma^\pm(0)$. Therefore, since $c$ is positive, we conclude that

$$c_1\phi_\gamma(0) - c_2\psi_\gamma^+(0) - c_3\psi_\gamma^-(0) = 0$$

for all $\gamma \in C_c^\infty \, (_0U_\mathrm{I})$, and now it follows from Lemma 19 (Section 6) that $c_1 = c_2 = c_3$.

**4. Proof of Theorem 2.** We shall now apply Theorem 1 to prove that the constants $c_1, \cdots, c_N$ of Lemma 29 of [4(e)] are all equal. But first we need one additional fact.

LEMMA 13. *Let $S$ denote the set of those elements in $\mathfrak{g}_0$ which are either regular or semiregular. Then $S$ is connected.*

Suppose $V_1$, $V_2$ are two closed subsets of $\mathfrak{g}_0$ such that $V_1 \cup V_2 = \mathfrak{g}_0$ and $V_1 \cap V_2 \cap S = \emptyset$. Then we have to show that $S$ is contained in one of the two sets $V_1$, $V_2$. Let $\mathfrak{h}_0$ be a Cartan subalgebra of $\mathfrak{g}_0$ such that $\theta(\mathfrak{h}_0) = \mathfrak{h}_0$. We use the notation of Section 2. For any root $\alpha \in P$, let $\sigma_\alpha$ denote the set of those $H \in \mathfrak{h}_0$ where $\alpha(H) = 0$. Moreover, let $U$ be the complement in $\mathfrak{h}_0$ of the union of the sets $\sigma_\alpha \cap \sigma_\beta$ $(\alpha, \beta \in P; \alpha \neq \beta)$. Then $U$ is connected (see the Corollary to Lemma 42 of [4(e)]), and it is obvious that $U \subset S$. Therefore $U$ is contained in one of the two sets $V_1$, $V_2$. But since $U$ is dense in $\mathfrak{h}_0$, we conclude that either $V_1$ or $V_2$ contains $\mathfrak{h}_0$.

Let $X$ be an element in $S$. Then since $G$ is connected, $\bigcup\limits_{x \in G} xX$ is a connected subset of $S$. Then if $X \in V_i$, it follows that $xX \in V_i$ for every $x \in G$ $(i = 1, 2)$. Now select $i$ such that $\mathfrak{h}_0 \subset V_i$. Then $x(\mathfrak{h}_0 \cap \mathfrak{g}_0') \subset V_i$ $(x \in G)$, and therefore since $\mathfrak{h}_0 \cap \mathfrak{g}_0'$ is dense in $\mathfrak{h}_0$, $x\mathfrak{h}_0 \subset V_i$.

Let $\mathfrak{a}_0$ be a fundamental Cartan subalgebra (see [4(e), §8]) of $\mathfrak{g}_0$ such that $\mathfrak{a}_0 = \theta(\mathfrak{a}_0)$. We may assume that $\mathfrak{a}_0 \subset V_1$. For any Cartan subalgebra $\Gamma_0$ of $\mathfrak{g}_0$, define $l_-(\Gamma_0)$ as in [4(e), §8]. We shall prove by induction on $l_-(\mathfrak{a}_0) - l_-(\mathfrak{h}_0)$ that $\mathfrak{h}_0 \subset V_1$. If $l_-(\mathfrak{a}_0) = l_-(\mathfrak{h}_0)$, $\mathfrak{h}_0$ is also fundamental and therefore conjugate to $\mathfrak{a}_0$ under $G$ (see the Corollary to Lemma 32 of [4(e)]). Hence it follows from what has been said above that $\mathfrak{h}_0 \subset V_1$. So let us now suppose that $l_-(\mathfrak{h}_0) < l_-(\mathfrak{a}_0)$. Since $K$ is compact and $\mathfrak{a}_0 \cap \mathfrak{k}_0$ is a maximal abelian subalgebra of $\mathfrak{k}_0$ (Corollary to Lemma 31 of [4(e)]), we may assume (by replacing $\mathfrak{h}_0$ by a conjugate Cartan subalgebra) that $\mathfrak{h}_0 \cap \mathfrak{k}_0 \subset \mathfrak{a}_0 \cap \mathfrak{k}_0$. Let $\mathfrak{q}_0$ be the centralizer of $\mathfrak{h}_0 \cap \mathfrak{k}_0$ in $\mathfrak{g}_0$. Then obviously $\theta(\mathfrak{q}_0) = \mathfrak{q}_0$ and therefore $\mathfrak{q}_0$ is reductive in $\mathfrak{g}_0$ (Lemma 10 of [4(e)]). Moreover, since $\mathfrak{h}_0$ is maximal abelian in $\mathfrak{g}_0$, it is clear that $\mathfrak{h}_0 \cap \mathfrak{p}_0$ is a maximal abelian subspace of $\mathfrak{q}_0 \cap \mathfrak{p}_0$. Let $L$ be the connected component of 1 in the centralizer of $\mathfrak{h}_0 \cap \mathfrak{k}_0$ in $K$. Then $L$ is compact and its Lie algebra is $\mathfrak{q}_0 \cap \mathfrak{k}_0$. Moreover, since $\mathfrak{q}_0$ is reductive and $\theta(\mathfrak{q}_0) = \mathfrak{q}_0$, any two maximal abelian subspaces of $\mathfrak{q}_0 \cap \mathfrak{p}_0$ are conjugate under $L$ (see Cartan [1] or Lemma 39 of

15

[4(b)]). Therefore since $\mathfrak{a}_0 \cap \mathfrak{p}_0 \subset \mathfrak{q}_0 \cap \mathfrak{p}_0$, we can select $k \in L$ such that $\mathfrak{a}_0 \cap \mathfrak{p}_0 \subset k(\mathfrak{h}_0 \cap \mathfrak{p}_0)$. Then by replacing $\mathfrak{h}_0$ by $k\mathfrak{h}_0$ we can assume that $\mathfrak{h}_0 \cap \mathfrak{p}_0 \supset \mathfrak{a}_0 \cap \mathfrak{p}_0$ and $\mathfrak{h}_0 \cap \mathfrak{k}_0 \subset \mathfrak{a}_0 \cap \mathfrak{k}_0$. Now since $\mathfrak{h}_0$ is not fundamental, there exists a root $\alpha_0 \in P$ which takes only real values on $\mathfrak{h}_0$ (see Lemma 33 of [4(e)]), and we can assume that $X_{\alpha_0}, X_{-\alpha_0} \in \mathfrak{g}_0$ and $X_{\alpha_0} - X_{-\alpha_0} \in \mathfrak{k}_0$ (Lemma 46 of [4(e)]). Then $\mathfrak{h}_1 = \sigma_{\alpha_0} + R(X_{\alpha_0} - X_{-\alpha_0})$ is a Cartan subalgebra of $\mathfrak{g}_0$ and $\theta(\mathfrak{h}_1) = \mathfrak{h}_1$. Since $\alpha_0$ is identically zero on $\mathfrak{h}_0 \cap \mathfrak{k}_0$, it follows that $\sigma_{\alpha_0} \supset \mathfrak{h}_0 \cap \mathfrak{k}_0$, and therefore $l_-(\mathfrak{h}_1) \geq l_-(\mathfrak{h}_0) + 1$. Hence $\mathfrak{h}_1 \subset V_1$ by our induction hypothesis. On the other hand, we can obviously choose a point $H \in \sigma_{\alpha_0}$ such that $\alpha(H) \neq 0$ for any root $\alpha \neq \alpha_0$ in $P$. Then $H$ is semiregular and therefore $\mathfrak{h}_0 \cap \mathfrak{h}_1 \cap S \neq \emptyset$. But since $V_1 \cap V_2 \cap S = \emptyset$, we conclude that $\mathfrak{h}_0$ must also be contained in $V_1$.

Now let $X$ be any element in $\mathfrak{g}_0'$ and let $\mathfrak{h}_X$ denote its centralizer in $\mathfrak{g}_0$. Then $\mathfrak{h}_X$ is a Cartan subalgebra of $\mathfrak{g}_0$ and some conjugate of $\mathfrak{h}_X$ (under $G$) is invariant under $\theta$ (see [4(c), Corollary to Lemma 1]). Hence it follows from our result that $\mathfrak{h}_X \subset V_1$. Therefore, since $X \in \mathfrak{h}_X$ and $\mathfrak{g}_0'$ is dense in $\mathfrak{g}_0$, we conclude that $V_1 = \mathfrak{g}_0$. This proves that $S \subset V_1$.

Define $\mathfrak{h}_0$ and $T'$ as in Lemma 27 of [4(e)]. Then $T'$ is a distribution on $\mathfrak{g}_0$.

LEMMA 14. *There exists an open neighborhood $V$ of $S$ in $\mathfrak{g}_0$ and a constant $c'$ such that $T' = c'$ on $V$.*

Since $S$ is connected, it would be sufficient to prove that $T'$ coincides with a constant on some open neighborhood (in $\mathfrak{g}_0$) of any point $X_0 \in S$. If $X_0$ is regular, this follows from Lemma 27 of [4(e)]. So now suppose that $X_0$ is semiregular. Then since $\mathrm{ad}\, X_0$ is semisimple, $X_0$ is contained in some Cartan subalgebra $\mathfrak{a}_0$ of $\mathfrak{g}_0$. Select $x \in G$ such that $\theta(x\mathfrak{a}_0) = x\mathfrak{a}_0$ (see the Corollary to Lemma 1 of [4(c)]), and put $H_0 = xX_0$, $x\mathfrak{a}_0 = \mathfrak{b}_0$. Then since $T'$ is invariant under $G$, it would be sufficient to prove that $T'$ coincides with a constant on some neighborhood of $H_0$. On the other hand, we know from Lemma 30 of [4(e)] that $\partial(\xi)T' = 0$ for any homogeneous element $\xi \in I(\mathfrak{g})$ of positive degree. Now select a neighborhood $N$ of $H_0$ corresponding to Lemmas 1 and 2, and let $\mathfrak{g}_i$ $(1 \leq i \leq k)$ be all the distinct connected components of $\mathfrak{g}_0'$ which meet $N$. Then $k \leq 3$ from Lemma 2. Moreover, it follows from Lemma 29 of [4(e)] that there exists an open neighborhood $N_0$ of $H_0$ in $N$, an integer $r \geq 0$ and complex numbers $c_i$ $(1 \leq i \leq k)$ such that

$$\langle \eta^r, \eta^r \rangle T'(f) = \sum_{1 \leq i \leq k} c_i \int_{N \cap \mathfrak{g}_i} \eta^r \partial(\eta^r) f \, dX$$

for all $f \in C_c^\infty(N_0)$. Now the set of singular elements of $\mathfrak{g}_0$ obviously has Euclidean measure zero. Therefore if $k = 1$, it is obvious that

$$\langle \eta^r, \eta^r \rangle T'(f) = c_1 \int_{\mathfrak{g}_0} \eta^r \partial(\eta^r) f \, dX = c_1 \langle \eta^r, \eta^r \rangle \int_{\mathfrak{g}_0} f \, dX,$$

since $\eta$ is a homogeneous polynomial of even degree. Moreover, $\langle \eta^r, \eta^r \rangle \neq 0$ (Corollary to Lemma 18 of $[4(\mathrm{d})]$). Therefore $T' = c_1$ on $N_0$ in this case. So let us now assume that $k \geq 2$. Then we may obviously suppose that $N$ is chosen in accordance with Lemma 8. Select $\xi$ corresponding to Lemma 9 and let us use the notation of Theorem 1. If $[\Xi : LB] = 2$, we know from Lemma 7 that there exists an element $x \in \Xi$ such that $x(X' - Y') = -(X' - Y')$. On the other hand, if $t$ is positive and sufficiently small, $H_0 + t(X' - Y') \in N_2$ while $x(H_0 + t(X' - Y')) = H_0 - t(X' - Y') \in N_3$. Moreover, since $G$ is connected, the points $y(H_0 + t(X' - Y'))$ $(y \in G)$ are all contained in the same connected component of $\mathfrak{g}_0'$. Therefore it is clear that both $N_2$ and $N_3$ lie in the same connected component of $\mathfrak{g}_0'$. Hence $k = 2$ in this case. This shows that

$$\langle \eta^r, \eta^r \rangle T'(f) = \sum_{1 \leq i \leq 3} c_i' \int_{N_i} \eta^r \partial(\eta^r) f \, dX \qquad (f \in C_c^\infty(N_0))$$

(in the notation of Theorem 1) where $c_1', c_2', c_3'$ are certain constants and $c_2' = c_3'$ in case $[\Xi : LB] = 2$. Furthermore, $\partial(\xi)T' = 0$ as we have already observed above. Hence it follows from Theorem 1 that $c_1' = c_2' = c_3'$, and therefore

$$\langle \eta^r, \eta^r \rangle T'(f) = c_1' \int_N \eta^r \partial(\eta^r) f \, dX = c_1' \langle \eta^r, \eta^r \rangle \int_{\mathfrak{g}_0} f \, dX$$

for all $f \in C_c^\infty(N_0)$. This proves that $T'$ coincides with $c_1'$ on $N_0$.

We can now prove the following theorem.

THEOREM 2. *Let $\mathfrak{h}_0$ be a Cartan subalgebra of $\mathfrak{g}_0$ such that*[3] $\theta(\mathfrak{h}_0) = \mathfrak{h}_0$ *and $\mathfrak{h}_1$ a connected component of $\mathfrak{h}_0' = \mathfrak{h}_0 \cap \mathfrak{g}_0'$. For any $f \in \mathscr{B}(\mathfrak{g}_0)$, put*

$$\phi_f(H) = \pi(H) \int_{G^*} f(x^* H) \, dx^* \qquad (H \in \mathfrak{h}_0')$$

*in the notation of Theorem 3 of $[4(\mathrm{e})]$. Then there exists a real number $c$ such that*

$$\lim_{H \to 0} \phi_f(H; \partial(\pi)) = cf(0) \qquad (H \in \mathfrak{h}_1)$$

*for all $f \in \mathscr{B}(\mathfrak{g}_0)$. Moreover $c = 0$ if $\mathfrak{h}_0$ is not fundamental.*

---

[3] In view of the Corollary to Lemma 1 of $[4(\mathrm{c})]$, the assumption $\theta(\mathfrak{h}_0) = \mathfrak{h}_0$ is obviously unnecessary. However, we make it here for convenience.

Since $\mathfrak{g}_0' \subset S$, it follows from Lemma 14 that $T' = c'$ on $\mathfrak{g}_0'$. Hence we conclude from the Corollary to Lemma 29 of [4(e)] that $T' = c'$ on $\mathfrak{g}_0$. In view of the definition of $T'$, this implies that

$$\underset{\mathfrak{h}_1}{\mathrm{Lim}} \, \phi_f(0 \, ; \partial(\pi)) = cf(0) \qquad\qquad (f \in \mathscr{E}(\mathfrak{g}_0)),$$

where $c$ is a complex number independent of $f$. We know that $\eta$ takes only real values on $\mathfrak{g}_0$ (see [4(e), §6]). Select an element $H_0 \in \mathfrak{h}_0'$ and put $\epsilon = |\pi(H_0)|/\pi(H_0)$. Then it is clear that $\epsilon\pi$ takes only real values on $\mathfrak{h}_0'$. Let us choose a real-valued function $f \in \mathscr{E}(\mathfrak{g}_0)$ such that $f(0) = 1$. Then obviously $\partial(\epsilon\pi)(\epsilon\phi_f) = \epsilon^2 \partial(\pi)\phi_f$ takes only real values on $\mathfrak{h}_0'$. Since $\epsilon^2 = \pm 1$, it follows that $c = \underset{\mathfrak{h}_1}{\mathrm{Lim}} \, \phi_f(0 \, ; \partial(\pi))$ is real. Finally we know from the Corollary to Lemma 34 of [4(e)] that $c = 0$ if $\mathfrak{h}_0$ is not fundamental.

**5. Proof of Theorem 3.** Our next object is to prove the following theorem.

THEOREM 3. *Let $\mathfrak{h}_0$ be a fundamental Cartan subalgebra of $\mathfrak{g}_0$ such that[3] $\theta(\mathfrak{h}_0) = \mathfrak{h}_0$. Let $\mathfrak{h}_{(1)}, \cdots, \mathfrak{h}_{(q)}$ be all the distinct connected components of $\mathfrak{h}_0' = \mathfrak{h}_0 \cap \mathfrak{g}_0'$ and $c_1, \cdots, c_q$ the corresponding real numbers of Theorem 2. Then $c_1 + c_2 + \cdots + c_q \neq 0$.*

If the ranks of $\mathfrak{g}$ and $\mathfrak{k}$ are equal, this follows immediately from Lemma 41 of [4(e)] and Theorem 2. So let us now suppose that $\mathrm{rank}\,\mathfrak{g} > \mathrm{rank}\,\mathfrak{k}$. We use the notation of [4(e), §5] and put $\mathfrak{g}_1 = \underset{x \in G}{\bigcup} x\mathfrak{h}_0'$. Then if the invariant measure $dx^*$ on $G/A$ is suitably normalized,

$$\int_{\mathfrak{g}_1} f(X)\,dX = \int_{G^* \times \mathfrak{h}_0'} f(x^*H) \, |\pi(H)|^2 \, dx^* dH$$

for $f \in C_c(\mathfrak{g}_1)$. Here $dX$ and $dH$ are the regular Euclidean measures on $\mathfrak{g}_0$ and $\mathfrak{h}_0$, respectively. On the other hand, it follows from the Corollary to Lemma 15 of [4(e)] that

$$\int_{G^*} f(x^*H)\,dx^* = |\pi_+(H)|^{-1} \int f(k(mH + Z))\,dk\,dm\,dZ$$

for $H \in \mathfrak{h}_0'$ and $f \in \mathscr{E}(\mathfrak{g}_0)$. Here $\pi_+ = \underset{\alpha \in P_+}{\prod} \alpha$, $dk$, $dm$ are the Haar measures on $K$ and $M$, respectively, and $dZ$ is the Euclidean measure on $\mathfrak{n}_0$. Moreover $\int_K dk = 1$. Since $\mathfrak{h}_0$ is fundamental, no root in $P$ can vanish identically on $\mathfrak{h}_0 \cap \mathfrak{k}_0$ (see Lemma 33 of [4(e)]). Therefore $\alpha \neq -\theta\alpha$ for $\alpha \in P_+$. On the other hand, since $\alpha$ and $-\theta\alpha$ coincide on $\mathfrak{h}_{\mathfrak{p}_0}$, it follows from the definition of $P$ and $P_+$ (see [4(e), §5]) that $-\theta\alpha \in P_+$ whenever $\alpha \in P_+$. Moreover, since $\mathrm{conj}\,\alpha(H) = -\alpha(\theta(H))$ ($\alpha \in P, H \in \mathfrak{h}_0$), it is clear that $\pi_+$ takes only nonnegative real values on $\mathfrak{h}_0$.

$A$ being the Cartan subgroup of $G$ corresponding to $\mathfrak{h}_0$, let $A'$ denote the normalizer of $A$ in $G$. Then if $W$ is the Weyl group of $\mathfrak{g}$ with respect to $\mathfrak{h}$, $W' = A'/A$ can be regarded as a subgroup[4] of $W$. Thus $W'$ is a finite group which operates on $G^* = G/A$ on the right and leaves the measure $dx^*$ invariant.[4] Now select two functions $\alpha_0$, $\beta_0$ in $C_c^\infty(G^*)$ and $C_c^\infty(\mathfrak{h}_0')$, respectively, such that

$$\int \alpha_0(x^*)\,dx^* = \int \beta_0(H)\,|\pi(H)|^2\,dH = 1.$$

Let $w'$ denote the order of $W'$ and put

$$\alpha(x^*) = (w')^{-1} \sum_{s \in W'} \alpha(x^*s), \qquad \beta = (w')^{-1} \sum_{s \in W'} \beta_0^s,$$

where $\beta_0^s(H) = \beta_0(s^{-1}H)$ ($H \in \mathfrak{h}_0$). Then it is clear (see [4(e), §11]) that there exists a function $F \in C_c^\infty(\mathfrak{g}_1)$ such that $F(x^*H) = \alpha(x^*)\beta(H)$ ($x^* \in G^*, H \in \mathfrak{h}_0'$). Moreover,

$$\int_{\mathfrak{g}_0} F(X)\,dX = \int_{G^* \times \mathfrak{h}_0'} F(x^*H)\,|\pi(H)|^2\,dx^*dH$$
$$= \int \alpha(x^*)\,dx^* \int \beta(H)\,|\pi(H)|^2\,dH = 1.$$

Let $\tilde{F}$ denote the Fourier transform of $F$ and put

$$\phi_{\tilde{F}}(H) = \pi(H) \int_{G^*} \tilde{F}(x^*H)\,dx^* \qquad\qquad (H \in \mathfrak{h}_0').$$

Then in view of what we have seen above,

$$\phi_{\tilde{F}}(H) = \pi'(H) \int \bar{F}(k(mH + Z))\,dk\,dm\,dZ,$$

where $\pi' = \prod_{\alpha \in P'} \alpha$ and $P' = P_0 \cup P_-$. Define $\zeta$, $I_0$ etc. as in Lemma 15 of [4(e)] and let $\mathfrak{h}_0''$ be the set of all points $H \in \mathfrak{h}_0$ where $\pi'(H) \neq 0$. Then it follows from the Corollary of Lemma 42 of [4(e)] that $\mathfrak{h}_0''$ and $\mathfrak{h}_0'$ have the same number of connected components. Let $\mathfrak{h}_{(i)}''$ denote the connected component of $\mathfrak{h}_0''$ containing $\mathfrak{h}_{(i)}$. Then $\mathfrak{h}_{(1)}'', \cdots, \mathfrak{h}_{(q)}''$ are all the distinct connected components of $\mathfrak{h}_0''$. For any $g \in \mathscr{E}(I_0)$, put

$$\psi_g(H) = \pi'(H) \int_M g(mH)\,dm \qquad\qquad (H \in \mathfrak{h}_0'').$$

---

[4] See Sections 6, 8 and 11 of [4(e)] for the proof of these facts.

Then it follows from Lemma 19 of [4(e)] that $\psi_g \in \mathscr{B}(\mathfrak{h}_0'')$ and $\partial(\pi_+)\psi_g = \psi_{\partial(\zeta)g}$. On the other hand, we know from Theorem 2 (applied to $\mathfrak{l}_0$) that there exist constants $c_i'$ $(1 \leq i \leq q)$ such that

$$\operatorname*{Lim}_{H \to 0} \psi_g(H; \partial(\pi')) = c_i'g(0) \qquad (H \in \mathfrak{h}_{(i)}'')$$

for every $g \in \mathscr{B}(\mathfrak{l}_0)$. Moreover, since $\mathfrak{h}_0 \cap \mathfrak{k}_0$ is maximal abelian in $\mathfrak{m}_0$, we conclude from Lemma 41 of [4(e)] (applied to $[\mathfrak{l}_0, \mathfrak{l}_0]$) that $c_1' + \cdots + c_q' \neq 0$. Now put

$$\tilde{F}_1(X) = \int_K \tilde{F}(kX)\,dk \quad (X \in \mathfrak{g}_0) \quad \text{and} \quad g_1(L) = \int_{\mathfrak{n}_0} \tilde{F}_1(L + Z)\,dZ \quad (L \in \mathfrak{l}_0).$$

Then $\phi_{\tilde{F}} = \psi_{g_1}$, and therefore $\partial(\pi)\phi_{\tilde{F}} = \partial(\pi')\psi_{\partial(\zeta)g_1}$ on $\mathfrak{h}_0'$. Hence

$$\operatorname*{Lim}_{H \to 0} \phi_{\tilde{F}}(H; \partial(\pi)) = \operatorname*{Lim}_{H \to 0} \psi_{\partial(\zeta)g_1}(H; \partial(\pi')) = c_i'g_1(0; \partial(\zeta)) \quad (H \in \mathfrak{h}_{(i)}).$$

Now put $F_1(X) = \int_K F(kX)\,dk$ $(X \in \mathfrak{g}_0)$. Then since $\mathfrak{g}_0$ is the orthogonal sum of $\mathfrak{n}_0$, $\mathfrak{l}_0$ and $\theta(\mathfrak{n}_0)$, we conclude from the theory of Fourier transforms that

$$g_1(L') = \int_{\mathfrak{n}_0} \tilde{F}_1(L' + Z)\,dZ$$

$$= \int_{\mathfrak{l}_0 \times \mathfrak{n}_0} \exp\{(-1)^{\frac{1}{2}}B(L', L)\}F_1(L + Z)\,dL\,dZ \qquad (L' \in \mathfrak{l}_0).$$

Here $dL$ is the (suitably normalized) Euclidean measure on $\mathfrak{l}_0$. Hence it follows easily that

$$g_1(0; \partial(\zeta)) = (-1)^{r_+/2} \int_{\mathfrak{l}_0 \times \mathfrak{n}_0} \zeta(L)F_1(L + Z)\,dL\,dZ,$$

where $r_+$ is the number of roots in $P_+$. Select $L \in \mathfrak{l}_0$ and $Z \in \mathfrak{n}_0$ such that $\zeta(L)F_1(L + Z) \neq 0$. Then it follows from Lemma 16 of [4(e)] that $nL = L + Z = xH$ for some $n \in N$, $x \in G$ and $H \in \mathfrak{h}_0'$. Hence $L = yH$ if $y = n^{-1}x$. Let $\mathfrak{h}_L$ denote the centralizer of $L$ in $\mathfrak{g}_0$. Then $\mathfrak{h}_L = y\mathfrak{h}_0$, and therefore $\mathfrak{h}_L$ is a fundamental Cartan subalgebra of $\mathfrak{g}_0$. On the other hand, since the ranks of $\mathfrak{l}_0$ and $\mathfrak{g}_0$ are the same, it is evident that $\mathfrak{h}_L \subset \mathfrak{l}_0$, and therefore, obviously, both $\mathfrak{h}_0$ and $\mathfrak{h}_L$ are fundamental Cartan subalgebras [5] of the reductive algebra $\mathfrak{l}_0$. Hence it follows from the Corollary to Lemma 32 of [4(e)] that $\mathfrak{h}_L = m\mathfrak{h}_0$ for some $m \in M$. Since $L$ is a regular element

---

[5] The notion of a fundamental Cartan subalgebra is extended to reductive Lie algebras over $R$ in the obvious way.

of $\mathfrak{h}_L$, we conclude that $L = mH_1$ for some $H_1 \in \mathfrak{h}_0{}'$, and therefore $\zeta(L) = \zeta(H_1) = \pi_*(H_1) > 0$. This shows that

$$g_1(0;\partial(\zeta)) = \epsilon \int |\zeta(L)| F_1(L+Z) \, dL \, dZ = \epsilon \int_{\mathfrak{l}_0 \times N} |\zeta(L)|^2 F_1(nL) \, dL \, dn$$

from Lemma 15 of [4(e)]. (Here $\epsilon = (-1)^{r_*/2}$.) Now put $\mathfrak{l}_1 = \bigcup_{m \in M} m(\mathfrak{h}_0{}')$. Then as we have seen above, $\zeta(L) F_1(nL) = 0$ $(n \in N, L \in \mathfrak{l}_0)$ unless $L \in \mathfrak{l}_1$. Moreover, it is clear that there exists a positive constant $a$ such that

$$\int_{\mathfrak{l}_1} f(L) \, dL = a \int_{M \times \mathfrak{h}_0{}'} f(mH) |\pi'(H)|^2 \, dm \, dH$$

for all $f \in C_c^\infty(\mathfrak{l}_1)$. Hence

$$g_1(0;\partial(\zeta)) = a\epsilon \int |\zeta(H)|^2 F_1(nmH) |\pi'(H)|^2 \, dm \, dH \, dn$$

$$= a\epsilon \int |\pi(H)|^2 F(knmH) \, dk \, dm \, dH \, dn$$

$$= a\epsilon \int |\pi(H)|^2 F(x^*H) \, dx^* \, dH$$

$$= a\epsilon \int F(X) \, dX = a\epsilon$$

from the Corollary to Lemma 14 of [4(e)]. This proves that

$$\lim_{H \to 0} \phi_{\tilde{F}}(H;\partial(\pi)) = a\epsilon c_i' \qquad (H \in \mathfrak{h}_{(i)}).$$

But since $\tilde{F}(0) = 1$, this implies that $c_i = a\epsilon c_i'$, and therefore $c_1 + \cdots + c_q = a\epsilon(c_1' + \cdots + c_q') \neq 0$.

**6. Appendix.** Let $\mathfrak{z}_0$ be a subalgebra of $\mathfrak{g}_0$ and $\Xi_0$ the analytic subgroup of $G$ corresponding to $\mathfrak{z}_0$. We assume that $\mathfrak{z}_0$ is reductive in $\mathfrak{g}_0$. Let $\Xi'$ be the normalizer of $\mathfrak{z}_0$ in $G$.

LEMMA 15. *Suppose* rank $\mathfrak{z}_0 = $ rank $\mathfrak{g}_0$. *Then the factor group* $\Xi'/\Xi_0$ *is finite.*

Since $\mathfrak{z}_0$ is reductive, we can select (see the Corollary to Lemma 2 of [4(c)]) a finite number of Cartan subalgebras $\mathfrak{h}_1, \cdots, \mathfrak{h}_r$ of $\mathfrak{z}_0$ such that every Cartan subalgebra of $\mathfrak{z}_0$ is of the form $x\mathfrak{h}_i$ for some $x \in \Xi_0$ and some $i$ $(1 \leq i \leq r)$. Suppose $\mathfrak{h}_1$ is conjugate to $\mathfrak{h}_i$ $(1 \leq i \leq k)$ under $\Xi'$ but not to $\mathfrak{h}_i$ $(k < i \leq r)$. Select $\xi_i \in \Xi'$ such that $\xi_i \mathfrak{h}_1 = \mathfrak{h}_i$ $1 \leq i \leq k$ $(\xi_1 = 1)$. Now let $x \in \Xi'$. Then it is clear that $x\mathfrak{h}_1 = y\mathfrak{h}_i = y\xi_i\mathfrak{h}_1$ for some $y \in \Xi_0$ and some $i$ $(1 \leq i \leq k)$. Since $\operatorname{rank} \mathfrak{z}_0 = \operatorname{rank} \mathfrak{g}_0$ and $\mathfrak{z}_0$ is reductive in $\mathfrak{g}_c$, $\mathfrak{h}_1$ is also a Cartan subalgebra of $\mathfrak{g}_0$. Let $A'$ be the normalizer of $\mathfrak{h}_1$ in $G$ and $A$ the Cartan subgroup of $G$ corresponding to $\mathfrak{h}_1$. Then $A'/A$ is a finite group (see [4(c), Lemma 10]). On the other hand, if $A_0$ is the analytic subgroup of $G$ corresponding to $\mathfrak{h}_1$, $A \cap \Xi_0 \supset A_0$. Moreover, from Lemma 7 of [4(e)], $A/A_0$ is both compact and discrete and hence finite. This proves that $(A' \cap \Xi')/A_0$ is a finite group. Select elements $a_j \in A' \cap \Xi'$ such that $A' \cap \Xi'$ is the union of $a_j A_0$ $(1 \leq j \leq N)$. Then it is clear that $x^{-1} y \xi_i \in a_j A_0$ for some $j$, and therefore $x \in y \xi_i A_0 a_j^{-1} \subset \xi_i a_j^{-1} \Xi_0$ since $\Xi_0$ is obviously normal in $\Xi'$. This shows that the order of the group $\Xi'/\Xi_0$ cannot exceed $rN$.

Let $\mathfrak{l}$ be the semisimple Lie **algebra of dimension 3 spanned over $R$ by the** three elements $H$, $X$, $Y$ which satisfy the relations $[H, X] = 2X$, $[H, Y] = -2Y$, $[X, Y] = H$. Let $\Omega$ denote the Casimir polynomial of $\mathfrak{l}$. Then

$$\Omega(tH + xX + yY) = 8(t^2 + xy) \qquad (t, x, y \in C).$$

Moreover, $\mathfrak{h} = CH$ is a Cartan subalgebra of $\mathfrak{l}$ and there are only two roots $\alpha$ and $-\alpha$ given by $\alpha(H) = 2$. **We take $\alpha$ as the positive root.** Then $\Omega$ coincides with $2\alpha^2$ on $\mathfrak{h}$. Hence an element $Z \in \mathfrak{l}$ is regular if and only if $\Omega(Z) \neq 0$. Let $\mathfrak{l}'$ denote the set of regular elements in $\mathfrak{l}$.

Consider the automorphism $\nu$ of $\mathfrak{l}$ given by [2]

$$\nu = \exp\{-(-1)^{\frac{1}{2}}(\pi/4)\operatorname{ad}(X + Y)\}.$$

A simple calculation shows that $\nu(H) = (-1)^{\frac{1}{2}}(X - Y)$, $\nu(X - Y) = (-1)^{\frac{1}{2}}H$ and $\nu(X + Y) = X + Y$. Therefore $RH + RX + RY$ and $\nu(RH + RX + RY) = R(-1)^{\frac{1}{2}}H + R(-1)^{\frac{1}{2}}(X - Y) + R(X + Y)$ are two isomorphic real forms of $\mathfrak{l}$. Since $\Omega(H) = 8 > 0$, they are not compact. On the other hand $R(-1)^{\frac{1}{2}}H + R(X - Y) + R(-1)^{\frac{1}{2}}(X + Y)$ is also a real form of $\mathfrak{l}$ which is compact since $\Omega$ is negative definite on it. We shall denote by $\mathfrak{l}_0$ either one of the two[6] real forms $RH + RX + RY$ and $R(-1)^{\frac{1}{2}}H + R(X - Y) + R(-1)^{\frac{1}{2}}(X + Y)$. Put $\mathfrak{l}_0' = \mathfrak{l}' \cap \mathfrak{l}_0$. If $\mathfrak{l}_0$ is compact, zero is the only singular point in $\mathfrak{l}_0$, and therefore $\mathfrak{l}_0'$ is connected. On the other hand, if $\mathfrak{l}_0$ is noncompact, the singular set of $\mathfrak{l}_0$ consists of all points $tH + xX + yY$ $(t, x, y \in R)$ satisfying the equation $t^2 + xy = 0$. In this case $\mathfrak{l}_0'$ has three connected components $\mathfrak{l}_1, \mathfrak{l}_2, \mathfrak{l}_3$ given as follows:

$$\mathfrak{l}_1 : t^2 + xy > 0; \quad \mathfrak{l}_2 : t^2 + xy < 0, x > y; \quad \mathfrak{l}_3 : t^2 + xy < 0, x < y.$$

Notice that $H \in \mathfrak{l}_1$ and $X - Y \in \mathfrak{l}_2$. Let $U$ be a neighborhood of zero in $\mathfrak{l}_0$.

---

[6] It is not difficult to see that apart from isomorphism these are the only two real forms of $\mathfrak{l}$. However, we do not need this fact.

We assume that $tZ \in U$ $(0 \leqq t \leqq 1)$ whenever $Z \in U$. Then it is obvious that $U \cap I_0'$ is connected if $I_0$ is compact and consists of three connected components $U \cap I_i$ $(i = 1, 2, 3)$ if $I_0$ is not compact.

From now on we consider only the noncompact case. The mapping $(tH + xX + yY) \to \begin{pmatrix} t & x \\ y & -t \end{pmatrix}$ $(t, x, y \in C)$ is an isomorphism of $I$ onto the Lie algebra of all $2 \times 2$ complex matrices with trace zero and we may identify the two under this isomorphism. Then $I_0$ becomes the real subalgebra of $I$ consisting of the real matrices. It is clear that $\Omega(Z) = -8 \det Z$ and the two eigenvalues of $Z$ are $\pm (-\det Z)^{\frac{1}{2}}$ $(Z \in I)$. Call an element $Z \in I_0'$ hyperbolic or elliptic according as these eigenvalues are real or pure imaginary. Then $I_1$ consists of hyperbolic and $I_2 \cup I_3$ of elliptic elements. Let $L$ be the (connected) adjoint group of $I_0$. It is easily seen that every hyperbolic element is conjugate under $L$ to $tH$ for some real $t \neq 0$. Moreover, since $\exp \frac{\pi}{2} \mathrm{ad}(X - Y)$ maps $H$ into $-H$, we can always assume that $t > 0$. Let $R^+$ and $R^-$ be the set of all positive and negative real numbers, respectively, and put $R' = R^+ \cup R^-$. Then we prove similarly that every element in $I_2$ and $I_3$, respectively, is conjugate to $\theta(X - Y)$ for some $\theta$ in $R^+$ and $R^-$. It is easy to check that there exists no element $x \in L$ such that $x(X - Y) = -(X - Y)$.

Let $dZ$ denote the Euclidean measure on $I_0$ given by [2] $dZ = (2\pi)^{-1} dt\,dx\,dy$, where $Z = tH + xX + yY$ $(t, x, y \in R)$. Consider the one parameter subgroup $K$ in $L$ corresponding to $(X - Y)$. Then $K$ is compact. Let $dk$ denote the normalized Haar measure on $K$ so that $\int_K dk = 1$.

LEMMA 16. *For any $f \in C_c(I_0)$ put*

$$\Phi_f(t) = \int_{-\infty}^{\infty} \bar{f}(tH + xX)\,dx \qquad (t \in R^+)$$

$$\Psi_f(\theta) = |\theta| \int_0^{\infty} \bar{f}(\theta(e^t X - e^{-t} Y))(e^t - e^{-t})\,dt \qquad (\theta \in R'),$$

*where $\bar{f}(Z) = \int f(kZ)\,dk$ $(Z \in I_0)$. Then*

$$\int_{I_1} f(Z)\,dZ = \int_0^{\infty} t\Phi_f(t)\,dt$$

$$\int_{I_2} f(Z)\,dZ = \int_0^{\infty} \theta\Psi_f(\theta)\,d\theta, \qquad \int_{I_3} f(Z)\,dZ = \int_{-\infty}^0 |\theta|\,\Psi_f(\theta)\,d\theta.$$

Since $f$ vanishes outside a compact set, it is seen without difficulty that $\Phi_f$ and $\Psi_f$ are well defined. Put $k_\theta = \exp\{\theta \operatorname{ad}(X - Y)\}$ $(\theta \in R)$ and consider the mapping $(\theta, t, x) \to (t', x', y')$ given by

$$k_\theta(tH + xX) = t'H + x'X + y'Y,$$

where $0 \leq \theta < \pi$, $t > 0$ and $x \in R$. It follows from general considerations (see [4(a), Lemma 9]) that

$$\left| \partial(t', x', y')/\partial(\theta, t, x) \right| = ct,$$

where $c$ is a constant. To obtain its value we compute the Jacobian at $\theta = 0$. Now

$$\{(\partial/\partial\theta) k_\theta(tH + xX)\}_{\theta=0} = [X - Y, tH + xX] = xH - 2tX - 2tY.$$

Hence

$$\left\{ \frac{\partial(t', x', y')}{\partial(\theta, t, x)} \right\}_{\theta=0} = \det \begin{pmatrix} x & -2t & -2t \\ 1 & 0 & 0 \\ 0 & 1 & 0 \end{pmatrix} = -2t.$$

Therefore $c = 2$. Let $A, N$ be the one-parameter subgroups of $L$ corresponding to $H$ and $X$, respectively. Then we know from general considerations (see Iwasawa [5]) that $L = KNA$. Therefore every element in $I_1$ can be written in the form $t(knH)$ $(k \in K, n \in N, t \in R^+)$. Since $k_\pi = 1$, it follows that every point of $I_1$ is obtained under our mapping $(\theta, t, x) \to k_\theta(tH + xX)$. Moreover, a direct computation shows that this mapping is univalent. Hence

$$\int_{I_1} f(Z)\, dZ = 2(2\pi)^{-1} \int_0^\pi d\theta \int_0^\infty t\, dt \int_{-\infty}^\infty f(k_\theta(tH + xX))\, dx$$

$$= \int_0^\infty t\, dt \int_{-\infty}^\infty \bar{f}(tH + xX)\, dx = \int_0^\infty t\Phi_f(t)\, dt.$$

Moreover, if $L^* = L/A$ and $dx^*$ is the invariant measure on $L^*$, it follows from the Corollary to Lemma 15 of [4(e)] that

$$\Phi_f(t) = t \int_{L^*} f(tx^*H)\, dx^* \qquad\qquad (f \in C_c(I_0), t \in R^+)$$

provided $dx^*$ is suitably normalized.

Now put $a_t = \exp\{t \operatorname{ad} H\}$ $(t \in R)$ and consider the mapping $(\phi, l, \theta) \to (t', x', y')$ given by

$$k_\phi a_t(\theta(X - Y)) = t'H + x'X + y'Y$$

$(0 \leqq \phi < \pi, t \in R^+, \theta \in R')$.   Then again it follows by general considerations [7] (or by actual computation) that

$$| \partial(t', x', y')/\partial(\phi, t, \theta) | = c\theta^2(e^{2t} - e^{-2t}),$$

where $c$ *is a constant.*   In order to obtain $c$ we evaluate the left side at $\phi = 0$. Then $\theta a_t(X - Y) = \theta(e^{2t}X - e^{-2t}Y)$.   Hence

$$[(\partial/\partial\phi)\{k_\phi a_t(\theta(X - Y))\}]_{\phi=0} = \theta[X - Y, e^{2t}X - e^{-2t}Y] = \theta(e^{2t} - e^{-2t})H.$$

Therefore

$$\left\{ \frac{\partial(t', x', y')}{\partial(\phi, t, \theta)} \right\}_{\phi=0} = \det \begin{pmatrix} \theta(e^{2t} - e^{-2t}) & 0 & 0 \\ 0 & 2\theta e^{2t} & 2\theta e^{-2t} \\ 0 & e^{2t} & -e^{-2t} \end{pmatrix}$$

$= -4\theta^2(e^{2t} - e^{-2t})$, and so $c = 4$. Let $A_+$ be the set of all $a_t$ with $t \geqq 0$. It is easy to see that $L = KA_+K$.   Therefore every element in $I_2(I_3)$ can be written in the form $k_\phi a_t(\theta(X - Y))$ with $0 \leqq \phi < \pi$, $t \geqq 0$, $\theta \in R^+$ ($\theta \in R^-$ respectively).   Moreover, if $t \neq 0$, this can be done only in one way since the normalizer of $R(X - Y)$ in $L$ is exactly $K$.   Hence

$$\int_{I_2} f(Z)\,dZ = 4(2\pi)^{-1} \int_0^\pi d\phi \int_0^\infty \theta^2 d\theta \int_0^\infty f(k_\phi a_t(\theta(X - Y)))(e^{2t} - e^{-2t})\,dt$$

$$= \int_0^\infty \theta\Psi_f(\theta)\,d\theta.$$

Similarly,

$$\int_{I_3} f(Z)\,dZ = 4(2\pi)^{-1} \int_0^\pi d\phi \int_0^\infty \theta^2 d\theta \int_0^\infty f(k_\phi a_t(-\theta(X - Y)))(e^{2t} - e^{-2t})\,dt$$

$$= \int_{-\infty}^0 |\theta|\,\Psi_f(\theta)\,d\theta.$$

On the other hand, if $dz$ is the Haar measure on $L$, it follows from [4(b), Lemma 38] that

$$\Psi_f(\theta) = |\theta| \int_L f(\theta z(X - Y))\,dz \qquad\qquad (f \in C_c(I_0))$$

if $dz$ is suitably normalized.

Now suppose $f \in C_c^\infty(I_0)$.   Then we can conclude from Theorem 3 of [4(e)] that $\Phi_f \in \mathcal{B}(R^+)$ and $\Psi_f \in \mathcal{B}(R')$.

---

[7] See [4(a), p. 501] and Lemma 38 of [4(b)].

LEMMA 17. *Let $f$ be a function in $C_c^\infty(I_0)$. Then*

$$\Phi_{\partial(\Omega)f}(t) = \tfrac{1}{8}\, d^2\Phi_f(t)/dt^2 \qquad\qquad (t \in R^+)$$

$$\Psi_{\partial(\Omega)f}(t) = -\tfrac{1}{8}\, d^2\Psi_f(\theta)/d\theta^2 \qquad\qquad (\theta \in R').$$

Since $\Omega(H) = 8$ and $\Omega(X - Y) = -8$, this is an immediate consequence of Theorem 3 of $[4(e)]$.

Put $\Phi_f(0) = \mathrm{Lim}_{t\to 0}\Phi_f(t)$ $(t \in R^+)$, $\Psi_f^+(0) = \mathrm{Lim}_{\theta\to 0}\Psi_f(\theta)$ $(\theta \in R^+)$ and $\Psi_f^-(0) = \mathrm{Lim}_{\theta\to 0}\Psi_f(\theta)$ $(\theta \in R^-)$ for $f \in C_c^\infty(I_0)$. (We know from Lemma 25 of $[4(e)]$ that all these limits exist). Then it is obvious that $\Phi_f(0) = \int_{-\infty}^{\infty} \bar{f}(xX)\,dx$. So now let us consider $\Psi_f^\pm(0)$. It is obvious that

$$\mathrm{Lim}_{\theta\to 0}\, \theta \int_0^\infty \bar{f}(\theta(e^t X - e^{-t}Y))e^{-t}\,dt = 0.$$

On the other hand, if $\theta \in R^+$,

$$\theta \int_0^\infty \bar{f}(\theta(e^t X - e^{-t}Y))e^t\,dt = \int_\theta^\infty \bar{f}(tX - \theta^2 t^{-1}Y)\,dt.$$

Moreover, if $M$ is an upper bound for $|\partial(Y)\bar{f}|$, it is clear that $|\bar{f}(tX + sY) - \bar{f}(tX)| \le |s|M$ $(s \in R)$, and since $\bar{f}$ vanishes outside a compact set, we can select $T \in R^+$ such that $\bar{f}(tX + sY) = 0$ for all $s \in R$ if $t \ge T$. Then if $\theta < T$, it follows that

$$\int_\theta^\infty |\bar{f}(tX - \theta^2 t^{-1}Y) - \bar{f}(tX)|\,dt \le \theta^2 \int_\theta^T M t^{-1}\,dt = M\theta^2 \log(T/\theta).$$

Hence

$$\mathrm{Lim}_{\theta\to 0} \int_0^\infty \bar{f}(tX - \theta^2 t^{-1}Y)\,dt = \int_0^\infty \bar{f}(tX)\,dt \qquad\qquad (\theta \in R^+).$$

This proves that $\Psi_f^+(0) = \int_0^\infty \bar{f}(tX)\,dt$. Similarly one shows that $\Psi_f^-(0) = \int_{-\infty}^0 \bar{f}(tX)\,dt$. Thus we have obtained the following result.

LEMMA 18. *For any $f \in C_c^\infty(I_0)$,*

$$\Phi_f(0) = \int_{-\infty}^\infty \bar{f}(tX)\,dt, \quad \Psi_f^+(0) = \int_0^\infty \bar{f}(tX)\,dt, \quad \Psi_f^-(0) = \int_{-\infty}^0 \bar{f}(tX)\,dt.$$

COROLLARY. *Put $f_m = \partial(\Omega^m)(\Omega^m f)$ $(f \in C_c^\infty(I_0), m \ge 0)$. Then*

$$\Phi_{f_m}(0) = c_m\Phi_f(0), \quad \Psi_{f_m}^+(0) = c_m\Psi_f^+(0), \quad \Psi_{f_m}^-(0) = c_m\Psi_f^-(0),$$

*where $c_m = 4^m m!\,\Gamma(m + \tfrac{1}{2})/\Gamma(\tfrac{1}{2})$.*

It is obvious that the differential operator $\partial(\Omega''') \circ \Omega'''$ is invariant under $L$ and therefore also under $K$. Moreover $\Omega(X) = 0$. Hence our assertion is an immediate consequence of the Corollary to Lemma 19 of [4(d)].

Let $U$ be an open neighborhood of zero in $\mathfrak{l}_0$.

LEMMA 19.  *Suppose $c_1, c_2, c_3$ are three constants and*

$$c_1 \Phi_f(0) = c_2 \Psi_f^+(0) + c_3 \Psi_f^-(0)$$

*for all $f \in C_c^\infty(U)$.  Then $c_1 = c_2 = c_3$.*

Let $\mathfrak{l}^+$ be the set of all points of the form $tH + xX + yY$ $(t, y \in R; x \in R^+)$. We can obviously choose a positive number $x_0$ and a real-valued nonnegative function $f \in C_c^\infty(U \cap \mathfrak{l}^+)$ such that $f(x_0 X) = 1$.  A simple calculation shows that

$$k_\theta X = (\cos\theta \sin\theta) H + (\cos\theta)^2 X - (\sin\theta)^2 Y \qquad (\theta \in R).$$

Hence it is clear from Lemma 18 that $\Phi_f(0) = \Psi_f^+(0) > 0$ while $\Psi_f^-(0) = 0$. Therefore it follows from our hypothesis that $c_1 = c_2$.  Similarly we prove that $c_1 = c_3$.

Let $E$ be a vector space over $R$ of dimension $n$ and let $d\mu$ denote the Euclidean measure on $E$.

LEMMA 20.  *Let $U$ be a nonempty open set in $E$ and $p_1, \cdots, p_r$ a finite number of continuous functions on $U$.  Suppose $p_1, \cdots, p_r$ are linearly independent over $C$.  Then we can select a function $f \in C_c^\infty(U)$ such that*

$$\int_U f p_i \, d\mu = \begin{cases} 1 & \text{if } i = 1, \\ 0 & \text{if } i \neq 1 \end{cases} \qquad (1 \leq i \leq r).$$

Let $\tau_i$ denote the distribution on $U$ given by

$$\tau_i(g) = \int_U g p_i \, d\mu \qquad (g \in C_c^\infty(U)).$$

Let $V$ be a vector space over $C$ of dimension $r$ and let $v_1, \cdots, v_r$ be a base for $V$.  For any $g \in C_c^\infty(U)$, put $\tau(g) = \sum_{1 \leq i \leq r} \tau_i(g) v_i$ and let $V_1$ be the sub-space of $V$ consisting of all elements of the form $\tau(g)$.  We claim $V_1 = V$. For otherwise, we could select a linear function $\lambda \neq 0$ on $V$ such that $\lambda(\tau(g)) = 0$ for all $g \in C_c^\infty(U)$.  Put $\lambda_i = \lambda(v_i)$ $1 \leq i \leq r$.  Then not every $\lambda_i$ is zero and

$$\lambda(\tau(g)) = \sum_{1 \leq i \leq r} \lambda_i \tau_i(g) = 0 \qquad (g \in C_c^\infty(U)).$$

Let $p = \sum_{1 \leq i \leq r} \lambda_i p_i$. Since $p_i$ are linearly independent, $p \neq 0$, and therefore it is obvious that $\int pg \, d\mu \neq 0$ for some $g \in C_c^\infty(U)$. However

$$\int pg \, d\mu = \sum_{1 \leq i \leq r} \lambda_i \tau_i(g) = \lambda(\tau(g)) = 0,$$

and so we get a contradiction. This proves that $V_1 = V$, and therefore we can choose $f \in C_c^\infty(U)$ such that $\tau(f) = v_1$. Then obviously $f$ fulfills the desired condition.

INSTITUTE FOR ADVANCED STUDY,
COLUMBIA UNIVERSITY.


## REFERENCES.

[1] E. Cartan, *Ann. Ecole Norm. Sup.*, vol. 44 (1927), pp. 345-467.

[2] C. Chevalley, *Theory of Lie groups*, Princeton University Press, 1946.

[3] L. Garding, *Math. Scand.*, vol. 1 (1953), pp. 55-72.

[4] Harish-Chandra, (a) *Trans. Amer. Math. Soc.*, vol. 76 (1954), pp. 485-528.
                (b) *Amer. Jour. Math.*, vol. 78 (1956), pp. 564-628.
                (c) *Trans. Amer. Math. Soc.*, vol. 83 (1956), pp. 98-163.
                (d) *Amer. Jour. Math.*, vol. 79 (1957), pp. 87-120.
                (e) *Amer. Jour. Math.*, vol. 79 (1957), pp. 193-257.
                (f) *Proc. Nat. Acad. Sci. U.S.A.*, vol. 42 (1956), pp. 538-540.

[5] K. Iwasawa, *Ann. of Math.*, vol. 50 (1949), pp. 507-557.

[6] F. John, *Proceedings of the symposium on spectral theory and differential problems*, Stillwater, Okla., (1951), pp. 113-175.

[7] L. Schwartz, *Théorie des distributions I*, Paris, Hermann, 1950.

[8] A. Weil, *L'intégration dans les groupes topologiques et ses applications*, Paris, Hermann, 1940.



Reprinted from
*Amer. J. of Math.*
**79** (1957), 653–686

# A FORMULA FOR SEMISIMPLE LIE GROUPS.*

By Harish-Chandra.

1. **Introduction.** Let $G$ be a connected semisimple Lie group and $\mathcal{E}$ the set of all equivalence classes of irreducible unitary representations of $G$. Let $T_\omega$ denote the character of any class $\omega \in \mathcal{E}$ (see [4(a)]). Then the Plancherel formula for $G$ is equivalent to the following statement. *There exists a unique positive measure $d\omega$ on $\mathcal{E}$ (called the Plancherel measure) such that*

$$f(1) = \int_{\mathcal{E}} T_\omega(f)\, d\omega$$

*for all*[1] $f \in C_c^\infty(G)$. Let $\delta$ denote the Dirac measure on $G$ corresponding to the unit mass at 1 so that $\delta(f) = f(1)$. Then the above formula may also be interpreted in the form

$$\delta = \int_{\mathcal{E}} T_\omega\, d\omega.$$

Let $T$ be a distribution and $D$ a differential operator on $G$. Then, for any $x, y \in G$, we define the distribution $T' = xTy$ and the differential operator $D' = xDy$ as follows. $T'(f) = T(f')$ and $D'f = Df'$ ($f \in C_c^\infty(G)$), where $f'(z) = f(xzy)$ ($z \in G$). Also, put $xT = xTy$, $Tx = yTx$, for $y = 1$. $T$ is called invariant if $xTx^{-1} = T$ for all $x \in G$. Let $\mathfrak{B}$ be the algebra of those differential operators $D$ for which $D = xDy$ ($x, y \in G$). We say that $T$ is an eigen-distribution of $\mathfrak{B}$ if it is an eigen-distribution for every $D$ in $\mathfrak{B}$. Let $Z$ be the center of $G$. Then $T$ is an eigen-distribution of $Z$ if $zT$ is a numerical multiple of $T$ for every $z \in Z$. It is known (see[4(a)]) that the characters $T_\omega$ ($\omega \in \mathcal{E}$) are invariant and that they are eigen-distributions of both $\mathfrak{B}$ and $Z$. Hence the Plancherel formula may be regarded as an "expansion" of $\delta$ in terms of such invariant eigen-distributions.

Although the existence and uniqueness of the Plancherel measure are known, so far no general method of computing it has been found. The object of this paper is to give a formula which can be regarded as a first step towards the determination of $d\omega$. Let $G'$ and $S$, respectively, denote the sets

---

* Received December 4, 1956.

[1] See footnote 1 of [4(g)] for certain notational conventions.

733

of regular and singular elements in $G$ (see [4(c)]). Consider a Cartan subgroup $A$ of $G$ and let $\mathfrak{H}$ denote the algebra of those differential operators on $A$ which are invariant under all translations of $A$. Then for any $f \in C_c^\infty(G)$, we shall define a function $\Psi_f$ on $A' = A \cap G'$ of class $C^\infty$ in such a way that the following conditions hold.[2]

(1)   If $f' = xfx^{-1}$ $(x \in G, f \in C_c^\infty(G))$, then $\Psi_f = \Psi_{f'}$.

(2)   There exists a homomorphism $\gamma$ of $\mathfrak{Z}$ into $\mathfrak{H}$ such that $\Psi_{zf} = \gamma(z)\Psi_f$ for all $z \in \mathfrak{Z}$ and $f \in C_c^\infty(G)$.

(3)   For every $u \in \mathfrak{H}$ and $f \in C_c^\infty(G)$, $u\Psi_f$ remains bounded on $A'$ and $\Psi_f$ vanishes outside some compact subset of $A$.

(4)   The mapping $f \to \Psi_f$ is continuous in a suitable sense (see Theorem 2).

Let $l$ be the rank of $G$. Then $A \cap S$ is the union of a finite number of closed subgroups of $A$ of dimension $l-1$. (3) implies that $u\Psi_f$ has no singularities (on $A$) except possible jumps across $A \cap S$. Let $\hat{A}$ be the character group of $A$ and $T_{\hat{a}}(f)$ the value of the Fourier transform of $\Psi_f$ at $\hat{a} \in \hat{A}$. Then it follows from (1) and (4) that $T_{\hat{a}}$ is an invariant distribution on $G$ (see Lemma 17). Moreover, since $Z \subset A$, it is clear that $T_{\hat{a}}$ is an eigen-distribution of $Z$. For any function $\phi \in C_c^\infty(A)$, let $\tilde{\phi}$ denote the Fourier transform of $\phi$. Then it is clear that, for any $u \in \mathfrak{H}$, there exists an analytic function $\hat{u}$ on $\hat{A}$ such that $(u\phi)^\sim = \hat{u}\tilde{\phi}$ $(\phi \in C_c^\infty(A))$. Let $\chi_{\hat{a}}(z)$ $(z \in \mathfrak{Z}, \hat{a} \in \hat{A})$ denote the value of $(\gamma(z))^\wedge$ at $\hat{a}$. Then if $\Psi_f$ could be extended to a function of class $C^\infty$ on $A$, it would follow from (2) that

$$T_{\hat{a}}(zf) = \chi_{\hat{a}}(z) \; T_{\hat{a}}(f).$$

Hence except for the above-mentioned jumps, $T_{\hat{a}}$ would have been an eigen-distribution of $\mathfrak{Z}$.

Let $l_-(A)$ denote the dimension of the maximal compact subgroup of $A/Z$. We say that $A$ is fundamental if $l_-(A)$ has the maximum possible value. If $A$ is fundamental, we can select an open neighborhood $B$ of $1$ in $A$ and a connected component $A_1$ of $B \cap A'$ such that $1$ lies in the closure of $A_1$ and the following additional condition holds.

(5)   There exists an element $u_0 \in \mathfrak{H}$ such that

$$(\mathrm{I}) \qquad\qquad f(1) = \lim_{a \to 1} (u_0\Psi_f)(a) \qquad\qquad (a \in A_1)$$

for all $f \in C_c^\infty(G)$.

---

[2] $\Psi_f = \epsilon_* F_f \prod_{\alpha \in P} |\eta_\alpha|^{\frac{1}{2}}$ in the notation of Lemma 18.

Let us agree, for a moment, to ignore the complications due to the jumps. Then (5) implies the formula

$$\text{(II)} \qquad f(1) = \int_{\hat{A}} \hat{u}_0(\hat{a})\, T_{\hat{a}}(f)\, d\hat{a} \qquad (f \in C_c^\infty(G)),$$

where $d\hat{a}$ is the (suitably normalized) Haar measure on $\hat{A}$. Therefore

$$\delta = \int_{\hat{A}} \hat{u}_0(\hat{a})\, T_{\hat{a}}\, d\hat{a},$$

and since $T_{\hat{a}}$ are invariant eigen-distributions of $Z$ and $\mathfrak{Z}$, we should expect this formula to be closely related to the Plancherel formula.

Let $r(G)$ be the maximum number of nonconjugate Cartan subgroups of $G$. In case $r(G) = 1$, it is actually true that $\Psi_f$ can be extended to a function of class $C^\infty$ on $A$ for every $f \in C_c^\infty(G)$. Hence the above formula holds in this case. Moreover, it is then possible to define a mapping $\hat{a} \to \omega(\hat{a})$ of $\hat{A}$ into $\mathcal{E}$ such that $T_{\hat{a}} = c T_{\omega(\hat{a})}$, where $c$ is a constant. Then

$$\delta = c \int_{\hat{A}} \hat{u}_0(\hat{a})\, T_{\omega(\hat{a})}\, d\hat{a}$$

and, in fact this is substantially the same as the Plancherel formula for $G$. Thus if $r(G) = 1$, the problem of the determination of the Plancherel measure is solved completely. This is so in particular if $G$ is either compact or complex (see [4(b)]).

Now in the general case $(r(G) > 1)$, the jumps of the function $\Psi_f$ (or its derivatives) cannot be ignored, and therefore $T_{\hat{a}}$ is no longer an eigen-distribution of $\mathfrak{Z}$. Nevertheless, there are good reasons to believe (see [4(c)]) that for suitable $\hat{a} \in \hat{A}$, $T_{\hat{a}}$ is intimately related to a character $T_\omega$ of $G$. Hence equation (I) may still be regarded as a first step towards the derivation of the Plancherel formula. The importance of such a relation has been clearly recognized by Gelfand and Graev [3].

Now suppose for a moment that $A$ is compact so that $\hat{A}$ is discrete. In view of equation (II) (which, however, is not quite exact due to the jumps of $\Psi_f$), one would expect to get a "discrete series" (see [4(f)]) in $\mathcal{E}$ parametrized by $\hat{A}$. More precisely, we should expect the existence of a mapping which assigns to each $\hat{a}$ lying in a suitable subset $\hat{A}_0$ of $\hat{A}$, a discrete class $\omega(\hat{a})$ in $\mathcal{E}$ such that the Plancherel measure of $\omega(\hat{a})$ is $c\hat{u}_0(\hat{a})$, where $c$ is a constant independent of $\hat{a}$. This expectation has actually been confirmed in a large number of cases (see [4(f)]).

If $G$ is simple, its first Betti number $b(G)$ is either 0 or 1. Assuming that $A$ is compact, we should expect to get a discrete series in both cases.

So far, a general method for constructing a discrete series has been obtained only when $b(G) = 1$ (see [4(d), (e), (f)]).

This paper is divided into two parts. The main results of Part I are contained in Theorem I and the Corollary to Lemma 12. We need them in Part II in order to carry over the results of [4(h), (i)] from its Lie algebra to the group $G$ itself. Sections 4 and 5 are devoted to the verification of the five conditions stated above. A short account of the principal results of this paper has appeared in [4(j)].

## Part I.

**2. Proof of Theorem 1.** Let $R$ and $C$ be the fields of real and complex numbers, respectively, and let $G$ be a connected semisimple Lie group with the Lie algebra $\mathfrak{g}_0$ over $R$. Let $\mathfrak{h}_0$ be a Cartan subalgebra of $\mathfrak{g}_0$ and $A$ the corresponding Cartan subgroup of $G$ (see [4(c), §2]). Our main object in this section is to prove the following theorem.

THEORM 1. *Let $a_0$ be an element in $A$ and $\Xi$ the centralizer of $a_0$ in $G$. Denote the natural mapping of $G$ on the factor space $G^* = G/\Xi$ by $x \to x^*$ ($x \in G$). Then we can find a neighborhood $B$ of $a_0$ in $A$ with the following property. Given any compact set $\omega$ in $G$, there exists a compact set $\Omega^*$ in $G^*$ satisfying the condition that, if $xax^{-1} \in \omega$ for some $a \in B$ and $x \in G$, then $x^* \in \Omega^*$.*

The proof is rather long and requires considerable preparation. First we need two results on nilpotent groups. Let $F$ be a field which is either $R$ or $C$, and let $N$ be a connected Lie group and $\mathfrak{n}$ its Lie algebra over $F$. (If $F = C$, we assume of course that $N$ is a complex analytic group.) Suppose $\mathfrak{n}$ is nilpotent and $\mathfrak{n}^{(k)}$ ($k = 1, 2, \cdots$) are ideals in $\mathfrak{n}$ such that $\mathfrak{n}^{(1)} = \mathfrak{n}$, $[\mathfrak{n}, \mathfrak{n}^{(k)}] \subset \mathfrak{n}^{(k+1)}$ ($k \geq 1$) and $\mathfrak{n}^{(k)} = \{0\}$ if $k$ is sufficiently large. Furthermore suppose $\mathfrak{n}$ can be written as a direct sum of two subspaces $\mathfrak{n}_1$, $\mathfrak{n}_2$ satisfying the following two conditions:

(1)    $\mathfrak{n}_1$ is a subalgebra of $\mathfrak{n}$,

(2)    $\mathfrak{n}^{(k)} = \mathfrak{n}_1 \cap \mathfrak{n}^{(k)} + \mathfrak{n}_2 \cap \mathfrak{n}^{(k)}$ for all $k \geq 1$.

Then we have the following result.

LEMMA 1. *Under the above conditions every element in $N$ can be written in the form $\exp X_2 \exp X_1$, where $X_1 \in \mathfrak{n}_1$ and $X_2 \in \mathfrak{n}_2$.*

Without loss of generality, we may assume that $N$ is simply connected. Then $\mathfrak{n}$ being nilpotent, the exponential mapping is a one-one regular mapping of $\mathfrak{n}$ onto $N$. We denote its inverse by $x \to \log x$ $(x \in N)$. $X$ being any element in $N$, we shall first show that for every integer $k \geq 0$, we can find $X_1 \in \mathfrak{n}_1$ such that

$$\log(\exp X \exp(-X_1)) \in \mathfrak{n}_2 + \mathfrak{n}^{(k+1)}.$$

We use induction on $k$. If $k = 0$ our assertion is true trivially. So let us suppose $k \geq 1$. By the induction hypothesis, we can choose $Y_1 \in \mathfrak{n}_1$ such that $Z = \log(\exp X \exp(-Y_1)) \in \mathfrak{n}_2 + \mathfrak{n}^{(k)}$. Select $Z' \in \mathfrak{n}^{(k)}$ such that $Z - Z' \in \mathfrak{n}_2$. Then $Z' = Z_1 + Z_2$, where $Z_i \in \mathfrak{n}_i \cap \mathfrak{n}^{(k)}$ $(i = 1, 2)$. Hence, by the Campbell-Hausdorff formula, we get

$$\log(\exp Z \exp(-Z_1)) \equiv Z - Z_1 \bmod \mathfrak{n}^{(k+1)}.$$

Moreover, $Z - Z_1 = (Z - Z') + Z_2 \in \mathfrak{n}_2$, and therefore

$$\log(\exp X \exp(-Y_1)\exp(-Z_1)) \in \mathfrak{n}_2 + \mathfrak{n}^{(k+1)}.$$

However, $\mathfrak{n}_1$ is a nilpotent algebra, and therefore $\exp Z_1 \exp Y_1 = \exp X_1$ for some $X_1 \in \mathfrak{n}_1$. Hence $\log(\exp X \exp(-X_1)) \in \mathfrak{n}_2 + \mathfrak{n}^{(k+1)}$.

Now taking $k$ sufficiently large, we get $X_2 = \log(\exp X \exp(-X_1)) \in \mathfrak{n}_2$, and so $\exp X = \exp X_2 \exp X_1$. This proves the lemma.

Let $X_1, \cdots, X_r$ be a base for $\mathfrak{n}$ over $F$ such that $[X_i, X_j] \in \sum_{k>j} F X_k$ $(1 \leq i \leq j \leq r)$. Put $\mathfrak{n}_{(0)} = \{0\}$, $\mathfrak{n}_{(j)} = \sum_{1 \leq i \leq j} F X_i$ $(1 \leq j \leq r)$, and let $\pi_j$ denote the projection of $\mathfrak{n}$ on $\mathfrak{n}_{(j)}$ given by $\pi_j X_i = X_i$ $(1 \leq i \leq j)$ and $\pi_j X_i = 0$ $(j < i \leq r)$. Let $_m\mathfrak{n}_{(j)}$ denote the Cartesian product of $\mathfrak{n}_{(j)}$ with itself $m$ times.

LEMMA 2. *Suppose $N$ is simply connected and $m$ is a positive integer. Then for each $j$ $(1 \leq j \leq r)$, there exists a polynomial mapping* [3] *$p_j$ of $_m\mathfrak{n}_{(j-1)}$ into $\mathfrak{n}_{(j)}$ such that*

$$\pi_j \log(\exp Y_1 \exp Y_2 \cdots \exp Y_m)$$
$$= \pi_j(Y_1 + \cdots + Y_m) + p_j(\pi_{j-1} Y_1, \pi_{j-1} Y_2, \cdots, \pi_{j-1} Y_m)$$

*for all $Y_1, \cdots, Y_m \in \mathfrak{n}$.*

We shall prove this by induction on $r = \dim \mathfrak{n}$. If $r = 1$, $\mathfrak{n}$ is abelian

---

[3] Let $U$, $V$ be two vector spaces over $F$ of finite dimension. A mapping $p$ of $U$ into $V$ is called a polynomial mapping if, for every linear function $\lambda$ on $V$, the function $u \to \lambda(p(u))$ $(u \in U)$ is a polynomial function on $U$.

and our statement is obvious. So now suppose $r \geqq 2$. Put $\mathfrak{n}^* = \mathfrak{n}/FX_r$ and $N^* = N/N_r$, where $N_r$ is the one-parameter subgroup of $N$ corresponding to $X_r$. Since $X_r$ lies in the center of $\mathfrak{n}$, it is clear that $\log(\exp X \exp Y) = X + Y$ $(X \in \mathfrak{n}, Y \in FX_r)$, and therefore the exponential mapping of $\mathfrak{n}^*$ onto $N^*$ is still one-one. Let $X \to X^*$ $(X \in \mathfrak{n})$ denote the natural mapping of $\mathfrak{n}$ on $\mathfrak{n}^*$, and let $\mathfrak{n}_{(j)}^*$ be the image of $\mathfrak{n}_{(j)}$ $(1 \leqq j \leqq r)$ under this mapping. Define the projection $\pi_j^*$ of $\mathfrak{n}^*$ on $\mathfrak{n}_{(j)}^*$ by $\pi_j^* X^* = (\pi_j X)^*$ $(X \in \mathfrak{n}, 1 \leqq j \leqq r)$. Then since $\dim \mathfrak{n}^* < \dim \mathfrak{n}$, our induction hypothesis is applicable to $\mathfrak{n}^*$. Hence for each $j$, there exists a polynomial mapping $q_j$ of $_m\mathfrak{n}_{(j-1)}$ into $\mathfrak{n}_j$ such that

$$\pi_j \log(\exp Y_1 \cdots \exp Y_m)$$
$$\equiv \pi_j(Y_1 + \cdots + Y_m) + q_j(\pi_{j-1}Y_1, \cdots, \pi_{j-1}Y_m) \bmod FX_r$$

for $Y_1, \cdots, Y_m \in \mathfrak{n}$. But if $j < r$, $\mathfrak{n}_{(j)} \cap (FX_r) = \{0\}$, and therefore

$$\pi_j \log(\exp Y_1 \cdots \exp Y_m) = \pi_j(Y_1 + \cdots + Y_m) + q_j(\pi_{j-1}Y_1, \cdots, \pi_{j-1}Y_m).$$

On the other hand, if $j = r$, $\pi_r$ is the identity mapping of $\mathfrak{n}$. Moreover, it is known that $(Y_1, \cdots, Y_m) \to \log(\exp Y_1 \cdots \exp Y_m)$ is a polynomial mapping of $_m\mathfrak{n} = {}_m\mathfrak{n}_{(r)}$ into $\mathfrak{n}$ (see Birkhoff [1]). Now put

$$p_r(Y_1, \cdots, Y_m) = \log(\exp Y_1 \cdots \exp Y_m) - (Y_1 + \cdots + Y_m).$$

Then since $X_r$ belongs to the center of $\mathfrak{n}$, it is clear that $p_r(Y_1, \cdots, Y_m) = p_r(\pi_{r-1}Y_1, \cdots, \pi_{r-1}Y_m)$. Hence

$$\log(\exp Y_1 \cdots \exp Y_m) = (Y_1 + \cdots + Y_m) + p_r(\pi_{r-1}Y_1, \cdots, \pi_{r-1}Y_m),$$

and this proves our assertion for $j = r$.

Let $\mathfrak{g}$, $\mathfrak{h}$ be the complexifications of $\mathfrak{g}_0$, $\mathfrak{h}_0$, respectively, and let $G_c$ be a complex analytic group with the Lie algebra $\mathfrak{g}$. From now on we shall use the notation of $[4(g), (h)]$ without further comment. In view of the Corollary of Lemma 1 of $[4(c)]$, we may obviously assume that $\theta(\mathfrak{h}_0) = \mathfrak{h}_0$. For each root $\alpha$ (of $\mathfrak{g}$ with respect to $\mathfrak{h}$), select an eleemnt $X_\alpha \neq 0$ in $\mathfrak{g}$ such that $[H, X_\alpha] = \alpha(H)X_\alpha$ for all $H \in \mathfrak{h}$. Define the set $P$ of positive roots as in $[4(h), \S 5]$, and put $\mathfrak{n} = \sum_{\alpha \in P} CX_\alpha$. Let $A_c$ be the analytic subgroup of $G_c$ corresponding to $\mathfrak{h}$. If $\alpha$ is a root, we denote by $\xi_\alpha$ the corresponding character of $A_c$ so that $\xi_\alpha(\exp H) = e^{\alpha(H)}$ $(H \in \mathfrak{h})$. Let $a_0$ be a fixed element in $A_c$ and $P_1$ the subset of those roots $\alpha \in P$ for which $\xi_\alpha(a_0) = 1$. Then if $P_2$

is the complement of $P_1$ on $P$, we put $\mathfrak{n}_1 = \sum\limits_{\alpha \in P_1} CX_\alpha,\ \mathfrak{n}_2 = \sum\limits_{\alpha \in P_2} CX_\alpha.$  Let $\alpha_1 < \alpha_2 < \cdots < \alpha_r$ be all the roots in $P$.  Put

$$\mathfrak{n}^{(k)} = \sum_{k \leq i \leq r} CX_{\alpha_i} \qquad\qquad (k \geq 1).$$

Then it is obvious that all the conditions of Lemma 1 are fulfilled.  Let $N_c$ and $N_1$ be the (complex) analytic subgroups of $G_c$ corresponding to $\mathfrak{n}$ and $\mathfrak{n}_1$, respectively.  Also, let $N_2$ denote the set of all elements in $N_c$ of the form $\exp X$ $(X \in \mathfrak{n}_2)$.  It is well known that $N_c$ is simply connected, and therefore the exponential mapping defines an isomorphism of the complex vector space $\mathfrak{n}$ onto $N_c$ (with respect to its structure as a complex manifold).  Hence $N_2$ is closed in $N_c$.

Choose a compact neighborhood $B_c$ of $a_0$ in $A_c$ such that, if $\alpha \in P_2$, then $\xi_\alpha$ never takes the value 1 on $B_c$.

LEMMA 3.  *Let $\Xi_c$ denote the centralizer of $a_0$ in $G_c$ and $x \to x^*$ the natural mapping of $G_c$ onto $G_c^* = G_c/\Xi_c$.  Then for any compact set $\omega_c$ in $G_c$, we can select a compact set $\Omega_c^*$ in $G_c^*$ satisfying the following condition. If $xax^{-1} \in \omega_c$ for some $x \in G_c$ and $a \in B_c$, then $x^* \in \Omega_c^*$.*

We shall derive Theorem 1 from the above lemma.  But in order to prove this lemma, we need some additional facts.

LEMMA 4.  *There exist compact sets $\omega_1,\ \omega_2$ in $N_1,\ N_2$, respectively, having the following property.  Suppose $n_2 a n_1 n_2^{-1} \in \omega_c$ for some $a \in B_c$, $n_1 \in N_1$ and $n_2 \in N_2$.  Then $n_1 \in \omega_1$ and $n_2 \in \omega_2$.*

It is well known that $A_c N_c$ is closed in $G_c$ and the mapping $(a, n) \to an$ $(a \in A_c, n \in N_c)$ of $A_c \times N_c$ into $G_c$ is topological (see Iwasawa [5]).  We define $\log n$ $(n \in N_c)$ as before and denote by $t_\alpha$ $(\alpha \in P)$ the Cartesian coordinates in the complex Euclidean space $\mathfrak{n}$ corresponding to the base $(X_\alpha)_{\alpha \in P}$.  Then it would be enough to show that, under our assumptions, $t_\alpha(\log n_1)$ and $t_\alpha(\log n_2)$ remain bounded for every $\alpha \in P$.  Put $\mathfrak{n}_\alpha = \sum\limits_{0 < \beta < \alpha} CX_\beta$, and let $\pi_\alpha$ denote the projection of $\mathfrak{n}$ on $\mathfrak{n}_\alpha$ given by $\pi_\alpha X_\beta = X_\beta$ $(0 < \beta < \alpha)$ and $\pi X_\beta = 0$ $(\beta \geq \alpha)$.  Then from Lemma 2, there exists a polynomial function $p_\alpha$ on $\mathfrak{n}_\alpha \times \mathfrak{n}_\alpha \times \mathfrak{n}_\alpha$ such that

$$t_\alpha(\log(\exp X_1 \exp X_2 \exp X_3)) = t_\alpha(X_1 + X_2 + X_3) + p_\alpha(\pi_\alpha X_1, \pi_\alpha X_2, \pi_\alpha X_3)$$

for $X_1, X_2, X_3 \in \mathfrak{n}$.  Now $n_2 a n_1 n_2^{-1} = a(a^{-1} n_2 a) n_1 n_2^{-1}$.  Since $\omega_c$ is compact, it follows from what we have said above that if $a'n' \in \omega_c$ $(a' \in A_c, n' \in N_c)$, then $t_\alpha(\log n')$ remains bounded for every $\alpha \in P$.  This shows that

$t_\alpha(\log(a^{-1}n_2an_1n^{-1}))$ must remain bounded for every $\alpha \in P$. Now suppose the assertion of the lemma is false. Then let $\beta$ be the lowest root in $P$ such that $|t_\beta(\log n_1)| + |t_\beta(\log n_2)|$ does not remain bounded when $n_1$, $n_2$ and $a$ vary in such a way as to fulfill our assumptions. Put $X_i = \log n_i$ $(i = 1, 2)$. Then [4] $\log(a^{-1}n_2a) = \mathrm{Ad}(a^{-1})X_2$ and so it is clear that [5]

$$t_\beta(\log(a^{-1}n_2an_1n_2^{-1})) = t_\beta((\mathrm{Ad}(a^{-1}) - 1)X_2 + X_1)$$
$$+ p_\beta(\pi_\beta\mathrm{Ad}(a^{-1})X_2, \pi_\beta X_1, -\pi_\beta X_2).$$

Therefore

$$(\xi_\beta(a^{-1}) - 1)t_\beta(X_2) + t_\beta(X_1)$$
$$= t_\beta(\log(a^{-1}n_2an_1n_2^{-1})) - p_\beta(\pi_\beta\mathrm{Ad}(a^{-1})X_2, \pi_\beta X_1, -\pi_\beta X_2).$$

Now it follows from the definition of $\beta$ and the compactness of $B_c$ that the right side remains bounded. Moreover, if $\beta \in P_1$, then $t_\beta(X_2) = 0$. On the other hand, if $\beta \in P_2$, then $t_\beta(X_1) = 0$ and $(\xi_\beta(a^{-1}) - 1)^{-1} = \xi_\beta(a)(1 - \xi_\beta(a))^{-1}$ remains bounded for $a \in B_c$. Therefore, in either case, we conclude from the above equation that $|t_\beta(X_1)| + |t_\beta(X_2)|$ remains bounded. As this contradicts the definition of $\beta$, the lemma is proved.

COROLLARY. *There exists a compact set $\omega^*$ in $G_c^*$ with the following property. If $nan^{-1} \in \omega_c$ for some $n \in N_c$ and $a \in B_c$, then $n^* \in \omega^*$.*

From Lemma 1, $n = n_2n_1$ $(n_2 \in N_2, n_1 \in N_1)$. Hence $nan^{-1} = n_2an_1'n_2^{-1}$. where $n_1' = a^{-1}n_1an_1^{-1}$. Choose compact sets $\omega_1$, $\omega_2$ corresponding to Lemma 4. Then $n_2 \in \omega_2$. Moreover, since $N_1 \subset \Xi_c$, it follows that $n^* = n_2^* \in \omega_2^*$. Hence we can take $\omega^* = \omega_2^*$.

Now we come to the proof of **Lemma 3**. **Let $U$ be the real analytic** subgroup of $G_c$ corresponding to the compact real form $\mathfrak{u} = \mathfrak{k}_0 + (-1)^{\frac{1}{2}}\mathfrak{p}_0$ of $\mathfrak{g}$. Then $U$ is compact and it is well known (see Iwasawa [5]) that $G = UA_cN_c = UN_cA_c$. This implies that $G_c = UN_2N_1A_c$. Moreover $\Xi_c \supset N_1A_c$. Now suppose $xax^{-1} \in \omega_c$ $(x \in G_c, a \in B_c)$. Then if $x = un_2n_1a'$ $(u \in U, n_2 \in N_2, n_1 \in N_1, a' \in A_c)$ and $n_1' = a^{-1}n_1an_1^{-1}$, we have

$$xax^{-1} = u(n_2an_1'n_2^{-1})u^{-1} \in \omega_c.$$

Therefore $n_2an_1'n_2^{-1} \in U\omega_cU$. Since $U\omega_cU$ is also compact, we can select sets $\omega_1$, $\omega_2$ in accordance with Lemma 4, corresponding to $U\omega_cU$. Then $n_2 \in \omega_2$ and $x \in U\omega_2N_1A_c \subset U\omega_2\Xi_c$. Hence $x^* \in \Omega_c^*$, where $\Omega_c^* = (U\omega_2)^*$. This completes the proof of Lemma 3.

---

[4] As usual $x \to \mathrm{Ad}(x)$ $(x \in G)$ denotes the adjoint representation.
[5] Here 1 stands for the identity mapping of $\mathfrak{g}$.

We now return to Theorem 1. Let $Z$ be the center of $G$. Since $Z \subset \Xi$, it is clear that, for the proof of this theorem, we may replace $G$ by any other connected group locally isomorphic to it. Let us therefore assume that $G$ is the real analytic subgroup of $G_c$ corresponding to $\mathfrak{g}_0$. Also, it will be convenient to suppose that $G_c$ is simply connected. Let $\eta$ be the conjugation of $\mathfrak{g}$ with respect to $\mathfrak{g}_0$. Then $\eta$ can be "extended" to a (real) automorphism of $G_c$ which we again denote by $\eta$. Since $G$ is the connected component of the subgroup consisting of those points of $G_c$ which are left fixed by $\eta$, $G$ is closed in $G_c$. Also $A = A_c \cap G$ (see $[4(c), \S 2]$) and $\Xi = G \cap \Xi_c$. Therefore $G^* = G/\Xi$ may be regarded as a subset of $G_c^*$. It is clear that Theorem 1 would follow immediately from Lemma 3 if we can show that $G^*$ is closed in $G_c^*$ and the inclusion mapping of $G^*$ into $G_c^*$ is topological. An elementary argument shows that this mapping is certainly continuous. Hence it is sufficient to prove the following result.

LEMMA 5. *Let $\{x_r{}^*\}_{r \geq 1}$ be a sequence in $G^*$ which converges in $G_c^*$ to some point $x^*$. Then $x^* \in G^*$ and $x_r{}^*$ converges to $x^*$ in $G^*$.*

Define the structure of a Hilbert space on $\mathfrak{g}$ by means of the positive-definite Hermitian form $\| X \|^2 = -B(X, \bar{\theta}(X))$ $(X \in \mathfrak{g})$ as in $[4(h), \S 5]$ For any linear transformation $T$ on $\mathfrak{g}$, put $\| T \|^2 = \mathrm{sp}(T^*T)$, where $T^*$ is the adjoint of $T$. Then $\| X \| = \| \mathrm{ad}\, X \|$ $(X \in \mathfrak{g})$. We write $\| x \| = \| \mathrm{Ad}(x) \|$ $(x \in G_c)$ and call a subset $\omega'$ of $G_c$ bounded if $\| x \|$ remains bounded for $x \in \omega'$. Since the center of $G_c$ is finite, it is clear that $\omega'$ is bounded if and only if its closure in $G_c$ is compact. Select points $x_r \in G$ and $x \in G_c$ lying in the cosets $x_r{}^*$ and $x^*$, respectively. Then obviously $x_r a_0 x_r^{-1} \to x a_0 x^{-1}$ in $G_c$, and therefore $\| x_r a_0 x_r^{-1} \|$ remains bounded. But as we shall show below (Lemma 6) this implies that $x_r{}^*$ are all contained in a compact subset of $G^*$, and from this, the assertion of Lemma 5 follows immediately. Thus it remains to prove the following lemma.

LEMMA 6. *Let $\omega$ be a compact set in $G$. Then there exists a compact subset $\Omega^*$ of $G^*$ satisfying the condition that if $x a_0 x^{-1} \in \omega$ $(x \in G)$, then $x^* \in \Omega^*$.*

Let $\mathfrak{z}_0$ be the centralizer of $a_0$ in $\mathfrak{g}_0$. Then it follows from Lemma 7 of $[4(c)]$ that $\theta(\mathfrak{z}_0) = \mathfrak{z}_0$, and therefore (see $[4(h)$, Lemma 10$]$) $\mathfrak{z}_0$ is reductive in $\mathfrak{g}_0$. Moreover $\mathfrak{h}_0 \subset \mathfrak{z}_0$, and therefore the ranks of $\mathfrak{g}_0$ and $\mathfrak{z}_0$ are the same. *Now it is convenient to drop our earlier notation and redefine $\mathfrak{h}_{\mathfrak{p}_0}$ and $\mathfrak{h}_0$ as follows.* Let $\mathfrak{h}_{\mathfrak{p}_0}$ be a maximal abelian subspace of $\mathfrak{z}_{\mathfrak{p}_0} = \mathfrak{z}_0 \cap \mathfrak{p}_0$. We extend it to a Cartan subalgebra $\mathfrak{h}_0$ of $\mathfrak{z}_0$. Then $\mathfrak{h}_0 = \mathfrak{h}_{\mathfrak{p}_0} + \mathfrak{h}_{\mathfrak{k}_0}$, where

$\mathfrak{h}_{\mathfrak{k}_0} = \mathfrak{h}_0 \cap \mathfrak{k}_0$, and since the ranks of $\mathfrak{g}_0$ and $\mathfrak{z}_0$ are equal and $\mathfrak{z}_0$ is reductive in $\mathfrak{g}_0$, $\mathfrak{h}_0$ is also a Cartan subalgebra of $\mathfrak{g}_0$. We now introduce the notation[6] of $[4(\mathrm{h}), \S 5]$ corresponding to the *new* $\mathfrak{h}_0$. In particular, $A$ is now the Cartan subgroup of $G$ corresponding to $\mathfrak{h}_0$ and $A_\mathfrak{p} = \exp \mathfrak{h}_{\mathfrak{p}_0}$. Since $\mathfrak{h}_0$ is contained in the centralizer of $a_0$, it is obvious that $a_0 \in A$. Let $a_0 = a_1 a_2$, where $a_1 \in A \cap K$ and $a_2 \in A_\mathfrak{p}$ (see $[4(\mathrm{c}), \text{Lemma } 7]$). Clearly $\mathfrak{m}_{\mathfrak{p}_0}$ and $\mathfrak{n}_0$ are invariant under $\mathrm{Ad}(a_0)$. Moreover, since $\mathfrak{h}_{\mathfrak{p}_0}$ is maximal abelian in $\mathfrak{z}_{\mathfrak{p}_0}$, $\mathfrak{z}_{\mathfrak{p}_0} \cap \mathfrak{m}_{\mathfrak{p}_0} = \{0\}$. Therefore $\mathrm{Ad}(a_0)$ can never take the eigenvalue 1 in $\mathfrak{m}_{\mathfrak{p}_0}$. Put $\mathfrak{n}_0' = \mathfrak{z}_0 \cap \mathfrak{n}_0$ and let $\mathfrak{n}_0''$ be the orthogonal complement of $\mathfrak{n}_0'$ in $\mathfrak{n}_0$. Since $\mathrm{Ad}(a_1)$ is unitary and $\mathrm{Ad}(a_2)$ is self-adjoint and they commute, it follows that both $\mathfrak{n}_0'$ and $\mathfrak{n}_0''$ are invariant under $\mathrm{Ad}(a_0)$ and no eigenvalue of $\mathrm{Ad}(a_0)$ in $\mathfrak{n}_0''$ can be 1. The same argument shows that $\mathfrak{z}_{\mathfrak{p}_0}$ and $\mathfrak{m}_{\mathfrak{p}_0}$ are mutually orthogonal.

Now let $\mathfrak{z}$, $\mathfrak{m}_\mathfrak{p}$, $\mathfrak{h}$, $\mathfrak{n}'$, $\mathfrak{n}''$, $\mathfrak{n}$, $\mathfrak{g}_\lambda$ ($\lambda \in \mathfrak{F}$) denote the complexifications of $\mathfrak{z}_0$, $\mathfrak{m}_{\mathfrak{p}_0}$, $\mathfrak{h}_0$, $\mathfrak{n}_0'$, $\mathfrak{n}_0''$, $\mathfrak{n}_0$, $\mathfrak{g}_{0\lambda}$, respectively, in $\mathfrak{g}$ and similarly for the others. Since every element in $\mathrm{ad}(\mathfrak{h}_{\mathfrak{p}_0})$ is self-adjoint, both $\mathfrak{n}'$, $\mathfrak{n}''$ are invariant under $\mathrm{ad}(\mathfrak{h}_\mathfrak{p})$. Put $\mathfrak{n}_\lambda' = \mathfrak{n}' \cap \mathfrak{g}_\lambda$, $\mathfrak{n}_\lambda'' = \mathfrak{n}'' \cap \mathfrak{g}_\lambda$ ($\lambda \in \mathfrak{F}$). Then $\mathfrak{n}_\lambda'$ and $\mathfrak{n}_\lambda''$ are mutually orthogonal and invariant under $\mathrm{Ad}(a_0)$, and $\mathfrak{n}' = \sum_{\lambda > 0} \mathfrak{n}_\lambda'$, $\mathfrak{n}'' = \sum_{\lambda > 0} \mathfrak{n}_\lambda''$, where both sums are orthogonal. Moreover, $\mathrm{Ad}(a_0)$ never takes the eigenvalue 1 on $\mathfrak{n}_\lambda''$, while its restriction on $\mathfrak{n}'$ is the identity. Choose a base for $\mathfrak{g}_\lambda$ ($\lambda > 0$) consisting of eigen-vectors of $\mathrm{Ad}(a_0)$. Then every element of this base lies either in $\mathfrak{n}_\lambda'$ or in $\mathfrak{n}_\lambda''$. Moreover, all these bases, put together for all $\lambda > 0$, form a base $(X_1, \cdots, X_q)$ for $\mathfrak{n}$. We number the elements $X_i$ in such a way that if $X_i \in \mathfrak{g}_\lambda$, $X_j \in \mathfrak{g}_\mu$ and $0 < \lambda < \mu$, then $i < j$. Then it is obvious that $[X_i, X_j] \in \sum_{k > j} C X_k$ ($1 \leq i < j \leq q$).

Let $N$, $N'$, respectively, be the analytic subgroups of $G$ corresponding to $\mathfrak{n}_0$, $\mathfrak{n}_0'$ and put $N'' = \exp \mathfrak{n}_0''$. Define $M$ and $M_\mathfrak{p} = \exp \mathfrak{m}_{\mathfrak{p}_0}$ as in $[4(\mathrm{h}),$ $\S 5]$. Then from Lemma 11 of $[4(\mathrm{h})]$, $G = K M_\mathfrak{p} N A_\mathfrak{p}$, and it is obvious that $\Xi \supset N' A_\mathfrak{p}$. Moreover, it follows from Lemma 1 that $N = N'' N'$. Finally, since $G$ is imbedded in the complex group $G_c$, the center of $G$ is finite, and therefore $K$ is compact. Therefore, we may assume without loss of generality that $\omega = K \omega K$. Now suppose $x = kmna$ ($k \in K, m \in M_\mathfrak{p}, n \in N, a \in A_\mathfrak{p}$) and $x a_0 x^{-1} \in \omega$. Then $(mna) a_0 (mna)^{-1} \in \omega$, and therefore

$$a_0^{-1} y a_0 y^{-1} = a_0^{-1} m n a_0 n^{-1} m^{-1} \in a_0^{-1} \omega,$$

---

where $y = mna$. On the other hand,

$$a_0^{-1}mna_0n^{-1}m^{-1} = (a_0^{-1}ma_0)\, m^{-1}m\, (a_0^{-1}na_0n^{-1})\, m^{-1}$$

and $(a_0^{-1}ma_0)m^{-1} \in M_\mathfrak{p}M_\mathfrak{p} \subset M \subset KM_\mathfrak{p}$ while $m(a_0^{-1}na_0n^{-1})m^{-1} \in mNm^{-1} = N$. Therefore, since $a_0^{-1}\omega$ is compact, it follows from Lemma 11 of [4(h)] that both $\|a_0^{-1}ma_0m^{-1}\|$ and $\|m(a_0^{-1}na_0n^{-1})m^{-1}\|$ must remain bounded (as $x$ varies in $G$ subject to the condition that $xa_0x^{-1} \in \omega$). However, as we shall see in Lemma 7, the boundedness of $\|a_0^{-1}ma_0m^{-1}\|$ implies that $m$ remains within a compact subset of $M_\mathfrak{p}$. Since $\|m^{-1}zm\| \leq \|m^{-1}\|\,\|z\|\,\|m\|$ for any $z \in G$, it follows that $\|a_0^{-1}na_0n^{-1}\|$ also remains bounded. Now let $n = n_2n_1$ $(n_1 \in N', n_2 \in N'')$. Then $a_0^{-1}na_0n^{-1} = a_0^{-1}n_2a_0n_2^{-1}$. But as we shall prove in Lemma 9, the boundedness of $\|a_0^{-1}n_2a_0n_2^{-1}\|$ implies the boundedness of $\|n_2\|$. Then if $x_1 = kmn_2$, we have $x^* = x_1^*$ and $\|x_1\| = \|mn_2\| \leq \|m\|\,\|n_2\|$. This shows that $x_1$ stays within a bounded subset of $G$, and therefore $x^* = x_1^*$ within a compact subset of $G^*$. Hence our assertion is now proved.

We have still to prove the two results which were used above.

LEMMA 7. *Suppose $m$ varies in $M_\mathfrak{p}$ in such a way that $\|ma_0m^{-1}\|$ remains bounded. Then $\|m\|$ also remains bounded.*

Define $a_1$, $a_2$ as above. Then $a_0 = a_1a_2$ and $a_2$ commutes with every element in $M_\mathfrak{p}$ since $[\mathfrak{h}_{\mathfrak{p}_0}, \mathfrak{m}_{\mathfrak{p}_0}] = \{0\}$. Therefore $ma_0m^{-1} = (ma_1m^{-1})a_2$, and so $\|ma_1m^{-1}\|$ remains bounded. Moreover, the exponential mapping being one-one and regular on $\mathfrak{p}_0$, we may denote by log its inverse on $\exp\mathfrak{p}_0$. Then it would be sufficient to show that $\|\log m\|$ remains bounded. For this we need the following lemma.

LEMMA 8. *Let $k$ be an element in $K$. Then if $n = \dim_C \mathfrak{g}$,*

$$\|pkp^{-1}\| \geq n^{\frac{1}{2}}\cosh(n^{-\frac{3}{2}}\|\log p - \mathrm{Ad}(k)(\log p)\|)$$

*for for $p \in \exp\mathfrak{p}_0$.*

Assuming this for a moment, Lemma 7 can now be proved as follows. We have seen that $\mathrm{Ad}(a_0)$ never takes the eigenvalue 1 in $\mathfrak{m}_{\mathfrak{p}_0}$ and $\mathrm{Ad}(a_0) = \mathrm{Ad}(a_1)$ on $\mathfrak{m}_{\mathfrak{p}_0}$. Hence we can find a positive number $c$ such that $\|X - \mathrm{Ad}(a_1)X\| \geq c\|X\|$ for all $X \in \mathfrak{m}_{\mathfrak{p}_0}$. Therefore, from Lemma 8,

$$\|ma_1m^{-1}\| \geq n^{\frac{1}{2}}\cosh(n^{-\frac{3}{2}}c\|\log m\|)$$

for all $m \in M_\mathfrak{p}$. But as we have seen above $\|ma_1m^{-1}\|$ remains bounded under the assumptions of Lemma 7, and so the same holds for $\|\log m\|$ and $\|m\|$.

Now in order to prove Lemma 8, we proceed as follows. Let $E$ be the space of all endomorphisms of $\mathfrak{g}$. Put $\langle S, T\rangle = \mathrm{sp}(S^*T)$ $(S, T \in E)$, where $S^*$ is the adjoint of $S$. This scalar product defines the structure of a Hilbert space on $E$. Let $d_X$ $(X \in \mathfrak{g})$ denote the linear mapping of $E$ given by $d_X T = [\mathrm{ad}\, X, T]$, where $[S, T] = ST - TS$ $(S, T \in E)$. If $X \in \mathfrak{p}_0$, $\mathrm{ad}\, X$ is self-adjoint, and therefore

$$\langle S, d_X T\rangle = \mathrm{sp}(S^*[\mathrm{ad}\, X, T]) = -\mathrm{sp}([\mathrm{ad}\, X, S^*]T)$$
$$= \mathrm{sp}((d_X S)^*T) = \langle d_X S, T\rangle.$$

This proves that $d_X$ is also self-adjoint. Moreover, if $p = \exp X \in \exp \mathfrak{p}_0$, it is obvious (by considering $E$ as a Lie algebra under the bracket operation $[S, T]$) that $\exp d_X$ is the mapping $\sigma_p \colon T \to \mathrm{Ad}(p)\, T\, \mathrm{Ad}(p^{-1})$ $(T \in E)$ of $E$. Clearly $\sigma_p$ is also self-adjoint. Hence we can choose an orthonormal base $(T_1, \cdots, T_r)$ of $E$ consisting of eigen-vectors of $d_X$ and $\sigma_p$. Let $\lambda_i$ be the eigenvalue of $d_X$ corresponding to $T_i$ $(1 \leq i \leq r)$. Then if $T = \sum_{1 \leq i \leq r} c_i T_i$ $(c_i \in C)$, it is clear that $\sigma_p T = \sum_i c_i e^{\lambda_i} T_i$, and therefore $\|\sigma_p T\|^2 = \sum_i |c_i|^2 e^{2\lambda_i}$. Now put $T = \mathrm{Ad}(k)$, where $k$ is some fixed element in $K$. Then this equation becomes

$$\| pkp^{-1}\|^2 = \sum_{1 \leq i \leq r} |c_i|^2 e^{2\lambda_i}.$$

On the other hand, since $\|Y\| = \|\theta(Y)\|$ $(Y \in \mathfrak{g})$, $\theta$ is a unitary transformation of $\mathfrak{g}$, and therefore $\|\theta S \theta^{-1}\| = \|S\|$ for any $S \in E$. Moreover, since $\theta(X) = -X$, it follows that $\theta(\mathrm{Ad}(pkp^{-1}))\theta^{-1} = \mathrm{Ad}(p^{-1}kp)$. Therefore

$$2\| pkp^{-1}\|^2 = \| pkp^{-1}\|^2 + \| p^{-1}kp\|^2 = \|\sigma_p T\|^2 + \|\sigma_{p^{-1}} T\|^2$$
$$= 2\sum_{1 \leq i \leq r} |c_i|^2 \cosh 2\lambda_i.$$

But $\cosh 2t = 1 + 2\sinh^2 t$ and $s(t) = \sinh t/t = \sum_{q \geq 0} t^{2q}/(2q+1)!$ is an increasing function of $t$ for $t \geq 0$. Therefore

$$\| pkp^{-1}\|^2 = \sum_{1 \leq i \leq r} |c_i|^2 + 2\sum_{1 \leq i \leq r} |c_i \lambda_i|^2 s(\lambda_i)^2.$$

Moreover, $\sum_i |c_i|^2 = \|k\|^2 = n$ since $\mathrm{Ad}(k)$ is unitary. Now choose an index $j$ $(1 \leq j \leq r)$ such that $|c_j \lambda_j| = \max_{1 \leq i \leq r} |c_i \lambda_i|$. Since $|c_j| \leq \|k\| = n^{\frac{1}{2}}$, it follows that $s(n^{-\frac{1}{2}}|c_j \lambda_j|) \leq s(|\lambda_j|)$, and therefore

$$\| pkp^{-1}\|^2 \geq n + 2|c_j \lambda_j|^2 s(\lambda_j)^2 \geq n + 2|c_j \lambda_j|^2 \{s(n^{-\frac{1}{2}}|c_j \lambda_j|)\}^2$$
$$= n \cosh(2n^{-\frac{1}{2}}|c_j \lambda_j|).$$

On the other hand, $\| d_X T \|^2 = \sum_{1 \leq i \leq r} | c_i \lambda_i |^2 \leq r \, | c_j \lambda_j |^2$ and $r = \dim E = n^2$.
Therefore $| c_j \lambda_j | \geq n^{-1} \| d_X T \|$ and

$$\| pkp^{-1} \| \geq n^{\frac{1}{2}} \cosh (n^{-\frac{3}{2}} \| d_X T \|)$$

since $\cosh t$ is an increasing function of $t$ for $t \geq 0$ and $\cosh 2t \geq \cosh^2 t$.
But $T = \mathrm{Ad}(k)$ is unitary, and therefore

$$\| d_X T \| = \| (d_X T) T^{-1} \| = \| (\mathrm{ad} X) - T(\mathrm{ad} X) T^{-1} \|$$

$$= \| \mathrm{ad} (X - \mathrm{Ad}(k) X) \| = \| X - \mathrm{Ad}(k) X \|.$$

This proves that

$$\| pkp^{-1} \| \geq n^{\frac{1}{2}} \cosh (n^{-\frac{3}{2}} \| X - \mathrm{Ad}(k) X \|)$$

which is equivalent to the statement of Lemma 8.

The second result which was required during the proof of Lemma 6, may be stated as follows.

LEMMA 9. *Suppose $n$ varies in $N''$ in such a way that $\| a_0^{-1} n a_0 n^{-1} \|$ remains bounded. Then $\| n \|$ itself remains bounded.*

Consider the base $(X_1, \cdots, X_q)$ of $\mathfrak{n}$ introduced during the proof of Lemma 6 and let $(t_1, \cdots, t_q)$ denote the complex Cartesian coordinates in $\mathfrak{n}$ corresponding to this base. For any $j$ $(1 \leq j \leq q)$, define the projection $\pi_j$ as in Lemma 2, and put $\pi_0 = 0$ and $\mathfrak{n}_j = \pi_j \mathfrak{n}$ $(0 \leq j \leq q)$. Then from Lemma 2, there exist polynomial functions $p_j$ on $\mathfrak{n}_{j-1} \times \mathfrak{n}_{j-1}$ such that

$$t_j (\log (\exp Y_1 \exp Y_2)) = t_j (Y_1 + Y_2) + p_j (\pi_{j-1} Y_1, \pi_{j-1} Y_2) \quad (1 \leq j \leq q)$$

for $Y_1, Y_2 \in \mathfrak{n}$. Now put $X = \log n$. Then it would be enough to prove that $t_j(X)$ $(1 \leq j \leq q)$ all remain bounded. Suppose this is false. Then consider the least index $j$ such that $| t_j(X) |$ does not stay bounded. Put

$$Z = \log (a_0^{-1} n a_0 n^{-1}) = \log (\exp (\mathrm{Ad}(a_0^{-1}) X) \exp (-X)).$$
Then

$$t_j(Z) = t_j (\mathrm{Ad}(a_0^{-1}) X) - t_j(X) + p_j (\pi_{j-1} \mathrm{Ad}(a_0^{-1}) X, -\pi_{j-1} X).$$

But since $X_i$ is an eigenvector of $\mathrm{Ad}(a_0)$, $\mathrm{Ad}(a_0) X_i = c_i X_i$ $(1 \leq i \leq q)$, where $c_i$ are nonzero complex numbers. Therefore $t_i(\mathrm{Ad}(a_0^{-1}) X) = c_i^{-1} t_i(X)$ and

$$(c_j^{-1} - 1) t_j(X) = t_j(Z) - p_j (\pi_{j-1} \mathrm{Ad}(a_0^{-1}) X, -\pi_{j-1} X).$$

In view of the hypothesis of the lemma and our choice of $j$, the right side remains bounded as $n$ varies. Moreover, $c_j \neq 1$ if $X_j \in \mathfrak{n}''$, and $t_j(X) = 0$ if $X_j \in \mathfrak{n}'$. Therefore, it follows from the above equation that $t_j(X)$ remains

3

bounded in either case. As this contradicts the definition of $j$, the lemma is proved.

The proof of Theorem 1 is now quite complete.

**3. The sets $V$ and $V_C$.** In order to prove our next result, we need a simple lemma. Let $W$ be a vector space over $C$ of finite dimension. We shall call an endomorphism $X$ of $W$ unipotent if $X-1$ is nilpotent. (Here 1 stands for the identity mapping of $W$.)

LEMMA 10. *Let $\mathfrak{N}$ and $\mathcal{E}$, respectively, be the sets of all nilpotent and unipotent endomorphisms of $W$. Then $X \to \exp X$ $(X \in \mathfrak{N})$ is a one-one mapping of $\mathfrak{N}$ onto $\mathcal{E}$.*

Since $\exp X - 1 = \sum_{k \geq 1} X^k/k!$, it is obvious that $\exp X \in \mathcal{E}$ for $X \in \mathfrak{N}$. For any $Y \in \mathcal{E}$, define $\log Y = \sum_{k \geq 1} (-1)^{k-1}(Y-1)^k/k$. Since $Y$ is unipotent, this series is actually finite, and it is clear that $\log Y \in \mathfrak{N}$. Our assertion now follows from the well-known identities (see Birkhoff [1]) $X = \log(\exp X)$ $(X \in \mathfrak{N})$ and $Y = \exp(\log Y)$ $(Y \in \mathcal{E})$.

Now $G$ being as in Theorem 1, we shall prove the following result.

LEMMA 11. *Let $V$ be the open set in the real Euclidean space $\mathfrak{g}_0$ defined as follows. An element $X$ of $\mathfrak{g}_0$ lies in $V$ if and only if the absolute value of every eigenvalue of $\operatorname{ad} X$ is less than[7] $\pi$. Then the exponential mapping of $\mathfrak{g}_0$ into $G$ is regular and univalent on $V$.*

If $X \in V$, it is clear that $\operatorname{ad} X$ can never have an eigenvalue of the form $2\pi(-1)^{\frac{1}{2}}n$ where $n$ is a nonzero integer. Therefore

$$(1 - \exp(-\operatorname{ad} X))/\operatorname{ad} X = \sum_{m \geq 0} (-1)^m (\operatorname{ad} X)^m/(m+1)!$$

is a nonsingular transformation of $\mathfrak{g}_0$. This shows (see Chevalley [2(a), p. 157]) that the exponential mapping is regular at $X$. Now suppose $\exp X_1 = \exp X_2$ $(X_1, X_2 \in V)$. Put $\sigma = (1 - \exp(-\operatorname{ad} X_1))/\operatorname{ad} X_1$. Then $\operatorname{Ad}(x) - 1 = \exp(\operatorname{ad} X_1) - 1 = \operatorname{Ad}(x)\sigma \operatorname{ad} X_1$, where $x = \exp X_1 = \exp X_2$. Since $\operatorname{Ad}(x)$ and $\sigma$ are both nonsingular, it follows that $\operatorname{Ad}(x)Y = Y$ for some $Y \in \mathfrak{g}$ if and only if $(\operatorname{ad} X_1)Y = 0$. Let $\mathfrak{c}_0$ denote the centralizer of $x$ in $\mathfrak{g}_0$. Then $\mathfrak{c}_0$ is also the centralizer of $X_1$ in $\mathfrak{g}_0$. But obviously $\operatorname{Ad}(x)X_2 = X_2$, and therefore $X_1$, $X_2$ commute. Hence $\exp(X_1 - X_2) = \exp X_1 \exp(-X_2) = 1$. Let $\lambda_{ji}$ $(1 \leq j \leq r_i)$ denote all the distinct eigenvalues of $\operatorname{ad} X_i$ $(i = 1, 2)$. Then every eigenvalue of $\operatorname{ad}(X_1 - X_2)$ must be

---

[7] Here $\pi$ denotes, as usual, the smallest positive root of the equation $\sin t = 0$.

of the form $\lambda_{j1} - \lambda_{k2}$ for some $j$ and $k$. Moreover it follows from the definition of $V$ that $|\lambda_{j1} - \lambda_{k2}| < 2\pi$. Hence $\exp(\lambda_{j1} - \lambda_{k2}) \neq 1$ unless $\lambda_{j1} = \lambda_{k2}$. Therefore since $\exp \mathrm{ad}(X_1 - X_2) = \mathrm{Ad}(\exp(X_1 - X_2)) = 1$, zero is the only eigenvalue of $\mathrm{ad}(X_1 - X_2)$. This implies that $\mathrm{ad}(X_1 - X_2)$ is nilpotent, and so it follows from Lemma 10 that $\mathrm{ad}(X_1 - X_2) = 0$. But $\mathfrak{g}$ being semisimple, we can now conclude that $X_1 = X_2$, and therefore the exponential mapping is univalent on $V$.

Let $V_G$ denote the image of $V$ in $G$ under the exponential mapping and let $Z$ denote the center of $G$. Obviously $V_G$ is open in $G$.

COROLLARY. *If $z \in Z$ then $V_G$ does not intersect $zV_G$ unless $z = 1$.*

This follows by applying Lemma 11 to $G/Z$ instead of $G$.

The following property of $V$ is important for our applications.

LEMMA 12. *Let $X_r$ $(r \geq 1)$ be a sequence in $V$. Then if $\| \exp X_r \|$ remains bounded, the same holds for $\| X_r \|$.*

For the purpose of this lemma we can obviously replace $G$ by any connected group locally isomorphic to it. Hence we may assume that $G \subset G_c$ (see §2). We shall now use the notation of the proofs of Lemmas 3 and 4. Put $\mathfrak{g}_1 = \bigcup_{u \in U} \mathrm{Ad}(u)(\mathfrak{h} + \mathfrak{n})$. Since $G = UA_cN_c$ and $\mathfrak{h} + \mathfrak{n}$ is invariant under $\mathrm{Ad}(A_cN_c)$, it is obvious that $\mathrm{Ad}(x)\mathfrak{g}_1 = \mathfrak{g}_1$ for $x \in G_c$. Since every regular element of $\mathfrak{g}$ is conjugate to some element in $\mathfrak{h}$ under $G_c$ (see Chevalley [2(b)]), $\mathfrak{g}_1$ contains the set of all regular elements in $\mathfrak{g}$. On the other hand, since $U$ is compact it is clear that $\mathfrak{g}_1$ is closed in $\mathfrak{g}$. Therefore, since regular elements are dense in $\mathfrak{g}$, $\mathfrak{g} = \mathfrak{g}_1$. Hence we can choose $u_r \in U$ and $Y_r \in \mathfrak{h} + \mathfrak{n}$ such that $X_r = \mathrm{Ad}(u_r)Y_r$. Since $\mathrm{Ad}(u_r)$ is unitary, $\| X_r \| = \| Y_r \|$ and $\| \exp X_r \| = \| \exp Y_r \|$. Let $Y_r = H_r + Z_r$ $(H_r \in \mathfrak{h}, Z_r \in \mathfrak{n})$. One proves without difficulty that the eigenvalues of $\mathrm{ad}\, Y_r$ and $\mathrm{ad}\, H_r$ are the same. Therefore since $X_r \in V$, it follows that $|\alpha(H_r)| < \pi$ for every root $\alpha \in P$. This shows that $\| H_r \|$ remains bounded. Now let $F(t)$ denote the entire function on the complex plane given by the series

$$F(t) = \sum_{m \geq 0} (-1)^m t^m / (m+1)! = (1 - e^{-t})/t.$$

Let $H_0$ be a fixed element in $\mathfrak{h}$ and let $t_\alpha$ $(\alpha \in P)$ denote the complex Cartesian coordinates in $\mathfrak{n}$ corresponding to the base $X_\alpha$ $(\alpha \in P)$. Then if $H \in \mathfrak{h}$ and $Z \in \mathfrak{n}$, it is obvious that $\mathrm{Ad}(\exp(H + Z))H_0 - H_0 \in \mathfrak{n}$. For any $\beta \in P$, consider the function

$$f_\beta(H, Z) = t_\beta(\mathrm{Ad}(\exp(H + Z))H_0 - H_0) + e^{\beta(H)}F(\beta(H))\beta(H_0)t_\beta(Z)$$

on $\mathfrak{h} + \mathfrak{n}$ ($H \in \mathfrak{h}, Z \in \mathfrak{n}$). Obviously, it is holomorphic. Moreover, if $\beta < \gamma$ ($\gamma \in P$), it is easily seen that $f_\beta(H, Z + tX_\gamma) = f_\beta(H, Z)$ ($t \in C$). Hence $f_\beta$ depends only on $H$ and $t_\alpha(Z)$ ($0 < \alpha \leq \beta$). Let us now consider $\{(d/dt)f_\beta(H, Z + tX_\beta)\}_{t=0}$. We know (see Chevalley [2(a), p. 157]) that

$$\{(d/dt)\operatorname{Ad}(\exp(H + Z + tX_\beta))H_0\}_{t=0} = \operatorname{Ad}(\exp(H + Z))(\operatorname{ad} Y)H_0,$$

where

$$Y = \{(1 - \exp(-\operatorname{ad}(H + Z)))/\operatorname{ad}(H + Z)\}X_\beta$$
$$= F(\beta(H))X_\beta \quad \mod \sum_{\gamma > \beta} CX_\gamma.$$

Therefore

$$\{(d/dt)f_\beta(H, Z + tX_\beta)\}_{t=0} = F(\beta(H))t_\beta(\operatorname{Ad}(\exp(H + Z))[X_\beta, H_0])$$
$$+ e^{\beta(H)}F(\beta(H))\beta(H_0).$$

But $[X_\beta, H_0] = -\beta(H_0)X_\beta$ and $\operatorname{Ad}(\exp(H + Z))X_\beta \equiv e^{\beta(H)}X_\beta \mod \sum_{\gamma > \beta} CX_\gamma$. Hence, it follows that $\{(d/dt)f_\beta(H + Z + tX_\beta)\}_{t=0} = 0$, and therefore $f_\beta$ depends only on $H$ and $t_\alpha(Z)$ ($0 < \alpha < \beta$).

Now suppose the assertion of the lemma is false. Then there exists a root $\beta \in P$ such that $t_\beta(Z_r)$ does not remain bounded. Select the least such root $\beta$. Then $\|H_r\|$ and $t_\alpha(Z_r)$ ($0 < \alpha < \beta$) all remain bounded and therefore $f_\beta(H_r, Z_r)$ also remains bounded as $r \to \infty$. On the other hand, by our hypothesis, $\|\exp Y_r\|$, and therefore, also, $\|\operatorname{Ad}(\exp(H_r + Z_r))\|$ stay bounded. Hence the same holds for

$$e^{\beta(H_r)}F(\beta(H_r))\beta(H_0)t_\beta(Z_r) = f_\beta(H_r, Z_r) - t_\beta(\operatorname{Ad}(\exp(H_r + Z_r))H_0 - H_0).$$

On the other hand, $|\beta(H_r)| < \pi$, and so $e^{-\beta(H_r)}\{F(\beta(H_r))\}^{-1}$ also remains bounded. Therefore by choosing $H_0$ in such a way that $\beta(H_0) \neq 0$, we conclude that $t_\beta(Z_r)$ also remains bounded. As this contradicts the definition of $\beta$, the lemma follows.

Let us say that a sequence $x_r$ ($r \geq 1$) in $G$ tends to infinity, if for every compact set $\omega$ of $G$, we can select a positive integer $r_0$ such that $x_r \notin \omega$ for $r \geq r_0$. Obviously if $x_r \to \infty$, the same holds for any subsequence of $\{x_r\}$.

COROLLARY. *Let $z_r$ ($r \geq 1$) be a sequence in $Z$ which tends to infinity. Then if $x_r \in z_r V_G$, $x_r$ also tends to infinity.*

For otherwise, by selecting a subsequence, we can arrange that $x_r$ converges to some $x \in G$. But $x_r = z_r \exp X_r$, where $X_r \in V$ and $\|x_r\| = \|\exp X_r\|$ remains bounded. Therefore by Lemma 12, $\|X_r\|$ also remains bounded and so again by selecting a subsequence, we may suppose that $X_r$ converges to some $X \in \mathfrak{g}_0$. Then $z_r = x_r \exp(-X_r) \to x \exp(-X)$, contradicting our hypothesis that $z_r \to \infty$.

## Part II.

**4. Proof of Theorem 2.** Let $A$ be a Cartan subgroup of $G$ (see [4(c)], §2]) and $\mathfrak{h}_0$ the Lie algebra of $A$. We assume that $\theta(\mathfrak{h}_0) = \mathfrak{h}_0$. Let $x \to x^*$ ($x \in G$) denote the natural mapping of $G$ on $G^* = G/A$ and put $h^{x^*} = xhx^{-1}$ ($x \in G, h \in A$). We shall call a subset of $G$ bounded if its closure is compact.

LEMMA 13. *Let $\omega$ be a bounded set in $G$ and $B$ the set of all $h \in A$ such that $h^{x^*} \in \omega$ for some $x^* \in G^*$. Then $B$ is also bounded.*

Let $h_r$ ($r \geq 1$) be a sequence of points in $B$. We have to show that some subsequence of $\{h_r\}$ is convergent. Choose $x_r \in G$ such that $h_r^{x_r^*} \in \omega$, and for any $y \in G$, define $\lambda(y)$ as follows. If $\lambda_1, \cdots, \lambda_s$ are all the distinct eigenvalues of $\mathrm{Ad}(y)$ and $m_i$ is the multiplicity of $\lambda_i$, then

$$\lambda(y) = m_1 |\lambda_1|^2 + \cdots + m_s |\lambda_s|^2.$$

Since $\omega$ is bounded, we can find a positive number $M$ such that $\lambda(y) \leq M$ for all $y \in \omega$. Hence $\lambda(h_r) = \lambda(h_r^{x_r^*}) \leq M$. But it is obvious that $\lambda(h) = \|h\|^2$ for $h \in A$ (in the notation of §2). Therefore $\|h_r\|^2 \leq M$, and so by selecting a subsequence, we can arrange that $\mathrm{Ad}(h_r)$ is convergent. Choose $h_\infty \in A$ such that $\mathrm{Ad}(h_r) \to \mathrm{Ad}(h_\infty)$. Then $\mathrm{Ad}(h_r'') \to 1$ if $h_r'' = h_r h_\infty^{-1}$, and $Z$ being the center of $G$, it is obvious that we can select $z_r \in Z$ such that $h_r' = z_r^{-1}h_r''$ converges to 1 in $G$. Since $h_r = z_r h_r' h_\infty$, it is enough to show that some subsequence of $z_r$ is convergent or, what is equivalent, that $z_r$ cannot tend to infinity. Let $\Xi_\infty$ denote the centralizer of $h_\infty$ in $G$ and $\bar{x}_r$ the coset $x_r \Xi_\infty$ in $G/\Xi_\infty$. Since $\mathrm{Ad}(h_r) \to \mathrm{Ad}(h_\infty)$ and $h_r^{x_r^*} \in \omega$, we can. by applying Theorem 1 to $\mathrm{Ad}(G) \cong G/Z$ at the point $\mathrm{Ad}(h_\infty)$, conclude that $\bar{x}_r$ are all contained in a compact subset of $G/\Xi_\infty$. But $z_r(h_r')^{x_r^*} = (h_r h_\infty^{-1})^{x_r^*}$. Therefore, since $h_r^{x_r^*} \in \omega$, the right side remains within a compact subset of $G$. Moreover $h_r' \to 1$, and so we can assume without loss of generality that $h_r' \in V_G$ (in the notation of §3). Then $(h_r')^{x_r^*}$ is also in $V_G$, and hence it follows from the Corollary of Lemma 12 that $z_r$ cannot tend to infinity. This proves the lemma.

From now on let us agree to use the notation of [4(h), §5]. For any $\alpha \in P$, let $\eta_\alpha$ denote the character of $A$ given by $\mathrm{Ad}(h).X_\alpha = \eta_\alpha(h).X_\alpha$ ($h \in A$). Put

$$\Delta'(h) = \prod_{\alpha \in P} (1 - \eta_\alpha(h^{-1})) \qquad (h \in A).$$

Let $A'$ be the set of those points $h \in A$ where $\Delta'(h) \neq 0$. Then $A'$ is exactly the set of regular elements in $A$ (see [4(c), §2]). Let $dx^*$ denote the invariant measure on $G^* = G/A$.

Lemma 14. Put[1]

$$F_f(h) = \Delta'(h) \int_{G^*} f(h^{x^*}) \, dx^* \qquad (h \in A', f \in C_c^\infty(G)).$$

*Then the above integral is convergent and $F_f$ is a function of class $C^\infty$ on $A'$.*

Fix $f \in C_c^\infty(G)$ and $h_0 \in A'$, and select a compact set $\omega$ in $G$ such that $f$ is zero outside $\omega$. Then if $\Xi$ is the centralizer of $h_0$ in $G$, it follows from the Corollary of Lemma 3 of [4(c)] that $\Xi/A$ is finite. Hence by Theorem 1, we can select an open neighborhood $B$ of $h_0$ in $A'$ and a compact set $\Omega^*$ in $G^*$ with the following property. If $xhx^{-1} \in \omega$ for some $h \in B$ and $x \in G$, then $x^* \in \Omega^*$. Therefore it is clear that

$$\int_{G^*} f(h^{x^*}) \, dx^* = \int_{\Omega^*} f(h^{x^*}) \, dx^* \qquad (h \in B),$$

and from this our lemma follows immediately.

Let $\mathfrak{B}$ be the universal enveloping algebra of $\mathfrak{g}$ and $\mathfrak{H}$ the subalgebra of $\mathfrak{B}$ generated by $(1, \mathfrak{h})$. Then we can regard the elements of $\mathfrak{B}$ and $\mathfrak{H}$ as left-invariant differential operators on $G$ and $A$, respectively (see [4(c), §4]). $S(\mathfrak{h})$ being the symmetric algebra over $\mathfrak{h}$, there exists a unique isomorphism of $S(\mathfrak{h})$ onto $\mathfrak{H}$ which preserves 1 and also every element in $\mathfrak{h}$. For any $q \in S(\mathfrak{h})$, we denote by $\partial(q)$ the image of $q$ under this isomorphism. Then $\partial(q)$ is a differential operator on $A$.

Let $A_1$ be an open subset of $A$. We regard it as an open submanifold of $A$ and consider the space $\mathscr{C}_0(A_1)$ of all complex-valued functions $g$ on $A_1$ of class $C^\infty$ satisfying the following two conditions.

(1) $g$ vanishes (on $A_1$) outside some bounded subset of $A_1$.

(2) For every[1] $v \in \mathfrak{H}$,

$$\tau_v(g) = \sup_{h \in A_1} |g(h; v)| < \infty.$$

Define a topology in $\mathscr{C}_0(A_1)$ by means of the collection of seminorms $\tau_v$ $(v \in \mathfrak{H})$. Then $\mathscr{C}_0(A_1)$ is a locally convex space and the same holds for $C_c^\infty(G)$ under its usual topology (see Schwartz [6, p. 67]). Our main object now is to prove the following theorem.

**THEOREM 2.** *Let*

$$F_f(h) = \Delta'(h) \int_{G^*} f(h^{x^*}) \, dx^* \qquad (h \in A', f \in C_c^\infty(G)).$$

*Then $F_f \in \mathscr{E}_0(A')$, and $f \to F_f$ is a continuous mapping of $C_c^\infty(G)$ into $\mathscr{E}_0(A')$. Moreover, for any bounded open subset $\omega$ of $G$, we can select a bounded subset $B$ of $A'$ such that $F_f$ is zero outside $B$ for every* [1] $f \in C_c^\infty(\omega)$.

We begin by first proving the following weaker result.

**LEMMA 15.** *There exists an open neighborhood $_0A$ of 1 in $A$ with the following property. Put $_0A' = {}_0A \cap A'$ and let $_0F_f$ denote the restriction of $F_f$ on $_0A'$ ($f \in C_c^\infty(G)$). Then $_0F_f \in \mathscr{E}_0(_0A')$, and $f \to {}_0F_f$ is a continuous mapping of $C_c^\infty(G)$ into $\mathscr{E}_0(_0A')$.*

Let $l$ be the rank of $\mathfrak{g}$ and $n$ the complex dimension of $\mathfrak{g}$. Then if $\lambda$ is an indeterminate and $I$ the identity mapping of $\mathfrak{g}$,

$$\det(\lambda I - \operatorname{ad} X) = \lambda^n + \sum_{l \leq r < n} (-1)^r p_r(X) \lambda^r \qquad (X \in \mathfrak{g}),$$

where [8] $p_r \in I(\mathfrak{g})$. Then if $\epsilon$ is a sufficiently small positive number, the inequalities $|p_r(X)| \leq \epsilon$ ($l \leq r < n$) imply that every eigenvalue of $\operatorname{ad} X$ is less that $\pi/2$ in absolute value. Choose a function $\phi(t)$ of class $C^\infty$ of $(n-l)$ real variables $t_r$ ($l \leq r < n$) such that $\phi(t) = 0$ unless $\max_r |t_r| \leq \epsilon$ and $\phi(t) = 1$ if $\max_r |t_r| \leq \epsilon/2$. Let $\Phi(X)$ denote the function on $\mathfrak{g}_0$ obtained from $\phi$ under the substitution $t_r = p_r(X)$ ($X \in \mathfrak{g}_0$). Define $V$ as in §3 and let $V_0$ be the set of those $X \in V$ where $\max_r |p_r(X)| < \epsilon/3$. Then obviously, $V_0$ is an open neighborhood of zero in $V$ and $\operatorname{Ad}(x) V_0 = V_0$, ($x \in G$). For a given $f \in C_c^\infty(G)$, consider the function $\bar{f}$ on $\mathfrak{g}_0$ defined by $\bar{f}(X) = f(\exp X) \Phi(X)$ ($X \in \mathfrak{g}_0$). It is obvious that the carrier of $\bar{f}$ is contained in $V$ and $\bar{f}(X) = f(\exp X)$ for ($X \in V_0$). Now define $x^*H = \operatorname{Ad}(x)H$ ($x \in G, H \in \mathfrak{h}$) and $\pi = \prod_{\alpha \in P} \alpha$. Let $\mathfrak{h}_0'$ be the set of those points $H \in \mathfrak{h}_0$ where $\pi(H) \neq 0$. The function $(1 - e^{-z})/z$ of the complex variable $z$, is holomorphic at $z = 0$ and takes the value 1 there. Therefore, since $\eta_\alpha(\exp H) = e^{\alpha(H)}$ ($H \in \mathfrak{h}_0$), it is clear that we can select an open neighborhood $\mathfrak{h}_2$ of zero in $\mathfrak{h}_0 \cap V_0$ and an analytic function $\zeta$ on $\mathfrak{h}_2$ such that

$$\Delta'(\exp H) = \zeta(H) \pi(H) \qquad (H \in \mathfrak{h}_2).$$

Choose another open neighborhood $\mathfrak{h}_1$ of zero in $\mathfrak{h}_0$ such that the closure of $\mathfrak{h}_1$ is compact and contained in $\mathfrak{h}_2$. Obviously, $\zeta(o) = 1$ and $\zeta$ is never

---

[8] We are using here the notation of [4(g)].

zero on $\mathfrak{h}_2$. Put $\mathfrak{h}_1' = \mathfrak{h}_1 \cap \mathfrak{h}_0'$ and $_0A = \exp \mathfrak{h}_1$. Then $_0A$ is an open neighborhood of 1 in $A$ and $\exp H \in {}_0A' = {}_0A \cap A'$ if $H \in \mathfrak{h}_1'$. For any $g \in C_c^\infty(\mathfrak{g}_0)$, put

$$\phi_g(H) = \pi(H) \int_{G^*} g(x^*H) \, dx^* \qquad\qquad (H \in \mathfrak{h}_1').$$

It follows from Theorem 3 of [4(h)] that the above integral is convergent and[8] $\phi_g \in \mathcal{B}(\mathfrak{h}_1')$. Moreover, it is obvious that $F_f(\exp H) = \zeta(H)\phi_{\bar{f}}(H)$ $(H \in \mathfrak{h}_1')$. But it follows from the definition of $\mathfrak{h}_1$, that[9] $\partial(q)\zeta$ remains bounded on $\mathfrak{h}_1$ for every $q \in S(\mathfrak{h})$. Therefore $\phi \to \zeta\phi$ $(\phi \in \mathcal{B}(\mathfrak{h}_1'))$ is a continuous mapping of $\mathcal{B}(\mathfrak{h}_1')$ into itself. In view of Theorem 3 of [4(h)], this proves that $g \to \zeta\phi_g$ $(g \in C_c^\infty(\mathfrak{g}_0))$ is a continuous mapping of $C_c^\infty(\mathfrak{g}_0)$ into $\mathcal{B}(\mathfrak{h}_1')$. On the other hand, $_0A' = \exp \mathfrak{h}_1'$ and, since the exponential mapping is one-one and regular on $\mathfrak{h}_2$ and the closure of $\mathfrak{h}_1$ is compact, it is clear that a function $\psi$ on $_0A'$ lies in $\mathcal{B}_0({}_0A')$ if and only if the function $\psi'(H) = \psi(\exp H)$ $(H \in \mathfrak{h}_1')$ lies in $\mathcal{B}(\mathfrak{h}_1')$. Moreover, it is obvious that $\psi \to \psi'$ is a topological mapping of $\mathcal{B}_0({}_0A')$ onto $\mathcal{B}(\mathfrak{h}_1')$. Put $F_f'(H) = F_f(\exp H)$ $(f \in C_c^\infty(G), H \in \mathfrak{h}_1')$. Then it would be enough to show that $F_f' \in \mathcal{B}(\mathfrak{h}_1')$ and $f \to F_f'$ is a continuous mapping of $C_c^\infty(G)$ into $\mathcal{B}(\mathfrak{h}_1')$. But we have just seen that $F_f' = \zeta\phi_{\bar{f}}$, and therefore, in view of what has been said above, it is sufficient to prove that $\sigma : f \to \bar{f}$ is a continuous mapping of $C_c^\infty(G)$ into $C_c^\infty(\mathfrak{g}_0)$. Let $\omega$ be a bounded open subset of $G$. Since $C_c^\infty(\mathfrak{g}_0)$ is a locally convex space, it would be enough to prove that $\sigma$ is continuous on $C_c^\infty(\omega)$ (see Schwartz [6, p. 69, Theorem II]). Let $\bar{\omega}$ denote the closure of $\omega$ and $W$ the carrier of $\Phi$. Then $W \subset V$, and it follows from Lemma 12 that $\exp W$ is closed in $G$. Hence $\bar{\omega} \cap \exp W$ is compact. Let $\Omega$ denote the complete inverse image of $\bar{\omega} \cap \exp W$ in $V$ under the exponential mapping. Then by Lemma 11, $\Omega$ is also compact. Select an open neighborhood $\Omega_1$ of $\Omega$ in $\mathfrak{g}_0$ such that its closure $\bar{\Omega}_1$ is compact and contained in $V$. Then for any $p \in S(\mathfrak{g})$, $\partial(p)\Phi$ remains bounded on $\Omega_1$. Since the exponential mapping defines an analytic isomorphism of $V$ with $V_G$, it is now clear that for any $q \in S(\mathfrak{g})$, we can choose a finite number of elements $b_1, \cdots, b_m \in \mathfrak{B}$ such that

$$|\bar{f}(X; \partial(q))| \leq \sum_{1 \leq i \leq m} |f(\exp X; b_i)|$$

for all $f \in C_c^\infty(\omega)$ and $X \in \Omega_1$. But since $\bar{f}$ is zero (on $\mathfrak{g}_0$) outside $\Omega$, this obviously implies that $\sigma$ is continuous on $C_c^\infty(\omega)$. Hence the lemma is proved.

Now fix a point $h_0$ in $A$ and let $\Xi$ and $\mathfrak{z}_0$ be the centralizers of $h_0$ in $G$

---

[9] We are using here the notation of [4(h), § 5].

and $\mathfrak{g}_0$, respectively. Then as we have already seen in §2, $\mathfrak{z}_0$ is reductive in $\mathfrak{g}_0$ and rank $\mathfrak{z}_0 =$ rank $\mathfrak{g}_0$. Let $\Xi_0$ be the analytic subgroup of $G$ corresponding to $\mathfrak{z}_0$. Then we know from Lemma 15 of [4(i)] that $\Xi/\Xi_0 Z$ is finite. Put $\bar{G} = G/\Xi$ and let $x \to \bar{x}$ $(x \in G)$ denote the natural mapping $G$ on $\bar{G}$. Since $\mathfrak{z}_0$ is reductive, there exists (see Weil [7, p. 45]) an invariant measure $d\bar{x}$ on $\bar{G}$. Similarly there exists a measure $d\xi^*$ on $\Xi^* = \Xi/A$ which is invariant under the operations of $\Xi$. It is well known (see Weil [7]) that $d\bar{x}$ and $d\xi^*$ can be so normalized that[1]

$$\int \gamma(x^*)\,dx^* = \int \bar{\gamma}(\bar{x})\,d\bar{x} \qquad\qquad (\gamma \in C_c(G^*)),$$

where

$$\bar{\gamma}(\bar{x}) = \int_{\Xi^*} \gamma(x\xi^*)\,d\xi^* \qquad\qquad (x \in G)$$

and $xy^* = (xy)^*$ $(x, y \in G)$. Select an open, connected and bounded neighborhood $B_0$ of $1$ in $A_0$ such that $B = h_0 B_0$ satisfies the condition of Theorem 1. We assume that $B_0$ is so small that $|\eta_\alpha(h) - 1| \geqq \frac{1}{2}|\eta_\alpha(h_0) - 1|$ for $h \in B$ and $\alpha \in P$. Let $\mathfrak{z}$ be the complexification of $\mathfrak{z}_0$ in $\mathfrak{g}$ and $P_1$ the set of all roots $\alpha \in P$ for which $X_\alpha \in \mathfrak{z}$. Also let $P_2$ the the complement of $P_1$ in $P$. Then if

$$\Delta_i' = \prod_{\alpha \in P_i} (1 - \eta_\alpha^{-1}) \qquad\qquad (i = 1, 2),$$

it is clear that $\Delta_2'$ is never zero on $B$ and $\Delta'(h_0 h) = \Delta_1'(h)\Delta_2'(h_0 h)$ $(h \in B_0)$. Hence $B_0' = h_0^{-1}(B \cap A')$ consists of all those $h \in B_0$ where $\Delta_1'(h) \neq 0$. Let $\Xi_0^*$ denote the image of $\Xi_0$ in $\Xi^*$ and put $\Xi_1 = \Xi_0 A$. Since $\Xi_0$ is normal in $\Xi$, $\Xi_1$ is a group and $\Xi_1 \supset \Xi_0 Z$. Therefore, in view of the finiteness of $\Xi/\Xi_0 Z$, we can select a finite number of elements $y_1 = 1, y_2, \cdots, y_r$ in $\Xi$ such that $\Xi$ is the disjoint union of the cosets $y_i \Xi_1$ $(1 \leqq i \leqq r)$. Then $\Xi^*$ is the disjoint union of the open sets $y_i \Xi_0^*$ $(1 \leqq i \leqq r)$ and $\Xi_0^* \cong \Xi_0/\Xi_0 \cap A$. Since $\mathfrak{z}_0$ is reductive, Lemma 15 is obviously applicable to $\Xi_0$ in place of $G$. Moreover, $B_0 \subset \Xi_0$ since $B_0$ is connected. Put

$$\psi_g(h) = \Delta_1'(h) \int_{\Xi_0^*} g(h^{\xi^*})\,d\xi^* \qquad\qquad (h \in B_0')$$

for $g \in C_c^\infty(\Xi_0)$. Then if $B_0$ is chosen sufficiently small, it follows from Lemma 15 (applied to $\Xi_0$) that $\psi_g \in \mathscr{B}_0(B_0')$ and $g \to \psi_g$ is a continuous mapping of $C_c^\infty(\Xi_0)$ into $\mathscr{B}_0(B_0')$.

Now if $f \in C_c^\infty(G)$, it is clear that

$$\int_{G^*} f(h^{x^*})\,dx^* = \int_{\bar{G}} f(h : \bar{x})\,d\bar{x} \qquad\qquad (h \in B \cap A'),$$

where

$$f(h:\bar{x}) = \int_{\Xi^*} f(xh^{\xi^*}x^{-1})\,d\xi^* = \sum_{1\le i\le r} \int_{\Xi_0^*} f(xy_ih^{\xi^*}y_i^{-1}x^{-1})\,d\xi^* \qquad (x\in G).$$

Let $\omega$ be a bounded open subset of $G$. Then it follows from our definition of $B$ that there exists a compact set $\bar{\Omega}$ in $\bar{G}$ with the following property. If $xhx^{-1}\in\omega$ for some $x\in G$ and $h\in B$, then $\bar{x}\in\bar{\Omega}$. Let $dx$ and $d\xi$ denote the Haar measures on $G$ and $\Xi$, respectively. We normalize them in such a way that

$$\int_G \gamma(x)\,dx = \int_{\bar{G}} \gamma'(\bar{x})\,d\bar{x} \qquad (\gamma\in C_c(G))$$

where $\gamma'(\bar{x}) = \int \gamma(x\xi)\,d\xi$ $(x\in G)$. Select a function $\alpha\in C_c(G)$ such that $\alpha' = 1$ on $\bar{\Omega}$ and put

$$g_f(\xi) = \sum_{1\le i\le r} \int_G \alpha(x)f(xy_ih_0\xi y_i^{-1}x^{-1})\,dx \qquad (\xi\in\Xi_0)$$

for any $f\in C_c^\infty(\omega)$. Then it is clear that

$$\int_{G^*} f((h_0h)^{x^*})\,dx^* = \int_{\bar{\Omega}} f(h_0h:\bar{x})\,d\bar{x} = \int_{\Xi_0^*} g_f(h^{\xi^*})\,d\xi^* \qquad (h\in B_0')$$

for $f\in C_c^\infty(\omega)$. Put

$$\beta'(h) = \Delta_2'(h)\beta(h_0^{-1}h) \qquad (h\in B\cap A')$$

for $\beta\in\mathcal{B}_0(B_0')$. It is obvious that $\Delta_2'$ is bounded away from zero on $B$. Therefore $\beta\to\beta'$ is a continuous mapping of $\mathcal{B}_0(B_0')$ onto $\mathcal{B}(B\cap A')$. On the other hand, if $f\in C_c^\infty(\omega)$,

$$F_f(h_0h) = \Delta'(h_0h)\int_{G^*} f((h_0h)^{x^*})\,dx^*$$

$$= \Delta_2'(h_0h)\psi_{g_f}(h) \qquad (h\in B_0'),$$

and therefore $F_f = \psi_{g_f}'$ on $B\cap A'$. The mapping $f\to g_f$ of $C_c^\infty(\omega)$ into $C_c^\infty(\Xi_0)$ is obviously continuous. Therefore $f\to\psi_{g_f}'$ is a continuous mapping of $C_c^\infty(\omega)$ into $\mathcal{B}_0(B\cap A')$. Thus we have obtained the following improved version of Lemma 15.

LEMMA 16. *Let $h_0$ be a point in $A$. Then there exists an open neighborhood $_0A$ of $h_0$ in $A$ with the following property. Put $_0A' = {_0A}\cap A'$ and let $_0F_f$ denote the restriction of $F_f$ on $_0A'$ $(f\in C_c^\infty(G))$. Then $_0F_f\in\mathcal{B}_0(_0A')$ and $f\to {_0F_f}$ is a continuous mapping of $C_c^\infty(G)$ into $\mathcal{B}_0(_0A')$.*

It is now easy to deduce Theorem 2. Let $\omega$ be a nonempty bounded open set in $G$. Define $B$ as in Lemma 13. Then it follows from Lemma 16 and the boundedness of $B$ that there exist a finite number of nonempty open subsets $A_i$ of $A$ $(1 \leq i \leq r)$ such that $B \subset \bigcup_{1 \leq i \leq r} A_i$ and the following conditions are fulfilled. Let $_iF_f$ denote the restriction of $F_f$ on $A_i$ for $f \in C_c^\infty(G)$. Then $_iF_f \in \mathscr{B}_0(A_i \cap A')$ and the mapping $f \to {}_iF_f$ of $C_c^\infty(G)$ into $\mathscr{B}_0(A_i \cap A')$ is continuous for every $i$. Now if $f \in C_c^\infty(\omega)$, $F_f$ is zero on $A'$ outside $B$. Hence it is clear that

$$\sup_{H \,\epsilon\, A'} |F_f(H;v)| \leq \sum_{1 \leq i \leq r} \sup_{H \,\epsilon\, A'_i} |F_f(H;v)| \qquad (v \in \mathfrak{H}),$$

where $A_i' = A_i \cap A'$. Therefore it follows that $F_f \in \mathscr{B}_0(A')$ and the mapping $f \to F_f$ of $C_c^\infty(\omega)$ into $\mathscr{B}_0(A')$ is continuous. All the statements of Theorem 2 are now obvious.

The following lemma is an immediate consequence of Theorem 2. Let $dh$ denote the Haar measure on $A$.

LEMMA 17. *Let $g$ be a measurable function on $A$ which is bounded on every compact set. Then the mapping*

$$f \to \int_A F_f(h) g(h) \, dh \qquad (f \in C_c^\infty(G))$$

*is a distribution on $G$.*

This holds in particular if $g$ is a character of $A$.

**5. The main theorems.** Let $\mathfrak{Z}$ denote the center of $\mathfrak{B}$. Then from Lemma 18 of $[4(c)]$ there exists an isomorphism $\gamma'$ of $\mathfrak{Z}$ into $S(\mathfrak{h})$ such that

$$z - \partial(\gamma'(z)) \in \sum_{\alpha \,\epsilon\, P} \mathfrak{B}X_\alpha \qquad (z \in \mathfrak{Z}).$$

THEOREM 3. *Define $F_f$ as in Theorem 2. Then*

$$F_{zf} = \partial(\gamma'(z)) F_f$$

*for $f \in C_c^\infty(G)$ and $z \in \mathfrak{Z}$.*

Fix a point $h_0 \in A'$ and let $\omega$ be a bounded open subset of $G$. Then as we saw during the proof of Lemma 14, there exists an open connected neighborhood $B$ of $h_0$ in $A'$ and a compact set $\Omega_0^*$ in $G^*$ with the following property. If $xhx^{-1} \in \omega$ for some $h \in B$ and $x \in G$, then $x^* \in \Omega_0^*$. Therefore

$$F_g(h) = \Delta'(h) \int_{\Omega_0^*} g(h^{x^*}) \, dx^* \qquad (h \in B, g \in C_c^\infty(\omega)).$$

Let $\Omega^*$ be a compact neighborhood of $\Omega_0^*$ in $G^*$. Normalize the Haar measures $dx$ and $dh$ on $G$ and $A$, respectively, in such a way that

$$\int_G \beta(x)\,dx = \int_{G^*} dx^* \int_A \beta(xh)\,dh \qquad (\beta \in C_c(G)),$$

and select $\alpha \in C_c^\infty(G)$ such that $\int \alpha(xh)\,dh = 1$ if $x^* \in \Omega^*$ $(x \in G)$. Then if $f \in C_c^\infty(\omega)$ and $z \in \mathfrak{Z}$, it is obvious that

$$F_{zf}(h) = \Delta'(h) \int_G \alpha(x)f(xhx^{-1};z)\,dx \qquad (h \in B).$$

Now we use the notation of $[4(c), \S 7]$ and put $f(x:h) = f(xhx^{-1})$ $(x \in G, h \in B)$. Then it follows from Lemma 23 of $[4(c)]$ that

$$z = \Gamma_h(1 \,\mathbf{X}\, \beta_h(z)) + \sum_{1 \leq i \leq N} a_i(h)\Gamma_h(b_i \,\mathbf{X}\, u_i) \qquad (h \in B),$$

where $b_i \in \mathfrak{B}\mathfrak{g}, u_i \in \mathfrak{H}$ and $a_i$ are analytic functions on $B$. Since $\mathfrak{B}\mathfrak{g}$ is the direct sum of $\mathfrak{S}'$ and $\mathfrak{B}\mathfrak{h}$ and $\Gamma_h(\mathfrak{B}\mathfrak{h} \,\mathbf{X}\, \mathfrak{H}) = \{0\}$, we may assume that $b_i \in \mathfrak{S}'$ and $u_i$ $(1 \leq i \leq N)$ are linearly independent. Then since $z^h = z$, it follows from Lemma 22 of $[4(c)]$ that $b_i{}^h = b_i$ $(h \in A)$. Therefore

$$f(xhx^{-1};z) = f(x: h;\beta_h(z)) + \sum_{1 \leq i \leq N} a_i(h)f(x;b_i: h;u_i) \qquad (h \in B, x \in G)$$

and

$$F_{zf}(h) = \Delta'(h) \int_G \alpha(x)f(x: h;\beta_h(z))\,dx$$

$$+ \sum_{1 \leq i \leq N} a_i(h)\Delta'(h) \int_G \alpha(x;b_i{}^*)f(x: h;u_i)\,dx$$

for $h \in B$. (Here $b_i{}^*$ is the adjoint of $b_i$). We now claim that

$$\int \alpha(x;b_i{}^*)f(x: h;u_i)\,dx = 0 \qquad (1 \leq i \leq N, h \in B).$$

It is clear that for a fixed $h, f(x: h;u_i)$ depends only on $x^*$. Hence it would be enough to prove that

$$\int_A \alpha(xh;b_i{}^*)\,dh = 0$$

if $x^* \in \Omega_0^*$. Choose a point $x_0 \in G$ such that $x_0^* \in \Omega_0^*$ and let $U$ be an open neighborhood of $1$ in $G$ such that $(x_0U)^* \subset \Omega^*$. Then it follows from the definition of $\alpha$ that

$$\int_A \alpha(x_0yh)\,dh = 1 \qquad (y \in U).$$

Put $\alpha(x:h) = \alpha(xh)$ $(x \in G, h \in A)$. It is obvious that $(b_i^*)^h = b_i^*$ $(h \in A)$ and therefore $\alpha(x; b_i^*: h) = \alpha(xh; b_i^*)$. Therefore

$$\int_A \alpha(x_0 h; b_i^*)\, dh = \int_A \alpha(x_0; b_i^*: h)\, dh = 0$$

since $b_i^* \in \mathfrak{B}\mathfrak{g}$ and $\int \alpha(x_0 y: h)\, dh = 1$ for $y \in U$. This proves that

$$F_{zf}(h) = \Delta'(h) \int_G \alpha(x) f(x: h; \beta_h(z))\, dx \qquad (h \in B).$$

On the other hand, Theorem 2 of $[4(c)]$ gives us a formula for $\beta_h(z)$. Define $d$ as in this theorem. It follows easily from Lemma 7 of $[4(c)]$ that $\operatorname{conj} \eta_\alpha(h^{-1}) = \eta_{\theta\alpha}(h)$ for $\alpha \in P$ and $h \in A$. Now if [9] $\alpha \in P_+$, the same holds for $-\theta\alpha$. Hence if $P' = P_0 \cup P_-$ (see $[4(h), \S 5]$ for notation), it is clear that

$$\operatorname{conj} \Delta'(h) = \prod_{\alpha \in P} (1 - \eta_{\theta\alpha}(h)) = (-1)^{r'} \prod_{\alpha \in P'} \eta_\alpha(h) \Delta'(h) \qquad (h \in A),$$

where $r'$ is the number of roots in $P'$. Therefore

$$|d(h)| = \left| \prod_{\alpha \in P} (\eta_\alpha(h^{-1}) - 1)(\eta_\alpha(h) - 1) \right|$$
$$= \left| \prod_{\alpha \in P_+} \eta_\alpha(h) \right| |\Delta'(h)|^2 = (-1)^{r'} \left| \prod_{\alpha \in P_+} \eta_\alpha(h) \right| \prod_{\alpha \in P'} \eta_\alpha(h) \Delta'(h)^2.$$

But

$$\operatorname{conj} \prod_{\alpha \in P_+} \eta_\alpha(h) = \prod_{\alpha \in P_+} \eta_{\theta\alpha}(h^{-1}) = \prod_{\alpha \in P_+} \eta_\alpha(h),$$

and so

$$|d(h)| = \pm \prod_{\alpha \in P} \eta_\alpha(h) \Delta'(h)^2.$$

Put $\rho = \frac{1}{2} \sum_{\alpha \in P} \alpha$ and select an open connected neighborhood $\mathfrak{h}_1$ of zero in $\mathfrak{h}_0$ such that $B_1 = h_0 \exp \mathfrak{h}_1 \subset B$. Then it is clear that

$$|d(h_0 \exp H)|^{\frac{1}{2}} = \epsilon e^{\rho(H)} \Delta'(h_0 \exp H) \qquad (H \in \mathfrak{h}_1),$$

where $\epsilon$ is a constant such that $\epsilon^4 = 1$. Therefore, it follows from Theorem 2 of $[4(c)]$ that $\beta(z) = (\Delta')^{-1} \partial(\gamma'(z)) \circ \Delta'$ on $B_1$ and so

$$F_{zf}(h) = \int \alpha(x) f(x: h; \partial(\gamma'(z)) \circ \Delta')\, dx$$
$$= F_f(h; \partial(\gamma'(z))) \qquad (h \in B_1).$$

This proves that $F_{zf}$ and $\partial(\gamma'(z)) F_f$ coincide at $h_0$, and hence the theorem is established.

**Put**

$$\Delta_+{}'(h) = \prod_{a \in P_+} (1 - \eta_a(h^{-1})), \qquad \Delta_0{}'(h) = \prod_{a \in P_0} (1 - \eta_a(h^{-1})) \qquad (h \in A).$$

Then $\Delta_+{}'(h)$ is real since

$$\operatorname{conj} \Delta_+{}'(h) = \prod_{a \in P_+} (1 - \eta_{-a}(h)) = \prod_{a \in P_+} (1 - \eta_a(h^{-1})) = \Delta_+{}'(h)$$

LEMMA 18. *Let $\epsilon_+(h)$ $(h \in A')$ denote the sign of $\Delta_+{}'(h)$ and $A''$ the set of all points $h \in A$ where $\Delta_0{}'(h) \neq 0$. Then for any $f \in C_c^\infty(G)$, $\epsilon_+ F_f$ can be extended to a function of class $C^\infty$ on $A''$.*

We use the notation of the proof of Lemma 15. Then

$$F_f(\exp H) = \zeta(H)\phi_f(H) \qquad\qquad (H \in \mathfrak{h}_1').$$

Put $\pi_+ = \prod_{a \in P_+} a \in S(\mathfrak{h})$. It is obvious that if $\mathfrak{h}_1$ is sufficiently small, $\Delta_+{}'(\exp H)/\pi_+(H)$ can be extended to an analytic function on $\mathfrak{h}_1$ which is nowhere zero and which takes the value 1 at $H = 0$. Therefore, since $\pi_+(H)$ is also real for $H \in \mathfrak{h}_0$, it has the same sign as $\Delta_+{}'(\exp H)$ for $H \in \mathfrak{h}_1'$. Hence it follows from Theorem 3 of [4(h)] that $\epsilon_+ F_f$ can be extended to a function of class $C^\infty$ on $B \cap A''$, where $B = \exp \mathfrak{h}_1$.

Now $h_0$ being a fixed point in $A''$, we use the notation of the proof of Lemma 16. Then $F_f = \psi_{g_f}{}'$ on $B \cap A'$. Since $\Delta_2'$ is never zero on $B$, it is clear that $\epsilon_+$ differs on $B$ from the sign of

$$\prod_{a \in P_1 \cap P_+} (1 - \eta_a(h^{-1}))$$

only by a constant factor. Therefore, by applying the above result to $\Xi_0$, we can conclude that there exists an open neighborhood $B_1$ of $h_0$ in $B$ such that $\epsilon_+ \psi_{g_f}{}'$ can be extended to a function of class $C^\infty$ on $B_1 \cap A''$. Since $\epsilon_+ F_f = \epsilon_+ \psi_{g_f}{}'$ on $B_1 \cap A'$ and $A'$ is dense in $A''$, the assertion of the lemma is now obvious.

Define $\mathfrak{h}_1$ and $\zeta$ as in the proof of Lemma 15 and put[8] $\pi' = \prod_{a \in P} \{\alpha + \langle \alpha, \rho \rangle\}$ and consider the differential operator $\partial(\pi') \circ \zeta$ on $\mathfrak{h}_1$. As usual we denote its local expression at zero by $(\partial(\pi') \circ \zeta)_0$.

LEMMA 19. $(\partial(\pi') \circ \zeta)_0 = \partial(\pi)$.

Since

$$\Delta'(\exp H) = e^{-\rho(H)} \prod_{a \in P} (e^{a(H)/2} - e^{-a(H)/2}) \qquad\qquad (H \in \mathfrak{h}_1)$$

and $e^{\rho}\partial(\pi') \circ e^{-\rho} = \partial(\pi)$, it would be enough to prove that $(\partial(\pi) \circ \zeta_1)_0 = \partial(\pi)$, where

$$\zeta_1(H) = \prod_{\alpha \in P} g(\alpha(H)) \qquad\qquad (H \in \mathfrak{h}_1)$$

and $g$ denotes the entire function $(e^{t/2} - e^{-t/2})/t$ of the complex variable $t$. Then

$$\zeta_1 = 1 + \sum_{k \geq 1} p_k,$$

where $p_k$ is a homogeneous polynomial in $S(\mathfrak{h})$ of degree $k$ and the above series converges to $\zeta_1$ in some neighborhood $\mathfrak{h}_2$ of zero in $\mathfrak{h}_1$. Let $W$ be the Weyl group of $\mathfrak{g}$ with respect to $\mathfrak{h}$. Then $\zeta_1$ can be regarded as a holomorphic function on some complex neighborhood $U$ of zero in $\mathfrak{h}$, and we may assume that $sU = U$ $(s \in W)$ and the above series converges to $\zeta_1$ on $U$. Then since $\zeta_1(sH) = \zeta_1(H)$ $(s \in W)$, it follows that $p_k{}^s = p_k$ $(s \in W, k \geq 1)$. Moreover, if $f$ is any homogeneous element in $S(\mathfrak{h})$, it is obvious that

$$f(0; \partial(\pi) \circ \zeta_1) = f(0; \partial(\pi)) + \sum_{1 \leq k \leq r} f(0; \partial(\pi) \circ p_k),$$

where $r$ is the degree of $\pi$. But it follows from Lemma 18 of $[4(g)]$ that $f(0; \partial(\pi) \circ p_k) = 0$, and so $f(0; \partial(\pi) \circ \zeta_1) = 'f(0; \partial(\pi))$. Hence, if $(\partial(\pi) \circ \zeta_1)_0 - \partial(\pi) = \partial(q)$ $(q \in S(\mathfrak{h}))$, then [8] $\langle p, q \rangle = 0$ for all $p \in S(\mathfrak{h})$, and therefore (see $[4(g), \S 2]$) $q = 0$. This proves the assertion of the lemma.

We say that the Cartan subgroup $A$ is fundamental if $\mathfrak{h}_0$ is fundamental in $\mathfrak{g}_0$ (see $[4(h), \S 8]$).

THEOREM 4. *We can select a neighborhood $B$ of $1$ in $A$ with the following property. Suppose $A_1$ is a connected component of $A' \cap B$ whose closure contains $1$. Then there exists a real number $c$ such that*

$$\mathrm{Lim}_{h \to 1} F_f(zh; \partial(\pi')) = cf(z) \qquad\qquad (h \in A_1)$$

*for all $f \in C_c^\infty(G)$ and $z \in Z$. Moreover, $c = 0$ if $A$ is not fundamental.*

If $z = 1$, our statement is an immediate consequence of Lemma 19 and the fact that

$$F_f(\exp H) = \zeta(H) \phi_f(H) \qquad\qquad (H \in \mathfrak{h}_1'),$$

if we take into account Theorem 2 of $[4(i)]$. For any arbitrary $z \in Z$, write $f_z(x) = f(zx) \cdot (x \in G)$. Then

$$\mathrm{Lim}_{h \to 1} F_f(zh; \partial(\pi')) = \mathrm{Lim}_{h \to 1} F_{f_z}(h; \partial(\pi')) = cf_z(1) = cf(z) \qquad (h \in A_1).$$

THEOREM 5. *Suppose $A$ is fundamental and $A_1, A_2, \cdots, A_r$ are all*

*the distinct connected components of $A' \cap B$ whose closures contain 1.
Let $c_i$ be the real number of Theorem 4 corresponding to $A_i$.    Then
$c_1 + c_2 + \cdots + c_r \neq 0$.*

This follows immediately from Theorem 3 of [4(i)].

*Note added on August 3, 1957.*  Recently I have been able to prove
that $c_1 = c_2 = \cdots = c_r$ in Theorem 5.    Since the normalization of the
measure $dx^*$ in Theorem 2 is arbitrary, only the sign of $c_1$ is of interest.
Assuming that $\mathfrak{h}_0$ is fundamental, let $p_+$ and $p_0$ denote the number of roots
in $P_+$ and $P_0$, respectively.    Then $p_+$ is even and, if $q = p_0 + \frac{1}{2}p_+$, the sign
of $c_1$ is $(-1)^q$.

Columbia University,
New York, N. Y.

---

## REFERENCES.


[1] G. Birkhoff, *Ann. of Math.*, vol. 38 (1937), pp. 526-532.
[2] C. Chevalley, (a) *Theory of Lie Groups*, Princeton University Press, 1946.
         (b) *Amer. Jour. Math.*, vol. 63 (1941), pp. 785-793.
[3] I. M. Gelfand and M. I. Graev, *Doklady Akad. Nauk. SSSR N. S.*, vol. 92 (1953),
         pp. 461-464.
[4] Harish-Chandra, (a) *Trans. Amer. Math. Soc.*, vol. 76 (1954), pp. 234-253.
         (b) ——, vol. 76 (1954), pp. 485-528.
         (c) ——, vol. 83 (1956), pp. 98-163.
         (d) *Amer. Jour. Math.*, vol. 77 (1955), pp. 743-777.
         (e) ——, vol. 78 (1956), pp. 1-41.
         (f) ——, vol. 78 (1956), pp. 564-628.
         (g) ——, vol. 79 (1957), pp. 87-120.
         (h) ——, vol. 79 (1957), pp. 193-257.
         (i) ——, vol. 79 (1957), pp. 653-686.
         (j) *Proc. Nat. Acad. Sci. U.S.A.*, vol. 42 (1956), pp. 538-540.
[5] K. Iwasawa, *Ann. of Math.*, vol. 50 (1949), pp. 507-557.
[6] L. Schwartz, *Theorie des distributions I*, (1950), Paris, Hermann.
[7] A. Weil, *L'integration dans les groupes topologiques et ses applications*, Paris,
    Hermann.



Reprinted from
*Amer. J. of Math.*
**79** (1957), 733-760

# Spherical Functions on a Semisimple Lie Group

HARISH-CHANDRA

DEPARTMENT OF MATHEMATICS, COLUMBIA UNIVERSITY
*Communicated by Paul A. Smith, March 29, 1957*

Let $R$ be the field of real numbers and $G$ a connected semisimple Lie group with the Lie algebra $\mathfrak{g}_0$ over $R$. We assume that the center of $G$ is finite. Let $K$ be a maximal compact subgroup of $G$. By a spherical function $f$ we mean a complex-valued function on $G$ such that $f(k_1 x k_2) = f(x)$ $(k_1, k_2 \in K;\ x \in G)$. Let $\mathfrak{k}_0$ be the Lie algebra of $K$. Define $\mathfrak{p}_0$, $\mathfrak{h}_{\mathfrak{p}_0}$, and $\mathfrak{n}_0$ as in an earlier paper,[1] and let $A$ and $N$ be the analytic subgroups of $G$ corresponding to $\mathfrak{h}_{\mathfrak{p}_0}$ and $\mathfrak{n}_0$, respectively. Then, for any $x \in G$, we denote by $H(x)$ the unique element in $\mathfrak{h}_{\mathfrak{p}_0}$ such that $x = k(\exp H(x))n$ for some $k \in K$ and $n \in N$. Introduce a linear function $\rho$ and a polynomial function $p$ on $\mathfrak{h}_{\mathfrak{p}_0}$ by means of the equations $e^{2\rho(H)} = \det(\mathrm{Ad}\ (\exp H))_{\mathfrak{n}_0}$ and $p(H) = \det(\mathrm{ad}\ H)_{\mathfrak{n}_0}(H \in \mathfrak{h}_{\mathfrak{p}_0})$, where the subscript indicates the restriction on $\mathfrak{n}_0$. Then $p$ is a product of real linear factors. Put $\pi = \alpha_1 \alpha_2 \cdots \alpha_r$, where $\alpha_1, \ldots, \alpha_r$ are all the distinct prime factors of $p$. Let $\mathfrak{F}$ denote the space of linear functions on $\mathfrak{h}_{\mathfrak{p}_0}$. Sometimes it would be convenient to identify $\mathfrak{F}$ with $\mathfrak{h}_{\mathfrak{p}_0}$ by means of the fundamental bilinear form on $\mathfrak{g}_0$. Then $\pi$ becomes a polynomial function also on $\mathfrak{F}$. Let $M$ be the centralizer and $M'$ the normalizer of $\mathfrak{h}_{\mathfrak{p}_0}$ in $G$. Then $W = M'/M$ is a finite group whose elements operate as linear transformations on $\mathfrak{h}_{\mathfrak{p}_0}$. Hence $W$ operates also on the ring of polynomial functions on $\mathfrak{h}_{\mathfrak{p}_0}$. It can be shown that $\pi^2$ is invariant under $W$, and therefore $\pi^s = \varepsilon(s)\pi$ $(s \in W)$, where $\varepsilon(s) = \pm 1$. Let $\mathfrak{h}_{\mathfrak{p}_0}'$ denote the set of those points $H \in \mathfrak{h}_{\mathfrak{p}_0}$, where $\pi(H) \neq 0$. There exists a unique connected component $\mathfrak{h}_{\mathfrak{p}_0}^+$ of $\mathfrak{h}_{\mathfrak{p}_0}'$ such that $\rho(H) \geqslant \rho(sH)$ for $s \in W$ and $H \in \mathfrak{h}_{\mathfrak{p}_0}^+$. Let $A_+$ denote the closure of $\exp(\mathfrak{h}_{\mathfrak{p}_0}^+)$. Then $G = K A_+ K$, and therefore a spherical function is completely determined by its restriction on $A_+$. We shall say that $H \to \infty$ $(H \in \mathfrak{h}_{\mathfrak{p}_0}^+)$ if $|\alpha_j(H)| \to \infty$ $(1 \leqq j \leqq r)$.

Put $\phi_\lambda(x) = \int_K \exp\{i\lambda(H(xk)) - \rho(H(xk))\}\, dk$ $(\lambda \in \mathfrak{F},\ x \in G)$, where $dk$ is the normalized Haar measure on $K$. Then $\phi_\lambda$ is spherical, and it is also an elementary function of the positive-definite type.[2]

**Theorem 1.** *There exists a unique analytic function $\beta$ on the real Euclidean space $\mathfrak{F}$ such that*

$$\lim_{H \to \infty} \left| \pi(\lambda) e^{\rho(H)} \phi_\lambda(\exp H) - \sum_{s \in W} \epsilon(s) \beta(s\lambda) \exp\left(i\lambda(s^{-1}H)\right) \right| = 0 \quad \left( H \in \mathfrak{h}_{\mathfrak{p}_0}{}^+ \right)$$

*for any $\lambda \in \mathfrak{F}$. Moreover, $|\beta(s\lambda)| = |\beta(\lambda)|$ for $s \in W$ and $\lambda \in \mathfrak{F}$.*

Extend the automorphism[1] $\theta$ of $\mathfrak{g}_0$ to $G$, and put $n' = \theta(n^{-1})$ $(n \in N)$.

**Theorem 2.** *It is possible to normalize the Haar measure $dn$ on $N$ in such a way that[3]*

$$\beta(\lambda) = \lim_{\epsilon \to 0} \pi(\lambda_\epsilon) \int_N \exp\{-i\lambda_\epsilon(H(n')) - \rho(H(n'))\} \, dn \quad (\epsilon > 0, \lambda \in \mathfrak{F}),$$

*where $\lambda_\epsilon = \lambda - i\epsilon\rho$. This normalization is characterized by the condition that $\int_N e^{-2\rho(H(n'))} \, dn = 1$.*

Define the space $\mathcal{C}(\mathfrak{F})$ as in a previous note,[4] and for any $a \in \mathcal{C}(\mathfrak{F})$, put $\phi_a(x) = \int_{\mathfrak{F}} \pi(\lambda) a(\lambda) \phi_\lambda(x) \, d\lambda$, where $d\lambda$ stands for the (suitably normalized) Euclidean measure on $\mathfrak{F}$. Then it can be shown that, for any $\mu \in \mathfrak{F}$, $\int_G |\phi_\mu(x) \phi_a(x)| dx < \infty$, where $dx$ denotes the Haar measure of $G$. Moreover, $\beta a \in \mathcal{C}(\mathfrak{F})$ for $a$ in $\mathcal{C}(\mathfrak{F})$.

**Theorem 3.** *$d\lambda$ can be so normalized that[5]*

$$\pi(\mu) \int_G \left(\operatorname{conj} \phi_\mu(x)\right) \phi_a(x) \, dx = |\beta(\mu)|^2 \sum_{s \in W} \epsilon(s) a(s\mu)$$

*for $\mu \in \mathfrak{F}$ and $a \in \mathcal{C}(\mathfrak{F})$.*

**Corollary 1.** *$\int_G |\phi_a(x)|^2 \, dx = w^{-1} \int_{\mathfrak{F}} |\beta(\lambda)|^2 |\Sigma_{s \in W} \epsilon(s) a(s\lambda)|^2 \, d\lambda \ (a \in \mathcal{C}(\mathfrak{F}))$, where $w$ is the order of $W$.*

It is possible to show that $\int_N |\phi_a(hn)| \, dn < \infty$ for $h \in A$ and $a \in \mathcal{C}(\mathfrak{F})$.

**Corollary 2.** *$e^{\rho(H)} \int_N \phi_a((\exp H)n) \, dn = \int_{\mathfrak{F}} |\beta(\lambda)|^2 \{\pi(\lambda)^{-1} \Sigma_{s \in W} \epsilon(s) a(s\lambda)\} e^{i\lambda(H)} d\lambda$ for $\lambda \in \mathfrak{F}$, $a \in \mathcal{C}(\mathfrak{F})$, and $H \in \mathfrak{h}_{\mathfrak{p}_0}$.*

In certain special cases it is possible to compute $\beta$ explicitly. For example, $\beta$ is a constant if $G$ is complex.[6]

We shall now indicate very briefly the central idea of our method of proof. Since $\phi_\lambda$ is spherical, it can be regarded as a function on the factor space $G/K$. Let $\mathfrak{Q}$ be the algebra of all differential operators on $G/K$ which are invariant under operations of $G$. It is known that $\phi_\lambda$ is an eigenfunction of every $D$ in $\mathfrak{Q}$. Let $\chi_\lambda(D)$ denote the corresponding eigenvalue. Then the above results can be obtained by a detailed study of the system of differential equations $D\phi_\lambda = \chi_\lambda(D)\phi_\lambda$ $(D \in \mathfrak{Q})$.

Let $L_2(G)$ denote the Hilbert space of all square-integrable functions on $G$, and $I_2(G)$ the closure of the subspace consisting of those continuous spherical functions which vanish outside a compact set. Then, if it could be shown that the functions $\phi_a$ ($a \in \mathcal{C}(\mathfrak{F})$) are dense in $I_2(G)$, the Plancherel formula[7] for functions on $G/K$ would follow in an explicit form from Corollary 1 of Theorem 3.

[1] *Trans. Am. Math. Soc.*, **75**, 187–188, 1953.
[2] See *Trans. Am. Math. Soc.*, **76**, 64, 1965, Theorem 5.
[3] Here we extend $\pi$ to a polynomial function on the complexification of $\mathfrak{F}$ in the obvious way.
[4] These PROCEEDINGS, **42**, 252–253, 1956.
[5] "conj $c$" denotes the conjugate of a complex number $c$.
[6] See *Trans. Am. Math. Soc.*, **76**, 253, 1954, Theorem 7.
[7] See these PROCEEDINGS, **40**, 203–204, 1954.

# SPHERICAL FUNCTIONS ON A SEMISIMPLE LIE GROUP, I.*

By Harish-Chandra.[1]

---

**1. Introduction.** Let $G$ be a connected semisimple Lie group with finite center and $K$ a maximal compact subgroup of $G$. A complex-valued function $f$ on $G$ is called a spherical function if $f(k_1 x k_2) = f(x)$ for $k_1, k_2 \in K$ and $x \in G$. Let $dk$ denote the normalized Haar measure of $K$. A spherical function $f \neq 0$ is said to be elementary (see Godement [4, p. 497]) if it is continuous and if

$$\int_K f(xky) \, dk = f(x) f(y)$$

for $x, y$ in $G$. It can be shown that every such function is analytic and, in fact, it is possible to give an alternative characterization of elementary spherical functions as follows. Let $\mathfrak{J}$ be the algebra of all differential operators on $G$ which are invariant under left translations by elements of $G$ and right translations by elements of $K$. Then a spherical function $f$ of class $C^\infty$ is elementary if and only if $f(1) = 1$ and $f$ is an eigenfunction of every differential operator in $\mathfrak{J}$. Finally, there exists a simple integral formula (see [5(d), Theorem 5]) for any such function.

It follows from the Plancherel formula for the factor space $G/K$ (see [5(m), p. 204]) that an "arbitrary" spherical function can be "expanded" in terms of the elementary ones. However the fundamental measure appearing in this formula had, so far, not been satisfactorily related to the group structure of $G$. If we agree to ignore certain technical complications, the situation may roughly be described as follows. A certain class of elementary spherical functions (of positive-definite type) can be parameterized by a space $E/W$. Here $E$ is a finite-dimensional real Euclidean space, $W$ is a finite group of linear transformations in $E$ and $E/W$ denotes the quotient space obtained by identifying points in $E$ which are congruent under $W$. Let $\phi_\lambda$ denote the elementary spherical function which corresponds to a point $\lambda \in E$ so that $\phi_{s\lambda} = \phi_\lambda$ ($s \in W$). For any continuous spherical function $f$ with compact support, put

$$\tilde{f}(\lambda) = \int f(x) \phi_\lambda(x^{-1}) \, dx$$

---

* Received July 16, 1957.
[1] John Simon Guggenheim Fellow.

241

where $dx$ is the Haar measure of $G$. Then the Plancherel formula asserts[2] the existence of a unique positive measure $d\mu$ on $E$ (which is invariant under $W$) such that

$$\int |f(x)|^2\, dx = \int |\tilde{f}(\lambda)|^2\, d\mu$$

for all such $f$. The problem is to determine this measure $d\mu$. Let $d\lambda$ denote the Euclidean measure on $E$. In this paper we shall give an asymptotic expansion for $\phi_\lambda$ on $G$. The leading terms of this expansion involve a certain coefficient $c(\lambda)$, which, considered as a function of $\lambda$, is analytic on $E$ except on certain hyperplanes (see Lemmas 37 and 52). In any case, the reciprocal $c^{-1}$ is analytic on $E$ and it will be shown in another paper that[3] $d\mu = |c(\lambda)|^{-2}\, d\lambda$ (if $dx$ and $d\lambda$ are suitably normalized). On the other hand the Fourier transform[2] of $c$ is a distribution on $E$ which is given by a simple formula in which the group structure of $G$ enters in a very direct manner (see Theorem 5).

The above outline shows that our problem can be divided into three more or less distinct parts: (1) the asymptotic formula for $\phi_\lambda$, (2) the investigation of the function $c$ and (3) the proof of the relation $d\mu = |c(\lambda)|^{-2}\, d\lambda$. Only the first two of these questions will be taken up in this paper. For (1) we consider the system of differential equations $D\phi_\lambda = \chi_\lambda(D)\phi_\lambda$ $(D \in \mathfrak{F})$ where $\chi_\lambda(D)$ is the eigenvalue corresponding to the operator $D$. Actually the equation $\omega\phi_\lambda = \chi_\lambda(\omega)\phi_\lambda$, corresponding to the Casimir operator $\omega$, plays a predominant role in this discussion. In fact, this single equation, together with some general properties of $\phi_\lambda$, permits us to derive the asymptotic formula for $\phi_\lambda$.

Now in order to obtain more information about $c$, we have to investigate $\phi_\lambda$ as a function of $\lambda$, in the neighborhood of infinity on $G$. It turns out that one can select a polynomial function $\pi$ on $E$ such that $b = \pi c$ is everywhere analytic on $E$. Moreover every derivative of $b$ is majorized on $E$ by a suitable polynomial function. We shall see in another paper that $|b(s\lambda)| = |b(\lambda)|$ for $s \in W$.

The contents of this paper are as follows. In Section 2 we collect some elementary facts which are obtained by considering the finite-dimensional representations of $G$. Section 3 is devoted to deriving the consequences of two unpublished lemmas of Chevalley. These results, which are of an

---

[2] This is a simplified version of the true picture and therefore is not entirely accurate.

[3] This is reminiscent of a result of Weyl [8(a), p. 266] on ordinary differential equations.

algebraic nature, will be used constantly during this and the next paper of this series. In Section 4 we define a homomorphism $\gamma$ of $\mathfrak{J}$ onto the algebra $J$ consisting of those polynomial functions on $E$ which are invariant under $W$. Let $l$ be the rank of symmetric Riemannian space $G/K$. Then it is possible to select a connected abelian Lie subgroup $A_{\mathfrak{p}}$ of $G$ of dimension $l$ such that $G = KA_{\mathfrak{p}}K$. Obviously a spherical function $f$ is completely determined by its restriction $\bar{f}$ on $A_{\mathfrak{p}}$. In Sections 5 and 6 we investigate the relationship[4] between $\overline{Df}$ and $\bar{f}$ for any $D \in \mathfrak{J}$. It turns out that one can define a differential operator $\delta'(D)$ on an open dense subset $A_{\mathfrak{p}}'$ of $A_{\mathfrak{p}}$ such that $\overline{Df} = \delta'(D)\bar{f}$ on $A_{\mathfrak{p}}$. Moreover there is an intimate connection between $\delta'(D)$ and $\gamma(D)$ and this permits us to prove (see the corollary of Theorem 2) that if $w$ is the order of the group $W$, there cannot exist more than $w$ linearly independent analytic functions on any open connected subset of $A_{\mathfrak{p}}'$, which are all eigenfunctions of $\delta'(D)$ with the same eigenvalues, for every $D \in \mathfrak{J}$. Therefore if we could somehow find $w$ such functions, corresponding to the eigenvalues $\chi_\lambda(D)$, on a connected component $A_{\mathfrak{p}}^+$ of $A_{\mathfrak{p}}'$, $\phi_\lambda$ would be expressible on $A_{\mathfrak{p}}^+$ as a linear combination of these. In order to do this we compute the operator $\delta'(\omega)$ and, starting from the equation $\delta'(\omega)\phi = \chi_\lambda(\omega)\phi$, give a method of constructing the required functions $\phi_\lambda^{(i)}$ $(1 \leq i \leq w)$. The asymptotic behaviour of the $\phi_\lambda^{(i)}$ is obvious from their construction and so in this way, we get an asymptotic formula for $\phi_\lambda$ on $A_{\mathfrak{p}}^+$.

In order to make further progress, it is necessary to make a closer study of the function $\phi_0$ corresponding to $\lambda = 0$. Theorem 3 contains the main result on $\phi_0$. It is possible to derive from this certain important consequences (see Lemma 45 and Theorems 4 and 5) and actually to obtain explicit formulae for $c$ and its Fourier transform. Section 12 is devoted to a deeper study of the function $c$ and the principal result is given in Lemma 52. In Section 13, we give some explicit calculations for the case $l = 1$. These will be required in the next paper of this series. The case when $G$ is complex is especially simple and therefore it is discussed separately in Section 14. Certain simple lemmas in analysis, which are often needed during this paper, are collected together in the Appendix (§ 15).

Some of the results of this paper have been announced in a short note [5(n)].


**2. Preliminary lemmas.** Let $R$ and $C$ be the fields of real and complex numbers respectively and $G$ a connected semisimple Lie group and $\mathfrak{g}_0$ its Lie algebra over $R$. Define $\mathfrak{k}_0$ and $\mathfrak{p}_0$ as usual (see [5(c), p. 187]) and let $K$ be

---

[4] The results of Sections 5 and 6 should be compared with Theorem 1 of [5(k)].

the analytic subgroup of $G$ corresponding to $\mathfrak{k}_0$. Since $K$ contains the center of $G$ (see [7]) and since we shall be concerned in this paper primarily with functions on $G/K$, $G$ can be replaced by any connected group locally isomorphic to it. Let $G_C$ be a simply connected complex-analytic group corresponding to the complexification $\mathfrak{g}$ of $\mathfrak{g}_0$. Then we may assume that $G$ is the real analytic subgroup of $G_C$ which corresponds to $\mathfrak{g}_0$. This permits us to identify the finite-dimensional representations of $G$ with those of $\mathfrak{g}_0$ and thus also with the complex representations of $G_C$ and $\mathfrak{g}$. If $\pi$ is such a representation on a vector space $V$, one can always introduce the structure of a Hilbert space in $V$ in such a way that $\pi(X)$ becomes skew-Hermitian for $X \in \mathfrak{u} = \mathfrak{k}_0 + (-1)^{\frac{1}{2}}\mathfrak{p}_0$. We shall always tacitly assume that such a structure has been defined. Also observe that $K$ is now compact.

Let $\mathfrak{h}_{\mathfrak{p}_0}$ be a maximal abelian subspace of $\mathfrak{p}_0$ and $\mathfrak{h}_0$ a Cartan subalgebra of $\mathfrak{g}_0$ containing $\mathfrak{h}_{\mathfrak{p}_0}$. Then $\mathfrak{h}_0 = \mathfrak{h}_{\mathfrak{p}_0} + \mathfrak{h}_{\mathfrak{k}_0}$ where $\mathfrak{h}_{\mathfrak{k}_0} = \mathfrak{h}_0 \cap \mathfrak{k}_0$. Complexify $\mathfrak{k}_0$, $\mathfrak{p}_0$, $\mathfrak{h}_0$, $\mathfrak{h}_{\mathfrak{p}_0}$, $\mathfrak{h}_{\mathfrak{k}_0}$ to $\mathfrak{k}$, $\mathfrak{p}$, $\mathfrak{h}$, $\mathfrak{h}_{\mathfrak{p}}$, $\mathfrak{h}_{\mathfrak{k}}$ respectively in $\mathfrak{g}$ and introduce compatible orders (see [4(1), §2]) in, the spaces of real-valued linear functions on $\mathfrak{h}_{\mathfrak{p}_0} + (-1)^{\frac{1}{2}}\mathfrak{h}_{\mathfrak{k}_0}$ and $\mathfrak{h}_{\mathfrak{p}_0}$. Let $P$ denote the set of all positive roots of $\mathfrak{g}$ (with respect to $\mathfrak{h}$) under this order. Consider the set $\Sigma$ of all linear functions $\lambda \neq 0$ on $\mathfrak{h}_{\mathfrak{p}}$ which are restrictions of some $\alpha$ in $P$. Then every element in $\Sigma$ is positive under the above order.

LEMMA 1. *Let $\pi$ be an irreducible finite-dimensional representation of $\mathfrak{g}$ such that the zero representation of $\mathfrak{k}$ occurs in the reduction[5] of $\pi$ with respect to $\mathfrak{k}$. Then if $\lambda$ is the highest weight of $\pi$, $\lambda$ is identically zero on $\mathfrak{h}_{\mathfrak{k}}$.*

Let $\psi_0 \neq 0$ be a vector in the representation space $V$ belonging to the weight $\lambda$. For any root $\alpha$, define $X_\alpha$ as usual (see [5(c), p. 188]) and put $\mathfrak{n} = \sum_{\alpha \in P_+} CX_\alpha$ where $P_+$ is the set of those roots $\alpha \in P$ whose restriction on $\mathfrak{h}_{\mathfrak{p}}$ is not zero. Let $A_{\mathfrak{p}}$ and $N$ be the analytic subgroups of $G$ corresponding to $\mathfrak{h}_{\mathfrak{p}_0}$ and $\mathfrak{n}_0 = \mathfrak{n} \cap \mathfrak{g}_0$ respectively. Then $G = KA_{\mathfrak{p}}N$ (see Iwasawa [6]). It is clear that $\pi(n)\psi_0 = \psi_0$ for $n \in N$ and therefore the vector space $C\psi_0$ is invariant under $\pi(A_{\mathfrak{p}}N)$. Hence $V$ is spanned by vectors of the form $\pi(k)\psi_0$ ($k \in K$). Put $E = \int_K \pi(k)\,dk$ where $dk$ is the normalized Haar measure of $K$. Then it follows that $EV = CE\psi_0$. Now we can, by hypothesis, select a unit vector $\psi$ in $V$ which is invariant under $K$. Then $\psi = E\psi = cE\psi_0$ where $c \in C$. But $\pi(X)E = E\pi(X) = 0$ if $X \in \mathfrak{k}$ and therefore $\lambda(H)E\psi_0 = E\pi(H)\psi_0 = 0$ for $H \in \mathfrak{h}_{\mathfrak{k}}$. Since $\psi \neq 0$, this implies that $\lambda(H) = 0$.

---

[5] It follows from Lemmas 2 and 3 of [5(c)] that $\pi(\mathfrak{k})$ is fully reducible.

It is well known that the exponential mapping is univalent and regular on $\mathfrak{p}_0$. Hence $A_\mathfrak{p}$ is simply connected and, for any $x \in G$, there exists a unique element $H(x) \in \mathfrak{h}_{\mathfrak{p}_0}$ such that $x \in K(\exp H(x))N$ (see [6]). Moreover $x \to H(x)$ is an analytic mapping of $G$ into $\mathfrak{h}_{\mathfrak{p}_0}$ (see [5(c), Lemma 26]).

Let $\mathfrak{F}_0$ denotes the set of all linear functions $\lambda$ of $\mathfrak{h}_\mathfrak{p}$ with the property that there exists a representation $\pi$ of $\mathfrak{g}$ satisfying the conditions of Lemma 1, whose highest weight coincides on $\mathfrak{h}_\mathfrak{p}$ with $\lambda$.

LEMMA 2. *Let $\Lambda$ be the highest weight of any irreducible finite-dimensional representation $\pi$ of $\mathfrak{g}$ and let $\lambda$ denote the restriction of $\Lambda$ on $\mathfrak{h}_\mathfrak{p}$. Then $2\lambda \in \mathfrak{F}_0$.*

Select a unit vector $\psi$ in the representation space $V$ belonging to the weight $\Lambda$ and put [6]

$$\phi(x) = \int_K |\pi(xk)\psi|^2 \, dk \qquad (x \in G).$$

Since $\pi(k)$ is unitary ($k \in K$) and $\pi(n)\psi = \psi$ ($n \in N$), it is clear that $|\pi(x)\psi| = e^{\Lambda(H(x))}$. Hence $\phi(x) = \int_K e^{2\Lambda(H(xk))} dk$. Also it follows from its definition that $\phi$ is spherical, that is, $\phi(k_1 x k_2) = \phi(x)$ ($k_1, k_2 \in K; x \in G$). Let $\phi_y$ ($y \in G$) denote the right translate of $\phi$ by $y$ so that $\phi_y(x) = \phi(xy)$ ($x \in G$). Select an orthonormal base $\psi_0 = \psi, \psi_1, \cdots, \psi_r$ for $V$ and let $\pi(x)\psi_i = \sum_{0 \leq j \leq r} \psi_j a_{ji}(x)$ ($0 \leq i \leq r$) where $a_{ji}$ are analytic functions on $G$. Then

$$\pi(xk)\psi = \sum_j \psi_j a_j(xk) = \sum_{j,i} \psi_j a_{ji}(x) a_i(k)$$

where $a_i = a_{i0}$. Therefore

$$|\pi(xk)\psi|^2 = \sum_j \left| \sum_i a_{ji}(x) a_i(k) \right|^2$$

and [7]

$$\phi(x) = \sum_{0 \leq i,j,m \leq r} a_{ji}(x)(\operatorname{conj} a_{jm}(x)) c_{im} \qquad (x \in G)$$

where $c_{im}$ are certain constants. Let $U'$ be the linear space spanned over $C$ by the $(r+1)^2$ functions $\sum_{0 \leq j \leq r} a_{ji}(\operatorname{conj} a_{jm})$ ($0 \leq i, m \leq r$). Then it is clear that $\phi_y \in U'$. Let $U$ be the subspace of $U'$ spanned by all $\phi_y$ ($y \in G$). We define a representation $\tau$ of $G$ on $U$ as follows. If $\phi' \in U$, $\tau(y)\phi'$ is the function $x \to \phi'(xy)$ on $G$. It is clear that $\tau(k)\phi = \phi_k = \phi$ ($k \in K$). Moreover $\phi \neq 0$ since $\phi(1) = 1$ and therefore the trivial representation of $K$

---

[6] We denote the scalar product of two elements $\phi_1, \phi_2$ in $V$ in the usual way by $(\phi_1, \phi_2)$. Similarly $|\phi_1|$ denotes the norm of $\phi_1$.

[7] $\operatorname{conj} c$ denotes the conjugate of a complex number $c$.

occurs in the reduction of $\tau$. Also as we shall see below (corollary to Lemma 3), $\tau$ is irreducible.

Now we may obviously assume that each $\psi_i$ belongs to a weight $\Lambda_i$ $(0 \leq i \leq r, \Lambda_0 = \Lambda)$ of $\pi$. Then $a_{ji}(x \exp H) = a_{ji}(x) e^{\Lambda_i(H)}$ $(H \in \mathfrak{h}_{\mathfrak{p}_0})$. Moreover the weights being real[8] linear functions,

$$\sum_j a_{ji}(xh)\operatorname{conj} a_{jm}(xh) = \sum_j a_{ji}(x)(\operatorname{conj} a_{jm}(x))\exp\{\Lambda_i(H) + \Lambda_m(H)\}$$

where $h = \exp H$ $(x \in G, H \in \mathfrak{h}_{\mathfrak{p}_0})$. Let $\mu$ be the highest weight of $\tau$. Then we conclude from the above formula that $\mu$ coincides on $\mathfrak{h}_{\mathfrak{p}_0}$ with $\Lambda_i + \Lambda_m$ for some $i$ and $m$. Since $\Lambda_i + \Lambda_m \leq 2\Lambda_0 = 2\Lambda$, it follows from the compatibility of our orders that $\bar\mu \leq 2\lambda$, where $\bar\mu$ is the restriction of $\mu$ on $\mathfrak{h}_{\mathfrak{p}}$. On the other hand[6]

$$\phi(x) = \int_K |\pi(xk)\psi|^2 \, dk$$

$$= \sum_{0 \leq i,j,m \leq r} a_{ji}(x)(\operatorname{conj} a_{jm}(x)) \int_K (\psi_i, \pi(k)\psi)\operatorname{conj}(\psi_m, \pi(k)\psi)\,dk.$$

Now[9] $a_{ji}(\exp H) = \delta_{ji} e^{\Lambda_j(H)}$ $(H \in \mathfrak{h}_{\mathfrak{p}_0})$. Hence $\phi(\exp H) = \sum_{0 \leq i \leq r} e^{2\Lambda_i(H)} c_i$ where

$$c_i = \int_K |(\psi_i, \pi(k)\psi)|^2 \, dk \geq 0.$$

Let $f_0, f_1, \cdots, f_p$ be a base for $U$ over $C$ such that each $f_i$ belongs to a weight $\mu_i$ of the representation $\tau$ and $\mu_0 = \mu$. Then $\phi = \sum_{0 \leq i \leq p} b_i f_i$ $(b_i \in C)$ and therefore

$$\tau(\exp H)\phi = \sum_{0 \leq i \leq p} b_i e^{\mu_i(H)} f_i \qquad\qquad (H \in \mathfrak{h}_{\mathfrak{p}_0}).$$

This shows that $\phi(\exp H) = \sum_i b_i f_i(1) e^{\mu_i(H)}$. Let $\bar\mu_i$ denote the restriction of $\mu_i$ on $\mathfrak{h}_{\mathfrak{p}}$. Then if we note that

$$c_0 = \int |(\psi, \pi(k)\psi)|^2 \, dk > 0$$

and recall that the exponentials of distinct linear functions on $\mathfrak{h}_{\mathfrak{p}}$ are linearly independent over $C$ (see [5(b), Lemma 41]), it follows from a comparison of the above two formulas for $\phi(\exp H)$ that $2\lambda = \bar\mu_i$ for some $i$. But $\mu \geq \mu_i$ and therefore $\bar\mu \geq \bar\mu_i = 2\lambda$. However we have seen above that $\bar\mu \leq 2\lambda$ and so $\bar\mu = 2\lambda$. This proves Lemma 2.

---

[8] A linear function on $\mathfrak{h}$ or $\mathfrak{h}_{\mathfrak{p}}$ is said to be real, if it takes only real values on $\mathfrak{h}_{\mathfrak{p}_0} + (-1)^{\frac{1}{2}}\mathfrak{h}_{\mathfrak{k}_0}$ or $\mathfrak{h}_{\mathfrak{p}_0}$ respectively.

[9] As usual $\delta_{ij} = 1$ or $0$ according as $i = j$ or not. Sometimes it will be convenient to write $\delta_i{}^j$ instead of $\delta_{ij}$.

Let $\mathfrak{B}$ be the universal enveloping algebra of $\mathfrak{g}$. We regard elements of $\mathfrak{B}$ as left-invariant differential operators on $G$ (see $[5(g), \S 4]$). From now on we shall make use of the notational conventions described in footnote 1 of $[5(k)]$. Define $\mathfrak{n} = \sum_{\alpha \in P_+} CX_\alpha$ as before and let $\mathfrak{H}_\mathfrak{p}$ be the subalgebra of $\mathfrak{B}$ generated by $(1, \mathfrak{h}_\mathfrak{p})$. We denote the centralizer of $\mathfrak{k}$ in $\mathfrak{B}$ by $I_\mathfrak{g}$.

LEMMA 3.   *For any $b \in \mathfrak{B}$, there exists a unique element $u_b \in \mathfrak{H}_\mathfrak{p}$ such that $b - u_b \in \mathfrak{k}\mathfrak{B} + \mathfrak{B}\mathfrak{n}$. The degree of $u_b$ does not exceed that of $b$. Moreover if $\nu$ is any linear function on $\mathfrak{h}_\mathfrak{p}$ and*

$$g(x) = \int_K e^{\nu(H(xk))} dk \qquad\qquad (x \in G),$$

*then $qg = \chi_\nu(u_q)g$ for $q \in I_\mathfrak{g}$. Here $\chi_\nu$ is the homomorphism of $\mathfrak{H}_\mathfrak{p}$ into $C$ such that $\chi_\nu(1) = 1$ and $\chi_\nu(H) = \nu(H)$ $(H \in \mathfrak{h}_\mathfrak{p})$.*

In order to prove the first statement, it is sufficient to show that $\mathfrak{B}$ is the direct sum of $\mathfrak{H}_\mathfrak{p}$ and $\mathfrak{k}\mathfrak{B} + \mathfrak{B}\mathfrak{n}$. Let $\mathfrak{X}$ and $\mathfrak{N}$ be the subalgebras of $\mathfrak{B}$ generated by $(1, \mathfrak{k})$ and $(1, \mathfrak{n})$ respectively. Consider the linear mapping of the tensor product $\mathfrak{X} \times \mathfrak{H}_\mathfrak{p} \times \mathfrak{N}$ into $\mathfrak{B}$ which maps $x \times h \times n$ onto $xhn$ $(x \in \mathfrak{X}, h \in \mathfrak{H}_\mathfrak{p}, n \in \mathfrak{N})$. Then our first two assertions follow immediately from Lemma 12 of $[5(c)]$ if we recall that $\mathfrak{g}$ is the direct sum of $\mathfrak{k}$, $\mathfrak{h}_\mathfrak{p}$ and $\mathfrak{n}$.

Now it is obvious from the definition of $H(x)$ that $H(xn) = H(x)$ $(n \in N)$ and [10] $H(xh) = H(x) + \log h$ $(h \in A_\mathfrak{p})$. Therefore if $F(x) = e^{\nu(H(x))}$ $(x \in G)$, it follows that $F(x; b) = 0$ for $b \in \mathfrak{B}\mathfrak{n}$ and $F(x; u) = \chi_\nu(u)F(x)$ $(u \in \mathfrak{H}_\mathfrak{p})$. Now put [11] $F(x : k) = F(xk)$ $(x \in G, k \in K)$. Then if $q \in I_\mathfrak{g}$, it is clear that $F(x; q : k) = F(xk; q)$. Hence

$$g(x; q) = \int_K F(x; q : k) dk = \int_K F(xk; q) dk.$$

But $q = b_1 + b_2 + u_q$ where $b_1 \in \mathfrak{k}\mathfrak{B}$ and $b_2 \in \mathfrak{B}\mathfrak{n}$. Hence

$$F(xk; q) = F(xk; b_1) + \chi_\nu(u_q)F(xk).$$

However if $b \in \mathfrak{B}$, it is obvious that $\int_K F(xkk_1; b) dk$ is actually independent of $k_1 \in K$. Therefore $\int_K F(xk; b') dk = 0$ for any $b' \in \mathfrak{k}\mathfrak{B}$ and hence

$$g(x; q) = \int_K F(xk; q) dk = \chi_\nu(u_q) \int_K F(xk) dk = \chi_\nu(u_q) g(x).$$

---

[10] $\log h$ denotes the unique element $H \in \mathfrak{h}_{\mathfrak{p}_0}$ such that $\exp H = h$.

[11] Here we follow the mode of writing introduced in $[5(k)], \S 2$.

COROLLARY. *The representation $\tau$ defined during the proof of Lemma 2 is irreducible.*

We denote the corresponding representation of $\mathfrak{B}$ on $U$ also by $\tau$. Then it is clear that $\tau(b)\phi' = b\phi'$ for any $\phi' \in U$ and $b \in \mathfrak{B}$. Hence $U$ consists of functions of the form $b\phi$ ($b \in \mathfrak{B}$). Let $E$ denote the projection $\int_K \tau(k)\,dk$. Then $EU = E\tau(\mathfrak{B})\phi$. However since $\tau(k)\phi = \phi$ ($k \in K$), it follows that $Eb\phi = q\phi$ where [12] $q = \int_K b^k\,dk \in I_{\mathfrak{G}}$ ($b \in \mathfrak{B}$). But

$$\phi(x) = \int_K e^{2\Lambda(H(xk))}\,dk$$

and therefore $EU = C\phi$ from Lemma 3. Hence the trivial representation of $K$ occurs exactly once in the reduction of $U$ under $\tau(K)$. Now let $U_1$ be any subspace of $U$ which is invariant and irreducible under $\tau$ and such that $EU_1 \neq \{0\}$. (It is obvious that such a space exists.) Then $\phi \in EU_1 \subset U_1$ and therefore $U = \tau(\mathfrak{B})\phi \subset U_1$. This proves that $U_1 = U$ and therefore $U$ is irreducible.

LEMMA 4. *Suppose $u$ is an element in $\mathfrak{H}_\mathfrak{p}$ such that $\chi_\lambda(u) = 0$ for all $\lambda \in \mathfrak{F}_0$. Then $u = 0$.*

Let $L = \dim \mathfrak{h}$. Then (see [5(b), Theorem 1]) there exist $L$ linearly independent linear functions $\Lambda_1, \cdots, \Lambda_L$ on $\mathfrak{h}$ with the property that $m_1\Lambda_1 + \cdots + m_L\Lambda_L$ is the highest weight of an irreducible finite-dimensional representation of $\mathfrak{g}$ whenever $m_1, \cdots, m_L$ are nonnegative integers. Let $\lambda_i$ denote the restriction of $2\Lambda_i$ on $\mathfrak{h}_\mathfrak{p}$. Then if $l = \dim \mathfrak{h}_\mathfrak{p}$, it is obvious that we can choose $l$ linearly independent elements, say $\lambda_1, \cdots, \lambda_l$, among $\lambda_1, \cdots, \lambda_L$. Then if $m_1, \cdots, m_l$ are nonnegative integers, it follows from Lemma 2 that $m_1\lambda_1 + \cdots + m_l\lambda_l \in \mathfrak{F}_0$. Our assertion is now an immediate consequence of Lemma 32 of [5(b)].

LEMMA 5. *Let $\pi$ be an irreducible representation of $G$ on a finite-dimensional vector space $V$. Suppose $\phi$ is a unit vector in $V$ such that $\pi(k)\phi = \phi$ for all $k \in K$. Then*

$$(\phi, \pi(x)\phi) = \int_K e^{\lambda(H(xk))}\,dk \qquad\qquad (x \in G)$$

*where $\lambda$ is the highest weight of $\pi$.*

---

[12] For any $x \in G_c$, $b \to b^x$ ($b \in \mathfrak{B}$) denotes the automorphism of $\mathfrak{B}$ which coincides with $\mathrm{Ad}(x)$ on $\mathfrak{g}$.

Let $\psi$ be a unit vector in $V$ belonging to the weight $\lambda$. Since $G = KA_\mathfrak{p}N$, it is clear that $V$ is spanned by vectors of the form $\pi(k)\psi$ $(k \in K)$. Put

$$E = \int_K \pi(k)\,dk.$$

Then $E\pi(k)\psi = E\psi$ and therefore $EV = CE\psi$. But $\phi = E\phi \in EV$ and so $\phi = cE\psi$ for some $c \in C$. It is obvious that $E$ is self-adjoint and $E^2 = E$. Hence

$$(\phi, \pi(x)\phi) = |c|^2 (\psi, E\pi(x)E\psi) \qquad (x \in G).$$

Now let $k \in K$ and $x \in G$. Then $xk = k'(\exp H(xk))n$ where $k' \in K$ and $n \in N$. Hence

$$E\pi(xk)\psi = e^{\lambda(H(xk))}E\pi(k')\psi = e^{\lambda(H(xk))}E\psi.$$

Therefore

$$E\pi(x)E\psi = \int_K E\pi(xk)\psi\,dk = \int_K e^{\lambda(H(xk))}dk\,E\psi.$$

Since $|c|^2 |E\psi|^2 = |\phi|^2 = 1$, this implies that

$$(\phi, \pi(x)\phi) = |c|^2 (\psi, E\pi(x)E\psi) = \int_K e^{\lambda(H(xk))}dk.$$

**3. Some results of Chevalley and their consequences.** Let $M$ be the centralizer and $M'$ the normalizer of $\mathfrak{h}_{\mathfrak{p}_0}$ in $K$. Then $W = M'/M$ is a finite group (see [5(j), p. 619]) which operates as a group of linear transformations on $\mathfrak{h}_\mathfrak{p}$ in the obvious way. This linear representation of $W$ being faithful, we can identify $W$ with the corresponding linear group. We shall call $W$ the little Weyl group of $\mathfrak{g}_0$ with respect to [13] $\mathfrak{h}_{\mathfrak{p}_0}$.

Put $B(X, Y) = \mathrm{sp}(\mathrm{ad}\,X\,\mathrm{ad}\,Y)$ $(X, Y \in \mathfrak{g})$ where $X \to \mathrm{ad}\,X$ denotes the adjoint representation of $\mathfrak{g}$. Then the quadratic form $B(X, X)$ is positive-definite on $\mathfrak{p}_0$. Therefore it defines a Euclidean metric on $\mathfrak{p}_0$ and hence also on $\mathfrak{h}_{\mathfrak{p}_0}$. Define $\Sigma$ as in the beginning of Section 2. Then for each $\alpha \in \Sigma$, there exists a unique element $H_\alpha \in \mathfrak{h}_{\mathfrak{p}_0}$ such that $B(H, H_\alpha) = \alpha(H)$ for every $H \in \mathfrak{h}_{\mathfrak{p}_0}$. Let $s_\alpha$ denote the linear transformation in $\mathfrak{h}_{\mathfrak{p}_0}$ which corresponds to the reflexion in the hyperplane $\alpha = 0$. Then $s_\alpha$ is given by $s_\alpha H = H - 2\{\alpha(H)/\alpha(H_\alpha)\}H_\alpha$ $(H \in \mathfrak{h}_\mathfrak{p})$. It is known (see Cartan [2]) that $s_\alpha \in W$. The following lemma has been proved by Chevalley.[14]

LEMMA 6 (Chevalley). *Let $H_0$ be any element in $\mathfrak{h}_\mathfrak{p}$ and $W'$ the sub-*

---

[13] It is seen without difficulty that $M'A_p$ and $MA_p$ respectively are the normalizer and centralizer of $\mathfrak{h}_{\mathfrak{p}_0}$ in $G$. Hence $W = M'A_p/MA_p$ is independent of our choice of $\mathfrak{k}_0$, so long as $\mathfrak{h}_{\mathfrak{p}_0}$ remains fixed.

[14] I am grateful to Professor Chevalley for showing me his unpublished results.

*group consisting of those elements $s \in W$ which leave $H_0$ fixed. Then $W'$ is generated by the reflexions $s_\alpha$ corresponding to those $\alpha \in \Sigma$ for which $\alpha(H_0) = 0$.*

COROLLARY. *$W$ is generated by $s_\alpha$ ($\alpha \in \Sigma$).*

This follows by taking $H_0 = 0$. (This corollary had been verified by Cartan [2] in the case of classical Lie algebras.)

Let $S(\mathfrak{p})$ and $S(\mathfrak{h}_\mathfrak{p})$ denote the symmetric algebras over $\mathfrak{p}$ and $\mathfrak{h}_\mathfrak{p}$ respectively. Then $S(\mathfrak{p}) \supset S(\mathfrak{h}_\mathfrak{p})$ and we can identify them with the algebras of polynomial functions on $\mathfrak{p}$ and $\mathfrak{h}_\mathfrak{p}$ respectively by means of the non-degenerate bilinear form $B(X, Y)$ (see [5(k), p. 93]). Moreover $\mathrm{Ad}(k)\mathfrak{p} = \mathfrak{p}$ ($k \in K$) and therefore $K$ operates on $\mathfrak{p}$ and therefore also on $S(\mathfrak{p})$. Similarly $W$ operates on $S(\mathfrak{h}_\mathfrak{p})$. Let $I(\mathfrak{p})$ and $I(\mathfrak{h}_\mathfrak{p})$ be the sub-algebras consisting of those elements of $S(\mathfrak{p})$ and $S(\mathfrak{h}_\mathfrak{p})$ which are invariant under $K$ and $W$ respectively.

LEMMA 7 (Chevalley [14]). *For any $p \in I(\mathfrak{p})$ let $\bar{p}$ denote the restriction of the polynomial function $p$ on $\mathfrak{h}_\mathfrak{p}$. Then $\bar{p} \in I(\mathfrak{h}_\mathfrak{p})$ and $p \to \bar{p}$ is an iso-morphism of $I(\mathfrak{p})$ onto $I(\mathfrak{h}_\mathfrak{p})$.*

Now fix an element $H_0$ in $\mathfrak{h}_\mathfrak{p}$ and define $W'$ as in Lemma 6. Put $J = I(\mathfrak{h}_\mathfrak{p})$ and let $J'$ denote the algebra of those elements in $S(\mathfrak{h}_\mathfrak{p})$ which are invariant under $W'$.

LEMMA 8. *Let [15] $r = [W : W']$. Then there exist $r$ homogeneous elements $v_1 = 1, v_2, \cdots, v_r$ in $J'$ such that $J' = \sum_{1 \leq i \leq r} J v_i$. Moreover $v_1, \cdots, v_r$ are linearly independent over the quotient field of $J$.*

Put $S = S(\mathfrak{h}_\mathfrak{p})$ and let $C(S)$, $C(J')$ and $C(J)$ denote the quotient fields of $S$, $J'$ and $J$ respectively. Then $C(S)/C(J)$ is normal and its Galois group is $W$ (see Chevalley [3(b), p. 781]). By the same argument $W'$ is the Galois group of $C(S)/C(J')$. Hence $[C(J') : C(J)] = [W : W'] = r$. Let $S_d$ denote the space of homogeneous elements in $S$ of degree $d$. Then it is obvious that $J = \sum_{d \geq 0} J_d$ and $J' = \sum_{d \geq 0} J_d'$ where $J_d = S_d \cap J$ and $J_d' = S_d \cap J'$. Put $J_+ = \sum_{d \geq 1} J_d$, $J_+' = \sum_{d \geq 1} J_d'$. We claim that $J' \cap (SJ_+) = J'J_+$. For if $u = \sum_{1 \leq i \leq m} p_i u_i \in J'$ ($p_i \in S, u_i \in J_+$), it is obvious that

$$u = u^s = \sum_i p_i^s u_i \qquad (s \in W')$$

and therefore $u = \sum_i q_i u_i \in J'J_+$ where $q_i = w'^{-1} \sum_{s \in W'} p_i^s$ and $w'$ is the order of $W'$. Hence

---

[15] $[W : W']$ stands for the index of $W'$ in $W$.

$$S/SJ_+ \supset (J' + SJ_+)/SJ_+ \cong J'/J' \cap SJ_+ = J'/J'J_+.$$

Moreover Chevalley has shown [3(b), Theorem (B)] that $\dim S/SJ_+ = w$ (where $w$ is the order of $W$) and the natural representation of $W$ on $S/SJ_+$ is equivalent to the regular representation of $W$. Hence in the reduction of this representation with respect to $W'$, the trivial representation of $W'$ occurs exactly $r = [W : W']$ times. Let $S'$ be the set of those elements $u \in S$ whose residue class mod $SJ_+$ is left fixed by $W'$. Then $\dim S'/SJ_+ = r$. Moreover $u - u^s \in SJ_+$ for any $u \in S'$ and $s \in W'$ and therefore $u \equiv w'^{-1} \sum_{s \in W'} u^s$ mod $SJ_+$. This shows that $S' = J' + SJ_+$ and hence $S'/SJ_+ \cong J'/J'J_+$. Therefore $\dim J'/J'J_+ = r$ and so we can select $r$ homogeneous elements $v_1 = 1, v_2, \cdots, v_r$ in $J'$ such that their residue classes mod $J'J_+$ form a base for $J'/J'J_+$ over $C$. Let $V = \sum_{1 \le i \le r} Cv_i$. We shall prove that $J' = JV$. It is clear that $J' = V + J_+J'$ and from this it follows by induction that $J' = JV + J_+^m J'$ for any integer $m \ge 1$. In particular $J_d' \subset JV + J_+^m J'$ for any $d \ge 0$. But then by choosing $m > d$, it is obvious that $J_d' \subset JV$ and therefore $J' = \sum_{d \ge 0} J_d' \subset JV$. This proves that $J' = JV$. Moreover since $C(J')/C(J)$ is a finite algebraic extension, it is clear that $C(J') = C(J)J' = C(J)V$. Therefore since $[C(J') : C(J)] = r$, $v_1, \cdots, v_r$ must be linearly independent over $C(J)$.

The above proof depended only on the fact that $W$ and $W'$ are finite groups generated by reflexions. Hence if we replace $(W, W')$ by $(W, 1)$ and $(W', 1)$ respectively, we get the following corollary.

COROLLARY. *There exist homogeneous elements* $u_1 = 1, u_2, \cdots, u_w$ *and* $u_1' = 1, u_2', \cdots, u_{w'}'$ *in $S$ such that*

$$S = \sum_{1 \le i \le w} Ju_i, \qquad S = \sum_{1 \le j \le w'} J'u_j'.$$

*Moreover* $(u_1, \cdots, u_w)$ *are linearly independent over $C(J)$ and* $(u_1', \cdots, u_{w'}')$ *over $C(J')$.*

Let $\mathfrak{m}$ denote the centralizer of $\mathfrak{h}_\mathfrak{p}$ in $\mathfrak{g}$ and $p(H)$ the determinant of the linear transformation on $\mathfrak{g}/\mathfrak{m}$ defined by $\operatorname{ad} H$ $(H \in \mathfrak{h}_\mathfrak{p})$. Then $p \in S$ and it is obvious that $p^s = p$ $(s \in W)$. Moreover $p(H) = \pm \prod_{\alpha \in P_+} \alpha(H)^2$. Hence if $\beta \in \Sigma$ and $s \in W$, it follows that either $s\beta$ or $-s\beta$ lies in $\Sigma$. Put $\langle H, H' \rangle = B(H, H')$ $(H, H' \in \mathfrak{h}_\mathfrak{p})$ and identify $\mathfrak{h}_\mathfrak{p}$ with its dual by means of the bilinear form $B$. We shall say that two nonzero linear functions $\lambda$, $\mu$ on $\mathfrak{h}_\mathfrak{p}$ are equivalent, if they are linearly dependent. Then if $\alpha$ and $\beta$ are equivalent elements in $\Sigma$, $\beta = c\alpha$ where $c$ is a positive real number.[16]

---

[16] Actually it is known (see Cartan [2]) that $c = \frac{1}{2}$, 1 or 2.

Let $l = \dim \mathfrak{h}_\mathfrak{p}$. Then we can select (see [5(1), Lemma 1]) $l$ linearly independent elements $\alpha_1, \cdots, \alpha_l \in \Sigma$ such that every $\alpha$ in $\Sigma$ can be written in the form $\alpha = m_1\alpha_1 + \cdots + m_l\alpha_l$ where $m_i$ are nonnegative integers. Let $q$ be the number of distinct equivalence classes in $\Sigma$. Extend $(\alpha_1, \cdots, \alpha_l)$ to a maximal set $(\alpha_1, \cdots, \alpha_q)$ of inequivalent elements in $\Sigma$ and put $\pi = \alpha_1\alpha_2 \cdots \alpha_q \in S$. Since $p^s = p$ $(s \in W)$, it is obvious that $\pi^s = \epsilon(s)\pi$ where $\epsilon(s)$ is a real number. But the mapping $s \to \epsilon(s)$ is obviously a homomorphism of the finite group $W$ into the multiplicative group of nonzero real numbers and therefore $\epsilon(s) = \pm 1$. We shall see presently that $\epsilon(s_\alpha) = -1$ for $\alpha \in \Sigma$.

Put $s_i = s_{\alpha_i}$ $1 \leq i \leq l$.

LEMMA 9. *$W$ is generated by $(s_1, \cdots, s_l)$. Moreover for any $\alpha \in \Sigma$, we can choose $s \in W$ and an index $i$ $(1 \leq i \leq l)$ such that $\alpha$ is equivalent to $s\alpha_i$.*

Let $W_1$ be the subgroup of $W$ generated by $(s_1, \cdots, s_l)$ and $\alpha$ a given element in $\Sigma$. Then $\alpha = m_1\alpha_1 + \cdots + m_l\alpha_l$ where $m_1, \cdots, m_l$ are nonnegative integers. Put $m(\alpha) = m_1 + \cdots + m_l$. We shall prove by induction on $m(\alpha)$ that $\alpha = cs\alpha_i$ for some $s \in W_1$, some $i$ $(1 \leq i \leq l)$ and a suitable positive number $c$. This is obvious if $m(\alpha) = 1$. So now suppose that $m(\alpha) \geq 2$ and $\alpha$ is not equivalent to any $\alpha_i$ $(1 \leq i \leq l)$. Since $\langle \alpha, \alpha \rangle$ is positive, $\langle \alpha, \alpha_j \rangle$ is positive for some $j$. Moreover since $\alpha$ is not equivalent to $\alpha_j$, $m_i > 0$ for some $i \neq j$. But $s_j\alpha = \alpha - 2\{\langle \alpha, \alpha_j \rangle / \langle \alpha_j, \alpha_j \rangle\}\alpha_j$ and $-s_j\alpha$ cannot lie in $\Sigma$ since $m_i > 0$. Hence $s_j\alpha \in \Sigma$ and $m(\alpha) > m(s_j\alpha)$. Our assertion now follows by applying the induction hypothesis to $s_j\alpha$. Now if $\alpha = cs\alpha_i$, it is clear that $s_\alpha = ss_is^{-1} \in W_1$. Since $s_\alpha$ $(\alpha \in \Sigma)$ generate $W$, this proves that $W_1 = W$.

LEMMA 10. *$\epsilon(s_\alpha) = -1$ for $\alpha \in \Sigma$.*

We first claim that $\pi^{s_i} = -\pi$ $(1 \leq i \leq l)$. It is sufficient to prove this for $i = 1$. Obviously $s_1\alpha_1, s_1\alpha_2, \cdots, s_1\alpha_q$ are all inequivalent and therefore $s_1\alpha_j = c_j\alpha_j'$ $(1 \leq j \leq q)$ where $\alpha_j \to \alpha_j'$ is a permutation of the set $(\alpha_1, \alpha_2, \cdots, \alpha_q)$ and $c_j$ are nonzero real numbers. Now suppose $j \neq 1$. Then if $\alpha_j = m_1\alpha_1 + \cdots + m_l\alpha_l$, $m_k > 0$ for some $k \neq 1$. On the other hand $s_1\alpha_j = \alpha_j + m\alpha_1$ for some integer $m$. Hence $-s_1\alpha_j$ cannot lie in $\Sigma$. Therefore $s_1\alpha_j \in \Sigma$ and $c_j > 0$. However $s_1\alpha_1 = -\alpha_1$ and so this proves that $\epsilon(s_1) = \prod_{1 \leq j \leq q} c_j < 0$. Hence $\epsilon(s_1) = -1$. Now let $\alpha \in \Sigma$. Then, as we have seen during the proof of Lemma 9, $s_\alpha = ss_is^{-1}$ for some $s \in W$ and some $i$ $(1 \leq i \leq l)$. Therefore $\epsilon(s_\alpha) = \epsilon(s_i) = -1$.

We shall call an element $u \in S$ skew (or skew-invariant) if $u^s = \epsilon(s)u$ for $s \in W$.

COROLLARY. *$J\pi$ is exactly the set of all skew-invariants in $S$.*

Let $S'$ be the set of all skew elements in $S$ . It is obvious that $J\pi \subset S'$. Now suppose $u \in S'$. Then $u^{s_\alpha} = -u$ ($\alpha \in \Sigma$) and therefore the polynomial function $u$ vanishes identically on the hyperplane $\alpha = 0$ in $\mathfrak{h}_{\mathfrak{p}}$. But this implies that $\alpha$ divides $u$ in $S$. This being true for every $\alpha \in \Sigma$, we conclude that $u = \pi v$ for some $v \in S$. But since both $u$ and $\pi$ are skew, it is clear that $v \in J$ and therefore $u \in J\pi$.

For any $x \in C(S)$, let $\mathrm{sp}_{S/J}x$ denote the relative trace of $x$ from $C(S)$ to $C(J)$. Let $\mathfrak{D}_{S/J}$ be the set of those elements $x \in C(S)$ which have the property that $\mathrm{sp}_{S/J}xy \in J$ for every $y$ in $S$.

LEMMA 11. *Let $x$ be an element in $C(S)$. Then $x\mathfrak{D}_{S/J} \subset S$ if and only if $x \in S\pi$.*

Let $y$ be any element in $S$. Since $W$ is the Galois group of $C(S)/C(J)$, it is clear that $\mathrm{sp}_{S/J}(y/\pi) = \sum_{s \in W} (y/\pi)^s = y_0/\pi$ where $y_0 = \sum_{s \in W} \epsilon(s)y^s$. But $y_0$ is obviously a skew-invariant and therefore, from the corollary to Lemma 10, it lies in $J\pi$. This proves that $\pi^{-1} \in \mathfrak{D}_{S/J}$. Hence if $x$ is an element in $C(S)$ such that $x\mathfrak{D}_{S/J} \subset S$, it follows that $x\pi^{-1} \in S$ and therefore $x \in S\pi$. The proof of the converse is however more complicated.

Select $u_i \in S$ $1 \leq i \leq w$ as in the corollary to Lemma 8 and let $u^i$ $1 \leq i \leq w$ denote the dual base of $C(S)/C(J)$ so that[9] $\mathrm{sp}_{S/J}(u_iu^j) = \delta_i{}^j$ ($1 \leq i,j \leq w$). Then it is easy to verify that $\mathfrak{D}_{S/J} = \sum_{1 \leq i \leq w} Ju^i$. Hence in order to complete the proof of our lemma, it is sufficient to show that $\pi u^i \in S$ ($1 \leq i \leq w$). Fix an element $\alpha$ in $\Sigma$. In view of Lemma 9, we may assume that $\alpha$ is equivalent to $s\alpha_1$ for some $s \in W$. Let $\beta_1, \cdots, \beta_k$ be all the elements among $(\alpha_1, \cdots, \alpha_q)$ which are equivalent to $t\alpha$ for some $t \in W$. Put $\pi_\alpha = \beta_1\beta_2 \cdots \beta_k$. Then it is clear that $\pi_\alpha{}^s = c(s)\pi_\alpha$ where $c(s) = \pm 1$ ($s \in W$). By the argument used in the proof of Lemma 10, we deduce easily that $c(s_1) = -1$ and therefore $c(s_\beta) = -1$ for $\beta = \beta_1, \beta_2, \cdots, \beta_k$. Now suppose $z \in J \cap S\alpha$. Then $z = y\alpha$ for some $y \in S$ and $z = z^{s_\alpha} = (y\alpha)^{s_\alpha} = -y^{s_\alpha}\alpha$. This shows that $y^{s_\alpha} = -y$ and therefore $\alpha$ divides $y$ in $S$. But then $\alpha^2$ divides $z$ (in $S$). However $z \in J$ and therefore $(s\alpha)^2$ divides $z$ for every $s \in W$. This shows that $z \in S\pi_\alpha{}^2$. Therefore since $\pi_\alpha{}^2 \in J$, we conclude, from the definition of $J$, that $z\pi_\alpha{}^{-2} \in S \cap C(J) = J$. This proves that $J \cap S\alpha = J\pi_\alpha{}^2$.

Let $S_1$ denote the set of all elements in $C(S)$ of the form $a/b$ $(a, b \in S)$ where $b$ and $\pi_a$ are relatively prime in $S$. It is obvious that $b^s$ and $\pi_a$ are also relatively prime for every $s \in W$. Hence if $b_0 = \prod_{s \in W} b^s$ and $b_1 = b_0/b$, it follows that $a/b = ab_1/b_0$. Notice that $b_0$ lies in $J$ but not in $J\pi_a^2$. Moreover $S\alpha$ is obviously a prime ideal in $S$ and therefore $J \cap S\alpha = J\pi_a^2$ is a prime ideal in $J$. Hence if $J_1$ denotes the quotient ring of $J$ with respect to $J\pi_a^2$, $b_0^{-1}$ lies in $J_1$ and this proves that $S_1 = SJ_1$. Finally if $a/b$ is in $C(J)$, it is obvious that $ab_1 \in J$. Therefore $S_1 \cap C(J) = J_1$.

By a result of Chevalley [3(b), Theorem (A)], we can select $\omega_1, \cdots, \omega_l \in J$ such that $J = C[\omega_1, \cdots, \omega_l]$. Then $\omega_1, \cdots, \omega_l$ are algebraically independent (over $C$). Obviously $\pi_a^2$ is also transcendental over $C$ and therefore we may assume that $(\pi_a^2, \omega_2, \cdots, \omega_l)$ are algebraically independent so that $C(J)$ is algebraic over $C(\pi_a^2, \omega_2, \cdots, \omega_l)$. Moreover since $\pi_a^2 \in J$, it can be expressed as a polynomial in $\omega_1, \omega_2, \cdots, \omega_l$ with coefficient in $C$. $\omega_1$ must actually appear in this polynomial because otherwise $\pi_a^2, \omega_2, \cdots, \omega_l$ would be algebraically dependent. Therefore $\pi_a^2$ cannot divide (in $J$) any nonzero element in $C[\omega_2, \cdots, \omega_l]$. Hence the field $C_1 = C(\omega_2, \cdots, \omega_l)$ is contained in the local ring $J_1$.

Now we regard $C(J)/C_1$ and $C(S)/C_1$ as fields of algebraic functions of one variable (see Chevalley [3(a)]). It is obvious that $\pi_a^2$ is a prime element of $J$ and therefore $J_1$ is a valuation ring in $C(J)$. Let $p$ be the corresponding place of $C(J)/C_1$. For any $j$ $(1 \leq j \leq k)$ and $x \in S$, let $v_j(x)$ denote the highest integer $m \geq 0$ such that $\beta_j^m$ divides $x$ in $S$. (If $x = 0$ we put $v_j(x) = \infty$.) Then it is seen without difficulty that $v_j$ can be extended (uniquely) to a valuation of $C(S)$. Obviously $J_1$ is contained in the valuation ring of $v_j$ and therefore $v_j$ defines a place $q_j$ of $C(S)$ lying above $p$. Since $C(S)$ is normal over $C(J)$, every place of $C(S)$ lying above $p$ is conjugate to $q_1$ under $W$ (see (see [3(a), p. 54])). But since $s\beta_1$ $(s \in W)$ is equivalent to $\beta_j$ for some $j$, it follows that $q_1, q_2, \cdots, q_k$ are all the distinct places of $C(S)$ above $p$. It is obvious that $S_1$ is exactly the set of those elements in $C(S)$ which are integral at $q_j$ for every $j$ $(1 \leq j \leq k)$. Moreover $S_1 = J_1 S = \sum_{1 \leq i \leq w} J_1 u_i$. Since $v_j(\pi_a^2) = 2$, it follows that the ramification index of $q_j$ with respect to $C(J)$ is 2. Therefore we conclude from Theorem 7 of [3(a), p. 69] that the differential exponent of $q_j$ is 1. Let $\mathfrak{D}_p$ be the set of all $x \in C(S)$ such that $sp_{S/J} x S_1 \subset J_1$. Then obviously $\mathfrak{D}_p = \sum_{1 \leq i \leq w} J_1 u^i$. Hence it follows from Lemma 4 of [3(a), p. 73] that $\min_{1 \leq i \leq w} v_j(u^i) = -1$.

Let $d$ be a nonzero element in $S$ of the lowest possible degree such that

$du^i \in S$ $1 \leq i \leq w$. Then the above result shows that $\pi_\alpha$ divides $d$ in $S$ and $\pi_\alpha$ and $d/\pi_\alpha$ are relatively prime in $S$. But since $\alpha$ was an arbitrary element in $\Sigma$, this implies that $d' = d/\pi$ lies in $S$ and it is relatively prime to $\pi$. In order to complete our proof, it only remains to show that $d' \in C$.

We shall now introduce some notation which will also be useful later. For any $H \in \mathfrak{h}_\mathfrak{p}$, let $S_H$ denote the set of all polynomial functions in $S$ which vanish at $H$ and put $J_H = J \cap S_H$. Obviously $S_H$ and $J_H$ are prime ideals in $S$ and $J$ respectively and $J_{sH} = J_H$ $(s \in W)$. Now suppose $\pi(H) \neq 0$. Then from Lemma 6, the $w$ elements $sH$ $(s \in W)$ are all distinct. Hence there exist $w$ distinct prime ideals $S_{sH}$ in $S$ lying above $J_H$. On the other hand $J = C + J_H$ and therefore $S = \sum_{1 \leq i \leq w} J u_i = \sum_{1 \leq i \leq w} C u_i + S J_H$. This proves that $\dim S/SJ_H \leq w$. But in view of what we have said above, the associative algebra $\bar{S} = S/SJ_H$ has $w$ distinct non-trivial homomorphisms into $C$. Therefore it must be semisimple and of dimension $w$ and $\bigcap_{s \in W} S_{sH} = SJ_H$. Let $u \to \bar{u}$ denote the natural homomorphism of $S$ on $\bar{S}$ and let $\mathrm{sp}\, \bar{u}$ denote the trace of $\bar{u}$ in the regular representation of $\bar{S}$. Then clearly $\mathrm{sp}\, \bar{u} = \sum_{s \in W} u(sH)$ $(u \in S)$. Moreover since $\bar{S}$ is semisimple, we can conclude that $\bar{u} = 0$ if $\mathrm{sp}\, \bar{u}_i \bar{u} = 0$ $(1 \leq i \leq w)$.

Now put $g_{ij} = \mathrm{sp}_{S/J}(u_i u_j)$ $1 \leq i, j \leq w$ and $g = \det(g_{ij})_{1 \leq i, j \leq w}$. Then $g \neq 0$. Let $(g^{ij})_{1 \leq i, j \leq w}$ denote the inverse of the matrix $(g_{ij})_{1 \leq i, j \leq w}$. It is obvious that $u^i = \sum_j g^{ij} u_j$ and therefore $d$ divides $g$ in $S$. We claim $g(H) \neq 0$. For otherwise we could choose complex numbers $c_1, \cdots, c_w$, not all zero, such that $\sum_j g_{ij}(H) c_j = 0$ $(1 \leq i \leq w)$. Put $u = \sum_j c_j u_j$. Then $\sum_j g_{ij} c_j = \mathrm{sp}_{S/J}(u_i u) = \sum_{s \in W} (u_i u)^s$. Hence $\mathrm{sp}\, \bar{u}_i \bar{u} = \sum_{s \in W} u_i(sH) u(sH) = 0$. But, as we have seen above, this implies that $\bar{u} = 0$. On the other hand $\bar{u}_1, \cdots, \bar{u}_w$ span $\bar{S}$ and so they must be linearly independent over $C$. Therefore $\bar{u} = \sum_j c_j \bar{u}_j \neq 0$ and we get a contradiction. Hence $g(H) \neq 0$ whenever $\pi(H) \neq 0$ $(H \in \mathfrak{h}_\mathfrak{p})$. This means that every prime factor of $g$ divides $\pi$. But we know that $\pi$ divides $d$ and $d$ divides $g$. Therefore $\pi$, $d$ and $g$ have exactly the same prime factors and this implies $d' \in C$. The proof of Lemma 11 is now complete.

Define $W'$ and $J'$ as before. We recall that $\pi = \alpha_1 \alpha_2 \cdots \alpha_q$. Let $\pi'$ denote the product of those elements among $\alpha_1, \cdots, \alpha_q$ which vanish at $H_0$. Define $\mathrm{sp}_{S/J'} x$ $(x \in C(S))$ to be the relative trace from $C(S)$ to $C(J')$.

Corollary. *Select $u_i$ and $u'_j$ $(1 \leq i \leq w, 1 \leq j \leq w')$ as in the corollary to Lemma 8 and define $u^i$ and $u'^j$ in $C(S)$ by the relations $\mathrm{sp}_{S/J}(u^i u_k) = \delta_k^i$*

$(1 \leq i, k \leq w)$ *and* $\mathrm{sp}_{S/J}(u'^j u'_m) = \delta_m{}^j$ $(1 \leq j, m \leq w')$ *respectively. Then* $\pi u^i$ *and* $\pi' u'^j$ *are in* $S$. *Moreover if* $d$ *and* $d'$ *are any two elements in* $S$ *such that* $du^i$ *and* $d'u'^j$ $(1 \leq i \leq w, 1 \leq j \leq w')$ *are all in* $S$, *then* $\pi$ *divides* $d$ *and* $\pi'$ *divides* $d'$ *in* $S$.

We have already seen this for $u^i$. The proof for $u'^j$ is entirely similar.

Choose $v_1, \cdots, v_r$ as in Lemma 8 and let $\mathrm{sp}_{J'/J} x$ $(x \in C(J'))$ denote the relative trace of $x$ from $C(J')$ to $C(J)$.

**LEMMA 12.** *Define* $v^i \in C(J')$ $1 \leq i \leq r$ *by the conditions* $\mathrm{sp}_{J'/J}(v^i v_j)$ $= \delta_j{}^i$ $(1 \leq i, j \leq r)$. *Then an element* $x \in S$ *is divisible by* $\pi/\pi'$ *in* $S$ *if and only if* $xv^i \in S$ $1 \leq i \leq r$.

Let $D$ be a nonzero element in $S$ of the least possible degree such that $Dv^i \in S$. Then the highest common factor (h. c. f.) of $Dv^1, \cdots, Dv^r$ (in $S$) is 1. Define $u'^j$ as above. Then

$$\mathrm{sp}_{S/J}(u'^i v^j u'_k v_m) = \mathrm{sp}_{J'/J}\mathrm{sp}_{S/J'}(u'^i u'_k v^j v_m) = \delta_k{}^i \mathrm{sp}_{J'/J}(v^j v_m) = \delta_k{}^i \delta_m{}^j$$

$(1 \leq i, k \leq w', 1 \leq j, m \leq r)$. Since $S = \sum\limits_{1 \leq k \leq w'} \sum\limits_{1 \leq m \leq r} J v_m u'_k$, it follows from the corollary to Lemma 11 that $\pi u'^i v^j \in S$. Since the h. c. f. of $Dv^j$ $(1 \leq j \leq r)$ is 1, this implies that $D^{-1}\pi u'^i \in S$. Therefore we conclude from the same corollary that $\pi/D\pi'$ is in $S$. Conversely $\pi' u'^i Dv^j \in S$ and therefore $\pi$ divides $\pi'D$ again by the same corollary. Hence $\pi'D/\pi$ lies in $C$ and the lemma follows.

Put $J_H' = J' \cap S_H$ for $H \in \mathfrak{h}_\mathfrak{p}$.

**LEMMA 13.** $\dim J'/J'J_H = [W : W']$, $\dim S/SJ_H = w$ *and* $\dim S/SJ_H'$ $= w'$ *for every* $H \in \mathfrak{h}_\mathfrak{p}$.

We prove only the first statement since the proofs of the other two are parallel. Suppose $\sum\limits_{1 \leq i \leq r} c_i v_i \in J'J_H$ $(c_i \in C)$. Then since $J'J_H = \sum\limits_i J_H v_i$, it follows from the linear independence of $v_1, \cdots, v_r$ over $C(J)$ (see Lemma 7) that $c_i \in J_H$. But obviously this implies that $c_i = 0$ $1 \leq l \leq r$. Hence the residue classes of $v_1, \cdots, v_r \bmod J'J_H$ form a base for $J'/J'J_H$ and therefore $\dim J'/J'J_H = r = [W : W']$.

Put $\sigma^i = \pi u^i$ $1 \leq i \leq w$. We have seen above that $\sigma^i \in S$.

**LEMMA 14.** *For any* $H \in \mathfrak{h}_\mathfrak{p}$, *let* $e_H = \sum\limits_{1 \leq i \leq w} \sigma^i(H) u_i$. *Then*

(1) $\quad pe_H \equiv p(H) e_H \bmod SJ_H$ *for any* $p \in S$,

(2) $\quad \sum\limits_{s \in W} \epsilon(s) e_{sH} \equiv \pi(H) \bmod SJ_H,$

(3)  $e_{sH}(tH) = \epsilon(s)\pi(sH)$ or zero according as $s = t$ or not $(s, t \in W)$.

We know that

$$\delta_j{}^i = \mathrm{sp}_{S/J}(u^i u_j) = \sum_{s \in W}(u^i u_j)^s = \pi^{-1}\sum_{s \in W}\epsilon(s)(\sigma^i u_j)^s.$$

Therefore $\sum_{s \in W}\epsilon(s)(\sigma^i u_j)^s = \pi\delta_j{}^i$ $(1 \leq i, j \leq w)$.  Now suppose $s_1, s_2, \cdots, s_w$ are all the distinct elements of $W$ and $\pi(H) \neq 0$.  Then this result implies that

$$\sum_k \epsilon(s_k)\sigma^i(s_k H)u_j(s_k H) = \pi(H)\delta_j{}^i.$$

Therefore if we regard

$$A = \{\sigma^i(s_k H)\}_{1 \leq i, k \leq w} \text{ and } B = \{u_j(s_k H)/\pi(H)\}_{1 \leq k, j \leq w}$$

as $w \times w$ matrices, it is clear that they are reciprocals of each other and therefore $BA$ is also the unit matrix.  Hence [9]

$$\sum_j \epsilon(s_i)\sigma^j(s_i H)u_j(s_k H) = \pi(H)\delta_{ik}.$$

This proves that $e_{s_i H}(s_k H) = \delta_{ik}\epsilon(s_i)\pi(H)$ and therefore $e_H \in S_{sH}$ if $s \neq 1$. Since $\pi(H) \neq 0$, we know (see the last part of the proof of Lemma 11) that $SJ_H = \bigcap_{s \in W} S_{sH}$.  Therefore $e_H S_H \subset SJ_H$ and hence $pe_H \equiv p(H)e_H \bmod SJ_H$ for $p \in S$.  Moreover it follows from the relation obtained above that $\sum_{s \in W}\epsilon(s)e_{sH}(tH) = \pi(H)$ for $t \in W$.  Therefore $\sum_{s \in W}\epsilon(s)e_H - \pi(H) \in \bigcap_{t \in W} S_{tH} = SJ_H$.  Thus all the statements of the lemma have been proved under the assumption that $\pi(H) \neq 0$.

Now we come to the general case.  Select $x_{ij}{}^k \in J$ such that $u_i u_j = \sum_k x_{ij}{}^k u_k$. Then $e_H u_j = \sum_i \sigma^i(H)u_i u_j = \sum_{i,j}\sigma^i(H)x_{ij}{}^k u_k \equiv \sum_{i,k}\sigma^i(H)x_{ij}{}^k(H)u_k \bmod SJ_H$. But we know from Lemma 13 that $u_1, \cdots, u_w$ are linearly independent $\bmod SJ_H$.  Therefore it is clear that the first statement of the lemma holds if and only $z_j{}^k(H) = 0$ $(1 \leq j, k \leq w)$ where $z_j{}^k = \sum_i \sigma^i x_{ij}{}^k - \sigma^k u_j$.  But then it follows from the above proof that $z_j{}^k(H) = 0$ whenever $\pi(H) \neq 0$.  Therefore since $z_j{}^k$ are polynomial functions on $\mathfrak{h}_\mathfrak{p}$, we conclude that they are all zero.  In the same way the second statement of the lemma is seen to be equivalent to the relations $\sum_{s \in W}\epsilon(s)\sigma^i(sH) = \pi(H)\delta_1{}^i$ $(1 \leq i \leq w)$.  Again since these equations hold when $\pi(H) \neq 0$, they hold in general.  The proof of (3) is entirely similar.

COROLLARY 1.  *If* $\pi(H) \neq 0$ *then* $e_{sH}$ $(s \in W)$ *are linearly independent* $\bmod SJ_H$.

2

This is an immediate consequence of the third statement of the above lemma.

COROLLARY 2. *Let $p$ be an element in $S$ and let $H \in \mathfrak{h}_\mathfrak{p}$. Consider the linear transformation $L$ in $S/SJ_H$ corresponding to multiplication by $p$. Then $\det(tI - L) = \prod_{s \in W} (t - p(sH))$ where $t$ is an indeterminate and $I$ is the identity mapping of $S/SJ_H$.*

Select $q_{ji} \in J$ such that $pu_i = \sum_j u_j q_{ji}$ $(1 \leq i \leq w)$. Then it is obvious that $pu_i \equiv \sum_j q_{ji}(H) u_j \bmod SJ_H$. Put

$$Q(t) = \det(t\delta_{ij} - q_{ij})_{1 \leq i, j \leq w}.$$

Then $Q(t)$ is a polynomial in $t$ with coefficients in $J$. Obviously it is sufficient to prove that $Q(t) = \prod_{s \in W} (t - p^s)$. Let $q_m$ and $p_m$ denote the coefficients of $t^m$ in $Q(t)$ and $\prod_{s \in W} (t - p^s)$ respectively. Put $Q(t:H) = \sum_{m \geq 0} q^m(H) t^m$. If $\pi(H) \neq 0$, $e_{sH}$ $(s \in W)$ form a base for $S \bmod SJ_H$ by Corollary 1 above and therefore it follows from the first statement of Lemma 14 that

$$Q(t:H) = \det(tI - L) = \prod_{s \in W} (t - p(sH)).$$

This shows that $q_m(H) = p_m(H)$ for every $m$, if $\pi(H) \neq 0$. But since $q_m$ and $p_m$ are polynomial functions on $\mathfrak{h}_\mathfrak{p}$, this implies that $q_m = p_m$.

Now put $D = \pi/\pi'$, $\tau^i = Dv^i$ $1 \leq i \leq r$ and select elements $s_1, \cdots, s_r$ in $W$ such that $W = \bigcup_{1 \leq i \leq r} W's_i$. (We recall that $r = [W : W']$.)

LEMMA 15. *Put $f_H = \sum_{1 \leq i \leq r} \tau^i(H) v_i$ for $H \in \mathfrak{h}_\mathfrak{p}$. Then*

(1)   $pf_H \equiv p(H) f_H \bmod SJ_H$ *for* $p \in J'$,

(2)   $\sum_{1 \leq i \leq r} \epsilon(s_i) \pi'(s_i H) f_{s_i H} \equiv \pi(H) \bmod SJ_H,$

(3)   $f_{s_i H}(s_j H) = \delta_{ij} D(s_i H)$ $(1 \leq i, j \leq r).$

*Finally, $f_{s_1 H}, \cdots, f_{s_r H}$ are linearly independent $\bmod SJ_H$ if $\pi(H) \neq 0$.*

The proof is similar to that of Lemma 14. Put $t_k = s_k^{-1}$ $(1 \leq k \leq r)$. Since $C(J')$ is the subfield of $C(S)$ consisting exactly of those elements which are left fixed by $W'$, we conclude from Galois theory that $t_1, \cdots, t_r$ define all the distinct isomorphisms of $C(J')/C(J)$ into $C(S)$. Therefore $\mathrm{sp}_{J'/J} v = \sum_{1 \leq i \leq r} v^{t_i}$ for $v \in C(J')$. This shows that

$$\delta_j{}^i = \mathrm{sp}_{J'/J}(v^i v_j) = \sum_{1 \leq k \leq r} (\tau^i v_j / D)^{t_k}.$$

Multiplying this equation by $\pi$, we get

$$\sum_{1 \leq k \leq r} \epsilon(t_k)\,(\pi'\tau^i v_j)\,t_k = \pi \delta_j{}^i \qquad (1 \leq i, j \leq r).$$

But this implies that

$$\sum_k \epsilon(s_k)\pi'(s_k H)\tau^i(s_k H)v_j(s_k H) = \pi(H)\delta_j{}^i.$$

Now first assume that $\pi(H) \neq 0$. Then the $r \times r$ matrices

$$A = \{\epsilon(s_k)\pi'(s_k H)\tau^i(s_k H)\}_{1 \leq i, k \leq r} \text{ and } B = \{v_j(s_k H)/\pi(H)\}_{1 \leq k, j \leq r}$$

are reciprocals of each other and therefore $BA$ is also the unit matrix. Hence

$$\sum_k \epsilon(s_i)\pi'(s_i H)\tau^k(s_i H)v_k(s_j H) = \pi(H)\delta_{ij}$$

and this proves that $f_{s_i H}(s_j H) = \delta_{ij} D(s_i H)$ $1 \leq i, j \leq r$. (By continuity this relation remains valid even when $\pi(H) = 0$.) Since $\pi(H) \neq 0$, it is clear that $D(sH) \neq 0$ for any $s \in W$. Therefore the linear independence of $f_{s_1 H}, \cdots, f_{s_r H} \bmod SJ_H$ is obvious. Furthermore

$$\sum_{1 \leq i \leq r} \epsilon(s_i)\pi'(s_i H)f_{s_i H}(s_j H) = \epsilon(s_j)\pi'(s_j H)D(s_j H) = \pi(H) \qquad (1 \leq i \leq r).$$

On the other hand $f_{s_i H} \in J'$ and therefore $f_{s_i H}(sH) = f_{s_i H}(s_j H)$ if $s \in W's_j$. Hence if we put $g = \sum_{1 \leq i \leq r} \epsilon(s_i)\pi'(s_i H)f_{s_i H} - \pi(H)$, it follows that $g(sH) = 0$ for all $s \in W$. But this implies that $g \in \bigcap_{s \in W} S_{sH} = SJ_H$ and so we get the second statement of the lemma. Now suppose $p \in J'$. Then $p(sH) = p(H)$ for $s \in W'$ and therefore $p - p(H) \in \bigcap_{s \in W'} S_{sH}$. On the other hand we have seen above that $f_H(sH) = 0$ if $s \notin W'$. Therefore $(p - p(H))f_H \in \bigcap_{s \in W} S_{sH} = SJ_H$ and this proves that $pf_H \equiv p(H)f_H \bmod SJ_H$.

In order to extend our results to the general case, we use the same method as before. For example, let us prove the first statement. It is enough to show that $v_i f_H \equiv v_i(H)f_H \bmod SJ_H$ $1 \leq i \leq r$. Since $w'v = \sum_{s \in W'} v^s$ for $v \in J' \cap (SJ_H)$, it follows that $J' \cap (SJ_H) = J'J_H$. Select $y_{ij}{}^k \in J$ such that $v_i v_j = \sum_k y_{ij}{}^k v_k$. Then due to Lemma 13, the required relations hold if and only if $\sum_j \tau^j(H)y_{ij}{}^k(H) = v_i(H)\tau^k(H)$ $(1 \leq i, k \leq r)$. The remaining argument is now the same as in the proof of Lemma 14.

COROLLARY. *Let $p$ be an element in $J'$ and let $H \in \mathfrak{h}_\mathfrak{p}$. Consider the linear transformation $L'$ in $J'/J'J_H$ corresponding to multiplication by $p$.*

*Then* $\det(tI' - L') = \prod\limits_{1 \leq i \leq r} (t - p(s_iH))$ *where $t$ is an indeterminate and $I'$ is the identity mapping of $J'/J'J_H$.*

This is proved in the same way as Corollary 2 to Lemma 14.

Define $J_+$ as in the proof of Lemma 8 and let $d_\pi$ and $d_i$ denote the degrees of $\pi$ and $u_i$ $(1 \leq i \leq w)$ respectively. Then the following result is of some interest, although it is not strictly necessary for our purpose. Therefore we state it here without proof.

**Lemma 16.** *Suppose $d_w = \max\limits_{1 \leq i \leq w} d_i$. Then $d_w = d_\pi$ and $d_i < d_\pi$ for $1 \leq i \leq w$. Moreover $u_w \equiv c\pi \bmod SJ_+$ where $c$ is a nonzero complex number.*

### 4. The homomorphism $\gamma$.

We now return to the notation of Section 2 and for any $q \in I_\mathfrak{g}$ define $u_q \in \mathfrak{H}_\mathfrak{p}$ as in Lemma 3. Put $\rho = \frac{1}{2} \sum\limits_{\alpha \in P_+} \bar{\alpha}$ where $\bar{\alpha}$ denotes the restriction of $\alpha$ on $\mathfrak{h}_\mathfrak{p}$. Also let $u \to {}^\backprime u$ denote the automorphism of $\mathfrak{H}_\mathfrak{p}$ such that ${}^\backprime H = H - \rho(H)$ $(H \in \mathfrak{h}_\mathfrak{p})$. We identify $S(\mathfrak{h}_\mathfrak{p})$ with $\mathfrak{H}_\mathfrak{p}$ under the isomorphism which preserves every element of $\mathfrak{h}_\mathfrak{p}$. Then $J = I(\mathfrak{h}_\mathfrak{p})$ becomes a subalgebra of $\mathfrak{H}_\mathfrak{p}$. Put $\gamma(q) = {}^\backprime u_q$ $(q \in I_\mathfrak{g})$.

**Theorem 1.** *$\gamma$ is a homomorphism of $I_\mathfrak{g}$ onto $I(\mathfrak{h}_\mathfrak{p})$ and its kernel is $I_\mathfrak{g} \cap \mathfrak{B}\mathfrak{k} = I_\mathfrak{g} \cap \mathfrak{k}\mathfrak{B}$.*

In order to show that $\gamma$ is a homomorphism, it is enough to verify that $u_{q_1q_2} = u_{q_1}u_{q_2}$ for $q_1, q_2 \in I_\mathfrak{g}$. Now $q_2 - u_{q_2} \in \mathfrak{k}\mathfrak{B} + \mathfrak{B}\mathfrak{n}$. Therefore $q_1(q_2 - u_{q_2}) \in \mathfrak{k}\mathfrak{B} + \mathfrak{B}\mathfrak{n}$ since $q_1\mathfrak{k} = \mathfrak{k}q_1$. But $q_1u_{q_2} = (q_1 - u_{q_1})u_{q_2} + u_{q_1}u_{q_2}$ and $(q_1 - u_{q_1})u_{q_2} \in (\mathfrak{k}\mathfrak{B} + \mathfrak{B}\mathfrak{n})u_{q_2} \subset \mathfrak{k}\mathfrak{B} + \mathfrak{B}\mathfrak{n}$ since $[\mathfrak{n}, \mathfrak{h}_\mathfrak{p}] \subset \mathfrak{n}$. Therefore $q_1q_2 - u_{q_1}u_{q_2} \in \mathfrak{k}\mathfrak{B} + \mathfrak{B}\mathfrak{n}$ and this proves that $u_{q_1q_2} = u_{q_1}u_{q_2}$.

Now fix $q$ in $I_\mathfrak{g}$. Then it follows from Lemma 4 that $\gamma(q) = 0$ if and only if $\chi_\nu(u_q) = 0$ for every $\nu \in \mathfrak{F}_0$. Define $\pi$ and $\phi$ as in Lemma 5 and put $g(x) = (\phi, \pi(x)\phi)$ for $x \in G$. Then if $\nu$ denotes the restriction on $\mathfrak{h}_\mathfrak{p}$ of the highest weight of $\pi$, it follows from Lemma 3 that $g(x;q) = \chi_\nu(u_q)g(x)$. Since $g(1) = 1$, this shows that $\chi_\nu(u_q) = g(1;q)$. But since $g$ is spherical, it is clear that $g(1;q) = 0$ if $q \in \mathfrak{k}\mathfrak{B} + \mathfrak{B}\mathfrak{k}$. Hence $I_\mathfrak{g} \cap (\mathfrak{k}\mathfrak{B} + \mathfrak{B}\mathfrak{k})$ is contained in the kernel of $\gamma$. On the other hand $\gamma(q) = 0$ implies that $g(x;q) = (\phi, \pi(x)\pi(q)\phi) = 0$. Since this holds for every $x$ in $G$ and since $\pi$ is irreducible, it follows that $\pi(q)\phi = 0$. Therefore by applying Lemma 1 of [5(d)] to $\mathfrak{Y} = \mathfrak{X}\mathfrak{k}$, we conclude that $q \in \mathfrak{B}\mathfrak{k}$. Combining this with the above result, we deduce that $I_\mathfrak{g} \cap \mathfrak{B}\mathfrak{k}$ is exactly the kernel of $\gamma$ and therefore $I_\mathfrak{g} \cap \mathfrak{B}\mathfrak{k} \supset I_\mathfrak{g} \cap \mathfrak{k}\mathfrak{B}$. Now consider the anti-automorphism $b \to b^*$ of $\mathfrak{B}$ such that $X^* = -X$ for every $X \in \mathfrak{g}$. It is obvious that $I_\mathfrak{g}^* = I_\mathfrak{g}$. Hence

$(I_\mathfrak{g} \cap \mathfrak{Bk})^* = I_\mathfrak{g} \cap \mathfrak{kB} \supset (I_\mathfrak{g} \cap \mathfrak{kB})^* = I_\mathfrak{g} \cap \mathfrak{Bk}$. This proves that $I_\mathfrak{g} \cap \mathfrak{Bk} = I_\mathfrak{g} \cap \mathfrak{kB}$.

In order to prove that $\gamma$ maps $I_\mathfrak{g}$ into $J$ we need some additional facts. For any $h \in A_\mathfrak{p}$ and $s \in W$, put $h^s = \exp(sH)$ where $H = \log h$ and let $dn$ denote the Haar measure on $N$.

LEMMA 17. *Put*

$$F_f(h) = e^{\rho(\log h)} \int_N f(hn)\, dn \qquad\qquad (h \in A_\mathfrak{p})$$

*for any function $f$ in* [17] *$C_c^\infty(G)$ such that $f(kxk^{-1}) = f(x)$ $(x \in G, k \in K)$. Then $F_f(h^s) = F_f(h)$ for $s \in W$ and $h \in A_\mathfrak{p}$.*

Let $x \to x^*$ denote the natural mapping of $G$ on the factor space $G^* = G/A_\mathfrak{p}$. Put $D(h) = \prod_{\alpha \in P_+} (e^{\alpha(H)/2} - e^{-\alpha(H)/2})$ and $h^{x^*} = xhx^{-1}$ $(h \in A_\mathfrak{p}, x \in G)$ where $H = \log h$. Denote the invariant measure on $G^*$ by $dx^*$ and consider the set $A_\mathfrak{p}'$ of those elements $h \in A_\mathfrak{p}$ where $D(h) \neq 0$. Then [18] the integral $\int_{G^*} g(h^{x^*})\, dx^*$ is convergent for $g \in C_c(G)$ and $h \in A_\mathfrak{p}'$. Moreover the measure $dn$ can be so normalized that

$$|D(h)| \int_{G^*} g(h^{x^*})\, dx^* = e^{\rho(\log h)} \int_{K \times N} g(khnk^{-1})\, dk\, dn \qquad (h \in A_\mathfrak{p}').$$

Let $s$ be an element in $W$ and select $k$ in $K$ such that $khk^{-1} = h^s$ for $h \in A_\mathfrak{p}$. Then for any $x \in G$, the coset $(xk)^*$ depends only on $x^*$ and $s$ and we denote it by $x^*s$. Since $W$ is a finite group, it follows without difficulty that the measure $dx^*$ is invariant under these operations of $W$ on $G^*$. Therefore

$$\int_{G^*} g((h^s)^{x^*})\, dx^* = \int_{G^*} g(h^{x^*})\, dx^*$$

for $g \in C_c(G)$ and $h \in A_\mathfrak{p}'$. If we apply this to $f$ and make use of the fact that $|D(h^s)| = |D(h)|$, it follows that $F_f(h^s) = F_f(h)$ for $h \in A_\mathfrak{p}'$. However it is obvious from its definition that $F_f \in C_c^\infty(A_\mathfrak{p})$. Therefore $F_f(h^s) = F_f(h)$ for all $h \in A_\mathfrak{p}$.

COROLLARY. *For any linear function $\lambda$ on $\mathfrak{h}_\mathfrak{p}$, put*

$$\phi_\lambda(x) = \int_K \exp\{(-1)^{\frac{1}{2}}\lambda(H(xk)) - \rho(H(xk))\}\, dk \qquad (x \in G).$$

*Then $\phi_{s\lambda} = \phi_\lambda$ for $s \in W$.*

---

[17] See footnote 1 of [5(k)] for notation.

[18] The facts stated here are quite well known. In any case their proofs offer no difficulty (see [5(f), p. 508]).

Obviously $\phi_\lambda$ is a spherical function of class $C^\infty$. Hence if $dx$ is the Haar meaure on $G$, it would be enough to show that for a fixed $s$,

$$\int \phi_\lambda(x)f(x)\,dx = \int \phi_{s\lambda}(x)f(x)\,dx$$

for every spherical function $f$ in $C_c^\infty(G)$. Let $dh$ denote the Haar measure on $A_\mathfrak{p}$. Then it is possible (see [5(c), Lemma 35]) to normalize it in such a way that $dx = e^{2\rho(\log h)}dk\,dh\,dn$ if $x = khn$ ($k \in K, h \in A_\mathfrak{p}, n \in N$). Therefore

$$\int \phi_\lambda(x)f(x)\,dx = \int \exp\{(-1)^{\frac{1}{2}}\lambda(H(x)) - \rho(H(x))\}f(x)\,dx$$

$$= \int \exp\{(-1)^{\frac{1}{2}}\lambda(\log h) + \rho(\log h)\}f(hn)\,dh\,dn$$

$$= \int \exp\{(-1)^{\frac{1}{2}}\lambda(\log h)\}F_f(h)\,dh.$$

Our assertion now follows immediately from the above lemma.

Now suppose $q \in I_\mathfrak{g}$. Then $\gamma(q)$, being an element of $S(\mathfrak{h}_\mathfrak{p})$, can be regarded as a polynomial function on $\mathfrak{h}_\mathfrak{p}$. We denote by $\gamma(q:H)$ the value of this function at $H \in \mathfrak{h}_\mathfrak{p}$. Also let us recall that $\mathfrak{h}_\mathfrak{p}$ has been identified with its dual under the bilinear form $B$. Hence the following result is an immediate consequence of Lemma 3 and the definition of $\gamma(q)$.

LEMMA 18. $\phi_\lambda(x;q) = \gamma(q:(-1)^{\frac{1}{2}}\lambda)\phi_\lambda(x)$ for $q \in I_\mathfrak{g}$, $x \in G$ and any linear function $\lambda$ on $\mathfrak{h}_\mathfrak{p}$.

It is obvious from this lemma and the corollary to Lemma 17 that $\gamma(q:H) = \gamma(q:sH)$ for $s \in W$ and $H \in \mathfrak{h}_\mathfrak{p}$. Hence $\gamma(q) = \gamma(q)^s$ and this proves that $\gamma(q) \in J$.

We still have to show that $\gamma$ maps $I_\mathfrak{g}$ onto $J$. Le $\lambda$ denote the canonical mapping (see [5(c), p. 192]) of [11] $S(\mathfrak{g})$ onto $\mathfrak{B}$ and put $\mathfrak{P} = \lambda(S(\mathfrak{p}))$ and $\mathfrak{P}_d = \lambda(S_d(\mathfrak{p}))$ where $S_d(\mathfrak{p})$ is the set of all homogeneous elements in $S(\mathfrak{p})$ of degree $d$. Since $\mathfrak{g}$ is the direct sum of $\mathfrak{k}$ and $\mathfrak{p}$, it follows [5(c), Lemma 12] that $\mathfrak{B} = \mathfrak{K}\mathfrak{P}$ and therefore $\mathfrak{B} = \mathfrak{P} + \mathfrak{k}\mathfrak{B}$, the sum being direct. Moreover if $X \in \mathfrak{k}$ it is obvious that both $\mathfrak{P}$ and $\mathfrak{k}\mathfrak{B}$ are invariant under the mapping [19] $b \to [X, b]$ ($b \in \mathfrak{B}$) of $\mathfrak{B}$. Hence $I_\mathfrak{g} = I_\mathfrak{g} \cap \mathfrak{P} + I_\mathfrak{g} \cap \mathfrak{k}\mathfrak{B}$ and so it is sufficient to prove the following lemma.

LEMMA 19. $\gamma$ defines a linear isomorphism of $I_\mathfrak{g} \cap \mathfrak{P}$ onto $J$.

Since $I_\mathfrak{g} \cap \mathfrak{k}\mathfrak{B}$ is the kernel of $\gamma$ and $\mathfrak{P} \cap \mathfrak{k}\mathfrak{B} = \{0\}$, it is clear that $\gamma$ is univalent on $I_\mathfrak{g} \cap \mathfrak{P}$. Also it follows from Lemma 11 of [5(c)] that

---

[19] $[a, b] = ab - ba$ for $a, b \in \mathfrak{B}$.

$I_\mathfrak{g} \cap \mathfrak{P} = \lambda(I(\mathfrak{p}))$ (in the notation of Lemma 7). Let $u$ be a homogeneous element in $J$ of degree $d$. We shall prove by induction on $d$ that there exists an element $q \in \sum_{e \leqq d} I_\mathfrak{g} \cap \mathfrak{P}_e$ such that $\gamma(q) = u$. This is obvious if $d = 0$. So suppose $d \geqq 1$. Let $\bar{p}$ denote the restriction on $\mathfrak{h}_\mathfrak{p}$ of any polynomial function $p \in S(\mathfrak{g})$. By Lemma 7, we can select a homogeneous element $p_0 \in I(\mathfrak{p})$ of degree $d$ such that $\bar{p}_0 = u$. Put $q_0 = \lambda(p_0)$. Then $q_0$ lies in $I_\mathfrak{g} \cap \mathfrak{P}$. We shall prove that $u - \gamma(q_0)$ is of degree lower than $d$ and hence, by induction hypothesis, we can choose $q_1 \in I_\mathfrak{g} \cap \mathfrak{P}$ of degree $< d$ such that $u - \gamma(q_0) = \gamma(q_1)$. Then $u = \gamma(q)$ where $q = q_0 + q_1$ and this would prove the lemma.

We know that $\mathfrak{g}$ is the direct sum of $\mathfrak{k}$, $\mathfrak{h}_\mathfrak{p}$ and $\mathfrak{n}$. Hence [11] $S(\mathfrak{g}) = S(\mathfrak{k})S(\mathfrak{h}_\mathfrak{p})S(\mathfrak{n})$. Now $\mathfrak{h}_\mathfrak{p}$ is orthogonal to $\mathfrak{k} + \mathfrak{n}$ under the bilinear form $B(X,Y)$ $(X,Y \in \mathfrak{g})$. Hence if $p \in S(\mathfrak{g})$, it is obvious that $p - \bar{p} \in \mathfrak{k}S(\mathfrak{g}) + S(\mathfrak{g})\mathfrak{n}$ (the products being taken in $S(\mathfrak{g})$). Therefore it is clear that $p_0 - u \in \mathfrak{k}S_{d-1}(\mathfrak{g}) + S_{d-1}(\mathfrak{g})\mathfrak{n}$, where $S_e(\mathfrak{g})$ denotes the space of homogeneous elements in $S(\mathfrak{g})$ of degree $e$. But then

$$q_0 - u \in \lambda(\mathfrak{k}S_{d-1}(\mathfrak{g}) + S_{d-1}(\mathfrak{g})\mathfrak{n}).$$

On the other hand if $f_1 \in S_{e_1}(\mathfrak{g})$, $f_2 \in S_{e_2}(\mathfrak{g})$, we know (see [5(a), p. 902]) that

$$\lambda(f_1 f_2) - \lambda(f_1)\lambda(f_2) \in \sum_{e < e_1 + e_2} \lambda(S_e(\mathfrak{g})).$$

Therefore

$$q_0 - u \in \mathfrak{k}\mathfrak{P} + \mathfrak{P}\mathfrak{n} + \sum_{e < d} \lambda(S_e(\mathfrak{g})).$$

Hence it follows from Lemma 3 and the definition of $\gamma$ that $\gamma(q_0) - u$ is of degree $< d$. The proof of Theorem 1 is now complete.[20]

The following result, which we note here for later use, has been obtained during the above proof.

LEMMA 20. *Let $\lambda$ denote the canonical mapping of $S(\mathfrak{g})$ onto $\mathfrak{P}$. Then if $p$ is any homogeneous element of degree $d$ in $I(\mathfrak{p})$, $\gamma(\lambda(p)) - \bar{p}$ is of degree $< d$. (Here $\bar{p}$ denotes the restriction of $p$ on $\mathfrak{h}_\mathfrak{p}$.)*

**5. The mapping $\mathfrak{d}'$.** Put $\Delta(H) = \prod_{\alpha \in P_+} (e^{\alpha(H)} - e^{-\alpha(H)})$ $(H \in \mathfrak{h}_\mathfrak{p})$ and $\Delta(h) = \Delta(\log h)$ $(h \in A_\mathfrak{p})$. We call a point $h$ in $A_\mathfrak{p}$ singular or regular according as $\Delta(h) = 0$ or not. Let $A_\mathfrak{p}'$ denote the set of regular points in $A_\mathfrak{p}$. We regard $\mathfrak{g}$ as a Hilbert space under the positive-definite Hermitian form

---

[20] Although our definition of $\gamma$ makes use of an order in the space of real linear functions on $\mathfrak{h}_{\mathfrak{p}_0}$, it can be shown that $\gamma$ is actually independent of the choice of this order (cf. [5(g), Lemma 20]).

$-B(X, \bar{\theta}(X))$ $(X \in \mathfrak{g})$ where $\bar{\theta}$ denotes the conjugation of $\mathfrak{g}$ with respect to its compact real form $\mathfrak{u} = \mathfrak{k}_0 + (-1)^{\frac{1}{2}}\mathfrak{p}_0$. Let $\mathfrak{m}$ be the centralizer of $\mathfrak{h}_\mathfrak{p}$ in $\mathfrak{g}$ and $\mathfrak{q}$ the orthogonal complement of $\mathfrak{m}$ in $\mathfrak{g}$. Then $\mathfrak{m} \cap \mathfrak{p} = \mathfrak{h}_\mathfrak{p}$ (because $\mathfrak{h}_\mathfrak{p}$ is maximal abelian in $\mathfrak{p}$) and $\mathfrak{q} = \mathfrak{q}_\mathfrak{k} + \mathfrak{q}_\mathfrak{p}$ where $\mathfrak{q}_\mathfrak{k} = \mathfrak{q} \cap \mathfrak{k}$ and $\mathfrak{q}_\mathfrak{p} = \mathfrak{q} \cap \mathfrak{p}$.

LEMMA 21. *Suppose* $h \in A_\mathfrak{p}'$. *Then* $\mathfrak{g} = \mathrm{Ad}(h^{-1})\mathfrak{q}_\mathfrak{k} + \mathfrak{h}_\mathfrak{p} + \mathfrak{k}$ *where the sum is direct.*

It is known [5(c), Lemma 4] that $\dim \mathfrak{q}_\mathfrak{k} = \dim \mathfrak{q}_\mathfrak{p}$. On the other hand $\dim \mathfrak{g} = \dim \mathfrak{k} + \dim \mathfrak{p}$ and $\mathfrak{p}$ is the orthogonal sum of $\mathfrak{h}_\mathfrak{p}$ and $\mathfrak{q}_\mathfrak{p}$. Hence $\dim \mathfrak{g} = \dim \mathfrak{k} + \dim \mathfrak{h}_\mathfrak{p} + \dim \mathfrak{q}_\mathfrak{p} = \dim \mathfrak{k} + \dim \mathfrak{h}_\mathfrak{p} + \dim \mathfrak{q}_\mathfrak{k}$. Therefore it is sufficient to prove that $\mathrm{Ad}(h^{-1})\mathfrak{q}_\mathfrak{k} \cap (\mathfrak{h}_\mathfrak{p} + \mathfrak{k}) = \{0\}$. Let $X$ be an element of $\mathfrak{q}_\mathfrak{k}$ such that $\mathrm{Ad}(h^{-1})X \in \mathfrak{h}_\mathfrak{p} + \mathfrak{k}$ and put $Y = (\mathrm{ad}\, H)^2 X$ where $H$ is some element in $\mathfrak{h}_{\mathfrak{p}_0}$. Then $Y \in \mathfrak{k}$ and $\mathrm{Ad}(h^{-1})Y \in \mathfrak{k}$. Therefore $\mathrm{Ad}(h)Y = \theta(\mathrm{Ad}(h^{-1})Y) = \mathrm{Ad}(h^{-1})Y$ where $\theta$ is defined as usual (see [5(c), p. 187]). But since $h \in A_\mathfrak{p}'$, $\mathfrak{m}$ is the centralizer of $h^2$ in $\mathfrak{g}$ and therefore $Y \in \mathfrak{m}$. This means that $(\mathrm{ad}\, H)^3 X = (\mathrm{ad}\, H)Y = 0$. However $\mathrm{ad}\, H$ is a self-adjoint operator on $\mathfrak{g}$ and so we conclude that $[H, X] = 0$. This being true for every $H \in \mathfrak{h}_{\mathfrak{p}_0}$, it follows that $X \in \mathfrak{m}$. But $\mathfrak{m} \cap \mathfrak{q} = \{0\}$, and therefore $X = 0$. This proves the lemma.

Let $\mathfrak{Q}_\mathfrak{k}$ be the image of [11] $S(\mathfrak{q}_\mathfrak{k})$ in $\mathfrak{B}$ under the canonical mapping.

COROLLARY 1. *The mapping* [12] $q \mathbf{X} v \mathbf{X} x \to q^{h^{-1}} v x$ $(q \in \mathfrak{Q}_\mathfrak{k}, v \in \mathfrak{H}_\mathfrak{p}, x \in \mathfrak{X})$ *defines a linear isomorphism of the tensor product* $\mathfrak{Q}_\mathfrak{k} \mathbf{X} \mathfrak{H}_\mathfrak{p} \mathbf{X} \mathfrak{X}$ *onto* $\mathfrak{B}$ *if* $h \in A_\mathfrak{p}'$.

This is an immediate consequence of the above result and Lemma 12 of [5(c)].

Put $\mathfrak{m}_\mathfrak{k} = \mathfrak{m} \cap \mathfrak{k}$ and let $\mathfrak{M}_\mathfrak{k}$ be the subalgebra of $\mathfrak{B}$ generated by $(1, \mathfrak{m}_\mathfrak{k})$. Since $\mathfrak{k} = \mathfrak{m}_\mathfrak{k} + \mathfrak{q}_\mathfrak{k}$, it follows from Lemma 12 of [5(c)] that $\mathfrak{X} = \mathfrak{Q}_\mathfrak{k} \mathfrak{M}_\mathfrak{k}$. Let $\mathfrak{Q}_\mathfrak{k}'$ denote the set of those elements in $\mathfrak{Q}_\mathfrak{k}$ whose homogeneous component [21] of degree zero is zero. Since $[\mathfrak{m}_\mathfrak{k}, \mathfrak{q}_\mathfrak{k}] \subset \mathfrak{q}_\mathfrak{k}$, it follows without difficulty that

$$\mathfrak{X}\mathfrak{k} = \mathfrak{k}\mathfrak{X} = \mathfrak{Q}_\mathfrak{k}'\mathfrak{M}_\mathfrak{k} + \mathfrak{Q}_\mathfrak{k}\mathfrak{M}_\mathfrak{k}\mathfrak{m}_\mathfrak{k}.$$

Hence in particular $\mathfrak{k}\mathfrak{Q}_\mathfrak{k} \subset \mathfrak{Q}_\mathfrak{k}'\mathfrak{M}_\mathfrak{k} + \mathfrak{Q}_\mathfrak{k}\mathfrak{M}_\mathfrak{k}\mathfrak{m}_\mathfrak{k}$. Moreover $\mathfrak{B} = (\mathfrak{Q}_\mathfrak{k})^{h^{-1}}\mathfrak{H}_\mathfrak{p}\mathfrak{X}$. from the above corollary and therefore $(\mathrm{Ad}(h^{-1})\mathfrak{k})\mathfrak{B} = (\mathfrak{k}\mathfrak{Q}_\mathfrak{k})^{h^{-1}}\mathfrak{H}_\mathfrak{p}\mathfrak{X}$. Since $[\mathfrak{m}, \mathfrak{h}_\mathfrak{p}] = \{0\}$, this implies that

---

[21] Define $\lambda$ and $S_d(\mathfrak{g})$ as during the proof of Lemma 19. Then $\mathfrak{B}$ is the direct sum of $\lambda(S_d(\mathfrak{g}))$ $(d \geq 0)$ and an element $b \in \mathfrak{B}$ is homogeneous of degree $d$ if $b \in \lambda(S_d(\mathfrak{g}))$.

$$\mathfrak{k}^{h-1}\mathfrak{B} \subset (\mathfrak{Q}_\mathfrak{t}')^{h-1}\mathfrak{H}_\mathfrak{p}\mathfrak{X} + \mathfrak{B}\mathfrak{k}$$

and therefore

$$\mathfrak{k}^{h-1}\mathfrak{B} + \mathfrak{B}\mathfrak{k} = (\mathfrak{Q}_\mathfrak{t}')^{h-1}\mathfrak{H}_\mathfrak{p}\mathfrak{X} + (\mathfrak{Q}_\mathfrak{t})^{h-1}\mathfrak{H}_\mathfrak{p}\mathfrak{X}\mathfrak{k}.$$

Combining this with Corollary 1 above, we get the following result.

COROLLARY 2. $\mathfrak{B}$ *is the direct sum of* $\mathfrak{H}_\mathfrak{p}$ *and* $\mathfrak{k}^{h-1}\mathfrak{B} + \mathfrak{B}\mathfrak{k}$ *if* $h \in A_\mathfrak{p}'$.

But this obviously implies the lemma below.

LEMMA 22. *Let* $h$ *be an element in* $A_\mathfrak{p}'$. *Then for any* $b \in \mathfrak{B}$ *there exists a unique element* $\delta_h'(b) \in \mathfrak{H}_\mathfrak{p}$ *such that*

$$b - \delta_h'(b) \in \mathfrak{k}^{h-1}\mathfrak{B} + \mathfrak{B}\mathfrak{k}.$$

The significance of $\delta_h'(b)$ is given by the following lemma.[22]

LEMMA 23. *Let* $\phi$ *be a spherical function on* $G$ *of class* $C^\infty$. *Then*[11]

$$\phi(h;b) = \phi(h;\delta_h'(b))$$

*for* $b \in \mathfrak{B}$ *and* $h \in A_\mathfrak{p}'$.

Since $\phi$ is spherical, it is obvious that $\phi(h;b') = 0$ for $b' \in \mathfrak{k}^{h-1}\mathfrak{B} + \mathfrak{B}\mathfrak{k}$. Therefore our assertion follows from the definition of $\delta_h'(b)$.

Now we regard $A_\mathfrak{p}'$ as an open submanifold (see [5(g), p. 110]) of the Lie group $A_\mathfrak{p}$. Then if $b \in \mathfrak{B}$, we intend to show that there exists a differential operator $\delta'(b)$ on $A_\mathfrak{p}'$ whose local expression (see [5(g), p. 112]) at any point $h \in A_\mathfrak{p}'$ coincides with $\delta_h'(b)$.

Define $\pi$ as in Lemma 11 and let $\mathfrak{h}_{\mathfrak{p}_0}'$ be the set of those points $H \in \mathfrak{h}_{\mathfrak{p}_0}$ where $\pi(H) \neq 0$. It is obvious that $\mathfrak{h}_{\mathfrak{p}_0}'$ is mapped onto $A_\mathfrak{p}'$ under the exponential mapping. Let $\mathfrak{R}$ be the ring of analytic functions on $A_\mathfrak{p}'$ generated[23] (over $C$) by the functions $g_\beta$ ($\beta \in P_+$) defined by

$$g_\beta(\exp H) = (e^{2\beta(H)} - 1)^{-1} \qquad (H \in \mathfrak{h}_{\mathfrak{p}_0}').$$

Note that $Hg_\beta = -2\beta(H)(g_\beta + g_\beta{}^2)$ for $H \in \mathfrak{h}_\mathfrak{p}$ and therefore $\mathfrak{R}$ is stable under the action of any differential operator in $\mathfrak{H}_\mathfrak{p}$. Let $\mathfrak{B}_d$ denote the space of all elements in $\mathfrak{B}$ of degree $\leq d$.

LEMMA 24. *Let* $b$ *be an element in* $\mathfrak{B}_d$. *Then we can choose* $u \in \mathfrak{H}_\mathfrak{p} \cap \mathfrak{B}_d$ *and a finite set of elements* $v_i \in \mathfrak{H}_\mathfrak{p} \cap \mathfrak{B}_{d-1}$ *and* $g_i \in \mathfrak{R}$ ($1 \leq i \leq r$) *such that*

---

[22] On account of our identification of $\mathfrak{H}_\mathfrak{p}$ with $S(\mathfrak{h}_\mathfrak{p})$, elements of $\mathfrak{H}_\mathfrak{p}$ can appear in two different roles, either as differential operators on $G$ (or $A_\mathfrak{p}$) or as polynomial functions on $\mathfrak{h}_\mathfrak{p}$. They should always be interpreted as differential operators unless some qualification to the contrary is made.

[23] Note that the constant function 1 is not included among the generators of $\mathfrak{R}$.

$$b - u - \sum_{1 \leq i \leq r} g_i(h) v_i \in \mathfrak{B}\mathfrak{k} + \mathfrak{k}^{h-1}\mathfrak{B}$$

*for all $h \in A_\mathfrak{p}'$.*

We shall use induction on $d$. Since $\mathfrak{g} = \mathfrak{k} + \mathfrak{h}_\mathfrak{p} + \theta(\mathfrak{n})$ and the sum is direct, it follows from [5(c), Lemma 12] that [11]

$$\mathfrak{B}_d = \sum_{p+q+r \leq d} \lambda(S_p(\theta(\mathfrak{n})))\lambda(S_q(\mathfrak{h}_\mathfrak{p}))\lambda(S_r(\mathfrak{k}))$$

where $\lambda$ is the canonical mapping of $S(\mathfrak{g})$ onto $\mathfrak{B}$ and $S_i$ refers to the space of homogeneous elements of degree $i$. Hence we can select $u \in \mathfrak{S}_\mathfrak{p} \cap \mathfrak{B}_d$ and $b_\beta \in \mathfrak{B}_{d-1}$ such that

$$b - u - \sum_{\beta \in P_+} \theta(X_\beta) b_\beta \in \mathfrak{B}\mathfrak{k}.$$

Now let $X_\beta = Y_\beta + Z_\beta$ $(Y_\beta \in \mathfrak{p}, Z_\beta \in \mathfrak{k})$. Then if $h \in A_\mathfrak{p}'$ and $H = \log h$,

$$\mathrm{Ad}(h^{-1})Z_\beta = \tfrac{1}{2}(\mathrm{Ad}(h^{-1})X_\beta + \theta(\mathrm{Ad}(h)X_\beta))$$
$$= -\tfrac{1}{2}(e^{\beta(H)} - e^{-\beta(H)})Y_\beta + \tfrac{1}{2}(e^{\beta(H)} + e^{-\beta(H)})Z_\beta.$$

Hence $Y_\beta \equiv (1 + 2g_\beta(h))Z_\beta \bmod \mathfrak{k}^{h-1}$. Therefore since $\theta(X_\beta) = -Y_\beta + Z_\beta$, we get the following result which will be useful later on.

LEMMA 25. *For any $\beta \in P_+$ let $X_\beta = Y_\beta + Z_\beta$ where $Y_\beta \in \mathfrak{p}$ and $Z_\beta \in \mathfrak{k}$. Then $\theta(X_\beta) \equiv -2g_\beta(h)Z_\beta \bmod \mathfrak{k}^{h-1}$ for $h \in A_\mathfrak{p}'$.*

This means that

$$b \equiv u - 2 \sum_{\beta \in P_+} g_\beta(h)[Z_\beta, b_\beta] \bmod (\mathfrak{B}\mathfrak{k} + \mathfrak{k}^{h-1}\mathfrak{B})$$

for all $h \in A_\mathfrak{p}'$. But $[Z_\beta, b_\beta] \in \mathfrak{B}_{d-1}$ and so the assertion of Lemma 24 follows immediately by induction hypothesis.

Now it is obvious from the definition of $\delta_h'(b)$ that $\delta_h'(b) = u + \sum_{1 \leq i \leq r} g_i(h)v_i$ for $h \in A_\mathfrak{p}'$, in the notation of Lemma 24. Therefore since $g_i$ are analytic functions on $A_\mathfrak{p}'$, $\delta'(b) = u + \sum_{1 \leq i \leq r} g_i v_i$ is an analytic differential operator on $A_\mathfrak{p}'$ whose local expression at $h$ is $\delta_h'(b)$.

**6. The relation between $\gamma$ and $\delta$.** Let $\mu$ be a linear function on $\mathfrak{h}_\mathfrak{p}$. Then $e^\mu$ is a holomorphic function on the complex Euclidean space $\mathfrak{h}_\mathfrak{p}$. We agree to denote the function $h \to e^{\mu(\log h)}$ on $A_\mathfrak{p}$ also by $e^\mu$. Now put [24] $\delta(b) = e^\rho \delta'(b) \circ e^{-\rho}$ $(b \in \mathfrak{B})$ where $\rho$ is defined as in Section 4. The following lemma establishes a connection between $\delta(q)$ and $\gamma(q)$ for $q \in I_\mathfrak{g}$.

---

[24] Whenever it is necessary to avoid confusion we denote the operator product of two differential operators $D$, $D'$ by $D \circ D'$. Since a function $f$ of class $C^\infty$ can also be regarded as a differential operator, one has to distinguish between $Df$ and $D \circ f$.

LEMMA 26. *Let $q$ be an element in $I_{\mathfrak{g}}$ of degree $d$. Then we can select a finite number of elements $v_i \in \mathfrak{H}_{\mathfrak{p}} \cap \mathfrak{B}_{d-1}$ and $g_i \in \mathfrak{R}$ $(1 \leqq i \leqq r)$ such that*

$$\delta(q) = \gamma(q) + \sum_{1 \leqq i \leqq r} g_i v_i$$

*on $A_{\mathfrak{p}}'$.*

We have seen in Section 4 that $I_{\mathfrak{g}} = I_{\mathfrak{g}} \cap \mathfrak{P} + I_{\mathfrak{g}} \cap \mathfrak{B}\mathfrak{k}$. Moreover it is easy to see that $\mathfrak{B}$ is the direct sum of $\mathfrak{P}$ and $\mathfrak{B}\mathfrak{k}$ and

$$\mathfrak{B}_d = (\mathfrak{B}_d \cap \mathfrak{P}) + (\mathfrak{B}_d \cap \mathfrak{B}\mathfrak{k}).$$

Therefore if $q_1$, $q_2$ are the components of $q$ in $I_{\mathfrak{g}} \cap \mathfrak{P}$ and $I_{\mathfrak{g}} \cap \mathfrak{B}\mathfrak{k}$ respectively, they are both of degree $\leqq d$. Also $\delta(q_2) = 0$, $\gamma(q_2) = 0$ and every homogeneous component of $q_1$ lies in $I_{\mathfrak{g}} \cap \mathfrak{P}$. Hence it would clearly be sufficient to consider the case when $q \in I_{\mathfrak{g}} \cap \mathfrak{P}$.

Let $b \to b^*$ denote the anti-automorphism of $\mathfrak{B}$ given by $X^* = -X$ $(X \in \mathfrak{g})$. Then if $p$ is a homogeneous polynomial in $S(\mathfrak{g})$ of degree $r$ and $\lambda$ the canonical mapping of $S(\mathfrak{g})$ into $\mathfrak{B}$, it is clear that $\lambda(p)^* = (-1)^r \lambda(p)$. But $X^* = \theta(X)$ for $X \in \mathfrak{p}$. Hence if we extend $\theta$ to an automorphism of $\mathfrak{B}$, it follows that $\theta(p) = p^*$ for $p \in \mathfrak{P}$. Let $\gamma'(q)$ denote the element in $\mathfrak{H}_{\mathfrak{p}}$ such that $q - \gamma'(q) \in \mathfrak{B}\mathfrak{n} + \mathfrak{k}\mathfrak{B}$ (see Lemma 3). Then $q^* - (\gamma'(q))^* \in \mathfrak{n}\mathfrak{B} + \mathfrak{B}\mathfrak{k}$ and therefore

$$q - \gamma'(q) = \theta(q^*) - \theta(\gamma'(q)^*) \in \theta(\mathfrak{n})\mathfrak{B} + \mathfrak{B}\mathfrak{k}.$$

Therefore we can select $b_\beta \in \mathfrak{B}_{d-1}$ such that

$$q - \gamma'(q) - \sum_{\beta \in P_+} \theta(X_\beta) b_\beta \in \mathfrak{B}\mathfrak{k}.$$

But then from Lemma 25,

$$q \equiv \gamma'(q) - \sum_{\beta \in P_+} g_\beta(h) b_\beta' \bmod (\mathfrak{B}\mathfrak{k} + \mathfrak{k}^{h-1}\mathfrak{B})$$

for $h \in A_{\mathfrak{p}}'$ where $b_\beta' = 2[Z_\beta, b_\beta] \in \mathfrak{B}_{d-1}$. Therefore by Lemma 24 we can select a finite number of elements $v_i' \in \mathfrak{H}_{\mathfrak{p}} \cap \mathfrak{B}_{d-1}$ and $g_i \in \mathfrak{R}$ $(1 \leqq i \leqq r)$ such that

$$q \equiv \gamma'(q) + \sum_{1 \leqq i \leqq r} g_i(h) v_i' \bmod (\mathfrak{B}\mathfrak{k} + \mathfrak{k}^{h-1}\mathfrak{B}) \qquad (h \in A_{\mathfrak{p}}').$$

But this means that $\delta'(q) = \gamma'(q) + \sum_i g_i v_i'$ and therefore

$$\delta(q) = \gamma(q) + \sum_{1 \leqq i \leqq r} g_i v_i$$

where [22] $v_i = e^\rho v_i' \circ e^{-\rho}$. Since it is obvious that $v_i \in \mathfrak{H}_{\mathfrak{p}}$, our assertion is proved.

Now select homogeneous elements $u_1 = 1, u_2, \cdots, u_w$ in $\mathfrak{H}_{\mathfrak{p}}$ in accordance with the corollary of Lemma 8, so that $\mathfrak{H}_{\mathfrak{p}} = \sum_{1 \leq i \leq w} J u_i$ where $J = I(\mathfrak{h}_{\mathfrak{p}})$. We know from Theorem 1 that $\gamma(I_{\mathfrak{g}}) = J$.

THEOREM 2. *Let $u$ be an element in $\mathfrak{H}_{\mathfrak{p}}$ and select $q_i \in I_{\mathfrak{g}}$ $(1 \leq i \leq w)$ such that $u = \sum_{1 \leq i \leq w} \gamma(q_i) u_i$. Then we can choose a finite number of elements $g_{ij} \in \mathfrak{R}$ and $q_{ij} \in I_{\mathfrak{g}}$ $(1 \leq i \leq w, 1 \leq j \leq r)$ such that*

$$u = \sum_{1 \leq i \leq w} u_i \circ \delta(q_i) + \sum_{1 \leq i \leq w} \sum_{1 \leq j \leq r} g_{ij} u_i \circ \delta(q_{ij})$$

*on $A_{\mathfrak{p}}'$.*

We shall use induction on the degree $d$ of $u$. Obviously $u$ can be assumed to be homogeneous. Put $I_{\mathfrak{p}} = I_{\mathfrak{g}} \cap \mathfrak{P}$. If we replace each $q_i$ by its component in $I_{\mathfrak{p}}$ (with respect to the direct sum $\mathfrak{B} = \mathfrak{P} + \mathfrak{Bf}$), neither $\gamma(q_i)$ nor $\delta(q_i)$ is affected. Hence we can suppose that $q_i \in I_{\mathfrak{p}}$. Let $d_i$ denote the degree of $u_i$. Then it follows from our assumptions (see the corollary of Lemma 8) that $\gamma(q_i)$ is homogeneous of degree $d - d_i$ if $d \geq d_i$ and $\gamma(q_i) = 0$ if $d_i > d$. Now fix $i$ and first suppose $d \geq d_i$. Then by Lemma 20, $q_i$ is of degree $d - d_i$ and $\gamma(q_i) u_i = u_i \circ \delta(q_i) - u_i \circ (\delta(q_i) - \gamma(q_i))$ on $A_{\mathfrak{p}}'$. Since $\mathfrak{R}$ is stable under the operations of $\mathfrak{H}_{\mathfrak{p}}$, it follows from Lemma 26 that $u_i \circ (\delta(q_i) - \gamma(q_i)) = \sum_{1 \leq j \leq r} g_j v_j$ where $g_j \in \mathfrak{R}$ and $v_j \in \mathfrak{H}_{\mathfrak{p}} \cap \mathfrak{B}_{d-1}$. On the other hand if $d < d_i$, $\gamma(q_i) = 0$ and therefore $q_i = 0$ by Lemma 19. Hence $\gamma(q_i) u_i = u_i \circ \delta(q_i) = 0$ in this case. This shows that we can select a finite set of elements $g_j \in \mathfrak{R}$ and $v_j \in \mathfrak{H}_{\mathfrak{p}} \cap \mathfrak{B}_{d-1}$ $(1 \leq j \leq s)$ such that

$$u = \sum_{1 \leq i \leq w} u_i \circ \delta(q_i) + \sum_{1 \leq j \leq s} g_j v_j$$

on $A_{\mathfrak{p}}'$. The statement of the theorem now follows immediately by applying the induction hypothesis to $v_j$.

The following consequence of this theorem will be important for our applications.

COROLLARY. *Let $U$ be a nonempty open connected subset of $A_{\mathfrak{p}}'$ and $\chi$ a homomorphism of $I_{\mathfrak{g}}$ into $C$. Let $E_{\chi}$ denote the space of all analytic functions $\phi$ on $U$ which satisfy the system of differential equations $\delta(q)\phi = \chi(q)\phi$ $(q \in I_{\mathfrak{g}})$. Then the dimension of $E_{\chi}$ over $C$ cannot exceed $w$.*

Choose a point $h_0$ in $U$. If $\dim E_{\chi} > w$, we can obviously find a function $\phi \neq 0$ in $E_{\chi}$ such that $\phi(h_0; u_i) = 0$ $(1 \leq i \leq w)$. But then it follows from the above theorem that $\phi(h_0; u) = 0$ for every $u \in \mathfrak{H}_{\mathfrak{p}}$. However since $U$ is connected and $\phi$ is analytic, this implies that $\phi = 0$ and so we get a contradiction. Hence $\dim E_{\chi} \leq w$.

**7. The operator $\delta'(\omega)$.** The Casimir operator $\omega$ of $\mathfrak{g}$ is an element lying in the center of $\mathfrak{B}$ which is defined as follows. Let $X_1, \cdots, X_n$ be a base for $\mathfrak{g}$ and put $g_{ij} = B(X_i, X_j)$ $1 \leq i, j \leq n$. Then the matrix $(g_{ij})_{1 \leq i, j \leq n}$ is nonsingular and if $(g^{ij})_{1 \leq i, j \leq n}$ denotes its inverse, $\omega = \sum\limits_{1 \leq i, j \leq n} g^{ij} X_i X_j$. (It is easy to check that $\omega$ does not depend on the choice of the base used in its definition.) We intend to compute $\delta'(\omega)$.

For any linear function $\mu$ on $\mathfrak{h}_\mathfrak{p}$, let $H_\mu$ denote the unique element in $\mathfrak{h}_\mathfrak{p}$ such that $B(H, H_\mu) = \mu(H)$ for all $H \in \mathfrak{h}_\mathfrak{p}$. Also put $\gamma'(q) = e^{-\rho} \gamma(q) \circ e^{\rho}$ for $q \in I_\mathfrak{g}$.

LEMMA 27. $\delta'(\omega) = \gamma'(\omega) + 2 \sum\limits_{\alpha \in P_+} (e^{2\bar{\alpha}} - 1)^{-1} H_{\bar{\alpha}}$ where $\bar{\alpha}$ denotes the restriction of $\alpha$ on $\mathfrak{h}_\mathfrak{p}$.

For each root $\alpha$, select $X_\alpha$ as before and normalize it in such a way that $B(X_\alpha, X_{-\alpha}) = 1$ for any $\alpha \in P$. Then $[X_\alpha, X_{-\alpha}] = H_\alpha$ where $H_\alpha$ is the element in $\mathfrak{h}$ such that $B(H, H_\alpha) = \alpha(H)$ for every $H \in \mathfrak{h}$. Choose bases $H_1, \cdots, H_l$ and $H_{l+1}, \cdots, H_m$ for $\mathfrak{h}_\mathfrak{p}$ and $\mathfrak{h}_\mathfrak{k}$ respectively such that[9] $B(H_i, H_j) = \delta_{ij}$ $(1 \leq i, j \leq m)$. Then $H_1, \cdots, H_m$ together with $X_\alpha$, $X_{-\alpha}$ $(\alpha \in P)$, form a base for $\mathfrak{g}$ and therefore it is clear that

$$\omega = H_1^2 + \cdots + H_m^2 + \sum\limits_{\alpha \in P} (X_\alpha X_{-\alpha} + X_{-\alpha} X_\alpha)$$

$$= H_1^2 + \cdots + H_m^2 + 2 \sum\limits_{\alpha \in P} X_{-\alpha} X_\alpha + \sum\limits_{\alpha \in P} H_\alpha$$

$$\equiv H_1^2 + \cdots + H_l^2 + 2 H_\rho + 2 \sum\limits_{\alpha \in P_+} X_{-\alpha} X_\alpha \bmod \mathfrak{k} \mathfrak{B} \cap \mathfrak{B} \mathfrak{k},$$

since $X_\alpha$, $X_{-\alpha}$ and $H_\alpha$ lie in $\mathfrak{k}$ if $\alpha$ vanishes identically on $\mathfrak{h}_\mathfrak{p}$. This shows that $\omega \equiv H_1^2 + \cdots + H_l^2 + 2 H_\rho \bmod (\mathfrak{k} \mathfrak{B} + \mathfrak{B} \mathfrak{n})$ and hence $\gamma'(\omega)$ $H_1^2 + \cdots + H_l^2 + 2 H_\rho$. Now for any root $\alpha$, which is not identically zero on $\mathfrak{h}_\mathfrak{p}$, let $X_\alpha = Y_\alpha + Z_\alpha$ $(Y_\alpha \in \mathfrak{p}, Z_\alpha \in \mathfrak{k})$ and put $g_\alpha(\exp H) = (e^{2\alpha(H)} - 1)^{-1}$ $(H \in \mathfrak{h}_{\mathfrak{p}_0}')$. We have seen in Section 5 that $Y_\alpha \equiv (1 + 2g_\alpha(h)) Z_\alpha \bmod \mathfrak{k}^{h-1}$ for $h \in A_\mathfrak{p}'$. Hence

$$X_\alpha X_{-\alpha} \equiv X_\alpha Y_{-\alpha} \equiv 2(1 + g_\alpha(h)) Z_\alpha Y_{-\alpha}$$

$$\equiv 2(1 + g_\alpha(h)) [Z_\alpha, Y_{-\alpha}] \bmod (\mathfrak{k}^{h-1} \mathfrak{B} + \mathfrak{B} \mathfrak{k}).$$

If we apply the automorphism $\theta$ to this congruence, we get

$$\theta(X_\alpha X_{-\alpha}) \equiv -2(1 + g_\alpha(h)) [Z_\alpha, Y_{-\alpha}] \bmod (\mathfrak{k}^h \mathfrak{B} + \mathfrak{B} \mathfrak{k}).$$

Replacing $\alpha$ and $h$ by $-\alpha$ and $h^{-1}$ respectively, we find that

$$\theta(X_{-\alpha} X_\alpha) \equiv -2(1 + g_\alpha(h)) [Z_{-\alpha}, Y_\alpha] \bmod (\mathfrak{k}^{h-1} \mathfrak{B} + \mathfrak{B} \mathfrak{k}).$$

This shows that

$$\tfrac{1}{2}\{X_\alpha X_{-\alpha} + \theta(X_{-\alpha}X_\alpha)\}$$
$$\equiv (1 + g_\alpha(h))\{[Z_\alpha, Y_{-\alpha}] - [Z_{-\alpha}, Y_\alpha]\} \bmod (\mathfrak{k}^{h-1}\mathfrak{B} + \mathfrak{B}\mathfrak{k}).$$

But $[X_\alpha, X_{-\alpha}] = H_\alpha$ and therefore

$$[Z_\alpha, Y_{-\alpha}] + [Y_\alpha, Z_{-\alpha}] = \tfrac{1}{2}(H_\alpha - \theta(H_\alpha)) = H_\alpha.$$

This proves that

$$\tfrac{1}{2}\{X_\alpha X_{-\alpha} + \theta(X_{-\alpha}X_\alpha)\} \equiv (1 + g_\alpha(h))H_{\bar\alpha} \bmod (\mathfrak{k}^{h-1}\mathfrak{B} + \mathfrak{B}\mathfrak{k})$$

for $h \in A_\mathfrak{p}'$. On the other hand it is easy to verify that $\theta(\omega) = \omega$. Therefore

$$\omega = \tfrac{1}{2}\{\omega + \theta(\omega)\}$$
$$\equiv (H_1^2 + \cdots + H_l^2) + \tfrac{1}{2}\sum_{\alpha \in P_+}\{X_\alpha X_{-\alpha} + X_{-\alpha}X_\alpha$$
$$+ \theta(X_\alpha X_{-\alpha} + X_{-\alpha}X_\alpha)\} \bmod \mathfrak{B}\mathfrak{k}.$$

But in view of the above result, this implies that

$$\omega \equiv (H_1^2 + \cdots + H_l^2) + \sum_{\alpha \in P_+}(1 + 2g_\alpha(h))H_{\bar\alpha} \bmod (\mathfrak{k}^{h-1}\mathfrak{B} + \mathfrak{B}\mathfrak{k})$$

since $g_\alpha - g_{-\alpha} = 1 + 2g_\alpha$. Hence

$$\omega \equiv \gamma'(\omega) + 2\sum_{\alpha \in P_+} g_\alpha(h)H_{\bar\alpha} \bmod (\mathfrak{k}^{h-1}\mathfrak{B} + \mathfrak{B}\mathfrak{k})$$

for $h \in A_\mathfrak{p}'$ and this proves the lemma.

COROLLARY 1. *Let $H_1, \cdots, H_l$ be any base for $\mathfrak{h}_\mathfrak{p}$ and let $(g^{ij})_{1 \le i, j \le l}$ denote the inverse of the matrix $g_{ij} = B(H_i, H_j)$ $(1 \le i, j \le l)$. Then*

$$\delta'(\omega) = \Delta^{-1} \sum_{1 \le i, j \le l} g^{ij} H_i \circ \Delta H_j.$$

It is easy to check that the right side is actually independent of the choice of the base $H_1, \cdots, H_l$. Hence it is sufficient to consider the base used in the proof of the above lemma. But then one finds by direct calculation that

$$\Delta^{-1} \sum_{1 \le i, j \le l} g^{ij} H_i \circ \Delta H_j = (H_1^2 + \cdots + H_l^2) + \sum_{\alpha \in P_+}\{(e^{\bar\alpha} + e^{-\bar\alpha})/(e^{\bar\alpha} - e^{-\bar\alpha})\}H_{\bar\alpha}$$
$$= \gamma'(\omega) + 2\sum_{\alpha \in P_+}(e^{2\bar\alpha} - 1)^{-1}H_{\bar\alpha}$$

and this proves our assertion.

Let $\bar\omega$ denote the restriction on $\mathfrak{h}_\mathfrak{p}$ of the Casimir polynomial of $\mathfrak{g}$ (see [5(k)], p. 98]).

COROLLARY 2. $\gamma(\omega) = \bar{\omega} - \langle \rho, \rho \rangle$.

We keep to the notation of the proof of Lemma 27. Since $\gamma'(\omega)$ $= (H_1{}^2 + \cdots + H_l{}^2) + 2H_\rho$ it is clear that

$$\gamma(\omega) = (H_1{}^2 + \cdots + H_l{}^2) - 2 \sum_{1 \leq i \leq l} \rho(H_i) H_i + \langle \rho, \rho \rangle + 2(H_\rho - \langle \rho, \rho \rangle)$$

$$= (H_1{}^2 + \cdots + H_l{}^2) - \langle \rho, \rho \rangle.$$

On the other hand $\bar{\omega} = H_1{}^2 + \cdots + H_l{}^2$ and so our assertion is obvious.


8. **Some consequences of Lemma 27.** Select $\alpha_1, \cdots, \alpha_l \in \Sigma$ as in Section 3 and let $L$ denote the set of all linear functions on $\mathfrak{h}_\mathfrak{p}$ of the form $m_1\alpha_1 + \cdots + m_l\alpha_l$ where $m_1, \cdots, m_l$ are nonnegative integers. Given two linear functions $\lambda, \lambda'$ on $\mathfrak{h}_\mathfrak{p}$, we write $\lambda << \lambda'$ (or $\lambda' >> \lambda$) if $\lambda \neq \lambda'$ and $\lambda' - \lambda \in L$. It is clear that for any two linear functions $\lambda_1, \lambda_2$ on $\mathfrak{h}_\mathfrak{p}$, there exist only a finite number of $\lambda$ such that $\lambda_1 << \lambda << \lambda_2$. Define $S = S(\mathfrak{h}_\mathfrak{p})$ as in Section 3 and let $Q = Q(\mathfrak{h}_\mathfrak{p})$ denote the quotient field of $S$. Then $Q$ is the field of rational functions on $\mathfrak{h}_\mathfrak{p}$. Introduce the scalar product $\langle p, q \rangle$ for $p, q \in S$ as in [5(k), p. 90] and put $m(\lambda) = m_1 + \cdots + m_l$ for $\lambda = m_1\alpha_1 + \cdots + m_l\alpha_l \in L$. Now for each $\lambda \in L$, we define an element $'\Gamma_\lambda \in Q$ by induction on $m(\lambda)$ as follows. $'\Gamma_0 = 1$ and [25]

$$\{\langle \lambda, \lambda - 2\rho \rangle - 2\lambda\}'\Gamma_\lambda = 2 \sum_{\alpha \in P_+} \sum_{k \geq 1} '\Gamma_{\lambda - 2k\bar{\alpha}} \{\langle \lambda - 2k\bar{\alpha}, \bar{\alpha} \rangle - \bar{\alpha}\},$$

where $\bar{\alpha}$ denotes the restriction of $\alpha$ on $\mathfrak{h}_\mathfrak{p}$ and $k$ runs over all positive integers such that $\lambda - 2k\bar{\alpha} \in L$. Let $L'$ denote the set of all elements $\lambda \neq 0$ in $L$. Consider the hyperplane $\sigma_\lambda'$ $(\lambda \in L')$ consisting of those points $H \in \mathfrak{h}_\mathfrak{p}$ where $2\lambda(H) = \langle \lambda, \lambda - 2\rho \rangle$. It is clear that any compact subset of $\mathfrak{h}_\mathfrak{p}$ meets $\sigma_\lambda'$ for only a finite number of $\lambda$ in $L'$. Let $\mathfrak{h}_\mathfrak{p}''$ denote the complement of $\bigcup_{\lambda \in L'} \sigma_\lambda'$ in $\mathfrak{h}_\mathfrak{p}$. Then $\mathfrak{h}_\mathfrak{p}''$ is an open, connected, dense subset of $\mathfrak{h}_\mathfrak{p}$. It is obvious that the rational function $'\Gamma_\lambda$ $(\lambda \in L)$ takes a well-defined value at any point $H \in \mathfrak{h}_\mathfrak{p}''$. We denote this value by $'\Gamma_\lambda(H)$.

LEMMA 28. *Let $U$ be a compact subset of $\mathfrak{h}_\mathfrak{p}''$. Then there exists a real number $c \geq 1$ such that $|'\Gamma_\lambda(H)| \leq c^{m(\lambda)}$ for $\lambda \in L$ and $H \in U$.*

Select positive numbers $c_1$ and $c_2$ such that

$$|\langle \lambda, \bar{\alpha} \rangle - \alpha(H)| \leq c_1(m(\lambda) + 1), \qquad |\langle \lambda, \lambda - 2\rho \rangle - 2\lambda(H)| \geq c_2 m(\lambda)^2$$

---

[25] As will become apparent during the proof of Lemma 29, the motivation for this definition comes from Lemma 27.

for every $\lambda \in L$, $\alpha \in P_+$ and $H \in U$. Obviously this is possible. Now if $\lambda' << \lambda$ $(\lambda' \in L)$, it is clear that $m(\lambda') < m(\lambda)$ and therefore

$$|\langle \lambda, \bar{\alpha} \rangle - \alpha(H)| \, |\langle \lambda, \lambda - 2\rho \rangle - 2\lambda(H)|^{-1} \leqq c_3/m(\lambda) \qquad (H \in U)$$

where $c_3 = 2c_1/c_2$. Hence it follows that

$$|{}'\Gamma_\lambda(H)| \leqq 2c_3 (m(\lambda))^{-1} \sum_{\alpha \in P_+} \sum_{k \geqq 1} |{}'\Gamma_{\lambda - 2k\bar{\alpha}}(H)| \qquad (H \in U, \lambda \in L').$$

Let $r$ be the number of roots in $P_+$ and put $c = \max\{rc_3, 1\}$. We shall prove by induction on $m(\lambda)$ that $|{}'\Gamma_\lambda(H)| \leqq c^{m(\lambda)}$ for $H \in U$ and $\lambda \in L$. If $\lambda = 0$, this is true. So now suppose $\lambda \neq 0$. Since $\lambda - 2k\bar{\alpha} \in L$, it is obvious that $k \leqq \frac{1}{2} m(\lambda)$. Hence it follows by induction hypothesis that

$$\sum_{k \geqq 1} |{}'\Gamma_{\lambda - 2k\bar{\alpha}}(H)| \leqq \tfrac{1}{2} m(\lambda) c^{m(\lambda)-1}$$

and therefore

$$|{}'\Gamma_\lambda(H)| \leqq c_3 r c^{m(\lambda)-1} \leqq c^{m(\lambda)}$$

For any number $M \geqq 0$, let $\mathfrak{h}_\mathfrak{p}(M)$ denote the set of those points $H \in \mathfrak{h}_\mathfrak{p}$ where the real part of $\alpha_i(H)$ is greater than $M$ for every $i$ $(1 \leqq i \leqq l)$.

COROLLARY 1. *We can choose $M \geqq 0$ such that the series*

$$\sum_{\lambda \in L} |{}'\Gamma_\lambda(H')e^{-\lambda(H)}|$$

*converges uniformly for $H \in \mathfrak{h}_\mathfrak{p}(M)$ and $H' \in U$.*

Obviously it is sufficient to choose $M$ so large that $c < e^M$.

COROLLARY 2. *Let $U'$ be an open subset of $\mathfrak{h}_\mathfrak{p}''$ whose closure (in $\mathfrak{h}_\mathfrak{p}''$) is compact. Then if $M$ is sufficiently large, the function $\phi'(H':H)$ $(H' \in U'$, $H \in \mathfrak{h}_\mathfrak{p}(M))$ defined by the series*

$$\phi'(H':H) = e^{\langle H', H \rangle} \sum_{\lambda \in L} {}'\Gamma_\lambda(H')e^{-\lambda(H)}$$

*is holomorphic on $U' \times \mathfrak{h}_\mathfrak{p}(M)$.*

This is an immediate consequence of Corollary 1.

We keep to the notation of Corollary 2 above and denote by $A_\mathfrak{p}(M)$ the set of those points $h \in A_\mathfrak{p}$ for which $\log h \in \mathfrak{h}_\mathfrak{p}(M)$. Also put $\Phi'(H':h) = \phi'(H':\log h)$ $(H' \in U', h \in A_\mathfrak{p}(M))$. Then $\Phi'$ is an analytic function on $U' \times A_\mathfrak{p}(M)$. The following lemma provides the justification for the above construction.

LEMMA 29. *Let $q$ be an element in $I_\mathfrak{g}$. Then* [11]

$$\Phi'(H':h;\delta'(q)) = \gamma'(q:H')\Phi'(H':h) \qquad (H' \in U, h \in A_\mathfrak{p}(M))$$

*where $\gamma'(q:H')$ denotes the value of the polynomial function* [26] *$\gamma'(q)$ at $H'$.*

---

[26] We recall that $\gamma'(q) = e_{-\rho}\gamma(q) \circ e_\rho$.

For otherwise suppose our assertion is false. Then from Lemma 26, $\delta'(q) = \gamma'(q) + \sum\limits_{1 \leq i \leq r} g_i v_i$ where $g_i \in \mathfrak{R}$ and $v_i \in \mathfrak{H}_\mathfrak{p}$. For any $v \in S(\mathfrak{h}_\mathfrak{p})$, let $\partial(v)$ denote the corresponding differential operator on $\mathfrak{h}_\mathfrak{p}$ (see [5(k), §2]). Then if $\psi$ is a holomorphic function on $\mathfrak{h}_\mathfrak{p}(M)$ and $\Psi(h) = \psi(\log h)$ $(h \in A_\mathfrak{p}(M))$, it is obvious that $\Psi(h;v) = \psi(\log h;\partial(v))$ for $v \in \mathfrak{H}_\mathfrak{p}$. Therefore in particular

$$\Psi(h; \sum_{1 \leq i \leq r} g_i v_i) = \sum_{1 \leq i \leq r} g_i(h)\psi(\log h;\partial(v_i)).$$

But $g_\alpha(\exp H) = (e^{2\alpha(H)} - 1)^{-1} = \sum\limits_{1 \leq k < \infty} e^{-2k\alpha(H)}$ for $\alpha \in P_+$ and $H \in \mathfrak{h}_\mathfrak{p}(M)$. Hence it is obvious that we can select an element $v_\lambda \in \mathfrak{H}_\mathfrak{p}$ for each $\lambda \in L'$, such that

$$\sum_{1 \leq i \leq r} g_i(\exp H)\psi(H;\partial(v_i)) = \sum_{\lambda \in L'} e^{-\lambda(H)}\psi(H;\partial(v_\lambda)) \qquad (H \in \mathfrak{h}_\mathfrak{p}(M))$$

for any holomorphic function $\psi$ on $\mathfrak{h}_\mathfrak{p}(M)$. Hence in particular

$$\Phi'(H':h;\delta'(q)) - \gamma'(q:H')\Phi'(H':h) = \phi_1(H':\log h) \qquad (h \in A_\mathfrak{p}(M))$$

where

$$\phi_1(H':H) = \phi'(H':H;\partial(\gamma'(q)) - \gamma'(q:H')) + \sum_{\lambda \in L'} e^{-\lambda(H)}\phi'(H':H;\partial(v_\lambda))$$

for $H' \in U'$ and $H \in \mathfrak{h}_\mathfrak{p}(M)$. Now put $\phi_0(H':H) = e^{-\langle H',H \rangle}\phi_1(H':H)$. Then it follows from our hypothesis that $\phi_0 \neq 0$. On the other hand it is obvious that

$$\phi'(H':H;\partial(v)) = e^{\langle H',H \rangle} \sum_{\lambda \in L} {}^{\backprime}\Gamma_\lambda(H')v(H'-H_\lambda)e^{-\lambda(H)}$$
$$(H' \in U', H \in \mathfrak{h}_\mathfrak{p}(M))$$

for $v \in \mathfrak{H}_\mathfrak{p}$. (Here $v(H'-H_\lambda)$ denotes the value of the polynomial function $v$ and $H'-H_\lambda$.) Hence

$$\phi_0(H':H) = \sum_{\lambda \in L} {}^{\backprime}\Gamma_\lambda(H')e^{-\lambda(H)}\{\gamma'(q:H'-H_\lambda) - \gamma'(q:H')\}$$
$$+ \sum_{\lambda \in L} \sum_{\mu \in L'} {}^{\backprime}\Gamma_\lambda(H')e^{-\lambda(H)-\mu(H)}v_\mu(H'-H_\lambda)$$
$$= \sum_{\lambda \in L} e^{-\lambda(H)}c_\lambda(H')$$

where

$$c_\lambda(H') = {}^{\backprime}\Gamma_\lambda(H')\{\gamma'(q:H'-H_\lambda) - \gamma'(q:H')\}$$
$$+ \sum_{0 < \mu < \lambda} {}^{\backprime}\Gamma_\mu(H')v_{\lambda-\mu}(H'-H_\mu).$$

Obviously $c_\lambda$ is a rational function on $\mathfrak{h}_\mathfrak{p}$ and since $\phi_0 \neq 0$, we can choose $\lambda_0 \in L$ such that $c_{\lambda_0} \neq 0$. We shall now show that this is actually impossible.

3

Let $\pi$ be an irreducible representation of $\mathfrak{g}$ on a finite-dimensional space $V_0$. We assume that there exists a unit vector $\xi$ in $V_0$ such that $\pi(\mathfrak{k})\xi = \{0\}$. As usual, we denote the corresponding representation of $G_c$ also by $\pi$ and put $\Xi(x) = (\xi, \pi(x)\xi)$ $(x \in G_c)$. Then $\Xi$ is a holomorphic function on $G_c$. Let $\Lambda_1' > \Lambda_2' > \cdots > \Lambda_s'$ be all the weights of $\pi$ and $\Lambda = \Lambda_1 > \Lambda_2 > \cdots > \Lambda_r$ all the linear functions on $\mathfrak{h}_\mathfrak{p}$ obtained by taking the restrictions of these weights on $\mathfrak{h}_\mathfrak{p}$. Then we know from [5(h), Lemma 2] that $\Lambda_i << \Lambda$ $(2 \leq i \leq r)$. Let $V_i$ be the subspace consisting of those elements $\eta \in V_0$ for which $\pi(H)\eta = \Lambda_i(H)\eta$ $(H \in \mathfrak{h}_\mathfrak{p})$ and let $E_i$ denote the orthogonal projection of $V_0$ on $V_i$. Then since $\pi(H)$ is self-adjoint for $H$ in $\mathfrak{h}_{\mathfrak{p}_0}$, it follows that $V_1, \cdots, V_r$ are mutually orthogonal and therefore

$$\Xi(\exp H) = \sum_{1 \leq i \leq r} |E_i\xi|^2 \, e^{\Lambda_i(H)} \qquad (H \in \mathfrak{h}_\mathfrak{p}).$$

On the other hand put $E = \int_K \pi(k)\,dk$ and let $\eta$ be a unit vector in $V_0$ belonging to the highest weight $\Lambda_1'$. Then we have seen during the proof of Lemma 5 that $\xi = cE\eta$ where $c$ is a nonzero complex number. This shows that $(\xi, \eta) = (E\xi, \eta) = (\xi, E\eta) = c^{-1} \neq 0$. But $\eta \in V_1$ and therefore $(\xi, \eta) = (\xi, E_1\eta) = (E_1\xi, \eta)$. Hence $E_1\xi \neq 0$ and so

$$\Xi(\exp H) = \sum_{1 \leq i \leq r} a_i' e^{\Lambda_i(H)} \qquad (H \in \mathfrak{h}_\mathfrak{p})$$

where $a_1' = |E_1\xi|^2 > 0$ and $a_i' = |E_i\xi|^2 \geq 0$ $(2 \leq i \leq r)$.

Let $\Xi_0$ denote the restriction of $\Xi$ on $G$. Then $\Xi_0$ is a spherical function and therefore $\Xi_0(h;b) = \Xi_0(h;\delta'(b))$ $(b \in \mathfrak{B}, h \in A_\mathfrak{p}')$ by Lemma 23. Moreover if $b \in I_\mathfrak{g}$, it follows from Lemmas 3 and 5 that $b\Xi_0 = \gamma'(b:H_\Lambda)\Xi_0$. Therefore

$$\Xi_0(h;\delta'(b)) - \gamma'(b:H_\Lambda)\Xi_0(h) = 0 \qquad (b \in I_\mathfrak{g}, h \in A_\mathfrak{p}').$$

Now put $\Xi_1(H) = (a_1')^{-1}e^{-\Lambda(H)}\Xi(\exp H)$ and $\Xi_1'(H) = \Xi(\exp H)$ $(H \in \mathfrak{h}_\mathfrak{p})$. Then $\Xi_1$ can be written in the form

$$\Xi_1(H) = 1 + \sum_{\lambda \in L'} a_\lambda e^{-\lambda(H)} \qquad (H \in \mathfrak{h}_\mathfrak{p})$$

where $a_\lambda$ are real numbers which are zero for all $\lambda$ in $L'$ except a finite number. Then by applying the above relation for $b = \omega$ and making use of Lemma 27, we find that

$$\Xi_1'(H;\partial(\gamma'(\omega))) + 2 \sum_{\alpha \in P_+} \sum_{1 \leq k < \infty} e^{-2k\alpha(H)}\, \Xi_1'(H;\partial(H_{\tilde\alpha}))$$

$$= \gamma'(\omega:H_\Lambda)\Xi_1'(H) \qquad (H \in \mathfrak{h}_\mathfrak{p}(M)).$$

Hence if $a_0 = 1$,

$$\sum_{\lambda \in L} a_\lambda \gamma'(\omega : H_\Lambda - H_\lambda) e^{-\lambda(H)} + 2 \sum_{\alpha \in P_+} \sum_{1 \le k < \infty} e^{-2k\alpha(H)} \sum_{\lambda \in L} e^{-\lambda(H)} a_\lambda \langle \bar\alpha, \Lambda - \lambda \rangle$$

$$= \gamma'(\omega : H_\Lambda) \sum_{\lambda \in L} a_\lambda e^{-\lambda(H)} \qquad (H \in \mathfrak{h}_\mathfrak{p}(M)).$$

But $\gamma'(\omega : H) = \langle H, H \rangle + 2\rho(H)$ $(H \in \mathfrak{h}_\mathfrak{p})$ from Corollary 2 to Lemma 27. Therefore if $E(M)$ stands for the set of all points $H \in \mathfrak{h}_\mathfrak{p}(2M)$ where[27] $\operatorname{Re} \alpha_i(H) > \frac{1}{2} |\alpha_j(H)|$ $(1 \le i, j \le l)$, it follows by applying the corollary of Lemma 57 of the Appendix (§ 15) to $V = E(M)$ that we can equate the coefficients of $e^{-\lambda}$ on both sides of the above equation and obtain

$$a_\lambda \langle \Lambda - \lambda, \Lambda - \lambda + 2\rho \rangle + 2 \sum_{\alpha \in P_+} \sum_{k \ge 1} a_{\lambda - 2k\bar\alpha} \langle \bar\alpha, \Lambda - \lambda + 2k\bar\alpha \rangle$$

$$= a_\lambda \langle \Lambda, \Lambda + 2\rho \rangle \qquad (\lambda \in L).$$

Here the sum is to be taken over those positive integers $k$ for which $\lambda - 2k\bar\alpha \in L$. This proves that

$$\{ \langle \lambda, \lambda - 2\rho \rangle - 2\lambda(H_\Lambda) \} a_\lambda$$

$$= 2 \sum_{\alpha \in P_+} \sum_{k \ge 1} a_{\lambda - 2k\bar\alpha} \{ \langle \lambda - 2k\bar\alpha, \bar\alpha \rangle - \bar\alpha(H_\Lambda) \}.$$

By comparing this with the recurrence relation for $'\Gamma_\lambda$, it is obvious that $a_\lambda = {}'\Gamma_\lambda(H_\Lambda)$ provided $H_\Lambda \notin \sigma_\mu'$ for any $\mu \in L'$ such that $\lambda - \mu \in L$. On the other hand it is clear from Lemma 4 that $\pi$ can be selected in such a way that (1) $H_\Lambda \notin \sigma_\mu'$ for $\mu = \lambda_0$ or $\mu << \lambda_0$ $(\mu \in L')$ and (2) the rational function $c_{\lambda_0}$ (which is then obviously defined at $H_\Lambda$) does not take the value zero at $H_\Lambda$. However the relation

$$\Xi_0(h ; \delta'(q)) - \gamma'(q : H_\Lambda) \Xi_0(h) = 0 \qquad (h \in A_\mathfrak{p}')$$

implies that

$$\sum_{\lambda \in L} a_\lambda e^{-\lambda(H)} \{ \gamma'(q : H_\Lambda - H_\lambda) - \gamma'(q : H_\Lambda) \}$$

$$+ \sum_{\lambda \in L} \sum_{\mu \in L'} a_\lambda e^{-\lambda(H) - \mu(H)} v_\mu(H_\Lambda - H_\lambda) = 0$$

for $H \in \mathfrak{h}_{\mathfrak{p}_0}(M) = \mathfrak{h}_{\mathfrak{p}_0} \cap \mathfrak{h}_\mathfrak{p}(M)$. Again by applying the corollary of Lemma 57 to $V = \mathfrak{h}_{\mathfrak{p}_0} \cap E(M)$, we can equate the coefficient of $e^{-\lambda}$ in the above expression, to zero. But it is obvious that this coefficient is $c_\lambda(H_\Lambda)$ for $\lambda = \lambda_0$ or $\lambda << \lambda_0$ $(\lambda \in L)$. Hence $c_{\lambda_0}(H_\Lambda) = 0$ and so we get a contradiction. This completes the proof of Lemma 29.

Let $\tau$ denote the mapping of $\mathfrak{h}_\mathfrak{p}$ into itself given by

$$\tau(H) = -(-1)^{\frac{1}{2}}(H + H_\rho) \qquad (H \in \mathfrak{h}_\mathfrak{p}).$$

---

[27] For any complex number $c$, $\operatorname{Re} c$ and $\operatorname{Im} c$ denote the real parts of $c$ and $-(-1)^{\frac{1}{2}} c$ respectively.

Then $\tau^{-1}(H) = (-1)^{\frac{1}{2}}H - H_\rho$. For any polynomial $p \in S(\mathfrak{h}_\mathfrak{p})$, let $p^\tau$ denote the function whose value at $H \in \mathfrak{h}_\mathfrak{p}$ is $p(\tau^{-1}(H))$. Then the mapping $p \to p^\tau$ can be extended to an automorphism of $Q(\mathfrak{h}_\mathfrak{p})$. It is clear that a rational function $p \in Q(\mathfrak{h}_\mathfrak{p})$ is defined at a point $H$ if and only if $p^\tau$ is defined at $\tau(H)$ and if this is so, $p(H) = p^\tau(\tau(H))$. Put $\sigma_\lambda = \tau(\sigma_\lambda')$ ($\lambda \in L'$) and $'\mathfrak{h}_\mathfrak{p} = \tau(\mathfrak{h}_\mathfrak{p}'')$. Then $\sigma_\lambda$ consists of those points $H \in \mathfrak{h}_\mathfrak{p}$ where $\langle \lambda, \lambda \rangle = 2(-1)^{\frac{1}{2}}\lambda(H)$. This shows that $\mathfrak{h}_{\mathfrak{p}_0} \subset {}'\mathfrak{h}_\mathfrak{p}$. Put $\Gamma_\lambda = ('\Gamma_\lambda)^\tau$ ($\lambda \in L$), $U = \tau(U')$ and

$$\phi(H_0:H) = e^{\rho(H)}\phi'(\tau^{-1}(H_0):H) = e^{(-1)^{\frac{1}{2}}\langle H_0, H \rangle} \sum_{\lambda \in L} \Gamma_\lambda(H_0) e^{-\lambda(H)}$$

for $H_0 \in U$ and $H \in \mathfrak{h}_\mathfrak{p}(M)$ (in the notation of Corollary 2 to Lemma 28). Moreover let

$$\Phi(H:h) = e^{\rho(\log h)}\Phi'(\tau^{-1}(H):h) \qquad\qquad (H \in U, h \in A_\mathfrak{p}(M)).$$

Then Lemma 29 can be restated as follows.

LEMMA 30.  *For any $q$ in $I_\mathfrak{g}$,*

$$\Phi(H:h;\delta(q)) = \gamma(q:(-1)^{\frac{1}{2}}H)\Phi(H:h)$$

*if $H \in U$ and $h \in A_\mathfrak{p}(M)$.*

Now put $\xi_\lambda(H_0:H) = \exp\{(-1)^{\frac{1}{2}}\langle H_0, H \rangle - \lambda(H)\}\Gamma_\lambda(H_0)$ for $H_0 \in {}'\mathfrak{h}_\mathfrak{p}$, $H \in \mathfrak{h}_\mathfrak{p}$ and $\lambda \in L$. Then if $u \in S(\mathfrak{h}_\mathfrak{p})$, it follows from the uniform convergence (see Corollary 1 to Lemma 28) of the series $\sum_{\lambda \in L} |\xi_\lambda(H_0:H)|$ on any compact subset $\Omega$ of $U \times \mathfrak{h}_\mathfrak{p}(M)$, that the series[11] $\sum_{\lambda \in L} \xi_\lambda(H_0;\partial(u):H)$ also converges uniformly on $\Omega$ to $\phi(H_0;\partial(u):H)$ (see Lemma 58 of the Appendix). Define $E(M)$ as above to be the set of all points $H \in \mathfrak{h}_\mathfrak{p}(2M)$ such that $\operatorname{Re}\alpha_i(H) > \frac{1}{2}|\alpha_j(H)|$ ($1 \leq i, j \leq l$). Then if $M$ is sufficiently large we have the following lemma.

LEMMA 31.  *Let $U_1$ be a compact subset of $U$. Then for any linear function $\mu$ on $\mathfrak{h}_\mathfrak{p}$ and $u \in S(\mathfrak{h}_\mathfrak{p})$, the series*

$$\sum_{\lambda \in L} |\xi_\lambda(H_0;\partial(u):H)e^{\mu(H)}|$$

*converges uniformly for $H_0 \in U_1$ and $H \in E(M)$.*

In view of Lemma 58 of the Appendix, it is sufficient to show that the series $\sum_{\lambda \in L} |\xi_\lambda(H_0:H)e^{\mu(H)}|$ converges uniformly for $H_0 \in U$ and $H \in E(M)$. Since the closure of $U$ is compact, we can find a real number $\nu$ such that

$$|\exp\{(-1)^{\frac{1}{2}}\langle H_0, H \rangle + \mu(H)\}| \leq \exp\{\nu \operatorname{Re}(\alpha_1(H) + \cdots + \alpha_l(H))\}$$

for all $H_0 \in U$ and $H \in E(M)$. Let $\beta(H) = \min_{1 \leq i \leq l} \mathrm{Re}\, \alpha_i(H)$. Then if $H \in E(M)$,
$\mathrm{Re}(\alpha_1(H) + \cdots + \alpha_l(H)) \leq 2l\beta(H)$ and $\mathrm{Re}\,\lambda(H) \geq m(\lambda)\beta(H)$ $(\lambda \in L)$.
Hence $|\xi_\lambda(H_0:H)e^{\mu(H)}| \leq |\Gamma_\lambda(H_0)| \exp\{(2\nu l - m(\lambda))2M\}$. Therefore if $m(\lambda) \geq 4\nu l$,

$$|\xi_\lambda(H_0:H)e^{\mu(H)}| \leq |\Gamma_\lambda(H_0)| \exp(-m(\lambda)M)$$

and our assertion follows from Lemma 28.

Now fix an element $H_0 \in U$. We now use the notation of Section 3 and define $W'$ and $J'$ corresponding to the element $H_0$. Let $V_{H_0}$ denote the subspace consisting of those elements $v \in S$ which satisfy the condition that $p(H_0; \partial(v)) = 0$ for every polynomial function $p \in SJ_{H_0}'$. Also let $J_+'$ be the subspace spanned by homogeneous elements in $J'$ of positive degree.

LEMMA 32. $S = V_{H_0} + SJ_+'$ and $\dim V_{H_0} = w'$.

Put $V = V_{H_0}$ and let $p \to p^\sigma$ $(p \in S)$ denote the automorphism of $S$ defined by $p^\sigma(H) = p(H + H_0)$ $(H \in \mathfrak{h}_\mathfrak{p})$. Since $H_0$ is left fixed by $W'$, it is clear that $(J')^\sigma = J'$. Moreover $p$ vanishes at $H_0$ if and only if $p^\sigma$ vanishes at zero. Hence $(J_{H_0}')^\sigma = J_+'$. Also it is obvious that $(\partial(H)p)^\sigma = \partial(H)p^\sigma$ $(H \in \mathfrak{h}_\mathfrak{p}, p \in S)$ and therefore $(\partial(u)p)^\sigma = \partial(u)p^\sigma$ $(u \in S)$. This shows that $V$ is exactly the set of those elements $v \in S$ which satisfy the condition $\langle v, p \rangle = 0$ for all $p \in SJ_+'$. For any $p \in S$, let $p_*$ denote the polynomial function on $\mathfrak{h}_\mathfrak{p}$ given by $p_*(H) = \mathrm{conj}\, p(H)$ $(H \in \mathfrak{h}_{\mathfrak{p}_0})$. Then $\langle p, p_* \rangle$ is a positive-definite Hermitian form on $S$ (see [5(k), p. 110]) and since $J_+'$ is invariant under the mapping $p \to p_*$, $V$ is the orthogonal complement of $SJ_+'$ in $S$ under this form. Hence $V \cap SJ_+' = \{0\}$. Moreover if $S_d$ denotes the space of homogeneous elements in $S$ of degree $d$, it is clear that $V \cap S_d$ is the orthogonal complement of $S_d \cap SJ_+'$ in $S_d$. Therefore since $\dim S_d$ is finite, it follows that $S_d = V \cap S_d + S_d \cap SJ_+'$ and so $S = V + SJ_+'$. But then $\dim V = \dim S/SJ_+' = w'$ from Lemma 13.

COROLLARY. *Suppose* $v \in V_{H_0}$ *and* $q \in I_\mathfrak{g}$. *Then*

$$\Phi(H_0; \partial(v):h; \delta(q)) = \gamma(q:(-1)^{\natural}H_0)\Phi(H_0; \partial(v):h)$$

*for* $h \in A_\mathfrak{p}(M)$.

Let $p$ denote the polynomial function $H \to \gamma(q:(-1)^{\natural}H)$ on $\mathfrak{h}_\mathfrak{p}$ and let $(\partial(v) \circ p)_H$ denote the local expression (see [5(k), p. 90]) at $H \in \mathfrak{h}_\mathfrak{p}$ of the differential operator [24] $\partial(v) \circ p$ on $\mathfrak{h}_\mathfrak{p}$. We know from Lemma 30 that

$$\Phi(H:h; \delta(q)) = p(H)\Phi(H:h) \qquad (H \in U, h \in A_\mathfrak{p}(M))$$

and therefore it is obvious that

$$\Phi(H;\partial(v):h;\delta(q)) = \Phi(H;(\partial(v)\circ p)_H:h)$$

Hence it would be sufficient to prove that $(\partial(v)\circ p)_{H_0} = p(H_0)\partial(v)$. So let us suppose that $D = (\partial(v)\circ p)_{H_0} - p(H_0)\partial(v) \neq 0$. Then we can choose $p_1 \in S$ such that $p_1(H_0;D) \neq 0$. Put $p_2 = (p - p(H_0))p_1$. Then $p_1(H_0;D) = p_2(H_0;\partial(v))$. On the other hand $p \in J$ and therefore $p - p(H_0) \in J_{H_0}{}'$ and $p_2 \in SJ_{H_0}{}'$. But since $v \in V_{H_0}$, this implies that $p_2(H_0;\partial(v)) = 0$. Hence we get a contradiction and so the corollary is proved.

Put $r = [W:W']$ and select $s_1 = 1, s_2, \cdots, s_r$ in $W$ such that $W = \bigcup_{1 \leq i \leq r} s_i W'$. Then the points $H_i = s_i H_0$ $(1 \leq i \leq r)$ are all distinct. We assume that $H_i \in {}'\mathfrak{h}_\mathfrak{p}$ $1 \leq i \leq r$ and $(-1)^{\frac{1}{2}}(H_i - H_j)$, regarded as a linear function on $\mathfrak{h}_\mathfrak{p}$, does not lie in $L$ for any pair of indices $i \neq j$ $(1 \leq i, j \leq r)$. Put $V_i = V_{H_i}$ $1 \leq i \leq r$ and for any $v \in V_i$, let $\psi_v{}^{(i)}$ denote the function $h \to \Phi(H_i;\partial(v):h)$ on $A_\mathfrak{p}(M)$ for $M$ sufficiently large.

LEMMA 33. *Select nonzero elements* $v_i \in V_i$ $(1 \leq i \leq r)$. *Then the functions* $\psi_{v_i}{}^{(i)}$ $1 \leq i \leq r$ *are linearly independent over* $C$.

It is clear that for fixed $H' \in {}'\mathfrak{h}_\mathfrak{p}$, $\lambda \in L$ and $v \in S$,

$$\xi_\lambda(H';\partial(v):H)\exp\{-(-1)^{\frac{1}{2}}\langle H',H\rangle + \lambda(H)\}$$

is obviously a polynomial function of $H$. We denote by $p_\lambda{}^{(i)}$ this polynomial corresponding to $H' = H_i$ and $v = v_i$. Then if $H \in E(M)$,

$$\phi(H_i;\partial(v_i):H) = \sum_{\lambda \in L} p_\lambda{}^{(i)}(H)\exp\{(-1)^{\frac{1}{2}}\langle H_i,H\rangle - \lambda(H)\}.$$

Now suppose $c_1, \cdots, c_r$ are complex numbers such that

$$\sum_i c_i\phi(H_i;\partial(v_i):H) = 0 \qquad\qquad (H \in E(M)).$$

Then

$$\sum_{1 \leq i \leq r}\sum_{\lambda \in L} c_i p_\lambda{}^{(i)}(H)\exp\{(-1)^{\frac{1}{2}}\langle H_i,H\rangle - \lambda(H)\} = 0$$

and it follows from Lemma 31, the corollary to Lemma 57 (§ 15) and our assumption that $(-1)^{\frac{1}{2}}(H_i - H_j) \notin L$ for $i \neq j$, that $c_i p_\lambda{}^{(i)} = 0$ for every $i$ and every $\lambda \in L$. On the other hand it is obvious that $p_0{}^{(i)}(H) = v_i((-1)^{\frac{1}{2}}H)$ $(H \in \mathfrak{h}_\mathfrak{p})$. Therefore since $v_i \neq 0$, it follows that $p_0{}^{(i)} \neq 0$ and so $c_i = 0$. This proves the lemma.

It follows from the corollary to Lemma 32 that

$$\delta(q)\psi_v{}^{(i)} = \gamma(q:(-1)^{\frac{1}{2}}H_0)\psi_v{}^{(i)} \qquad (v \in V_i, q \in I_\mathfrak{g}, 1 \leq i \leq r).$$

Moreover $\dim V_i = w'$ by Lemma 29. Hence if $r_{ij}$ $1 \leq j \leq w'$ is a base for $V_i$, we conclude from Lemma 33, that the $w$ functions $\psi_{ij} = \psi_{v_{ij}}{}^{(i)}$ $(1 \leq i \leq r,$

$1 \leq j \leq w'$) are linearly independent and $\delta(q)\psi_{ij} = \gamma(q:(-1)^{\delta}H_0)\psi_{ij}$ ($q \in I_{\mathfrak{g}}$). Therefore in view of Lemmas 18 and 23, we get the following result from the corollary of Theorem 2.

LEMMA 34. *Under the above assumptions there exist unique complex numbers $c_{ij}$ such that*

$$e^{\rho(\log h)} \int_K \exp\{(-1)^{\delta}\langle H_0, H(hk)\rangle - \rho(H(hk))\}dk = \sum_{1 \leq i \leq r} \sum_{1 \leq j \leq w'} c_{ij}\psi_{ij}(h)$$

*for all $h \in A_{\mathfrak{p}}(M)$.*

In particular let us consider the case when $H_0 = 0$. Then $W' = W$ and $V = V_1$ is the set of all $v \in S(\mathfrak{h}_{\mathfrak{p}})$ such that $\langle v, p \rangle = 0$ for $p \in SJ_+$. Moreover since $\mathfrak{h}_{\mathfrak{p}_0} \subset {}'\mathfrak{h}_{\mathfrak{p}}$, all the required conditions for $H_0$ hold in this case. Therefore we obtain the following corollary.

COROLLARY. *There exists an element $v \in V$ and a number $M \geq 0$ such that*

$$e^{\rho(\log h)} \int_K e^{-\rho(H(hk))}dk = \Phi(0;\partial(v):h)$$

*for all $h \in A_{\mathfrak{p}}(M)$.*

**9. An important inequality.** Define $\tilde{\theta}$ as in Section 5 and put $\|X\|^2 = -B(X, \tilde{\theta}(X))$ ($X \in \mathfrak{g}$). As before, we regard $\mathfrak{g}$ as a Hilbert space under the norm $\|X\|$. Let $\mathfrak{h}_{\mathfrak{p}_0}^+$ be the set of those points $H \in \mathfrak{h}_{\mathfrak{p}_0}$ where $\alpha(H) > 0$ for every $\alpha \in \Sigma$. We put $A_{\mathfrak{p}}^+ = \exp \mathfrak{h}_{\mathfrak{p}_0}^+$. Our main object in this section is to prove the following result which will play an important role later.

THEOREM 3. *There exists an integer $d \geq 0$ and a positive number $a$ such that*

$$\int_K e^{-\rho(H(hk))}dk \leq a(1 + \|\log h\|)^d e^{-\rho(\log h)}$$

*for all $h \in A_{\mathfrak{p}}^+$.*

For the proof we need some auxiliary results. Let ${}^+\mathfrak{h}_{\mathfrak{p}_0}$ denote the set of all $H \in \mathfrak{h}_{\mathfrak{p}_0}$ such that $\langle H, H' \rangle \geq 0$ for every $H'$ in $\mathfrak{h}_{\mathfrak{p}_0}^+$. Also let $\mathrm{Cl}(A_{\mathfrak{p}}^+)$ denote the closure of $A_{\mathfrak{p}}^+$ in $A_{\mathfrak{p}}$.

LEMMA 35. *Define $\mathfrak{F}_0$ as in Lemma 2. An element $H' \in \mathfrak{h}_{\mathfrak{p}_0}$ lies in ${}^+\mathfrak{h}_{\mathfrak{p}_0}$ if and only if $\lambda(H') \geq 0$ for every $\lambda \in \mathfrak{F}_0$. Moreover $\mathfrak{h}_{\mathfrak{p}_0}^+ \subset {}^+\mathfrak{h}_{\mathfrak{p}_0}$ and $\log h - H(hk) \in {}^+\mathfrak{h}_{\mathfrak{p}_0}$ for $h \in \mathrm{Cl}(A_{\mathfrak{p}}^+)$ and $k \in K$.*

Select $\alpha_1, \cdots, \alpha_l \in \Sigma$ as in Section 3 and choose $H_i \in \mathfrak{h}_{\mathfrak{p}_0}$ such that[9] $\alpha_i(H_j) = \delta_{ij}$ $(1 \leq i, j \leq l)$. Then it is obvious that $\mathfrak{h}_{\mathfrak{p}_0}{}^+$ consists of all elements of the form $t_1 H_1 + \cdots + t_l H_l$ where $t_1, \cdots, t_l$ are positive numbers. Therefore $H'$ lies in ${}^+\mathfrak{h}_{\mathfrak{p}_0}$ if and only if $\langle H', H_i \rangle \geq 0$ for every $i$. For any $\beta \in P$ define $H_\beta$ as during the proof of Lemma 27 and, for a fixed $i$, consider the linear function $\mu_i$ on $\mathfrak{h}$ given by $\mu_i(H) = B(H_i, H)$ $(H \in \mathfrak{h})$. It is clear that $\mu_i(H_\beta) = \beta(H_i) \geq 0$ for $\beta \in P$. Since $\beta(H_\beta)$ are positive rational numbers, we can choose a positive integer $m$ such that $2m\mu_i(H_\beta)/\beta(H_\beta)$ is an integer for every $\beta \in P$ and every $i$ $(1 \leq i \leq l)$. Let $\lambda_i$ denote the restriction of $2m\mu_i$ on $\mathfrak{h}_{\mathfrak{p}}$. Then it follows from Theorem 1 of [5(b)] and Lemma 2 that $\lambda_i \in \mathfrak{F}_0$. Now suppose $H' \in \mathfrak{h}_{\mathfrak{p}_0}$. Then $\lambda_i(H') = 2m\langle H', H_i \rangle$ and $\langle \lambda_i, \alpha_j \rangle = 2m\alpha_j(H_i) = 2m\delta_{ji}$. Therefore if $\lambda$ is any linear function on $\mathfrak{h}_{\mathfrak{p}}$, $\lambda = \sum_i c_i \lambda_i$ where $c_i = \langle \lambda, \alpha_i \rangle / 2m$. But $\lambda \geq s_{\alpha_i} \lambda$ if $\lambda \in \mathfrak{F}_0$ and so $2mc_i = \langle \lambda, \alpha_i \rangle \geq 0$. Hence $\langle H', H_i \rangle \geq 0$ $(1 \leq i \leq l)$ if and only if $\lambda(H') \geq 0$ for every $\lambda \in \mathfrak{F}_0$. This proves the first statement of the lemma.

Now let $\pi$ be an irreducible representation of $\mathfrak{g}$ on a finite-dimensional space $V$ and let $\phi$ be a unit vector in $V$ belonging to the highest weight $\Lambda$ of $\pi$. Then (see [5(h), Lemma 2]) every other weight of $\pi$ is of the form $\Lambda - (m_1\beta_1 + \cdots + m_r\beta_r)$ where $\beta_1, \cdots, \beta_r \in P$ and $m_1, \cdots, m_r$ are positive integers. Hence if $H \in \mathfrak{h}_{\mathfrak{p}_0}{}^+$, it follows that $\Lambda(H) \geq \Lambda'(H)$ for every weight $\Lambda'$ of $\pi$. But since $\pi(H)$ is a self-adjoint operator of trace zero, this implies that $\Lambda(H) \geq 0$. Similarly if $h \in \mathrm{Cl}(A_\mathfrak{p}{}^+)$, $e^{\Lambda(\log h)}$ is the largest eigenvalue of the self-adjoint operator $\pi(h)$. Therefore

$$| \pi(hk)\phi | \leq e^{\Lambda(\log h)} | \pi(k)\phi | = e^{\Lambda(\log h)}$$

for $k \in K$. However it is obvious that $| \pi(hk)\phi | = e^{\Lambda(H(hk))}$ and so $\Lambda(\log h - H(hk)) \geq 0$. This shows in particular that $\lambda(H) \geq 0$ and $\lambda(\log h - H(hk)) \geq 0$ for $\lambda \in \mathfrak{F}_0$. Therefore by the criterion established above, $H$ and $\log h - H(hk)$ lie in ${}^+\mathfrak{h}_{\mathfrak{p}_0}$. Thus the lemma is proved.

COROLLARY 1. *Let $H$ and $H'$ be two elements in $\mathfrak{h}_{\mathfrak{p}_0}{}^+$. Then $\langle H', H \rangle > \langle H', sH \rangle$ for $s \neq 1$ in $W$.*

Fix an element $s \neq 1$ in $W$ and select $k \in K$ such that $\mathrm{Ad}(k)H = sH$. Then $\pi(sH) = \pi(k)\pi(H)\pi(k^{-1})$ in the above notation. Therefore since $\Lambda(H)$ is the greatest eigenvalue of $\pi(H)$, it follows that $\Lambda(sH) \leq \Lambda(H)$. But in view of Lemma 35, this implies that $H - sH \in {}^+\mathfrak{h}_{\mathfrak{p}_0}$. Now $\lambda_1, \cdots, \lambda_l$ are linearly independent and by Lemma 6 $H \neq sH$ since $H \in \mathfrak{h}_{\mathfrak{p}_0}{}^+$. Therefore $\lambda_i(H - sH) \neq 0$ for some $i$. Moreover we have seen that if $\lambda$ is any linear function on $\mathfrak{h}_{\mathfrak{p}}$, $\lambda = \sum_i c_i \lambda_i$ where $c_i = \langle \lambda, \alpha_i \rangle / 2m$. Therefore $\langle H', H - sH \rangle$

$= (2m)^{-1} \sum_{1 \leq i \leq l} \alpha_i(H')\lambda_i(H - sH)$. But from Lemma 35, $\lambda_i(H - sH)$ are non-negative and $\alpha_i(H') > 0$ $(1 \leq i \leq l)$ since $H' \in \mathfrak{h}_{\mathfrak{p}_0}{}^+$. Therefore $\langle H', H - sH \rangle > 0$.

From [5(j), Lemma 37] we can choose $s_0 \in W$ such that $s_0 \mathfrak{h}_{\mathfrak{p}_0}{}^+ = -\mathfrak{h}_{\mathfrak{p}_0}{}^+$. Put $H^* = -s_0 H$ for any $H \in \mathfrak{h}_{\mathfrak{p}}$. It is obvious that the mapping $\alpha \to -s_0 \alpha$ $(\alpha \in \Sigma)$ is a permutation on $\Sigma$ and therefore $\rho(H^*) = \rho(H)$.

COROLLARY 2. $(\log h)^* + H(hk) \in {}^+\mathfrak{h}_{\mathfrak{p}_0}$ for $h \in \mathrm{Cl}(A_{\mathfrak{p}}{}^+)$ and $k \in K$.

Fix $h$ and $k$ and let $hk = k' \exp H(hk) n$ $(k' \in K, n \in N)$. Then $H(hk) = -H(h^{-1}k')$ (see [5(c), Lemma 36]). Now select $k_0 \in K$ such that $\mathrm{Ad}(k_0)H' = s_0 H'$ for every $H'$ in $\mathfrak{h}_{\mathfrak{p}_0}$. Then if $h^* = \exp(\log h)^*$, it is obvious that $h^* = k_0 h^{-1} k_0{}^{-1}$ and therefore $H(h^* k_0 k') = H(h^{-1}k') = -H(hk)$. But from Lemma 35, this implies that

$$\log h^* + H(hk) = \log h^* - H(h^* k_0 k') \in {}^+\mathfrak{h}_{\mathfrak{p}_0}.$$

COROLLARY 3. $\rho(\log h) \geq |\rho(H(hk))|$ for $h \in \mathrm{Cl}(A_{\mathfrak{p}}{}^+)$ and $k \in K$.

It follows from the definition of $\rho$ that $s_\alpha \rho < \rho$ and therefore $\langle \rho, \alpha \rangle > 0$ for any $\alpha \in \Sigma$. Hence $H_\rho \in \mathfrak{h}_{\mathfrak{p}_0}{}^+$. Moreover $\rho(\log h^*) = \rho(\log h)$ as we have seen above. Hence our assertion follows from the fact that $\log h - H(hk)$ and $\log h^* + H(hk)$ are both in ${}^+\mathfrak{h}_{\mathfrak{p}_0}$.

Now put $\psi(h) = e^{\rho(\log h)} \int_K e^{-\rho(H(hk))} dk$ $(h \in A_{\mathfrak{p}})$.

LEMMA 36. $\psi(h) \leq \psi(hh_1)$ for $h \in A_{\mathfrak{p}}$ and $h_1 \in \mathrm{Cl}(A_{\mathfrak{p}}{}^+)$.

Fix $h$ and $h_1$ and for any $k \in K$, let $k'$ denote the unique element in $K$ such that $h_1 k \in k' A_{\mathfrak{p}} N$. Then $k \to k'$ is a homeomorphism of $K$, $H(hh_1 k) = H(hk') + H(h_1 k)$ and $dk' = e^{-2\rho(H(h_1 k))} dk$ (see [5(c), pp. 240-241]). Hence

$$\int_K e^{-\rho(H(hk))} dk = \int_K e^{-\rho(H(hk'))} dk' = \int_K \exp\{-\rho(H(hk')) - 2\rho(H(h_1 k))\} dk$$

$$= \int_K \exp\{-\rho(H(hh_1 k)) - \rho(H(h_1 k))\} dk \leq e^{\rho(\log h_1)} \int_K e^{-\rho(H(hh_1 k))} dk$$

from Corollary 3 to Lemma 35. This implies that $\psi(h) \leq \psi(hh_1)$.

Now we come to the proof of Theorem 3. Select $H_0 \in \mathfrak{h}_{\mathfrak{p}_0}{}^+$ such that $\alpha_i(H_0) = 1$ $(1 \leq i \leq l)$ and put $F(t) = \psi(\exp tH_0)$ $(t \in R)$. Then in the notation of Section 8, $F(t) = \phi(0; \partial(v) : tH_0)$ for $t > M$, where $v$ and $M$ are defined as in the corollary to Lemma 34. Let $d$ be the degree of $v$. Then it follows from Lemma 31 that

$$F(t) = \sum_{\lambda \in L} p_\lambda(t) e^{-t\lambda(H_0)} \qquad (t > M),$$

where $p_\lambda$ are polynomials of degree $\leq d$ and the series $\sum_{\lambda \in L} | p_\lambda(t) e^{-t\lambda(H_0)} |$ converges uniformly for $t > M$ (if $M$ is chosen sufficiently large). Hence we can find a finite number of distinct elements $\lambda_0 = 0, \lambda_1, \cdots \lambda_r \in L$ such that

$$| F(t) - \sum_{0 \leq i \leq r} p_{\lambda_i}(t) e^{-t\lambda_i(H_0)} | \leq 1$$

for $t > M$. But since $\lambda(H_0) = m(\lambda)$ (in the notation of Lemma 28), it is obvious that

$$\operatorname*{Lim}_{t \to +\infty} \sum_{1 \leq i \leq r} p_{\lambda_i}(t) e^{-t\lambda_i(H_0)} = 0.$$

Therefore $F(t) \leq 2 + p_0(t)$ for all sufficiently large positive values of $t$. Since the degree of $p_0$ is $\leq d$, this means that we can select a positive constant $a$ such that $F(t) \leq a(1 + t)^d$ for all $t \geq 0$.

Let $\mathrm{Cl}(\mathfrak{h}_{\mathfrak{p}_0}^+)$ denote the closure of $\mathfrak{h}_{\mathfrak{p}_0}^+$ in $\mathfrak{h}_{\mathfrak{p}_0}$ and put $\beta(H) = \max_{1 \leq i \leq l} \alpha_i(H)$ for $H \in \mathfrak{h}_{\mathfrak{p}_0}$. Then it is obvious that $\beta(H)H_0 - H \in \mathrm{Cl}(\mathfrak{h}_{\mathfrak{p}_0}^+)$. Hence it follows from Lemma 36 that $\psi(\exp H) \leq F(\beta(H)) \leq a(1 + \beta(H))^d$ for $H \in \mathrm{Cl}(\mathfrak{h}_{\mathfrak{p}_0}^+)$. On the other hand $\| H \|^2 = \mathrm{sp}(\operatorname{ad} H)^2 \geq \sum_{1 \leq i \leq l} \alpha_i(H)^2 \geq \beta(H)^2$. Therefore

$$\psi(\exp H) \leq a(1 + \| H \|)^d$$

and this proves the theorem.

*Remark.* Let $d_\pi$ denote the degree of the polynomial function $\pi$ of Section 3 and define $V$ as in the corollary to Lemma 34. Then it follows from Lemma 16 that no nonzero element in $V$ can have a degree greater than $d_\pi$. This shows that $d \leq d_\pi$.

**10. The function $c$.** For any $\lambda \in L'$ define $\sigma_\lambda$ (as in Section 8) to be the hyperplane consisting of those points $H \in \mathfrak{h}_{\mathfrak{p}}$ where $\langle \lambda, \lambda \rangle = 2(-1)^{\frac{1}{2}}\lambda(H)$ and let $\tau_\lambda(s, t)$ $(s, t \in W)$ denote the set of all $H \in \mathfrak{h}_{\mathfrak{p}}$ for which $(-1)^{\frac{1}{2}}(sH - tH) = H_\lambda$. Then if $\mathfrak{S}$ is the collection of all hyperplanes of the form $s\sigma_\lambda$ or $\tau_\lambda(s.t)$ $(s. t \in W; \lambda \in L')$, it is obvious that any compact subset of $\mathfrak{h}_{\mathfrak{p}}$ meets only a finite number of hyperplanes in $\mathfrak{S}$. Let $*\mathfrak{h}_{\mathfrak{p}}$ be the complement (in $\mathfrak{h}_{\mathfrak{p}}$) of the union of all hyperplanes in $\mathfrak{S}$. Then $*\mathfrak{h}_{\mathfrak{p}}$ is an open, connected and dense subset of $\mathfrak{h}_{\mathfrak{p}}$. Moreover it follows from its definition that $*\mathfrak{h}_{\mathfrak{p}}$ is invariant under $W$. Define $\pi$ as in Lemma 11 and let $*\mathfrak{h}_{\mathfrak{p}}'$ be the set of those $H \in *\mathfrak{h}_{\mathfrak{p}}$ where $\pi(H) \neq 0$. Since the zeros of $\pi$ consist of a finite number of hyperplanes, $*\mathfrak{h}_{\mathfrak{p}}'$ is also connected. We shall now use the notation of Section 8.

LEMMA 37. *There exists a holomorphic function $c$ on $*\mathfrak{h}_{\mathfrak{p}}'$ with the* *following property. For any compact subset $U$ of $*\mathfrak{h}_\mathfrak{p}$, we can select a number $M \geq 0$ such that*

$$e^{\rho(\log h)} \int_K \exp\{(-1)^{\frac{1}{2}}\langle H_0, H(hk)\rangle - \rho(H(hk))\}dk$$
$$= \sum_{s \in W} c(sH_0)\Phi(sH_0:h)$$

*for $H_0 \in U \cap *\mathfrak{h}_\mathfrak{p}'$ and $h \in A_\mathfrak{p}(M)$. (Here $\Phi$ has the same meaning as in Lemma 30.)*

Put

$$\psi(H_0:h) = e^{\rho(\log h)} \int_K \exp\{(-1)^{\frac{1}{2}}\langle H_0, H(hk)\rangle - \rho(H(hk))\}dk$$

for $H_0 \in \mathfrak{h}_\mathfrak{p}$ and $h \in A_\mathfrak{p}$. It is obvious that for fixed $h \in A_\mathfrak{p}$ and[22] $u \in \mathfrak{H}_\mathfrak{p}$, $\psi(H_0:h;u)$ is a holomorphic function of $H_0$. Moreover if $H_0 \in *\mathfrak{h}_\mathfrak{p}'$, the $w$ elements $sH_0$ $(s \in W)$ are all distinct. Therefore it follows from Lemma 34 that there exist unique complex numbers $c_s(H_0)$ such that

$$\psi(H_0:h) = \sum_{s \in W} c_s(H_0)\Phi(sH_0:h)$$

for $h \in A_\mathfrak{p}(M)$ provided $M$ is sufficiently large. In view of Corollary 1 of Lemma 28, it is sufficient to show that $c_s$ $(s \in W)$, regarded as functions of $H_0$, are holomorphic on $*\mathfrak{h}_\mathfrak{p}'$ and $c_s(H_0) = c_1(sH_0)$ $(H_0 \in *\mathfrak{h}_\mathfrak{p}')$.

Let $U_0$ be a nonempty open set in $*\mathfrak{h}_\mathfrak{p}$ which is invariant under $W$. We assume that the closure of $U_0$ in $*\mathfrak{h}_\mathfrak{p}$ is compact. Select $M$ so large that for any $H \in U_0, \Phi(H:h)$ is defined for $h \in A_\mathfrak{p}(M)$ and choose $u_1, \cdots, u_w \in \mathfrak{H}_\mathfrak{p}$ as in the corollary to Lemma 8. Also put $U_0' = U_0 \cap *\mathfrak{h}_\mathfrak{p}'$.

LEMMA 38. *Let $s_1, s_2, \cdots, s_w$ be all the elements of $W$. Then $\det\{\Phi(s_iH_0:h_0;u_j)\}_{1 \leq i,j \leq w} \neq 0$ for any $H_0 \in U_0'$ and $h_0 \in A_\mathfrak{p}(M)$.*

For otherwise we can select complex numbers $a_s$ $(s \in W)$, not all zero, such that $\sum_{s \in W} a_s\Phi(sH_0:h_0;u_j) = 0$ $(1 \leq j \leq w)$. Put $f(h) = \sum_{s \in W} a_s\Phi(sH_0:h)$ for $h \in A_\mathfrak{p}(M)$. Then $f$ is an analytic function on $A_\mathfrak{p}(M)$ and it follows from Lemma 30 that $\delta(q)f = \gamma(q:(-1)^{\frac{1}{2}}H_0)f$ for $q \in I_\mathfrak{g}$. Therefore since $f(h_0;u_j) = 0$ $(1 \leq j \leq w)$, we conclude from Theorem 2 that $f(h_0;u) = 0$ for every $u \in \mathfrak{H}_\mathfrak{p}$. But since $f$ is analytic and $A_\mathfrak{p}(M)$ is connected, this implies that $f = 0$. However this contradicts Lemma 33 since $W' = \{1\}$ in the present case.

Now in order to complete the proof of Lemma 37, fix $h_0 \in A_\mathfrak{p}(M)$. Then by Lemma 38 there exist holomorphic functions $a_{si}$ on $U_0'$ $(s \in W, 1 \leq i \leq w)$ such that $\sum_{1 \leq i \leq w} a_{si}(H)\Phi(s'H:h_0;u_i) = 1$ or $0$ according as $s = s'$

or not $(s, s' \in W, H \in U_0')$. Therefore $c_s(H) = \sum\limits_{1 \leq i \leq w} a_{si}(H)\psi(H : h_0 ; u_i)$ $(s \in W, H \in U_0')$ and this proves that $c_s$ is holomorphic on $U_0'$. On the other hand we know from the corollary to Lemma 17 that $\psi(sH : h_0) = \psi(H : h_0)$ for $s \in W$ and $H \in \mathfrak{h}_\mathfrak{p}$. This implies that $c_s(H) = c(sH)$ $(s \in W, H \in U_0)$ where $c = c_1$. Hence Lemma 37 is now proved completely.

**11. A formula for $c$.** We shall now derive an explicit formula for $c$ which will be valid on a suitable subdomain of $*\mathfrak{h}_\mathfrak{p}'$. For any root $\alpha \in P$, define $X_\alpha$ and $X_{-\alpha}$ as in Section 7 and put $\mathfrak{n}^+ = \sum\limits_{\alpha \in P} CX_\alpha, \mathfrak{n}^- = \sum\limits_{\alpha \in P} CX_{-\alpha}$. Let $N_c{}^+$, $N_c{}^-$ and $A_c$ denote the complex-analytic subgroups of $G_c$ corresponding to $\mathfrak{n}^+$, $\mathfrak{n}^-$ and $\mathfrak{h}$ respectively and put $G_c' = N_c{}^- A_c N_c{}^+$. Since $\mathfrak{g}$ is the direct sum of $\mathfrak{n}^-$, $\mathfrak{h}$ and $\mathfrak{n}^+$, it follows easily that $G_c'$ is open in $G_c$. We regard $G_c'$ as an open submanifold of $G_c$. Then (see [5(i), Lemma 1]) the mapping $(n_1, a, n_2) \to n_1 a n_2$ $(n_1 \in N_c{}^-, a \in A_c, n_2 \in N_c{}^+)$ is a one-one holomorphic mapping of $N_c{}^- \times A_c \times N_c{}^+$ onto $G_c'$ which is regular everywhere. For any $z \in G_c'$, let $a(z)$ denote the unique element in $A_c$ such that $z \in N_c{}^- a(z) N_c{}^+$. Then $z \to a(z)$ is a holomorphic mapping of $G_c'$ onto $A_c$.

LEMMA 39. *Let $n_r', a_r, n_r$ $(r \geq 1)$ be three sequences in $N_c{}^-$, $A_c$ and $N_c{}^+$ respectively such that $a_r$ and $n_r' a_r n_r$ converge in $G_c$. Then $n_r'$ and $n_r$ are also convergent.*

Obviously it would be enough to prove that $n_r'$ is convergent. Since $A_c$ is closed in $G_c$, $a_r$ converges to an element $a_0 \in A_c$. Let $\pi$ be an irreducible representation of $\mathfrak{g}$ (and therefore also of $G_c$) on a finite-dimensional vector space $V$ and $\xi$ a unit vector belonging to the highest weight $\Lambda$ of $\pi$. For any root $\alpha$ define $H_\alpha$ as in Section 7. Then by choosing $\pi$ suitably, we can assume (see [5(b), Theorem 1]) that $\Lambda(H_\alpha) > 0$ for every $\alpha \in P$. Let $\chi$ denote the character of $A_c$ such that $\pi(a)\xi = \chi(a)\xi$ for $a \in A_c$. Then if $n_r' a_r n_r$ converges to $z$ in $G_c$, it is clear that

$$\pi(n_r')\xi = \chi(a_r{}^{-1})\pi(n_r' a_r n_r)\xi \to \chi(a_0{}^{-1})\pi(z)\xi.$$

Since $\mathfrak{n}^-$ is a nilpotent Lie algebra, we can select $Y_r \in \mathfrak{n}^-$ such that $\exp Y_r = n_r'$. Then it would be sufficient to prove that $Y_r$ converges in $\mathfrak{n}^-$. Thus it remains to prove the following result.

LEMMA 40. *Suppose $Y$ varies in $\mathfrak{n}^-$ in such a way that $\pi(\exp Y)\xi$ converges in $V$. Then $Y$ itself converges in $\mathfrak{n}^-$.*

Let $\beta_1 < \beta_2 < \cdots < \beta_r$ be all the roots in $P$. Then $Y = \sum\limits_{1 \leq i \leq r} t_i(Y)X_{-\beta_i}$

$(t_i(Y) \in C)$ and it would be sufficient to prove that $t_i(Y)$ converges in $C$ for every $i$. Hence suppose that this is false and let $j$ be the least index such that $t_j(Y)$ does not converge. Let $V_j$ be the subspace of $V$ consisting of all vectors belonging to the weight $\Lambda - \beta_j$ and let $E_j$ denote the orthogonal projection of $V$ on $V_j$. Then

$$E_j \pi (\exp Y) \xi = \sum_{m \geq 0} E_j \pi (Y^m) \xi / m\,!.$$

But if $Y' = \sum_{1 \leq i < j} t_i(Y) X_{-\beta_i}$, it is obvious that $E_j \pi (Y^m) \xi = E_j \pi (Y'^m) \xi$ for $m > 1$. Moreover $E_j \pi (Y') \xi = 0$ while $E_j \pi (Y) \xi = t_j(Y) \pi (X_{-\beta_j}) \xi$. Hence

$$E_j \pi (\exp Y) \xi = t_j(Y) \pi (X_{-\beta_j}) \xi + E_j \pi (\exp Y') \xi.$$

On the other hand, in view of our hypothesis and the definition of $j$, both $\pi (\exp Y) \xi$ and $\pi (\exp Y') \xi$ converge in $V$. Therefore the same holds for $t_j(Y) \pi (X_{-\beta_j}) \xi$. But $\pi (X_{-\beta_j}) \xi \neq 0$ since $\Lambda (H_{\beta_j}) > 0$ (see [5(h), Lemma 1]). Hence $t_j(Y)$ also converges in $C$. As this contradicts the definition of $j$, our assertion follows.

COROLLARY 1. *Let $z$ be an element in $G_c$. Then $z \in G_c'$ if and only if* $(\xi, \pi(z)\xi) \neq 0$.

Put $f(z) = (\xi, \pi(z)\xi)$ and let $G_c''$ be the set of all $z \in G_c$ where $f(z) \neq 0$. Since $f$ is a holomorphic function on $G_c$ and $f(1) = 1$, its set of zeros is a complex subvariety of $G_c$ of one complex dimension less. Therefore $G_c''$ is an open connected subset of $G_c$. Moreover if $z \in G_c'$, it is obvious that $f(z) = \chi(a(z)) \neq 0$. Therefore $G_c' \subset G_c''$. Hence it would be sufficient to show that $G_c'$ is closed in $G_c''$. Let $z_r$ $(r \geq 1)$ be a sequence in $G_c'$ which converges to $z \in G_c''$. Suppose $z_r = n_r' a_r n_r$ $(n_r' \in N_c^-, a_r \in A_c, n_r \in N_c^+)$. Then $\chi(a_r) = f(z_r) \to f(z) \neq 0$ and therefore $\pi(n_r')\xi = \chi(a_r^{-1})\pi(z_r)\xi \to f(z)^{-1}\pi(z)\xi$. But then by the above lemma, $n_r'$ converges to an element $n'$ in $N_c^-$. However this implies that $a_r n_r = n_r'^{-1} z_r \to n'^{-1} z$. On the other hand $A_c N_c^+$ is closed in $G_c$ and hence $n'^{-1} z \in A_c N_c^+$. This proves that $z \in G_c'$ and therefore $G_c'$ is closed in $G_c''$.

COROLLARY 2. *Put $K' = K \cap G_c'$. Then $K'$ is an open dense subset of $K$ whose complement (in $K$) is of measure zero with respect to the Haar measure of $K$.*

Let $g$ be the restriction on $K$ of the function $f$ defined above. Then $g$ is an analytic function which is not identically zero since $g(1) = 1$. Moreover $K'$ is the set of all points $k \in K$ where $g(k) \neq 0$. From this our statement follows immediately.

Since $G_c$ is simply connected, we can extend $\theta$ to an automorphism of $G_c$. Put $\mathfrak{n} = \sum_{\alpha \in P_+} C X_\alpha$ and $\bar{\mathfrak{n}} = \theta(\mathfrak{n}) = \sum_{\alpha \in P_+} C X_{-\alpha}$ and let $N_c$, $\bar{N}_c$ denote the complex analytic subgroups of $G_c$ corresponding to $\mathfrak{n}$, $\bar{\mathfrak{n}}$ respectively. Also let $\Xi_c$ and $\mathfrak{z}$ denote the centralizers of $\mathfrak{h}_\mathfrak{p}$ in $G_c$ and $\mathfrak{g}$ respectively. Since $\mathfrak{n}^+ + \mathfrak{h}_\mathfrak{p} \subset \mathfrak{n} + \mathfrak{z}$, it follows easily that $\bar{N}_c \Xi_c N_c \supset N_c^- A_c N_c^+ = G_c'$.

LEMMA 41. *Suppose $Y \in \bar{\mathfrak{n}}$ and $\exp(\operatorname{ad} Y)$ maps $\mathfrak{n}$ into itself. Then $Y = 0$.*

Otherwise suppose that $Y \neq 0$. Let $\beta_1 < \beta_2 < \cdots < \beta_r$ be all the roots in $P_+$. Then $Y = \sum_i c_i X_{-\beta_i}$ where $c_i \in C$. Let $k$ be the least integer such that $c_k \neq 0$. Then it is obvious that $\exp(\operatorname{ad} Y) X_{\beta_k} - X_{\beta_k} \equiv - c_k H_{\beta_k} \bmod \mathfrak{n}^-$. Therefore since the sum $\mathfrak{h} + \mathfrak{n}^+ + \mathfrak{n}^-$ is direct, it follows that $\exp(\operatorname{ad} Y) X_{\beta_k}$ does not lie in $\mathfrak{n}^+ + \mathfrak{n}^-$. But this contradicts our hypothesis that $\exp(\operatorname{ad} Y)$ maps $\mathfrak{n}$ into itself. Hence the lemma.

COROLLARY. $\bar{N}_c \cap (\Xi_c N_c) = \{1\}$ *and* $\Xi_c \cap N_c = \{1\}$.

Suppose $y \in \bar{N}_c \cap (\Xi_c N_c)$. Then $y = \exp Y$ for some $Y \in \bar{\mathfrak{n}}$ and $\operatorname{Ad}(y)$ leaves $\mathfrak{n}$ invariant. Hence $Y = 0$ and so $y = 1$. Now suppose $x \in \Xi_c \cap N_c$. Then $x = \exp X$ for some $X \in \mathfrak{n}$. Put $Y = \theta(X)$. Since $x \in \Xi_c$, $\operatorname{Ad}(x)$ leaves $\bar{\mathfrak{n}}$ invariant and therefore $\exp(\operatorname{ad} Y)$ leaves $\mathfrak{n}$ invariant. Hence $Y = 0$. This proves that $x = 1$.

Define $N$ as in Section 2 and let $M$ be the centralizer of $A_\mathfrak{p}$ in $K$. Put $S = M A_\mathfrak{p} N$. Obviously $S$ is a subgroup of $G$.

LEMMA 42. $G \cap (\bar{N}_c \Xi_c N_c) = \bar{N} S$ *where* $\bar{N} = \theta(N)$.

Suppose $x \in G \cap (\bar{N}_c \Xi_c N_c)$. Then $x = \bar{n} \xi n$ $(\bar{n} \in \bar{N}_c, \xi \in \Xi_c, n \in N_c)$. Let $\eta$ denote the conjugation of $\mathfrak{g}$ with respect to $\mathfrak{g}_0$. We extend it to a real automorphism of $G_c$. Then $x = \eta(x) = \eta(\bar{n})\eta(\xi)\eta(n)$. Since $\eta$ maps $\mathfrak{n}$ and $\bar{\mathfrak{n}}$ into themselves, it follows from the corollary to Lemma 41 that $\eta(\bar{n}) = \bar{n}$ and $\eta(n) = n$. Choose $X \in \mathfrak{n}$ and $Y \in \bar{\mathfrak{n}}$ such that $\exp X = n$, $\exp Y = \bar{n}$. Since the exponential mapping is univalent on both $\mathfrak{n}$ and $\bar{\mathfrak{n}}$, we conclude that $X$ and $Y$ are left fixed by $\eta$. Hence $X \in \mathfrak{n} \cap \mathfrak{g}_0$, $Y \in \bar{\mathfrak{n}} \cap \mathfrak{g}_0$ and therefore $n \in N$, $\bar{n} \in \bar{N}$ and $\xi \in G \cap \Xi_c$. This shows that $G \cap (\bar{N}_c \Xi_c N_c) = \bar{N}(G \cap \Xi_c) N$. But since $M A_\mathfrak{p}$ is the centralizer[13] of $A_\mathfrak{p}$ in $G$, it follows that $G \cap \Xi_c = M A_\mathfrak{p}$ and therefore $G \cap (\bar{N}_c \Xi_c N_c) = \bar{N} S$.

COROLLARY. *Put $K_1 = K \cap (\bar{N} S)$. Then $K_1 \supset K'$.*

We have seen above that $\bar{N}_c \Xi_c N_c \supset G_c'$ and therefore $K_1 = K \cap (\bar{N}_c \Xi_c N_c) \supset K \cap G_c' = K'$.

LEMMA 43. *Let $\bar{n} \in \bar{N}$ and $h \in \mathrm{Cl}(A_{\mathfrak{p}}^{+})$. Then both $H(\bar{n})$ and $H(\bar{n})$* *$- H(h\bar{n}h^{-1})$ lie in* $^{+}\mathfrak{h}_{\mathfrak{p}_0}$.

Let $\pi$ be an irreducible representation of $\mathfrak{g}$ on a finite-dimensional space and let $\phi$ be a unit vector belonging to the highest weight $\Lambda$ of $\pi$. Then $|\pi(\bar{n})\phi| = e^{\Lambda(H(\bar{n}))}$. Moreover since $h \in \mathrm{Cl}(A_{\mathfrak{p}}^{+})$, $e^{\Lambda(\log h)}$ is the greatest eigenvalue of the self-adjoint operator $\pi(h)$ (see the proof of Lemma 35). Hence

$$e^{\Lambda(H(h\bar{n}h^{-1}))} = |\pi(h\bar{n}h^{-1})\phi| \leq e^{\Lambda(\log h)}|\pi(\bar{n}h^{-1})\phi| = |\pi(\bar{n})\phi| = e^{\Lambda(H(\bar{n}))}.$$

This proves that $\Lambda(H(\bar{n})) - \Lambda(H(h\bar{n}h^{-1})) \geq 0$. Now choose $Y \in \bar{\mathfrak{n}}$ such that $\bar{n} = \exp Y$. Then $\pi(\exp Y)\phi - \phi$ is obviously a sum of vectors belonging to weights lower than $\Lambda$. Hence $\phi$ and $\pi(\bar{n})\phi - \phi$ are mutually orthogonal and therefore $|\pi(\bar{n})\phi|^2 = |\phi|^2 + |\pi(\bar{n})\phi - \phi|^2 \geq 1$. This shows that $\Lambda(H(\bar{n})) \geq 0$ and the required statements now follow from Lemma 35.

COROLLARY. $\rho(H(\bar{n})) \geq 0$ *for* $\bar{n} \in \bar{N}$.

Since $H_\rho \in \mathfrak{h}_{\mathfrak{p}_0}^{+}$, this is an immediate consequence of the above lemma.

For any $\bar{n} \in \bar{N}$, let $k(\bar{n})$ denote the unique element in $K$ such that $\bar{n} \in k(\bar{n})A_{\mathfrak{p}}N$. We denote by $C(K)$ the space of continuous functions on $K$ and by $dm$ the normalized Haar measure on $M$.

LEMMA 44. *The Haar measure $d\bar{n}$ on $\bar{N}$ can be so normalized that*

$$\int_K f(k)\,dk = \int_{M \times \bar{N}} f(mk(\bar{n}))\,e^{-2\rho(H(\bar{n}))}\,dm\,d\bar{n}$$

*for any $f \in C(K)$. This normalization is characterized by the condition that*

$$\int_{\bar{N}} e^{-2\rho(H(\bar{n}))}\,d\bar{n} = 1.$$

Put $S_0 = A_{\mathfrak{p}}N$ and let $ds_0$ denote the left-invariant measure on $S_0$. Similarly let $ds$ denote the left-invariant measure on $S = MS_0$. We may assume that $ds = dm\,ds_0$ $(s = ms_0, m \in M, s_0 \in S_0)$. Let $dx$ denote the Haar measure on $G$ and $D(s)$ $(s \in S)$ the absolute value of the determinant of the restriction of $\mathrm{Ad}(s)$ on $\mathfrak{z} + \mathfrak{n}$. Since $G = KS_0$, a simple calculation (see [5(c), Lemma 35]) shows that

$$\int_G F(x)\,dx = \int_{K \times S_0} F(ks_0)D(s_0)\,dk\,ds_0 \qquad (F \in C_c(G)),[17]$$

provided $dx$ is suitably normalized. Now $\bar{N}S = K_1S_0$ and therefore we conclude from Corollary 2 to Lemma 40 and the corollary to Lemma 42 that the complement of $\bar{N}S$ in $G$ is of measure zero. On the other hand the mapping $(\bar{n},s) \to \bar{n}s$ of $\bar{N} \times S$ into $G$ is univalent from the corollary to Lemma 41. Put $\mathfrak{n}_0 = \mathfrak{n} \cap \mathfrak{g}_0$ and $\mathfrak{z}_0 = \mathfrak{z} \cap \mathfrak{g}_0$. Then $\mathfrak{g}_0 = \theta(\mathfrak{n}_0) + \mathfrak{z}_0$ and from this it follows easily that the above mapping is everywhere regular. Moreover a straightforward calculation shows that $dx = D(s)\,d\bar{n}\,ds$ $(x = \bar{n}s)$. Therefore

$$\int_{K \times S_0} F(ks_0) D(s_0)\,dk\,ds_0 = \int_{\bar{N} \times S} F(\bar{n}s) D(s)\,d\bar{n}\,ds.$$

Now $\bar{n} = k(\bar{n})s(\bar{n})$ where $s(\bar{n}) \in S_0$. Therefore it follows from the left-invariance of $ds$ that

$$\int_{\bar{N} \times S} F(\bar{n}s) D(s)\,d\bar{n}\,ds = \int_{\bar{N} \times S} F(k(\bar{n})s) D(s) D(s(\bar{n}))^{-1}\,d\bar{n}\,ds.$$

But $D(s(\bar{n})) = e^{2\rho(H(\bar{n}))}$. Hence

$$\int_{K \times S_0} F(ks_0) D(s_0)\,dk\,ds_0 = \int F(k(\bar{n})ms_0) D(s_0) e^{-2\rho(H(\bar{n}))}\,dm\,d\bar{n}\,ds_0$$

because $D(m) = 1$. Now select a function [17] $g \in C_c(S_0)$ such that

$$\int g(s_0) D(s_0)\,ds_0 = 1$$

and define $F$ by $F(ks_0) = f(k)g(s_0)$ $(k \in K, s_0 \in S_0)$ where $f$ is a given function in $C(K)$. Then

$$\int F(ks_0) D(s_0)\,dk\,ds_0 = \int_K f(k)\,dk$$

while

$$\int F(k(\bar{n})ms_0) D(s_0) e^{-2\rho(H(\bar{n}))}\,dm\,d\bar{n}\,ds_0 = \int f(k(\bar{n})m) e^{-2\rho(H(\bar{n}))}\,d\bar{n}\,dm.$$

But for a fixed $m \in M$, the measure $d\bar{n}$ is invariant under the transformation $\bar{n} \to \bar{n}' = m^{-1}\bar{n}m$ and $k(\bar{n}') = m^{-1}k(\bar{n})m$, $H(\bar{n}') = H(\bar{n})$. Therefore

$$\int_K f(k)\,dk = \int f(mk(\bar{n})) e^{-2\rho(H(\bar{n}))}\,dm\,d\bar{n}.$$

In particular if we take $f = 1$, we get the relation $\int e^{-2\rho(H(\bar{n}))}\,d\bar{n} = 1$.

COROLLARY 1. *Let $\nu$ be a linear function on $\mathfrak{h}_\mathfrak{p}$ and put $\nu_+ = \nu + \rho$, $\nu_- = \nu - \rho$. Then*

$$\int_K \exp\{\nu(H(hk)) - \rho(H(hk))\}\,dk$$
$$= e^{\nu\text{-}(\log h)} \int_{\bar{N}} \exp\{\nu_-(H(h\bar{n}h^{-1})) - \nu_+(H(\bar{n}))\}\,d\bar{n}$$

*for $h \in A_\mathfrak{p}$.*

Let $\psi(h)$ denote the left side. Then it follows from the above lemma that

$$\psi(h) = \int \exp\{\nu_-(H(hmk(\bar{n}))) - 2\rho(H(\bar{n}))\}dm\,d\bar{n}.$$

But $m$ commutes with $h$ and therefore $H(hmk(\bar{n})) = H(hk(\bar{n}))$. Also $\bar{n} \in k(\bar{n})\exp H(\bar{n})N$ and therefore $h\bar{n}\exp(-H(\bar{n})) \in hk(\bar{n})N$. But this implies that $H(hk(\bar{n})) = H(h\bar{n}) - H(\bar{n}) = \log h + H(h\bar{n}h^{-1}) - H(\bar{n})$. Therefore

$$\psi(h) = e^{\nu_-(\log h)} \int \exp\{\nu_-(H(h\bar{n}h^{-1})) - \nu_+(H(\bar{n}))\}d\bar{n}.$$

COROLLARY 2.

$$e^{\rho(\log h)} \int_K e^{-\rho(H(hk))}\,dk = \int_{\bar{N}} \exp\{-\rho(H(h\bar{n}h^{-1})) - \rho(H(\bar{n}))\}d\bar{n}$$

for $h \in A_{\mathfrak{p}}$.

This follows by putting $\nu = 0$ in Corollary 1.

The following lemma is of decisive significance for our purpose.

LEMMA 45. *Let $\epsilon$ be a positive number and $d$ the integer of Theorem 3. Then*

$$\int_{\bar{N}} e^{-\rho(H(\bar{n}))}\{1 + \rho(H(\bar{n}))\}^{-d-\epsilon}\,d\bar{n} < \infty.$$

Select an element $H_0 \in \mathfrak{h}_{\mathfrak{p}_0}{}^+$ and put $h_t = \exp tH_0$ $(t \in R)$. Then it follows from Theorem 3 and Corollary 2 above that

$$\int_{\bar{N}} \exp\{-\rho(H(h_t\bar{n}h_t^{-1})) - \rho(H(\bar{n}))\}d\bar{n} \leq c(1+t)^d \qquad (t \geq 0)$$

for a suitable positive number $c$. Put $\Lambda = \frac{1}{2}\sum_{\alpha \in P}\alpha$ and let $\sigma$ denote the irreducible finite-dimensional representation of $\mathfrak{g}$ on a space $V$ with the highest weight [28] $\Lambda$. Choose a unit vector $\phi_0$ in $V$ belonging to the weight $\Lambda$. Then

$$|\sigma(h\bar{n}h^{-1})\phi_0|^2 = \exp 2\rho(H(h\bar{n}h^{-1}))$$

for $h \in A_{\mathfrak{p}}$ and $\bar{n} \in \bar{N}$. Let $\phi_0, \phi_1, \cdots, \phi_p$ be an orthonormal base for $V$ such that $\phi_i$ belongs to the weight $\Lambda_i$ $(\Lambda_0 = \Lambda)$. Then it is obvious that $\pi(\bar{n})\phi_0 - \phi_0$ is orthogonal to $\phi_0$ and therefore

$$|\sigma(h\bar{n}h^{-1})\phi_0|^2 = 1 + \sum_{1 \leq i \leq p}\exp\{2\Lambda_i(\log h) - 2\Lambda(\log h)\}\,|(\phi_i, \pi(\bar{n})\phi_0)|^2.$$

---

[28] It is known (see Weyl [8(c)]) that such a representation exists.

4

Let $\beta_1, \cdots, \beta_q$ be all the distinct roots in $P_+$ and select $X \in \bar{\mathfrak{n}}_0 = \theta(\mathfrak{n}_0)$ such that $\bar{n} = \exp X$. Then $\pi(\bar{n})\phi_0 - \phi_0 = \sum_{r \geq 1} \sigma(X^r)\phi_0/r!$ and if $i \neq 0$, $(\phi_i, \pi(\bar{n})\phi_0) = 0$ unless there exist nonnegative integers $m_1, \cdots, m_q$, not all zero, such that $\Lambda_i = \Lambda - (m_1\beta_1 + \cdots + m_q\beta_q)$ on $\mathfrak{h}_\mathfrak{p}$. Hence if $a = \min_{1 \leq i \leq q} \beta_i(H_0)$, it follows that

$$|\sigma(h_t \bar{n} h_t^{-1})\phi_0|^2 \leq 1 + \sum_{1 \leq i \leq p} e^{-2at} \, |(\phi_i, \pi(\bar{n})\phi_0)|^2$$

$$\leq 1 + e^{-2at} \, |\pi(\bar{n})\phi_0|^2 = 1 + \exp\{2\rho(H(\bar{n})) - 2at\} \qquad (t \geq 0).$$

This proves that

$$\exp\{2\rho(H(h_t \bar{n} h_t^{-1}))\} \leq 1 + \exp\{2\rho(H(\bar{n})) - 2at\}$$

and therefore

$$\int e^{-\rho(H(\bar{n}))} \{1 + e^{-2at + 2\rho(H(\bar{n}))}\}^{-\frac{1}{2}} \, d\bar{n} \leq c(1+t)^d \qquad (t \geq 0).$$

Now for any integer $r \geq 0$, let $\bar{N}_r$ denote the set of all points $\bar{n} \in \bar{N}$ where $\rho(H(\bar{n})) \leq 2^r$. It is obvious from its definition that $a$ is positive. Therefore if $t = 2^r/a$ and $\bar{n} \in \bar{N}_r$, it is clear that $e^{-2at} e^{2\rho(H(\bar{n}))} \leq 1$ and therefore

$$\int_{\bar{N}_r} e^{-\rho(H(\bar{n}))} 2^{-\frac{1}{2}} \, dn \leq c(1 + a^{-1}2^r)^d.$$

Hence

$$\int_{\bar{N}_r} e^{-\rho(H(\bar{n}))} \, d\bar{n} \leq c_1 2^{rd} \qquad (r \geq 0)$$

where $c_1 = 2^{\frac{1}{2}}c(1 + a^{-1})^d$. Let $\bar{N}_{r+1} - \bar{N}_r$ denote the complement of $\bar{N}_r$ in $\bar{N}_{r+1}$. Then if $\epsilon > 0$,

$$\int_{\bar{N}_{r+1} - \bar{N}_r} e^{-\rho(H(\bar{n}))} \{1 + \rho(H(\bar{n}))\}^{-d-\epsilon} \, d\bar{n} \leq 2^{-r(d+\epsilon)} \int_{\bar{N}_{r+1}} e^{-\rho(H(\bar{n}))} \, d\bar{n} \leq c_2 2^{-r\epsilon}$$

where $c_2 = 2^d c_1$. Therefore

$$\int_{\bar{N} - \bar{N}_0} e^{-\rho(H(\bar{n}))} \{1 + \rho(H(\bar{n}))\}^{-d-\epsilon} \, d\bar{n} \leq c_2 \sum_{r \geq 0} 2^{-r\epsilon} < \infty.$$

Hence in order to complete the proof, it would be sufficient to verify that $\bar{N}_0$ is compact. But this is an immediate consequence of Lemma 40.

    COROLLARY. *Let $\epsilon$ be a positive number. Then*

$$\int_{\bar{N}} \exp\{-(1+\epsilon)\rho(H(\bar{n}))\} \, d\bar{n} < \infty.$$

This is obvious because $e^{-\epsilon t}(1+t)^{d+1}$ remains bounded for $t \geqq 0$.

For any $H \in \mathfrak{h}_\mathfrak{p}$ we denote by $\operatorname{Re} H$ and $\operatorname{Im} H$ the elements $H_1$, $H_2$ respectively in $\mathfrak{h}_{\mathfrak{p}_0}$ such that $H = H_1 + (-1)^{\frac{1}{2}} H_2$.

**THEOREM 4.** *Define the function $c$ as in Lemma 37 and normalize the Haar measure $d\bar{n}$ on $\bar{N}$ by the condition*

$$\int_{\bar{N}} e^{-2\rho(H(\bar{n}))}\, d\bar{n} = 1.$$

*Then if $H'$ is an element in $*\mathfrak{h}_\mathfrak{p}'$ such that $-\operatorname{Im} H' \in \mathfrak{h}_{\mathfrak{p}_0}{}^+$, we have the relation*

$$c(H') = \int_{\bar{N}} \exp\{-(-1)^{\frac{1}{2}}\langle H', H(\bar{n})\rangle - \rho(H(\bar{n}))\}\, d\bar{n}.$$

Put $\nu(H) = (-1)^{\frac{1}{2}}\langle H', H\rangle$ for $H \in \mathfrak{h}_\mathfrak{p}$. Then it follows from Lemma 37 and Corollary 1 of Lemma 44 that

$$\sum_{s \in W} c(sH') e^{-\nu(\log h)} \Phi(sH' : h) = \int_{\bar{N}} \exp\{\nu_-(H(h\bar{n}h^{-1})) - \nu_+(H(\bar{n}))\}\, d\bar{n}$$

for $h \in A_\mathfrak{p}(M)$. Select a point $H_0 \in \mathfrak{h}_{\mathfrak{p}_0}{}^+$ and put $h_t = \exp tH_0$ $(t \geqq 0)$. Then it follows from Corollary 1 of Lemma 35 that the real part of $(-1)^{\frac{1}{2}}\langle sH' - H', H_0\rangle$ is negative if $s \neq 1$ $(s \in W)$. Therefore it is obvious from Lemma 31 that $\underset{t \to +\infty}{\operatorname{Lim}} e^{-t\nu(H_0)}\Phi(sH' : h_t) = 1$ or $0$ according as $s = 1$ or not. Hence

$$c(H') = \underset{t \to +\infty}{\operatorname{Lim}} \int_{\bar{N}} \exp\{\nu_-(H(h_t\bar{n}h_t^{-1})) - \nu_+(H(\bar{n}))\}\, d\bar{n}.$$

On the other hand we can evidently choose a positive number $\epsilon < 1$ such that $-\operatorname{Im} H' - \epsilon H_\rho \in \mathfrak{h}_{\mathfrak{p}_0}{}^+$. Put $\nu' = \nu - \epsilon\rho$. Then $\nu_- = \nu' - (1-\epsilon)\rho$, $\nu_+ = \nu' + (1+\epsilon)\rho$ and $\operatorname{Re}\nu'(H) \geqq 0$ for $H \in {}^+\mathfrak{h}_{\mathfrak{p}_0}$. On the other hand

$$\nu_-(H(h_t\bar{n}h_t^{-1})) - \nu_+(H(\bar{n})) = \nu'(H(h_t\bar{n}h_t^{-1}) - H(\bar{n})) - (1-\epsilon)\rho(H(h_t\bar{n}h_t^{-1}))$$
$$- (1+\epsilon)\rho(H(\bar{n})).$$

Therefore it follows from Lemma 43 that

$$\operatorname{Re}\{\nu_-(H(h_t\bar{n}h_t^{-1})) - \nu_+(H(\bar{n}))\} \leqq -(1+\epsilon)\rho(H(\bar{n}))$$

for $t \geqq 0$. Hence we conclude from the corollary of Lemma 45 that

$$c(H') = \int_{\bar{N}} \underset{t \to +\infty}{\operatorname{Lim}} \exp\{\nu_-(H(h_t\bar{n}h_t^{-1})) - \nu_+(H(\bar{n}))\}\, d\bar{n}.$$

Now for a fixed $\bar{n} \in \bar{N}$, select $X \in \bar{\mathfrak{n}}$ such that $\exp X = \bar{n}$. Then $h_t\bar{n}h_t^{-1}$

$= \exp(\mathrm{Ad}(h_t)X)$ and if $X = \sum\limits_{\beta \in P_+} a_\beta X_{-\beta}$ $(a_\beta \in C)$, it is obvious that $\mathrm{Ad}(h_t)X = \sum\limits_{\beta} a_\beta e^{-t\beta(H_0)} X_{-\beta}$. Since $\beta(H_0) > 0$ for every $\beta \in P_+$, it follows that $h_t \bar{n} h_t^{-1} \to 1$ as $t \to +\infty$ and therefore

$$c(H') = \int_{\bar{N}} \exp\{-\nu_+(H(\bar{n}))\}d\bar{n}.$$

This proves the theorem.

We recall that $\mathfrak{h}_{\mathfrak{p}_0}'$ is the set of those points $H \in \mathfrak{h}_{\mathfrak{p}_0}$ where $\alpha(H) \neq 0$ for every $\alpha \in P_+$. We have seen in Section 10 that $\mathfrak{h}_{\mathfrak{p}_0}' \subset {}^*\mathfrak{h}_{\mathfrak{p}_0}'$ and therefore $c$ is defined on $\mathfrak{h}_{\mathfrak{p}_0}'$.

COROLLARY. *Let $H_0$ be any point in $\mathfrak{h}_{\mathfrak{p}_0}'$. Then*

$$c(H_0) = \operatorname*{Lim}_{\epsilon \to 0} \int_{\bar{N}} \exp\{-(-1)^{\frac{1}{2}}\langle H_0, H(\bar{n})\rangle - (1+\epsilon)\rho(H(\bar{n}))\}d\bar{n} \quad (\epsilon > 0).$$

Put $H_\epsilon = H_0 - (-1)^{\frac{1}{2}}\epsilon H_\rho$ where $\epsilon$ is a positive number. It is obvious that $H_\epsilon \in {}^*\mathfrak{h}_\mathfrak{p}'$ and $-\operatorname{Im} H_\epsilon \in \mathfrak{h}_{\mathfrak{p}_0}^+$ if $\epsilon$ is sufficiently small. Moreover $c(H_0) = \operatorname*{Lim}_{\epsilon \to 0} c(H_\epsilon)$ since $c$ is holomorphic at $H_0$. Hence the corollary follows immediately from Theorem 4.

We shall now give another interpretation of the above corollary. Let $\mathfrak{D}(\mathfrak{h}_\mathfrak{p})$ be the algebra of all polynomial differential operators [5(k), §2] on $\mathfrak{h}_\mathfrak{p}$ and let $\mathscr{B}(\mathfrak{h}_{\mathfrak{p}_0})$ denote the space of all functions on $\mathfrak{h}_{\mathfrak{p}_0}$ of class $C^\infty$ such that $Da$ remains bounded on $\mathfrak{h}_{\mathfrak{p}_0}$ for every $D \in \mathfrak{D}(\mathfrak{h}_{\mathfrak{p}_0})$. We topologize $\mathscr{B}(\mathfrak{h}_{\mathfrak{p}_0})$ in the usual way and consider a $\mathscr{B}$-distribution $\tau$ on $\mathfrak{h}_{\mathfrak{p}_0}$ (see [5(k), §2]). Let $dH$ denote the regular Euclidean measure [5(k), p. 91] on $\mathfrak{h}_{\mathfrak{p}_0}$ and for any $a \in \mathscr{B}(\mathfrak{h}_{\mathfrak{p}_0})$, put

$$a'(H_0) = \int_{\mathfrak{h}_{\mathfrak{p}_0}} \exp\{-(-1)^{\frac{1}{2}}\langle H_0, H\rangle\}a(H)dH \qquad (H_0 \in \mathfrak{h}_{\mathfrak{p}_0}).$$

Then the Fourier transform of $\tau$ is the $\mathscr{B}$-distribution $\bar{\tau}$ given by $\bar{\tau}(a) = \tau(a')$ $(a \in \mathscr{B}(\mathfrak{h}_{\mathfrak{p}_0}))$.

THEOREM 5. *Let $\bar{\tau}$ denote the Fourier transform of the distribution $\tau$ defined by*

$$\tau(a) = \int_{\bar{N}} a(H(\bar{n}))e^{-\rho(H(\bar{n}))}\,d\bar{n} \qquad (a \in \mathscr{B}(\mathfrak{h}_{\mathfrak{p}_0})).$$

*Then $\bar{\tau}$ coincides on $\mathfrak{h}_{\mathfrak{p}_0}'$ with the analytic function $c$.*

We know from Lemma 45 that

$$c = \int_{\bar{N}} e^{-\rho(H(\bar{n}))}\{1 + \rho(H(\bar{n}))\}^{-r}\,d\bar{n} < \infty,$$

if $r$ is a suitable positive integer. Hence

$$\int_{\bar{N}} |\,a(H(\bar{n}))\,|\, e^{-\rho(H(\bar{n}))}\, d\bar{n} \leqq c \sup_{H \in \mathfrak{h}_{\mathfrak{p}_0}} |\,(1+\rho(H))^r a(H)\,| \qquad (a \in \mathscr{C}\,(\mathfrak{h}_{\mathfrak{p}_0}))$$

and therefore $\tau$ is a $\mathscr{C}$-distribution. Moreover this inequality implies that

$$\tau(a) = \operatorname*{Lim}_{\epsilon \to 0} \int_{\bar{N}} a(H(\bar{n}))\, e^{-(1+\epsilon)\rho(H(\bar{n}))}\, d\bar{n} \qquad\qquad (\epsilon > 0).$$

Now fix an element [17] $a \in C_c^\infty\,(\mathfrak{h}_{\mathfrak{p}_0}')$. Then

$$\bar{\tau}(a) = \operatorname*{Lim}_{\epsilon \to 0} \int_{\bar{N}} a'(H(\bar{n}))\, e^{-(1+\epsilon)\rho(H(\bar{n}))}\, d\bar{n}.$$

On the other hand

$$\int_{\bar{N}} a'(H(\bar{n}))\, e^{-(1+\epsilon)\rho(H(\bar{n}))}\, d\bar{n}$$
$$= \int_{\bar{N}} e^{-(1+\epsilon)\rho(H(\bar{n}))}\, d\bar{n} \int_{\mathfrak{h}_{\mathfrak{p}_0}} \exp\{-(-1)^{\frac{1}{2}}\langle H(\bar{n}), H\rangle\} a(H)\, dH$$

and it follows from the corollary of Lemma 45 that

$$\int_{\bar{N}} e^{-(1+\epsilon)\rho(H(\bar{n}))}\, d\bar{n} \int_{\mathfrak{h}_{\mathfrak{p}_0}} |\,a(H)\,|\, dH < \infty.$$

Therefore by Fubini's Theorem

$$\bar{\tau}(a) = \operatorname*{Lim}_{\epsilon \to 0} \int_{\mathfrak{h}_{\mathfrak{p}_0}} a(H)\,dH \int_{\bar{N}} \exp\{-(-1)^{\frac{1}{2}}\langle H(\bar{n}), H\rangle - (1+\epsilon)\rho(H(\bar{n}))\}\,d\bar{n}.$$

Hence if $H_\epsilon = H - (-1)^{\frac{1}{2}}\epsilon H_\rho$, we can conclude from Theorem 4 that

$$\bar{\tau}(a) = \operatorname*{Lim}_{\epsilon \to 0} \int_{\mathfrak{h}_{\mathfrak{p}_0}} a(H)\, c(H_\epsilon)\, dH.$$

Now let $\Omega$ denote the carrier of $a$. Then $\Omega$ is a compact subset of $\mathfrak{h}_{\mathfrak{p}_0}'$ and therefore $c$ is holomorphic on some *complex* neighborhood of $\Omega$ in $\mathfrak{h}_\mathfrak{p}$. Hence it is obvious that

$$\bar{\tau}(a) = \int_{\mathfrak{h}_{\mathfrak{p}_0}} a(H)\, c(H)\, dH.$$

This proves that $\bar{\tau}$ coincides with $c$ on $\mathfrak{h}_{\mathfrak{p}_0}'$.

**12. Further study of the function $c$.** We shall now investigate the function $c$ a little more closely on $\mathfrak{h}_{\mathfrak{p}_0}'$. Let $\mathfrak{F}$ be the space of all linear functions [29] on $\mathfrak{h}_\mathfrak{p}$ and $\mathfrak{F}_R$ be the subspace (over $R$) consisting of the real [8]

---

[29] Although $\mathfrak{F}$ and $\mathfrak{h}_\mathfrak{p}$ have been identified under the bilinear form $B$, it is sometimes convenient to distinguish between the two.

functions. For any $\lambda \in \mathfrak{F}$, let $\lambda_R$ and $\lambda_I$ denote the real and imaginary parts of $\lambda$ so that $\lambda = \lambda_R + (-1)^{\frac{1}{2}}\lambda_I$ $(\lambda_R, \lambda_I \in \mathfrak{F}_R)$. Then the following lemma establishes an important inequality.

LEMMA 46. *Put*

$$\psi(\lambda : x) = \int_K \exp\{(-1)^{\frac{1}{2}}\lambda(H(xk)) - \rho(H(xk))\}dk$$

*for* $\lambda \in \mathfrak{F}$ *and* $x \in G$. *Then for any* $b \in \mathfrak{B}$, *we can select an integer* $d \geq 0$ *and a positive constant* $a$ *such that* [30]

$$|\psi(\lambda : x ; b)| \leq a(1 + \|\lambda\|)^d \psi((-1)^{\frac{1}{2}}\lambda_I : x)$$

*for all* $x \in G$ *and* $\lambda \in \mathfrak{F}$.

Let $\beta_1, \cdots, \beta_r$ be a fundamental system of roots in $P$. Select linear functions $\Lambda_1, \cdots, \Lambda_r$ on $\mathfrak{h}$ such that [9] $\Lambda_i(H_{\beta_j}) = 2\beta_j(H_{\beta_j})\delta_{ij}$ $(1 \leq i, j \leq r)$ where, for any root $\beta$, $H_\beta$ is defined in the same way as in Section 7. Then, by Theorem 1 of [5(b)], there exists an irreducible representation $\pi_i$ of $\mathfrak{g}$ on a finite-dimensional space $V_i$ with the highest weight $\Lambda_i$. Select a unit vector $\xi_i$ in $V_i$ belonging to $\Lambda_i$. Extend $\lambda$ and $\rho$ to linear functions on $\mathfrak{h}$ by defining them to be zero on $\mathfrak{h}_{\mathfrak{k}}$. Then $\lambda = \sum_{1 \leq i \leq r} \lambda_i \Lambda_i$ and $\rho = \sum_{1 \leq i \leq r} \rho_i \Lambda_i$ where $\lambda_i \in C$ and $\rho_i \in R$. First we shall prove the following result.

LEMMA 47. *Suppose* $\lambda \in \mathfrak{F}$ *and* $x \in G$. *Then*

$$\psi(\lambda : x) = \psi(-\lambda : x^{-1}) = \int_K \prod_{1 \leq i \leq r} |\pi_i(x^{-1}k)\xi_i|^{-\nu_i} dk$$

*where* $\nu_i = (-1)^{\frac{1}{2}}\lambda_i + \rho_i$ $1 \leq i \leq r$.

For any $k \in K$, let $k_x$ denote the unique element in $K$ such that $xk \in k_x A_\mathfrak{p} N$. Then $k \to k_x$ is a topological mapping of $K$ onto itself and $dk = \exp\{2\rho(H(xk))\}dk_x$ (see [5(c), p. 241]). Hence

$$\psi(\lambda : x) = \int_K \exp\{(-1)^{\frac{1}{2}}\lambda(H(xk)) + \rho(H(xk))\}dk_x.$$

Replacing $k$ by $k_{x^{-1}}$ in this integral and making use of the relation $H(xk_{x^{-1}}) = -H(x^{-1}k)$ (see [5(c), Lemma 36]), we get

$$\psi(\lambda : x) = \int_K \exp\{-(-1)^{\frac{1}{2}}\lambda(H(x^{-1}k)) - \rho(H(x^{-1}k))\}dk = \psi(-\lambda : x^{-1}).$$

---

[30] In view of the identification of $\mathfrak{F}$ and $\mathfrak{h}_\mathfrak{p}$, $\|\lambda\|$ is well-defined $(\lambda \in \mathfrak{F})$.

But $|\pi_i(x^{-1}k)\xi_i| = \exp\{\Lambda_i(H(x^{-1}k))\}$. Hence

$$\psi(-\lambda:x^{-1}) = \int_K \prod_{1 \leq i \leq r} |\pi_i(x^{-1}k)\xi_i|^{-\nu_i} \, dk.$$

Now we come to the proof of Lemma 46. Select a base $X_1, \cdots, X_n$ for $\mathfrak{g}_0$ over $R$ and for any $x \in G$ and $t = (t_1, \cdots, t_n) \in R^n$, put $x_t = x \exp(t_1 X_1 + \cdots + t_n X_n)$. For any ordered set $M = (m_1, \cdots, m_n)$ of nonnegative integers, we write $|M| = m_1 + \cdots + m_n$, $t^M = t_1^{m_1} \cdots t_n^{m_n}$ and denote by $X(M)$ the coefficient (in $\mathfrak{B}$) of $t^M$ in $(|M|!)^{-1} (t_1 X_1 + \cdots + t_n X_n)^{|M|}$. Also put $|t| = \max_{1 \leq i \leq n} |t_i|$ and $M + M' = (m_1 + m_1', \cdots, m_n + m_n')$ if $M' = (m_1', \cdots, m_n')$. Let $E_j$ be the space of all endomorphisms of $V_j$ $(1 \leq j \leq r)$. For any $T \in E_j$, put $|T| = \sup_{|v| \leq 1} |Tv|$ $(v \in V_j)$. Then $E_j$ is a Banach space under this norm and

$$\pi_j(\exp\{-(t_1 X_1 + \cdots + t_n X_n)\}) = \sum_M t^M (-1)^{|M|} \pi_j(X(M)),$$

the series converging absolutely and uniformly in $E_j$ (see [5(c), §5]) provided $|t|$ remains bounded. Define $\bar{\theta}$ as in Section 5 and let $b \to b^*$ $(b \in \mathfrak{B})$ be the anti-automorphism of $\mathfrak{B}$ over $R$ which coincides with $-\bar{\theta}$ on $\mathfrak{g}$. Then it is clear that $\pi_j(b^*)$ is the adjoint of $\pi_j(b)$ (in the sense of Hilbert space theory). Put

$$b_M = \sum_{M_1 + M_2 = M} (-1)^{|M_1|} (X(M_1))^* X(M_2)$$

for any $M$. Then it is obvious that

$$|\pi_j(x_t^{-1}k)\xi_j|^2 = |\pi_j(x^{-1}k)\xi_j|^2 + \sum_{|M| \geq 1} t^M (\pi_j(x^{-1}k)\xi_j, \pi_j(b_M)\pi_j(x^{-1}k)\xi_j)$$

for all $k$, $x$ and $t$ $(1 \leq j \leq r)$. Put

$$\Psi_{M,j}(x) = |\pi_j(x)\xi_j|^{-2} (\pi_j(x)\xi_j, \pi_j(b_M)\pi_j(x)\xi_j) \qquad (x \in G).$$

Then $|\Psi_{M,j}(x)| \leq |\pi_j(b_M)|$. Hence $\Psi_{M,j}$ is a bounded analytic function on $G$ and

$$|\pi_j(x_t^{-1}k)\xi_j|^2 = |\pi_j(x^{-1}k)\xi_j|^2 \{1 + \sum_{|M| \geq 1} t^M \Psi_{M,j}(x^{-1}k)\}.$$

Obviously this series converges uniformly with respect to $x$, $k$ and $t$ provided $x$ varies within a compact subset of $G$ and $|t|$ remains bounded. Therefore by the binomial theorem,

$$\exp\{-(-1)^{\frac{1}{2}}\lambda(H(x_t^{-1}k)) - \rho(H(x_t^{-1}k))\} = \prod_{1 \leq j \leq r} |\pi_j(x_t^{-1}k)\xi_j|^{-\nu_j}$$

$$= \exp\{-(-1)^{\frac{1}{2}}\lambda(H(x^{-1}k)) - \rho(H(x^{-1}k))\} \sum_M t^M \Psi_M(x^{-1}k:\lambda)$$

provides $|t|$ is sufficiently small. Here $\Psi_M(x:\lambda)$ is a function on $G \times \mathfrak{F}$ which can be written as a polynomial in $\nu_j$ and $\Psi_{M',j}$ $(1 \leq j \leq r, |M'| \leq |M|)$ with constant coefficients. Therefore it is clear that there exists a positive number $a_M$ and an integer $d_M \geq 0$ such that

$$|\Psi_M(x:\lambda)| \leq a_M(1 + \|\lambda\|)^{d_M} \qquad\qquad (x \in G, \lambda \in \mathfrak{F}).$$

Moreover it is obvious that for a fixed $x$, the above series converges uniformly with respect to $k$ and $t$ provided $|t|$ is sufficiently small. Therefore by integrating over $K$, we get an expansion of $\psi(-\lambda:x_t^{-1}) = \psi(\lambda:x_t)$ in powers of $t_1, \cdots, t_n$. This shows that if $M = (m_1, \cdots, m_n)$,

$$\psi(\lambda:x;X(M)) = (m_1! \cdots m_n!)^{-1}\{\partial^{m_1+\cdots+m_n}/\partial t_1{}^{m_1} \cdots \partial t_n{}^{m_n})\psi(\lambda:x_t)\}_{t=0}$$

$$= \int_K \Psi_M(x^{-1}k:\lambda)\exp\{-(-1)^{\frac{1}{2}}\lambda(H(x^{-1}k)) - \rho(H(x^{-1}k))\}dk.$$

Hence

$$|\psi(\lambda:x;X(M))| \leq a_M(1 + \|\lambda\|)^{d_M}\psi((-1)^{\frac{1}{2}}\lambda_I:x)$$

and the assertion of the lemma follows from the fact that the elements $X(M)$ form a base for $\mathfrak{B}$.

The following result will be needed in another paper.

LEMMA 48. *Let $\Omega$ be a compact set in $G$. Then we can select a positive number $c$ such that $\psi(0:xy) \leq c\psi(0:x)$ for all $x \in G$ and $y \in \Omega$.*

It is clear that

$$e^{-\rho(H(y^{-1}x^{-1}k))} = \prod_{1 \leq j \leq r} |\pi_j(y^{-1}x^{-1}k)\xi_j|^{-\rho_j}$$

and

$$|\pi_j(y)|^{-1}|\pi_j(x^{-1}k)\xi_j| \leq |\pi_j(y^{-1}x^{-1}k)\xi_j| \leq |\pi_j(y^{-1})||\pi_j(x^{-1}k)\xi_j|.$$

Therefore if we put

$$c = \sup_{y \in \Omega} \prod_{1 \leq j \leq r} \{|\pi_j(y)|^{\rho_j} + |\pi_j(y^{-1})|^{-\rho_j}\},$$

it follows that

$$e^{-\rho(H(y^{-1}x^{-1}k))} \leq ce^{-\rho(H(x^{-1}k))}$$

for $y \in \Omega$. In view of Lemma 47 our assertion is now obvious.

Now for any positive number $\epsilon$, let $\mathfrak{F}_\epsilon$ denote the open neighborhood of $\mathfrak{F}_R$ in $\mathfrak{F}$ consisting of those points $\lambda \in \mathfrak{F}$ for which [27] $|\operatorname{Im}\lambda_j| < \epsilon/2$ $(1 \leq j \leq r)$. Then it follows from Lemmas 2 and 35 and Corollary 2 of Lemma 35 that

$$| \lambda_I(H(hk))| \leq \tfrac{1}{2}\epsilon \sum_{1 \leq j \leq r} | \Lambda_j(H(hk))|$$

$$\leq \tfrac{1}{2}\epsilon \sum_{1 \leq j \leq r} \Lambda_j(\log h + \log h^*) = \epsilon\rho(\log h) \qquad (h \in \mathrm{Cl}(A_{\mathfrak{p}}^+)),$$

since [31] $\sum_{1 \leq j \leq r} \Lambda_j = \tfrac{1}{2}\sum_{\beta \in P} \beta$ and $\rho(\log h^*) = \rho(\log h)$. Therefore we get the following result.

LEMMA 49. *Suppose* $\lambda \in \mathfrak{F}_\epsilon$ *and* $h \in \mathrm{Cl}(A_{\mathfrak{p}}^+)$. *Then*

$$\psi((-1)^{\frac{1}{2}}\lambda_I : h) \leq e^{\epsilon\rho(\log h)}\psi(0:h).$$

Lemmas 46 and 49, together with Theorem 3, give us an estimate for $\psi(\lambda:h;b)$ which will be of great value in the discussion below.

Select $u_1, \cdots, u_w \in \mathfrak{H}_{\mathfrak{p}}$ as in the corollary of Lemma 8 and put

$$\psi_i(\lambda:H) = e^{\rho(H)}\psi(\lambda:\exp H ; u_i') \qquad (\lambda \in \mathfrak{F}, H \in \mathfrak{h}_{\mathfrak{p}_0}, 1 \leq i \leq w)$$

where $u_i' = e^\rho u_i \circ e^{-\rho}$. Fix an element $H_0 \in \mathfrak{h}_{\mathfrak{p}_0}^+$ and let $\Psi(\lambda:t)$ $(t \in R)$ denote the one-column matrix $\psi_i(\lambda:tH_0)$ $1 \leq i \leq w$. Let us consider $d\Psi/dt$. Choose $q_{ij} \in I_{\mathfrak{g}}$ such that $H_0 u_j = \sum_{1 \leq i \leq w} u_i \gamma(q_{ij})$ $(1 \leq j \leq w)$. Then if $h_t = \exp tH_0$ $(t \in R)$, it follows from Theorem 2 and Lemmas 18 and 23 that

$$d\psi_j(\lambda:tH_0)/dt = \sum_{1 \leq i \leq w} \psi_i(\lambda:tH_0)\gamma(q_{ij}:(-1)^{\frac{1}{2}}\lambda) + \theta_j(\lambda:h_t) \qquad (t \geq 1)$$

where

$$\theta_j(\lambda:h) = \sum_{1 \leq i \leq w} \sum_{1 \leq k \leq s} g_{jik}(h)\psi_i(\lambda:\log h)\gamma(q_{jik}:(-1)^{\frac{1}{2}}\lambda) \qquad (\lambda \in \mathfrak{F}, h \in A_{\mathfrak{p}}^+),$$

$g_{jik} \in \mathfrak{R}$, $q_{jik} \in I_{\mathfrak{g}}$ and $s$ is a positive integer. Put $\beta(H_0 = \min_{\alpha \in P_+} \alpha(H_0)$. Then it is obvious that for any $g \in \mathfrak{R}$, there exists a positive number $c$ such that $| g(h_t)| \leq ce^{-2t\beta(H_0)}$ $(t \geq 1)$. Let $\Theta(\lambda:t)$ denote the one-column matrix $\theta_i(\lambda:h_t)$ $1 \leq i \leq w$ and put $\| a \| = (\sum_{1 \leq i \leq m} \sum_{1 \leq j \leq n} | a_{ij}|^2)^{\frac{1}{2}}$ for any $m \times n$ matrix $a = (a_{ij})$ with complex coefficients. Then it follows from Lemmas 46 and 49 and Theorem 3 that

$$\| \Theta(\lambda:t)\| \leq c_1 \exp\{\epsilon t\rho(H_0) - 2t\beta(H_0)\}(1 + \| \lambda \|)^{d_1}t^{d_1}$$

for $t \geq 1$ and $\lambda \in \mathfrak{F}_\epsilon$. Here $c_1$ is a positive number and $d_1$ a nonnegative integer. Let $\Xi(\lambda)$ denote the $w \times w$ matrix given by $\Xi_{ij}(\lambda) = \gamma(q_{ji}:(-1)^{\frac{1}{2}}\lambda)$

---

[31] Put $\Lambda = \tfrac{1}{2}\sum_{\beta \in P} \beta$. Then $2\Lambda(H_{\beta_j}) = \beta_j(H_{\beta_j})$ for $1 \leq j \leq r$ (see Weyl [8(c)]). Hence $\Lambda = \Lambda_1 + \cdots + \Lambda_r$.

$(1 \leq i, j \leq w)$. Then the above equation can be written in matrix notation as follows:

$$d\Psi(\lambda:t)/dt = \Xi(\lambda)\Psi(\lambda:t) + \Theta(\lambda:t) \qquad (t \geq 1).$$

Now put $\Psi_0(\lambda:t) = \exp(-t\Xi(\lambda))\Psi(\lambda:t)$ and $\Theta_0(\lambda:t) = \exp(-t\Xi(\lambda))\Theta(\lambda:t)$. Then it is clear that

$$d\Psi_0(\lambda:t)/dt = \Theta_0(\lambda:t) \qquad (t \geq 1).$$

We shall now estimate $\| \Theta_0(\lambda:t) \|$. It is obvious from the definition of $\Xi(\lambda)$ that

$$H_0 u_j \equiv \sum_{1 \leq i \leq w} \Xi_{ji}(\lambda) u_i \bmod SJ_\lambda.$$

in the notation of Section 3, if we put $\lambda^* = (-1)^{\frac{1}{2}}\lambda$. Hence we conclude from Corollary 2 of Lemma 14 that $\lambda^*(sH_0)$ $(s \in W)$ are all the eigenvalues of $\Xi(\lambda)$. Put $\tau(\lambda) = 2w \max_{s \in W} |\lambda_I(sH_0)|$. Then it follows from Lemma 60 of the Appendix that

$$\| \exp(-t\Xi(\lambda)) \| \leq e^{t\tau(\lambda)w^{\frac{1}{2}}} \sum_{0 \leq k < w} t^k \| \Xi(\lambda) \|^k \qquad (t > 0).$$

Therefore since $\| \Theta_0(\lambda:t) \| \leq \| \exp(-t\Xi(\lambda)) \| \, \| \Theta(\lambda:t) \|$, it is obvious that there exists a positive number $c_2$ and an integer $d_2 \geq 0$ such that

$$\| \Theta_0(\lambda:t) \| \leq c_2(1 + \| \lambda \|)^{d_2} t^{d_2} \exp\{\tau(\lambda)t + \epsilon t\rho(H_0) - 2t\beta(H_0)\}$$

for $\lambda \in \mathfrak{F}_\epsilon$ and $t \geq 1$. Moreover by choosing $\epsilon$ sufficiently small, we can obviously arrange that $\beta(H_0) \geq \tau(\lambda) + \epsilon\rho(H_0)$ for all $\lambda \in \mathfrak{F}_\epsilon$. Then in this case

$$\| \Theta_0(\lambda:t) \| \leq c_2(1 + \| \lambda \|)^{d_2} t^{d_2} e^{-t\beta(H_0)} \qquad (t \geq 1, \lambda \in \mathfrak{F}_\epsilon).$$

This shows that the integral $\int_1^\infty \| \Theta_0(\lambda:t) \| \, dt$ converges uniformly with respect to $\lambda$ as $\lambda$ varies in a compact subset of $\mathfrak{F}_\epsilon$. Moreover there exists a positive number $c_3$ such that

$$\int_1^\infty \| \Theta_0(\lambda:t) \| \, dt \leq c_3(1 + \| \lambda \|)^{d_2} \qquad (\lambda \in \mathfrak{F}_\epsilon).$$

But since $d\Psi_0(\lambda:t)/dt = \Theta_0(\lambda:t)$, this implies that $\Psi_0(\lambda:t)$ tends to a limit as $t \to +\infty$. If we denote this limit by $\Psi_0(\lambda:\infty)$, we have the relation

$$\Psi_0(\lambda:\infty) = \Psi_0(\lambda:1) + \int_1^\infty \Theta_0(\lambda:t) \, dt \qquad (\lambda \in \mathfrak{F}_\epsilon).$$

In view of the uniform convergence of the integral, it is clear that $\Psi_0(\lambda:\infty)$ is a holomorphic (matrix-) function of $\lambda$ on $\mathfrak{F}_\epsilon$. Moreover since $\|\Psi_0(\lambda:1)\| \leq \|\exp(-\Xi(\lambda))\| \|\Psi(\lambda:1)\|$, it follows without difficulty (by making use of Lemmas 46 and 49 and the above estimate for $\|\exp(-\Xi(\lambda))\|$) that $\|\Psi_0(\lambda:1)\| \leq c_4(1+\|\lambda\|)^{d_4}$ for suitable positive numbers $c_4$, $d_4$. Therefore $\|\Psi_0(\lambda:\infty)\| \leq c_5(1+\|\lambda\|)^{d_5}$ $(\lambda \in \mathfrak{F}_\epsilon)$ for an appropriate choice of $c_5$, $d_5$. Hence if we apply Lemma 58 of the Appendix to $V = \mathfrak{F}_{\epsilon/2}$, we get the following result.

LEMMA 50. *If $\epsilon$ is sufficiently small, there exists an integer $d \geq 0$ with the following property. For any $u \in S(\mathfrak{h}_\mathfrak{p})$, the function*

$$(1+\|\lambda\|)^{-d}\|\Psi_0(\lambda;\partial(u):\infty)\|$$

*remains bounded as $\lambda$ varies in $\mathfrak{F}_\epsilon$.*

Now define $\pi$ as in Lemma 11 and let $\mathfrak{F}_\epsilon'$ denote the set of those elements $\lambda \in \mathfrak{F}_\epsilon$ where $\pi(\lambda) \neq 0$. We may assume that $\epsilon$ is so small that $\|\lambda_I\| < \|\Lambda\|/2$ for any $\lambda \in \mathfrak{F}_\epsilon$ and $\Lambda \in L'$ (see Section 8 for the definition of $L'$). Then it follows without difficulty that $\mathfrak{F}_\epsilon \subset {}^*\mathfrak{h}_\mathfrak{p}$ and therefore $\mathfrak{F}_\epsilon' \subset {}^*\mathfrak{h}_\mathfrak{p}'$. Hence if $\lambda \in \mathfrak{F}_\epsilon'$, we conclude from Lemma 37 that

$$\psi_i(\lambda:tH_0) = \sum_{s \in W} c(s\lambda)\Phi(s\lambda:h_t;u_i) \qquad (1 \leq i \leq w, t > M)$$

where $M$ is a suitable positive number. Define $e_{sH} \in S(\mathfrak{h}_\mathfrak{p})$ $(s \in W, H \in \mathfrak{h}_\mathfrak{p})$ as in Lemma 14 and put $e_s = e_{s\lambda^*}$. Moreover define the $w \times w$ matrix $E(s)$ as follows:

$$e_s u_i \equiv \sum_{1 \leq j \leq w} E_{ij}(s)u_j \bmod SJ_{\lambda^*}. \qquad (E_{ij}(s) \in C).$$

Then it follows from Lemma 14 that $\sum_{s \in W} \epsilon(s)E(s) = \pi(\lambda^*)I$ where $I$ is the unit matrix. Moreover $e_s H_0 = H_0 e_s \equiv \lambda^*(s^{-1}H_0)e_s \bmod SJ_{\lambda^*}$ by the same lemma and therefore

$$E(s)\Xi(\lambda) = \Xi(\lambda)E(s) = \lambda^*(s^{-1}H_0)E(s).$$

Hence

$$\pi(\lambda^*)\Psi_0(\lambda:t) = \sum_{s \in W} \epsilon(s)E(s)\exp(-t\Xi(\lambda))\Psi(\lambda:t)$$

$$= \sum_{s \in W} \epsilon(s)\exp\{-\lambda^*(s^{-1}H_0)t\}E(s)\Psi(\lambda:t).$$

Now put $\xi(\mu:h) = \exp\{\mu^*(\log h)\}$ $(\mu \in \mathfrak{F}, h \in A_\mathfrak{p})$ where $\mu^* = (-1)^{\frac{1}{2}}\mu$. Then it is obvious that $\xi(\mu:h;u) = u(\mu^*)\xi(\mu:h)$ for $u \in \mathfrak{H}_\mathfrak{p}$ and therefore $\xi(\mu:h;u) = 0$ for $u \in SJ_{\mu^*}$. Also

$$\sum_{1\leq j\leq w} E_{ij}(s)u_j \equiv e_s u_i \equiv u_i(s\lambda^*)e_s \bmod SJ_\lambda. \qquad (1\leq i\leq w)$$

from Lemma 14. Therefore

$$\sum_{1\leq j\leq w} E_{ij}(s)\xi(s'\lambda:h;u_j) = u_i(s\lambda^*)e_s(s'\lambda^*)\xi(s'\lambda:h) \qquad (s,s'\in W).$$

Hence it follows from the third statement of Lemma 14 that

$$\sum_j E_{ij}(s)\xi(s'\lambda:h;u_j) = \begin{cases} 0 \text{ if } s\neq s' \\ u_i(s\lambda^*)\epsilon(s)\pi(\lambda^*)\xi(s\lambda:h) \text{ if } s=s'. \end{cases}$$

Let $\psi_{0j}(\lambda:\infty)$ $(1\leq j\leq w)$ denote the coefficients of the one-column matrix $\Psi_0(\lambda:\infty)$. Then

$$\pi(\lambda^*)\psi_{0i}(\lambda:\infty) = \operatorname*{Lim}_{t\to+\infty} \sum_{s,s'\in W}\sum_j \epsilon(s)\xi(s\lambda:h_t^{-1})E_{ij}(s)c(s'\lambda)\Phi(s'\lambda:h_t;u_j).$$

But since $\beta(H_0)\geq \tau(\lambda)+\epsilon\rho(H_0)$, it follows that $|\lambda_I(sH_0)|-\beta(H_0)$ $\leq -\epsilon\rho(H_0)$ for $s\in W$. Therefore it is obvious that

$$\operatorname*{Lim}_{t\to+\infty} |\Phi(s'\lambda:h_t;u_j)-\xi(s'\lambda:h_t;u_j)| = 0 \qquad (s'\in W)$$

and hence

$$\begin{aligned}
\pi(\lambda^*)\psi_{0i}(\lambda:\infty) &= \operatorname*{Lim}_{t\to+\infty} \sum_{s,s'\in W}\sum_j \epsilon(s)\xi(s\lambda:h_t^{-1})E_{ij}(s)c(s'\lambda)\xi(s'\lambda:h_t;u_j) \\
&= \pi(\lambda^*)\operatorname*{Lim}_{t\to+\infty}\sum_{s\in W} \xi(s\lambda:h_t^{-1})c(s\lambda)u_i(s\lambda^*)\xi(s\lambda:h_t) \\
&= \pi(\lambda^*)\sum_{s\in W} u_i(s\lambda^*)c(s\lambda)
\end{aligned}$$

in view of what we have said above. Thus we have obtained the following result.

LEMMA 51.   $\psi_{0i}(\lambda:\infty) = \sum_{s\in W} u_i(s\lambda^*)c(s\lambda)$ *for any* $\lambda\in\mathfrak{F}_\epsilon'$.

Define $\sigma^i$ $(1\leq i\leq w)$ as in Lemma 14.

COROLLARY.   $\sum_{1\leq i\leq w} \sigma^i(\lambda^*)\psi_{0i}(\lambda:\infty) = \pi(\lambda^*)c(\lambda)$ *for* $\lambda\in\mathfrak{F}_\epsilon'$.

This follows immediately from the above lemma and the third statement of Lemma 14.

We have seen that $\Psi_0(\lambda:\infty)$ is a holomorphic function of $\lambda$ on $\mathfrak{F}_\epsilon$. Therefore in view of the above corollary and Lemma 50, we have the following result.

LEMMA 52. *Suppose* $\epsilon$ *is sufficiently small. Then there exists a holomorphic function* $\boldsymbol{b}$ *on* $\mathfrak{F}_\epsilon$ *such that* $\boldsymbol{b}(\lambda) = \pi(\lambda)c(\lambda)$ *for* $\lambda\in\mathfrak{F}_\epsilon'$. *Moreover* *we can select an integer $d \geqq 0$ with the property that for any $u \in S(\mathfrak{h}_\mathfrak{p})$, $(1 + \| \lambda \|)^{-d} | \boldsymbol{b}(\lambda; \partial(u)) |$ remains bounded as $\lambda$ varies in $\mathfrak{F}_\epsilon$.*

Define the $\mathscr{E}$-distributions $\tau$ and $\bar{\tau}$ on $\mathfrak{h}_{\mathfrak{p}_0}$ as in Theorem 5.

COROLLARY. *The distribution $\pi\bar{\tau}$ coincides on $\mathfrak{h}_{\mathfrak{p}_0}$ with $\boldsymbol{b}$.*

We use the notation of the proof of Theorem 5. It is sufficient to show that

$$\bar{\tau}(\pi a) = \int_{\mathfrak{h}_{\mathfrak{p}_0}} \boldsymbol{b}(H) a(H) dH$$

for any $a \in C_c^\infty(\mathfrak{h}_{\mathfrak{p}_0})$. Put $H_\epsilon = H - (-1)^{\frac{1}{2}}\epsilon H_\rho$ $(H \in \mathfrak{h}_{\mathfrak{p}_0})$ where $\epsilon$ is a positive number. Then it is obvious that $\pi(H_\epsilon) \neq 0$. Hence if $\epsilon$ is sufficiently small, $H_\epsilon \in {}^*\mathfrak{h}_\mathfrak{p}'$ and therefore $\boldsymbol{c}(H_\epsilon)$ is well-defined for all $H \in \mathfrak{h}_{\mathfrak{p}_0}$. Moreover if we use the method of proof of Theorem 5, it follows without difficulty that

$$\bar{\tau}(\pi a) = \operatorname*{Lim}_{\epsilon \to 0} \int_{\mathfrak{h}_{\mathfrak{p}_0}} a(H) \pi(H) \boldsymbol{c}(H_\epsilon) dH.$$

On the other hand it is obvious that $| \alpha(H) | \leqq | \alpha(H_\epsilon) |$ for $\alpha \in \Sigma$ and $H \in \mathfrak{h}_{\mathfrak{p}_0}$ and therefore $| \pi(H) \boldsymbol{c}(H_\epsilon) | \leqq | \boldsymbol{b}(H_\epsilon) |$. Let $\Omega$ be the carrier of $a$. Then since $\Omega$ is compact and since $\boldsymbol{b}$ is holomorphic on a complex neighborhood of $\mathfrak{h}_{\mathfrak{p}_0}$ in $\mathfrak{h}_\mathfrak{p}$, it is obvious that $| \boldsymbol{b}(H_\epsilon) |$ remains uniformly bounded for all sufficiently small values of $\epsilon$ as $H$ varies in $\Omega$. Hence we conclude that

$$\bar{\tau}(\pi a) = \int_{\mathfrak{h}_{\mathfrak{p}_0}} \operatorname*{Lim}_{\epsilon \to 0} \{ a(H) \pi(H) \boldsymbol{c}(H_\epsilon) \} dH.$$

But it is obvious that $\operatorname*{Lim}_{\epsilon \to 0} \pi(H) \boldsymbol{c}(H_\epsilon) = \boldsymbol{b}(H)$ if $\pi(H) \neq 0$. Therefore

$$\bar{\tau}(\pi a) = \int_{\mathfrak{h}_{\mathfrak{p}_0}} a(H) \boldsymbol{b}(H) dH$$

and this proves the corollary.

**13. The case $l = 1$.** We assume in this section that [32] $\dim \mathfrak{h}_\mathfrak{p} = 1$. Then [16] we can select $\alpha \in \Sigma$ such that $2\alpha$ is the only other possible element in $\Sigma$. Let $p$ and $q$ denote the number of roots in $P_+$ which coincide on $\mathfrak{h}_\mathfrak{p}$ with $\alpha$ and $2\alpha$ respectively. Then $\rho = \frac{1}{2}(p + 2q)\alpha$. Let $H$ be the element in $\mathfrak{h}_\mathfrak{p}$ such that $\alpha(H) = 1$. Then $\langle H, H \rangle = 2 \sum_{\beta \in P_+} \beta(H)^2 = 2(p + 4q)$. This

---

[32] A special case of this type has been discussed by Bargmann [1]. We follow the same method.

implies that $H_\alpha = (2p + 8q)^{-1}H$ and $H_\rho = \frac{1}{2}(p + 2q)H_\alpha$. Hence it follows from Lemma 27 that

$$2(p + 4q)\delta'(\omega) = H^2 + (p + 2q)H + \{2p(e^{2\alpha} - 1)^{-1} + 4q(e^{4\alpha} - 1)^{-1}\}H.$$

Put $t(h) = \alpha(\log h)$ for $h \in A_\mathfrak{p}$. Then $t$ can be regarded as the coordinate function on the one-dimensional Lie group $A_\mathfrak{p}$ and so it is clear that

$$2(p + 4q)\delta'(\omega) = d^2/dt^2 + \{p \coth t + 2q \coth 2t\}d/dt$$

$$= d^2/dt^2 + \{(p + q)\coth t + q \tanh t\}d/dt,$$

since $\coth 2t = \frac{1}{2}(\coth t + \tanh t)$. Now put $z = -(\sinh t)^2$. Then the above relation becomes

$$\tfrac{1}{2}(p + 4q)\delta'(\omega) = z(z - 1)d^2/dz^2 + \tfrac{1}{2}\{(p + 2q + 2)z - (p + q + 1)\}d/dz.$$

On the other hand if $\lambda$ is any linear function on $\mathfrak{h}_\mathfrak{p}$, we know from Corollary 2 of Lemma 27 that

$$\gamma(\omega : (-1)^{\frac{1}{2}}\lambda) = -\langle\lambda, \lambda\rangle - \langle\rho, \rho\rangle = -\tfrac{1}{2}(p + 4q)^{-1}\{\lambda(H)^2 + \rho(H)^2\}.$$

Therefore if we put $\phi_\lambda(h) = \psi(\lambda : h)$ $(h \in A_\mathfrak{p})$, it follows from Lemma 18 that

$$z(z - 1)d^2\phi_\lambda/dz^2$$

$$+ \tfrac{1}{2}\{(p + 2q + 2)z - (p + q + 1)\}d\phi_\lambda/dz + \tfrac{1}{4}\{\lambda(H)^2 + (\tfrac{1}{2}p + q)^2\}\phi_\lambda = 0.$$

Now let

$$a = \{p + 2q + 2\lambda^*(H)\}/4, \quad b = \{p + 2q - 2\lambda^*(H)\}/4, \quad c = (p + q + 1)/2$$

where $\lambda^* = (-1)^{\frac{1}{2}}\lambda$. Then the above equation becomes

$$z(z - 1)d^2\phi_\lambda/dz^2 + \{(a + b + 1)z - c\}d\phi_\lambda/dz + ab\phi_\lambda = 0.$$

Now $z$ can be regarded as a coordinate function on $A_\mathfrak{p}^+$. Therefore $\phi_\lambda = g(z)$ where $g$ is analytic function on the interval $-\infty < z < 0$. Moreover $W = \{1, s\}$ where $sH = -H$. Therefore $\psi_\lambda(h) = \psi_\lambda(h^s) = \psi_\lambda(h^{-1})$ $(h \subset A_\mathfrak{p})$ and so $\phi_\lambda$ can be expanded in powers of $t^2$ in the neighborhood of the point $h = 1$. But this implies that $g$ is actually analytic also at $z = 0$. Moreover $g(0) = \phi_\lambda(1) = 1$. Therefore since $c > 0$, it follows by a direct substitution of the power series for $g$ around the origin, in the above differential equation, that $g(z)$ is the hypergeometric function $F(a, b, c, z)$. But it is known (see Bargmann [1, p. 627]) that

$$F(a, b, c, z) = |z|^{-a}\Gamma(c)\Gamma(b-a)\{\Gamma(b)\Gamma(c-a)\}^{-1}F(a, 1+a-c, 1+a-b, z^{-1})$$
$$+ |z|^{-b}\Gamma(c)\Gamma(a-b)\{\Gamma(a)\Gamma(c-b)\}^{-1}F(b, 1+b-c, 1+b-a, z^{-1})$$
$$(z < 0)$$

provided $a - b = \lambda^*(H)$ is not an integer. (Here $\Gamma$ denotes the classical Gamma function.) Therefore if $\lambda$ is real, it is clear that

$$\lim_{t' \to +\infty} | e^{t'\rho(H)}\phi_\lambda(\exp t'H) - \boldsymbol{c}(\lambda)e^{t'\lambda^*(H)} - \boldsymbol{c}(-\lambda)e^{-t'\lambda^*(H)} | = 0$$

where

$$\boldsymbol{c}(\lambda) = 2^{-\lambda^*(H)+\rho(H)}\,\Gamma(\tfrac{1}{2}(p+q+1))\Gamma(\lambda^*(H))$$
$$\times \{\Gamma(4^{-1}(p+2q+2\lambda^*(H)))\Gamma(4^{-1}(p+2+2\lambda^*(H)))\}^{-1}.$$

Moreover in the present case $\pi = \alpha$ and therefore

$$\pi(\lambda) = \lambda(H_\alpha) = (2p + 8q)^{-1}\lambda(H).$$

Hence

$$(-1)^{\frac{1}{2}}\boldsymbol{b}(\lambda) = 2^{-\lambda^*(H)+\rho(H)}(2p + 8q)^{-1}\Gamma((p+q+1)/2)\Gamma(1+\lambda^*(H))$$
$$\times \{\Gamma((p+2q+2\lambda^*(H))/4)\Gamma((p+2+2\lambda^*(H))/4)\}^{-1},$$

and by analytic continuation this formula holds for all $\lambda \in \mathfrak{F}_\epsilon$ if $\epsilon$ is sufficiently small. Moreover if $\lambda \in \mathfrak{F}_R$, it is obvious that $|\boldsymbol{b}(-\lambda)| = |\boldsymbol{b}(\lambda)|$. Hence we have obtained the following result.

LEMMA 53. *If $\epsilon$ is sufficiently small, $\boldsymbol{b}$ is never zero on $\mathfrak{F}_\epsilon$. Moreover $|\boldsymbol{b}(s'\lambda)| = |\boldsymbol{b}(\lambda)|$ for $s' \in W$ and $\lambda \in \mathfrak{F}_R$.*

We shall see in another paper that a similar result holds when $l > 1$.

**14. The complex case.**[33] In this section we assume that $G$ is a *complex* group. Then it follows easily (see [5(f), p. 513]) that no root $\alpha \in P$ can vanish identically either on $\mathfrak{h}_\mathfrak{p}$ or on $\mathfrak{h}_\mathfrak{t}$. Moreover the restrictions on $\mathfrak{h}_\mathfrak{p}$ of two distinct roots $\alpha, \beta \in P$, are linearly independent unless $\beta = -\theta\alpha$. Hence $P = P_+$ and we can select a subset $Q$ of $P$ such that exactly one of the two roots $\alpha, -\theta\alpha$ lies in $Q$ for any $\alpha \in P$. Put $d(H) = \prod_{\alpha \in Q} (e^{\alpha(H)} - e^{-\alpha(H)})$ $(H \in \mathfrak{h}_\mathfrak{p})$. It is obvious that $\pi(H) = \prod_{\alpha \in Q} \alpha(H)$ and therefore $d(sH) = \epsilon(s)d(H)$

---

[33] After this paper had been written, two recent notes of F. A. Beresin (Doklady Akad. Nauk. SSSR N. S., vol. 107 (1956), pp. 9-12, vol. 110 (1956), pp. 897-900) came to my attention. The results of this section are also contained in these notes.

$(s \in W)$. Moreover $\rho(H) = \frac{1}{2} \sum_{\alpha \in P} \alpha(H) = \sum_{\alpha \in Q} \alpha(H)$. Hence it follows by elementary considerations of divisibility (see Weyl [8(b)]) that

$$d(H) = \sum_{s \in W} \epsilon(s) e^{\rho(s^{-1}H)} \qquad (H \in \mathfrak{h}_\mathfrak{p}).$$

Now put $D(h) = d(\log h)$ for $h \in A_\mathfrak{p}$ and let us use the notation of Section 7. Then it is obvious [22] that $\varpi D = \langle \rho, \rho \rangle D$. If we take this fact and Lemma 27 into account, a simple calculation gives the following result.

LEMMA 54. $\delta'(\omega) = D^{-1}\gamma(\omega) \circ D$ in the present case.

Fix $H' \in \mathfrak{h}_\mathfrak{p}$ and consider the function

$$F(H':h) = D(h)^{-1} \exp\{\rho(\log h) + \langle H', \log h \rangle\}$$

$(h \in A_\mathfrak{p}^+)$ on $A_\mathfrak{p}^+$. In view of the above lemma, $F$ is an eigenfunction of the operator $\delta'(\omega)$ and the corresponding eigenvalue is $\gamma'(\omega:H')$. Moreover $D(h)^{-1}e^{\rho(\log h)} = \prod_{\alpha \in Q} \{1 - e^{-2\alpha(\log h)}\}^{-1}$. Therefore it follows easily from the reasoning of Section 8 that $\Phi' = F$ in the notation of Lemma 29. Now put

$$\phi_\lambda(x) = \int_K \exp\{(-1)^{\frac{1}{2}}\lambda(H(xk)) - \rho(H(xk))\}dk \qquad (x \in G, \lambda \in \mathfrak{F})$$

and fix $\lambda \in \mathfrak{F}_R' = \mathfrak{h}_{\mathfrak{p}_0}'$. Then it follows from Lemma 37 that [34]

$$D(h)\phi_\lambda(h) = \sum_{s \in W} c(s\lambda) \exp\{s\lambda^*(\log h)\} \qquad (h \in A_\mathfrak{p}(M))$$

where $\lambda^* = (-1)^{\frac{1}{2}}\lambda$. However since both sides are analytic functions on $A_\mathfrak{p}$ this equation must hold for all $h \in A_\mathfrak{p}$. On the other hand $\phi_\lambda(h^s) = \phi_\lambda(h)$ while $D(h^s) = \epsilon(s)D(h)$. Therefore since the points $s\lambda$ $(s \in W)$ are all distinct, it follows (see [5(b), Lemma 41]) that $c(s\lambda) = \epsilon(s)c(\lambda)$. Hence

$$D(h)\phi_\lambda(h) = c(\lambda) \sum_{s \in W} \epsilon(s) \exp\{s\lambda^*(\log h)\} \qquad (h \in A_\mathfrak{p}).$$

Now $\pi$, being an element in $\mathfrak{H}_\mathfrak{p}$, can be considered as a differential operator on $A_\mathfrak{p}$. In view of the expression for $D$ obtained earlier, it is clear that $D(1;\pi) = w\pi(\rho)$ where $w$ is the order of $W$. Hence if we apply the operator $\pi$ to the above equation and put $h = 1$, we get

$$w\pi(\rho) = wc(\lambda)\pi(\lambda^*)$$

since $\phi_\lambda(1) = 1$. Hence

$$\pi(\lambda)D(h)\phi_\lambda(h)$$
$$= (-1)^{-q/2}\pi(\rho) \sum_{s \in W} \epsilon(s) \exp\{(-1)^{\frac{1}{2}}s\lambda(\log h)\} \qquad (h \in A_\mathfrak{p})$$

---

[34] $s\mu(H)$ denotes the value of $s\mu$ at $H$ $(s \in W, \mu \in \mathfrak{F}, H \in \mathfrak{h}_\mathfrak{p})$.

where $q$ is the number of roots in $Q$. However both sides are holomorphic functions of $\lambda$ on $\mathfrak{F}$ and so the above equation must hold for all $\lambda \in \mathfrak{F}$. Thus we have obtained another proof of an earlier result (see [5(e), Theorem 7]). Also $\boldsymbol{b}(\lambda) = (-1)^{-q/2}\pi(\rho)$. This shows that $\boldsymbol{b}$ is a constant in this case.

It is interesting to note that from the above results one can deduce the following extension of Lemma 54, which we state here without proof.[35]

LEMMA 55. $\delta'(q) = D^{-1}\gamma(q) \circ D$ for every $q \in I_{\mathfrak{G}}$.

**15. Appendix.** LEMMA 56. *Let $t$ be a real variable and $k_1, \cdots, k_r$ distinct real numbers. Then*

$$\operatorname*{Lim\,sup}_{t \to +\infty} \Big| \sum_{1 \le i \le r} c_i \exp\{(-1)^{\frac{1}{2}}k_i t\} \Big| \ge \Big\{ \sum_{1 \le i \le r} |c_i|^2 \Big\}^{\frac{1}{2}}$$

*for any complex numbers $c_1, \cdots, c_r$.*

Put $f(t) = \sum_i c_i \exp\{(-1)^{\frac{1}{2}}k_i t\}$. Then it follows by direct computation that

$$\operatorname*{Lim}_{T \to +\infty} T^{-1} \int_0^T |f(t)|^2 \, dt = \sum_i |c_i|^2.$$

On the other hand if $a = \operatorname*{Lim\,sup}_{t \to +\infty} |f(t)|$, it is obvious that

$$\operatorname*{Lim}_{T \to +\infty} T^{-1} \int_0^T |f(t)|^2 \, dt \le a^2.$$

Therefore $a^2 \ge \sum_i |c_i|^2$.

COROLLARY. *Let $k_1, \cdots, k_r$ be nonzero complex numbers and $p_0, p_1, \cdots, p_r$ polynomials in $t$ with complex coefficients. Suppose*

$$\operatorname*{Lim\,sup}_{t \to +\infty} |p_0(t) + p_1(t)e^{k_1 t} + \cdots + p_r(t)e^{k_r t}| \le a$$

*for some real number $a$. Then $p_0$ is a constant and $|p_0| \le a$.*

Let $l_i$ denote the real part of $k_i$. If $l_i < 0$, it is obvious that $\operatorname*{Lim}_{t \to +\infty} |p_i(t)e^{k_i t}| = 0$. Therefore if we suppose that $l_i \ge 0$ for $1 \le i \le s$ while $l_i < 0$ for $i > s$, it follows that

$$\operatorname*{Lim\,sup}_{t \to +\infty} \Big| p_0(t) + \sum_{1 \le i \le s} p_i(t)e^{k_i t} \Big| \le a.$$

---

[35] Cf. Theorem 2 of [5(g)]

5

Hence without loss of generality we may assume that $s = r$ and $k_1, \cdots, k_r$ are distinct and $p_i \neq 0$ $(0 \leq i \leq r)$. We shall now first show that $l_1 = l_2 = \cdots = l_r = 0$. For otherwise suppose $l = \max(l_1, \cdots, l_r) > 0$. We may assume that $l_1 = l_2 = \cdots = l_j = l$ while $l_i < l$ for $i > j$. Let $n$ be the highest among the degrees of $p_1, \cdots, p_j$ and put $c_i' = \operatorname*{Lim}_{t \to +\infty} p_i(t)/t^n$. Then $c_1', \cdots, c_j'$ cannot all be zero and it is clear that

$$0 = \operatorname*{Lim\,sup}_{t \to +\infty} t^{-n} e^{-lt} \left| p_0(t) + p_1(t)e^{k_1 t} + \cdots + p_r(t)e^{k_r t} \right|$$

$$= \operatorname*{Lim\,sup}_{t \to +\infty} \left| c_1' e^{(k_1 - l)t} + c_2' e^{(k_2 - l)t} + \cdots + c_j' e^{(k_j - l)t} \right|.$$

But since $k_i - l$ $(1 \leq i \leq j)$ are distinct pure imaginary numbers, we get a contradiction with Lemma 56. This proves that $l = 0$.

Now let $m$ be the maximum of the degrees of $p_0, p_1, \cdots, p_r$. We claim $m = 0$. For otherwise since $l_i = 0$ $1 \leq i \leq r$, it follows that

$$\operatorname*{Lim}_{t \to +\infty} t^{-m} \left| p_0(t) + \sum_{1 \leq i \leq r} p_i(t)e^{k_i t} \right| = 0.$$

This implies that

$$\operatorname*{Lim}_{t \to +\infty} \left| c_0 + \sum_{1 \leq i \leq r} c_i e^{k_i t} \right| = 0$$

where $c_i = \operatorname*{Lim}_{t \to +\infty} p_i(t)/t^m$ $(0 \leq i \leq r)$. But in view of the definition of $m, c_0, \cdots, c_r$ cannot all be zero and so again we get a contradiction with the above lemma. This shows that $m = 0$ and hence $p_0, \cdots, p_r$ are all constants. Therefore again by the above lemma, $|p_0| \leq a$.

Let $E$ be a vector space over $R$ of finite dimension. We shall say that a subset $F$ of $E$ is *full* if $tH \in F$ whenever $H \in F$ and $t \geq 1$ $(t \in R)$.

LEMMA 57. *Let $k_i \neq 0$ $(1 \leq i \leq r)$ be a finite set of linear functions*[36] *and $p_0, \cdots, p_r$ polynomial functions on $E$. Suppose $V$ is a nonempty, open and full subset of $E$ and $a$ a real number such that*

$$\left| p_0(H) + p_1(H)e^{k_1(H)} + \cdots + p_r(H)e^{k_r(H)} \right| \leq a$$

*for all $H \in V$. Then $p_0$ is a constant and $|p_0| \leq a$.*

Let $m$ be the degree of $p_0$. Since $V$ is open, we can select $H_0 \in V$ in such a way that $k_i(H_0) \neq 0$ $(1 \leq i \leq r)$ and $q_0(t) = p_0(tH_0)$ is a polynomial of degree $m$ in the real variable $t$. Put $q_i(t) = p_i(tH_0)$ and $k_i' = k_i(H_0)$ $1 \leq i \leq r$. Then since $tH_0 \in V$ for $t \geq 1$, it follows that

$$\left| q_0(t) + q_1(t)e^{k_1' t} + \cdots + q_r(t)e^{k_r' t} \right| \leq a \qquad (t \geq 1).$$

---

[36] We permit linear or polynomial functions on $E$ to take complex values.

Hence we conclude from the corollary of Lemma 56 that $q_0$ is a constant and $|q_0| \leq a$. This proves that $m = 0$ and therefore $|p_0| = |q_0| \leq a$.

COROLLARY. *Let $k_i$ $(i \geq 1)$ be a sequence of distinct linear functions and $p_i$ $(i \geq 1)$ a sequence of polynomial functions on $E$ and $V$ a nonempty, open and full subset of $E$. Suppose the following two conditions hold.*

(1) *For each linear function $k$ on $E$ the series*

$$\sum_{1 \leq i < \infty} |p_i(H)\exp(k(H) + k_i(H))|$$

*converges uniformly for $H \in V$.*

(2) $\sum_{1 \leq i < \infty} p_i(H)e^{k_i(H)} = 0$ *for $H \in V$.*

*Then $p_i = 0$ for every $i$.*

For otherwise select an index $j$ such that $p_j \neq 0$. For a given $\epsilon > 0$, choose an integer $N \geq j$ such that

$$\sum_{i > N} |p_i(H)\exp(k_i(H) - k_j(H))| \leq \epsilon$$

for all $H \in V$. This is possible by condition (1). Moreover it is obvious from condition (2) that

$$|\sum_{1 \leq i \leq N} p_i(H)\exp(k_i(H) - k_j(H))| \leq \epsilon \qquad (H \in V).$$

On the other hand $k_i - k_j \neq 0$ for $i \neq j$ and so we conclude from the above lemma that $p_j$ is a constant and $|p_j| \leq \epsilon$. But $\epsilon$ being arbitrary, this implies that $p_j = 0$. As this contradicts the choice of $j$, our assertion is proved.

Let $(z_1, \cdots, z_l)$ denote the Cartesian coordinates of a point $z$ in the complex Euclidean space $C^l$ of dimension $l$. We regard $C^l$ as a vector space over $C$ in the usual way and put $|z| = \max_{1 \leq i \leq l} |z_i|$. The distance between two sets $U$ and $V$ in $C^l$ is defined to be $\inf |z - \zeta|$ $(z \in U, \zeta \in V)$.

LEMMA 58. *Let $U$ be an open set in $C^l$ and $V$ a subset of $U$ and let $\epsilon$ denote the distance between $V$ and the complement of $U$ in $C^l$. Then if $\xi$ is any holomorphic function on $U$,*

$$\sup_{z \in V} |(\partial^{k_1 + \cdots + k_l}/\partial z_1^{k_1} \cdots \partial z_l^{k_l})\xi(z)|$$
$$\leq k_1! k_2! \cdots k_l! (2\pi/\epsilon)^{k_1 + \cdots + k_l} \sup_{z \in U} |\xi(z)|$$

*for any integers $k_1, \cdots, k_l \geq 0$.*

We may obviously assume that $\epsilon > 0$. Let $\epsilon'$ be any positive number less than $\epsilon$. Then if $z \in V$ and $|z - \zeta| \leq \epsilon'$, it follows that $\zeta \in U$. Hence

$$(\partial^{k_1 + \cdots + k_l}/\partial z_1^{k_1} \cdots \partial z_l^{k_l}) \xi(z)$$

$$= k_1! \cdots k_l! \{2\pi(-1)^{\frac{1}{2}}\}^{-1} \oint_1 \cdots \oint_l \xi(\zeta) \{ (\zeta_1 - z_1)^{k_1+1} \cdots (\zeta_l - z_l)^{k_l+1} \}^{-1}$$
$$\times d\zeta_1 \cdots d\zeta_l$$

where $\oint_i$ denotes complex integration with respect to $\zeta_i$ on the circle $|z_i - \zeta_i| = \epsilon'$. Therefore it is obvious that

$$\left|(\partial^{k_1 + \cdots + k_l}/\partial z_1^{k_1} \cdots \partial z_l^{k_l}) \xi(z)\right| \leq k_1! \cdots k_l! \sup_{\zeta \in U} |\xi(\zeta)| \, \epsilon'^{-k} (2\pi)^{k-1}$$

where $k = k_1 + \cdots + k_l$. This being true for every $\epsilon' < \epsilon$, our assertion now follows immediately.

The following lemma and its corollary will be needed in another paper.

LEMMA 59. *Suppose $\epsilon$ is positive in Lemma 57 and $\eta = z_1 \xi$. Then if $\sup_{z \in U} (1 + |z|)^{-d} |\eta(z)| < \infty$ for some integer $d \geq 0$, we can conclude that $\sup_{z \in V} (1 + |z|)^{-d} |\xi(z)| < \infty$.*

Select a positive number $\epsilon' < \min(1/6, \epsilon/3)$ and let $V'$ be the set of those points $z \in V$ where $|z_1| < \epsilon'$. Obviously it would be enough to verify that $\sup_{z \in V'} (1 + |z|)^{-d} |\xi(z)| < \infty$. Now fix a point $z^0 = (z_1^0, z_2^0, \cdots, z_l^0)$ in $V'$. Then

$$\xi(z^0) = (2\pi(-1)^{\frac{1}{2}})^{-1} \oint z_1^{-1} \eta(z_1, z_2^0, \cdots z_l^0) (z_1 - z_1^0)^{-1} dz_1$$

where $\oint$ denotes complex integration over the circle $|z_1| = 2\epsilon'$ in the $z_1$-plane. Now let $z' = (z_1, z_2^0, \cdots, z_l^0)$ where $|z_1| = 2\epsilon'$. Then it is clear that

$$1 + |z^0| \geq 1 + |z'| - 3\epsilon' \geq \tfrac{1}{2}(1 + |z'|).$$

Hence it follows that

$$(1 + |z^0|)^{-d} |\xi(z^0)| \leq (2^d/\epsilon') \sup_{z \in U} (1 + |z|)^{-d} |\eta(z)|.$$

This proves the lemma.

COROLLARY. *Let $p$ be a polynomial function on $C^l$ which can be written as a product of linear factors. Then the above lemma also holds if $\eta = p\xi$.*

This follows by an easy induction on the degree of $p$.

For any $m \times m$ matrix $\Lambda = (\lambda_{ij})_{1 \leq i, j \leq m}$ (with complex coefficients) put
$\| \Lambda \| = (\sum_{i,j} | \lambda_{ij} |^2)^{\frac{1}{2}}$.

LEMMA 60. *Let $\Lambda$ be an $m \times m$ matrix and let $\lambda_1, \cdots, \lambda_m$ denote all the eigenvalues of $\Lambda$. Put $\nu = \max_{1 \leq i \leq m} | \operatorname{Re} \lambda_i |$. Then*

$$\| e^{\Lambda} \| \leq m^{\frac{1}{2}} e^{2m\nu} \sum_{0 \leq k < m} \| \Lambda \|^k.$$

It is well known that we can find a unitary matrix $U$ such that $\Lambda' = U\Lambda U^{-1}$ has zeros everywhere below the diagonal. Since $\Lambda$ and $\Lambda'$ have the same eigenvalues and $\| \Lambda' \| = \| \Lambda \|$, $\| e^{\Lambda'} \| = \| e^{\Lambda} \|$, we may replace $\Lambda$ by $\Lambda'$ in our problem. Therefore we can assume that $\Lambda = A + N$ where $A$ is a diagonal and $N$ a supertriangular [37] matrix. Now consider $e^{t\Lambda} = e^{t(A+N)}$ $(t \in R)$ and put $n(t) = e^{-tA} e^{t(A+N)}$. Then it is clear that

$$dn(t)/dt = e^{-tA}(-A)e^{t(A+N)} + e^{-tA}(A+N)e^{t(A+N)}$$

$$= e^{-tA}Ne^{t(A+N)} = e^{-tA}Ne^{tA}n(t).$$

This differential equation can be solved by the method of iteration and one finds that

$$n(t) = 1 + \sum_{k \geq 1} \int_{0 \leq \tau_1 \leq \tau_2 \leq \cdots \leq \tau_k \leq t} N(\tau_k)N(\tau_{k-1}) \cdots N(\tau_1) \, d\tau_1 \cdots d\tau_k$$

for $t \geq 0$. Here 1 stands for the unit matrix and $N(\tau) = e^{-\tau A}Ne^{\tau A}$ $(\tau \in R)$. But since $N(\tau)$ is also supertriangular, the product $N(\tau_k) \cdots N(\tau_1)$ is zero if $k \geq m$. Moreover $A = A_1 + (-1)^{\frac{1}{2}}A_2$ where $A_1$, $A_2$ are both diagonal and real. Then since $e^{(-1)^{\frac{1}{2}}\tau A_2}$ is unitary, it follows that

$$\| N(\tau) \| = \| e^{-\tau A_1}Ne^{\tau A_1} \| \qquad\qquad (\tau \in R).$$

But the diagonal elements of $A$ are exactly the eigenvalues of $\Lambda$. Therefore if $| \tau | \leq 1$, no coefficient of $e^{\tau A_1}$ can exceed $e^{\nu}$ in absolutee value. Hence $\| N(\tau) \| \leq e^{2\nu} \| N \|$ for $| \tau | \leq 1$. This shows that

$$\int_{0 \leq \tau_1 \leq \cdots \leq \tau_k \leq t} \| N(\tau_k) \cdots N(\tau_1) \| \, d\tau_1 \cdots d\tau_k \leq \| N \|^k e^{2k\nu}$$

for $0 \leq t \leq 1$ and $k \geq 1$. Therefore

$$\| n(1) \| \leq \| 1 \| + \sum_{1 \leq k < m} \| N \|^k e^{2k\nu}.$$

---

[37] This means that all the coefficients of $N$ on or below the diagonal are zero.

310                HARISH-CHANDRA.

But $e^{\Lambda} = e^{A}n(1)$ and so $\| e^{\Lambda} \| \leqq e^{\nu} \| n(1) \|$. Moreover $\| \Lambda \|^{2} = \| A \|^{2} + \| N \|^{2} \geqq \| N \|^{2}$. Therefore

$$\| e^{\Lambda} \| \leqq e^{\nu} \| n(1) \| \leqq m^{\frac{1}{2}} e^{2m\nu} \sum_{0 \leqq k < m} \| N \|^{k}$$

$$\leqq m^{\frac{1}{2}} e^{2m\nu} \sum_{0 \leqq k < m} \| \Lambda \|^{k}$$

since $\| 1 \| = m^{\frac{1}{2}}$.

COLUMBIA UNIVERSITY.

---

## REFERENCES.


[1] V. Bargmann, *Ann. of Math.*, vol. 48 (1947), pp. 568-640.

[2] E. Cartan, *Ann. École Norm. Sup.*, vol. 44 (1927), pp. 345-467.

[3] C. Chevalley, a) *Introduction to the theory of algebraic functions of one variable*, American Mathematical Society, 1951.
      b) *Amer. Jour. Math.*, vol. 77 (1955), pp. 778-782.

[4] R. Godement, *Trans. Amer. Math. Soc.*, vol. 73 (1953), pp. 496-556.

[5] Harish-Chandra, a) *Ann. of Math.*, vol. 50 (1949), pp. 900-915.
      b) *Trans. Amer. Math. Soc.*, vol. 70 (1951), pp. 28-96.
      c) " " " " vol. 75 (1953), pp. 185-243.
      d) " " " " vol. 76 (1954), pp. 26-65.
      e) " " " " vol. 76 (1954), pp. 234-253.
      f) " " " " vol. 76 (1954), pp. 485-528.
      g) " " " " vol. 83 (1956), pp. 98-163.
      h) *Amer. Jour. Math.*, vol. 77 (1955), pp. 743-777.
      i) " " " vol. 78 (1956), pp. 1-41.
      j) " " " vol. 78 (1956), pp. 564-628.
      k) " " " vol. 79 (1957), pp. 87-120.
      l) " " " vol. 79 (1957), pp. 193-257.
      m) *Proc. Nat. Acad. Sci. U.S.A.*, vol. 40 (1954), pp. 200-204.
      n) " " " " " vol. 43 (1957), pp. 408-409.

[6] K. Iwasawa, *Ann. of Math.*, vol. 50 (1949), pp. 507-557.

[7] G. D. Mostow, *Bull. Amer. Math. Soc.*, vol. 55 (1949), pp. 969-980.

[8] H. Weyl, a) *Math. Ann.*, vol. 68 (1910), pp. 220-269.
      b) *Math. Zeit.*, vol. 24 (1925), pp. 328-395.
      c) *The structure and representations of continuous groups*, Princeton, The Institute for Advanced Study, 1935.



Reprinted from
*Amer. J. of Math.*
**80** (1958), 241–310

# SPHERICAL FUNCTIONS ON A SEMISIMPLE LIE GROUP II.*

By Harish-Chandra.[1]

---

**1. Introduction.** Define $G$ and $K$ as in $[2(i), \S 1]$ and let $I_2(G)$ be the Hilbert space of all square-integrable spherical functions on $G$. Our main problem is to give an explicit "expansion" of a function $f$ in $I_2(G)$, in terms of the elementary spherical functions. We shall construct a subspace $I_0(G)$ of $I_2(G)$ and give the required expansion for elements in $I_0(G)$. It seems likely that $I_0(G)$ is actually dense in $I_2(G)$. If this could be proved, our problem would be solved completely.

We have seen in $[2(i), \S 1]$ that a certain class of elementary spherical functions are parameterized by the space $E/W$. Fix a point $\lambda$ in $E$. Then in $[2(i)]$, we have studied the asymptotic behaviour of $\phi_\lambda$ on [2] $A_\mathfrak{p}$ at points which are far away from the singular [3] set. In this paper we shall consider the singular points as well. The main object here is to show that $\phi_\lambda(x)$ goes to zero very fast as $x$ approaches infinity on $G$ (see Theorem 2 of Section 12). Now let $d\lambda$ denote the Euclidean measure on $E$ and put

$$\phi_a{}'(x) = \int_E \pi(\lambda) a(\lambda) \phi_\lambda(x) \, d\lambda \qquad (x \in G, a \in C_c{}^\infty(E)),$$

where $\pi$ has the same meaning as in $[2(i), \S 3]$. Let $I_0{}'(G)$ stand for the space of all such functions $\phi_a{}'$. Then we shall prove that for $f \in I_0{}'(G)$, the integral

$$\bar{f}(\lambda) = \int_G f(x) \phi_\lambda(x^{-1}) \, dx \qquad (\lambda \in E)$$

is well defined and, in fact, it is possible to normalize $dx$ in such a way that [2]

$$\int_G |f(x)|^2 \, dx = \int_E |\bar{f}(\lambda)|^2 \, |\mathbf{c}(\lambda)|^{-2} \, d\lambda$$

for all $f$ in $I_0{}'(G)$. In order to extend this formula to functions on $I_2(G)$,

---

* Received July 16, 1957.

[1] John Simon Guggenheim fellow.

[2] As far as possible we keep to the notation of $[2(i)]$. Hence any symbol which is not explicitly defined afresh should be given the same meaning as in $[2(i)]$.

[3] We recall that a point $h \in A\mathfrak{p}$ is called singular if $\Delta(h) = 0$.

553

one would have to show that $I_0'(G)$ is dense in $I_2(G)$. This however still remains to be done.

The central idea of this paper can be described as follows. We give a method of reducing questions about spherical functions on $G$ to similar questions on a proper subgroup of $G$. This process enables us to use induction. Again, it is the detailed consideration of certain differential equations on $G$, which permits us to effect this reduction. Since we are primarily interested in the asymptotic behaviour of the corresponding solutions, we have to undertake a study of the asymptotic behaviour of the coefficients of these equations. This is done in Sections 2 to 5. Lemma 17 and its corollary bring these equations into a manageable form and we then obtain (in Section 6) some rather rough estimates concerning the behaviour of the corresponding solutions at infinity. These, in turn, enable us to prove certain results on convergence in Sections 7 and 8 and to simplify the equations by replacing them by their asymptotic form (see Lemma 24). It is this step which brings about the reduction of the problem from $G$ to a proper subgroup (see Lemmas 27, 30 and 35).

Now we combine the above reduction with induction and first use it to make a detailed study of the function $\dot{c}$ in Section 9. In particular, we show that $|c(s\lambda)| = |c(\lambda)|$ for $s \in W$ and $c^{-1}$ is analytic on $E$. The same method is then applied in Sections 11 and 12 to study $\phi_\lambda$ and $\phi_a'$ ($\lambda \in E$, $a \in C_c^\infty(E)$). The main results are contained in Theorems 1, 2 and 3.

In Section 13 we show that the mapping $a \to \int^{\cdot} \phi_a'(x)\phi_\lambda(x^{-1})\,dx$ ($a \in C_c^\infty(E)$), for a fixed $\lambda \in E$, is a distribution on $E$ which is given by a simple formula (see the corollary of Lemma 41). However this formula involves an unknown coefficient $\beta(\lambda)$ and the rest of the paper is devoted to establishing a relationship between $\beta$ and $c$. This requires a considerable amount of fresh work. First we prove an inequality (Lemma 42) in order to ensure the convergence of some integrals taken over a certain nilpotent subgroup $\bar{N}$ of $G$. The corollary of Lemma 43 expresses the value of these integrals in terms of the function $\beta$. On the other hand we show in Section 15 (Theorem 4) that the same integrals can also be expressed by means of $c$ and this gives the required connection between $\beta$ and $c$. However, the proof of Theorem 4 is rather long. It depends upon a detailed study of certain differential equations on the solvable group $\bar{N}A_\mathfrak{p}$.

Finally in Section 16, we state two conjectures and show how one can obtain an explicit Plancherel formula for spherical functions on $G$, by assuming their validity.

The principal results of this paper have been announced in a short note [2(k)].

**2. The ring $\mathfrak{R}$.** Define $\mathfrak{h}_{\mathfrak{p}_0}^+$ as in [2(i), §9] and fix an element $H_0 \neq 0$ in [2] $Cl(\mathfrak{h}_{\mathfrak{p}_0}^+)$. Let $\mathfrak{m}_0$ be the centralizer of $H_0$ in $\mathfrak{g}_0$. Then if $\theta$ is the automorphism of $\mathfrak{g}$ given by $\theta(X+Y) = X-Y$ $(X \in \mathfrak{k}, Y \in \mathfrak{p})$, it is obvious that $\theta(\mathfrak{m}_0) = \mathfrak{m}_0$. This implies (see [2(h), Lemma 10]) that $\mathfrak{m}_0$ is reductive in $\mathfrak{g}_0$. Let $\mathfrak{m}$ be the complexification of $\mathfrak{m}_0$ in $\mathfrak{g}$. We regard $\mathfrak{g}$ as a Hilbert space in the usual way [2(i), §9] and consider the orthogonal complement $\mathfrak{q}$ of $\mathfrak{m}$ in $\mathfrak{g}$. Then $\theta(\mathfrak{q}) = \mathfrak{q}$ and $[\mathfrak{m}, \mathfrak{q}] \subset \mathfrak{q}$. Let $I$ be the centralizer of $\mathfrak{m}$ in $\mathfrak{h}_\mathfrak{p}$ and put $I_0 = I \cap \mathfrak{h}_{\mathfrak{p}_0}$.

As before let $P$ denote the set of all positive roots of $\mathfrak{g}$ (with respect to $\mathfrak{h}$). Then $\alpha(H_0) \geq 0$ for $\alpha \in P$. Let $Q$ be the set of those $\alpha \in P$ for which $\alpha(H_0) > 0$. Put $\mathfrak{q}_+ = \sum_{\alpha \in Q} CX_\alpha$, $\mathfrak{q}_- = \sum_{\alpha \in Q} CX_{-\alpha}$ where $X_\alpha$ has the usual meaning (see [2(i), §7]). Then $\mathfrak{q} = \mathfrak{q}_+ + \mathfrak{q}_-$ and $\mathfrak{m} = \mathfrak{h} + \sum_{\alpha \in P'} (CX_\alpha + CX_{-\alpha})$, $P'$ being the complement of $Q$ in $P$. It is now obvious that $[\mathfrak{m}, \mathfrak{q}_+] \subset \mathfrak{q}_+$, $[\mathfrak{m}, \mathfrak{q}_-] \subset \mathfrak{q}_-$.

Let $M$ and $M_K$ be the analytic subgroups of $G$ corresponding to $\mathfrak{m}_0$ and $\mathfrak{m}_{\mathfrak{k}_0} = \mathfrak{m}_0 \cap \mathfrak{k}_0$ respectively, Put $\mathfrak{m}_{\mathfrak{p}_0} = \mathfrak{m}_0 \cap \mathfrak{p}_0$ and let $M_\mathfrak{p}$ be the set of all elements in $M$ of the form $\exp Y$ $(Y \in \mathfrak{m}_{\mathfrak{p}_0})$. Also put [4]

$$\Delta_M(m) = \det(\mathrm{Ad}(m) - \mathrm{Ad}(\theta(m)))_{\mathfrak{q}_+} \qquad (m \in M)$$

where the subscript denotes restriction on $\mathfrak{q}_+$. If $m = kp$ $(k \in M_K, p \in M_\mathfrak{p})$, it is obvious that $\Delta_M(m) = \det(\mathrm{Ad}(k))_{\mathfrak{q}_+}\Delta_M(p)$. Let $M'$ be the set of those points $m \in M$ where $\Delta_M(m) \neq 0$. Then it follows that $M' = M_K M' M_K$ and $(M')^{-1} = M'$. Moreover

$$\Delta_M(\exp H) = \prod_{\alpha \in Q} (e^{\alpha(H)} - e^{-\alpha(H)}) \qquad (H \in \mathfrak{h}_{\mathfrak{p}_0}).$$

Let $m$ be an element in $M'$. Since $\theta(\mathfrak{q}_+) = \mathfrak{q}_-$, it is clear that $X \to \{\mathrm{Ad}(m^{-1}) - \mathrm{Ad}(\theta(m^{-1}))\}X$ $(X \in \mathfrak{q})$ is a nonsingular linear mapping of $\mathfrak{q}$. We denote its inverse by $A(m)$.

LEMMA 1. *Put $\mathfrak{q}_{\mathfrak{k}} = \mathfrak{q} \cap \mathfrak{k}$ and $\mathfrak{q}_\mathfrak{p} = \mathfrak{q} \cap \mathfrak{p}$. Then if $m \in M'$, $A(m)\mathfrak{q}_{\mathfrak{k}} = \mathfrak{q}_\mathfrak{p}$ and $A(m)\mathfrak{q}_\mathfrak{p} = \mathfrak{q}_{\mathfrak{k}}$.*

Since $\theta(\mathrm{Ad}(x)Y) = \mathrm{Ad}(\theta(x))\theta(Y)$ $(x \in G, Y \in \mathfrak{g})$, it is obvious that under the mapping $\mathrm{Ad}(x) - \mathrm{Ad}(\theta(x))$, $\mathfrak{k}$ is mapped into $\mathfrak{p}$ and $\mathfrak{p}$ into $\mathfrak{k}$. From this our assertion follows immediately.

---

[4] Here we have extended $\theta$ to an automorphism of $G$.

For any $m \in M'$, define a linear transformation $c(m)$ on $\mathfrak{q}$ by $c(m)X = \mathrm{Ad}(m^{-1})A(m)X$ $(X \in \mathfrak{q})$. Then both $\mathfrak{q}_+$ and $\mathfrak{q}_-$ are invariant under $c(m)$. Let

$$c(m)X_\alpha = \sum_{\beta \in Q} X_\beta c_{\beta\alpha}(m) \qquad (\alpha \in Q, \, c_{\beta\alpha}(m) \in C).$$

Then $c_{\beta\alpha}(m)$, regarded as a function of $m$, is analytic on $M'$. Let $\mathfrak{R}_0$ be the ring of analytic functions on $M'$ generated[5] over $C$ by $c_{\beta\alpha}$ $(\beta, \alpha \in Q)$. Let $k \in M_K$ and $m \in M'$. Then it is easy to verify that $c(km) = c(m)$ and $c(mk)X = \mathrm{Ad}(k^{-1})c(m)\mathrm{Ad}(k)X$ $(X \in \mathfrak{q})$. Moreover

$$\theta(c(m)X) = \mathrm{Ad}(\theta(m^{-1}))\theta(A(m)X) = -\mathrm{Ad}\,\theta(m^{-1})A(m)\theta(X)$$
$$= \theta(X) - c(m)\theta(X),$$

and therefore

$$c(m)\theta(X_\alpha) = \theta(X_\alpha) - \theta(c(m)X_\alpha)$$
$$= \theta(X_\alpha) - \sum_{\beta \in Q} \theta(X_\beta)c_{\beta\alpha}(m) \qquad (\alpha \in Q).$$

Hence we get the following result.

LEMMA 2. *Let $k$ be an element in $M_K$ and $g$ in $\mathfrak{R}_0$. Then $g(km) = g(m)$ $(m \in M')$ and the function $m \rightarrow g(mk)$ is also in $\mathfrak{R}_0$. Moreover the matrix coefficients of $c(m)$ (with respect to any fixed base of $\mathfrak{q}$), considered as functions on $M'$, lie in the ring $C + \mathfrak{R}_0$.*

Now for any $x \in G$, put $x^\dagger = \theta(x^{-1})$.

LEMMA 3. $(1 - \mathrm{Ad}(m^\dagger m))c(m)X = X$ *for $X \in \mathfrak{q}$ and $m \in M'$.*

For $c(m)X = \mathrm{Ad}(m^{-1})A(m)X$ and therefore $(1 - \mathrm{Ad}(m^\dagger m))c(m)X = (\mathrm{Ad}(m^{-1}) - \mathrm{Ad}(m^\dagger))A(m)X = X$.

Let $U$ be any linear subspace of $\mathfrak{g}$ containing $\mathfrak{q}_+$ and let $T$ be a linear transformation in $U$ which leaves $\mathfrak{q}_+$ invariant. Then we set $|T|_+ = \sup \|TX\|$ where $X$ runs over all elements in $\mathfrak{q}_+$ with $\|X\| \leq 1$. Let $M_+$ be the set of those points $m \in M$ where $|\mathrm{Ad}(m^{-1})|_+ < 1$. It is obvious that $M_+ \subset M'$, $A_\mathfrak{p}^+ \subset M_+$ and $M_K M_+ M_K = M_+$. Moreover if $m_1, m_2$ are in $M_+$, the same holds for $m_1 m_2$.

LEMMA 4. *Let $m$ be an element in $M_+$. Then*

$$|c(m)|_+ \leq |\mathrm{Ad}(m^{-1})|_+^2 \{1 - |\mathrm{Ad}(m^{-1})|_+^2\}^{-1}.$$

---

[5] This means that $\mathfrak{R}_0$ consists of those functions on $M'$ which can be written as polynomials in $c_{\beta\alpha}$ (with complex coefficients), the constant term of the polynomial being zero.

It is clear that $\mathrm{Ad}(m^\dagger)$ is the adjoint of $\mathrm{Ad}(m)$ (in the sense of Hilbert space theory). Hence $|\mathrm{Ad}(m^\dagger m)^{-1}|_+ \leqq |\mathrm{Ad}(m^{-1})|_+^2 < 1$. Therefore since

$$(1 - \mathrm{Ad}((m^\dagger m)^{-1}))c(m)X = - \mathrm{Ad}((m^\dagger m)^{-1})X \qquad (X \in \mathfrak{q}_+)$$

from Lemma 3, it is clear that

$$c(m)X = - \sum_{1 \leqq r < \infty} \mathrm{Ad}((m^\dagger m)^{-r})X.$$

Hence $|c(m)|_+ \leqq \sum_{1 \leqq r < \infty} |\mathrm{Ad}(m^{-1})|_+^{2r} = |\mathrm{Ad}(m^{-1})|_+^2 \{1 - |\mathrm{Ad}(m^{-1})|_+^2\}^{-1}.$

As usual let $\mathfrak{B}$ denote the universal enveloping algebra of $\mathfrak{g}$ and consider the subalgebra $\mathfrak{M}$ of $\mathfrak{B}$ generated by $(1, \mathfrak{m})$. We can regard any element $\mu$ of $\mathfrak{M}$ as a left-invariant differential operator on $M$. Applying this operator to each coefficient of the matrix of $c(m)$ (corresponding to some fixed base of $\mathfrak{q}$), we get a linear transformation $c(m;\mu)$ on $\mathfrak{q}$ $(m \in M')$.

LEMMA 5. *Let $\mu$ be any element in $\mathfrak{M}$ of degree $r$. Then there exists a positive number $a$ such that*

$$|c(m;\mu)|_+ \leqq a\,|\mathrm{Ad}(m^{-1})|_+^2 \{1 - |\mathrm{Ad}(m^{-1})|_+^2\}^{-(r+1)}$$

*for all $m \in M_+$.*

Let $\sigma$ denote the representation of $\mathfrak{M}$ on $\mathfrak{q}_+$ defined by $\sigma(Y)X = [Y, X]$ $(Y \in \mathfrak{m}, X \in \mathfrak{q}_+)$. We shall first prove the following result.

LEMMA 6. *Let $\mu$ be an element in $\mathfrak{M}$ of degree $r$. Then we can select a finite number of elements $\mu_{ij} \in \mathfrak{M}$ $(1 \leqq i \leqq 5, 1 \leqq j \leqq s)$ with the following properties. $\mu_{3j}$ are of degree lower than $r$ and*

$$(1 - \mathrm{Ad}(m^\dagger m)^{-1})c(m;\mu)X = \sum_{1 \leqq j \leqq s} \sigma(\mu_{1j})\mathrm{Ad}((m^\dagger m)^{-1})\sigma(\mu_{2j})c(m;\mu_{3j})X$$

$$+ \sum_{1 \leqq j \leqq s} \sigma(\mu_{4j})\mathrm{Ad}((m^\dagger m)^{-1})\sigma(\mu_{5j})X$$

*for all $X \in \mathfrak{q}_+$ and $m \in M'$.*

If $r = 0$, our statement follows from Lemma 3. So now suppose $r \geqq 1$ and use induction. It is clearly enough to consider the case when $\mu = Y_1 Y_2 \cdots Y_r$ $(Y_i \in \mathfrak{m}_0)$. Put $\mu_0 = Y_2 \cdots Y_r$ and $m_t = m \exp tY_1$ for $m \in M$ and $t \in R$. Then if $X \in \mathfrak{q}_+$, $m \in M'$, it is obvious that

$$[(d/dt)\{(1 - \mathrm{Ad}(m_t^\dagger m_t)^{-1})c(m_t;\mu_0)X\}]_{t=0}$$

$$= (1 - \mathrm{Ad}(m^\dagger m)^{-1})c(m;\mu)X + \sigma(Y_1)\mathrm{Ad}(m^\dagger m)^{-1}c(m;\mu_0)X$$

$$- \mathrm{Ad}(m^\dagger m)^{-1}\sigma(\theta(Y_1))c(m;\mu_0)X.$$

On the other hand our induction hypothesis is applicable to $\mu_0$ and so we can choose elements $v_{ij} \in \mathfrak{M}$ $(1 \leq i \leq 5, 1 \leq j \leq q)$ such that the degree of $v_{3j}$ is lower than $r-1$ and

$$(1 - \mathrm{Ad}(m^\dagger m)^{-1}) c(m;\mu_0) X = \sum_{1 \leq j \leq q} \sigma(v_{1j}) \mathrm{Ad}(m^\dagger m)^{-1} \sigma(v_{2j}) c(m;v_{3j}) X$$
$$+ \sum_{1 \leq j \leq q} \sigma(v_{4j}) \mathrm{Ad}(m^\dagger m)^{-1} \sigma(v_{5j}) X$$

for $m \in M'$ and $X \in \mathfrak{q}_+$. Therefore

$$[(d/dt)\{(1 - \mathrm{Ad}(m_t{}^\dagger m_t)^{-1}) c(m_t;\mu_0) X]_{t=0}$$
$$= \sum_{1 \leq i \leq q} \{ -\sigma(v_{1j} Y_1) \mathrm{Ad}(m^\dagger m)^{-1} \sigma(v_{2j}) c(m;v_{3j})$$
$$+ \sigma(v_{1j}) \mathrm{Ad}(m^\dagger m)^{-1} \sigma(\theta(Y_1) v_{2j}) c(m;v_{3j})$$
$$+ \sigma(v_{1j}) \mathrm{Ad}(m^\dagger m)^{-1} \sigma(v_{2j}) c(m;Y_1 v_{3j}) - \sigma(v_{4j} Y_1) \mathrm{Ad}(m^\dagger m)^{-1} \sigma(v_{5j})$$
$$+ \sigma(v_{4j}) \mathrm{Ad}(m^\dagger m)^{-1} \sigma(\theta(Y_1) v_{5j}) \} X.$$

By equating the two expressions for

$$[(d/dt)\{(1 - \mathrm{Ad}(m_t{}^\dagger m_t)^{-1}) c(m_t;\mu_0) X\}]_{t=0}$$

obtained above, the assertion of the lemma follows.

Now we come to the proof of Lemma 5. If $r=0$, our statement is true in consequence of Lemma 4. Hence we may suppose that $r \geq 1$ and use induction on $r$. But we know from Lemma 6 that

$$c(m;\mu) X = \sum_{1 \leq j \leq s} \sum_{0 \leq k < \infty} \mathrm{Ad}(m^\dagger m)^{-k} \{ \sigma(\mu_{1j}) \mathrm{Ad}(m^\dagger m)^{-1} \sigma(\mu_{2j}) c(m;\mu_{3j}) X$$
$$+ \sigma(\mu_{4j}) \mathrm{Ad}(m^\dagger m)^{-1} \sigma(\mu_{5j}) X \}$$

for $X \in \mathfrak{q}_+$ and $m \in M_+$. Moreover

$$| \mathrm{Ad}(m^\dagger m)^{-k} \sigma(\mu_{1j}) \mathrm{Ad}(m^\dagger m)^{-1} \sigma(\mu_{2j}) c(m;\mu_{3j}) |_+$$
$$\leq | \mathrm{Ad}(m^{-1}) |_+^{2(k+1)} | \sigma(\mu_{1j}) |_+ | \sigma(\mu_{2j}) |_+ | c(m;\mu_{3j}) |_+$$

and the degree of $\mu_{3j}$ is less than $r$. Therefore if we apply the induction hypothesis to $\mu_{3j}$ and observe that

$$| \mathrm{Ad}(m^\dagger m)^{-k} \sigma(\mu_{4j}) \mathrm{Ad}(m^\dagger m)^{-1} \sigma(\mu_{5j}) |_+ \leq | \mathrm{Ad}(m^{-1}) |_+^{2(k+1)} | \sigma(\mu_{4j}) |_+ | \sigma(\mu_{5j}) |_+,$$

the required result follows immediately.

Let $\mathfrak{R}$ be the ring of analytic functions on $M'$ generated over $C$ by all functions of the form $\mu c_{\alpha\beta}$ $(\mu \in \mathfrak{M}; \alpha, \beta \in Q)$. It is obvious that if $g \in \mathfrak{R}$ and $\mu \in \mathfrak{M}$, then $\mu g$ is also in $\mathfrak{R}$. The following result is an immediate consequence of Lemma 5.

LEMMA 7. *Let $g$ be an element in $\Re$. Then there exists a positive number $a$ and an integer $r \geq 1$ such that*

$$| g(m) | \leq a | \operatorname{Ad}(m^{-1}) |_+^2 \{1 - | \operatorname{Ad}(m^{-1}) |_+^2\}^{-r}$$

*for all $m \in M_+$.*

**3. The mapping $\mu_0$.** Let $q$ be the number of roots in $Q$. Since $\mathfrak{g}$ is the orthogonal sum of $\mathfrak{m}$ and $\mathfrak{q}$, $\dim \mathfrak{g} = \dim \mathfrak{m} + \dim \mathfrak{q} = \dim \mathfrak{m} + 2q$. On the other hand, the restriction of $\operatorname{ad} H_0$ on $\mathfrak{q}$ is obviously nonsingular and therefore it defines a linear isomorphism of $\mathfrak{q}_{\mathfrak{k}}$ onto $\mathfrak{q}_{\mathfrak{p}}$. Hence $\dim \mathfrak{q}_{\mathfrak{k}} = \frac{1}{2} \dim \mathfrak{q} = q$. Let $X_\alpha = Y_\alpha + Z_\alpha$ $(Y_\alpha \in \mathfrak{p}, Z_\alpha \in \mathfrak{k})$ for $\alpha \in P$. Then $\theta(X_\alpha) = -Y_\alpha + Z_\alpha$. Since $X_\alpha, \theta(X_\alpha)$ $(\alpha \in Q)$ span $\mathfrak{q}$, it follows that $Z_\alpha$ $(\alpha \in Q)$ span $\mathfrak{q}_{\mathfrak{k}}$. Therefore $Z_\alpha$ $(\alpha \in Q)$ must be linearly independent over $C$. Put $\mathfrak{m}_{\mathfrak{k}} = \mathfrak{m} \cap \mathfrak{k}$ and $\mathfrak{m}_{\mathfrak{p}} = \mathfrak{m} \cap \mathfrak{p}$.

LEMMA[6] 8. *Suppose $m$ is an element in $M'$. Then $\mathfrak{g} = \operatorname{Ad}(m^{-1}) \mathfrak{q}_{\mathfrak{k}} + \mathfrak{m}_{\mathfrak{p}} + \mathfrak{k}$ where the sum is direct.*

Since $\mathfrak{m}_{\mathfrak{p}} + \mathfrak{k} = \mathfrak{m} + \mathfrak{q}_{\mathfrak{k}}$ and $\dim \mathfrak{g} = \dim \mathfrak{m} + 2 \dim \mathfrak{q}_{\mathfrak{k}}$, it would be enough to prove that $\operatorname{Ad}(m^{-1}) \mathfrak{q}_{\mathfrak{k}} \cap (\mathfrak{m}_{\mathfrak{p}} + \mathfrak{k}) = \{0\}$. So let us suppose that $X \in \mathfrak{q}_{\mathfrak{k}}$ and $\operatorname{Ad}(m^{-1}) X \in \mathfrak{m}_{\mathfrak{p}} + \mathfrak{k}$. Then $(\operatorname{ad} H_0)^2 \operatorname{Ad}(m^{-1}) X \in \mathfrak{k}$ and therefore

$$(\operatorname{ad} H_0)^2 \operatorname{Ad}(m^{-1}) X = \theta((\operatorname{ad} H_0)^2 \operatorname{Ad}(m^{-1}) X) = (\operatorname{ad} H_0)^2 \operatorname{Ad}(m^\dagger) X.$$

This shows that $(\operatorname{ad} H_0)^2 (\operatorname{Ad} m^{-1} - \operatorname{Ad} m^\dagger) X = 0$. But since $\operatorname{ad} H_0$ defines a nonsingular linear transformation on $\mathfrak{q}$ and since $m \in M'$, it is now clear that $X = 0$. This proves that $\operatorname{Ad}(m^{-1}) \mathfrak{q}_{\mathfrak{k}} \cap (\mathfrak{m}_{\mathfrak{p}} + \mathfrak{k}) = \{0\}$.

Let $\mathfrak{X}$ be the subalgebra of $\mathfrak{B}$ generated by $(1, \mathfrak{k})$ and let $\mathfrak{Q}_{\mathfrak{k}}$ and $\mathfrak{M}_{\mathfrak{p}}$ be the images in $\mathfrak{B}$ of[7] $S(\mathfrak{q}_{\mathfrak{k}})$ and $S(\mathfrak{m}_{\mathfrak{p}})$ respectively under the canonical mapping (see [2(b), p. 192]) of $S(\mathfrak{g})$ into $\mathfrak{B}$. Moreover for any $x \in G$, let $b \to b^x$ $(b \in \mathfrak{B})$ denote the automorphism of $\mathfrak{B}$ which coincides with $\operatorname{Ad}(x)$ on $\mathfrak{g}$. Consider the tensor product $V = \mathfrak{Q}_{\mathfrak{k}} \mathbf{X} \mathfrak{M}_{\mathfrak{p}} \mathbf{X} \mathfrak{X}$. Then it follows from the above result and Lemma 12 of [2(b)] that for any $m \in M'$, there exists a linear isomorphism of $V$ onto $\mathfrak{B}$ which maps $q \mathbf{X} \mu \mathbf{X} x$ on $q^{m^{-1}} \mu x$ $(q \in \mathfrak{Q}_{\mathfrak{k}}, \mu \in \mathfrak{M}_{\mathfrak{p}}, x \in \mathfrak{X})$. Hence $\mathfrak{B} = (\mathfrak{Q}_{\mathfrak{k}})^{m^{-1}} \mathfrak{M}_{\mathfrak{p}} \mathfrak{X}$. Let $\mathfrak{Q}_{\mathfrak{k}}'$ denote the subspace consisting of those elements in $\mathfrak{Q}_{\mathfrak{k}}$ whose homogeneous component (see [2(i), Footnote 21]) of degree zero is zero. Then since $V$ is the direct sum of $\mathfrak{Q}_{\mathfrak{k}} \mathbf{X} \mathfrak{M}_{\mathfrak{p}} \mathbf{X} (\mathfrak{X}\mathfrak{k})$, $\mathfrak{Q}_{\mathfrak{k}}' \mathbf{X} \mathfrak{M}_{\mathfrak{p}} \mathbf{X} 1$ and $1 \mathbf{X} \mathfrak{M}_{\mathfrak{p}} \mathbf{X} 1$, the following corollary is obvious.

---

[6] Cf. [2(i), Lemma 21].

[7] Here we follow the notation introduced in [2(g), §2].

COROLLARY. *Suppose $m$ lies in $M'$. Then*

$$\mathfrak{B} = (\mathfrak{Q}_{\mathfrak{k}})^{m-1}\mathfrak{M}_{\mathfrak{p}}\mathfrak{X} = \mathfrak{B}\mathfrak{k} + (\mathfrak{Q}_{\mathfrak{k}}')^{m-1}\mathfrak{M}_{\mathfrak{p}} + \mathfrak{M}_{\mathfrak{p}}$$

*where the sum is direct.*

For any finite set of elements $X_1, \cdots, X_d \in \mathfrak{g}$, put

$$\sigma(X_1, X_2, \cdots, X_d) = (d!)^{-1} \sum X_{i_1} X_{i_2} \cdots X_{i_d} \in \mathfrak{B},$$

where the sum is over all permutations $(i_1, \cdots, i_d)$ of $(1, 2, \cdots, d)$. More-over let $\mathfrak{B}_d$ denote the set of all element in $\mathfrak{B}$ of degree $\leqq d$.

LEMMA 9. *Let $b$ be a given element in $\mathfrak{B}_d$. Then we can choose $\mu_0$ in $\mathfrak{M}_{\mathfrak{p}} \cap \mathfrak{B}_d$ and a finite number of elements $\mu_i \in \mathfrak{M}_{\mathfrak{p}} \cap \mathfrak{B}_{d-1}$, $g_i \in \mathfrak{R}_0$ $(1 \leqq i \leqq r)$ such that*

$$b - \mu_0 - \sum_{1 \leqq i \leqq r} g_i(m)\mu_i \in \mathfrak{B}\mathfrak{k} + (\mathfrak{Q}_{\mathfrak{k}}')^{m-1}\mathfrak{M}_{\mathfrak{p}}$$

*for all $m \in M'$.*

We shall use the induction on $d$. Since $\mathfrak{g} = \mathfrak{q}_{\mathfrak{p}} + \mathfrak{m}_{\mathfrak{p}} + \mathfrak{k}$, it is obviously enough to consider the case when $b = \sigma(Y_1, \cdots, Y_e)\mu$ and $Y_j \in \mathfrak{q}_{\mathfrak{p}}$, $\mu \in \mathfrak{M}_{\mathfrak{p}}$ and $\mu$ is of degree $d - e$ $(e \geqq 1)$. Now for any $Y \in \mathfrak{q}_{\mathfrak{p}}$ and $m \in M'$,

$$c(m)Y = \text{Ad}(m^{-1})A(m)Y = \tfrac{1}{2}Y + \tfrac{1}{2}(\text{Ad}(m^{-1}) + \text{Ad}\,\theta(m^{-1}))A(m)Y.$$

Since $A(m)Y \in \mathfrak{q}_{\mathfrak{k}} \subset \mathfrak{k}$ from Lemma 1, it is clear that $c(m)Y - \tfrac{1}{2}Y \in \mathfrak{k}$. Put

$$U_i(m) = 2A(m)Y_i, \qquad V_i(m) = 2c(m)Y_i - Y_i \qquad (1 \leqq i \leqq e).$$

Then $U_i(m)$, $V_i(m)$ are in $\mathfrak{k}$ and $\text{Ad}(m^{-1})U_i(m) = Y_i + V_i(m)$. Let $(i_1, i_2, \cdots, i_e)$ be any permutation of $(1, 2, \cdots, e)$. Then

$$\begin{aligned}
(U_{i_1}(m) \cdots U_{i_e}(m))^{m-1}\mu &= (Y_{i_1} + V_{i_1}(m)) \cdots (Y_{i_e} + V_{i_e}(m))\mu \\
&= (L_{i_1} + R_{i_1} - D_{i_1}) \cdots (L_{i_e} + R_{i_e} - D_{i_e})\mu \\
&\equiv (L_{i_1} - D_{i_1}) \cdots (L_{i_e} - D_{i_e})\mu \bmod \mathfrak{B}\mathfrak{k}.
\end{aligned}$$

Here $L_i$, $R_i$ and $D_i$ respectively denote the linear mappings $a \to Y_i a$, $a \to aV_i(m)$ and $a \to [V_i(m), a] = V_i(m)a - aV_i(m)$ $(a \in \mathfrak{B})$ of $\mathfrak{B}$ into itself. Since $[X, a] \in \mathfrak{B}_r$ for $X \in \mathfrak{g}$ and $a \in \mathfrak{B}_r$, it follows from Lemma 2 that we can select $b_1, \cdots, b_s \in \mathfrak{B}_{d-1}$ and $g_1', \cdots, g_s' \in C + \mathfrak{R}_0$ such that

$$(\sigma(U_1(m), \cdots, U_e(m)))^{m-1}\mu \equiv b + \sum_{1 \leqq i \leqq s} g_i'(m)b_i \bmod \mathfrak{B}\mathfrak{k}$$

for all $m \in M'$. But $\sigma(U_1(m), \cdots, U_e(m)) \in \mathfrak{Q}_\mathfrak{k}'$ since $U_j(m) \in \mathfrak{q}_\mathfrak{k}$ and therefore

$$b \equiv - \sum_{1 \leq i \leq s} g_i'(m) b_i \bmod \{\mathfrak{B}\mathfrak{k} + (\mathfrak{Q}_\mathfrak{k}')^{m-1}\mathfrak{M}_\mathfrak{p}\}.$$

Our assertion now follows immediately by the induction hypothesis.

Let $\mathfrak{R}_0\mathfrak{M}_\mathfrak{p}$ denote the set of all differential operators on $M'$ which are of the form $\sum_{1 \leq i \leq r} g_i\mu_i$ $(g_i \in \mathfrak{R}_0, \mu_i \in \mathfrak{M}_\mathfrak{p})$. Then the above lemma shows that for any $b \in \mathfrak{B}$, we can select $\mu_0 \in \mathfrak{M}_\mathfrak{p}$ and $\mu \in \mathfrak{R}_0\mathfrak{M}_\mathfrak{p}$, such that

$$b - \mu_0 - \mu_m \in \mathfrak{B}\mathfrak{k} + (\mathfrak{Q}_\mathfrak{k}')^{m-1}\mathfrak{M}_\mathfrak{p}$$

for all $m \in M'$. (Here $\mu_m$ denotes, as usual, the local expression [2(d), p. 112] of $\mu$ at $m$.) Moreover it follows from the Corollary of Lemma 8 that the differential operator $\mu_0 + \mu$ on $M'$ is uniquely determined by $b$. Now suppose $\mu_0 + \mu = \mu_0' + \mu'$ $(\mu_0' \in \mathfrak{M}_\mathfrak{p}, \mu' \in \mathfrak{R}_0\mathfrak{M}_\mathfrak{p})$. Then $\mu_0 - \mu_0' = \mu' - \mu$ and therefore $\mu_0 - \mu_0' = (\mu' - \mu)_m$ for every $m \in M'$. In particular we can take $m = \exp tH_0$ $(t > 0)$. Then if $g \in \mathfrak{R}_0$, it follows from Lemma 7 that $\underset{t \to +\infty}{\mathrm{Lim}}\, g(\exp tH_0) = 0$. Therefore $\mu_0 - \mu_0' = 0$. This proves that both $\mu_0$ and $\mu$ are uniquely determined by $b$. We shall denote them by $\mu_0(b)$ and $\mu(b)$ respectively.

LEMMA 10. *Let $k$ be an element in $M_K$. Then $(\mu_0(b))^k = \mu_0(b^k)$ and*[8] $\mu(b)^k = \mu(b^k)$ *for $b \in \mathfrak{B}$.*

Since $b - \mu_0(b) - \mu_m(b) \in \mathfrak{B}\mathfrak{k} + (\mathfrak{Q}_\mathfrak{k}')^{m-1}\mathfrak{M}_\mathfrak{p}$ $(m \in M')$, it follows that

$$b^k - \mu_0(b)^k - (\mu_m(b))^k \in \mathfrak{B}\mathfrak{k} + (\mathfrak{Q}_\mathfrak{k}')^{km-1}\mathfrak{M}_\mathfrak{p}.$$

Therefore replacing $m$ by $mk$, we find that

$$b^k - \mu_0(b)^k - (\mu_{mk}(b))^k \in \mathfrak{B}\mathfrak{k} + (\mathfrak{Q}_\mathfrak{k}')^{m-1}\mathfrak{M}_\mathfrak{p}.$$

Now $(\mu(b)^k)_m = (\mu(b)_{k^{-1}mk})^k$ from Lemma 14 of [2(d)]. But it is clear from Lemma 2 that the differential operator $\mu(b)^k$ lies in $\mathfrak{R}_0\mathfrak{M}_\mathfrak{p}$ and $\mu(b)_{k^{-1}mk} = \mu(b)_{mk}$. Therefore

$$b^k - \mu_0(b)^k - (\mu(b)^k)_m \in \mathfrak{B}\mathfrak{k} + (\mathfrak{Q}_\mathfrak{k}')^{m-1}\mathfrak{M}_\mathfrak{p}$$

and this proves that $\mu_0(b^k) = \mu_0(b)^k$ and $\mu(b^k) = \mu(b)^k$.

Let $I_\mathfrak{g}$ be the centralizer of $\mathfrak{k}$ in $\mathfrak{B}$ and $I_\mathfrak{m}$ the centralizer of $\mathfrak{m}_\mathfrak{k}$ in $\mathfrak{M}$.

COROLLARY. *If $b \in I_\mathfrak{g}$ then $\mu_0(b) \in I_\mathfrak{m}$.*

This is obvious from Lemma 10.

---

[8] See [2(d), p. 118] for the definition of $\mu(b)^k$.

For any $b \in \mathfrak{B}$, define the differential operator $\delta'(b)$ on[2] $A_\mathfrak{p}'$ as in [2(i), §5]. It is obvious that $A_\mathfrak{p}' \subset M'$.

LEMMA 11. *Let $b$ be an element in $\mathfrak{B}$ and select $g_i \in \mathfrak{R}_0, \mu_i \in \mathfrak{M}_\mathfrak{p}$ $(1 \leq i \leq r)$ such that*

$$b - \mu_0(b) - \sum_{1 \leq i \leq r} g_i(m)\mu_i \in \mathfrak{B}\mathfrak{k} + (\mathfrak{Q}_\mathfrak{k}')^{m-1}\mathfrak{M}_\mathfrak{p}$$

*for all $m \in M'$. Then*

$$\delta_h'(b) = \delta_h'(\mu_0(b)) + \sum_{1 \leq i \leq r} g_i(h)\delta_h'(\mu_i)$$

*for $h \in A_\mathfrak{p}'$.*

Clearly $\mathfrak{Q}_\mathfrak{k}' \subset \mathfrak{k}\mathfrak{B}$ and therefore

$$b - \mu_0(b) - \sum_{1 \leq i \leq r} g_i(h)\mu_i \in \mathfrak{B}\mathfrak{k} + \mathfrak{k}^{h-1}\mathfrak{B} \qquad (h \in A_\mathfrak{p}').$$

Also $a - \delta_h'(a) \in \mathfrak{B}\mathfrak{k} + \mathfrak{k}^{h-1}\mathfrak{B}$ for any $a \in \mathfrak{B}$. Therefore[2]

$$\delta_h'(b) - \delta_h'(\mu_0(b)) - \sum_{1 \leq i \leq r} g_i(h)\delta_h'(\mu_i) \in \{\mathfrak{B}\mathfrak{k} + \mathfrak{k}^{h-1}\mathfrak{B}\} \cap \mathfrak{H}_\mathfrak{p} = \{0\}$$

from Corollary 2 to Lemma 21 of [2(i)].

**4. The connection between $\mu_0$ and $\gamma$.** Define the homomorphism $\gamma$ of $I_\mathfrak{g}$ onto $I(\mathfrak{h}_\mathfrak{p})$ as in Theorem 1 of [2(i)] and consider the subgroup $W'$ consisting of those elements[2] $s \in W$ which leave $H_0$ fixed. Now $\mathfrak{m}_0$ is reductive in $\mathfrak{g}_0$ and $\mathfrak{h}_{\mathfrak{p}_0}$ is a maximal abelian subspace of $\mathfrak{m}_{\mathfrak{p}_0}$. Therefore since $\mathfrak{m}_0$ is the centralizer of $H_0$ in $\mathfrak{g}_0$, it is clear that $W'$ can also be considered as the little Weyl group[9] of $\mathfrak{m}_0$ with respect to $\mathfrak{h}_{\mathfrak{p}_0}$. Let $J'$ be the ring of those elements in $\mathfrak{H}_\mathfrak{p}$ which are invariant under $W'$. Then corresponding to Theorem 1 of [2(i)], we have a homomorphism $\gamma_\mathfrak{m}$ of $I_\mathfrak{m}$ onto $J'$. Define the linear function $\sigma$ on $\mathfrak{h}_\mathfrak{p}$ by $\sigma(H) = \frac{1}{2} \sum_{\beta \in P'} \beta(H)$ where $P'$ is the subset of those roots in $P$ which vanish at $H_0$.

LEMMA 12. $\gamma(q) = e^{\rho - \sigma}\gamma_\mathfrak{m}(\mu_0(q)) \circ e^{-(\rho - \sigma)}$ *for $q \in I_\mathfrak{g}$.*

Define $\delta(b)$ $(b \in \mathfrak{B})$ as in Lemma 26 of [2(i)]. Then it follows from Lemma 11 that $\delta_h(q) = \delta_h(\mu_0(q)) + \delta_h(\mu_h(q))$ for $h \in A_\mathfrak{p}'$. On the other hand

$$\delta(\mu_0(q)) = e^\rho \delta'(\mu_0(q)) \circ e^{-\rho} = e^{\rho - \sigma}\{e^\sigma \delta'(\mu_0(q)) \circ e^{-\sigma}\} \circ e^{-\rho + \sigma}.$$

Now put $h_t = \exp tH$ $(t \in R)$ where $H$ is some fixed element in $\mathfrak{h}_{\mathfrak{p}_0}^+$. Then

---

[9] Here we are extending the notion of the little Weyl group [2(i), §3] to reductive Lie algebras in the obvious way (see [2(d), p. 118]).

$h_t \in A_\mathfrak{p}'$ if $t > 0$. Hence if $d$ is the degree of $q$ it follows from Lemma 26 of [2(i)] by taking limits in the finite-dimensional space $\mathfrak{B}_d$, that

$$\operatorname*{Lim}_{t \to +\infty} \delta_{h_t}(q) = \gamma(q), \qquad \operatorname*{Lim}_{t \to +\infty} \delta_{h_t}(\mu_{h_t}(q)) = 0.$$

(Here we have to make use of Lemma 7 and the definition of $\mu(q)$.) Therefore $\gamma(q) = \operatorname*{Lim}_{t \to +\infty} \delta_{h_t}(\mu_0(q))$. On the other hand we can apply Corollary 2 to Lemma 21 of [2(i)] to $\mathfrak{m}$ instead of $\mathfrak{g}$. Hence for any $b \in \mathfrak{M}$ and $h \in A_\mathfrak{p}'$, there exists a unique element $u$ in $\mathfrak{H}_\mathfrak{p}$ such that $b - u \in (\mathfrak{m}_\mathfrak{f})^{h-1}\mathfrak{M} + \mathfrak{M}\mathfrak{m}_\mathfrak{f}$. But then it follows from the definition of $\delta'(b)$ that $u = \delta_h'(b)$. Now put $\delta''(b) = e^\sigma \delta'(b) \circ e^{-\sigma}$ and apply Lemma 26 of [2(i)] to $\mathfrak{m}$. Then it follows that $\operatorname*{Lim}_{t \to +\infty} \delta_{h_t}''(b) = \gamma_\mathfrak{m}(b)$ provided $b \in I_\mathfrak{m}$. By the Corollary of Lemma 10, this holds in particular if $b = \mu_0(q)$. But since

$$\delta(\mu_0(q)) = e^{\rho-\sigma}\delta''(\mu_0(q)) \circ e^{-\rho+\sigma},$$

it is obvious that

$$\gamma(q) = \operatorname*{Lim}_{t \to +\infty} \delta_{h_t}(\mu_0(q)) = e^{\rho-\sigma}\gamma_\mathfrak{m}(\mu_0(q)) \circ e^{-\rho+\sigma}.$$

This proves the lemma.

COROLLARY. *Suppose $q_1$, $q_2$ are two elements in $I_\mathfrak{g}$. Then*

$$\mu_0(q_1)\mu_0(q_2) \equiv \mu_0(q_2)\mu_0(q_1) \equiv \mu_0(q_1 q_2) \bmod \mathfrak{M}\mathfrak{m}_\mathfrak{f}.$$

Since $\gamma(q_1 q_2) = \gamma(q_1)\gamma(q_2)$, it follows from Lemma 12 that $\gamma_\mathfrak{m}$ maps $\mu_0(q_1)\mu_0(q_2)$, $\mu_0(q_2)\mu_0(q_1)$ and $\mu_0(q_1 q_2)$ into the same element $\gamma(q_1)\gamma(q_2)$. The required result now follows from the fact that the kernel of $\gamma_\mathfrak{m}$ is contained in $\mathfrak{M}\mathfrak{m}_\mathfrak{f}$ (see [2(i), Theorem 1]).

**5. Some consequences of Lemma 12.** For any linear function $\lambda$ on $\mathfrak{h}_\mathfrak{p}$, let $H_\lambda$ denote the corresponding element in $\mathfrak{h}_\mathfrak{p}$ so that $\langle H, H_\lambda \rangle = \lambda(H)$ for all $H \in \mathfrak{h}_\mathfrak{p}$. Also define $I$ and $I_0$ as in the beginning of Section 2.

LEMMA 13. *$\rho - \sigma$ is invariant under $W'$ and $H_{\rho-\sigma} \in I_0$.*

Let $(\operatorname{ad} H)_{\mathfrak{q}_+}$ denote the restriction of $\operatorname{ad} H$ on $\mathfrak{q}_+$ for $H \in \mathfrak{h}_\mathfrak{p}$. It is obvious that $\operatorname{sp}(\operatorname{ad} H)_{\mathfrak{q}_+} = \rho(H) - \sigma(H)$. Since $W'$ is the little Weyl group of $\mathfrak{m}$, we can, for any given $s \in W'$, choose $k \in M_K$ such that $\operatorname{Ad}(k)H = sH$. Since $\mathfrak{q}_+$ is invariant under $k$, it follows that $\operatorname{sp}(\operatorname{ad} H)_{\mathfrak{q}_+} = \operatorname{sp}(\operatorname{ad}(sH))_{\mathfrak{q}_+}$ and this shows that $\rho - \sigma$ is invariant under $W'$. Now in order to prove that $H_{\rho-\sigma} \in I_0$, it is obviously sufficient to verify that $\beta(H_{\rho-\sigma}) = 0$ for any root $\beta \in P$ for which $\beta(H_0) = 0$. Let $\alpha$ denote the restriction of $\beta$ on $\mathfrak{h}_\mathfrak{p}$. We may assume that $\alpha \neq 0$. Consider the corresponding reflexion $s_\alpha \in W$ (see $[2(\mathrm{i}), \S 3]$). It is obvious that $s_\alpha$ lies in $W'$ and therefore it must leave $H_{\rho-\sigma}$ fixed. Hence $\beta(H_{\rho-\sigma}) = \alpha(H_{\rho-\sigma}) = 0$.

Let $\omega$ denote the Casimir operator of $\mathfrak{g}$ (see $[2(\mathrm{i}), \S 7]$). We wish to compute $\mu_0(\omega)$.

LEMMA 14. *Let* $X_1, \cdots, X_q$ *be an orthonormal base for* $\mathfrak{m}_{\mathfrak{p}_0}$ *over* $R$. *Then* $\mu_0(\omega) = X_1^2 + \cdots + X_q^2 + 2H_{\rho-\sigma}$.

Put $\eta_0 = X_1^2 + \cdots + X_q^2$. It is easy to check that $\eta_0 \in I_\mathfrak{m}$ and if we apply Corollary 2 of Lemma 27 of $[2(\mathrm{i})]$ to $\mathfrak{m}$ (in place of $\mathfrak{g}$), it follows that $\gamma_\mathfrak{m}(\eta_0) = \bar{\omega} - \langle \sigma, \sigma \rangle$. However it is obvious that

$$e^{-\rho+\sigma}\bar{\omega} \circ e^{\rho-\sigma} = \bar{\omega} + 2H_{\rho-\sigma} + \langle \rho - \sigma, \rho - \sigma \rangle.$$

Therefore if $\eta = \eta_0 + 2H_{\rho-\sigma} - 2\langle \rho - \sigma, \sigma \rangle$, it follows again by the same corollary that

$$e^{\rho-\sigma}\gamma_\mathfrak{m}(\eta) \circ e^{-\rho+\sigma} = \bar{\omega} - \langle \rho, \rho \rangle = \gamma(\omega).$$

But in view of Lemma 12, this implies that $\gamma_\mathfrak{m}(\eta - \mu_0(\omega)) = 0$. However $\eta - \mu_0(\omega)$ lies in $\mathfrak{M}_\mathfrak{p} \cap I_\mathfrak{m}$. Therefore it follows from $[2(\mathrm{i}), \text{Lemma } 19]$ that $\eta = \mu_0(\omega)$. But we have seen above that $\beta(H_{\rho-\sigma}) = 0$ if $\beta(H_0) = 0$ $(\beta \in P)$. Therefore $\langle \rho - \sigma, \sigma \rangle = 0$ and this proves our assertion.

Put $\gamma_\mathfrak{m}'(\nu) = e^{\rho-\sigma}\gamma_\mathfrak{m}(\nu) \circ e^{-\rho+\sigma}$ $(\nu \in I_\mathfrak{m})$. Since $\rho - \sigma$ is invariant under $W'$, it is clear that $J' = \gamma_\mathfrak{m}(I_\mathfrak{m}) = \gamma_\mathfrak{m}'(I_\mathfrak{m})$. Put $r = [W:W']$ and choose elements $\nu_1 = 1, \nu_2, \cdots, \nu_r$ in $I_\mathfrak{m}$ such that $\gamma_\mathfrak{m}'(\nu_i)$ $(1 \leq i \leq r)$ are homogeneous and[2] $J' = \sum_{1 \leq i \leq r} J\gamma_\mathfrak{m}'(\nu_i)$. From Lemma 8 of $[2(\mathrm{i})]$ this is possible.

LEMMA 15. $I_\mathfrak{m} \subset \sum_{1 \leq i \leq r} \nu_i\mu_0(I_\mathfrak{g}) + \mathfrak{M}\mathfrak{m}_\mathfrak{k}$.

Let $\nu$ be a given element in $I_\mathfrak{m}$. Then $\gamma_\mathfrak{m}'(\nu) \in J'$ and therefore $\gamma_\mathfrak{m}'(\nu) = \sum_{1 \leq i \leq r} \gamma(q_i)\gamma_\mathfrak{m}'(\nu_i)$ for suitable $q_i \in I_\mathfrak{g}$ since $\gamma(I_\mathfrak{g}) = J$. But $\gamma(q_i) = \gamma_\mathfrak{m}'(\mu_0(q_i))$ from Lemma 12 and therefore $\gamma_\mathfrak{m}'(\nu - \sum_{1 \leq i \leq r} \nu_i\mu_0(q_i)) = 0$. However the kernels of $\gamma_\mathfrak{m}$ and $\gamma_\mathfrak{m}'$ are the same and so by applying Theorem 1 of $[2(\mathrm{i})]$ to $\mathfrak{m}$, we conclude that $\nu - \sum_i \nu_i\mu_0(q_i) \in \mathfrak{M}\mathfrak{m}_\mathfrak{k}$. This proves the lemma.

COROLLARY. *Suppose* $\sum_{1 \leq i \leq r} \nu_i\mu_0(q_i) \in \mathfrak{M}\mathfrak{m}_\mathfrak{k}$ *for some* $q_i \in I_\mathfrak{g}$. *Then* $q_i \in \mathfrak{Bf}$ $1 \leq i \leq r$.

Put $\nu = \sum_i \nu_i\mu_0(q_i)$. Then $\nu \in I_\mathfrak{m} \cap (\mathfrak{M}\mathfrak{m}_\mathfrak{k})$ and therefore $\gamma_\mathfrak{m}(\nu) = 0$.

This implies that $\gamma_{\mathfrak{m}}'(\nu) = \sum_i \gamma_{\mathfrak{m}}'(\nu_i)\gamma(q_i) = 0$. But $\gamma_{\mathfrak{m}}'(\nu_i)$ $1 \leq i \leq r$ are linearly independent over $C(J)$ (see [2(i), Lemma 8]) and therefore $\gamma(q_i) = 0$ $1 \leq i \leq r$. Our assertion now follows from Theorem 1 of [2(i)].

The following lemma contains the main result of this section.

LEMMA 16. *Let $\nu$ be an element in $I_{\mathfrak{m}}$ and suppose $\nu - \sum\limits_{1 \leq i \leq r} \nu_i \mu_0(q_i)$*
*$\in \mathfrak{M}\mathfrak{m}_{\mathfrak{l}}$ $(q_i \in I_{\mathfrak{g}})$. Then we can select a finite number of elements $g_j \in \mathfrak{R}$ and $\tau_j \in \mathfrak{M}$ $(1 \leq j \leq s)$ with the following property. If $\phi$ is any spherical function on $G$ of class $C^\infty$,*

$$\phi(m\,;\nu) = \sum_{1 \leq i \leq r} \phi(m\,;\nu_i q_i) + \sum_{1 \leq j \leq s} g_j(m)\phi(m\,;\tau_j)$$

*for all $m \in M'$.*

It is obvious that $\phi(m\,;v) = 0$ for $v \in \mathfrak{M}\mathfrak{m}_{\mathfrak{l}}$. Therefore $\phi(m\,;\nu)$ $= \sum\limits_i \phi(m\,;\nu_i \mu_0(q_i))$. On the other hand since

$$q_i - \mu_0(q_i) - \mu_{\mathfrak{m}}(q_i) \in \mathfrak{B}\mathfrak{k} + (\mathfrak{Q}_{\mathfrak{l}}')^{m-1}\mathfrak{M}_{\mathfrak{p}},$$

it is clear that

$$\phi(m\,;\mu_0(q_i)) = \phi(m\,;q_i) - \phi(m\,;\mu(q_i)) \qquad (m \in M').$$

Therefore

$$\phi(m\,;\nu_i\mu_0(q_i)) = \phi(m\,;\nu_i q_i) - \phi(m\,;\nu_i \circ \mu(q_i))$$

and

$$\phi(m\,;\nu) = \sum_{1 \leq i \leq r} \phi(m\,;\nu_i q_i) - \sum_{1 \leq i \leq r} \phi(m\,;\nu_i \circ \mu(q_i)).$$

But since $\mu(q_i) \in \mathfrak{R}_0\mathfrak{M}_{\mathfrak{p}}$, it is obvious that $\sum\limits_i \nu_i \circ \mu(q_i) = -\sum\limits_{1 \leq j \leq s} g_j\tau_j$ for suitable $g_j \in \mathfrak{R}$ and $\tau_j \in \mathfrak{M}$. This proves our assertion.

Let $l_0 = \dim I$ and let $I_0{}^+$ denote the set of those elements $H \in I_0$ for which $\alpha(H) > 0$ for every $\alpha \in Q$ (see Section 2 for the definition of $Q$). Then $H_0 \in I_0{}^+$. Select a base $H_p$ $(1 \leq p \leq l_0)$ for $I_0$ over $R$ and choose $q_{ijp} \in I_{\mathfrak{g}}$ $(1 \leq i, j \leq r, 1 \leq p \leq l_0)$ such that

$$H_p\nu_j \equiv \sum_{1 \leq i \leq r} \nu_i\mu_0(q_{ijp}) \bmod \mathfrak{M}\mathfrak{m}_{\mathfrak{l}} \qquad (1 \leq j \leq r, 1 \leq p \leq l_0).$$

Then $q_{ijp}$ are unique mod $(I_{\mathfrak{g}} \cap \mathfrak{B}\mathfrak{k})$ from the Corollary to Lemma 15. Moreover

$$H_s H_p\nu_j \equiv \sum_{1 \leq i \leq r} H_s\nu_i\mu_0(q_{ijp}) \equiv \sum_{1 \leq i,k \leq r} \nu_k\mu_0(q_{kis})\mu_0(q_{ijp}) \bmod \mathfrak{M}\mathfrak{m}_{\mathfrak{l}},$$

since $\mu_0(q_{ijp}) \in I_{\mathfrak{m}}$. But $H_s H_p = H_p H_s$ and so we conclude from the corollaries of Lemmas 12 and 15 that

$$(1) \qquad \sum_{1 \leq i \leq r} (q_{kis}q_{ijp} - q_{kip}q_{ijs}) \equiv 0 \bmod \mathfrak{B}\mathfrak{k} \qquad (1 \leq k, j \leq r; 1 \leq p, s \leq l_0).$$

Moreover from Lemma 16, we can select a positive integer $N$ and elements $g_{pja} \in \mathfrak{R}$, $\tau_{pja} \in \mathfrak{M}$ $(1 \leqq p \leqq l_0, 1 \leqq j \leqq r, 1 \leqq a \leqq N)$ such that

$$(2) \qquad \phi(m; H_p \nu_j) = \sum_{1 \leqq i \leqq r} \phi(m; \nu_i q_{ijp})$$

$$+ \sum_{1 \leqq a \leqq N} g_{pja}(m) \phi(m; \tau_{pja}) \qquad (m \in M')$$

for any spherical function $\phi$ on $G$ of class $C^\infty$ $(1 \leqq j \leqq r, 1 \leqq p \leqq l_0)$.

**6. Some rough estimates.** Let $\mathfrak{F}$ be the space of all linear functions on $\mathfrak{h}_{\mathfrak{p}}$. For any $\lambda \in \mathfrak{F}$, put

$$\phi_\lambda(x) = \phi(\lambda:x) = \int_K \exp\{(-1)^{\frac{1}{2}}\lambda(H(xk)) - \rho(H(xk))\}dk$$

$$(x \in G)$$

in the notation of Lemma 3 of $[2(\mathrm{i})]$. For $m \in M$ and $H \in \mathfrak{h}_{\mathfrak{p}_0}$ we write

$$\Phi_j(\lambda:m:H) = e^{\rho(H)}\phi(\lambda:m \exp H; \nu_j) \qquad (1 \leqq j \leqq r)$$

and

$$\Psi_{pj}(\lambda:m:H) = e^{\rho(H)} \sum_{1 \leqq a \leqq N} g_{pja}(m \exp H)\phi(\lambda:m \exp H; \tau_{pja})$$

$$(1 \leqq j \leqq r, 1 \leqq p \leqq l_0)$$

where $g_{pja}$ and $\tau_{jpa}$ have the same meaning as in equation (2) of Section 5 and $m, H$ are so chosen that $m \exp H \in M'$. Now let $\Phi(\lambda:m:H)$ and $\Psi_p(\lambda:m:H)$ denote the corresponding one-column matrices and define the $r \times r$ matrices $\Xi_p(\lambda)$ $(1 \leqq p \leqq l_0)$ by $\Xi_{pij}(\lambda) = \gamma(q_{jip}:(-1)^{\frac{1}{2}}\lambda)$ $(1 \leqq i, j \leqq r)$. Then it follows from equation (1) that $\Xi_p(\lambda)$ $(1 \leqq p \leqq l_0)$ commute with each other. For any $H' \in I$, put

$$\Psi_{H'}(\lambda:m:H) = \sum_p c_p \Psi_p(\lambda:m:H), \qquad \Xi(\lambda:H') = \sum_p c_p \Xi_p(\lambda) + \rho(H')I$$

where $H' = \sum_p c_p H_p$ $(c_p \in C)$ and $I$ is the unit matrix. If $m \in M_+$ and $H \in I_0^+$, it is obvious that $m \exp tH \in M_+$ for all $t \geqq 0$. Therefore from Lemma 18 of $[2(\mathrm{i})]$ and equation (2), we find that

$$d\Phi(\lambda:m:tH)/dt = \Xi(\lambda:H)\Phi(\lambda:m:tH) + \Psi_H(\lambda:m:tH) \qquad (t \geqq 0).$$

The following lemma is now obvious.

LEMMA 17.

$$d\Phi(\lambda:m;\eta:tH)/dt = \Xi(\lambda:H)\Phi(\lambda:m;\eta:tH) + \Psi_H(\lambda:m;\eta:tH)$$

for $\lambda \in \mathfrak{F}$, $m \in M_+$, $H \in I_0^+$, $\eta \in \mathfrak{M}$ and $t \geqq 0$.

Now put $\Phi^0(\lambda:m:H) = \exp(-\Xi(\lambda:H))\Phi(\lambda:m:H)$ and

$$\Psi_{H'}{}^0(\lambda:m:H) = \exp(-\Xi(\lambda:H))\Psi_{H'}(\lambda:m:H) \qquad (H' \in \mathfrak{I}).$$

COROLLARY. $d\Phi^0(\lambda:m;\eta:tH)/dt = \Psi_H{}^0(\lambda:m;\eta:tH)$ *under the conditions of Lemma* 17.

This is an immediate consequence of the above lemma.

For any $p \times q$ matrix $u = (u_{ij})$ with complex coefficients, put $\|u\| = \{\sum_{i,j} |u_{ij}|^2\}^{\frac{1}{2}}$. We now intend to obtain an estimate for $\|\Psi_H{}^0(\lambda:m;\eta:tH)\|$. It follows from Lemma 46 of [2(i)] and Lemma 7 that for any fixed $\eta$, we can choose an integer $k \geq 0$ and a positive number $a'$ such that[2]

$$\|\Psi_p(\lambda:m;\eta:H)\|$$

$$\leq a'(1 + \|\lambda\|)^k |\operatorname{Ad}(h^{-1}m^{-1})|_+^2 \{1 - |\operatorname{Ad}(h^{-1}m^{-1})|_+^2\}^{-(k+1)} e^{\rho(H)} \phi((-1)^{\frac{1}{2}}\lambda_I:mh)$$

for $m \in M_+$, $H \in \mathfrak{I}_0{}^+$, $\lambda \in \mathfrak{F}$ and $1 \leq p \leq l_0$. (Here $h = \exp H$.) Now put $\beta(H) = \min_{\alpha \in Q} \alpha(H)$ and $\Lambda_I{}^+ = \exp \mathfrak{I}_0{}^+$. Then

$$|\operatorname{Ad}(h^{-1}m^{-1})|_+ \leq |\operatorname{Ad}(m^{-1})|_+ e^{-\beta(\log h)} \leq e^{-\beta(\log h)}$$

for $m \in M_+$ and $h \in A_I{}^+$. Hence we get the following result immediately.

LEMMA 18. *For any* $\eta \in \mathfrak{M}$, *we can choose an integer* $k \geq 0$ *and a positive number* $a$ *such that*

$$\|\Psi_H(\lambda:mh;\eta:tH)\|$$

$$\leq a(1 + \|\lambda\|)^k \|H\| \exp\{t\rho(H) - t\beta(H) - \beta(\log h)\} \phi((-1)^{\frac{1}{2}}\lambda_I:mh \exp tH)$$

*for* $\lambda \in \mathfrak{F}$, $m \in M_+$, $h \in A_I{}^+$, $H \in \mathfrak{I}_0{}^+$ *and* $t \in R$ *provided* $t\beta(H) + \beta(\log h) \geq 1$.

On the other hand

$$\|\Psi_H{}^0(\lambda:m;\eta:tH)\| \leq \|\exp(-t\Xi(\lambda:H))\| \|\Psi_H(\lambda:m;\eta:tH)\|.$$

So now we have to estimate $\|\exp(-t\Xi(\lambda:H))\|$. Put $\lambda^* = (-1)^{\frac{1}{2}}\lambda$ for $\lambda \in \mathfrak{F}$.

LEMMA 19. *Select* $r$ *elements* $s_1, s_2, \cdots, s_r$ *in* $W$ *such that* $W = \bigcup_{1 \leq i \leq r} W's_i$. *Then if* $T$ *is an indeterminate and* $I$ *the* $r \times r$ *unit matrix,*

$$\det(TI - \Xi(\lambda:H)) = \prod_{1 \leq i \leq r} (T - \lambda^*(s_iH))$$

*for* $\lambda \in \mathfrak{F}$ *and* $H \in \mathfrak{I}$.

Fix $\lambda$ and $H$ and select $q_{ij} \in I_{\mathfrak{G}}$ $(1 \leq i, j \leq r)$ such that $Hv_j \equiv \sum_i v_i \mu_0(q_{ij})$

mod $\mathfrak{M}\mathfrak{m}_{\mathfrak{k}}$. Since I is contained in the center of $\mathfrak{m}$, it follows by applying the homomorphism $\gamma_{\mathfrak{m}}$ to this relation that $Hv_j{}' = \sum_i v_i{}' \gamma_{\mathfrak{m}}(\mu_0(q_{ij}))$ where $v_i{}' = \gamma_{\mathfrak{m}}(v_i)$. Moreover $e^{\rho-\sigma} H \circ e^{-\rho+\sigma} = H - \rho(H)$ since $\sigma$ is zero on I. So if $v_i = e^{\rho-\sigma} v_i{}' \circ e^{-\rho+\sigma}$, it follows from Lemma 12 that $Hv_j = \rho(H)v_j + \sum_i v_i \gamma(q_{ij})$. Now if we recall the definition of the matrix $\Xi(\lambda:H)$ and denote its coefficients by $\xi_{ij}$, it becomes clear that

$$Hv_j \equiv \sum_i \xi_{ji} v_i \bmod J' J_\lambda.$$

in the notation of Lemma 13 of [2(i)]. But it follows from our definition of $v_1, \cdots, v_r$ and this lemma that $v_1, \cdots, v_r$ are linearly independent $\bmod J' J_\lambda$. Hence our assertion is an immediate consequence of the corollary of Lemma 15 of [2(i)].

COROLLARY. *There exists a positive number $a$ and an integer $k \geqq 0$ such that*

$$\| \exp(t\Xi(\lambda:H)) \| \leqq a(1 + \| \lambda \|)^k (1 + |t|)^r (1 + \| H \|)^r$$
$$\times \exp\{2r |t| \max_{1 \leqq i \leqq r} | \lambda_I(s_i H)| \}$$

*for $H \in I_0$, $t \in R$ and $\lambda \in \mathfrak{F}$.*

It follows from the definition of $\Xi(\lambda:H)$ that we can choose a positive number $b$ and an integer $d \geqq 0$ such that

$$\| \Xi(\lambda:H) \| \leqq b(1 + \| \lambda \|)^d \| H \| \qquad\qquad (\lambda \in \mathfrak{F}, H \in I).$$

Moreover it follows from the above lemma and [2(i), Lemma 60] that

$$\| \exp(t\Xi(\lambda:H)) \|$$
$$\leqq r^{\frac{1}{2}} \exp(2r |t| \max_{1 \leqq i \leqq r} | \lambda_I(s_i H)|) \sum_{0 \leqq k \leqq r} |t|^k \| \Xi(\lambda:H) \|^k.$$

Therefore the statement of the corollary is now obvious.

For a positive number $\epsilon$, let $\mathfrak{F}_\epsilon$ denote the set of all $\lambda \in \mathfrak{F}$ such that $\| \lambda_I \| < \epsilon$.

LEMMA 20. *Let $V$ be a compact set in $I_0{}^+$. Then there exists a positive number $\epsilon$ with the following property. Let $\eta$ be an element in $\mathfrak{M}$. Then we can choose an integer $k \geqq 0$ and, for any compact set $\omega$ in $M_+$, positive numbers $a_1$, $a_2$ and $t_0$ such that*

$$\| \Psi_H{}^0(\lambda:mh;\eta:tH) \| \leqq a_1(1 + \| \lambda \|)^k t^k e^{-a_2 t}$$

*for $\lambda \in \mathfrak{F}_\epsilon$, $m \in \omega$, $h \in A_I{}^+$, $H \in V$ and $t \geqq t_0$.*

Our proof depends on Lemma 18. But first we need an estimate for $\phi((-1)^{\frac{1}{2}}\lambda_I : mh)$.

LEMMA 21. *Let $\epsilon'$ be a positive number $< 1$. Then we can select an integer $d \geqq 0$ and positive numbers $\epsilon$ and $a$ such that*

$$\phi((-1)^{\frac{1}{2}}\lambda_I : mh) \leqq a(1 + \| \log h \|)^d e^{-(1-\epsilon')\rho(\log h)}$$

*for $m \in \omega$, $h \in A_I^+$ and $\lambda \in \mathfrak{F}_\epsilon$. Moreover $d$ and $\epsilon$ may be so chosen that they are independent of $\omega$.*

From Theorem 3 and Lemma 49 of [2(i)], we can select $d$, $\epsilon$ and a positive number $a'$ such that

$$\phi((-1)^{\frac{1}{2}}\lambda_I : \exp H) \leqq a'(1 + \| H \|)^d e^{-(1-\epsilon')\rho(H)}$$

if $H \in \mathrm{Cl}(\mathfrak{h}_{\mathfrak{p}_0}{}^+)$ and $\lambda \in \mathfrak{F}_\epsilon$. On the other hand since $\phi_\mu$ is a spherical function for any $\mu \in \mathfrak{F}$, it is obvious that $\phi((-1)^{\frac{1}{2}}\lambda_I : \exp H) = \phi((-1)^{\frac{1}{2}}\lambda_I : \exp sH)$ $(s \in W, H \in \mathfrak{h}_{\mathfrak{p}_0})$. Moreover $\rho(H) \geqq \rho(sH)$ for $s \in W$ and $H \in \mathfrak{h}_{\mathfrak{p}_0}{}^+$. Therefore since $\mathfrak{h}_{\mathfrak{p}_0} = \bigcup_{s \in W} s(\mathrm{Cl}(\mathfrak{h}_{\mathfrak{p}_0}{}^+))$ (see [2(f), Lemma 37]), we conclude that the above inequality actually holds for all $H \in \mathfrak{h}_{\mathfrak{p}_0}$.

On the other hand $M = M_K A_{\mathfrak{p}} M_K$ (see [2(f), Lemmas 31 and 33]) and since $\omega$ is compact, we can select a positive number $c$ such that [10] $\| \log | m | \| \leqq c$ for $m \in \omega$. For a fixed $m \in \omega$, choose $h_1 \in A_{\mathfrak{p}}$ such that $m \in M_K h_1 M_K$. Then $\| \log h_1 \| = \| \log | m | \| \leqq c$ and

$$\phi((-1)^{\frac{1}{2}}\lambda_I : mh) = \phi((-1)^{\frac{1}{2}}\lambda_I : h_1 h)$$
$$\leqq a'(1 + \| \log(h_1 h) \|)^d \exp\{-(1-\epsilon')\rho(\log(h_1 h))\}$$

if $h \in A_I^+$ and $\lambda \in \mathfrak{F}_\epsilon$. On the other hand we can select a positive number $a''$ such that

$$(1 + \| H \|)^d e^{-(1-\epsilon')\rho(H)} \leqq a''$$

for all elements $H \in \mathfrak{h}_{\mathfrak{p}_0}$ with $\| H \| \leqq c$. Then if $a = a'a''$, it follows that

$$\phi((-1)^{\frac{1}{2}}\lambda_I : mh) \leqq a(1 + \| \log h \|)^d e^{-(1-\epsilon')\rho(\log h)}$$

for $m \in \omega$, $h \in A_I^+$ and $\lambda \in \mathfrak{F}_\epsilon$.

Now we come to the proof of Lemma 20. Since $V$ is compact, we can choose a positive number $a_2$ such that $\beta(H) \geqq 2a_2$ for $H \in V$. Put $t_0 = \max(1, a_2^{-1})$ and select $\epsilon$ and $\epsilon'$ as in Lemma 21. We can assume that $\epsilon$ and $\epsilon'$ are so small that $\epsilon'\rho(H) \leqq a_2/2$ and $2r | \lambda_I(sH) | \leqq a_2/2$ for $H \in V$,

---

[10] For any $x \in G$, $| x |$ denotes the unique element $p \in \exp \mathfrak{p}_0$ such that $xp^{-1} \in K$. Moreover $\log | x | = X$ where $| x | = \exp X$ $(X \in \mathfrak{p}_0)$.

$s \in W$ and $\lambda \in \mathfrak{F}_\epsilon$. Therefore it follows from Lemma 18 and the corollary of Lemma 19 that we can select an integer $k_1 \geqq 0$ and a positive number $a'$ such that

$$\| \Psi_H{}^0(\lambda : m ; \eta : tH) \|$$
$$\leqq a'(1 + \| \lambda \|)^{k_1} t^r \exp\{t\rho(H) - 3ta_2/2\} \phi((-1)^{\frac{1}{2}}\lambda_I : m \exp tH)$$

if $m \in M_+$, $H \in V$, $t \geqq t_0$ and $\lambda \in \mathfrak{F}_\epsilon$. On the other hand by Lemma 21,

$$\phi((-1)^{\frac{1}{2}}\lambda_I : mh \exp tH)$$
$$\leqq a(1 + t \| H \|)^d e^{-(1-\epsilon')t\rho(H)} (1 + \| \log h \|)^d e^{-(1-\epsilon')\rho(\log h)}$$

for $m \in \omega$, $h \in A_I^+$ and $t \geqq t_0$. Moreover it is obvious that there exists a positive number $c$ such that $\rho(H') \geqq c \| H' \|$ for $H' \in \mathrm{Cl}(\mathfrak{h}_{\mathfrak{p}_0}{}^+)$. Therefore since $(1 + t)^d e^{-(1-\epsilon')ct} \to 0$ as $t \to +\infty$, we can select a number $a'' > 0$ such that

$$\phi((-1)^{\frac{1}{2}}\lambda_I : mh \exp tH) \leqq a'' t^d e^{-(1-\epsilon')t\rho(H)}$$

for $m \in \omega$, $h \in A_I^+$, $t \geqq t_0$ and $\lambda \in \mathfrak{F}_\epsilon$. This shows that

$$\| \Psi_H{}^0(\lambda : mh ; \eta : tH) \| \leqq a'a''(1 + \| \lambda \|)^{k_1} t^{r+d} \exp\{\epsilon' t\rho(H) - 3ta_2/2\}$$

under the same conditions. Therefore if we put $a_1 = a'a''$, $k = \max(k_1, r + d)$ and recall that $\epsilon'\rho(H) \leqq a_2/2$ for $H \in V$, the statement of Lemma 20 follows immediately.

**7. The function $\Theta_H$.** Now fix an element $H$ in $I_0^+$. Then if $\epsilon$ is sufficiently small, it follows from Lemma 20 and the corollary to Lemma 17 that for any $\eta \in \mathfrak{M}$, the integral

$$\int_0^\infty \| d\Phi^0(\lambda : m ; \eta : tH)/dt \| \, dt$$

converges uniformly as $m$ and $\lambda$ vary within compact subsets of $M_+$ and $\mathfrak{F}_\epsilon$ respectively. This shows that $\mathrm{Lim}_{t \to +\infty} \Phi^0(\lambda : m : tH)$ exists and for a fixed $m$, it represents a holomorphic (matrix-) function of $\lambda$ on $\mathfrak{F}_\epsilon$. We shall denote this limit by $\Theta_H(\lambda : m)$. Moreover in view of the uniform convergence of the above integral, the following facts are obvious. For a fixed $\lambda$ in $\mathfrak{F}_\epsilon$, $\Theta_H(\lambda : m)$ is a $C^\infty$ function of $m$ on $M_+$ and $\Theta_H(\lambda : m ; \eta) = \mathrm{Lim}_{t \to +\infty} \Phi^0(\lambda : m ; \eta : tH)$ for $\eta \in \mathfrak{M}$. Now suppose $\omega$ is a compact subset of $M$. Then we can obviously select a positive number $t_0$ such that $m \exp t_0 H \in M_+$ for $m \in \omega$. Also it follows from the definition of $\Phi^0$ that

$$\Phi^0(\lambda : m : tH + t_0 H) = e^{t_0\rho(H)} \exp\{-t_0 \Xi(\lambda : H)\} \Phi^0(\lambda : m \exp t_0 H : tH).$$

Therefore if $\lambda \in \mathfrak{F}_\epsilon$ and $m \in \omega$,

$$\operatorname*{Lim}_{t \to +\infty} \Phi^0(\lambda:m:tH) = e^{t_0\rho(H)} \exp\{- t_0 \Xi(\lambda:H)\}\Theta_H(\lambda:m \exp t_0 H).$$

Hence we can extend $\Theta_H$ on $\mathfrak{F}_\epsilon \times M$ by setting

$$\Theta_H(\lambda:m) = \operatorname*{Lim}_{t \to +\infty} \Phi^0(\lambda:m:tH) \qquad (\lambda \in \mathfrak{F}_\epsilon, m \in M).$$

The following result is now obvious.

LEMMA 22. $\Theta_H(\lambda:m \exp tH) = e^{-t\rho(H)} \exp(t\Xi(\lambda:H))\Theta_H(\lambda:m)$ for $t \in R$, $m \in M$ and $\lambda \in \mathfrak{F}_\epsilon$.

This lemma shows that for a fixed $\lambda$ in $\mathfrak{F}_\epsilon$, $\Theta_H$ is a $C^\infty$ function on $M$ and for a fixed $m$ in $M$, it is a holomorphic function on $\mathfrak{F}_\epsilon$.

LEMMA 23. *For any* $\eta \in \mathfrak{M}$ *we can select an integer* $k$ *such that* $(1 + \|\lambda\|)^{-k} \|\Theta_H(\lambda:m;\eta)\|$ *remains bounded for any fixed* $m \in M$ *as* $\lambda$ *varies in* $\mathfrak{F}_\epsilon$.

Select $t_0 > 0$ such that $m_0 = m \exp t_0 H \in M_+$. Then

$$\Theta_H(\lambda:m;\eta) = e^{t_0\rho(H)} \exp\{- t_0 \Xi(\lambda: H)\}\Theta_H(\lambda:m_0;\eta).$$

Hence in view of the corollary to Lemma 19, it is sufficient to consider the case when $m \in M_+$. But then

$$\Theta_H(\lambda:m;\eta) = \Phi^0(\lambda:m;\eta:TH) + \int_T^\infty \Psi_H^0(\lambda:m;\eta:tH)\,dt$$

for $T \geqq 0$. Therefore

$$\|\Theta_H(\lambda:m;\eta)\| \leqq \|\exp(-T\Xi(\lambda:H))\| \, \|\Phi(\lambda:m;\eta:TH)\|$$
$$+ \int_T^\infty \|\Psi_H^0(\lambda:m;\eta:tH)\|\,dt.$$

On the other hand it follows from Lemma 46 of [2(i)] that

$$\|\Phi(\lambda:m;\eta:TH)\| \leqq a_1(1 + \|\lambda\|)^{k_1}\phi((-1)^{\frac{1}{2}}\lambda_I:m \exp TH)e^{T\rho(H)}$$

$$(\lambda \in \mathfrak{F}_\epsilon, T \geqq 0)$$

where $a_1$ is a positive number and $k_1$ a nonnegative integer and the latter depends only on $\eta$. Hence if we apply Lemmas 20 and 21 and take into account the corollary of Lemma 19, our assertion becomes obvious.

Choose $\epsilon' > 0$ such that $\epsilon'\rho(H) < \beta(H)/3$ and $\epsilon' < 1$. We assume that $\epsilon$ is so small that the pair $(\epsilon, \epsilon')$ fulfills the condition of Lemma 21 and $6r\epsilon \|H\| < \beta(H)$.

LEMMA 24. *Let $q$ be an element in $I_{\mathfrak{g}}$. Then*

$$\Theta_H(\lambda:m;\mu_0(q)) = \gamma(q:(-1)^{\frac{1}{2}}\lambda)\Theta_H(\lambda:m)$$

*for $\lambda \in \mathfrak{F}_\epsilon$ and $m \in M$.*

In view of Lemma 22, we may assume that $m \in M_+$. Now

$$\phi(\lambda:m;\nu_i\mu_0(q)) = \phi(\lambda:m;\nu_iq) - \phi(\lambda:m;\nu_i\circ\mu(q)) \qquad (1 \leqq i \leqq r)$$

in the notation of Section 5. But $\phi(\lambda:m;\nu_iq) = \gamma(q:(-1)^{\frac{1}{2}}\lambda)\phi(\lambda:m;\nu_i)$ and $\nu_i\mu_0(q) - \mu_0(q)\nu_i \in \mathfrak{M}m_t$ by Theorem 1 of [2(i)] since $\mu_0(q) \in I_{\mathfrak{m}}$. Therefore

$$\phi(\lambda:m;\mu_0(q)\nu_i) = \gamma(q:(-1)^{\frac{1}{2}}\lambda)\phi(\lambda:m;\nu_i) - \phi(\lambda:m;\nu_i\circ\mu(q)).$$

Now fix $m$ and $\lambda$. Then it follows from Lemmas 7 and 21 and [2(i), Lemma 46] that

$$|\phi(\lambda:m\exp tH;\nu_i\circ\mu(q))| \leqq a(1+t)^d \exp\{-t\beta(H) - (1-\epsilon')t\rho(H)\}$$

$$(t \geqq 0)$$

where $a$ is a positive constant. Hence we can select a constant $a_1$ such that

$$\|\Phi(\lambda:m;\mu_0(q):tH) - \gamma(q:(-1)^{\frac{1}{2}}\lambda)\Phi(\lambda:m:tH)\|$$

$$\leqq a_1(1+t)^d e^{-2t\beta(H)/3} \qquad (t \geqq 0).$$

Therefore

$$\|\Phi^0(\lambda:m;\mu_0(q):tH) - \gamma(q:(-1)^{\frac{1}{2}}\lambda)\Phi^0(\lambda:m:tH)\|$$

$$\leqq a_1(1+t)^d e^{-2t\beta(H)/3} \|\exp(-t\Xi(\lambda:H))\|.$$

In view of the inequality $6r\epsilon\|H\| \leqq \beta(H)$, it follows from the corollary of Lemma 19 that

$$\lim_{t\to+\infty} \|\Phi^0(\lambda:m;\mu_0(q):tH) - \gamma(q:(-1)^{\frac{1}{2}}\lambda)\Phi^0(\lambda:m:tH)\| = 0.$$

This proves that $\Theta_H(\lambda:m;\mu_0(q)) = \gamma(q:(-1)^{\frac{1}{2}}\lambda)\Theta_H(\lambda:m)$.

COROLLARY. *For a fixed $\lambda$ in $\mathfrak{F}_\epsilon$, $\Theta_H(\lambda:m)$ is an analytic function of $m$ on $M$.*

Let $Y_1, \cdots, Y_s$ be an orthonormal base for $\mathfrak{m}_{1_0}$ over $R$. Then it is easy to verify that $Y_1{}^2 + \cdots + Y_s{}^2 \in I_{\mathfrak{m}}$ and if $\omega$ is the Casimir operator of $\mathfrak{g}$, it follows from Lemma 14 that $\eta = \mu_0(\omega) + Y_1{}^2 + \cdots + Y_s{}^2$ is an elliptic differential operator (see Garding [1] for the definition of ellipticity) on $M$. On the other hand it is obvious from the definition of $\Theta_H$ that $\Theta_H(\lambda:k_1mk_2)$

$= \Theta_H(\lambda:m)$ for $k_1, k_2 \in M_K$. Therefore $\Theta_H(\lambda:m;\eta) = \Theta_H(\lambda:m;\mu_0(\omega))$ $= \gamma(\omega:(-1)^{\frac{1}{2}}\lambda)\Theta_H(\lambda:m)$. This shows that for a fixed $\lambda$, every coefficient of the matrix $\Theta_H$ is an eigenfunction of $\eta$. Therefore since $\eta$ is an analytic differential operator on $M$ of the elliptic type, we can conclude (see John [3]) that all these coefficients are also analytic in $m$.

We shall now try to determine $\Theta_H$ explicitly.

LEMMA 25. *Let $\theta'$ be an analytic function on $M$ such that $\theta'(k_1 m k_2)$ $= \theta'(m)$ for $k_1, k_2 \in M_K$ and $m \in M$ and suppose $\theta'$ is an eigenfunction of $\mu_0(q)$ for every $q \in I_{\mathfrak{g}}$. Then the equations $\theta'(1;v_i) = 0$ $(1 \leqq i \leqq r)$ imply that $\theta'$ is identically zero.*

Let $\eta$ be any element in $\mathfrak{M}$. Put $\eta_0 = \int_{M_K} \eta^{k'} \, dk'$ where $dk'$ denotes the normalized Haar measure on $M_K$. It is clear that $\theta'(1;\eta) = \theta'(1;\eta_0)$. But $\eta_0 \in I_{\mathfrak{m}}$ and $I_{\mathfrak{m}} \subset \sum_i v_i \mu_0(I_{\mathfrak{g}}) + \mathfrak{M} \mathfrak{m}_{\mathfrak{l}}$ from Lemma 15. Hence it follows from our assumptions that $\theta'(1;\eta_0) = 0$. This shows that $\theta'(1;\eta) = 0$ for every $\eta \in \mathfrak{M}$. But since $\theta'$ is analytic, this means that $\theta' = 0$.

For any $\lambda \in \mathfrak{F}$, put

$$\theta_\lambda(m) = \theta(\lambda:m) = \int_{M_K} \exp\{(-1)^{\frac{1}{2}}\lambda(H(mk')) - \rho(H(mk'))\} dk'$$

$$(m \in M)$$

where $dk'$ has the same meaning as above. Define $\gamma_{\mathfrak{m}}'(\eta) = e^{\rho-\sigma}\gamma_{\mathfrak{m}}(\eta) \circ e^{-\rho+\sigma}$ for $\eta \in I_{\mathfrak{m}}$ as in Section 5. Then by applying Lemma 18 of [2(i)] to $\mathfrak{m}$ (in place of $\mathfrak{g}$), it follows without difficulty that

$$\theta(\lambda:m;\eta) = \gamma_{\mathfrak{m}}'(\eta:(-1)^{\frac{1}{2}}\lambda)\theta(\lambda:m) \qquad (\eta \in I_{\mathfrak{g}}),$$

where $\gamma_{\mathfrak{m}}'(\eta:(-1)^{\frac{1}{2}}\lambda)$ denotes the value of the polynomial function $\gamma_{\mathfrak{m}}'(\eta)$ $\in S(\mathfrak{h}_{\mathfrak{p}})$ at $(-1)^{\frac{1}{2}}\lambda$. Therefore

$$\theta(\lambda:m;\mu_0(q)) = \gamma(q:(-1)^{\frac{1}{2}}\lambda)\theta(\lambda:m) \qquad (q \in I_{\mathfrak{g}})$$

from Lemma 12.

Put $v_i = \gamma_{\mathfrak{m}}'(v_i)$ $1 \leqq i \leqq r$ and define $\pi$, $\pi'$, $D = \pi/\pi'$ and $\tau^i$ $(1 \leqq i \leqq r)$ as in Lemma 15 of [2(i)]. Also put $\lambda^* = (-1)^{\frac{1}{2}}\lambda$ $(\lambda \in \mathfrak{F})$.

LEMMA 26. *Select $r$ elements $s_1 = 1, s_2, \cdots, s_r$ in $W$ such that $W = \bigcup_{1 \leqq i \leqq r} W' s_i$. Then*

$$\pi(\lambda^*)\Theta_H(\lambda:m) = \sum_{1 \leqq i, j \leqq r} \epsilon(s_i)\pi'(s_i\lambda^*)\tau^j(s_i\lambda^*)\Theta_H(\lambda:1;v_j)\theta(s_i\lambda:m)$$

*for $\lambda \in \mathfrak{F}_\epsilon$ and $m \in M$.*

For a fixed $m$, both sides are holomorphic functions of $\lambda$. Hence it is enough to prove the above relation under the assumption that $\pi(\lambda) \neq 0$. Put $\lambda_i = s_i \lambda$. Then the rational functions $v^k = \tau^k/D$ on $\mathfrak{h}_\mathfrak{p}$ $(1 \leq k \leq r)$ are all defined at $\lambda_i^*$ and we have to show that

$$f(\lambda:m) = \Theta_H(\lambda:m) - \sum_{1 \leq i,j \leq r} v^j(\lambda_i^*)\Theta_H(\lambda:1;v_j)\theta(\lambda_i:m)$$

is zero. On the other hand $\theta(\lambda_i:m;v_k) = v_k(\lambda_i^*)\theta(\lambda_i:m)$ and therefore

$$f(\lambda:1;v_k) = \Theta_H(\lambda:1;v_k) - \sum_{1 \leq i,j \leq r} v^j(\lambda_i^*)\Theta_H(\lambda:1;v_j)v_k(\lambda_i^*).$$

But since $\mathrm{sp}_{J'/J}(v^j v_k) = \delta_k^j$ (see $[2(\mathrm{i}), \S 3]$), it follows that

$$\sum_i v^j(\lambda_i^*)v_k(\lambda_i^*) = \delta_k^j \qquad\qquad (1 \leq j, k \leq r)$$

and therefore $f(\lambda:1;v_k) = 0$ $(1 \leq k \leq r)$. Hence our assertion is a consequence of Lemma 25.

**8. The function $\Theta$.** We shall now show that if $\lambda \in \mathfrak{F}_R$, $\Theta_H(\lambda:m)$ is actually independent of $H$. Let $H'$ be another element in $I_0^+$ and for a fixed $m \in M$, choose a positive number $T_0'$ such that $m \exp T_0' H' \in M_+$. Then

$$\Phi^0(\lambda:m:TH+T'H')$$
$$= \Phi^0(\lambda:m:T'H') + \int_0^T \{d\Phi^0(\lambda:m:tH+T'H')/dt\}dt$$

for $T, T' \geq 0$. On the other hand

$$\Phi^0(\lambda:m:tH+T'H') = e^{T'\rho(H')}\exp(-T'\Xi(\lambda:H'))\Phi^0(\lambda:m\exp T'H':tH)$$

from the definition of $\Phi^0$. Therefore

$$\| d\Phi^0(\lambda:m:tH+T'H')/dt \|$$
$$\leq e^{T'\rho(H')} \| \exp(-T'\Xi(\lambda:H')) \| \, \| \Psi_H^0(\lambda:m\exp T'H':tH) \|$$

for $t \geq 0$ and $T' \geq T_0'$. Now fix $\lambda$ and $m$ and assume that $T_0'$ is sufficiently large. Then from Lemma 18 we can choose a constant $a$ such that

$$\| d\Phi^0(\lambda:m:tH+T'H')/dt \|$$
$$\leq ae^{T'\rho(H')} \| \exp(-T'\Xi(\lambda:H')) \| \, \| \exp(-t\Xi(\lambda:H)) \|$$
$$\cdot \times \exp\{t\rho(H) - t\beta(H) - T'\beta(H')\}\phi(0:\exp(T'H'+tH))$$

for $T' \geq T_0'$ and $t \geq 0$. Hence in view of Theorem 3 of $[2(\mathrm{i})]$ and the corollary of Lemma 19, we can find a positive number $c$ and an integer $k \geq 0$ such that

$$\| d\Phi^0(\lambda:m:tH+T'H')/dt \| \leq c(1+t)^k T'^k \exp\{-t\beta(H) - T'\beta(H')\}$$
$$(T' \geq T_0', t \geq 0).$$

But this implies that

$$\operatorname*{Lim}_{T'\to+\infty} \int_0^\infty \| \, d\Phi^0(\lambda:m:tH+T'H')/dt \, \| \, dt = 0$$

and therefore $\Phi^0(\lambda:m:TH+T'H')$ converges to $\operatorname*{Lim}_{t'\to+\infty} \Phi^0(\lambda:m:t'H')$
$= \Theta_{H'}(\lambda:m)$ as $T'\to+\infty$, uniformly with respect to $T$. Hence

$$\Theta_{H'}(\lambda:m) = \operatorname*{Lim}_{T,T'\to+\infty} \Phi^0(\lambda:m:TH+T'H').$$

But since the right side is symmetric with respect to $H$ and $H'$, we conclude that $\Theta_{H'}(\lambda:m) = \Theta_H(\lambda:m)$.

Now put $\Theta(\lambda:m) = \Theta_H(\lambda:m)$ $(\lambda\in\mathfrak{F}_R, m\in M)$ for any $H\in\mathfrak{I}_0{}^+$. We propose to obtain an estimate for the difference $\Phi^0 - \Theta$.

LEMMA 27. *For any $\eta\in\mathfrak{M}$, we can choose an integer $k\geq 0$ and a positive constant $a$ such that*

$$\| \, \Phi^0(\lambda:h;\eta:H) - \Theta(\lambda:h;\eta) \, \|$$
$$\leq a(1+\|\lambda\|)^k(1+\|H\|)^k e^{-\beta(H)}(1+\|\log h\|)^d e^{-\rho(\log h)}$$

*for $\lambda\in\mathfrak{F}_R$, $h\in A_\mathfrak{p}{}^+$ and $H\in\mathfrak{I}_0{}^+$, provided $\beta(H)\geq 1$. (Here $d$ has the same meaning as in [2(i), Theorem 3].)*

We know that

$$\Theta(\lambda:h;\eta) = \Phi^0(\lambda:h;\eta:H) + \int_1^\infty \{d\Phi^0(\lambda:h;\eta:tH)/dt\}dt.$$

Moreover

$$\| \, d\Phi^0(\lambda:h;\eta:tH)/dt \, \| = \| \, \Psi_H{}^0(\lambda:h;\eta:tH) \, \|$$
$$\leq \| \exp(-t\Xi(\lambda:H)) \| \, \| \Psi_H(\lambda:h;\eta:tH) \|.$$

Therefore it follows from Lemma 18 and the corollary to Lemma 19, that we can select integers $k_1, k_2\geq 0$ and a positive number $c_1$ such that

$$\| \, d\Phi^0(\lambda:h;\eta:tH)/dt \, \|$$
$$\leq c_1(1+\|\lambda\|)^{k_1}t^{k_2}(1+\|H\|)^{k_2}\exp\{t\rho(H) - t\beta(H)\}\phi(0:h\exp tH)$$

for $\lambda\in\mathfrak{F}_R$, $h\in A_\mathfrak{p}{}^+$, $H\in\mathfrak{I}_0{}^+$ and $t\geq 1$ provided $\beta(H)\geq 1$. On the other hand by Theorem 3 of [2(i)],

$$\phi(0:h\exp tH) \leq c_2(1+\|\log h\|)^d(1+\|H\|)^d\exp\{-t\rho(H) - \rho(\log h)\}$$

where $c_2$ is a constant. Therefore

$$\| \, d\Phi^0(\lambda:h;\eta:tH)/dt \, \| \leq c_1 c_2(1+\|\lambda\|)^{k_1}t^{k_2}(1+\|H\|)^{k_2+d}$$
$$\times \exp\{-t\beta(H) - \rho(\log h)\}(1+\|\log h\|)^d.$$

But

$$\int_1^\infty t^{k_2} e^{-t\beta(H)} dt = e^{-\beta(H)} \int_0^\infty (1+t)^{k_2} e^{-t\beta(H)} dt$$

$$\leq e^{-\beta(H)} \int_0^\infty (1+t)^{k_2} e^{-t} dt,$$

since $\beta(H) \geqq 1$. Therefore our assertion is now obvious.

COROLLARY. *For a given* $\eta \in \mathfrak{M}$, *we can chose an integer* $k_1 \geqq 0$ *and a positive number* $a_1$ *such that*

$$\| \Phi(\lambda:h;\eta:H) - \exp\Xi(\lambda:H)\Theta(\lambda:h;\eta) \|$$
$$\leqq a_1(1 + \| \lambda \|)^{k_1}(1 + \| H \|)^{k_1} e^{-\beta(H)}(1 + \| \log h \|)^d e^{-\rho(\log h)}$$

*for* $\lambda \in \mathfrak{F}_R$, $h \in A_\mathfrak{p}^+$ *and* $H \in I_0^+$ *provided* $\beta(H) \geqq 1$.

This is an immediate consequence of the above lemma and the corollary to Lemma 19.

**9. The functions $b$ and $b_0$.** Define $b$ as in Lemma 52 of $[2(i)]$. We intend to show that $| b(s\lambda) | = | b(\lambda) |$ for $s \in W$ and $\lambda \in \mathfrak{F}_R$. Let us say that $h \to \infty$ $(h \in A_\mathfrak{p}^+)$ if $\alpha(\log h) \to +\infty$ for every $\alpha \in P_+$. Moreover if $f$ and $g$ are two functions on $A_\mathfrak{p}^+$, we write $f(h) \sim g(h)$ if $\underset{h \to \infty}{\mathrm{Lim}} | f(h) - g(h) | = 0$.

We now use the notation of Sections 7 and 8 and assume that $\epsilon$ is sufficiently small. Let $\mathfrak{F}_\epsilon'$ denote the set of those points $\lambda \in \mathfrak{F}_\epsilon$ where $\pi(\lambda) \neq 0$. Then it follows from Lemma 37 of $[2(i)]$ (applied to $\mathfrak{m}$ instead of $\mathfrak{g}$) that there exists a holomorphic function $c'$ on $\mathfrak{F}_\epsilon'$ such that [11]

$$e^{\rho(\log h)}\theta(\lambda:h) \sim \sum_{s \in W} c'(s\lambda)\exp\{s\lambda^*(\log h)\} \qquad (h \in A_\mathfrak{p}^+)$$

for $\lambda \in \mathfrak{F}_\epsilon'$. Now let $\mathfrak{F}_R' = \mathfrak{F}_\epsilon' \cap \mathfrak{F}_R$ and fix $\lambda \in \mathfrak{F}_R'$ and $H \in I_0^+$. Put $\lambda_i = s_i\lambda$ and $f_i = f_{\lambda_i}$ $(1 \leqq i \leqq r)$ in the notation of $[2(i), \text{Lemma } 15]$. Define the $r \times r$ matrix $F(i)$ with complex coefficients by

$$f_i v_j = \sum_k F_{jk}(i) v_k \bmod SJ_{\lambda^*}.$$

Then if $I$ is the unit matrix of degree $r$, it follows from Lemma 15 of $[2(i)]$ that $\sum_i \epsilon(s_i)\pi'(\lambda_i^*)F(i) = \pi(\lambda^*)I$ and $F(i)\Xi(\lambda:h) = \lambda_i^*(H)F(i)$ $1 \leqq i \leqq r$. Hence

$$\pi(\lambda^*)\Phi^0(\lambda:h:tH) = \pi(\lambda^*)\exp(-t\Xi(\lambda:H))\Phi(\lambda:h:tH)$$
$$= \sum_i \epsilon(s_i)\pi'(\lambda_i^*)e^{-t\lambda_i^*(H)}F(i)\Phi(\lambda:h:tH) \qquad (t \in R, h \in A_\mathfrak{p}).$$

---

[11] If $s \in W$ and $\mu \in \mathfrak{F}$, $s\mu(H')$ denotes the value of the linear function $s\mu$ at $H' \in \mathfrak{hp}$.

Therefore

$$\pi(\lambda^*)\Theta(\lambda:h) = \operatorname*{Lim}_{t\to+\infty} \sum_i \epsilon(s_i)\pi'(\lambda_i^*)e^{-t\lambda_i^*(H)}F(i)\Phi(\lambda:h:tH).$$

Now put $\phi'(\lambda:h) = e^{\rho(\log h)}\phi(\lambda:h)$. Since $v_i = \gamma_m'(v_i)$ $(1\leq i\leq r)$, it follows without difficulty by applying Lemma 26 of [2(i)] to $\mathfrak{m}$ (in place of $\mathfrak{g}$) that

$$\operatorname*{Lim}_{t\to+\infty} |\Phi_j(\lambda:h:tH) - \phi'(\lambda:h\exp tH;v_j)| = 0$$

for $h \in A_\mathfrak{p}^+$. Therefore

$$\pi(\lambda^*)e^{\rho(\log h)}\Theta_k(\lambda:h)$$
$$= \operatorname*{Lim}_{t\to+\infty} \sum_{1\leq i,j\leq r} \epsilon(s_i)\pi_i'(\lambda_i^*)e^{-t\lambda_i^*(H)}F_{kj}(i)\phi'(\lambda:h\exp tH;v_j).$$

On the other hand it follows from Lemma 37 of [2(i)] that

$$\operatorname*{Lim}_{\substack{t\to+\infty\\ h\to\infty}} |\phi'(\lambda:h\exp tH;v_j) - \sum_{s\in W} c(s\lambda)v_j(s\lambda^*)\exp\{s\lambda^*(\log h + tH)\}| = 0.$$

But we know from [2(i), Lemma 15] that $f_iv_k \equiv v_k(\lambda_i^*)f_i \bmod SJ_\lambda$. and therefore

$$\sum_j F_{kj}(i)v_j(\lambda_m^*) = v_k(\lambda_i^*)f_i(\lambda_m^*) = v_k(\lambda_i^*)\delta_{im}D(\lambda_i^*) \qquad (1\leq i,m\leq r).$$

Moreover $H$ is invariant under $W'$, since it lies in the center of $\mathfrak{m}$. Hence

$$\sum_j F_{kj}(i)\sum_{s\in W} c(s\lambda)v_j(s\lambda^*)e^{s\lambda^*(\log h + tH)}$$
$$= \sum_{j,m} F_{kj}(i)v_j(\lambda_m^*)\sum_{s'\in W'} c(s'\lambda_m)e^{s'\lambda_m^*(\log h + tH)}$$
$$= e^{t\lambda_i^*(H)}v_k(\lambda_i^*)D(\lambda_i^*)\sum_{s'\in W'} c(s'\lambda_i)e^{s'\lambda_i^*(\log h)}.$$

This proves that

$$\pi(\lambda^*)e^{\rho(\log h)}\Theta_k(\lambda:h) \sim \pi(\lambda^*)\sum_{1\leq i\leq r} v_k(\lambda_i^*)\sum_{s'\in W'} c(s'\lambda_i)e^{s'\lambda_i^*(\log h)}$$
$$(1\leq k\leq r).$$

Now put (in the notation of Lemma 26)[12]

$$`\Gamma_{ki}(\lambda) = (-1)^{-d/2}\epsilon(s_i)\sum_{1\leq j\leq r}\tau^j(\lambda_i^*)\Theta_k(\lambda:1;v_j),$$

where $d$ is the degree of $D = \pi/\pi'$. Then it follows from Lemma 26 that

$$\pi(\lambda)e^{\rho(\log h)}\Theta_k(\lambda:h) \sim \sum_{1\leq i\leq r} `\Gamma_{ki}(\lambda)\pi'(\lambda_i)\sum_{s'\in W'} c'(s'\lambda_i)e^{s'\lambda_i^*(\log h)}.$$

This shows that

$$\operatorname*{Lim}_{h\to\infty} |\sum_i \sum_{s'\in W'} \{v_k(\lambda_i^*)\pi(\lambda)c(s'\lambda_i) - `\Gamma_{ki}(\lambda)\pi'(\lambda_i)c'(s'\lambda_i)\}e^{s'\lambda_i^*(\log h)}| = 0.$$

---

[12] For any integer $k$, $(-1)^{k/2} = ((-1)^i)^k$.

But since $\pi(\lambda) \neq 0$, the $w$ elements $s'\lambda_i$ $(s' \in W', 1 \leq i \leq r)$ are all distinct. Hence we conclude from Lemma 57 of $[2(\mathrm{i})]$ that

$$v_k(\lambda_i^*)\pi(\lambda)c(s'\lambda_i) = {}'\Gamma_{ki}(\lambda)\pi'(\lambda_i)c'(s'\lambda_i) \qquad (s' \in W', 1 \leq i, k \leq r).$$

Putting $i = k = 1$ and ${}'\Gamma_{11} = \Gamma$, this becomes $\pi(\lambda)c(s'\lambda) = \Gamma(\lambda)\pi'(\lambda)c'(s'\lambda)$ $(\lambda \in \mathfrak{F}_R')$. Now if $\epsilon$ is sufficiently small, it follows from the results of Section 7, that ${}'\Gamma_{ij}$ can be extended to holomorphic functions on $\mathfrak{F}_\epsilon$. Moreover there exist (see $[2(\mathrm{i}), \text{Lemma } 52]$) holomorphic functions $b$ and $b'$ on $\mathfrak{F}_\epsilon$ such that $b(\lambda) = \pi(\lambda)c(\lambda)$ and $b'(\lambda) = \pi'(\lambda)c'(\lambda)$ for $\lambda \in \mathfrak{F}_\epsilon'$. The above result shows that $b(s'\lambda) = \Gamma(\lambda)b'(s'\lambda)$ for $s' \in W'$ and $\lambda \in \mathfrak{F}_R'$. Hence by analytic continuation we obtain the following lemma.

LEMMA 28.  $b(s'\lambda) = \Gamma(\lambda)b'(s'\lambda)$ *for* $s' \in W'$ *and* $\lambda \in \mathfrak{F}_\epsilon$.

COROLLARY 1.  $|b(s\lambda)| = |b(\lambda)|$ *for* $s \in W$ *and* $\lambda \in \mathfrak{F}_R$.

It is obvious that $\phi(-\lambda:h) = \operatorname{conj}\phi(\lambda:h)$ for $\lambda \in \mathfrak{F}_R$ and $h \in A_\mathfrak{p}$. Hence it follows from Lemmas 37 and 57 of $[2(\mathrm{i})]$ that $c(-\lambda) = \operatorname{conj}c(\lambda)$ for $\lambda \in \mathfrak{F}_R'$. Therefore if $d$ is the degree of $\pi$, $\operatorname{conj}b(\lambda) = (-1)^d b(-\lambda)$ $(\lambda \in \mathfrak{F}_R)$. Let $l = \dim \mathfrak{h}_\mathfrak{p}$. If $l = 1$, $W$ consists of only two elements. Hence if $s \neq 1$, $s\lambda = -\lambda$ and therefore our assertion is obvious. So now assume that $l \geq 2$ and use induction. Define $\Sigma$ as in $[2(\mathrm{i}), \S 3]$. Since the reflexions $s_\alpha$ $(\alpha \in \Sigma)$ generate $W$, it would be enough to prove that $|b(s_\alpha\lambda)| = |b(\lambda)|$ for $\alpha \in \Sigma$ and $\lambda \in \mathfrak{F}_R$. Now fix $\alpha \in \Sigma$ and select the element $H_0$ of Section 2 in such a way that $\alpha(H_0) = 0$. Then $s_\alpha \in W'$ and it follows by the induction hypothesis (applied to $\mathfrak{m}' = [\mathfrak{m}, \mathfrak{m}]$) that $|b'(s'\lambda)| = |b'(\lambda)|$ for $s' \in W'$ and $\lambda \in \mathfrak{F}_R$. But then by Lemma 27, $|b(s_\alpha\lambda)| = |\Gamma(\lambda)b'(s_\alpha\lambda)| = |\Gamma(\lambda)b'(\lambda)| = |b(\lambda)|$. This proves our assertion.

COROLLARY 2.  $b(\lambda)b(-\lambda) = b(s\lambda)b(-s\lambda)$ *for* $s \in W$ *and* $\lambda \in \mathfrak{F}_\epsilon$.

First suppose $\lambda \in \mathfrak{F}_R$. Then we have seen above that

$$b(-\lambda) = (-1)^d \operatorname{conj}b(\lambda)$$

and therefore

$$b(\lambda)b(-\lambda) = (-1)^d |b(\lambda)|^2 = (-1)^d |b(s\lambda)|^2 = b(s\lambda)b(-s\lambda)$$

by Corollary 1. Hence the same identity holds for all $\lambda \in \mathfrak{F}_\epsilon$ by analytic continuation.

COROLLARY 3.  *Suppose* $\epsilon$ *is sufficiently small. Then if* $b(\lambda) = 0$ *for some* $\lambda \in \mathfrak{F}_\epsilon$, *it follows that* $b(s\lambda) = 0$ *for every* $s \in W$.

If $l=1$, our statement is true since $\boldsymbol{b}(\lambda)$ is never zero for $\lambda \in \mathfrak{F}_\epsilon$ (see Lemma 53 of $[2(i)]$). So let us assume that $l \geqq 2$ and use induction. Fix an element $\alpha \in \Sigma$. Obviously it would be enough to prove that $\boldsymbol{b}(s_\alpha \lambda) = 0$. Again choose $H_0$ in such a way that $s_\alpha \in W'$. If $\boldsymbol{b}'(\lambda) = 0$, it follows by induction hypothesis that $\boldsymbol{b}'(s_\alpha \lambda) = 0$. On the other hand if $\boldsymbol{b}'(\lambda) \neq 0$, $\Gamma(\lambda) = \boldsymbol{b}(\lambda)/\boldsymbol{b}'(\lambda) = 0$. Therefore in either case $\boldsymbol{b}(s_\alpha \lambda) = \Gamma(\lambda)\boldsymbol{b}'(s_\alpha \lambda) = 0$.

Now we regard $\mathfrak{F}_\epsilon$ as a complex-analytic manifold and consider the divisor $\boldsymbol{d}$ of the function $\boldsymbol{b}$ on $\mathfrak{F}_\epsilon$. Select $\alpha_1, \cdots, \alpha_q \in \Sigma$ such that $\pi = \alpha_1 \alpha_2 \cdots \alpha_q$ and let $\boldsymbol{d}_i$ denote the divisor of $\alpha_i$. Then it is obvious that $\boldsymbol{d}_i$ $(1 \leqq i \leqq q)$ are distinct prime divisors. Consider an element $\lambda \in \mathfrak{F}_\epsilon'$. We claim that $\boldsymbol{b}(\lambda) \neq 0$. For otherwise it follows from Corollary 3 of Lemma 28 that $\boldsymbol{b}(s\lambda) = 0$ for every $s \in W$. But since $\pi(\lambda) \neq 0$, this implies that $\boldsymbol{c}(s\lambda) = 0$ for all $s \in W$. However in view of $[2(i), \text{Lemma } 37]$, this means that the analytic function $\phi_\lambda$ on $A_\mathfrak{p}$ must be identically zero. But this is impossible since $\phi(\lambda : 1) = 1$. Hence $\boldsymbol{b}(\lambda)$ is never zero unless $\pi(\lambda) = 0$ $(\lambda \in \mathfrak{F}_\epsilon)$ and therefore $\boldsymbol{d} = k_1 \boldsymbol{d}_1 + \cdots + k_q \boldsymbol{d}_q$ where $k_1, \cdots, k_q$ are certain nonnegative integers. The divisors $\boldsymbol{d}_i$ are obviously invariant under the transformation $\lambda \to -\lambda$ of $\mathfrak{F}_\epsilon$. Hence $2\boldsymbol{d}$ is the divisor of the function $\boldsymbol{b}(\lambda)\boldsymbol{b}(-\lambda)$. But then by Corollary 2 of Lemma 28, $\boldsymbol{d}$ must be invariant under $W$. We claim that $k_i \leqq 1$ $(1 \leqq i \leqq q)$. For otherwise suppose $k_1 \geqq 2$ and select a point $\lambda_0 \in \mathfrak{F}_R$ such that $\alpha_1(\lambda_0) = 0$ while $\alpha_i(\lambda_0) \neq 0$ $(2 \leqq i \leqq q)$. Then $\boldsymbol{c} = \boldsymbol{b}/\pi$, considered as a meromorphic function on $\mathfrak{F}_\epsilon$, is holomorphic around $\lambda_0$ and $\boldsymbol{c}(\lambda_0) = 0$. Since the divisor of $\boldsymbol{c}$ is invariant under $W$, it follows that $\boldsymbol{c}$ is holomorphic around $s\lambda_0$ and $\boldsymbol{c}(s\lambda_0) = 0$ for every $s \in W$. Let $U$ be a sufficiently small compact neighborhood of $\lambda_0$ in $\mathfrak{F}_R$. We now use the notation of $[2(i), \S 10]$. Then if $N$ is a sufficiently large positive number, $\Phi(s\lambda : h)$ is defined for $\lambda \in U$, $s \in W$ and $h \in A_\mathfrak{p}(N)$ and it follows from $[2(i), \text{Lemma } 37]$ that

$$e^{\rho(\log h)}\phi(\lambda : h) = \sum_{s \in W} \boldsymbol{c}(s\lambda)\Phi(s\lambda : h) \qquad (h \in A_\mathfrak{p}(N))$$

for $\lambda \in U' = U \cap \mathfrak{F}_R'$. Hence if we fix $h \in A_\mathfrak{p}(N)$ and make $\lambda$ tend to $\lambda_0$ it becomes clear (see $[2(i), \text{Corollary 2 of Lemma 28}]$) that

$$e^{\rho(\log h)}\phi(\lambda_0 : h) = \sum_{s \in W} \boldsymbol{c}(s\lambda_0)\Phi(s\lambda_0 : h) = 0 \qquad (h \in A_\mathfrak{p}(N)).$$

However this is impossible because the analytic function $\phi_{\lambda_0}$ on $A_\mathfrak{p}$ is not identically zero. In fact $\phi(\lambda_0 : 1) = 1$. This proves that $k_i = 0$ or 1 $(1 \leqq i \leqq q)$. Hence we may assume that the divisor of $\boldsymbol{c}^{-1}$ is $\boldsymbol{d}_1 + \boldsymbol{d}_2 + \cdots + \boldsymbol{d}_p$ for some integer $p$ $(0 \leqq p \leqq q)$ . Put $\pi_0 = \alpha_1 \alpha_2 \cdots \alpha_p$. $(\pi_0 = 1$ if $p = 0$.) Since

$d_1 + \cdots + d_p$ is invariant under $W$, it is clear that $\pi_0{}^s = \epsilon_0(s)\pi_0$ $(s \in W)$ where $\epsilon_0(s) = \pm 1$. Put $b_0 = \pi_0 c$. Then the divisor of $b_0$ is zero. Hence we have obtained the following result.

LEMMA 29. *The functions $c^{-1}$ and $b_0 = \pi_0 c$ are everywhere holomorphic on $\mathfrak{F}_\epsilon$. Moreover $b_0$ is nowhere zero on $\mathfrak{F}_\epsilon$.*

Define $b_0'$ and $\pi_0'$ in an analogous way (corresponding to $\mathfrak{m}$). Then $b_0' = \pi_0' c'$ and $b_0'$, $1/b_0'$ are both holomorphic on $\mathfrak{F}_\epsilon$. Put $\pi_1 = \pi/\pi_0$, $\pi_1' = \pi'/\pi_0'$. Then $b_0 = \Gamma \pi_1' b_0'/\pi_1$. But since $b_0$ and $b_0'$ have no zeros, it follows that $\Gamma_0 = \Gamma \pi_1'/\pi_1$ and its reciprocal are both holomorphic on $\mathfrak{F}_\epsilon$. Moreover we saw during the proof of Lemma 28 that

$$v_k(\lambda_i{}^*)\pi(\lambda)c(\lambda_i) = {}^{\backprime}\Gamma_{ki}(\lambda)\pi'(\lambda_i)c'(\lambda_i) \qquad (1 \leq i, k \leq r)$$

for $\lambda \in \mathfrak{F}_R'$. But $\pi(\lambda)c(\lambda_i) = \epsilon(s_i)b(\lambda_i) = \epsilon(s_i)\Gamma(\lambda_i)b'(\lambda_i)$. Also $b'(\lambda_i) \neq 0$ since $\pi(\lambda) \neq 0$ (Corollary 3 of Lemma 28). Therefore

$$'\Gamma_{ki}(\lambda) = \epsilon(s_i)v_k(s_i\lambda^*)\Gamma(s_i\lambda) \qquad (1 \leq i, k \leq r)$$

and so, by analytic continuation, this relation holds for all $\lambda \in \mathfrak{F}_\epsilon$. Put $\Gamma_{ki}(\lambda) = \epsilon_0(s_i)v_k(s_i\lambda^*)\Gamma_0(s_i\lambda)$ $(1 \leq i, k \leq r, \lambda \in \mathfrak{F}_\epsilon)$. Then it follows from Lemma 26 and the definition of $'\Gamma_{ki}(\lambda)$ that

$$\pi_0(\lambda)\Theta_k(\lambda:m) = \sum_{1 \leq i \leq r} \Gamma_{ki}(\lambda)\pi_0'(\lambda_i)\theta(\lambda_i:m) \qquad (m \in M)$$

for $\lambda \in \mathfrak{F}_R'$. Therefore since both sides are analytic functions of $\lambda$, this holds for all $\lambda \in \mathfrak{F}_R$.

LEMMA 30. *There exist analytic functions $\Gamma_{ij}$ $1 \leq i, j \leq r$ on $\mathfrak{F}_R$ with the following properties.*

(1) $\quad \pi_0(\lambda)\Theta_i(\lambda:m) = \sum_{1 \leq j \leq r} \Gamma_{ij}(\lambda)\pi_0'(s_j\lambda)\theta(s_j\lambda:m)$

$$(1 \leq i \leq r, \lambda \in \mathfrak{F}_R, m \in M).$$

(2) *We can find an integer $d \geq 0$ such that*

$$(1 + \|\lambda\|)^{-d} \sum_{i,j} |\Gamma_{ij}(\lambda; \partial(v))|$$

*remains bounded for any fixed $v \in S(\mathfrak{h}_\mathfrak{p})$ as $\lambda$ varies in $\mathfrak{F}_R$.*

In order to verify the second property we proceed as follows. In view of Lemma 58 of [2(i)] and the definition of $\Gamma_{ij}$, it would be enough to show that $\sup_{\lambda \in \mathfrak{F}_{\epsilon'}} (1 + \|\lambda\|)^{-d} |\Gamma_0(\lambda)| < \infty$ for some $\epsilon' > 0$ and $d \geq 0$. Since $\Gamma_0 = \Gamma \pi_1'/\pi_1$ and $\Gamma_0$ has no zeros on $\mathfrak{F}_\epsilon$, it follows that $\pi_1'$ divides $\pi_1$ in

$S(\mathfrak{h}_\mathfrak{p})$. Therefore by the corollary of Lemma 59 of $[2(i)]$, it would be sufficient to verify that $\sup_{\lambda \in \mathfrak{F}_\epsilon} (1 + \| \lambda \|)^{-d} | \Gamma(\lambda) | < \infty$ for some $d \geq 0$. But $\Gamma(\lambda) = '\Gamma_{11}(\lambda)$ and so this follows immediately from Lemma 23 and the definition of $'\Gamma_{11}$.

If we take into account Lemma 52 of $[2(i)]$, we get the following result by a similar argument.

LEMMA 31. *We can select an integer $d \geq 0$ with the following property. For any $u \in S(\mathfrak{h}_\mathfrak{p})$, $\sup_{\lambda \in \mathfrak{F}_R} (1 + \| \lambda \|)^{-d} | \boldsymbol{b}_0(\lambda; \partial(u)) |$ is finite.*

The following lemma will be needed in Section 13.

LEMMA 32. *Let $p$ be an element in $S(\mathfrak{h}_\mathfrak{p})$ such that $p^s = \epsilon_0(s)p$ $(s \in W)$. Then $p = \pi_0 q$ for some $q \in I(\mathfrak{h}_\mathfrak{p})$.*

Fix an element $\alpha \in \Sigma$. It follows easily (see the proof of Lemma 10 of $[2(i)]$) that $\epsilon_0(s_\alpha) = -1$ or 1 according as $\alpha$ divides $\pi_0$ or not. Now suppose $\alpha$ divides $\pi_0$. Then $p^{s_\alpha} = -p$ and therefore $\alpha$ divides $p$. This shows that $\pi_0$ divides $p$ in $S(\mathfrak{h}_\mathfrak{p})$ and our assertion is now obvious.

**10. The relationship between $\varphi$ and $\theta$.** Lemmas 27 and 30 give us an approximate connection between the functions $\phi$ and $\theta$. Since $I$ lies in the center of $\mathfrak{m}$, it is obvious that $\theta(\lambda : m \exp H) = \exp\{\lambda^*(H) - \rho(H)\}\theta(\lambda : m)$ for $H \in I_0$, $\lambda \in \mathfrak{F}$ and $m \in M$. Therefore it follows from Lemma 30 that

$$\pi_0(\lambda) e^{\rho(H)} \Theta_i(\lambda : m \exp(H)) = \sum_{1 \leq j \leq r} \Gamma_{ij}(\lambda) e^{\lambda_j^*(H)} \pi_0'(\lambda_j) \theta(\lambda_j : m)$$
$$(1 \leq i \leq r).$$

On the other hand

$$\Phi_i(\lambda : h ; \eta : H) = e^{\rho(H)} \phi(\lambda : h \exp H ; \eta \nu_i) \qquad (1 \leq i \leq r)$$

for $\eta \in \mathfrak{M}$. Therefore we get the following result from the corollary of Lemma 27.

LEMMA 33. *For a given $\eta \in \mathfrak{M}$, we can choose an integer $k \geq 0$ and a positive number $a$ such that*

$$\left| \pi_0(\lambda) e^{\rho(\log h + H)} \phi(\lambda : h \exp H ; \eta \nu_i) - \sum_{1 \leq j \leq r} \Gamma_{ij}(\lambda) e^{\lambda_j^*(H)} \pi_0'(\lambda_j) \theta(\lambda_j : h ; \eta) e^{\rho(\log h)} \right|$$

$$\leq a(1 + \| \lambda \|)^k (1 + \| H \|)^k e^{-\beta(H)} (1 + \| \log h \|)^d \qquad (1 \leq i \leq r)$$

*for $\lambda \in \mathfrak{F}_R, h \in A_\mathfrak{p}^+$ and $H \in I_0^+$ provided $\beta(H) \geq 1$. (Here $d$ has the same meaning as in $[2(i), \text{Theorem } 3]$.)*

Now fix an element $b \in \mathfrak{B}$ and consider the differential operators $\mu_0(b)$ and $\mu(b)$ of Lemma 10. Then $\mu(b) = \sum\limits_{1 \leq i \leq s} g_i \eta_i$ where $g_i \in \mathfrak{R}_0$ and $\eta_i \in \mathfrak{M}_\mathfrak{p}$. From Lemma 7, we can select a positive number $a_1$ such that $|g_i(h \exp H)| \leq a_1 e^{-\beta(H)}$ for $h \in A_\mathfrak{p}^+$ and $H \in \mathfrak{I}_0^+$ provided $\beta(H) \geq 1$. Hence

$$|\phi(\lambda : h \exp H ; \mu(b))| \leq a_1 e^{-\beta(H)} \sum_{1 \leq j \leq s} |\phi(\lambda : h \exp H ; \eta_j)|.$$

On the other hand by Lemma 46 of $[2(i)]$, there exists an integer $k_1 \geq 0$ and a constant $a_2 > 0$ such that

$$\sum_{1 \leq j \leq s} |\phi(\lambda : h \exp H ; \eta_j)|$$
$$\leq a_2 (1 + \|\lambda\|)^{k_1} (1 + \|H\|)^d (1 + \|\log h\|)^d e^{-\rho(H + \log h)}$$

for all $\lambda \in \mathfrak{F}_R$. This shows that

$$e^{\rho(H + \log h)} |\phi(\lambda : h \exp H ; b) - \phi(\lambda : h \exp H ; \mu_0(b))|$$
$$= e^{\rho(H + \log h)} |\phi(\lambda : h \exp H ; \mu(b))|$$
$$\leq a_1 a_2 (1 + \|\lambda\|)^{k_1} e^{-\beta(H)} (1 + \|H\|)^d (1 + \|\log h\|)^d \quad (h \in A_\mathfrak{p}^+, H \in \mathfrak{I}_0^+)$$

provided $\beta(H) \geq 1$. Since $\mu_0(b) \in \mathfrak{M}$ and $\nu_1 = 1$, we can now apply Lemma 33 to $\eta = \mu_0(b)$ and thus obtain the following result. Put $\Gamma_j(\lambda) = \Gamma_{1j}(\lambda)$ $(1 \leq j \leq r)$.

LEMMA 34. *For any $b \in \mathfrak{B}$, we can choose an integer $k \geq 0$ and a positive number $a$ such that*

$$|\pi_0(\lambda) e^{\rho(\log h + H)} \phi(\lambda : h \exp H ; b)$$
$$- \sum_{1 \leq j \leq r} \Gamma_j(\lambda) e^{\lambda_j^*(H)} \pi_0'(\lambda_j) \theta(\lambda_j : h ; \mu_0(b)) e^{\rho(\log h)}|$$
$$\leq a (1 + \|\lambda\|)^k (1 + \|H\|)^k e^{-\beta(H)} (1 + \|\log h\|)^d$$

*for $\lambda \in \mathfrak{F}_R$, $h \in A_\mathfrak{p}^+$ and $H \in \mathfrak{I}_0^+$ provided $\beta(H) \geq 1$.*

We recall that the element $H_0$ of Section 2 has been kept fixed throughout our discussion. It is obvious that $\rho(H_0) > 0$. Select a number $\epsilon$ such that $0 < \epsilon < 1$ and let $U$ be the set of all points $H \in \mathfrak{h}_{\mathfrak{p}_0}$ which satisfy the conditions $\alpha(H) \geq (1 - \epsilon)\alpha(H_0)$ $(\alpha \in P)$ and $\|H\| \leq (1 + \epsilon)\|H_0\|$. Then $U$ is a neighborhood of $H_0$ in $\mathrm{Cl}(\mathfrak{h}_{\mathfrak{p}_0}^+)$. Put $H_\epsilon = H - (1 - \epsilon)H_0$ for $H \in U$. Then $H_\epsilon \in \mathrm{Cl}(\mathfrak{h}_{\mathfrak{p}_0}^+)$ and $\exp tH = \exp t(1 - \epsilon)H_0 \exp tH_\epsilon$ $(t \in R)$. Hence if $t \geq t_0 = \{(1 - \epsilon)\beta(H_0)\}^{-1}$, it follows from Lemma 34 that

$$|\pi_0(\lambda) e^{t\rho(H)} \phi(\lambda : \exp tH ; b)$$
$$- \sum_{1 \leq i \leq r} \Gamma_i(\lambda) \exp\{t(1 - \epsilon)\lambda_i^*(H_0)\} \pi_0'(\lambda_i) \theta(\lambda_i : \exp tH_\epsilon ; \mu_0(b)) e^{t\rho(H_\epsilon)}|$$
$$\leq a (1 + \|\lambda\|)^k (1 + \|tH_0\|)^k (1 + \|tH_\epsilon\|)^d e^{-t/t_0}$$

for $H \in U$ and $\lambda \in \mathfrak{F}_R$. Moreover since $H_0$ lies in the center of $\mathfrak{m}$, it is clear that

$$\theta(\lambda_i : \exp tH ; \mu_0(b))$$
$$= \exp\{t(1-\epsilon)\lambda_i^*(H_0) - t(1-\epsilon)\rho(H)\}\theta(\lambda_i : \exp tH_\epsilon ; \mu_0(b)).$$

Also $\|H_\epsilon\| \leqq \|H\| + (1-\epsilon)\|H_0\| \leqq 2\|H_0\|$. Therefore if we put $a_1 = 2^d(1 + \|H_0\|)^{k+d}a$, it follows that

$$\left| \pi_0(\lambda)e^{t\rho(H)}\phi(\lambda : \exp tH ; b) - \sum_i \Gamma_i(\lambda)\pi_0'(\lambda_i)e^{t\rho(H)}\theta(\lambda_i : \exp tH ; \mu_0(b)) \right|$$
$$\leqq a_1(1 + \|\lambda\|)^k(1 + t)^{k+d}e^{-t/t_0}$$

for $H \in U$, $\lambda \in \mathfrak{F}_R$ and $t \geqq t_0$. On the other hand by applying Lemma 46 and Theorem 3 of $[2(\mathrm{i})]$ both to $\phi(\lambda : \exp tH ; b)$ and $\theta(\lambda_i : \exp tH ; \mu_0(b))$ and making use of the second statement of Lemma 30, we conclude that

$$\left| \pi_0(\lambda)\phi(\lambda : h ; b) \right| + \sum_i \left| \Gamma_i(\lambda)\pi_0'(\lambda_i)\theta(\lambda_i : h ; \mu_0(b)) \right|$$
$$\leqq a_2(1 + \|\lambda\|)^{k_2}e^{-\rho(\log h)}(1 + \|\log h\|)^{d_2}$$

for suitable constants $a_2, k_2, d_2 \geqq 0$ and $h \in \mathrm{Cl}(A_\mathfrak{p}^+)$, $\lambda \in \mathfrak{F}_R$. Since $\underset{t \to +\infty}{\mathrm{Lim}}\, t^q e^{-t/2t_0} = 0$ for any $q \geqq 0$, we thus obtain the following result.

LEMMA 35. *There exists a neighborhood $U$ of $H_0$ in $\mathrm{Cl}(\mathfrak{h}_{\mathfrak{p}_0}^+)$ and a positive number $t_0$ with the following property. If $b$ is any given element in $\mathfrak{B}$, we can select an integer $k \geqq 0$ and a constant $a > 0$ such that*

$$\left| \pi_0(\lambda)e^{t\rho(H)}\phi(\lambda : \exp tH ; b) - \sum_i \Gamma_i(\lambda)\pi_0'(\lambda_i)e^{t\rho(H)}\theta(\lambda_i : \exp tH ; \mu_0(b)) \right|$$
$$\leqq a(1 + \|\lambda\|)^k e^{-t/2t_0}$$

*for $H \in U$, $\lambda \in \mathfrak{F}_R$ and $t \geqq 0$.*

**11. Proof of Theorem 1.** We shall now deduce an important result from Lemma 35.

THEOREM 1. *For any $b \in \mathfrak{B}$, we can choose an integer $k \geqq 0$ such that $(1 + \|\lambda\|)^{-k}|\pi_0(\lambda)\phi(\lambda : h ; b)e^{\rho(\log h)}|$ remains bounded as $\lambda$ and $h$ vary in $\mathfrak{F}_R$ and $\mathrm{Cl}(A_\mathfrak{p}^+)$ respectively.*

We shall prove this theorem by induction on $\dim_R \mathfrak{g}_0$. Let $S_+$ be the set of all points $H \in \mathrm{Cl}(\mathfrak{h}_{\mathfrak{p}_0}^+)$ with $\|H\| = 1$. Then $S_+$ is compact. Therefore it is obviously sufficient to prove the following lemma.

LEMMA 36. *For a given* $b \in \mathfrak{B}$ *and* $H_0 \in S_+$, *we can choose a neighborhood* $V$ *of* $H_0$ *in* $S_+$, *an integer* $k_1 \geqq 0$ *and a positive number* $a_1$ *such that*

$$\left| \pi_0(\lambda) e^{t\rho(H)} \phi(\lambda : \exp tH ; b) \right| \leqq a_1 (1 + \| \lambda \|)^{k_1}$$

*for* $\lambda \in \mathfrak{F}_R$, $H \in V$ *and* $t \geqq 0$.

We now construct $\mathfrak{m}_0$ corresponding to this $H_0$ and use the notation of Lemma 35. Since $\mathfrak{m}_0$ is reductive, its derived algebra $\mathfrak{m}_0' = [\mathfrak{m}_0, \mathfrak{m}_0]$ is semi-simple and $\dim_R \mathfrak{m}_0' < \dim_R \mathfrak{g}_0$ because $H_0$ lies in the center of $\mathfrak{m}_0$. Therefore our induction hypothesis is applicable to $\mathfrak{m}_0'$. Define $\sigma$ as in Lemma 12 and put

$$\theta'(\lambda : m) = \int_{M_K} \exp\{\lambda^*(H(mk')) - \sigma(H(mk'))\} dk' \qquad (\lambda \in \mathfrak{F}_R, m \in M),$$

where $dk'$ denotes, as before, the normalized Haar measure of $M_K$. Now $\mathfrak{l}_0$ lies in the center of $\mathfrak{m}_0$ and $\mathfrak{m}_0 = \mathfrak{l}_0 + \mathfrak{m}_0' + \mathfrak{m}_{\mathfrak{k}_0}$. Hence by applying the induction hypothesis to $\mathfrak{m}_0'$, it is easily seen that for any $\eta \in \mathfrak{M}$, we can select an integer $k_2 \geqq 0$ and a positive number $a_2$ such that

$$\left| \pi_0'(\lambda) e^{\sigma(H)} \theta'(\lambda : \exp H ; \eta) \right| \leqq a_2 (1 + \| \lambda \|)^{k_2}$$

for all $\lambda \in \mathfrak{F}_R$ and $H \in \mathrm{Cl}(\mathfrak{h}_{\mathfrak{p}_0}{}^+)$. Also it is obvious that

$$\theta(\lambda : h) = \exp(\sigma(\log h) - \rho(\log h)) \theta'(\lambda : h)$$

for $h \in A_\mathfrak{p}$. Let $\mu_0(b)'$ denote the image of $\mu_0(b)$ under the automorphism of $\mathfrak{M}$ over $C$ which leaves every element of $\mathfrak{m}_0' + \mathfrak{m}_{\mathfrak{k}_0}$ fixed and maps $H$ on $H - \rho(H)$ for $H \in \mathfrak{l}$. Then clearly

$$\theta(\lambda : h ; \mu_0(b)) = \exp(\sigma(\log h) - \rho(\log h)) \theta'(\lambda : h ; \mu_0(b)')$$

and therefore if we take $\eta = \mu_0(b)'$, it follows that

$$\left| \pi_0'(\lambda) e^{\rho(H)} \theta(\lambda : \exp H ; \mu_0(b)) \right| \leqq a_2 (1 + \| \lambda \|)^{k_2}$$

for $\lambda \in \mathfrak{F}_R$ and $H \in \mathrm{Cl}(\mathfrak{h}_{\mathfrak{p}_0}{}^+)$. But then from the second statement of Lemma 30, we can conclude that

$$\left| \sum_i \pi_0'(\lambda_i) \Gamma_i(\lambda) e^{\rho(H)} \theta(\lambda_i : \exp H ; \mu_0(b)) \right| \leqq a_3 (1 + \| \lambda \|)^{k_3}$$
$$(\lambda \subset \mathfrak{F}_R, H \subset \mathrm{Cl}(\mathfrak{h}_{\mathfrak{p}_0}{}^+))$$

for suitable constants $a_3, k_3 \geqq 0$. Hence it follows from Lemma 35 that we can choose constants $a_1, k_1 \geqq 0$ and a neighborhood $V$ of $H_0$ in $S_+$ such that

$$\left| \pi_0(\lambda) e^{t\rho(H)} \phi(\lambda : \exp tH ; b) \right| \leqq a_1 (1 + \| \lambda \|)^{k_1}$$

for $H \in V$, $\lambda \in \mathfrak{F}_R$ and $t \geqq 0$. This proves Lemma 36 and hence also Theorem 1.

**12. The space $I(G)$.** Let $x$ be an element in $G$. Then $x$ can be written uniquely (see [2(f), Lemma 31]) in the form $k \exp X$ ($k \in K, X \in \mathfrak{p}_0$). Put $D(x) = \{\det(\sinh \operatorname{ad} X/\operatorname{ad} X)_\mathfrak{p}\}^{\frac{1}{2}}$, where $(\sinh \operatorname{ad} X/\operatorname{ad} X)_\mathfrak{p}$ denotes the restriction on $\mathfrak{p}$ of the linear transformation

$$\sinh \operatorname{ad} X/\operatorname{ad} X = \sum_{q \geq 0} (\operatorname{ad} X)^{2q}/(2q+1)!$$

of $\mathfrak{g}$. Then $D$ is an analytic function on $G$ which is obviously spherical. Put [10] $D_q(x) = (1 + \| \log |x| \|)^q D(x)$.

DEFINITION. *Let $I(G)$ denote the set of all complex-valued functions $\phi$ on $G$ of class $C^\infty$ satisfying the following two conditions:*

(1)   *$\phi$ is spherical.*

(2)   *For any $b \in \mathfrak{B}$ and any integer $q \geq 0$,*

$$\sup_{x \in G} D_q(x) | \phi(x;b) | < \infty.$$

Put $\tau_{b,q}(\phi) = \sup D_q(x) | \phi(x;b)|$. Then $\tau_{b,q}$ is a seminorm on $I(G)$ and the collection of these seminorms, for all $b \in \mathfrak{B}$ and $q \geq 0$, defines a topology on $I(G)$ in the usual way so that $I(G)$ becomes a locally convex space.

We note that $(e^t - e^{-t})/2t = \sum_{q \geq 0} t^{2q}/(2q+1)!$ is an increasing function of $t$ for $t \geq 0$. Hence it follows easily that $D(x) \geq D(1) = 1$ ($x \in G$). Moreover $(1 - e^{-2t})/2t$ remains bounded for $t \geq 0$ and therefore $D(h)e^{-\rho(\log h)}$ also remains bounded as $h$ varies in $\operatorname{Cl}(A_\mathfrak{p}^+)$. We shall need these facts presently.

THEOREM 2. *For any $b \in \mathfrak{B}$, we can select an integer $q \geq 0$ and a constant $c > 0$ such that*

$$D(x) | \pi_0(\lambda)\phi(\lambda:x;b) | \leq c(1 + \| \lambda \|)^q$$

*for all $x \in G$ and $\lambda \in \mathfrak{F}_R$.*

Obviously we can choose a finite number of elements $b_1, \cdots, b_r \in \mathfrak{B}$ and analytic functions $a_1, \cdots, a_r$ on $K$ such that $b^k = \sum_{1 \leq i \leq r} a_i(k)b_i$ for $k \in K$. Now suppose $h \in \operatorname{Cl}(A_\mathfrak{p}^+)$ and $k \in K$. Then

$$\phi(\lambda:hk;b) = \phi(\lambda:h;b^k) = \sum_i a_i(k)\phi(\lambda:h;b_i)$$

and therefore

$$| \phi(\lambda:hk;b)| \leq a \sum_i | \phi(\lambda:h;b_i)|$$

where $a = \sup_{k \in K} (| a_1(k)|, \cdots, | a_r(k)|)$. On the other hand $D(hk) = D(h)$

3

and $D(h)e^{-\rho(\log h)}$ remains bounded as $h$ varies in $\mathrm{Cl}(A_\mathfrak{p}^+)$. Therefore by Theorem 1, we can choose an integer $q \geqq 0$ and a constant $c > 0$ such that

$$D(h) |\pi_0(\lambda)\phi(\lambda:hk;b)| \leqq c(1 + \|\lambda\|)^q$$

for all $\lambda \in \mathfrak{F}_R$, $h \in \mathrm{Cl}(A_\mathfrak{p}^+)$ and $k \in K$. But $G = K(\mathrm{Cl}(A_\mathfrak{p}^+))K$ (see [2(f), §12]) and $D$ and $\phi_\lambda$ are spherical functions. Hence it is clear that

$$D(x) |\pi_0(\lambda)\phi(\lambda:x;b)| \leqq c(1 + \|\lambda\|)^q$$

for $x \in G$ and $\lambda \in \mathfrak{F}_R$. This proves the theorem.

Now $\mathfrak{F}_R$ being a real Euclidean space, we can define the locally convex space $\mathscr{C}(\mathfrak{F}_R)$ as usual (see [2(g), §2]). It consists of all complex-valued $C^\infty$ functions $a$ on $\mathfrak{F}_R$ such that $\sup_{\lambda \in \mathfrak{F}_R} (1 + \|\lambda\|)^q |a(\lambda; \partial(p)| < \infty$ for $p \in S(\mathfrak{h}_\mathfrak{p})$ and any integer $q \geqq 0$. Let $d\lambda$ denote the Euclidean measure on $\mathfrak{F}_R$. Then if $a \in \mathscr{C}(\mathfrak{F}_R)$, it follows from Theorem 2 that the integral

$$\int_{\mathfrak{F}_R} |\pi_0(\lambda)a(\lambda)\phi(\lambda:x;b)| \, d\lambda \qquad (b \in \mathfrak{B})$$

converges uniformly with respect to $x \in G$. This shows that

$$\phi_a(x) = \int_{\mathfrak{F}_R} \pi_0(\lambda)a(\lambda)\phi(\lambda:x) \, d\lambda \qquad (x \in G)$$

is a $C^\infty$ function on $G$ and

$$\phi_a(x;b) = \int \pi_0(\lambda)a(\lambda)\phi(\lambda:x;b) \, d\lambda \qquad (b \in \mathfrak{B}).$$

THEOREM 3. *For any* $a \in \mathscr{C}(\mathfrak{F}_R)$, *put*

$$\phi_a(x) = \int_{\mathfrak{F}_R} \pi_0(\lambda)a(\lambda)\phi(\lambda:x) \, d\lambda \qquad (x \in G).$$

*Then* $\phi_a \in I(G)$ *and* $a \to \phi_a$ *is a continuous mapping of* $\mathscr{C}(\mathfrak{F}_R)$ *into* $I(G)$. *Also*

$$\phi_a(x;b) = \int_{\mathfrak{F}_R} \pi_0(\lambda)a(\lambda)\phi(\lambda:x;b) \, d\lambda \qquad (x \in G, b \in \mathfrak{B}).$$

It is obvious from its definition that $\phi_a$ is spherical. Moreover if $x = k_1 h k_2$ $(k_1, k_2 \in K; h \in \mathrm{Cl}(A_\mathfrak{p}^+))$, it is clear that

$$D_r(x)\phi_a(x;b) = D_r(h)\phi_a(h;b^{k_2})$$

for $b \in \mathfrak{B}$ and any integer $r \geqq 0$. Let $\mathfrak{D}(\mathfrak{h}_\mathfrak{p})$ denote the algebra of polynomial differential operators on $\mathfrak{h}_\mathfrak{p}$ (see [2(g), §2]). Then as we saw during the proof of Theorem 2, it would be sufficient to prove the following lemma.

Lemma 37. *Let $b$ be an element in $\mathfrak{B}$ and $s$ an integer $\geq 0$. Then we can select a finite number of elements $d_0, d_1, \cdots, d_N$ in $\mathfrak{D}(\mathfrak{h}_\mathfrak{p})$ such that*

$$(1 + \|\log h\|)^s e^{\rho(\log h)} |\phi_a(h; b)| \leq \sum_{0 \leq i \leq N} \sup_{\lambda \in \mathfrak{F}_R} |a(\lambda; d_i)|$$

*for $a \in \mathscr{B}(\mathfrak{F}_R)$ and $h \in \mathrm{Cl}(A_\mathfrak{p}^+)$.*

We shall use induction on $\dim \mathfrak{h}_\mathfrak{p}$. The case $\mathfrak{h}_\mathfrak{p} = \{0\}$ is obviously trivial. So let us assume that $\dim \mathfrak{h}_\mathfrak{p} \geq 1$. Define $S_+$ as in Section 11. Since $S_+$ is compact, it would be enough to prove the following statement.

Lemma 38. *Fix $b$ and $s$ as above. Then for a given $H_0 \in S_+$, we can select a neighborhood $V$ of $H_0$ in $S_+$ and a finite number of elements $d_0, \cdots, d_N \in \mathfrak{D}(\mathfrak{h}_\mathfrak{p})$ such that*

$$(1 + t)^s e^{t\rho(H)} |\phi_a(\exp tH; b)| \leq \sum_{0 \leq i \leq N} \sup_{\lambda \in \mathfrak{F}_R} |a(\lambda; d_i)|$$

*for $a \in \mathscr{B}(\mathfrak{F}_R)$, $H \in V$ and $t \geq 0$.*

We use the notation of Lemma 35 (corresponding to the element $H_0$ selected above). Then by this lemma, we can choose a neighborhood $V$ of $H_0$ in $S_+$, an integer $k \geq 0$ and two constants $c, t_0 > 0$ such that

$$\left| \pi_0(\lambda) e^{t\rho(H)} \phi(\lambda : \exp tH ; b) - \sum_{1 \leq i \leq r} \Gamma_i(\lambda) \pi_0'(\lambda_i) \theta(\lambda_i : \exp tH ; \mu_0(b)) e^{t\rho(H)} \right|$$

$$\leq c(1 + \|\lambda\|)^k e^{-t/t_0}$$

for $H \in V$, $\lambda \in \mathfrak{F}_R$ and $t \geq 0$. Put

$$f_a(m) = \int_{\mathfrak{F}_R} \pi_0'(\lambda) a(\lambda) \theta(\lambda : m) d\lambda \qquad (m \in M, a \in \mathscr{B}(\mathfrak{F}_R)).$$

Then by induction hypothesis, Lemma 37 is applicable to $\mathfrak{m}_0' = [\mathfrak{m}_0, \mathfrak{m}_0]$ instead of $\mathfrak{g}_0$ and so it follows easily that we can choose a finite number of elements $d_1', \cdots, d_p' \in \mathfrak{D}(\mathfrak{h}_\mathfrak{p})$ such that

$$(1 + \|\log h\|)^s e^{\rho(\log h)} |f_a(h; \mu_0(b))| \leq \sum_{1 \leq i \leq p} \sup_{\lambda \in \mathfrak{F}_R} |a(\lambda; d_i')|$$

for $a \in \mathscr{B}(\mathfrak{F}_R)$ and $h \in \mathrm{Cl}(A_\mathfrak{p}^+)$. Now put $a_i(\lambda) = \Gamma_i(s_i^{-1}\lambda) a(s_i^{-1}\lambda)$ $(1 \leq i \leq r)$ for $a \in \mathscr{B}(\mathfrak{F}_R)$ and $\lambda \in \mathfrak{F}_R$. Then it follows from the second statement of Lemma 30 that $a_i$ are also in $\mathscr{B}(\mathfrak{F}_R)$. Moreover in view of the inequality stated above, it is obvious that

$$\left| e^{t\rho(H)} \phi_a(\exp tH ; b) \right.$$

$$\left. - \sum_{1 \leq i \leq r} f_{a_i}(\exp tH ; \mu_0(b)) e^{t\rho(H)} \right| \leq c e^{-t/t_0} \int (1 + \|\lambda\|)^k |a(\lambda)| d\lambda$$

for $H \in V$ and $t \geq 0$. On the other hand we can choose $d_0 \in S(\mathfrak{h}_\mathfrak{p})$ such that

$$c(1+t)^s e^{-t/t_0} \int (1+\| \lambda \|)^k |a(\lambda)| \, d\lambda \leq \sup_{\lambda \in \mathfrak{F}_R} |d_0(\lambda)a(\lambda)|$$

for $t \geq 0$ and $a \in \mathcal{B}(\mathfrak{F}_R)$. Furthermore

$$(1+t)^s e^{t\rho(H)} |f_{a_i}(\exp tH ; \mu_0(b))| \leq \sum_{1 \leq j \leq p} \sup_{\lambda \in \mathfrak{F}_R} |a(\lambda; d_j')|$$

for $H \in V$ and $t \geq 0$ and, in view of the second statement of Lemma 30, we can select $d_1, \cdots, d_N \in \mathfrak{D}(\mathfrak{h}_\mathfrak{p})$ such that

$$\sum_{1 \leq i \leq r} \sum_{1 \leq j \leq p} \sup_{\lambda \in \mathfrak{F}_R} |a_i(\lambda; d_j')| \leq \sum_{1 \leq i \leq N} \sup_{\lambda \in \mathfrak{F}_R} |a(\lambda; d_i)|$$

for $a \in \mathcal{B}(\mathfrak{F}_R)$. Therefore the assertion of Lemma 38 is now obvious. This completes the proof of Theorem 3.

### 13. The distribution $T_\mu$.

Let $dx$ denote the Haar measure on $G$.

LEMMA 39. *Fix an element $\mu$ in $\mathfrak{F}_R$. Then for any $a \in \mathcal{B}(\mathfrak{F}_R)$, the integral*

$$\int_G (\operatorname{conj} \phi(\mu : x)) \phi_a(x) \, dx$$

*is well defined and the mapping*

$$T_\mu : a \to \int (\operatorname{conj} \phi(\mu : x)) \phi_a(x) \, dx \qquad (a \in \mathcal{B}(\mathfrak{F}_R))$$

*is a $\mathcal{B}$-distribution on $\mathfrak{F}_R$.*

This is an immediate consequence of Theorem 3 and the following result.

LEMMA 40. *Let $f \in I(G)$, $\mu \in \mathfrak{F}_R$ and $b_1, b_2 \in \mathfrak{B}$. Then the integral*

$$\int_G |\phi(\mu : xy ; b_1)f(x ; b_2)| \, dx \qquad (y \in G)$$

*converges uniformly with respect to $y$ on every compact subset of $G$. Moreover the mapping*

$$f \to \int_G |\phi(\mu : x ; b_1)f(x ; b_2)| \, dx \qquad (f \in I(G))$$

*is continuous on $I(G)$.*

From Lemma 46 of [2(i)] we can select a positive number $c_1$ such that $|\phi(\mu : x ; b_1)| \leq c_1 \phi(0 : x)$ for $x \in G$. Moreover if $\Omega$ is a compact subset of $G$, there exists a positive number $c_2$ such that $\phi(0 : xy) \leq c_2 \phi(0 : x)$ for

$x \in G$ and $y \in \Omega$ (see [2(i), Lemma 48]). Therefore it would be sufficient to prove that

$$\int |\phi(0:x)f(x;b_2)|\,dx < \infty$$

and the mapping

$$f \to \int |\phi(0:x)f(x;b_2)|\,dx \qquad (f \in I(G))$$

is continuous on $I(G)$. Put $\Delta(\exp H) = \prod_{\alpha \in P_+} (e^{\alpha(H)} - e^{-\alpha(H)})$ $(H \in \mathfrak{h}_{\mathfrak{p}_0})$ as in [2(i), §5]. Then if the Haar measure $dh$ on $A_{\mathfrak{p}}$ is suitably normalized, we have

$$\int g(x)\,dx = \int_{A_{\mathfrak{p}}^+} \Delta(h)\,dh \int_{K \times K} g(k_1 h k_2)\,dk_1 dk_2$$

for any $g \in C_c(G)$ (see [2(f), §12]). Therefore

$$\int |\phi(0:x)f(x;b_2)|\,dx = \int_{A_{\mathfrak{p}}^+} \Delta(h)\phi(0:h)\,dh \int_K |f(h;b_2^k)|\,dk.$$

We can obviously select a finite number of linearly independent elements $b_3, \cdots, b_N$ in $\mathfrak{B}$ and analytic functions $g_3, \cdots, g_N$ on $K$ such that $b_2^k = \sum_{3 \le i \le N} g_i(k) b_i$ $(k \in K)$. Then

$$|f(h;b_2^k)| \le \sum_{3 \le i \le N} |g_i(k)|\,|f(h;b_i)|.$$

Since $K$ is compact, the $g_i$ remain bounded on $K$ and so it would be sufficient to verify that for any $b \in \mathfrak{B}$,

$$\int_{A_{\mathfrak{p}}^+} \Delta(h)\phi(0:h)|f(h;b)|\,dh < \infty$$

and the mapping

$$f \to \int_{A_{\mathfrak{p}}^+} \Delta(h)\phi(0:h)|f(h;b)|\,dh \qquad (f \in I(G))$$

is continuous on $I(G)$. Choose an integer $q_1 \ge 0$ such that

$$\int_{A_{\mathfrak{p}}} (1 + \|\log h\|)^{-q_1}\,dh < \infty$$

Since $|\Delta(h)|^{\frac{1}{2}} e^{-\rho(\log h)}$ obviously remains bounded for $h \in A_{\mathfrak{p}}^+$, we can, from Theorem 3 of [2(i)], select an integer $q_2 \ge 0$ such that

$$(1 + \|\log h\|)^{-q_2} |\Delta(h)|^{\frac{1}{2}} \phi(0:h)$$

is bounded on $A_{\mathfrak{p}}^+$. On the other hand it follows from the definition of $D$ that $|\Delta(h)|^{\frac{1}{2}}(1+\|\log h\|)^{-q_3}D(h)^{-1}$ remains bounded on $A_\mathfrak{p}$ for a suitable integer $q_3 \geqq 0$. Hence there exists a positive number $c$ such that $\Delta(h)\phi(0:h) \leqq cD_{q_2+q_3}(h)$ for $h \in A_\mathfrak{p}^+$. Therefore if we put $q = q_1 + q_2 + q_3$, it is clear that

$$\int_{A_\mathfrak{p}^+} \Delta(h)\phi(0:h)|f(h;b)|\,dh \leqq c\tau_{b,q}(f) \int_{A_\mathfrak{p}} (1+\|\log h\|)^{q_1}\,dh$$

in the notation of Section 12. This proves our assertion.

COROLLARY 1. *Let $\Omega$ be a compact subset of $G \times G$. Then if $f \in I(G)$, the integral* $\int_G |\phi(\mu:xy;b_1)f(xz;b_2)|\,dx$ *converges uniformly for $(y,z) \in \Omega$.*

Without loss of generality we may assume that $\Omega = \omega_1 \times \omega_2$ where $\omega_1, \omega_2$ are compact subsets of $G$. Let $\omega$ be the image of $\Omega$ in $G$ under the mapping $(y,z) \to z^{-1}y$. Then $\omega$ is also compact. Given any $\epsilon > 0$ we can, by Lemma 40, select a compact set $V$ in $G$ such that

$$\int_{cV} |\phi(\mu:xy;b_1)f(x;b_2)|\,dx < \epsilon$$

for $y \in \omega$. (As usual $^cV$ denotes the complement of $V$ in $G$.) Put $U = V\omega_2^{-1}$. Then $U$ is also compact and $^cUz \subset {}^cV$ for $z \in \omega_2$. Hence if $y \in \omega_1$ and $z \in \omega_2$, it is clear that

$$\int_{cU} |\phi(\mu:xy;b_1)f(xz;b_2)|\,dx = \int_{cUz} |\phi(\mu:xz^{-1}y;b_1)f(x;b_2)|\,dx < \epsilon$$

from the right-invariance of the measure $dx$. This proves the corollary.

Now put

$$F(y:z) = \int_G \{\operatorname{conj}\phi(\mu:xy)\}f(xz)\,dx \qquad (y,z \in G).$$

Then the above corollary implies that $F$ is a $C^\infty$-function on $G \times G$ and

$$F(y;b_1:z;b_2) = \int_G (\operatorname{conj}\phi(\mu:xy;b_1))f(xz;b_2)\,dx$$

for $b_1, b_2 \in \mathfrak{B}$. On the other hand $F(y;b_1:z;b_2) = F(z^{-1}y;b_1:1;b_2)$ by the right-invariance of $dx$. Hence if we put $y = 1$ and $z = \exp tX$ ($X \in \mathfrak{g}_0, t \in R$) and differentiate with respect to $t$ at $t = 0$, we get $F(1;b_1:1;Xb_2) = -F(1;Xb_1:1;b_2)$. Let $b \to b^*$ denote the anti-automorphism of $\mathfrak{B}$ over $R$ such that $(X + (-1)^{\frac{1}{2}}Y)^* = -X + (-1)^{\frac{1}{2}}Y$ ($X, Y \in \mathfrak{g}_0$). Then the above result obviously implies the following corollary.

COROLLARY 2. *For any $b \in \mathfrak{B}$, $f \in I(G)$ and $\mu \in \mathfrak{F}_R$,*

$$\int_G (\operatorname{conj}\phi(\mu:x))f(x;b)\,dx = \int_G (\operatorname{conj}\phi(\mu:x;b^*))f(x)\,dx.$$

Moreover since $|\phi(\mu:x)| \leqq \phi(0:x)$ and the integral $\displaystyle\int \phi(0:x)|f(x)|\,dx$ is convergent, the following result is also obvious.

COROLLARY 3. *Fix an element $f \in I(G)$. Then the mapping*

$$\mu \to \int_G (\operatorname{conj}\phi(\mu:x))f(x)\,dx \qquad\qquad (\mu \in \mathfrak{F}_R)$$

*is continuous on $\mathfrak{F}_R$.*

We shall now try to determine the distribution $T_\mu$ of Lemma 39 explicitly. Define the homomorphism $\gamma$ of $I_\mathfrak{g}$ into $I(\mathfrak{h}_\mathfrak{p})$ as in Section 4. Then we know from Lemma 18 of [2(i)] that $\phi(\lambda:x;q) = \gamma(q:(-1)^{\mathfrak{i}}\lambda)\phi(\lambda:x)$ $(x \in G,$ $q \in I_\mathfrak{g}, \lambda \in \mathfrak{F})$. Let $p_q$ denote the polynomial function on $\mathfrak{F}$ whose value at $\lambda$ is $\gamma(q:(-1)^{\mathfrak{i}}\lambda)$. Then it follows from Theorem 3 that $q\phi_a = \phi_{p_q a}$ for $a \in \mathcal{B}(\mathfrak{F}_R)$ and we conclude from Corollary 2 of Lemma 40 that

$$T_\mu(p_q a) = \int (\operatorname{conj}\phi(\mu:x))\phi_a(x;q)\,dx = \int (\operatorname{conj}\phi(\mu:x;q^*))\phi_a(x)\,dx.$$

On the other hand $\operatorname{conj}\phi(\mu:x) = \phi(\mu:x^{-1})$ (see [2(i), Lemma 47]). Therefore it is obvious that $\operatorname{conj}\phi(\mu:1;b^*) = \phi(\mu:1;b)$ for any $b \in \mathfrak{B}$. On the other hand if $b$ lies in $I_\mathfrak{g}$, the same holds for $b^*$ and $\phi(\mu:1;b) = \gamma(b:(-1)^{\mathfrak{i}}\mu)$. Therefore

$$\operatorname{conj}\gamma(q^*:(-1)^{\mathfrak{i}}\mu) = \operatorname{conj}\phi(\mu:1;q^*) = \phi(\mu:1;q) = \gamma(q:(-1)^{\mathfrak{i}}\mu).$$

This proves that $T_\mu(p_q a) = p_q(\mu)T_\mu(a)$. Now let $p$ be any element in $I(\mathfrak{h}_\mathfrak{p})$. Then by Theorem 1 of [2(i)], $p = p_q$ for some $q \in I_\mathfrak{g}$ and so we get the following result.

LEMMA 41. $pT_\mu = p(\mu)T_\mu$ *for any* $p \in I(\mathfrak{h}_\mathfrak{p})$ *and* $\mu \in \mathfrak{F}_R$.

COROLLARY. *There exists a continuous function $\mathfrak{\beta}$ on $\mathfrak{F}_R$ such that*

$$\pi_0(\mu)T_\mu(a) = \mathfrak{\beta}(\mu)\sum_{s \in W}\epsilon_0(s)a(s\mu)$$

*for all $a \in \mathcal{B}(\mathfrak{F}_R)$ and $\mu \in \mathfrak{F}_R$. Moreover $\mathfrak{\beta}$ is unique and $\mathfrak{\beta}(s\mu) = \mathfrak{\beta}(\mu)$ $(s \in W)$.*

Fix a point $\mu_0 \in \mathfrak{F}_R$. Then if $\mu \in \mathfrak{F}_R$ and $\mu \neq s\mu_0$ for any $s \in W$, it is clear that we can choose $p_0 \in I(\mathfrak{h}_\mathfrak{p})$ such that $p_0(\mu) \neq p_0(\mu_0)$. Put $p = p_0 - p_0(\mu_0)$. Then it follows from Lemma 41 that $pT_{\mu_0} = 0$. On the other hand $p(\mu) \neq 0$ and so we conclude that $\mu$ does not lie in the carrier of $T_{\mu_0}$. This shows that this carrier is contained in the finite set of points $s\mu_0$ $(s \in W)$.

For any function $a$ on $\mathfrak{F}_R$ and $s \in W$ define the function $a^s$ by $a^s(\lambda)$ $= a(s^{-1}\lambda)$ $(\lambda \in \mathfrak{F}_R)$. Then if $a \in \mathscr{B}(\mathfrak{F}_R)$, it is obvious that $a^s$ also lies in $\mathscr{B}(\mathfrak{F}_R)$ and $\phi_{a^s} = \epsilon_0(s)\phi_a$. Hence $T_\mu(a^s) = \epsilon_0(s)T_\mu(a)$. Also $T_\mu = T_{s\mu}$ since $\phi_{s\mu} = \phi_\mu$ (see the corollary of Lemma 17 of [2(i)]). Let $U$ be an open neighborhood of $\mu_0$ in $\mathfrak{F}_R$ such that its closure is compact and $sU = U$ $(s \in W)$. Select a function $a_0 \in C_c^\infty(\mathfrak{F}_R)$ such that $a_0 = 1$ on $U$. Then if $\mu \in U$, it is clear that $\sum_{s \in W} \epsilon_0(s)\pi_0(s\mu)a_0(s\mu) = w\pi_0(\mu)$ where $w$ is the order of $W$. Hence the uniqueness of $\boldsymbol{\beta}$ follows from its continuity and the equation $\pi_0(\mu)T_\mu(\pi_0 a_0) = w\boldsymbol{\beta}(\mu)\pi_0(\mu)$ $(\mu \in U)$. Therefore $\boldsymbol{\beta}(s\mu) = \boldsymbol{\beta}(\mu)$ since $T_{s\mu} = T_\mu$. Thus only the existence of $\boldsymbol{\beta}$ requires proof. Fix an integer $k \geqq 0$. Then for any given $a \in \mathscr{B}(\mathfrak{F}_R)$ and $\mu \in U$, we can select a polynomial $p \in S(\mathfrak{h}_\mathfrak{p})$ such that all derivatives of $a - p$ of order $\leqq k$ are zero at the points $s\mu$ $(s \in W)$. Hence if $k$ is chosen sufficiently large, $T_\mu(a) = T_\mu(pa_0)$ (see Schwartz [4, p. 99]). But $a_0{}^s - a_0$ is identically zero on $U$ and so it is obvious that $T_\mu(pa_0) = \epsilon_0(s)T_\mu(p^s a_0)$. Hence $T_\mu(a) = T_\mu(qa_0)$ where $q = w^{-1}\sum_{s \in W}\epsilon_0(s)p^s$. But from Lemma 32 we can choose $p_0 \in I(\mathfrak{h}_\mathfrak{p})$, such that $q = \pi_0 p_0$. Therefore

$$T_\mu(a) = T_\mu(p_0\pi_0 a_0) = p_0(\mu)T_\mu(\pi_0 a_0) \qquad (\mu \in U).$$

from Lemma 41. On the other hand $\pi_0(\mu)p_0(\mu) = q(\mu) = w^{-1}\sum_{s \in W}\epsilon_0(s)p(s\mu)$ $= w^{-1}\sum_{s \in W}\epsilon_0(s)a(s\mu)$. Therefore

$$\pi_0(\mu)T_\mu(a) = w^{-1}T_\mu(\pi_0 a_0)\sum_{s \in W}\epsilon_0(s)a(s\mu) \qquad (\mu \in U).$$

But from Corollary 3 of Lemma 40, $T_\mu(\pi_0 a_0)$ is a continuous function of $\mu$ and so our assertion is proved.

Define $\boldsymbol{b}_0$ as in Section 9. We shall see in Section 15 that if the measure $d\lambda$ on $\mathfrak{F}_R$ is suitably normalized, $\boldsymbol{\beta}(\lambda) = |\boldsymbol{b}_0(\lambda)|^2$ for $\lambda \in \mathfrak{F}_R$. This, in fact, is one of the main results of this paper.

**14. An inequality and its consequences.** Put $\mathfrak{n} = \sum_{\alpha \in P_+} CX_\alpha$ as in [2(i), §2] and $\bar{\mathfrak{n}} = \theta\mathfrak{n} = \sum_{\alpha \in P_+} CX_{-\alpha}$ where $\theta$ is the automorphism of $\mathfrak{g}$ defined in Section 2. Let $N$ and $\bar{N}$ be the analytic subgroups of $G$ corresponding to $\mathfrak{n} \cap \mathfrak{g}_0$ and $\bar{\mathfrak{n}} \cap \mathfrak{g}_0$ respectively. We normalize their Haar measures $dn$ and $d\bar{n}$ in such a way (see [2(i), Lemma 44]) that

$$\int_N e^{-2\rho(H(n'))}dn = \int_{\bar{N}} e^{-2\rho(H(\bar{n}))}d\bar{n} = 1$$

where $n' = \theta(n^{-1})$ $(n \in N)$.

LEMMA 42. *Let $h$ and $\bar{n}$ be two elements in $A_{\mathfrak{p}}$ and $\bar{N}$ respectively and select $h' \in \mathrm{Cl}(A_{\mathfrak{p}}^+)$ such that $\bar{n}h \in Kh'K$. Then $\rho(\log h') \geq \rho(H(\bar{n})) + \rho(\log h)$ and therefore $\|\log h'\| \geq \|\rho\|^{-1}\{\rho(H(\bar{n})) + \rho(\log h)\}$.*

Put $\Lambda = \frac{1}{2}\sum\limits_{\alpha \in P} \alpha$ and let $\sigma$ denote the irreducible representation of $\mathfrak{g}$ on a finite-dimensional space $V$ with the highest weight $\Lambda$ (see the proof of Lemma 45 of $[2(\mathrm{i})]$). Choose a unit vector $\psi_0$ in $V$ belonging to the weight $\Lambda$. Then if $\bar{n}h = k_1 h' k_2$ $(k_1, k_2 \in K)$, it follows that $|\sigma(\bar{n}h)\psi_0| = |\sigma(h'k_2)\psi_0|$. But since $\Lambda$ coincides with $\rho$ on $\mathfrak{h}_{\mathfrak{p}}$, this shows that $\rho(H(\bar{n})) + \rho(\log h) = \rho(H(h'k_2))$. However $h' \in \mathrm{Cl}(A_{\mathfrak{p}}^+)$ and therefore we conclude from Corollary 3 of Lemma 35 of $[2(\mathrm{i})]$ that $\rho(H(h'k_2)) \leq \rho(\log h')$. Hence the lemma.

COROLLARY 1. *Let $\Omega$ be a subset of $A_{\mathfrak{p}}$ such that $\rho(\log h)$ remains bounded from below as $h$ varies in $\Omega$. Then for any integer $k \geq 0$, we can select a positive number $c_1$ and an integer $k_1 \geq 0$ such that*

$$D_{k_1}(\bar{n}h) \geq c_1\{1 + \rho(H(\bar{n}))\}^k \exp\{\rho(H(\bar{n})) + |\rho(\log h)|\}$$

*for all $\bar{n} \in \bar{N}$ and $h \in \Omega$.*

It is easily seen that we can select an integer $d \geq 0$ and a positive number $c_2$ such that $D(h_0)e^{-\rho(\log h_0)}(1 + \|\log h_0\|)^d \geq c_2$ for $h_0 \in \mathrm{Cl}(A_{\mathfrak{p}}^+)$. Therefore if $h'$ is defined as above, it follows that

$$D_d(\bar{n}h) = D_d(h') \geq c_2 e^{\rho(\log h')} \geq c_2 \exp\{\rho(H(\bar{n})) + \rho(\log h)\}.$$

Hence in view of Lemma 42 and our assumption on $\Omega$, the statement of the corollary is obvious.

COROLLARY 2. *Let $U$ be a compact subset of $\bar{N}$ and fix $f$ and $b$ in $I(G)$ and $\mathfrak{B}$ respectively. Then the integral $\int_{\bar{N}} |f(\bar{n}\bar{n}_0 h; b)| d\bar{n}$ converges uniformly as $\bar{n}_0$ and $h$ vary in $U$ and $\Omega$ respectively. Moreover, we can find an integer $r \geq 0$ and a positive number $c$ such that*

$$\int_{\bar{N}} |f(\bar{n}h; b)| d\bar{n} \leq c \sup_{x \in G} D_r(x)|f(x; b)|$$

*for all $h \in \Omega$, $f \in I(G)$ and $b \in \mathfrak{B}$.*

From Lemma 45 of $[2(\mathrm{i})]$, we can select an integer $k \geq 0$ so large that

$$c_2 = \int_{\bar{N}} \{1 + \rho(H(\bar{n}))\}^{-k} e^{-\rho(H(\bar{n}))} d\bar{n} < \infty.$$

Choose $k_1$ and $c_1$ corresponding to Corollary 1 above. Then

$$|f(\bar n h;b)| \leq c_1^{-1}\tau_b(f)e^{-\rho(H(\bar n))}\{1+\rho(H(\bar n))\}^{-k}$$

for $\bar n \in \bar N$, $h \in \Omega$, $b \in \mathfrak{B}$ and $f \in I(G)$, where

$$\tau_b(f) = \sup_{x \in G} D_{k_1}(x)|f(x;b)|.$$

Hence it is sufficient to choose $r = k_1$ and $c = c_1^{-1}c_2$. Moreover for any $\epsilon > 0$, we can select a compact set $V_0$ in $\bar N$ such that

$$\int_{{}^c V_0} \{1+\rho(H(\bar n))\}^{-k} e^{-\rho(N(\bar n))}d\bar n < \epsilon.$$

($^c V_0$ is the complement of $V_0$ in $\bar N$.) Put $V = V_0 U^{-1}$. Then $V$ is also compact and

$$\int_{{}^c V} \{1+\rho(H(\bar n \bar n_0))\}^{-k}\exp\{-\rho(H(\bar n \bar n_0))\}d\bar n < \epsilon$$

for $\bar n_0 \in U$. From this the first assertion of the corollary follows immediately.

Put $\bar{\mathfrak{N}}_1 = \bar{\mathfrak{N}}\bar n$ where $\bar{\mathfrak{N}}$ is the subalgebra of $\mathfrak{B}$ generated by $(1,\bar n)$. Moreover for any function $F$ on $G$ write $F(\bar n:h) = F(\bar n h)$ $(\bar n \in \bar N, h \in A_\mathfrak{p})$.

COROLLARY 3. *Suppose $\eta \in \bar{\mathfrak{N}}_1$ and $u \in \mathfrak{H}_\mathfrak{p}$. Then*

$$\int_{\bar N} f(\bar n;\eta:h;u)d\bar n = 0$$

*for $f \in I(G)$ and $h \in A_\mathfrak{p}$.*

Fix $h$ and put $\eta' = (\eta)^{h^{-1}}$ for any $\eta \in \bar{\mathfrak{N}}$. Then $f(\bar n;\eta:h;u) = f(\bar n h;\eta'u)$ and therefore, by Corollary 2, the integral

$$\int_{\bar N} |f(\bar n \bar n_0;\eta:h;u)|\,d\bar n$$

converges uniformly as $\bar n_0$ varies within a compact subset of $\bar N$. Put

$$F(\bar n_0) = \int f(\bar n \bar n_0;\eta:h;u)d\bar n \qquad\qquad (\bar n_0 \in \bar N).$$

Then it follows from the above remarks that $F$ is of class $C^\infty$ on $\bar N$ and

$$F(\bar n_0;\eta_1) = \int f(\bar n \bar n_0;\eta_1\eta:h;u)d\bar n$$

for $\eta_1 \in \bar{\mathfrak{N}}$. However the Haar measure on the nilpotent group $\bar N$ is both left- and right-invariant. Therefore $F(\bar{n}_0) = F(1)$ and hence $F(\bar{n}_0;X) = 0$ for $X \in \bar{\mathfrak{n}}$. This shows that

$$\int f(\bar{n};X_\eta:h;u)\,d\bar{n} = 0$$

for $X \in \bar{\mathfrak{n}}$ and $\eta \in \mathfrak{R}$ and now our assertion is obvious.

Let $\mathscr{B}(A_\mathfrak{p})$ denote the set of all functions $F$ on $A_\mathfrak{p}$ of class $C^\infty$ such that

$$\sigma_{q,u}(F) = \sup_{h \in A_\mathfrak{p}} (1 + \|\log h\|)^q \,|\,F(h;u)\,| < \infty$$

for any integer $q \geqq 0$ and $u \in \mathfrak{H}_\mathfrak{p}$. We define a topology in $\mathscr{B}(A_\mathfrak{p})$ by means of the seminorms $\sigma_{q,u}$ and thus turn it into a locally convex space. Let $I(A_\mathfrak{p})$ denote the subspace of $\mathscr{B}(A_\mathfrak{p})$ consisting of those elements which are invariant under $W$. Obviously $I(A_\mathfrak{p})$ is closed in $\mathscr{B}(A_\mathfrak{p})$.

COROLLARY 4. *For any* $f \in I(G)$, *put* $F_f(h) = e^{\rho(\log h)} \int_{\bar{N}} f(\bar{n}h)\,d\bar{n}$ $(h \in A_\mathfrak{p})$. *Then* $F_f \in I(A_\mathfrak{p})$ *and* $f \to F_f$ *is a continuous mapping of* $I(G)$ *into* $I(A_\mathfrak{p})$. *Moreover*

$$F_f(h;u) = e^{\rho(\log h)} \int_{\bar{N}} f(\bar{n}h;u')\,d\bar{n} \qquad (h \in A_\mathfrak{p}, u \in \mathfrak{H}_\mathfrak{p})$$

*where* $u' = e^{-\rho}u \circ e^{\rho}$.

It is obvious from Corollary 2 that $F_f$ is of class $C^\infty$ and

$$F_f(h;u) = e^{\rho(\log h)} \int_{\bar{N}} f(\bar{n}h;u')\,d\bar{n}$$

for $u \in \mathfrak{H}_\mathfrak{p}$. Moreover $f(x) = f(\theta(x^{-1}))$ $(x \in G)$, since $f$ is spherical. Therefore

$$F_f(h) = e^{\rho(\log h)} \int_N f(hn)\,dn.$$

We can now apply to $f$ the proof of Lemma 17 of $[2(\mathrm{i})]$ and conclude that $F_f(h^s) = F_f(h)$ $(s \in W)$.

Select $k$ as in the proof of Corollary 2. Since $\rho(\log h) \geqq 0$ for $h \in \mathrm{Cl}(A_\mathfrak{p}{}^+)$, we can take $\Omega = \mathrm{Cl}(A_\mathfrak{p}{}^+)$. Now choose $k_1$ as in Corollary 1. Then it is obvious from Lemma 42, that for any integer $q \geqq 0$, there exists a positive constant $c_q$ such that

$$D_{k_1+q}(\bar{n}h) \geqq c_q(1 + \|\log h\|)^q \{1 + \rho(H(\bar{n}))\}^k \exp\{\rho(\log h) + \rho(H(\bar{n}))\}$$

for $h \in \mathrm{Cl}(A_\mathfrak{p}{}^+)$ and $\bar{n} \in \bar{N}$. This shows that

$$(1 + \|\log h\|)^q \,|\,F_f(h;u)\,| \leqq cc_q^{-1} \sup_{x \in G} D_{k_1+q}(x)\,|\,f(x;u')\,|$$

for $h \in \mathrm{Cl}(A_\mathfrak{p}^{+})$, $u \in \mathfrak{H}_\mathfrak{p}$ and $f \in I(G)$, where

$$c = \int_{\bar{N}} e^{-\rho(H(\bar{n}))} \{1 + \rho(H(\bar{n}))\}^{-k} \, d\bar{n}.$$

Let $u \to u^s$ $(s \in W)$ denote the automorphism of $\mathfrak{H}_\mathfrak{p}$ which coincides with $s$ on $\mathfrak{h}_\mathfrak{p}$. Then obviously $F_f(h \, ; u) = F_f(h^s \, ; u^s)$. Therefore since $A_\mathfrak{p} = \bigcup_{s \in W} s(\mathrm{Cl}(A_\mathfrak{p}^{+}))$, it follows that

$$\sup_{h \in A_\mathfrak{p}} (1 + \| \log h \|)^q \, | F_f(h \, ; u)| \leqq c c_q^{-1} \sum_{s \in W} \sup_{x \in G} D_{k_1 + q}(x) | f(x \, ; (u^s)')|$$

for $f \in I(G)$, $u \in \mathfrak{H}_\mathfrak{p}$ and $q \geqq 0$. This proves the corollary.

We normalize the Haar measure $dh$ on $A_\mathfrak{p}$ in such a way (see [2(b), Lemma 35]) that $dx = e^{2\rho(\log h)} dk \, dh \, dn$ for $x = khn$ $(k \in K, h \in A_\mathfrak{p}, n \in N)$.

LEMMA 43. *For any* $\lambda \in \mathfrak{F}_R$ *and* $f \in I(G)$,

$$\int_G \phi_\lambda(x) f(x) \, dx = \int_{A_\mathfrak{p}} F_f(h) \exp\{(-1)^{\frac{1}{2}} \lambda(\log h)\} \, dh.$$

We keep to the notation of the proof of Corollary 4 of Lemma 42. Select an integer $q \geqq 0$ so large that $\int_{A_\mathfrak{p}} (1 + \| \log h \|)^{-q} \, dh < \infty$. Fix $f \in I(G)$ and put

$$g(h) = e^{\rho(\log h)} \int_N | f(hn)| \, dn = e^{\rho(\log h)} \int_{\bar{N}} | f(\bar{n}h)| \, d\bar{n}$$

for $h \in A_\mathfrak{p}$. One proves without difficulty (see [2(i), Lemma 17]) that $g(h^s) = g(h)$ for $s \in W$. Hence

$$\int_G | f(x)| \, e^{-\rho(H(x))} dx = \int_{A_\mathfrak{p}} g(h) \, dh = w \int_{A_\mathfrak{p}^{+}} g(h) \, dh$$

where $w$ is the order of $W$. On the other hand

$$e^{-\rho(H(\bar{n}))} \{1 + \rho(H(\bar{n}))\}^{-k} \sup_{x \in G} D_{k_1 + q}(x) | f(x)|$$

$$\geqq c_q (1 + \| \log h \|)^q e^{\rho(\log h)} | f(\bar{n}h)|$$

for $\bar{n} \in \bar{N}$ and $h \in \mathrm{Cl}(A_\mathfrak{p}^{+})$ and therefore

$$g(h) \leqq c c_q^{-1} (1 + \| \log h \|)^{-q} \sup_{x \in G} D_{k_1 + q}(x) | f(x)| \qquad (h \in A_\mathfrak{p}^{+}).$$

This shows that $\displaystyle\int | f(x)| \, e^{-\rho(H(x))} dx < \infty$. Hence it follows immediately that

$$\int_G \phi_\lambda(x)f(x)dx = \int_G f(x)\exp\{(-1)^{\frac{1}{2}}\lambda(H(x)) - \rho(H(x))\}dx$$

$$= \int_{A_\mathfrak{p}} \exp\{(-1)^{\frac{1}{2}}\lambda(\log h) + \rho(\log h)\}dh \int_N f(hn)dn$$

$$= \int_{A_\mathfrak{p}} F_f(h)\exp\{(-1)^{\frac{1}{2}}\lambda(\log h)\}dh.$$

Let us now assume that $d\lambda$ is the *regular* Euclidean measure (see [2(g), §2]) on $\mathfrak{F}_R$ and the Haar measure $dh$ on $A_\mathfrak{p}$ is so normalized that

$$\int_{A_\mathfrak{p}} dh \int_{\mathfrak{F}_R} a(\lambda)\exp\{(-1)^{\frac{1}{2}}\lambda(\log h)\}d\lambda = a(0)$$

for any $a \in \mathcal{B}(\mathfrak{F}_R)$. In view of the relation $dx = e^{2\rho(\log h)}dkdhdn$, this fixes the normalization of $dx$ as well. We shall call these the regular normalizations of $d\lambda$, $dx$ and $dh$. Define $\boldsymbol{\beta}$ as in the corollary of Lemma 41.

COROLLARY. *Let a be any element in $\mathcal{B}(\mathfrak{F}_R)$. Then*

$$e^{\rho(H)}\int_{\bar{N}} \phi_a(\bar{n}\exp H)d\bar{n} = \int_{\mathfrak{F}_R} \boldsymbol{\beta}(\lambda)\{\pi_0(\lambda)^{-1}\sum_{s\in W}\epsilon_0(s)a(s\lambda)\}e^{(-1)^{\frac{1}{2}}\lambda(H)}d\lambda$$

*for $H \in \mathfrak{h}_{\mathfrak{p}_0}$.*

Put $f = \phi_a$. Then from Theorem 3, $f$ lies in $I(G)$ and therefore by Lemma 43,

$$\int F_f(h)\exp\{-(-1)^{\frac{1}{2}}\lambda(\log h)\}dh = \int \phi_a(x)\mathrm{conj}\,\phi_\lambda(x)dx \quad (\lambda \in \mathfrak{F}_R).$$

Now apply the corollary of Lemma 41. Then

$$\pi_0(\lambda)\int F_f(h)\exp\{-(-1)^{\frac{1}{2}}\lambda(\log h)\}dh = \boldsymbol{\beta}(\lambda)\sum_{s\in W}\epsilon_0(s)a(s\lambda).$$

Therefore our assertion follows immediately by the Fourier inversion formula.

We shall see in the next section that the above relation also holds if $\boldsymbol{\beta}(\lambda)$ is replaced by $|\boldsymbol{b}_0(\lambda)|^2$. This would prove that $\boldsymbol{\beta}(\lambda) = |\boldsymbol{b}_0(\lambda)|^2$ for $\lambda \in \mathfrak{F}_R$.

**15. Determination of the function $\boldsymbol{\beta}$.** Extend $\theta$ to an automorphism of $\mathfrak{B}$ and put $\mathfrak{N} = \theta(\bar{\mathfrak{N}})$, $\mathfrak{N}_1 = \theta(\bar{\mathfrak{N}}_1)$ and $\gamma'(q) = e^{-\rho}\gamma(q) \circ e^\rho$ $(q \in I_\mathfrak{g})$. Also define $\mathfrak{B}_d$ as in Lemma 9.

LEMMA 44. *Let q be an element in $I_\mathfrak{g}$ and let d denote the degree of $\gamma(q)$. Then $q - \gamma'(q) \in \mathfrak{B}\mathfrak{k} + (\bar{\mathfrak{N}}_1\mathfrak{H}_\mathfrak{p}) \cap \mathfrak{B}_d$.*

Define $\mathfrak{P}$ as in Lemma 19 of $[2(\mathrm{i})]$. Then we have seen in $[2(\mathrm{i}), \S 4]$ that $I_\mathfrak{g} = I_\mathfrak{g} \cap \mathfrak{P} + I_\mathfrak{g} \cap (\mathfrak{k}\mathfrak{B})$. On the other hand $I_\mathfrak{g} \cap (\mathfrak{k}\mathfrak{B}) = I_\mathfrak{g} \cap (\mathfrak{B}\mathfrak{k})$ is the kernel of $\gamma$ (see $[2(\mathrm{i}), \text{Theorem } 1]$). So it would be sufficient to consider the case when $q \in I_\mathfrak{g} \cap \mathfrak{P}$. But then it follows from $[2(\mathrm{i}), \text{Lemma } 20]$ that $q$ itself is of degree $d$. Now $\mathfrak{g}$ is the direct sum of $\mathfrak{k}$, $\mathfrak{h}_\mathfrak{p}$ and $\mathfrak{n}$. Hence $\mathfrak{B}_d \subset \mathfrak{k}\mathfrak{B} + (\mathfrak{H}_\mathfrak{p}\mathfrak{N}) \cap \mathfrak{B}_d$ from $[2(\mathrm{b}), \text{Lemma } 12]$. Therefore it follows from the definition of $\gamma$ (see $[2(\mathrm{i}), \S 4]$) that $q - \gamma'(q) \in \mathfrak{k}\mathfrak{B} + (\mathfrak{H}_\mathfrak{p}\mathfrak{N}_1) \cap \mathfrak{B}_d$. Let $b \to b^\dagger$ $(b \in \mathfrak{B})$ denote the anti-automorphism of $\mathfrak{B}$ such that $X^\dagger = -\theta(X)$ $(X \in \mathfrak{g})$. It is clear that every element of $\mathfrak{p}$ (and therefore also of $\mathfrak{P}$) is left fixed by this anti-automorphism. Hence

$$q - \gamma'(q) = (q - \gamma'(q))^\dagger \in \mathfrak{B}\mathfrak{k} + (\bar{\mathfrak{N}}_1\mathfrak{H}_\mathfrak{p}) \cap \mathfrak{B}_d.$$

This proves the lemma.

Select $u_1 = 1, u_2, \cdots, u_w \in \mathfrak{H}_\mathfrak{p}$ as in the corollary of Lemma 8 of $[2(\mathrm{i})]$ and let $u \to u'$ denote the automorphism of $\mathfrak{H}_\mathfrak{p}$ given by $u' = e^{-\rho}u \circ e^\rho$ $(u \in \mathfrak{H}_\mathfrak{p})$.

LEMMA 45. *Suppose* $u = \sum\limits_{1 \leq i \leq w} u_i\gamma(q_i)$ $(q_i \in I_\mathfrak{g})$. *Then we can choose a finite number of elements* $q_{ki} \in I_\mathfrak{g}$ *and* $\eta_{ki} \in \bar{\mathfrak{N}}_1$ $(1 \leq k \leq r, 1 \leq i \leq w)$ *such that*

$$u' \equiv \sum\limits_{1 \leq i \leq w} u_i'q_i + \sum\limits_{1 \leq k \leq r} \sum\limits_{1 \leq i \leq w} \eta_{ki}u_i'q_{ki} \bmod \mathfrak{B}\mathfrak{k}.$$

We shall use induction on the degree $d$ of $u$. Obviously we may assume that $u$ is homogeneous and $d \geq 1$. Then $u_i\gamma(q_i)$ $(1 \leq i \leq w)$ are also homogeneous of degree $d$ (see the corollary of Lemma 8 of $[2(\mathrm{i})]$) and it follows from Lemma 44 that

$$\sum\limits_i u_i'q_i - \sum\limits_i u_i'\gamma'(q_i) \in \mathfrak{B}\mathfrak{k} + (\bar{\mathfrak{N}}_1\mathfrak{H}_\mathfrak{p}) \cap \mathfrak{B}_d.$$

But this implies that

$$u' \equiv \sum\limits_i u_i'q_i + \sum\limits_{1 \leq k \leq s} \eta_k v_k' \bmod \mathfrak{B}\mathfrak{k}$$

where $\eta_k \in \bar{\mathfrak{N}}_1$ and $v_k \in \mathfrak{H}_\mathfrak{p} \cap \mathfrak{B}_{d-1}$ $(1 \leq k \leq s)$. On the other hand $\mathfrak{H}_\mathfrak{p} = \sum\limits_{1 \leq i \leq w} Ju_i$ where $J = I(\mathfrak{h}_\mathfrak{p}) = \gamma(I_\mathfrak{g})$. Therefore our assertion follows by applying the induction hypothesis to $v_k$.

Select $\alpha_1, \cdots, \alpha_l \in \Sigma$ as in $[2(\mathrm{i}), \S 3]$ and consider the additive group $L$ generated by $\Sigma$. Then $(\alpha_1, \cdots, \alpha_l)$ is a set of independent generators of $L$. We say that an element $b \in \mathfrak{B}$ is of rank $\beta$ $(\beta \in L)$ if $[H, b] = Hb - bH = \beta(H)b$ for every $H \in \mathfrak{h}_\mathfrak{p}$. Let $L_+$ be the set of all elements $\beta \in L$ of the form $\beta = m_1\alpha_1 + \cdots + m_l\alpha_l$ where $m_1, \cdots, m_l$ are nonnegative integers which are not all zero.  Let $(\bar{\mathfrak{N}}_1)_{-\beta}$ $(\beta \in L_+)$ denote the set of all elements in $\bar{\mathfrak{N}}_1$ of rank $-\beta$.  Then it is easy to see that $\bar{\mathfrak{N}}_1$ is the direct sum of $(\bar{\mathfrak{N}}_1)_{-\beta}$ for $\beta \in L_+$.  Therefore it is obvious that the elements $\eta_{ki}$ of Lemma 45 can be so chosen that $\eta_{ki}$ is of the rank $-\beta_{ki}$ for some $\beta_{ki} \in L_+$.  Now we use the notation of Section 14.

COROLLARY.  *Assume that $\eta_{ki}$ is of rank $-\beta_{ki}$ $(\beta_{ki} \in L_+)$ in Lemma 45. Then*

$$\phi_\lambda(\bar{n}:h;u') = \sum_{1 \le i \le w} \gamma(q_i:(-1)^{\frac{1}{2}}\lambda)\phi_\lambda(\bar{n}:h;u_i')$$
$$+ \sum_{1 \le k \le r}\sum_{1 \le i \le w} e^{-\beta_{ki}(\log h)}\gamma(q_{ki}:(-1)^{\frac{1}{2}}\lambda)\phi_\lambda(\bar{n};\eta_{ki}:h;u_i')$$

*for $\lambda \in \mathfrak{F}$, $\bar{n} \in \bar{N}$ and $h \in A_\mathfrak{p}$.*

It follows from Lemma 45 and [2(i), Lemma 18] that

$$\phi_\lambda(x;u') = \sum_i \gamma(q_i:(-1)^{\frac{1}{2}}\lambda)\phi_\lambda(x;u_i') + \sum_{k,i} \gamma(q_{ki}:(-1)^{\frac{1}{2}}\lambda)\phi_\lambda(x;\eta_{ki}u_i')$$

for $x \in G$.  Now if we put $x = \bar{n}h$ and observe that

$$\phi_\lambda(\bar{n}h;\eta_{ki}u_i') = e^{-\beta_{ki}(\log h)}\phi_\lambda(\bar{n};\eta_{ki}:h;u_i'),$$

our assertion follows immediately.

Now put

$$\Phi_j(\lambda:\bar{n}:h:H) = \pi_0(\lambda)e^{\rho\,(\log h + H)}\phi_\lambda(\bar{n}:h\exp H;u_j') \qquad (1 \le j \le w)$$

for $\lambda \in \mathfrak{F}$, $\bar{n} \in \bar{N}$, $h \in A_\mathfrak{p}$ and $H \in \mathfrak{h}_{\mathfrak{p}_0}$.  Suppose $Hu_i = \sum_j u_j \gamma(q_{ji})$ where $q_{ji} \in I_\mathfrak{g}$. Then we put $\Xi_{ij}(\lambda:H) = \gamma(q_{ji}:(-1)^{\frac{1}{2}}\lambda)$ $(1 \le i,j \le w)$.  It is clear that $\Xi_{ij}(\lambda:H)$ depend on $H$ but not on the particular choice of $q_{ji}$.  Now fix an element $H_0 \in \mathfrak{h}_{\mathfrak{p}_0}^+$.  By Lemma 45 we can choose elements $q_{ji}$, $q_{kji}$ in $I_\mathfrak{g}$ and $\eta_{kij}$ in $\bar{\mathfrak{N}}_1$ $(1 \le i,j \le w, 1 \le k \le r)$ such that

$$H_0 u_i' \equiv \sum_{1 \le j \le w} u_j' q_{ji} + \sum_{1 \le k \le r}\sum_{1 \le j \le w} \eta_{kij}u_j' q_{kji} \bmod \mathfrak{B}\mathfrak{k} \qquad (1 \le i \le w).$$

Moreover we may assume that $\eta_{kij}$ is of rank $-\beta_{kij}$ for suitable $\beta_{kij} \in L_+$. Then it follows from the corollary of Lemma 45 that

$$d\Phi_i(\lambda:\bar{n}:h:tH_0)/dt = \sum_{1 \le j \le w} \gamma(q_{ji}:(-1)^{\frac{1}{2}}\lambda)\Phi_j(\lambda:\bar{n}:h:tH_0)$$
$$+ \sum_{1 \le k \le r}\sum_{1 \le j \le w} \gamma(q_{kji}:(-1)^{\frac{1}{2}}\lambda)\Phi_j(\lambda:\bar{n};\eta_{kij}:h:tH_0)$$
$$\times \exp\{-\beta_{kij}(\log h + tH_0)\}.$$

In matrix notation, this equation can be written as follows.

$$d\Phi(\lambda:\bar{n}:h:tH_0)/dt = \Xi(\lambda:H_0)\Phi(\lambda:\bar{n}:h:tH_0) + \Psi(\lambda:\bar{n}:h:t)$$

where

$$\Psi_i(\lambda:\bar{n}:h:t)$$
$$= \sum_{1\leq k\leq r}\sum_{1\leq j\leq w}\gamma(q_{kji}:(-1)^i\lambda)\Phi_j(\lambda:\bar{n};\eta_{kij}:h:tH_0)$$
$$\times \exp\{-\beta_{kij}(\log h + tH_0)\}$$

$(1\leq i\leq w)$. Put $\Phi(\lambda:\bar{n}:h:t)=\Phi(\lambda:\bar{n}:h:tH_0),\Phi(\lambda:\bar{n}:h)=\Phi(\lambda:\bar{n}:h:0)$,

$$\Phi^0(\lambda:\bar{n}:h:t)=\exp(-t\Xi(\lambda:H_0))\Phi(\lambda:\bar{n}:h:t),$$
$$\Psi^0(\lambda:\bar{n}:h:t)=\exp(-t\Xi(\lambda:H_0))\Psi(\lambda:\bar{n}:h:t) \qquad (t\in R).$$

Then the following lemma is obvious.

LEMMA 46. $d\Phi^0(\lambda:\bar{n}:h:t)/dt=\Psi^0(\lambda:\bar{n}:h:t)$ and therefore

$$\Phi^0(\lambda:\bar{n}:h:T)=\Phi(\lambda:\bar{n}:h) + \int_0^T \Psi^0(\lambda:\bar{n}:h:t)dt$$

for $T\geq 0$.

Put $\beta = \min_{\alpha\in P_+} \alpha(H_0)$ and let $\omega$ denote a compact subset of $\mathfrak{F}_R$. It follows from Corollary 2 of Lemma 14 of $[2(i)]$ that $(-1)^i\lambda(sH_0)$ $(s\in W)$ are all the eigenvalues of $\Xi(\lambda:H_0)$. Hence by Lemma 60 of $[2(i)]$, there exists a positive number $c_1$ such that $\|\exp(-t\Xi(\lambda:H_0))\|\leq c_1(1+t)^w$ for $t\geq 0$ and $\lambda\in\omega$. Put $x^\dagger=\theta(x^{-1})$ as in Section 2 $(x\in G)$. Then $x^\dagger\in KxK$ and therefore $\phi_\lambda(x^\dagger)=\phi_\lambda(x)$. Let $b\to b^\dagger$ denote the anti-automorphism of $\mathfrak{B}$ such that $X^\dagger=-\theta(X)$ $(X\in\mathfrak{g})$. Then if $\eta\in\mathfrak{N}$ and $u\in\mathfrak{H}_\mathfrak{p}$, it follows without difficulty that

$$\phi(\lambda:\bar{n};\eta:h;u)=\phi_\lambda(hn;b)$$

where $n=(\bar{n})^\dagger$ and $b=((\eta u)^{\bar{n}})^\dagger$. Therefore since $D(x)\geq 1$ $(x\in G)$, we get the following result from Theorem 2.

LEMMA 47. *Let $U$ and $\omega$ denote two compact subsets of $\bar{N}$ and $\mathfrak{F}_R$ respectively. Then for any $h\in A_\mathfrak{p}$, we can select a positive constant $c$ such that*

$$\|\Psi^0(\lambda:\bar{n}:h:t)\|\leq c(1+t)^w e^{-\beta t}$$

*for $\lambda\in\omega$, $\bar{n}\in U$ and $t\geq 0$.*

It is obvious from Lemmas 46 and 47 that $\Phi^0(\lambda:\bar{n}:h:t)$ tends to a limit as $t\to+\infty$ if $\lambda\in\mathfrak{F}_R$. We denote this limit by $\Phi^0(\lambda:\bar{n}:h:\infty)$.

COROLLARY. *For a fixed $h$ in $A_\mathfrak{p}$, $\Phi^0(\lambda:\bar{n}:h:\infty)$ is a continuous function of $(\lambda, \bar{n})$ on $\mathfrak{F}_R \times \bar{N}$.*

It follows from Lemma 47 that the integral $\displaystyle\int_0^\infty \|\Psi^0(\lambda:\bar{n}:h:t)\|\,dt$ converges uniformly as $\lambda$ and $\bar{n}$ vary in $\omega$ and $U$ respectively. The corollary is an immediate consequence of this fact.

As before let $\mathfrak{F}_R'$ be the set of those points $\lambda$ in $\mathfrak{F}_R$ where $\pi(\lambda) \neq 0$. Fix an open subset $\omega$ of $\mathfrak{F}_R'$ such that its closure in $\mathfrak{F}_R'$ is compact. We also assume that $s\omega = \omega$ for $s \in W$. Select an element $p$ in $S(\mathfrak{h}_\mathfrak{p})$ and for any $a \in C_c^\infty(\omega)$, put $a_t(\lambda) = a(\lambda)\pi(\lambda)^{-1}p(\lambda)\exp\{-(-1)^{\frac{1}{2}}t\lambda(H_0)\}$ $(t \in R, \lambda \in \mathfrak{F}_R)$. Obviously $a_t \in C_c^\infty(\omega)$ for any $t$. Now choose an integer $k_0 \geq 0$ such that

$$\int_{\bar{N}} e^{-\rho(H(\bar{n}))}\{1 + \rho(H(\bar{n}))\}^{-k_0}\,d\bar{n} < \infty.$$

LEMMA 48. *Define $\Omega$ as in Corollary 1 of Lemma 42 and let $b$ be an element in $\mathfrak{B}$. Then there exists an integer $m \geq 0$ with the following property. For any $a \in C_c^\infty(\omega)$, we can select a positive number $c_a$ such that*

$$|\phi_{a_t}(\bar{n}h;b)| \leq c_a(1+t)^m\exp\{-|\rho(\log h)|-\rho(H(\bar{n}))\}\{1+\rho(H(\bar{n}))\}^{-k_0}$$

*for $\bar{n} \in \bar{N}$, $h \in \Omega$ and $t \geq 0$.*

From Corollary 1 of Lemma 42, we can choose an integer $k_1 \geq 0$ and a positive number $c_1$ such that

$$|f(\bar{n}h;b)|$$

$$\leq c_1\exp\{-|\rho(\log h)|-\rho(H(\bar{n}))\}\{1+\rho(H(\bar{n}))\}^{-k_0}\sup_{x \in G}D_{k_1}(x)|f(x;b)|$$

for any $f \in I(G)$. On the other hand by Theorem 3 we can select a finite number of differential operators $d_1, \cdots, d_r$ in $\mathfrak{D}(\mathfrak{h}_\mathfrak{p})$ such that

$$\sup_{x \in G}D_{k_1}(x)|\phi_a(x;b)| \leq \sum_{1 \leq i \leq r}\sup_{\lambda \in \omega}|a(\lambda;d_i)|$$

for $a \in C_c^\infty(\omega)$. Let $m_i$ denote the order of $d_i$. Then if $m = \max(m_1, \cdots, m_r)$, it is obvious that for any $a \in C_c^\infty(\omega)$, there exists a positive constant $c'$ such that

$$\sum_{1 \leq i \leq r}\sup_{\lambda \in \omega}|a_t(\lambda;d_i)| \leq c'(1+t)^m$$

for all $t \geq 0$. From this our assertion follows immediately.

COROLLARY 1. *Let $u$ be an element in $\mathfrak{H}_\mathfrak{p}$ and $\eta$ an element in $\bar{\mathfrak{N}}_1$ of rank $-\beta_0$ $(\beta_0 \in L_+)$. Then we can select an integer $m \geq 0$ and, for any $a \in C_c^\infty(\omega)$ and $h \in A_\mathfrak{p}$, a positive constant $c(a,h)$ such that*

4

$$e^{\rho(\log h + tH_0)} \left| \phi_{a_i}(\bar{n}; \eta : h \exp tH_0 ; u) \right| e^{-\beta_0(\log h + tH_0)}$$

$$\leq c(a, h)(1 + t)^m e^{-\rho(H(\bar{n}))} \{1 + \rho(H(\bar{n}))\}^{-k_0}$$

*for $\bar{n} \in \bar{N}$ and $t \geq 0$.*

This is obvious from the above lemma if we note that

$$\phi_{a_i}(\bar{n}; \eta : h \exp tH_0 ; u) e^{-\beta_0(\log h + tH_0)} = \phi_{a_i}(\bar{n}h \exp tH_0 ; \eta u).$$

For any $a \in C_c^\infty(\omega)$, put

$$\Phi^0(a : \bar{n} : h : t) = \int_{\mathfrak{F}_R} a(\lambda) \Phi^0(\lambda : \bar{n} : h : t) d\lambda,$$

$$\Psi^0(a : \bar{n} : h : t) = \int_{\mathfrak{F}_R} a(\lambda) \Psi^0(\lambda : \bar{n} : h : t) d\lambda$$

and $\Phi(a : \bar{n} : h) = \Phi^0(a : \bar{n} : h : 0)$.

COROLLARY 2. *There exists an integer $m \geq 0$ with the following property. For any $a \in C_c^\infty(\omega)$ and $h \in A_\mathfrak{p}$, we can select a positive number $c(a, h)$ such that*

$$\| \Phi^0(a : \bar{n} : h : t) \| \leq c(a, h)(1 + t)^m e^{-\rho(H(\bar{n}))} \{1 + \rho(H(\bar{n}))\}^{-k_0}$$

*for all $\bar{n} \in \bar{N}$ and $t \geq 0$.*

We use the notation of [2(i), Lemma 14]. Put $\lambda^* = (-1)^{\frac{1}{2}}\lambda$ and and $e_{\lambda} = \sum_{1 \leq i \leq w} \sigma^i(\lambda^*) u_i$ $(\lambda \in \mathfrak{F}_R)$. Select $v_{kij} \in J$ such that $u_k u_i = \sum_j v_{kij} u_j$ $(1 \leq k, i \leq w)$ and put $E_{ij} = \sum_k \sigma^k v_{kij} \in S$. Then it is obvious that

$$e_{\lambda} \cdot u_i \equiv \sum_{1 \leq j \leq w} E_{ij}(\lambda^*) u_j \bmod SJ_{\lambda} . \qquad (1 \leq i \leq w).$$

Let $E(\lambda^*)$ denote the $w \times w$ matrix with the coefficients $E_{ij}(\lambda^*)$. Then we have seen during the proof of Lemma 51 of [2(i)] that [11]

$$E(s\lambda^*)\Xi(\lambda : H_0) = \Xi(\lambda : H_0)E(s\lambda^*) = s\lambda^*(H_0)E(s\lambda^*) \qquad (s \in W)$$

and $\sum_{s \in W} \epsilon(s) E(s\lambda^*) = \pi(\lambda^*) I$ where $I$ is the unit matrix. Hence

$$\pi(\lambda^*)\exp\{-t\Xi(\lambda : H_0)\} = \sum_{s \in W} \epsilon(s)\exp\{-ts\lambda^*(H_0)\}E(s\lambda^*)$$

for $t \in R$ and $\lambda \in \mathfrak{F}_R$. Therefore

$$\Phi^0(a : \bar{n} : h : t) = \sum_{s \in W} \epsilon(s) \int_{\mathfrak{F}_R} a(\lambda) \pi(\lambda^*)^{-1}$$

$$\times \exp\{-ts\lambda^*(H_0)\}E(s\lambda^*)\Phi(\lambda : \bar{n} : h : t) d\lambda.$$

But since $E_{ij} \in S$, the statement of the corollary now follows from Lemma 48.

LEMMA 49. *The integral* $\int_{\bar{N}} \| \Phi^0(a:\bar{n}:h:t) \| \, d\bar{n}$ *is convergent for* $a \in C_c^\infty(\omega)$, $h \in A_{\mathfrak{p}}$ *and* $t \geqq 0$. *Moreover*

$$\int_{\bar{N}} \Phi(a:\bar{n}:h)\,d\bar{n} = \int_{\bar{N}} \Phi^0(a:\bar{n}:h:t)\,d\bar{n} \qquad (t \geqq 0).$$

The first statement is obvious from Corollary 2 of Lemma 48. Moreover it follows from Lemmas 46 and 47 that

$$\Phi^0(a:\bar{n}:h:T) = \Phi(a:\bar{n}:h) + \int_0^T \Psi^0(a:\bar{n}:h:t)\,dt$$

for $T \geqq 0$. On the other hand

$$\Psi^0(a:\bar{n}:h:t) = \sum_{s \in W} \epsilon(s) \int_{\mathfrak{F}_R} a(\lambda)\pi(\lambda^*)^{-1}$$

$$\times \exp\{-ts\lambda^*(H_0)\} E(s\lambda^*)\Psi(\lambda:\bar{n}:h:t)\,d\lambda.$$

Therefore if we take into account the definition of $\Psi(\lambda:\bar{n}:h:t)$, it follows from Corollary 1 of Lemma 48 that

$$\int \| \Psi^0(a:\bar{n}:h:t\|)\,d\bar{n} \leqq c'(1+t)^m \qquad (t \geqq 0),$$

where $c'$ and $m$ are independent of $t$. This shows that

$$\int_{\bar{N}} d\bar{n} \int_0^T \| \Psi^0(a:\bar{n}:h:t) \| \, dt < \infty.$$

Hence by Fubini's Theorem,

$$\int_{\bar{N}} \Phi^0(a:\bar{n}:h:T)\,d\bar{n} = \int_{\bar{N}} \Phi(a:\bar{n}:h)\,d\bar{n} + \int_0^T dt \int_{\bar{N}} \Psi^0(a:\bar{n}:h:t)\,d\bar{n}.$$

So in order to prove the second statement of the lemma, it is enough to verify that $\int_{\bar{N}} \Psi^0(a:\bar{n}:h:t)\,d\bar{n} = 0$ for any $t \geqq 0$. But in view of the definition of $\Psi(\lambda:\bar{n}:h:t)$ and the expression for $\Psi^0(a:\bar{n}:h:t)$ in terms of $\Psi(\lambda:\bar{n}:h:t)$ given above, it would be sufficient to show that $\int \phi_a(\bar{n};\eta:h;u)\,d\bar{n}$ $=0$ for $h \in A_{\mathfrak{p}}$, $\eta \in \bar{\mathfrak{N}}_1$, $u \in \mathfrak{H}_{\mathfrak{p}}$ and $a \in C_c^\infty(\omega)$. But this is a consequence of Theorem 3 and Corollary 3 of Lemma 42.

Put $\Phi^0(a:\bar{n}:h:\infty) = \int a(\lambda)\Phi^0(\lambda:\bar{n}:h:\infty)\,d\lambda$ $(a \in C_c^\infty(\omega))$. Then it follows from Lemma 47 that $\Phi^0(a:\bar{n}:h:\infty) = \underset{t \to +\infty}{\mathrm{Lim}}\,\Phi(a:\bar{n}:h:t)$. On the other hand

$$\int_{\bar{N}} \Phi(a:\bar{n}:h)\,d\bar{n} = \operatorname*{Lim}_{t\to+\infty} \int_{\bar{N}} \Phi^0(a:\bar{n}:h:t)\,d\bar{n}$$

by Lemma 49. From these facts we intend to deduce that

$$\int_{\bar{N}} \Phi_1(a:\bar{n}:h)\,d\bar{n} = \int_{\bar{N}} \Phi_1{}^0(a:\bar{n}:h:\infty)\,d\bar{n}.$$

(Here $\Phi_1$ and $\Phi_1{}^0$ stand for the first coefficients of the one-column matrices $\Phi$ and $\Phi^0$ respectively.) However first let us compute $\Phi_1{}^0(\lambda:\bar{n}:h:\infty)$. In order to do this we need the following lemma.

LEMMA 50. *Fix* $\bar{n}\in\bar{N}$ *and for* $h\in A_\mathfrak{p}$ *let* $h'$ *denote some element in* $\mathrm{Cl}(A_\mathfrak{p}{}^+)$ *such that* $\bar{n}h\in Kh'K$. *Then* $\|\log h'-\log h-H(\bar{n})\|\to 0$ *as* $\min_{\alpha\in P_+}\alpha(\log h)\to +\infty$.

Let $\sigma$ be an irreducible finite-dimensional representation of $G$ on a vector space $V$ and let $\Lambda$ denote the highest weight of $\sigma$. Select an orthonormal base $\psi_0,\psi_1,\cdots,\psi_p$ for $V$ such that $\psi_i$ belongs to the weight $\Lambda_i$ $0\leq i\leq p$ $(\Lambda_0=\Lambda)$. Put $\|T\|=(\sum_{0\leq i\leq p}|T\psi_i|^2)^{\frac{1}{2}}$ for any linear transformation $T$ on $V$. Then

$$\|\sigma(h')\|^2 = \|\sigma(\bar{n}h)\|^2 = \sum_{0\leq i\leq p}|\sigma(\bar{n}h)\psi_i|^2.$$

Let $\beta_i$ $(1\leq i\leq p)$ denote the restriction of $\Lambda-\Lambda_i$ on $\mathfrak{h}_\mathfrak{p}$. Then it follows from [2(e), Lemma 2] that $\beta_i\in L_+$ if $\beta_i\neq 0$. Suppose $\beta_i=0$ for $1\leq i< r$ and $\beta_i\in L_+$ for $r\leq i\leq p$. Then the above relation implies that

$$e^{2\Lambda(\log h')}\{r+\sum_{r\leq i\leq p}e^{-2\beta_i(\log h')}\}$$
$$= e^{2\Lambda(\log h)}\{\sum_{0\leq i< r}|\sigma(\bar{n})\psi_i|^2 + \sum_{r\leq i\leq p}|\sigma(\bar{n})\psi_i|^2\, e^{-2\beta_i(\log h)}\}.$$

Now suppose $i< r$ and $\alpha\in P_+$. It is obvious that

$$\sigma(H)\sigma(X_\alpha)\psi_i = \{\Lambda(H)+\alpha(H)\}\sigma(X_\alpha)\psi_i$$

for $H\in\mathfrak{h}_\mathfrak{p}$. On the other hand $\alpha$ cannot coincide on $\mathfrak{h}_\mathfrak{p}$ with $-\beta$ for any $\beta\in L_+$. Therefore $\Lambda+\alpha$ cannot coincide on $\mathfrak{h}_\mathfrak{p}$ with $\Lambda_j$ for any $j$ $(0\leq j\leq p)$. This implies that $\sigma(X_\alpha)\psi_i=0$. Hence $\sigma(n)\psi_i=\psi_i$ for $n\in N$ and so $|\sigma(\bar{n})\psi_i|=e^{\Lambda(H(\bar{n}))}$ $(0\leq i< r)$. Now put $H'(h)=\log h'-\log h-H(\bar{n})$. Then the above equation becomes

$$r+\sum_{r\leq i\leq p}|\sigma(\bar{n})\psi_i|^2\exp\{-2\beta_i(\log h)-2\Lambda(H(\bar{n}))\}$$
$$= e^{2\Lambda(H'(h))}\{r+\sum_{r\leq i\leq p}e^{-2\beta_i(\log h')}\}.$$

Now let $h \to \infty$ in the terminology of Section 9.   Then $\beta_i(\log h) \to +\infty$ $(r \leq i \leq p)$ and hence we get

$$r = \operatorname*{Lim}_{h \to \infty} e^{2\Lambda(H'(h))} \{ r + \sum_{r \leq i \leq p} e^{-2\beta_i(\log h')} \} \geq r \operatorname*{Lim\,sup}_{h \to \infty} e^{2\Lambda(H'(h))}.$$

Since $r \geq 1$, this shows that $\operatorname*{Lim\,sup}_{h \to \infty} \Lambda(H'(h)) \leq 0$.   On the other hand $\bar{n}h = k_1 h' k_2$ $(k_1, k_2 \in K)$ and therefore $H(\bar{n}) + \log h = H(h'k_2)$.   But it follows from Lemma 35 of $[2(\mathrm{i})]$ that $\Lambda(\log h') \geq \Lambda(H(h'k_2))$ and hence $\Lambda(H'(h)) \geq 0$.   This proves that $\operatorname*{Lim}_{h \to \infty} \Lambda(H'(h)) = 0$.   This being true for every highest weight $\Lambda$, we conclude from Theorem 1 of $[2(\mathrm{a})]$ that

$$\operatorname*{Lim}_{h \to \infty} \| H'(h) \| = 0.$$

Define the function $\boldsymbol{b}_0$ on $\mathfrak{F}_R$ as in Section 9.

LEMMA 51.[11]

$$\Phi_1^0(\lambda : \bar{n} : h : \infty) = e^{-\rho(H(\bar{n}))} \sum_{s \in W} \epsilon_0(s) \boldsymbol{b}_0(s\lambda) \exp\{ (-1)^{\frac{1}{2}} s\lambda(\log h + H(\bar{n})) \}$$

for $\lambda \in \mathfrak{F}_R$, $\bar{n} \in \bar{N}$ and $h \in A_\mathfrak{p}$.

Let $\mathfrak{F}_R''$ be the set of all points $\lambda \in \mathfrak{F}_R$ such that $\lambda(s_1 H_0) \neq \lambda(s_2 H_0)$ for any two distinct elements $s_1, s_2$ in $W$.   Then $\mathfrak{F}_R'' \subset \mathfrak{F}_R'$ and since $H_0 \in \mathfrak{h}_{\mathfrak{p}_0}{}^+$, it is obvious that $\mathfrak{F}_R''$ is dense in $\mathfrak{F}_R$.   Now, for fixed $\bar{n}$ and $h$, both sides of the above equation are continuous functions of $\lambda$ (see the corollary of Lemma 47).   Therefore it is enough to prove their equality for $\lambda \in \mathfrak{F}_R''$.

Since $\Phi(\lambda : \bar{n} : h : t) = \exp\{ t\Xi(\lambda : H_0) \} \Phi^0(\lambda : \bar{n} : h : t)$, it is clear that

$$\pi(\lambda^*) \Phi(\lambda : \bar{n} : h : t) = \sum_{s \in W} \epsilon(s) \exp\{ ts\lambda^*(H_0) \} E(s\lambda^*) \Phi^0(\lambda : \bar{n} : h : t)$$

for $t \in R$.   (Here the notation is the same as during the proof of Corollary 2 to Lemma 48.)   Hence

$$\pi(\lambda^*) \Phi_1(\lambda : \bar{n} : h : t) - \sum_{s \in W} \epsilon(s) \exp\{ ts\lambda^*(H_0) \} g_s(\lambda : \bar{n} : h) \to 0$$

as $t \to +\infty$, where

$$g_s(\lambda : \bar{n} : h) = \sum_{1 \leq j \leq w} E_{1j}(s\lambda^*) \Phi_j^0(\lambda : \bar{n} : h : \infty) \qquad (s \in W).$$

On the other hand

$$\Phi_1(\lambda : \bar{n} : h : t) = \pi_0(\lambda) e^{\rho(\log h + tH_0')} \phi_\lambda(\bar{n}h \exp tH_0).$$

Now fix $\lambda$, $h$ and $\bar{n}$ $(\lambda \in \mathfrak{F}_R'')$ and for $t \geq 0$, select an element $h_t' \in \mathrm{Cl}(A_\mathfrak{p}{}^+)$ such that $\bar{n}h \exp tH_0 \in Kh_t'K$.   Also put $H'(t) = \log h_t' - \log h - tH_0 - H(\bar{n})$.

Then it follows from Lemma 50 that $\|H'(t)\| \to 0$ and therefore $h_{t'} \to \infty$ (in the termininology of Section 9) as $t \to +\infty$. But

$$e^{\rho(\log h + tH_0)}\phi_\lambda(\bar{n}h\exp tH_0) = \exp\{-\rho(H(\bar{n})) - \rho(H'(t))\}e^{\rho(\log h_{t'})}\phi_\lambda(h_{t'})$$

and

$$\operatorname*{Lim}_{h' \to \infty}\,\Big|\,e^{\rho(\log h')}\phi_\lambda(h') - \sum_{s \in W}\boldsymbol{c}(s\lambda)\exp\{s\lambda^*(\log h')\}\Big| = 0$$

from Lemma 37 of $[2(\mathrm{i})]$. Therefore

$$\Phi_1(\lambda:\bar{n}:h:t) - e^{-\rho(H(\bar{n}))}\sum_{s \in W}\epsilon_0(s)\boldsymbol{b}_0(s\lambda)\exp\{s\lambda^*(\log h + tH_0 + H(\bar{n}))\} \to 0$$

as $t \to +\infty$. This implies that

$$\sum_{s \in W} e^{s\lambda^*(tH_0)}\epsilon(s)g_s(\lambda:\bar{n}:h)$$
$$- e^{-\rho(H(\bar{n}))}\sum_{s \in W}\pi(\lambda^*)\epsilon_0(s)\boldsymbol{b}_0(s\lambda)\exp\{s\lambda^*(\log h + tH_0 + H(\bar{n}))\} \to 0$$

as $t \to +\infty$. But since $\lambda \in \mathfrak{F}_R''$, it follows from Lemma 56 of $[2(\mathrm{i})]$ that

$$g_s(\lambda:\bar{n}:h) = \pi(\lambda^*)\epsilon(s)\epsilon_0(s)\boldsymbol{b}_0(s\lambda)\,e^{-\rho(H(\bar{n}))}\exp\{s\lambda^*(\log h + H(\bar{n}))\} \quad (s \in W).$$

On the other hand $\sum_{s \in W}\epsilon(s)E(s\lambda^*) = \pi(\lambda^*)I$ and therefore

$$\sum_{s \in W}\epsilon(s)g_s(\lambda:\bar{n}:h) = \pi(\lambda^*)\Phi_1{}^0(\lambda:\bar{n}:h:\infty).$$

Hence

$$\pi(\lambda^*)\Phi_1{}^0(\lambda:\bar{n}:h:\infty)$$
$$= \pi(\lambda^*)e^{-\rho(H(\bar{n}))}\sum_{s \in W}\epsilon_0(s)\boldsymbol{b}_0(s\lambda)\exp\{s\lambda^*(\log h + H(\bar{n}))\}.$$

The statement of the lemma is now obvious since $\pi(\lambda^*) \neq 0$.

COROLLARY. *The integral* $\displaystyle\int_{\bar{N}}|\,\Phi_1{}^0(a:\bar{n}:h:\infty)|\,d\bar{n}$ *converges if* $a \in C_c^\infty(\omega)$ *and* $h \in A_\mathfrak{p}$.

Fix $h$ and $a$. Then it is obvious that for any $s \in W$,

$$\sup_{H \in \mathfrak{h}_{\mathfrak{p}0}}(1 + \|H\|)^{k_0}\Big|\int_{\mathfrak{F}_R}\boldsymbol{b}_0(s\lambda)a(\lambda)\exp\{s\lambda^*(\log h + H)\}d\lambda\Big| < \infty.$$

Hence there exists a positive number $c$ such that

$$|\,\Phi_1{}^0(a:\bar{n}:h:\infty)| \leq c(1 + \|H(\bar{n})\|)^{-k_0}e^{-\rho(H(\bar{n}))} \qquad (\bar{n} \in \bar{N})$$

and from this the corollary follows immediately.

Now we regard the Lie group $\bar{N}$ as an analytic manifold. For any $\nu \in C_c^\infty(\bar{N})$ and $a \in C_c^\infty(\omega)$, put

$$\Phi^0(a:\nu:\bar{n}_0:h:t) = \int_{\bar{N}} \nu(\bar{n})\Phi^0(a:\bar{n}_0\bar{n}:h:t)\,d\bar{n} \qquad (\bar{n}_0 \in \bar{N}, h \in A_\mathfrak{p}, t \geqq 0)$$

and $\Phi(a:\nu:\bar{n}_0:h) = \Phi^0(a:\nu:\bar{n}_0:h:0)$. It follows from the corollary of Lemma 47 that the integral

$$\Phi^0(a:\nu:\bar{n}_0:h:\infty) = \int_{\bar{N}} \nu(\bar{n})\Phi^0(a:\bar{n}_0\bar{n}:h:\infty)\,d\bar{n}$$

is also well defined.

LEMMA 52. *The integrals*

$$\int_{\bar{N}} \Phi(a:\nu:\bar{n}:h)\,d\bar{n}, \qquad \int_{\bar{N}} \Phi^0(a:\nu:\bar{n}:h:\infty)\,d\bar{n}$$

*are both well defined for $a \in C_c^\infty(\omega)$, $\nu \in C_c^\infty(\bar{N})$, $h \in A_\mathfrak{p}$ and they are equal.*

In view of Lemma 49 and the right-invariance of the Haar measure of the nilpotent group $\bar{N}$, it is clear that

$$\int \| \Phi(a:\nu:\bar{n}:h)\|\,d\bar{n} \leqq \int |\nu(\bar{n})|\,d\bar{n} \int \| \Phi(a:\bar{n}:h)\|\,d\bar{n} < \infty.$$

Moreover it follows from Lemmas 46 and 47 that

$$\Phi^0(a:\nu:\bar{n}:h:\infty) = \Phi(a:\nu:\bar{n}:h) + \int_0^\infty \Psi^0(a:\nu:\bar{n}:h:t)\,dt$$

where

$$\Psi^0(a:\nu:\bar{n}_0:h:t) = \int \nu(\bar{n})\Psi^0(a:\bar{n}_0\bar{n}:h:t)\,d\bar{n} \qquad (\bar{n}_0 \in \bar{N}).$$

We claim that

$$\int_{\bar{N}} d\bar{n} \int_0^\infty \| \Psi^0(a:\nu:\bar{n}:h:t)\|\,dt < \infty.$$

If we recall the definition of $\Psi^0(\lambda:\bar{n}:h:t)$ $(\lambda \in \mathfrak{F}_R)$, it becomes clear (see the proof of Lemma 49) that for this it would be sufficient to prove the following lemma. Put $\phi_a(\bar{n}:h) = \phi_a(\bar{n}h)$ $(a \in C_c^\infty(\omega))$ and use the notation of Corollary 1 of Lemma 48.

LEMMA 53. *For any $a \in C_c^\infty(\omega)$ and $\nu \in C_c^\infty(\bar{N})$, put*

$$\phi_a(\nu:\bar{n}_0:h) = \int_{\bar{N}} \nu(\bar{n})\phi_a(\bar{n}_0\bar{n}:h)\,d\bar{n} \qquad (\bar{n}_0 \in \bar{N}, h \in A_\mathfrak{p}).$$

*Then for any $u \in \mathfrak{H}_{\mathfrak{p}}$ and $h \in A_{\mathfrak{p}}$, the integral*

$$\int_{\bar{N}} d\bar{n} \int_0^\infty \left| \phi_{a_i}(v : \bar{n} ; \eta : h \exp tH_0 ; u) \right|$$
$$\times \exp\{\rho(\log h + tH_0) - \beta_0(\log h + tH_0)\} dt$$

*is finite.*

Let $\xi \to \xi'$ denote the anti-automorphism of $\mathfrak{N}$ such that $X' = -X$ ($X \in \bar{\mathfrak{n}}$). Then in view of the left-invariance of the measure $d\bar{n}$, it is clear that

$$\phi_a(v : \bar{n} ; \eta : h ; u) = \phi_a(v' : \bar{n} : h ; u),$$

where $v'(\bar{n}^{-1}) = v(\bar{n}^{-1} ; (\eta')^{\bar{n}})$ ($\bar{n} \in \bar{N}$). Hence $\phi_{a_i}(v : \bar{n} ; \eta : h \exp tH_0 ; u)$ $= \phi_{a_i}(v' : \bar{n} : h \exp tH_0 ; u)$. Now fix $v$, $h$ and $a$. Then from Lemma 48, we can choose a positive constant $c$ and an integer $m \geqq 0$ such that

$$\left| \phi_{a_i}(v' : \bar{n}_0 : h \exp tH_0 ; u) \right| e^{\rho(\log h + tH_0)}$$
$$\leqq c(1 + t)^m \int \left| v'(\bar{n}) \right| e^{-\rho(H(\bar{n}_0\bar{n}))} \{1 + \rho(H(\bar{n}_0\bar{n}))\}^{-k_0} d\bar{n}$$

for all $\bar{n}_0 \in \bar{N}$ and $t \geqq 0$. Since $\beta_0(H_0) > 0$, the assertion of the lemma is now obvious.

This proves that

$$\int \| \Phi^0(a : v : \bar{n} : h : \infty) \| d\bar{n}$$
$$\leqq \int \| \Phi(a : v : \bar{n} : h) \| d\bar{n} + \int d\bar{n} \int_0^\infty \| \Psi^0(a : v : \bar{n} : h : t) \| dt < \infty.$$

Therefore it follows by Fubini's Theorem that

$$\int \Phi^0(a : v : \bar{n} : h : \infty) d\bar{n} = \int \Phi(a : v : \bar{n} : h) d\bar{n}$$
$$+ \int_0^\infty dt \int \Psi^0(a : v : \bar{n} : h : t) d\bar{n}.$$

Hence in order to prove Lemma 52, it would be sufficient to verify that

$$\int \Psi^0(a : v : \bar{n} : h : t) d\bar{n} = 0$$

for any $t \geqq 0$. But we have seen during the proof of Lemma 49 that the integral $\int \| \Psi^0(a : \bar{n} : h : t) \| d\bar{n}$ is finite and $\int \Psi^0(a : \bar{n} : h : t) d\bar{n} = 0$. Hence it is obvious that

$$\int \Psi^0(a : v : \bar{n} : h : t) d\bar{n} = \int v(\bar{n}) d\bar{n} \int \Psi^0(a : \bar{n} : h : t) d\bar{n} = 0.$$

This completes the proof of Lemma 52.

Now select $\nu$ in such a way that $\int \nu(\bar{n})d\bar{n} = 1$. Then it follows from Lemmas 52 and 49 and the corollary of Lemma 51 that

$$\int \Phi_1{}^0(a:\bar{n}:h:\infty)\,d\bar{n} = \int \Phi_1{}^0(a:\nu:\bar{n}:h:\infty)\,d\bar{n}$$
$$= \int \Phi_1(a:\nu:\bar{n}:h)\,d\bar{n} = \int \Phi_1(a:\bar{n}:h)\,d\bar{n}$$

for $a \in C_c{}^\infty(\omega)$ and $h \in A_{\mathfrak{p}}$. Now fix $a$ and $h$ and put

$$a_s{}'(H) = \int \boldsymbol{b}_0(s\lambda)a(\lambda)\exp\{s\lambda^*(\log h + H)\}d\lambda$$

for $s \in W$ and $H \in \mathfrak{h}_{\mathfrak{p}_0}$. Then it follows from the theory of Fourier transforms that $\sup\limits_{H \in \mathfrak{h}_{\mathfrak{p}_0}}(1 + \|H\|)^{k_0}|a_s{}'(H)| < \infty$. On the other hand,

$$\Phi_1{}^0(a:\bar{n}:h:\infty) = e^{-\rho(H(\bar{n}))}\sum_{s\in W}\epsilon_0(s)a_s{}'(H(\bar{n}))$$

by Lemma 51. Therefore it is obvious that

$$\int \Phi_1{}^0(a:\bar{n}:h:\infty)\,d\bar{n} = \sum_{s\in W}\epsilon_0(s)\int a_s{}'(H(\bar{n}))e^{-\rho(H(\bar{n}))}\,d\bar{n}.$$

Now let $\delta$ be a sufficiently small positive number. By the corollary of Lemma 45 of [2(i)], $\int \exp\{-(1+\delta)\rho(H(\bar{n}))\}d\bar{n} < \infty$ and therefore, by Fubini's Theorem,

$$\int a_s{}'(H(\bar{n}))\exp\{-(1+\delta)\rho(H(\bar{n}))\}d\bar{n}$$
$$= \int \boldsymbol{b}_0(s\lambda)a(\lambda)e^{s\lambda^*(\log h)}\,d\lambda \int \exp\{s\lambda^*(H(\bar{n})) - (1+\delta)\rho(H(\bar{n}))\}d\bar{n}$$
$$= \int \boldsymbol{b}_0(\lambda)a(s^{-1}\lambda)e^{\lambda^*(\log h)}\operatorname{conj}\boldsymbol{c}(\lambda_\delta)\,d\lambda$$

from Theorem 4 of [2(i)], where $\lambda_\delta = \lambda - (-1)^{\frac{1}{2}}\delta\rho$. But since $\omega \subset \mathfrak{F}_R'$, $\boldsymbol{c}$ is a holomorphic function on a complex neighborhood of $\omega$ in $\mathfrak{F}$. Therefore if we recall that $\omega$ is invariant under $W$, it becomes obvious that

$$\int a_s{}'(H(\bar{n}))e^{-\rho(H(\bar{n}))}d\bar{n} = \operatorname*{Lim}_{\delta\to 0}\int a_s{}'(H(\bar{n}))e^{-(1+\delta)\rho(H(\bar{n}))}d\bar{n}$$
$$= \int \boldsymbol{b}_0(\lambda)a(s^{-1}\lambda)e^{\lambda^*(\log h)}\operatorname{conj}\boldsymbol{c}(\lambda)\,d\lambda$$
$$= \int |\boldsymbol{b}_0(\lambda)|^2 \pi_0(\lambda)^{-1}a(s^{-1}\lambda)e^{\lambda^*(\log h)}d\lambda.$$

Hence

$$\int \Phi_1^{\,0}(a:\bar n:h:\infty)\,d\bar n = \int |b_0(\lambda)|^2 \{\pi_0(\lambda)^{-1}\sum_{s\in W}\epsilon_0(s)a(s\lambda)\}e^{\lambda^*(\log h)}d\lambda.$$

But we have seen above that

$$\int \Phi_1^{\,0}(a:\bar n:h:\infty)\,d\bar n = \int \Phi_1(a:\bar n:h)\,d\bar n = e^{\rho(\log h)}\int \phi_a(\bar n h)\,d\bar n.$$

Therefore it follows easily from the corollary of Lemma 43 that $\beta(\lambda) = |b_0(\lambda)|^2$ for $\lambda\in\omega$. But since this holds for any choice of $\omega$ and since $\beta$ and $b_0$ are both continuous on $\mathfrak{F}_R$, it is obvious that this relation actually holds for all $\lambda\in\mathfrak{F}_R$.

THEOREM 4. *Let a be an element in* $\mathscr{B}(\mathfrak{F}_R)$. *Then*

$$e^{\rho(H)}\int_{\bar N} \phi_a(\bar n\exp H)\,d\bar n = \int_{\mathfrak{F}_R} |b_0(\lambda)|^2\{\pi_0(\lambda)^{-1}\sum_{s\in W}\epsilon_0(s)a(s\lambda)\}e^{(-1)^{\frac{1}{2}}\lambda(H)}d\lambda$$

*for* $H\in\mathfrak{h}_{\mathfrak{p}_0}$.

In view of the result obtained above, this is an immediate consequence of the corollary of Lemma 43, if we use the regular normalization of $d\lambda$. On the other hand the statement of the theorem is obviously independent of the normalization of $d\lambda$. Hence it holds in general.

THEOREM 5. *Let* $a\in\mathscr{B}(\mathfrak{F}_R)$ *and* $\lambda\in\mathfrak{F}_R$. *Then*

$$\pi_0(\lambda)\int_G \phi_a(x)\operatorname{conj}\phi_\lambda(x)\,dx = |b_0(\lambda)|^2\sum_{s\in W}\epsilon_0(s)a(s\lambda)$$

*if we use the regular normalizations of dx and d$\lambda$.*

This is merely a restatement of the corollary of Lemma 41.

COROLLARY 1. *Let w denote the order of W. Then under the above normalizations,*

$$\int_G |\phi_a(x)|^2\,dx = w^{-1}\int_{\mathfrak{F}_R} |b_0(\lambda)|^2|\sum_{s\in W}\epsilon_0(s)a(s\lambda)|^2\,d\lambda$$

*for* $a\in\mathscr{B}(\mathfrak{F}_R)$.

This would follow immediately from Theorem 5 by Fubini's Theorem, as soon as we can verify that

$$\int \mid \pi_0(\lambda)a(\lambda)\phi_\lambda(x)\phi_a(x)\mid dxd\lambda < \infty.$$

This is done as follows. We know (see $[2(f), \S 12]$) that if the Haar measure $dh$ on $A_\mathfrak{p}$ is suitably normalized,

$$\int f(x)dx = \int_{A_\mathfrak{p}^+} f(h)\Delta(h)dh$$

for any spherical function $f$ in $C_c(G)$. (Here $\Delta$ has the same meaning as in Section 13.) Now select an integer $q \geqq 0$ such that

$$c_1 = \int_{A_\mathfrak{p}} (1 + \parallel \log h \parallel)^{-q} dh < \infty.$$

Then

$$\mid \pi_0(\lambda)\mid \int \mid \phi_\lambda(x)\phi_a(x)\mid dx = \mid \pi_0(\lambda)\mid \int_{A_\mathfrak{p}^+} \mid \phi_\lambda(h)\phi_a(h)\mid \Delta(h)dh$$

$$\leqq c_1 \sup_{h \in A_\mathfrak{p}^+} \Delta(h)(1 + \parallel \log h \parallel)^q \mid \pi_0(\lambda)\phi_\lambda(h)\phi_a(h)\mid.$$

Hence it follows from Theorems 1 and 3 that there exists an integer $k \geqq 0$ and, for a fixed $a$, a positive number $c$ such that

$$\mid \pi_0(\lambda)\mid \int \mid \phi_\lambda(x)\phi_a(x)\mid dx \leqq c(1 + \parallel \lambda \parallel)^k$$

for all $\lambda \in \mathfrak{F}_R$. Therefore

$$\int \mid \pi_0(\lambda)a(\lambda)\phi_\lambda(x)\phi_a(x)\mid dxd\lambda \leqq c\int (1 + \parallel \lambda \parallel)^k \mid a(\lambda)\mid d\lambda < \infty.$$

Let $I_0(G)$ denote the subspace of $I(G)$ consisting of all elements of the form $\phi_a$ ($a \in \mathcal{B}(\mathfrak{F}_R)$).

COROLLARY 2. *For any $f$ in $I(G)$, put*

$$\bar{f}(\lambda) = \int_G f(x)\mathrm{conj}\,\phi_\lambda(x)dx \qquad (\lambda \in \mathfrak{F}_R).$$

*Then*

$$\int_G \mid f(x)\mid^2 dx = w^{-1} \int_{\mathfrak{F}_R} \mid c(\lambda)\mid^{-2} \mid \bar{f}(\lambda)\mid^2 d\lambda$$

*and*

$$f(x) = w^{-1} \int_{\mathfrak{F}_R} |c(\lambda)|^{-2} \tilde{f}(\lambda) \phi_\lambda(x) \, d\lambda$$

*for $f \in I_0(G)$ and $x \in G$, if $dx$ and $d\lambda$ are regularly normalized.*

This is an immediate consequence of Theorems 5 and Corollary 1 above. We have seen (Lemma 29) that $c^{-1}$ is analytic on $\mathfrak{F}_R$. Moreover Theorem 5 of [2(i)] gives a geometric interpretation of $c$.

**16. Two conjectures.** Let $I_c(G)$ denote the subspace consisting of those elements of $I(G)$ which have compact support. Also let $L_2(G)$ denote the Hilbert space of square-integrable functions on $G$ and $I_2(G)$ the closure of $I_c(G)$ in $L_2(G)$. It is seen without difficulty that $I_c(G) \subset I(G) \subset I_2(G)$. Therefore if the statements of Corollary 2 of Theorem 5 could be extended to all functions $f \in I(G)$, we would immediately get an explicit Plancherel formula (see [2(j)]) for $I_2(G)$. Moreover Theorem 5 of [2(i)] would then provide a simple interpretation for the Plancherel measure. For this, it would obviously be enough to show that $I_c(G) \subset I_0(G)$. However this appears to be a rather difficult problem which I have not yet been able to solve. Its solution seems to depend on the following two unproved statements.

(1)   *There exists a polynomial function $p \in S(\mathfrak{h}_\mathfrak{p})$ such that $|\,b_0(\lambda)p(\lambda)\,|$
       $\geqq 1$ for all $\lambda \in \mathfrak{F}_R$.*

(2)   *For any $f \in I(G)$, define $F_f$ as in Corollary 4 to Lemma 42. Then
       $F_f \neq 0$ unless $f = 0$.*

Assuming these two conjectures, we shall show that $I_c(G) \subset I_0(G)$. Let $f$ be any element in $I_c(G)$. Then it follows from Lemma 43 and Corollary 4 of Lemma 42 that $\tilde{f} \in \mathscr{B}(\mathfrak{F}_R)$. On the other hand if we take Lemma 31 into account, it follows from the first conjecture that the function $a(\lambda) = w^{-1}\pi_0(\lambda)|\,b_0(\lambda)\,|^{-2}\tilde{f}(\lambda)$ $(\lambda \in \mathfrak{F}_R)$ also lies in $\mathscr{B}(\mathfrak{F}_R)$. Now put $g = \phi_a$. We know from the corollary of Lemma 17 of [2(i)] that $\phi_{s\lambda} = \phi_\lambda (s \in W)$ and therefore $\tilde{f}(s\lambda) = \tilde{f}(\lambda)$ and $\pi_0(\lambda)\tilde{g}(\lambda) = \pi_0(\lambda)\tilde{f}(\lambda)$ $(\lambda \in \mathfrak{F}_R)$ by Theorem 5. Moreover $\tilde{g}$ and $\tilde{f}$ are both continuous functions on $\mathfrak{F}_R$ (see Corollary 3 of Lemma 40). Hence $\tilde{g} = \tilde{f}$. But then by Lemma 43, $F_f = F_g$ and therefore $f = g = \phi_a$ by the second conjecture. This proves that $f \in I_0(G)$ and therefore $I_c(G) \subset I_0(G)$.

Although there seems to be considerable evidence in support of these conjectures, 1 have at present no clear idea as to how to proceed to prove them.

COLUMBIA UNIVERSITY.

REFERENCES.

[1] L. Garding, *Math. Scand.*, vol. 1 (1953), pp. 55-72.

[2] Harish-Chandra, (a) *Trans. Amer. Math. Soc.*, vol. 70 (1951), pp. 28-96.
    (b) "  "  "  "  vol. 75 (1953), pp. 185-243.
    (c) "  "  "  "  vol. 76 (1954), pp. 234-253.
    (d) "  "  "  "  vol. 83 (1956), pp. 98-163.
    (e) *Amer. Jour. Math.*, vol. 77 (1955), pp. 743-777.
    (f) "  "  "  vol. 78 (1956), pp. 564-628.
    (g) "  "  "  vol. 79 (1957), pp. 87-120.
    (h) "  "  "  vol. 79 (1957), pp. 193-257.
    (i) "  "  "  vol. 80 (1958), pp. 241-310.
    (j) *Proc. Nat. Acad. Sci. U.S.A.*, vol. 40 (1954), pp. 200-204.
    (k) "  "  "  "  "  vol. 43 (1957), pp. 408-409.

[3] F. John, *Proceedings of the Symposium on Spectral Theory and Differential Problems*, Stillwater, Okla., 1951, pp. 113-175.

[4] L. Schwartz, *Théorie des distributions I*, Paris, Hermann, 1950.

Reprinted from
*Amer. J. of Math.*
**80** (1958), 553–613

If you have any concerns about our products,
you can contact us on
ProductSafety@springernature.com

In case Publisher is established outside the EU,
the EU authorized representative is:
Springer Nature Customer Service Center GmbH
Europaplatz 3, 69115 Heidelberg, Germany

Printed by Libri Plureos GmbH
in Hamburg, Germany